Special Volume:
Numerical Methods in Electromagnetics
Guest Editors: W.H.A. Schilders and E.J.W. ter Maten

Handbook of Numerical Analysis

General Editor:

P.G. Ciarlet

Laboratoire Jacques-Louis Lions
Université Pierre et Marie Curie
4 Place Jussieu
75005 PARIS, France

and

Department of Mathematics
City University of Hong Kong
Tat Chee Avenue
KOWLOON, Hong Kong

ELSEVIER
NORTH
HOLLAND

Amsterdam • Boston • Heidelberg • London • New York • Oxford • Paris
San Diego • San Francisco • Singapore • Sydney • Tokyo

Volume XIII

Special Volume: Numerical Methods in Electromagnetics

Guest Editors:

W.H.A. Schilders

Philips Research Laboratories, IC Design
Prof. Holstlaan 4, 5656 AA, Eindhoven
The Netherlands

E.J.W. ter Maten

Philips Research Laboratories
Electronic Design & Tools/Analogue Simulation
Prof. Holstlaan 4, 5656 AA, Eindhoven
The Netherlands

2005

ELSEVIER
NORTH
HOLLAND

Amsterdam • Boston • Heidelberg • London • New York • Oxford • Paris
San Diego • San Francisco • Singapore • Sydney • Tokyo

ELSEVIER B.V.
Radarweg 29
P.O. Box 211, 1000 AE Amsterdam
The Netherlands

ELSEVIER Inc.
525 B Street, Suite 1900
San Diego, CA 92101-4495
USA

ELSEVIER Ltd
The Boulevard, Langford Lane
Kidlington, Oxford OX5 1GB
UK

ELSEVIER Ltd
84 Theobalds Road
London WC1X 8RR
UK

First edition 2005

Library of Congress Cataloging in Publication Data
A catalog record is available from the Library of Congress.

British Library Cataloguing in Publication Data
A catalogue record is available from the British Library.

ISBN: 0-444-51375-2
ISSN (Series): 1570-8659

⊗ The paper used in this publication meets the requirements of ANSI/NISO Z39.48-1992 (Permanence of Paper).

Printed in the Netherlands

General Preface

In the early eighties, when Jacques-Louis Lions and I considered the idea of a *Handbook of Numerical Analysis*, we carefully laid out specific objectives, outlined in the following excerpts from the "General Preface" which has appeared at the beginning of each of the volumes published so far:

> During the past decades, giant needs for ever more sophisticated mathematical models and increasingly complex and extensive computer simulations have arisen. In this fashion, two indissociable activities, *mathematical modeling* and *computer simulation*, have gained a major status in all aspects of science, technology and industry.
>
> In order that these two sciences be established on the safest possible grounds, mathematical rigor is indispensable. For this reason, two companion sciences, *Numerical Analysis* and *Scientific Software*, have emerged as essential steps for validating the mathematical models and the computer simulations that are based on them.
>
> *Numerical Analysis* is here understood as the part of *Mathematics* that describes and analyzes all the numerical schemes that are used on computers; its objective consists in obtaining a clear, precise, and faithful, representation of all the "information" contained in a mathematical model; as such, it is the natural extension of more classical tools, such as analytic solutions, special transforms, functional analysis, as well as stability and asymptotic analysis.
>
> The various volumes comprising the *Handbook of Numerical Analysis* will thoroughly cover all the major aspects of Numerical Analysis, by presenting accessible and in-depth surveys, which include the most recent trends.
>
> More precisely, the Handbook will cover the *basic methods of Numerical Analysis*, gathered under the following general headings:
>
> – Solution of Equations in $\mathbb{R}^n$,
> – Finite Difference Methods,
> – Finite Element Methods,
> – Techniques of Scientific Computing.

It will also cover the *numerical solution of actual problems of contemporary interest in Applied Mathematics*, gathered under the following general headings:

– Numerical Methods for Fluids,
– Numerical Methods for Solids.

In retrospect, it can be safely asserted that Volumes I to IX, which were edited by both of us, fulfilled most of these objectives, thanks to the eminence of the authors and the quality of their contributions.

After Jacques-Louis Lions' tragic loss in 2001, it became clear that Volume IX would be the last one of the type published so far, i.e., edited by both of us and devoted to some of the general headings defined above. It was then decided, in consultation with the publisher, that each future volume will instead be devoted to a single "*specific application*" and called for this reason a "*Special Volume*". "*Specific applications*" will include Mathematical Finance, Meteorology, Celestial Mechanics, Computational Chemistry, Living Systems, Electromagnetism, Computational Mathematics etc. It is worth noting that the inclusion of such "specific applications" in the *Handbook of Numerical Analysis* was part of our initial project.

To ensure the continuity of this enterprise, I will continue to act as Editor of each Special Volume, whose conception will be jointly coordinated and supervised by a Guest Editor.

P.G. CIARLET
July 2002

Preface

The electronics industry has shown extremely rapid advances over the past 50 years, and it is largely responsible for the economic growth in that period. It all started with the invention of the bipolar transistor based on silicon at the end of the 1940s, and since then the industry has caused another evolution for mankind. It is hard to imagine a world without all the achievements of the electronics industry.

In order to be able to continue these rapid developments, it is absolutely necessary to perform virtual experiments rather than physical experiments. Simulations are indispensable in the electronics industry nowadays. Current electronic circuits are extremely complex, and its production requires hundreds of steps that altogether take several months of fabrication time. The adagio is "first time right", and this has its repercussions for the way designers work in the electronics industry. Nowadays, they make extensive use of software tools embedded in virtual design environments. The so-called "virtual fab" has made an entry, and it is foreseen that its importance will only grow in the future.

Numerical methods are a key ingredient of a simulation environment, whence it is not surprising that the electronics industry has become one of the most fertile working environments for numerical mathematicians. Since the 1970s, there is a strong demand for efficient and robust software tools for electronic circuit simulation. Initially, this development started with the analysis of large networks of resistors, capacitors and inductors, but soon other components such as bipolar transistors and diodes were added. Specialists made models for these components, but the problems associated with the extreme nonlinearities introduced by these models had to be tackled by numerical analysts. It was one of the first serious problems that were encountered in the field, and it initiated research into damped Newton methods for extremely nonlinear problems. In the past 30 years, electronic circuit simulation has become a very mature subject, with many beautiful results (both from the engineering and the mathematical point of view), and it still is a very active area of mathematical research. Nowadays, hot topics are the research into differential algebraic equations and the efficient calculation of (quasi-)periodic steady states.

Although circuit simulation was one of the first topics to be addressed by numerical mathematicians in the electronics industry, the simulation of semiconductor devices quickly followed at the end of the 1970s. Transistors rapidly became more complex, and a multitude of different devices was discovered. Transistors of the MOS-type (metal-oxide-semiconductor) became much more popular, and are now mainly responsible for the rapid developments in the industry. In order to be able to simulate the behavior of these devices, research into semiconductor device simulation was carried out. Soon it became clear that this was a very demanding problem from the numerical point of view, and it took many years and many conferences before some light was seen at the end of the tunnel. Applied mathematicians analyzed the famous drift-diffusion problem, and numerical mathematicians developed algorithms for its discretization and solution. During the 1990s, extended models were introduced for the modelling of semiconductor devices (hydrodynamic models, quantum effects), and nowadays this development is still continuing.

Parallel to these developments in the area of electronic circuits and devices, the more classical electromagnetics problems were also addressed. Design of magnets for loud-speakers and magnet design for MRI (magnetic resonance imaging) were important tasks, for which we can also observe a tendency towards heavy usage of simulation tools and methods. The field also generated many interesting mathematical and numerical results, whereas the role of the numerical mathematician was again indispensable in this area.

Whether it is by coincidence or not, the fields of circuit/device simulation and the more classical electromagnetics simulation, have come very close to each other in recent years. Traditionally, researchers working in the two areas did not communicate much, and separate conferences were organized with a minimum of cross-fertilization. However, owing to the increased operating frequencies of devices and the shrinking dimensions of electronics circuits, electromagnetic effects have started to play an important role. These effects influence the behavior of electronic circuits, and it is foreseen that these effects may be dramatic in the future if they are not understood well and precautions are taken. Hence, recent years show an increased interest in combined simulations of circuit behavior with electromagnetics that, in turn, has led to new problems for numerical mathematicians. One of these new topics is model order reduction, which is the art of reducing large discrete systems to a much smaller model that nevertheless exhibits behavior similar to the large system. Model order reduction is a topic at many workshops and conferences nowadays, with a multitude of applications also outside the electronics industry.

From the foregoing, it is clear that the electronics industry has always been, and still is, a very fruitful area for numerical mathematics. On the one hand, numerical mathematicians have played an important role in enabling the set-up of virtual design environments. On the other hand, many new methods have been developed as a result of the work in this specialist area. Often, the methods developed to solve the electronics problems can also be applied in other application areas. Therefore, the reason for this special volume is twofold. The first aim is to give insight in the way numerical methods are being used to solve the wide variety of problems in the electronics industry. The second aim is to give researchers from other fields of application the opportunity to benefit from the results, which have been obtained in the electronics industry.

This special volume of the Handbook of Numerical Analysis gives a broad overview of the use of numerical methods in the electronics industry. Since it is not assumed that all readers are familiar with the concepts being used in the field, Chapter 1 gives a detailed overview of models being used. The starting point is the set of Maxwell equations, and from this all models can be derived. The chapter serves as the basis for the other chapters, so that readers can always go back to Chapter 1 for a physical explanation, or a derivation of the models.

The remaining chapters discuss the use of numerical methods for different applications within the electronics industry. We have attempted to organize the book in the same way as numerical analysis is performed in practice: modelling, followed by discretization, followed by solution of nonlinear and linear systems. Unfortunately, our attempts to obtain a chapter on nonlinear solution strategies have failed in the end. The corresponding chapter would have been a very interesting one, with results on damped Newton methods and nonlinear variable transformations. These methods will now be discussed in a separate book, and the reader is referred to this or to the extensive literature on the subject. Fortunately, all other aspects of numerical analysis are present in this volume, and in the following we give a short summary of the remaining chapters.

Chapter 2 is devoted to the more classical form of electromagnetics simulations, but as can be seen from the chapter, the field leads to beautiful mathematical results. Chapter 3 also discusses methods for discretising the Maxwell equations, using the finite difference time domain method that is extremely popular nowadays. The authors of this chapter have widespread experience in applying the method to practical problems, and the chapter discusses a multitude of related topics. Chapters 4 and 5 are devoted to the simulation of the behavior of semiconductor devices, with an emphasis again on discretization methods. Chapter 4 discusses the well known drift-diffusion model and some extensions, whereas Chapter 5 concentrates on extended models.

Circuit simulation is the topic discussed in Chapter 6, where both the modelling and the discretization of these problems is addressed. The concept of differential-algebraic equations is discussed extensively, together with its importance for the analysis of circuits. Furthermore, time discretization and the solution of periodic steady-state problems can be found in this chapter. In Chapter 7, the first step towards coupled circuit/device simulations with electromagnetic effects is made by considering the problem of analyzing the electromagnetic behavior of printed circuit boards. The chapter discusses in detail the efficient evaluation of the interaction integrals, and shows the use of some numerical techniques that are not very well known.

Chapters 9 and 10 are of a more theoretical character, which does not mean that their contents are less important. On the contrary, the solution techniques for linear systems discussed in Chapter 9 are at the core of all simulation software, and hence it is extremely important to perform the solution of linear systems as efficiently as possible. The model order reduction methods discussed in Chapter 10 are equally important, since they provide a sound basis for enabling the coupled simulations required in present-day design environments. Strangely enough, it turns out that the techniques used in the area of model order reduction, are intimately related to the solution methods for linear systems. In this respect, the last two chapters are closely related, though very different in character.

We hope that this volume will inspire readers, and that the presentation given in the various chapters is of interest to a large community of researchers and engineers. It is also hoped that the volume reflects the importance of numerical mathematics in the electronics industry. In our experience, we could attach tags to almost all electronic products with the statement: "Mathematics inside". Let this be an inspiration for young people to not only benefit from the developments of the electronics industry, but also contribute physically to the developments in the future by becoming an enthusiastic numerical mathematician!

Eindhoven, June 2004

Wil Schilders
Jan ter Maten

Contents of Volume XIII

SPECIAL VOLUME: NUMERICAL METHODS IN ELECTROMAGNETICS

GENERAL PREFACE v

PREFACE vii

Introduction to Electromagnetism, *W. Magnus, W. Schoenmaker* 3
Discretization of Electromagnetic Problems: The "Generalized Finite Differences" Approach, *A. Bossavit* 105
Finite-Difference Time-Domain Methods, *S.C. Hagness, A. Taflove, S.D. Gedney* 199
Discretization of Semiconductor Device Problems (I), *F. Brezzi, L.D. Marini, S. Micheletti, P. Pietra, R. Sacco, S. Wang* 317
Discretization of Semiconductor Device Problems (II), *A.M. Anile, N. Nikiforakis, V. Romano, G. Russo* 443
Modelling and Discretization of Circuit Problems, *M. Günther, U. Feldmann, J. ter Maten* 523
Simulation of EMC Behaviour, *A.J.H. Wachters, W.H.A. Schilders* 661
Solution of Linear Systems, *O. Schenk, H.A. van der Vorst* 755
Reduced-Order Modelling, *Z. Bai, P.M. Dewilde, R.W. Freund* 825

SUBJECT INDEX 897

Contents of the Handbook

VOLUME I

FINITE DIFFERENCE METHODS (PART 1)

Introduction, *G.I. Marchuk* 3
Finite Difference Methods for Linear Parabolic Equations, *V. Thomée* 5
Splitting and Alternating Direction Methods, *G.I. Marchuk* 197

SOLUTION OF EQUATIONS IN $\mathbb{R}^n$ (PART 1)

Least Squares Methods, *Å. Björck* 465

VOLUME II

FINITE ELEMENT METHODS (PART 1)

Finite Elements: An Introduction, *J.T. Oden* 3
Basic Error Estimates for Elliptic Problems, *P.G. Ciarlet* 17
Local Behavior in Finite Element Methods, *L.B. Wahlbin* 353
Mixed and Hybrid Methods, *J.E. Roberts and J.-M. Thomas* 523
Eigenvalue Problems, *I. Babuška and J. Osborn* 641
Evolution Problems, *H. Fujita and T. Suzuki* 789

VOLUME III

TECHNIQUES OF SCIENTIFIC COMPUTING (PART 1)

Historical Perspective on Interpolation, Approximation and Quadrature, *C. Brezinski* 3
Padé Approximations, *C. Brezinski and J. van Iseghem* 47
Approximation and Interpolation Theory, *Bl. Sendov and A. Andreev* 223

NUMERICAL METHODS FOR SOLIDS (PART 1)

Numerical Methods for Nonlinear Three-Dimensional Elasticity, *P. Le Tallec* 465

SOLUTION OF EQUATIONS IN $\mathbb{R}^n$ (PART 2)

Numerical Solution of Polynomial Equations, *Bl. Sendov, A. Andreev and N. Kjurkchiev* 625

VOLUME IV

FINITE ELEMENT METHODS (PART 2)

Origins, Milestones and Directions of the Finite Element Method – A Personal View, *O.C. Zienkiewicz* 3
Automatic Mesh Generation and Finite Element Computation, *P.L. George* 69

NUMERICAL METHODS FOR SOLIDS (PART 2)

Limit Analysis of Collapse States, *E. Christiansen* 193
Numerical Methods for Unilateral Problems in Solid Mechanics, *J. Haslinger, I. Hlaváček and J. Nečas* 313
Mathematical Modelling of Rods, *L. Trabucho and J.M. Viaño* 487

VOLUME V

TECHNIQUES OF SCIENTIFIC COMPUTING (PART 2)

Numerical Path Following, *E.L. Allgower and K. Georg* 3
Spectral Methods, *C. Bernardi and Y. Maday* 209
Numerical Analysis for Nonlinear and Bifurcation Problems, *G. Caloz and J. Rappaz* 487
Wavelets and Fast Numerical Algorithms, *Y. Meyer* 639
Computer Aided Geometric Design, *J.-J. Risler* 715

VOLUME VI

NUMERICAL METHODS FOR SOLIDS (PART 3)

Iterative Finite Element Solutions in Nonlinear Solid Mechanics, *R.M. Ferencz and T.J.R. Hughes* 3
Numerical Analysis and Simulation of Plasticity, *J.C. Simo* 183

NUMERICAL METHODS FOR FLUIDS (PART 1)

Navier–Stokes Equations: Theory and Approximation, *M. Marion and R. Temam* 503

VOLUME VII

SOLUTION OF EQUATIONS IN $\mathbb{R}^n$ (PART 3)

Gaussian Elimination for the Solution of Linear Systems of Equations, *G. Meurant* 3

TECHNIQUES OF SCIENTIFIC COMPUTING (PART 3)

The Analysis of Multigrid Methods, *J.H. Bramble and X. Zhang* 173
Wavelet Methods in Numerical Analysis, *A. Cohen* 417
Finite Volume Methods, *R. Eymard, T. Gallouët and R. Herbin* 713

VOLUME VIII

SOLUTION OF EQUATIONS IN $\mathbb{R}^n$ (PART 4)

Computational Methods for Large Eigenvalue Problems, *H.A. van der Vorst* 3

TECHNIQUES OF SCIENTIFIC COMPUTING (PART 4)

Theoretical and Numerical Analysis of Differential–Algebraic Equations, *P.J. Rabier and W.C. Rheinboldt* 183

NUMERICAL METHODS FOR FLUIDS (PART 2)

Mathematical Modeling and Analysis of Viscoelastic Fluids of the Oldroyd Kind, *E. Fernández-Cara, F. Guillén and R.R. Ortega* 543

VOLUME IX

NUMERICAL METHODS FOR FLUIDS (PART 3)

Finite Element Methods for Incompressible Viscous Flow, *R. Glowinski* 3

VOLUME X

SPECIAL VOLUME: COMPUTATIONAL CHEMISTRY

Computational Quantum Chemistry: A Primer, *E. Cancès, M. Defranceschi, W. Kutzelnigg, C. Le Bris, Y. Maday* 3
The Modeling and Simulation of the Liquid Phase, *J. Tomasi, B. Mennucci, P. Laug* 271
An Introduction to First-Principles Simulations of Extended Systems, *F. Finocchi, J. Goniakowski, X. Gonze, C. Pisani* 377
Computational Approaches of Relativistic Models in Quantum Chemistry, *J.P. Desclaux, J. Dolbeault, M.J. Esteban, P. Indelicato, E. Séré* 453

Quantum Monte Carlo Methods for the Solution of the Schrödinger Equation for Molecular Systems, *A. Aspuru-Guzik, W.A. Lester, Jr.* 485
Linear Scaling Methods for the Solution of Schrödinger's Equation, *S. Goedecker* 537
Finite Difference Methods for Ab Initio Electronic Structure and Quantum Transport Calculations of Nanostructures, *J.-L. Fattebert, M. Buongiorno Nardelli* 571
Using Real Space Pseudopotentials for the Electronic Structure Problem, *J.R. Chelikowsky, L. Kronik, I. Vasiliev, M. Jain, Y. Saad* 613
Scalable Multiresolution Algorithms for Classical and Quantum Molecular Dynamics Simulations of Nanosystems, *A. Nakano, T.J. Campbell, R.K. Kalia, S. Kodiyalam, S. Ogata, F. Shimojo, X. Su, P. Vashishta* 639
Simulating Chemical Reactions in Complex Systems, *M.J. Field* 667
Biomolecular Conformations Can Be Identified as Metastable Sets of Molecular Dynamics, *C. Schütte, W. Huisinga* 699
Theory of Intense Laser-Induced Molecular Dissociation: From Simulation to Control, *O. Atabek, R. Lefebvre, T.T. Nguyen-Dang* 745
Numerical Methods for Molecular Time-Dependent Schrödinger Equations – Bridging the Perturbative to Nonperturbative Regime, *A.D. Bandrauk, H.-Z. Lu* 803
Control of Quantum Dynamics: Concepts, Procedures and Future Prospects, *H. Rabitz, G. Turinici, E. Brown* 833

VOLUME XI

SPECIAL VOLUME: FOUNDATIONS OF COMPUTATIONAL MATHEMATICS

On the Foundations of Computational Mathematics, *B.J.C. Baxter, A. Iserles* 3
Geometric Integration and its Applications, *C.J. Budd, M.D. Piggott* 35
Linear Programming and Condition Numbers under the Real Number Computation Model, *D. Cheung, F. Cucker, Y. Ye* 141
Numerical Solution of Polynomial Systems by Homotopy Continuation Methods, *T.Y. Li* 209
Chaos in Finite Difference Scheme, *M. Yamaguti, Y. Maeda* 305
Introduction to Partial Differential Equations and Variational Formulations in Image Processing, *G. Sapiro* 383

VOLUME XII

SPECIAL VOLUME: COMPUTATIONAL MODELS FOR THE HUMAN BODY

Mathematical Modelling and Numerical Simulation of the Cardiovascular System, *A. Quarteroni, L. Formaggia* 3

Computational Methods for Cardiac Electrophysiology, *M.E. Belik, T.P. Usyk, A.D. McCulloch* 129
Mathematical Analysis, Controllability and Numerical Simulation of a Simple Model of Avascular Tumor Growth, *J.I. Díaz, J.I. Tello* 189
Human Models for Crash and Impact Simulation, *E. Haug, H.-Y. Choi, S. Robin, M. Beaugonin* 231
Soft Tissue Modeling for Surgery Simulation, *H. Delingette, N. Ayache* 453
Recovering Displacements and Deformations from 3D Medical Images Using Biomechanical Models, *X. Papademetris, O. Škrinjar, J.S. Duncan* 551
Methods for Modeling and Predicting Mechanical Deformations of the Breast under External Perturbations, *F.S. Azar, D.N. Metaxas, M.D. Schnall* 591

VOLUME XIII

SPECIAL VOLUME: NUMERICAL METHODS IN ELECTROMAGNETICS

Introduction to Electromagnetism, *W. Magnus, W. Schoenmaker* 3
Discretization of Electromagnetic Problems: The "Generalized Finite Differences" Approach, *A. Bossavit* 105
Finite-Difference Time-Domain Methods, *S.C. Hagness, A. Taflove, S.D. Gedney* 199
Discretization of Semiconductor Device Problems (I), *F. Brezzi, L.D. Marini, S. Micheletti, P. Pietra, R. Sacco, S. Wang* 317
Discretization of Semiconductor Device Problems (II), *A.M. Anile, N. Nikiforakis, V. Romano, G. Russo* 443
Modelling and Discretization of Circuit Problems, *M. Günther, U. Feldmann, J. ter Maten* 523
Simulation of EMC Behaviour, *A.J.H. Wachters, W.H.A. Schilders* 661
Solution of Linear Systems, *O. Schenk, H.A. van der Vorst* 755
Reduced-Order Modelling, *Z. Bai, P.M. Dewilde, R.W. Freund* 825

Special Volume: Numerical Methods in Electromagnetics

Introduction to Electromagnetism

Wim Magnus

IMEC, Silicon Process and Device Technology Division (SPDT), Quantum Device Modeling Group (QDM), Kapeldreef 75, Flanders, B-3001 Leuven, Belgium
E-mail address: wim.magnus@imec.be

Wim Schoenmaker

MAGWEL N.V., Kapeldreef 75, B-3001 Leuven, Belgium

List of symbols

$\mathbf{A}$	vector potential
$\mathbf{A}_{\text{EX}}$	external vector potential
$\mathbf{A}_{\text{IN}}$	induced vector potential
$\mathbf{B}$, $\mathbf{B}_{\text{IN}}$	magnetic induction
C	capacitance
c	speed of light, concentration
$\mathbf{dr}$	line element
$\mathbf{dS}$	surface element
$\mathrm{d}s$	elementary distance in Riemannian geometry
$\mathrm{d}\tau$	volume element
D_{n}	electron diffusion coefficient
D_{p}	hole diffusion coefficient
$\mathbf{D}$	electric displacement vector
E	energy
E_{F}	Fermi energy
$E_{\alpha\mathbf{k}}(W)$	electron energy

Numerical Methods in Electromagnetics
Special Volume (W.H.A. Schilders and E.J.W. ter Maten, Guest Editors) of
HANDBOOK OF NUMERICAL ANALYSIS, VOL. XIII
P.G. Ciarlet (Editor)

ISSN 1570-8659
DOI 10.1016/S1570-8659(04)13001-9

e	elementary charge
$\mathbf{E}$	electric field
$\mathbf{E}_{\mathrm{C}}$	conservative electric field
$\mathbf{E}_{\mathrm{EX}}$	external electric field
$\mathbf{E}_{\mathrm{IN}}$	induced electric field
$\mathbf{E}_{\mathrm{NC}}$	non-conservative electric field
$\mathbf{e}_{\mathrm{z}}$	unit vector along z-axis
$\mathbf{e}_{\phi}$	azimuthal unit vector
$F_{\mu\nu}$	electromagnetic field tensor
$f, f_{\mathrm{n}}, f_{\mathrm{p}}$	(Boltzmann) distribution function
G	conductance, generation rate
G_{Q}	quantized conductance
$g_{\mu\nu}$	metric tensor
H	Hamiltonian
$H_{\mathbf{p}'\mathbf{p}}$	Hamiltonian scattering matrix element
h	Planck's constant
$\hbar$	reduced Planck constant ($h/2\pi$)
I	electric current
i	imaginary unit
J_{G}	gate leakage current
$\mathbf{J}, \mathbf{J}_{\mathrm{n}}, \mathbf{J}_{\mathrm{p}}$	electric current density
$\mathbf{H}$	magnetic field intensity
k	wavenumber
k_{B}	Boltzmann's constant
$\mathbf{k}$	electron wave vector
L	inductance, Lagrangian, length
L_x, L_y	length
$\mathbf{L}$	total angular momentum
l	subband index, angular momentum quantum number, length
m	angular momentum quantum number
m, m_n	charge carrier effective mass
m_0	free electron mass
$m_{\mathrm{n}}, m_{\mathrm{g}\alpha x}, m_{\mathrm{g}\alpha y}, m_{\mathrm{g}\alpha z}, m_{1,\mathrm{ox},\alpha}, m_{2,\mathrm{ox},\alpha}, m_{3,\mathrm{ox},\alpha}, \ldots, m_{\alpha x}, m_{\alpha y}, m_{\alpha z}$	electron effective mass
m_{p}	hole effective mass
$\mathbf{M}$	magnetization vector
$\mathbf{m}$	magnetic moment
N	number of particles, coordinates or modes
N_A	acceptor doping density
n	electron concentration
$\mathbf{n}$	unit vector
$p, p_i, \mathbf{p}, \mathbf{p}_i, P, P_i, \mathbf{P}, \mathbf{P}_i, \ldots$	generalized momenta

p	hole concentration
$\mathbf{P}$	total momentum, electric polarization vector
$\mathbf{p}$	momentum, electric dipole moment
$q, q_i, \mathbf{q}, \mathbf{q}_i, Q,$ $Q_i, \mathbf{Q}, \mathbf{Q}_i, \ldots$	generalized coordinates
Q	electric charge
q_n	carrier charge
Q_{A}	electric charge residing in active area
R	resistance, recombination rate
R_{H}	Hall resistance
R_{K}	von Klitzing resistance
R_{L}	lead resistance
R_{Q}	quantized resistance
$R^{\mu}_{\rho\lambda\sigma}$	Riemann tensor
$\mathbb{R}$	set of real numbers
(r, θ, ϕ)	spherical coordinates
$\mathbf{r}, \mathbf{r}_n$	position vector
S	action, entropy
$S(\mathbf{p}, \mathbf{p}')$	transition rate
$\mathbf{S}$	Poynting vector
$\mathbf{S}_{\mathrm{n}}, \mathbf{S}_{\mathrm{p}}$	energy flux vector
t	time
T	lattice temperature
T_{n}	electron temperature
T_{p}	hole temperature
$\mathbf{T}, \mathbf{T}_{\alpha\beta}$	EM energy momentum tensor
U_{E}	electric energy
U_{M}	magnetic energy
U_{EM}	EM energy
$U(y), U(z)$	potential energy
u_{EM}	EM energy density
V	scalar electric potential
V_{H}	Hall voltage
V_{G}	gate voltage
$\mathbf{v}_n$	carrier velocity
$\mathbf{v}_{\mathrm{n}}, \mathbf{v}_{\mathrm{p}}$	drift velocity
$\mathbf{v}$	drift velocity, velocity field
$W, W_{\alpha l}(W)$	subband energy
$w, w_{\mathrm{n}}, w_{\mathrm{p}}$	carrier energy density
(x, y, z)	Cartesian coordinates
Y	admittance
Z	impedance
α	summation index, valley index, variational parameters
β	summation index, $1/k_{\mathrm{B}}T$

$\partial\Omega$	boundary surface of Ω
$\partial\Omega_\infty$	boundary surface of Ω_∞
ε_0	electric permittivity of vacuum
ε	electric permittivity
$\varepsilon_\mathrm{r}, \varepsilon_\mathrm{S}$	relative electric permittivity
Γ	closed curve inside a circuit
$\Gamma_{\alpha l}$	resonance width
$\Gamma^\alpha_{\mu\nu}$	affine connection
κ	wavenumber
$\kappa_\mathrm{n}, \kappa_\mathrm{p}$	thermal conductivity
Λ_EM	EM angular momentum density
μ	magnetic permeability, chemical potential
μ_0	magnetic permeability of vacuum
μ, μ_n, μ_p	carrier mobility
μ_r	relative magnetic permeability
Ω	connected subset of $\mathbb{R}^3$, volume, circuit region
Ω_∞	all space
ω	angular frequency
$\boldsymbol{\pi}_\mathrm{EM}$	EM momentum density
ρ	electric charge density
(ρ, ϕ, z)	cylindrical coordinates
σ	electrical conductivity, spin index
$\tau, \tau_0, \tau_\mathrm{e}, \tau_\mathbf{p}$, $\tau_\mathrm{en}, \tau_\mathrm{ep}$	relaxation time
$\tau_{\alpha l}$	resonance lifetime
χ	gauge function
χ_e	electric susceptibility
$\chi_k(y)$	wave function
χ_m	magnetic susceptibility
Φ_D	electric flux (displacement)
Φ_E	electric flux (electric field)
Φ_ex	external magnetic flux
Φ, Φ_M	magnetic flux
$\psi(\mathbf{r}), \psi_\alpha(\mathbf{r}, z)$, $\psi_{\alpha\mathbf{k}}(\mathbf{r}, z)$, $\phi_\alpha(W, z), \psi(x, y)$	wave function
$\boldsymbol{\nabla}$	gradient
$\boldsymbol{\nabla}\cdot$	divergence
$\boldsymbol{\nabla}\times$	curl
$\boldsymbol{\nabla}^2$	vectorial Laplace operator
∇^2	Laplace operator
$\mathcal{L}$	Lagrange density, inductance per unit length
V_ε	electromotive force

1. Preface

Electromagnetism, formulated in terms of the Maxwell equations, and quantum mechanics, formulated in terms of the Schrödinger equation, constitute the physical laws by which the bulk of natural experiences are described. Apart from the gravitational forces, nuclear forces and weak decay processes, the description of the physical facts starts with these underlying microscopic theories. However, knowledge of these basic laws is only the beginning of the process to apply these laws in realistic circumstances and to determine their quantitative consequences. With the advent of powerful computer resources, it has become feasible to extract information from these basic laws with unprecedented accuracy. In particular, the complexity of realistic systems manifests itself in the non-trivial boundary conditions, such that without computers, reliable calculation are beyond reach.

The ambition of physicists, chemists and engineers, to provide tools for performing calculations, does not only boost progress in technology but also has a strong impact on the formulation of the equations that represent the physics knowledge and hence provides a deeper understanding of the underlying physics laws. As such, computational physics has become a cornerstone of theoretical physics and we may say that without a computational recipe, a physics law is void or at least incomplete. Contrary to what is sometimes claimed, that after having found the unifying theory for gravitation and quantum theory, there is nothing left to investigate, we believe that physics has just started to flourish and there are wide fields of research waiting for exploration.

This volume is dedicated to the study of electrodynamic problems. The Maxwell equations appear in the form

$$\Delta(\text{field}) = \text{source}, \tag{1.1}$$

where Δ describes the near-by field variable correlation of the field that is induced by a source or field disturbance. Near-by correlations can be mathematically expressed by differential operators that probe changes going from one location the a neighboring one. It should be emphasized that "near-by" refers to space and time.

One could "easily" solve these equations by construction the inverse of the differential operator. Such an inverse is usually known as a Green function.

There are two main reasons that prevent a straightforward solution of the Maxwell equations. First of all, realistic structure boundaries may be very irregular, and therefore the corresponding boundary conditions cannot be implemented analytically. Secondly, the sources themselves may depend on the values of the fields and will turn the problem in a highly non-linear one, as may be seen from Eq. (1.1) that should be read as

$$\Delta(\text{field}) = \text{source}(\text{field}). \tag{1.2}$$

The bulk of this volume is dedicated to find solutions to equations of this kind. In particular, Chapters II, III, IV and V are dealing with above type of equations. A considerable amount of work deals with obtaining the details of the right-hand side of Eq. (1.2), namely how the source terms, being charges and currents depend in detail on the values of the field variables.

Whereas, the microscopic equation describe the physical processes in great detail, i.e., at every space–time point field and source variables are declared, it may be profitable to collect a whole bunch of these variables into a single basket and to declare for each basket a few representative variables as the appropriate values for the fields and the sources. This kind of reduction of parameters is the underlying strategy of circuit modeling. Here, the Maxwell equations are replaced by Kirchhoff's network equations. This is the starting point for Chapter VI.

The "basket" containing a large collection of fundamental degrees of freedom of field and source variables, should not be filled at random. Physical intuition suggests that we put together in one basket degrees of freedom that are "alike". Field and source variables at near-by points are candidates for being grabbed together, since physical continuity implies that a all elements in the basket will have similar values.[1]

The baskets are not only useful for simplifying the continuous equations. They are vital to the discretization schemes. Since any computer has only a finite memory storage, the continuous or infinite collection of degrees of freedom must be mapped onto a finite subset. This may be accomplished by appropriately positioning and sizing of all the baskets. This procedure is named "grid generation" and the construction of a good grid is often of great importance to obtain accurate solutions.

After having mapped the continuous problem onto a finite grid one may establish a set of algebraic equations connecting the grid variables (basket representatives) and explicitly reflecting the non-linearity of the original differential equations. The solution of large systems of non-linear algebraic equations is based on Newton's iterative method. To find the solution of the set of non-linear equations $\mathbf{F}(\mathbf{x}) = \mathbf{0}$, an initial guess is made: $\mathbf{x} = \mathbf{x}_{\text{init}} = \mathbf{x}_0$. Next the guess is (hopefully) improved by looking at the equation:

$$\mathbf{F}(\mathbf{x}_0 + \Delta\mathbf{x}) \simeq \mathbf{F}(\mathbf{x}_0) + \mathbf{A} \cdot \Delta\mathbf{x}, \tag{1.3}$$

where the matrix $\mathbf{A}$ is

$$\mathbf{A}_{ij} = \left(\frac{\partial F_i(\mathbf{x})}{\partial x_j}\right)_{\mathbf{x}_0}. \tag{1.4}$$

In particular, by assuming that the correction brings us close to the solution, i.e., $\mathbf{x}_1 = \mathbf{x}_0 + \Delta\mathbf{x} \simeq \mathbf{x}^*$, where $\mathbf{F}(\mathbf{x}^*) = 0$, we obtain that

$$\begin{aligned} 0 &= \mathbf{F}(\mathbf{x}_0) + \mathbf{A} \cdot \Delta\mathbf{x} \quad \text{or} \\ \Delta\mathbf{x} &= -\mathbf{A}^{-1} \cdot \mathbf{F}(\mathbf{x}_0). \end{aligned} \tag{1.5}$$

Next we repeat this procedure, until convergence is reached. A series of vectors, $\mathbf{x}_{\text{init}} = \mathbf{x}_0, \mathbf{x}_1, \mathbf{x}_2, \ldots, \mathbf{x}_{n-1}, \mathbf{x}_n = \mathbf{x}_{\text{final}}$, is generated, such that $|\mathbf{F}(\mathbf{x}_{\text{final}})| < \varepsilon$, where ε is some prescribed error criterion. In each iteration a large linear matrix problem of the type $\mathbf{A}|\mathbf{x}\rangle = |\mathbf{b}\rangle$ needs to be solved.

---

[1]It should be emphasized that such a picture works at the classical level. Quantum physics implies that near-by field point may take any value and the continuity of fields is not required.

2. The microscopic Maxwell equations

2.1. The microscopic Maxwell equations in integral and differential form

In general, any electromagnetic field can be described and characterized on a microscopic scale by two vector fields $\mathbf{E}(\mathbf{r}, t)$ and $\mathbf{B}(\mathbf{r}, t)$ specifying respectively the electric field and the magnetic induction in an arbitrary space point $\mathbf{r}$ at an arbitrary time t. All dynamical features of these vector fields are contained in the well-known Maxwell equations (MAXWELL [1954a], MAXWELL [1954b], JACKSON [1975], FEYNMAN, LEIGHTON and SANDS [1964a])

$$\nabla \cdot \mathbf{E} = \frac{\rho}{\varepsilon_0}, \tag{2.1}$$

$$\nabla \cdot \mathbf{B} = 0, \tag{2.2}$$

$$\nabla \times \mathbf{E} = -\frac{\partial \mathbf{B}}{\partial t}, \tag{2.3}$$

$$\nabla \times \mathbf{B} = \mu_0 \mathbf{J} + \varepsilon_0 \mu_0 \frac{\partial \mathbf{E}}{\partial t}. \tag{2.4}$$

They describe the spatial and temporal behavior of the electromagnetic field vectors and relate them to the sources of electric charge and current that may be present in the region of interest. Within the framework of a microscopic description, the electric charge density ρ and the electric current density $\mathbf{J}$ are considered spatially localized distributions residing in vacuum. As such they represent both mobile charges giving rise to macroscopic currents in solid-state devices, chemical solutions, plasmas, etc., and bound charges that are confined to the region of an atomic nucleus. In turn, the Maxwell equations in the above presented form explicitly refer to the values taken by $\mathbf{E}$ and $\mathbf{B}$ in vacuum and, accordingly, the electric permittivity ε_0 and the magnetic permeability μ_0 appearing in Eqs. (2.1) and (2.4) correspond to vacuum.

From the mathematical point of view, the solution of the differential equations (2.1)–(2.4) together with appropriate boundary conditions in space and time, should in principle unequivocally determine the fields $\mathbf{E}(\mathbf{r}, t)$ and $\mathbf{B}(\mathbf{r}, t)$. In practice however, analytical solutions may be achieved only in a limited number of cases and, due to the structural and geometrical complexity of modern electronic devices, one has to adopt advanced numerical simulation techniques to obtain reliable predictions of electromagnetic field profiles. In this light, the aim is to solve Maxwell's equations on a discrete set of mesh points using suitable discretization techniques which are often taking advantage of integral form of Maxwell's equations. The latter may be derived by a straightforward application of Gauss' and Stokes' theorems. In particular, one may integrate Eqs. (2.1) and (2.1) over a simply connected region $\Omega \in \mathbb{R}^3$ bounded by a closed surface $\partial\Omega$ to obtain

$$\int_{\partial\Omega} \mathbf{E}(\mathbf{r}, t) \cdot \mathbf{dS} = \frac{1}{\varepsilon_0} Q(t), \tag{2.5}$$

$$\int_{\partial\Omega} \mathbf{B}(\mathbf{r}, t) \cdot \mathbf{dS} = 0, \tag{2.6}$$

where $Q(t)$ denotes the instantaneous charge residing in the volume Ω, i.e.,

$$Q(t) = \int_{\Omega} \rho(\mathbf{r}, t)\, \mathrm{d}\tau. \tag{2.7}$$

Eq. (2.5) is nothing but Gauss' law stating that the total outward flux of the electric field threading the surface $\partial\Omega$ equals the total charge contained in the volume Ω up to a factor ε_0 whereas Eq. (2.6) reflects the absence of magnetic monopoles.

Similarly, introducing an arbitrary, open and simply connected surface Σ bounded by a simple, closed curve Γ, one may extract the induction law of Faraday and Ampère's law by integrating respectively Eqs. (2.3) and (2.4) over Σ:

$$\oint_{\Gamma} \mathbf{E}(\mathbf{r}, t) \cdot \mathbf{dr} = -\frac{\mathrm{d}\Phi_{\mathrm{M}}(t)}{\mathrm{d}t}, \tag{2.8}$$

$$\oint_{\Gamma} \mathbf{B}(\mathbf{r}, t) \cdot \mathbf{dr} = \mu_0 \left(I(t) + \varepsilon_0 \frac{\mathrm{d}\Phi_{\mathrm{E}}(t)}{\mathrm{d}t} \right). \tag{2.9}$$

The variables $\Phi_{\mathrm{E}}(t)$ and $\Phi_{\mathrm{M}}(t)$ are representing the time-dependent electric and magnetic fluxes piercing the surface Σ and are defined as:

$$\Phi_{\mathrm{E}}(t) = \int_{\Sigma} \mathbf{E}(\mathbf{r}, t) \cdot \mathbf{dS}, \tag{2.10}$$

$$\Phi_{\mathrm{M}}(t) = \int_{\Sigma} \mathbf{B}(\mathbf{r}, t) \cdot \mathbf{dS}, \tag{2.11}$$

while the circulation of the electric field around Γ is the instantaneous electromotive force $V_{\varepsilon}(t)$ along Γ is:

$$V_{\varepsilon}(t) = \oint_{\Gamma} \mathbf{E}(\mathbf{r}, t) \cdot \mathbf{dr}. \tag{2.12}$$

The right-hand side of Eq. (2.9) consists of the total current flowing through the surface Σ

$$I(t) = \int_{\Sigma} \mathbf{J}(\mathbf{r}, t) \cdot \mathbf{dS} \tag{2.13}$$

and the so-called displacement current which is proportional to the time derivative of the electric flux. The sign of the above line integrals depends on the orientation of the closed loop Γ, the positive traversal sense of which is uniquely defined by the orientation of the surface Σ imposed by the vectorial surface element $\mathbf{dS}$. Apart from this restriction it should be noted that the surface Σ can be chosen freely so as to extract meaningful physical information from the corresponding Maxwell equation. In particular, though being commonly labeled by the symbol Σ, the surfaces appearing in Faraday's and Ampère's laws (Eqs. (2.8)–(2.9)) will generally be chosen in a different way as can be illustrated by the example of a simple electric circuit. In the case of Faraday's law, one usually wants $\Phi_{\mathrm{M}}(t)$ to be the magnetic flux threading the circuit and therefore Σ would be chosen to "span" the circuit while Γ would be located in the interior of the circuit area. On the other hand, in order to exploit Ampère's law, the surface Σ should be pierced by the current density in the circuit in order to make $I(t)$ the current flowing through the circuit.

2.2. *Conservation laws*

Although a complete description of the electromagnetic field requires the full solution of the Maxwell equations in their differential form, one may extract a number of conservation laws may by simple algebraic manipulation. The differential form of the conservation laws takes the generic form

$$\nabla \cdot \mathbf{F} + \frac{\partial \mathbf{G}}{\partial t} = \mathbf{K}, \tag{2.14}$$

where $\mathbf{F}$ is the generalized flow tensor associated with the field $\mathbf{G}$ and $\mathbf{K}$ is related to any possible external sources or sinks.

2.2.1. *Conservation of charge – the continuity equation*

Taking the divergence of Eq. (2.4) and the time derivative of Eq. (2.1) and combining the resulting equations, one easily obtains the charge-current continuity equation expressing the conservation of electric charge:

$$\nabla \cdot \mathbf{J} + \frac{\partial \rho}{\partial t} = 0. \tag{2.15}$$

Integration over a closed volume Ω yields

$$\int_{\partial\Omega} \mathbf{J} \cdot \mathbf{dS} = -\frac{\partial}{\partial t} \int_{\Omega} \rho \, \mathrm{d}\tau, \tag{2.16}$$

which states that the total current flowing through the bounding surface $\partial\Omega$ equals the time rate of change of all electric charge residing within Ω.

2.2.2. *Conservation of energy – Poynting's theorem*

The electromagnetic energy flow generated by a time dependent electromagnetic field is most adequately represented by the well-known Poynting vector given by

$$\mathbf{S} = \frac{1}{\mu_0} \mathbf{E} \times \mathbf{B}. \tag{2.17}$$

Calculating the divergence of $\mathbf{S}$ and using the Maxwell equations, one may relate the Poynting vector to the electromagnetic energy density u_{EM} through the energy conservation law

$$\nabla \cdot \mathbf{S} + \frac{\partial u_{\mathrm{EM}}}{\partial t} = -\mathbf{J} \cdot \mathbf{E}, \tag{2.18}$$

which is also known as the Poynting theorem. The energy density u_{EM} is given by

$$u_{\mathrm{EM}} = \frac{1}{2}\left(\varepsilon_0 E^2 + \frac{B^2}{\mu_0}\right). \tag{2.19}$$

The energy conservation expressed in Eq. (2.18) refers to the total energy of the electromagnetic field and all charged particles contributing to the charge and current distributions. In particular, denoting the mechanical energy of the charged particles residing in the volume Ω by E_{MECH} one may derive for both classical and quantum mechanical systems that the work done per unit time by the electromagnetic field on the charged volume is given by

$$\frac{dE_{MECH}}{dt} = \int_{\Omega} \mathbf{J} \cdot \mathbf{E}\, d\tau. \tag{2.20}$$

Introducing the total electromagnetic energy associated with the volume Ω as $E_{EM} = \int_{\Omega} u_{EM}\, d\tau$ one may integrate Poynting's theorem to arrive at

$$\frac{d}{dt}(E_{MECH} + E_{EM}) = -\int_{\partial\Omega} \mathbf{S} \cdot \mathbf{dS}. \tag{2.21}$$

It should be emphasized that the above result also covers most of the common situations where the energy of the charged particles is relaxed to the environment through dissipative processes. The latter may be accounted for by invoking appropriate constitutive equations expressing the charge and current densities as linear or non-linear responses to the externally applied electromagnetic fields and other driving force fields. As an example, we mention Ohm's law, proposing a linear relation between the macroscopic electric current density and the externally applied electric field in a non-ideal conductor:

$$\mathbf{J}_M = \sigma \mathbf{E}_{EXT}. \tag{2.22}$$

Here, the conductivity σ is assumed to give an adequate characterization of all microscopic elastic and inelastic scattering processes that are responsible for the macroscopically observable electric resistance. The derivation of constitutive equations will be discussed in greater detail in Section 4.

2.3. Conservation of linear momentum – the electromagnetic field tensor

In an analogous way, an appropriate linear momentum density π_{EM} may be assigned to the electromagnetic field, which differs from the Poynting vector merely by a factor $\varepsilon_0\mu_0 = 1/c^2$:

$$\pi_{EM} = \varepsilon_0 \mathbf{E} \times \mathbf{B} = \frac{1}{c^2}\mathbf{S}. \tag{2.23}$$

The time evolution of π_{EM} is not only connected to the rate of change of the mechanical momentum density giving rise to the familiar Lorentz force term, but also involves the divergence of a second rank tensor $\mathbf{T}$ which is usually called the Maxwell stress tensor (JACKSON [1975], LANDAU and LIFSHITZ [1962]). The latter is defined most easily by its Cartesian components

$$\mathbf{T}_{\alpha\beta} = \varepsilon_0\left(\frac{1}{2}|\mathbf{E}|^2\delta_{\alpha\beta} - E_\alpha E_\beta\right) + \frac{1}{\mu_0}\left(\frac{1}{2}|\mathbf{B}|^2\delta_{\alpha\beta} - B_\alpha B_\beta\right) \tag{2.24}$$

with $\alpha, \beta = x, y, z$.

A straightforward calculation yields:

$$\frac{\partial \pi_{EM}}{\partial t} = -\rho\mathbf{E} - \mathbf{J} \times \mathbf{B} - \nabla \cdot \mathbf{T}. \tag{2.25}$$

2.3.1. Angular momentum conservation

The angular momentum density of the electromagnetic field and its corresponding flux may be defined respectively by the relations

$$\boldsymbol{\Lambda}_{\mathrm{EM}} = \mathbf{r} \times \boldsymbol{\pi}_{\mathrm{EM}}, \qquad \boldsymbol{\Gamma} = \mathbf{r} \times \mathbf{T}. \tag{2.26}$$

The conservation law that governs the angular momentum, reads

$$\frac{\partial \boldsymbol{\Lambda}_{\mathrm{EM}}}{\partial t} = -\mathbf{r} \times (\rho \mathbf{E} + \mathbf{J} \times \mathbf{B}) - \nabla \cdot \boldsymbol{\Gamma}. \tag{2.27}$$

3. Potentials and fields, the Lagrangian

Not only the Maxwell equations themselves but also all related conservation laws have been expressed with the help of two key observables describing the microscopic electromagnetic field, namely $\mathbf{E}$ and $\mathbf{B}$. Strictly speaking, all relevant physics involving electromagnetic phenomena can be described correctly and completely in terms of the variables $\mathbf{E}$ and $\mathbf{B}$ solely, and from this point of view there is absolutely no need of defining auxiliary potentials akin to $\mathbf{E}$ and $\mathbf{B}$. Nevertheless, it proves quite beneficial to introduce the scalar potential $V(\mathbf{r}, t)$ and the vector potential $\mathbf{A}(\mathbf{r}, t)$ as alternative electrodynamical degrees of freedom.

3.1. The scalar and vector potential

From the Maxwell equation $\nabla \cdot \mathbf{B} = 0$ and Helmholtz' theorem it follows that, within a simply connected region Ω, there exists a regular vector field $\mathbf{A}$ – called vector potential – such that

$$\mathbf{B} = \nabla \times \mathbf{A}, \tag{3.1}$$

which allows us to rewrite Faraday's law (2.8) as

$$\nabla \times \left(\mathbf{E} + \frac{\partial \mathbf{A}}{\partial t}\right) = 0. \tag{3.2}$$

The scalar potential V emerges from the latter equation and Helmholtz' theorem stating that, in a simply connected region Ω there must exist a regular scalar function V such that

$$\mathbf{E} = -\nabla V - \frac{\partial \mathbf{A}}{\partial t}. \tag{3.3}$$

Although V and $\mathbf{A}$ do not add new physics, there are at least three good reasons to introduce them anyway. First, it turns out that (JACKSON [1975], FEYNMAN, LEIGHTON and SANDS [1964a]) the two potentials greatly facilitate the mathematical treatment of classical electrodynamics in many respects. For instance, the choice of an appropriate gauge [2] allows one to convert the Maxwell equations into convenient wave equations for V and $\mathbf{A}$ for which analytical solutions can be derived occasionally. Moreover, the

[2] Gauge transformations will extensively be treated in Section 7.

scalar potential V provides an natural link to the concept of macroscopic potential differences that are playing a crucial role in conventional simulations of electric circuits.

Next, most quantum mechanical treatments directly invoke the "potential" picture to deal with the interaction between a charged particle and an electromagnetic field. In particular, adopting the path integral approach, one accounts for the presence of electric and magnetic fields by correcting the action functional S related to the propagation from $(\mathbf{r}_0, t_0)$ to $(\mathbf{r}_1, t_1)$ along a world line, according to

$$S[V, \mathbf{A}] = S[0, \mathbf{0}] + q\left(\int_{\mathbf{r}_1}^{\mathbf{r}_2} \mathbf{A} \cdot \mathbf{dr} - \int_{t_0}^{t_1} \mathrm{d}t\, V(\mathbf{r}, t)\right), \tag{3.4}$$

while the field-dependent Hamiltonian term appearing in the non-relativistic, one-particle Schrödinger equation $\mathrm{i}\hbar(\partial\psi/\partial t) = H\psi$, takes the form

$$H = \frac{1}{2m}(\mathbf{p} - q\mathbf{A})^2 + qV \tag{3.5}$$

with $\mathbf{p} = -\mathrm{i}\hbar\boldsymbol{\nabla}$. Furthermore, the canonical quantization of the electromagnetic radiation field leads to photon modes corresponding to the quantized transverse modes of the vector potential.

Finally, the third motivation for adopting scalar and vector potentials lies in the perspective of developing new numerical simulation techniques. For example, it was observed recently (SCHOENMAKER, MAGNUS and MEURIS [2002]) that the magnetic field generated by a steady current distribution may alternatively be extracted from the fourth Maxwell equation (Ampère's law),

$$\boldsymbol{\nabla} \times \boldsymbol{\nabla} \times \mathbf{A} = \mu_0 \mathbf{J} \tag{3.6}$$

by assigning discretized vector potential variables to the *links* connecting adjacent nodes. This will be discussed in Section 8.

3.2. *Gauge invariance*

In contrast to the electric field and the magnetic induction, neither the scalar nor the vector potential are uniquely defined. Indeed, performing a so-called gauge transformation

$$\begin{aligned} \mathbf{A}'(\mathbf{r}, t) &= \mathbf{A}(\mathbf{r}, t) + \boldsymbol{\nabla}\chi(\mathbf{r}, t), \\ V'(\mathbf{r}, t) &= V(\mathbf{r}, t) - \frac{\partial\chi(\mathbf{r}, t)}{\partial t}, \end{aligned} \tag{3.7}$$

where the gauge field $\chi(\mathbf{r}, t)$ is an arbitrary regular, real scalar field, one clearly observes that the potentials are modified while the electromagnetic fields $\mathbf{E}(\mathbf{r}, t)$ and $\mathbf{B}(\mathbf{r}, t)$ remain unchanged. Similarly, any quantum mechanical wave function $\psi(\mathbf{r}, t)$ transforms according to

$$\begin{aligned} \psi'(\mathbf{r}, t) &= \psi(\mathbf{r}, t) \exp\bigl(\mathrm{i}q\chi(\mathbf{r}, t)\bigr), \\ \psi'^*(\mathbf{r}, t) &= \psi^*(\mathbf{r}, t) \exp\bigl(-\mathrm{i}q\chi(\mathbf{r}, t)\bigr), \end{aligned}$$

whereas the quantum mechanical probability density $|\psi(\mathbf{r}, t)|^2$ and other observable quantities are invariant under a gauge transformation, as required.

3.3. Lagrangian for an electromagnetic field interacting with charges and currents

While the Maxwell equations are the starting point in the so-called *inductive approach*, one may alternatively adopt the *deductive approach* and try to "derive" the Maxwell equations from a proper variational principle. As a matter of fact it is possible indeed to postulate a Lagrangian density $\mathcal{L}(\mathbf{r}, t)$ and an action functional $S[\mathcal{L}, t_0, t_1] = \int_{t_0}^{t_1} \mathcal{L}(\mathbf{r}, t)\, \mathrm{d}\tau$ such that the Maxwell equations emerge as the Euler–Lagrange equations that make the action

$$\delta S = 0 \tag{3.8}$$

stationary. While such a "derivation" is of utmost importance for the purpose of basic understanding from the theoretical point of view, the Lagrangian and Hamiltonian formulation of electromagnetism may look redundant when it comes to numerical computations. However, we have quoted the Lagrangian density of the electromagnetic field not only for the sake of completeness but also to illustrate the numerical potential of the underlying variational principle.

The Lagrangian density for the interacting electromagnetic field is conventionally postulated as a quadratic functional of the scalar and vector potential and their derivatives:

$$\mathcal{L} = \frac{1}{2}\varepsilon_0 \left| \nabla V + \frac{\partial \mathbf{A}}{\partial t} \right|^2 - \frac{1}{2\mu_0} |\nabla \times \mathbf{A}|^2 + \mathbf{J} \cdot \mathbf{A} - \rho V, \tag{3.9}$$

where the field variables V and $\mathbf{A}$ are linearly coupled to the charge and current distribution ρ and $\mathbf{J}$.

It is now straightforward to obtain the Maxwell equations as the Euler–Lagrange equations corresponding to Eq. (3.9) provided that the set of field variables is chosen to be either V or A_α. The first possibility gives rise to

$$\sum_{\beta=x,y,z} \frac{\partial}{\partial x_\beta} \left[\frac{\partial \mathcal{L}}{\partial \left(\frac{\partial V}{\partial x_\beta} \right)} i \right] + \frac{\partial}{\partial t} \left[\frac{\partial \mathcal{L}}{\partial \left(\frac{\partial V}{\partial t} \right)} \right] = \frac{\partial \mathcal{L}}{\partial V}. \tag{3.10}$$

Inserting all non-zero derivatives, we arrive at

$$\varepsilon_0 \sum_\beta \frac{\partial}{\partial x_\beta} \left(\frac{\partial V}{\partial x_\beta} + \frac{\partial A_\beta}{\partial t} \right) = -\rho, \tag{3.11}$$

which clearly reduces to the first Maxwell equation

$$\varepsilon_0 \nabla \cdot \mathbf{E} = \rho \quad \text{(Gauss' law)}. \tag{3.12}$$

Similarly, the three Euler–Lagrange equations

$$\sum_{\beta=x,y,z} \frac{\partial}{\partial x_\beta} \left[\frac{\partial \mathcal{L}}{\partial \left(\frac{\partial A_\alpha}{\partial x_\beta} \right)} \right] + \frac{\partial}{\partial t} \left[\frac{\partial \mathcal{L}}{\partial \left(\frac{\partial A_\alpha}{\partial t} \right)} \right] = \frac{\partial \mathcal{L}}{\partial A_\alpha}, \quad \alpha = x, y, z \tag{3.13}$$

lead to the fourth Maxwell equation

$$\frac{1}{\mu_0} \left(\nabla \times \mathbf{B} - \varepsilon_0 \frac{\partial \mathbf{E}}{\partial t} \right) = \mathbf{J} \quad \text{(Ampère–Faraday's law)}. \tag{3.14}$$

It should be noted that, within the deductive approach, the electric and magnetic field vectors are *defined* by the equations

$$\mathbf{E} = -\nabla V - \frac{\partial \mathbf{A}}{\partial t}, \qquad \mathbf{B} = \nabla \times \mathbf{A}, \tag{3.15}$$

whereas the latter are directly resulting from the Maxwell equations in the inductive approach. Mutatis mutandis, the two remaining Maxwell equations $\nabla \cdot \mathbf{B} = 0$ and $\nabla \times \mathbf{E} = -\partial \mathbf{B}/\partial t$ are a direct consequence of the operation of the vector identities (A.34) and (A.35) on Eqs. (3.15). It should also be noted that the Lagrangian density may be written as

$$\mathcal{L} = \frac{1}{2}\varepsilon_0 \mathbf{E}^2 - \frac{1}{2\mu_0}\mathbf{B}^2. \tag{3.16}$$

So far, we have considered the Maxwell equations from the perspective that the charge and the current densities are given and the fields should be determined. However, as was already mentioned in the introduction, the charge and current densities may also be influenced by the fields. In order to illustrate the opposite cause–effect relation, we consider the Lagrangian of N charged particles moving in an electromagnetic field. The Lagrangian is

$$L = \sum_{n=1}^{N} \frac{1}{2} m_n v_n^2 + \frac{1}{2}\int \mathrm{d}\tau \left(\varepsilon_0 \mathbf{E}^2 - \frac{1}{\mu_0}\mathbf{B}^2\right) - \int \mathrm{d}\tau\, \rho V + \int \mathrm{d}\tau\, \mathbf{J} \cdot \mathbf{A}, \tag{3.17}$$

where we defined the charge and current densities as

$$\begin{aligned} \rho(\mathbf{r}, t) &= \sum_{n=1}^{N} q_n \delta(\mathbf{r} - \mathbf{r}_n), \\ \mathbf{J}(\mathbf{r}, t) &= \sum_{n=1}^{N} q_n \mathbf{v}_n \delta(\mathbf{r} - \mathbf{r}_n) \end{aligned} \tag{3.18}$$

and the particles' velocities as $\mathbf{v}_n = \mathrm{d}\mathbf{r}_n/\mathrm{d}t$. Applying the Euler–Lagrange equations:

$$\frac{\mathrm{d}}{\mathrm{d}t}\left(\frac{\partial L}{\partial \mathbf{v}_n}\right) - \frac{\partial L}{\partial \mathbf{r}_n} = 0, \tag{3.19}$$

gives

$$m_n \frac{\mathrm{d}^2 \mathbf{r}_n}{\mathrm{d}t^2} = q_n \mathbf{E}(\mathbf{r}_n, t) + q_n \mathbf{v}_n \times \mathbf{B}(\mathbf{r}_n, t). \tag{3.20}$$

The last term is recognized as the Lorentz force.

3.4. Variational calculus

Although the numerical implementation of the variational principle leading to the Maxwell equations is not a common practice in numerical analysis, it may nevertheless turn out to be a useful approximation technique for particular classes of problems.

The exact solution of the Euler–Lagrange equations determines an extremum of the action functional which becomes stationary with respect to *any arbitrary* variations of the field functions that meet the boundary conditions invoked. On the other hand, being inspired by physical intuition or analogy with similar problems, one may be able to propose a class of trial functions satisfying the boundary conditions and exhibiting the expected physical behavior. If these trial functions can be characterized by one or more adjustable parameters $\alpha_1, \ldots, \alpha_n$, then one may calculate the values of $\alpha_1, \ldots, \alpha_n$ for which the action integral becomes stationary. Although the corresponding numerical value of the action will generally differ from the true extremum that is attained by the exact solution, the resulting trial function may surprisingly lead to rather accurate estimates of the physical quantities of interest. A nice example of this phenomenon is given in FEYNMAN, LEIGHTON and SANDS [1964a] (Part II, Chapter 19) where a variational calculation of the capacitance of a cylindrical coaxial cable is presented and compared with the exact formula for various values of the inner and outer radii of the cable.

As an illustration, we have worked out the case of a long coaxial cable with a square cross section, for which the inductance is estimated within the framework of variational calculus.

Consider an infinitely long coaxial cable centered at the z-axis, consisting of a conducting core, a magnetic insulator and a conducting coating layer. Both the core and the coating layer have a square cross section of sizes a and b, respectively. The core carries a current I in the z-direction which is flowing back to the current source through the coating layer, thereby closing the circuit as depicted in Fig. 3.1. Neglecting skin effects we assume that the current density is strictly localized at the outer surface of the core and the inner surface of the coating layer, respectively. Moreover, the translational symmetry in the z-direction reduces the solution of Maxwell's equations essentially to a two-dimensional problem whereas the square symmetry of the cable allows us to divide an arbitrary cable cross-section into four identical triangles and to work out the solution for just one triangular area. In particular, we will focus on the region Δ (see Fig. 3.2) bounded by

$$x \geqslant 0; \quad -x \leqslant y \leqslant x. \tag{3.21}$$

FIG. 3.1. Infinitely long coaxial cable carrying a stationary surface current.

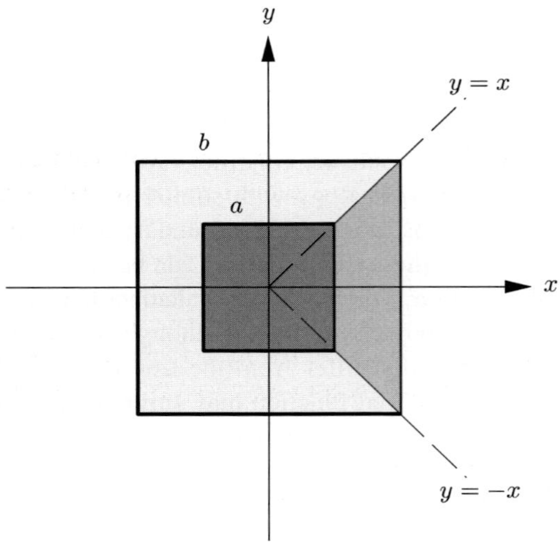

FIG. 3.2. Cross section of the coaxial cable.

Within this region, the current density takes the form

$$\begin{aligned} \mathbf{J}(x, y) &= J_z(x)\mathbf{e}_z, \\ J_z(x) &= \frac{I}{4}\left[\frac{1}{a}\delta\left(x - \frac{a}{2}\right) - \frac{1}{b}\delta\left(x - \frac{b}{2}\right)\right], \end{aligned} \tag{3.22}$$

where the factor 4 indicates that the region Δ accounts for only a quarter of the total current flowing through the cable's cross-section. The particular shape of the current density reflects the presence of perfect shielding requiring that the magnetic field be vanishing for $x < a/2$ and $x > b/2$ whereas B_y should abruptly jump[3] to a non-zero value at $x = a/2 + \varepsilon$ and $x = b/2 - \varepsilon$ where $\varepsilon \to 0^+$. The non-zero limiting values of B_y are used to fix appropriate boundary for B_y simply by integrating the z-component of the Maxwell equation $\boldsymbol{\nabla} \times \mathbf{B} = \mathbf{0}$ over the intervals $[a/2 - \varepsilon, a/2 + \varepsilon]$ and $[b/2 - \varepsilon, b/2 + \varepsilon]$, respectively. For instance, from

$$\int_{a/2-\varepsilon}^{a/2+\varepsilon} \mathrm{d}x \left[\frac{\partial B_y(x, y)}{\partial x} - \frac{\partial B_x(x, y)}{\partial y}\right] = \frac{\mu_0 I}{4a} \tag{3.23}$$

and

$$B_y(x, y) = 0 \quad \text{for } x < a/2, \tag{3.24}$$

it follows that

$$\lim_{x \to 1/2a^+} B_y(x, y) = \frac{\mu_0 I}{4a} \tag{3.25}$$

[3] If the current density were smeared out, the magnetic field would gradually tend to zero inside the core and the coating layer.

and similarly

$$\lim_{x \to 1/2b^-} B_y(x, y) = \frac{\mu_0 I}{4b}. \tag{3.26}$$

Finally, the boundary conditions reflecting the connection of adjacent triangular areas are directly dictated by symmetry considerations requiring that the magnetic field vector be orthogonal to the segments $y^2 = x^2$:

$$B_y(x, \pm x) = \mp B_x(x, \pm x) \quad \text{for } \frac{a}{2} < x < \frac{b}{2}. \tag{3.27}$$

Next, we propose a set of trial functions for B_x and B_y that meet the above boundary conditions as well as the symmetry requirement that B_x change sign at $y = 0$:

$$B_x(x, y) = \frac{-\mu_0 I}{8} y \left[\frac{1}{x^2} + \alpha \frac{x^2 - y^2}{a^4} \right], \tag{3.28}$$

$$B_y(x, y) = \frac{\mu_0 I}{8x}, \tag{3.29}$$

if (x, y) lies inside the trapezoid $a/2 < x < b/2$, $|y| \leqslant x$ and $B_x = B_y = 0$ elsewhere. The parameter α is a variational parameter that will be chosen such that the action functional attains a minimum with respect to the class of trial functions generated by Eqs. (3.28) and (3.29). Since no dynamics is involved in the present problem, the time integral occurring in the action integral becomes irrelevant and the least action principle amounts to the minimization of the magnetic energy stored in the insulator.

Anticipating the discussions of Chapter VI, we may calculate the inductance L of an electric circuit by equating $1/2LI^2$ to the magnetic energy stored in the circuit:

$$\begin{aligned} \frac{1}{2} L I^2 = U_M &= \frac{1}{2\mu_0} \int_\Omega \mathrm{d}\tau \, |\mathbf{B}|^2 \\ &= \frac{4l}{2\mu_0} \int_{a/2}^{b/2} \mathrm{d}x \int_{-x}^{x} \mathrm{d}y \left[B_x^2(x, y) + B_y^2(x, y) \right], \end{aligned} \tag{3.30}$$

where Ω refers to the volume of the insulator and l is the length of the cable and the pre-factor 4 accounts for the identical contributions from the four identical trapezoidal areas. From Eq. (3.30) we obtain the following expression for $\mathcal{L}$, the inductance per unit length:

$$\mathcal{L} \equiv \frac{L}{l} = \frac{4}{\mu_0 I^2} \int_{a/2}^{b/2} \mathrm{d}x \int_{-x}^{x} \mathrm{d}y \left[B_x^2(x, y) + B_y^2(x, y) \right]. \tag{3.31}$$

Since the trial functions defined in Eqs. (3.28) and (3.29) are chosen to meet the boundary conditions, the variational problem is reduced to the minimization of U_M, or equivalently, $\mathcal{L}$ with respect to α. The calculation of $\mathcal{L}(\alpha)$ is elementary and here we only quote the final result:

$$\frac{\mathcal{L}(\alpha)}{\mu_0} = \frac{1}{6} \log u + \frac{(u^4 - 1)}{215040} \left[112\alpha + (u^4 + 1)\alpha^2 \right] \tag{3.32}$$

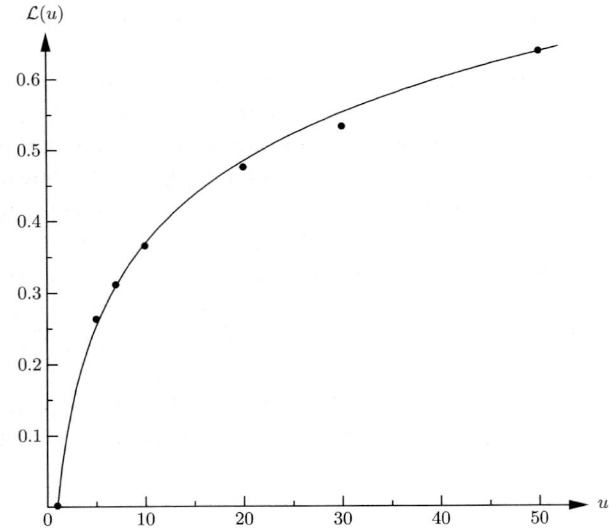

FIG. 3.3. Inductance per unit length: variational estimate (full line) versus numerical evaluation (•).

with $u = b/a$. Clearly, the required minimum corresponding to $\partial\mathcal{L}(\alpha)/\partial\alpha = 0$, is obtained for

$$\alpha = -\frac{56}{1+u^4}. \tag{3.33}$$

Finally, inserting the above result into Eq. (3.31), we obtain the inductance per length as follows:

$$\frac{\mathcal{L}}{\mu_0} = \frac{1}{6}\log u - \frac{7}{480}\frac{(u^4-1)}{(u^4+1)}. \tag{3.34}$$

The variational result is plotted against the "exact" numerical evaluation of the inductance in Fig. 3.3. Being a variational estimate, Eq. (3.34) provides a rigorous upper bound for the true inductance.

4. The macroscopic Maxwell equations

4.1. Constitutive equations

The Maxwell equations contain source terms being the charge densities and the currents. In this section we will present the physics behind these terms and derive their precise form. We will see that the charge and current formulas depend very much on the medium in which these charges and currents are present. For solid media we can distinguish between insulators, semiconductors and conductors. The corresponding expressions differ considerably for the different materials. Furthermore in the gas phase or the liquid phase again other expressions will be found. In the latter case we enter the realm of plasma physics and magnetohydrodynamics. These topics are beyond the present scope.

Before starting to derive the constitutive equations we need to address another machinery, namely statistical physics. From a philosophical point of view, statistical physics is a remarkable part of natural science. It does not contribute to a deeper understanding of the fundamental forces of nature, yet it introduces a fundamental constant of nature, the Boltzmann constant $k_B = 1.3805 \times 10^{23}$ J/K. Furthermore, there has been a discussion over several generations of physicists, debating the reality of irreversibility. The dispute in a nutshell is whether the idea of entropy increase is a sensible one, considering the fact that the microscopic dynamics is time-reversal invariant. As has been demonstrated in MAGNUS and SCHOENMAKER [1993] the time reversal invariance is broken in the limit of infinitely many degrees of freedom. In practice, 'infinity' is already reached for 30 degrees of freedom in the study of MAGNUS and SCHOENMAKER [1993]. Therefore, we believe that the dispute is settled and statistical physics is 'solid as a rock'.

4.2. Boltzmann transport equation

In this section we will consider the assumptions that lead to the Boltzmann transport equation. This equation serves as the starting point for deriving the formules for the constitutive equation for the currents in metals, semiconductors and insulators.

When describing the temporal evolution of many particles, one it not interested in the detailed trajectory of each individual particle in space and time. First of all, the particles are identical and therefore their trajectories are interchangeable. Secondly, the individual trajectories exhibit stochastic motion on a short time scale that is irrelevant on a larger time scale. In a similar way, the detailed knowledge at a short length scale is also not of interest for understanding the behavior at larger length scales. Thus we must obtain a procedure for eliminating the short-distance fluctuations from the description of the many particle system. In fact, to arrive at a manageable set of equations such a procedure should also reduce the number of variables for which the evolution equations need to be formulated.

There are a number of schemes that allow for such a reduction. All methods apply some kind of coarse graining, i.e., a number of microscopic variables are bundled and are represented by a single effective variable. In this section, we discuss the method that is due to Boltzmann and that leads to the Boltzmann transport equation.

Consider N particles with generalized coordinates $\mathbf{q}_i$, $i = 1, \ldots, N$, and generalized momenta $\mathbf{p}_i$, $i = 1, \ldots, N$. Each particle can be viewed as a point of the so-called μ-space, a six-dimensional space, spanned by the coordinates $\mathbf{q}$, $\mathbf{p}$. In this light, the N particles will trace out N curves in phase space as time evolves. Let us now subdivide the phase space into cells of size $\Delta\Omega = \Delta q^3 \Delta p^3$. Each cell can be labeled by a pair of coordinates $\mathbf{Q}_i$ and momenta $\mathbf{P}_i$. The number of particles that is found in the cell Ω_i is given by $f(\mathbf{P}_i, \mathbf{Q}_i, t)$. We can illustrate the role of the cell size setting $\Delta\Omega$. The function $f(\mathbf{P}_i, \mathbf{Q}_i, t)$ is given by

$$f(\mathbf{P}_i, \mathbf{Q}_i, t) = \sum_{i=1}^{N} \int_{\Delta\Omega} \mathrm{d}^3 p \, \mathrm{d}^3 q \, \delta\big(\mathbf{p} - \mathbf{p}_i(t)\big)\delta\big(\mathbf{q} - \mathbf{q}_i(t)\big). \tag{4.1}$$

We can illustrate the role of the coarse-graining scaling parameter $\Delta\Omega$. If we take the size of the cell arbitrary small then we will occasionally find a particle in the cell. Such a choice of $\Delta\Omega$ corresponds to a fully microscopic description of the mechanical system and we will not achieve a reduction in degrees of freedom.

On the other hand, if we choose $\Delta\Omega$ arbitrary large, then all degrees of freedom are represented by one (static) point f, and we have lost all knowledge of the system. Therefore $\Delta\Omega$ must be chosen such that it acts as the "communicator" between the microscopic and macroscopic worlds. This connection can be obtained by setting the size of the cell large enough such that each cell contains a number of particles. Within each cell the particles are considered to be in a state of thermal equilibrium. Thus for each cell a temperature T_i and a chemical potential μ_i can be given. The (local) thermal equilibrium is realized if there occurs a thermalization, i.e., within the cell collisions should occur within a time interval Δt. Therefore, the cell should be chosen such that its size exceeds at least a few mean-free path lengths.

On the macroscopic scale, the cell labels $\mathbf{P}_i$ and $\mathbf{Q}_i$ are smooth variables. The cell size is the denoted by the differential $\mathrm{d}\Omega = \mathrm{d}^3 p \, \mathrm{d}^3 q$. Then we may denote the distribution functions as $f(\mathbf{P}, \mathbf{Q}, t) \equiv f(\mathbf{p}, \mathbf{q}, t)$. From the distribution function $f(\mathbf{p}, \mathbf{q}, t)$, the particle density function can be obtained from

$$\int \mathrm{d}^3 p \, f(\mathbf{p}, \mathbf{q}, t) = \rho(\mathbf{q}, t). \tag{4.2}$$

As time progresses from t to $t + \delta t$, all particles in a cell at $\mathbf{p}$, $\mathbf{q}$ will be found in a cell at $\mathbf{p}'$, $\mathbf{q}'$, provided that no collisions occurred. Hence

$$f(\mathbf{p}, \mathbf{q}, t) \, \mathrm{d}^3 p \, \mathrm{d}^3 q = f(\mathbf{p} + \mathbf{F}\delta t, \mathbf{q} + \mathbf{v}\delta t, t + \delta t) \, \mathrm{d}^3 p' \, \mathrm{d}^3 q'. \tag{4.3}$$

According to Liouville's theorem (FOWLER [1936], HUANG [1963]), the two volume elements $\mathrm{d}^3 p \, \mathrm{d}^3 q$ and $\mathrm{d}^3 p' \, \mathrm{d}^3 q'$ are equal, which may appear evident if there are no external forces. If there are forces that do not explicitly depend on time, any cubic element deforms into a parallelepiped but with the same volume as the original cube. Taking also into account the effect of collisions that may kick particles in or out of the cube in the time interval δt, we arrive at the following equation for the distribution function

$$\left(\frac{\partial}{\partial t} + \frac{\mathbf{p}}{m} \cdot \nabla_{\mathbf{q}} + \mathbf{F} \cdot \nabla_{\mathbf{p}}\right) f(\mathbf{p}, \mathbf{q}, t) = \left(\frac{\partial f}{\partial t}\right)_{\mathrm{c}}, \tag{4.4}$$

where the "collision term" $(\partial f/\partial t)_{\mathrm{c}}$ *defines* the effects of scattering. A quantitative estimate of this term is provided by studying the physical mechanisms that contribute to this term. As carriers traverse, their motion is frequently disturbed by scattering due to collisions with impurity atoms, phonons, crystal defects, other carriers or even with foreign particles (cosmic rays). The frequency at which such events occur can be estimated by assuming that these events take place in an uncorrelated way; in other words two such events are statistically independent. Each physical mechanism is described by an interaction Hamiltonian or potential function, $U_S(\mathbf{r})$ that describes the details of the scattering process. The matrix element that describes the transition from a carrier in a state with momentum $|\mathbf{p}\rangle$ to a state with momentum $|\mathbf{p}'\rangle$ is

$$H_{\mathbf{p}'\mathbf{p}} = \frac{1}{\Omega}\int \mathrm{d}\tau\, \mathrm{e}^{-\frac{\mathrm{i}}{\hbar}\mathbf{p}'\cdot\mathbf{r}} U_S(\mathbf{r}) \mathrm{e}^{\frac{\mathrm{i}}{\hbar}\mathbf{p}\cdot\mathbf{r}}, \tag{4.5}$$

where Ω is a box that is used to count the number of momentum states. This box is of the size Δq^3 as defined above.

The evaluation of the transition amplitude relies on Fermi's Golden Rule. The transition rate then becomes

$$S(\mathbf{p}', \mathbf{p}) = \frac{2\pi}{\hbar} |H_{\mathbf{p}'\mathbf{p}}|^2 \delta\big(E(\mathbf{p}') - E(\mathbf{p}) - \Delta E\big), \tag{4.6}$$

where ΔE is the change in energy related to the transition. If $\Delta E = 0$, the collision is *elastic* The collision term is the result of the balance between kick-in and kick-out of the transitions that take place per unit time:

$$\left(\frac{\partial f}{\partial t}\right)_{\mathrm{c}} = \sum_{\mathbf{p}'} \big(S(\mathbf{p}', \mathbf{p}) f(\mathbf{q}, \mathbf{p}', t) - S(\mathbf{p}, \mathbf{p}') f(\mathbf{q}, \mathbf{p}, t)\big). \tag{4.7}$$

Once more it should be emphasized that although this balance picture is heuristic, looks reasonable and leads to a description of irreversibility it does not explain the latter. The collision term can be further fine-tuned to mimic the consequences of Pauli's exclusion principle by suppression of multiple occupation of states:

$$\begin{aligned}\left(\frac{\partial f}{\partial t}\right)_{\mathrm{c}} = \sum_{\mathbf{p}'} \big[&S(\mathbf{p}', \mathbf{p}) f(\mathbf{q}, \mathbf{p}', t)\big(1 - f(\mathbf{q}, \mathbf{p}, t)\big) \\ &- S(\mathbf{p}', \mathbf{p}) f(\mathbf{q}, \mathbf{p}, t)\big(1 - f(\mathbf{q}, \mathbf{p}', t)\big)\big].\end{aligned} \tag{4.8}$$

4.3. Currents in metals

In many materials, the conduction current that flows due to the presence of an electric field, $\mathbf{E}$, is proportional to $\mathbf{E}$, so that

$$\mathbf{J} = \sigma \mathbf{E}, \tag{4.9}$$

where the electrical conductivity σ is a material parameter. In metallic materials, Ohm's law, Eq. (4.9) is accurate. However, a fast generalization should be allowed for anisotropic conducting media. Moreover, the conductivity may depend on the frequency mode such that we arrive at

$$\mathbf{J}_i(\omega) = \boldsymbol{\sigma}_{ij}(\omega) \mathbf{E}_j(\omega) \tag{4.10}$$

and $\boldsymbol{\sigma}$ is a second-rank tensor. The derivation of Ohm's law from the Boltzmann transport equation was initiated by Drude. In Drude's model (DRUDE [1900a], DRUDE [1900b]), the electrons move as independent particles in the metallic region suffering from scattering during their travel from the cathode to the anode. The distribution function is assumed to be of the following form:

$$f(\mathbf{q}, \mathbf{p}, t) = f_0(\mathbf{q}, \mathbf{p}, t) + f_\mathrm{A}(\mathbf{q}, \mathbf{p}, t), \tag{4.11}$$

where f_0 is the equilibrium distribution function, being symmetric in the momentum variable $\mathbf{p}$, and f_A is a perturbation due to an external field that is anti-symmetric in the momentum variable. The collision term in Drude's model is crudely approximated by the following assumptions:

- only kick-out,
- all $S(\mathbf{p}, \mathbf{p}')$ are equal,
- no Pauli exclusion principle,
- no carrier heating, i.e., low-field transitions.

The last assumption implies that only the anti-symmetric part participates in the collision term (LUNDSTROM [1999]). Defining a characteristic time $\tau_\mathbf{p}$, the momentum-relaxation time, we find that

$$\left(\frac{\partial f}{\partial t}\right)_\mathrm{c} = -\frac{f_\mathrm{A}}{\tau_\mathbf{p}} \quad \text{and} \quad \frac{1}{\tau_\mathbf{p}} = \sum_{\mathbf{p}'} S(\mathbf{p}, \mathbf{p}'). \tag{4.12}$$

Furthermore, assuming a constant electric field $\mathbf{E}$ and a spatially uniform charge electron distribution, the Boltzmann transport equation becomes

$$-q\mathbf{E} \cdot \nabla(f_0 + f_\mathrm{A}) = -\frac{f_\mathrm{A}}{\tau_\mathbf{p}}. \tag{4.13}$$

Finally, if we assume that $f \simeq f_0 \propto \exp(-p^2/2mk_\mathrm{B}T)$ then

$$f_\mathrm{A} = q\tau_\mathbf{p}\mathbf{E} \cdot \nabla_\mathbf{p} f_0 = \frac{q\tau_\mathbf{p}}{k_\mathrm{B}T}\mathbf{E} \cdot \mathbf{v} f_0. \tag{4.14}$$

Another way of looking at this result is to consider $f = f_0 + f_A$ as a Taylor series for f_0:

$$f(\mathbf{p}) = f_0(\mathbf{p}) + (q\tau_\mathbf{p}\mathbf{E}) \cdot \nabla_\mathbf{p} f_0(\mathbf{p}) + \cdots = f_0(\mathbf{p} + q\tau_\mathbf{p}\mathbf{E}). \tag{4.15}$$

This is a *displaced* Maxwellian distribution function in the direction opposite to the applied field $\mathbf{E}$. The current density is $\mathbf{J} = qn\mathbf{v}$ follows from the averaged velocity

$$\mathbf{J} = qn\frac{\int \mathrm{d}^3p\,(\mathbf{p}/m) f(\mathbf{p})}{\int \mathrm{d}^3p\, f(\mathbf{p})} = \frac{q^2\tau_\mathbf{p}}{m} n\mathbf{E}. \tag{4.16}$$

The electron mobility, μ_n, is defined as the proportionality constant in the constitutive relation $\mathbf{J} = q\mu_n n\mathbf{E}$, such that

$$\mu_n = \frac{q\tau_\mathbf{p}}{m}. \tag{4.17}$$

So we have been able to "deduce" Ohm's law from the Boltzmann transport equation.

It is a remarkable fact that Drude's model is quite accurate, given the fact that no reference was made to Pauli's exclusion principle and the electron waves do not scatter while traveling in a perfect crystal lattice. Indeed, it was recognized by Sommerfeld that ignoring these effects will give rise to errors in the calculation of the order of 10^2, but both these errors cancel. Whereas Drude's model explains the existence of resistance, more advanced models are needed to accommodate for the non-linear current-voltage characteristics, the frequency dependence and the anisotropy of the conductance for some materials. A "modern" approach to derive conductance properties was initiated by KUBO [1957]. His theory naturally leads to the inclusion of anisotropy, non-linearity and frequency dependence. Kubo's approach also serves as the starting point to calculate transport properties in the quantum theory of many particles at finite temperature (MAHAN [1981]). These approaches start from the quantum-Liouville equation and the Gibb's theory of assembles on phase space. The latter has a more transparent generalization to the many-particle Hilbert space of quantum states.

Instead of reproducing here text book presentations of these various domains of physics, we intend to give the reader some sense of alertness, that the validity of some relations is limited. In order to push back the restrictions, one needs to re-examine the causes of the limitations. Improved models can be *guessed* by widening the defining expression as in the foregoing case where the scalar σ was substituted by the conductivity tensor $\boldsymbol{\sigma}$. The consequences of these guesses can be tested in simulation experiments. Therefore, simulation plays an important role to obtain improved models.

In the process of purchasing model improvements a few guidelines will be of help. First of all, the resulting theory should respect some fundamental physical principles. The *causality* principle is an important example. It states that there is a retarded temporal relation between cause and effect. The causality principle is a key ingredient to derive the Kramers–Kronig relations, that put severe limitations on the real and imaginary parts of the material parameters. Yet these relationships are not sufficient to determine the models completely, but one needs to include additional physical models.

4.4. *Charges in metals*

Metallic materials are characterized as having an appreciable conductivity. Any excess free charge distribution in the metal will decay exponentially to zero in a small time. Combining Gauss' law with the current continuity equation

$$\nabla \cdot (\varepsilon \mathbf{E}) = \rho, \qquad \nabla \cdot (\sigma \mathbf{E}) = \frac{\partial \rho}{\partial t} \tag{4.18}$$

and considering ε and σ constant, we find

$$\frac{\partial \rho}{\partial t} = -\frac{\sigma}{\varepsilon}\rho, \qquad \rho = \rho_0 \exp\left(-\frac{\sigma}{\varepsilon}t\right). \tag{4.19}$$

In metallic materials, the decay time $\tau = \varepsilon/\sigma$ is of the order of 10^{-18} s, such that $\rho = 0$ at any instant.

For conducting materials one usually assumes $\nabla \cdot \mathbf{D} = 0$ and for constant ε and ρ, the electric field $\mathbf{E}$ and current density $\mathbf{J}$ are constant (COLLIN [1960]). A subtlety arises when ε and ρ are varying in space. Considering the steady-state version of above set of equations, we obtain

$$\nabla \cdot (\varepsilon \mathbf{E}) = \rho, \qquad \nabla \cdot (\sigma \mathbf{E}) = 0. \tag{4.20}$$

The field **E** should simultaneously obey two equations. Posed as a boundary-value problem for the scalar potential, V, we may determine V from the second equation and determine ρ as a "post-processing" result originating from the first equation.

4.5. Semiconductors

Intrinsic semiconductors are insulators at zero temperature. This is because the band structure of semiconductors consists of bands that are either filled or empty. At zero temperature, the chemical potential falls between the highest filled band which is called the valence band and the lowest empty band which is named the conduction band. The separation of the valance and conduction band is sufficiently small such that at some temperature, there is an appreciable amount of electrons that have an energy above the conduction band onset. As a consequence these electron are mobile and will contribute to the current if a voltage drop is put over the semiconducting material. The holes in the valance band act as positive charges with positive effective mass and therefore they also contribute to the net current. Intrinsic semiconductors are rather poor conductors but their resistance is very sensitive to the temperature ($\sim\exp(-A/T)$). By adding dopants to the intrinsic semiconductor, the chemical potential of the electrons and holes may by shifted up or down with respect to the band edges. Before going into further descriptions of dopant distributions, we would like to emphasize the following fact: *Each thermodynamic system in thermal equilibrium has constant intensive conjugated variables*. In particular, the temperature, T, conjugated to the internal energy of the system and the chemical potential, μ, conjugated to the number of particles in the systems are constant for a system in equilibrium. Therefore, if the dopant distribution varies in the device and the distance between the chemical potential and the band edges is modulated, then for the device being in equilibrium, the band edges must vary in accordance with the dopant variations, as illustrated in Fig. 4.1.

4.6. Currents in semiconductors

Whereas in metals the high conductivity prevents local charge accumulation at an detectable time scale, the situation in semiconductors is quite different. In uniformly doped semiconductors, the decay of an excess charge spot occurs by a diffusion process, that takes place on much longer time scale. In non-uniformly doped semiconductors, there

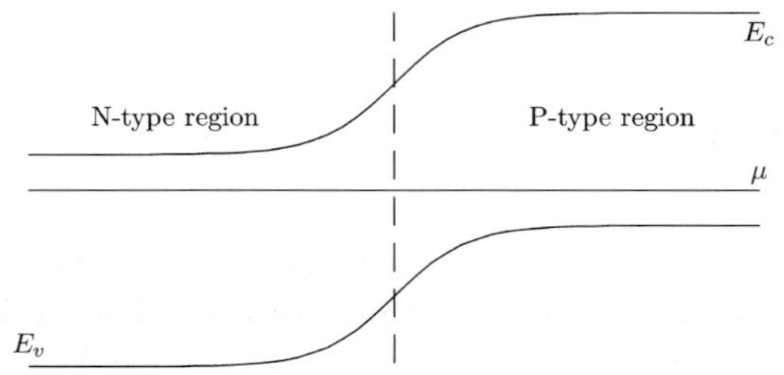

FIG. 4.1. Band edge modulation by doping.

are depletion layers, or accumulation layers of charges that permanently exists even in thermal equilibrium.

The charge and current densities in semiconductors follow also from the general Boltzmann transport theory, but this theory needs to be complemented with specific details such as the band gap, the dopant distribution, and the properties related to the interfaces to other materials.

Starting from the Boltzmann transport equation, the *moment expansion* considers variables that are averaged quantities as far as the momentum dependence is concerned. The generic expression for the moment expansion is

$$\frac{1}{\Omega}\sum_{\mathbf{p}} Q(\mathbf{p})\left(\frac{\partial}{\partial t} + \frac{\mathbf{p}}{m}\cdot\nabla_{\mathbf{q}} + \mathbf{F}\cdot\nabla_{\mathbf{p}}\right) f(\mathbf{p},\mathbf{q},t) = \frac{1}{\Omega}\sum_{\mathbf{p}} Q(\mathbf{p})\left(\frac{\partial f}{\partial t}\right)_{\mathrm{c}}, \tag{4.21}$$

where $Q(\mathbf{p})$ is an polynomial in the components of $\mathbf{p}$ and the normalization $1/\Omega$ allows for a smooth transition to integrate over all momentum states in the Brillouin zone

$$\frac{1}{\Omega}\sum_{\mathbf{p}} \to \frac{1}{4\pi^3}\int_{\mathrm{BZ}} \mathrm{d}^3k. \tag{4.22}$$

The zeroth order expansion gives (LUNDSTROM [1999])

$$\begin{aligned} &\frac{\partial n}{\partial t} - \frac{1}{q}\nabla\cdot\mathbf{J}_{\mathrm{n}} = -U, \\ &\frac{\partial p}{\partial t} + \frac{1}{q}\nabla\cdot\mathbf{J}_{\mathrm{p}} = U \end{aligned} \tag{4.23}$$

and where the various variables are:

electrons *holes*

$$\begin{aligned} &n(\mathbf{r},t) = \frac{1}{\Omega}\sum_{\mathbf{p}} f_{\mathrm{n}}(\mathbf{p},\mathbf{r},t), && p(\mathbf{r},t) = \frac{1}{\Omega}\sum_{\mathbf{p}} f_{\mathrm{p}}(\mathbf{p},\mathbf{r},t), \\ &\mathbf{J}_{\mathrm{n}}(\mathbf{r},t) = -qn(\mathbf{r},t)\mathbf{v}_{\mathrm{n}}(\mathbf{r},t), && \mathbf{J}_{\mathrm{p}}(\mathbf{r},t) = qp(\mathbf{r},t)\mathbf{v}_{\mathrm{p}}(\mathbf{r},t), \\ &\mathbf{v}_{\mathrm{n}}(\mathbf{r},t) = \frac{1}{\Omega}\sum_{\mathbf{p}} \frac{\mathbf{p}}{m} f_{\mathrm{n}}(\mathbf{p},\mathbf{r},t), && \mathbf{v}_{\mathrm{p}}(\mathbf{r},t) = \frac{1}{\Omega}\sum_{\mathbf{p}} \frac{\mathbf{p}}{m} f_{\mathrm{p}}(\mathbf{p},\mathbf{r},t) \end{aligned} \tag{4.24}$$

and

$$U = \frac{1}{\Omega}\sum_{\mathbf{p}}\left(\frac{\partial f}{\partial t}\right)_{\mathrm{c}} = R - G. \tag{4.25}$$

The particle velocities give an expression for the current densities but by choosing $Q(\mathbf{p}) = \mathbf{p}$, we obtain the first moment of the expansion that can be further approximated to give alternative expressions for the current densities. Defining the momentum relaxation time τ_{p} as a characteristic time for the momentum to reach thermal equilibrium from a non-equilibrium state and the electron and hole temperature tensors

(FORGHIERI, GUERRRI, CIAMPOLINI, GNUDI and RUDAN [1988])

$$\frac{1}{2}nk_{\mathrm{B}}T_{\mathrm{n},ij}(\mathbf{r},t)=\frac{1}{\Omega}\sum_{\mathbf{p}}\frac{1}{2m}(p_i-mv_{\mathrm{n},i})(p_j-mv_{\mathrm{n},j})f_{\mathrm{n}}(\mathbf{p},\mathbf{r},t)$$
$$=\frac{1}{2}nk_{\mathrm{B}}T_{\mathrm{n}}(\mathbf{r},t)\delta_{ij},$$
$$\frac{1}{2}pk_{\mathrm{B}}T_{\mathrm{p},ij}(\mathbf{r},t)=\frac{1}{\Omega}\sum_{\mathbf{p}}\frac{1}{2m}(p_i-mv_{\mathrm{p},i})(p_j-mv_{\mathrm{p},j})f_{\mathrm{p}}(\mathbf{p},\mathbf{r},t)$$
$$=\frac{1}{2}pk_{\mathrm{B}}T_{\mathrm{p}}(\mathbf{r},t)\delta_{ij}, \tag{4.26}$$

where the last equality follows from assuming an isotropic behavior, then one arrives at the following constitutive equation for the currents in semiconducting materials

$$\mathbf{J}_{\mathrm{n}}+n\tau_{\mathrm{pn}}\frac{\mathrm{d}}{\mathrm{d}t}\left(\frac{\mathbf{J}_{\mathrm{n}}}{n}\right)=q\mu_{\mathrm{n}}n\left(\mathbf{E}+\frac{k_{\mathrm{B}}}{q}\nabla T_{\mathrm{n}}\right)+qD_{\mathrm{n}}\nabla n,$$
$$\mathbf{J}_{\mathrm{p}}+p\tau_{\mathrm{pp}}\frac{\mathrm{d}}{\mathrm{d}t}\left(\frac{\mathbf{J}_{\mathrm{p}}}{p}\right)=q\mu_{\mathrm{p}}p\left(\mathbf{E}-\frac{k_{\mathrm{B}}}{q}\nabla T_{\mathrm{p}}\right)-qD_{\mathrm{p}}\nabla p. \tag{4.27}$$

The momentum relaxation times, the electron and hole mobilities and the electron and hole diffusivities are related through the Einstein relations

$$D=\frac{k_{\mathrm{B}}T}{q}\mu=\frac{k_{\mathrm{B}}T}{m}\tau. \tag{4.28}$$

The second terms on the left-hand sides of Eq. (4.27) are the *convective currents*. The procedure of taking moments of the Boltzmann transport equation always involves a truncation, i.e., the nth order equation in the expansion demands information of the $(n+1)$th order moment to be supplied. For the second-order moment, one thus needs to provide information on the third moment

$$\frac{1}{\Omega}\sum_{\mathbf{p}}p_ip_jp_kf(\mathbf{p},\mathbf{r},t). \tag{4.29}$$

In the above scheme the second-order expansion leads to the *hydrodynamic model* (FORGHIERI, GUERRRI, CIAMPOLINI, GNUDI and RUDAN [1988]). In this model the carrier temperatures are determined self-consistently with the carrier densities. The closure of the system of equations is achieved by assuming a model for the term (4.29) that only contains lower order variables. The thermal flux Q, being the energy that gets transported through thermal conductance can be expressed as

$$\mathbf{Q}=\frac{1}{\Omega}\sum_{\mathbf{p}}\frac{1}{2m}|\mathbf{p}-m\mathbf{v}|^2\left(\frac{\mathbf{p}}{m}-\mathbf{v}\right)=-\kappa\nabla T, \tag{4.30}$$

where $\kappa=\kappa_{\mathrm{n}},\kappa_{\mathrm{p}}$ are the thermal conductivities.

Besides the momentum flux, a balance equation is obtained for the energy flux:

$$\begin{aligned}
&\frac{\partial(nw_\mathrm{n})}{\partial t} + \nabla\cdot\mathbf{S}_\mathrm{n} = \mathbf{E}\cdot\mathbf{J}_\mathrm{n} + n\left(\frac{\partial w_\mathrm{n}}{\partial t}\right)_\mathrm{c},\\
&\frac{\partial(pw_\mathrm{p})}{\partial t} + \nabla\cdot\mathbf{S}_\mathrm{p} = \mathbf{E}\cdot\mathbf{J}_\mathrm{p} + p\left(\frac{\partial w_\mathrm{p}}{\partial t}\right)_\mathrm{c}.
\end{aligned} \tag{4.31}$$

The energy flux is denoted as $\mathbf{S}$ and w is the energy density. In the isotropic approximation, the latter reads

$$w_\mathrm{n} = \frac{3}{2}k_\mathrm{B}T_\mathrm{n} + \frac{1}{2}m_\mathrm{n}v_\mathrm{n}^2, \qquad w_\mathrm{p} = \frac{3}{2}k_\mathrm{B}T_\mathrm{p} + \frac{1}{2}m_\mathrm{p}v_\mathrm{p}^2. \tag{4.32}$$

The energy flux can be further specified as

$$\begin{aligned}
&\mathbf{S}_\mathrm{n} = \kappa_\mathrm{n}\nabla T_\mathrm{n} - (w_\mathrm{n} + k_\mathrm{B}T_\mathrm{n})\frac{\mathbf{J}_\mathrm{n}}{q},\\
&\mathbf{S}_\mathrm{p} = \kappa_\mathrm{p}\nabla T_\mathrm{p} + (w_\mathrm{p} + k_\mathrm{B}T_\mathrm{p})\frac{\mathbf{J}_\mathrm{p}}{q}.
\end{aligned} \tag{4.33}$$

Just as for the momentum, one usually assumes a characteristic time, τ_e, for a non-equilibrium energy distribution to relax to equilibrium. Then the collision term in the energy balance equation becomes

$$\begin{aligned}
&n\left(\frac{\partial w_\mathrm{n}}{\partial t}\right)_\mathrm{c} = -n\frac{w_\mathrm{n} - w^*}{\tau_\mathrm{en}} - Uw_\mathrm{n},\\
&p\left(\frac{\partial w_\mathrm{p}}{\partial t}\right)_\mathrm{c} = -p\frac{w_\mathrm{p} - w^*}{\tau_\mathrm{ep}} - Uw_\mathrm{p}
\end{aligned} \tag{4.34}$$

and w^* is the carrier mean energy at the lattice temperature. In order to complete the hydrodynamic model the thermal conductivities are given by the Wiedemann–Franz law for thermal conductivity

$$\kappa = \left(\frac{k_\mathrm{B}}{q}\right)^2 T\sigma(T)\Delta(T). \tag{4.35}$$

Herein is $\Delta(T)$ a value obtained from evaluating the steady-state Boltzmann transport equation for uniform electric fields and $\sigma(T) = q\mu c$ the electrical conductivity ($c = n, p$). If a power-law dependence for the energy relaxation times can be assumed, i.e.,

$$\tau_\mathrm{e} = \tau_0\left(\frac{w}{k_\mathrm{B}T^*}\right)^\nu, \tag{4.36}$$

then $\Delta(T) = 5/2 + \nu$. Occasionally, ν is considered to be a constant ($\nu = 0.5$). However, this results into too restrictive an expression for the $\tau_\mathrm{e}(w)$. Therefore $\Delta(T)$ is often tuned towards Monte-Carlo data.

Comparing the present elaboration on deriving constitutive equations from the Boltzmann transport equation with the derivation of the currents in metals we note that we did not refer to a displaced Maxwellian distribution. Such a derivation is also possible for semiconductor currents. The method was used by STRATTON [1962]. A difference pops up in the diffusion term of the carrier current. For the above results we obtained

$$\mathbf{J}(\text{diffusive part}) \propto \mu \nabla T. \tag{4.37}$$

In Stratton's model one obtains

$$\mathbf{J}(\text{diffusive part}) \propto \nabla(\mu T), \tag{4.38}$$

the difference being a term

$$\xi = \frac{\partial \log \mu(T)}{\partial \log(T)}. \tag{4.39}$$

Stratton's model is usually referred to as the *energy transport* model.

For the semiconductor environment, the Scharfetter–Gummel scheme provides a means to discretize the current equations on a grid (SCHARFETTER and GUMMEL [1969]). In the case that no carrier heating effects are considered (T is constant) the diffusion equations are

$$\mathbf{J} = q\mu c\mathbf{E} \pm kT\mu \nabla c, \tag{4.40}$$

where the plus (minus) sign refers to negatively (positively) charged particles and c denotes the corresponding carrier density. It is assumed that both the current $\mathbf{J}$ and the electric field $\mathbf{E}$ are constant along a link and that the potential V varies linearly along the link. Adopting a local coordinate axis u with $u = 0$ corresponding to node i, and $u = h_{ij}$ corresponding to node j, we may integrate Eq. (4.40) along the link ij to obtain

$$J_{ij} = q\mu_{ij} c\left(\frac{V_i - V_j}{h_{ij}}\right) \pm kT\mu_{ij}\frac{\mathrm{d}c}{\mathrm{d}u}, \tag{4.41}$$

which is a first-order differential equation in c. The latter is solved using the aforementioned boundary conditions and gives rise to a non-linear carrier profile. The current J_{ij} can then be rewritten as

$$\frac{J_{ij}}{\mu_{ij}} = -\frac{\alpha}{h_{ij}} B\left(\frac{-\beta_{ij}}{\alpha}\right) c_i + \frac{\alpha}{h_{ij}} B\left(\frac{\beta_{ij}}{\alpha}\right) c_j, \tag{4.42}$$

using the Bernoulli function

$$B(x) = \frac{x}{\mathrm{e}^x - 1}. \tag{4.43}$$

Furthermore, we used $\alpha = \pm kT$ and $\beta_{ij} = q(V_i - V_j)$.

Before turning to the consideration of insulating materials, we briefly discuss the influence of strong magnetic fields on the currents. These fields will bend the trajectories due to the Lorentz force. In the derivation of the macroscopic current densities from the Boltzmann transport equation, we should include this force. The result is that in the constitutive current expression we must make the replacement: $\mathbf{E} \rightarrow \mathbf{E} + q\mathbf{v} \times \mathbf{B}$. Since $\mathbf{J} = qc\mathbf{v}$, we arrive at the following *implicit* relation for $\mathbf{J}$:

$$\mathbf{J} = \sigma\mathbf{E} + \mu\mathbf{J} \times \mathbf{B}, \tag{4.44}$$

where $\sigma = q\mu c$ is the conductivity and μ is the mobility. This relation can be made *explicit* by solving the following set of linear equations:

$$\begin{bmatrix} 1 & -\mu B_z & \mu B_y \\ \mu B_z & 1 & -\mu B_x \\ -\mu B_y & \mu B_x & 1 \end{bmatrix} \cdot \begin{bmatrix} J_x \\ J_y \\ J_z \end{bmatrix} = \begin{bmatrix} \sigma E_x \\ \sigma E_y \\ \sigma E_x \end{bmatrix} \tag{4.45}$$

of which the solution is:

$$\mathbf{J} = \left[\sigma\mathbf{E} + \mu\sigma\mathbf{E} \times \mathbf{B} + \mu^2\sigma(\mathbf{E}\cdot\mathbf{B})\mathbf{B}\right]/(1+\mu^2B^2). \tag{4.46}$$

Above considerations are required for the description of Hall sensors. Here we will not further elaborate on this extension, nor will we consider the consequences of anisotropic conductivity properties.

4.7. Insulators

So far, we have been rather sloppy in classifying materials as being an insulator, semiconductor or metal. We have referred to the reader's qualitative awareness of the conduction quality of a material under consideration. For the time being we will sustain in this practice and define insulators as having a negligible conductivity. Therefore, in an insulating material there are no conduction currents. The constitutive equation for **J** becomes trivial.

$$\mathbf{J} = \mathbf{0}. \tag{4.47}$$

Recently, there is an increased interest in currents in insulating materials. The gate dielectric material SiO_2 that has been used in mainstream CMOS technology has a band gap of 3.9 eV and therefore acts as a perfect insulator for normal voltage operation conditions around 3 V and using 60 Å thick oxides. However, the continuous down scaling of the transistor architecture requires that the oxides thicknesses are also reduced. With the current device generation (100 nm gate length), the oxide thickness should be less than 20 Å. For these thin layers, direct tunneling through the layer barrier becomes a dominating current leakage in integrated CMOS devices.

4.7.1. Subband states and resonances

A planar p-type silicon metal-insulator-semiconductor (MIS) capacitor consisting of a gate electrode, a gate stack and a silicon substrate is considered. The gate stack has a thickness T_{ox} ranging from 15 to 40 Å and contains N_{ox} layers of insulating material such as SiO_2, Si_3Ni_5, etc. When a positive gate voltage V_G is applied to the gate electrode, the electrons residing in the electron inversion layer formed near the Si/insulator interface, are coupled to both the gate and the gate stack through non-vanishing tunneling amplitudes. As a result, measurable tunneling currents are observed that involve a net migration of electrons from the leaky inversion layer to the gate electrode.

In this section, we have summarized the approach followed in MAGNUS and SCHOENMAKER [2000a] and MAGNUS and SCHOENMAKER [2002] to calculate these tunneling currents.

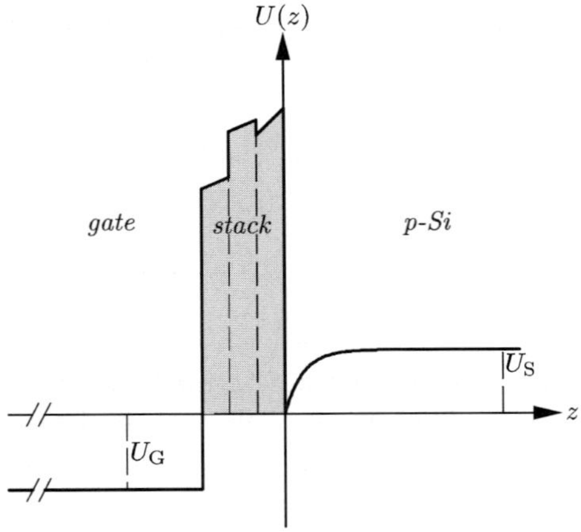

FIG. 4.2. Conduction band profile of a MIS capacitor. (Figure reproduced by permission of the Americal Institute of Physics and Springer Verlag.)

The z-axis is chosen to be perpendicular to the SiO_2-interface that is taken to be the (x, y)-plane. The gate, gate stack and semiconductor region are defined by $-\infty \leqslant z < t_{ox}$, $-t_{ox} \leqslant z < 0$ and $0 \leqslant z \leqslant +\infty$, respectively, as depicted in Fig. 4.2.

All electron energies including the chemical potential, are measured with respect to the edge of the conduction band at the Si/insulator interface. The potential energy takes a uniform value in the gate region whereas it approaches the limit U_S in the bulk substrate.

The whole MIS capacitor can be treated as a single quantum mechanical entity for which the Schrödinger equation needs to be solved. Adopting the effective mass approximation for the electrons in the different valleys, and the Hartree approximation to describe the electron–electron interaction in the inversion layer, the three-dimensional time-independent Schrödinger equation for the semiconductor region takes the form

$$-\frac{\hbar^2}{2}\left(\frac{1}{m_{\alpha x}}\frac{\partial^2}{\partial x^2}+\frac{1}{m_{\alpha y}}\frac{\partial^2}{\partial y^2}+\frac{1}{m_{\alpha z}}\frac{\partial^2}{\partial z^2}\right)\psi_\alpha(\mathbf{r}, z)+\left[U(z)-E\right]\psi_\alpha(\mathbf{r}, z)=0, \tag{4.48}$$

where $\mathbf{r} = (x, y)$, α is a valley index and $m_{\alpha x}$, $m_{\alpha y}$ and $m_{\alpha z}$ denote the components of the effective mass tensor along the principle directions of the silicon valleys. The same equation applies to the other regions upon insertion of appropriate effective masses. Assuming translational invariance in the lateral directions, one may write each one-electron wave function as a plane wave modulated by a one-dimensional envelope wave function $\phi_\alpha(W, z)$ and the corresponding one-electron eigenenergy $E_{\alpha\mathbf{k}}(W)$ as follows:

$$\begin{aligned}\psi_{\alpha k}(W, \mathbf{r}, z) &= \frac{1}{\sqrt{L_x L_y}}\mathrm{e}^{\mathrm{i}\mathbf{k}\cdot\mathbf{r}}\phi_\alpha(W, z),\\ E_{\alpha\mathbf{k}}(W) &= \frac{\hbar^2}{2}\left(\frac{k_x^2}{m_{\alpha x}}+\frac{k_y^2}{m_{\alpha y}}\right)+W,\end{aligned} \tag{4.49}$$

where $\mathbf{k} = (k_x, k_y)$ and $\phi_\alpha(W, z)$ is an eigenfunction of the one-dimensional Schrödinger equation

$$-\frac{\hbar^2}{2m_{\alpha z}}\frac{\mathrm{d}^2\phi_\alpha(W, z)}{\mathrm{d}z^2} + \big[U(z) - W\big]\phi_\alpha(W, z) = 0 \tag{4.50}$$

corresponding to the energy eigenvalue W.

Since the size of the whole system is assumed to be large in all directions, the energy spectrum will be dense and in particular the eigenvalues W can take all real values exceeding U_G. Moreover, the complete set of wave functions solving Eq. (4.50) constitutes an orthogonal, continuous basis for which a proper delta-normalization is invoked:

$$\big\langle\phi_\alpha(W')\big|\phi_\alpha(W)\big\rangle \equiv \int_{-\infty}^{\infty} \mathrm{d}z\, \phi_\alpha^*(W, z)\phi_\alpha(W, z) = \delta(W' - W). \tag{4.51}$$

Although the insulating layers are relatively thin, the energy barriers separating the inversion layer from the gate electrode are generally high enough to prevent a flood of electrons leaking away into the gate. In other words, in most cases of interest the potential well, hosting the majority of inversion layer electrons, will be coupled only weakly to the gate region. It follows from ordinary quantum mechanics (FLUEGGE [1974]) that the relative probability of finding an electron in the inversion layer well should exhibit sharply peaked maxima for a discrete set of W-values. The latter are the resonant energies corresponding to a set of virtually bound states, also called quasi-bound states, that may be regarded as the subband states of the coupled system. This becomes intuitively clear when the thickness of the barrier region is arbitrarily increased so that the coupling between the gate electrode and the semiconductor region vanishes. In this limiting case, the resonant energies will coincide with the true subband energies of the isolated potential well while the resonant wave functions drop to zero at the interface plane $z = 0$. Similarly, the spectral widths of the resonant wave functions tend to zero and the resonance peaks turn into genuine delta functions of W.

The above picture provides a way to investigate the subband structure of an inversion layer. By applying a transfer matrix approach to a piecewise constant potential profile and tracing the maxima of the squared wave function amplitudes as a function of W the continuous wave functions can be calculated. Once the sequence of resonant subband energies $\{W_{\alpha l} \mid l = 1, 2, \ldots\}$ and the corresponding wave functions are found, one may analytically determine the spectral widths that are directly related to the second derivative of the wave functions, with respect to W, evaluated at the resonant energies.

Within the Hartree approximation, the potential energy profile $U(z)$ needs to be determined by solving self-consistently the above mentioned Schrödinger equation (4.50) and the one-dimensional Poisson equation

$$\frac{\mathrm{d}^2U(z)}{\mathrm{d}z^2} = -\frac{e^2}{\varepsilon_\mathrm{S}}\big[n(z) - p(z) + N_\mathrm{A}(z)\big], \tag{4.52}$$

where $n(z)$, $p(z)$, $N_\mathrm{A}(z)$ and ε_S denote, respectively, the electron, hole and acceptor concentrations and the permittivity in the silicon part of the structure. In the present work we have not treated the occurrence of free charges in the gate and the gate stack.

On the other hand, charges trapped by interface states are incorporated through a surface charge density D_{it}.

The potential energy is modeled by a piecewise constant profile defined on a one-dimensional mesh reflecting the gate stack layers and a user-defined number of substrate layers. In this light the self-consistent link between $n(z)$ and $U(z)$ is not provided for each point in the inversion layer but rather for their averages over the subsequent cells of the mesh. This approach is adequate whenever the number of cells is sufficiently large and it has been successfully employed in the past (JOOSTEN, NOTEBORN and LENSTRA [1990], NOTEBORN, JOOSTEN, LENSTRA and KASKI [1990]). In the following however, we focus on the procedure to extract the resonant energies and spectral widths.

The solutions to the Schrödinger equation for the layered structure can now compactly be written as linear combinations of u_1 and u_2, being generic basis functions in each cell.

In order to trace the resonance peaks and spectral widths, a numerically stable probability function scanning the presence of an electron in the inversion layer as a function of W, needs to be determined. Rewriting the gate and substrate wave functions as

$$\phi_\alpha(W,z) = \begin{cases} C_{\text{g},\alpha}\sin(k_{\text{g},\alpha}(z+t_{\text{ox}})+\theta_\alpha) & \text{for } z < -t_{\text{ox}}, \\ C_{\text{s},\alpha}\exp(-k_{\text{s},\alpha}(z-a)) & \text{for } z > a, \end{cases} \tag{4.53}$$

one obtains the relative probability of an electron for being in the inversion layer:

$$P_\alpha(W) \equiv \left|\frac{C_{\text{s},\alpha}(W)}{C_{\text{g},\alpha}(W)}\right|^2. \tag{4.54}$$

Emerging as resonance energies in the continuous energy spectrum, the subband energies $W_{\alpha l}$ correspond to distinct and sharply peaked maxima of the $P_\alpha(W)$, or well defined minima of $P_\alpha^{-1}(W)$, even for oxide thicknesses as low as 10 Å. As a consequence, expanding $P_\alpha^{-1}(W)$ in a Taylor series around $W = W_{\alpha l}$, we may replace $P_\alpha(W)$ by a sum of Lorentz-shaped functions:

$$P_\alpha(W) \to \sum_l P_\alpha(W_{\alpha l})\frac{\Gamma_{\alpha l}^2}{(W-W_{\alpha l})^2+\Gamma_{\alpha l}^2}, \tag{4.55}$$

where the resonance widths $\Gamma_{\alpha l}^2$ are related to the second derivative of $P_\alpha^{-1}(W)$ through

$$\Gamma_{\alpha l}^2 = 2P_\alpha^{-1}(W_{\alpha l})\left[\frac{\partial^2 P_\alpha^{-1}}{\partial W^2}(W_{\alpha l})\right]^{-1} \tag{4.56}$$

and can be directly extracted from the transmission matrices and their derivatives, evaluated at $W = W_{\alpha l}$.

4.7.2. *Tunneling gate currents*

The subband structure of a p-type inversion layer channel may be seen to emerge from an enumerable set of sharp resonances appearing in the continuous energy spectrum of the composed system consisting of the gate contact, the gate stack (insulating layers), the inversion layer and the substrate contact. In particular, the discreteness of the subband states is intimately connected with the presence of energy barriers in the gate stack that restrict the coupling between the channel and the gate regions and therefore the amplitude for electrons tunneling through the barriers (see Fig. 4.2). Clearly, the smallness of the above mentioned coupling is reflected in the size of the resonance width – or equivalently, the resonance lifetime $\tau_{\alpha l} = \hbar/2\Gamma_{\alpha l}$ – as compared to the resonance energy.

It is tempting to identify the gate leakage current as a moving ensemble of electrons originating from decaying subband states. However, before such a link can be established, a conceptual problem should be resolved. Although intuition obviously suggests that an electron residing in a particular subband $\rangle\ \alpha l$ should contribute an amount $-e/\tau_{\alpha l}$ to the gate current, this is apparently contradicted by the observation that *the current density corresponding to each individual subband wave function identically vanishes*. The latter is due to the nature of the resonant states. Contrary to the case of the doubly degenerate running wave states having energies above the bottom of the conduction band in the substrate, the inversion layer resonant states are non-degenerate and virtually bound, and the wave functions are rapidly decaying into the substrate area. As a consequence, all wave functions are real (up to an irrelevant phase factor) and the diagonal matrix elements of the current density operator vanishes. The vanishing of the current for the envelope wave functions was also noted in SUNE, OLIVIO and RICCO [1991], MAGNUS and SCHOENMAKER [1999]. Therefore, we need to establish a sound physical model (workaround) resolving the current paradox and connecting the resonance lifetimes to the gate current. Since we do not adopt a plane-wave hypothesis for the inversion layer electrons in the perpendicular direction, our resolution of the paradox differs from the one that is proposed in SUNE, OLIVIO and RICCO [1991].

The paradox can be resolved by noting that the resonant states, though diagonalizing the electron Hamiltonian in the presence of the gate bias, are constituting a *non-equilibrium* state of the whole system which is not necessarily described by a Gibbs-like statistical operator, even not when the steady state is reached. There are at least two alternatives to solve the problem in practice.

The most rigorous approach aims at solving the full time dependent problem starting from a MIS capacitor that is in thermal equilibrium ($V_G = 0$) until some initial time $t = 0$. Before $t = 0$, the potential profile is essentially determined by the gate stack barriers and, due to the absence of an appreciable inversion layer potential well, all eigen solutions of the time independent Schrödinger equation are linear combinations of transmitted and reflected waves. In other words, almost all states are carrying current, although the thermal average is of course zero (equilibrium). However, it should be possible to calculate the time evolution of the creation and annihilation operators related to the unperturbed states. The perturbed resonant states, defining the subband structure for $V_G > 0$, would serve as a set of intermediate states participating in all transitions between the unperturbed states caused by the applied gate voltage. Although such an approach is conceptually straightforward, it is probably rather cumbersome to be carried out in practice.

One may consider a strategy that is borrowed from the theory of nuclear decay (MERZBACHER [1970], LANDAU and LIFSHITZ [1958]). The resulting model leads to a concise calculation scheme for the gate current. Under the assumption that the resonance widths of the virtual bound states are much smaller than their energies, the corresponding real wave functions can be extended to the complex plane if the resonance energies and the corresponding resonance widths are combined to form complex energy eigenvalues of the Schrödinger equation (MAGNUS and SCHOENMAKER [2000a]). Such an extension enables us to mimic both the supply (creation) and the decay (disintegration) of particles in a resonant bound state by studying the wave functions in those regions of space where the real, i.e., non-complex, wave functions would be standing waves either asymptotically or exactly.

Within the scope of the this work, scattering by phonons, or any other material dependent interactions is neglected. Moreover, electron–electron interaction is treated in the Hartree approximation that, in practice, amounts to a self-consistent solution of the one-particle Schrödinger equation and Poisson's equation. Therefore, bearing in mind that normal transport through the gate stack is limited by tunneling events, the time-reversal symmetry breaking between decaying and loading states can be inserted through the boundary conditions for the statistical operator corresponding to the non-interacting Liouville equation. Consequently, the gate current density is given by

$$J_{\mathrm{G}} = -\frac{e}{\pi \hbar^2 \beta} \sum_{\alpha l} \frac{\sqrt{m_{\alpha x} m_{\alpha y}}}{\tau_{\alpha l}} \log \frac{1 + \exp(\beta(E_{\mathrm{F}} - W_{\alpha l} - eV_{\mathrm{G}}))}{1 + \exp(\beta(E_{\mathrm{F}} - W_{\alpha l}))}. \tag{4.57}$$

It is clear from Eq. (4.57) that the resonance lifetimes are the key quantities building up the new formula for the gate leakage current. These variables apparently replace the familiar transmission coefficients that would emerge from traveling states contributing to the current in accumulation mode. This feature reflects the scope of nuclear decay theory which is a fair attempt to resolve the leakage current paradox. Although the latter theory produces a dynamical evolution of the one-particle wave functions, one can eventually insert a time independent, yet non-equilibrium, statistical operator to calculate the averages. It would be desirable to verify the success of this procedure on the grounds of sound time-dependent non-equilibrium theory. The same recommendation can be made regarding a more systematic investigation of the agreement between the results of the present calculation and the simulations based on Bardeen's approach (BARDEEN [1961]).

The above considerations have been used to evaluate the gate current numerically (MAGNUS and SCHOENMAKER [2000a], MAGNUS and SCHOENMAKER [2000b]). In Fig. 4.3 the simulation results are compared with a gate current characteristic that was obtained from measurements on a large MIS transistor with a NO insulator and grounded source and drain contacts. The latter serve as huge electron reservoirs capable of replacing the channel electrons (inversion) that participate in the gate tunneling current, such that the assumption on instantaneous injection or absorption compensating for migrating electrons is justified.

The following parameters are used: $T = 300$ K, $T_{\mathrm{ox}} = 25$ Å, $m_{g\alpha x} = m_{g\alpha y} = m_{g\alpha z} = 0.32 m_0$, $N_{\mathrm{ox}} = 3$, $m_{1,\mathrm{ox},\alpha} = m_{2,\mathrm{ox},\alpha} = m_{3,\mathrm{ox},\alpha} = 0.42 m_0$. The barrier height and the dielectric constant of the NO layer are taken to be 3.15 and 3.9 eV, respectively, while the acceptor concentration N_{A} is 4×10^{17} cm^{-3}. Fig. 4.4 shows typical current-voltage

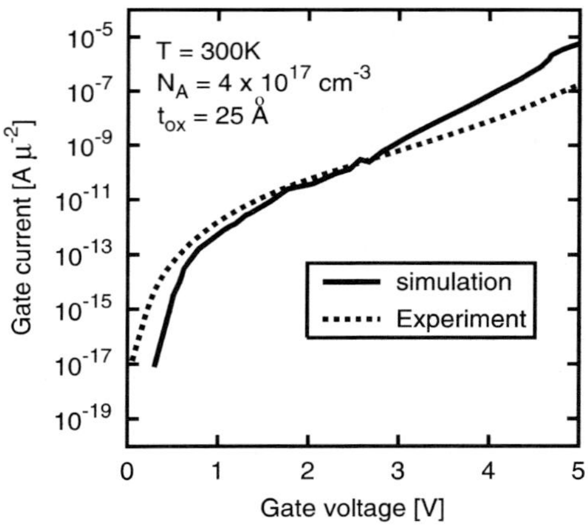

FIG. 4.3. Gate tunneling current vs. gate voltage for a NO layer with thickness of 25 Å. The doping is 4×10^{17} cm^{-3} and $T = 300$ K. (Figure reproduced by permission of The American Institute of Physics and Springer Verlag.)

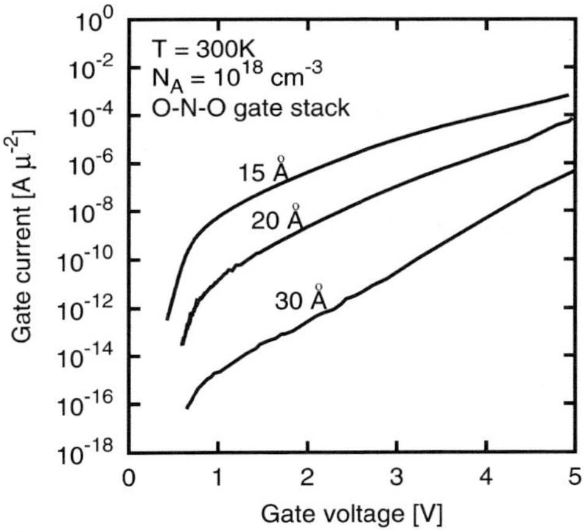

FIG. 4.4. Gate tunneling current vs. gate voltage for a NO layer with thickness of 15, 20 and 30 Å, the substrate doping being 10^{18} cm^{-3}. All other parameters are the same as in Fig. 4.3. (Figure reproduced by permission of The American Institute of Physics and Springer Verlag.)

characteristics for oxide thicknesses of 15, 20 and 30 Å and $N_{\mathrm{A}} = 10^{18}$ cm^{-3}. The simulation results show a good agreement with the experimental data in the range 1–4 V. It should be noted that the results are based on a set of "default" material parameters (BRAR, WILK and SEABAUGH [1996], DEPAS, VANMEIRHAEGHE, LAFLERE and CARDON [1994]). In particular for the effective electron mass in SiO_2, we used the results from Brar et al. The latter ones were obtained by measurements on *accumulation* layers. We suspect that the overestimation of the gate leakage currents for higher voltages is partly caused by the depletion layer in the poly-crystalline gate region ("poly-depletion") such that a shift in the gate potential at the gate/insulator interface occurs. Another origin of the discrepancy may be found in the approximations that are used in the method. The evaluation of the resonance lifetimes of the states using the Breit–Wigner expansion (BREIT and WIGNER [1936]) becomes less accurate if the overlap increases.

4.8. *Charges in insulators*

Although there are no mobile charges in perfect insulators, static charges may be present. Physically, these charges could be trapped during the processing of the insulator, or caused by radiation damage or stressing conditions. In the simulation of charges in insulators one first has to determine which time scale one is interested in. On the time scale of the operation of device switching characteristics, one may safely assume that the charges in insulators are immobile. However, on the time scale of the device lifetime or accelerated stressing condition, one must consider tunneling currents and trap generations that definitely can be traced to mobile charges.

4.9. *Dielectric media*

A dielectric material increases the storage capacity of a condenser or a capacitor by neutralizing charges at the electrodes that would otherwise contribute to the external field. Faraday identified this phenomenon as dielectric polarization. The polarization is caused by a microscopic alignment of dipole charges with respect to the external field. Looking at the macroscopic scale, we may introduce a polarization vector field, **P**.

In order to give an accurate formulation of dielectric polarization we first consider an arbitrary charge distribution localized around the origin. The electric potential in some point **r**, is

$$V(\mathbf{r}) = \frac{1}{4\pi\varepsilon_0}\int \frac{\rho(\mathbf{r}')}{|\mathbf{r}-\mathbf{r}'|}\,\mathrm{d}\tau'. \tag{4.58}$$

Now let **r** be a point outside the localization region of the charge distribution, i.e., $|\mathbf{r}| > |\mathbf{r}'|$. From the completeness of the series of the spherical harmonics, $Y_{lm}(\theta,\phi)$, one obtains

$$\frac{1}{|\mathbf{r}-\mathbf{r}'|} = 4\pi\sum_{l=0}^{\infty}\sum_{m=-l}^{l}\frac{1}{2l+1}\frac{|\mathbf{r}'|^l}{|\mathbf{r}|^{l+1}}Y_{lm}^*(\theta',\phi')Y_{lm}(\theta,\phi), \tag{4.59}$$

where

$$Y_{lm}(\theta,\phi) = \sqrt{\frac{2l+1}{4\pi}\frac{(l-m)!}{(l+m)!}}\,P_l^m(\cos\theta)\mathrm{e}^{\mathrm{i}m\phi} \tag{4.60}$$

and

$$P_l^m(x) = (-1)^m(1-x^2)^{m/2}\frac{\mathrm{d}^m}{\mathrm{d}x^m}P_l(x) \tag{4.61}$$

are the associated Legendre polynomials. Using above expansion, the potential of the charge distribution can be written as:

$$V(\mathbf{r}) = \frac{1}{4\pi\varepsilon_0}\sum_{l=0}^{\infty}\sum_{m=-l}^{l}\frac{4\pi}{2l+1}q_{lm}\frac{Y_{lm}(\theta,\phi)}{r^{l+1}} \tag{4.62}$$

and

$$q_{lm} = \int Y^*_{lm}(\theta',\phi')(r')^l\rho(\mathbf{r}')\,\mathrm{d}\tau' \tag{4.63}$$

are the *multipole moments* of the charge distribution. The zeroth-order expansion coefficient

$$q_{00} = \frac{1}{4\pi}\int\rho(\mathbf{r})\,\mathrm{d}\tau = \frac{Q}{4\pi} \tag{4.64}$$

corresponds to total charge of the localized charge distribution. The total charge can be referred to as the electric *monopole* moment. The electric dipole moment

$$\mathbf{p} = \int \mathbf{r}\rho(\mathbf{r})\,\mathrm{d}\tau \tag{4.65}$$

and the first order expansion coefficients are related according to

$$\begin{aligned} q_{1,1} &= -\sqrt{\frac{3}{8\pi}}(p_x - \mathrm{i}p_y), \\ q_{1,-1} &= \sqrt{\frac{3}{8\pi}}(p_x + \mathrm{i}p_y), \\ q_{1,0} &= \sqrt{\frac{3}{4\pi}}p_z. \end{aligned} \tag{4.66}$$

The higher-order moments depend on the precise choice of the origin inside the charge distribution and therefore their usage is mainly restricted to cases where a preferred choice of the origin is dictated by the physical systems.[4] The potential of the charge distribution, ignoring second and higher order terms is

$$V(\mathbf{r}) = \frac{1}{4\pi\varepsilon_0}\left(\frac{q}{r} + \frac{\mathbf{p}\cdot\mathbf{r}}{r^3}\right) \tag{4.67}$$

and the electric field of a dipole $\mathbf{p}$ located at the origin is

$$\mathbf{E}(\mathbf{r}) = \frac{3\hat{\mathbf{n}}(\mathbf{p}\cdot\hat{\mathbf{n}}) - \mathbf{p}}{4\pi\varepsilon_0 r^3} \tag{4.68}$$

and $\hat{\mathbf{n}} = \mathbf{r}/|\mathbf{r}|$. This formula is correct provided that $\mathbf{r} \neq \mathbf{0}$. An idealized dipole sheet at $x = 0$ is described by a charge distribution

$$\rho(\mathbf{r}) = \frac{\sigma}{4\pi\varepsilon_0}\delta'(x), \tag{4.69}$$

[4] For example, the center of a nucleus provides a preferred choice of the origin. The quadrupole moment of a nucleus is an important quantity in describing the nuclear structure.

where δ' is the derivative of the delta function. The corresponding electric field is

$$\mathbf{E}(\mathbf{r}) = -\frac{\sigma}{4\pi\varepsilon_0}\delta(x). \tag{4.70}$$

We will now consider the polarization of dielectric media and derive the macroscopic version of Gauss' law. If an electric field is applied to a medium consisting of a large number of atoms and molecules, the molecular charge distribution will be distorted. In the medium an electric polarization is produced. The latter can be quantitatively described as a macroscopic variable or cell variable such as $\mathbf{P} = \Delta\mathbf{p}/\Delta V$, i.e., as the dipole moment per unit volume. On a macroscopic scale, we may consider the polarization as a vector field, i.e., $\mathbf{P}(\mathbf{r})$. The potential $V(\mathbf{r})$ can be constructed by linear superposition of the contributions from each volume element $\Delta\Omega$ located at $\mathbf{r}'$. Each volume element gives a contribution originating from the net charge and a contributions arising from the dipole moment.

$$\Delta V(\mathbf{r}) = \frac{1}{4\pi\varepsilon_0}\left(\frac{\rho(\mathbf{r}')}{|\mathbf{r}-\mathbf{r}'|}\Delta\Omega + \frac{\mathbf{P}(\mathbf{r}')\cdot(\mathbf{r}-\mathbf{r}')}{|\mathbf{r}-\mathbf{r}'|^3}\right). \tag{4.71}$$

Adding all contributions and using the fact that

$$\nabla'\left(\frac{1}{|\mathbf{r}-\mathbf{r}'|}\right) = \frac{\mathbf{r}-\mathbf{r}'}{|\mathbf{r}-\mathbf{r}'|^3}, \tag{4.72}$$

we obtain

$$V(\mathbf{r}) = \frac{1}{4\pi\varepsilon_0}\int \mathrm{d}\tau' \frac{1}{|\mathbf{r}-\mathbf{r}'|}\big(\rho(\mathbf{r}') - \nabla'\cdot\mathbf{P}(\mathbf{r}')\big). \tag{4.73}$$

This corresponds to the potential of a charge distribution $\rho - \nabla\cdot\mathbf{P}$. Since the microscopic equation $\nabla\times\mathbf{E} = 0$ does apply also on the macroscopic scale, we conclude that $\mathbf{E}$ is still derivable from a potential field, $\mathbf{E} = -\nabla V$, and

$$\nabla\cdot\mathbf{E} = \frac{1}{\varepsilon_0}(\rho - \nabla\cdot\mathbf{P}). \tag{4.74}$$

This result can be easily confirmed by using

$$\nabla^2\left(\frac{1}{|\mathbf{r}-\mathbf{r}'|}\right) = -4\pi\delta(\mathbf{r}-\mathbf{r}'). \tag{4.75}$$

The electric displacement, $\mathbf{D}$, is defined as

$$\mathbf{D} = \varepsilon_0\mathbf{E} + \mathbf{P} \tag{4.76}$$

and the first Maxwell equation becomes

$$\nabla\cdot\mathbf{D} = \rho. \tag{4.77}$$

If the response of the medium to the electric field is linear and isotropic then the coefficient of proportionality is the electric susceptibility, χ_e and the polarization reads

$$\mathbf{P} = \varepsilon_0\chi_e\mathbf{E}. \tag{4.78}$$

and consequently,

$$\mathbf{D} = \varepsilon_0(1 + \chi_e)\mathbf{E} = \varepsilon_0\varepsilon_r\mathbf{E}. \tag{4.79}$$

This is a *constitutive* relation connecting **D** and **E**, necessary to solve the field equations. Here we have limited ourselves to consider an elementary connection. However, in general the connection can be non-linear and anisotropic, such that $\mathbf{P} = \mathbf{P}(\mathbf{E})$ will involve a non-trivial expression.

It is instructive to apply above terminology to a parallel-plate capacitor. The storage capacity C of two electrodes with charges $\pm Q$ in vacuum is $C = Q/V$, where V is the voltage drop. Filling the volume between the plates with a dielectric material results into a voltage drop

$$V = \frac{Q/\varepsilon_r}{C}. \tag{4.80}$$

This equation may be interpreted as stating that of the total charge Q, the *free* charge Q/ε_r contributes to the voltage drop, whereas the *bound* charge $(1 - 1/\varepsilon_r)Q$, is neutralized by the polarization of the dielectric material. The electric susceptibility, χ_e emerges as the ratio of the bound charge and the free charge:

$$\chi_e = \frac{(1 - 1/\varepsilon_r)Q}{Q/\varepsilon_r} = \varepsilon_r - 1. \tag{4.81}$$

The displacement and the polarization both have the dimension [charge/area]. These variables correspond to electric flux densities. Given an infinitesimal area element **dS** on an electrode, the normal component of **D** corresponds to the charge $\mathrm{d}Q = \mathbf{D} \cdot \mathbf{dS}$ on the area element and the normal component of **P** represents the bound charge $(1 - 1/\varepsilon_r)\,\mathrm{d}Q$ on the area element. Finally, the normal component of $\varepsilon_0\mathbf{E}$ corresponds to the free charge $\mathrm{d}Q/\varepsilon_r$ residing on the area element. The question arises how the displacement **D**, the polarization **P** and $\varepsilon_0\mathbf{E}$ can be associated to flux densities while there is no flow. In fact, the terminology is justified by analogy or mathematical equivalence with real flows. Consider for instance a stationary flow of water in $\mathbb{R}^3$. There exists a one-parameter family of maps $\phi_t : \mathbb{R}^3 \to \mathbb{R}^3$ that takes the molecule located at the position $\mathbf{r}_0$ at t_0 to the position $\mathbf{r}_1$ at t_1. Associated to the flow there exists a flux field

$$\mathbf{J}(\mathbf{r}) = \frac{\mathrm{d}\mathbf{r}}{\mathrm{d}t}. \tag{4.82}$$

The velocity field describes the streamlines of the flow. For an incompressible stationary flow we have that for any volume Ω

$$\oint_{\partial\Omega} \mathbf{J} \cdot \mathbf{dS} = 0 \quad \text{or} \quad \nabla \cdot \mathbf{J} = 0. \tag{4.83}$$

The number of water molecules that enter a volume exactly balances the number of water molecules that leave the volume. Now suppose that it is possible that water molecules are created or annihilated, e.g., by a chemical reaction $2H_2O \leftrightarrow O_2 + 2H_2$ in some volume. This process corresponds to a source/sink, Σ in the balance equation

$$\nabla \cdot \mathbf{J}(\mathbf{r}) = \Sigma(\mathbf{r}). \tag{4.84}$$

Comparing this equation with the first Maxwell equation, we observe the mathematical equivalence. The charge density ρ acts as a source/sink for the flux field $\mathbf{D}$.

4.10. Magnetic media

A stationary current density, $\mathbf{J}(\mathbf{r})$, generates a magnetic induction given by

$$\mathbf{B}(\mathbf{r}) = \frac{\mu_0}{4\pi}\int \mathrm{d}\tau'\, \mathbf{J}(\mathbf{r}') \times \frac{\mathbf{r}-\mathbf{r}'}{|\mathbf{r}-\mathbf{r}'|^3}. \tag{4.85}$$

This result is essentially the finding of Biot, Savart and Ampère. With the help of Eq. (4.72) we may write (4.85) as

$$\mathbf{B}(\mathbf{r}) = \frac{\mu_0}{4\pi}\nabla \times \int \mathrm{d}\tau'\, \frac{\mathbf{J}(\mathbf{r}')}{|\mathbf{r}-\mathbf{r}'|}. \tag{4.86}$$

An immediate consequence is $\nabla \cdot \mathbf{B} = 0$. Using the identity $\nabla \times \nabla \times \mathbf{A} = \nabla(\nabla \cdot \mathbf{A}) - \nabla^2\mathbf{A} = 0$, and the fact that $\mathbf{J} = 0$, as well as Eq. (4.75) one obtains that

$$\nabla \times \mathbf{B} = \mu_0 \mathbf{J}. \tag{4.87}$$

Helmholtz' theorem implies that there will be a vector field $\mathbf{A}$ such that $\mathbf{B} = \nabla \times \mathbf{A}$ and a comparison with Eq. (4.86) shows that

$$\mathbf{A}(\mathbf{r}, t) = \frac{\mu_0}{4\pi}\int \mathrm{d}\tau'\, \frac{\mathbf{J}(\mathbf{r}')}{|\mathbf{r}-\mathbf{r}'|} + \nabla\chi(\mathbf{r}, t), \tag{4.88}$$

where χ is an arbitrary scalar function. The arbitrariness in the solution (4.88) for $\mathbf{A}$ illustrates the freedom to perform gauge transformations. This freedom however is lifted by fixing a gauge condition, i.e., by inserting an additional constraint that the component of $\mathbf{A}$ should obey, such that not all components are independent anymore. A particular choice is the Coulomb gauge, $\nabla \times \mathbf{A} = \mathbf{0}$. In that case, χ is a solution of Laplace's equation $\nabla^2\chi = 0$. Provided that there are no sources at infinity and space is unbounded, the unique solution for χ is a constant, such that

$$\mathbf{A}(\mathbf{r}, t) = \frac{\mu_0}{4\pi}\int \mathrm{d}\tau'\, \frac{\mathbf{J}(\mathbf{r}')}{|\mathbf{r}-\mathbf{r}'|}. \tag{4.89}$$

We will now consider a localized current distribution around some origin, $\mathbf{0}$. Then we may expand Eq. (4.89) for $|\mathbf{r}| > |\mathbf{r}'|$ using

$$\frac{1}{|\mathbf{r}-\mathbf{r}'|} = \frac{1}{|\mathbf{r}|} + \frac{\mathbf{r}\cdot\mathbf{r}'}{|\mathbf{r}|^3} + \cdots \tag{4.90}$$

as

$$\mathbf{A}(\mathbf{r}) = \frac{\mu_0}{4\pi r}\int \mathrm{d}\tau'\, \mathbf{J}(\mathbf{r}') + \frac{\mu_0}{4\pi r^3}\int \mathrm{d}\tau'\, (\mathbf{r}\cdot\mathbf{r}')\mathbf{J}(\mathbf{r}'). \tag{4.91}$$

The first integral is zero, i.e., $\int \mathrm{d}\tau\, \mathbf{J}(\mathbf{r}) = 0$, whereas the second integral gives

$$\mathbf{A}(\mathbf{r}) = \frac{\mu_0}{4\pi}\frac{\mathbf{m}\times\mathbf{r}}{r^3}, \quad \mathbf{m} = \frac{1}{2}\int \mathrm{d}\tau\, \mathbf{r}\times\mathbf{J}(\mathbf{r}). \tag{4.92}$$

The variable $\mathbf{m}$ is the *magnetic moment* of the current distribution. Following a similar reasoning as was done for the dielectric media, we consider the macroscopic effects of magnetic materials. Since $\nabla \cdot \mathbf{B} = 0$ at the microscopic scale, this equation also is valid at macroscopic scale. Therefore, Helmholtz' theorem is still applicable. By dividing space into volume elements ΔV, we can assign to each volume element a magnetic moment

$$\Delta \mathbf{m} = \mathbf{M}(\mathbf{r})\Delta V, \tag{4.93}$$

where $\mathbf{M}$ is the magnetization or magnetic moment density. For a substance consisting of k different atoms or molecules with partial densities ρ_i $(i = 1, \ldots, k)$ and with magnetic moment $\mathbf{m}_i$ for the ith atom or molecule, the magnetization is

$$\mathbf{M}(\mathbf{r}) = \sum_{i=1}^{k} \rho_i(\mathbf{r})\mathbf{m}_i. \tag{4.94}$$

The free-charge current density and the magnetization of the volume element ΔV at location $\mathbf{r}'$, give rise to a contribution to the vector potential at location $\mathbf{r}$ being

$$\Delta \mathbf{A}(\mathbf{r}) = \frac{\mu_0}{4\pi}\frac{\mathbf{J}(\mathbf{r}')}{|\mathbf{r}-\mathbf{r}'|}\Delta V + \frac{\mu_0}{4\pi}\frac{\mathbf{M}(\mathbf{r}') \times (\mathbf{r}-\mathbf{r}')}{|\mathbf{r}-\mathbf{r}'|^3}\Delta V. \tag{4.95}$$

Adding all contributions

$$\mathbf{A}(\mathbf{r}) = \frac{\mu_0}{4\pi}\int \mathrm{d}\tau' \frac{\mathbf{J}(\mathbf{r}') + \nabla \times' \mathbf{M}(\mathbf{r}')}{|\mathbf{r}-\mathbf{r}'|}. \tag{4.96}$$

This corresponds to the vector potential of a current distribution $\mathbf{J} + \nabla \times \mathbf{M}$ and therefore

$$\nabla \times \mathbf{B} = \mu_0(\mathbf{J} + \nabla \times \mathbf{M}). \tag{4.97}$$

The magnetic *field* is defined as

$$\mathbf{H} = \frac{1}{\mu_0}\mathbf{B} - \mathbf{M}. \tag{4.98}$$

Then the stationary macroscopic equations become

$$\nabla \times \mathbf{H} = \mathbf{J}, \qquad \nabla \cdot \mathbf{B} = 0. \tag{4.99}$$

If we follow a strict analogy with the discussion on electrical polarization we should adopt a linear relation between the magnetization $\mathbf{M}$ and the induction $\mathbf{B}$ in order to obtain a constitutive relation between $\mathbf{H}$ and $\mathbf{B}$. However, historically it has become customary to define the *magnetic susceptibility* χ_m as the ratio of the magnetization and the magnetic field

$$\mathbf{M} = \chi_\mathrm{m}\mathbf{H}. \tag{4.100}$$

Then we obtain

$$\mathbf{B} = \mu_0(\mathbf{H} + \mathbf{M}) = \mu_0(1 + \chi_\mathrm{m})\mathbf{H} = \mu_0\mu_\mathrm{r}\mathbf{H} = \mu\mathbf{H}. \tag{4.101}$$

In here, μ is the *permeability* and μ_r is the *relative* permeability.

Just as is the case for electrical polarization, the constitutive relation, $\mathbf{B} = \mathbf{B}(\mathbf{H})$, can be anisotropic and non-linear. In fact, the $\mathbf{B}(\mathbf{H})$ relation may be multiple-valued depending on the history of the preparation of the material or the history of the applied magnetic fields (hysteresis).

In deriving the macroscopic field equations, we have so far been concerned with stationary phenomena. Both the charge distributions and the current distributions were assumed to be time-independent. The resulting equations are

$$\nabla \times \mathbf{E} = 0, \tag{4.102}$$

$$\nabla \cdot \mathbf{B} = 0, \tag{4.103}$$

$$\nabla \cdot \mathbf{D} = \rho, \tag{4.104}$$

$$\nabla \times \mathbf{H} = \mathbf{J}. \tag{4.105}$$

Faraday's law that was obtained from experimental observation, relates the circulation of the electric field to the time variation of the magnetic flux

$$\oint \mathbf{E} \cdot \mathbf{dr} = -\frac{\mathrm{d}}{\mathrm{d}t} \int \mathbf{B} \cdot \mathbf{dS}, \tag{4.106}$$

or

$$\nabla \times \mathbf{E} + \frac{\partial \mathbf{B}}{\partial t} = 0. \tag{4.107}$$

Magnetic monopoles have never been observed nor mimiced by time-varying fields. Therefore, the equation $\nabla \cdot \mathbf{B} = 0$ holds in all circumstances. Maxwell observed that the simplest generalization of Eqs. (4.104) and (4.105) that apply to time-dependent situations and that are consistent with charge conservation, are obtained by substituting $\mathbf{J}$ in Eq. (4.105) by $\mathbf{J} + \partial \mathbf{D}/\partial t$, since using the charge conservation and Gauss' law gives

$$\nabla \cdot \left(\mathbf{J} + \frac{\partial \mathbf{D}}{\partial t} \right) = 0, \tag{4.108}$$

such that the left- and right-hand side of

$$\nabla \times \mathbf{H} = \mathbf{J} + \frac{\partial \mathbf{D}}{\partial t} \tag{4.109}$$

are both divergenceless. Eqs. (4.103), (4.107), (4.104) and (4.109) are referred to as the (macroscopic) *Maxwell equations*. From a theoretical point of view, the Maxwell equations (4.103) and (4.107) found their proper meaning within the geometrical interpretation of electrodynamics, where they are identified as the Bianci identities for the curvature (see Section 8).

5. Wave guides and transmission lines

An important application of the Maxwell theory concerns the engineering of physical devices that are capable of transporting electromagnetic energy. This transport takes place in a wave-like manner. The static limit does not take into account the wave behavior of the Maxwell equations. The easiest way to implement this feature is by confining the field in two dimensions, allowing it to move freely along the third dimension (i.e., longitudinal sections are much larger than transversal directions). In this way, guided waves are recovered. A particular case of this model is the transmission line.

The wave guide consists of boundary surfaces that are good conductors. In practical realizations these surfaces are metallic materials such that the ohmic losses will be low. In the description of wave guides one usually assumes that the surfaces are perfectly conducting in a first approximation and that for large but finite conductivity, the ohmic losses can be calculated by perturbative methods. Besides the (idealized) boundary surfaces, the wave guide consists of a dielectric medium with no internal charges ($\rho = 0$), no internal currents ($\mathbf{J} = \mathbf{0}$). Furthermore, for an idealized description it is assumed that the conductivity of the dielectric medium vanishes ($\sigma = 0$). Finally, a wave guide is translational invariant in one direction. It has become customary, to choose the z-axis parallel to this direction.

In order to solve the Maxwell equations for wave guides, one considers harmonic fields (modes). The generic solution may be obtained as a superposition of different modes. The physical fields $\mathbf{E}(\mathbf{r}, t)$ and $\mathbf{H}(\mathbf{r}, t)$ are obtained from

$$\mathbf{E}(\mathbf{r}, t) = \Re\big(\mathbf{E}(\mathbf{r})\mathrm{e}^{\mathrm{i}\omega t}\big), \qquad \mathbf{H}(\mathbf{r}, t) = \Re\big(\mathbf{H}(\mathbf{r})\mathrm{e}^{\mathrm{i}\omega t}\big), \tag{5.1}$$

where $\mathbf{E}(\mathbf{r})$ and $\mathbf{H}(\mathbf{r})$ are complex phasors. The Maxwell equations governing these phasors are

$$\begin{aligned} &\nabla \cdot \mathbf{E} = 0, && \nabla \cdot \mathbf{H} = 0, \\ &\nabla \times \mathbf{E} = -\mathrm{i}\omega\mu\mathbf{H}, && \nabla \times \mathbf{H} = \mathrm{i}\omega\varepsilon\mathbf{E}. \end{aligned} \tag{5.2}$$

Defining $\omega\mu = k\zeta$ and $\omega\varepsilon = k/\zeta$ then $k = \omega\sqrt{\mu\varepsilon}$ and $\zeta = \sqrt{\mu/\varepsilon}$. From Eqs. (5.2) it follows that the phasors satisfy the following equation:

$$(\nabla^2 + k^2)\begin{Bmatrix} \mathbf{E} \\ \mathbf{H} \end{Bmatrix} = 0. \tag{5.3}$$

The translational invariance implies that if $\mathbf{E}(\mathbf{r}), \mathbf{H}(\mathbf{r})$ is a solution of Eq. (5.3), then $\mathbf{E}(\mathbf{r} + \boldsymbol{a})$, $\mathbf{H}(\mathbf{r} + \boldsymbol{a})$ with $\boldsymbol{a} = a\mathbf{e}_z$, is also a solution of Eq. (5.3). We may therefore introduce a shift operator, $\hat{S}(\boldsymbol{a})$ such that

$$\hat{S}(\boldsymbol{a})\begin{Bmatrix} \mathbf{E}(\mathbf{r}) \\ \mathbf{H}(\mathbf{r}) \end{Bmatrix} = \begin{Bmatrix} \mathbf{E}(\mathbf{r} + \boldsymbol{a}) \\ \mathbf{H}(\mathbf{r} + \boldsymbol{a}) \end{Bmatrix}. \tag{5.4}$$

Performing a Taylor series expansion gives

$$\mathbf{E}(\mathbf{r} + \boldsymbol{a}) = \sum_{n=0}^{\infty} \frac{a^n}{n!}\frac{\partial^n}{\partial z^n}\mathbf{E}(\mathbf{r}) = \exp\left(a\frac{\partial}{\partial z}\right)\mathbf{E}(\mathbf{r}) \tag{5.5}$$

and therefore $\hat{S}(\boldsymbol{a}) = \exp(a\frac{\partial}{\partial z}) = \exp(\mathrm{i}a\hat{k})$ with $\hat{k} = -\mathrm{i}\frac{\partial}{\partial z}$. The Helmholtz operator $\hat{H} = \nabla^2 + k^2$ commutes with $\hat{k}$, i.e., $[\hat{H}, \hat{k}] = 0$. As a consequence we can write the solutions of Eq. (5.3) in such a way that they are simultaneously eigenfunctions of $\hat{H}$ and $\hat{k}$. The eigenfunctions of $\hat{k}$ are easily found to be

$$f(z) = \mathrm{e}^{\mathrm{i}\kappa z}, \tag{5.6}$$

since

$$-\mathrm{i}\frac{\mathrm{d}}{\mathrm{d}z}f(z)=\kappa f(z). \tag{5.7}$$

Thus from the translational invariance we may conclude that it suffices to consider solutions for $\mathbf{E}$ and $\mathbf{H}$ of the form $\mathbf{E}(x,y)\mathrm{e}^{\mathrm{i}\kappa z}$ and $\mathbf{H}(x,y)\mathrm{e}^{\mathrm{i}\kappa z}$. Defining explicitly the transversal and the longitudinal components of the fields

$$\begin{aligned}\mathbf{E}(x,y)&=\mathbf{E}_{\mathrm{T}}(x,y)+\mathbf{E}_{\mathrm{L}}(x,y), & \mathbf{E}_{\mathrm{L}}(x,y)&=E_z(x,y)\mathbf{e}_{\mathrm{z}},\\ \mathbf{H}(x,y)&=\mathbf{H}_{\mathrm{T}}(x,y)+\mathbf{H}_{\mathrm{L}}(x,y), & \mathbf{H}_{\mathrm{L}}(x,y)&=H_z(x,y)\mathbf{e}_{\mathrm{z}},\end{aligned} \tag{5.8}$$

and

$$\nabla^2=\nabla_{\mathrm{T}}^2+\frac{\partial}{\partial z^2}=\nabla_{\mathrm{T}}^2-\kappa^2, \tag{5.9}$$

where the subscript T stands for a transverse field in the x–y-plane, while the subscript L denotes the longitudinal fields along the z-axis, we obtain

$$\begin{aligned}&\left(\nabla_{\mathrm{T}}^2+k^2-\kappa^2\right)\begin{Bmatrix}\mathbf{E}_{\mathrm{T}}(x,y)\\ \mathbf{H}_{\mathrm{T}}(x,y)\end{Bmatrix}=0,\\ &\left(\nabla_{\mathrm{T}}^2+k^2-\kappa^2\right)\begin{Bmatrix}E_z(x,y)\\ H_z(x,y)\end{Bmatrix}=0.\end{aligned} \tag{5.10}$$

The transverse equations correspond to an eigenvalue problem with fields vanishing at the boundaries in the transverse directions. The characteristic equations that need to be solved are the Helmholtz equations resulting into eigenvalue problems, where the eigenvalues are $p^2=k^2-\kappa^2$. The boundary conditions for the fields on the boundary surfaces are

$$\mathbf{n}\times\mathbf{E}=0, \qquad \mathbf{n}\cdot\mathbf{H}=0. \tag{5.11}$$

For the transverse components, going back to the full Maxwell equations, we get from Eq. (5.2)

$$\boldsymbol{\nabla}_{\mathrm{T}}E_z-\frac{\partial}{\partial z}\mathbf{E}_{\mathrm{T}}=-\mathrm{i}\omega\mu\mathbf{e}_{\mathrm{z}}\times\mathbf{H}_{\mathrm{T}} \tag{5.12}$$

and

$$\boldsymbol{\nabla}_{\mathrm{T}}H_z-\frac{\partial}{\partial z}\mathbf{H}_{\mathrm{T}}=\mathrm{i}\omega\varepsilon\mathbf{e}_{\mathrm{z}}\times\mathbf{E}_{\mathrm{T}}. \tag{5.13}$$

Combining (5.12) and (5.13), gives

$$\begin{aligned}p^2\mathbf{E}_{\mathrm{T}}&=\mathrm{i}\omega\mu\mathbf{e}_{\mathrm{z}}\times\boldsymbol{\nabla}_{\mathrm{T}}H_z+\mathrm{i}\kappa\boldsymbol{\nabla}_{\mathrm{T}}E_z,\\ p^2\mathbf{H}_{\mathrm{T}}&=-\mathrm{i}\omega\varepsilon\mathbf{e}_{\mathrm{z}}\times\boldsymbol{\nabla}_{\mathrm{T}}E_z+\mathrm{i}\kappa\boldsymbol{\nabla}_{\mathrm{T}}H_z.\end{aligned}$$

We may define the transversal fields as

$$\mathbf{E}_{\mathrm{T}}\propto V(z)\boldsymbol{e}_{\mathrm{t}}^{(1)}, \qquad \mathbf{H}_{\mathrm{T}}\propto I(z)\boldsymbol{e}_{\mathrm{t}}^{(2)}, \tag{5.14}$$

where $\boldsymbol{e}_{\mathrm{t}}^{(1)}$ and $\boldsymbol{e}_{\mathrm{t}}^{(2)}$ are transversal vectors independent of z.

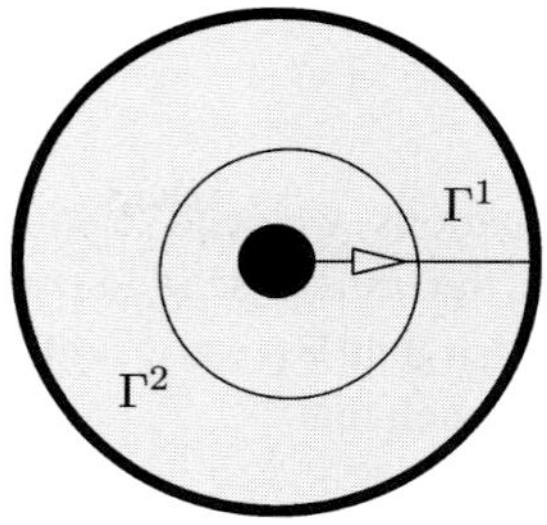

FIG. 5.1. Contours for evaluating voltage drops and currents of a two-conductor system in a TEM mode.

5.1. TEM modes

Inspired by waves in free space, we might look for modes that have a transverse behavior for both electric as magnetic field component, i.e., $E_z = H_z = 0$. These solutions are the transverse electromagnetic or TEM modes.

$$[\nabla_T^2 + p^2]\mathbf{E}_T = \mathbf{0}, \qquad [\nabla_T^2 + p^2]\mathbf{H}_T = \mathbf{0}. \tag{5.15}$$

For the TEM mode, the Maxwell equations result into $\kappa = k$. As a consequence Eqs. (5.15) are void. However, one also obtains from the Maxwell equations (5.12) and (5.13) that

$$\nabla \times \mathbf{E}_T = 0, \qquad \nabla \cdot \mathbf{E}_T = 0, \qquad \mathbf{H}_T = \frac{1}{\zeta}\mathbf{e}_z \times \mathbf{E}_T. \tag{5.16}$$

Therefore the TEM modes are as in an infinite medium. Since $E_z = 0$, the surfaces are equipotential boundaries and therefore at least two surfaces are needed to carry the wave. Since in any plane with constant z, we have a static potential, we can consider an arbitrary path going from one conductor to another. The voltage drop will be

$$V(z) = \int_{\Gamma^1} \mathbf{E}_T \cdot \mathbf{dr}. \tag{5.17}$$

The current in one conductor can be evaluated by taking a closed contour around the conductor and evaluate the field circulation. This is illustrated in Fig. 5.1.

$$I(z) = \oint_{\Gamma^2} \mathbf{H}_T \cdot \mathbf{dr}. \tag{5.18}$$

5.2. TM modes

When we look at solutions for which the longitudinal magnetic field vanishes ($H_z = 0$ everywhere), the magnetic field is always in the transverse direction. These solutions are the transverse magnetic or TM modes.

$$[\nabla_T^2 + p^2]E_z = \mathbf{0}, \tag{5.19}$$

$$p^2\mathbf{E}_\mathrm{T} = \mathrm{i}\kappa\nabla_\mathrm{T}E_z, \tag{5.20}$$

$$p^2\mathbf{H}_\mathrm{T} = -\mathrm{i}\omega\varepsilon\mathbf{e}_\mathrm{z} \times \nabla_\mathrm{T}E_z. \tag{5.21}$$

To find the solution of these equations, we need to solve a Helmholtz equation for H_z, and from Eqs. (5.20) and (5.21), the transverse field components are derived. Eq. (5.20) implies that $\nabla_\mathrm{T} \times \mathbf{E}_\mathrm{T} = 0$ and also that $\nabla_\mathrm{T} \times \boldsymbol{e}_T^{(1)} = \mathbf{0}$. Therefore, we may introduce a (complex) transverse potential ϕ such that

$$\boldsymbol{e}_\mathrm{t}^{(1)} = -\nabla_\mathrm{T}\phi. \tag{5.22}$$

This potential is proportional to E_z, i.e.,

$$E_z = -\frac{p^2}{\mathrm{i}\kappa}V(z)\phi. \tag{5.23}$$

Substitution of the (5.14) and (5.14) into (5.13) gives that $\boldsymbol{e}_\mathrm{t}^{(2)} = \mathbf{e}_\mathrm{z} \times \boldsymbol{e}_\mathrm{t}^{(1)}$ and $V(z) = -(\kappa/\omega\varepsilon)I(z)$.

5.3. TE modes

Similarly, when we look at solutions for which the longitudinal electric field vanishes ($E_z = 0$ everywhere), the electric field is always in the transverse direction. These solutions are the transverse electric or TE modes.

$$\left[\nabla_\mathrm{T}^2 + p^2\right]B_z = 0, \tag{5.24}$$

$$p^2\mathbf{E}_\mathrm{T} = \mathrm{i}\omega\mu\mathbf{e}_\mathrm{z} \times \nabla_\mathrm{T}H_z, \tag{5.25}$$

$$p^2\mathbf{H}_\mathrm{T} = \mathrm{i}\kappa\nabla_\mathrm{T}H_z. \tag{5.26}$$

To find the solution of these equations, we need to solve a Helmholtz equation for B_z, and from Eqs. (5.25) and (5.26), the transverse field components are derived. Since in this case $\nabla_\mathrm{T} \times \mathbf{H}_\mathrm{T} = \mathbf{0}$ there exists a scalar potential ψ such that

$$\boldsymbol{e}_\mathrm{t}^{(2)} = -\nabla_\mathrm{T}\psi. \tag{5.27}$$

Following a similar reasoning as above we obtain that

$$H_z = \frac{p^2}{\mathrm{i}k\zeta}V(z)\psi. \tag{5.28}$$

Furthermore, we find that $\boldsymbol{e}_\mathrm{t}^{(1)} = -\mathbf{e}_\mathrm{z} \times \boldsymbol{e}_\mathrm{t}^{(2)}$ and $V(z) = -(\omega\mu/\kappa)I(z)$.

5.4. Transmission line theory – S parameters

The structure of the transverse components of the electric and magnetic fields gives rise to an equivalent-circuit description. In order to show this, we will study the TM mode, but the TE description follows the same reasoning. By assuming the generic transmission-line solutions

$$V(z) = V_+ e^{-i\kappa z} + V_- e^{i\kappa z}, \tag{5.29}$$

$$I(z) = \frac{1}{Z_c}\left(V_+ e^{-i\kappa z} - V_- e^{i\kappa z}\right), \tag{5.30}$$

where Z_c is the characteristic impedance of the transmission line or the "telegraph" equations

$$\frac{dV(z)}{dz} = -ZI(z), \tag{5.31}$$

$$\frac{dI(z)}{dz} = -YV(z). \tag{5.32}$$

In these equations, the series impedance is denoted by Z and Y is the shunt admittance of the equivalent transmission line model. Each propagating mode corresponds to an eigenvalue p and we find that

$$Z = \frac{p^2 - k^2}{i\omega\varepsilon}, \qquad Y = i\omega\varepsilon. \tag{5.33}$$

From these expressions, the resulting equivalent circuit can be constructed.

6. From macroscopic field theory to electric circuits

6.1. Kirchhoff's laws

Electronic circuits consist of electronic components or devices integrated in a network. The number of components may range form a few to several billion. In the latter case the network is usually subdivided in functional blocks and each block has a unique functional description. The hierarchal approach is vital to the progress of electronic design and reuse of functional blocks (sometimes referred to as intellectual property) determines the time-to-market of new electronic products. Besides the commercial value of the hierarchical approach, there is also a scientific benefit. It is not possible to design advanced electronic circuits by solving the Maxwell equations using the boundary conditions that are imposed by the circuit. The complexity of the problem simply does not allow such an approach taking into account the available compute power and the constraints that are imposed on the design time. Moreover, a full solution of the Maxwell equations is often not very instructive in obtaining insight into the operation of the circuit. In order to understand the operation or input/output response of a circuit, it is beneficial to describe the circuit in effective variables. These coarse-grained variables (in the introduction we referred to these variables as "baskets") should be detailed enough such that a physical meaning can be given to them, whereas on the other hand they should be sufficiently "coarse" so as to mask details that are not relevant for understanding the circuit properties. The delicate balancing between these two requirements has resulted into "electronic circuit theory". The latter is based on the physical laws that are expressed by Maxwell's equations, and the laws of energy and charge conservation. The purpose of this section is to analyze how the circuit equations may be extracted from these microscopic physical laws. It should be emphasized that the extraction is not a rigorous derivation in the mathematical sense but relies on the validity of a number of approximations and assumptions reflecting the ideal behavior of electric circuits. These assumptions should be critically revised if one wishes to apply the circuit equations in areas that are outside the original scope of circuit theory. A simple example is a capacitor consisting of two large, conducting parallel plates separated by a relatively thin insulating layer: its capacity may be a suitable, characteristic variable for describing its impact in a circuit at low and moderately high frequencies. However, at extremely large frequencies the same device may act as a wave guide or an antenna, partly radiating the stored electromagnetic energy.

Being aware of such pitfalls, we continue our search for effective formulations of the circuit equations. In fact, the underlying prescriptions are given by the following (plausible) statements:

- A circuit can be represented by a topological network that consists of branches and nodes.
- **Kirchhoff's voltage law** (KVL) – The algebraic sum of all voltages along any arbitrary loop of the network equals zero at every instant of time.
- **Kirchhoff's current law** (KCL) – The algebraic sum of all currents entering or leaving any particular network node equals zero at every instant of time.

In order to make sense out of these statements we first need to have a clear understanding of the various words that were encountered; in particular, we must explain what is meant by a node, a branch, a voltage and a current. For that purpose we consider the most elementary circuit: a battery and a resistor that connects the poles of the battery. The circuit is depicted in Fig. 6.1. We have explicitly taken into account the finite resistance of the leads. In fact, a more realistic drawing is presented in Fig. 6.2, where we account for the fact that the leads have a finite volume. In particular, we have divided the full circuit volume into four different regions: (1) the battery region Ω_{B}, (2) the left lead region $\Omega_{1\mathrm{L}}$, (3) the right lead region $\Omega_{2\mathrm{L}}$, and (4) the resistor region Ω_{A}.

We will now consider the power supplied by the battery to the circuit volume. The work done by the electromagnetic field on all charges in the circuit volume per unit time

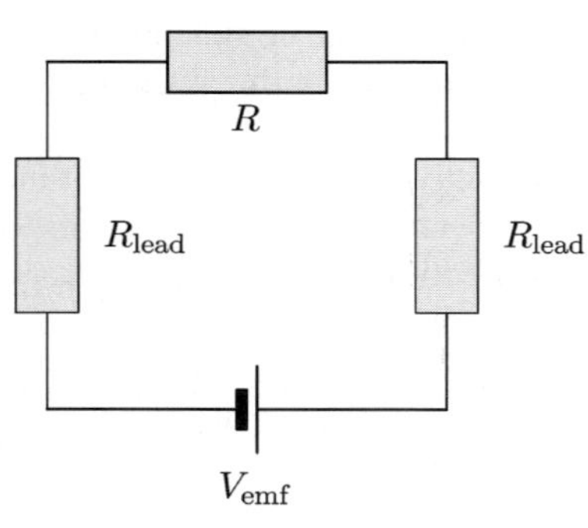

FIG. 6.1. Closed electric circuit containing a resistor connected to a DC power supply through two resistive leads.

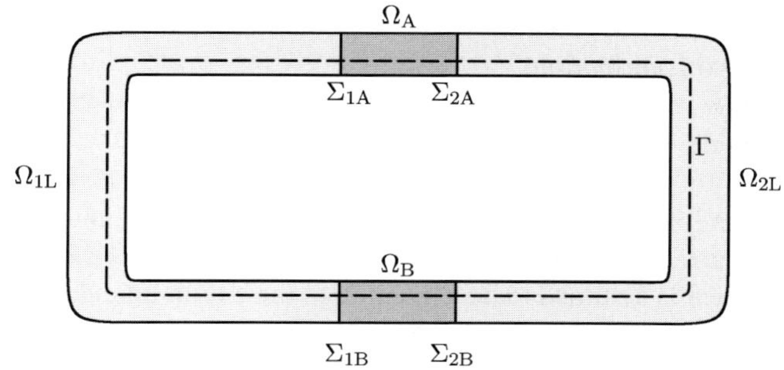

FIG. 6.2. The electric circuit of Fig. 6.1, taking into account the spatial extension of the leads. Γ is a circuit loop, i.e., an internal, closed loop encircling the "hole" of the circuit. (Figure reproduced by permission of the American Physical Society and Springer Verlag.)

is given by

$$\frac{\mathrm{d}E_{\mathrm{MECH}}}{\mathrm{d}t} = \int_{\Omega} \mathbf{J} \cdot \mathbf{E} \, \mathrm{d}\tau. \tag{6.1}$$

This corresponds to the dissipated power in steady-state conditions for which $(\partial\rho/\partial t) = 0$. As a consequence, $\nabla \cdot \mathbf{J} = 0$ and therefore we may apply the $\mathbf{J} \cdot \mathbf{E}$ theorem (see Appendix A). We obtain:

$$\int_{\Omega} \mathbf{J} \cdot \mathbf{E} \, \mathrm{d}\tau = \left(\oint_{\Sigma} \mathbf{J} \cdot \mathbf{dS} \right) \left(\oint_{\Gamma} \mathbf{E} \cdot \mathbf{dr} \right), \tag{6.2}$$

where Σ is an arbitrary cross section of the circuit and Γ is a circuit loop, i.e., an arbitrary closed path inside the circuit region. We identify the first integral of the right-hand side of Eq. (6.2) as the *current* in the circuit. The second integral of the right-hand side of Eq. (6.2) is identified as the *electromotive force* (EMF) or the *voltage* that is supplied by the battery, V_{ε}. The latter is nothing but the work done per unit charge by the electric field when the charge has made one full revolution around the circuit. Note the integral $\oint_{\Gamma} \mathbf{E} \cdot \mathbf{dr}$ is *non-zero*, although $\nabla \times \mathbf{E} = \mathbf{0}$. This is possible because the circuit is not a simply connected region in $\mathbb{R}^3$. More precisely, the topology of the circuit is that of a manifold of genus one, say a torus or a toroidal region with one "hole". We may now consider the left-hand side of Eq. (6.2) and consider the contributions to Eq. (6.2). For region (2) we obtain:

$$\int_{\Omega_{1L}} \mathbf{J} \cdot \mathbf{E} \, \mathrm{d}\tau = -\int_{\Omega_{1L}} (\nabla V) \cdot \mathbf{J} \, \mathrm{d}\tau = -\int_{\Omega_{1L}} \nabla \cdot (V\mathbf{J}) \, \mathrm{d}\tau + \int_{\Omega_{1L}} V \nabla \cdot \mathbf{J} \, \mathrm{d}\tau. \tag{6.3}$$

The first equality is valid since $\mathbf{E} = -\nabla V$, in a simply-connected region such as Ω_{1L}, Ω_{2L}, Ω_A or Ω_B. The last integral is equal to zero, since $\nabla \cdot \mathbf{J} = 0$ and the one-but-last integral is

$$-\int_{\Omega_{1L}} \nabla \cdot (V\mathbf{J}) \, \mathrm{d}\tau = -\oint_{\partial\Omega_{1L}} V\mathbf{J} \cdot \mathbf{dS}. \tag{6.4}$$

If we now *assume* that the potential is constant on a cross-section of the circuit, then this integral has two contributions:

$$-\oint_{\partial\Omega_{1L}} V\mathbf{J}\cdot\mathbf{dS} = -V_{\Sigma_{1B}}\int_{\Sigma_{1B}}\mathbf{J}\cdot\mathbf{dS} - V_{\Sigma_{1A}}\int_{\Sigma_{1A}}\mathbf{J}\cdot\mathbf{dS}. \tag{6.5}$$

Using Gauss' theorem we may identify the two remaining surface integrals can be identified as the total current I. Indeed, in the steady state regime ($\partial\rho/\partial t = 0$) the divergence of $\mathbf{J}$ vanishes while $\mathbf{J}$ is assumed to be tangential to the circuit boundary $\partial\Omega$. Therefore, the vanishing volume integral of $\nabla\cdot\mathbf{J}$ over Ω_{1L} reduces to

$$0 = \int_{\partial\Omega_{1L}}\mathbf{J}\cdot\mathbf{dS} = \int_{\Sigma_{1A}}\mathbf{J}\cdot\mathbf{dS} - \int_{\Sigma_{1B}}\mathbf{J}\cdot\mathbf{dS}, \tag{6.6}$$

which justifies the identification

$$\int_{\Sigma_{1A}}\mathbf{J}\cdot\mathbf{dS} = \int_{\Sigma_{1B}}\mathbf{J}\cdot\mathbf{dS} \equiv I \tag{6.7}$$

whence

$$-\oint_{\partial\Omega_{1L}} V\mathbf{J}\cdot\mathbf{dS} = I(V_{\Sigma_{1A}} - V_{\Sigma_{1B}}). \tag{6.8}$$

The regions (3) and (4) can be evaluated in a similar manner. As a consequence we obtain:

$$I(V_{\Sigma_{2B}} - V_{\Sigma_{2A}}) + I(V_{\Sigma_{2A}} - V_{\Sigma_{1A}}) + I(V_{\Sigma_{1A}} - V_{\Sigma_{1B}}) + \int_{\Omega_B}\mathbf{J}\cdot\mathbf{E}\,d\tau = IV_{\varepsilon}. \tag{6.9}$$

The final integral that applies to the battery region, is also equal to zero. This is because the electric field consists of two components: a conservative component and a non-conservative component, i.e., $\mathbf{E} = \mathbf{E}_C + \mathbf{E}_{NC}$. The purpose of the ideal[5] battery is to cancel the conservative field, such that after a full revolution around the circuit a net energy supply is obtained from the electric field. Then we finally arrive at the following result:

$$V_{\varepsilon} = V_{\Sigma_{2B}} - V_{\Sigma_{1B}}. \tag{6.10}$$

Eq. (6.10) is not a trivial result: having been derived from energy considerations, it relates the EMF of the battery, arising from a non-conservative field, to the potential difference at its terminals, i.e., a quantity characterizing a conservative field. Physically, it reflects the concept that an ideal battery is capable of maintaining a constant potential difference at its terminals even if a current is flowing through the circuit. This example illustrates how Kirchhoff's laws can be extracted from the underlying physical laws. It should be emphasized that we achieved more than what is provided by Kirchhoff's laws. Often Kirchhoff's voltage law is presented as a *trivial* identity, i.e., by putting N nodes on a closed path, as we have done by selecting a series of cross sections, it is

[5]The internal resistance of a real battery is neglected here.

always true that

$$(V_1 - V_2) + (V_2 - V_3) + \cdots + (V_{N-1} - V_N) + (V_N - V_1) = 0. \tag{6.11}$$

Physics enters this identity (turning it into a useful equation) by relating the potential differences to their physical origin. In the example above, the potential difference, $V_{\Sigma_{2B}} - V_{\Sigma_{1B}}$, is the result of a power supply.

By the in-depth discussion of the simple circuit, we have implicitly provided a detailed understanding of what is understood to be a voltage, a current, a node and a branch in a Kirchhoff network. The nodes are geometrically idealized regions of the circuit to which network branches can be attached. The nodes can be electrically described by a single voltage value. A branch is also a geometrical idealization. Knowledge of the current *density* inside the branch is not required. All that counts is the total current in the branch. We also have seen that at some stages only progress could be made by making simplifying assumptions and finally that all variables are time independent. The last condition is a severe limitation. In the next section we will discuss the consequences of eliminating this restriction. We can insert more physics in the network description. So far, we have not exploited Ohm's law, $\mathbf{J} = \sigma\mathbf{E}$. For a resistor with length L, cross sectional area A and constant resistivity σ, we find that

$$\int_{\Omega_A} \mathbf{J} \cdot \mathbf{E}\,d\tau = \sigma \int (\nabla V)^2\,d\tau = \sigma L \cdot A\left(\frac{V_{\Sigma_{2A}} - V_{\Sigma_{1A}}}{L}\right)^2 = I(V_{\Sigma_{2A}} - V_{\Sigma_{1A}}). \tag{6.12}$$

As a result, we can "define" the resistance as the ratio of the potential difference and the current:

$$V_{\Sigma_{2A}} - V_{\Sigma_{1A}} = RI, \qquad R = \frac{L}{\sigma A}. \tag{6.13}$$

6.2. Circuit rules

In the foregoing section we have considered DC steady-state currents, for which $\nabla \cdot \mathbf{J} = 0$ and $\partial\mathbf{B}/\partial t = \mathbf{0}$, such that the $\mathbf{J} \cdot \mathbf{E}$ theorem could be applied. In general, these conditions are not valid and the justification of using the Kirchhoff's laws becomes more difficult. Nevertheless, the guiding principles remain unaltered, i.e., the conservation of charge and energy will help us in formulating the circuit equations. On the other hand, as was already mentioned in the previous section, the idealization of a real circuit involves a number of approximations and assumptions that are summarized below in a – non-exhaustive – list of circuit rules:

(1) An electric circuit, or more generally, a circuit network, is a manifold of genus $N \geqslant 1$, i.e., a multiply connected region with N holes. The branches of this manifold consist of distinct circuit segments or devices, mainly active and passive components, interconnecting conductors and seats of EMF.
(2) The active components typically include devices that are actively processing signals, such as transistors, vacuum tubes, operational amplifiers, A/D converters.

(3) Passive components refer to ohmic resistors, capacitors and inductors or coils, diodes, tunneling junctions, Coulomb blockade islands, etc. They are representing energy dissipation, induction effects, quantum mechanical tunneling processes and many other phenomena.

(4) The seats of EMF include both DC and AC power supplies, i.e., chemical batteries, EMFs induced by externally applied magnetic fields, all different kinds of current and voltages sources and generators, etc. The electromagnetic power supplied by the EMF sources is dissipated entirely in the circuit. No energy is released to the environment of the circuit through radiation or any other mechanism.

(5) In compliance with the previous rule, all circuit devices are assumed to behave in an ideal manner. First, all conductors are taken to be perfect conductors. Considering perfect conduction as the infinite conductivity limit of realistic conduction ($\mathbf{J} = \sigma \mathbf{E}$), it is clear that no electric fields can survive inside a perfect conductor which therefore can be considered an equipotential volume. Clearly, from $\nabla \cdot \mathbf{E} = \rho/\varepsilon$ it follows that the charge density also vanishes inside the conductor. Furthermore, a perfect conductor is perfectly shielded from any magnetic field. Strictly speaking, this is not a direct consequence of Maxwell's third equation, since $\nabla \times \mathbf{E} = \mathbf{0}$ would only imply $\partial \mathbf{B}/\partial t = \mathbf{0}$ but the effect of static magnetic fields on the circuit behavior will not be considered here. It should also be noted that a perfect conductor is not the same as a superconductor. Although for both devices the penetration of magnetic fields is restricted to a very narrow boundary layer, called penetration depth, only the superconductor hosts a number of "normal" electrons (subjected to dissipative transport) and will even switch entirely to the normal state when the supercurrent attains its critical value. Furthermore, a supercurrent can be seen as a coherent, collective motion of so-called Cooper pairs of electrons, i.e., *bosons* while perfect conduction is carried by unpaired electrons or holes, i.e., *fermions*. Next, all energy dissipation exclusively takes place inside the circuit resistors. This implies that all capacitors and inductors are assumed to be made of perfect conductors. Inside the windings of an inductor and the plates of a capacitor, no electric or magnetic fields are present. The latter exist only in the cores of the inductors[6] while the corresponding vector potential and induced electric field are localized in the inductor. Similarly, the electric charge on the plates of a capacitor are residing in a surface layer and the corresponding, conservative electric field is strictly localized between the plates while all stray fields are ignored. Finally, the ideal behavior of the seats of EMF is reflected in the absence of internal resistances and the strict localization of the non-conservative electric fields that are causing the EMFs.

(6) The current density vector $\mathbf{J}$ defines a positive orientation of the circuit loop Γ. It corresponds the motion of a positive charge moving from the anode to the cathode outside the EMF seat and from cathode to anode inside the EMF seat.

[6]Topologically, the cores are not part of the circuit region Ω.

6.3. Inclusion of time dependence

The previous set of rules will guide us towards the derivation of the final circuit equations. However, before turning to the latter, it is worth to have a second look at Eq. (6.11). In the continuum, this identity can be given in the following way:

$$\oint_{\Gamma} \mathbf{dr} \cdot \nabla V(\mathbf{r}, t) = 0, \tag{6.14}$$

where Γ is an arbitrary closed loop. Note that above equation includes time-dependent fields $V(\mathbf{r}, t)$. In order to validate the first Kirchhoff law (KVL), we insert into Eq. (6.14) the potential that corresponds to

$$\nabla V = -\mathbf{E} - \frac{\partial \mathbf{A}}{\partial t}. \tag{6.15}$$

Of course, if we were to plug this expression into Eq. (6.14), we would just arrive at Faraday's law. The transition to the circuit equations is realized by cutting the loop into discrete segments (rule 1) and assigning to each segment appropriate lumped variables. To illustrate this approach we revisit the circuit of Fig. 6.1, where we have now folded the resistor of the left lead into a helix and, according to the circuit rules, its resistance is taken to be zero whereas the top resistor is replaced by a capacitor. The resulting, idealized circuit depicted in Fig. 6.3 has four segments. The battery region, that now may produce a time-dependent EMF, and the right-lead region can be handled as was done in the foregoing section. According to the circuit rules, it is assumed that all resistance is concentrated in the resistor located between node 3 and node 4, while both the inductor and the capacitor are made of perfect conductors and no leakage current is flowing between the capacitor plates. Starting from the identities

$$V_1 - V_2 + V_2 - V_3 + V_3 - V_4 + V_4 - V_1 = 0, \tag{6.16}$$

$$\oint_{\Gamma} \mathbf{E} \cdot \mathbf{dr} + \frac{\partial}{\partial t} \oint_{\Gamma} \mathbf{A} \cdot \mathbf{dr} = 0, \tag{6.17}$$

we decompose the electric field into a conservative, an external and induced component:

$$\mathbf{E} = \mathbf{E}_{\mathrm{C}} + \mathbf{E}_{\mathrm{EX}} + \mathbf{E}_{\mathrm{IN}}, \tag{6.18}$$

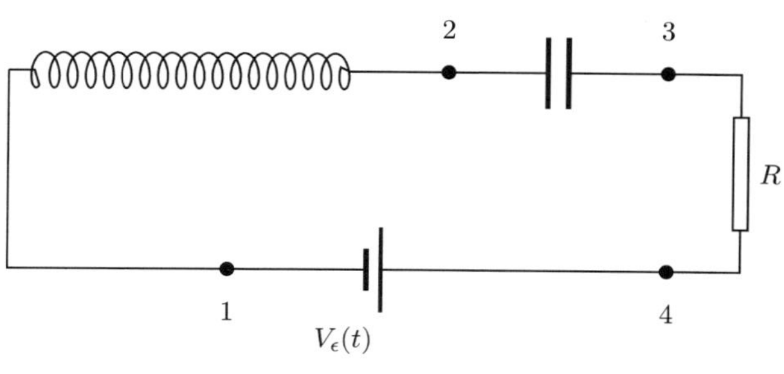

FIG. 6.3. The electric circuit of Fig. 6.2 with a helix-shaped "resistor".

where

$$\mathbf{A} = \mathbf{A}_{\mathrm{EX}} + \mathbf{A}_{\mathrm{IN}}, \qquad \mathbf{E}_{\mathrm{C}} = -\nabla V,$$
$$\mathbf{E}_{\mathrm{EX}} = -\frac{\partial}{\partial t}\mathbf{A}_{\mathrm{EX}}, \qquad \mathbf{E}_{\mathrm{IN}} = -\frac{\partial}{\partial t}\mathbf{A}_{\mathrm{IN}}. \tag{6.19}$$

Since the battery and the inductor are perfect conductors, the total electric field in these devices is identically zero:

$$\int_1^4 \mathbf{dr} \cdot \mathbf{E} = 0, \tag{6.20}$$

$$\int_2^1 \mathbf{dr} \cdot \mathbf{E} = 0. \tag{6.21}$$

Following the circuit rules, we assume that the induced electric field and the external field are only present in the inductor region and the battery region, respectively. Then Eq. (6.21) can be evaluated as

$$\int_2^1 \mathbf{dr} \cdot \mathbf{E} = \int_2^1 \mathbf{dr} \cdot (\mathbf{E}_{\mathrm{C}} + \mathbf{E}_{\mathrm{IN}}) = V_2 - V_1 + \int_2^1 \mathbf{dr} \cdot \mathbf{E}_{\mathrm{IN}} = 0 \tag{6.22}$$

and therefore

$$V_1 - V_2 = \int_2^1 \mathbf{dr} \cdot \mathbf{E}_{\mathrm{IN}}. \tag{6.23}$$

For the battery region we obtain:

$$\int_1^4 \mathbf{dr} \cdot \mathbf{E} = \int_1^4 \mathbf{dr} \cdot (\mathbf{E}_{\mathrm{C}} + \mathbf{E}_{\mathrm{EX}}) = V_1 - V_4 + \int_1^4 \mathbf{dr} \cdot \mathbf{E}_{\mathrm{EX}} = 0 \tag{6.24}$$

and therefore

$$V_4 - V_1 = \int_1^4 \mathbf{dr} \cdot \mathbf{E}_{\mathrm{EX}} = \oint \mathbf{dr} \cdot \mathbf{E}_{\mathrm{EX}} = V_\varepsilon. \tag{6.25}$$

Inside the capacitor, the induced and external fields are zero, and therefore we obtain

$$\int_3^2 \mathbf{dr} \cdot \mathbf{E} = \int_3^2 \mathbf{dr} \cdot \mathbf{E}_{\mathrm{C}} = V_3 - V_2. \tag{6.26}$$

On the other hand, the potential difference between the capacitor is assumed to be proportional to charge Q stored on one of the plates, i.e., $Q = CV$, where C is the *capacitance*. The resistor is treated in an analogous manner:

$$V_4 - V_3 = \int_4^3 \mathbf{dr} \cdot \mathbf{E}_{\mathrm{C}} = IR. \tag{6.27}$$

Insertion of all these results into Eq. (6.16) gives:

$$\int_1^2 \mathbf{E}_{\mathrm{IN}} \cdot \mathbf{dr} = -V_\varepsilon + IR + \frac{Q}{C}, \tag{6.28}$$

where we anticipated that the electric field between the capacitor plates is given by $Q/(Cd)$ and d is the thickness of the dielectric. The integral at the left-hand side of Eq. (6.28) can be obtained by using Faraday's law once again:

$$\int_2^1 \mathbf{E}_{\mathrm{IN}} \cdot \mathbf{dr} \simeq \oint_\Gamma \mathbf{E}_{\mathrm{IN}} \cdot \mathbf{dr} = -\frac{\partial}{\partial t} \oint_\Gamma \mathbf{A}_{\mathrm{IN}} \cdot \mathbf{dr} = -\frac{\partial}{\partial t} \int_{S(\Gamma)} \mathbf{B}_{\mathrm{IN}} \cdot \mathbf{dS}, \tag{6.29}$$

where $S(\Gamma)$ is the area enclosed by the loop Γ. Since the magnetic field $\mathbf{B}$ is only appreciably different from zero inside the core of the inductor, the integral may be identified as the magnetic self-flux Φ_{M} of the inductor. This flux is proportional to the circuit current I that also flows through the windings of the coil. Hence, $\Phi_{\mathrm{M}} = LI$, where L is the *inductance* of the inductor and therefore Eq. (6.23) becomes:

$$V_1 - V_2 = -L\frac{\mathrm{d}I}{\mathrm{d}t}. \tag{6.30}$$

We are now in the position to write down the circuit equation for the simple circuit of Fig. 6.3. Starting from the identity of Eq. (6.16), we find

$$-L\frac{\mathrm{d}I}{\mathrm{d}t} + V_\varepsilon - IR - \frac{Q}{C} = 0. \tag{6.31}$$

So far, we have not considered energy conservation for the time-dependent circuit equations. However, this conservation law is important for determining explicit expressions for the inductances and capacitances. Integrating the electromagnetic energy density u_{EM} over an arbitrarily large volume Ω_∞ with a boundary surface $\partial\Omega_\infty$, we obtain the total energy content of the electromagnetic field:

$$U_{\mathrm{EM}} = \frac{1}{2}\int_{\Omega_\infty} \mathrm{d}\tau \left(\varepsilon E^2 + \frac{B^2}{\mu}\right) = \frac{1}{2}\int_{\Omega_\infty} \mathrm{d}\tau\, (\mathbf{E}\cdot\mathbf{D} + \mathbf{B}\cdot\mathbf{H}). \tag{6.32}$$

Replacing $\mathbf{E}$ and $\mathbf{B}$ by $-\nabla V - \partial\mathbf{A}/\partial t$ and $\nabla \times \mathbf{A}$, respectively, we may rewrite Eq. (6.32) as

$$U_{\mathrm{EM}} = \frac{1}{2}\int_{\Omega_\infty} \mathrm{d}\tau \left[-\left(\nabla V + \frac{\partial \mathbf{A}}{\partial t}\right)\cdot\mathbf{D} + \mathbf{H}\cdot\nabla\times\mathbf{A}\right]. \tag{6.33}$$

Next, exploiting the vector identity (A.40), we applying Gauss' theorem to the volume Ω_∞ thereby neglecting all fields at the outer surface $\partial\Omega_\infty$, i.e.,

$$\int_{\partial\Omega_\infty} \mathbf{dS}\cdot(V\mathbf{D}) = 0, \tag{6.34}$$

we obtain:

$$-\int_{\Omega_\infty} \mathrm{d}\tau\, \nabla V \cdot \mathbf{D} = \int_{\Omega_\infty} \mathrm{d}\tau\, V\nabla\cdot\mathbf{D} = \int_{\Omega_\infty} \mathrm{d}\tau\, \rho V, \tag{6.35}$$

where the last equality follows from the first Maxwell equation $\nabla\cdot\mathbf{D} = \rho$.

Similarly, using the identity (A.39) and inserting the fourth Maxwell equation, we may convert the volume integral of $\mathbf{H}\cdot\nabla\times\mathbf{A}$ appearing in Eq. (6.33):

$$\int_{\Omega_\infty} \mathrm{d}\tau\, \mathbf{H}\cdot\nabla\times\mathbf{A} = \int_{\Omega_\infty} \mathrm{d}\tau\, \mathbf{A}\cdot\left(\mathbf{J} + \frac{\partial\mathbf{D}}{\partial t}\right). \tag{6.36}$$

Putting everything together, we may express the total electromagnetic energy as follows:

$$U_{\mathrm{EM}} = \frac{1}{2}\int_{\Omega_\infty} \mathrm{d}\tau \left[\rho V - \frac{\partial \mathbf{A}}{\partial t}\cdot\mathbf{D} + \mathbf{A}\cdot\left(\mathbf{J} + \frac{\partial \mathbf{D}}{\partial t}\right)\right] \tag{6.37}$$

$$= \frac{1}{2}\int_{\Omega} \mathrm{d}\tau \left[\rho V - \frac{\partial \mathbf{A}}{\partial t}\cdot\mathbf{D} + \mathbf{A}\cdot\left(\mathbf{J} + \frac{\partial \mathbf{D}}{\partial t}\right)\right], \tag{6.38}$$

where the last integral is restricted to the circuit region Ω in view of the circuit rules stating that all electromagnetic fields are vanishing outside the circuit region. It is easy to identify in Eq. (6.38) the "electric" and "magnetic" contributions respectively referring to E^2 and B^2 in Eq. (6.32):

$$U_{\mathrm{EM}} = U_{\mathrm{E}} + U_{\mathrm{M}}, \tag{6.39}$$

$$U_{\mathrm{E}} = \frac{1}{2}\int_{\Omega} \mathrm{d}\tau \left(\rho V - \frac{\partial \mathbf{A}}{\partial t}\cdot\mathbf{D}\right), \tag{6.40}$$

$$U_{\mathrm{M}} = \frac{1}{2}\int_{\Omega} \mathrm{d}\tau\, \mathbf{A}\cdot\left(\mathbf{J} + \frac{\partial \mathbf{D}}{\partial t}\right). \tag{6.41}$$

Neglecting the magnetic field inside the ideal circuit conductors according to the circuit rules, we take $\nabla\times\mathbf{A}$ to be zero inside the circuit. Moreover, bearing in mind that the identity

$$\nabla\cdot\left(\mathbf{J} + \frac{\partial \mathbf{D}}{\partial t}\right) = 0 \tag{6.42}$$

is generally valid, we may now apply the $\mathbf{J}\cdot\mathbf{E}$ theorem to the combination $\mathbf{A}\cdot(\mathbf{J} + \partial\mathbf{D}/\partial t)$:

$$U_{\mathrm{M}} = \frac{1}{2}\left(\oint_{\Gamma} \mathbf{dr}\cdot\mathbf{A}\right)\left(\int_{\Sigma} \mathbf{dS}\cdot\left(\mathbf{J} + \frac{\partial \mathbf{D}}{\partial t}\right)\right). \tag{6.43}$$

The loop integral clearly reduces to the total magnetic flux, which consists of the self-flux Φ_{M} and the external flux Φ_{ex}. Furthermore, due to Eq. (6.42), the surface integral of Eq. (6.41) can be calculated for any cross-section Σ that does not contain accumulated charge. Taking Σ in a perfectly conducting lead, we have $\mathbf{D} = \mathbf{0}$ and the integral reduces to the total current $I = \int_{\Sigma} \mathbf{dS}\cdot\mathbf{J}$. On the other hand, if we were choosing Σ to cross the capacitor dielectric, the current density would vanish and the integral would be equal to $\mathrm{d}\Phi_{\mathrm{D}}(t)/\mathrm{d}t$ where

$$\Phi_{\mathrm{D}}(t) = \int_{\Sigma} \mathbf{dS}\cdot\mathbf{D}(\mathbf{r}, t) \tag{6.44}$$

is the flux of the displacement vector. Since both choices of Σ should give rise to identical results, we conclude that

$$I(t) = \frac{\mathrm{d}\Phi_{\mathrm{D}}(t)}{\mathrm{d}t} \tag{6.45}$$

which confirms the observation that the circuit of Fig. 6.3 where the capacitor is in series with the other components, can only carry charging and discharging currents. In any case, we are left with

$$U_{\mathrm{M}} = \tfrac{1}{2}(\Phi_{\mathrm{M}} + \Phi_{\mathrm{ex}})I \tag{6.46}$$

or, reusing the "definition" of inductance, i.e., $\Phi_{\mathrm{M}} = LI$,

$$U_{\mathrm{M}} = \tfrac{1}{2}LI^2 + \tfrac{1}{2}\Phi_{\mathrm{ex}}I, \tag{6.47}$$

where $(1/2)LI^2$ is the familiar expression for the magnetic energy stored in the core of the inductor.

The electric energy may be rewritten in terms of capacitances in an analogous manner. The contribution of $\partial \mathbf{A}/\partial t \cdot \mathbf{D}$ in Eq. (6.40) vanishes because $\partial \mathbf{A}/\partial t$, representing the non-conservative electric field, is non-zero only inside the inductor and the generator regions, where the total electric field reduces to zero. On the other hand, for perfectly conducting leads that are also equipotential domains, the first term gives:

$$U_{\mathrm{E}} = \frac{1}{2}\sum_n Q_n V_n, \tag{6.48}$$

where V_n generally denotes the potential of the nth (ideal) conductor, containing a charge Q_n. Being expressed in terms of bare potentials, the result of Eq. (6.48) seems to be gauge dependent at a first glimpse. It should be noted however, that Eq. (6.48) has been derived within the circuit approximation, which implies that the charged conductors are not arbitrarily distributed in space, but are all part of the – localized – circuit. In particular, the charges Q_n are assumed to be stored on the plates of the capacitors of the circuit, and as such the entire set $\{Q_n\}$ can be divided into pairs of opposite charges $\{(Q_j, -Q_j)\}$. Hence, Eq. (6.48) should be read

$$U_{\mathrm{E}} = \frac{1}{2}\sum_j Q_j(V_{1j} - V_{2j}) = \frac{1}{2}\sum_j C_j(V_{1j} - V_{2j})^2, \tag{6.49}$$

where $V_{1j} - V_{2j}$ is the gauge-invariant potential difference between the plates of the jth capacitor.

The second Kirchhoff law (KCL), follows from charge conservation. The branches in the network can not store charge, unless capacitors are included. The integrated charge is denoted by Q_n and

$$\frac{\mathrm{d}Q_j}{\mathrm{d}t} = -\int \mathbf{J} \cdot \mathbf{dS} = \sum_k I_{jk}, \tag{6.50}$$

where the surface integral is over a surface around charge-storage domain and I_{jk} is the current flowing from the charge-storage region j into the jth circuit branch. As in the steady-state case, the Kirchhoff laws, in particular the expressions for the various voltage differences could only be obtained if some simplifying assumptions are made. For the inductor it was assumed that the induced magnetic field is only different from zero inside the core. For the capacitor, in a similar way it was assumed that the energy of storing the charge is localized completely between the plates. These assumptions need to be carefully checked before applying the network equations. As an illustration of this remark we emphasize that we ignored the volume integrals that are not parts of the circuits. In particular, the integral of the electric energy outside the circuit is the kinetic part of the radiation energy:

$$U_{\mathrm{E}}^{\mathrm{rad}} = -\frac{1}{2}\int_{\Omega_\infty\backslash\Omega} \mathrm{d}\tau\, \frac{\partial \mathbf{A}}{\partial t}\cdot\mathbf{D} = \frac{1}{2}\varepsilon\int_{\Omega_\infty\backslash\Omega} \mathrm{d}\tau\, \frac{\partial \mathbf{A}}{\partial t}\cdot\frac{\partial \mathbf{A}}{\partial t}, \tag{6.51}$$

and the potential energy of the radiation field:

$$U_{\mathrm{M}}^{\mathrm{rad}} = -\frac{1}{2\mu}\int_{\Omega_\infty\backslash\Omega} \mathrm{d}\tau\, (\nabla\times\mathbf{A})\cdot(\nabla\times\mathbf{A}) \tag{6.52}$$

are not considered at the level of circuit modeling.

7. Gauge conditions

The Maxwell theory of electrodynamics describes the interaction between radiation and charged particles. The electromagnetic fields are described by six quantities, the vector components of **E** and **B**. The sources of the radiation fields are represented by the charge density ρ and the current density **J**. If the sources are prescribed functions $\rho(\mathbf{r},t)$ and $\mathbf{J}(\mathbf{r},t)$, then the evolution of $\mathbf{E}(\mathbf{r},t)$ and $\mathbf{B}(\mathbf{r},t)$ is completely determined. The fields **E** and **B** may be obtained from a scalar potential V and a vector potential **A** such that

$$\mathbf{E} = -\nabla V - \frac{\partial \mathbf{A}}{\partial t}, \qquad \mathbf{B} = \nabla\times\mathbf{A}. \tag{7.1}$$

As was mentioned already in Section 3, the potentials $(V,\mathbf{A})$ are not unique. The choice

$$V \to V' = V - \frac{\partial \Lambda}{\partial t}, \qquad \mathbf{A}\to\mathbf{A}' = \mathbf{A} + \nabla\Lambda \tag{7.2}$$

gives rise two the same fields **E** and **B**. A change in potential according to Eq. (7.2) is a gauge transformation. The Lagrangian density

$$\mathcal{L} = \frac{1}{2}\varepsilon_0\left(\nabla V + \frac{\partial \mathbf{A}}{\partial t}\right)^2 - \frac{1}{2\mu_0}(\nabla\times\mathbf{A})^2 + \mathbf{J}\cdot\mathbf{A} - \rho V \tag{7.3}$$

gives rise to an action integral

$$S = \int \mathrm{d}t \int \mathrm{d}^3 r\, \mathcal{L}(\mathbf{r},t) \tag{7.4}$$

that is gauge invariant under the transformation (7.2). The gauge invariance of the Maxwell equations has been found a posteriori. It was the outcome of a consistent theory for numerous experimental facts. In modern physics invariance principles play a key role in order to classify experimental results. One often postulates some symmetry or some gauge invariance and evaluates the consequences such that one can decide whether the supposed symmetry is capable of correctly ordering the experimental data.

The equations of motion that follow from the variation of the action S are

$$-\varepsilon_0\left(\nabla^2 V + \nabla\cdot\frac{\partial \mathbf{A}}{\partial t}\right) = \rho, \tag{7.5}$$

$$\frac{1}{\mu_0}\nabla\times\nabla\times\mathbf{A} = \mathbf{J} - \varepsilon_0\frac{\partial}{\partial t}\left(\nabla V + \frac{\partial \mathbf{A}}{\partial t}\right). \tag{7.6}$$

These equations may be written as

$$M * \begin{bmatrix} V \\ \mathbf{A} \end{bmatrix} = \begin{bmatrix} \rho \\ \mathbf{J} \end{bmatrix}, \tag{7.7}$$

where the matrix operator M is defined as

$$M = \begin{bmatrix} -\varepsilon_0\nabla^2 & -\varepsilon_0\nabla\cdot\frac{\partial}{\partial t} \\ \varepsilon_0\nabla\cdot\frac{\partial}{\partial t} & \varepsilon_0\frac{\partial^2}{\partial t^2} + \frac{1}{\mu_0}\nabla\times\nabla\times \end{bmatrix}. \tag{7.8}$$

This operator is *singular*, i.e., there exist non-zero fields $(X, \mathbf{Y})$ such that

$$M * \begin{bmatrix} X \\ \mathbf{Y} \end{bmatrix} = \begin{bmatrix} 0 \\ \mathbf{0} \end{bmatrix}. \tag{7.9}$$

An example is the pair $(X, \mathbf{Y}) = (-\partial\Lambda/\partial t, \nabla\Lambda)$, where $\Lambda(\mathbf{r}, t)$ is an arbitrary scalar field.

The matrix M corresponds to the second variation of the action integral and therefore $\mathcal{L}$ corresponds to a singular Lagrangian density. The singularity of M implies that there does not exist an unique inverse matrix M^{-1} and therefore, Eq. (7.7) cannot be solved for the fields $(V, \mathbf{A})$ for given sources $(\rho, \mathbf{J})$. The singularity of the Lagrangian density also implies that not all the fields $(V, \mathbf{A})$ are independent. In particular, the canonical momentum conjugated to the generalized coordinate $V(\mathbf{r}, t)$ vanishes

$$\frac{\partial\mathcal{L}}{\partial\left(\frac{\partial V}{\partial t}\right)} = 0.$$

In fact, Gauss' law can be seen as a constraint for the field degrees of freedom and we are forced to restrict the set of field configurations by a gauge condition.

A gauge condition breaks the gauge invariance but it should not effect the theory such that the physical outcome is sensitive to it. In different words: the gauge condition should not influence the results of the calculation of the fields $\mathbf{E}$ and $\mathbf{B}$ and, furthermore, it must not make any field configurations of $\mathbf{E}$ and $\mathbf{B}$ "unreachable". Finally, the gauge condition should result into a non-singular Lagrangian density such that the potentials can be uniquely determined from the source distributions. We will now discuss a selection of gauge conditions that can be found in the physics literature.

7.1. The Coulomb gauge

The Coulomb gauge is a constraint on the components of the vector potential such

$$C[\mathbf{A}] \equiv \nabla\cdot\mathbf{A} = 0. \tag{7.10}$$

The constraint can be included in the action, S, by adding a term to the Lagrangian that explicitly breaks the gauge invariance of the action. The new action becomes "gauge-conditioned". We set:

$$S \to S_{\text{g.c.}} = S_0 + \frac{\lambda}{\mu_0} \int \mathrm{d}t\, \mathrm{d}\tau\, C^2[\mathbf{A}], \tag{7.11}$$

where $S_{\text{g.c.}}$ it the gauge-conditioned action, S_0 is the gauge-invariant action and λ is a dimensionless parameter. Then the equations for the potentials are

$$-\varepsilon_0 \left(\nabla^2 V + \nabla \cdot \frac{\partial \mathbf{A}}{\partial t} \right) = \rho, \tag{7.12}$$

$$\frac{1}{\mu_0} \nabla \times \nabla \times \mathbf{A} - 2\frac{\lambda}{\mu_0} \nabla (\nabla \cdot \mathbf{A}) = \mathbf{J} - \varepsilon_0 \frac{\partial}{\partial t} \left(\nabla V + \frac{\partial \mathbf{A}}{\partial t} \right). \tag{7.13}$$

The parameter λ, can be chosen freely. Exploiting the constraint in Eqs. (7.10) and (7.12), we obtain

$$-\varepsilon_0 \nabla^2 V = \rho, \tag{7.14}$$

$$\left(\varepsilon_0 \frac{\partial^2}{\partial t} - \frac{1}{\mu_0} \nabla^2 \right) \mathbf{A} = \mathbf{J} - \varepsilon_0 \frac{\partial}{\partial t} (\nabla V), \tag{7.15}$$

$$\nabla \cdot \mathbf{A} = 0. \tag{7.16}$$

Eq. (7.14) justifies the name of this gauge: the scalar potential is the instantaneous Coulomb potential of the charge distribution.

Eqs. (7.14) and (7.15) can be formally solved by Green functions. In general, a Green function corresponding to a differential operator Δ is the solution of the following equation:

$$\Delta * G(\mathbf{r}, \mathbf{r}') = \delta(\mathbf{r} - \mathbf{r}'). \tag{7.17}$$

We have already seen that the Coulomb problem can be solved by the Green function $G(\mathbf{r}, \mathbf{r}') = -(1/4\pi)\delta(\mathbf{r} - \mathbf{r}')$. It should be emphasized that the Green function is not only determined by the structure of the differential operator but also by the boundary conditions. The wave equation (7.15) can also be formally solved by a Green function obeying

$$\left(\frac{1}{c^2} \frac{\partial^2}{\partial t^2} - \nabla^2 \right) G(\mathbf{r}, t, \mathbf{r}', t') = \delta(\mathbf{r} - \mathbf{r}')\delta(t - t'), \tag{7.18}$$

such that

$$\mathbf{A}(\mathbf{r}, t) = \int_{-\infty}^{\infty} \mathrm{d}t' \int \mathrm{d}\tau'\, G(\mathbf{r}, t, \mathbf{r}', t') \left(\mathbf{J}(\mathbf{r}', t') - \varepsilon \frac{\partial}{\partial t} \nabla V \right). \tag{7.19}$$

In free space the Green function is easily found by carrying out a Fourier expansion

$$G(\mathbf{r}, t, \mathbf{r}', t') = \frac{1}{(2\pi)^4} \int_{-\infty}^{\infty} \mathrm{d}\omega \int \mathrm{d}^3\mathbf{k}\, G(\omega, \mathbf{k}) \exp\bigl[\mathrm{i}\bigl(\omega(t - t') - \mathbf{k} \cdot (\mathbf{r} - \mathbf{r}')\bigr)\bigr]. \tag{7.20}$$

Defining $k^2 = (\omega/c)^2 - |\mathbf{k}|^2$, the Green function is $G(\omega, \mathbf{k}) = k^{-2}$. In order to respect physical causality the $(\omega, \mathbf{k})$ – integration should be done in such a way that the *retarded* Green function is obtained. This can be done by adding an infinitesimal positive shift to the poles of the Green function or propagator in the momentum representation, i.e., $G(\omega, \mathbf{k}) = 1/(k^2 - \mathrm{i}\varepsilon)$. The ω-integral then generates a step function in the difference of the time arguments

$$\frac{1}{2\pi}\int_{-\infty}^{\infty} \mathrm{d}\omega \frac{\mathrm{e}^{\mathrm{i}\omega(t-t')}}{\omega - \omega_0 - \mathrm{i}\varepsilon} = \mathrm{i}\theta(t - t')\mathrm{e}^{\mathrm{i}\omega_0(t-t')}. \tag{7.21}$$

7.2. *The Lorenz gauge*

The next most commonly used gauge condition is the Lorenz gauge. In this gauge the scalar potential and vector potential are treated on an equal footing. The condition reads

$$C[\mathbf{A}, V] \equiv \nabla \cdot \mathbf{A} + \frac{1}{c^2}\frac{\partial V}{\partial t} = 0, \tag{7.22}$$

where $c^{-1} = \sqrt{\mu_0 \varepsilon_0}$ is the (vacuum) speed of light. The generic equations of motion (7.5) and (7.6) then lead to

$$\left(\frac{1}{c^2}\frac{\partial^2}{\partial t^2} - \nabla^2\right)V = \frac{\rho}{\varepsilon_0}, \tag{7.23}$$

$$\left(\frac{1}{c^2}\frac{\partial^2}{\partial t^2} - \nabla^2\right)\mathbf{A} = \mu_0 \mathbf{J}. \tag{7.24}$$

The Lorenz gauge is very suitable for performing calculations in the radiation regime. First of all, the similar treatment of all potentials simplifies the calculations and next, the traveling time intervals of the waves are not obscured by the "instantaneous" adaption of the fields to the sources as is done in the Coulomb gauge. This point is not manifest for free-field radiation, since for sourceless field solutions the absence of charges leads to $\nabla \cdot \mathbf{E} = 0$ which is solved by $V(\mathbf{r}, t) = 0$. Therefore the Coulomb gauge is suitable to handle plane electromagnetic waves. These waves have two transverse polarization modes. In the case of extended charge distributions, Gauss' law gets modified and as a consequence the scalar potential cannot be taken identically equal to zero anymore. In the Lorenz gauge, there are four fields participating in the free-field solution. Definitely two of these fields are fictitious and, as such, they are called "ghost" fields. The longitudinal polarization of an electromagnetic wave corresponds to a ghost field. Care must be taken that these unphysical fields do not have an impact on the calculation of the physical quantities **E** and **B**.

7.3. *The Landau gauge*

Various derivations of the integer quantum Hall effect (IQHE) are based on the Landau gauge. The IQHE that was discovered by VON KLITZING, DORDA and PEPPER [1980] may generally occur in two-dimensional conductors with a finite width, such as the conduction channel in the inversion layer of a metal-oxide-semiconductor field-effect transistor (MOSFET) or the potential well of a semiconductor heterojunction.

Consider a two-dimensional electron gas (2DEG) confined to a ribbon $0 \leqslant x \leqslant L$, $|y| \leqslant W/2$, $z = 0$ carrying an electron current I in the x-direction. When a homogeneous magnetic field $\mathbf{B}$ is applied perpendicularly to the strip, the electrons are deflected by the Lorentz force $-e\mathbf{v} \times \mathbf{B}$ and start piling up at one side of the strip leaving a positive charge at the other side. As a result, a transverse Hall voltage V_{H} arises and prevents any further lateral transfer of deflected electrons. This phenomenon is of course nothing but the classical Hall effect for which the Hall field is probed by the Hall resistance being defined as the ratio of the Hall voltage and the longitudinal current I:

$$R_{\mathrm{H}} = \frac{V_{\mathrm{H}}}{I}. \tag{7.25}$$

However, if the ribbon is cooled down to cryogenic temperatures and the density of the 2DEG is systematically increased by changing the gate voltage, one may observe subsequent plateaus in the Hall resistance, corresponding to a series of quantized values

$$R_{\mathrm{H}} = \frac{h}{2e^2\nu} = \frac{R_{\mathrm{K}}}{\nu}, \tag{7.26}$$

where $R_{\mathrm{K}} = h/2e^2 = 25812.8\ \Omega$ is the von Klitzing resistance and ν is a positive integer.

Moreover, each time the Hall resistance attains a plateau, the longitudinal resistance of the ribbon drops to zero, which is a clear indication of ballistic, scattering free transport. For extensive discussions on the theory of the quantum Hall effect, we refer to BUTCHER, MARCH and TOSI [1993], DATTA [1995], DITTRICH, HAENGGI, INGOLD, KRAMER, SCHOEN and ZWERGER [1997], EZAWA [2000] and all references therein. Here we would merely like to sketch how the choice of a particular gauge may facilitate the description of electron transport in terms of spatially separated, current carrying states (edge states).

The one-electron Hamiltonian reads

$$H = \frac{(\mathbf{p} + e\mathbf{A})^2}{2m} + U(y), \tag{7.27}$$

where $\mathbf{A}$ is the vector potential incorporating the external magnetic field and $U(y)$ describes the confining potential in the lateral direction. In view of the longitudinal, macroscopic current, it is quite natural to inquire whether the eigensolutions of $H\psi(x, y, z) = E\psi(x, y, z)$ are modulated by plane waves propagating along the x-direction, i.e.,

$$\psi(x, y) = \frac{1}{\sqrt{L}} \mathrm{e}^{ikx} \chi_k(y), \tag{7.28}$$

where the wave number k would be an integer multiple of $2\pi/L$ to comply with periodic boundary conditions. Clearly, the establishment of full translational invariance for the Hamiltonian proposed in Eq. (7.27) is a prerequisite and so we need to construct a suitable gauge such that the non-zero components of $\mathbf{A}$ do not depend on x. The simplest gauge meeting this requirement is the Landau gauge, which presently takes the form

$$\mathbf{A} = (-By, 0, 0), \tag{7.29}$$

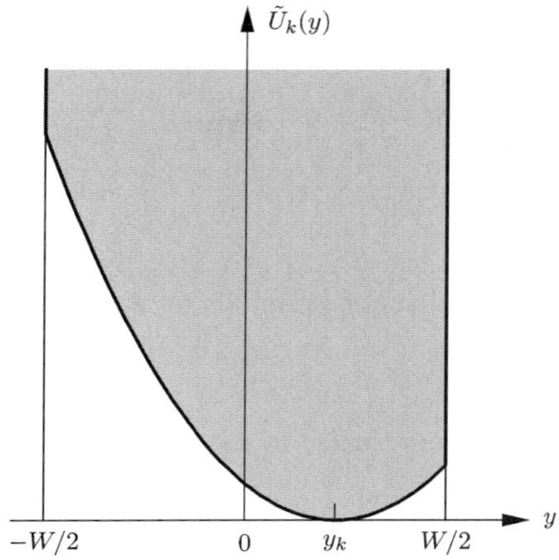

FIG. 7.1. Effective confinement potential in a Hall bar (shaded area). The bare confinement is invoked by a "hard wall" restricting the lateral motion to the interval $|y| < W/2$.

thereby giving rise to the correct magnetic field $\nabla \times \mathbf{A} = B\mathbf{e}_z$. Combining Eqs. (7.27), (7.28) and (7.29), we obtain an effective Schrödinger equation for the "transverse" wave functions $\chi_k(y)$:

$$-\frac{\hbar^2}{2m}\frac{\mathrm{d}^2\chi_k(y)}{\mathrm{d}y^2} + \left[\tilde{U}_k(y) - E\right]\chi_k(y) = 0 \tag{7.30}$$

with

$$\tilde{U}_k(y) = U(y) + \frac{1}{2}m\omega_c^2(y - y_k)^2. \tag{7.31}$$

$\tilde{U}_k(y)$ acts as an effective confinement potential, centered around its minimum at $y = y_k$ (see Fig. 7.1) where

$$y_k = \frac{\hbar k}{eB} \tag{7.32}$$

and $\omega_c = eB/m$ is the cyclotron frequency. For strong magnetic fields, the eigenfunctions of Eq. (7.30) corresponding to a given wave number k are strongly peaked around $y = y_k$ where the probability of finding an electron outside the effective potential well falls off very rapidly. In particular, when $|k|$ increases, y_k will become of the same order of magnitude as the ribbon half-width or get even larger, so that the corresponding eigenstates – the so-called "edge states" – are strongly localized near the edges of the Hall bar while states with positive momenta $\hbar k$ have no significant lateral overlap with states having negative momenta. The spatial separation of edge states with different propagation directions and the resulting reduction of scattering matrix elements is crucial for the occurrence of the quantized Hall plateaus and can obviously be investigated most conveniently by adopting the Landau gauge since the latter ensures translational invariance in the direction of the current. It should be noted however that a full analytical solution cannot be given in terms of the familiar harmonic oscillator functions

(Hermite functions) because of the edge-related boundary condition

$$\chi_k\left(\pm\frac{W}{2}\right)=0. \tag{7.33}$$

7.4. The temporal gauge

The temporal gauge is given by the condition that the scalar field V vanish identically.

$$V(\mathbf{r},t)=0. \tag{7.34}$$

The electric field is then solely represented by the time derivative of the vector potential.

$$\mathbf{E}(\mathbf{r},t)=-\frac{\partial\mathbf{A}(\mathbf{r},t)}{\partial t}. \tag{7.35}$$

In particular, this implies that for a static field the vector potential grows unboundedly in time. This gauge has the nice property that from a Lagrangian point of view the electric field is just the canonical momentum conjugated to the vector field variables, i.e.,

$$\mathcal{L}=\frac{1}{2}\varepsilon_0\left(\frac{\partial\mathbf{A}}{\partial t}\right)^2-\frac{1}{2\mu_0}(\boldsymbol{\nabla}\times\mathbf{A})^2. \tag{7.36}$$

7.5. The axial gauge

The axial gauge is a variation of the theme above. In this gauge one component of the vector potential, e.g., A_z is set identically equal to zero.

$$A_z=0. \tag{7.37}$$

This gauge may be exploited if a cylindrical symmetry is present. This symmetry can be inserted by setting

$$\mathbf{A}(\rho,\phi,z)=\big(A_\rho(\rho,\phi),A_\phi(\rho,\phi),0\big) \tag{7.38}$$

in cylindrical coordinates (ρ,ϕ,z). Then

$$\mathbf{B}=\boldsymbol{\nabla}\times\mathbf{A}=\mathbf{e}_z\frac{1}{\rho}\left(\frac{\partial}{\partial\rho}(\rho A_\phi)-\frac{\partial A_\rho}{\partial\phi}\right). \tag{7.39}$$

An infinitely thin solenoid along the z-axis corresponds to a magnetic field distribution like a “needle”, i.e., $\mathbf{B}=\Phi\delta(x)\delta(y)\mathbf{e}_z$. Such a field can be represented by the following vector potential:

$$\mathbf{A}=\frac{\Phi}{2\pi\rho}\mathbf{e}_\phi, \tag{7.40}$$

where Φ denotes the magnetic flux generated by the solenoid.

7.6. *The 't Hooft gauge*

The selection of a gauge should be done by first identifying the problem that one wants to solve. Experience has shown that a proper selection of the gauge condition is essential to handle a particular issue. At all times it should be avoided that in the process of constructing the solution one should jump ad-hoc from one gauge condition to another. There can be found examples in the literature, where this is done, e.g., a sudden jump is taken from the Coulomb gauge to the temporal gauge, without defining the transition function that accompanies such a gauge transformation. Moreover, the demonstration that the physical results are insensitive to such transitions is often neither given. The gauge fixing method due to 'T HOOFT [1971] carefully takes the above considerations into account. It illustrates the freedom in choosing a gauge condition as well as the sliding in going from one gauge condition to another. Whereas 't Hooft's original work deals with the theory of weak interactions, the ideas can also be applied to condensed matter physics. Suppose that the physical system consists of the electromagnetic fields $(V, \mathbf{A})$ and some charged scalar field ϕ. For the latter, there is a Lagrangian density

$$\mathcal{L}_{\text{scalar}} = \frac{1}{2}\mathrm{i}\hbar\left(\phi^*\frac{\partial\phi}{\partial t} - \phi\frac{\partial\phi^*}{\partial t}\right) - \frac{\hbar^2}{2m}(\nabla\phi^*)\cdot(\nabla\phi) - W(\phi^*\phi). \tag{7.41}$$

The potential W describes the (massive) mode of this scalar field and possible self-interactions. If this potential has the form

$$W(\phi^*\phi) = c_2|\phi|^2 + c_3|\phi|^3 + c_4|\phi|^4 \tag{7.42}$$

with c_2 a positive number the field ϕ then this Lagrangian density describes massive scalar particles and the vacuum corresponds to $\phi = 0$. On the other hand, if $c_2 < 0$ then the minimum of W occurs at $|\phi| \equiv \phi_0 \neq 0$. In condensed matter physics, the ground state of a superconductor has non-zero expectation value for the presence of Cooper pairs. These Cooper pairs can be considered as a new particle having zero spin, i.e., it is a boson and its charge is $2e$. The corresponding field for these bosons can be given by ϕ as above, and the ground state is characterized by some non-zero value of ϕ. This can be realized by setting $c_2 < 0$. The interaction of this scalar field with the electromagnetic field is provided by the minimal substitution procedure and leads to the following Lagrangian

$$\begin{aligned}
&\mathcal{L} = \mathcal{L}_{\text{EM}} + \mathcal{L}_{\text{scalar}} + \mathcal{L}_{\text{int}},\\
&\mathcal{L}_{\text{int}} = \mathbf{J}\cdot\mathbf{A} - \rho V - \frac{e}{m}\rho A^2,\\
&\rho = -e\phi^*\phi,\\
&\mathbf{J} = \frac{\mathrm{i}e\hbar}{2m}\left[\phi^*\nabla\phi - (\nabla\phi^*)\phi\right] + \frac{e}{m}\rho\mathbf{A}.
\end{aligned} \tag{7.43}$$

The complex field $\phi = \phi_1 + \mathrm{i}\phi_2$ can now be expanded around the vacuum expectation value $\phi = \phi_0 + \chi + \mathrm{i}\phi_2$. The interaction Lagrangian will contain terms being quadratic in the fields that mix the electromagnetic potentials with the scalar fields. Such terms can be eliminated by choosing the gauge condition in such way that these terms cancel, i.e., by properly selecting the constants α_1 and α_2 in

$$C[\mathbf{A}, V, \chi, \phi_2] \equiv \nabla \cdot \mathbf{A} + \frac{1}{c}\frac{\partial V}{\partial t} + \alpha_1 \chi + \alpha_2 \phi_2 = 0. \tag{7.44}$$

8. The geometry of electrodynamics

Electrodynamics was discovered as a phenomenological theory. Starting from early experiments with amber, permanent magnets and conducting wires, one finally arrived after much effort at Gauss' law. Biot–Savart's law and Faraday's law of induction. Only Maxwell's laws were obtained by theoretical reasoning being confirmed experimentally later on by Herz. Maxwell's great achievement was later equalized by Einstein who proposed in the general theory of relativity that

gravity = curvature.

Ever since Einstein's achievement of describing gravity in terms of non-Euclidean geometry, theoretical physics has witnessed a stunning development based on geometrical reasoning. Nowadays it is generally accepted that the standard model of matter, based on gauge theories, is the correct description (within present-day experimental accessibility) of matter and its interaction. These gauge theories have a geometrical interpretation very analogous to Einstein's theory of gravity. In fact, we may widen our definition of "geometry" such that gravity (coordinate covariance) and the standard theory (gauge covariance) are two realizations of the same mechanism. Electrodynamics is the low-energy part of the standard model. Being a major aspect of this book, it deserves special attention and in this interpretation. Besides the esthetic beauty that results from these insights, there is also pragmatic benefit. Solving electrodynamic problems on the computer, guided by the geometrical meaning of the variables has been a decisive factor for the success of the calculation. This was already realized by WILSON [1974] when he performed computer calculations of the quantum aspects of gauge theories. In order to perform computer calculations of the classical fields, geometry plays an important role as is discussed in Chapter II. However, the classical fields **E** and **B** as well as the sources ρ and **J** are invariant under gauge transformations and therefore their deeper geometrical meaning is hidden. In fact, we can identify the proper geometric character for these variables, such as scalars (zero-forms), force fields (one-forms), fluxes (two-forms) or volume densities (three-forms) as can be done for any other fluid dynamic system, but this can be done without making any reference to the geometric nature of electrodynamics in the sense that **E** and **B** represent the *curvature* in the geometrical interpretation of electrodynamics. Therefore, in this section we will consider the scalar potential and vector potential fields that do depend on gauge transformations and as such will give access to the geometry of electrodynamics.

8.1. Gravity as a gauge theory

The history of the principle of gauge invariance begins with the discovery of the principle of general covariance in general relativity. According to this principle the physical laws should maintain their form for all coordinate systems. In 1918, Hermann Weyl made an attempt to unify electrodynamics with gravity in WEYL [1918]. According to the general theory of relativity, the gravitational field corresponds to curvature of space–time, and therefore, if a vector is parallel transported along a closed loop, the angle between the starting vector and the final vector will differ from zero. Furthermore, this angle is a measure for the curvature in space. Weyl extended the Riemann geometry in such a way that not only the angle changes but also the *length* of the vector. The relative change in length is described by an anti-symmetric tensor and this tensor is invariant under changing the "unit of length". This invariance is closely related to charge conservation. Weyl called this "Maszstab Invarianz". The theory turned out to be contradictory and was abandoned, but the term "Maszstab Invarianz" survived (Maszstab = measure = gauge). With the arrival of quantum mechanics the principle of gauge invariance obtained its final interpretation: gauge invariance should refer to the phase transformations that may be applied on the wave functions. In particular, the phase transformation may be applied with different angles for different points in space and time.

$$\psi(\mathbf{r},t) \to \psi'(\mathbf{r},t) = \exp\left(\frac{ie}{\hbar}\chi(\mathbf{r},t)\right)\psi(\mathbf{r},t). \tag{8.1}$$

At first sight it looks as if we have lost the geometrical connection and the link is only historical. However, a closer look at gravity shows that the link is still present.

Starting from the idea that all coordinate systems are equivalent, we may consider a general coordinate transformation

$$x^\mu \to x'^\mu = x'^\mu(x^\nu). \tag{8.2}$$

The transformation rule for coordinate differentials is

$$\mathrm{d}x'^\mu = \frac{\partial x'^\mu}{\partial x^\nu}\,\mathrm{d}x^\nu. \tag{8.3}$$

An ordered set of functions transforming under a change of coordinates in the same way as the coordinate differentials is defined to be a *contravariant vector*

$$V'^\mu = \frac{\partial x'^\mu}{\partial x^\nu} V^\nu. \tag{8.4}$$

A *scalar* transforms in an invariant way, i.e.,

$$\phi(x) \to \phi'(x') = \phi(x). \tag{8.5}$$

The derivatives of a scalar transform as

$$V'_\mu = \frac{\partial x^\nu}{\partial x'^\mu} V_\nu. \tag{8.6}$$

Any ordered set of functions transforming under a change of coordinates as the derivatives of a scalar function is a *covariant vector*. In general, *tensors* transform according to a multiple set of pre-factors, i.e.,

$$V'^{\alpha_1\alpha_2\ldots}_{\mu_1\mu_2\ldots} = \frac{\partial x'^{\alpha_1}}{\partial x^{\beta_1}}\frac{\partial x'^{\alpha_2}}{\partial x^{\beta_2}}\frac{\partial x^{\nu_1}}{\partial x'^{\mu_1}}\frac{\partial x^{\nu_2}}{\partial x'^{\mu_2}}\cdots V^{\beta_1\beta_2\ldots}_{\nu_1\nu_2\ldots}. \tag{8.7}$$

The principle of general coordinate covariance can be implemented by claiming that all physical laws should be expressed as tensor equations. Since left- and right-hand sides will transform with equal sets of pre-factors, the form invariance is guaranteed.

So far, we have only been concerned with the change from one arbitrary coordinate system to another. One might argue that this will just hide well-known results in a thick shell of notational complexity. In order to peal off these shells and to find the physical implications one must refer to the *intrinsic* properties of the geometric structure. Occasionally, the intrinsic structure is simple, e.g., flat space time, and the familiar relations are recovered. It was Einstein's discovery that space–time is *not* flat in the presence of matter and therefore the physical laws are more involved.

Riemann geometry is a generalization of Euclidean geometry in the sense that locally one can still find coordinate systems $\xi^\mu = (ict, \mathbf{x})$, such that the distance between two near-by points is given by Pythagoras' theorem, i.e.,

$$\mathrm{d}s^2 = \delta_{\mu\nu}\,\mathrm{d}\xi^\mu\,\mathrm{d}\xi^\nu. \tag{8.8}$$

In an arbitrary coordinate system the distance is given by

$$\mathrm{d}s^2 = g_{\mu\nu}(x)\,\mathrm{d}x^\mu\,\mathrm{d}x^\nu, \tag{8.9}$$

where

$$g_{\mu\nu}(x) = \frac{\partial\xi^\alpha}{\partial x^\mu}\frac{\partial\xi^\beta}{\partial x^\nu}\delta_{\alpha\beta} \tag{8.10}$$

is the metric tensor of the coordinate system.

In the local coordinate system, ξ, the equation of motion of a freely falling particle is given by

$$\frac{\mathrm{d}^2\xi^\mu}{\mathrm{d}s^2} = 0. \tag{8.11}$$

In an arbitrary coordinate system, this equation becomes

$$\frac{\mathrm{d}}{\mathrm{d}s}\left(\frac{\partial\xi^\mu}{\partial x^\alpha}\frac{\mathrm{d}x^\alpha}{\mathrm{d}s}\right) = 0. \tag{8.12}$$

This can be evaluated to

$$\frac{\mathrm{d}^2x^\alpha}{\mathrm{d}s^2} + \Gamma^\alpha_{\mu\nu}\frac{\mathrm{d}x^\mu}{\mathrm{d}s}\frac{\mathrm{d}x^\nu}{\mathrm{d}s} = 0, \tag{8.13}$$

where $\Gamma^\alpha_{\mu\nu}$ is the *affine connection*

$$\Gamma^\alpha_{\mu\nu} = \frac{\partial x^\alpha}{\partial\xi^\beta}\frac{\partial^2\xi^\beta}{\partial x^\mu\partial x^\nu}. \tag{8.14}$$

The affine connection transform under general coordinate transformations as

$$\Gamma'^\alpha_{\mu\nu} = \frac{\partial x'^\alpha}{\partial x^\rho}\frac{\partial x^\tau}{\partial x'^\mu}\frac{\partial x^\sigma}{\partial x'^\nu}\Gamma^\rho_{\tau\sigma} + \frac{\partial x'^\alpha}{\partial x^\rho}\frac{\partial^2 x^\rho}{\partial x'^\mu\partial x'^\nu}. \tag{8.15}$$

The second term destroys the covariance of the affine connection, i.e., the affine connection is *not* a tensor.

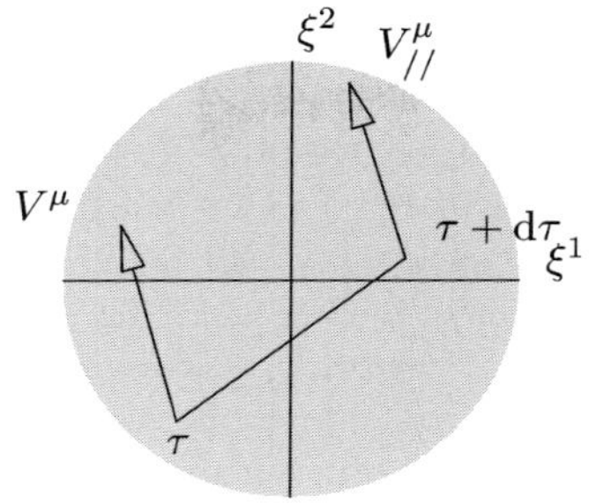

FIG. 8.1. Parallel displacement in the locally Euclidean coordinate system.

The metric tensor $g_{\mu\nu}(x)$ contains information on the local curvature of the Riemann geometry. Now consider a vector $V^{\mu}(\tau)$ along a curve $x^{\mu}(\tau)$. In the locally Euclidean coordinate system (ξ), the change of the vector along the curve is $\mathrm{d}V^{\mu}/\mathrm{d}\tau$. In another coordinate system (x'), we find from the transformation rule (8.4)

$$\frac{\mathrm{d}V'^{\mu}}{\mathrm{d}\tau} = \frac{\partial x'^{\mu}}{\partial x^{\nu}}\frac{\mathrm{d}V^{\nu}}{\mathrm{d}\tau} + \frac{\partial^2 x'^{\mu}}{\partial x^{\nu}\partial x^{\lambda}}\frac{\partial x^{\lambda}}{\partial \tau}V^{\nu}(\tau). \tag{8.16}$$

The second derivative in the second term is an inhomogeneous term in the transformation rule that prevents $\mathrm{d}V^{\mu}/\mathrm{d}\tau$ from being a vector and contains the key to curvature. This term is directly related to the affine connection. The combination

$$\frac{DV^{\mu}}{D\tau} = \frac{\mathrm{d}V^{\mu}}{\mathrm{d}\tau} + \Gamma^{\mu}_{\nu\lambda}\frac{\mathrm{d}x^{\lambda}}{\mathrm{d}\tau}V^{\nu} \tag{8.17}$$

transforms as a vector and is called the *covariant* derivative along the curve. In the restricted region where we can use the Euclidean coordinates, ξ, we may apply Euclidean geometrical methods, and in particular we can shift a vector over an infinitesimal distance from one base point to another and keep the initial and final vector parallel. This is depicted in Fig. 8.1. The component of the vector do not alter by the shift operation: $\delta V^{\mu} = 0$. Furthermore, in the local frame $x^{\mu} = \xi^{\mu}_{x(\tau)}$, the affine connection vanishes, i.e., $\Gamma^{\alpha}_{\mu\nu} = 0$. Therefore, the conventional operation of parallelly shifting a vector in the locally Euclidean coordinate system can be expressed by the equation $DV^{\mu}/D\tau = 0$. Being a tensor equation, this it true in all coordinate systems. A vector, whose covariant derivative along a curve vanishes is said to be *parallel* transported along the curve. The coordinates satisfy the following first-order differential equations:

$$\frac{\mathrm{d}V^{\mu}}{\mathrm{d}\tau} = -\Gamma^{\mu}_{\nu\lambda}\frac{\mathrm{d}x^{\lambda}}{\mathrm{d}\tau}V^{\nu}. \tag{8.18}$$

The parallel transport of a vector V^{μ} over a small distance $\mathrm{d}x^{\nu}$ changes the components of the vector by amounts

$$\delta V^{\mu} = -\Gamma^{\mu}_{\nu\lambda}V^{\nu}\delta x^{\lambda}. \tag{8.19}$$

In general, if we want to perform the differentiation of a tensor field with respect to the coordinates, we must compare tensors in two nearby points. In fact, the comparison corresponds to subtraction, but a subtraction is only defined if the tensors are

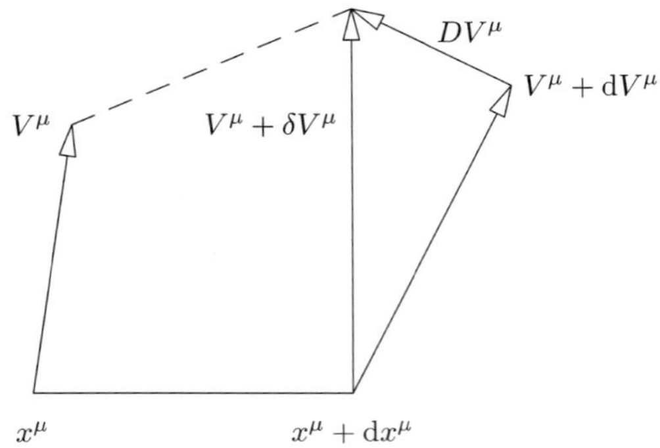

FIG. 8.2. The covariant derivative of a vector field.

anchored to the same point. (In different points, we have different local coordinate systems.) Therefore we must first parallel transport the initial tensor to the nearby point before the subtraction can be performed. This is illustrated in Fig. 8.2. For example, the covariant differential of a vector field is

$$DV^\mu = \mathrm{d}V^\mu - \delta V^\mu = \left(\frac{\partial V^\mu}{\partial x^\lambda} + \Gamma^\mu_{\lambda\kappa} V^\kappa\right) \mathrm{d}x^\lambda = D_\lambda V^\mu \, \mathrm{d}x^\lambda. \tag{8.20}$$

So far, the general coordinate systems include both accelerations originating from non-uniform boosts of the coordinate systems as well as acceleration that may be caused by gravitational field due to the presence of matter. In the first case, space–time is not really curved. In the second case space–time is curved. In order to find out whether gravitation is present one must extract information about the intrinsic properties of space–time. This can be done by the parallel transport of a vector field along a closed loop. If the initial and final vector differ, one can conclude that gravity is present. The difference that a closed loop (see Fig. 8.3) transport generates is given by

$$\Delta V^\mu = V^\mu_{\text{via B}} - V^\mu_{\text{via D}} = R^\mu_{\rho\lambda\sigma} V^\rho \delta x^\lambda \delta x^\sigma, \tag{8.21}$$

where

$$R^\mu_{\rho\lambda\sigma} = \frac{\partial \Gamma^\mu_{\rho\lambda}}{\partial x^\sigma} - \frac{\partial \Gamma^\mu_{\rho\sigma}}{\partial x^\lambda} + \Gamma^\eta_{\rho\lambda}\Gamma^\mu_{\sigma\eta} - \Gamma^\eta_{\rho\sigma}\Gamma^\mu_{\lambda\eta} \tag{8.22}$$

is the *curvature* tensor or Riemann tensor. This tensor describes the intrinsic curvature in a point.

We are now prepared to consider the geometrical basis of electrodynamics and other gauge theories but we will first summarize a few important facts:

- in each space–time point a local frame may be erected,
- the affine connection is a path-dependent quantity,
- the affine connection does not transform as a tensor,
- the field strength (curvature) may be obtained by performing a parallel transport along a closed loop.

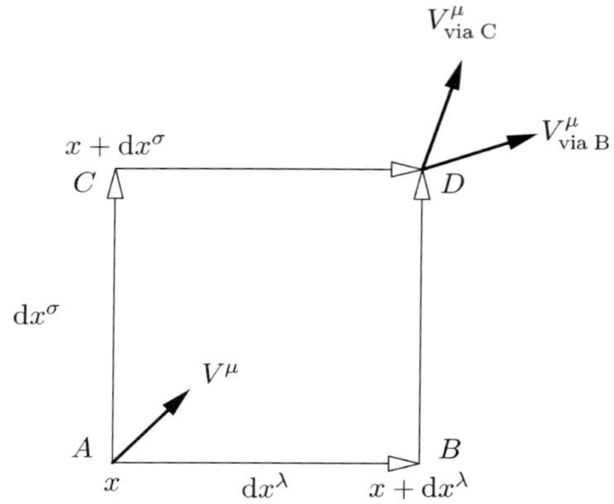

FIG. 8.3. Determination of the curvature from a round trip along a closed loop.

8.2. *The geometrical interpretation of electrodynamics*

As for the local Euclidean coordinate systems, we will consider the possibility of setting up in each space–time point a local frame for fixing the phase of the complex wave function $\psi(\mathbf{r}, t)$ (see Fig. 8.4). Since the choice of such a local frame (gauge) is not unique we may rotate the frame without altering the physical content of a frame fixing.

We can guarantee the latter by demanding appropriate transformation properties (see the above section about tensors) of the variables. Changing the local frame for the phase of a wave function amounts to

$$\begin{aligned} \psi'(\mathbf{r}, t) &= \exp\left(\frac{ie}{\hbar}\chi(\mathbf{r}, t)\right)\psi(\mathbf{r}, t), \\ \psi'^*(\mathbf{r}, t) &= \exp\left(-\frac{ie}{\hbar}\chi(\mathbf{r}, t)\right)\psi^*(\mathbf{r}, t). \end{aligned} \tag{8.23}$$

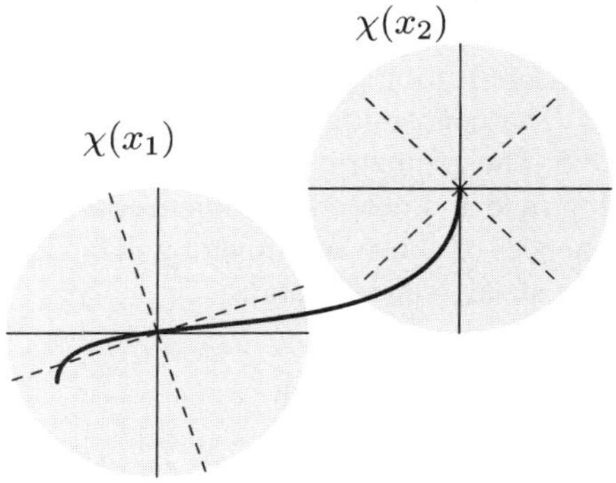

FIG. 8.4. Local frames for the phase of a wave function.

These transformation rules are similar to the contravariant and covariant transformation rules for vectors in the foregoing section. We can similarly construct a "scalar" by taking $\psi^*\psi$. The derivative of the wave function transforms as

$$\frac{\partial \psi'}{\partial x^\mu} = \exp\left(\frac{ie}{\hbar}\chi\right)\frac{\partial \psi}{\partial x^\mu} + \frac{ie}{\hbar}\frac{\partial \chi}{\partial x^\mu}\exp\left(\frac{ie}{\hbar}\chi\right)\psi. \tag{8.24}$$

The second term prevents the derivative of ψ from transforming as a "vector" under the change of gauge. However, geometry will now be of help to construct gauge covariant variables from derivatives. We must therefore postulate an "affine connection", such that a covariant derivative can be defined. For that purpose a connection, A_μ, is proposed that transforms as

$$A_\mu = A_\mu + \frac{\partial \chi}{\partial x^\mu}. \tag{8.25}$$

The covariant derivative is

$$D_\mu = \frac{\partial}{\partial x^\mu} + \frac{ie}{\hbar}A_\mu. \tag{8.26}$$

Similar to the gravitational affine connection, the field A_μ can be used to construct "parallel" transport. Therefore, the field A_μ must be assigned to the *paths* along which the transport takes place. The curvature of the connection can also be constructed by making a complete turn around a closed loop. The result is

$$F_{\mu\nu}\delta x^\mu \delta x^\nu = \oint \mathrm{d}x^\mu A_\mu, \tag{8.27}$$

where

$$F_{\mu\nu} = \frac{\partial A_\mu}{\partial x^\nu} - \frac{\partial A_\nu}{\partial x^\mu} \tag{8.28}$$

is the electromagnetic field tensor.

In order to perform numerical computations starting from the fields A_μ it is necessary to introduce a discretization grid. The simulation of a finite space or space–time domain requires that each grid point be separated by a finite distance from its neighboring points. The differential operators that appear in the continuous field equations must be translated to the discretization grid by properly referencing to the geometrical meaning of the variables. The connections A_μ should be assigned to the links of the grid, as depicted in Fig. 8.5. The geometrical interpretation suggests that this is the only correct scheme for solving field and potential problems on the computer.

The numerical consequences of above assignment will be considered in the following example. We will solve the steady-state equation

$$\begin{aligned} &\nabla \times \mathbf{B} = \mu_0 \mathbf{J}, \qquad \mathbf{B} = \nabla \times \mathbf{A}, \\ &\mathbf{J} = \sigma \mathbf{E}, \qquad \mathbf{E} = -\nabla V, \end{aligned} \tag{8.29}$$

by discretizing the set of equations on a regular Cartesian grid having N nodes in each direction. The total number of nodes in D dimensions is $M_{\text{nodes}} = N^D$. To each node

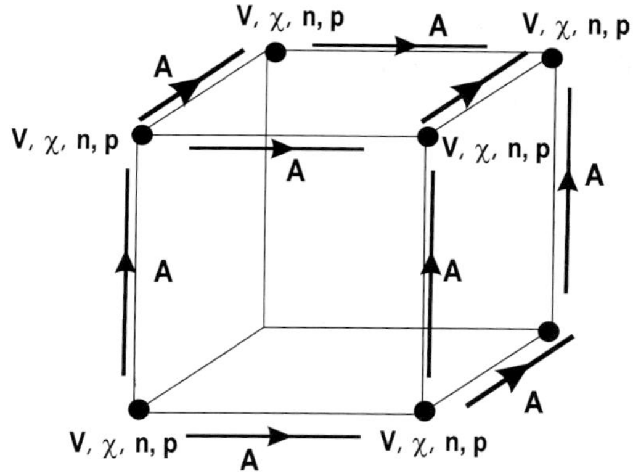

FIG. 8.5. The fundamental variables on the Cartesian grid.

we may associate D links along the positive directions, and therefore the grid has approximately DN^D links. There are $2D$ sides with each a number of $N^{(D-1)}$ nodes. Half the fraction of side nodes will not contribute a link in the positive direction. Therefore, the precise number of links in the lattice is $M_{\text{links}} = DN^D(1 - \frac{1}{N})$.

As far as the description of the electromagnetic field is concerned, the counting of unknowns for the full lattice results into M_{links} variables (A_{ij}) for the links, and M_{nodes} variables (V_i) for the nodes. Since each link (node) gives rise to one equation, the naive counting is consistent. However, we have not yet implemented the gauge condition. The conventional Coulomb gauge $\nabla \cdot \mathbf{A} = 0$, constraints the link degrees of freedom and therefore not all link fields are independent. There are $3N^3(1 - \frac{1}{N})$ link variables and $3N^3(1 - \frac{1}{N}) + N^3$ equations, including the constraints. As a consequence, at first sight it seems that we are confronted with an overdetermined system of equations, since each node provides an extra equation for $\mathbf{A}$. However, the translation of the Maxwell–Ampère equation on the lattice leads to a singular matrix, i.e., not all rows are independent. The rank of the corresponding matrix is $3N^3(1 - \frac{1}{N})$, whereas there are $3N^3(1 - \frac{1}{N}) + N^3$ rows and $3N^3(1 - \frac{1}{N})$ columns. Such a situation is highly inconvenient for solving non-linear systems of equations, where the non-linearity stems from the source terms being explicitly dependent on the fields. The application of the Newton–Raphson method requires that the matrices in the related Newton equation be non-singular and square. In fact, the non-singular and square form of the Newton–Raphson matrix can be recovered by introducing the more general gauge $\nabla \cdot \mathbf{A} + \nabla^2\chi = 0$, where an additional field χ, i.e., one unknown per node, is introduced. In this way the number of unknowns and the number of equations match again. In the continuum limit ($N \to \infty$), the field χ and one component of $\mathbf{A}$ can be eliminated. Though being irrelevant for theoretical understanding, the auxiliary field χ is essential for obtaining numerical stability on a discrete, finite lattice. In other words, our specific gauge solely serves as a tool to obtain a discretization scheme that generates a regular Newton–Raphson matrix, as explained in MEURIS, SCHOENMAKER and MAGNUS [2001].

It should be emphasized that the inclusion of the gauge-fixing field χ should not lead to unphysical currents. As a consequence, the χ-field should be a solution of $\nabla\chi = 0$. To summarize, instead of solving the problem

$$\begin{aligned} \nabla \times \nabla \times \mathbf{A} &= \mu_0 \mathbf{J}(\mathbf{A}), \\ \nabla \cdot \mathbf{A} &= 0, \end{aligned} \tag{8.30}$$

we solve the equivalent system of equations

$$\begin{aligned} \nabla \times \nabla \times \mathbf{A} - \gamma \nabla \chi &= \mu_0 \mathbf{J}(\mathbf{A}), \\ \nabla \cdot \mathbf{A} + \nabla^2 \chi &= 0. \end{aligned} \tag{8.31}$$

The equivalence of both sets of Eqs. (8.30) and (8.31) can be demonstrated by considering the action integral

$$S = -\frac{1}{2\mu_0} \int \mathrm{d}\tau \, |\nabla \times \mathbf{A}|^2 + \int \mathrm{d}\tau \, \mathbf{J} \cdot \mathbf{A}. \tag{8.32}$$

Functional differentiation with respect to $\mathbf{A}$ yields the field equations

$$\frac{\delta S}{\delta \mathbf{A}} = -\frac{1}{\mu_0} \nabla \times \nabla \times \mathbf{A} + \mathbf{J} = 0. \tag{8.33}$$

The constraint corresponding to the Coulomb gauge can be taken into account by adding a Lagrange multiplier term to the action integral

$$S = -\frac{1}{2\mu_0} \int \mathrm{d}\tau \, |\nabla \times \mathbf{A}|^2 + \int \mathrm{d}\tau \, \mathbf{J} \cdot \mathbf{A} + \gamma \int \mathrm{d}\tau \, \chi \nabla \cdot \mathbf{A} \tag{8.34}$$

and perform the functional differentiation with respect to χ

$$\frac{\delta S}{\delta \chi} = \nabla \cdot \mathbf{A} = 0. \tag{8.35}$$

Finally, the Lagrange multiplier field χ becomes a dynamical variable by adding a free-field part to the action integral

$$S = -\frac{1}{2\mu_0} \int \mathrm{d}\tau \, |\nabla \times \mathbf{A}|^2 + \int \mathrm{d}\tau \, \mathbf{J} \cdot \mathbf{A} + \gamma \int \mathrm{d}\tau \, \chi \nabla \cdot \mathbf{A} - \frac{1}{2}\gamma \int \mathrm{d}\tau \, |\nabla \chi|^2 \tag{8.36}$$

and functional differentiation with respect to $\mathbf{A}$ and χ results into the new system of equations. Physical equivalence is guaranteed provided that $\nabla\chi$ does not lead to an additional current source. Therefore, it is required that $\nabla\chi = 0$. In fact, acting with the divergence operator on the first equation of (8.31) gives Laplace's equation for χ. The solution of the Laplace equation is identically zero if the solution vanishes at the boundary.

We achieved to implement the gauge condition resulting into a unique solution and simultaneously to arrive at a system containing the same number of equations and unknowns. Hence a square Newton–Raphson matrix is guaranteed while solving the full set of non-linear equations.

8.3. Differential operators in Cartesian grids

Integrated over a test volume ΔV_i surrounding a node i, the divergence operator, acting on vector potential $\mathbf{A}$, can be discretized as a combination of 6 neighboring links

$$\int_{\Delta V_i} \nabla \cdot \mathbf{A}\, \mathrm{d}\tau = \int_{\partial(\Delta V_i)} \mathbf{A} \cdot \mathbf{dS} \sim \sum_{k}^{6} S_{ik} A_{ik}. \tag{8.37}$$

The symbol $\sim$ represents the conversion to the grid formulation and $\partial(\Delta V_i)$ denotes the boundary of ΔV_i.

Similarly, the gradient operator acting on the ghost field χ or any scalar field V, can be discretized for a link ij using the nodes i and j. Integration over a surface S_{ij} perpendicular to the link ij gives

$$\int_{\Delta S_{ij}} \nabla \chi \cdot \mathbf{dS} \sim \frac{\chi_j - \chi_i}{h_{ij}} S_{ij}, \tag{8.38}$$

where h_{ij} denotes the length of the link between the nodes i and j.

The gradient operator for a link ij, integrated along the link ij, is given by

$$\int_{\Delta L_{ij}} \nabla \chi \cdot \mathbf{dr} \sim \chi_j - \chi_i. \tag{8.39}$$

The *curl–curl* operator can be discretized for a link ij using a combination of 12 neighboring links and the link ij itself. As indicated in Fig. 8.6, the field $\mathbf{B}_i$ in the center of the "wing" i, can be constructed by taking the circulation of the vector potential $\mathbf{A}$

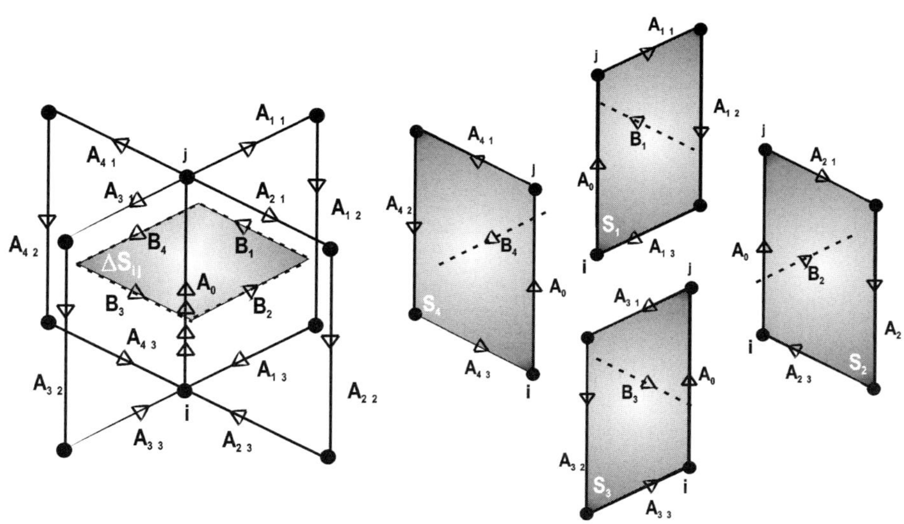

FIG. 8.6. The assembly of the $\nabla \times \nabla \times$-operator using 12 contributions of neighboring links.

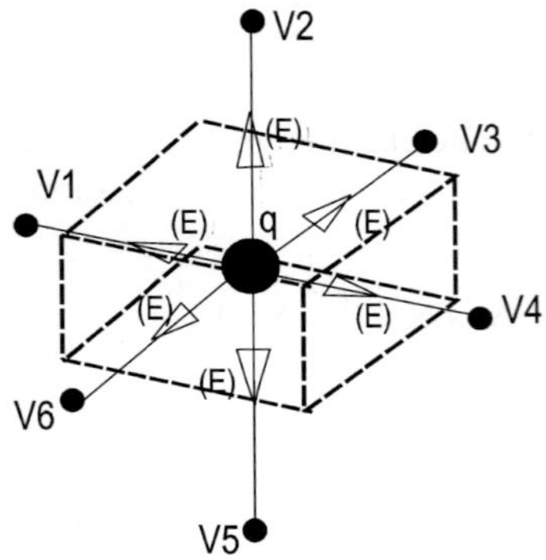

FIG. 8.7. The assembly of the $\nabla \cdot \nabla$-operator using 6 contributions of neighboring nodes.

around the wing i $(i = 1, 4)$

$$\mathbf{B}_i S_i = \sum_{j=1}^{3} \boldsymbol{A}_{ij} h_{ij} + \boldsymbol{A}_0 h_0, \tag{8.40}$$

where h_α is the length of the corresponding link α. Integration over a surface S_{ij} perpendicular to the link ij yields a linear combination of different $\boldsymbol{A}_{ij}$'s, the coefficients of which are denoted by Λ_{ij}.

$$\begin{aligned}\int_{\Delta S_{ij}} \nabla \times \nabla \times \mathbf{A} \cdot \mathbf{dS} = \int_{\partial(\Delta S_{ij})} \nabla \times \mathbf{A} \cdot \mathbf{dr} = \int_{\partial(\Delta S_{ij})} \mathbf{B} \cdot \mathbf{dr} \\ \sim \Lambda_{ij} A_{ij} + \sum_{kl}^{12} \Lambda_{ij}^{kl} A_{kl}.\end{aligned} \tag{8.41}$$

The div-grad (Laplacian) operator can be discretized (see Fig. 8.7) being integrated over a test volume ΔV_i surrounding a node i as a combination of 6 neighboring nodes and the node i itself.

$$\int_{\Delta V_i} \nabla \cdot (\nabla \chi)\, \mathrm{d}\tau = \int_{\partial(\Delta V_i)} \nabla \chi \cdot \mathbf{dS} \sim \sum_{k}^{6} S_{ik} \frac{\chi_k - \chi_i}{h_{ik}}. \tag{8.42}$$

8.4. Discretized equations

The fields $(\mathbf{A}, \chi)$ need to be solved throughout the simulation domain, i.e., for conductors, semiconducting regions as well as for the dielectric regions. The discretization of these equations by means of the box/surface-integration method gives

$$\int_{\Delta S} (\nabla \times \nabla \times \mathbf{A} - \gamma \nabla \chi - \mu_0 \mathbf{J}) \cdot \mathbf{dS} = 0, \tag{8.43}$$

$$\int_{\Delta V} \nabla \cdot \mathbf{J}\, \mathrm{d}\tau = 0, \tag{8.44}$$

$$\int_{\Delta V} (\nabla \cdot \mathbf{A} + \nabla^2 \chi)\, \mathrm{d}\tau = 0 \tag{8.45}$$

leading for the independent variables $\mathbf{A}$, χ to

$$\Lambda_{ij} A_{ij} + \sum_{kl}^{12} \Lambda_{ij}^{kl} A_{kl} - \mu_0 S_{ij} J_{ij} - \gamma S_{ij} \frac{\chi_j - \chi_i}{h_{ij}} = 0, \tag{8.46}$$

$$\sum_k^6 S_{ik} J_{ik} = 0, \tag{8.47}$$

$$\sum_k^6 S_{ik} \left(A_{ik} + \frac{\chi_k - \chi_i}{h_{ik}} \right) = 0. \tag{8.48}$$

Depending on the region under consideration, the source terms $(Q_i, \mathbf{J}_{ij})$ differ. In a conductor we implement Ohm's law, $\mathbf{J} = \sigma \mathbf{E}$ on a link ij:

$$J_{ij} = -\sigma_{ij} \left(\frac{V_j - V_i}{h_{ij}} \right) \tag{8.49}$$

and Q_i is determined by charge conservation.

For the semiconductor environment we follow the Scharfetter–Gummel scheme (SCHARFETTER and GUMMEL [1969]). In this approach, the diffusion equations

$$\mathbf{J} = q\mu c \mathbf{E} \pm kT \mu \nabla c, \tag{8.50}$$

where the plus (minus) sign refers to negatively (positively) charged particles and c denotes the corresponding carrier density. It is assumed that both the current $\mathbf{J}$ and vector potential $\mathbf{A}$ are constant along a link and that the potential V and the gauge field χ vary linearly along the link. Adopting a local coordinate axis u with $u = 0$ corresponding to node i, and $u = h_{ij}$ corresponding to node j, we may integrate Eq. (8.50) along the link ij to obtain

$$J_{ij} = q\mu_{ij} c \left(\frac{V_i - V_j}{h_{ij}} \right) \pm k_\mathrm{B} T \mu_{ij} \frac{\mathrm{d}c}{\mathrm{d}u} \tag{8.51}$$

which is a first-order differential equation in c. The latter is solved using the aforementioned boundary conditions and gives rise to a non-linear carrier profile. The current J_{ij} can then be rewritten as

$$\frac{J_{ij}}{\mu_{ij}} = -\frac{\alpha}{h_{ij}} B\left(\frac{-\beta_{ij}}{\alpha} \right) c_i + \frac{\alpha}{h_{ij}} B\left(\frac{\beta_{ij}}{\alpha} \right) c_j, \tag{8.52}$$

where $B(x)$ is the Bernoulli function

$$B(x) = \frac{x}{\mathrm{e}^x - 1} \tag{8.53}$$

and

$$\alpha = \pm k_\mathrm{B} T, \tag{8.54}$$

$$\beta_{ij} = q(V_i - V_j). \tag{8.55}$$

8.5. Examples

We present a few examples demonstrating that the proposed potential formulation in terms of the Poisson scalar field V, the vector potential field $\mathbf{A}$ and the ghost field χ, is a viable method to solve the Maxwell field problem. All subtleties related to that formulation, i.e., the positioning of the vector potential on links, and the introduction of the ghost field χ, are already encountered in constructing the solutions of the static equations (SCHOENMAKER and MEURIS [2002]).

8.5.1. Crossing wires

The first example concerns two crossing wires and thereby addresses the three-dimensional features of the solver. The structure is depicted in Fig. 8.8 and has four

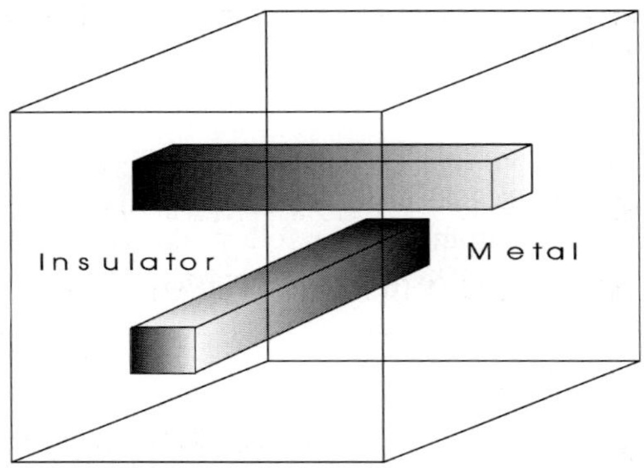

FIG. 8.8. Layout of two crossing wires in insulating environment.

TABLE 8.1
Some characteristic results for two crossing wires

Electric energy (J)		Magnetic energy (J)	
$\frac{1}{2}\varepsilon_0 \int_\Omega \mathrm{d}\tau\, E^2$	1.03984×10^{-18}	$\frac{1}{2\mu_0} \int_\Omega \mathrm{d}\tau\, B^2$	2.89503×10^{-11}
$\frac{1}{2} \int_\Omega \mathrm{d}\tau\, \rho\phi$	1.08573×10^{-18}	$\frac{1}{2} \int_\Omega \mathrm{d}\tau\, \mathbf{J}\cdot\mathbf{A}$	2.92924×10^{-11}

TABLE 8.2
Some characteristic results for a square coaxial cable

a	b	b/a	L	
μm	μm		(cylindrical) (nH)	(square) (nH)
2	6	3	220	255
1	5	5	322	329
1	7	7	389	390
1	10	10	461	458

ports. In the simulation we put one port at 0.1 V and kept the other ports grounded. The current is 4 A. The simulation domain is $10 \times 10 \times 14\ \mu m^3$. The metal lines have a perpendicular cross section of $2 \times 2\ \mu m^2$. The resistivity is $10^{-8}\ \Omega\,m$. In Tables 8.1–8.3, some typical results are presented. The energies have been calculated in two different ways and good agreement is observed. This confirms that the new methods underlying the field solver are trustworthy. The χ-field is zero within the numerical accuracy, i.e., $\chi \sim O(10^{-14})$.

8.5.2. *Square coaxial cable*

To show that also inductance calculations are adequately addressed, we calculate the inductance per unit length (L) of a square coaxial cable as depicted in Fig. 8.9. The inductance of such a system with inner dimension a and outer dimension b, was calculated from

$$l \times \frac{1}{2} L I^2 = \frac{1}{2\mu_0} \int_{\Omega} B^2\,d\tau = \frac{1}{2} \int_{\Omega} d\tau\, \mathbf{J} \cdot \mathbf{A} \tag{8.56}$$

with l denoting the length of the cable. As expected, for large values of the ratio $r = b/a$, the numerical result for the square cable approaches the analytical result for a cylindrical cable, $L = (\mu_0/2\pi)\ln(b/a)$.

TABLE 8.3
Some characteristic results for the spiral inductor

Electric energy (J)		Magnetic energy (J)	
$\frac{1}{2}\varepsilon_0 \int_{\Omega} d\tau\, E^2$	2.2202×10^{-18}	$\frac{1}{2\mu_0} \int_{\Omega} d\tau\, B^2$	3.8077×10^{-13}
$\frac{1}{2} \int_{\Omega} d\tau\, \rho\phi$	2.3538×10^{-18}	$\frac{1}{2} \int_{\Omega} d\tau\, \mathbf{J} \cdot \mathbf{A}$	3.9072×10^{-13}

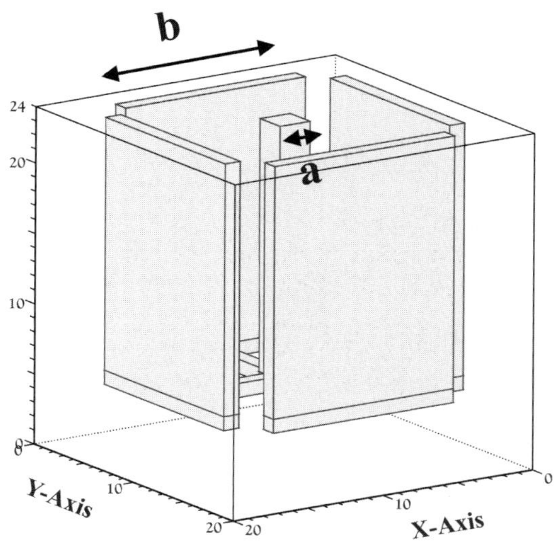

FIG. 8.9. Layout of the square coax structure.

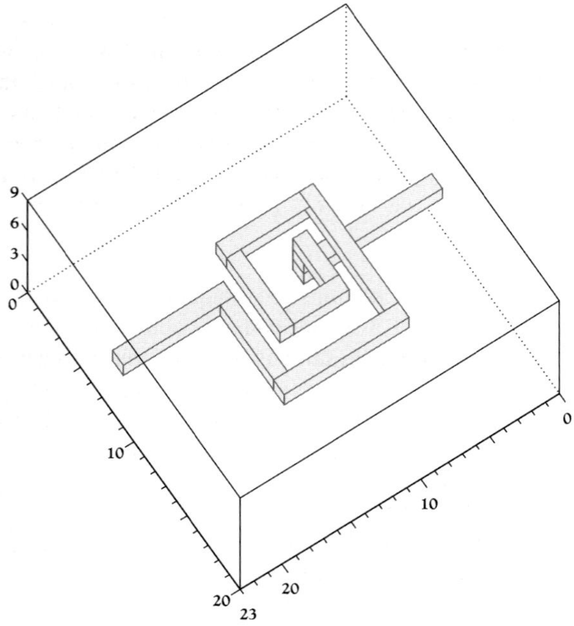

FIG. 8.10. Layout of the spiral inductor structure.

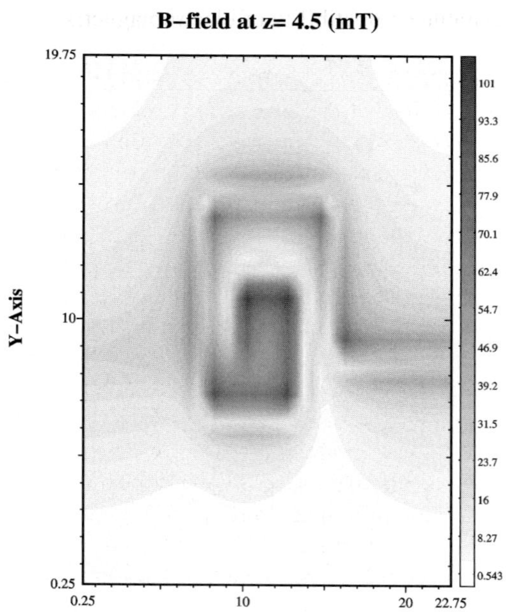

FIG. 8.11. Magnetic field strength in the plane of the spiral inductor.

8.5.3. *Spiral inductor*

A spiral inductor, as shown in Fig. 8.10 was simulated. This structure also addresses the three-dimensional features of the solver. The cross-section of the different lines is 1 μm × 1 μm. The overall size of the structure is 8 μm × 8 μm and the simulation domain is 23 × 20 × 9 μm^3. The resistance is evaluated as $R = V/I$ and equals 0.54 Ω. In Fig. 8.11, the intensity of the magnetic field is shown at height 4.5 μm. From the results in Table 8.3 we obtain that the inductance of the spiral inductor is 4.23×10^{-11} Henry.

9. Outlook

The preceding sections have been meant to offer the reader a glimpse of the achievements and the present activities in the field of numerical modeling of electromagnetic problems within the framework of 19th century classical electromagnetism that was physically founded by MAXWELL [1954a], MAXWELL [1954b], Faraday, Lenz, Lorentz and many others, and mathematically shaped by the upcoming vector calculus of those days (MORSE and FESHBACH [1953]).

The enormous predictive power of the resulting, "classical" electromagnetic theory and the impressive technological achievements that have emerged from it, may create the false impression that, from the physics point of view, electromagnetism has come to a dead end where no new discoveries should be expected and all remaining questions are reduced to the numerical solubility of the underlying mathematical problems.

Truly, after the inevitable compatibility of electromagnetism with the theory of relativity (EINSTEIN, LORENTZ, MINKOWSKI and WEYL [1952]) had been achieved and the theory of quantum electrodynamics (QED) (SCHWINGER [1958]) had been successfully established in the first half of the 20th century, neither new fundamental laws nor extensions of the old Maxwell theory have been proposed ever since.

Nevertheless, as was pointed out already in Section 8, modern concepts borrowed from the theory of differential geometry turn out to provide exciting alternatives to formulate the laws of electromagnetism and may gain new insights similar to the understanding of the intimate link between gravity and geometrical curvature of the Minkowski space. Moreover, recent technological developments in the fabrication of nanometer-sized semiconductor structures and mesoscopic devices (DATTA [1995]) have raised new as well as unanswered old questions concerning the basic quantum mechanical features of carrier transport in solids and its relation to both externally applied and induced electromagnetic fields. The topology of electric circuits such as mesoscopic rings carrying persistent currents, mesoscopic devices with macroscopic leads, including quantum wires, quantum dots, quantum point contacts, Hall bars, etc. appears to be a major component determining the transport properties. In particular, the spatial localization of both the electromagnetic fields and the carrier energy dissipation plays an essential role in the quantum theory that governs carrier transport.

This section addresses just a few topics of the above mentioned research domain in order to illustrate that quantum mechanics is invoked not only to provide a correct description of the particles participating in the electric current but also to extend the theory of the electromagnetic field beyond the framework of Maxwell's equations and QED. As the corresponding research area is still being established and theoretical understanding is often still premature, several statements presented in the remainder of this chapter, should be regarded as possible but not final answers to existing problems, thereby mainly reflecting the personal view of the authors. A more detailed treatment of the topics considered below can be found in MAGNUS and SCHOENMAKER [2000c] and MAGNUS and SCHOENMAKER [2002].

9.1. *Quantum mechanics, electric circuits and topology*

Quantization of the electric conductance of quantum point contacts (QPC) is a striking example of a transport phenomenon that cannot be accounted for by combining classical electrodynamics with conventional transport theory that inherently neglects preservation of phase coherence. A typical QPC consists of a two-dimensional electron gas (2DEG) residing at in a high-mobility semiconductor structure near the interface of, say an AlGaAs/GaAs heterojunction, whereas a negatively biased gate provides a narrow constriction hampering the electron flow in the direction perpendicular to the gate arms (see Fig. 9.1). While the length of the gate arms (along the propagation direction may be of the order of 1 μm, the width is usually smaller than 250 nm. Experimentally, conductance quantization was originally observed by the groups of WHARAM, THORNTON, NEWBURY, PEPPER, AHMED, FROST, HASKO, PEACOCK, RITCHIE and JONES [1988] and VAN WEES, VAN HOUTEN, BEENAKKER, WILLIAMSON, KOUWENHOVEN, VAN DER MAREL and FOXON [1988] by connecting the QPC to an external power source (V) through a couple of conducting leads as sketched in Fig. 9.2. While the total resistance R of the circuit was determined by measuring its ohmic response to a given bias voltage V, the resistance R_Q associated with the very QPC was obtained by subtracting the resistance R_L of the two leads:

$$R_\mathrm{Q} = R - 2R_\mathrm{L}. \tag{9.1}$$

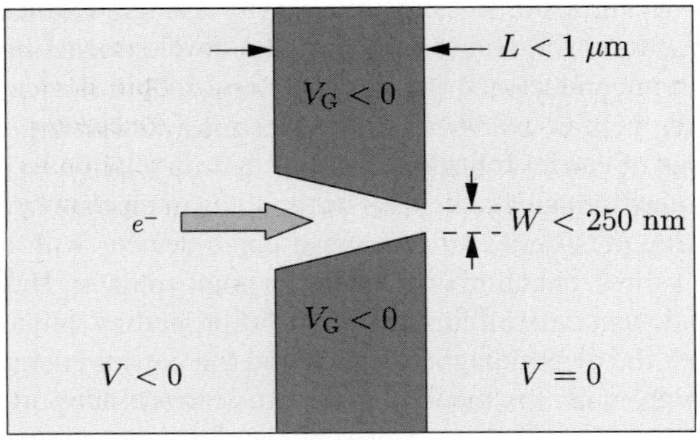

FIG. 9.1. Quantum point contact with length L and width W, considered as a two-terminal device. The source contact (left) is kept on a negative potential V with respect to the drain contact (right).

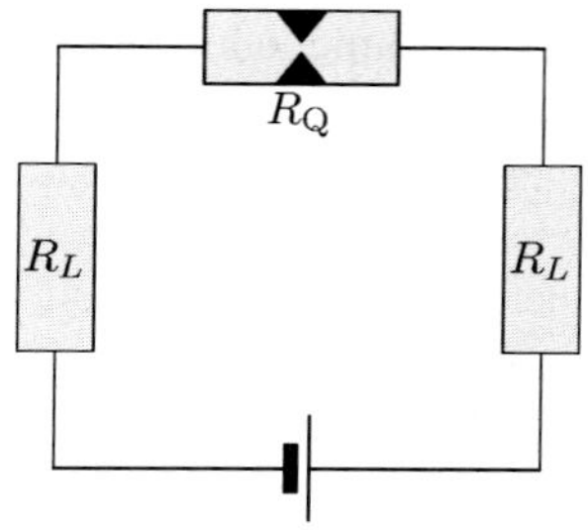

FIG. 9.2. Closed electric circuit containing a QPC connected to a DC power supply through two resistive leads.

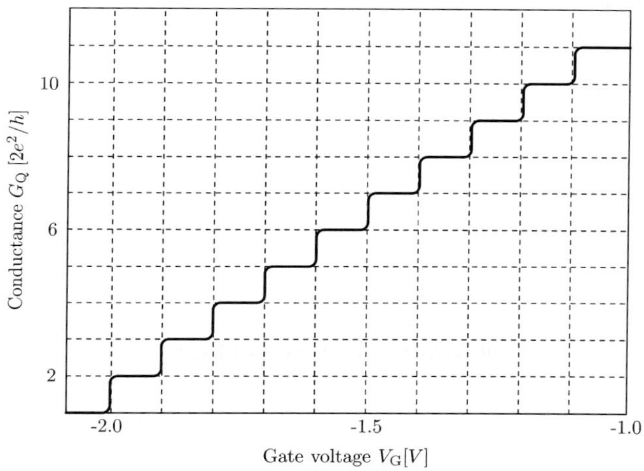

FIG. 9.3. Quantized conductance of a quantum point contact under cryogenic conditions.

After they had cooled down the QPC below 4 K, the experimentalists of both the Delft and Cambridge groups measured the circuit current I as a function of the gate voltage V_G for a fixed bias voltage. As a result, they obtained a staircase-like pattern in the profile of the electric conductance $G_Q = R_Q^{-1}$ associated with the QPC, as indicated schematically in Fig. 9.3. From this observation it follows that the conductance G_Q is quantized in units of $R_K^{-1} = 2e^2/h$ where $R_K = h/e^2 = 25128\ \Omega$ denotes von Klitzing's resistance. A quantitative description is provided by the well-known Landauer–Büttiker formula

$$G_Q = \frac{2e^2}{h} N, \tag{9.2}$$

where N is the number of "conduction channels" that are open for ballistic electron transport through the QPC, given a particular value of the gate voltage V_G. Eq. (9.2) is a special case of a formula that was proposed by LANDAUER [1957], LANDAUER [1970] to describe electron propagation through disordered materials, while it was recovered by BUETTIKER [1986] to cope with semiconductors with mesoscopic active areas. For a two-terminal device the generalized conductance formula reads

$$G = \frac{2e^2}{h} \sum_{n=1}^{N} T_n, \tag{9.3}$$

where the transmission probabilities $\{T_n\}$ reduce 1 for purely ballistic transport. Although conductance quantization in a QPC does not reach the degree of exactness suggested by the idealized drawing of Fig. 9.3, the stair-case profile has been repeatedly observed by many other researchers in the field and, consequently one should be confident in the experimental verification of this phenomenon. On the other hand, it is the strong belief of the authors that, presently – when this manuscript is being written – the commonly accepted theoretical explanation of conductance quantization runs far behind its experimental realization. It is commonly accepted that the absence of energy dissipation and other decoherence effects and, correspondingly, the preservation of the phase of the electron wave functions over a mesoscopic distance are major keys for understanding the mechanism of quantum transport. Nevertheless, numerous questions concerning the localization of energy dissipation are left unanswered by the underlying theories and a generalized, unifying transport theory connecting the macroscopic models based on the Drude–Lorenz model on one hand and the Landauer–Büttiker picture on the other hand, is still lacking. For a more extensive discussion on common models and theories leading to the Landauer–Büttiker formula, we refer to DATTA [1995], BUTCHER, MARCH and TOSI [1993], STONE and SZAFER [1988], LENSTRA and SMOKERS [1988], LENSTRA, VAN HAERINGEN and SMOKERS [1990], STONE [1992], IMRY and LANDAUER [1999]. Here we would like to summarize briefly the main results of conventional theory and discuss an alternative approach which has been proposed recently by the authors in MAGNUS and SCHOENMAKER [2000c].

In the case of conventional conductors one can easily trace back the macroscopic, electric resistance to dissipation of energy and decoherence effects that are due to various elastic and inelastic scattering mechanisms. On the other hand, the question arises why a mesoscopic, ballistic conductor the active region of which is supposed to be free of scattering, can still have a non-zero resistance. Moreover, as one may conclude from Eq. (9.3), this resistance merely depends on the fundamental constants e and $\hbar$ and a set of quantum mechanical transmission coefficients. The latter are usually extracted from the single-electron Schrödinger equation, i.e., under the assumption that many-particle interactions such as electron–electron and even electron–phonon scattering can be neglected. Consequently, the resistance of a ballistic conductor appears to be expressible in quantities that are not referring to neither decoherence nor energy dissipation. As discussed extensively in the above references, a common explanation for this phenomenon is provided by the concept of so-called contact resistance. The underlying picture considers the ballistic conductor as being connected on the "left" and the "right" to two huge reservoirs that are kept on two different chemical potentials μ_L, μ_R so as to maintain between the reservoirs a net current of electrons propagating through one or more channels of the ballistic conductor (such as a QPC or a quantum dot). Due to the mismatch of the huge, macroscopic leads and the mesoscopic active area, two interface regions separating the active area from the "bulk" of the leads. Assuming further that electrons are entering and leaving the active area without undergoing any quantum mechanical reflections in the interface regions, the latter emerge as the missing spots where the phase coherence characterizing the transport in the ballistic region is broken. In other words, the resistance associated with a mesoscopic active areas should be considered localized, being realized in the interface or "contact" regions while the main electrostatic potential drop is still falling over the active area. Even when the notion of non-local resistance is rather conceivable in a medium where phase-coherent transport along nanometer-sized paths may demand that Ohm's law $\mathbf{J}(\mathbf{r}) = \sigma \mathbf{E}(\mathbf{r})$ be generalized to $\mathbf{J}(\mathbf{r}) = \int d^3r' \sigma(\mathbf{r}, \mathbf{r}')\mathbf{E}(\mathbf{r}')$, we feel that the reservoir picture does not satisfactorily explain the phenomenon of quantized conductance. First, to the best of our knowledge, there is neither an unambiguous way of defining the contact regions interfacing between an active area and a reservoir nor a trace of experimental evidence for it. Next, invoking the chemical potentials μ_L and μ_R and the corresponding local thermal equilibria states for the two reservoirs already silently postulates the existence of a finite current without providing explicitly a current limiting mechanism. Moreover, the equation $eV_{app} = \mu_L - \mu_R$ relating the applied bias to the chemical potential difference as a crucial step in conventional treatments, is simply taken for granted (sometimes even taken as a definition of bias voltage!) while FENTON [1992], FENTON [1994] already pointed out that it should be rigorously derived from quantum mechanical first principles. Finally, the topology of an electric circuit containing a ballistic conductor or any mesoscopic device is not reflected in the reservoir concept that treats the circuit as a simply connected, open-ended region. The latter has severe consequences for the description of the driving electric field existing in the active area as will be discussed in the following lines.

For the sake of simplicity, we will consider a DC power source providing the electric circuit with the energy required to maintain a steady current of electrons flowing through a toroidal (doughnut-shaped, torus-like) circuit Ω. In addition, we will assume that no external magnetic field is applied in the circuit region so that the only magnetic field existing in the torus is the self-induced one which is constant in time. According to the third Maxwell equation, the total electric field acting on the electrons in the circuit, should therefore be irrotational, i.e.,

$$\nabla \times \mathbf{E} = \mathbf{0}. \tag{9.4}$$

In spite of Eq. (9.4), the electric field $\mathbf{E}$ is *not* conservative. Indeed, the electromotive force or EMF characterizing the strength of the DC power source, is nothing but the non-vanishing loop integral of $\mathbf{E}$ around any closed curve Γ lying in the interior of the torus and encircling the hole of the torus once and only once (winding number = 1). According to Stokes' theorem for multiply connected regions the curve Γ is arbitrary as long as it is located in a region where $\nabla \times \mathbf{E}$ vanishes, so any internal curve of Ω will do. Physically, the EMF represents the work done by the electric field on a unit charge that makes one complete turn around the circuit (moving along Γ). As an immediate consequence, we need to be most careful when dealing with innocent looking quantities such as electrostatic potential and the notion of potential difference. While an irrotational field $\mathbf{E}(\mathbf{r})$ can always be derived from a scalar potential $V(\mathbf{r})$ in any *simply-*connected subset of the torus (see the Helmholtz theorem), there exists no such scalar potential doing the job along the entire circuit. Mathematically speaking, one could of course imagine a brute force definition for such a potential anyway, namely the line integral of the field **E** along a subset of Γ connecting some reference point $\mathbf{r}_0$ with the field point **r**. However, since the circulation of **E** is non-zero when **r** travels all around Γ, the value of such a potential would unlimitedly increase (or decrease) when **r** keeps on traveling around the circuit. This would give rise to a potential function $V(x_1, x_2, x_3)$ that would be multivalued in the cyclic coordinate, say x_3. Such a function would clearly be unacceptable from the physical point of view which requires all physically meaningful functions to be periodic in x_3. It goes without saying that the concept of EMF is hardly conceivable in a theory describing the electric circuit as an open-ended region. Such a simply-connected region exclusively leads to conservative, irrotational electric fields that cannot give rise to a steady energy supply. The latter is therefore emulated by introducing position dependent chemical potentials artificially keeping the lead reservoirs on different levels of electron supply.

It should be noted at this point that the above topology considerations have already given rise to at least two major conceptual differences between open-ended conductors and closed electric circuits.

First, electrons entering the active area coming from one lead and moving to the other are never seen to return to their "origin" except when they are reflected.[7] As such, the open-ended conductor is very similar to a system of two large water buckets, one of them being emptied into the other through a narrow tube. Although the water flow resembles a steady flow after the initial and before the final transient regime, the water is not being pumped back into the first bucket and the flow trivially stops when the first bucket is empty. On the contrary, although quantum mechanics does not allow an accurate localization of electrons in the transport direction when they reside in delocalized, current carrying states, the electrons are confined to the interior of the circuit region and will make a huge number of turns when a steady-state current is maintained on a macroscopic time scale. Next, in most cases the open-ended conductor model leads to an artificial, spatial division of the circuit into a finite active area and two infinite lead regions. Indeed, position dependence is not only inferred for the chemical potential, in various treatments such as DATTA [1995] one also assigns separate sets of energy spectra and their corresponding quantum states to the three distinct regions: two continuous energy spectra representing the huge and wide leads and a discrete spectrum providing a small number of conduction channels (referred to as N in the Landauer–Büttiker formula). Moreover, at both interfaces emerges a mismatch between the enumerable discrete spectrum and the two continuous spectra and this very mismatch is even considered the origin of the so-called "contact resistance" explaining the phenomenon of conductance quantization.

However, it is known from elementary quantum mechanics that energy and position, being represented by non-commuting operators cannot be simultaneously measured. In other words, there is no physical ground for setting up different quantum mechanical treatments of distinct spatial areas (unless they are completely isolated from each other thereby preventing any exchange of particles, which is obviously not the case). Treating

[7] In principle electrons may undergo quantum mechanical reflections at the interfaces between the lead and the active part of the device, but these reflections are explicitly ignored in most of the conventional theories.

the complete circuit – including power source, conducting leads and mesoscopic active area – as a single quantum mechanical entity, a single spectrum of allowed energies and corresponding eigenstates is to be assigned to the entire circuit, not to parts of it. Clearly, unless we are discussing isolated microcircuits such as mesoscopic rings carrying persistent currents, the circuit inevitably becomes huge, due to the presence of the huge leads. Consequently, the single energy spectrum turns out to be a continuous one, consisting virtually of all energies that are accessible by the circuit system. On the other hand, the influence of the active area with either its narrow spatial confinement (QPC) or its huge potential barriers is reflected in the occurrence of a discrete set of sharply peaked resonances emerging in the quantum mechanical transmission coefficient as a function of energy. The corresponding states are genuine "conduction channel states" allowing an appreciable transmission of electrons, while the latter is negligible for any other state. In this picture however, there is no "mismatch" between quantum states, since all states simply pertain to the entire system and only the wave functions (not the energies) depend on position. Consequently, the notion of contact resistance relying on the existence of a mismatch of states, looses its meaning and the basis question remains: what causes the resistance of a mesoscopic active area embedded in a closed electric circuit and why does it take the form of Eq. (9.3)?

Being inspired by the experimental setup, we propose to consider the simplest possible, closed circuit, i.e., a torus-shaped region Ω consisting of a DC power source ("battery" region Ω_B), two ideally conducting leads Ω_{1L} and Ω_{2L} connecting the active area Ω_A, as depicted in Fig. 9.4. In general, the electric field in the circuit region may be decomposed into a conservative and non-conservative part:

$$\mathbf{E}(\mathbf{r}) = \mathbf{E}_C(\mathbf{r}) + \mathbf{E}_{NC}(\mathbf{r}), \tag{9.5}$$

where the conservative component $\mathbf{E}_C$ is derived from an appropriate scalar potential which is periodic along any interior, closed loop Γ (see Fig. 9.5),

$$\mathbf{E}_C(\mathbf{r}) = -\nabla V(\mathbf{r}) \tag{9.6}$$

with

$$\oint_\Gamma \mathbf{E}_C(\mathbf{r}) \cdot \mathbf{dr} = 0, \tag{9.7}$$

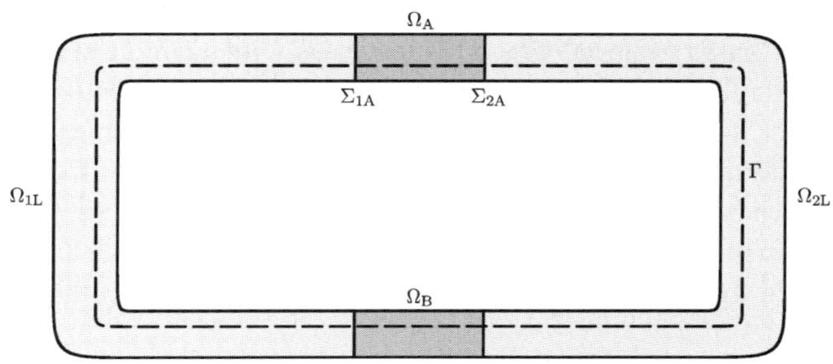

FIG. 9.4. Toroidal electric circuit. (Figure reproduced by permission of the American Physical Society and Springer Verlag.)

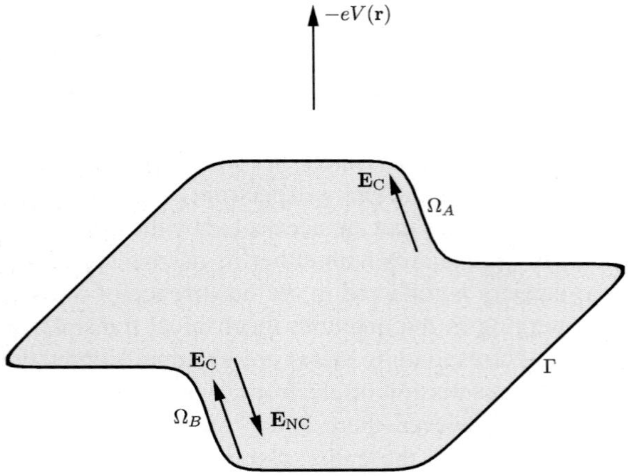

FIG. 9.5. Electrostatic potential energy profile along Γ. (Figure reproduced by permission of the American Physical Society and Springer Verlag.)

whereas the EMF is entirely due to the non-conservative component $\mathbf{E}_{\rm NC}$:

$$V_\varepsilon = \oint_\Gamma \mathbf{E}_{\rm NC}(\mathbf{r}) \cdot \mathbf{dr}. \tag{9.8}$$

Taking the leads to be ideal, dissipationless conductors (which corresponds to the subtraction of the lead resistances of the experimental result setup), we implicitly require that the total electric field vanishes in the leads:

$$\mathbf{E}(\mathbf{r}) = \mathbf{0} \quad \text{for } \mathbf{r} \in \Omega_{\rm 1L} \text{ or } \Omega_{\rm 2L}. \tag{9.9}$$

Furthermore, as we are looking for a universal mechanism that is able to limit the current in a mesoscopic circuit, we have explicitly omitted any source of incidental inelastic scattering and hence neglected all energy dissipation in the circuit, including the internal resistance of the power source. For the sake of simplicity we have also assumed that the non-conservative electric field component $\mathbf{E}_{\rm NC}$ is strictly localized in the seat of the EMF, i.e., in the "battery region" $\Omega_{\rm B}$. This leaves us with a circuit where free electrons can pile up only in the active region due to electrostatic confinement or the presence of a potential barrier, while the leads appear to be equipotential volumes. Since the power source has no internal resistance, the non-conservative component $\mathbf{E}_{\rm NC}$ is pumping all electrons that arrived on the positive pole back to the negative pole at no energy cost. In other words, within the "battery region" $\mathbf{E}_{\rm NC}$ exactly counteracts the effect of the conservative field that would decelerate all electrons climbing up the potential hill in $\Omega_{\rm B}$ (see Fig. 9.5):

$$\mathbf{E}_{\rm NC}(\mathbf{r}) = \begin{cases} -\mathbf{E}_{\rm C}(\mathbf{r}) & \text{for } \mathbf{r} \in \Omega_{\rm B}, \\ \mathbf{0} & \text{elsewhere.} \end{cases} \tag{9.10}$$

From Eqs. (9.5)–(9.10) it follows that

$$V_{\mathcal{E}} = \oint_{\Gamma} \mathbf{E} \cdot \mathbf{dr} = \int_{\Sigma_{1\mathrm{A}}}^{\Sigma_{2\mathrm{A}}} \mathbf{E}_{\mathrm{C}}(\mathbf{r}) \cdot \mathbf{dr} = V_1 - V_2. \tag{9.11}$$

In view of the permanently available power supply and the absence of energy dissipation, one would expect the circuit current to grow unlimitedly. Indeed, the counteracting electromotive force arising from the self-induced magnetic field generated by the current, though initially delaying the current increase because of Lenz' law, would not be capable of slowing down the electron flow in the long term. The latter of course follows directly from elementary, classical mechanics but also from the equation of an L–R-circuit where the circuit resistance R tends to zero:

$$I(t) = \frac{V_{\mathcal{E}}}{R}\left(1 - \mathrm{e}^{-Rt/L}\right) \xrightarrow{R\to 0} \frac{V_{\mathcal{E}}}{L} t. \tag{9.12}$$

Clearly, this simple result does not hold if the current should become so large that radiation losses can no longer be neglected. However, the corresponding radiation resistance is typically of the order of the vacuum impedance (see JACKSON [1975]) $Z_0 = \mu_0 c \approx 120\pi\ \Omega$, which is not only smaller than von Klitzing's resistance by roughly two orders of magnitude, but also does not inherently contain the constants e and h. We therefore believe that radiation resistance is not the appropriate mechanism to explain conductance quantization.

Although the idealized circuit under consideration should not be regarded as a superconductor, we might be inspired by the phenomenon of flux quantization governing the electromagnetic response of type-I superconductors, as explained in various textbooks by many authors, such as KITTEL [1976], KITTEL [1963] and FEYNMAN, LEIGHTON and SANDS [1964b]. In type-I superconducting rings with an appreciable thickness (exceeding the coherence length), flux quantization emerges from the Meissner effect according to which all magnetic field lines are expelled from the interior of the ring, and the requirement that the wave function describing Cooper pairs in the superconducting state be single-valued when a virtual turn along an interior closed curve is made. More precisely, as stated in SAKURAI [1976], the deflection of the magnetic field causes the vector potential to be irrotational inside the ring which, in turn, allows one to fully absorb the vector potential into the phase of the wave function:[8]

$$\psi(\mathbf{r}) = \psi_0(\mathbf{r}) \exp\left(\frac{2ie}{\hbar} \int_P \mathbf{A} \cdot \mathbf{dr}\right). \tag{9.13}$$

The fields $\psi(\mathbf{r})$ and $\psi_0(\mathbf{r})$ respectively denote the wave function in the presence and absence of an irrotational vector potential and P represents an internal path connecting an arbitrary reference point with the point $\mathbf{r}$. Moving $\mathbf{r}$ all around the ring turns the line integral of $\mathbf{A}$ into the magnetic flux $\Phi = \oint \mathbf{A} \cdot \mathbf{dr}$ trapped by some closed loop Γ. Obviously, $\psi(\mathbf{r})$ becomes multi-valued unless the flux Φ equals an integer multiple of the London flux quantum $\Phi_{\mathrm{L}} = h/2e$. In the case of our circuit however, we do not consider external magnetic fields and the only magnetic field that may pierce the circuit region Ω is the self-induced magnetic field $\mathbf{B} = \nabla \times \mathbf{A}$ generated by the current

[8]The factor 2 in the phase factor reflects the charge $-2e$ of a Cooper pair.

flowing through the circuit. Though not vanishing everywhere inside Ω, **B** is circulating around the current density vector **J** representing the current distribution in the circuit. As a consequence, the azimuthal component of **B** (along **J**) will generally vanish, while each transverse component changes sign in the region where **J** is non-zero, i.e., inside the circuit region. In other words, there exists a closed, internal curve Γ_0 along which $\mathbf{B} = \mathbf{0}$ and **A** is irrotational. Hence, provided the point **r** is close enough to the curve Γ_0, we may repeat the above argument and approximately absorb **A** into the phase of the electron wave functions. Similarly, approximate flux quantization may be invoked, provided that the flux is now strictly defined as the loop integral of **A** around Γ_0 and the flux quantum is taken to be the double of the previous one, i.e., the Dirac flux quantum $\Phi_0 = h/e$. Complying with the flux quantization constraint means that any increase of the induced magnetic flux caused by an increase of the circuit current should be stepwise. Within the scope of a semi-classical picture, one could propose that an electron cannot extract energy from the power supply, unless the time slot during which it is exposed to the external electric field, is large enough to generate one quantum of induced magnetic flux. Indeed, if the energy extraction were continuous, the induced magnetic flux could be raised by an arbitrary small amount, thereby violating the (approximate) flux quantization constraint. The characteristic time τ_0 required to add one flux quantum, can easily be estimated by comparing the electron energy $\Delta E_{\mathrm{MECH},n}$ gained from the external field during a time interval $[t_n - \frac{1}{2}\tau_0, t_n + \frac{1}{2}\tau_0]$ with the corresponding magnetic energy increase ΔU_{M} of the circuit, where a flux jump occurs at $t = t_n$. Integrating the energy rate equation (2.20) from $t_n - \frac{1}{2}\tau_0$ to $t_n + \frac{1}{2}\tau_0$, we may express $\Delta E_{\mathrm{MECH},n}$ as follows:

$$\Delta E_{\mathrm{MECH},n} = \int_{t_n - \frac{1}{2}\tau_0}^{t_n + \frac{1}{2}\tau_0} \mathrm{d}t \int_{\Omega} \mathrm{d}\tau\, \mathbf{J}(\mathbf{r}, t) \cdot \mathbf{E}(\mathbf{r}, t). \tag{9.14}$$

During $[t_n - \frac{1}{2}\tau_0, t_n + \frac{1}{2}\tau_0]$, the charge density remains unchanged before and after the jump at $t = t_n$ and consequently, the current density is solenoidal, while the external electric field is irrotational. Hence, according to the recently derived $\mathbf{J} \cdot \mathbf{E}$ integral theorem for multiply connected regions (see Appendix A.1 and MAGNUS and SCHOENMAKER [1998]), we may disentangle the right-hand side of Eq. (9.14):

$$\int_{t_n - \frac{1}{2}\tau_0}^{t_n + \frac{1}{2}\tau_0} \mathrm{d}t \int_{\Omega} \mathrm{d}\tau\, \mathbf{J}(\mathbf{r}, t) \cdot \mathbf{E}(\mathbf{r}, t) = \tfrac{1}{2}[I_{n-1} + I_n] V_{\varepsilon} \tau_0, \tag{9.15}$$

where $I_n = \int_{\Sigma_{1A}} \mathbf{J}(\mathbf{r}, t_n) \cdot \mathbf{dS}$ is the net current entering the cross section Σ_{1A} at a time t_n. On the other hand the flux change $\Delta\Phi_n$ associated with the jump $\Delta I_n \equiv I_n - I_{n-1}$, reads

$$\Delta\Phi_n = L \Delta I_n, \tag{9.16}$$

where L is the inductance of the circuit. Since $\Delta\Phi_n$ is to be taken equal to Φ_0, we obtain the increased magnetic energy of the circuit:

$$\Delta U_{\mathrm{M}} = \tfrac{1}{2} L I_n^2 - \tfrac{1}{2} L I_{n-1}^2 = \tfrac{1}{2}(I_{n-1} + I_n)\Phi_0. \tag{9.17}$$

Combining Eqs. (9.14), (9.15) and (9.17) and putting $\Delta U_M = \Delta E_{MECH,n}$, we derive the following result:

$$\tau_0 = \frac{\Phi_0}{V_\varepsilon}. \tag{9.18}$$

If an electron has been sufficiently accelerated such that the time it is exposed to the localized electric field becomes smaller than τ_0, energy extraction is stopped and the one-electron current will never exceed e/τ_0. For an electron ensemble carrying spin and being distributed over N ballistic channels, the total current predicted by the Landauer–Büttiker formula (9.2) is therefore recovered:

$$I = 2N\frac{e}{\tau_0} = \frac{2e^2}{h} N V_\varepsilon. \tag{9.19}$$

In spite of the naive calculation leading to Eq. (9.19), it is shown that the interplay between circuit topology, flux quantization and the localized electric field may lead to a kind of "selection rule" prohibiting the unlimited extraction of energy from a power supply, even if all dissipative mechanisms are (artificially) turned off.

9.2. *Quantum circuit theory*

On the other hand, it goes without saying that a sound theory is required not only to support and to refine the concept of flux quantization for non-superconducting circuits, but also to bridge the gap between the rigorous, microscopic transport description and the global circuit model that is to reflect the quantum mechanical features of coherent transport through the electric circuit or part of it. Such a theory which could be called "quantum circuit theory" (QCT) might emerge as an extension of the good old theory of QED that would generalize the quantization of the electromagnetic field on two levels: not only should one address non-trivial topologies such as toroidal regions in which finite currents may flow and finite charges may be induced, but also an appropriate set of conjugate observables describing the global circuit properties should be defined. In view of the previous considerations regarding the magnetic flux trapped by the circuit, a natural pair of variables could be the flux of the electric displacement field $\mathbf{D}$ through a cross section Σ_0 crossing the circuit in the interior of the active region and the magnetic flux threaded by the loop Γ_0:

$$\Phi_D = \int_{\Sigma_0} \mathbf{D} \cdot \mathbf{dS}, \tag{9.20}$$

$$\Phi_M = \oint_{\Gamma_0} \mathbf{A} \cdot \mathbf{dr}. \tag{9.21}$$

Taking the electric displacement field instead of the electric field itself to construct a "partner" for Φ_M has mainly to do with the requirement that the product of two conjugate variables have the dimension of an action ($\propto \hbar$). Assuming that $\mathbf{D}$ vanishes outside the active region Ω_A, one may consider the latter as a leaky capacitor the plates of which are separated by Σ_0 such that, according to Gauss' law, Φ_D would equal the charge accumulated on one plate, say Q_A (see Fig. 9.6). Canonical quantization would

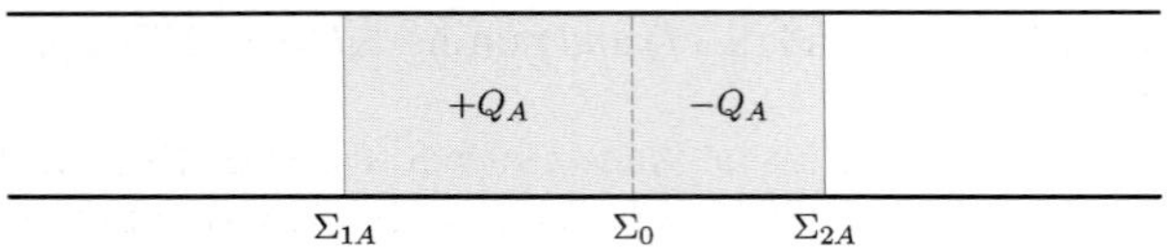

FIG. 9.6. Cross section Σ_0 separating positive and negative charges in the active region Ω_{A}.

then impose

$$[\Phi_{\mathrm{D}}, \Phi_{\mathrm{M}}] = \mathrm{i}\hbar, \qquad [\Phi_{\mathrm{D}}, \Phi_{\mathrm{D}}] = [\Phi_{\mathrm{M}}, \Phi_{\mathrm{M}}] = 0. \tag{9.22}$$

It is now tempting to propose a phenomenological expression like

$$H = \frac{\Phi_{\mathrm{D}}^2}{2C} + \frac{\Phi_{\mathrm{M}}^2}{2L} + \Phi_{\mathrm{D}} V_\varepsilon - \Phi_{\mathrm{M}} I \tag{9.23}$$

for a circuit Hamiltonian describing the interaction between the electromagnetic field variables $\{\Phi_{\mathrm{D}}, \Phi_{\mathrm{M}}\}$ and the electron current operator $I = \int_{\Sigma_0} \mathbf{J} \cdot \mathbf{dS}$ under the constraint $Q_{\mathrm{A}} = \langle \Phi_{\mathrm{D}} \rangle$, and to derive the corresponding Heisenberg equations of motion with the help of the commutation relations (9.22):

$$\frac{\mathrm{d}\Phi_{\mathrm{D}}(t)}{\mathrm{d}t} = -\frac{\mathrm{i}}{\hbar}\big[\Phi_{\mathrm{D}}(t), H\big] = \frac{\Phi_{\mathrm{M}}(t)}{L} - I(t), \tag{9.24}$$

$$\frac{\mathrm{d}\Phi_{\mathrm{M}}(t)}{\mathrm{d}t} = -\frac{\mathrm{i}}{\hbar}\big[\Phi_{\mathrm{M}}(t), H\big] = -\frac{\Phi_{\mathrm{D}}(t)}{C} - V_\varepsilon. \tag{9.25}$$

At first sight, the above equations are satisfied by meaningful steady-state solutions that may be obtained by setting the long-time averages $\langle \ldots \rangle = \lim_{t\to\infty} \langle \ldots \rangle_t$ of $\mathrm{d}\Phi_{\mathrm{D}}(t)/\mathrm{d}t$ and $\mathrm{d}\Phi_{\mathrm{M}}(t)/\mathrm{d}t$ equal to zero. Indeed, the resulting equations

$$\langle I \rangle = \frac{\langle \Phi_{\mathrm{M}} \rangle}{L}, \tag{9.26}$$

$$\frac{Q_{\mathrm{A}}}{C} = \frac{\langle \Phi_{\mathrm{D}} \rangle}{C} = -V_\varepsilon \tag{9.27}$$

are restating the familiar result that the steady-state of the circuit is determined by a current that is proportional to the magnetic flux, while the capacitor voltage tends to the externally applied electromotive force.

However, in order to investigate whether the quantum dynamics generated by the proposed Hamiltonian eventually leads to the Landauer–Büttiker formula or not, would require us to give a meaningful definition of the inductance and capacitance coefficients L and C as well as a recipe to calculate the statistical averages in a straightforward manner. Clearly, this can only be accomplished if a full microscopic investigation of the circuit is performed including both the self-consistent solution of the one-electron Schrödinger equation and the fourth Maxwell equation, and a rigorous evaluation of the dynamical, quantum-statistical ensemble averages. As such, this is quite an elaborate task which, however, may open new perspectives in the boundary region between electromagnetism and quantum mechanics.

Acknowledgements

We gratefully acknowledge Herman Maes (former Director Silicon Technology and Device Integration Division at IMEC) for giving us the opportunity to contribute to this book. We owe special thanks to our colleagues at IMEC for their willingness to discuss issues of electromagnetism and device physics and to give valuable comments and remarks. In particular, we would like to thank Peter Meuris as a co-worker realizing the geometrical picture of the magnetic vector potential and its related ghost field including its numerical implementation. Stefan Kubicek is gratefully acknowledged for stimulating discussions on differential geometry and its applications to engineering. Finally, we would like to thank the editors of this special volume to give us the opportunity to present our views concerning the simulation of electromagnetic fields and quantum transport phenomena.

Appendix A.1. Integral theorems

Integral theorems borrowed from the differential geometry of curves, surfaces and connected regions (MORSE and FESHBACH [1953], MAGNUS and SCHOENMAKER [1998]) turn out to be useful and perhaps even indispensible for a thorough understanding of elementary electromagnetic theory. Not only are they quite helpful in converting the differential form of Maxwell's equations into their equivalent integral form, but they also offer a convenient tool to define a discretized version of the field variables in the framework of numerical simulation. Moreover, they naturally bridge the gap between the microscopic interaction of the electromagnetic fields and charges in a solid-state conductor and the global circuit models envisaged on the macroscopic level.

The first three integral theorems that are summarized below, are extensively referred to in Section 2. The fourth one is the Helmholtz theorem, which allows one to decompose any well-behaved vector field into a longitudinal and a transverse part.

THEOREM A.1 (Stokes' theorem). *Let Σ be an open, orientable, multiply connected surface in $\mathbb{R}^3$ bounded by an outer, closed curve $\partial\Sigma_0$ and n inner, closed curves $\partial\Sigma_1$, ..., $\partial\Sigma_n$ defining n holes. If Σ is oriented by a surface element* **dS** *and if* **A** *is a differentiable vector field defined on Σ, then*

$$\int_\Sigma \nabla \times \mathbf{A} \cdot \mathbf{dS} = \oint_{\partial\Sigma_0} \mathbf{A} \cdot \mathbf{dr} - \sum_{j=1}^{n} \oint_{\partial\Sigma_j} \mathbf{A} \cdot \mathbf{dr}, \tag{A.1}$$

where the orientation of all boundary curves is uniquely determined by the orientation of **dS**.

THEOREM A.2 (Gauss' theorem). *Let Ω be a closed, orientable, multiply connected subset of $\mathbb{R}^3$ bounded by an outer, closed surface $\partial\Omega_0$ and n inner, closed surfaces defining n holes. If* **E** *is a differentiable vector field defined on Ω then*

$$\int_\Omega \nabla \cdot \mathbf{E}\, \mathrm{d}\tau = \int_{\partial\Omega_0} \mathbf{E} \cdot \mathbf{dS} - \sum_{j=1}^{n} \int_{\partial\Omega_j} \mathbf{E} \cdot \mathbf{dS} \tag{A.2}$$

and

$$\int_{\Omega} \nabla \times \mathbf{E}\, \mathrm{d}\tau = \int_{\partial\Omega_0} \mathbf{dS} \times \mathbf{E} - \sum_{j=1}^{n} \int_{\partial\Omega_j} \mathbf{dS} \times \mathbf{E}, \tag{A.3}$$

where all boundary surfaces have the same orientation as the outward pointing surface element of the outer surface.

The scalar Gauss theorem (A.2) reduces to *Green's Theorem* when the vector field takes the form $\mathbf{E} = f\nabla g - g\nabla f$

$$\begin{aligned}\int_{\Omega} (f\nabla^2 g - g\nabla^2 f)\, \mathrm{d}\tau &= \int_{\partial\Omega_0} (f\nabla g - g\nabla f) \cdot \mathbf{dS} \\ &\quad - \sum_{j=1}^{n} \int_{\partial\Omega_j} (f\nabla g - g\nabla f) \cdot \mathbf{dS}, \end{aligned} \tag{A.4}$$

where the scalar fields f and g are differentiable on Ω.

THEOREM A.3 ($\mathbf{J} \cdot \mathbf{E}$ theorem). *Let Ω be a closed, multiply connected, bounded subset of $\mathbb{R}^3$ with one hole and boundary surface $\partial\Omega$. If $\mathbf{J}$ and $\mathbf{E}$ are two differentiable vector fields on Ω, circulating around the hole and satisfying the conditions*

$$\nabla \cdot \mathbf{J} = 0, \tag{A.5}$$

$$\nabla \times \mathbf{E} = \mathbf{0}, \tag{A.6}$$

$$\mathbf{J} \parallel \partial\Omega \quad \text{or} \quad \mathbf{J} = \mathbf{0} \quad \text{in each point of } \partial\Omega, \tag{A.7}$$

then

$$\int_{\Omega} \mathbf{J} \cdot \mathbf{E}\, \mathrm{d}\tau = \left(\int_{\Sigma} \mathbf{J} \cdot \mathbf{dS}\right)\left(\oint_{\Gamma} \mathbf{E} \cdot \mathbf{dr}\right), \tag{A.8}$$

where Σ is an arbitrary cross section, intersecting Ω only once and Γ is a simple closed curve, encircling the hole and lying within Ω but not intersecting $\partial\Omega$. The orientation of Σ is uniquely determined by the positive orientation of Γ.

PROOF. Without any loss of generality one may define curvilinear coordinates (x^1, x^2, x^3) and a corresponding set of covariant basis vectors $(\boldsymbol{a}_1, \boldsymbol{a}_2, \boldsymbol{a}_3)$ and its contravariant counterpart, which are compatible with the topology of the toroidal (torus-like) region Ω. More precisely, x^1, x^2 and x^3 may be chosen such that the boundary surface $\partial\Omega$ coincides with one of the coordinate surfaces $\mathrm{d}x^1 = 0$ while the curves $\mathrm{d}x^1 = \mathrm{d}x^2 = 0$ are closed paths encircling the hole only once and x^3 is a cyclic coordinate. Then the inner volume contained within Ω may be conveniently parametrized by restricting the range of (x^1, x^2, x^3) to some rectangular interval $[c^1, d^1] \times [c^2, d^2] \times [c^3, d^3]$. Since Ω is multiply connected, the irrotational vector field $\mathbf{E}$ cannot generally be derived from a scalar potential for the whole region Ω. However, for the given topology of Ω, it is always possible to assign such a potential to the "transverse" components of $\mathbf{E}$ only:

$$E_1(x^1,x^2,x^3) = -\frac{\partial V(x^1,x^2,x^3)}{\partial x^1}, \tag{A.9}$$

$$E_2(x^1,x^2,x^3) = -\frac{\partial V(x^1,x^2,x^3)}{\partial x^2}, \tag{A.10}$$

but

$$E_3(x^1,x^2,x^3) \neq -\frac{\partial V(x^1,x^2,x^3)}{\partial x^3}, \tag{A.11}$$

where $V(x^1,x^2,x^3)$ can be constructed straightaway by invoking the first two components of $\nabla \times \mathbf{E} = \mathbf{0}$:

$$V(x^1,x^2,x^3) = V(c^1,c^2,x^3) - \int_{c^1}^{x^1} \mathrm{d}s\, E_1(s,x^2,x^3) - \int_{c^2}^{x^2} \mathrm{d}t\, E_2(c^1,t,x^3). \tag{A.12}$$

The potential term $V(c^1,c^2,x^3)$ naturally arises as an integration constant which, depending on x^3 only, may be absorbed in the definition of $V(x^1,x^2,x^3)$ and will therefore be omitted. Eqs. (A.9) and (A.10) are now easily recovered by taking the derivative of (A.12) with respect to x^1 and x^2, and inserting the third component of $\nabla \times \mathbf{E} = \mathbf{0}$.

Finally, taking also the derivative with respect to x^3, one obtains:

$$E_3(x^1,x^2,x^3) = -\frac{\partial V(x^1,x^2,x^3)}{\partial x^3} + E_3(c^1,c^2,x^3). \tag{A.13}$$

From Eqs. (A.9), (A.10) and (A.13) arises a natural decomposition of $\mathbf{E}$ into a conservative vector field $\mathbf{E}_\mathrm{C}$ and a non-conservative field $\mathbf{E}_\mathrm{NC}$ that is oriented along $\boldsymbol{a}_3$, thereby depending only on the cyclic coordinate x^3:

$$\mathbf{E} = \mathbf{E}_\mathrm{C} + \mathbf{E}_\mathrm{NC} \tag{A.14}$$

with

$$\mathbf{E}_\mathrm{C}(x^1,x^2,x^3) = -\nabla V(x^1,x^2,x^3), \tag{A.15}$$

$$\mathbf{E}_\mathrm{NC}(x^1,x^2,x^3) = E_3(c^1,c^2,x^3)\boldsymbol{a}^3. \tag{A.16}$$

The conservative part of $\mathbf{E}$ does not contribute to the volume integral of $\mathbf{J} \cdot \mathbf{E}$. Indeed, from (A.15) it follows

$$\int_\Omega \mathbf{J} \cdot \mathbf{E}_\mathrm{C}\, \mathrm{d}\tau = \int_\Omega \mathbf{J} \cdot \nabla V\, \mathrm{d}\tau = \int_\Omega \nabla \cdot (V\mathbf{J})\, \mathrm{d}\tau - \int_\Omega V \nabla \cdot \mathbf{J}\, \mathrm{d}\tau. \tag{A.17}$$

With the help of Gauss' theorem – which is also valid for multiply connected regions – the first term of the right-hand side of Eq. (A.17) can be rewritten as a surface integral of $V\mathbf{J}$ which is seen to vanish as $\mathbf{J}$ is assumed to be tangential to the surface $\partial\Omega$ in all of its points. Clearly, the second integral in the right-hand side of (A.17) is identically zero due to $\nabla \cdot \mathbf{J} = 0$ and one is therefore lead to the conclusion

$$\int_\Omega \mathbf{J} \cdot \mathbf{E}_\mathrm{C}\, \mathrm{d}\tau = 0. \tag{A.18}$$

On the other hand, the contribution of $\mathbf{E}_{\mathrm{NC}}$ can readily be evaluated in terms of the curvilinear coordinates. Denoting the Jacobian determinant by $g(x^1, x^2, x^3)$ one may express the volume integral as a threefold integral over the basic interval $[c^1, d^1] \times [c^2, d^2] \times [c^3, d^3]$, thereby exploiting the fact that the non-conservative contribution merely depends on x^3:

$$\begin{aligned}
&\int_\Omega \mathbf{J}\cdot\mathbf{E}\,\mathrm{d}\tau \\
&\quad = \int_\Omega \mathbf{J}\cdot\mathbf{E}_{\mathrm{NC}}\,\mathrm{d}\tau \\
&\quad = \int_{c^3}^{d^3} \mathrm{d}x^3\, E_3(c^1, c^2, x^3) \int_{c^1}^{d^1} \mathrm{d}x^1 \int_{c^2}^{d^2} \mathrm{d}x^2\, g(x^1, x^2, x^3) J^3(x^1, x^2, x^3).
\end{aligned} \tag{A.19}$$

The last integral can conveniently be interpreted as the flux of $\mathbf{J}$ through the single cross section $\Sigma(x^3)$ defined by

$$\Sigma(x^3) = \big\{(x^1, x^2, x^3) \mid c^1 \leqslant x^1 \leqslant d^1;\ c^2 \leqslant x^2 \leqslant d^2;\ x^3 \text{ fixed}\big\}. \tag{A.20}$$

Indeed, expanding the Jacobian determinant as a mixed product of the three basis vectors, i.e.,

$$g = \boldsymbol{a}_1 \times \boldsymbol{a}_2 \cdot \boldsymbol{a}_3 \tag{A.21}$$

and identifying the two-form $\boldsymbol{a}_1 \times \boldsymbol{a}_2\, \mathrm{d}x^1\, \mathrm{d}x^2$ as a generic surface element $\mathbf{dS}$ perpendicular to $\Sigma(x^3)$, one easily arrives at

$$\begin{aligned}
&\int_{c^1}^{d^1} \mathrm{d}x^1 \int_{c^2}^{d^2} \mathrm{d}x^2\, g(x^1, x^2, x^3) J^3(x^1, x^2, x^3) \\
&\quad = \int_{c^1}^{d^1} \mathrm{d}x^1 \int_{c^2}^{d^2} \mathrm{d}x^2\, \boldsymbol{a}_1 \times \boldsymbol{a}_2 \cdot \mathbf{J}(x^1, x^2, x^3) = \int_{\Sigma(x^3)} \mathbf{dS}\cdot\mathbf{J} \equiv I(x^3)
\end{aligned} \tag{A.22}$$

and

$$\int_\Omega \mathbf{J}\cdot\mathbf{E}\,\mathrm{d}\tau = \int_{c^3}^{d^3} \mathrm{d}x^3\, E_3(c^1, c^2, x^3) I(x^3). \tag{A.23}$$

The sign of the flux $I(x^3)$ obviously depends on the orientation of $\Sigma(x^3)$, which is unequivocally determined by the surface element $\mathbf{dS} = \boldsymbol{a}_1 \times \boldsymbol{a}_2\, \mathrm{d}x^1\, \mathrm{d}x^2$. As long as only positive body volumes are concerned, one may equally require that each infinitesimal volume element $\mathrm{d}\tau = g\, \mathrm{d}x^1\, \mathrm{d}x^2\, \mathrm{d}x^3$ be positive for positive incremental values $\mathrm{d}x^1$, $\mathrm{d}x^2$ and $\mathrm{d}x^3$. Moreover, since $\mathbf{dr} = \mathrm{d}x^3 \boldsymbol{a}_3$ is the elementary tangent vector of the coordinate curve $\Gamma(x^1, x^2) = \{(x^1, x^2, x^3) \mid x^1, x^2 \text{ fixed};\ c^3 \leqslant x^3 \leqslant d^3\}$ orienting $\Gamma(x^1, x^2)$ in a positive traversal sense through increasing x^3, one easily arrives at

$$\mathrm{d}\tau = \mathbf{dS}\cdot\mathbf{dr} > 0. \tag{A.24}$$

In other words, the orientation of $\Sigma(x^3)$ is completely fixed by the positive traversal sense of $\Gamma(x^1, x^2)$. However, since $\mathbf{J}$ is solenoidal within Ω as well as tangential to $\partial\Omega$, one may conclude from Gauss' theorem that the value of the flux $I(x^3)$ does not depend on the particular choice of the cross section $\Sigma(x^3)$ which may thus be replaced by any other single cross section Σ provided that the orientation is preserved. Consequently, $I(x^3)$ reduces to a constant value I and may be taken out of the integral of Eq. (A.23) which now simplifies to:

$$\int_\Omega \mathbf{J}\cdot\mathbf{E}\,\mathrm{d}\tau = I\int_{c^3}^{d^3} \mathrm{d}x^3\, E_3(c^1,c^2,x^3). \tag{A.25}$$

The remaining integral turns out to be the line integral of **E** along the coordinate curve $\Gamma(c^1,c^2)$:

$$\int_\Omega \mathbf{J}\cdot\mathbf{E}\,\mathrm{d}\tau = I V_\varepsilon(c^1,c^2) \tag{A.26}$$

with

$$V_\varepsilon(c^1,c^2) = \oint_{\Gamma(c^1,c^2)} \mathbf{E}\cdot\mathbf{dr}. \tag{A.27}$$

Since **E** is irrotational, according to Stokes' theorem its circulation does not depend on the particular choice of the circulation curve as was already discussed in more detail in the previous section. Consequently, $\Gamma(c^1,c^2)$ may be replaced by any other interior closed curve Γ encircling the hole region and sharing the traversal sense with $\Gamma(c^1,c^2)$:

$$V_\varepsilon(c^1,c^2) = V_\varepsilon \equiv \oint_\Gamma \mathbf{E}\cdot\mathbf{dr}. \tag{A.28}$$

Hence,

$$\int_\Omega \mathbf{J}\cdot\mathbf{E}\,\mathrm{d}\tau = I V_\varepsilon. \tag{A.29}$$

This completes the proof. □

THEOREM A.4 (Helmholtz' theorem). *Let Ω be a simply connected, bounded subset of $\mathbb{R}^3$. Then, any finite, continuous vector field* **F** *defined on Ω can be derived from a differentiable vector potential* **A** *and a differentiable scalar potential χ such that*

$$\mathbf{F} = \mathbf{F}_\mathrm{L} + \mathbf{F}_\mathrm{T}, \tag{A.30}$$

$$\mathbf{F}_\mathrm{L} = \nabla\chi, \tag{A.31}$$

$$\mathbf{F}_\mathrm{T} = \nabla\times\mathbf{A}. \tag{A.32}$$

Due to the obvious properties

$$\begin{aligned} &\nabla\times\mathbf{F}_\mathrm{L} = \mathbf{0},\\ &\nabla\cdot\mathbf{F}_\mathrm{T} = 0. \end{aligned} \tag{A.33}$$

$\mathbf{F}_\mathrm{L}$ and $\mathbf{F}_\mathrm{T}$ are respectively called the longitudinal and transverse components of **F**.

Appendix A.2. Vector identities

Let f, $\mathbf{A}$ and $\mathbf{B}$ represent a scalar field and two vector fields defined on a connected subset Ω of $\mathbb{R}^3$, all being differentiable on Ω. Then the following (non-exhaustive) list of identities may be derived using familiar vector calculus:

$$\nabla \cdot (\nabla \times \mathbf{A}) \equiv 0, \tag{A.34}$$

$$\nabla \times (\nabla f) \equiv \mathbf{0}, \tag{A.35}$$

$$\begin{aligned}\nabla(\mathbf{A}\cdot\mathbf{B}) = \mathbf{A}(\nabla\cdot\mathbf{B}) + \mathbf{B}(\nabla\cdot\mathbf{A}) + (\mathbf{A}\cdot\nabla)\mathbf{B} \\ + (\mathbf{B}\cdot\nabla)\mathbf{A} + \mathbf{A}\times(\nabla\times\mathbf{B}) + \mathbf{A}\times(\nabla\times\mathbf{B}),\end{aligned} \tag{A.36}$$

$$\begin{aligned}\nabla\times(\mathbf{A}\times\mathbf{B}) = -\mathbf{A}(\nabla\cdot\mathbf{B}) + \mathbf{B}(\nabla\cdot\mathbf{A}) - (\mathbf{A}\cdot\nabla)\mathbf{B} \\ + (\mathbf{B}\cdot\nabla)\mathbf{A} - \mathbf{A}\times(\nabla\times\mathbf{B}) + \mathbf{A}\times(\nabla\times\mathbf{B}),\end{aligned} \tag{A.37}$$

$$\nabla\times(f\mathbf{A}) = f\nabla\times\mathbf{A} + \nabla f\times\mathbf{A}, \tag{A.38}$$

$$\nabla\cdot(\mathbf{A}\times\mathbf{B}) = \mathbf{B}\cdot\nabla\times\mathbf{A} - \mathbf{A}\cdot\nabla\times\mathbf{B}, \tag{A.39}$$

$$\nabla\cdot(f\mathbf{A}) = f\nabla\cdot\mathbf{A} + \nabla f\cdot\mathbf{A}, \tag{A.40}$$

$$\nabla\times(\nabla\times\mathbf{A}) = \nabla(\nabla\cdot\mathbf{A}) - \nabla^2\mathbf{A}. \tag{A.41}$$

It should be noted that Eq. (A.41) should be considered as a definition of the vectorial Laplace operator ("Laplacian"), rather than a vector identity. Clearly, if one expands the left-hand side of Eq. (A.41) in Cartesian coordinates, one may straightforwardly obtain

$$\left[\nabla\times(\nabla\times\mathbf{A})\right]_x = \frac{\partial}{\partial x}\nabla\cdot\mathbf{A} - \nabla^2 A_x, \tag{A.42}$$

etc., which does indeed justify the identification $\nabla^2\mathbf{A} = (\nabla^2 A_x, \nabla^2 A_y, \nabla^2 A_z)$ for Cartesian coordinates, but not for an arbitrary system of curvilinear coordinates.

References

BARDEEN, J. (1961). Tunneling from a many-particle point of view. *Phys. Rev. Lett.* **6**, 57–59.

BRAR, B., WILK, G., SEABAUGH, A. (1996). Direct extraction of the electron tunneling effective mass in ultrathin SiO_2. *Appl. Phys. Lett.* **69**, 2728–2730.

BREIT, G., WIGNER, E.P. (1936). Capture of slow neutrons. *Phys. Rev.* **49**, 519–531.

BUETTIKER, M. (1986). Role of quantum coherence in series resistors. *Phys. Rev. B* **33**, 3020–3026.

BUTCHER, P., MARCH, N.H., TOSI, M.P. (eds.) (1993). *Physics of Low-Dimensional Semiconductor Structures* (Plenum Press, New York).

COLLIN, R. (1960). *Field Theory of Guided Waves* (Mc-Graw Hill, New York).

DATTA, S. (1995). *Electronic Transport in Mesoscopic Systems* (Cambridge University Press, UK).

DEPAS, M., VANMEIRHAEGHE, R., LAFLERE, W., CARDON, F. (1994). Electrical characteristics of Al/SiO_2/n-Si tunnel-diodes with an oxide layer grown by rapid thermal-oxidation. *Solid-State Electron.* **37**.

DITTRICH, T., HAENGGI, P., INGOLD, G.-L., KRAMER, B., SCHOEN, G., ZWERGER, W. (1997). *Quantum Transport and Dissipation* (Wiley-VCH, Weinheim, Germany).

DRUDE, P. (1900a). Zur Elektronentheorie der Metalle, I. Teil. *Ann. Phys.* **1**, 566–613.

DRUDE, P. (1900b). Zur Elektronentheorie der Metalle, II. Teil. *Ann. Phys.* **3**, 369–402.

EINSTEIN, A., LORENTZ, H.A., MINKOWSKI, H., WEYL, H. (1952). *The Principle of Relativity, Collected Papers* (Dover, New York).

EZAWA, Z.F. (2000). *Quantum Hall Effects – Field Theoretical Approach and Related Topics* (World Scientific Publishing, Singapore).

FENTON, E.W. (1992). Electric-field conditions for Landauer and Boltzmann–Drude conductance equations. *Phys. Rev. B* **46**, 3754–3770.

FENTON, E.W. (1994). Electrical and chemical potentials in a quantum-mechanical conductor. *Superlattices Microstruct.* **16**, 87–91.

FEYNMAN, R.P., LEIGHTON, R.B., SANDS, M. (1964a). *The Feynman Lectures on Physics, vol. 2* (Addison-Wesley, New York).

FEYNMAN, R.P., LEIGHTON, R.B., SANDS, M. (1964b). *The Feynman Lectures on Physics, vol. 3* (Addison-Wesley, New York).

FLUEGGE, S. (1974). *Practical Quantum Mechanics* (Springer, New York).

FORGHIERI, A., GUERRRI, R., CIAMPOLINI, P., GNUDI, A., RUDAN, M. (1988). A new discretization strategy of the semiconductor equations comprising momentum and energy balance. *IEEE Trans. Computer-Aided Design* **7**, 231–242.

FOWLER, R.H. (1936). *Statistical Mechanics* (MacMillan, New York).

HUANG, K. (1963). *Statistical Mechanics* (John Wiley & Sons, New York).

IMRY, Y., LANDAUER, R. (1999). Conductance viewed as transmission. *Rev. Mod. Phys.* **71**, S306–S312.

JACKSON, J.D. (1975). *Classical Electrodynamics* (John Wiley & Sons, New York).

JOOSTEN, H., NOTEBORN, H., LENSTRA, D. (1990). Numerical study of coherent tunneling in a double-barrier structure. *Thin Solid Films* **184**, 199–206.

KITTEL, C. (1963). *Quantum Theory of Solids* (John Wiley, New York).

KITTEL, C. (1976). *Introduction to Solid State Physics* (John Wiley, New York).

KUBO, R. (1957). Statistical-mechanical theory of irreversible processes. I. General theory and simple applications to magnetic and conduction problems. *J. Phys. Soc. Japan* **12**, 570.

LANDAU, L.D., LIFSHITZ, E.M. (1958). *Quantum Mechanics (Non-Relativistic Theory)* (Pergamon Press, London).

LANDAU, L., LIFSHITZ, E.M. (1962). *The Classical Theory of Fields* (Addison-Wesley, Reading, MA).

LANDAUER, R. (1957). Spatial variation of currents and fields due to localized scatterers in metallic conduction. *IBM J. Res. Dev.* **1**, 223–231.

LANDAUER, R. (1970). Electrical resistance of disordered one-dimensional lattices. *Philos. Mag.* **21–25**, 863.

LENSTRA, D., SMOKERS, T.M. (1988). Theory of nonlinear quantum tunneling resistance in one-dimensional disordered systems. *Phys. Rev. B* **38**, 6452–6460.

LENSTRA, D., VAN HAERINGEN, W., SMOKERS, T.M. (1990). Carrier dynamics in a ring, Landauer resistance and localization in a periodic system. *Physica A* **162**, 405–413.

LUNDSTROM, M. (1999). *Fundamentals of Carrier Transport*, second ed. (Cambridge University Press, Cambridge).

MAGNUS, W., SCHOENMAKER, W. (1993). Dissipative motion of an electron–phonon system in a uniform electric field: an exact solution. *Phys. Rev. B* **47**, 1276–1281.

MAGNUS, W., SCHOENMAKER, W. (1998). On the use of a new integral theorem for the quantum mechanical treatment of electric circuits. *J. Math. Phys.* **39**, 6715–6719.

MAGNUS, W., SCHOENMAKER, W. (1999). Full quantum mechanical treatment of charge leakage in MOS capacitors with ultra-thin oxide layers. In: *Proc. 29th European Solid-State Device Research Conference (ESSDERC'99), Leuven, Editions Frontières, 13–15 September 1999*, pp. 248–251.

MAGNUS, W., SCHOENMAKER, W. (2000a). Full quantum mechanical model for the charge distribution and the leakage currents in ultrathin metal-insulator-semiconductor capacitors. *J. Appl. Phys.* **88**, 5833–5842.

MAGNUS, W., SCHOENMAKER, W. (2000b). On the calculation of gate tunneling currents in ultra-thin metal-insulator-semiconductor capacitors. *Microelectronics and Reliability* **41**, 31–35.

MAGNUS, W., SCHOENMAKER, W. (2000c). Quantized conductance, circuit topology, and flux quantization. *Phys. Rev. B* **61**, 10883–10889.

MAGNUS, W., SCHOENMAKER, W. (2002). *Quantum Transport in Submicron Devices: A Theoretical Introduction* (Springer, Berlin, Heidelberg).

MAHAN, G.D. (1981). *Many-Particle Physics* (Plenum Press, New York).

MAXWELL, J.C. (1954a). *A Treatise on Electricity and Magnetism, vol. 1* (Dover Publications, New York).

MAXWELL, J.C. (1954b). *A Treatise on Electricity and Magnetism, vol. 2* (Dover Publications, New York).

MERZBACHER, E. (1970). *Quantum Mechanics* (John Wiley & Sons, New York).

MEURIS, P., SCHOENMAKER, W., MAGNUS, W. (2001). Strategy for electromagnetic interconnect modeling. *IEEE Trans. Computer-Aided Design of Circuits and Integrated Systems* **20** (6), 753–762.

MORSE, P.M., FESHBACH, H. (1953). *Methods of Theoretical Physics, Part I* (McGraw-Hill Book Company, Inc).

NOTEBORN, H., JOOSTEN, H., LENSTRA, D., KASKI, K. (1990). Selfconsistent study of coherent tunneling through a double-barrier structure. *Phys. Scripta T* **33**, 219–226.

SAKURAI, J.J. (1976). *Advanced Quantum Mechanics* (Addison-Wesley, Reading, MA).

SCHARFETTER, D., GUMMEL, H. (1969). Large-signal analysis of a silicon read diode oscillator. *IEEE Trans. Electron Devices* **ED-16**, 64–77.

SCHOENMAKER, W., MAGNUS, W., MEURIS, P. (2002). Ghost fields in classical gauge theories. *Phys. Rev. Lett.* **88** (18), 181602-01–181602-04.

SCHOENMAKER, W., MEURIS, P. (2002). Electromagnetic interconnects and passives modeling: software implementation issues. *IEEE Trans. Computer-Aided Design of Circuits and Integrated Systems* **21** (5), 534–543.

SCHWINGER, J. (1958). *Selected Papers on Quantum Electrodynamics* (Dover Publications, New York).

STONE, A.D., SZAFER, A. (1988). What is measured when you measure a resistance – the Landauer formula revisited. *IBM J. Res. Dev.* **32**, 384–413.

STONE, D. (1992). Physics of nanostructures. In: Davies, J.H., Long, A.R. (eds.), *Proceedings of the 38th Scottish Universities Summer School in Physics, St Andrews, July–August 1991* (IOP Publishing Ltd., London), pp. 65–76.

STRATTON, R. (1962). Diffusion of hot and cold electrons in semiconductor devices. *Phys. Rev.* **126**, 2002–2014.

SUNE, J., OLIVIO, P., RICCO, B. (1991). Self-consistent solution of the Poisson and Schrödinger-equations in accumulated semiconductor-insulator interfaces. *J. Appl. Phys.* **70**, 337–345.

'T HOOFT, G. (1971). Renormalizable Lagrangians for massive Yang–Mills fields. *Nucl. Phys. B* **35**, 167–188.

VAN WEES, B.J., VAN HOUTEN, H., BEENAKKER, C.W.J., WILLIAMSON, J.G., KOUWENHOVEN, L.P., VAN DER MAREL, D., FOXON, C.T. (1988). Quantized conductance of point contacts in a two-dimensional electron-gas. *Phys. Rev. Lett.* **60**, 848–850.

VON KLITZING, K., DORDA, G., PEPPER, M. (1980). New method for high-accuracy determination of the fine-structure constant based on quantized Hall resistance. *Phys. Rev. Lett.* **45**, 494–497.

WEYL, H. (1918). Gravitation und Elektrizität. *Sitzungsber. Preussischen Akad. Wiss.* **26**, 465–480.

WHARAM, D.A., THORNTON, T.J., NEWBURY, R., PEPPER, M., AHMED, H., FROST, J.E.F., HASKO, D.G., PEACOCK, D.C., RITCHIE, D.A., JONES, G.A.C. (1988). One-dimensional transport and the quantization of the ballistic resistance. *J. Phys. C* **21**, L209–L214.

WILSON, K. (1974). Confinement of quarks. *Phys. Rev. D* **10**, 2445–2459.

Discretization of Electromagnetic Problems: The "Generalized Finite Differences" Approach

Alain Bossavit

Laboratoire de Génie Électrique de Paris,
11 Rue Joliot-Curie,
91192 Gif-sur-Yvette Cedex,
France
E-mail address: Bossavit@lgep.supelec.fr

Numerical Methods in Electromagnetics
Special Volume (W.H.A. Schilders and E.J.W. ter Maten, Guest Editors) of
HANDBOOK OF NUMERICAL ANALYSIS, VOL. XIII
P.G. Ciarlet (Editor)

ISSN 1570-8659
DOI 10.1016/S1570-8659(04)13002-0

Contents

CHAPTER I 109

1. Affine space 110
2. Piecewise smooth manifolds 113
3. Orientation 115
4. Chains, boundary operator 121
5. Metric notions 123

CHAPTER II 127

6. Integration: Circulation, flux, etc. 127
7. Differential forms, and their physical relevance 130
8. The Stokes theorem 134
9. The magnetic field, as a 2-form 136
10. Faraday and Ampère 138
11. The Hodge operator 139
12. The Maxwell equations: Discussion 140

CHAPTER III 147

13. A model problem 147
14. Primal mesh 149
15. Dual mesh 152
16. A discretization kit 155
17. Playing with the kit: Full Maxwell 159
18. Playing with the kit: Statics 161
19. Playing with the kit: Miscellanies 164

CHAPTER IV 167

20. Consistency 168
21. Stability 172
22. The time-dependent case 173
23. Whitney forms 174

24. Higher-degree forms 181
25. Whitney forms for other shapes than simplices 184

REFERENCES 193

Further reading 196

CHAPTER I

Preliminaries: Euclidean Space

What we shall do in this preliminary chapter (Sections 1–5, out of a total of 25) can be described as "deconstructing Euclidean space". Three-dimensional Euclidean space, denoted by E_3 here, is a relatively involved mathematical structure, made of an affine 3D space (more on this below), equipped with a metric and an orientation. By taking the Cartesian product of that with another Euclidean space, one-dimensional and meant to represent Time, one gets the mathematical framework in which most of classical physics is described. This framework is often taken for granted, and should not.

By this we do not mean to challenge the separation between space and (absolute) time, which would be getting off to a late start, by a good century. Relativity is not our concern here, because we won't deal with moving conductors, which makes it all right to adopt a privileged reference frame (the so-called laboratory frame) and a unique chronometry. The problem we perceive is with E_3 itself, too rich a structure in several respects. For one thing, orientation of space is *not* necessary. (How could it be? How could physical phenomena depend on this social convention by which we class right-handed and left-handed helices, such as shells or staircases?) And yet, properties of the cross product, or of the curl operator, so essential tools in electromagnetism, crucially depend on orientation. As for metric (i.e., the existence of a dot product, from which norms of vectors and distances between points are derived), it also seems to be involved in the two main equations, $\partial_t \mathrm{B} + \mathrm{rot}\,\mathrm{E} = 0$ (Faraday's law) and $-\partial_t \mathrm{D} + \mathrm{rot}\,\mathrm{H} = \mathrm{J}$ (Ampère's theorem), since the definition of rot depends on the metric. We shall discover that it plays no role there, actually, because a change of metric, in the description of some electromagnetic phenomenon, would change *both* rot *and* the vector fields E, B, etc., in such a way that the equations would stay unchanged. Metric is no less essential for that, but its intervention is limited to the expression of constitutive laws, that is, to what will replace in our notation the standard $\mathrm{B} = \mu \mathrm{H}$ and $\mathrm{D} = \varepsilon \mathrm{E}$.[1]

Our purpose, therefore, is to separate the various layers present in the structure of E_3, in view of using exactly what is needed, and nothing more, for each subpart of the Maxwell system of equations. That this can be done is no news: As reported by POST [1972], the metric-free character of the two main Maxwell equations was pointed out by Cartan, as early as 1924, and also by KOTTLER [1922] and VAN DANTZIG [1934]. But the exploitation of this remark in the design of numerical schemes is

[1] We shall most often ignore Ohm's law here, for shortness, and therefore, treat the current density J as a data. It would be straightforward to supplement the equations by the relation $\mathrm{J} = \sigma \mathrm{E} + \mathrm{J}^s$, where only the "source current" J^s is known in advance.

a contemporary thing, which owes much to (again, working independently) TONTI [2001], Tonti (see TONTI [1996], MATTIUSSI [2000]) and Weiland (see EBELING, KLATT, KRAWCZYK, LAWINSKY, WEILAND, WIPF, STEFFEN, BARTS, BROWMAN, COOPER, DEAVEN and RODENZ [1989], WEILAND [1996]). See also SORKIN [1975], HYMAN and SHASHKOV [1997], TEIXEIRA and CHEW [1999]. Even more recent (BOSSAVIT and KETTUNEN [1999], MATTIUSSI [2000]) is the realization that such attention to the underlying geometry would permit to soften the traditional distinctions between finite-difference, finite-element, and finite-volume approaches. In particular, it will be seen here that a common approach to error analysis applies to the three of them, which does rely on the existence of finite elements, but not on the variational methods that are often considered as foundational in finite element theory. These finite elements, moreover, are not of the Lagrange (node based) flavor. They are differential geometric objects, created long ago for other purposes, the Whitney forms (WHITNEY [1957]), whose main characteristic is the interpretation they suggest of degrees of freedom (DoF) as integrals over geometric elements (edges, facets, ...) of the discretization mesh.

As a preparation to this deconstruction process, we need to recall a few notions of geometry and algebra which do not seem to get, in most curricula, the treatment they deserve. First on this agenda is the distinction between vector space and affine space.

1. Affine space

A *vector space*[2] on the reals is a set of objects called *vectors*, which one can (1) add together (in such a way that they form an Abelian group, the neutral element being the null vector) and (2) multiply by real numbers. No need to recall the axioms which harmonize these two groups of features. Our point is this: The three-dimensional vector space (for which our notation will be V_3) makes an awkward model of physical space,[3] unless one deals with situations with a privileged point, such as for instance a center of mass, which allows one to identify a spatial point x with the translation vector that sends this privileged point to x. Otherwise, the idea to add points, or to multiply them by a scalar, is ludicrous. On the other hand, taking the midpoint of two points, or more generally, barycenters, makes sense, and is an allowed operation in affine space, as will follow from the definition.

An *affine space* is a set on which a vector space, considered as an additive group, acts effectively, transitively and regularly. Let's elaborate.

A group G *acts* on a set X if for each $g \in G$ there is a map from X to X, that we shall denote by a_g, such that a_1 is the identity map, and $a_{gh} = a_g a_h$. (Symbol 1 denotes

[2]Most definitions will be implicit, with the defined term set, on first appearance, in *italics* style. The same style is also used, occasionally, for emphasis.

[3]Taking $\mathbb{R}^3$, the set of triples of real numbers, with all the topological and metric properties inherited from $\mathbb{R}$, is even worse, for this implies that some basis $\{\partial_1, \partial_2, \partial_3\}$ has been selected in V_3, thanks to which a vector v writes as $v = \sum_i v^i \partial_i$, hence the identification between v and the triple $\{v^i\}$ of components (or coordinates of the point v stands for). In most situations which require mathematical modelling, no such basis imposes itself. There may exist privileged directions, as when the device to be modelled has some kind of translational invariance, but even this does not always mandate a choice of basis.

the neutral element, and will later double for the group made of this unique element.) The action is *effective* if $a_g = 1$ implies $g = 1$, that is to say, if all nontrivial group elements "do something" to X. The *orbit* of x under the action is the set $\{a_g(x)\colon\ g \in G\}$ of transforms of x. Belonging to the same orbit is an equivalence relation between points. One says the action is *transitive* if all points are thus equivalent, i.e., if there is a single orbit. The *isotropy group* (or stabilizer, or little group) of x is the subgroup $G_x = \{g \in G\colon\ a_g(x) = x\}$ of elements of G which fix x. In the case of a transitive action, little groups of all points are conjugate (because $g_{xy} G_y = G_x g_{xy}$, where g_{xy} is any group element whose action takes x to y), and thus "the same" in some sense. A transitive action is *regular* (or *free*) if it has no fixed point, that is, if $G_x = 1$ for all x. If so is the case, X and G are in one-to-one correspondence, so they look very much alike. Yet they should not be identified, for they have quite distinctive structures. Hence the concept of *homogeneous space*: A set, X here, on which some group acts transitively and effectively. (A standard example is given by the two-dimensional sphere S_2 under the action of the group SO_3 of rotations around its center.) If, moreover, the little group is trivial (regular action), the only difference between the homogeneous space X and the group G lies in the existence of a distinguished element in G, the neutral one. Selecting a point 0 in X (the origin) and then identifying $a_g(0)$ with g (and hence 0 in X with the neutral element of G) provides X with a group structure, but the isomorphism with G thus established is not canonical, and this group structure is most often irrelevant, just like the vector-space structure of 3D space.

Affine space is a case in point. Intuitively, take the n-dimensional vector space V_n, and forget about the origin: What remains is A_n, the affine space of dimension n. More rigorously, a vector space V, considered as an additive group, acts on itself (now considered as just a set, which we acknowledge by calling its elements *points*, instead of vectors) by the mappings[4] $a_v = x \to x + v$, called *translations*. This action is transitive, because for any pair of points $\{x, y\}$, there is a vector v such that $y = x + v$, and regular, because $x + v \neq x$ if $v \neq 0$, whatever x. The structure formed by V as a set equipped with this group action is called the *affine space A associated with V*. Each vector of V has thus become a point of A, but there is nothing special any longer with the vector 0, as a point in A. Reversing the viewpoint, one can say that an affine space A is a homogeneous space with respect to the action of some vector space V, considered as an additive group. (Points of A will be denoted x, y, etc., and $y - x$ will stand, by a natural notational abuse, for the vector that carries x to y.) The most common example is obtained by considering as equivalent, in some vector space V, two vectors u and v such that $u - v$ belong to some fixed vector subspace W. Each equivalence class has an obvious affine structure (W acts on it regularly by $v \to v + w$). Such a class is called an *affine subspace* of V, *parallel to* W[5] (see Fig. 1.1) Of course, no vector in such an

[4]We'll find it convenient to denote a map f by $x \to \mathrm{Expr}(x)$, where Expr is the defining expression, and to link name and definition by writing $f = x \to \mathrm{Expr}(x)$. (The arrow is a "stronger link" than the equal sign in this expression.) In the same spirit, $X \to Y$ denotes the set of all maps "of type $X \to Y$", that is, maps from X to Y, not necessarily defined over all X. Points x for which f is defined form its *domain* $\mathrm{dom}(f) \subset X$, and their images form the *codomain* $\mathrm{cod}(f) \subset Y$, also called the *range* of f.

[5]Notice how the set of all affine subspaces parallel to W also constitutes an affine space under the action of V, or more pointedly – because then the action is regular – of the quotient space V/W. A "point", there, is a whole affine subspace.

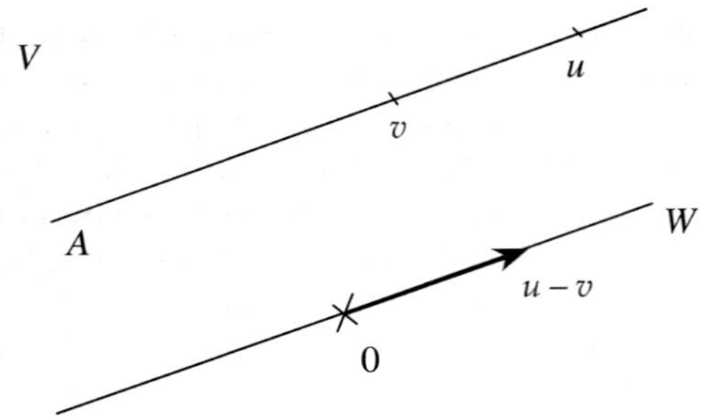

FIG. 1.1. No point in the affine subspace A, parallel to W, can claim the role of "origin" there.

affine subspace qualifies more than any other as origin, and calling its elements "points" rather than "vectors" is therefore appropriate.

At this stage, we may introduce the *barycenter* of points x and y, with weights λ and $1-\lambda$, as the translate $x+\lambda(y-x)$ of x by the vector $\lambda(y-x)$, and generalize to any number of points. The concepts of affine independence, dimension of the affine space, and affine subspaces follow from the similar ones about the vector space. *Barycentric coordinates*, with respect to $n+1$ affinely independent points $\{a_0, \ldots, a_n\}$ in A_n are the weights $\lambda^i(x)$ such that $\sum_i \lambda^i(x) = 1$ and $\sum_i \lambda^i(x)(x - a_i) = 0$, which we shall feel free to write $x = \sum_i \lambda^i(x) a_i$. *Affine maps* on A_n are those that are linear with respect to the barycentric coordinates. If x is a point in affine space A, vectors of the form $y - x$ are called *vectors at* x. They form of course a vector space isomorphic to the associate V, called the *tangent space at* x, denoted T_x. (I will call *free* vectors the elements of V, as opposed to vectors "at" some point, dubbed *bound* (or *anchored)* vectors. Be aware that this usage is not universal.) The tangent space to a curve or a surface which contains x is the subspace of T_x formed by vectors at x tangent to this curve or surface.[6] Note that vector fields are maps of type *POINT* → *BOUND_VECTOR*, actually, subject to the restriction that the value of v at x, notated $v(x)$, is a vector at x. The distinction between this and a *POINT* → *FREE_VECTOR* map, which may seem pedantic when the point spans ordinary space, must obviously be maintained in the case of tangent vector fields defined over a surface or a curve.

Homogeneous space is a key concept: Here is the mathematical construct by which we can best model humankind's *physical* experience of spatial homogeneity. Translating from a spatial location to another, we notice that similar experiments give similar results, hence the concept of invariance of the structure of space with respect to the group of such motions. By taking as mathematical model of space a homogeneous space relative to the action of this group (in which we recognize V_3, by observing how translations compose), we therefore acknowledge an essential *physical* property of the space we live in.

REMARK 1.1. In fact, translational invariance is only approximately verified, so one should perhaps approach this basic modelling issue more cautiously: Imagine space as

[6]For a piecewise smooth manifold (see below), such a subspace may fail to exist at some points, which will not be a problem.

a seamless assembly (via smooth transition functions) of patches of affine space, each point covered by at least one of them, which is enough to capture the idea of *local* translational invariance of physical space. This idea gets realized with the concept of smooth manifold (see below) of dimension 3. What we shall eventually recognize as the metric-free part of the Maxwell's system (Ampère's and Faraday's laws) depends on the manifold structure only. Therefore, postulating an affine structure is a *modelling decision*, one that goes a trifle beyond what would strictly be necessary to account for the homogeneity of space, but will make some technical discussions easier when (about Whitney forms) barycentric coordinates will come to the fore.

There is no notion of distance in affine space, but this doesn't mean no topology: Taking the preimages of neighborhoods of $\mathbb{R}^n$ under any one-to-one affine map gives a system of neighborhoods, hence a topology – the same for all such maps. (So we shall talk loosely of a "ball" or a "half ball" in reference to an affine one-to-one image of $B = \{\xi \in \mathbb{R}^n\colon \sum_i (\xi^i)^2 < 1\}$ or of $B \cap \{\xi\colon \xi^1 \geqslant 0\}$.) Continuity and differentiability thus make sense for a function f of type $A_p \to A_n$. In particular, the derivative of f at x is the linear map $\mathrm{D}f(x)$, from V_p to V_n, such that $|f(x+v) - f(x) - \mathrm{D}f(x)(v)|/|v| = \mathrm{o}(|v|)$, if such a map exists, which does not depend on which norms $|\,|$ on V_p and V_n are used to check the property. The same symbol, $\mathrm{D}f(x)$, will be used for the *tangent map* that sends a vector v anchored at x to the vector $\mathrm{D}f(x)(v)$ anchored at $f(x)$.

2. Piecewise smooth manifolds

We will do without a formal treatment of manifolds. Most often, we shall just use the word as a generic term for lines, surfaces, or regions of space ($p = 1, 2, 3$, respectively), piecewise smooth (as defined in a moment), connected or not, with or without a boundary. A 0-manifold is a collection of isolated points.

For the rare cases when the general concept is evoked, suffice it to say that a p-dimensional manifold is a set M equipped with a set of maps of type $M \to \mathbb{R}^p$, called *charts*, which make M look, for all purposes, but only locally, like $\mathbb{R}^p$ (and hence, like p-dimensional affine space). *Smooth* manifolds are those for which the so-called *transition functions* $\varphi \circ \psi^{-1}$, for any pair $\{\varphi, \psi\}$ of charts, are smooth, i.e., possess derivatives of all orders. (So-called C^k manifolds obtain when continuous derivatives exist up to order k.) Then, if some property P makes sense for functions of type $\mathbb{R}^p \to X$, where X is some target space, f from M to X is reputed to have property P if all composite functions $f \circ \varphi^{-1}$, now of type $\mathbb{R}^p \to X$, have it. A manifold M *with boundary* has points where it "looks, locally, like" a closed half-space of $\mathbb{R}^p$; these points form, taken together, a (boundaryless) $(p-1)$-manifold ∂M, called the *boundary* of M. Connectedness is not required: A manifold can be in several pieces, all of the same dimension p.

In practice, our manifolds will be glued assemblies of *cells*, as follows.

First, let us define "reference cells" in $\mathbb{R}^p$, as illustrated on Fig. 2.1. These are bounded convex polytopes of the form

$$K_p^\alpha = \left\{\xi \in \mathbb{R}^p\colon \xi^l \geqslant 0\ \forall l = 1, \ldots, p,\ \sum_{j=1}^{p} \alpha^i_j \xi^j \leqslant 1\ \forall i = 1, \ldots, k\right\}, \tag{2.1}$$

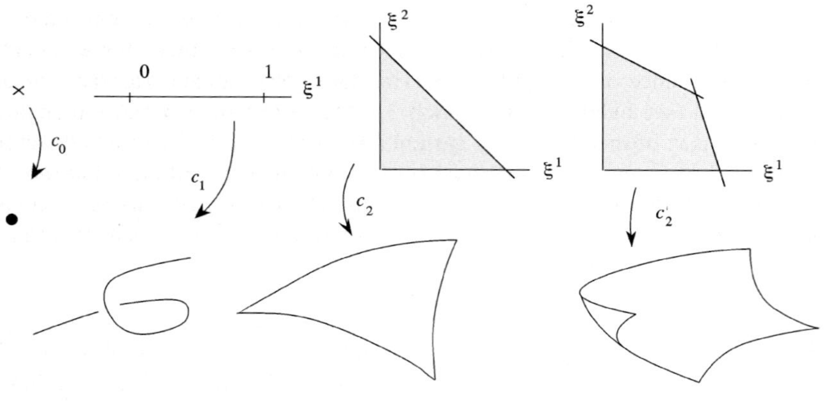

FIG. 2.1. Some cells in A_3, of dimensions 0, 1, 2.

where the α^i_j's form a rectangular $(k \times p)$-matrix with nonnegative entries, and no redundant rows.

Now, a *p-cell* in A_n, with $0 \leqslant p \leqslant n$, is a smooth map c from some K^α_p into A_n, one-to-one, and such that the derivative $\mathrm{D}c(\xi)$ has rank p for all ξ in K^α_p. (These restrictions, which qualify c as an *embedding*, are meant to exclude double points, and cusps, pleats, etc., which smoothness alone is not enough to warrant.) The same symbol c will serve for the map and for the image $c(K^\alpha_p)$. The *boundary* ∂c of the cell is the image under c of the topological boundary of K^α_p, i.e., of points ξ for which at least one equality holds in (2.1). Remark that ∂c is an assembly of $(p-1)$-cells, which themselves intersect, if they do, along parts of their boundaries.

Thus, a 0-cell is just a point. A 1-cell, or "path", is a simple parameterized curve. The simplest 2-cell is the triangular "patch", a smooth embedding of the triangle $\{\xi\colon \xi^1 \geqslant 0,\ \xi^2 \geqslant 0,\ \xi^1 + \xi^2 \leqslant 1\}$. The definition is intended to leave room for polygonal patches as well, and for three-dimensional "blobs", i.e., smooth embeddings of convex polyhedra.

We shall have use for the *open* cell corresponding to a cell c (then called a *closed* cell for contrast), defined as the restriction of c to the interior of its reference cell.

A subset M of A_n will be called a *piecewise smooth* p-manifold if (1) there exists a finite family $\mathcal{C} = \{c_i\colon i = 1, \dots, m\}$ of p-cells whose union is M, (2) the open cell corresponding to c_i intersects no other cell, (3) intersections $c_i \cap c_j$ are piecewise smooth $(p-1)$-manifolds (the recursive twist in this clause disentangles at $p = 0$), (4) the cells are properly joined at their boundaries,[7] i.e., in such a way that each point of M has a neighborhood in M homeomorphic to either a p-ball or half a p-ball.

Informally, therefore, piecewise smooth manifolds are glued assemblies of cells, obtained by topological identification of parts of their respective boundaries. (Surface S in Fig. 4.1, below, is typical.)

[7]This is regrettably technical, but it can't be helped, if M is to be a manifold. The assembly of *three* curves with a common endpoint, for instance, is not a manifold. See also HENLE [1994] for examples of 3D-spaces obtained by identification of facets of some polyhedra, which fail to be manifolds. Condition (2) forbids self-intersections, which is overly drastic and could be avoided, but will not be too restrictive in practice.

Having introduced this category of objects – which we shall just call manifolds, from now on – we should, as it is the rule and almost a reflex in mathematical work, deal with maps between such objects, called *morphisms*, that preserve their relevant structures. About cells, first: A map between two images of the same reference cell which is bijective and smooth (in both directions) is called a *diffeomorphism*. Now, about our manifolds: There is a *piecewise smooth diffeomorphism* between two of them (and there too, we shall usually dispense with the "piecewise smooth" qualifier) if they are homeomorphic and can both be chopped into sets of cells which are, two by two, diffeomorphic.

3. Orientation

To get oneself oriented, in the vernacular, consists in knowing where is South, which way is uptown, etc. To orient a map, one makes its upper side face North. Pigeons, and some persons, have a sense of orientation. And so forth. *Nothing* of this kind is implied by the mathematical concept of orientation – which may explain why so simple a notion may be so puzzling to many. Not that mathematical orientation has no counterpart in everyday's life, it has, but in something else: When entering a roundabout or a circle with a car, you know whether you should turn clockwise or counterclockwise. *That* is orientation, as regards the ground's surface. Notice how it depends on customs and law. For the spatial version of it, observe what "right-handed" means, as applied to a staircase or a corkscrew.

3.1. Oriented spaces

Now let us give the formal definition. A *frame* in V_n is an ordered n-tuple of linearly independent vectors. Select a basis (which is thus a frame among others), and for each frame, look at the determinant of its n vectors, as expressed in this basis, hence a *FRAME* $\rightarrow$ *REAL* function. This function is basis-dependent, but the equivalence relation defined by "$f \equiv f'$ if and only if frames f and f' have determinants of the same sign" does not depend on the chosen basis, and is thus intrinsic to the structure of V_n. There are two equivalence classes with respect to this relation. Orienting V_n consists in designating one of them as the class of "positively oriented" frames. This amounts to defining a function, which assigns to each frame a label, either *direct* or *skew*, two equivalent frames getting the same label. There are two such functions, therefore two possible orientations. An *oriented vector space* is thus a pair $\{V, Or\}$, where Or is one of the two orientation classes of V. (Equivalently, one may define an oriented vector space as a pair {*vector space*, *privileged basis*}, provided it's well understood that this basis plays no other role than specifying the orientation.) We shall find convenient to extend the notion to a vector space of dimension 0 (i.e., one reduced to the single element 0), to which also correspond, by convention, two oriented vector spaces, labelled $+$ and $-$.

REMARK 3.1. Once a vector space has been oriented, there are direct and skew *frames*, but there is no such thing as direct or skew *vectors*, except, one may concede, in dimension 1. A vector does not acquire new features just because the space where it belongs has been oriented! Part of the confusion around the notion of "axial" (vs. "polar") vectors stems from this semantic difficulty (BOSSAVIT [1998a, p. 296]). As axial vectors will not be used here, the following description should be enough to deal with the issue. Let's agree that, if Or is one of the orientation classes of V, the expression $-Or$ denotes the other class. Now, form pairs $\{v, Or\}$, where v is a vector and Or any orientation class of V, and consider two pairs $\{v, Or\}$ and $\{v', Or'\}$ as equivalent when $v' = -v$ and $Or' = -Or$. *Axial vectors* are, by definition, the equivalence classes of such pairs. (*Polar* vectors is just a redundant name, inspired by a well-minded sense of equity, for vectors of V.) Notice that axial *scalars* can be defined the same way: substitute a real number for v. Hence axial vector fields and axial functions (more often called "pseudo-functions" in physics texts). The point of defining such objects is to become able to express Maxwell's equations in *non*-oriented Euclidean space, i.e., V_3 with a dot product but no specific orientation. See BOSSAVIT [1998b] or [1999] for references and a discussion.

An affine space, now, is oriented by orienting its vector associate: a *bound frame* at x in A_n, i.e., a set of n independent vectors at x, is direct (respectively skew) if these n vectors form a direct (respectively skew) frame in V_n.

Vector subspaces of a given vector space (or affine subspaces of an affine space[8]) can have their own orientation. Orienting a line, in particular, means selecting a vector parallel to it, called a *director* vector for the line, which specifies the "forward" direction along it.

Such orientations of different subspaces are a priori unrelated. Orienting 3D space by the corkscrew rule, for instance, does not imply any orientation in a given plane. This remark may hurt common sense, for we are used to think of the standard orientation of space and of, say, a horizontal plane, as somehow related. And they are, indeed, but only because we think of vertical lines as oriented, bottom up. This is the convention known as *Ampère's rule*. To explain what happens there, suppose space is oriented, and some privileged straightline is oriented too, on its own. Then, any plane *transverse* to this line (i.e., thus placed that the intersection reduces to a single point) inherits an orientation, as follows: To know whether a frame in the plane is direct or skew, make a list of vectors composed of, in this order, (1) the line's director, (2) the vectors of the planar frame; hence an enlarged spatial frame, which is either direct or skew, which tells us about the status of the plane frame.

More generally, there is an interplay between the orientations of complementary subspaces and those of the encompassing space. Recall that two subspaces U and W of V are *complementary* if their *span* is all V (i.e., each v in V can be decomposed as $v = u + w$, with u in U and w in W) and if they are *transverse* ($U \cap W = \{0\}$, which makes the decomposition unique). We shall refer to V as the "ambient" space, and write $V = U + W$. If both U and W have orientation, this orients V, by the following convention: the frame obtained by listing the vectors of a direct frame in U first, then those of a direct frame in W, is direct. Conversely, if both U and V are oriented, one may orient W as follows: to know whether a given frame in W is direct or skew, list its vectors behind those of a direct frame of U, and check whether the enlarged frame thus obtained is direct or skew in V. This is a natural generalization of Ampère's rule.

[8] An affine subspace is oriented by orienting the parallel vector subspace. A point, which is an affine subspace parallel to $\{0\}$, can therefore be oriented, which we shall mark by apposing a sign to it, $+$ or $-$.

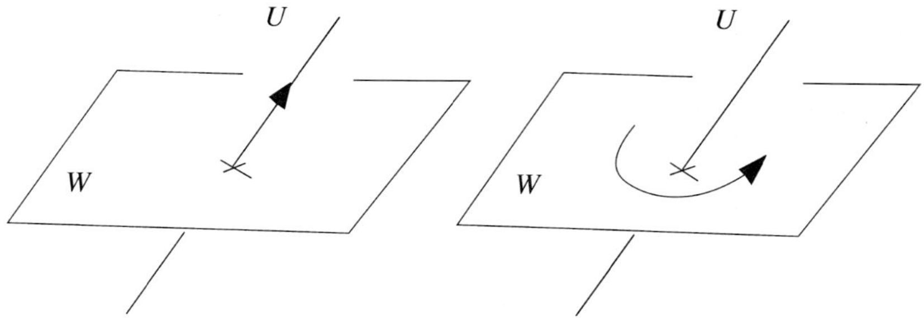

FIG. 3.1. Left: Specifying a "crossing direction" through a plane W by inner-orienting a line U transverse to it. Right: Outer-orienting U, i.e., giving a sense of going around it, by inner-orienting W.

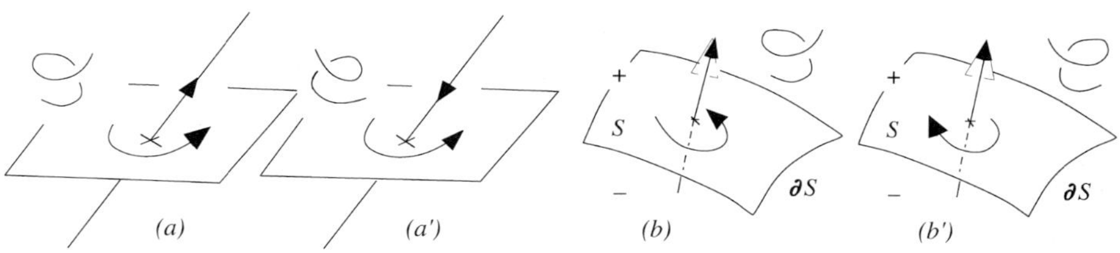

FIG. 3.2. Left: How an externally oriented line acquires inner orientation, depending on the orientation of ambient space. (Alternative interpretation: if one knows both orientations, inner and outer, for a line, one knows the ambient orientation.) Right: Assigning to a surface a crossing direction (here from region "−" below to region "+" above) will not by itself imply an inner orientation. But it does if ambient space is oriented, as seen in (b) and (b′). Figs. 3.2(a) and 3.2(b) can be understood as an explanation of Ampère's rule, in which the ambient orientation is, by convention, the one shown here by the "right corkscrew" icon.

Now what if U is oriented, but ambient space is not? Is U's orientation of any relevance to the complement W? Yes, as Fig. 3.1 suggests (left): For instance, if W has dimension $n-1$, an orientation of the one-dimensional complement U can be interpreted as a crossing direction relative to W, an obviously useful notion. (Flow of something through a surface, for instance, presupposes a crossing direction.) Hence the concept of *external*, or *outer orientation* of subspaces of V: Outer orientation of a subspace is, by definition, an orientation of one[9] of its complements. Outer orientation of V itself is thus a sign, + or −. (For contrast and clarity, we shall call *inner* orientation what was simply "orientation" up to this point.) The notion (which one can trace back to Veblen (VEBLEN and WHITEHEAD [1932]), cf. VAN DANTZIG [1954] and SCHOUTEN [1989]) passes to affine subspaces of an affine space the obvious way.

Note that *if* ambient space is oriented, outer orientation determines inner orientation (Fig. 3.2). But otherwise, the two kinds of orientation are independent. As we shall see, they cater for different needs in modelling.

[9]Nothing ambiguous in that. There is a canonical linear map between two complements W_1 and W_2 of the same subspace U, namely, the "affine projection" π_U along U, thus defined: for v in W_1, set $\pi_U(v) = v + u$, where u is the unique vector in U such that $v + u \in W_2$. Use π_U to transfer orientation from W_1 to W_2.

3.2. *Oriented manifolds*

Orientation can be defined for other figures than linear subspaces. Connected parts of affine subspaces, such as polygonal facets, or line segments, can be oriented by orienting the supporting subspace (i.e., the smallest one containing them). Smooth lines and surfaces as a whole are oriented by attributing orientations to all their tangents or tangent planes in a consistent way.

"Consistent"? Let's explain what that means, in the case of a surface. First, subspaces parallel to the tangent planes at all points in the neighborhood $N(x)$ of a given surface point x have, if $N(x)$ is taken small enough, a common complement, characterized by a director $n(x)$ (not the "normal" vector, since we have no notion of orthogonality at this stage, but the idea is the same). Then $N(x)$ is consistently oriented if all these orientations correspond via the affine projection along $n(x)$ (cf. Note 9). But this is only *local* consistency, which can always be achieved, and one wants more: *global* consistency, which holds if the surface can be covered by such neighborhoods, with consistent orientation in each non-empty intersection $N(x) \cap N(y)$. This may not be feasible, as in the case of a Möbius band, hence the distinction between (internally) orientable and non-orientable manifolds.

Cells, as defined above, are inner orientable, thanks to the fact that Dc does not vanish. For instance (cf. Fig. 3.3), for a path c, i.e., a smooth embedding $t \to c(t)$ from $[0, 1]$ to A_n, the tangent vectors $\partial_t c(t)$ determine consistent orientations of their supporting lines, hence an orientation of the path. (The other orientation would be obtained by starting from the "reverse" path, $t \to c(1-t)$.) Same with a patch $\{s, t\} \to S(s, t)$ on the triangle $T = \{\{s, t\}\colon 0 \leqslant s,\ 0 \leqslant t,\ s + t \leqslant 1\}$: The vectors $\partial_s S(s, t)$ and $\partial_t S(s, t)$, in this order, form a basis at $S(s, t)$ which orients the tangent plane, and these orientations are consistent.

As for piecewise smooth manifolds, finally, the problem is at points x where cells join, for a tangent subspace may not exist there. But according to our conventions, there must be a neighborhood homeomorphic to a ball or half-ball, which *is* orientable, hence a way to check whether tangent subspaces at regular points in the vicinity of x have consistent orientations, and therefore, to check whether the manifold as a whole is or is not orientable.

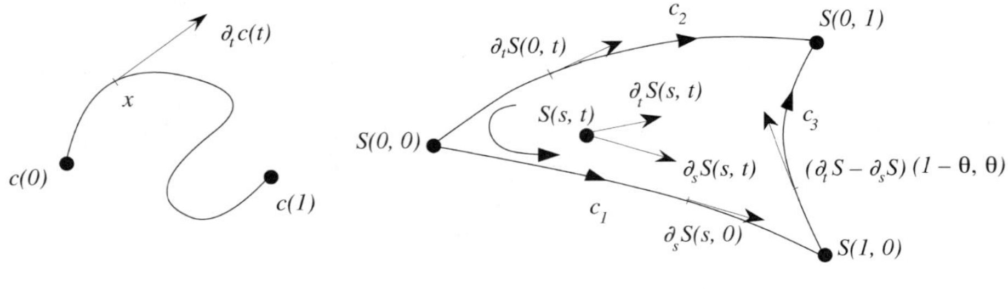

FIG. 3.3. A path and a patch, with natural inner orientations. Observe how their boundaries are themselves assemblies of cells: $\partial c = c(0) - c(1)$ and $\partial S = c_1 - c_2 + c_3$, with a notation soon to be introduced more formally. Paths c_i are $c_1 = s \to S(s, 0)$, $c_2 = t \to S(0, t)$, and $c_3 = \theta \to S(1 - \theta, \theta)$, each with its natural inner orientation.

Similar considerations hold for external orientation. Outer-orienting a surface consists in giving a (globally consistent) crossing direction through it. For a line, it's a way of "turning around" it, or "gyratory sense" (Fig. 3.1, right). For a point, it's an orientation of the space in its neighborhood. For a connected region of space, it's just a sign, $+$ or $-$.

3.3. Induced orientation

Surfaces which enclose a volume V (which one may suppose connected, though the boundary ∂V itself need not be) can always be outer oriented, because the "inside out" crossing direction is always globally consistent. Let us, by convention, take this direction as defining the canonical outer orientation of ∂V. No similarly canonical *inner* orientation of the surface results, as could already be seen on Fig. 3.2, since there are, in the neighborhood of each boundary point, two eligible orientations of ambient space. But if V is inner oriented, this orientation can act in conjunction with the outer one of ∂V to yield a natural inner orientation of V's boundary about this point. For example, on the left of Fig. 3.4, the 2-frame $\{v_1, v_2\}$ in the tangent plane of a boundary point is taken as direct because, by listing its vectors behind an outward directed vector ν, one gets the direct 3-frame $\{\nu, v_1, v_2\}$. Consistency of these orientations stems from the consistency of the crossing direction. Hence V's inner orientation *induces* one on each part of its boundary.

The same method applies to manifolds of lower dimension p, by working inside the affine p-subspace tangent to each boundary point. See Fig. 3.4(b) for the case $p = 2$. The p-manifold, thus, serves as ambient space with respect to its own boundary, for the purpose of inducing orientation.

In quite a similar way (Fig. 3.5), *outer* orientation of a manifold induces an *outer* orientation of each part of its boundary. (For a volume V, the induced outer orientation of ∂V is the inside-out or outside-in direction, depending on the outer orientation, $+$ or $-$, of V.)

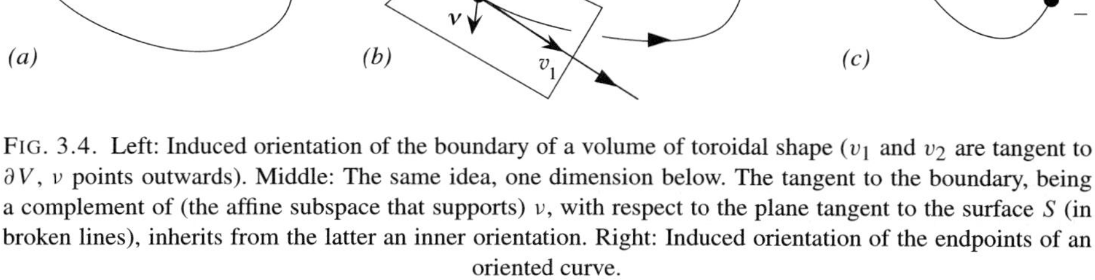

FIG. 3.4. Left: Induced orientation of the boundary of a volume of toroidal shape (v_1 and v_2 are tangent to ∂V, ν points outwards). Middle: The same idea, one dimension below. The tangent to the boundary, being a complement of (the affine subspace that supports) ν, with respect to the plane tangent to the surface S (in broken lines), inherits from the latter an inner orientation. Right: Induced orientation of the endpoints of an oriented curve.

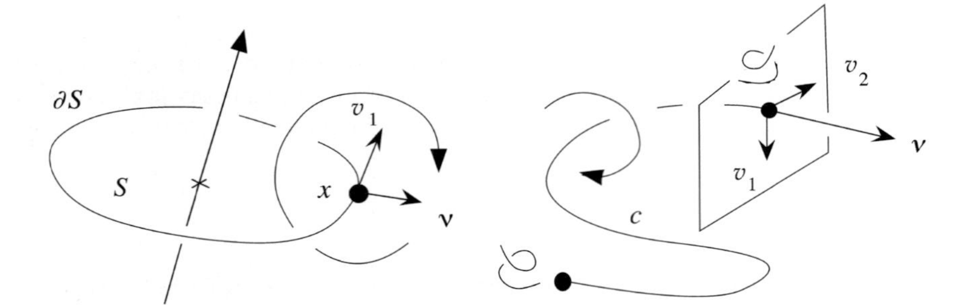

FIG. 3.5. Left: To outer-orient ∂S is to (consistently) inner-orient complements of the tangent, one at each boundary point x. For this, take as direct the frame $\{v_1, \nu\}$, where $\{v_1\}$ is a direct frame in the complement of the plane tangent to S at x, and ν an outward directed vector tangent to S. That $\{v_1\}$ is direct is known from the outer orientation of S. Right: Same idea about the boundary points of line c. Notice that ν is now appended *behind* the list of frame vectors. Consistency stems from the consistency of ν, the inside-out direction with respect to S. The icons near the endpoints are appropriate, since outer orientation of a point is inner orientation of the space in its vicinity.

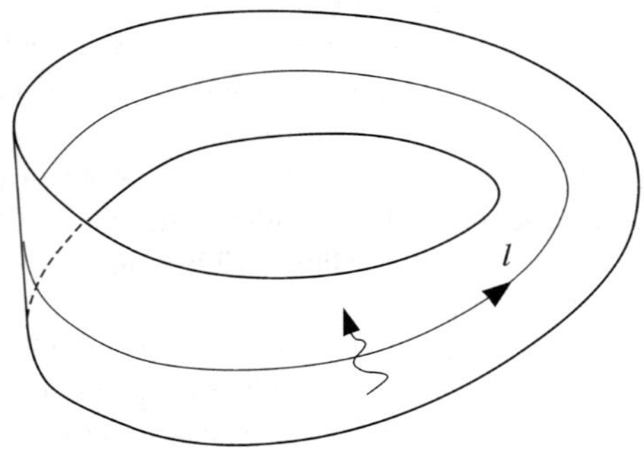

FIG. 3.6. Möbius band, not orientable. As the middle line l does not separate two regions, it cannot be assigned any consistent crossing direction, so it has no outer orientation with respect to the "ambient" band.

3.4. Inner vs outer orientation of submanifolds

We might (but won't, as the present baggage is enough) extend these concepts to submanifolds of ambient manifolds other that A_3, including non-orientable ones. A two-dimensional example will give the idea (Fig. 3.6): Take as ambient manifold a Möbius band M, and forget about the 3-dimensional space it is embedded in for the sake of the drawing. Then it's easy to find in M a line which (being a line) is inner orientable, but cannot consistently be outer oriented. Note that the band by itself, i.e., considered as its own ambient space, can be outer oriented, by giving it a sign: Indeed, outer orientation of the tangent plane at each point of M, being inner orientation of this point, is such a sign, so consistent orientation means attributing the same sign to all points. (By the same token, any manifold is outer orientable, with respect to itself as ambient space.)

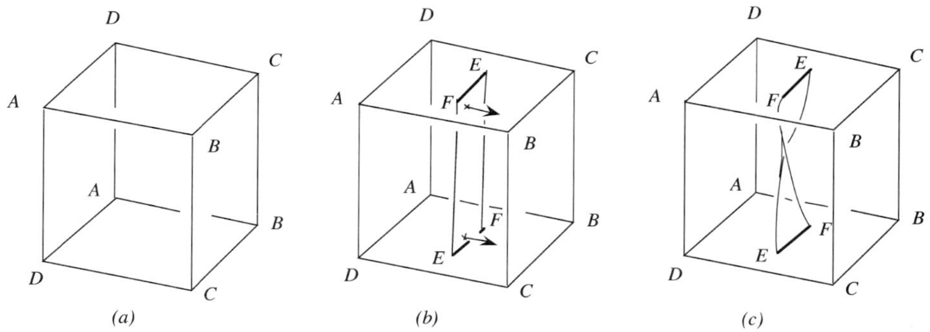

FIG. 3.7. Left: Non-orientable 3-manifold with boundary: Identify top and bottom by matching upper A with lower A, etc. Middle: Embedded Möbius band, with a globally consistent crossing direction. Right: Embedded ribbon.

For completeness, let us give another example (Fig. 3.7), this time of an outer-orientable surface without inner orientation, owing to non-orientability of the ambient manifold. The latter (whose boundary is a Klein bottle) is made by sticking together the top and bottom sides of a vertical cube, according to the rule of Fig. 3.7(a). The ribbon shown in (b) is topologically a Möbius band, a non-(inner)orientable surface. Yet, it plainly has a consistent set of transverse vectors. (Follow the upper arrow as its anchor point goes up and reenters at the bottom, and notice that the arrow keeps pointing in the direction of AB in the process. So it coincides with the lower arrow when this passage has been done.) Contrast with the ordinary ribbon in (c), orientable, but not outer orientable with respect to this ambient space.

The two concepts of orientation are therefore essentially different.

In what follows, we shall use the word "twisted" (as opposed to "straight") to connote anything that is to do with outer (as opposed to inner) orientation.

4. Chains, boundary operator

It may be convenient at times to describe a manifold M as an assembly of several manifolds, even if M is connected. Think for example of the boundary of a triangle, as an assembly of three edges, and more generally of a piecewise smooth assembly of cells. But it may happen – so will be the case here, later – that these various manifolds have been *independently* oriented, with orientations which may or may not coincide with the desired one for M. This routinely occurs with boundaries, in particular. The concept of chain will be useful to deal with such situations.

A *p-chain* is a finite family $\mathcal{M} = \{M_i\colon i = 1, \ldots, k\}$ of oriented connected p-manifolds,[10] to which we shall loosely refer below as the "components" of the chain, each loaded with a weight μ^i belonging to some ring of coefficients, such as $\mathbb{R}$ or $\mathbb{Z}$ (say $\mathbb{R}$ for definiteness, although weights will be signed integers in most of our examples). Such a chain is conveniently denoted by the "formal" sum $\sum_i \mu^i M_i \equiv \mu^1 M_1 + \cdots + \mu^k M_k$,

[10]For instance, cells. But we don't request that. Each M_i may be a piecewise smooth manifold already.

thus called because the + signs do not mean "add" in any standard way. On the other hand, chains themselves, as whole objects, can be added, and there the notation helps: To get the sum $\sum_i \mu^i M_i + \sum_j \nu^j N_j$, first merge the two families $\mathcal{M}$ and $\mathcal{N}$, then attribute weights by adding the weights each component has in each chain, making use of the convention that $\mu M'$ is the same chain as $-\mu M$ when M' is the same manifold as M with opposite orientation. If all weights are zero, we have the *null chain*, denoted 0. All this amounts, as one sees, to handling chains according to the rules of algebra, when they are represented via formal sums, which is the point of such a notation. *Twisted* chains are defined the same way, except that all orientations are external. (Twisted and straight chains are not to be added, or otherwise mixed.)

If M is an oriented piecewise smooth manifold, all its cells c_i inherit this orientation, but one may have had reasons to orient them on their own, independently of M. (The same cell may well be part of several piecewise smooth manifolds, for instance.) Then, it is natural to associate with M the chain $\sum_i \pm c_i$, also denoted by M, with ith weight -1 when the orientations of M and c_i differ. (Refer back to Fig. 3.3 for simple examples.)

Now, the boundary of an oriented piecewise smooth $(p+1)$-manifold M is an assembly of p-manifolds, each of which we assume has an orientation of its own. Let us assign each of them the weight ± 1, according to whether its orientation coincides with the one inherited from M. (We say the two orientations *match* when this coincidence occurs.) Hence a chain, also denoted ∂M. By linearity, the operator ∂ extends to chains: $\partial(\sum_i \mu^i M_i) = \sum_i \mu^i \partial M_i$. A chain with null boundary is called a *cycle*. A chain which is the boundary of another chain is called, appropriately, a *boundary*. Boundaries are cycles, because of the fundamental property

$$\partial \circ \partial = 0, \tag{4.1}$$

i.e., the boundary of a boundary is the null chain. A concrete example, as in Fig. 4.1, will be more instructive here than a formal proof.

REMARK 4.1. Beyond its connection with assemblies of oriented cells, no too definite intuitive interpretation of the concept of chain should be looked for. Perhaps, when $p = 1$, one can think of the chain $\sum_i \gamma_i c_i$, with integer weights, as "running along each c_i, in turn, $|\gamma_i|$ times, in the direction indicated by c_i's orientation, or in the reverse direction, depending on the sign of γ_i". But this is a bit contrived. Chains are better conceived as algebraic objects, based on geometric ones in a useful way – as the example in Fig. 4.1 should suggest, and as we shall see later. However, we shall indulge in language abuse, and say that a closed curve "is" a 1-cycle, or that a closed surface "is" a 2-cycle, with implicit reference to the associated chain.

So boundaries are cycles, after (4.1). Whether the converse is true is an essential question. In affine space, the answer is positive: A closed surface encloses a volume, a closed curve (even if knotted) is the boundary of some surface (free of self-intersections, amazing as this may appear), called a Seifert surface (SEIFERT and THRELFALL [1980], ARMSTRONG [1979, p. 224]). But in some less simple ambient manifolds, a cycle need not bound. In the case of a solid torus, for instance, a meridian circle is a boundary, but a parallel circle is not, because none of the disks it bounds in A_3 is entirely contained in

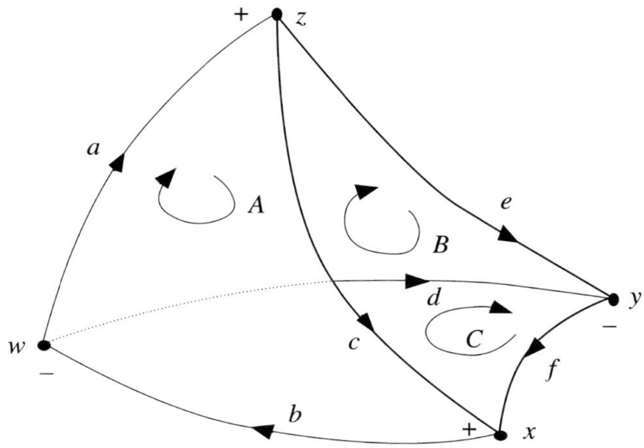

FIG. 4.1. Piecewise smooth surface S, inner oriented (its orientation is taken to be that of the curved triangle in the fore, marked A), represented as the chain $A - B - C$ based on the oriented curved triangles A, B, C. (Note the minus signs: B's and C's orientations don't match that of S.) One has $\partial A = a + b + c$, $\partial B = e + a - d$, $\partial C = b + d + f$, where a, b, c, d, e, f are the boundary curves, arbitrarily oriented as indicated. Now, $\partial S = \partial(A - B - C) = c - e - f$: Observe how the "seams" a, b, c automatically receive null weights in this 1-chain, whatever their orientation, because they appear twice with opposite signs. Next, since $\partial c = x - z$, $\partial e = -y - z$, and $\partial f = x + y$, owing to the (arbitrary) orientations assigned to points w, x, y, z, one has $\partial\partial S = \partial(c - e - f) = 0$, by the same process of cancellation by pairs. The reader is invited to work out a similar example involving twisted chains instead of straight ones.

the torus. Whether cycles are or aren't boundaries is therefore an issue when investigating the global topological properties of a manifold. Chains being algebraic objects then becomes an asset, for it makes possible to harness the power of algebra to the study of topology. This is the gist of *homology* (HENLE [1994], HILTON and WYLIE [1965]), and of algebraic topology in general.

5. Metric notions

Now, let us equip V_n with a dot product: $u \cdot v$ is a real number, linearly depending on vectors u and v, with symmetry ($u \cdot v = v \cdot u$) and strict positive-definiteness ($u \cdot u > 0$ if $u \neq 0$). Come from this, first the notions of orthogonality and angle, next a norm $|u| = (u \cdot u)^{1/2}$ on V_n, then a distance $d(x, y) = |y - x|$, translation-invariant by construction, between points of the affine associate A_n.

DEFINITION 5.1. Euclidean space, E_n, is the structure composed of A_n, plus a dot product on its associate V_n, plus an orientation.

Saying "the" structure implies that two realizations of it (with two different dot products and/or orientations) are isomorphic in some substantial way. This is so: For any other dot product, "•" say, there is an invertible linear transform L such that

$u \cdot v = Lu \cdot Lv$. Moreover,[11] one may have L "direct", in the sense that it maps a frame to another frame of the same orientation class, or "skew". Therefore, two distinct Euclidean structures on A_n are linked by some L. In the language of group actions, the linear group GL_n, composed of the above L's, acts transitively on Euclidean structures, i.e., with a unique orbit, which is our justification for using the singular. (These structures are said to be *affine equivalent*,[12] a concept that will recur.) The point can vividly be made by using the language of group actions: the isotropy group of $\{\cdot, Or\}$ "cannot be any larger". (More precisely, it is maximal, as a subgroup, in the group of direct linear transforms.)

In dimension 3,[13] dot product and orientation conspire in spawning the *cross product*: $u \times v$ is characterized by the equality

$$|u \times v|^2 + (u \cdot v)^2 = |u|^2 |v|^2 \tag{5.1}$$

and the fact that vectors u, v and $u \times v$ form, in this order, a direct frame. The 3-*volume* of the parallelotope built on vectors u, v, w, defined by $\mathrm{vol}(u, v, w) = (u \times v) \cdot w$, is equal, up to sign, to the above volumic measure, with equality if the frame is direct.[14] Be well aware that $\times$ doesn't make any sense in *non*-oriented three-space.

We shall have use for the related notion of *vectorial area* of an outer oriented triangle T, defined as the vector $\vec{\mathrm{T}} = \mathrm{area}(\mathrm{T})n$, where n is the normal unit vector that provides the crossing direction. (If an ambient orientation exists, two vectors u and v can be laid along two of the three sides, in such a way that $\{u, v, n\}$ is a direct frame. Then, $\vec{\mathrm{T}} = \frac{1}{2} u \times v$. Fig. 6.1 gives an example.) More generally, an outer oriented surface of E_3 has a vectorial area: Chop the surface into small adjacent triangular patches, add the vectorial areas of these, and pass to the limit. (This yields 0 for a closed surface.)

For later use, we state the relations between the structures induced by $\{\cdot, Or\}$ and $\{\bullet, \mathbf{Or}\}$, where $\mathbf{Or} = \pm Or$, the sign being that of $\det(L)$. (There is no ambiguity about "$\det(L)$", understood as the determinant of the matrix representation of L: its value is the same in any basis.) The norm $(u \bullet u)^{1/2}$ will be denoted by $\|u\|$. The corresponding cross product $\boldsymbol{\times}$ (boldface) is defined by $\|u \boldsymbol{\times} v\|^2 + (u \bullet v)^2 = \|u\|^2 \|v\|^2$ as in (5.1) (plus the request that $\{u, v, u \boldsymbol{\times} v\}$ be $\mathbf{Or}$-direct), and the new volume is $\mathbf{vol}(u, v, w) = (u \boldsymbol{\times} v) \bullet w$. It's a simple exercise to show that

$$\|u\| = |Lu|, \quad L(u \boldsymbol{\times} v) = Lu \times Lv, \quad \mathbf{vol}(u, v, w) = \det(L)\,\mathrm{vol}(u, v, w). \tag{5.2}$$

(It all comes from the equality $\det(Lu, Lv, Lw) = \det(L) \det(u, v, w)$, when u, v, w, and L are represented in some basis, a purely affine formula.) Notice that, for any w,

[11] L is not unique, since UL, for any *unitary* U (i.e., such that $|Uv| = |v|\ \forall v$), will work as well. In particular, one might force L to be self-adjoint, but we won't take advantage of that.

[12] Such equivalence is what sets Euclidean norms apart among all conceivable norms on V_n, like for instance $|v| = \sum_i |v^i|$. As argued at more length in BOSSAVIT [1998a], choosing to work in a Euclidean framework is an acknowledgment of another observed symmetry of the world we live in: its *isotropy*, in addition to its homogeneity.

[13] A binary operation with the properties of the cross product can exist only in dimensions 3 and 7 (SHAW and YEADON [1989], ECKMANN [1999]).

[14] An n-volume could directly be defined on V_n, as a map $\{v_1, \ldots, v_n\} \to \mathrm{vol}(v_1, \ldots, v_n)$, multilinear and null when two vectors of the list are equal. Giving an n-volume implies an orientation (direct frames are those with positive n-volumes), but no metric (unless $n = 1$).

one has $L^a L(u \mathbf{\times} v) \cdot w = L(u \mathbf{\times} v) \cdot Lw = \det(L)(u \times v) \cdot w$, where L^a denotes the *adjoint* of L (defined by $Lu \cdot v = u \cdot L^a v$ for all u, v), hence an alternative formula:

$$u \mathbf{\times} v = \det(L)(L^a L)^{-1}(u \times v). \tag{5.3}$$

As for the vectorial area, denoted $\vec{\mathbf{T}}$ in the "bold" metric, one will see that

$$\vec{\mathbf{T}} = \left|\det(L)\right|(L^a L)^{-1}\vec{\mathrm{T}}, \tag{5.4}$$

with a factor $|\det(L)|$, not $\det(L)$, because $\vec{\mathrm{T}}$ and $\vec{\mathbf{T}}$, both going along the crossing direction, point towards the same side of T.

We shall also need a topology on the space of p-chains, in order to define differential forms as *continuous* linear functionals on this space. As we shall argue later, physical observables such as electromotive force, flux, and so forth, can be conceived as the values of functionals of this kind, the chain operand being the idealization of some measuring device. Such values don't change suddenly when the measurement apparatus is slightly displaced, which is the rationale for continuity. But to make precise what "slightly displaced" means, we need a notion of "nearness" between chains – a topology.[15]

First thing, nearness between manifolds. Let us define the distance $d(M, N)$ between two of them as the greatest lower bound (the infimum) of $d_\phi(M, N) = \sup\{x \in M: |x - \phi(x)|\}$ with respect to all orientation-preserving piecewise smooth diffeomorphisms (OPD) ϕ that exist between M and N. There may be no such OPD, in which case we take the distance as infinite, but otherwise there is symmetry between M and N (consider ϕ^{-1} from N to M), positivity, d can't be zero if $M \neq N$, and the triangle inequality holds. (*Proof*: Take M, N, P, select OPDs ϕ and ψ from P to M and N, and consider x in P. Then $|\phi(x) - \psi(x)| \leqslant |\phi(x) - x| + |x - \psi(x)|$, hence $d_{\psi \circ \phi^{-1}}(M, N) \leqslant d_\phi(M, P) + d_\psi(N, P)$, then minimize with respect to ϕ and ψ.) Nearness of two manifolds, in this sense, does account for the intuitive notion of "slight displacement" of a line, a surface, etc. The topology thus obtained does not depend on the original dot product, although d does.

Next, on to chains. The notion of convergence we want to capture is clear enough: a sequence of chains $\{c_n = \sum_{i=1,\ldots,k} \mu_n^i M_{i,n}: n \in \mathbb{N}\}$ should certainly converge towards the chain $c = \sum_{i=1,\ldots,k} \mu^i M_i$ when the sequences of components $\{M_{i,n}: n \in \mathbb{N}\}$ all converge, in the sense of the previous distance, to M_i, while the weights $\{\mu_n^i: n \in \mathbb{N}\}$ converge too, towards μ^i. But knowing some convergent sequences is not enough to know the topology. (For that matter, even the knowledge of *all* convergent sequences would not suffice, see GELBAUM and OLMSTED [1964, p. 161].) On the other hand, the finer the topology, i.e., the more open sets it has, the more difficult it is for a sequence to converge, which tells us what to do: Define the desired topology as the finest one which (1) is compatible with the vector space structure of p-chains (in particular, each neighborhood of 0 should contain a convex neighborhood) (2) makes all sequences of the above kind converge.

---

[15] What follows is an attempt to bypass, rather than to face, this difficult problem, to which Harrison's work on "chainlet" spaces (nested Banach spaces which include chains and their limits with respect to various norms, HARRISON [1998]), provides a much more satisfactory solution.

The space of straight [respectively twisted] p-chains, as equipped with this topology, will be denoted by $\mathcal{C}_p$ [respectively $\tilde{\mathcal{C}}_p$]. Both spaces are purely affine constructs, independent of the Euclidean structure, which only played a transient role in their definition.

It now makes sense to ask whether the linear map ∂ is continuous from $\mathcal{C}_p$ to $\mathcal{C}_{p-1}$. The answer is by the affirmative, thanks to the linearity of ∂ and the inequality $d(\partial M, \partial N) \leqslant d(M, N)$. [*Proof*: The restriction to ∂M of an OPD ϕ is an OPD which sends it to ∂N, so $d(\partial M, \partial N) \leqslant \inf_\phi \sup\{x \in \partial M\colon |\phi(x) - x|\} \leqslant \inf_\phi \sup\{x \in M\colon |\phi(x) - x|\} = d(M, N)$.]

CHAPTER II

Rewriting the Maxwell Equations

Deconstruction calls for reconstruction: We now resettle the Maxwell system in the environment just described, paying attention to what makes use of the metric structure and what does not. In the process, differential forms will displace vector fields as basic entities.

6. Integration: Circulation, flux, etc.

Simply said, differential forms are, among mathematical objects, those meant to be integrated. So let us revisit Integration.

In standard integration theory (HALMOS [1950], RUDIN [1973], YOSIDA [1980]), one has a set X equipped with a measure dx. Then, to a pair $\{A, f\}$, where A is a part of X and f a function, integration associates a number, denoted $\int_A f(x)\,dx$ (or simply $\int_A f$, if there is no doubt on the underlying measure), with additivity and continuity with respect to both arguments, A and f. In what follows, we operate a slight change of viewpoint: Instead of leaving the measure dx in background of a stage on which the two objects of interest would be A and f, we consider the whole integrand $f(x)\,dx$ as a single object (later to be given its proper name, "differential form"), and A as some piecewise smooth manifold of A_3. This liberates integration from its dependence on the metric structure: The integral becomes a map of type *MANIFOLD* $\times$ *DIFFERENTIAL_FORM* $\rightarrow$ *REAL* (by linearity, *CHAIN* will eventually replace *MANIFOLD* there), which we shall see is the right approach as far as Electromagnetics is concerned. The transition will be in two steps, one in which the Euclidean structure is used, one in which we get rid of it.

The dot product of E_n induces measures on its submanifolds: By definition, the Euclidean measure of the parallelotope built on p vectors $\{v_1, \ldots, v_p\}$ anchored at x, i.e., of the set $\{x + \sum_i \lambda^i v_i\colon\ 0 \leqslant \lambda^i \leqslant 1,\ i = 1, \ldots, p\}$, is the square-root of the so-called Gram determinant of the v_i's, whose entries are the dot products $v_i \cdot v_j$, for all i, j from 1 to p. One can build from this, by the methods of classical measure theory (HALMOS [1950]), the p-dimensional measures, i.e., the lineal, areal, volumic, etc., measures of a (smooth, bounded) curve, surface, volume, etc. (what Whitney and his followers call its "mass", WHITNEY [1957]). For $p = 0$ not to stand out as an exception there, we attribute to an isolated point the measure 1. (This is the so-called *counting measure*, for which the measure of a set of points is the number of its elements.)

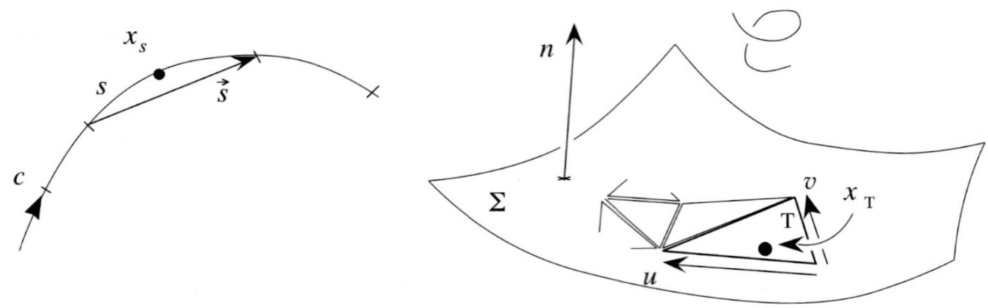

FIG. 6.1. Forming the terms of Riemann sums. Left: generic "curve segment" s, with associated sampling point x_s and vector $\vec{s}$. Right: generic triangular small patch T, with sampling point x_T. Observe how, with the ambient orientation indicated by the icon, the vectorial area of T happens to be $\frac{1}{2}u \times v$.

We shall consider, corresponding to the four dimensions $p = 0, \dots, 3$ of manifolds in E_3, four kinds of integrals which are constantly encountered in Physics. Such integrals will be defined on cells first, then extended by linearity to chains, which covers the case of piecewise smooth manifolds.

First, $p = 0$, a point, x say. The integral of a smooth function φ is then[16] $\varphi(x)$. If the point is inner oriented, i.e., if it bears a sign $\varepsilon(x) = \pm 1$, the integral is by convention $\varepsilon(x)\varphi(x)$.

Next ($p = 1$), let c be a 1-cell. At point $x = c(t)$, define the *unit tangent vector* $\tau(x)$ as the vector at x equal to $\partial_t c(t)/|\partial_t c(t)|$, which inner-orients c. Given a smooth vector field u, the dot product $\tau \cdot u$ defines a real-valued function on the image of c. We call *circulation* of u, along c thus oriented, the integral $\int_c \tau \cdot u$ of this function with respect to the Euclidean measure of lengths.

REMARK 6.1. Integrals (of smooth enough functions) are limits of Riemann sums. In the present case, such a sum can be obtained as suggested by Fig. 6.1, left: Chop the curve into a finite family $\mathcal{S}$ of adjacent curve segments s, pick a point x_s in each of them, and let $\vec{s}$ be the vector, oriented along c, that joins the extremities of s. The Riemann sum associated with $\mathcal{S}$ is then $\sum_{s \in \mathcal{S}} \vec{s} \cdot u(x_s)$, and converges towards $\int_c \tau \cdot u$ when $\mathcal{S}$ is properly refined.

Further up ($p = 2$), let Σ be a 2-cell, to which a crossing direction has been assigned, and choose the parameterization $\{s, t\} \to \Sigma(s, t)$ in such a way that vectors $\eta(s, t) = \partial_s \Sigma(s, t) \times \partial_t \Sigma(s, t)$ point in this direction. Then set $n(x) = \eta(s, t)/|\eta(s, t)|$, at point $x = \Sigma(s, t)$, to obtain the outer-orienting *unit normal field*. Given a smooth vector field u, we define the *flux* through Σ, thus outer oriented, as the integral $\int_\Sigma n \cdot u$ of the real-valued function $n \cdot u$ with respect to, this time, the Euclidean measure of

[16]This is also its integral over the set $\{x\}$, with respect to the counting measure, in the sense of Integration Theory. The integral over a *finite* set $\{x_1, \dots, x_k\}$, in this sense, would be $\sum_i \varphi(x_i)$. Notice the difference between this and what we are busy defining right now, the integral on a 0-chain, which will turn out to be a weighted sum of the reals $\varphi(x_i)$.

areas. (No ambiguity on this point, since the status of Σ as a surface has been made clear.)

REMARK 6.2. For Riemann sums, dissect Σ into a family $\mathcal{T}$ of small triangular patches T, whose vectorial areas are $\vec{\mathrm{T}}$, pick a point x_{T} in each of them, and consider $\sum_{\mathrm{T}\in\mathcal{T}} \vec{\mathrm{T}} \cdot u(x_{\mathrm{T}})$.

Last, for $p = 3$, and a 3-cell V with outer orientation $+$, the integral of a function f is the standard $\int_V f$, integral of f over the image of V with respect to the Lebesgue measure. This is consistent with the frequent physical interpretation of $\int_V f$ as the quantity, in V, of something (mass, charge, ...) present with density f in V. With outer orientation $-$, the integral is $-\int_V f$. Thus, outer orientation helps fix bookkeeping conventions when f is a rate of variation, like for instance, heat production or absorption. The inner orientation of V is irrelevant here.

Now, let us extend the notion to chains based on oriented cells. In dimension 0, where an oriented point is a point-cum-sign pair $\{x, \varepsilon\}$, a 0-chain m is a finite collection $\{\{x_i, \varepsilon_i\}\colon i = 1, \dots, k\}$ of such pairs, each with a weight μ^i. The integral $\int_m \varphi$ is then defined as $\sum_i \mu^i \varepsilon_i \varphi(x_i)$.[17] In dimension 1, the circulation along the 1-chain $c = \sum_i \mu^i c_i$ is $\int_c \tau \cdot u = \sum_i \mu^i \int_{c_i} \tau \cdot u$. The flux $\int_\Sigma n \cdot u$ through the *twisted* (beware!) chain $\Sigma = \sum_i \mu^i \Sigma_i$ is defined as $\sum_i \mu^i \int_{\Sigma_i} n \cdot u$. As for dimension 3, a twisted chain manifold V is a finite collection[18] $\{\{V_i, \varepsilon_i\}\colon i = 1, \dots, k\}$ of 3D blobs-with-sign, with weights μ^i, and $\int_V f$ is, by definition, $\sum_i \mu^i \varepsilon_i \int_{V_i} f$.

Note that we have implicitly defined integrals on piecewise smooth manifolds there, since these can be considered as cell-based chains with "orientation matching weights" (1 if the cell's orientation and the manifold's match, -1 if they don't).

Thus the most common ways[19] to integrate things in three-space lead to the definition of integrals over *inner* oriented manifolds or chains in cases $p = 0$ and 1 and *outer* oriented ones[20] in cases $p = 2$ and 3. An unpleasant asymmetry. But since we work in *oriented* Euclidean space, where one may, as we have seen, derive outer from inner orientation, and the other way round, this restores the balance, hence finally *eight* kinds of integrals, depending on the dimension and on the nature (internal or external) of the orientation of the underlying chain.

Thus we have obtained a series of maps of type *CHAIN* $\to$ *REAL*, but in a pretty awkward way, one must admit. Could there be an underlying unifying concept that would make it all simpler?

---

[17]One might think, there, that orientation-signs and weights do double duty. Indeed, a convention could be made that all points are positively oriented, and this would dispose of the ε_is. We won't do this, for the sake of uniformity of treatment with respect to dimension.

[18]Again, one might outer-orient such elementary volumes by giving them all a $+$ sign, reducing the redundancy, and we refrain to do so for the same reason.

[19]Others reduce to one of these. For instance, when using Cartesian coordinates x–y–z, $\int_c f(x, y, z)\,\mathrm{d}x$ is simply the circulation along c, in the sense we have defined above, of the field of x-directed basis vectors magnified by the scalar factor f.

[20]A tradition initiated in FIRESTONE [1933] distinguishes between so-called "across" and "through" physical quantities (KOENIG and BLACKWELL [1960], BRANIN [1961]), expressible by circulations and fluxes, respectively. As we shall see, this classification is not totally satisfying.

7. Differential forms, and their physical relevance

Indeed, these maps belong to a category of objects that can be defined without recourse to the Euclidean structure, and have thus a purely affine nature:

DEFINITION 7.1. A straight [respectively twisted] differential form of degree p, or p-form, is a real-valued map ω over the space of straight [respectively twisted] p-chains, linear with respect to chain addition, and continuous in the sense of the above-defined topology of chains (end of Section 5).

Differential forms, thus envisioned, are dual objects with respect to chains, which prompts us to mobilize the corresponding machinery of functional analysis (YOSIDA [1980]): Call $\mathcal{F}^p$ [respectively $\widetilde{\mathcal{F}}^p$] the space of straight [respectively twisted] p-forms, as equipped with its so-called "strong" topology.[21] Then $\mathcal{C}_p$ and $\mathcal{F}^p$ [respectively $\widetilde{\mathcal{C}}_p$ and $\widetilde{\mathcal{F}}^p$] are *in duality* via the bilinear bicontinuous map $\{c, \omega\} \to \int_c \omega$, of type p-*CHAIN* $\times$ p-*FORM* $\to$ *REAL*. A common notation for such duality products being $\langle c; \omega\rangle$, we shall use that as a convenient alternative[22] to $\int_c \omega$. A duality product should be *non-degenerate*, i.e., $\langle c'; \omega\rangle = 0\ \forall c'$ implies $\omega = 0$, and $\langle c; \omega'\rangle = 0\ \forall \omega'$ forces $c = 0$. The former property holds true by definition, and the latter is satisfied because, if $c \neq 0$, one can construct an ad hoc smooth vector field or function with nonzero integral, hence a nonzero form ω such that $\langle c; \omega\rangle \neq 0$.

The above eight kinds of integrals, therefore, are instances of differential forms, which we shall denote (in their order of appearance) by ${}^0\varphi$, ${}^1u$ (circulation of u), ${}^2\tilde{u}$ (flux of u), ${}^3\tilde{\varphi}$, and ${}^0\tilde{\varphi}$, ${}^1\tilde{u}$, ${}^2u$, ${}^3\varphi$. This is of course ad hoc notation, to be abandoned as soon as the transition from fields to forms is achieved. Note the use of the pre-superscript p, accompanied or not by the tilde as the case may be, as an *operator*, that transforms functions or vector fields into differential forms (twisted ones, if the tilde is there). This operator, being relative to a specific Euclidean structure is as a rule metric- and orientation-dependent. (We'll use $^{\mathbf{p}}$, and $\tilde{}$, versus ^{p}, and $\tilde{}$, to distinguish[23] the $\{\bullet, \mathbf{Or}\}$ and the $\{\cdot, Or\}$ structure.) For instance, the 2 in ${}^2u$ means that, given the straight 2-chain S, one uses both the inner orientation of each of its components

[21] Differential forms converge, in this topology, if their integrals converge uniformly on bounded sets of chains. (A *bounded* set B is one that is *absorbed* by any neighborhood V of 0, i.e., such that $\lambda B \subset V$ for some $\lambda > 0$.) We won't have to invoke such technical notions in the sequel. (Again, see HARRISON [1998] for *norms* on (Banach) spaces of differential forms.) Note the generic use of "differential form" here: Whether an object qualifies as differential form depends on the chosen topology on chain spaces.

[22] In line with the convention of Note 4, we shall denote by ω the map $c \to \langle c; \omega\rangle$, and feel free to write $\omega = c \to \langle c; \omega\rangle$. Of course, the symmetric construct $c = \omega \to \langle c; \omega\rangle$ is just as valid. Maps of the latter kind, from forms to reals, were called *currents* in DE RHAM [1960]. (See DE RHAM [1936, p. 220], for the physical justification of the term.) There are, a priori, much more currents than chains (or even chainlets, HARRISON [1998]), and one should not be fooled by the expression "in duality" into thinking that the dual of $\mathcal{F}^p$, i.e., the bidual of $\mathcal{C}_p$, is $\mathcal{C}_p$ itself.

[23] This play on styles is only a temporary device, not to be used beyond the present Chapter. Later we shall revert to the received "musical" notation, which assumes a single, definite metric structure in background, and cares little about ambiguity: $\sharp u$ denotes the vector proxy of form u, and $\flat$U is the form represented by the vector field U.

and the ambient orientation to define a crossing direction, then the metric in order to build a normal vector field n in this direction, over each component of the chain. Then, $\langle S;\ {}^2u\rangle = \int_S n \cdot u$ defines ${}^2u$, a straight 2-form indeed. (Notice that $\langle S;\ {}^2u\rangle$ does *not* depend on the ambient orientation.)

REMARK 7.1. In the foregoing example, it would be improper to describe $\langle S;\ {}^2u\rangle$ as the flux of u "through" S, since the components of S, a straight chain, didn't come equipped with crossing directions. These were derived from the ambient orientation, part of the Euclidean structure, instead of being given as an attribute of S's components. To acknowledge this difference, we shall refer to $\int_S n \cdot u$ as the flux "embraced by" S. This is not mere fussiness, as will be apparent when we discuss magnetic flux.

One may wonder, at this point, whether substituting the single concept of differential form for those of point-value, circulation, flux, etc., has gained us any real generality, besides the obvious advantage of conceptual uniformity. Let us examine this point carefully, because it's an essential part of the deconstruction of Euclidean space we have undertaken.

On the one hand, the condition that differential forms should be continuous with respect to deformations of the underlying manifolds doesn't leave room, in dimension 3, for other kinds of differential forms than the above eight. First, it eliminates many obvious linear functionals from consideration. (For instance, γ being an outer-oriented curve, the *intersection number*, defined as the number of times γ crosses S, counted algebraically (i.e., with sign – if orientations do not match), provides a linear map $S \to S \wedge \gamma$, which is not considered as a bona fide differential form. Indeed, it lacks continuity.) Second, it allows one, by using the Riesz representation theorem, to build vector fields or functions that reduce the given form to one of the eight types: For instance, given a 1-form ω, there is[24] a vector field Ω such that $\langle c;\ \omega\rangle = \int_c \tau \cdot \Omega$, which is our first example of what will later be referred to as a "proxy" field: A scalar or vector field that stands for a differential form. For other degrees, forms in 3D are representable by vector fields ($p = 1$ and 2) or by functions ($p = 0$ and 3).

However, the continuity condition requires less regularity from the proxy fields than the smoothness we have assumed up to now. Not to the point of allowing them to be only piecewise smooth: What is required lies in between, and should be clear from Fig. 7.1, which revisits a well known topic from the present viewpoint. As one sees, the contrived "transmission conditions", about tangential continuity of this or normal continuity of that, are implied by the very definition of forms as continuous maps.

Last, the generalization is genuine in spatial dimensions higher than 3: A two-form in 4-space, for instance, has no vector proxy, as a rule.

So, although differential forms do extend a little the scope of integration, this is but a marginal improvement, at least in the 3D context. The real point lies elsewhere, and will

[24] The proof is involved. From a vector field v, build a 1-chain $\sum_i \mu_i s_i$, akin to the graphic representation of v by arrows, i.e., s_i is an oriented segment that approximates v in a region of volume μ_i. Apply ω to this chain, go to the limit. The real-valued linear map thus generated is then shown, thanks to the continuity of ω, to be continuous with respect to the L^2 norm on vector fields. Hence a Riesz vector field Ω, which turns out to be a proxy for ω.

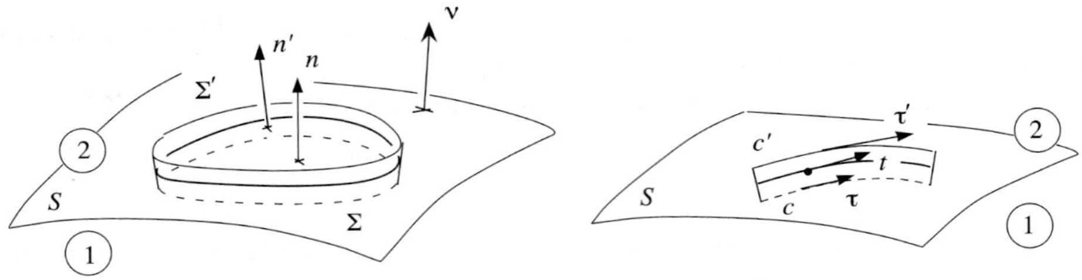

FIG. 7.1. The interface S, equipped with the unit normal field ν, separates two regions where the vector field u is supposed to be smooth, except for a possible discontinuity across S. Suppose Σ or c, initially below S, is moved up a little, thus passing into region 2. Under such conditions, the flux of u through Σ (left) and circulation of u along c (right) can yet be *stable*, i.e., vary continuously with deformations of c and Σ, provided u has some partial regularity: As is well known, and easily proven thanks to the Stokes theorem, *normal* continuity (zero jump $[\nu \cdot u]$ of the normal component across the interface) ensures continuity of the flux $\int_\Sigma n \cdot u$ with respect to Σ (left), while *tangential* continuity of u (zero jump $[u_S]$ of the tangential component across the interface) is required for continuity of the circulation $\int_c \tau \cdot u$ (right) with respect to c. Forms ${}^0\varphi$ and ${}^0\tilde{\varphi}$ require a continuous φ. Piecewise continuity of the proxy function φ is enough for ${}^3\varphi$ and ${}^3\tilde{\varphi}$.

now be argued: Which differential form is built from a given (scalar or vector) field depends on the Euclidean structure, *but the physical entity one purports to model via this field does not*, as a rule. Therefore, the entity of physical significance is the form, conceived as an affine object, and not the field. Two examples will suffice to settle this point.

Consider an electric charge, Q coulombs strong, which is made to move along an oriented smooth curve c, in the direction indicated by the tangent vector field τ. We mean a *test* charge, with Q small enough to leave the ambient electromagnetic field $\{\mathrm{E}, \mathrm{B}\}$ undisturbed, and a *virtual* motion, which allows us to consider the field as frozen at its present value. The work involved in this motion is Q times the quantity $\int_c \tau \cdot \mathrm{E}$, called the *electromotive force* (e.m.f.) *along* c, and expressed in volts (i.e., joules per coulomb). No unit of length is invoked in this description.

Then why is E expressed in volts *per meter* (or whatever unit one adopts)? Only because a vector v such that $|v| = 1$ is one meter long, which makes $\mathrm{E} \cdot v$, and the integral $\int_c \tau \cdot \mathrm{E}$ as well, a definite amount of *volts*, indeed. This physical data, of course, only depends on the field and the curve, not on the metric structure. Yet, change the dot product, from $\cdot$ to $\bullet$ (recall that $u \bullet v = Lu \cdot Lv$), which entails a change in the measure of lengths (hence a rescaling of the unitary vector, now $\boldsymbol{\tau}$ instead of τ), and the circulation of E is now[25] $\boldsymbol{\int_c} \boldsymbol{\tau} \bullet \mathrm{E} = \int_c \tau \cdot L^a L\mathrm{E}$, a different (and physically meaningless) number. On the other hand, there *is* a field $\mathbf{E}$ such that $\boldsymbol{\int_c} \boldsymbol{\tau} \bullet \mathbf{E} = \int_c \tau \cdot \mathrm{E}$, namely $\mathbf{E} = (L^a L)^{-1}\mathrm{E}$. Conclusion: *Which vector field encodes the physical data* (here, e.m.f.'s along all curves) *depends on the chosen metric, although the data themselves do not.* This metric-dependence of E is the reason to call it a vector *proxy*: It merely *stands* for

[25]On the left of the equal sign, the integral and the symbols $\bullet$ and $\boldsymbol{\tau}$ are boldface. (One should see the difference, unless something is amiss in the visualization chain.) So the circulation of E is with respect to the "bold" measure of lengths on the left. The easiest way to verify this equality (and others like it to come) is to work on the above Riemann sums $\sum_s v_s \bullet \mathrm{E}(x_s)$ of the "bold" circulation of E: One has, for each term (omitting the subscript), $v \bullet \mathrm{E} = Lv \cdot L\mathrm{E} = v \cdot L^a L\mathrm{E}$, hence the result.

the real thing, which is the mapping $c \to \langle$e.m.f. along $c\rangle$, i.e., a differential form of degree 1, which we shall from now on denote by e.

Thus, summoning all the equivalent notations introduced so far,

$$e = {}^1\mathrm{E} = {}^1\mathbf{E} = c \to \langle c; e\rangle, \quad \text{where } \langle c; e\rangle \equiv \int_c e = \int_c \tau \cdot \mathrm{E} = \int_c \boldsymbol{\tau} \boldsymbol{\cdot} \mathbf{E}. \tag{7.1}$$

This (straight) 1-form is the right mathematical object by which to represent the electric field, for it tells all about it: Electromotive forces along curves are, one may argue (TONTI [1996]), all that can be observed as regards the electric field.[26] To the point that one can get rid of all the vector-field-and-metric scaffolding, and introduce e directly, by reasoning as follows: The *1-CHAIN* $\to$ *REAL* map we call e.m.f. depends linearly and continuously, *as can experimentally be established*, on the chain over which it is measured. But this is the very definition of a 1-form. Hence e is the minimal, necessary and sufficient, mathematical description of the (empirical) electric field.

REMARK 7.2. The chain/form duality, thus, takes on a neat physical meaning: While the form e models the field, chains are abstractions of the *probes*, of more or less complex structure, that one may place here and there in order to measure it.

The electric field is not the whole electromagnetic field: it only accounts for forces (and their virtual work) exerted on non-moving electric charges. We shall deal later with the magnetic field, which gives the motion-dependent part of the Lorentz force, and recognize it as a 2-form. But right now, an example involving a *twisted* 2-form will be more instructive.

So consider current density, classically a vector field J, whose purpose is to account for the quantity of electric charge, $\int_\Sigma n \cdot \mathrm{J}$, that traverses, per unit of time, a surface Σ in the direction of the unit normal field n that outer-orients it. (Note again this quantity is in ampères, whereas the dimension of the proxy field J is A/m^2.) This map, $\Sigma \to \langle$intensity through $\Sigma\rangle$, a twisted 2-form (namely, ${}^2\tilde{\mathrm{J}}$), is what we can measure and know about the electric current, and the metric plays no role there. Yet, change $\cdot$ to $\boldsymbol{\cdot}$, which affects the measure of areas, and the flux of J becomes[27] $\int_\Sigma \mathbf{n} \boldsymbol{\cdot} \mathrm{J} = |\det(L)| \int_\Sigma n \cdot \mathrm{J}$. The "bold" vector proxy, therefore, should be $\mathbf{J} = |\det(L)|^{-1}\mathrm{J}$, and then ${}^2\tilde{\mathbf{J}} = {}^2\tilde{\mathrm{J}}$. Again, different vector proxies, but the same twisted 2-form, which thus appears as the invariant and physically meaningful object. It will be denoted by j.

This notational scheme will be systematized: Below, we shall call e, h, d, b, j, a, etc., the differential forms that the traditional vector fields E, H, D, B, J, A, etc., represent.

[26] Pointwise values cannot directly be measured, which is why they are somewhat downplayed here, but of course they do make sense, at points of regularity of the field: Taking for c the segment $[x, x+v]$, where v is a vector at x that one lets go to 0, generates at the limit a linear map $v \to \omega_x(v)$. This map, an element of the dual of T_x, is called a *covector* at x. A 1-form, therefore, can be conceived as a (smooth enough) field of covectors. In coordinates, covectors such as $v \to v^i$, where v^i is the ith component of v at point x, form a basis for covectors at x. (They are what is usually denoted by $\mathrm{d}x^i$; but d^i makes better notation, that should be used instead, on a par with ∂_i for basis vectors.)

[27] Same trick, with Riemann sums of the form $\sum_{\mathrm{T}} \vec{\mathbf{T}} \boldsymbol{\cdot} \mathrm{J}(x_{\mathrm{T}})$. After (5.2) and (5.4), $\vec{\mathbf{T}} \boldsymbol{\cdot} \mathrm{J} = L\vec{\mathrm{T}} \cdot L\mathrm{J} = L^a L\vec{\mathrm{T}} \cdot \mathrm{J} = |\det(L)|\vec{\mathrm{T}} \cdot \mathrm{J}$. Hence $\int_\Sigma \mathbf{n} \boldsymbol{\cdot} \mathrm{J} = |\det(L)| \int_\Sigma n \cdot \mathrm{J}$.

8. The Stokes theorem

The Stokes "theorem" hardly deserves such a status in the present approach, for it reduces to a mere

DEFINITION 8.1. The exterior derivative $\mathrm{d}\omega$ of the $(p-1)$-form ω is the p-form $c \to \int_{\partial c} \omega$.

In plain words: To integrate $\mathrm{d}\omega$ over the p-chain c, integrate ω over its boundary ∂c. (This applies to straight or twisted chains and forms equally. Note that d is well defined, thanks to the continuity of ∂ from $\mathcal{C}_{p-1}$ to $\mathcal{C}_p$.) In symbols: $\int_{\partial c} \omega = \int_c \mathrm{d}\omega$, which is the common form of the theorem, or equivalently,

$$\langle \partial c; \omega \rangle = \langle c; \mathrm{d}\omega \rangle \quad \forall c \in \mathcal{C}_p \text{ and } \omega \in \mathcal{F}^{p-1} \tag{8.1}$$

(put tildes over $\mathcal{C}$ and $\mathcal{F}$ for twisted chains and forms), which better reveals what is going on: d is the *dual* of ∂ (YOSIDA [1980]). As a corollary of (4.1), one has

$$\mathrm{d} \circ \mathrm{d} = 0. \tag{8.2}$$

A form ω is *closed* if $\mathrm{d}\omega = 0$, and *exact* if $\omega = \mathrm{d}\alpha$ for some form α. (Synonyms, perhaps more mnemonic, are *cocycle* and *coboundary*. The integral of a cocycle over a boundary, or of a coboundary over a cycle, vanishes.)

REMARK 8.1. In A_n, all closed forms are exact: this is known as the *Poincaré Lemma* (see, e.g., SCHUTZ [1980, p. 140]). But closed forms need not be exact in general manifolds: this is the dual aspect of the "not all cycles bound" issue we discussed earlier. Studying forms, consequently, is another way, dual to homology, to investigate topology. The corresponding theory is called *cohomology* (JÄNICH [2001], MADSEN and TORNEHAVE [1997]).

In three dimensions, the d is the affine version of the classical differential operators, grad, rot, and div, which belong to the Euclidean structure. Let's review this.

First, the gradient: Given a smooth function φ, we define $\operatorname{grad}\varphi$ as the vector field such that, for any 1-cell c with unit tangent field τ,

$$\int_c \tau \cdot (\operatorname{grad}\varphi) = \int_{\partial c} \varphi, \tag{8.3}$$

the latter quantity being of course $\varphi(c(1)) - \varphi(c(0))$. By linearity, this extends to any 1-chain. One recognizes (8.1) there. The relation between gradient and d, therefore, is ${}^1(\operatorname{grad}\varphi) = \mathrm{d}^0\varphi \equiv \mathrm{d}\varphi$, the third term being what is called the *differential* of φ. (The zero superscript can be dropped, because there is only one way to turn a function into a 0-form, whatever the metric.) The vector field $\operatorname{grad}\varphi$ is a proxy for the 1-form $\mathrm{d}\varphi$.

Thus defined, $\operatorname{grad}\varphi$ depends on the metric. If the dot product is changed from "·" to "•", the vector field whose circulation equals the right-hand side of (8.3) is a different proxy, $\mathbf{grad}\,\varphi$, which relates to the first one, as one will see using (5.2), by $\operatorname{grad}\varphi = L^a L\, \mathbf{grad}\,\varphi$.

Up in degree, rot and div are defined in similar fashion. Thus, all in all,

$$^1(\mathrm{grad}\,\varphi) = \mathrm{d}^0\varphi, \quad ^2(\mathrm{rot}\,u) = \mathrm{d}^1 u, \quad ^3(\mathrm{div}\,v) = \mathrm{d}^2 v. \tag{8.4}$$

Be well aware that all forms here are *straight*. Yet their proxies may behave in confusing ways with respect to orientation, as we shall presently see.

About curl, (8.4) says that the curl of a smooth field u, denoted $\mathrm{rot}\,u$, is the vector field such that, for any inner oriented surface S,

$$\int_S n \cdot \mathrm{rot}\,u = \int_{\partial S} \tau \cdot u. \tag{8.5}$$

Here, τ corresponds to the induced orientation of ∂S, and n is obtained by the Ampère rule. So the ambient orientation is explicitly used. Changing it reverses the sign of $\mathrm{rot}\,u$. The curl behaves like the cross product in this respect. If, moreover, the dot product is changed, the bold curl and the meager one relate as follows:

PROPOSITION 8.1. *With* $u \boldsymbol{\cdot} v = Lu \cdot Lv$ *and* $\mathbf{Or} = \mathrm{sign}(\det(L))Or$, *one has*

$$\mathbf{rot}\,u = \big(\det(L)\big)^{-1}\,\mathrm{rot}(L^a Lu). \tag{8.6}$$

PROOF. Because of the hybrid character of (8.5), with integration over an outer oriented surface on the left, and over an inner oriented line on the right, the computation is error prone, so let's be careful. On the one hand (Note 25), $\boldsymbol{\int}_{\partial S} \boldsymbol{\tau} \boldsymbol{\cdot} u = \int_{\partial S} \tau \cdot L^a Lu = \int_S n \cdot \mathrm{rot}(L^a Lu)$. On the other hand (Note 27), setting $\mathbf{J} = \mathbf{rot}\,u$, we know that $\boldsymbol{\int}_S \mathbf{n} \boldsymbol{\cdot} \mathbf{J} = |\det(L)| \int_S n \cdot \mathbf{J}$, hence ... but wait! In Note 27, we had both normals n and $\mathbf{n}$ on the same side of the surface, but here (see Fig. 3.2, left), they may point to opposite directions if $\mathbf{Or} \neq Or$. The correct formula is thus $\boldsymbol{\int}_S \mathbf{n} \boldsymbol{\cdot} \mathbf{rot}\,u = \det(L) \int_S n \cdot \mathbf{rot}\,u \equiv \int_S n \cdot \mathrm{rot}(L^a Lu)$, hence (8.6). □

As for the divergence, (8.4) defines $\mathrm{div}\,v$ as the function such that, for any volume V with outgoing normal n on ∂V,

$$\int_V \mathrm{div}\,v = \int_{\partial V} n \cdot v. \tag{8.7}$$

No vagaries due to orientation this time, because both integrals represent the same kind of form (twisted). Moreover, $\mathbf{div}\,v = \mathrm{div}\,v$, because the same factor $|\det(L)|$ pops up on both sides of $\boldsymbol{\int}_V \mathbf{div}\,v = \boldsymbol{\int}_{\partial V} \mathbf{n} \boldsymbol{\cdot} v$. (These integrals, as indicated by the boldface summation sign, are with respect to the "bold" measure. For the one on the left, it's the 3D measure $|\mathbf{vol}|$, and $\mathbf{vol} = \det(L)\,\mathrm{vol}$ after (5.2).)

REMARK 8.2. The invariance of div is consistent with its physical interpretation: if v is the vector field of a fluid mass, its divergence is the rate of change of the volume occupied by this mass, and though volumes depend on the metric, volume *ratios* do not, again after (5.2).

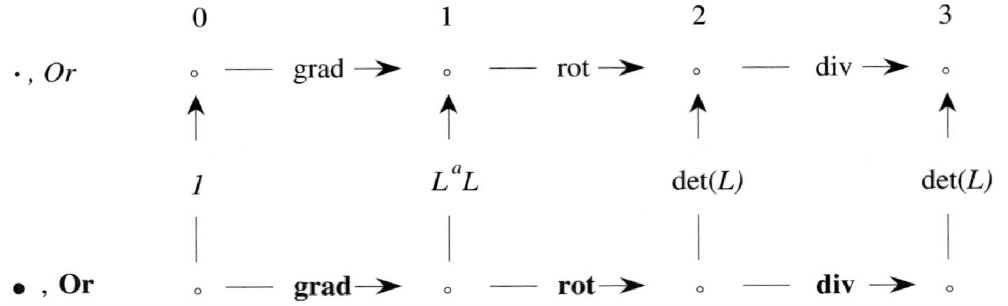

FIG. 8.1. Vertical arrows show how to relate vector or scalar proxies that correspond to the *same* straight form, of degree 0 to 3, in two different Euclidean structures. For *twisted* forms, use the same diagram, but with $|\det(L)|$ substituted for $\det(L)$.

For reference, Fig. 8.1 gathers and displays the previous results. This is a commutative diagram, from which transformation formulas about the differential operators can be read off.[28]

As an illustration of how such a diagram can be used, let us prove something the reader has probably anticipated: the invariance of Faraday's law with respect to a change of metric and orientation. Let two vector fields E and B be such that $\partial_t \mathrm{B} + \mathrm{rot}\,\mathrm{E} = 0$, and set $\mathbf{B} = \mathrm{B}/\det(L)$, $\mathbf{E} = (L^a L)^{-1}\mathrm{E}$, which represent the same differential forms (call them b and e) in the $\{\bullet, \mathbf{Or}\}$ framework, as B and E in the $\{\cdot, Or\}$ one. Then $\partial_t \mathbf{B} + \mathbf{rot}\,\mathbf{E} = 0$. We now turn to the significance of the single physical law underlying these two relations.

9. The magnetic field, as a 2-form

Electromagnetic forces on moving charges, i.e., currents, will now motivate the introduction of the magnetic field. Consider a current loop, I ampères strong, which is made to move – virtual move, again – so as to span a surface S (Fig. 9.1). The virtual work involved is then I times $\int_S n \cdot \mathrm{B}$ ("cut flux" rule), as explained in the caption. Experience establishes the linearity and continuity of the factor $\int_S n \cdot \mathrm{B}$, called the *induction flux*, as a function of S. Hence a 2-form, again the minimal description of the (empirical) magnetic field, which we denote by b and call *magnetic induction*.

In spite of the presence of n in the formula, b is not a twisted but a straight 2-form, as it should, since ambient orientation cannot influence the sign of the virtual work in any way. Indeed, what is relevant is the direction of the current along the loop, which inner-orients c, and the inner orientation of S is the one that matches the orientation of the chain $c' - c$ ("final position minus initial position" in the virtual move). The intervention of a normal field, therefore, appears as the result of the will to represent b with help of a vector, the traditional B such that $b = {}^2\mathrm{B}$. No surprise, then, if this vector

[28] It should be clear that L might depend on the spatial position x, so this diagram is more general than what we contracted for. It gives the correspondence between differential operators relative to different Riemannian structures on the same 3D manifold.

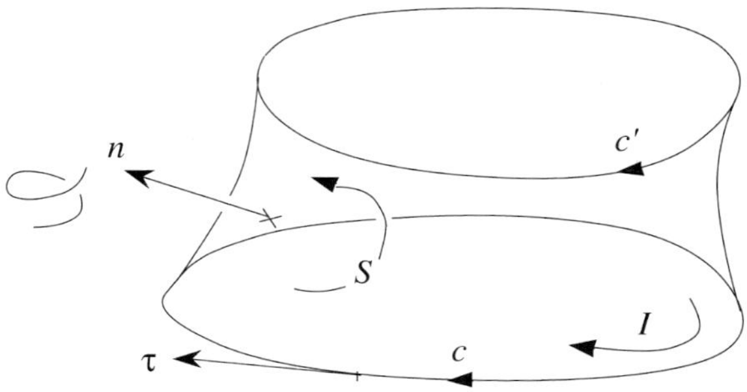

FIG. 9.1. Conventions for the virtual work due to B on a current loop, in a virtual move from position c to position c'. The normal n is the one associated, by Ampère's rule, with the inner orientation of S, a surface such that $\partial S = c' - c$. The virtual work of the $\mathrm{J} \times \mathrm{B}$ force, with $\mathrm{J} = I\tau$, is then I times the flux $\int_S n \cdot \mathrm{B}$.

Nature of the proxy	*for a* straight	*or* twisted	*DF of degree*
function	polar	axial	*0*
vector field	polar	axial	*1*
vector field	axial	polar	*2*
function	axial	polar	*3*

FIG. 9.2. Nature of the proxies in *non*-oriented 3D space with dot product.

proxy "changes sign" with ambient orientation! Actually, it cannot do its job, that is, represent b, without an ambient orientation.

If one insists on a proxy that can act to this effect in autonomy, this object has to carry on its back, so to speak, an orientation of ambient space, i.e., it must be a field of *axial* vectors. Even so, the dependence on metric is still there, so the benefit of using such objects is tiny. Yet, why not, if one is aware that (polar) vector field and axial vector field are just mathematical *tools*,[29] which may be more or less appropriate, depending on the background structures, to represent a given physical entity. In this respect, it may be useful to have a synoptic guide (Fig. 9.2).

We can fully appreciate, now, the difference between j and b, between current flow and magnetic flux. Current density, the twisted 2-form j, is meant to be integrated over surfaces Σ with crossing direction: its proxy J is independent of the ambient orientation. Magnetic induction, the straight 2-form b, is meant to be integrated over surfaces S with inner orientation: its proxy B changes sign if ambient orientation is changed. Current, clearly, flows through a surface, so intensity is one of these "through variables" of

[29]Thus axiality or polarity is by no means a property of the physical objects. But the way physicists write about it doesn't help clarify this. For instance (BAEZ and MUNIAIN [1994, p. 61]): "In physics, the electric field E is called a vector, while the magnetic field B is called an axial vector, because E changes sign under parity transformation, while B does not". Or else (ROSEN [1973]): "It is well known that under the space inversion transformation, $P:(x,y,z) \to (-x,-y,-z)$, the electric field transforms as a polar vector, while the magnetic field transforms as an axial vector, $P:\{E \to -E, B \to B\}$". This may foster confusion, as some passages in BALDOMIR and HAMMOND [1996] demonstrate.

Note 20. But thinking of the magnetic flux as going *through* S is misleading. Hence the expression used here, flux *embraced by* a surface.[30]

10. Faraday and Ampère

We are now ready to address Faraday's famous experiment: variations of the flux embraced by a conducting loop create an electromotive force. A mathematical statement meant to express this law with maximal economy will therefore establish a link between the integral of b over a fixed surface S and the integral of e over its boundary ∂S. Here it is: one has

$$\partial_t \int_S b + \int_{\partial S} e = 0 \quad \forall S \in \mathcal{C}_2, \tag{10.1}$$

i.e., for any straight 2-chain, and in particular, any inner oriented surface S. Numbers in (10.1) have dimension: webers for the first integral, and volts (i.e., Wb/s) for the second one. *Inner* orientation of ∂S (and hence, of S itself) makes lots of physical sense: it corresponds to selecting one of the two ways a galvanometer can be inserted in the circuit idealized by ∂S. Applying the Stokes theorem – or should we say, the definition of d – we find the local, infinitesimal version of the global, integral law (10.1), as this:

$$\partial_t b + \mathrm{d}e = 0, \tag{10.2}$$

the metric- and orientation-free version of $\partial_t \mathrm{B} + \mathrm{rot}\,\mathrm{E} = 0$.

As for Ampère's theorem, the expression is similar, except that twisted forms are now involved:

$$-\partial_t \int_\Sigma d + \int_{\partial\Sigma} h = \int_\Sigma j \quad \forall \Sigma \in \widetilde{\mathcal{C}}_2, \tag{10.3}$$

i.e., for any twisted 2-chain, and in particular, any outer oriented surface Σ. Its infinitesimal form is

$$-\partial_t d + \mathrm{d}h = j, \tag{10.4}$$

again the purely affine version of $-\partial_t \mathrm{D} + \mathrm{rot}\,\mathrm{H} = \mathrm{J}$. Since j is a twisted form, d must be one, and h as well,[31] which suggests that its proxy H will not behave like E under a change of the background Euclidean structure. Indeed, one has $\mathbf{H} = \mathrm{sign}(\det(L))(L^a L)^{-1}\mathrm{H}$ in the now familiar notation. In non-oriented space with metric, the proxy H would be an axial vector field, on a par with B. Vector proxies D and J would be polar, like E.

At this stage, we may announce the strategy that will lead to a discretized form of (10.1) and (10.3): Instead of requesting their validity for *all* chains S or Σ, we shall be

[30] This exposes the relative inadequacy of the "across vs. through" concept, notions which roughly match those of straight 1-form and twisted 2-form (BRANIN [1961]). Actually, between lines and surfaces on the one hand, and inner or outer orientation on the other hand, it's *four* different "vectorial" entities one may have to deal with, and the vocabulary may not be rich enough to cope.

[31] A *magnetomotive force* (m.m.f.), therefore, is a real value (in ampères) attached to an *outer* oriented line γ, namely the integral $\int_\gamma h$.

content with enforcing them for a *finite* family of chains, those generated by the 2-cells of an appropriate finite element mesh, hence a system of differential equations. But first, we must deal with the constitutive laws linking b and d to h and e.

11. The Hodge operator

For it seems a serious difficulty exists there: Since b and h, or d and e, are objects of different types, simple proportionality relations between them, such as $b = \mu h$ and $d = \varepsilon e$, won't make sense if μ and ε are mere scalar factors. To save this way of writing, as it is of course desirable, we must properly redefine μ and ε as *operators*, of type 1-*FORM* $\to$ 2-*FORM*, one of the forms twisted, the other one straight.

So let's try to see what it takes to go from e to d. It consists in being able to determine $\int_\Sigma d$ over any given outer oriented surface Σ, knowing two things: the form e on the one hand, i.e., the value $\int_c e$ for any inner oriented curve c, and the relation $\mathrm{D} = \varepsilon \mathrm{E}$ between the proxies, on the other hand. (Note that ε can depend on position. We shall assume it's piecewise smooth.) How can that be done?

The answer is almost obvious if Σ is a small[32] piece of plane. Build, then, a small segment c meeting Σ orthogonally at a point x where ε is smooth. Associate with c the vector $\vec{c}$ of same length that points along the crossing direction through Σ, and let this vector also serve to inner-orient c. Let $\vec{\Sigma}$ stand for the vectorial area of Σ, and take note that $\vec{\Sigma}/\operatorname{area}(\Sigma) = \vec{c}/\operatorname{length}(c)$. Now dot-multiply this equality by D on the left, εE on the right. The result is

$$\int_\Sigma d = \varepsilon(x)\frac{\operatorname{area}(\Sigma)}{\operatorname{length}(c)}\int_c e, \tag{11.1}$$

which does answer the question.

How to lift the restrictive hypothesis that Σ be small? Riemann sums, again, are the key. Divide Σ into small patches T, as above (Fig. 6.1, right), equip each of them with a small orthogonal segment c_{T}, meeting it at x_{T}, and such that $\vec{c}_{\mathrm{T}} = \vec{\mathrm{T}}$. Next, define $\int_\Sigma d$ as the limit of the Riemann sums[33] $\sum_{\mathrm{T}} \varepsilon(x_{\mathrm{T}}) \int_{c\mathrm{T}} e$. One may then define the *operator* ε, with reuse of the symbol, as the map $e \to d$ just constructed, from $\mathcal{F}^1$ to $\widetilde{\mathcal{F}}^2$. A similar definition holds for μ, of type $\widetilde{\mathcal{F}}^1 \to \mathcal{F}^2$, and for the operators ε^{-1} and μ^{-1} going in the other direction. (Later, we shall substitute ν for μ^{-1}.)

REMARK 11.1. We leave aside the anisotropic case, with a (symmetric) tensor ε^{ij} instead of the scalar ε. In short: Among the variant "bold" metrics, there is one in which ε^{ij} reduces to unity. Then apply what precedes, with "orthogonality", "length", and "area" understood in the sense of this modified metric. (The latter may depend on position, however, so this stands a bit outside our present framework. Details are given in BOSSAVIT [2001b].)

[32] To make up for the lack of rigor which this word betrays, one should treat c and Σ as "p-vectors" ($p = 1$ and 2 respectively), which are the infinitesimal avatars of p-chains. See BOSSAVIT [1998b] for this approach.

[33] Singular points of ε, at which $\varepsilon(x_{\mathrm{T}})$ is not well defined, can always be avoided in such a process, unless Σ coincides with a surface of singularities, like a material interface. But then, move Σ a little, and extend d to such surfaces by continuity.

REMARK 11.2. When the scalar ε or μ equals 1, what has just been defined is the classical *Hodge operator* of differential geometry (BURKE [1985], SCHUTZ [1980]), usually denoted by $*$, which maps p-forms, straight or twisted, to $(n-p)$-forms of the other kind, with $** = \pm 1$, depending on n and p. In dimension $n = 3$, it's a simple exercise to show that the above construction then reduces to $*\,{}^1u = {}^2\tilde{u}$, which prompts the following definition: $*\,{}^0\varphi = {}^3\tilde{\varphi}$, $*\,{}^1u = {}^2\tilde{u}$, $*\,{}^2u ={}^1\tilde{u}$, ${}^3\varphi = *\,{}^0\tilde{\varphi}$. Note that $** = 1$ for all p in 3D.

The metric structure has played an essential role in this definition: areas, lengths, and orthogonality depend on it. So we now distinguish, in the Maxwell equations, the two metric-free main ones,

$$\partial_t b + \mathrm{d}e = 0, \tag{10.2}$$

$$-\partial_t d + \mathrm{d}h = j, \tag{10.4}$$

and the metric-dependent constitutive laws

$$b = \mu h, \tag{11.2}$$

$$d = \varepsilon e, \tag{11.3}$$

where μ and ε are operators of the kind just described. To the extent that no metric element is present in these equations, except for the operators μ and ε, from which one can show the metric can be inferred (BOSSAVIT [2001b]), one may even adopt the radical point of view (DI CARLO and TIERO [1991]) that μ and ε *encode* the metric information.

12. The Maxwell equations: Discussion

With initial conditions on e and h at time $t = 0$, and conditions about the "energy" of the fields to which we soon return, the above system makes a well-posed problem. Yet a few loose ends must be tied.

First, recall that j is supposed to be known. But reintroducing Ohm's law at this stage would be no problem: replace j in (10.4) by $j^s + \sigma e$, where j^s is a given twisted 2-form (the source current), and σ a third Hodge-like operator on the model of ε and μ.

12.1. Boundary conditions, transmission conditions

Second, boundary conditions, if any. Leaving aside artificial "absorbing" boundary conditions (MITTRA, RAMAHI, KHEBIR, GORDON and KOUKI [1989]), not addressed here, there are essentially four basic ones, as follows.

Let's begin with "electric walls", i.e., boundaries of perfect conductors, inside which $\mathrm{E} = 0$, hence the standard $n \times \mathrm{E} = 0$ on the boundary. In terms of the form e, it means that $\int_c e = 0$ for all curves c contained in such a surface. This motivates the following definition, stated in dimension n for generality: S being an $(n-1)$-manifold, call $\mathcal{C}_p(S)$ the space of p-chains whose components are all supported in S; then,

DEFINITION 12.1. The trace $\mathrm{t}_S\omega$ of the p-form ω is the restriction of ω to $\mathcal{C}_p(S)$, i.e., the map $c \to \int_c \omega$ restricted to p-chains based on components which are contained in S.

Of course this requires $p < n$. So the boundary condition at an electric wall S^e is $\mathrm{t}_{S^e} e = 0$, which we shall rather write, for the sake of clarity, as "$\mathrm{t}e = 0$ on S^e". Symmetrically, the condition $\mathrm{t}h = 0$ on S^h corresponds to a magnetic wall S^h.

The Stokes theorem shows that d, and t, commute: $\mathrm{dt}\omega = \mathrm{td}\omega$ for any ω of degree not higher than $n - 2$. Therefore $\mathrm{t}e = 0$ implies $\mathrm{td}e = 0$, hence $\partial_t(\mathrm{t}b) = 0$ by (10.2), that is, $\mathrm{t}b = 0$ if one starts from null fields at time 0. For the physical interpretation of this, observe that $\mathrm{t}b = 0$ on S^b means $\int_S b = 0$ for any surface piece S belonging to S^b, or else, in terms of the vector proxy, $\int_S n \cdot \mathrm{B} = 0$, which implies $n \cdot \mathrm{B} = 0$ on all S^b: a "no-flux" surface, called a "magnetic barrier" by some. We just proved anew, in the present language, that electric walls are impervious to magnetic flux. One will see in the same manner that $\mathrm{t}j = 0$ corresponds to "insulating boundaries" ($n \cdot \mathrm{J} = 0$) and $\mathrm{t}d = 0$ to "dielectric barriers" ($n \cdot \mathrm{D} = 0$). If j is given with $\mathrm{t}j = 0$ at the boundary of the domain of interest (which is most often the case) then $\mathrm{t}h = 0$ on S^h implies $\mathrm{t}d = 0$ there. (In eddy current problems, where d is neglected, but j is only partially given, $\mathrm{t}h = 0$ on S^h implies $\mathrm{t}j = 0$, i.e., no current through the surface.)

Conditions $\mathrm{t}b = 0$ or $\mathrm{t}d = 0$ being thus weaker than $\mathrm{t}e = 0$ or $\mathrm{t}h = 0$, one may well want to enforce them independently. Many combinations are thereby possible. As a rule (but there are exceptions in non-trivial topologies, see BOSSAVIT [2000]), well-posedness in a domain D bounded by surface S obtains if S can be subdivided as $S = S^e \cup S^h \cup S^{eh}$, with $\mathrm{t}e = 0$ on S^e (electric wall), $\mathrm{t}h = 0$ on S^h (magnetic wall), and *both* conditions $\mathrm{td}e = 0$ and $\mathrm{td}h = 0$ on S^{eh}, which corresponds to $\mathrm{t}b = 0$ and $\mathrm{t}d = 0$ taken together (boundary which is both a magnetic and a dielectric barrier, or, in the case of eddy-current problems, an insulating interface).

REMARK 12.1. It may come as a surprise that the standard Dirichlet/Neumann opposition is not relevant here. It's because a Neumann condition is just a Dirichlet condition composed with the Hodge and the trace operators (BOSSAVIT [2001c]): Take for instance the standard $n \times \mu^{-1} \operatorname{rot} \mathrm{E} = 0$, which holds on magnetic walls in the E formulation. This is (up to an integration with respect to time) the proxy form of $\mathrm{t}h = 0$, i.e., of the *Dirichlet* condition $n \times \mathrm{H} = 0$. In short, Neumann conditions on e are Dirichlet conditions on h, and the other way round. They only become relevant when one eliminates either e or h in order to formulate the problem in terms of the other field exclusively, thus breaking the symmetry inherent in Maxwell's equations (which we have no intention to do unless forced to!).

Third point, what about the apparently missing equations, $\operatorname{div} \mathrm{D} = \mathrm{Q}$ and $\operatorname{div} \mathrm{B} = 0$ in their classical form (Q is the density of electric charge)? These are not equations, actually, but relations implied by the Maxwell equations, or at best, constraints that initial conditions should satisfy, as we now show.

Let's first define q, the electric charge, of which the above Q is the proxy scalar field. Since j accounts for its flow, charge conservation implies $d_t \int_V q + \int_{\partial V} j = 0$ for all volumes V, an integral law the infinitesimal form of which is

$$\partial_t q + \mathrm{d}j = 0. \tag{12.1}$$

Suppose both q and j were null before time $t = 0$. Later, then, $q(t) = -\int_0^t (\mathrm{d}j)(s)\,ds$. Note that q, like $\mathrm{d}j$, is a *twisted* 3-form, as should be the case for something that accounts for the density of a substance. (Twisted forms are often called "densities", by the way, as in BURKE [1985].)

Now, if one accepts the physical premise that no electromagnetic field exists until its sources (charges and their flow, i.e., q and j) depart from zero, all fields are null at $t = 0$, and in particular, after (10.4), $d(t) = d(0) + \int_0^t [(\mathrm{d}h)(s) - j(s)]\,ds$, hence, by using (8.2), $\mathrm{d}d(t) = -\int_0^t (\mathrm{d}j)(s)\,ds \equiv q(t)$, at all times, hence the derived relation $\mathrm{d}d = q$. As for b, the same computation shows that $\mathrm{d}b = 0$.

So-called "transmission conditions", classically $[n \times \mathrm{E}] = 0$, $[n \cdot \mathrm{B}] = 0$, etc., at material interfaces, can be evoked at this juncture, for these too are not equations, in the sense of additional constraints that the unknowns e, b, etc., would have to satisfy. They *are* satisfied from the outset, being a consequence of the very definition of differential forms (cf. Fig. 7.1).

12.2. Wedge product, energy

Fourth point, the notion of energy. The physical significance of such integrals as $\int \mathrm{B} \cdot \mathrm{H}$ or $\int \mathrm{J} \cdot \mathrm{E}$ is well known, and it's easy to show, using the relations displayed on Fig. 8.1, that both are metric-independent. So they should be expressible in non-metric terms. This is so, thanks to the notion of *wedge product*, an operation which creates a $(p+q)$-form $\omega \wedge \eta$ (straight when both factors are of the same kind, twisted otherwise) out of a p-form ω and a q-form η. We shall only describe this in detail in the case of a 2-form b and a 1-form h, respectively straight and twisted.

The result, a twisted 3-form $b \wedge h$, is known if integrals $\int_V b \wedge h$ are known for all volumes V. In quite the same way as with the Hodge map, the thing is easy when V is a small parallelepiped, as shown in Fig. 12.1. Observe that, if $b = {}^2\mathrm{B}$ and $h = {}^1\widetilde{\mathrm{H}}$, then $\int_V b \wedge h = \mathrm{B} \cdot \mathrm{H}\,\mathrm{vol}(V)$, if one follows the recipe of Fig. 12.1, confirming the soundness of the latter. The extension to finite-size volumes is made by constructing Riemann sums, as usual.

REMARK 12.2. Starting from the equality $\int b \wedge h' = \int \mathrm{B} \cdot \mathrm{H}'$, setting $b = \mu h$ yields $\int \mu h \wedge h' = \int \mu \mathrm{H} \cdot \mathrm{H}' = \int \mu \mathrm{H}' \cdot \mathrm{H} = \int \mu h' \wedge h$, a *symmetry* property of the Hodge operator to which we didn't pay attention till now. Note also that $\int \mu h \wedge h = \int \mu |\mathrm{H}|^2 > 0$, unless $h = 0$. Integrals such as $\int \mu h \wedge h'$, or $\int \nu b \wedge b'$, etc., can thus be understood as *scalar products* on spaces of forms, which can thereby be turned (after due completion) into Hilbert spaces. The corresponding norms, i.e., the square roots of $\int \mu h \wedge h$, of $\int \nu b \wedge b$, and other similar constructs on e or d, will be denoted by $|h|_\mu$, $|b|_\nu$, etc.

Other possible wedge products are ${}^0\varphi \wedge \omega = {}^0(\varphi\omega)$ (whatever the degree of ω), ${}^1u \wedge {}^1v = {}^2(u \times v)$, ${}^2u \wedge {}^1v = {}^3(u \cdot v)$. (If none or both factors are straight forms, the product is straight.) It's an instructive exercise to work out the exterior derivative of

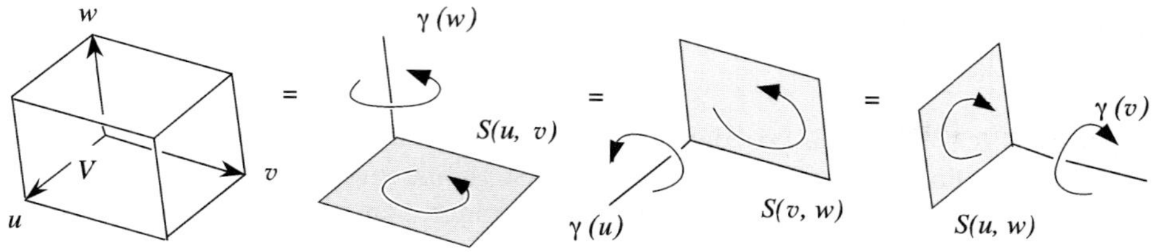

FIG. 12.1. There are three ways, as shown, to see volume V, built on u, v, w, as the extrusion of a surface S along a line segment γ. A natural definition of the integral of $b \wedge h$ is then $\int_V b \wedge h = (\int_{S(u,v)} b)(\int_{\gamma(w)} h) + (\int_{S(v,w)} b)(\int_{\gamma(u)} h) + (\int_{S(u,w)} b)(\int_{\gamma(v)} h)$. Note the simultaneous inner and outer orientations of S and γ, which should match (if the outer orientation of V is $+$, as assumed), but are otherwise arbitrary.

such products, using the Stokes theorem, and to look for the equivalents of the standard integration by parts formulas, such as

$$\int_\Omega (\mathrm{H} \cdot \operatorname{rot} \mathrm{E} - \mathrm{E} \cdot \operatorname{rot} \mathrm{H}) = \int_{\partial\Omega} n \cdot (\mathrm{E} \times \mathrm{H}),$$

$$\int_\Omega (\mathrm{D} \cdot \operatorname{grad} \Psi + \Psi \operatorname{div} \mathrm{D}) = \int_{\partial\Omega} \Psi n \cdot \mathrm{D}.$$

They are, respectively,

$$\int_\Omega (\mathrm{d}e \wedge h - e \wedge \mathrm{d}h) = \int_{\partial\Omega} e \wedge h, \tag{12.2}$$

$$\int_\Omega (\mathrm{d}\psi \wedge d + \psi\, \mathrm{d}d) = \int_{\partial\Omega} \psi\, d. \tag{12.3}$$

Now, let us consider a physically admissible field, that is, a quartet of forms b, h, e, d, which may or may not satisfy Maxwell's equations when taken together, but are each of the right degree and kind in this respect.

DEFINITION 12.2. The following quantities:

$$\frac{1}{2}\int \mu^{-1} b \wedge b, \quad \frac{1}{2}\int \mu h \wedge h, \quad \frac{1}{2}\int \varepsilon e \wedge e, \quad \frac{1}{2}\int \varepsilon^{-1} d \wedge d, \tag{12.4}$$

are called, respectively, *magnetic energy*, *magnetic coenergy*, *electric energy*, and *electric coenergy* of the field. The integral $\int j \wedge e$ is the *power* released by the field.

The latter definition, easily derived from the expression of the Lorentz force, is a statement about field–matter energy exchanges from which the use of the word "energy" could rigorously be justified, although we shall not attempt that here (cf. BOSSAVIT [1990a]). The definition entails the following relations:

$$\frac{1}{2}\int \mu^{-1} b \wedge b + \frac{1}{2}\int \mu h \wedge h \geqslant \int b \wedge h,$$

$$\frac{1}{2}\int \varepsilon^{-1} d \wedge d + \frac{1}{2}\int \varepsilon\, e \wedge e \geqslant \int d \wedge e,$$

with equality if and only if $b = \mu h$ and $d = \varepsilon e$. One may use this as a way to set up the constitutive laws.

REMARK 12.3. The well-posedness evoked earlier holds if one restricts the search to fields with finite energy. Otherwise, of course, nonzero solutions to (10.2), (10.4), (11.2), (11.3) with $j = 0$ do exist (such as, for instance, plane waves).

The integrals in (12.4) concern the whole space, or at least, the whole region of existence of the field. One may wish to integrate on some domain Ω only, and to account for the energy balance. This is again an easy exercise:

PROPOSITION 12.1 (Poynting's theorem). *If the field* $\{b, h, e, d\}$ *does satisfy the Maxwell equations* (10.2), (10.4), (11.2), (11.3), *one has*

$$d_t\left[\frac{1}{2}\int_\Omega \mu^{-1} b \wedge b + \frac{1}{2}\int_\Omega \varepsilon e \wedge e\right] + \int_{\partial\Omega} e \wedge h = -\int_\Omega j \wedge e$$

for any fixed domain Ω.

PROOF. "Wedge multiply" (10.2) and (10.4), from the right, by e and $-h$, add, use (12.2) and Stokes. □

As one sees, all equalities and inequalities on which a variational approach to Maxwell's theory can be based do have their counterparts with differential forms. We shall not follow this thread any further, since what comes ahead is not essentially based on variational methods. Let's rather close this section with a quick review of various differential forms in Maxwell's theory and how they relate.

12.3. *The "Maxwell house"*

To the field quartet and the source pair $\{q, j\}$, one may add the *electric potential* ψ and the *vector potential* a, a straight 0-form and 1-form respectively, such that $b = \mathrm{d}a$ and $e = -\partial_t a + \mathrm{d}\psi$. Also, the *magnetic potential* φ (twisted 0-form) and the twisted 1-form τ such that $h = \tau + \mathrm{d}\varphi$, whose proxy is the T of Carpenter's "T–Ω" method (CARPENTER [1977]). None of them is as fundamental as those in (10.2), (10.4), but each can be a useful auxiliary at times. The *magnetic current* k and *magnetic charge* m can be added to the list for the sake of symmetry (Fig. 12.2), although they don't seem to represent any real thing (GOLDHABER and TROWER [1990]).

For easier reference, Fig. 12.2 displays all these entities as an organized whole, each one "lodged" according to its degree and nature as a differential form. Since primitives in time may have to be considered, we can group the differential forms of electromagnetism in four similar categories, shown as vertical pillars on the figure. Each pillar symbolizes the structure made by spaces of forms of all degrees, linked together by the d operator. Straight forms are on the left and twisted forms on the right. Differentiation or integration with respect to time links each pair of pillars (the front one and the rear one) forming the sides of the structure. Horizontal beams symbolize constitutive laws.

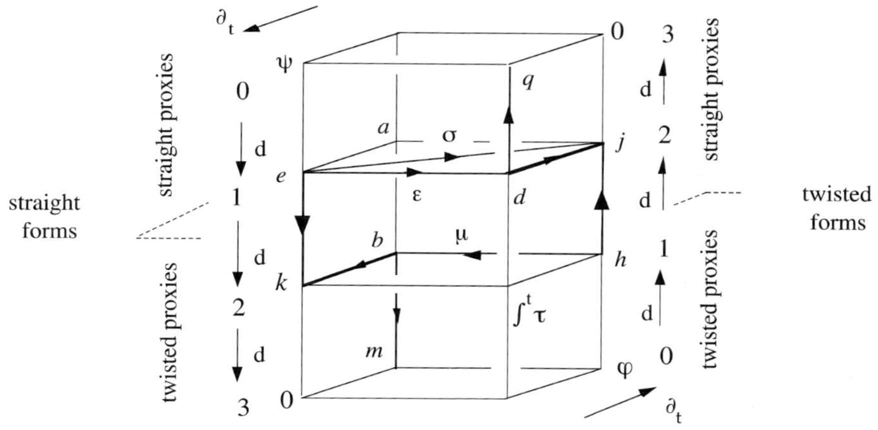

FIG. 12.2. Structures underlying the Maxwell system of equations. For more emphasis on their symmetry, Faraday's law is here taken to be $\partial_t b + \mathrm{d}e = -k$, with $k = 0$. (The straight 2-form k would stand for the flow of magnetic charge, if such a thing existed. Then, one would have $\mathrm{d}b = m$, where the straight 3-form m represents magnetic charge, linked with its current by the conservation law $\partial_t m + \mathrm{d}k = 0$.)

As one can see, each object has its own room in the building: b, a 2-form, at level 2 of the "straight" side, the 1-form a such that $b = \mathrm{d}a$ just above it, etc. Occasional asymmetries (e.g., the necessity to time-integrate τ before lodging it, the bizarre layout of Ohm's law ...) point to weaknesses which are less those of the diagram than those of the received nomenclature or (more ominously) to some hitch about Ohm's law (BOSSAVIT [1996]). Relations mentioned up to now can be directly read off from the diagram, up to sporadic sign inversions. An equation such as $\partial_t b + \mathrm{d}e = -k$, for instance, is obtained by gathering at the location of k the contributions of all adjacent niches, including k's, in the direction of the arrows. Note how the rules of Fig. 9.2, about which scalar- or vector-proxies must be twisted or straight, are in force.

But the most important thing is probably the neat separation, in the diagram, between "vertical" relations, of purely affine nature, and "horizontal" ones, which depend on metric. If this was not drawing too much on the metaphor, one could say that a change of metric, as encoded in ε and μ (due for instance to a change in their local values, because of a temperature modification or whatever) would shake the building horizontally but leave the vertical panels unscathed.

This suggests a method for *discretizing* the Maxwell equations: The orderly structure of Fig. 12.1 should be preserved, if at all possible, in numerical simulations. Hence in particular the search for finite elements *which fit differential forms*, which will be among our concerns in the sequel.

CHAPTER III

Discretizing

It's a good thing to keep in mind a representative of the family of problems one wishes to model. Here, we shall have wave-propagation problems in view, but heuristic considerations will be based on the much simpler case of static fields. The following example can illustrate both things, depending on whether the exciting current, source of the field, is transient or permanent, and lends itself to other useful variations.

13. A model problem

In a closed cavity with metallic walls (Fig. 13.1), which has been free from any electromagnetic activity till time $t = 0$, suppose a flow of electric charge is created in an enclosed antenna after this instant, by some unspecified agency. An electromagnetic field then develops, propagating at the speed of light towards the walls which, as soon as they are reached by the wavefront, begin to act as secondary antennas. Dielectric or magnetizable bodies inside the cavity, too, may scatter waves. Hence a complex evolution, which one may imagine simulating by numerical means. (How else?)

For the sake of generality, let's assume a symmetry plane, and a symmetrically distributed current. (In that case, the plane acts as a magnetic wall.) The computation will thus be restricted to a spatial domain D coinciding with one half of the cavity, on the left of the symmetry plane, say. Calling S^e and Σ^h, as Fig. 13.1 shows, the two parts

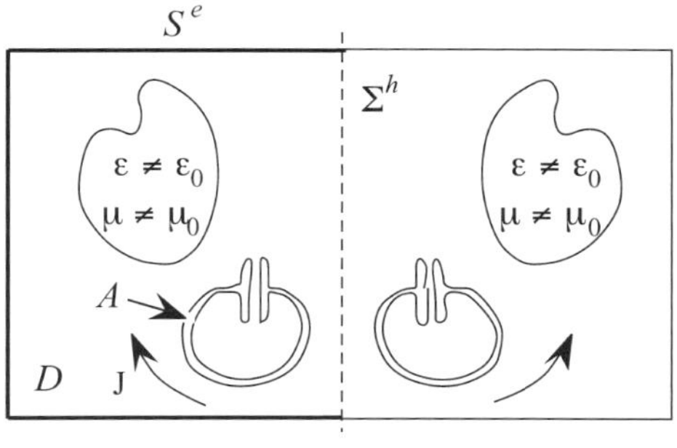

FIG. 13.1. Situation and notation (dimension 3). Region D is the left half of the cavity. Its boundary S has a part S^e in the conductive wall and a part Σ^h in the symmetry plane. Region A, the left "antenna", is the support of the given current density J (mirrored on the right), for which some generator, not represented and not included in the modelling, is responsible.

of its surface, an electric wall and a magnetic wall respectively, we write the relevant equations in D as

$$\begin{aligned} &\partial_t b + \mathrm{d}e = 0, && -\partial_t d + \mathrm{d}h = j, \\ &d = \varepsilon e, && b = \mu h, \\ &\mathrm{t}e = 0 \quad \text{on } S^e, && \mathrm{t}h = 0 \quad \text{on } \Sigma^h. \end{aligned} \tag{13.1}$$

The coefficients ε and μ which generate their Hodge namesakes are real, constant in time, but not necessarily equal to their vacuum values ε_0 and μ_0, and may therefore depend on x. (They could even be tensors, as observed earlier.) The current density j is given, and assumed to satisfy $j(t) = 0$ for $t \leqslant 0$. All fields, besides j, are supposed to be null before $t = 0$, hence initial conditions $e(0) = 0$ and $h(0) = 0$. Notice that $\mathrm{d}j = 0$ is *not* assumed: some electric charge may accumulate at places in the antenna, in accordance with the charge-conservation equation (12.1).

Proving this problem well-posed[34] is not our concern. Let's just recall that it is so, under reasonable conditions on j, when all fields e and h are constrained to have finite energy.

Two further examples will be useful. Suppose j has reached a steady value for so long that all fields are now time-independent. The magnetic part of the field, i.e., the pair $\{b, h\}$, can then be obtained by solving, in domain D,

$$\begin{aligned} &\mathrm{d}b = 0, && \mathrm{d}h = j, \\ & & b = \mu h, & \\ &\mathrm{t}b = 0 \quad \text{on } S^e, && \mathrm{t}h = 0 \quad \text{on } \Sigma^h. \end{aligned} \tag{13.2}$$

This is also a well-posed problem (magnetostatics), provided $\mathrm{d}j = 0$. As for the electric part of the field, which has no reason to be zero since the asymptotic charge density $q = q(\infty) = -\int_0^\infty \mathrm{d}j(t)\,dt$ does not vanish, as a rule, one will find it by solving

$$\begin{aligned} &\mathrm{d}d = q, && \mathrm{d}e = 0, \\ & & d = \varepsilon e, & \\ &\mathrm{t}e = 0 \quad \text{on } S^e, && \mathrm{t}d = 0 \quad \text{on } \Sigma^h \end{aligned} \tag{13.3}$$

(electrostatics). The easy task of justifying the boundary conditions in (13.2) and (13.3) is left to the reader. One should recognize in (13.3), thinly veiled behind the present notation, the most canonical example there is of elliptic boundary-value problem.[35]

Finally, let's give an example of eddy-current problem in harmonic regime, assuming a conductivity $\sigma \geqslant 0$ in D and $\sigma = 0$ in A. This time, all fields are of the form $u(t, x) =$

[34]Its physical relevance has been challenged (by SMYTH and SMYTH [1977]), on the grounds that assuming a given current density (which is routinely done in such problems) neglects the reaction of the antenna to its own radiated field. This is of course true – and there are other simplifications that one might discuss – but misses the point of what *modelling* is about. See UMAN [1977] and BOSSAVIT [1998b, p. 153], for a discussion of this issue.

[35]Mere changes of symbols would yield the stationary heat equation, the equation of steady flow in porous media, etc. Notice in particular how the steady current equation, with Ohm's law, can be written as $\mathrm{d}j = 0$, $j = \sigma e$, $\mathrm{d}e = 0$, plus boundary conditions (non-homogeneous, to include source terms).

$\mathrm{Re}[\exp(i\omega t)\,\mathrm{U}(x)]$, with U complex-valued (SMALL CAPITALS will denote such fields). The given current in A, now denoted J^s (s for "source"), is solenoidal, displacement currents are neglected, and Ohm's law $\mathrm{J} = \sigma \mathrm{E} + \mathrm{J}^s$ is in force, where σ is of course understood as a Hodge-like operator, but positive semi-definite only. The problem is then, with the same boundary conditions as above,

$$\mathrm{dH} = \sigma \mathrm{E} + \mathrm{J}^s, \qquad \mathrm{H} = \nu \mathrm{B}, \qquad \mathrm{dE} = -i\omega \mathrm{B},$$

and B and H can be eliminated, hence a second-order equation in terms of E:

$$i\omega\sigma \mathrm{E} + \mathrm{d}\nu\,\mathrm{dE} = -i\omega \mathrm{J}^s, \tag{13.4}$$

with boundary conditions $\mathrm{tE} = 0$ on S^e and $\mathrm{t}\nu\,\mathrm{dE} = 0$ on Σ^h.

Nothing forbids σ and μ there to be complex-valued too. (Let's however request them to have Hermitian symmetry.) A complex μ can sometimes serve as a crude but effective way to model ferromagnetic hysteresis. And since the real σ can be replaced by $\sigma + i\omega\varepsilon$, we are not committed to drop out displacement currents, after all. Hence, (13.4) can well be construed as the general version of the Maxwell equations in harmonic regime, at angular frequency ω, with dissipative materials possibly present. In particular, (13.4) can serve as a model for the "microwave oven" problem. Note that what we have here is a Fredholm equation: Omitting the excitation term J^s and replacing σ by $i\omega\varepsilon$ gives the "resonant cavity problem" in D, namely, to find frequencies ω at which $\mathrm{d}\nu\,\mathrm{dE} = \omega^2\varepsilon \mathrm{E}$ has a nonzero solution E.

14. Primal mesh

Let's define what we shall call a "cellular paving". This is hardly different from a finite-element mesh, just a bit more general, but we need to be more fussy than is usual about some details. We pretend to work in n-dimensional Euclidean space E_n, but of course $n = 3$ is the case in point. The cells we use here are those introduced earlier[36] (Fig. 2.1), with the important caveat that they are all "open" cells, in the sense of Section 2, i.e., do not include their boundaries. (The only exception is for $p = 0$, nodes, which are both open and closed.) The corresponding closed cell will be denoted with an overbar (also used for the topological closure).

This being said, a *cellular paving* of some region R of space is a finite set of open p-cells such that (1) two distinct cells never intersect, (2) the union of all cells is R, (3) if the closures of two cells c and c' meet, their intersection is the closure of some (unique) cell c''. It may well happen that c'' is c, or c'. In such a case, e.g., if $\overline{c} \cap \overline{c'} = \overline{c}$, we say that c is *a face of* c'. For instance, on Fig. 14.1, left, c_3 is a face of c_4. If c is a face of c' which itself is a face of c'', then c is a face of c''. Cells in ambient dimension 3 or lower will be called *nodes, edges, facets*, and *volumes*, with symbols n, e, f, v to match.

We'll say we have a *closed* paving if R is closed. (Fig. 14.1, left, gives a two-dimensional example, where $R = \overline{D}$.) But it need not be so. Closed pavings are not

[36] Topologically simple *smooth* cells, therefore. But the latter condition is not strict and we shall relax it to *piecewise* smooth, in the sequel, without special warning.

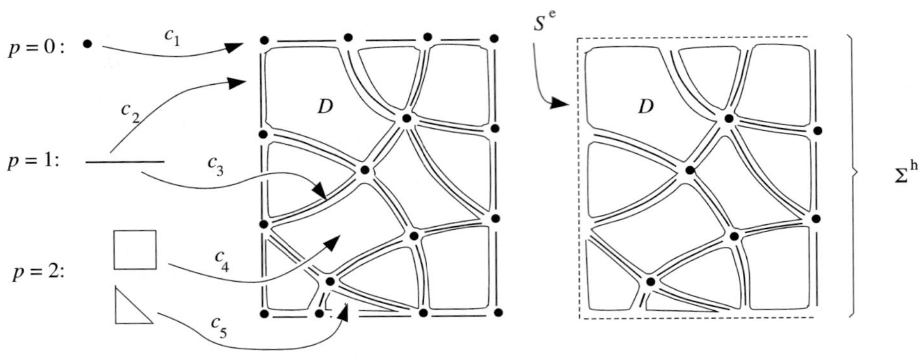

FIG. 14.1. Left: A few p-cells, contributing to a closed cellular paving of D. (This should be imagined in dimension 3.) Right: A culled paving, now "closed relative to" S^e. This is done in anticipation of the modelling we have in mind, in which cells of S^e would carry null degrees of freedom, so they won't be missed.

necessarily what is needed in practice, as one may rather wish to discard some cells in order to deal with boundary conditions. Hence the relevance of the following notion of "relative closedness": C being a closed part of R, we shall say that a paving of R is *closed modulo C* if it can be obtained by removing, from some closed paving, all the cells which map into C. The case we shall actually need, of a paving of $R = \overline{D} - S^e$ which is closed modulo S^e, is displayed on the right of Fig. 14.1. Informally said, "pave $\overline{D}$ first, then remove all cells from the electric boundary".

Each cell has its own inner orientation. These orientations are arbitrary and independent. In three dimensions, we shall denote by $\mathcal{N}$, $\mathcal{E}$, $\mathcal{F}$, $\mathcal{V}$, the sets of oriented p-cells of the paving, and by N, E, F, V the number of cells in each of these sets. (The general notation, rarely required, will be $\mathcal{S}_p$ for the set of p-cells and S_p for the number of such cells.)

Two cells σ and c, of respective dimensions p and $p+1$, are assigned an *incidence number*, equal to ± 1 if σ is a face of c, and to 0 otherwise. As for the sign, recall that each cell orients its own boundary (Section 4), so this orientation may or may not coincide with the one attributed to σ. If orientations match, the sign is $+$, else it's $-$. Fig. 14.2 illustrates this point. (Also refer back to Fig. 4.1.)

Collecting these numbers in arrays, we obtain rectangular matrices $\mathbf{G}$, $\mathbf{R}$, $\mathbf{D}$, called *incidence matrices* of the tessellation. For instance (Fig. 14.2), the incidence number for edge e and facet f is denoted $\mathbf{R}^e_f$, and makes one entry in matrix $\mathbf{R}$, whose rows and columns are indexed over facets and edges, respectively. The entry $\mathbf{G}^n_e$ of $\mathbf{G}$ is -1 in the case displayed, because n, positively oriented, is at the start of edge e (cf. Fig. 3.4(c)). And so on. Symbols $\mathbf{G}$, $\mathbf{R}$, $\mathbf{D}$ are of course intentionally reminiscent of grad, rot, div, but we still have a long way to go to fully understand the connection. Yet, one thing should be conspicuous already: contrary to grad, rot, div, the incidence matrices are *metric-independent* entities, so the analogy cannot be complete. Matrices $\mathbf{G}$, $\mathbf{R}$, $\mathbf{D}$ are more akin to the (metric-independent) operator d from this viewpoint, and the generic symbol $\mathbf{d}$, indexed by the dimension p if needed, will make cleaner notation in spatial

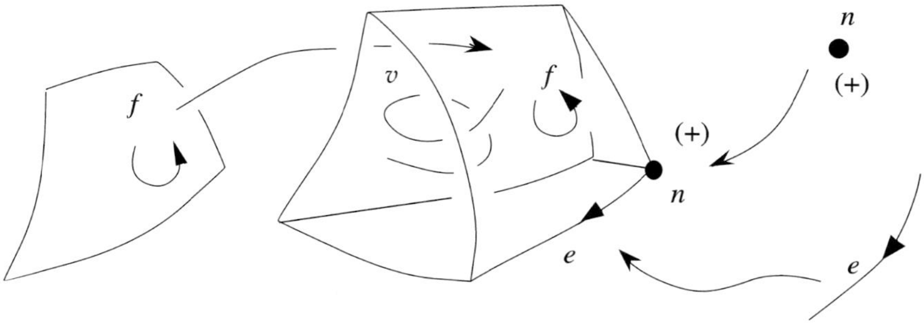

FIG. 14.2. Sides: Individual oriented cells. Middle: The same, plus a 3-cell, as part of a paving, showing respective orientations. Those of v and f match, those of f and e, or of e and n, don't. So $\mathbf{G}_e^n = -1$, $\mathbf{R}_f^e = -1$, and $\mathbf{D}_v^f = 1$.

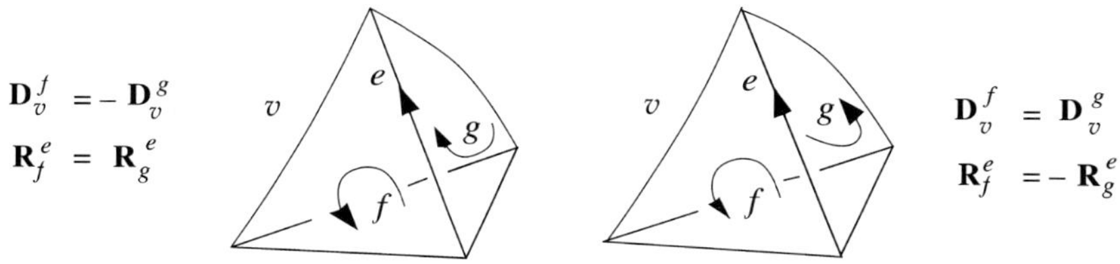

FIG. 14.3. Relation $\mathbf{DR} = 0$, and how it doesn't depend on the cells' individual orientations: In both cases, one has $\mathbf{D}_v^f \mathbf{R}_f^e + \mathbf{D}_v^g \mathbf{R}_g^e = 0$.

dimensions higher than 3, with $\mathbf{d}_0 = \mathbf{G}$, $\mathbf{d}_1 = \mathbf{R}$, $\mathbf{d}_2 = \mathbf{D}$. The mnemonic value of $\mathbf{G}$, $\mathbf{R}$, $\mathbf{D}$, however, justifies keeping them in use.

Just as $\text{rot} \circ \text{grad} = 0$ and $\text{div} \circ \text{rot} = 0$, one has $\mathbf{RG} = 0$ and $\mathbf{DR} = 0$. Indeed, for an edge e and a volume v, the $\{v, e\}$-entry of $\mathbf{DR}$ is $\sum_{f \in \mathcal{F}} \mathbf{D}_v^f \mathbf{R}_f^e$. Nonzero terms occur, in this sum over facets, only for those which both contain e and are a face of v, which happens only if e belongs to $\overline{v}$. In that case, there are exactly two facets f and g of v meeting along e (Fig. 14.3), and hence two nonzero terms. As Fig. 14.3 shows, they have opposite signs, whatever the orientations of the individual cells, hence the result, $\mathbf{DR} = 0$. By a similar argument, $\mathbf{RG} = 0$, and more generally, $\mathbf{d}_{p+1}\mathbf{d}_p = 0$.

REMARK 14.1. The answer to the natural question, "then, is the kernel of $\mathbf{R}$ equal to the range of $\mathbf{G}$?", is "yes" here, because $\overline{D} - S^e$ has simple topology. (See the remark at the end of Section 4 about homology. This time, going further would lead us into *co*homology.) For the same reason, $\ker(\mathbf{D}) = \text{cod}(\mathbf{R})$. This will be important below.

It is no accident if this proof of $\mathbf{d} \circ \mathbf{d} = 0$ evokes the one about $\partial \circ \partial = 0$ in Section 4, and the caption of Fig. 4.1. The same basic observation, "the boundary of a boundary is zero" (TAYLOR and WHEELER [1992], KHEYFETS and WHEELER [1986]), underlies all proofs of this kind. In fact, the above incidence matrices can be used to find the boundaries, chainwise, of each cell. For instance, f being understood as the 2-chain based on facet f with weight 1, one has $\partial f = \sum_{e\in\mathcal{E}} \mathbf{R}^e_f e$. So if S is the straight 2-chain $\sum_f w^f f$ with weights w^f (which we shall call a *primal* 2-*chain*, or "m-surface", using m as a mnemonic for the underlying mesh), its boundary[37] is the 1-chain

$$\partial S = \sum_{e\in\mathcal{E}} \sum_{f\in\mathcal{F}} \mathbf{R}^e_f w^f e.$$

More generally, let's write $\boldsymbol{\partial}_p$, boldface,[38] for the transpose of the above matrix $\mathbf{d}_{p-1}$. Then, if $c = \sum_{\sigma\in\mathcal{S}_p} w^\sigma \sigma$ is a p-chain, its boundary is $\partial c = \sum\{s \in \mathcal{S}_{p-1}\colon (\boldsymbol{\partial}_p \mathbf{w})^s s\}$, where $\mathbf{w}$ stands for the vector of weights. Thus, $\boldsymbol{\partial}$ is to ∂ what $\mathbf{d}$ is to d. Moreover, the duality between d and ∂ is matched by a similar duality between their finite-dimensional counterparts $\mathbf{d}$ and $\boldsymbol{\partial}$.

15. Dual mesh

A *dual* mesh, with respect to m, is also a cellular paving, though not of the same region exactly, and with *outer* orientation of cells. Let's explain.

To each p-cell c of the primal mesh, we assign an $(n-p)$-cell, called the *dual* of c and denoted $\tilde{c}$, which meets c at a single point x_c. (Ways to build $\tilde{c}$ will soon be indicated.) Hence a one-to-one correspondence between cells of complementary dimensions. Thus, for instance, facet f is pierced by the dual edge $\tilde{f}$ (a line), node n is inside the dual volume $\tilde{n}$, and so forth. Since the tangent spaces at x_c to c and $\tilde{c}$ are complementary, the inner orientation of c provides an outer orientation for $\tilde{c}$ (Fig. 15.1). Incidence matrices $\widetilde{\mathbf{G}}$, $\widetilde{\mathbf{R}}$, $\widetilde{\mathbf{D}}$ can then be defined, as above, the sign of each nonzero entry depending on whether outer orientations match or not.

Moreover, it is required that, when c is a face of c', the dual $\tilde{c}'$ be a face of $\tilde{c}$, and the other way round. This has two consequences. First, we don't really need new names for the dual incidence matrices. Indeed, consider for instance edge e and facet f, and suppose $\mathbf{R}^e_f = 1$, i.e., e is a face of f and their orientations match: Then the dual edge $\tilde{f}$ is a face of the dual facet $\tilde{e}$, whose outer orientations match, too. So what we would otherwise denote $\widetilde{\mathbf{R}}^{\tilde{f}}_{\tilde{e}}$ is equal to $\mathbf{R}^e_f$. Same equality if $\mathbf{R}^e_f = -1$, and same reasoning for

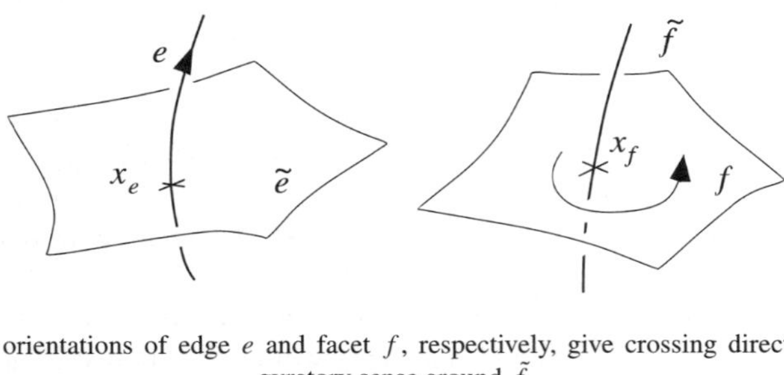

FIG. 15.1. Inner orientations of edge e and facet f, respectively, give crossing direction through $\tilde{e}$ and gyratory sense around $\tilde{f}$.

[37]More accurately, its boundary *relative to* Σ^h.

[38]Boldface, from now on, connotes mesh-related things, such as DoF arrays, etc.

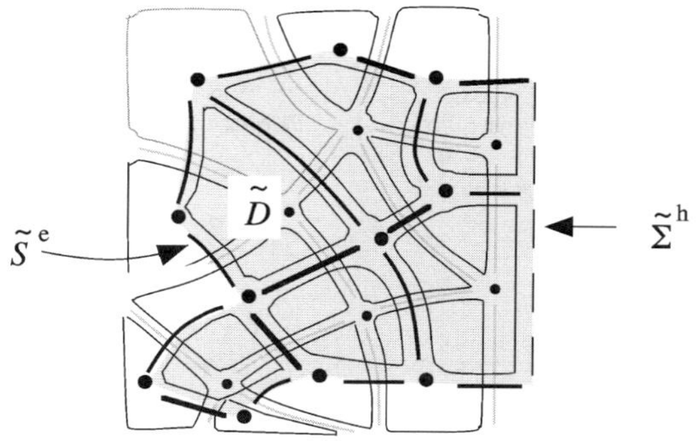

FIG. 15.2. A dual paving, overlaid on the primal one.

other kinds of cells, from which we conclude that the would-be dual incidence matrices $\widetilde{\mathbf{G}}$, $\widetilde{\mathbf{R}}$, $\widetilde{\mathbf{D}}$ are just the transposes $\mathbf{D}^t$, $\mathbf{R}^t$, $\mathbf{G}^t$ of the primal ones.

Second consequence, there is no gap between dual cells, which thus form a cellular paving of a connected region $\widetilde{R}$, the interior $\widetilde{D}$ of which is nearly D, but not quite (Fig. 15.2). A part of its boundary is paved by dual cells: We name it $\widetilde{S}^e$, owing to its kinship with S^e (not so obvious on our coarse drawing! but the finer the mesh, the closer $\widetilde{S}^e$ and S^e will get). The other part is denoted $\widetilde{\Sigma}^h$. So the cellular paving we now have is closed modulo $\widetilde{\Sigma}^h$, whereas the primal one was closed modulo S^e.

Given the mesh m, all its conceivable duals have the same *combinatorial* structure (the same incidence matrices), but can differ as regards *metric*, which leaves much leeway to construct dual meshes. Two approaches are noteworthy, which lead to the "barycentric dual" and the "Voronoi–Delaunay dual". We shall present them as special cases of two slightly more general procedures, the "star construction" and the "orthogonal construction" of meshes in duality. For this we shall consider only *polyhedral* meshes (those with polyhedral 3-cells), which is not overly restrictive in practice.

The orthogonal construction consists in having each dual cell orthogonal to its primal partner. (Cf. Figs. 15.3 and 15.5, left.) A particular case is the Voronoi–Delaunay tessellation (DIRICHLET [1850]), under the condition that dual nodes should be inside primal volumes. Alas, as Fig. 15.4 shows, orthogonality can be impossible to enforce, if the primal mesh is imposed. If one starts from a simplicial primal for which all circumscribed spheres have their center inside the tetrahedron, and all facets are acute triangles, all goes well. (One then takes these circumcenters as dual nodes.) But this property, desirable on many accounts, is not so easily obtained, and certainly not warranted by common mesh generators.

Hence the usefulness of the star construction, more general, because it applies to any primal mesh with star-shaped cells. A part A of A_n is *star-shaped* if it contains a point a, that we shall call a *center*, such that the whole segment $[a, x]$ belongs to A when x belongs to A. Now, pick such a center in each primal cell (the center of a primal node is itself), and join it to centers of all faces of the cell. This way, *simplicial* subcells are obtained (tetrahedra and their faces, in 3D). One gets the dual mesh by rearranging them, as follows: for each primal cell c, build its dual by putting together all k-subcells,

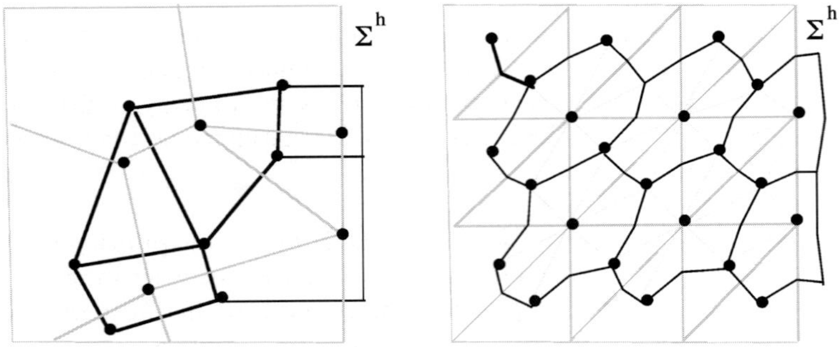

FIG. 15.3. Left: Orthogonal dual mesh. (Same graphic conventions as in Fig. 15.2, slightly simplified.) Right: Star construction of a dual mesh (close enough, here, to a barycentric mesh, but not quite the same). Notice the isolated dual edge, and the arbitrariness in shaping dual cells beyond Σ^h.

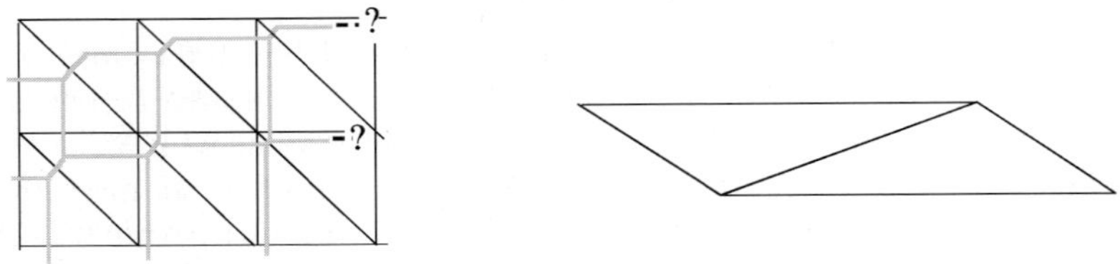

FIG. 15.4. Left: How hopeless the orthogonal construction can become, even with a fairly regular primal mesh. Right: Likely the simplest example of a 2D mesh without any orthogonal dual.

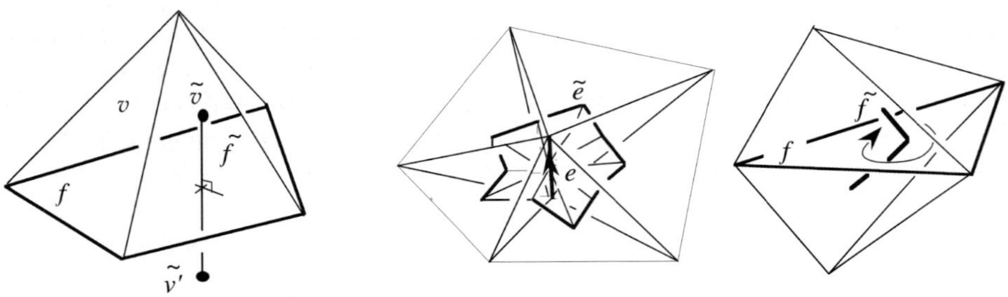

FIG. 15.5. Left: A facet f and its dual edge $\tilde{f}$ in the orthogonal construction ($\tilde{v}$ and $\tilde{v}'$ are the dual nodes which lie inside the volumes v and v' just above and just below f). From $\tilde{v}$, all boundary facets of v can directly be seen at right angle, but we don't require more: $\tilde{v}$ is neither v's barycenter nor the center of its circumscribed sphere, if there is such a sphere. Right: A dual facet and a dual edge, in the case of a simplicial primal mesh and of its barycentric dual. Observe the orientations.

$k \leqslant n - p$, which have one of their vertices at c's center, and other vertices at centers of cells incident on c. Figs. 15.3 and 15.5, right, give the idea. If all primal cells are simplices to start with, taking the barycenters of their faces as centers will give the *barycentric* dual mesh evoked a bit earlier.

REMARK 15.1. The recipe is imprecise about cells dual to those of Σ^h, whose shape outside D can be as one fancies (provided the requirements about duality are satisfied). Nothing there to worry about: Such choices are just as arbitrary as the selection of the centers of cells. It's all part of the unavoidable approximation error, which can be reduced at will by refinement.[39]

REMARK 15.2. If, as suggested above ("pave $\overline{D}$ first ..."), the primal mesh has been obtained by culling from a closed one, subcells built from the latter form a refinement of *both* the primal mesh and the dual mesh. The existence of this common "underlying simplicial complex" will be an asset when designing finite elements.

16. A discretization kit

We are ready, now, to apply the afore-mentioned strategy: Satisfy the balance equations (10.1) and (10.3) for a selected *finite* family of surfaces.

Let's first adopt a finite, approximate representation of the fields. Consider b, for instance. As a 2-form, it is meant to be integrated over inner oriented surfaces. So one may consider the integrals $\int_f b$, denoted $\mathbf{b}_f$, for all facets f, as a kind of "sampling" of b, and take the array of such "degrees of freedom" (DoF), $\{\mathbf{b} = \mathbf{b}_f\colon\ f \in \mathcal{F}\}$, indexed over primal facets, as a finite representation of b. This does not tell us about the *value* of the field at any given point, of course. But is that the objective? Indeed, all we know about a field is what we can measure, and we don't measure point values. These are abstractions. What we do measure is, indirectly, the *flux* of b, embraced by the loop of a small enough magnetic probe, by reading off the induced e.m.f. The above sampling thus consists in having each facet of the mesh play the role of such a probe, and the smaller the facets, the better we know the field. Conceivably, the mesh may be made so fine that the $\mathbf{b}_f$'s are *sufficient information* about the field, in practice. (Anyway, we'll soon see how to compute an approximation of the flux for any surface, knowing the $\mathbf{b}_f$'s, hence an approximation of b.) So one may be content with a method that would yield the four meaningful arrays of degrees of freedom, listing

- the edge e.m.f.'s, $\mathbf{e} = \{\mathbf{e}_e\colon\ e \in \mathcal{E}\}$,
- the facet fluxes, $\mathbf{b} = \{\mathbf{b}_f\colon\ f \in \mathcal{F}\}$,
- the dual-edge m.m.f.'s, $\mathbf{h} = \{\mathbf{h}_f\colon\ f \in \mathcal{F}\}$,
- and the dual-facet displacement currents, $\mathbf{d} = \{\mathbf{d}_e\colon\ e \in \mathcal{E}\}$,

all that from a similar sampling, across dual facets, of the given current j, encoded in the DoF array $\mathbf{j} = \{\mathbf{j}_e\colon\ e \in \mathcal{E}\}$.

In this respect, considering the integral form (10.1) and (10.3) of the basic equations will prove much easier than dealing with so-called "weak forms" of the infinitesimal equations (10.2) and (10.4). In fact, this simple shift of emphasis (which is the gist of Weiland's "finite integration theory", WEILAND [1992], and of Tonti's "cell method", TONTI [2001], MATTIUSSI [2000]) will so to speak *force on us* the right and unique discretization, as follows.

[39] A *refinement* of a paving is another paving of the same region, which restricts to a proper cellular paving of each original cell.

16.1. Network equations, discrete Hodge operator

Suppose the chain S in (10.1) is the simplest possible in the present context, that is, a *single* primal facet, f. The integral of e along ∂f is the sum of its integrals along edges that make ∂f, with proper signs, which are precisely the signs of the incidence numbers, by their very definition. Therefore, Eq. (10.1) applied to f yields

$$\partial_t \mathbf{b}_f + \sum_{e \in \mathcal{E}} \mathbf{R}_f^e \mathbf{e}_e = 0.$$

There is one equation like this for each facet of the primal mesh, that is – thanks for having discarded facets in S^e, for which the flux is known to be 0 – one for each genuinely unknown facet-flux of b. Taken together, in matrix form,

$$\partial_t \mathbf{b} + \mathbf{R}\mathbf{e} = 0, \tag{16.1a}$$

they form the first group of our *network* differential *equations*.

The same reasoning about each dual facet $\tilde{e}$ (the simplest possible outer-oriented surface that Σ in (10.3) can be) yields

$$-\partial_t \mathbf{d}_e + \sum_{f \in \mathcal{F}} \mathbf{R}_f^e \mathbf{h}_f = \mathbf{j}_e,$$

for all e in $\mathcal{E}$, i.e., in matrix form,

$$-\partial_t \mathbf{d} + \mathbf{R}^t \mathbf{h} = \mathbf{j}, \tag{16.1b}$$

the second group of network equations.

To complete this system, we need discrete counterparts to $b = \mu h$ and $d = \varepsilon e$, i.e., *network constitutive laws*, of the form

$$\mathbf{b} = \boldsymbol{\mu}\mathbf{h}, \qquad \mathbf{d} = \boldsymbol{\varepsilon}\mathbf{e}, \tag{16.2}$$

where $\boldsymbol{\varepsilon}$ and $\boldsymbol{\mu}$ are appropriate square symmetric matrices. Understanding how such matrices can be built is our next task. It should be clear that no *canonical* construction can exist – for sure, nothing comparable to the straightforward passage from (10.1), (10.3) to (16.1a), (16.1b) – because the metric of both meshes must intervene (Eq. (11.1) gives a clue in this respect). Indeed, the exact equivalent of (16.1), up to notational details, can be found in most published algorithms (including those based on the Galerkin method, see, e.g., LEE and SACKS [1995]), whereas a large variety of proposals exist as regards $\boldsymbol{\varepsilon}$ and $\boldsymbol{\mu}$. These "discrete Hodge operators" are the real issue. Constructing "good" ones, in a sense we still have to discover, is the central problem.

Our approach will be as follows: First – just not to leave the matter dangling too long – we shall give *one* solution, especially simple, to this problem, which makes $\boldsymbol{\varepsilon}$ and $\boldsymbol{\mu}$ *diagonal*, a feature the advantages of which we shall appreciate by working out a few examples. Later (in Section 20), a generic error analysis method will be sketched, from which a *criterion* as to what makes a good $\boldsymbol{\varepsilon}$–$\boldsymbol{\mu}$ pair will emerge. Finite elements will enter the stage during this process, and help find other solutions to the problem, conforming to the criterion.

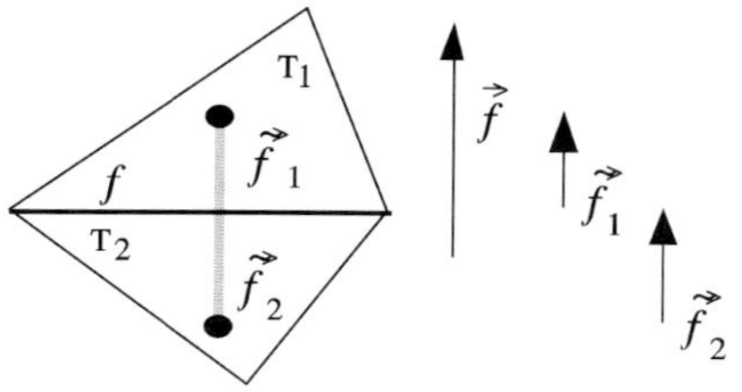

FIG. 16.1. The case of a discontinuous permeability (μ_1 and μ_2 in primal volumes T_1 and T_2, separated by facet f). We denote by $\vec{f}$ the vectorial area of f and by $\vec{\tilde{f}}_1$, $\vec{\tilde{f}}_2$, the vectors along both parts of $\tilde{f}$. Let u and v be arbitrary vectors, respectively normal and tangent to f, and let $\mathbf{H}_1 = u + v$ in T_1. Transmission conditions across f determine a unique uniform field $\mathrm{B}_2 = \mu_1 u + \mu_2 v$ in T_2. Then $\mathbf{b}_f = \mu_1 \vec{f} \cdot u$ and $\mu_2 \mathbf{h}_f = \mu_2 \vec{\tilde{f}}_1 \cdot u + \mu_1 \vec{\tilde{f}}_2 \cdot u$. As $\vec{f}$, $\vec{\tilde{f}}_1$, and $\vec{\tilde{f}}_2$ are collinear, u disappears from the quotient $\mathbf{b}_f/\mathbf{h}_f$, yielding (16.4).

The simple solution is available if one has been successful in building a dual mesh by the orthogonal construction (Figs. 15.3 and 15.5, left). Then, in the case when ε and μ are uniform,[40] one sets $\boldsymbol{\varepsilon}^{ee'} = 0$ if $e \neq e'$, $\boldsymbol{\mu}^{ff'} = 0$ if $f \neq f'$, and (cf. (11.1))

$$\boldsymbol{\varepsilon}^{ee} = \varepsilon \frac{\mathrm{area}(\tilde{e})}{\mathrm{length}(e)}, \qquad \boldsymbol{\mu}^{ff} = \mu \frac{\mathrm{area}(f)}{\mathrm{length}(\tilde{f})}, \tag{16.3}$$

which does provide diagonal matrices $\boldsymbol{\varepsilon}$ and $\boldsymbol{\mu}$. (The inverse of $\boldsymbol{\mu}$ will be denoted by $\boldsymbol{\nu}$.) The heuristic justification (TONTI [2001]) is that *if* the various fields happened to be piecewise constant (relative to the primal mesh), formulas (16.3) would exactly correspond to the very definition (11.1) of the Hodge operator. (Section 20 will present a stronger argument.) In the case of non-uniform coefficients, formulas such as

$$\boldsymbol{\mu}^{ff} = \frac{\mu_1 \mu_2 \,\mathrm{area}(f)}{\mu_2 \,\mathrm{length}(\tilde{f}_1) + \mu_1 \,\mathrm{length}(\tilde{f}_2)}, \tag{16.4}$$

where $\tilde{f}_1$ and $\tilde{f}_2$ are the parts of $\tilde{f}$ belonging to the two volumes adjacent to f, apply instead (Fig. 16.1). Observe the obvious intervention of metric elements (lengths, areas, angles) in these constructions.

REMARK 16.1. Later, when edge elements w^e and facet elements w^f will enrich the toolkit, we shall consider another solution, that consists in setting $\boldsymbol{\varepsilon}^{ee'} = \int_D \varepsilon w^e \wedge w^{e'}$ and $\boldsymbol{\nu}^{ff'} = \int_D \mu^{-1} w^f \wedge w^{f'}$. For reference, let's call this the "Galerkin approach" to the problem. We shall use loose expressions such as "the Galerkin $\boldsymbol{\varepsilon}$", or "the diagonal hodge", to refer to various brands of discrete Hodge operators.

16.2. The toolkit

At this stage, we have obtained discrete counterparts (Fig. 16.2) to most features of the "Maxwell building" of Fig. 12.2, but time differentiation and wedge product still miss

[40] I'll use "uniform" and "steady" for "constant in space" and "constant in time", respectively.

Primal-based DoF arrays

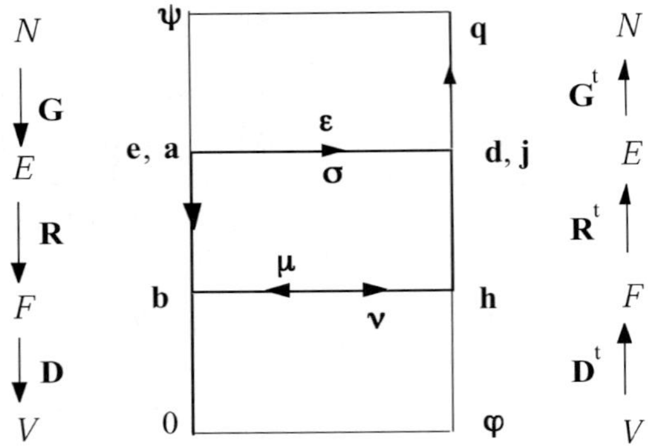

Dual-based DoF arrays

FIG. 16.2. A "discretization toolkit" for Maxwell's equations.

theirs. Some thought about how the previous ideas would apply in four dimensions should quickly suggest the way to deal with time derivatives: δt being the time step, call $\mathbf{b}^k$, $\mathbf{h}^k$, the values of $\mathbf{b}$, $\mathbf{h}$ at time $k\delta t$, for $k = 0, 1, \ldots$, call $\mathbf{j}^{k+1/2}$, $\mathbf{d}^{k+1/2}$, $\mathbf{e}^{k+1/2}$ those of $\mathbf{j}$, $\mathbf{d}$, $\mathbf{e}$ at time $(k + 1/2)\delta t$, and approximate $\partial_t \mathbf{b}$, at time $(k + 1/2)\delta t$, by $(\mathbf{b}^{k+1} - \mathbf{b}^k)/\delta t$, and similarly, $\partial_t \mathbf{d}$, now at time $k\delta t$, by $(\mathbf{d}^{k+1/2} - \mathbf{d}^{k-1/2})/\delta t$.

As for the wedge product, to $\int_D b \wedge h$ corresponds the sum $\sum_{f\in\mathcal{F}} \mathbf{b}_f \mathbf{h}_f$, which we shall denote by $(\mathbf{b}, \mathbf{h})$, with bold parentheses. Similarly, $\int_D d \wedge e$ corresponds to $\sum_{e\in\mathcal{E}} \mathbf{d}_e \mathbf{e}_e$, also denoted $(\mathbf{d}, \mathbf{e})$. Hence we may define "discrete energy" quadratic forms, $1/2(\boldsymbol{\nu}\mathbf{b}, \mathbf{b})$, $1/2(\boldsymbol{\mu}\mathbf{h}, \mathbf{h})$, $1/2(\boldsymbol{\varepsilon}\mathbf{e}, \mathbf{e})$, and $1/2(\boldsymbol{\varepsilon}^{-1}\mathbf{d}, \mathbf{d})$, all quantities with, indeed, the physical dimension of energy (but be aware that $(\mathbf{j}, \mathbf{e})$ is a power instead, like $\int_D j \wedge e$). Some notational shortcuts: Square roots such as $(\boldsymbol{\nu}\mathbf{b}, \mathbf{b})^{1/2}$, or $(\boldsymbol{\varepsilon}\mathbf{e}, \mathbf{e})^{1/2}$, etc., will be denoted by $|\mathbf{b}|_\nu$, or $|\mathbf{e}|_\varepsilon$, in analogy with the above $|b|_\nu$, or $|e|_\varepsilon$, and serve as various, physically meaningful *norms* on the vector spaces of DoF arrays. We'll say the "ν-norm", the "ε-norm", etc., for brevity.

PROPOSITION 16.1. *If Eqs.* (16.1)–(16.2) *are satisfied, one has*

$$d_t\left[\frac{1}{2}(\boldsymbol{\nu}\mathbf{b}, \mathbf{b}) + \frac{1}{2}(\boldsymbol{\varepsilon}\mathbf{e}, \mathbf{e})\right] = -(\mathbf{j}, \mathbf{e}). \tag{16.5}$$

PROOF. Take the bold scalar product of (16.1a) and (16.1b) by $\mathbf{h}$ and $-\mathbf{e}$, add, and use the equality $(\mathbf{R}\mathbf{e}, \mathbf{h}) = (\mathbf{e}, \mathbf{R}^t\mathbf{h})$. □

REMARK 16.2. The analogue of $\int_S h \wedge e$, when S is some m-surface, is

$$\sum_{f\in\mathcal{F}(S),\, e\in\mathcal{E}} \mathbf{R}^e_f \mathbf{h}_f \mathbf{e}_e,$$

where $\mathcal{F}(S)$ stands for the subset of facets which compose S. (Note how this sum vanishes if S is the domain's boundary.) By exploiting this, the reader will easily modify (16.5) in analogy with the Poynting theorem. In spite of such formal correspondences, energy and discrete energy have, a priori, no relation. To establish one, we shall need "interpolants", such as finite elements, enabling us to pass from degrees of freedoms to fields. For instance, facet elements will generate a mapping $\mathbf{b} \to b$, with $b = \sum_f \mathbf{b}_f w^f$. If $\boldsymbol{\nu}$ is the Galerkin hodge, then $\int_D \nu b \wedge b = (\boldsymbol{\nu}\mathbf{b}, \mathbf{b})$. Such built-in equality between energy and discrete energy is an exception, a distinctive feature of the Ritz–Galerkin approach. With other discrete hodges, even *convergence* of discrete energy, as the mesh is refined, towards the true one, should not be expected.

17. Playing with the kit: Full Maxwell

Now we have enough to discretize any model connected with Maxwell's equations. Replacing, in (13.1), rot by $\mathbf{R}$ or $\mathbf{R}^t$, ε and μ by $\boldsymbol{\varepsilon}$ and $\boldsymbol{\mu}$, and ∂_t by the integral or half-integral differential quotient, depending on the straight or twisted nature of the differential form in consideration, we obtain this:

$$\frac{\mathbf{b}^{k+1} - \mathbf{b}^k}{\delta t} + \mathbf{R}\mathbf{e}^{k+1/2} = 0, \qquad -\boldsymbol{\varepsilon}\frac{\mathbf{e}^{k+1/2} - \mathbf{e}^{k-1/2}}{\delta t} + \mathbf{R}^t\boldsymbol{\nu}\mathbf{b}^k = \mathbf{j}^k \tag{17.1}$$

(where $\mathbf{j}^k$ is the array of intensities through dual facets, at time[41] $k\delta t$), with initial conditions

$$\mathbf{b}^0 = 0, \qquad \mathbf{e}^{-1/2} = 0. \tag{17.2}$$

In the simplest case where the primal and dual mesh are plain rectangular staggered grids, (17.1) and (17.2) is the well known Yee scheme (YEE [1966]). So what we have here is the closest thing to Yee's scheme in the case of *cellular* meshes.

A similar numerical behavior can therefore be expected. Indeed,

PROPOSITION 17.1. *The scheme* (17.1) *and* (17.2) *is stable for* δt *small enough, provided both* $\boldsymbol{\varepsilon}$ *and* $\boldsymbol{\nu}$ *are symmetric positive definite.*

PROOF. For such a proof, one may assume $\mathbf{j} = 0$ and nonzero initial values in (17.2), satisfying $\mathbf{D}\mathbf{b}^0 = 0$. Eliminating $\mathbf{e}$ from (17.1), one finds that

$$\mathbf{b}^{k+1} - 2\mathbf{b}^k + \mathbf{b}^{k-1} + (\delta t)^2\mathbf{R}\boldsymbol{\varepsilon}^{-1}\mathbf{R}^t\boldsymbol{\nu}\mathbf{b}^k = 0. \tag{17.3}$$

Since $\mathbf{DR} = 0$, the "loop invariant" $\mathbf{D}\mathbf{b}^k = 0$ holds, so one may work in the corresponding subspace, ker($\mathbf{D}$). Let's introduce the (generalized) eigenvectors $\mathbf{v}_i$ such that $\mathbf{R}\boldsymbol{\varepsilon}^{-1}\mathbf{R}^t\mathbf{v}_i = \lambda_i\boldsymbol{\mu}\mathbf{v}_i$, which satisfy $(\boldsymbol{\mu}\mathbf{v}_i, \mathbf{v}_j) = 1$ if $i = j$, 0 if $i \neq j$. In this "μ-orthogonal" basis, $\mathbf{b}^k = \boldsymbol{\mu}\Sigma_i\eta_i^k\mathbf{v}_i$, and (17.3) becomes

$$\eta_i^{k+1} - \left(2 - \lambda_i(\delta t)^2\right)\eta_i^k + \eta_i^{k-1} = 0$$

for all i. The η_i^ks, and hence the $\mathbf{b}^k$s, stay bounded if the characteristic equation of each of these recurrences has imaginary roots, which happens (Fig. 17.1) if $0 < \lambda_j\delta t < 2$ for all j. □

[41] For easier handling of Ohm's law, $\mathbf{j}(k\delta t)$ may be replaced by $(\mathbf{j}^{k+1/2} + \mathbf{j}^{k-1/2})/2$.

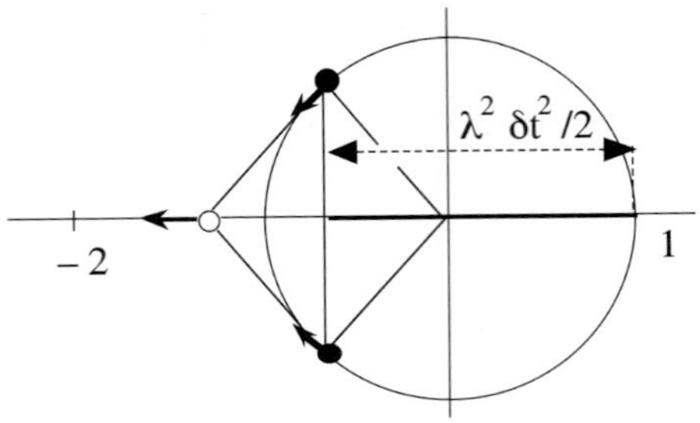

FIG. 17.1. The white spot lies at the sum of roots of the characteristic equation $r^2 - (2 - \lambda_i(\delta t)^2)r + 1 = 0$. Stability is lost if it leaves the interval $[-2, 2]$.

In the case of the original Yee scheme, eigenvalues could explicitly be found, hence the well-known relation (YEE [1966]) between the maximum possible value of δt and the lengths of the cell sides. For general grids, we have no explicit formulas, but the thumbrule is the same: δt should be small enough for a signal travelling at the speed of light (in the medium under study) not to cross more than one cell during this lapse of time.

This stringent stability condition makes the scheme unattractive if not fully explicit, or nearly so: $\boldsymbol{\varepsilon}$ should be *diagonal*, or at the very least, block-diagonal with most blocks of size 1 and a few small-size ones, and $\boldsymbol{\nu}$ should be sparse. If so is the case, each time step will only consist in a few matrix–vector products plus, perhaps, the resolution of a few small linear systems, which makes up for the large number of time steps. Both conditions are trivially satisfied with the orthogonal construction (cf. (16.3), (16.4)), but we have already noticed the problems this raises. Hence the sustained interest for so-called "mass-lumping" procedures, which aim at replacing the Galerkin $\boldsymbol{\varepsilon}$ by a diagonal matrix without compromising convergence: see COHEN, JOLY and TORDJMAN [1993], ELMKIES and JOLY [1997], HAUGAZEAU and LACOSTE [1993] (a coordinate-free reinterpretation of which can be found in BOSSAVIT and KETTUNEN [1999]).

REMARK 17.1. Obviously, there is another version of the scheme, in **h** and **d**, for which what is relevant is sparsity of $\boldsymbol{\varepsilon}^{-1}$ and diagonality of $\boldsymbol{\mu}$, i.e., of $\boldsymbol{\nu}$. Unfortunately, the diagonal lumping procedure that worked for edge elements fails when applied to the Galerkin $\boldsymbol{\nu}$, i.e., to the mass-matrix of facet elements (BOSSAVIT and KETTUNEN [1999]).

There are of course other issues than stability to consider, but we shall not dwell on them right now. For *convergence* (to be treated in detail later, but only in statics), cf. MONK and SÜLI [1994], NICOLAIDES and WANG [1998], BOSSAVIT and KETTUNEN [1999]. On *dispersion* properties, little can be said unless the meshes have some translational symmetry, at least locally, and this is beyond our scope. As for *conservation* of some quantities, it would be nice to be able to say, in the case when $\mathbf{j} = 0$, that "total discrete energy is conserved", but this is only almost true. Conserved quantities, as one will easily verify, are $\frac{1}{2}(\boldsymbol{\mu}\mathbf{h}^{k+1}, \mathbf{h}^k) + \frac{1}{2}(\boldsymbol{\varepsilon}\mathbf{e}^{k+1/2}, \mathbf{e}^{k+1/2})$ and

$\frac{1}{2}(\boldsymbol{\mu}\mathbf{h}^k, \mathbf{h}^k) + \frac{1}{2}(\boldsymbol{\varepsilon}\mathbf{e}^{k-1/2}, \mathbf{e}^{k+1/2})$, both independent of k. So their half-sum, which can suggestively be written as

$$W_k = \tfrac{1}{2}(\boldsymbol{\mu}\mathbf{h}^{k+1/2}, \mathbf{h}^k) + \tfrac{1}{2}(\boldsymbol{\varepsilon}\mathbf{e}^k, \mathbf{e}^{k+1/2}),$$

if one agrees on $\mathbf{h}^{k+1/2}$ and $\mathbf{e}^k$ as shorthands for $[\mathbf{h}^k + \mathbf{h}^{k+1}]/2$ and $[\mathbf{e}^{k-1/2} + \mathbf{e}^{k+1/2}]/2$, is conserved: *Not* the discrete energy, definitely, however close.

18. Playing with the kit: Statics

Various discrete models can be derived from (17.1) by the usual maneuvers (neglect the displacement current term $\boldsymbol{\varepsilon}\mathbf{e}$, omit time-derivatives in static situations), but it may be more instructive to obtain them from scratch. Take the magnetostatic model (13.2), for instance: Replace forms b and h by the DoF arrays $\mathbf{b}$ and $\mathbf{h}$, the d by the appropriate matrix, as read off from Fig. 16.2, and obtain

$$\mathbf{D}\mathbf{b} = 0, \qquad \mathbf{h} = \boldsymbol{\nu}\mathbf{b}, \qquad \mathbf{R}^t\mathbf{h} = \mathbf{j}, \tag{18.1}$$

which automatically includes the boundary conditions, thanks for having discarded[42] "passive" boundary cells. Observe that $\mathbf{G}^t\mathbf{j} = 0$ must hold for a solution to exist: But this is the discrete counterpart, as Fig. 16.2 shows, of $\mathrm{d}j = 0$, i.e., of $\operatorname{div} \mathrm{J} = 0$ in vector notation.

In the next section, we shall study the convergence of (18.1). When it holds, all schemes equivalent to (18.1) that can be obtained by algebraic manipulations are thereby equally valid – and there are lots of them. First, let $\mathbf{h}^{\mathbf{j}}$ be one of the facet-based arrays[43] such that $\mathbf{R}^t\mathbf{h}^{\mathbf{j}} = \mathbf{j}$. Then $\mathbf{h}$ in (18.1) must be of the form $\mathbf{h} = \mathbf{h}^{\mathbf{j}} + \mathbf{D}^t\boldsymbol{\varphi}$. Hence (18.1) becomes

$$\mathbf{D}\boldsymbol{\mu}\mathbf{D}^t\boldsymbol{\varphi} = -\mathbf{D}\boldsymbol{\mu}\mathbf{h}^{\mathbf{j}}. \tag{18.2}$$

This, which corresponds to $-\operatorname{div}(\mu(\operatorname{grad}\Phi + \mathrm{H}^j)) = 0$, the scalar potential formulation of magnetostatics, is not interesting unless $\boldsymbol{\nu}$ is diagonal, or nearly so, since $\boldsymbol{\mu}$ is full otherwise. So it requires the orthogonal construction, and is not an option in the case of the Galerkin $\boldsymbol{\nu}$. It's a well-studied scheme (cf. BANK and ROSE [1987], COURBET and CROISILLE [1998], GALLOUET and VILA [1991], HEINRICH [1987], HUANG and XI [1998], SÜLI [1991]), called "block-centered" in other sectors of numerical engineering (KAASSCHIETER and HUIJBEN [1992], WEISER and WHEELER [1988]), because degrees of freedom, assigned to the *dual* nodes, appear as lying inside the primal volumes,

---

[42] Alternatively (and this is how non-homogeneous boundary conditions can be handled), one may work with enlarged incidence matrices $\mathbf{R}$ and $\mathbf{D}$ and enlarged DoF arrays, taking all cells into account, then assign boundary values to passive cells, and keep only active DoFs on the left-hand side.

[43] There are such arrays, owing to $\mathbf{G}^t\mathbf{j} = 0$, because $\ker(\mathbf{G}^t) = \operatorname{cod}(\mathbf{R}^t)$, by transposition of $\operatorname{cod}(\mathbf{G}) = \ker(\mathbf{R})$, in the simple situation we consider. Finding one is an easy task, which does not require solving a linear system. Also by transposition of $\operatorname{cod}(\mathbf{R}) = \ker(\mathbf{D})$, one has $\ker(\mathbf{R}^t) = \operatorname{cod}(\mathbf{D}^t)$, and hence $\mathbf{R}^t(\mathbf{h} - \mathbf{h}^{\mathbf{j}}) = 0$ implies $\mathbf{h} = \mathbf{h}^{\mathbf{j}} + \mathbf{D}^t\boldsymbol{\varphi}$.

or “blocks”. Uniqueness of $\boldsymbol{\varphi}$ is easily proved,[44] which implies the uniqueness – not so obvious, a priori – of $\mathbf{h}$ and $\mathbf{b}$ in (18.1).

Symmetrically, there is a scheme corresponding to the vector potential formulation (i.e., $\mathrm{rot}(\nu\,\mathrm{rot}\,\mathrm{A}) = \mathrm{J}$):

$$\mathbf{R}^t \boldsymbol{\nu} \mathbf{R} \mathbf{a} = \mathbf{j}, \tag{18.3}$$

obtained by setting $\mathbf{b} = \mathbf{R}\mathbf{a}$, where the DoF array $\mathbf{a}$ is indexed over (active) edges. (If $\boldsymbol{\nu}$ is the Galerkin hodge, (18.3) is what one obtains when using edge elements to represent the vector potential.) Existence in (18.3) stems from $\mathbf{G}^t\mathbf{j} = 0$. No uniqueness this time, because $\ker(\mathbf{R})$ does not reduce to 0, but all solutions $\mathbf{a}$ give the same $\mathbf{b}$, and hence the same $\mathbf{h} = \boldsymbol{\nu}\mathbf{b}$.

REMARK 18.1. Whether to “gauge” $\mathbf{a}$ in this method, that is, to impose a condition that would select a unique solution, such as $\mathbf{G}^t\boldsymbol{\varepsilon}\mathbf{a} = 0$ for instance, remains to these days a contentious issue. It depends on which method is used to solve (18.3), and on how well the necessary condition $\mathbf{G}^t\mathbf{j} = 0$ is implemented. With iterative methods such as the conjugate gradient and its variants, and if one takes care to use $\mathbf{R}^t\mathbf{h}^{\mathbf{j}}$ instead of $\mathbf{j}$ in (18.3), then it’s better *not* to gauge (REN [1996]).

This is not all. If we refrain to eliminate $\mathbf{h}$ in the reduction from (18.1) to (18.3), but still set $\mathbf{b} = \mathbf{R}\mathbf{a}$, we get an intermediate two-equation system,

$$\begin{pmatrix} -\boldsymbol{\mu} & \mathbf{R} \\ \mathbf{R}^t & 0 \end{pmatrix} \begin{pmatrix} \mathbf{h} \\ \mathbf{a} \end{pmatrix} = \begin{pmatrix} 0 \\ \mathbf{j} \end{pmatrix}, \tag{18.4}$$

often called a *mixed* algebraic system (ARNOLD and BREZZI [1985]). (Again, little interest if $\boldsymbol{\mu}$ is full, i.e., unless $\boldsymbol{\nu}$ was diagonal from the outset.) The same manipulation in the other direction (eliminating $\mathbf{h}$ by $\mathbf{h} = \mathbf{h}^{\mathbf{j}} + \mathbf{D}^t\boldsymbol{\varphi}$, but keeping $\mathbf{b}$) gives

$$\begin{pmatrix} -\boldsymbol{\nu} & \mathbf{D}^t \\ \mathbf{D} & 0 \end{pmatrix} \begin{pmatrix} \mathbf{b} \\ \boldsymbol{\varphi} \end{pmatrix} = \begin{pmatrix} -\mathbf{h}^{\mathbf{j}} \\ 0 \end{pmatrix}. \tag{18.5}$$

We are not yet through. There is an interesting variation on (18.5), known as the mixed-hybrid approach. It’s a kind of “maximal domain decomposition”, in the sense that all volumes are made independent by “doubling” the degrees of freedom of $\mathbf{b}$ and $\mathbf{h}$ (two distinct values on sides of each facet not in Σ^h). Let’s redefine the enlarged arrays and matrices accordingly, and call them $\overline{\mathbf{b}}$, $\overline{\mathbf{h}}$, $\overline{\boldsymbol{\nu}}$, $\overline{\mathbf{D}}$, $\overline{\mathbf{R}}$. Constraints on $\overline{\mathbf{b}}$ (equality of up- and downstream fluxes) can be expressed as $\mathbf{N}\overline{\mathbf{b}} = 0$, where $\mathbf{N}$ has very simple structure (one 1×2 block, with entries 1 and -1, for each facet). Now, introduce an array $\boldsymbol{\lambda}$ of facet-based Lagrange multipliers, and add $(\boldsymbol{\lambda}, \mathbf{N}\overline{\mathbf{b}})$ to the underlying Lagrangian of (18.5). This gives a new discrete formulation (still equivalent to (18.1), if one derives $\mathbf{b}$

[44] It stems from $\ker(\mathbf{D}^t) = 0$. Indeed, $\mathbf{D}^t\boldsymbol{\psi} = 0$ means that $\Sigma_v \mathbf{D}_v^f \boldsymbol{\psi}_v = 0$ for all primal facets f. For some facets (those in Σ^h), there is but *one* volume v such that $\mathbf{D}_v^f \neq 0$, which forces $\boldsymbol{\psi}_v = 0$ for this v. Remove all such volumes v, and repeat the reasoning and the process, thus spreading the value 0 to all $\boldsymbol{\psi}_v$s.

and $\mathbf{h}$ from $\overline{\mathbf{b}}$ and $\overline{\mathbf{h}}$ the obvious way):

$$\begin{pmatrix} -\overline{\nu} & \overline{\mathbf{D}}^t & \mathbf{N}^t \\ \overline{\mathbf{D}} & 0 & 0 \\ \mathbf{N} & 0 & 0 \end{pmatrix} \begin{pmatrix} \overline{\mathbf{b}} \\ \varphi \\ \boldsymbol{\lambda} \end{pmatrix} = \begin{pmatrix} -\overline{\mathbf{h}}^{\mathbf{j}} \\ 0 \\ 0 \end{pmatrix}.$$

Remark that the enlarged $\overline{\nu}$ is block-diagonal (as well as its inverse $\overline{\mu}$), hence easy elimination of $\overline{\mathbf{b}}$. What then remains is a symmetric system in φ and $\boldsymbol{\lambda}$:

$$\begin{pmatrix} \overline{\mathbf{D}}\,\overline{\mu}\,\overline{\mathbf{D}}^t & \overline{\mathbf{D}}\,\overline{\mu}\,\mathbf{N}^t \\ \mathbf{N}\,\overline{\mu}\,\overline{\mathbf{D}}^t & \mathbf{N}\,\overline{\mu}\,\mathbf{N}^t \end{pmatrix} \begin{pmatrix} \varphi \\ \boldsymbol{\lambda} \end{pmatrix} = -\begin{pmatrix} \overline{\mathbf{D}}\,\overline{\mu}\,\overline{\mathbf{h}}^{\mathbf{j}} \\ \mathbf{N}\,\overline{\mu}\,\overline{\mathbf{h}}^{\mathbf{j}} \end{pmatrix}.$$

The point of this manipulation is that $\overline{\mathbf{D}}\overline{\mu}\overline{\mathbf{D}}^t$ is *diagonal*, equal to $\mathbf{K}$, say. So we may again eliminate φ, which leads to a system in terms of only $\boldsymbol{\lambda}$:

$$\mathbf{N}[\overline{\mu} - \overline{\mu}\,\overline{\mathbf{D}}^t\mathbf{K}^{-1}\overline{\mathbf{D}}\,\overline{\mu}]\mathbf{N}^t\boldsymbol{\lambda} = \mathbf{N}[\overline{\mu}\,\overline{\mathbf{D}}^t\mathbf{K}^{-1}\overline{\mathbf{D}}\,\overline{\mu} - \overline{\mu}]\overline{\mathbf{h}}^{\mathbf{j}}. \tag{18.6}$$

Contrived as it may look, (18.6) is a quite manageable system, with a sparse symmetric matrix. (The bracketed term on the left is block-diagonal, like $\overline{\mu}$.)

REMARK 18.2. In $(\boldsymbol{\lambda}, \mathbf{N}\overline{\mathbf{b}})$, each $\boldsymbol{\lambda}_f$ multiplies a term $(\mathbf{N}\overline{\mathbf{b}})_f$ which is akin to a magnetic charge. Hence the $\boldsymbol{\lambda}_f$s should be interpreted as facet-DoFs of a magnetic potential, which assumes the values necessary to reestablish the equality between fluxes that has been provisionally abandoned when passing from $\mathbf{b}$ to the enlarged (double size) flux vector $\overline{\mathbf{b}}$. This suggests a way to "complementarity" (obtaining bilateral estimates of some quantities) which is explored in BOSSAVIT [2003].

There is a dual mixed-hybrid approach, starting from (18.4), where *dual* volumes are made independent, hence (in the case of a simplicial primal mesh) three DoFs per facet, for both $\mathbf{b}$ and $\mathbf{h}$, and two Lagrange multipliers to enforce their equality. This leads to a system similar to (18.6) – but with twice as many unknowns, which doesn't make it attractive.

Systems (18.2), (18.3), (18.4), (18.5) and (18.6) all give the same solution pair $\{\mathbf{b}, \mathbf{h}\}$ – the solution of (18.1). Which one effectively to solve, therefore, is uniquely a matter of algorithmics, in which size, sparsity, and effective conditioning should be considered. The serious contenders are the one-matrix semi-definite systems, i.e., (18.2), (18.3), and (18.6). An enumeration of the number of off-diagonal terms (which is a fair figure of merit when using conjugate gradient methods on such matrices), shows that (18.6) rates better than (18.3), as a rule. The block-centered scheme (18.2) outperforms both (18.3) and (18.6), but is not available[45] with the Galerkin hodge. Hence the enduring interest (CHAVENT and ROBERTS [1991], KAASSCHIETER and HUIJBEN [1992], MOSÉ, SIEGEL, ACKERER and CHAVENT [1994], HAMOUDA, BANDELIER and RIOUX-DAMIDAU [2001]) for the "mixed-hybrid" method (18.6).

Each of the above schemes could be presented as the independent discretization of a specific mixed or mixed-hybrid variational formulation, and the literature is replete

[45] Unless one messes up with the computation of the terms of the mass-matrix, by using ad-hoc approximate integration formulas. This is precisely one of the devices used in mass-lumping.

with sophisticated analyses of this kind. Let's reemphasize that all these schemes are *algebraically* equivalent, as regards **b** and **h**. Therefore, an error analysis of one of them applies to all: For instance, if $\boldsymbol{\nu}$ is the Galerkin hodge, the standard variational convergence proof for (18.3), or if $\boldsymbol{\mu}$ is the diagonal hodge of (16.4), the error analysis we shall perform next section, on the symmetrical system (18.1).

19. Playing with the kit: Miscellanies

The advantage of working at the discrete level from the outset is confirmed by most examples one may tackle. For instance, the discrete version of the eddy-current problem (13.4) is, without much ado, found to be

$$i\omega\boldsymbol{\sigma}\mathbf{E} + \mathbf{R}^t\boldsymbol{\nu}\mathbf{R}\mathbf{E} = -i\omega\mathbf{J}^s. \tag{19.1}$$

As a rule, σ vanishes outside of a closed region $C = D - \Delta$ of the domain, C for "conductor". (Assume, then, that A, which is $\mathrm{supp}(\mathrm{J}^s)$, is contained in Δ.) The system matrix then has a non-trivial null space, $\ker(\boldsymbol{\sigma}) \cap \ker(\mathbf{R})$, and uniqueness of $\mathbf{E}$ is lost. It can be restored by enforcing the constraint $\mathbf{G}^t\boldsymbol{\varepsilon}_\Delta\mathbf{E} = 0$, where $\boldsymbol{\varepsilon}_\Delta$ is derived from $\boldsymbol{\varepsilon}$ by setting to zero all rows and columns which correspond to edges borne by C. Physically, this amounts to assume a zero electric charge density outside the conductive region $C = \mathrm{supp}(\sigma)$. (Beware, the electric field obtained this way can be seriously wrong about A, where this assumption is not warranted, in general. However, the electric field in C is correct.) Mathematically, the effect is to limit the span of the unknown $\mathbf{E}$ to a subspace over which $i\omega\boldsymbol{\sigma} + \mathbf{R}^t\boldsymbol{\nu}\mathbf{R}$ is regular.

In some applications, however, the conductivity is nonzero in all D, but may assume values of highly different magnitudes, and the above matrix, though regular, is ill-conditioned. One then will find in the kit the right tools to "regularize" such a "stiff" problem. See CLEMENS and WEILAND [1999] for an example of the procedure, some aspects of which are studied in BOSSAVIT [2001a]. Briefly, it consists in adding to the left-hand side of (19.1) a term, function of $\mathbf{E}$, that vanishes when $\mathbf{E}$ is one of the solutions of (19.1), which supplements the $\mathbf{R}^t\boldsymbol{\nu}\mathbf{R}$ matrix by, so to speak, what it takes to make it regular (and hence, to make the whole system matrix well conditioned, however small σ can be at places). The modified system is

$$i\omega\boldsymbol{\sigma}\mathbf{E} + \mathbf{R}^t\boldsymbol{\nu}\mathbf{R}\mathbf{E} + \boldsymbol{\sigma}\mathbf{G}\boldsymbol{\delta}\mathbf{G}^t\boldsymbol{\sigma}\mathbf{E} = -i\omega\mathbf{J}^s, \tag{19.2}$$

where $\boldsymbol{\delta}$ is a Hodge-like matrix, node based, diagonal, whose entries are $\delta^{nn} = \int_{\tilde{n}} 1/\mu\sigma^2$. A rationale for this can be found in BOSSAVIT [2001a]: In a nutshell, the idea is to "load the null space" of $\mathbf{R}^t\boldsymbol{\nu}\mathbf{R}$, and dimensional considerations motivate the above choice of $\boldsymbol{\delta}$. Our sole purpose here is to insist that all this can be done at the discrete level.

REMARK 19.1. One *might* motivate this procedure by starting from the following equation, here derived from (19.2) by simply using the toolkit in the other direction ("discrete" to "continuous"):

$$i\omega\sigma\mathrm{E} + \mathrm{rot}(\nu\,\mathrm{rot}\,\mathrm{E}) - \sigma\,\mathrm{grad}\left(\frac{1}{\mu\sigma^2}\,\mathrm{div}(\sigma\mathrm{E})\right) = -i\omega\mathrm{J}^s, \tag{19.3}$$

but which can be seen as a natural regularization of (13.4). (We revert to vector proxies here to call attention on the use of a variant of the $-\Delta = \text{rot} \circ \text{rot} - \text{grad} \circ \text{div}$ formula, which is relevant when both μ and σ are uniform in (19.3).) This is a time-honored idea (LEIS [1968]). Part of its present popularity may stem from its allowing standard discretization via *node-based* vector-valued elements (the discrete form is then of course quite different[46] from (19.2)), because E in (19.3) has more a priori regularity than E in (13.4). Even if one has reasons to prefer using such elements, the advantage is only apparent, because the discrete solution may converge towards something else than the solution of (13.4) in some cases (e.g., reentrant corners, cf. COSTABEL and DAUGE [1997]), where the solution of (19.3) has *too much* regularity to satisfy (13.4). This should make one wary of this approach.

Many consider the nullspace of $\mathbf{R}^t \boldsymbol{\nu} \mathbf{R}$ as a matter of concern, too, as regards the eigenmode problem,

$$\mathbf{R}^t \boldsymbol{\nu} \mathbf{R} \mathbf{E} = \omega^2 \boldsymbol{\varepsilon} \mathbf{E}, \tag{19.4}$$

because $\omega = 0$ is an eigenvalue of multiplicity N (the number of active nodes). Whether the concern is justified is debatable, but again, there are tools in the kit to address it. First, regularization, as above:

$$[\mathbf{R}^t \boldsymbol{\nu} \mathbf{R} + \boldsymbol{\varepsilon} \mathbf{G} \boldsymbol{\delta} \mathbf{G}^t \boldsymbol{\varepsilon}] \mathbf{E} = \omega^2 \boldsymbol{\varepsilon} \mathbf{E}, \tag{19.5}$$

with $\delta^{nn} = \int_{\tilde{n}} 1/\mu\varepsilon^2$ this time. Zero is not an eigenvalue any longer, but new eigenmodes appear, those of $\boldsymbol{\varepsilon} \mathbf{G} \boldsymbol{\delta} \mathbf{G}^t \boldsymbol{\varepsilon} \mathbf{E} = \omega^2 \boldsymbol{\varepsilon} \mathbf{E}$ under the restriction $\mathbf{E} = \mathbf{G} \boldsymbol{\psi}$. As remarked by WHITE and KONING [2000], we have here (again, assuming uniform coefficients) a phenomenon of "spectral complementarity" between the operators $\text{rot} \circ \text{rot}$ and $-\text{grad} \circ \text{div}$. The new modes, or "ghost modes" as they are called in WEILAND [1985], have to be sifted out, which is in principle easy[47] (evaluate the norm $|\mathbf{G}^t \boldsymbol{\varepsilon} \mathbf{E}|_\delta$), or "swept to the right" by inserting an appropriate scalar factor in front of the regularizing term. Second solution (TRAPP, MUNTEANU, SCHUHMANN, WEILAND and IOAN [2002]): Restrict the search of $\mathbf{E}$ to a complement of $\ker(\mathbf{R}^t \boldsymbol{\nu} \mathbf{R})$, which one can do by so-called "tree-cotree" techniques (ALBANESE and RUBINACCI [1988], MUNTEANU [2002]). This verges on the issue of *discrete Helmholtz decompositions*, another important tool in the kit, which cannot be given adequate treatment here (see RAPETTI, DUBOIS and BOSSAVIT [2002]).

[46]When $\boldsymbol{\sigma}$ and $\boldsymbol{\nu}$ are the Galerkin hodges, (19.2) corresponds to the edge-element discretization of (19.3).

[47]These ghost modes are *not* the (in)famous "spurious modes" which were such a nuisance before the advent of edge elements (cf. BOSSAVIT [1990b]). Spurious modes occur when one solves the eigenmode problem $\text{rot}(\nu \, \text{rot} \, \text{E}) = \omega^2 \varepsilon \text{E}$ by using *nodal vectorial* elements. Then (barring exceptional boundary conditions) the $\text{rot}(\nu \, \text{rot})$ matrix is regular (because the approximation space does not contain gradients, contrary to what happens with edge elements), but also – and for the same reason, as explained in BOSSAVIT [1998a] – poorly conditioned, which is the root of the evil. It would be wise *not* to take "ghost modes" and "spurious modes" as synonyms, in order to avoid confusion on this tricky point.

CHAPTER IV

Finite Elements

We now tackle the convergence analysis of the discrete version of problem (13.2), magnetostatics:

$$\mathbf{D}\mathbf{b} = 0, \qquad \mathbf{h} = \nu\mathbf{b}, \qquad \mathbf{R}^t\mathbf{h} = \mathbf{j}. \tag{18.1}$$

A preliminary comment on what that means is in order.

A few notational points before: The mesh is denoted m, the dual mesh is $\widetilde{m}$, and we shall subscript by m, when necessary, all mesh-related entities. For instance, the largest diameter of all p-cells, $p \geqslant 1$, primal and dual, will be denoted γ_m (with a mild abuse, since it also depends on the metric of the dual mesh), and called the "grain" of the pair of meshes. The computed solution $\{\mathbf{b}, \mathbf{h}\}$ will be $\{\mathbf{b}_m, \mathbf{h}_m\}$ when we wish to stress its dependence on the mesh-pair. And so on.

A first statement of our purpose is "study $\{\mathbf{b}_m, \mathbf{h}_m\}$ when γ_m tends to 0". Alas, this lacks definiteness, because how the *shapes* of the cells change in the process does matter a lot. In the case of triangular 2D meshes, for instance, there are well-known counter-examples (BABUŠKA and AZIZ [1976]) showing that, if one tolerates too much "flattening" of the triangles as the grain tends to 0, convergence may fail to occur. Hence the following definition: A family $\mathcal{M}$ of (pairs of interlocked) meshes is *uniform* if there is a *finite* catalogue of "model cells" such that any cell in any m or $\widetilde{m}$ of the family is the transform by similarity of one of them. The notation "$m \to 0$" will then refer to a sequence of meshes, all belonging to some definite uniform family, and such that their γ_ms tend to zero. Now we redefine our objective: Show that the error incurred by taking $\{\mathbf{b}_m, \mathbf{h}_m\}$ as a substitute for the real field $\{b, h\}$ tends to zero when $m \to 0$.

The practical implications of achieving this are well known. If, for a given m, the computed solution $\{\mathbf{b}_m, \mathbf{h}_m\}$ is not deemed satisfactory, one must *refine* the mesh and redo the computation, again and again. If the refinement rule guarantees that all meshes such a process can generate belong to some definite uniform family, then the convergence result means "you may get as good an approximation as you wish by refining this way", a state of affairs we are more or less happy to live with.

Fortunately, such refinement rules do exist (this is an active area of research: BÄNSCH [1991], BEY [1995], DE COUGNY and SHEPHARD [1999], MAUBACH [1995]). Given a pair of coarse meshes to start with, there are ways to subdivide the cells so as to keep bounded the number of different cell-shapes that appear in the process, hence a potential infinity of refined meshes, which do constitute a uniform family. (A refinement process for tetrahedra is illustrated by Fig. 20.1. As one can see, at most five different shapes

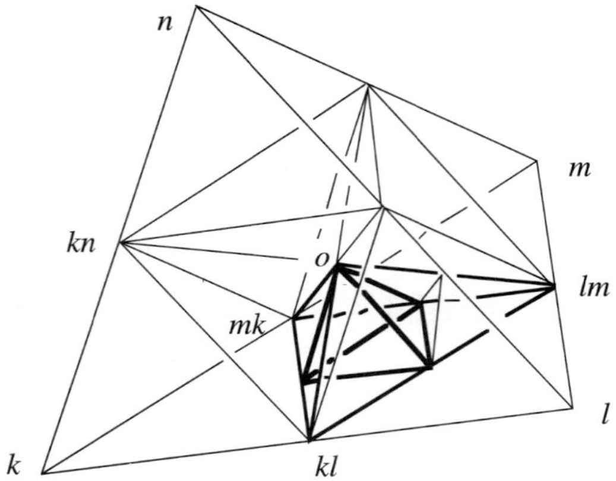

FIG. 20.1. Subdivision rule for a tetrahedron $\mathrm{T} = \{k, l, m, n\}$. (Mid-edges are denoted kl, lm, etc., and o is the barycenter.) A first halving of edges generates four small tetrahedra and a core octahedron, which itself can be divided into eight "octants" such as $\mathrm{O} = \{o, kl, lm, mk\}$, of at most four different shapes. Now, octants like O should be subdivided as follows: divide the facet in front of o into four triangles, and join to o, hence a tetrahedron similar to T, and three peripheral tetrahedra. These, in turn, are halved, as shown for the one hanging from edge $\{o, lm\}$. Its two parts are similar to O and to the neighbor octant $\{o, kn, kl, mk\}$ respectively.

can occur, for each tetrahedral shape present in the original coarse mesh. In practice, not all volumes get refined simultaneously, so junction dissection schemes are needed, which enlarges the catalogue of shapes, but the latter is bounded nonetheless.)

For these reasons, we shall feel authorized to assume uniformity in this sense. We shall also posit that the hodge entries, whichever way they are built, only depend (up to a multiplicative factor) on the *shapes* of the cells contributing to them. Although stronger than necessary, these assumptions will make some proofs easier, and thus help focus on the main ideas.

20. Consistency

Back to the comparison between $\{\mathbf{b}_m, \mathbf{h}_m\}$ and $\{b, h\}$, a natural idea is to compare the computed DoF arrays, $\mathbf{b}_m$ and $\mathbf{h}_m$, with arrays of the same kind, $r_m b = \{\int_f b \colon f \in \mathcal{F}\}$ and $r_m h = \{\int_{\tilde{f}} h \colon f \in \mathcal{F}\}$, composed of the fluxes and m.m.f.'s of the (unknown) solution $\{b, h\}$ of the original problem (13.2). This implicitly defines two operators with the same name, r_m: one that acts on 2-forms, giving an array of facet-fluxes, one that acts on twisted 1-forms, giving an array of dual-edge m.m.f.'s. (No risk of confusion, since the name of the operand, b or h, reveals its nature.)

Since $\mathrm{d}b = 0$, the flux of b embraced by the boundary of any primal 3-cell v must vanish, therefore the sum of facet fluxes $\sum_f \mathbf{D}_v^f \int_f b$ must vanish for all v. Similarly, $\mathrm{d}h = j$ yields the relation $\sum_f \mathbf{R}_f^e \int_{\tilde{f}} h = \int_{\tilde{e}} j$, by integration over a dual 2-cell. In matrix form, all this becomes

$$\mathbf{D} r_m b = 0, \qquad \mathbf{R}^t r_m h = \mathbf{j}, \tag{20.1}$$

since the entries of $\mathbf{j}$ are precisely the intensities across dual facets. Comparing with (18.1), we obtain

$$\mathbf{D}(\mathbf{b}_m - r_m b) = 0, \qquad \mathbf{R}^t(\mathbf{h}_m - r_m h) = 0, \tag{20.2}$$

and

$$(\mathbf{h}_m - r_m h) - \boldsymbol{\nu}(\mathbf{b}_m - r_m b) = (\boldsymbol{\nu} r_m - r_m \nu)b \equiv \boldsymbol{\nu}(r_m \mu - \boldsymbol{\mu} r_m)h. \tag{20.3}$$

Let us compute the μ-norm of both sides of (20.3). (For this piece of algebra, we shall use the notation announced in last chapter: $(\mathbf{b}, \mathbf{h})$ for a sum such as $\sum_{f \in \mathcal{F}} \mathbf{b}_f \mathbf{h}_f$, and $|\mathbf{h}|_\mu$ for $(\boldsymbol{\mu}\mathbf{h}, \mathbf{h})^{1/2}$, the μ-norm of $\mathbf{h}$, and other similar constructs.)

As this is done, "square" and "rectangle" terms appear. The rectangle term for the left-hand side is $-2(\mathbf{b}_m - r_m b, \mathbf{h}_m - r_m h)$, but since $\mathbf{D}(\mathbf{b}_m - r_m b) = 0$ implies the existence of some $\mathbf{a}$ such that $\mathbf{b}_m - r_m b = \mathbf{R}\mathbf{a}$, we have

$$(\mathbf{b}_m - r_m b, \mathbf{h}_m - r_m h) = (\mathbf{R}\mathbf{a}, \mathbf{h}_m - r_m h) = \big(\mathbf{a}, \mathbf{R}^t(\mathbf{h}_m - r_m h)\big) = 0,$$

after (20.2). Only square terms remain, and we get

$$\begin{aligned} &|\mathbf{h}_m - r_m h|_\mu^2 + |\mathbf{b}_m - r_m b|_\nu^2 \\ &\quad = \big|(\boldsymbol{\nu} r_m - r_m \nu)b\big|_\mu^2 \equiv \big|(\boldsymbol{\mu} r_m - r_m \mu)h\big|_\nu^2 \equiv (\boldsymbol{\nu} r_m b - r_m h, r_m b - \boldsymbol{\mu} r_m h). \end{aligned} \tag{20.4}$$

On the left-hand side, which has the dimension of an energy, we spot two plausible estimators for the error incurred by taking $\{\mathbf{b}_m, \mathbf{h}_m\}$ as a substitute for the real field $\{b, h\}$: the "error in (discrete) energy" [respectively coenergy], as regards $\mathbf{b}_m - r_m b$ [respectively $\mathbf{h}_m - r_m h$]. Components of $\mathbf{b}_m - r_m b$ are what can be called the "residual fluxes" $\mathbf{b}_f - \int_f b$, i.e., the difference between the computed flux embraced by facet f and the genuine (but unknown) flux $\int_f b$. Parallel considerations apply to h, with m.m.f.'s along $\tilde{f}$ instead of fluxes. It makes sense to try and *bound* these error terms by some function of γ_m. So let us focus on the right-hand side of (20.4), for instance on its second expression, the one in terms of h.

By definition of r_m, the f-component of $r_m(\mu h)$ is the flux of $b = \mu h$ embraced by f. On the other hand, the flux array $\boldsymbol{\mu} r_m h$ is the result of applying the discrete Hodge operator to the m.m.f. array $r_m h$, so the compound operators $r_m \mu$ and $\boldsymbol{\mu} r_m$ will not be equal: they give different fluxes when applied to a generic h. This contrasts with the equalities $(\mathbf{D} r_m - r_m \mathrm{d})b = 0$ and $(\mathbf{R}^t r_m - r_m \mathrm{d})h = 0$, which stem from the Stokes theorem. The mathematical word to express such equalities is "conjugacy": $\mathbf{D}$ and d are conjugate via r_m, and so are $\mathbf{R}^t$ and d, too. Thus, μ and $\boldsymbol{\mu}$ are *not* conjugate via r_m – and this is, of course, the reason why discretizing entails some error.

Yet, it may happen that $r_m \mu$ and $\boldsymbol{\mu} r_m$ *do* coincide for *some* hs. This is so, for instance, with piecewise constant fields, when $\boldsymbol{\mu}$ is the diagonal hodge of (16.3) and (16.4): in fact, these formulas were motivated by the desire to achieve this coincidence for such fields. Also, as we shall prove later, $r_m \nu$ and $\boldsymbol{\nu} r_m$ coincide on facet-element approximations of b, i.e., on divergence-free fields of the form $\sum_{f \in \mathcal{F}} \mathbf{b}_f w^f$ (which are meshwise constant), when $\boldsymbol{\nu}$ is the Galerkin hodge. Since all piecewise smooth fields differ from such special fields by some small residual, and the finer the mesh the smaller, we may

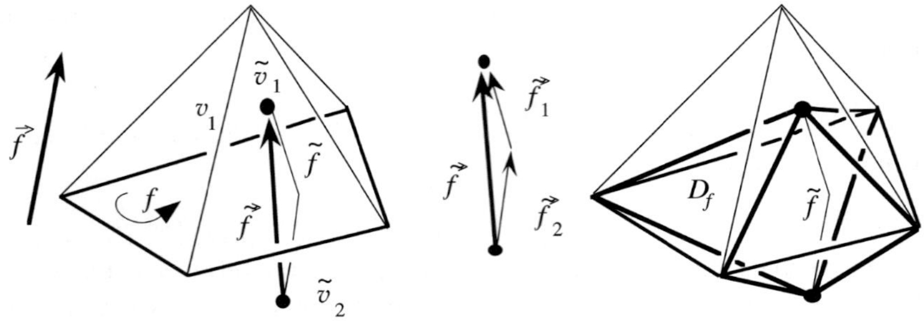

FIG. 20.2. As in Fig. 16.1, $\vec{f}$ denotes the vectorial area of facet f: the vector of magnitude area(f), normal to f, that points away from f in the direction derived from f's inner orientation by Ampère's rule. By $\overset{\rightsquigarrow}{f}$ we denote the vector that joins the end points of the associated dual edge $\tilde{f}$. (An ambient orientation is assumed here. One could do without it by treating both $\vec{f}$ and $\overset{\rightsquigarrow}{f}$ as axial vectors.) In case ν is not the same on both sides of f, understand $\nu \overset{\rightsquigarrow}{f}$ as $\nu_2 \overset{\rightsquigarrow}{f_2} + \nu_1 \overset{\rightsquigarrow}{f_1}$, where $\overset{\rightsquigarrow}{f_2}$ and $\overset{\rightsquigarrow}{f_1}$ are as suggested. Region D_f is the volume enclosed by the "tent" determined by the extremities of $\tilde{f}$ and the boundary of f. Note that $\vec{f}$ and $\nu \overset{\rightsquigarrow}{f}$ always cross f in the same direction, but only in the orthogonal construction are they parallel (cf. Fig. 16.1): In that case, (20.6) can be satisfied by a *diagonal* hodge – cf. (16.3) and (16.4).

in such cases expect "asymptotic conjugacy", in the sense that the right-hand side of (20.4) will tend to 0 with m, for a piecewise smooth b or h. This property, which we rewrite informally but suggestively as

$$\boldsymbol{\nu} r_m - r_m \nu \to 0 \quad \text{when } m \to 0, \qquad \boldsymbol{\mu} r_m - r_m \mu \to 0 \quad \text{when } m \to 0 \tag{20.5}$$

(two equivalent statements), is called *consistency* of the approximation of μ and ν by $\boldsymbol{\mu}$ and $\boldsymbol{\nu}$. Consistency, thus, implies asymptotic vanishing of the error in (discrete) energy, after (20.4).

Let's now take a heuristic step. (We revert to vector proxies for this. Fig. 20.2 explains about $\vec{f}$ and $\overset{\rightsquigarrow}{f}$, and n and τ are normal and tangent unit vector fields, as earlier. The norm of an ordinary vector is $|\;|$.) Remark that the right-hand side of (20.4) is, according to its rightmost avatar, a sum of terms, one for each f, of the form

$$\left[\sum_{f'} \boldsymbol{\nu}^{ff'} \int_{f'} n \cdot \mathrm{B} - \int_{\tilde{f}} \nu\tau \cdot \mathrm{B}\right]\left[\int_{f} \mu n \cdot \mathrm{H} - \sum_{f''} \boldsymbol{\mu}^{ff''} \int_{\tilde{f}''} \tau \cdot \mathrm{H}\right],$$

which we'll abbreviate as [B, f][H, f]. Each should be made as small as possible for the sum to tend to 0. Suppose ν is uniform, and that boundary conditions are such that B and H are uniform. Then $[\mathrm{B}, f] = \mathrm{B} \cdot (\sum_{f'} \boldsymbol{\nu}^{ff'} \vec{f}' - \nu \overset{\rightsquigarrow}{f})$. This term vanishes if

$$\sum_{f' \in \mathcal{F}} \boldsymbol{\nu}^{ff'} \vec{f}' = \nu \overset{\rightsquigarrow}{f}. \tag{20.6}$$

(This implies $\sum_{f' \in \mathcal{F}} \boldsymbol{\mu}^{ff'} \nu \overset{\rightsquigarrow}{f'} = \vec{f}$, and hence, cancellation of [H, f], too.) We therefore adopt this geometric compatibility condition as a *criterion* about $\boldsymbol{\nu}$. Clearly, the diagonal hodge of (16.4) passes this test. But on the other hand, no diagonal $\boldsymbol{\nu}$ can satisfy (20.6) unless $\vec{f}$ and $\nu\, \overset{\leadsto}{f}$ are collinear.

PROPOSITION 20.1. *If $\boldsymbol{\nu}$ is diagonal, with $\boldsymbol{\nu}^{ff}\,\vec{f} = \nu\,\overset{\leadsto}{f}$, as required by the criterion, there is consistency.*

PROOF. (All C's, from now on, denote constants, not necessarily the same each time, possibly depending on the solution, but not on the mesh.) This time, the solution proxy B is only piecewise smooth, and possibly discontinuous if ν is not uniform, but its component parallel to $\vec{f}$, say $\mathcal{B}$, satisfies $|\mathcal{B}(x) - \mathcal{B}(y)| \leqslant C|x-y|$ in the region D_f of Fig. 20.2. One has[48] $\int_f n \cdot \mathrm{B} = \mathrm{area}(f)\mathcal{B}(x_f)$ and $\int_{\tilde{f}} \nu\tau \cdot \mathrm{B} = \mathrm{length}(\nu\,\overset{\leadsto}{f})\mathcal{B}(x_{\tilde{f}})$, for some averaging points x_f and $x_{\tilde{f}}$, the distance of which doesn't exceed γ_m, hence $[\mathrm{B}, f] \leqslant C\gamma_m \boldsymbol{\nu}^{ff}\,\mathrm{area}(f)$, by factoring out $\boldsymbol{\nu}^{ff}\,\mathrm{area}(f) \equiv \mathrm{length}(\nu\,\overset{\leadsto}{f})$, and similarly, $[\mathrm{H}, f] \leqslant C\gamma_m \boldsymbol{\mu}^{ff}\,\mathrm{length}(\nu\,\overset{\leadsto}{f})$. Noticing that $\mathrm{area}(f)\,\mathrm{length}(\nu\,\overset{\leadsto}{f}) = 3\int_{D_f} \nu$, and summing up with respect to f, one finds that

$$|\mathbf{h}_m - r_m h|_\mu^2 + |\mathbf{b}_m - r_m b|_\nu^2 \leqslant C\gamma_m^2, \tag{20.7}$$

the consistency result. □

Going back to (20.4), we conclude that both the ν-norm of the residual flux array and the μ-norm of the residual m.m.f. array tend to 0 as fast as γ_m, or faster,[49] a result we shall exploit next.

One may wonder whether the proof can be carried out in the case of a non-diagonal hodge, assuming (20.6). The author has not been able to do so on the basis of (20.6) only. The result is true under stronger hypotheses (stronger than necessary, perhaps): When the construction of $\boldsymbol{\nu}$ is a local one, i.e., $\boldsymbol{\nu}^{ff'} = 0$ unless facets f and f' belong to a common volume, and when the *infimum* δ_m of all cell diameters verifies $\delta_m \geqslant \beta\gamma_m$, with β independent of m. Then $\boldsymbol{\nu}$ has a band structure, and its terms behave in γ_m^{-1}, which makes it easy to prove that $[\mathrm{B}, f]$ is in $\mathrm{O}(\gamma_m^2)$. Handling $[\mathrm{H}, f]$ is more difficult, because $\boldsymbol{\mu}$ is full, and the key argument about averaging points not being farther apart than γ_m breaks down. On the other hand, owing to the band structure of $\boldsymbol{\nu}$, and its positive-definite character, $\boldsymbol{\mu}^{ff'}$ is small for distant f and f', which allows one to also bound $[\mathrm{H}, f]$ by $C\gamma_m^2$. The number of faces being in γ_m^{-3}, consistency ensues.

There is some way to go to turn such an argument into a proof, but this is enough to confirm (20.6) in its status of criterion as regards $\boldsymbol{\nu}$, a criterion which is satisfied, by construction (Fig. 16.1), in FIT (WEILAND [1996]) and in the cell method (TONTI

---

[48] In case ν is not the same on both sides of f, understand $\mathrm{length}(\nu\,\overset{\leadsto}{f})$ as $\nu_1\,\mathrm{length}(\overset{\leadsto}{f_1}) + \nu_2\,\mathrm{length}(\overset{\leadsto}{f_2})$. The underlying measure of lengths is not the Euclidean one, but the one associated with the metric induced by the Hodge operator ν.

[49] Convergence in γ_m^2 is in fact often observed when the meshes have some regularity, such as crystal-like symmetries, which may cancel out some terms in the Taylor expansions implicit in the above proof. For instance, the distance between points x_f and $x_{\tilde{f}}$ may well be in γ_m^2 rather than γ_m. This kind of phenomenon is commonplace in Numerical Analysis (SCHATZ, SLOAN and WAHLBIN [1996]).

[2001]), but allows a much larger choice. We'll see in a moment how and why it is satisfied in the Galerkin approach.

21. Stability

So, the left-hand side of (20.4) tends to 0. Although this is considered by many as sufficient in practice, one cannot be satisfied with such "discrete energy" estimates. Fields should be compared with fields. To really prove convergence, one should build from the DoF arrays $\mathbf{b}_m$ and $\mathbf{h}_m$ an approximation $\{b_m, h_m\}$ of the pair of differential forms $\{b, h\}$, and show that the discrepancies $b_m - b$ and $h_m - h$ tend to 0 with m in some definite sense. So we are after some map, that we shall denote by p_m, that would transform a flux array $\mathbf{b}$ into a 2-form $p_m \mathbf{b}$ and an m.m.f. array $\mathbf{h}$ into a twisted 1-form $p_m \mathbf{h}$. The map should behave naturally with respect to r_m, i.e.,

$$r_m p_m \mathbf{b} = \mathbf{b}, \qquad r_m p_m \mathbf{h} = \mathbf{h}, \tag{21.1}$$

as well as

$$|p_m r_m b - b|_\nu \to 0 \quad \text{and} \quad |p_m r_m h - h|_\mu \to 0 \quad \text{when } m \to 0 \tag{21.2}$$

(asymptotic vanishing of the "truncation error" $p_m r_m - 1$). A satisfactory result, then, would be that both $|b - p_m \mathbf{b}_m|_\nu$ and $|h - p_m \mathbf{h}_m|_\mu$ tend to 0 with m (convergence "in energy"). As will now be proved, this is warranted by the following inequalities:

$$\alpha |p_m \mathbf{b}|_\nu \leqslant |\mathbf{b}|_\nu, \qquad \alpha |p_m \mathbf{h}|_\mu \leqslant |\mathbf{h}|_\mu \tag{21.3}$$

for all $\mathbf{b}$ and $\mathbf{h}$, where the constant $\alpha > 0$ does not depend on m. Since $|\mathbf{b}|_\nu$ and $|\mathbf{h}|_\mu$ depend on the discrete hodge, (21.3) is a property of the approximation scheme, called *stability*.

PROPOSITION 21.1. *Consistency* (20.5) *being assumed,* (21.2) *and* (21.3) *entail convergence.*

PROOF. By consistency, the right-hand side of (20.4) tends to 0, whence $|\mathbf{b}_m - r_m b|_\nu \to 0$, and $|p_m \mathbf{b}_m - p_m r_m b|_\nu \to 0$ by (21.3). Therefore $p_m \mathbf{b}_m \to b$, "in energy", thanks to (21.2). Same argument about h. □

This is Lax's celebrated folk theorem (LAX and RICHTMYER [1956]): *consistency + stability = convergence.*

Below, we shall find a systematic way to construct p_m, the so-called *Whitney map.* But if we don't insist right now on generality, there is an easy way to find a suitable such map in the case of a simplicial primal mesh and of DoF arrays $\mathbf{b}$ that satisfy $\mathbf{D}\mathbf{b} = 0$ (luckily, only these do matter in magnetostatics). The idea is to find a vector proxy $\overline{\mathrm{B}}$ uniform inside each tetrahedron with facet fluxes $\overline{\mathrm{B}} \cdot \vec{f}$ equal to $\mathbf{b}_f$. (Then, $\operatorname{div} \overline{\mathrm{B}} = 0$ all over D.) This, which would not be possible with cells of arbitrary shapes, can be done with tetrahedra, for there are, for each tetrahedral volume v, three unknowns (the components of $\overline{\mathrm{B}}$) to be determined from four fluxes linked by one linear relation, $\sum_f \mathbf{D}_v^f \mathbf{b}_f = 0$, so the problem has a solution, which we take as $p_m \mathbf{b}$.

Then,[50] $p_m r_m b \to b$. As for the stability condition (21.3), one has $|p_m \mathbf{b}|_\nu^2 = \int_D \nu |\overline{\mathrm{B}}|^2$, a quadratic form with respect to the facet fluxes, which we may therefore denote by $(\mathbf{b}, \mathbf{N}\mathbf{b})$, with $\mathbf{N}$ some positive definite matrix. Now, suppose first a *single* tetrahedron in the mesh m, and consider the Rayleigh-like quotient $(\mathbf{b}, \boldsymbol{\nu}\mathbf{b})/(\mathbf{b}, \mathbf{N}\mathbf{b})$. Its lower bound, strictly positive, depends only on the *shape* of the tetrahedron, not on its size. Then, uniformity of the family of meshes, and of the construction of $\boldsymbol{\nu}$, allows us to take for α in (21.3) the smallest of these lower bounds, which is strictly positive and independent of m. We may thereby conclude that $p_m \mathbf{b}_m$ converges towards b in energy.

No similar construction on the side of h is available, but this is not such a handicap: if $p_m \mathbf{b}_m \to b$, then $\nu p_m \mathbf{b}_m \to h$. This amounts to setting p_m on the dual side equal to $\nu p_m \boldsymbol{\mu}$. The problem with that is, $p_m \mathbf{h}$ fails to have the continuity properties we expect from a magnetic field: its vector proxy H is not tangentially continuous across facets, so one cannot take its curl. But never mind, since this "non-conformal" approximation converges in energy.

Yet, we need a more encompassing p_m map, if only because $\mathbf{D}\mathbf{b} = 0$ was just a happy accident. Before turning to that, which will be laborious, let's briefly discuss convergence in the full Maxwell case.

22. The time-dependent case

Here is a sketch of the convergence proof for the generalized Yee scheme (17.1) and (17.2) of last chapter.

First, linear interpolation in time between the values of the DoF arrays, as output by the scheme, provides DoF-array-valued functions of time which converge, when δt tends to zero, towards the solution of the "spatially discretized" equations (16.1) and (16.2). This is not difficult.

Next, linearity of the equations allows one to pass from the time domain to the frequency domain, via a Laplace transformation. Instead of studying (16.1) and (16.2), therefore, one may examine the behavior of the solution of

$$-p\mathbf{D} + \mathbf{R}^t \mathbf{H} = \mathbf{J}, \qquad p\mathbf{B} + \mathbf{R}\mathbf{E} = 0, \tag{22.1}$$

$$\mathbf{D} = \boldsymbol{\varepsilon}\mathbf{E}, \qquad \mathbf{B} = \boldsymbol{\mu}\mathbf{H}, \tag{22.2}$$

when $m \to 0$. Here, $p = \xi + i\omega$, with $\xi > 0$, and small capitals denote Laplace transforms, which are arrays of *complex*-valued DoFs. If one can prove uniform convergence with respect to ω (which the requirement $\xi > 0$ makes possible), convergence of the solution of (16.1) and (16.2) will ensue, by inverse Laplace transformation. The main problem, therefore, is to compare $\mathbf{E}$, $\mathbf{B}$, $\mathbf{H}$, $\mathbf{D}$, as given by (22.1) and (22.2), with $r_m \mathrm{E}$, $r_m \mathrm{B}$, $r_m \mathrm{H}$, $r_m \mathrm{D}$, where small capitals, again, denote Laplace transforms, but of differential forms this time.

[50] This is an exercise, for which the following hints should suffice. Start from b, piecewise smooth, such that $\mathrm{d}b = 0$, set $\mathbf{b} = r_m b$, get $\overline{\mathrm{B}}$ as above, and aim at finding an upper bound for $|\mathrm{B} - \overline{\mathrm{B}}|$, where B is the proxy of b, over a tetrahedron T. For this, evaluate $\nabla\lambda \cdot \int_{\mathrm{T}} (\mathrm{B} - \overline{\mathrm{B}})$, where λ is an affine function such that $|\nabla\lambda| = 1$. Integrate by parts, remark that $\int_f \lambda n \cdot \overline{\mathrm{B}} = \lambda(x_f)\mathbf{b}_f$, where x_f is the barycenter of f. Taylor-expand $n \cdot \mathrm{B}$ about x_f, hence a bound in $C\gamma_m^4$ for $\int_{\partial\mathrm{T}} \lambda n \cdot (\mathrm{B} - \overline{\mathrm{B}})$, from which stems $|\int_{\mathrm{T}} (\mathrm{B} - \overline{\mathrm{B}})| \leqslant C\gamma_m^4$. Use uniformity to conclude that $|\mathrm{B} - \overline{\mathrm{B}}| \leqslant C\gamma_m$.

The approach is similar to what we did in statics. First establish that

$$p\boldsymbol{\mu}(\mathbf{H} - r_m \mathrm{H}) + \mathbf{R}(\mathbf{E} - r_m \mathrm{E}) = p(r_m \mu - \boldsymbol{\mu} r_m)\mathrm{H}, \tag{22.3}$$

$$-p\boldsymbol{\varepsilon}(\mathbf{E} - r_m \mathrm{E}) + \mathbf{R}^t(\mathbf{H} - r_m \mathrm{H}) = -p(r_m \varepsilon - \boldsymbol{\varepsilon} r_m)\mathrm{E}. \tag{22.4}$$

Then, right-multiply (22.3) (in the sense of (,)) by $(\mathbf{H} - r_m \mathrm{H})^*$ and the complex conjugate of (22.4) by $-(\mathbf{E} - r_m \mathrm{E})$, add. The middle terms (in $\mathbf{R}$ and $\mathbf{R}^t$) cancel out, and energy estimates follow. The similarity between the right-hand sides of (20.3), on the one hand, and (22.3), (22.4), on the other hand, shows that no further consistency requirements emerge. Stability, thanks to $\xi > 0$, holds there if it held in statics. What is a good discrete hodge in statics, therefore, is a good one in transient situations. Let's tentatively promote this remark to the rank of heuristic principle:

As regards discrete constitutive laws, *what makes a convergent scheme for* static *problems will, as a rule, make one for the Maxwell evolution equations.*

At this stage, we may feel more confident about the quality of the toolkit: If the discrete hodges and the meshes are compatible in the sense of (20.6), so that consistency can be achieved, if there is a way to pass from DoFs to fields which binds energy and discrete energy tightly enough for stability (21.3) to hold, then convergence will ensue. So we need the p_m operator. We would need it, anyway, to determine fluxes, e.m.f.'s, etc., at a finer scale than what the mesh provides – motivation enough to search for interpolants, but not the most compelling reason to do so: Field reconstruction from the DoFs is needed, basically, *to assess stability*, in the above sense, and thereby, the validity of the method. Whitney forms, which will now enter the scene, provide this mechanism.

23. Whitney forms

Let's summarize the requirements about the generic map p_m. It should map each kind of DoF array to a differential form of the appropriate kind: $p_m \mathbf{e}$, starting from an edge-based DoF array $\mathbf{e}$, should be a 1-form; $p_m \mathbf{b}$, obtained from a facet-based $\mathbf{b}$, should be a 2-form, and so forth. Properties (21.1) and (21.2) should hold for all kinds, too, so in short,

$$r_m p_m = 1, \quad p_m r_m \to 1 \quad \text{when } m \to 0. \tag{23.1}$$

The stability property (21.3) will automatically be satisfied in the case of a uniform family of meshes. Moreover, we expect $\mathrm{d}b = 0$ when $\mathbf{Db} = 0$, $\mathrm{d}e = 0$ when $\mathbf{Re} = 0$, etc. More generally, $\mathbf{Ra} = \mathbf{b}$ should entail $\mathrm{d}a = b$, and so forth. These are desirable features of the toolkit. The neatest way to secure them is to enforce the structural property

$$\mathrm{d}p_m = p_m \mathbf{d}, \tag{23.2}$$

at all levels (Fig. 23.1): d and **d** should be conjugate, via p_m, or said differently, Fig. 23.1 should be a *commutative diagram*. Remarkably, these prescriptions will prove sufficient to generate interpolants in an essentially unique way. Such interpolants are known as *Whitney forms* (WHITNEY [1957]), and we shall refer to the structure they constitute as the *Whitney complex*.

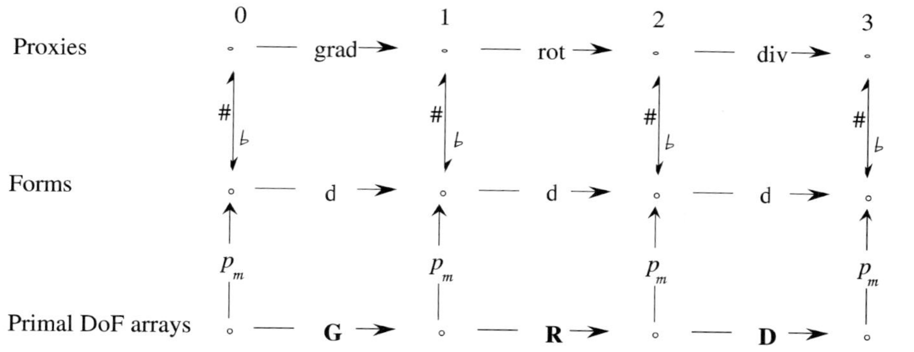

FIG. 23.1. Diagrammatic rendering of (23.2), with part of Fig. 8.1 added. Flat and sharp symbols represent the isomorphism between differential forms and their scalar or vector proxies.

23.1. *Whitney forms as a device to approximate manifolds*

We address the question by taking a detour, to see things from a viewpoint consistent with our earlier definition of differential forms as maps from manifolds to numbers. A differential form, say, for definiteness, b, maps a p-manifold S to the number $\int_S b$, with $p = 2$ here. Suppose we are able to approximate S by a p-chain, i.e., here, a chain based on facets, $p_m^t S = \sum_{f \in \mathcal{F}} \mathbf{c}^f f$. Then a natural approximation to $\int_S b$ is $\int_{p_m^t S} b$. But this number we know, by linearity: since $\int_f r_m b = \mathbf{b}_f$, it equals the sum $\sum_f \mathbf{c}^f \mathbf{b}_f$, that we shall denote $\boldsymbol{\langle} \mathbf{c}; \mathbf{b} \boldsymbol{\rangle}$ (with boldface brackets). Hence an approximate knowledge of the field b, i.e., of all its measurable attributes – the fluxes – from the DoF array $\mathbf{b}$. In particular, fluxes embraced by *small* surfaces (small with respect to the grain of the mesh) will be computable from $\mathbf{b}$, which meets our expectations about interpolating to local values of b. The question has thus become "how best to represent S by a 2-chain?". Fig. 23.2 (where $p = 1$, so a curve c replaces S) gives the idea.

Once we know about the manifold-to-chain map p_m^t, we know about Whitney forms: For instance, the one associated with facet f is, like the field b itself, a map from surfaces to numbers, namely the map $S \to \mathbf{c}^f$ that assigns to S its weight with respect to f. We denote this map by w^f and its value at S by $\int_S w^f$ or by $\langle S; w^f \rangle$ as we have done earlier. (The notational redundancy will prove useful.) Note that $\boldsymbol{\langle} p_m^t S; \mathbf{b} \boldsymbol{\rangle} = \int_S \sum_f \mathbf{b}_f w^f = \int_S p_m \mathbf{b} \equiv \langle S; p_m \mathbf{b} \rangle$, which justifies the "$p_m^t$" notation: A transposition is indeed involved.

23.2. *A generating formula*

Now, let's enter the hard core of it. A simplicial primal mesh will be assumed until further notice. (We shall see later how to lift this restriction.) Results will hold for any spatial dimension n and all simplicial dimensions $p \leqslant n$, but will be stated as if n was 3 and $p = 1$ or 2 (edge and facet elements). So we shall also write proofs, even recursive ones that are supposed to move from p to $p + 1$ (see, e.g., Proposition 23.1), as if p had a specific value (1 or 2), and thereby prefer $\mathbf{R}, \mathbf{D}$, or $\mathbf{R}^t$, $\mathbf{D}^t$, to $\mathbf{d}$ or $\boldsymbol{\partial}$. That the proof has general validity notwithstanding should be obvious each time.

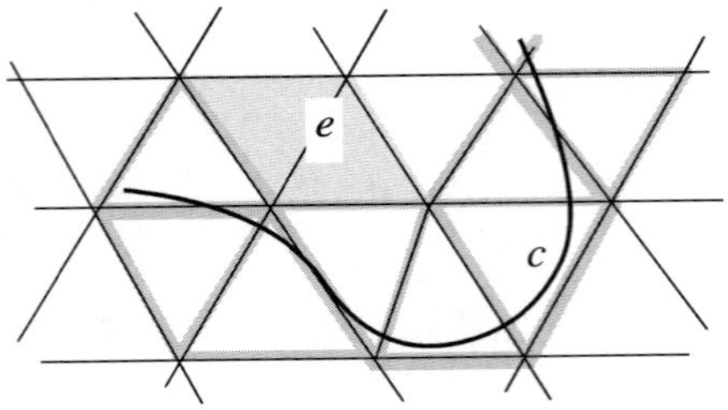

FIG. 23.2. Representing curve c by a weighted sum of mesh-edges, i.e., by a 1-chain. Graded thicknesses of the edges are meant to suggest the respective weights assigned to them. Edges such as e, whose "control domain" (shaded) doesn't intersect c, have zero weight. (A weight can be negative, if the edge is oriented backwards with respect to c.) Which weights thus to assign is the central issue in our approach to Whitney forms.

We use $\lambda^n(x)$ for the barycentric weight of point x with respect to node n, when x belongs to one of the tetrahedra which share node n (otherwise, $\lambda^n(x) = 0$). We'll soon see that $w^n = \lambda^n$ is the natural choice for nodal 0-forms, and again this dual notation will make some formulas more readable. We define $\lambda^e = \lambda^m + \lambda^n$, when edge $e = \{m, n\}$, as well as $\lambda^f = \lambda^l + \lambda^m + \lambda^n$ for facet $f = \{l, m, n\}$, etc. When $e = \{m, n\}$ and $f = \{l, m, n\}$, we denote node l by $f - e$. Thus λ^{f-e} refers to (in that case) λ^l, and equals $\lambda^f - \lambda^e$. The oriented segment from point x to point y is xy, the oriented triangle formed by points x, y, z, in this order, is xyz. And although node n and its location x_n should not be confused, we shall indulge in writing, for instance, ijx for the triangle based on points x_i, x_j, and x, when i and j are node labels.

The weights in the case of a "small manifold", such as a point, a segment, etc.,[51] will now be constructed, and what to use for non-small ones, i.e., the maps w^e, w^f, etc., from lines, surfaces, etc., to reals, will follow by linearity. The principle of this construction is to enforce the following commutative diagram property:

$$\boldsymbol{\partial} p_m^t = p_m^t \partial, \tag{23.3}$$

which implies, by transposition, $\mathrm{d}p_m = p_m \mathbf{d}$, the required structural property (23.2).[52] We shall not endeavor to prove, step by step, that our construction does satisfy (23.3), although that would be an option. Rather, we shall let (23.3) inspire the definition that follows, and then, directly establish that $\mathrm{d}p_m = p_m \mathbf{d}$. This in turn will give (23.3) by transposition.

DEFINITION 23.1. Starting from $w^n = \lambda^n$, the simplicial Whitney forms are

$$w^e = \sum_{n \in \mathcal{N}} \mathbf{G}_e^n \lambda^{e-n} \,\mathrm{d}w^n, \quad w^f = \sum_{e \in \mathcal{E}} \mathbf{R}_f^e \lambda^{f-e} \,\mathrm{d}w^e, \quad w^v = \sum_{f \in \mathcal{F}} \mathbf{D}_v^f \lambda^{v-f} \,\mathrm{d}w^f \tag{23.4}$$

(and so on, recursively, to higher dimensions).

[51] The proper underlying concept, not used here, is that of *multivector* at point x.

[52] If moreover $\ker(\boldsymbol{\partial}_p) = \mathrm{cod}(\boldsymbol{\partial}_{p+1})$, i.e., in the case of a trivial topology, then $\ker(\mathrm{d}_p) = \mathrm{cod}(\mathrm{d}_{p-1})$, just as, by transposition, $\ker(\mathbf{d}_p) = \mathrm{cod}(\mathbf{d}_{p-1})$. One says the Whitney spaces of forms, as linked by the d_p, form an *exact sequence*.

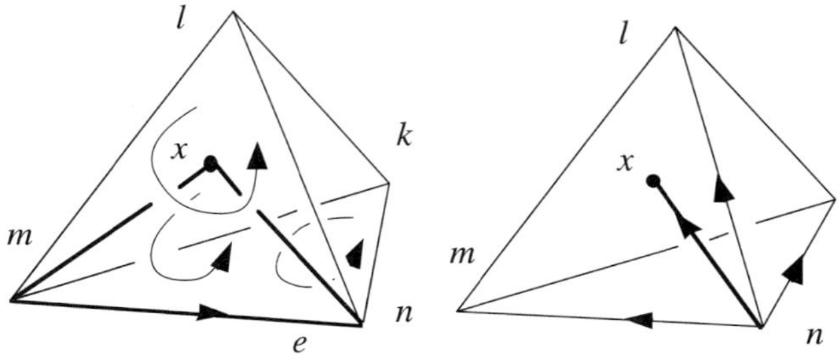

FIG. 23.3. Left: With edge $e = \{m, n\}$ and facets $\{m, n, k\}$ and $\{m, n, l\}$ oriented as shown, the 2-chain to associate with the "join" $x \vee e$, alias mnx, can only be $\lambda^k(x)mnk + \lambda^l(x)mnl$. This is what (23.5) says. Right: Same relation between the join $x \vee n$ and the 1-chain $\lambda^k(x)nk + \lambda^l(x)nl + \lambda^m(x)nm$.

Let us justify this statement, by showing how compliance with (23.3) suggests these formulas. The starting point comes from finite element interpolation theory, which in our present stand consists in expressing a point x as a weighted sum of nodes, the weights $w^n(x)$ being the barycentric ones, $\lambda^n(x)$. (Note how the standard p_m for nodal DoFs, $p_m\varphi = \sum_n \varphi_n w^n$, comes from $p^t_m x = \sum_n w^n(x)n$ by transposition.) Recursively, suppose we know the proper weights for a segment yz, i.e., the bracketed terms in the sum $p^t_m yz = \sum_e \langle yz;\, w^e\rangle e$, and let us try to find $p^t_m xyz$. By linearity, $p^t_m xyz = \sum_e \langle yz;\, w^e\rangle p^t_m(x \vee e)$, where the "join" $x \vee e$ is the triangle displayed in Fig. 23.3, left. So the question is: which 2-chain best represents $x \vee e$? As suggested by Fig. 23.3, the only answer consistent with (23.3) is

$$p^t_m(x \vee e) = \sum_{f \in \mathcal{F}} \mathbf{R}^e_f \lambda^{f-e}(x) f. \tag{23.5}$$

Indeed, this formula expresses $x \vee e$ as the average of mnk and mnl (the only two facets f for which $\mathbf{R}^e_f \neq 0$), with weights that depend on the relative proximity of x to them. So $p^t_m xyz = \sum_{e,f} \mathbf{R}^e_f \lambda^{f-e}(x)\langle yz;\, w^e\rangle f \equiv \sum_f \langle xyz;\, w^f\rangle f$, hence

$$\langle xyz;\, w^f\rangle = \sum_e \mathbf{R}^e_f \lambda^{f-e}(x)\langle yz;\, w^e\rangle. \tag{23.6}$$

On the other hand, since a degenerate triangle such as xzx should get zero weights, we expect $0 = \langle xzx;\, w^f\rangle = \sum_e \mathbf{R}^e_f \lambda^{f-e}(x)\langle zx;\, w^e\rangle$, and the same for $\langle xxy;\, w^f\rangle$. From this (which will come out true after Proposition 23.1 below), we get

$$\begin{aligned}\langle xyz;\, w^f\rangle &= \sum_e \mathbf{R}^e_f \lambda^{f-e}(x)\langle yz + zx + xy;\, w^e\rangle \\ &= \sum_e \mathbf{R}^e_f \lambda^{f-e}(x)\big\langle \partial(xyz);\, w^e\big\rangle = \sum_e \mathbf{R}^e_f\, \lambda^{f-e}(x)\langle xyz;\, \mathrm{d}w^e\rangle\end{aligned}$$

for any small triangle xyz, by Stokes, and hence $w^f = \sum_e \mathbf{R}^e_f \lambda^{f-e}\, \mathrm{d}w^e$.

Thus, formulas (23.4) – which one should conceive as the unfolding of a unique formula – are forced on us, as soon as we accept (23.5) as the right way, amply suggested by Fig. 23.3, to pass from the weights for a simplex s to those for the join $x \vee s$. The reader will easily check that (23.4) describes the Whitney forms as they are more widely known, that is, on a tetrahedron $\{k, l, m, n\}$,

$$w^n = \lambda^n$$

for node n,

$$w^e = \lambda^m \,\mathrm{d}\lambda^n - \lambda^n \,\mathrm{d}\lambda^m$$

for edge $e = \{m, n\}$,

$$w^f = 2(\lambda^l \,\mathrm{d}\lambda^m \wedge \mathrm{d}\lambda^n + \lambda^m \,\mathrm{d}\lambda^n \wedge \mathrm{d}\lambda^l + \lambda^n \,\mathrm{d}\lambda^l \wedge \mathrm{d}\lambda^m)$$

for facet $f = \{l, m, n\}$, and

$$w^v = 6(\lambda^k \,\mathrm{d}\lambda^l \wedge \mathrm{d}\lambda^m \wedge \mathrm{d}\lambda^n + \lambda^l \,\mathrm{d}\lambda^m \wedge \mathrm{d}\lambda^n \wedge \mathrm{d}\lambda^k + \lambda^m \,\mathrm{d}\lambda^n \wedge \mathrm{d}\lambda^k \wedge \mathrm{d}\lambda^l \\ + \lambda^n \,\mathrm{d}\lambda^k \wedge \mathrm{d}\lambda^l \wedge \mathrm{d}\lambda^m)$$

for volume $v = \{k, l, m, n\}$. In higher dimensions (WHITNEY [1957]), the Whitney form of a p-simplex $s = \{n_0, n_1, \ldots, n_p\}$, with inner orientation implied by the order of the nodes, is

$$w^s = p! \sum_{i=0,\ldots,p} (-1)^i w^{n_i} \,\mathrm{d}w^{n_0} \wedge \cdots \langle i \rangle \cdots \wedge \mathrm{d}w^{n_p},$$

where the $\langle i \rangle$ means "omit the term $\mathrm{d}w^{n_i}$".

From now on, we denote by W^p the finite-dimensional subspaces of $\mathcal{F}^p$ generated by these basic forms.

REMARK 23.1. To find the vector proxies of w^e and w^f, substitute ∇ and $\times$ to d and $\wedge$. The scalar proxy of w^v is simply the function equal to $1/\operatorname{vol}(v)$ on v, 0 elsewhere. The reader is invited to establish the following formulas:

$$w^{mn}(x) = (kl \times kx)/6 \operatorname{vol}(klmn), \qquad w^{mnk}(x) = xl/3 \operatorname{vol}(v),$$

very useful when it comes to actual coding. (Other handy formulas, at this stage, are $\operatorname{rot}(x \to v \times ox) = 2v$ and $\operatorname{div}(x \to ox) = 3$, where o is some origin point and v a fixed vector. As an exercise, one may use this to check on Proposition 23.3 below.)

REMARK 23.2. One may recognize in (23.6) the development of the 3×3 determinant of the array of barycentric coordinates of points x, y, z, with respect to nodes l, m, n, hence the geometrical interpretation of the weights displayed in Fig. 23.4.

23.3. *Properties of Whitney forms*

Thus in possession of a rationale for (23.4), we now derive from it a few formulas, for their own sake and as a preparation for the proof of the all important $\mathrm{d}p_m = p_m\mathbf{d}$ result, Proposition 23.3 below.

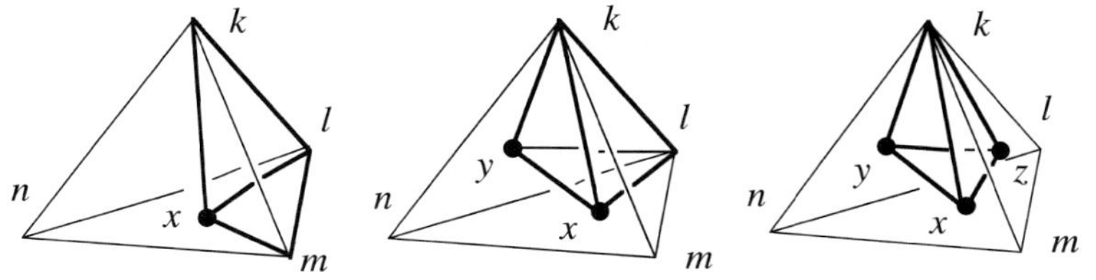

FIG. 23.4. Just as the barycentric weight of point x with respect to node n is $\mathrm{vol}(klmx)$, if one takes $\mathrm{vol}(klmn)$ as unit, the weight of the segment xy with respect to edge $\{m, n\}$ is $\mathrm{vol}(klxy)$, and the weight of the triangle xyz with respect to facet $\{l, m, n\}$ is $\mathrm{vol}(kxyz)$.

PROPOSITION 23.1. *For each p-simplex, there is one linear relation between Whitney forms associated with $(p-1)$-faces of this simplex. For instance, for each f,*

$$\sum_{e \in \mathcal{E}} \mathbf{R}^e_f \lambda^{f-e} w^e = 0.$$

PROOF. By (23.4), $\sum_e \mathbf{R}^e_f \lambda^{f-e} w^e = \sum_{e,n} \lambda^{f-e} \lambda^{e-n} \mathbf{R}^e_f \mathbf{G}^n_e w^n = 0$, thanks to the relation $\mathbf{RG} = 0$, because $\lambda^{f-e}\lambda^{e-n}$, which is the same for all e in ∂f, can be factored out. □

As a corollary, and by using $\mathrm{d}(\lambda\omega) = \mathrm{d}\lambda \wedge \omega + \lambda\, \mathrm{d}\omega$, we have

$$w^f = -\sum_{e \in \mathcal{E}} \mathbf{R}^e_f \, \mathrm{d}\lambda^{f-e} \wedge w^e,$$

and other similar alternatives to (23.4).

PROPOSITION 23.2. *For each p-simplex s, one has*

$$\text{(i)} \quad \lambda^s \, \mathrm{d}w^s = (p+1)\, \mathrm{d}\lambda^s \wedge w^s, \qquad \text{(ii)} \quad \mathrm{d}\lambda^s \wedge \mathrm{d}w^s = 0. \tag{23.7}$$

PROOF. This is true for $p = 0$. Assume it for $p = 1$. Then

$$\mathrm{d}w^f = \sum_e \mathbf{R}^e_f \, \mathrm{d}\lambda^{f-e} \wedge \mathrm{d}w^e = \sum_e \mathbf{R}^e_f \, \mathrm{d}\lambda^f \wedge \mathrm{d}w^e \equiv \mathrm{d}\lambda^f \wedge \sum_e \mathbf{R}^e_f \, \mathrm{d}w^e$$

by (ii), hence $\mathrm{d}\lambda^f \wedge \mathrm{d}w^f = 0$. Next,

$$\begin{aligned} \lambda^f \, \mathrm{d}w^f &= \lambda^f \Big(\sum_e \mathbf{R}^e_f \, \mathrm{d}\lambda^f \wedge \mathrm{d}w^e \Big) = \mathrm{d}\lambda^f \wedge \Big(\sum_e \mathbf{R}^e_f \lambda^f \, \mathrm{d}w^e \Big) \\ &= \mathrm{d}\lambda^f \wedge \Big(w^f + \sum_e \mathbf{R}^e_f \lambda^e \, \mathrm{d}w^e \Big), \end{aligned}$$

which thanks to (i) equals

$$\begin{aligned} \mathrm{d}\lambda^f \wedge \Big(w^f + 2 \sum_e \mathbf{R}^e_f \, \mathrm{d}\lambda^e \wedge w^e \Big) &= \mathrm{d}\lambda^f \wedge w^f - 2\mathrm{d}\lambda^f \wedge \sum_e \mathbf{R}^e_f \, \mathrm{d}\lambda^{f-e} \wedge w^e \\ &= 3\mathrm{d}\lambda^f \wedge w^f, \end{aligned}$$

which proves (i) for $p = 2$. Hence (ii) for $p = 2$ by taking the d. □

Next, yet another variant of (23.4), but without summation this time. For any edge e such that $\mathbf{R}^e_f \neq 0$, one has

$$\mathbf{R}^e_f w^f = \lambda^{f-e}\,\mathrm{d}w^e - 2\,\mathrm{d}\lambda^{f-e} \wedge w^e. \tag{23.8}$$

This is proved by recursion, using $\mathbf{G}^n_{e'} w^{e'} = \lambda^{e'-n}\,\mathrm{d}w^n - \mathrm{d}\lambda^{e'-n} w^n$, where $n = e \cap e'$, and the identity $\mathbf{G}^n_{e'}\mathbf{G}^n_e = -\mathbf{R}^{e'}_f \mathbf{R}^e_f$. We may now conclude with the main result about structural properties (cf. Fig. 23.1):

PROPOSITION 23.3. *One has*

$$\mathrm{d}w^e = \sum_{f \in \mathcal{F}} \mathbf{R}^e_f w^f,$$

and hence, by linearity, $\mathrm{d}p_m = p_m \mathbf{d}$.

PROOF. Since both sides vanish out of the "star" of e, i.e., the union $\mathrm{st}(e)$ of volumes containing it, one may do as if $\mathrm{st}(e)$ were the whole meshed region. Note that $\sum_f \mathbf{R}^e_f \lambda^f = 1 - \lambda^e$ on $\mathrm{st}(e)$. Then,

$$\sum_f \mathbf{R}^e_f w^f = \sum_f [\lambda^{f-e}\,\mathrm{d}w^e - 2\,\mathrm{d}\lambda^{f-e} \wedge w^e] = (1-\lambda^e)\,\mathrm{d}w^e - 2\,\mathrm{d}(1-\lambda^e) \wedge w^e$$
$$= (1-\lambda^e)\,\mathrm{d}w^e + \lambda^e \wedge \mathrm{d}w^e \equiv \mathrm{d}w^e,$$

by using (i). Now, $\mathrm{d}(p_m \mathbf{a}) = \mathrm{d}(\sum_e \mathbf{a}_e w^e) = \sum_{e,f} \mathbf{R}^e_f \mathbf{a}_e w^f = \sum_f (\mathbf{Ra})_f w^f = p_m(\mathbf{da})$. □

As a corollary, $\mathrm{d}W^{p-1} \subset W^p$, and if $\ker(\mathbf{d}_p) = \mathrm{cod}(\mathbf{d}_{p-1})$, then $\ker(\mathrm{d}; W^p) = \mathrm{d}W^{p-1}$, the *exact sequence* property of Whitney spaces in case of trivial topology.

23.4. "Partition of unity"

For what comes now, we revert to the standard vector analysis framework, where w^f denotes the proxy vector field (i.e., $2(\lambda^l \nabla\lambda^m \times \nabla\lambda^n + \cdots)$) of the Whitney form w^f.

Recall that barycentric functions sum to 1, thus forming a "partition of unity": $\sum_{n \in \mathcal{N}} w^n = 1$. We shall drop the ugly arrows in what follows, and use symbol f not only as a label, but also for the vectorial area of f (Fig. 20.2). Same dual use of $\tilde{f}$. Same convention for xyz, to be understood as a triangle or as its vectorial area, according to the context.

PROPOSITION 23.4. *At all points x, for all vectors v,*

$$\sum_{f \in \mathcal{F}} \big(w^f(x) \cdot v\big) f = v. \tag{23.9}$$

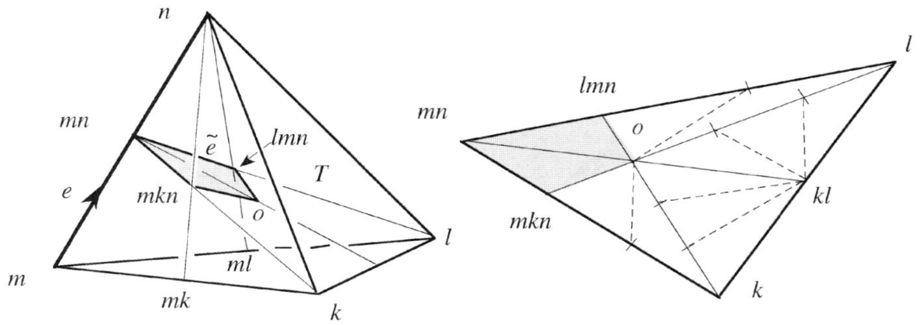

FIG. 23.5. Why $\int_T w^e = \tilde{e}$ in the barycentric construction of the dual mesh. First, the length of the altitude from n is $1/|\nabla w^n|$, therefore $\int_T \nabla w^n = klm/3$. Next, the average of w^n or w^m is 1/4. So $\int_T w^e \equiv \int_T [w^m \nabla w^n - w^n \nabla w^m]$ is a vector equal to $(klm/3 + kln/3)/4$. As the figure shows (all twelve triangles on the right have the same area), this is precisely the vectorial area of $\tilde{e}$.

This is a case of something true of all simplices, and a consequence of the above construction in which the weights $\langle xyz;\, w^f(x)\rangle$ were assigned in order to have $xyz = \sum_f \langle xyz;\, w^f(x)\rangle\, f$. Replacing there w^f by its proxy, and xyz and f by their vectorial areas, we do find (23.9). As a corollary (replace f by g, v by $\nu w^f(x)$, and integrate in x), the entries $\boldsymbol{\nu}^{fg}$ of the Galerkin facet elements mass matrix satisfy

$$\sum_{g \in \mathcal{F}} \boldsymbol{\nu}^{fg} g = \nu \tilde{f},$$

where $\nu \tilde{f}$ is as explained on Fig. 20.2, but with the important specification that here, we are dealing with the *barycentric* dual mesh. That $\int \nu w^f = \nu \tilde{f}$ is an exercise in elementary geometry, and a similar formula holds for all Whitney forms (Fig. 23.5). Now, compare this with (20.6), the compatibility condition that was brought to light by the convergence analysis: We have proved, at last, that the Galerkin hodges do satisfy it.

24. Higher-degree forms

Let's sum up: Whitney forms were built in such a way that the partition of unity property (23.9) ensues. This property makes the mass matrix $\boldsymbol{\nu}$ of facet elements satisfy, with respect to the mesh and its barycentric dual, a compatibility criterion, (20.6), which we earlier recognized as a requisite for consistency. Therefore, we may assert that *Whitney forms of higher polynomial degree*, too, *should satisfy* (23.9), and take this as heuristic guide in the derivation of such forms.

Being a priori more numerous, higher-degree forms will make a finer partition. But we have a way to refine the partition (23.9): Multiply it by the λ^ns, which themselves form a partition of unity. This results in

$$\sum_{f \in \mathcal{F},\, n \in \mathcal{N}} \big(\lambda^n w^f(x) \cdot v\big) f = v,$$

hence the recipe: Attach to edges, facets, etc., the products $\lambda^n w^e$, $\lambda^n w^f$, etc., where n spans $\mathcal{N}$. Instead of the usual Whitney spaces W^p, with forms of polynomial degree

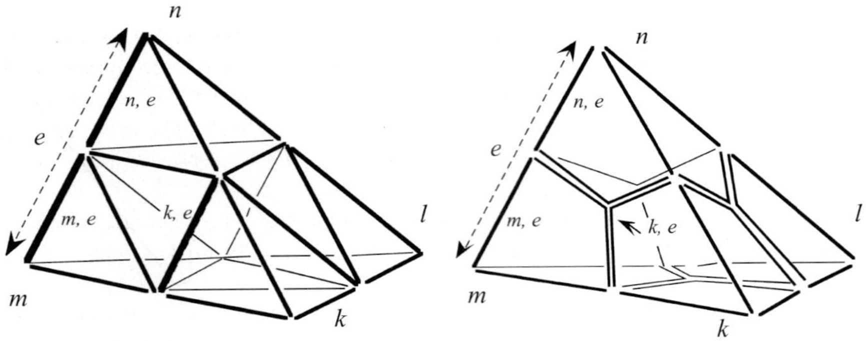

FIG. 24.1. Left: "Small" edges, in one-to-one correspondence with the forms $\lambda^n w^e$, and how they are labelled. Right: A variant where some small edges, such as $\{k, e\}$, are broken lines. These three crooked small edges, with proper signs, add up to the null chain, hence the compatibility condition of Note 53 is built in.

1 at most, we thus obtain larger spaces W_2^p, with forms of polynomial degree 2 at most. (For consistency, W^p may now be denoted W_1^p.) As we shall prove in a moment (under the assumption of trivial topology, but this is no serious restriction), the complex they constitute enjoys the exact sequence property: If for instance $b = \sum_{n,f} \mathbf{b}_{nf} \lambda^n w^f$ satisfies $\mathrm{d}b = 0$ (which means it has a divergence-free proxy) then there are DoFs $\mathbf{a}_{ne}$ such that $b = \mathrm{d}(\sum_{n,e} \mathbf{a}_{ne} \lambda^n w^e)$. (How to define W_k^p, for polynomial degrees $k = 3, \ldots,$ should now be obvious.)

Note however that, because of Proposition 23.1, these new forms are not linearly independent. For instance, the span of the $\lambda^n w^e$s, over a tetrahedron, has dimension 20 instead of the apparent 24, because Proposition 23.1 imposes one linear relation per facet. Over the whole mesh, with N nodes, E edges, F facets, the two products $\lambda^m w^e$ and $\lambda^n w^e$ for each edge $e = \{m, n\}$, and the three products $\lambda^{f-e} w^e$ for each facet f, make a total of $2E + 3F$ generators for W_2^1. But with one relation per facet, the dimension of W_2^1 is only $2(E + F)$. (The spans of the $\lambda^n w^n$s, the $\lambda^n w^f$s, and the $\lambda^n w^v$s, have respective dimensions $N + E$, $3(F + V)$, and $4V$. The general formula is $\dim(W_2^p) = (p + 1)(S_p + S_{p+1})$, where S_p is the number of p-simplices. Note that $\sum_p (-1)^p \dim(W_2^p) = \sum_p (-1)^p S_p \equiv \chi$, the Euler–Poincaré constant of the meshed domain.)

Owing to this redundancy, the main problem with these forms is, how to interpret the DoFs. With standard edge elements, the DoF $\mathbf{a}_{e'}$ is the integral of the 1-form $a = \sum_e \mathbf{a}_e w^e$ over edge e'. In different words, the square matrix of the circulations $\langle e'; w^e \rangle$ is the identity matrix: edges and edge elements are *in duality* in this precise sense (just like the basis vectors and covectors ∂_i and d^j of Note 26). Here, we cannot expect to find a family of 1-chains in such duality with the $\lambda^n w^e$s. The most likely candidates in this respect, the "small edges" denoted $\{n, e\}$, etc., on Fig. 24.1, left, don't pass, because the matrix of the $\langle \{n', e'\}; \lambda^n w^e \rangle$ is not the identity matrix. If at least this matrix was regular, finding chains in duality with the basis forms, or the other way round, would be straightforward. But regular it is not, because of the relations of Proposition 23.1. We might just omit one small edge out of three on each facet, but this is an ugly solution. Better to reason in terms of *blocks* of DoF of various dimensions, and to be content with a rearrangement of chains that makes the matrix block-diagonal: Blocks of size 1 for small edges which are part of the "large" ones, blocks of size three for small edges inside the facets. Each of these 3-blocks corresponds to a subspace of dimension *two*, owing to Proposition 23.1, be it the subspace of forms or of chains. The triple of degrees of freedom, therefore, is up to an additive constant. Yet, the circulations[53] do determine the *form*, if not the DoF, uniquely ("unisolvence" property).

The reader will easily guess about "small facets" (16 of them on a single tetrahedron, for a space of dimension $3(F+T)=3(4+1)=15$) and "small volumes" (four), in both variants.

Which leaves us with the task of proving the exact sequence property, that is to say, the validity of Poincaré's Lemma in the complex of the W_2^p: Show that $\mathrm{d}b=0$ for $b\in W_2^p$ implies the existence, locally at least, of $a\in W_2^{p-1}$ such that $b=\mathrm{d}a$. We'll treat the very case this notation suggests, i.e., $p=2$, and assume trivial topology ("contractible" meshed domain), which does no harm since only a local result is aimed at. We use rot and div rather than d for more clarity. First, two technical points:

LEMMA 24.1. *If* $\sum_{n\in\mathcal{N}}\beta_n\lambda^n(x)=\beta_0$ *for all* x, *where the* β*s are real numbers, then* $\beta_n=\beta_0$ *for all nodes* $n\in\mathcal{N}$.

PROOF. Clear, since $\sum_n\lambda^n=1$ is the only relation linking the $\lambda^n(x)$s. □

LEMMA 24.2. *If* $a\in W^1$, *then* $2\,\mathrm{rot}(\lambda^n a)-3\lambda^n\,\mathrm{rot}\,a\in W^2$.

PROOF. If $a=w^e$ and $n=f-e$, this results from (23.8). If n is one of the end points of e, e.g., $e=\{m,n\}$, a direct computation, inelegant as it may be, will do: $2\,\mathrm{d}\lambda^n\wedge(\lambda^m\,\mathrm{d}\lambda^n-\lambda^n\,\mathrm{d}\lambda^m)=-2\lambda^n\,\mathrm{d}\lambda^n\wedge\mathrm{d}\lambda^m=\lambda^n\,\mathrm{d}w^e$. □

Now,

PROPOSITION 24.1. *If the* W_1^p *sequence is exact, the* W_2^p *sequence is exact.*

PROOF (at level $p=2$). Suppose $b=b_0+\sum_{n\in\mathcal{N}}\lambda^n b_n$, with b_0 and all the b_n in W^2, and $\mathrm{div}\,b=0$. Taking the divergence of the sum and applying Lemma 24.1 in each volume, one sees that $\mathrm{div}\,b_n$ is the same field for all n. So there is some common $\bar{b}$ in W^2 such that $\mathrm{div}(b_n-\bar{b})=0$ for all n, and since the W^p complex is exact, there is an a_n in W^1 such that $b_n=\bar{b}+\mathrm{rot}\,a_n$. Hence, $b=b_0+\bar{b}+\sum_n\lambda^n\,\mathrm{rot}\,a_n$. By Lemma 24.2, there is therefore some $\hat{b}$ in W^2 such that $b=\hat{b}+\frac{2}{3}\mathrm{rot}(\sum_n\lambda^n a_n)$. Since $\mathrm{div}\,\hat{b}=0$, the solenoidal b in W_2^2 we started from is indeed the curl of some element of W_2^1. □

Very little is needed to phrase the proof in such a way that the contractibility assumption becomes moot. Actually, the complexes W_1^p and W_2^p have *the same cohomology*,

[53] Since the matrix has no maximal rank, small-edge circulations must satisfy compatibility conditions for the form to exist. (Indeed, one will easily check that any element of W_2^1 has a null circulation along the chain made by the boundary of a facet minus four times the boundary of the small facet inside it.) This raises a minor problem with the r_m map, whose images need not satisfy this condition. The problem is avoided with a slightly different definition of the small edges (KAMEARI [1999]), as suggested on the right of Fig. 24.1.

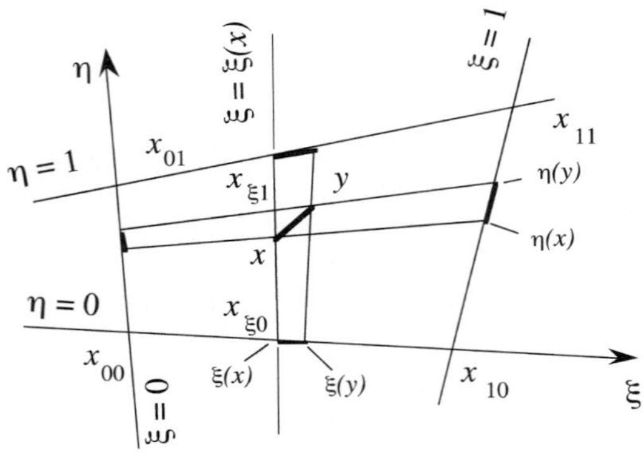

FIG. 25.1. The system of projections, in dimension 2.

whatever the topology of the domain and the culling of passive simplices (i.e., those bearing a null DoF) implied by the boundary conditions.

25. Whitney forms for other shapes than simplices

This simple idea, *approximate p-manifolds by p-chains based on p-cells of the mesh*, is highly productive, as we presently see.

25.1. Hexahedra

First example, the well-known isoparametric element (ERGATOUDIS, IRONS and ZIENKIEWICZ [1968]) on hexahedra can thus be understood. A 2D explanation (Fig. 25.1) will suffice, the generalization being easy. Let us take a convex quadrangle based on points $x_{00}, x_{10}, x_{01}, x_{11}$, and wonder about which weights $w^n(x)$ should be assigned to them (label n designates the generic node) in order to have $x = \sum_{n \in \{00,10,10,11\}} w^n(x) x_n$ in a sensible way. The weights are obvious if x lies on the boundary. For instance, if $x = (1-\xi)x_{00} + \xi x_{10}$, a point we shall denote by $x_{\xi 0}$, weights are $\{1-\xi, \xi, 0, 0\}$. Were it $x \equiv x_{\xi 1} = (1-\xi)x_{01} + \xi x_{11}$, we would take $\{0, 0, 1-\xi, \xi\}$. Now, each x is part of some segment $[x_{\xi 0} x_{\xi 1}]$, for a *unique* value $\xi(x)$ of the weight ξ, in which case $x = (1-\eta)x_{\xi 0} + \eta x_{\xi 1}$, for some $\eta = \eta(x)$, hence it seems natural to distribute the previous weights in the same proportion:

$$x = \big(1-\eta(x)\big)\big(1-\xi(x)\big)x_{00} + \big(1-\eta(x)\big)\xi(x)x_{10} + \eta(x)\big(1-\xi(x)\big)x_{01} + \eta(x)\xi(x)\,x_{11}, \tag{25.1}$$

and we are staring at the basis functions. They form, obviously, a partition of unity.

Looking at what we have done, and generalizing to dimension 3 or higher, we notice a *system of projections*, associated with a trilinear [54] *chart*, $x \to \{\xi(x), \eta(x), \zeta(x)\}$, from

[54] Thus called because ξ, η, and ζ, though cubic polynomials in terms of the Cartesian coordinates of x, are affine functions of each of them, taken separately.

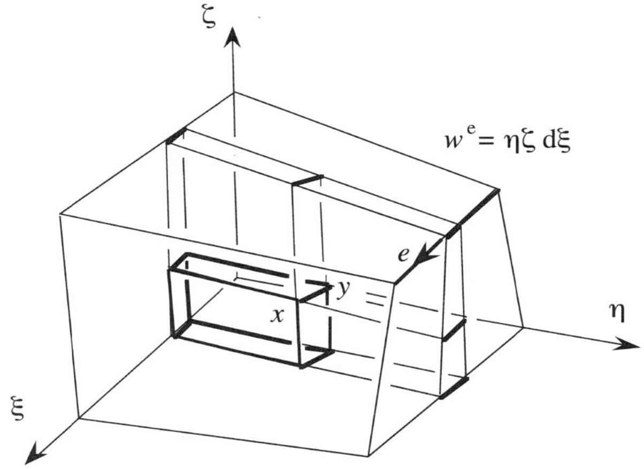

FIG. 25.2. Weight $w^e(xy)$ is the $\xi\eta\zeta$-volume of the "hinder region" of xy with respect to edge e.

a hexahedron to the unit cube in $\xi\eta\zeta$-space. The successive projections (which can be performed in any order) map a point $x \equiv x_{\xi\eta\zeta}$ to its images $x_{0\eta\zeta}$ and $x_{1\eta\zeta}$ on opposite facets[55] $\xi = 0$ and $\xi = 1$, then, recursively, send these images to points on opposite edges, etc., until eventually a node n is reached. In the process, the weight $\langle x;\, w^n\rangle$ of x is recursively determined by formulas such as (assuming for the sake of the example that n belongs to the facet $\xi = 0$)

$$\langle x_{\xi\eta\zeta};\, w^n\rangle = (1 - \xi)\langle x_{0\eta\zeta};\, w^n\rangle.$$

The final weight of x with respect to n is thus the product of factors, such as here $(1-\xi)$, collected during the projection process. (They measure the relative proximity of each projection to the face towards which next projection will be done.) The last factor in this product is 1, obtained when the projection reaches n. Observe the fact, essential of course, that whatever the sequence of projections, the partial weights encountered along the way are the same, only differently ordered, and hence the weight of x with respect to node n is a well-defined quantity.

The viewpoint thus adopted makes the next move obvious. Now, instead of a point x, we deal with a vector v at x, small enough for the segment xy (where $y = x + v$) to be contained in a single hexahedron. The above projections send x and y to facets, edges, etc. Ending the downward recursion one step higher than previously, at the level of edges, we get projections $x_e y_e$ of xy onto all edges e. The weight $\langle xy;\, w^e\rangle$ is the product of weights of x collected along the way, but the last factor is now the algebraic ratio $x_e y_e/e$ (which makes obvious sense) instead of 1. Hence the analytical expression of the corresponding Whitney form, for instance, in the case of Fig. 25.2, $w^e = \eta\zeta\, \mathrm{d}\xi$. (Notice the built-in "partition of unity" property: $xy = \Sigma_e \langle xy;\, w^e\rangle e$.) The proxies, $w^e = \eta\zeta\nabla\xi$ in this example, were proposed as edge elements for hexahedra by VAN WELIJ [1985].

[55] Be aware that p-faces need not be "flat", i.e., lie within an affine p-subspace for $p > 1$, in dimension higher than 2. To avoid problems this would raise, we assume here a mesh generation which enforces this extra requirement.

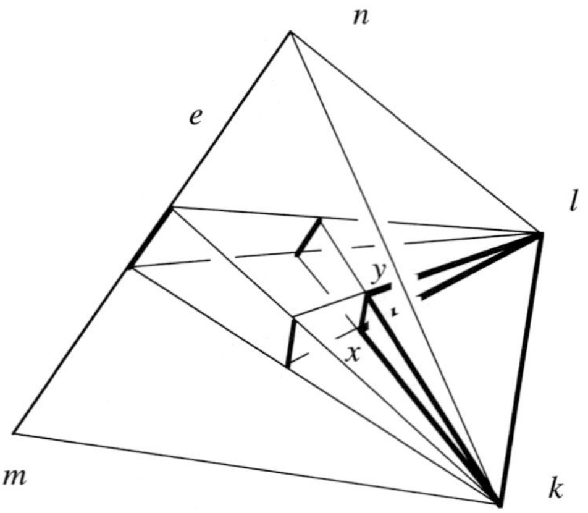

FIG. 25.3. There too, weight $w^e(xy)$ is the relative volume of the hinder region.

One may wonder whether weights such as $\langle xy; w^e \rangle$ have a geometric interpretation there too. They do: $\langle xy; w^e \rangle$ is the relative volume, in the *reference hexahedron*[56] $H = \{\xi, \eta, \zeta\}$: $0 \leqslant \xi \leqslant 1$, $0 \leqslant \eta \leqslant 1$, $0 \leqslant \zeta \leqslant 1$, of the "hinder region" of Fig. 25.2, made of points "behind" xy with respect to edge e. This may seem fairly different from the situation in Fig. 23.4, middle, but a suitable reinterpretation of the system of projections in the tetrahedron (Fig. 25.3) shows the analogy.

A similar reasoning gives facet elements: the last weight, for a small triangle xyz, is $x_f y_f z_f / f$, which again makes sense: Take the ratio of the areas (an affine notion) of the images of these surfaces in the reference cube, with sign + if orientations of $x_f y_f z_f$ and f match, − otherwise. Whitney forms such as $w^f = \xi \, \mathrm{d}\eta \, \mathrm{d}\zeta$ (when f is the facet $\xi = 1$) result. The proxy of that particular one is $\xi \nabla\eta \times \nabla\zeta$.

25.2. *Prisms*

So, Cartesian coordinates and barycentric coordinates provide two systems of projections which make obvious the weight allocation. These systems can be mixed: one of them in use for $p < n$ dimensions, the other one for the $n - p$ remaining dimensions. In dimension 3, this gives only one new possibility, the prism (Fig. 25.4).

Such a variety of shapes makes the mesh generation more flexible (DULAR, HODY, NICOLET, GENON and LEGROS [1994]). Yet, do the elements of a given degree, edge elements say, fit together properly when one mixes tetrahedra, hexahedra, and prisms? Yes, because of the recursivity of the weight allocation: If a segment xy lies entirely in the facet common to two volumes of different kind, say a tetrahedron and a prism, the weights $\langle xy; w^e \rangle$ for edges belonging to this facet only depend on what happens in the facet, i.e., they are the same as evaluated with both formulas for w^e, the one valid in the tetrahedron, the one valid in the prism. This is enough to guarantee the *tangential continuity* of such composite edge elements.

[56] Recall that all tetrahedra are affine equivalent, which is why we had no need for a reference one. The situation is different with hexahedra, which form several orbits under the action of the affine group.

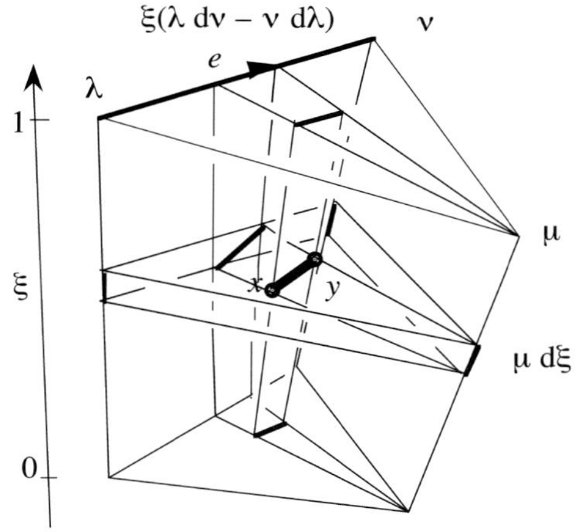

FIG. 25.4. Projective system and edge elements for a prism. Observe the commutativity of the projections.

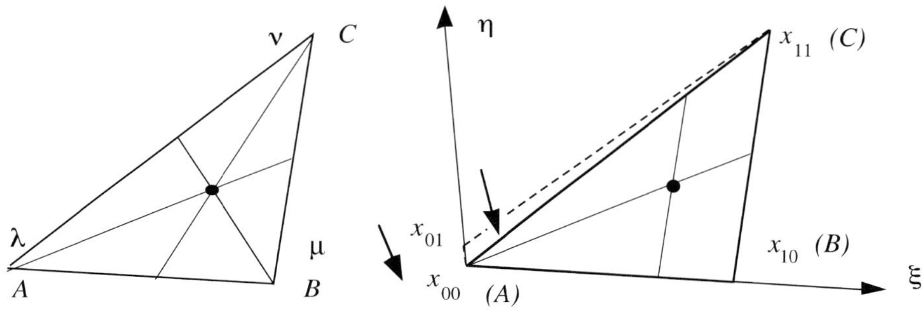

FIG. 25.5. Projective systems for the same triangle, in the barycentric coordinates on the left, and by degeneracy of the quadrilateral system on the right.

25.3. *"Degeneracies"*

Yet one may yearn for even more flexibility, and edge elements for *pyramids* have been proposed (COULOMB, ZGAINSKI and MARÉCHAL [1997], GRADINARU and HIPTMAIR [1999]). A systematic way to proceed, in this respect, is to recourse to "degenerate" versions of the hexahedron or the prism, obtained by fusion of one or more pair of nodes and or edges.

To grasp the idea, let's begin with the case of the degenerated quadrilateral, in two dimensions (Fig. 25.5). With the notations of the figure, where $\{\lambda, \mu, \nu\}$ are the barycentric coordinates in the left triangle, the map $\{\mu, \nu\} \to \{\eta, \xi\}$, where $\eta = \nu/(\mu + \nu)$ and $\xi = \mu + \nu$, sends the interior of the triangle to the interior of the right quadrilateral. When, by deformation of the latter, x_{10} merges with x_{00}, the projective system of the quadrilateral generates a new projective system on the triangle.

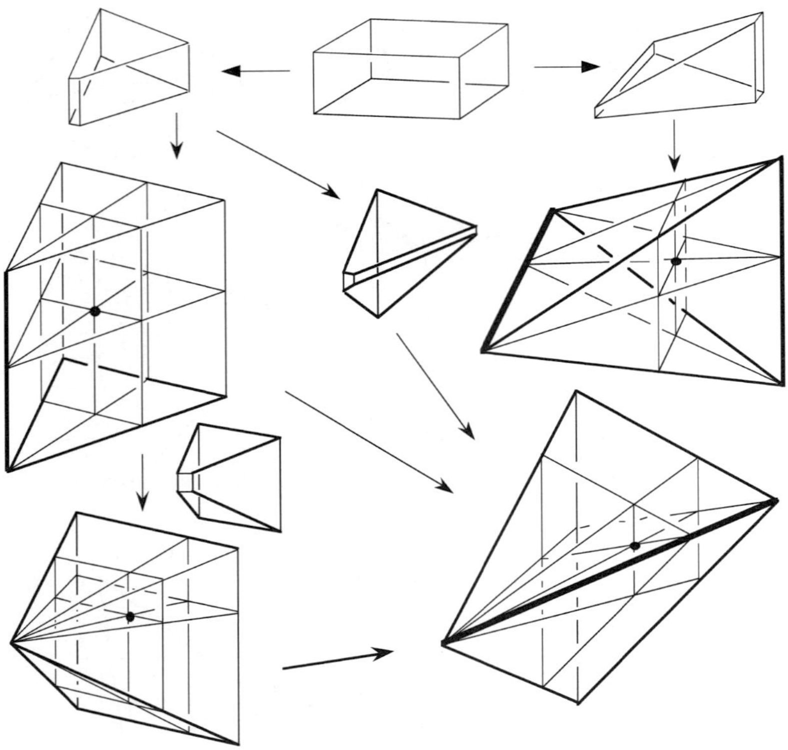

FIG. 25.6. Projective systems in four degenerations of the hexahedron. Thick lines indicate the merged edges.

The weights assigned to the nodes, and hence the nodal elements, are the same in both systems, for $\xi\eta = \nu$ for point C (cf. (25.1)), $\xi(1-\eta) = \mu$ for B, and the sum $(1-\xi)(1-\eta) + (1-\xi)\eta$, attributed to A by adding the loads of x_{00} and x_{01}, does equal λ. But the edge elements differ: For AC, $\eta\,\mathrm{d}\xi \equiv -(1-\lambda)^{-1}\mu\,\mathrm{d}\lambda$ on the right instead of $\lambda\,\mathrm{d}\nu - \nu\,\mathrm{d}\lambda$ on the left, $-(1-\lambda)^{-1}\mu\,\mathrm{d}\lambda$ for AB, and $\mathrm{d}\nu + (1-\lambda)^{-1}\nu\,\mathrm{d}\lambda$ for BC. (The singularity of shape functions at point A is never a problem, because integrals where they appear always converge.)

In dimension 3, the principle is the same: When two edges merge, by degeneration of a hexahedron or of a prism, the Whitney form of the merger is the sum of the Whitney forms of the two contributors, which one may wish to rewrite in a coordinate system adapted to the degenerate solid. Figs. 25.6 and 25.7 show seven degeneracies, all those that one can obtain from a hexahedron or a prism with plane facets under the constraint of not creating curved facets in the process. As one sees, the only novel shape is the pyramid, while the prism is retrieved once and the tetrahedron four times.

But, as was predictible from the 2-dimensional case, it's *new* Whitney forms, on these solids, that are produced by the merging, because the projection systems are different. In particular, we have now *five* distinct projective systems on the tetrahedron (and two on the pyramid and the prism), and the equality of traces is not automatic any longer. One

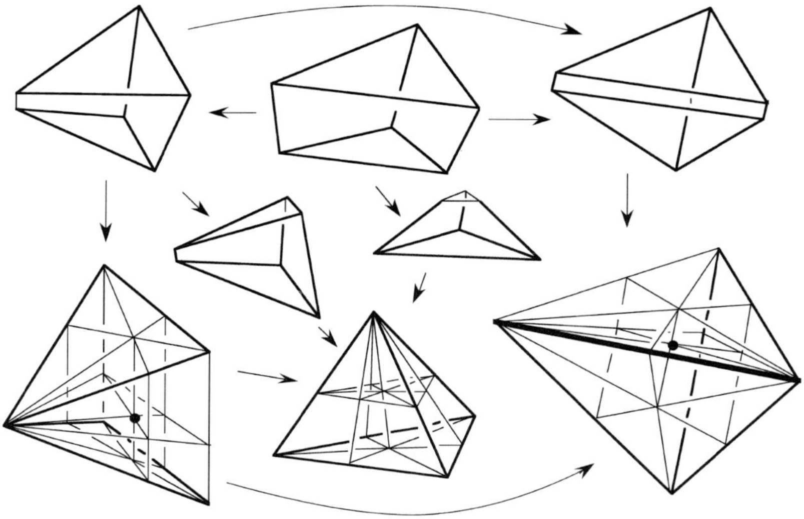

FIG. 25.7. Projective systems in three degenerations of the prism. Note how the pyramid has two ways to degenerate towards the tetrahedron.

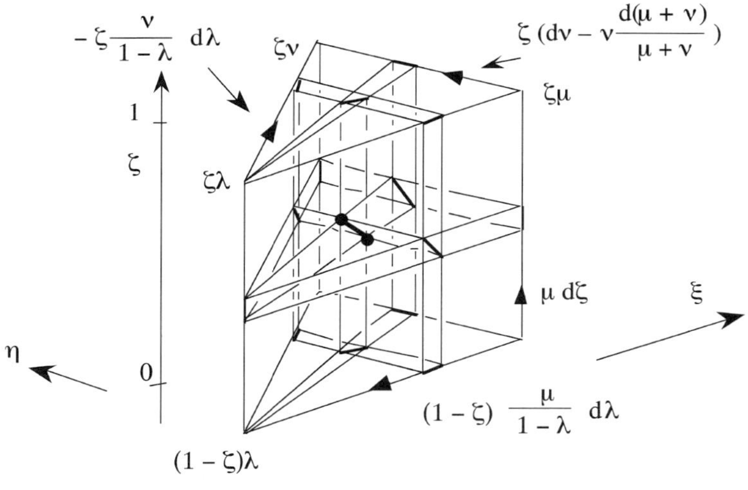

FIG. 25.8. Nodal and edge elements for the projective system of Fig. 25.5. One passes from the previous coordinate system $\{\xi, \eta, \zeta\}$ to the prism-adapted $\{\zeta, \lambda, \mu, \nu\}$ system by the formulas $\xi = \mu + \nu$, $\eta = \nu/(\mu + \nu)$, with $\lambda + \mu + \nu = 1$.

must therefore care about correct assembly, in order to get the same projection system on each facet.

The advantage of having the pyramid available is thus marred by the necessity of an extended shape-functions catalogue (on at least two triangular facets of a pyramid, the projection system cannot match the tetrahedron's one), and by the existence of cum-

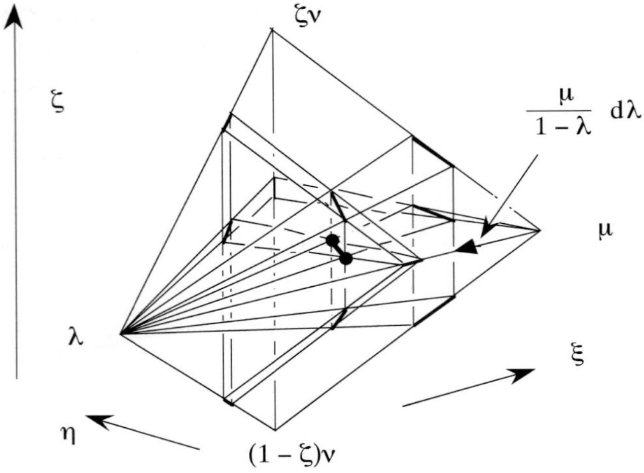

FIG. 25.9. Degeneration of the prism of Fig. 25.8. Two edges disappear, and a new edge element, $\mu(1-\lambda)^{-1}\,\mathrm{d}\lambda$ is created by the merging. The coordinate system is the same here as in Fig. 25.8, so $\{\lambda, \mu, \nu\}$ should not be confused with barycentric coordinates of this tetrahedron. Denoting the latter by $\{\bar\kappa, \bar\lambda, \bar\mu, \bar\nu\}$, and using the formulas $\nu = \bar\nu + \bar\kappa$ and $\zeta = \bar\nu/(\bar\nu + \bar\kappa)$, one has $\xi = \bar\mu + \bar\nu + \bar\kappa = 1 - \bar\lambda$, $\eta = (\bar\nu + \bar\kappa)/(1 - \bar\lambda)$. Thus, for instance, the shape function $\mu(1-\lambda)^{-1}\,\mathrm{d}\lambda$ rewrites as $\bar\mu(1-\bar\lambda)^{-1}\,\mathrm{d}\bar\lambda$ in barycentric coordinates.

bersome assembly rules. Yet, finding the new shape-functions is not too difficult, as exemplified by Figs. 25.8 and 25.9.

25.4. Star-shaped cells, dual cells

Let's end all this by an indication on how to build Whitney forms on any star-shaped polyhedron.

Suppose each p-cell of the mesh m, for all p, has been provided with a "center", in the precise sense of Section 15, i.e., a point with respect to which the cell is star-shaped. Then, join the centers in order to obtain a simplicial refinement, $\overline{m}$ say, where the new sets of p-simplices are $\overline{\mathcal{S}}_p$, the old sets of cells being $\mathcal{S}_p$. In similar style, let $\mathbf{u}$ and $\overline{\mathbf{u}}$ stand for DoF arrays indexed over $\mathcal{S}_p$ and $\overline{\mathcal{S}}_p$ respectively, with the compatibility relation $\mathbf{u}_s = \Sigma_{s'} \pm \overline{\mathbf{u}}_{s'}$ for all s in $\mathcal{S}_p$, the sum running over all small simplices in the refinement of cell s, and the signs taking care of relative orientations. To define $p_m\mathbf{u}$, knowing what $p_{\overline{m}}\overline{\mathbf{u}}$ is, we just take the *smallest*, in the energy norm, of the $p_{\overline{m}}\overline{\mathbf{u}}$'s, with respect to all $\overline{\mathbf{u}}$'s compatible with $\mathbf{u}$.

The family of interpolants thus obtained is to the cellular mesh, for all purposes, what Whitney forms were to a simplicial mesh. Whether they deserve to be called "Whitney forms" is debatable, however, because they are metric-dependent, unlike the standard Whitney forms. The same construction on the dual side provides similar pseudo-Whitney forms on the dual mesh. (More precisely, there is, as we have observed at the end of Section 15, a common simplicial refinement of both m and $\widetilde{m}$. The process just defined constructs forms on both, but it's easy to check that the pseudo-Whitneys on the *primal* mesh are just the Whitney forms.) This fills a drawer in the toolkit, the emptiness of which we took some pain to hide until now, although it was conspicuous at places, on Fig. 23.1, for instance.

References

ALBANESE, R., RUBINACCI, G. (1988). Integral formulations for 3-D eddy-currents computation using edge-elements. *IEE Proc. A* **135**, 457–462.

ARMSTRONG, M.A. (1979). *Basic Topology* (McGraw-Hill, London).

ARNOLD, D.N., BREZZI, F. (1985). Mixed and non-conforming finite element methods: implementation, postprocessing and error estimates. M^2AN **19**, 7–32.

BABUŠKA, I., AZIZ, A.K. (1976). On the angle condition in the finite element method. *SIAM J. Numer. Anal.* **13**, 214–226.

BAEZ, J., MUNIAIN, J.P. (1994). *Gauge Fields, Knots and Gravity* (World Scientific, Singapore).

BALDOMIR, D., HAMMOND, P. (1996). *Geometry of Electromagnetic Systems* (Oxford Univ. Press, Oxford).

BANK, R.E., ROSE, D.J. (1987). Some error estimates for the box method. *SIAM J. Numer. Anal.* **24**, 777–787.

BÄNSCH, E. (1991). Local mesh refinement in 2 and 3 dimensions. *Impact Comput. Sci. Engrg.* **3**, 181–191.

BEY, J. (1995). Tetrahedral grid refinement. *Computing* **55**, 355–378.

BOSSAVIT, A. (1990a). Eddy-currents and forces in deformable conductors. In: Hsieh, R.K.T. (ed.), *Mechanical Modellings of New Electromagnetic Materials: Proc. IUTAM Symp., Stockholm, April 1990* (Elsevier, Amsterdam), pp. 235–242.

BOSSAVIT, A. (1990b). Solving Maxwell's equations in a closed cavity, and the question of spurious modes. *IEEE Trans. Magn.* **26**, 702–705.

BOSSAVIT, A. (1996). A puzzle. *ICS Newsletter* **3** (2), 7; **3** (3), 14; **4** (1) (1997) 17–18.

BOSSAVIT, A. (1998a). *Computational Electromagnetism* (Academic Press, Boston).

BOSSAVIT, A. (1998b). Computational electromagnetism and geometry. *J. Japan Soc. Appl. Electromagn. Mech.* **6**, 17–28, 114–123, 233–240, 318–326; **7** (1999) 150–159, 249–301, 401–408; **8** (2000) 102–109, 203–209, 372–377.

BOSSAVIT, A. (1999). On axial and polar vectors. *ICS Newsletter* **6**, 12–14.

BOSSAVIT, A. (2000). Most general 'non-local' boundary conditions for the Maxwell equations in a bounded region. *COMPEL* **19**, 239–245.

BOSSAVIT, A. (2001a). 'Stiff' problems in eddy-current theory and the regularization of Maxwell's equations. *IEEE Trans. Magn.* **37**, 3542–3545.

BOSSAVIT, A. (2001b). On the notion of anisotropy of constitutive laws: some implications of the 'Hodge implies metric' result. *COMPEL* **20**, 233–239.

BOSSAVIT, A. (2001c). On the representation of differential forms by potentials in dimension 3. In: van Rienen, U., Günther, M., Hecht, D. (eds.), *Scientific Computing in Electrical Engineering* (Springer-Verlag, Berlin), pp. 97–104.

BOSSAVIT, A. (2003). Mixed-hybrid methods in magnetostatics: complementarity in one stroke. *IEEE Trans. Magn.* **39**, 1099–1102.

BOSSAVIT, A., KETTUNEN, L. (1999). Yee-like schemes on a tetrahedral mesh, with diagonal lumping. *Int. J. Numer. Modelling* **12**, 129–142.

BRANIN JR., F.H. (1961). An abstract mathematical basis for network analogies and its significance in physics and engineering. *Matrix and Tensor Quarterly* **12**, 31–49.

BURKE, W.L. (1985). *Applied Differential Geometry* (Cambridge Univ. Press, Cambridge).

DI CARLO, A., TIERO, A. (1991). The geometry of linear heat conduction. In: Schneider, W., Troger, H., Ziegler, F. (eds.), *Trends in Applications of Mathematics to Mechanics* (Longman, Harlow), pp. 281–287.

CARPENTER, C.J. (1977). Comparison of alternative formulations of 3-dimensional magnetic-field and eddy-current problems at power frequencies. *Proc. IEE* **124**, 1026–1034.

CHAVENT, G., ROBERTS, J.E. (1991). A unified presentation of mixed, mixed-hybrid finite elements and standard finite difference approximations for the determination of velocities in waterflow problems. *Adv. Water Resources* **14**, 329–348.

CLEMENS, M., WEILAND, T. (1999). Transient eddy-current calculation with the FI-method. *IEEE Trans. Magn.* **35**, 1163–1166.

COHEN, G., JOLY, P., TORDJMAN, N. (1993). Construction and analysis of higher order elements with mass-lumping for the wave equation. In: *Mathematical Aspects of Wave Propagation Phenomena* (SIAM, Philadelphia), pp. 152–160.

COSTABEL, M., DAUGE, M. (1997). Singularités des équations de Maxwell dans un polyèdre. *C. R. Acad. Sci. Paris I* **324**, 1005–1010.

DE COUGNY, H.L., SHEPHARD, M.S. (1999). Parallel refinement and coarsening of tetrahedral meshes. *Int. J. Numer. Meth. Engng.* **46**, 1101–1125.

COULOMB, J.L., ZGAINSKI, F.X., MARÉCHAL, Y. (1997). A pyramidal element to link hexahedral, prismatic and tetrahedral edge finite elements. *IEEE Trans. Magn.* **33**, 1362–1365.

COURBET, B., CROISILLE, J.P. (1998). Finite volume box schemes on triangular meshes. M^2AN **32**, 631–649.

VAN DANTZIG, D. (1934). The fundamental equations of electromagnetism, independent of metrical geometry. *Proc. Cambridge Phil. Soc.* **30**, 421–427.

VAN DANTZIG, D. (1954). On the geometrical representation of elementary physical objects and the relations between geometry and physics. *Nieuw. Archief vor Wiskunde* **3**, 73–89.

DIRICHLET, G.L. (1850). Über die Reduktion der positiven quadratischen Formen mit drei unbestimmten ganzen Zahlen. *J. Reine Angew. Math.* **40**, 209.

DULAR, P., HODY, J.-Y., NICOLET, A., GENON, A., LEGROS, W. (1994). Mixed finite elements associated with a collection of tetrahedra, hexahedra and prisms. *IEEE Trans. Magn.* **30**, 2980–2983.

EBELING, F., KLATT, R., KRAWCZYK, F., LAWINSKY, E., WEILAND, T., WIPF, S.G., STEFFEN, B., BARTS, T., BROWMAN, M.J., COOPER, R.K., DEAVEN, H., RODENZ, G. (1989). The 3-D MAFIA group of electromagnetic codes. *IEEE Trans. Magn.* **25**, 2962–2964.

ECKMANN, B. (1999). Topology, algebra, analysis – relations and missing links. *Notices AMS* **46**, 520–527.

ELMKIES, A., JOLY, P. (1997). Éléments finis d'arête et condensation de masse pour les équations de Maxwell: le cas de dimension 3. *C. R. Acad. Sci. Paris Sér. I* **325**, 1217–1222.

ERGATOUDIS, J.G., IRONS, B.M., ZIENKIEWICZ, O.C. (1968). Curved, isoparametric, 'quadrilateral' elements for finite element analysis. *Int. J. Solids Struct.* **4**, 31–42.

FIRESTONE, F.A. (1933). A new analogy between mechanical and electrical systems. *J. Acoust. Soc. Am.* **0**, 249–267.

GALLOUET, T., VILA, J.P. (1991). Finite volume element scheme for conservation laws of mixed type. *SIAM J. Numer. Anal.* **28**, 1548–1573.

GELBAUM, B.R., OLMSTED, J.M.H. (1964). *Counterexamples in Analysis* (Holden-Day, San Francisco).

GOLDHABER, A.S., TROWER, W.P. (1990). Resource letter MM-1: Magnetic monopoles. *Am. J. Phys.* **58**, 429–439.

GRADINARU, V., HIPTMAIR, R. (1999). Whitney elements on pyramids. *ETNA* **8**, 154–168.

HALMOS, P.R. (1950). *Measure Theory* (Van Nostrand, Princeton).

HAMOUDA, L., BANDELIER, B., RIOUX-DAMIDAU, F. (2001). Mixed formulation for magnetostatics. In: *Proc. Compumag* (paper PE4–11).

HARRISON, J. (1998). Continuity of the integral as a function of the domain. *J. Geometric Anal.* **8**, 769–795.

HAUGAZEAU, Y., LACOSTE, P. (1993). Condensation de la matrice masse pour les éléments finis mixtes de $H(\mathrm{rot})$. *C. R. Acad. Sci. Paris I* **316**, 509–512.

HEINRICH, B. (1987). *Finite Difference Methods on Irregular Networks* (Akademie-Verlag, Berlin).

HENLE, A. (1994). *A Combinatorial Introduction to Topology* (Dover, New York).

HILTON, P.J., WYLIE, S. (1965). *Homology Theory, An Introduction to Algebraic Topology* (Cambridge Univ. Press, Cambridge).

HUANG, J., XI, S. (1998). On the finite volume element method for general self-adjoint elliptic problems. *SIAM J. Numer. Anal.* **35**, 1762–1774.

HYMAN, J.M., SHASHKOV, M. (1997). Natural discretizations for the divergence, gradient, and curl on logically rectangular grids. *Comput. Math. Appl.* **33**, 81–104.

JÄNICH, K. (2001). *Vector Analysis* (Springer, New York).

KAASSCHIETER, E.F., HUIJBEN, A.J.M. (1992). Mixed-hybrid finite elements and streamline computation for the potential flow problem. *Numer. Meth. PDE* **8**, 221–266.

KAMEARI, A. (1999). Symmetric second order edge elements for triangles and tetrahedra. *IEEE Trans. Magn.* **35**, 1394–1397.

KHEYFETS, A., WHEELER, J.A. (1986). Boundary of a boundary and geometric structure of field theories. *Int. J. Theor. Phys.* **25**, 573–580.

KOENIG, H.E., BLACKWELL, W.A. (1960). Linear graph theory: a fundamental engineering discipline. *IRE Trans. Edu.* **3**, 42–49.

KOTTLER, F. (1922). Maxwell'sche Gleichungen und Metrik. *Sitzungber. Akad. Wien IIa* **131**, 119–146.

LAX, P.D., RICHTMYER, R.D. (1956). Survey of the stability of linear finite difference equations. *Comm. Pure Appl. Math.* **9**, 267–293.

LEE, J.-F., SACKS, Z. (1995). Whitney elements time domain (WETD) methods. *IEEE Trans. Magn.* **31**, 1325–1329.

LEIS, R. (1968). Zur Theorie elektromagnetischer Schwingungen in anisotropen inhomogenen Medien. *Math. Z.* **106**, 213–224.

MADSEN, I., TORNEHAVE, J. (1997). *From Calculus to Cohomology* (Cambridge Univ. Press, Cambridge).

MATTIUSSI, C. (2000). The finite volume, finite element, and finite difference methods as numerical methods for physical field problems. *Adv. Imag. Electron Phys.* **113**, 1–146.

MAUBACH, J.M. (1995). Local bisection refinement for N-simplicial grids generated by reflection. *SIAM J. Sci. Stat.* **16**, 210–227.

MITTRA, R., RAMAHI, O., KHEBIR, A., GORDON, R., KOUKI, A. (1989). A review of absorbing boundary conditions for two and three-dimensional electromagnetic scattering problems. *IEEE Trans. Magn.* **25**, 3034–3038.

MONK, P., SÜLI, E. (1994). A convergence analysis of Yee's scheme on nonuniform grids. *SIAM J. Numer. Anal.* **31**, 393–412.

MOSÉ, R., SIEGEL, P., ACKERER, P., CHAVENT, G. (1994). Application of the mixed hybrid finite element approximation in a groundwater flow model: Luxury or necessity?. *Water Resources Res.* **30**, 3001–3012.

MUNTEANU, I. (2002). Tree-cotree condensation properties. *ICS Newsletter* **9**, 10–14.

NICOLAIDES, R., WANG, D.-Q. (1998). Convergence analysis of a covolume scheme for Maxwell's equations in three dimensions. *Math. Comp.* **67**, 947–963.

POST, E.J. (1972). The constitutive map and some of its ramifications. *Ann. Phys.* **71**, 497–518.

RAPETTI, F., DUBOIS, F., BOSSAVIT, A. (2002). Integer matrix factorization for mesh defect detection. *C. R. Acad. Sci. Paris* **334**, 717–720.

REN, Z. (1996). Autogauging of vector potential by iterative solver – numerical evidence. In: *3d Int. Workshop on Electric and Magnetic Fields, A.I.M. (31 Rue St-Gilles, Liège)*, pp. 119–124.

DE RHAM, G. (1936). Relations entre la topologie et la théorie des intégrales multiples. *L'Enseignement Math.* **35**, 213–228.

DE RHAM, G. (1960). *Variétés différentiables* (Hermann, Paris).

ROSEN, J. (1973). Transformation properties of electromagnetic quantities under space inversion, time reversal, and charge conjugation. *Am. J. Phys.* **41**, 586–588.

RUDIN, W. (1973). *Functional Analysis* (McGraw-Hill, New York).

SCHATZ, A.H., SLOAN, I.H., WAHLBIN, L.B. (1996). Superconvergence in finite element methods and meshes that are locally symmetric with respect to a point. *SIAM J. Numer. Anal.* **33**, 505–521.

SCHOUTEN, J.A. (1989). *Tensor Analysis for Physicists* (Dover, New York).

SCHUTZ, B. (1980). *Geometrical Methods of Mathematical Physics* (Cambridge Univ. Press, Cambridge).

SEIFERT, H., THRELFALL, W. (1980). *A Textbook of Topology* (Academic Press, Orlando) (first German ed., 1934).

SHAW, R., YEADON, F.J. (1989). On $(a \times b) \times c$. *Am. Math. Monthly* **96**, 623–629.

SMYTH, J.B., SMYTH, D.C. (1977). Critique of the paper 'The electromagnetic radiation from a finite antenna'. *Am. J. Phys.* **45**, 581–582.

SORKIN, R. (1975). The electromagnetic field on a simplicial net. *J. Math. Phys.* **16**, 2432–2440.

SÜLI, E. (1991). Convergence of finite volume schemes for Poisson's equation on nonuniform meshes. *SIAM J. Numer. Anal.* **28**, 1419–1430.

TAYLOR, E.F., WHEELER, J.A. (1992). *Spacetime Physics* (Freeman, New York).

TEIXEIRA, F.L., CHEW, W.C. (1999). Lattice electromagnetic theory from a topological viewpoint. *J. Math. Phys.* **40**, 169–187.

TONTI, E. (1996). On the geometrical structure of electromagnetism. In: Ferrarese, G. (ed.), *Gravitation, Electromagnetism and Geometrical Structures* (Pitagora, Bologna), pp. 281–308.

TONTI, E. (2001). A direct formulation of field laws: the cell method. *CMES* **2**, 237–258.

TRAPP, B., MUNTEANU, I., SCHUHMANN, R., WEILAND, T., IOAN, D. (2002). Eigenvalue computation by means of a tree–cotree filtering technique. *IEEE Trans. Magn.* **38**, 445–448.

UMAN, M.A. (1977). Reply to Smyth and Smyth. *Am. J. Phys.* **45**, 582.

VEBLEN, O., WHITEHEAD, J.H.C. (1932). *The Foundations of Differential Geometry* (Cambridge Univ. Press, Cambridge).

WEILAND, T. (1992). Maxwell's grid equations. In: *Proc. URSI Int. Symp. Electromagnetic Theory, Sydney*, pp. 37–39.

WEILAND, T. (1996). Time domain electromagnetic field computation with finite difference methods. *Int. J. Numer. Modelling* **9**, 295–319.

WEILAND, T. (1985). Three dimensional resonator mode computation by finite difference methods. *IEEE Trans. Magn.* **21**, 2340–2343.

WEISER, A., WHEELER, M.F. (1988). On convergence of block-centered finite differences for elliptic problems. *SIAM J. Numer. Anal.* **25**, 351–375.

VAN WELIJ, J.S. (1985). Calculation of eddy currents in terms of H on hexahedra. *IEEE Trans. Magn.* **21**, 2239–2241.

WHITE, D.A., KONING, J.M. (2000). A novel approach for computing solenoidal eigenmodes of the vector Helmholtz equation. CEFC'00 (p. 328 of the "digest of abstracts").

WHITNEY, H. (1957). *Geometric Integration Theory* (Princeton Univ. Press, Princeton).

YEE, K.S. (1966). Numerical solution of initial boundary value problems involving Maxwell's equations in isotropic media. *IEEE Trans. AP* **14**, 302–307.

YOSIDA, K. (1980). *Functional Analysis* (Springer-Verlag, Berlin) (first ed., 1965).

Further reading

DODZIUK, J. (1976). Finite-difference approach to the Hodge theory of harmonic forms. *Amer. J. Math.* **98**, 79–104.

FRANKEL, T. (1997). *The Geometry of Physics, An Introduction* (Cambridge Univ. Press, Cambridge).

HIPTMAIR, R. (2001). Discrete Hodge operators. *Progress in Electromagnetics Research* **32**, 122–150.

KOTIUGA, P.R. (1984). Hodge decompositions and computational electromagnetics. Thesis (Department of Electrical Engng., McGill University, Montréal).

MAXWELL, J.C. (1864). On reciprocal figures and diagrams of forces. *Phil. Mag. Ser.* **4**, 250–261.

VON MISES, R. (1952). On network methods in conformal mapping and in related problems. In: Appl. Math. Series **18** (US Department of Commerce, NBS), pp. 1–6.

MÜLLER, W. (1978). Analytic torsion and R-torsion of Riemannian manifolds. *Adv. Math.* **28**, 233–305.

NEDELEC, J.C. (1980). Mixed finite elements in $\mathbb{R}^3$. *Numer. Math.* **35**, 315–341.

POST, E.J. (1979). Kottler–Cartan–van Dantzig (KCD) and noninertial systems. *Found. Phys.* **9**, 619–640.

POST, E.J. (1984). The metric dependence of four-dimensional formulations of electromagnetism. *J. Math. Phys.* **25**, 612–613.

REN, Z., IDA, N. (2002). High-order elements of complete and incomplete bases in electromagnetic field computation. *IEE Proc. Science, Measurement and Technology* **149**, 147–151.

SILVESTER, P., CHARI, M.V.K. (1970). Finite element solution of saturable magnetic field problems. *IEEE Trans. PAS* **89**, 1642–1651.

TAFLOVE, A. (1995). *Computational Electromagnetics: The Finite-Difference Time-Domain Method* (Artech House, Boston).

XIANG, YOU-QING, ZHOU, KE-DING, LI, LANG-RU (1989). A new network-field model for numerical analysis of electromagnetic field. In: Shunnian, Ding (ed.), *Electromagnetic Fields in Electrical Engineering: Proc. BISEF'88, October 19–21, Beijing* (Pergamon Press, Oxford), pp. 391–398.

YIOULTSIS, T.V., TSIBOUKIS, T.D. (1997). Development and implementation of second and third order vector finite elements. *IEEE Trans. Magn.* **33**, 1812–1815.

Finite-Difference Time-Domain Methods

Susan C. Hagness

University of Wisconsin-Madison, College of Engineering, Department of Electrical and Computer Engineering, 3419 Engineering Hall, 1415 Engineering Drive, Madison, WI 53706, USA
E-mail address: hagness@engr.wisc.edu

Allen Taflove

Northwestern University, Computational Electromagnetics Lab (NUEML), Department of Electrical and Computer Engineering, 2145 Sheridan Road, Evanston, IL 60208-3118, USA
E-mail address: taflove@ece.northwestern.edu

Stephen D. Gedney

University of Kentucky, College of Engineering, Department of Electrical and Computer Engineering, 687C F. Paul Anderson Tower Lexington, KY 40506-0046, USA
E-mail address: gedney@engr.uky.edu

1. Introduction

1.1. Background

Prior to about 1990, the modeling of electromagnetic engineering systems was primarily implemented using solution techniques for the sinusoidal steady-state Maxwell's equations. Before about 1960, the principal approaches in this area involved closed-form and infinite-series analytical solutions, with numerical results from these analyses obtained using mechanical calculators. After 1960, the increasing availability of programmable electronic digital computers permitted such frequency-domain approaches to rise

Numerical Methods in Electromagnetics
Special Volume (W.H.A. Schilders and E.J.W. ter Maten, Guest Editors) of
HANDBOOK OF NUMERICAL ANALYSIS, VOL. XIII
P.G. Ciarlet (Editor)

ISSN 1570-8659
DOI 10.1016/S1570-8659(04)13003-2

markedly in sophistication. Researchers were able to take advantage of the capabilities afforded by powerful new high-level programming languages such as Fortran, rapid random-access storage of large arrays of numbers, and computational speeds orders of magnitude faster than possible with mechanical calculators. In this period, the principal computational approaches for Maxwell's equations included the high-frequency asymptotic methods of KELLER [1962] and KOUYOUMJIAN and PATHAK [1974] and the integral-equation techniques of HARRINGTON [1968].

However, these frequency-domain techniques have difficulties and trades-off. For example, while asymptotic analyses are well suited for modeling the scattering properties of electrically large complex shapes, such analyses have difficulty treating nonmetallic material composition and volumetric complexity of a structure. While integral equation methods can deal with material and structural complexity, their need to construct and solve systems of linear equations limits the electrical size of possible models, especially those requiring detailed treatment of geometric details within a volume, as opposed to just the surface shape.

While significant progress has been made in solving the ultra-large systems of equations generated by frequency-domain integral equations (see, for example, SONG and CHEW [1998]), the capabilities of even the latest such technologies are exhausted by many volumetrically complex structures of engineering interest. This also holds for frequency-domain finite-element techniques, which generate sparse rather than dense matrices. Further, the very difficult incorporation of material and device nonlinearities into frequency-domain solutions of Maxwell's equations poses a significant problem as engineers seek to design active electromagnetic/electronic and electromagnetic/quantum-optical systems such as high-speed digital circuits, microwave and millimeter-wave amplifiers, and lasers.

1.2. Rise of finite-difference time-domain methods

During the 1970s and 1980s, a number of researchers realized the limitations of frequency-domain integral-equation solutions of Maxwell's equations. This led to early explorations of a novel alternative approach: direct time-domain solutions of Maxwell's differential (curl) equations on spatial grids or lattices. The finite-difference time-domain (FDTD) method, introduced by YEE [1966], was the first technique in this class, and has remained the subject of continuous development (see TAFLOVE and HAGNESS [2000]).

There are seven primary reasons for the expansion of interest in FDTD and related computational solution approaches for Maxwell's equations:

(1) FDTD uses no linear algebra. Being a fully explicit computation, FDTD avoids the difficulties with linear algebra that limit the size of frequency-domain integral-equation and finite-element electromagnetics models to generally fewer than 10^6 field unknowns. FDTD models with as many as 10^9 field unknowns have been run. There is no intrinsic upper bound to this number.

(2) FDTD is accurate and robust. The sources of error in FDTD calculations are well understood and can be bounded to permit accurate models for a very large variety of electromagnetic wave interaction problems.

(3) FDTD treats impulsive behavior naturally. Being a time-domain technique, FDTD directly calculates the impulse response of an electromagnetic system. Therefore, a single FDTD simulation can provide either ultrawideband temporal waveforms or the sinusoidal steady-state response at any frequency within the excitation spectrum.
(4) FDTD treats nonlinear behavior naturally. Being a time-domain technique, FDTD directly calculates the nonlinear response of an electromagnetic system.
(5) FDTD is a systematic approach. With FDTD, specifying a new structure to be modeled is reduced to a problem of mesh generation rather than the potentially complex reformulation of an integral equation. For example, FDTD requires no calculation of structure-dependent Green's functions.
(6) Computer memory capacities are increasing rapidly. While this trend positively influences all numerical techniques, it is of particular advantage to FDTD methods which are founded on discretizing space over a volume, and therefore inherently require a large random access memory.
(7) Computer visualization capabilities are increasing rapidly. While this trend positively influences all numerical techniques, it is of particular advantage to FDTD methods which generate time-marched arrays of field quantities suitable for use in color videos to illustrate the field dynamics.

An indication of the expanding level of interest in FDTD Maxwell's equations solvers is the hundreds of papers currently published in this area worldwide each year, as opposed to fewer than ten as recently as 1985 (see SHLAGER and SCHNEIDER [1998]). This expansion continues as engineers and scientists in non-traditional electromagnetics-related areas such as digital systems and integrated optics become aware of the power of such direct solution techniques for Maxwell's equations.

1.3. Characteristics of FDTD and related space-grid time-domain techniques

FDTD and related space-grid time-domain techniques are direct solution methods for Maxwell's curl equations. These methods employ no potentials. Rather, they are based upon volumetric sampling of the unknown electric and magnetic fields within and surrounding the structure of interest, and over a period of time. The sampling in space is at sub-wavelength resolution set by the user to properly sample the highest near-field spatial frequencies thought to be important in the physics of the problem. Typically, 10–20 samples per wavelength are needed. The sampling in time is selected to ensure numerical stability of the algorithm.

Overall, FDTD and related techniques are marching-in-time procedures that simulate the continuous actual electromagnetic waves in a finite spatial region by sampled-data numerical analogs propagating in a computer data space. Time-stepping continues as the numerical wave analogs propagate in the space lattice to causally connect the physics of the modeled region. For simulations where the modeled region must extend to infinity, absorbing boundary conditions (ABCs) are employed at the outer lattice truncation planes which ideally permit all outgoing wave analogs to exit the region with negligible reflection. Phenomena such as induction of surface currents, scattering and multiple scattering, aperture penetration, and cavity excitation are modeled time-step by time-step by the action of the numerical analog to the curl equations. Self-consistency of these modeled phenomena is generally assured if their spatial and temporal variations are well resolved by the space and time sampling process. In fact, the goal is to provide a self-consistent model of the mutual coupling of all of the electrically small volume cells constituting the structure and its near field, even if the structure spans tens of wavelengths in three dimensions and there are hundreds of millions of space cells.

Time-stepping is continued until the desired late-time pulse response is observed at the field points of interest. For linear wave interaction problems, the sinusoidal response at these field points can be obtained over a wide band of frequencies by discrete Fourier transformation of the computed field-versus-time waveforms at these points. Prolonged "ringing" of the computed field waveforms due to a high Q-factor or large electrical size of the structure being modeled requires a combination of extending the computational window in time and extrapolation of the windowed data before Fourier transformation.

1.4. Classes of algorithms

Current FDTD and related space-grid time-domain algorithms are fully explicit solvers employing highly vectorizable and parallel schemes for time-marching the six components of the electric and magnetic field vectors at each of the space cells. The explicit nature of the solvers is usually maintained by employing a leapfrog time-stepping scheme. Current methods differ primarily in how the space lattice is set up. In fact, gridding methods can be categorized according to the degree of structure or regularity in the mesh cells:

(1) Almost completely structured. In this case, the space lattice is organized so that its unit cells are congruent wherever possible. The most basic example of such a mesh is the pioneering work of YEE [1966], who employed a uniform Cartesian grid having rectangular cells. Staircasing was used to approximate the surface of structural features not parallel to the grid coordinate axes. Later work showed that it is possible to modify the size and shape of the space cells located immediately adjacent to a structural feature to conformally fit its surface (see, for example, JURGENS, TAFLOVE, UMASHANKAR and MOORE [1992] and DEY and MITTRA [1997]). This is accurate and computationally efficient for large structures because the number of modified cells is proportional to the surface area of the structure. Thus, the number of modified cells becomes progressively smaller relative to the number of regular cells filling the structure volume as its size increases. As a result, the computer resources needed to implement a fully conformal model approximate those required for a staircased model. However, a key disadvantage of this technique is that special mesh-generation software must be constructed.

(2) Surface-fitted. In this case, the space lattice is globally distorted to fit the shape of the structure of interest. The lattice can be divided into multiple zones to accommodate a set of distinct surface features (see, for example, SHANKAR, MOHAMMADIAN and HALL [1990]). The major advantage of this approach is

that well-developed mesh-generation software of this type is available. The major disadvantage is that, relative to the Yee algorithm, there is substantial added computer burden due to:

(a) memory allocations for the position and stretching factors of each cell;

(b) extra computer operations to implement Maxwell's equations at each cell and to enforce field continuity at the interfaces of adjacent cells.

Another disadvantage is the possible presence of numerical dissipation in the time-stepping algorithm used for such meshes. This can limit the range of electrical size of the structure being modeled due to numerical wave-attenuation artifacts.

(3) Completely unstructured. In this case, the space containing the structure of interest is completely filled with a collection of lattice cells of varying sizes and shapes, but conforming to the structure surface (see, for example, MADSEN and ZIOLKOWSKI [1990]). As for the case of surface-fitted lattices, mesh-generation software is available and capable of modeling complicated three-dimensional shapes possibly having volumetric inhomogeneities. A key disadvantage of this approach is its potential for numerical inaccuracy and instability due to the unwanted generation of highly skewed space cells at random points within the lattice. A second disadvantage is the difficulty in mapping the unstructured mesh computations onto the architecture of either parallel vector computers or massively parallel machines. The structure-specific irregularity of the mesh mandates a robust pre-processing algorithm that optimally assigns specific mesh cells to specific processors.

At present, the best choice of computational algorithm and mesh remains unclear. For the next several years, we expect continued progress in this area as various groups develop their favored approaches and perform validations.

1.5. Predictive dynamic range

For computational modeling of electromagnetic wave interaction structures using FDTD and related space-grid time-domain techniques, it is useful to consider the concept of predictive dynamic range. Let the power density of the primary (incident) wave in the space grid be P_0 W/m^2. Further, let the minimum observable power density of a secondary (scattered) wave be P_S W/m^2, where "minimum observable" means that the accuracy of the field computation degrades due to numerical artifacts to poorer than n dB (some desired figure of merit) at lower levels than P_S. Then, we can define the predictive dynamic range as $10\log_{10}(P_0/P_S)$ dB.

This definition is well suited for FDTD and other space-grid time-domain codes for two reasons:

- It squares nicely with the concept of a "quiet zone" in an experimental anechoic chamber, which is intuitive to most electromagnetics engineers;
- It succinctly quantifies the fact that the desired numerical wave analogs propagating in the lattice exist in an additive noise environment due to nonphysical propagating wave analogs caused by the imperfect ABCs.

In addition to additive noise, the desired physical wave analogs undergo gradual progressive deterioration while propagating due to accumulating numerical dispersion artifacts, including phase velocity anisotropies and inhomogeneities within the mesh.

In the 1980s, researchers accumulated solid evidence for a predictive dynamic range on the order of 40–50 dB for FDTD codes. This value is reasonable if one considers the additive noise due to imperfect ABCs to be the primary limiting factor, since the analytical ABCs of this era (see, for example, MUR [1981]) provided outer-boundary reflection coefficients in the range of about 0.3–3% (−30 to −50 dB).

The 1990s saw the emergence of powerful, entirely new classes of ABCs including the perfectly matched layer (PML) of BERENGER [1994]; the uniaxial anisotropic PML (UPML) of SACKS, KINGSLAND, LEE and LEE [1995] and GEDNEY [1996]; and the complementary operator methods (COM) of RAMAHI [1997], RAMAHI [1998]. These ABCs were shown to have effective outer-boundary reflection coefficients of better than −80 dB for impinging pulsed electromagnetic waves having ultrawideband spectra. Solid capabilities were demonstrated to terminate free-space lattices, multimoding and dispersive waveguiding structures, and lossy and dispersive materials.

However, for electrically large problems, the overall dynamic range may not reach the maximum permitted by these new ABCs because of inaccuracies due to accumulating numerical-dispersion artifacts generated by the basic grid-based solution of the curl equations. Fortunately, by the end of the 1990s, this problem was being attacked by a new generation of low-dispersion algorithms. Examples include the wavelet-based multi-resolution time-domain (MRTD) technique introduced by KRUMPHOLZ and KATEHI [1996] and the pseudo-spectral time-domain (PSTD) technique introduced by LIU, Q.H. [1996], LIU, Q.H. [1997]. As a result of these advances, there is emerging the possibility of FDTD and related space-grid time-domain methods demonstrating predictive dynamic ranges of 80 dB or more in the first decade of the 21st century.

1.6. Scaling to very large problem sizes

Using FDTD and related methods, we can model electromagnetic wave interaction problems requiring the solution of considerably more than 10^8 field-vector unknowns. At this level of complexity, it is possible to develop detailed, three-dimensional models of complete engineering systems, including the following:

- Entire aircraft and missiles illuminated by radar at 1 GHz and above;
- Entire multilayer circuit boards and multichip modules for digital signal propagation, crosstalk, and radiation;
- Entire microwave and millimeter-wave amplifiers, including the active and passive circuit components and packaging;
- Entire integrated-optical structures, including lasers, waveguides, couplers, and resonators.

A key goal for such large models is to achieve algorithm/computer-architecture scaling such that for N field unknowns to be solved on M processors, we approach an order(N/M) scaling of the required computational resources.

We now consider the factors involved in determining the computational burden for the class of FDTD and related space-grid time-domain solvers.

(1) *Number of volumetric grid cells*, N. The six vector electromagnetic field components located at each lattice cell must be updated at every time step. This yields by itself an order(N) scaling.
(2) *Number of time steps*, $n_{\max}$. A self-consistent solution in the time domain mandates that the numerical wave analogs propagate over time scales sufficient to causally connect each portion of the structure of interest. Therefore, $n_{\max}$ must increase as the maximum electrical size of the structure. In three dimensions, it can be argued that $n_{\max}$ is a fractional power function of N such as $N^{1/3}$. Further, $n_{\max}$ must be adequate to step through "ring-up" and "ring-down" times of energy storage features such as cavities. These features vary from problem to problem and cannot be ascribed a dependence relative to N.
(3) *Cumulative propagation errors*. Additional computational burdens may arise due to the need for either progressive mesh refinement or progressively higher-accuracy algorithms to bound cumulative positional or phase errors for propagating numerical modes in progressively enlarged meshes. Any need for progressive mesh refinement would feed back to factor 1.

For most free-space problems, factors 2 and 3 are weaker functions of the size of the modeled structure than factor 1. This is because geometrical features at increasing electrical distances from each other become decoupled due to radiative losses by the electromagnetic waves propagating between these features. Further, it can be shown that replacing second-order accurate algorithms by higher-order versions sufficiently reduces numerical dispersion error to avoid the need for progressive mesh refinement for object sizes up to the order of 100 wavelengths. Overall, a computational burden of order($N \cdot n_{\max}$) = order($N^{4/3}$) is estimated for very large FDTD and related models.

2. Maxwell's equations

In this section, we establish the fundamental equations and notation for the electromagnetic fields used in the remainder of this chapter.

2.1. Three-dimensional case

Using MKS units, the time-dependent Maxwell's equations in three dimensions are given in differential and integral form by

Faraday's Law:

$$\frac{\partial \vec{B}}{\partial t} = -\nabla \times \vec{E} - \vec{M}, \tag{2.1a}$$

$$\frac{\partial}{\partial t} \iint_A \vec{B} \cdot \mathrm{d}\vec{A} = -\oint_\ell \vec{E} \cdot \mathrm{d}\vec{\ell} - \iint_A \vec{M} \cdot \mathrm{d}\vec{A}. \tag{2.1b}$$

Ampere's Law:

$$\frac{\partial \vec{D}}{\partial t} = \nabla \times \vec{H} - \vec{J}, \tag{2.2a}$$

$$\frac{\partial}{\partial t}\iint_A \vec{D}\cdot \mathrm{d}\vec{A} = \oint_\ell \vec{H}\cdot \mathrm{d}\vec{\ell} - \iint_A \vec{J}\cdot \mathrm{d}\vec{A}. \tag{2.2b}$$

Gauss' Law for the electric field:

$$\nabla \cdot \vec{D} = 0, \tag{2.3a}$$

$$\oiint_A \vec{D}\cdot \mathrm{d}\vec{A} = 0. \tag{2.3b}$$

Gauss' Law for the magnetic field:

$$\nabla \cdot \vec{B} = 0, \tag{2.4a}$$

$$\oiint_A \vec{B}\cdot \mathrm{d}\vec{A} = 0. \tag{2.4b}$$

In (2.1)–(2.4), the following symbols (and their MKS units) are defined:

$\vec{E}$: electric field (volts/meter)
$\vec{D}$: electric flux density (coulombs/meter^2)
$\vec{H}$: magnetic field (amperes/meter)
$\vec{B}$: magnetic flux density (webers/meter^2)
A: arbitrary three-dimensional surface
$\mathrm{d}\vec{A}$: differential normal vector that characterizes surface A (meter^2)
ℓ: closed contour that bounds surface A
$\mathrm{d}\vec{\ell}$: differential length vector that characterizes contour ℓ (meters)
$\vec{J}$: electric current density (amperes/meter^2)
$\vec{M}$: equivalent magnetic current density (volts/meter^2)

In linear, isotropic, nondispersive materials (i.e., materials having field-independent, direction-independent, and frequency-independent electric and magnetic properties), we can relate $\vec{D}$ to $\vec{E}$ and $\vec{B}$ to $\vec{H}$ using simple proportions:

$$\vec{D} = \varepsilon \vec{E} = \varepsilon_r \varepsilon_0 \vec{E}; \qquad \vec{B} = \mu \vec{H} = \mu_r \mu_0 \vec{H}, \tag{2.5}$$

where

ε: electrical permittivity (farads/meter)
ε_r: relative permittivity (dimensionless scalar)
ε_0: free-space permittivity (8.854×10^{-12} farads/meter)
μ: magnetic permeability (henrys/meter)
μ_r: relative permeability (dimensionless scalar)
μ_0: free-space permeability ($4\pi \times 10^{-7}$ henrys/meter)

Note that $\vec{J}$ and $\vec{M}$ can act as *independent sources* of E- and H-field energy, $\vec{J}_{\mathrm{source}}$ and $\vec{M}_{\mathrm{source}}$. We also allow for materials with isotropic, nondispersive electric and magnetic losses that attenuate E- and H-fields via conversion to heat energy. This yields:

$$\vec{J} = \vec{J}_{\text{source}} + \sigma \vec{E}; \qquad \vec{M} = \vec{M}_{\text{source}} + \sigma^* \vec{H}, \tag{2.6}$$

where

σ: electric conductivity (siemens/meter)

σ^*: equivalent magnetic loss (ohms/meter)

Finally, we substitute (2.5) and (2.6) into (2.1a) and (2.2a). This yields Maxwell's curl equations in linear, isotropic, nondispersive, lossy materials:

$$\frac{\partial \vec{H}}{\partial t} = -\frac{1}{\mu} \nabla \times \vec{E} - \frac{1}{\mu} \big(\vec{M}_{\text{source}} + \sigma^* \vec{H} \big), \tag{2.7}$$

$$\frac{\partial \vec{E}}{\partial t} = \frac{1}{\varepsilon} \nabla \times \vec{H} - \frac{1}{\varepsilon} \big(\vec{J}_{\text{source}} + \sigma \vec{E} \big). \tag{2.8}$$

We now write out the vector components of the curl operators of (2.7) and (2.8) in Cartesian coordinates. This yields the following system of six coupled scalar equations:

$$\frac{\partial H_x}{\partial t} = \frac{1}{\mu} \left[\frac{\partial E_y}{\partial z} - \frac{\partial E_z}{\partial y} - \big(M_{\text{source}_x} + \sigma^* H_x \big) \right], \tag{2.9a}$$

$$\frac{\partial H_y}{\partial t} = \frac{1}{\mu} \left[\frac{\partial E_z}{\partial x} - \frac{\partial E_x}{\partial z} - \big(M_{\text{source}_y} + \sigma^* H_y \big) \right], \tag{2.9b}$$

$$\frac{\partial H_z}{\partial t} = \frac{1}{\mu} \left[\frac{\partial E_x}{\partial y} - \frac{\partial E_y}{\partial x} - \big(M_{\text{source}_z} + \sigma^* H_z \big) \right], \tag{2.9c}$$

$$\frac{\partial E_x}{\partial t} = \frac{1}{\varepsilon} \left[\frac{\partial H_z}{\partial y} - \frac{\partial H_y}{\partial z} - \big(J_{\text{source}_x} + \sigma E_x \big) \right], \tag{2.10a}$$

$$\frac{\partial E_y}{\partial t} = \frac{1}{\varepsilon} \left[\frac{\partial H_x}{\partial z} - \frac{\partial H_z}{\partial x} - \big(J_{\text{source}_y} + \sigma E_y \big) \right], \tag{2.10b}$$

$$\frac{\partial E_z}{\partial t} = \frac{1}{\varepsilon} \left[\frac{\partial H_y}{\partial x} - \frac{\partial H_x}{\partial y} - \big(J_{\text{source}_z} + \sigma E_z \big) \right]. \tag{2.10c}$$

The system of six coupled partial differential equations of (2.9) and (2.10) forms the basis of the FDTD numerical algorithm for electromagnetic wave interactions with general three-dimensional objects. The FDTD algorithm need not explicitly enforce the Gauss' Law relations indicating zero free electric and magnetic charge, (2.3) and (2.4). This is because these relations are theoretically a direct consequence of the curl equations, as can be readily shown. However, the FDTD space grid must be structured so that the Gauss' Law relations are *implicit* in the positions of the E- and H-field vector components in the grid, and in the numerical space-derivative operations upon these components that model the action of the curl operator. This will be discussed later in the context of the Yee mesh.

Before proceeding with the introduction of the Yee algorithm, it is instructive to consider simplified two-dimensional cases for Maxwell's equations. These cases demonstrate important electromagnetic wave phenomena and can yield insight into the analytical and algorithmic features of the general three-dimensional case.

2.2. *Reduction to two dimensions*

Let us assume that the structure being modeled extends to infinity in the z-direction with no change in the shape or position of its transverse cross section. If the incident wave is also uniform in the z-direction, then all partial derivatives of the fields with respect to z must equal zero. Under these conditions, the full set of Maxwell's curl equations given by (2.9) and (2.10) reduces to two modes, the *transverse-magnetic mode with respect to* z (TM_z) and the *transverse-electric mode with respect to* z (TE_z). The reduced sets of Maxwell's equations for these modes are as follows.

TM_z mode (involving only H_x, H_y, and E_z)

$$\frac{\partial H_x}{\partial t} = \frac{1}{\mu}\left[-\frac{\partial E_z}{\partial y} - \left(M_{\text{source}_x} + \sigma^* H_x\right)\right], \tag{2.11a}$$

$$\frac{\partial H_y}{\partial t} = \frac{1}{\mu}\left[\frac{\partial E_z}{\partial x} - \left(M_{\text{source}_y} + \sigma^* H_y\right)\right], \tag{2.11b}$$

$$\frac{\partial E_z}{\partial t} = \frac{1}{\varepsilon}\left[\frac{\partial H_y}{\partial x} - \frac{\partial H_x}{\partial y} - (J_{\text{source}_z} + \sigma E_z)\right]. \tag{2.11c}$$

TE_z mode (involving only E_x, E_y, and H_z)

$$\frac{\partial E_x}{\partial t} = \frac{1}{\varepsilon}\left[\frac{\partial H_z}{\partial y} - (J_{\text{source}_x} + \sigma E_x)\right], \tag{2.12a}$$

$$\frac{\partial E_y}{\partial t} = \frac{1}{\varepsilon}\left[-\frac{\partial H_z}{\partial x} - (J_{\text{source}_y} + \sigma E_y)\right], \tag{2.12b}$$

$$\frac{\partial H_z}{\partial t} = \frac{1}{\mu}\left[\frac{\partial E_x}{\partial y} - \frac{\partial E_y}{\partial x} - (M_{\text{source}_z} + \sigma^* H_z)\right]. \tag{2.12c}$$

The TM_z and TE_z modes contain no common field vector components. Thus, these modes can exist simultaneously with *no* mutual interactions for structures composed of isotropic materials or anisotropic materials having no off-diagonal components in the constitutive tensors.

Physical phenomena associated with these two modes can be very different. The TE_z mode can support propagating electromagnetic fields bound closely to, or guided by, the surface of a metal structure (the "creeping wave" being a classic example for curved metal surfaces). On the other hand, the TM_z mode sets up an E-field which must be negligible at a metal surface. This diminishes or eliminates bound or guided near-surface propagating waves for metal surfaces. The presence or absence of surface-type waves can have important implications for scattering and radiation problems.

3. The Yee algorithm

3.1. *Basic ideas*

YEE [1966] originated a set of finite-difference equations for the time-dependent Maxwell's curl equations of (2.9) and (2.10) for the lossless materials case $\sigma = 0$ and

$\sigma^* = 0$. This section summarizes Yee's algorithm, which forms the basis of the FDTD technique. Key ideas underlying the robust nature of the Yee algorithm are as follows:

(1) The Yee algorithm solves for both electric and magnetic fields in time and space using the coupled Maxwell's curl equations rather than solving for the electric field alone (or the magnetic field alone) with a wave equation.
   - This is analogous to the combined-field integral equation formulation of the method of moments, wherein both $\vec{E}$ and $\vec{H}$ boundary conditions are enforced on the surface of a material structure.
   - Using both $\vec{E}$ and $\vec{H}$ information, the solution is more robust than using either alone (i.e., it is accurate for a wider class of structures). Both electric and magnetic material properties can be modeled in a straightforward manner. This is especially important when modeling radar cross section mitigation.
   - Features unique to each field such as tangential $\vec{H}$ singularities near edges and corners, azimuthal (looping) $\vec{H}$ singularities near thin wires, and radial $\vec{E}$ singularities near points, edges, and thin wires can be individually modeled if both electric and magnetic fields are available.

(2) As illustrated in Fig. 3.1, the Yee algorithm centers its $\vec{E}$ and $\vec{H}$ components in three-dimensional space so that every $\vec{E}$ component is surrounded by four circulating $\vec{H}$ components, and every $\vec{H}$ component is surrounded by four circulating $\vec{E}$ components.

   This provides a beautifully simple picture of three-dimensional space being filled by an interlinked array of Faraday's Law and Ampere's Law contours. For example, it is possible to identify Yee $\vec{E}$ components associated with displacement current flux linking $\vec{H}$ loops, as well as $\vec{H}$ components associated with magnetic flux linking $\vec{E}$ loops. In effect, the Yee algorithm simultaneously simulates the pointwise differential form *and* the macroscopic integral form of

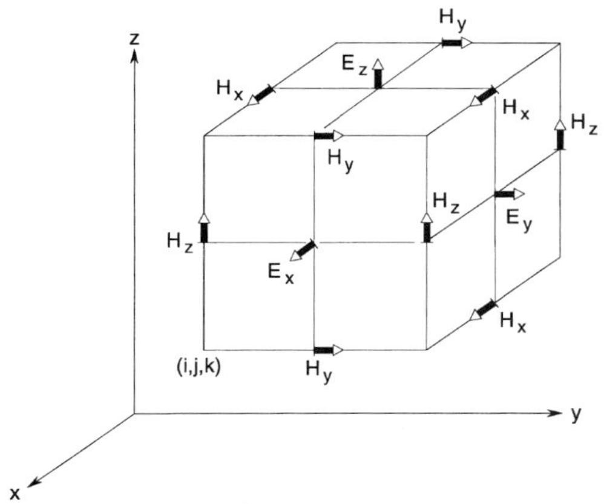

FIG. 3.1. Position of the electric and magnetic field vector components about a cubic unit cell of the Yee space lattice. *After*: K.S. Yee, *IEEE Trans. Antennas and Propagation*, Vol. 14, 1966, pp. 302–307, © 1966 IEEE.

Maxwell's equations. The latter is extremely useful in specifying field boundary conditions and singularities.

In addition, we have the following attributes of the Yee space lattice:

- The finite-difference expressions for the space derivatives used in the curl operators are central-difference in nature and second-order accurate.
- Continuity of tangential $\vec{E}$ and $\vec{H}$ is naturally maintained across an interface of dissimilar materials if the interface is parallel to one of the lattice coordinate axes. For this case, there is no need to specially enforce field boundary conditions at the interface. At the beginning of the problem, we simply specify the material permittivity and permeability at each field component location. This yields a stepped or "staircase" approximation of the surface and internal geometry of the structure, with a space resolution set by the size of the lattice unit cell.
- The location of the $\vec{E}$ and $\vec{H}$ components in the Yee space lattice and the central-difference operations on these components implicitly enforce the two Gauss' Law relations (see Section 3.6.9). Thus, the Yee mesh is divergence-free with respect to its E- and H-fields in the absence of free electric and magnetic charge.

(3) As illustrated in Fig. 3.2, the Yee algorithm also centers its $\vec{E}$ and $\vec{H}$ components in time in what is termed a leapfrog arrangement. All of the $\vec{E}$ computations in the modeled space are completed and stored in memory for a particular time point using previously stored $\vec{H}$ data. Then all of the $\vec{H}$ computations in the space are completed and stored in memory using the $\vec{E}$ data just computed. The cycle begins again with the recomputation of the $\vec{E}$ components based on the newly obtained $\vec{H}$. This process continues until time-stepping is concluded.

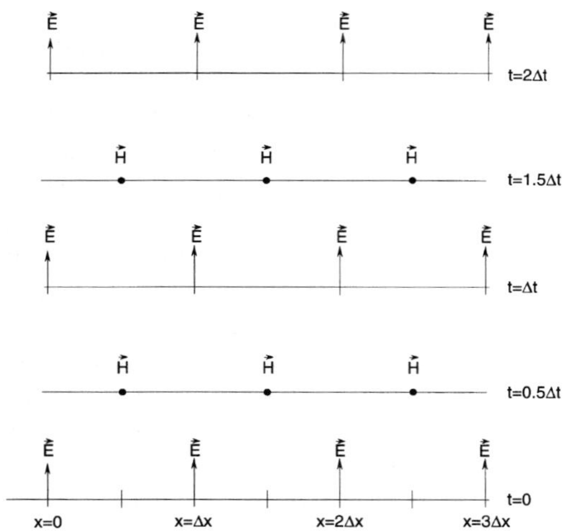

FIG. 3.2. Space–time chart of the Yee algorithm for a one-dimensional wave propagation example showing the use of central differences for the space derivatives and leapfrog for the time derivatives. Initial conditions for both electric and magnetic fields are zero everywhere in the grid.

- Leapfrog time-stepping is fully explicit, thereby avoiding problems involved with simultaneous equations and matrix inversion.
- The finite-difference expressions for the time derivatives are central-difference in nature and second-order accurate.
- The time-stepping algorithm is nondissipative. That is, numerical wave modes propagating in the mesh do not spuriously decay due to a nonphysical artifact of the time-stepping algorithm.

3.2. *Finite differences and notation*

YEE [1966] introduced the following notation for space points and functions of space and time. A space point in a uniform, rectangular lattice is denoted as

$$(i, j, k) = (i\Delta x, j\Delta y, k\Delta z). \tag{3.1}$$

Here, Δx, Δy, and Δz are, respectively, the lattice space increments in the x, y, and z coordinate directions, and i, j, and k are integers. Further, we denote any function u of space and time evaluated at a discrete point in the grid and at a discrete point in time as

$$u(i\Delta x, j\Delta y, k\Delta z, n\Delta t) = u^n_{i,j,k}, \tag{3.2}$$

where Δt is the time increment, assumed uniform over the observation interval, and n is an integer.

Yee used centered finite-difference (central-difference) expressions for the space and time derivatives that are both simply programmed and second-order accurate in the space and time increments. Consider his expression for the first partial space derivative of u in the x-direction, evaluated at the fixed time $t_n = n\Delta t$:

$$\frac{\partial u}{\partial x}(i\Delta x, j\Delta y, k\Delta z, n\Delta t) = \frac{u^n_{i+1/2,j,k} - u^n_{i-1/2,j,k}}{\Delta x} + \mathrm{O}\big[(\Delta x)^2\big]. \tag{3.3}$$

We note the $\pm 1/2$ increment in the i subscript (x-coordinate) of u, denoting a space finite-difference over $\pm 1/2\Delta x$. Yee's goal was second-order accurate central differencing, but it is apparent that he desired to take data for his central differences to the right and left of his observation point by only $\Delta x/2$, rather than a full Δx.

Yee chose this notation because he wished to interleave his $\vec{E}$ and $\vec{H}$ components in the space lattice at intervals of $\Delta x/2$. For example, the difference of two adjacent $\vec{E}$ components, separated by Δx and located $\pm 1/2\Delta x$ on either side of an $\vec{H}$ component, would be used to provide a numerical approximation for $\partial E/\partial x$ to permit stepping the $\vec{H}$ component in time. For completeness, it should be added that a numerical approximation analogous to (3.3) for $\partial u/\partial y$ or $\partial u/\partial z$ can be written simply by incrementing the j or k subscript of u by $\pm 1/2\Delta y$ or $\pm 1/2\Delta z$, respectively.

Yee's expression for the first time partial derivative of u, evaluated at the fixed space point (i, j, k), follows by analogy:

$$\frac{\partial u}{\partial t}(i\Delta x, j\Delta y, k\Delta z, n\Delta t) = \frac{u^{n+1/2}_{i,j,k} - u^{n-1/2}_{i,j,k}}{\Delta t} + \mathrm{O}\big[(\Delta t)^2\big]. \tag{3.4}$$

Now the $\pm 1/2$ increment is in the n superscript (time coordinate) of u, denoting a time finite-difference over $\pm 1/2\Delta t$. Yee chose this notation because he wished to interleave his $\vec{E}$ and $\vec{H}$ components in time at intervals of $1/2\Delta t$ for purposes of implementing a leapfrog algorithm.

3.3. Finite-difference expressions for Maxwell's equations in three dimensions

We now apply the above ideas and notation to achieve a finite-difference numerical approximation of the Maxwell's curl equations in three dimensions given by (2.9) and (2.10). We begin by considering as an example the E_x field-component equation (2.10a). Referring to Figs. 3.1 and 3.2, a typical substitution of central differences for the time and space derivatives in (2.10a) at $E_x(i, j+1/2, k+1/2, n)$ yields the following expression:

$$\frac{E_x|^{n+1/2}_{i,j+1/2,k+1/2} - E_x|^{n-1/2}_{i,j+1/2,k+1/2}}{\Delta t} = \frac{1}{\varepsilon_{i,j+1/2,k+1/2}} \cdot \left(\begin{array}{c} \frac{H_z|^n_{i,j+1,k+1/2} - H_z|^n_{i,j,k+1/2}}{\Delta y} - \frac{H_y|^n_{i,j+1/2,k+1} - H_y|^n_{i,j+1/2,k}}{\Delta z} \\ - J_{\text{source}_x}|^n_{i,j+1/2,k+1/2} - \sigma_{i,j+1/2,k+1/2} E_x|^n_{i,j+1/2,k+1/2} \end{array} \right). \tag{3.5}$$

Note that all field quantities on the right-hand side are evaluated at time-step n, including the electric field E_x appearing due to the material conductivity σ. Since E_x values at time-step n are not assumed to be stored in the computer's memory (only the previous values of E_x at time-step $n - 1/2$ are assumed to be in memory), we need some way to estimate such terms. A very good way is as follows, using what we call a *semi-implicit approximation*:

$$E_x|^n_{i,j+1/2,k+1/2} = \frac{E_x|^{n+1/2}_{i,j+1/2,k+1/2} + E_x|^{n-1/2}_{i,j+1/2,k+1/2}}{2}. \tag{3.6}$$

Here E_x values at time-step n are assumed to be simply the arithmetic average of the stored values of E_x at time-step $n - 1/2$ and the yet-to-be computed new values of E_x at time-step $n + 1/2$. Substituting (3.6) into (3.5) and collecting terms yields the following explicit time-stepping relation for E_x (which is numerically stable for values of σ from zero to infinity):

$$\begin{aligned} E_x|^{n+1/2}_{i,j+1/2,k+1/2} &= \left(\frac{1 - \frac{\sigma_{i,j+1/2,k+1/2}\Delta t}{2\varepsilon_{i,j+1/2,k+1/2}}}{1 + \frac{\sigma_{i,j+1/2,k+1/2}\Delta t}{2\varepsilon_{i,j+1/2,k+1/2}}} \right) E_x|^{n-1/2}_{i,j+1/2,k+1/2} \\ &\quad + \left(\frac{\frac{\Delta t}{\varepsilon_{i,j+1/2,k+1/2}}}{1 + \frac{\sigma_{i,j+1/2,k+1/2}\Delta t}{2\varepsilon_{i,j+1/2,k+1/2}}} \right) \cdot \left(\begin{array}{c} \frac{H_z|^n_{i,j+1,k+1/2} - H_z|^n_{i,j,k+1/2}}{\Delta y} \\ - \frac{H_y|^n_{i,j+1/2,k+1} - H_y|^n_{i,j+1/2,k}}{\Delta z} \\ - J_{\text{source}_x}|^n_{i,j+1/2,k+1/2} \end{array} \right). \end{aligned} \tag{3.7a}$$

Similarly, we can derive finite-difference expressions based on Yee's algorithm for the E_y and E_z field components given by Maxwell's equations (2.10b) and (2.10c).

Referring again to Fig. 3.1, we have:

$$E_y\Big|_{i-1/2,j+1,k+1/2}^{n+1/2} = \left(\frac{1-\frac{\sigma_{i-1/2,j+1,k+1/2}\Delta t}{2\varepsilon_{i-1/2,j+1,k+1/2}}}{1+\frac{\sigma_{i-1/2,j+1,k+1/2}\Delta t}{2\varepsilon_{i-1/2,j+1,k+1/2}}}\right) E_y\Big|_{i-1/2,j+1,k+1/2}^{n-1/2} + \left(\frac{\frac{\Delta t}{\varepsilon_{i-1/2,j+1,k+1/2}}}{1+\frac{\sigma_{i-1/2,j+1,k+1/2}\Delta t}{2\varepsilon_{i-1/2,j+1,k+1/2}}}\right)\cdot\left(\begin{array}{c}\frac{H_x|_{i-1/2,j+1,k+1}^{n}-H_x|_{i-1/2,j+1,k}^{n}}{\Delta z} \\ -\frac{H_z|_{i,j+1,k+1/2}^{n}-H_z|_{i-1,j+1,k+1/2}^{n}}{\Delta x} \\ -J_{\text{source}_y}|_{i-1/2,j+1,k+1/2}^{n}\end{array}\right), \tag{3.7b}$$

$$E_z\Big|_{i-1/2,j+1/2,k+1}^{n+1/2} = \left(\frac{1-\frac{\sigma_{i-1/2,j+1/2,k+1}\Delta t}{2\varepsilon_{i-1/2,j+1/2,k+1}}}{1+\frac{\sigma_{i-1/2,j+1/2,k+1}\Delta t}{2\varepsilon_{i-1/2,j+1/2,k+1}}}\right) E_z\Big|_{i-1/2,j+1/2,k+1}^{n-1/2} + \left(\frac{\frac{\Delta t}{\varepsilon_{i-1/2,j+1/2,k+1}}}{1+\frac{\sigma_{i-1/2,j+1/2,k+1}\Delta t}{2\varepsilon_{i-1/2,j+1/2,k+1}}}\right)\cdot\left(\begin{array}{c}\frac{H_y|_{i,j+1/2,k+1}^{n}-H_y|_{i-1,j+1/2,k+1}^{n}}{\Delta x} \\ -\frac{H_x|_{i-1/2,j+1,k+1}^{n}-H_x|_{i-1/2,j,k+1}^{n}}{\Delta y} \\ -J_{\text{source}_z}|_{i-1/2,j+1/2,k+1}^{n}\end{array}\right). \tag{3.7c}$$

By analogy we can derive finite-difference equations for (2.9a)–(2.9c) to time-step H_x, H_y, and H_z. Here $\sigma^* H$ represents a magnetic loss term on the right-hand side of each equation, which is estimated using a semi-implicit procedure analogous to (3.6). Referring again to Figs. 3.1 and 3.2, we have for example the following time-stepping expressions for the H components located about the unit cell:

$$H_x\Big|_{i-1/2,j+1,k+1}^{n+1} = \left(\frac{1-\frac{\sigma^*_{i-1/2,j+1,k+1}\Delta t}{2\mu_{i-1/2,j+1,k+1}}}{1+\frac{\sigma^*_{i-1/2,j+1,k+1}\Delta t}{2\mu_{i-1/2,j+1,k+1}}}\right) H_x\Big|_{i-1/2,j+1,k+1}^{n} + \left(\frac{\frac{\Delta t}{\mu_{i-1/2,j+1,k+1}}}{1+\frac{\sigma^*_{i-1/2,j+1,k+1}\Delta t}{2\mu_{i-1/2,j+1,k+1}}}\right)\cdot\left(\begin{array}{c}\frac{E_y|_{i-1/2,j+1,k+3/2}^{n+1/2}-E_y|_{i-1/2,j+1,k+1/2}^{n+1/2}}{\Delta z} \\ -\frac{E_z|_{i-1/2,j+3/2,k+1}^{n+1/2}-E_z|_{i-1/2,j+1/2,k+1}^{n+1/2}}{\Delta y} \\ -M_{\text{source}_x}|_{i-1/2,j+1,k+1}^{n+1/2}\end{array}\right), \tag{3.8a}$$

$$H_y\Big|_{i,j+1/2,k+1}^{n+1} = \left(\frac{1-\frac{\sigma^*_{i,j+1/2,k+1}\Delta t}{2\mu_{i,j+1/2,k+1}}}{1+\frac{\sigma^*_{i,j+1/2,k+1}\Delta t}{2\mu_{i,j+1/2,k+1}}}\right) H_y\Big|_{i,j+1/2,k+1}^{n} + \left(\frac{\frac{\Delta t}{\mu_{i,j+1/2,k+1}}}{1+\frac{\sigma^*_{i,j+1/2,k+1}\Delta t}{2\mu_{i,j+1/2,k+1}}}\right)\cdot\left(\begin{array}{c}\frac{E_z|_{i+1/2,j+1/2,k+1}^{n+1/2}-E_z|_{i-1/2,j+1/2,k+1}^{n+1/2}}{\Delta x} \\ -\frac{E_x|_{i,j+1/2,k+3/2}^{n+1/2}-E_x|_{i,j+1/2,k+1/2}^{n+1/2}}{\Delta z} \\ -M_{\text{source}_y}|_{i,j+1/2,k+1}^{n+1/2}\end{array}\right), \tag{3.8b}$$

$$H_z\big|_{i,j+1,k+1/2}^{n+1} = \left(\frac{1-\dfrac{\sigma^*_{i,j+1,k+1/2}\Delta t}{2\mu_{i,j+1,k+1/2}}}{1+\dfrac{\sigma^*_{i,j+1,k+1/2}\Delta t}{2\mu_{i,j+1,k+1/2}}}\right) H_z\big|_{i,j+1,k+1/2}^{n} + \left(\frac{\dfrac{\Delta t}{\mu_{i,j+1,k+1/2}}}{1+\dfrac{\sigma^*_{i,j+1,k+1/2}\Delta t}{2\mu_{i,j+1,k+1/2}}}\right) \cdot \left(\begin{array}{c} \dfrac{E_x\big|_{i,j+3/2,k+1/2}^{n+1/2} - E_x\big|_{i,j+1/2,k+1/2}^{n+1/2}}{\Delta y} \\ -\dfrac{E_y\big|_{i+1/2,j+1,k+1/2}^{n+1/2} - E_y\big|_{i-1/2,j+1,k+1/2}^{n+1/2}}{\Delta x} \\ -M_{\text{source}_z}\big|_{i,j+1,k+1/2}^{n+1/2} \end{array}\right). \tag{3.8c}$$

With the systems of finite-difference expressions of (3.7) and (3.8), the new value of an electromagnetic field vector component at any space lattice point depends only on its previous value, the previous values of the components of the other field vector at adjacent points, and the known electric and magnetic current sources. Therefore, at any given time step, the computation of a field vector can proceed either one point at a time, or, if p parallel processors are employed concurrently, p points at a time.

3.4. Field updating coefficients

To implement the finite-difference systems of (3.7) and (3.8) for a region having a continuous variation of material properties with spatial position, it is desirable to define and store the following updating coefficients for each field vector component:

Updating coefficients at the general E-field component location (i, j, k):

$$C_a\big|_{i,j,k} = \left(1 - \frac{\sigma_{i,j,k}\Delta t}{2\varepsilon_{i,j,k}}\right) \Big/ \left(1 + \frac{\sigma_{i,j,k}\Delta t}{2\varepsilon_{i,j,k}}\right), \tag{3.9a}$$

$$C_{b_1}\big|_{i,j,k} = \left(\frac{\Delta t}{\varepsilon_{i,j,k}\Delta_1}\right) \Big/ \left(1 + \frac{\sigma_{i,j,k}\Delta t}{2\varepsilon_{i,j,k}}\right), \tag{3.9b}$$

$$C_{b_2}\big|_{i,j,k} = \left(\frac{\Delta t}{\varepsilon_{i,j,k}\Delta_2}\right) \Big/ \left(1 + \frac{\sigma_{i,j,k}\Delta t}{2\varepsilon_{i,j,k}}\right). \tag{3.9c}$$

Updating coefficients at the general H-field component location (i, j, k):

$$D_a\big|_{i,j,k} = \left(1 - \frac{\sigma^*_{i,j,k}\Delta t}{2\mu_{i,j,k}}\right) \Big/ \left(1 + \frac{\sigma^*_{i,j,k}\Delta t}{2\mu_{i,j,k}}\right), \tag{3.10a}$$

$$D_{b_1}\big|_{i,j,k} = \left(\frac{\Delta t}{\mu_{i,j,k}\Delta_1}\right) \Big/ \left(1 + \frac{\sigma^*_{i,j,k}\Delta t}{2\mu_{i,j,k}}\right), \tag{3.10b}$$

$$D_{b_2}\big|_{i,j,k} = \left(\frac{\Delta t}{\mu_{i,j,k}\Delta_2}\right) \Big/ \left(1 + \frac{\sigma^*_{i,j,k}\Delta t}{2\mu_{i,j,k}}\right). \tag{3.10c}$$

In (3.9) and (3.10), Δ_1 and Δ_2 denote the two possible lattice space increments used for the finite differences in each field-component calculation. For a cubic lattice, $\Delta x = \Delta y = \Delta z = \Delta$ and thus $\Delta_1 = \Delta_2 = \Delta$. For this case, $C_{b_1} = C_{b_2}$ and $D_{b_1} = D_{b_2}$, reducing the storage requirement to two updating coefficients per field vector component. Here, the approximate total computer storage needed is $18N$, where N is the number of space cells in the FDTD lattice. The finite-difference expressions of (3.7) and (3.8) can now be rewritten more simply. For example, to update E_x we have:

$$\begin{aligned} E_x\big|_{i,j+1/2,k+1/2}^{n+1/2} &= C_{a,E_x}\big|_{i,j+1/2,k+1/2} E_x\big|_{i,j+1/2,k+1/2}^{n-1/2} \\ &\quad + C_{b,E_x}\big|_{i,j+1/2,k+1/2} \cdot \begin{pmatrix} H_z\big|_{i,j+1,k+1/2}^{n} - H_z\big|_{i,j,k+1/2}^{n} + H_y\big|_{i,j+1/2,k}^{n} \\ - H_y\big|_{i,j+1/2,k+1}^{n} - J_{\text{source}_x}\big|_{i,j+1/2,k+1/2}^{n}\Delta \end{pmatrix}. \end{aligned} \tag{3.11}$$

Similarly, to update H_x we have:

$$\begin{aligned} H_x\big|_{i-1/2,j+1,k+1}^{n+1} &= D_{a,H_x}\big|_{i-1/2,j+1,k+1} H_x\big|_{i-1/2,j+1,k+1}^{n} \\ &\quad + D_{b,H_x}\big|_{i-1/2,j+1,k+1} \cdot \begin{pmatrix} E_y\big|_{i-1/2,j+1,k+3/2}^{n+1/2} - E_y\big|_{i-1/2,j+1,k+1/2}^{n+1/2} \\ + E_z\big|_{i-1/2,j+1/2,k+1}^{n+1/2} - E_z\big|_{i-1/2,j+3/2,k+1}^{n+1/2} \\ - M_{\text{source}_x}\big|_{i-1/2,j+1,k+1}^{n+1/2}\Delta \end{pmatrix}. \end{aligned} \tag{3.12}$$

For a space region with a finite number of media having distinct electrical properties, the computer storage requirement can be further reduced. This can be done by defining an integer array, MEDIA(i, j, k), for each set of field vector components. This array stores an integer "pointer" at each location of such a field component in the space lattice, enabling the proper algorithm coefficients to be extracted. For example, to update E_x we have:

$$\begin{aligned} m &= \text{MEDIA}_{E_x}\big|_{i,j+1/2,k+1/2}, \\ E_x\big|_{i,j+1/2,k+1/2}^{n+1/2} &= C_a(m) E_x\big|_{i,j+1/2,k+1/2}^{n-1/2} + C_b(m) \cdot \big(H_z\big|_{i,j+1,k+1/2}^{n} - H_z\big|_{i,j,k+1/2}^{n} \\ &\quad + H_y\big|_{i,j+1/2,k}^{n} - H_y\big|_{i,j+1/2,k+1}^{n} - J_{\text{source}_x}\big|_{i,j+1/2,k+1/2}^{n}\Delta\big). \end{aligned} \tag{3.13}$$

Similarly, to update H_x we have:

$$\begin{aligned} m &= \text{MEDIA}_{H_x}\big|_{i-1/2,j+1,k+1}, \\ H_x\big|_{i-1/2,j+1,k+1}^{n+1} &= D_a(m) H_x\big|_{i-1/2,j+1,k+1}^{n} + D_b(m) \cdot \big(E_y\big|_{i-1/2,j+1,k+3/2}^{n+1/2} - E_y\big|_{i-1/2,j+1,k+1/2}^{n+1/2} \\ &\quad + E_z\big|_{i-1/2,j+1/2,k+1}^{n+1/2} - E_z\big|_{i-1/2,j+3/2,k+1}^{n+1/2} - M_{\text{source}_x}\big|_{i-1/2,j+1,k+1}^{n+1/2}\Delta\big). \end{aligned} \tag{3.14}$$

We note that the coefficient arrays $C_a(m)$, $C_b(m)$, $D_a(m)$, and $D_b(m)$ each contain only M elements, where M is the number of distinct material media in the FDTD space lattice. Thus, if separate MEDIA(i, j, k) integer pointer arrays are provided for each field vector component, the approximate total computer storage needed is reduced to $12N$, where N is the number of space cells in the FDTD lattice. This reduction in computer storage comes at some cost, however, since additional computer instructions must be executed at each field vector location to obtain the pointer integer m from the associated MEDIA array and then extract the $C(m)$ or $D(m)$ updating coefficients.

Taking advantage of the integer nature of the MEDIA arrays, further reduction in computer storage can be achieved by bitwise packing of these integers. For example, a 64-bit word can be divided into sixteen 4-bit pointers. Such a composite pointer could specify up to $2^4 = 16$ distinct media at each of 16 locations in the grid. This provides the means to reduce the overall computer storage for the MEDIA arrays by a factor of $15/16$ (94%).

3.5. *Space region with nonpermeable media*

Many electromagnetic wave interaction problems involve nonpermeable media ($\mu = \mu_0$, $\sigma^* = 0$) and can be implemented on a uniform cubic-cell FDTD space lattice. For such problems, the field updating expressions can be further simplified by defining the proportional $\vec{E}$ and $\vec{M}$ vectors:

$$\hat{\vec{E}} = (\Delta t/\mu_0\Delta)\vec{E}; \tag{3.15a}$$

$$\hat{\vec{M}} = (\Delta t/\mu_0)\vec{M}, \tag{3.15b}$$

where $\Delta = \Delta x = \Delta y = \Delta z$ is the cell size of the space lattice. Assuming that $\hat{E}_x$, $\hat{E}_y$, and $\hat{E}_z$ are stored in the computer memory, and further defining a scaled E-field updating coefficient $\hat{C}_b(m)$ as

$$\hat{C}_b(m) = (\Delta t/\mu_0\Delta)C_b(m) \tag{3.16}$$

we can rewrite (3.13) as:

$$\begin{aligned} m &= \text{MEDIA}_{E_x}|_{i,j+1/2,k+1/2}, \\ \hat{E}_x\big|^{n+1/2}_{i,j+1/2,k+1/2} &= C_a(m)\hat{E}_x\big|^{n-1/2}_{i,j+1/2,k+1/2} + \hat{C}_b(m)\cdot\big(H_z\big|^n_{i,j+1,k+1/2} - H_z\big|^n_{i,j,k+1/2} \\ &\quad + H_y\big|^n_{i,j+1/2,k} - H_y\big|^n_{i,j+1/2,k+1} - J_{\text{source}_x}\big|^n_{i,j+1/2,k+1/2}\Delta\big). \end{aligned} \tag{3.17}$$

Finite-difference expression (3.14) can now be rewritten very simply as:

$$\begin{aligned} H_x\big|^{n+1}_{i-1/2,j+1,k+1} &= H_x\big|^n_{i-1/2,j+1,k+1} + \hat{E}_y\big|^{n+1/2}_{i-1/2,j+1,k+3/2} - \hat{E}_y\big|^{n+1/2}_{i-1/2,j+1,k+1/2} \\ &\quad + \hat{E}_z\big|^{n+1/2}_{i-1/2,j+1/2,k+1} - \hat{E}_z\big|^{n+1/2}_{i-1/2,j+3/2,k+1} - \hat{M}_{\text{source}_x}\big|^{n+1/2}_{i-1/2,j+1,k+1}. \end{aligned} \tag{3.18}$$

This technique eliminates the multiplications previously needed to update the H components, and requires storage of MEDIA arrays only for the E components. At the end of the run, the desired values of the unscaled E-fields can be obtained simply by multiplying the stored E values by the reciprocal of the scaling factor of (3.15a).

3.6. *Reduction to the two-dimensional* TM_z *and* TE_z *modes*

The finite-difference systems of (3.7) and (3.8) can be reduced for the decoupled, two-dimensional TM_Z and TE_Z modes summarized in Section 2.2. For convenience and consistency, we again consider the field vector components in the space lattice represented by the unit cell of Fig. 3.1. Assuming now that all partial derivatives of the fields with respect to z are equal to zero, the following conditions hold:

(1) The sets of (E_z, H_x, H_y) components located in each lattice cut plane k, $k+1$, etc. are identical and can be completely represented by any one of these sets, which we designate as the TM_Z mode.

(2) The sets of (H_z, E_x, E_y) components located in each lattice cut plane $k+1/2$, $k+3/2$, etc. are identical and can be completely represented by any one of these sets, which we designate as the TE_Z mode.

The resulting finite-difference systems for the TM_Z and TE_Z modes are as follows:

TM_Z *mode, corresponding to the system of* (2.11)

$$H_x\big|_{i-1/2,j+1}^{n+1} = \left(\frac{1-\dfrac{\sigma^*_{i-1/2,j+1}\Delta t}{2\mu_{i-1/2,j+1}}}{1+\dfrac{\sigma^*_{i-1/2,j+1}\Delta t}{2\mu_{i-1/2,j+1}}}\right) H_x\big|_{i-1/2,j+1}^{n} + \left(\frac{\dfrac{\Delta t}{\mu_{i-1/2,j+1}}}{1+\dfrac{\sigma^*_{i-1/2,j+1}\Delta t}{2\mu_{i-1/2,j+1}}}\right)$$
$$\times \left(\frac{E_z\big|_{i-1/2,j+1/2}^{n+1/2} - E_z\big|_{i-1/2,j+3/2}^{n+1/2}}{\Delta y} - M_{\mathrm{source}_x}\big|_{i-1/2,j+1}^{n+1/2}\right), \tag{3.19a}$$

$$H_y\big|_{i,j+1/2}^{n+1} = \left(\frac{1-\dfrac{\sigma^*_{i,j+1/2}\Delta t}{2\mu_{i,j+1/2}}}{1+\dfrac{\sigma^*_{i,j+1/2}\Delta t}{2\mu_{i,j+1/2}}}\right) H_y\big|_{i,j+1/2}^{n} + \left(\frac{\dfrac{\Delta t}{\mu_{i,j+1/2}}}{1+\dfrac{\sigma^*_{i,j+1/2}\Delta t}{2\mu_{i,j+1/2}}}\right)$$
$$\times \left(\frac{E_z\big|_{i+1/2,j+1/2}^{n+1/2} - E_z\big|_{i-1/2,j+1/2}^{n+1/2}}{\Delta x} - M_{\mathrm{source}_y}\big|_{i,j+1/2}^{n+1/2}\right), \tag{3.19b}$$

$$E_z\big|_{i-1/2,j+1/2}^{n+1/2} = \left(\frac{1-\dfrac{\sigma_{i-1/2,j+1/2}\Delta t}{2\varepsilon_{i-1/2,j+1/2}}}{1+\dfrac{\sigma_{i-1/2,j+1/2}\Delta t}{2\varepsilon_{i-1/2,j+1/2}}}\right) E_z\big|_{i-1/2,j+1/2}^{n-1/2} + \left(\frac{\dfrac{\Delta t}{\varepsilon_{i-1/2,j+1/2}}}{1+\dfrac{\sigma_{i-1/2,j+1/2}\Delta t}{2\varepsilon_{i-1/2,j+1/2}}}\right)$$
$$\times \left(\begin{array}{l}\dfrac{H_y\big|_{i,j+1/2}^{n} - H_y\big|_{i-1,j+1/2}^{n}}{\Delta x} + \dfrac{H_x\big|_{i-1/2,j}^{n} - H_x\big|_{i-1/2,j+1}^{n}}{\Delta y} \\ \quad - J_{\mathrm{source}_z}\big|_{i-1/2,j+1/2}^{n}\end{array}\right). \tag{3.19c}$$

TE$_z$ *mode, corresponding to the system of* (2.12)

$$E_x\big|_{i,j+1/2}^{n+1/2} = \left(\frac{1-\frac{\sigma_{i,j+1/2}\Delta t}{2\varepsilon_{i,j+1/2}}}{1+\frac{\sigma_{i,j+1/2}\Delta t}{2\varepsilon_{i,j+1/2}}}\right) E_x\big|_{i,j+1/2}^{n-1/2} + \left(\frac{\frac{\Delta t}{\varepsilon_{i,j+1/2}}}{1+\frac{\sigma_{i,j+1/2}\Delta t}{2\varepsilon_{i,j+1/2}}}\right) \times \left(\frac{H_z\big|_{i,j+1}^{n} - H_z\big|_{i,j}^{n}}{\Delta y} - J_{\text{source}_x}\big|_{i,j+1/2}^{n}\right), \tag{3.20a}$$

$$E_y\big|_{i-1/2,j+1}^{n+1/2} = \left(\frac{1-\frac{\sigma_{i-1/2,j+1}\Delta t}{2\varepsilon_{i-1/2,j+1}}}{1+\frac{\sigma_{i-1/2,j+1}\Delta t}{2\varepsilon_{i-1/2,j+1}}}\right) E_y\big|_{i-1/2,j+1}^{n-1/2} + \left(\frac{\frac{\Delta t}{\varepsilon_{i-1/2,j+1}}}{1+\frac{\sigma_{i-1/2,j+1}\Delta t}{2\varepsilon_{i-1/2,j+1}}}\right) \times \left(\frac{H_z\big|_{i-1,j+1}^{n} - H_z\big|_{i,j+1}^{n}}{\Delta x} - J_{\text{source}_y}\big|_{i-1/2,j+1}^{n}\right), \tag{3.20b}$$

$$H_z\big|_{i,j+1}^{n+1} = \left(\frac{1-\frac{\sigma^*_{i,j+1}\Delta t}{2\mu_{i,j+1}}}{1+\frac{\sigma^*_{i,j+1}\Delta t}{2\mu_{i,j+1}}}\right) H_z\big|_{i,j+1}^{n} + \left(\frac{\frac{\Delta t}{\mu_{i,j+1}}}{1+\frac{\sigma^*_{i,j+1}\Delta t}{2\mu_{i,j+1}}}\right) \times \left(\begin{array}{l} \frac{E_x|_{i,j+3/2}^{n+1/2} - E_x|_{i,j+1/2}^{n+1/2}}{\Delta y} \\ \quad + \frac{E_y|_{i-1/2,j+1}^{n+1/2} - E_y|_{i+1/2,j+1}^{n+1/2}}{\Delta x} - M_{\text{source}_z}\big|_{i,j+1}^{n+1/2} \end{array}\right). \tag{3.20c}$$

3.7. *Interpretation as Faraday's and Ampere's Laws in integral form*

The Yee algorithm for FDTD was originally interpreted as a direct approximation of the pointwise derivatives of Maxwell's time-dependent curl equations by numerical central differences. Although this interpretation is useful for understanding how FDTD models wave propagation away from material interfaces, it sheds little light on what algorithm modifications are needed to properly model the electromagnetic field physics of fine geometrical features such as wires, slots, and curved surfaces requiring subcell spatial resolution.

The literature indicates that FDTD modeling can be extended to such features by departing from Yee's original pointwise derivative thinking (see, for example, TAFLOVE, UMASHANKAR, BEKER, HARFOUSH and YEE [1988] and JURGENS, TAFLOVE, UMASHANKAR and MOORE [1992]). As shown in Fig. 3.3, the idea involves starting with a more macroscopic (but still local) combined-field description based upon Ampere's Law and Faraday's Law in *integral* form, implemented on an array of electrically small, spatially orthogonal contours. These contours mesh (intersect) in the manner of links in a chain, providing a geometrical interpretation of the coupling of these two laws. This meshing results in the filling of the FDTD modeled space by a three-dimensional "chain-link" array of intersecting orthogonal contours. The presence of wires, slots, and curved surfaces can be modeled by incorporating appropriate field behavior into the contour and surface integrals used to implement Ampere's and Faraday's Laws at selected meshes, and by deforming contour paths as required to conform with surface curvature.

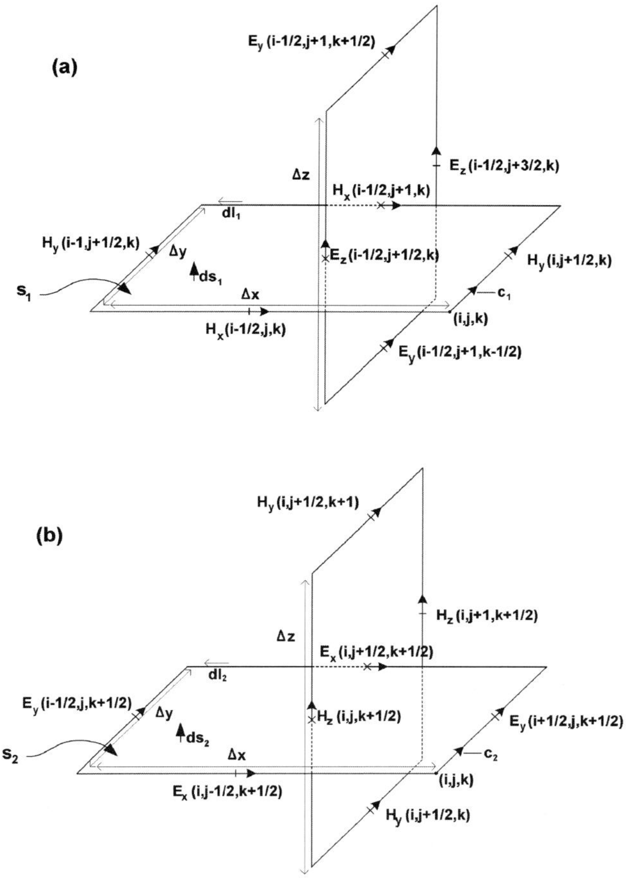

FIG. 3.3. Examples of chain-linked orthogonal contours in the free-space Yee mesh. (a) Ampere's Law for time-stepping E_z; (b) Faraday's Law for time-stepping H_z. *Adapted from*: A. Taflove et al., *IEEE Trans. Antennas and Propagation*, 1988, pp. 247–257, © 1988 IEEE.

This approach is intuitively satisfying to an electrical engineer since it permits the FDTD numerical model to deal with physical quantities such as:

- Electromotive forces (EMFs) and magnetomotive forces (MMFs) developed when completing one circuit about a Faraday's or Ampere's Law contour path;
- Magnetic flux and electric displacement current when performing the surface integrations on the patches bounded by the respective contours.

In this section, we demonstrate the equivalence of the Yee and contour-path interpretations for the free-space case. For simplicity, FDTD time-stepping expressions are developed for only one E and one H field component. Extension to all the rest is straightforward. We further assume lossless free space with no electric or magnetic current sources. Applying Ampere's Law along contour C_1 in Fig. 3.3(a), and assuming that the field value at a midpoint of one side of the contour equals the average value of that field component along that side, we obtain

$$\frac{\partial}{\partial t}\int S_1\vec{D}\cdot \mathrm{d}\vec{S}_1 = \oint_{C_1} \vec{H}\cdot \mathrm{d}\vec{\ell}_1, \tag{3.21a}$$

$$\begin{aligned}\frac{\partial}{\partial t}\int S_1\varepsilon_0 E_z|_{i-1/2,j+1/2,k}\,\mathrm{d}S_1 \cong{}& H_x|_{i-1/2,j,k}\Delta x + H_y|_{i,j+1/2,k}\Delta y\\ &- H_x|_{i-1/2,j+1,k}\Delta x - H_y|_{i-1,j+1/2,k}\Delta y.\end{aligned} \tag{3.21b}$$

Now further assume that $E_z|_{i-1/2,j+1/2,k}$ equals the average value of E_z over the surface patch S_1 and that the time derivative can be numerically realized by using a central-difference expression. Then (3.21b) yields

$$\begin{aligned}&\varepsilon_0\Delta x\Delta y\left(\frac{E_z|^{n+1/2}_{i-1/2,j+1/2,k} - E_z|^{n-1/2}_{i-1/2,j+1/2,k}}{\Delta t}\right)\\ &\quad = \left(H_x|^n_{i-1/2,j,k} - H_x|^n_{i-1/2,j+1,k}\right)\Delta x + \left(H_y|^n_{i,j+1/2,k} - H_y|^n_{i-1,j+1/2,k}\right)\Delta y.\end{aligned} \tag{3.21c}$$

Multiplying both sides by $\Delta t/(\varepsilon_0\Delta x\Delta y)$ and solving for $E_z|^{n+1/2}_{i-1/2,j+1/2,k}$ provides

$$\begin{aligned}E_z|^{n+1/2}_{i-1/2,j+1/2,k} ={}& E_z|^{n-1/2}_{i-1/2,j+1/2,k} + \left(H_x|^n_{i-1/2,j,k} - H_x|^n_{i-1/2,j+1,k}\right)\Delta t/(\varepsilon_0\Delta y)\\ &+ \left(H_y|^n_{i,j+1/2,k} - H_y|^n_{i-1,j+1/2,k}\right)\Delta t/(\varepsilon_0\Delta x).\end{aligned} \tag{3.22}$$

Eq. (3.22) is simply the free-space version of (3.7c), the Yee time-stepping equation for E_z that was obtained directly from implementing the curl $\vec{H}$ equation with finite differences. The only difference is that (3.22) is evaluated at $(i-1/2, j+1/2, k)$ whereas (3.7c) is evaluated at $(i-1/2, j+1/2, k+1)$ shown in Fig. 3.1.

In an analogous manner, we can apply Faraday's Law along contour C_2 in Fig. 3.3(b) to obtain

$$\frac{\partial}{\partial t}\int S_2\vec{B}\cdot \mathrm{d}\vec{S}_2 = -\oint_{C_2} \vec{E}\cdot \mathrm{d}\vec{\ell}_2, \tag{3.23a}$$

$$\begin{aligned}\frac{\partial}{\partial t}\int S_2\mu_0 H_z|_{i,j,k+1/2}\,\mathrm{d}S_2 \cong{}& -E_x|_{i,j-1/2,k+1/2}\Delta x - E_y|_{i+1/2,j,k+1/2}\Delta y\\ &+ E_x|_{i,j+1/2,k+1/2}\Delta x + E_y|_{i-1/2,j,k+1/2}\Delta y,\end{aligned} \tag{3.23b}$$

$$\begin{aligned}&\mu_0\Delta x\Delta y\left(\frac{H_z|^{n+1}_{i,j,k+1/2} - H_z|^n_{i,j,k+1/2}}{\Delta t}\right)\\ &\quad = \left(E_x|^{n+1/2}_{i,j+1/2,k+1/2} - E_x|^{n+1/2}_{i,j-1/2,k+1/2}\right)\Delta x\\ &\qquad + \left(E_y|^{n+1/2}_{i-1/2,j,k+1/2} - E_y|^{n+1/2}_{i+1/2,j,k+1/2}\right)\Delta y.\end{aligned} \tag{3.23c}$$

Multiplying both sides by $\Delta t/(\mu_0\Delta x\Delta y)$ and solving for $H_z|^{n+1/2}_{i,j,k+1/2}$ provides

$$\begin{aligned}H_z|^{n+1}_{i,j,k+1/2} ={}& H_z|^n_{i,j,k+1/2} + \left(E_x|^{n+1/2}_{i,j+1/2,k+1/2} - E_x|^{n+1/2}_{i,j-1/2,k+1/2}\right)\Delta t/(\mu_0\Delta y)\\ &+ \left(E_y|^{n+1/2}_{i-1/2,j,k+1/2} - E_y|^{n+1/2}_{i+1/2,j,k+1/2}\right)\Delta t/(\mu_0\Delta x).\end{aligned} \tag{3.24}$$

Eq. (3.24) is simply the free-space version of (3.8c), the Yee time-stepping expression for H_z that was obtained directly from implementing the curl $\vec{E}$ equation with finite differences. The only difference is that (3.24) is evaluated at $(i, j, k + 1/2)$ whereas (3.8c) is evaluated at $(i, j + 1, k + 1/2)$ shown in Fig. 3.1.

3.8. Divergence-free nature

We now demonstrate that the Yee algorithm satisfies Gauss' Law for the electric field, Eq. (2.3), and hence is divergence-free in source-free space. We first form the time derivative of the total electric flux over the surface of a single Yee cell of Fig. 3.1:

$$\begin{aligned}
&\frac{\partial}{\partial t} \oiint_{\text{Yee cell}} \vec{D} \cdot \mathrm{d}\vec{S} \\
&= \underbrace{\varepsilon_0 \frac{\partial}{\partial t}\left(E_x|_{i,j+1/2,k+1/2} - E_x|_{i-1,j+1/2,k+1/2}\right)}_{\text{Term 1}} \Delta y \Delta z \\
&\quad + \underbrace{\varepsilon_0 \frac{\partial}{\partial t}\left(E_y|_{i-1/2,j+1,k+1/2} - E_y|_{i-1/2,j,k+1/2}\right)}_{\text{Term 2}} \Delta x \Delta z \\
&\quad + \underbrace{\varepsilon_0 \frac{\partial}{\partial t}\left(E_z|_{i-1/2,j+1/2,k+1} - E_z|_{i-1/2,j+1/2,k}\right)}_{\text{Term 3}} \Delta x \Delta y.
\end{aligned} \tag{3.25}$$

Using the Yee algorithm time-stepping relations for the E-field components according to (3.7), we substitute appropriate H-field spatial finite differences for the E-field time derivatives in each term:

$$\begin{aligned}
&\text{Term 1} \\
&= \left(\frac{H_z|_{i,j+1,k+1/2} - H_z|_{i,j,k+1/2}}{\Delta y} - \frac{H_y|_{i,j+1/2,k+1} - H_y|_{i,j+1/2,k}}{\Delta z}\right) \\
&\quad - \left(\frac{H_z|_{i-1,j+1,k+1/2} - H_z|_{i-1,j,k+1/2}}{\Delta y} - \frac{H_y|_{i-1,j+1/2,k+1} - H_y|_{i-1,j+1/2,k}}{\Delta z}\right),
\end{aligned} \tag{3.26a}$$

$$\begin{aligned}
&\text{Term 2} \\
&= \left(\frac{H_x|_{i-1/2,j+1,k+1} - H_x|_{i-1/2,j+1,k}}{\Delta z} - \frac{H_z|_{i,j+1,k+1/2} - H_z|_{i-1,j+1,k+1/2}}{\Delta x}\right) \\
&\quad - \left(\frac{H_x|_{i-1/2,j,k+1} - H_x|_{i-1/2,j,k}}{\Delta z} - \frac{H_z|_{i,j,k+1/2} - H_z|_{i-1,j,k+1/2}}{\Delta x}\right),
\end{aligned} \tag{3.26b}$$

$$\begin{aligned}
&\text{Term 3} \\
&= \left(\frac{H_y|_{i,j+1/2,k+1} - H_y|_{i-1,j+1/2,k+1}}{\Delta x} - \frac{H_x|_{i-1/2,j+1,k+1} - H_x|_{i-1/2,j,k+1}}{\Delta y}\right) \\
&\quad - \left(\frac{H_y|_{i,j+1/2,k} - H_y|_{i-1,j+1/2,k}}{\Delta x} - \frac{H_x|_{i-1/2,j+1,k} - H_x|_{i-1/2,j,k}}{\Delta y}\right).
\end{aligned} \tag{3.26c}$$

For all time steps, this results in

$$\frac{\partial}{\partial t}\oiint_{\text{Yee cell}} \vec{D}\cdot \mathrm{d}\vec{S} = (\text{Term } 1)\Delta y\Delta z + (\text{Term } 2)\Delta x\Delta z + (\text{Term } 3)\Delta x\Delta y \\ = 0. \tag{3.27}$$

Assuming zero initial conditions, the constant zero value of the time derivative of the net electric flux leaving the Yee cell means that this flux never departs from zero:

$$\oiint_{\text{Yee cell}} \vec{D}(t)\cdot \mathrm{d}\vec{S} = \oiint_{\text{Yee cell}} \vec{D}(t=0)\cdot \mathrm{d}\vec{S} = 0. \tag{3.28}$$

Therefore, the Yee cell satisfies Gauss' Law for the E-field in charge-free space and thus is divergence-free with respect to its E-field computations. The proof of the satisfaction of Gauss' Law for the magnetic field, Eq. (2.4), is by analogy.

4. Nonuniform Yee grid

4.1. Introduction

The FDTD algorithm is second-order-accurate by nature of the central-difference approximations used to realize the first-order spatial and temporal derivatives. This leads to a discrete approximation for the fields based on a uniform space lattice. Unfortunately, structures with fine geometrical features cannot always conform to the edges of a uniform lattice. Further, it is often desirable to have a refined lattice in localized regions, such as near sharp edges or corners, to accurately model the local field phenomena.

A quasi-nonuniform grid FDTD algorithm was introduced by SHEEN [1991]. This method is based on reducing the grid size by exactly one-third. By choosing the subgrid to be exactly one-third, the spatial derivatives of the fields at the interface between the two regions can be expressed using central-difference approximations, resulting in a second-order-accurate formulation. This technique was successfully applied to a number of microwave circuit and antenna problems (see, for example, SHEEN [1991] and TULINTSEFF [1992]). However, this method is limited to specific geometries that conform to this specialized grid.

It is clear that more general geometries could be handled by a grid with arbitrary spacing. Unfortunately, central differences can no longer be used to evaluate the spatial derivatives of the fields for such a grid, leading to first-order error. However, it was demonstrated by MONK and SULI [1994] and MONK [1994] that, while this formulation does lead to first-order error locally, it results in second-order error globally. This is known as *supraconvergence* (see also MANTEUFFEL and WHITE [1986] and KREISS, MANTEUFFEL, SCHWARTZ, WENDROFF and WHITE [1986]).

4.2. Supraconvergent FDTD algorithm

This section presents the supraconvergent FDTD algorithm based on nonuniform meshing that was discussed by GEDNEY and LANSING [1995]. Following their notation, a three-dimensional nonuniform space lattice is introduced. The vertices of the lattice are defined by the general one-dimensional coordinates:

$$\{x_i;\, i=1, N_x\}; \qquad \{y_j;\, j=1, N_y\}; \qquad \{z_k;\, k=1, N_z\}. \tag{4.1}$$

The edge lengths between vertices are also defined as

$$\begin{aligned}
&\{\Delta x_i = x_{i+1} - x_i;\ i=1,\ N_x-1\};\\
&\{\Delta y_j = y_{j+1} - y_j;\ j=1, N_y-1\};\\
&\{\Delta z_k = z_{k+1} - z_k;\ k=1, N_z-1\}.
\end{aligned} \tag{4.2}$$

Within the nonuniform space, a reduced notation is introduced, defining the cell and edge centers:

$$x_{i+1/2} = x_i + \Delta x_i/2; \qquad y_{j+1/2} = y_j + \Delta y_j/2; \qquad z_{k+1/2} = z_k + \Delta z_k/2. \tag{4.3}$$

A set of dual edge lengths representing the distances between the edge centers is then introduced:

$$\begin{aligned}
&\left\{h_i^x = (\Delta x_i + \Delta x_{i-1})/2;\ i=2, N_x\right\};\\
&\left\{h_j^y = (\Delta y_j + \Delta y_{j-1})/2;\ j=2, N_y\right\};\\
&\left\{h_k^z = (\Delta z_k + \Delta z_{k-1})/2;\ k=2, N_z\right\}.
\end{aligned} \tag{4.4}$$

Finally, the E- and H-fields in the discrete nonuniform grid are denoted as in the following examples:

$$E_x\big|_{i+1/2,j,k}^{n} \equiv E_x(x_{i+1/2}, y_j, z_k, n\Delta t), \tag{4.5a}$$

$$H_x\big|_{i,j+1/2,k+1/2}^{n+1/2} \equiv H_x\big(x_i, y_{j+1/2}, z_{k+1/2}, (n+1/2)\Delta t\big). \tag{4.5b}$$

The nonuniform FDTD algorithm is based on a discretization of Maxwell's equations in their integral form, specifically, Faraday's Law and Ampere's Law:

$$\oint_C \vec{E}\cdot \mathrm{d}\vec{\ell} = -\frac{\partial}{\partial t}\iint_S \vec{B}\cdot \mathrm{d}\vec{s} - \iint_S \vec{M}\cdot \mathrm{d}\vec{s}, \tag{4.6}$$

$$\oint_{C'} \vec{H}\cdot \mathrm{d}\vec{\ell} = \frac{\partial}{\partial t}\iint_{S'} \vec{D}\cdot \mathrm{d}\vec{s} + \iint_{S'} \sigma\vec{E}\cdot \mathrm{d}\vec{s} + \iint_{S'} \vec{J}\cdot \mathrm{d}\vec{s}. \tag{4.7}$$

The surface integral in (4.6) is performed over a lattice cell face, and the contour integral is performed over the edges bounding the face, as illustrated in Fig. 4.1(a). Similarly, the surface integral in (4.7) is performed over a dual-lattice cell face.

Evaluating (4.6) and (4.7) over the cell faces using (4.5), and evaluating the time derivatives using central-differencing leads to

$$\begin{aligned}
&E_x\big|_{i+1/2,j+1,k}^{n}\Delta x_i - E_x\big|_{i+1/2,j,k}^{n}\Delta x_i - E_y\big|_{i+1,j+1/2,k}^{n}\Delta y_j + E_y\big|_{i,j+1/2,k}^{n}\Delta y_j\\
&\quad = -\left[\mu_{i+1/2,j+1/2,k}\left(\frac{H_z\big|_{i+1/2,j+1/2,k}^{n+1/2} - H_z\big|_{i+1/2,j+1/2,k}^{n-1/2}}{\Delta t}\right) + M_z\big|_{i+1/2,j+1/2,k}^{n+1/2}\right]\\
&\qquad \times \Delta x_i \Delta y_j,
\end{aligned} \tag{4.8}$$

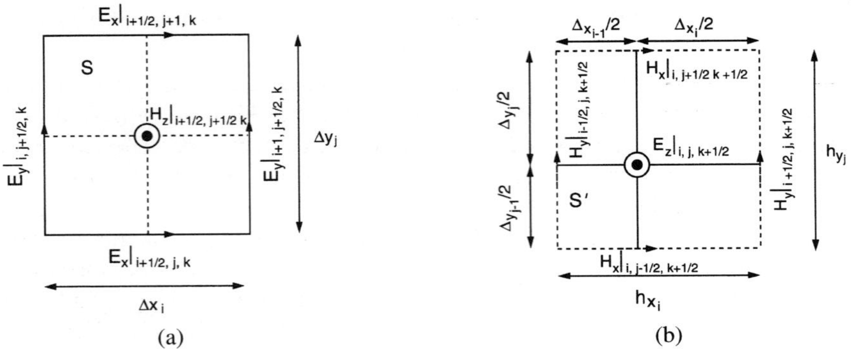

FIG. 4.1. Lattice faces bounded by lattice edges defining surfaces of integration bounded by closed contours. (a) Lattice cell face bounded by grid edges, showing a dual-lattice edge passing through its center. (b) The dual-lattice face bounded by dual edges. *Source:* GEDNEY and LANSING [1995].

$$
\begin{aligned}
&H_x\big|_{i,j+1/2,k+1/2}^{n+1/2} h_{x_i} - H_x\big|_{i,j-1/2,k+1/2}^{n+1/2} h_{x_i} - H_y\big|_{i+1/2,j,k+1/2}^{n+1/2} h_{y_j} + H_y\big|_{i-1/2,j,k+1/2}^{n+1/2} h_{y_j} \\
&\quad = \left[\begin{array}{l} \varepsilon_{i,j,k+1/2}\left(\dfrac{E_z\big|_{i,j,k+1/2}^{n+1} - E_z\big|_{i,j,k+1/2}^{n}}{\Delta t}\right) \\ \quad + \dfrac{\sigma_{i,j,k+1/2}}{2}\left(\dfrac{E_z\big|_{i,j,k+1/2}^{n+1} + E_z\big|_{i,j,k+1/2}^{n}}{\Delta t}\right) + J_z\big|_{i,j,k+1/2}^{n+1/2} \end{array} \right] h_{x_i} h_{y_j},
\end{aligned} \tag{4.9}
$$

where $\varepsilon_{i,j,k+1/2}$, $\sigma_{i,j,k+1/2}$, and $\mu_{i+1/2,j+1/2,k}$ are the averaged permittivity, conductivity, and permeability, respectively, about the grid edges. Subsequently, this leads to an explicit update scheme:

$$
\begin{aligned}
&H_z\big|_{i+1/2,j+1/2,k}^{n+1/2} \\
&\quad = H_z\big|_{i+1/2,j+1/2,k}^{n-1/2} - \frac{\Delta t}{\mu_{i+1/2,j+1/2,k}} \\
&\qquad \times \left[\begin{array}{l} \frac{1}{\Delta y_j}\left(E_x\big|_{i+1/2,j+1,k}^{n} - E_x\big|_{i+1/2,j,k}^{n}\right) \\ \quad - \frac{1}{\Delta x_i}\left(E_y\big|_{i+1,j+1/2,k}^{n} - E_y\big|_{i,j+1/2,k}^{n}\right) + M_z\big|_{i+1/2,j+1/2,k}^{n+1/2} \end{array} \right],
\end{aligned} \tag{4.10}
$$

$$
\begin{aligned}
&E_z\big|_{i,j,k+1/2}^{n+1} \\
&\quad = \left(\frac{2\varepsilon_{i,j,k+1/2} - \sigma_{i,j,k+1/2}\Delta t}{2\varepsilon_{i,j,k+1/2} + \sigma_{i,j,k+1/2}\Delta t}\right) E_z\big|_{i,j,k+1/2}^{n} + \left(\frac{2\Delta t}{2\varepsilon_{i,j,k+1/2} + \sigma_{i,j,k+1/2}\Delta t}\right) \\
&\qquad \times \left[\begin{array}{l} \frac{1}{h_{y_j}}\left(H_x\big|_{i,j+1/2,k+1/2}^{n+1/2} - H_x\big|_{i,j-1/2,k+1/2}^{n+1/2}\right) \\ \quad - \frac{1}{h_{x_i}}\left(H_y\big|_{i+1/2,j,k+1/2}^{n+1/2} - H_y\big|_{i-1/2,j,k+1/2}^{n+1/2}\right) - J_z\big|_{i,j,k+1/2}^{n+1/2} \end{array} \right].
\end{aligned} \tag{4.11}
$$

Similar updates for the remaining field components are easily derived by permuting the indices in (4.10) and (4.11) in a right-handed manner.

4.3. Demonstration of supraconvergence

The explicit updates for the H-fields in (4.10) are second-order-accurate in both space and time, since the vertices of the dual lattice are assumed to be located at the cell centers of the primary lattice. On the other hand, the explicit updates for the E-fields in (4.11) are only first-order-accurate in space. This results in local first-order error in regions where the grid is nonuniform.

However, via a numerical example, GEDNEY and LANSING [1995] showed that this method is supraconvergent, i.e., it converges with a higher order accuracy than the local error mandates. They considered calculation of the resonant frequencies of a fixed-size rectangular cavity having perfect electric conductor (PEC) walls. A random, nonuniform grid spacing for x_i, y_j, and z_k was assumed within the cavity such that

$$\{x_i = (i-1)\Delta x + 0.5\Re\Delta x;\ i = 1, N_x\}, \tag{4.12a}$$

$$\{y_j = (j-1)\Delta y + 0.5\Re\Delta y;\ j = 1, N_y\}, \tag{4.12b}$$

$$\{z_k = (k-1)\Delta z + 0.5\Re\Delta z;\ k = 1, N_z\}, \tag{4.12c}$$

where $-1/2 \leqslant \Re \leqslant 1/2$ denotes a random number. The interior of the cavity was excited with a Gaussian-pulsed, z-directed magnetic dipole placed off a center axis:

$$\vec{M}(t) = \hat{z}\mathrm{e}^{-(t-t_0)^2/T^2}. \tag{4.13}$$

The calculated time-varying E-field was probed off a center axis (to avoid the nulls of odd resonant modes), and the cavity resonant frequencies were extracted using a fast Fourier transform (FFT). Spectral peaks resulting from this procedure corresponded to the resonant frequencies of the cavity modes. Subsequently, the average grid cell size h was reduced, and the entire simulation was run again.

Fig. 4.2 graphs the results of four such runs for the error of the nonuniform grid FDTD model in calculating the resonant frequency of the TE_{110} mode relative to the exact solution, as well as a generic order(h^2) accuracy slope. We see that the convergence of the resonant frequency is indeed second-order.

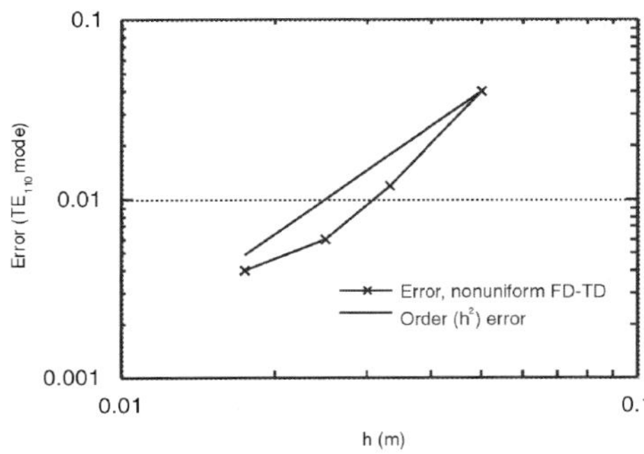

FIG. 4.2. Error convergence of the resonant frequency of the TE_{110} mode of a rectangular PEC cavity computed using the nonuniform FDTD algorithm. *Source*: GEDNEY and LANSING [1995].

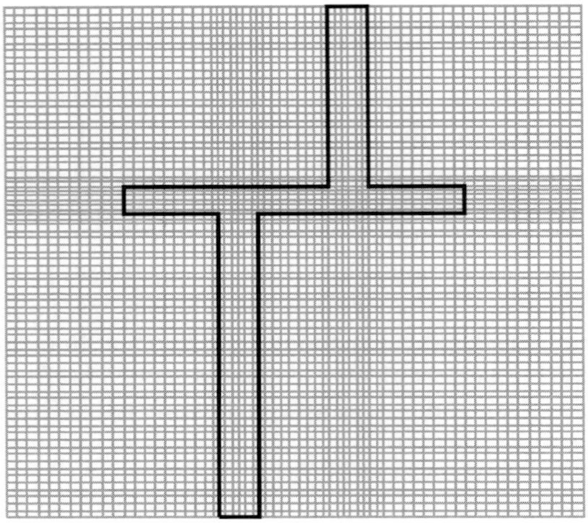

FIG. 4.3. Typical nonuniform grid used for a planar microwave circuit.

The nonuniform FDTD method is well suited for the analysis of planar microwave circuits. The geometrical details of such circuits are typically electrically small, leading to small cell sizes. Further, microwave circuits are often located in an unbounded medium, requiring absorbing boundaries to be placed a sufficient distance from the circuit to avoid nonphysical reflections. For uniform meshing, these two characteristics can combine to produce very large space lattices. With nonuniform meshing, the local cell size can be refined such that the circuit trace size, shape, and field behavior are accurately modeled, while coarser cells are used in regions further from the metal traces. Fig. 4.3 illustrates a typical nonuniform grid used for a microstrip circuit.

5. Alternative finite-difference grids

Thus far, this chapter has considered several fundamental aspects of the uniform Cartesian Yee space lattice for Maxwell's equations. Since 1966, this lattice and its associated staggered leapfrog time-stepping algorithm have proven to be very flexible, accurate, and robust for a wide variety of engineering problems. However, Yee's staggered, uncollocated arrangement of electromagnetic field components is but one possible alternative in a Cartesian coordinate system (see, for example, LIU, Y. [1996]). In turn, a Cartesian grid is but one possible arrangement of field components in two and three dimensions. Other possibilities include hexagonal grids in two dimensions and tetradecahedron/dual-tetrahedron meshes in three dimensions (see again LIU, Y. [1996]).

It is important to develop criteria for the use of a particular space lattice and time-stepping algorithm to allow optimum selection for a given problem. A key consideration is the capability of rendering the geometry of the structure of interest within the space lattice with sufficient accuracy and detail to obtain meaningful results. A second fundamental consideration is the accuracy by which the algorithm simulates the propagation of electromagnetic waves as they interact with the structure.

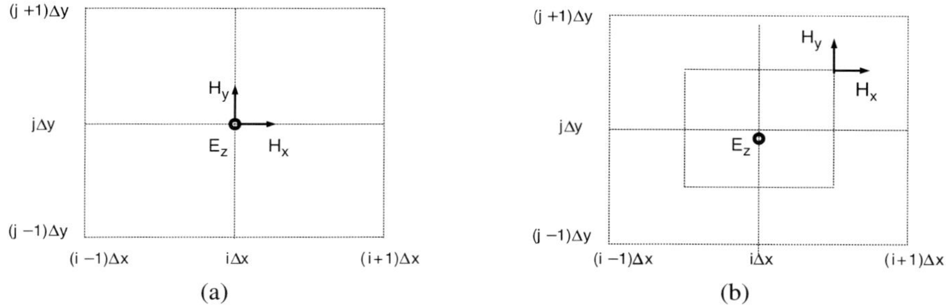

FIG. 5.1. Two Cartesian grids that are alternatives to Yee's arrangement (illustrated in two dimensions for the TM_Z case). (a) Unstaggered, collocated grid. (b) Staggered, collocated grid. *Source*: Y. Liu, *J. Computational Physics*, 1996, pp. 396–416.

5.1. Cartesian grids

Fig. 5.1 illustrates two Cartesian grids that are alternatives to Yee's arrangement in two dimensions for the TM_Z case, as discussed by LIU, Y. [1996]: (a) the unstaggered, collocated grid, in which all E- and H-components are collocated at a single set of grid-cell vertices; and (b) the staggered, collocated grid, in which all E-components are collocated at a distinct set of grid-cell vertices that are spatially interleaved with a second distinct set of vertices where all H-components are collocated.

Upon applying second-order-accurate central space differences to the TM_Z mode equations of (2.11) for the unstaggered, collocated grid of Fig. 5.1(a) (with a lossless material background assumed for simplicity), we obtain as per LIU, Y. [1996]:

$$\frac{\partial H_x|_{i,j}}{\partial t} = -\frac{1}{\mu_{i,j}} \cdot \left(\frac{E_z|_{i,j+1} - E_z|_{i,j-1}}{2\Delta y}\right), \tag{5.1a}$$

$$\frac{\partial H_y|_{i,j}}{\partial t} = \frac{1}{\mu_{i,j}} \cdot \left(\frac{E_z|_{i+1,j} - E_z|_{i-1,j}}{2\Delta x}\right), \tag{5.1b}$$

$$\frac{\partial E_z|_{i,j}}{\partial t} = \frac{1}{\varepsilon_{i,j}} \cdot \left(\frac{H_y|_{i+1,j} - H_y|_{i-1,j}}{2\Delta x} - \frac{H_x|_{i,j+1} - H_x|_{i,j-1}}{2\Delta y}\right). \tag{5.1c}$$

Similarly, applying second-order-accurate central space differences to the TM_Z mode equations of (2.11) for the staggered, collocated grid of Fig. 5.1(b) yields:

$$\frac{\partial H_x|_{i+1/2,j+1/2}}{\partial t} = -\frac{0.5}{\mu_{i+1/2,j+1/2}} \cdot \left[\frac{(E_z|_{i,j+1} + E_z|_{i+1,j+1}) - (E_z|_{i,j} + E_z|_{i+1,j})}{\Delta y}\right], \tag{5.2a}$$

$$\frac{\partial H_y|_{i+1/2,j+1/2}}{\partial t} = \frac{0.5}{\mu_{i+1/2,j+1/2}} \cdot \left[\frac{(E_z|_{i+1,j} + E_z|_{i+1,j+1}) - (E_z|_{i,j} + E_z|_{i,j+1})}{\Delta x}\right], \tag{5.2b}$$

$$\frac{\partial E_z|_{i,j}}{\partial t} = \frac{0.5}{\varepsilon_{i,j}} \cdot \left[\begin{array}{l} \frac{(H_y|_{i+1/2,j-1/2}+H_y|_{i+1/2,j+1/2})-(H_y|_{i-1/2,j-1/2}+H_y|_{i-1/2,j+1/2})}{\Delta x} \\ -\frac{(H_x|_{i-1/2,j+1/2}+H_x|_{i+1/2,j+1/2})-(H_x|_{i-1/2,j-1/2}+H_x|_{i+1/2,j-1/2})}{\Delta y} \end{array} \right]. \tag{5.2c}$$

LIU, Y. [1996] analyzed the Yee grid and the alternative Cartesian grids of Figs. 5.1(a) and 5.1(b) for a key source of error: the numerical phase-velocity anisotropy. This error, discussed in Section 6, is a nonphysical variation of the speed of a numerical wave within an empty grid as a function of its propagation direction. To limit this error to less than 0.1%, LIU, Y. [1996] showed that we require a resolution of 58 points per free-space wavelength λ_0 for the grid of Fig. 5.1(a), 41 points per λ_0 for the grid of Fig. 5.1(b), and only 29 points per λ_0 for the Yee grid. Thus, Yee's grid provides more accurate modeling results than the two alternatives of Fig. 5.1.

5.2. *Hexagonal grids*

LIU, Y. [1996] proposed using regular hexagonal grids in two dimensions to reduce the numerical phase-velocity anisotropy well below that of Yee's Cartesian mesh. Here, the primary grid is composed of equilateral hexagons of edge length Δs. Each hexagon can be considered to be the union of six equilateral triangles. Connecting the centroids of these triangles yields a second set of regular hexagons that comprises a dual grid.

Fig. 5.2 illustrates for the TM_z case in two dimensions the two principal ways of arranging E and H components in hexagonal grids. Fig. 5.2(a) shows the unstaggered, collocated grid in which Cartesian E_z, H_x, and H_y components are collocated at the vertices of the equilateral triangles. No dual grid is used. Fig. 5.2(b) shows the field arrangement for the staggered, uncollocated grid and its associated dual grid, the latter indicated by the dashed line segments. Here, only E_z components are defined at the vertices of the equilateral triangles, which are the centroids of the hexagonal faces of the dual grid. Magnetic field components H_1, H_2, H_3, etc. are defined to be tangential to, and centered on, the edges of the dual-grid hexagons. These magnetic components

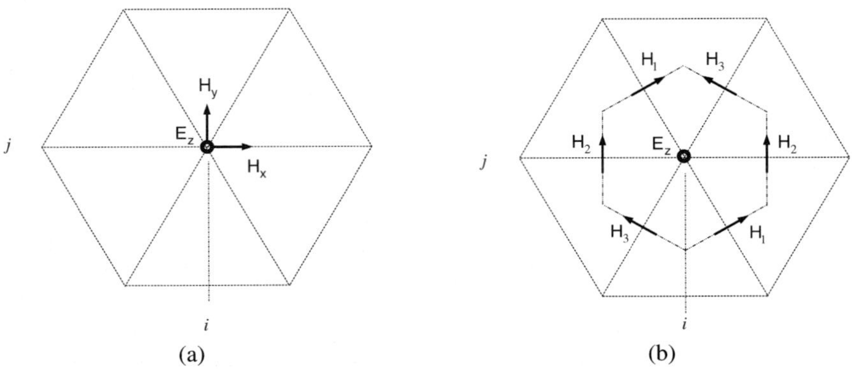

FIG. 5.2. Two central-difference hexagonal grids that are alternatives to Yee's arrangement (illustrated in two dimensions for the TM_Z case). (a) Unstaggered, collocated grid, with no dual grid. (b) Staggered, uncollocated grid and its associated dual grid. *Source*: Y. Liu, *J. Computational Physics*, 1996, pp. 396–416.

are also perpendicular to, and centered on, the edges of the primary-grid triangles. We note that the grid of Fig. 5.2(b) is a direct extension of Yee's interleaved E and H component topology from rectangular to hexagonal cells.

Upon applying second-order-accurate central space differences to the $\mathrm{TM_z}$ mode equations of (2.11) for the unstaggered, collocated hexagonal grid of Fig. 5.2(a) (with a lossless material background assumed for simplicity), we obtain as per LIU, Y. [1996]:

$$\frac{\partial H_x|_{i,j}}{\partial t} = -\frac{\sqrt{3}}{\mu_{i,j}6\Delta s}\begin{pmatrix} E_z|_{i-1/2,j+0.5\sqrt{3}} + E_z|_{i+1/2,j+0.5\sqrt{3}} \\ - E_z|_{i-1/2,j-0.5\sqrt{3}} - E_z|_{i+1/2,j-0.5\sqrt{3}} \end{pmatrix}, \tag{5.3a}$$

$$\begin{aligned}\frac{\partial H_y|_{i,j}}{\partial t} &\\ = \frac{1}{\mu_{i,j}6\Delta s}&\begin{pmatrix} 2E_z|_{i+1,j} - 2E_z|_{i-1,j} + E_z|_{i+1/2,j+0.5\sqrt{3}} \\ - E_z|_{i-1/2,j+0.5\sqrt{3}} + E_z|_{i+1/2,j-0.5\sqrt{3}} - E_z|_{i-1/2,j-0.5\sqrt{3}} \end{pmatrix},\end{aligned} \tag{5.3b}$$

$$\begin{aligned}\frac{\partial E_z|_{i,j}}{\partial t} &\\ = \frac{1}{\varepsilon_{i,j}6\Delta s}&\begin{pmatrix} 2H_y|_{i+1,j} - 2H_y|_{i-1,j} + H_y|_{i+1/2,j+0.5\sqrt{3}} \\ - H_y|_{i-1/2,j+0.5\sqrt{3}} + H_y|_{i+1/2,j-0.5\sqrt{3}} - H_y|_{i-1/2,j-0.5\sqrt{3}} \\ - \sqrt{3}H_x|_{i+1/2,j+0.5\sqrt{3}} + \sqrt{3}H_x|_{i+1/2,j-0.5\sqrt{3}} \\ - \sqrt{3}H_x|_{i-1/2,j+0.5\sqrt{3}} + \sqrt{3}H_x|_{i-1/2,j-0.5\sqrt{3}} \end{pmatrix}.\end{aligned} \tag{5.3c}$$

Similarly, applying second-order-accurate central space differences to the $\mathrm{TM_z}$ mode equations for the staggered, uncollocated grid of Fig. 5.2(b) yields:

$$\frac{\partial H1|_{i+1/4,j-0.25\sqrt{3}}}{\partial t} = \frac{1}{\mu_{i+1/4,j-0.25\sqrt{3}}\Delta s}\left(E_z|_{i+1/2,j-0.5\sqrt{3}} - E_z|_{i,j}\right), \tag{5.4a}$$

$$\frac{\partial H2|_{i+1/2,j}}{\partial t} = \frac{1}{\mu_{i+1/2,j}\Delta s}\left(E_z|_{i+1,j} - E_z|_{i,j}\right), \tag{5.4b}$$

$$\frac{\partial H3|_{i+1/4,j+0.25\sqrt{3}}}{\partial t} = \frac{1}{\mu_{i+1/4,j+0.25\sqrt{3}}\Delta s}\left(E_z|_{i+1/2,j+0.5\sqrt{3}} - E_z|_{i,j}\right), \tag{5.4c}$$

$$\begin{aligned}\frac{\partial E_z|_{i,j}}{\partial t} &= \frac{2}{\varepsilon_{i,j}3\Delta s} \\ &\times \begin{pmatrix} H1|_{i+1/4,j-0.25\sqrt{3}} + H2|_{i+1/2,j} + H3|_{i+1/4,j+0.25\sqrt{3}} \\ - H1|_{i-1/4,j+0.25\sqrt{3}} - H2|_{i-1/2,j} - H3|_{i-1/4,j-0.25\sqrt{3}} \end{pmatrix}.\end{aligned} \tag{5.4d}$$

We note that the total number of field unknowns for the staggered, uncollocated grid of Fig. 5.2(b) is 33% more than that for the unstaggered grid of Fig. 5.2(a), but the discretization is simpler and the number of total operations is less by about 50%.

LIU, Y. [1996] showed that the numerical velocity anisotropy errors of the hexagonal grids of Figs. 5.2(a) and 5.2(b) are 1/200th and 1/1200th, respectively, that of the rectangular Yee grid for a grid sampling density of 20 points per free-space wavelength. This represents a large potential advantage in computational accuracy for the hexagonal grids. Additional details are provided in Section 6.

5.3. Tetradecahedron/dual-tetrahedron mesh in three dimensions

In three dimensions, the uniform Cartesian Yee mesh consists of an ordered array of hexahedral unit cells ("bricks"), as shown in Fig. 3.1. This simple arrangement is attractive since the location of every field component in the mesh is easily and compactly specified, and geometry generation can be performed in many cases with paper and pencil.

However, from the discussion of Section 5.2, it is clear that constructing a uniform mesh in three dimensions using shapes other than rectangular "bricks" may lead to superior computational accuracy with respect to the reduction of the velocity-anisotropy error. Candidate shapes for unit cells must be capable of assembly in a regular mesh to completely fill space. In addition to the hexahedron, space-filling shapes include the tetradecahedron (truncated octahedron), hexagonal prism, rhombic dodecahedron, and elongated rhombic dodecahedron (see LIU, Y. [1996]). We note that the three-dimensional lattice corresponding to the two-dimensional, staggered, uncollocated hexagonal grid of Fig. 5.2(b) is the tetradecahedron/dual-tetrahedron configuration shown in Fig. 5.3. Here, the primary mesh is comprised of tetradecahedral units cells having 6 square faces and 8 regular hexagonal faces. The dual mesh is comprised of tetrahedral cells having isosceles-triangle faces with sides in the ratio of $\sqrt{3}$ to 2.

LIU, Y. [1996] reports a study of the extension of Yee's method to the staggered tetradecahedron/dual-tetrahedron mesh of Fig. 5.3. The algorithm uses a centered finite-difference scheme involving 19 independent unknown field components, wherein 12

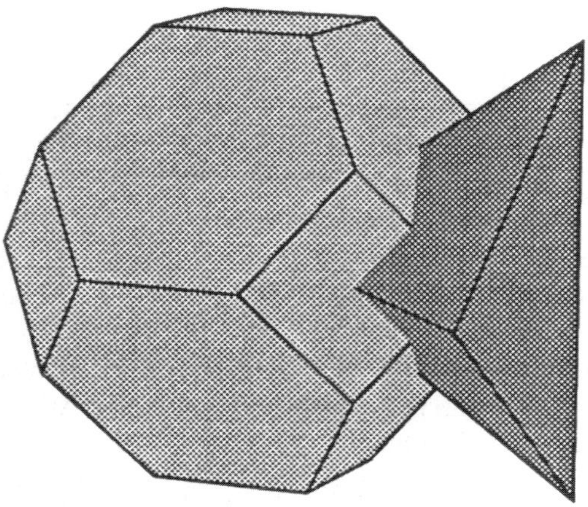

FIG. 5.3. Tetradecahedron and dual-tetrahedron unit cells for the extension of Yee's method to a regular non-Cartesian mesh in three dimensions. This mesh has very favorable numerical wave-velocity anisotropy characteristics relative to the Cartesian arrangement of Fig. 3.1. *Source*: Y. Liu, *J. Computational Physics*, 1996, pp. 396–416.

are defined on the edges of tetradecahedra, and 7 are defined on the edges of the dual tetrahedra. Similar to the staggered, uncollocated hexagonal grid of Fig. 5.2(b), this mesh has very favorable numerical wave-velocity anisotropy characteristics relative to its Yee Cartesian counterpart, shown in Fig. 3.1.

Despite this advantage, the usage of the tetradecahedron/dual-tetrahedron mesh by the FDTD community has been very limited. This is due to the additional complexity in its mesh generation relative to Yee's Cartesian space lattice.

6. Numerical dispersion

6.1. Introduction

The FDTD algorithms for Maxwell's curl equations reviewed in Sections 3–5 cause nonphysical dispersion of the simulated waves in a free-space computational lattice. That is, the phase velocity of numerical wave modes can differ from c by an amount varying with the wavelength, direction of propagation in the grid, and grid discretization. An intuitive way to view this phenomenon is that the FDTD algorithm embeds the electromagnetic wave interaction structure of interest in a tenuous "numerical aether" having properties very close to vacuum, but not quite. This "aether" causes propagating numerical waves to accumulate delay or phase errors that can lead to nonphysical results such as broadening and ringing of pulsed waveforms, imprecise cancellation of multiple scattered waves, anisotropy, and pseudorefraction. Numerical dispersion is a factor that must be accounted to understand the operation of FDTD algorithms and their accuracy limits, especially for electrically large structures.

This section reviews the numerical dispersion characteristics of Yee's FDTD formulation. Section 7 will review proposed low-dispersion FDTD methods, not necessarily based on Yee's space grid and/or the use of explicit central differences.

6.2. Two-dimensional wave propagation, Cartesian Yee grid

We begin our discussion of numerical dispersion with an analysis of the Yee algorithm for the two-dimensional $\mathrm{TM_z}$ mode, (3.19a)–(3.19c), assuming for simplicity no electric or magnetic loss. It can be easily shown that the dispersion relation obtained is valid for any two-dimensional TM or TE mode in a Cartesian Yee grid. The analysis procedure involves substitution of a plane, monochromatic, sinusoidal traveling-wave mode into (3.19a)–(3.19c). After algebraic manipulation, an equation is derived that relates the numerical wavevector components, the wave frequency, the time step, and the grid space increments. This equation, the numerical dispersion relation, can be solved for a variety of grid discretizations, wavevectors, and wave frequencies to illustrate the principal nonphysical results associated with numerical dispersion.

Initiating this procedure, we assume the following plane, monochromatic, sinusoidal traveling wave for the $\mathrm{TM_z}$ mode:

$$E_z|_{I,J}^n = E_{z0}\mathrm{e}^{\mathrm{j}(\omega n\Delta t - \tilde{k}_x I\Delta x - \tilde{k}_y J\Delta y)}, \tag{6.1a}$$

$$H_x|_{I,J}^n = H_{x0}\mathrm{e}^{\mathrm{j}(\omega n\Delta t - \tilde{k}_x I\Delta x - \tilde{k}_y J\Delta y)}, \tag{6.1b}$$

$$H_y|_{I,J}^n = H_{y0} e^{j(\omega n \Delta t - \tilde{k}_x I \Delta x - \tilde{k}_y J \Delta y)}, \tag{6.1c}$$

where $\tilde{k}_x$ and $\tilde{k}_y$ are the x- and y-components of the numerical wavevector and ω is the wave angular frequency. Substituting the traveling-wave expressions of (6.1) into the finite-difference equations of (3.19) yields, after simplification, the following relations for the lossless material case:

$$H_{x0} = \frac{\Delta t E_{z0}}{\mu \Delta y} \cdot \frac{\sin(\tilde{k}_y \Delta y/2)}{\sin(\omega \Delta t/2)}, \tag{6.2a}$$

$$H_{y0} = -\frac{\Delta t E_{z0}}{\mu \Delta x} \cdot \frac{\sin(\tilde{k}_x \Delta x/2)}{\sin(\omega \Delta t/2)}, \tag{6.2b}$$

$$E_{z0} \sin\left(\frac{\omega \Delta t}{2}\right) = \frac{\Delta t}{\varepsilon}\left[\frac{H_{x0}}{\Delta y} \sin\left(\frac{\tilde{k}_y \Delta y}{2}\right) - \frac{H_{y0}}{\Delta x} \sin\left(\frac{\tilde{k}_x \Delta x}{2}\right)\right]. \tag{6.2c}$$

Upon substituting H_{x0} of (6.2a) and H_{y0} of (6.2b) into (6.2c), we obtain

$$\left[\frac{1}{c\Delta t} \sin\left(\frac{\omega \Delta t}{2}\right)\right]^2 = \left[\frac{1}{\Delta x} \sin\left(\frac{\tilde{k}_x \Delta x}{2}\right)\right]^2 + \left[\frac{1}{\Delta y} \sin\left(\frac{\tilde{k}_y \Delta y}{2}\right)\right]^2, \tag{6.3}$$

where $c = 1/\sqrt{\mu\varepsilon}$ is the speed of light in the material being modeled. Eq. (6.3) is the general numerical dispersion relation of the Yee algorithm for the TM_z mode.

We shall consider the important special case of a square-cell grid having $\Delta x = \Delta y = \Delta$. Then, defining the *Courant stability factor* $S = c\Delta t/\Delta$ and the *grid sampling density* $N_\lambda = \lambda_0/\Delta$, we rewrite (6.3) in a more useful form:

$$\frac{1}{S^2} \sin^2\left(\frac{\pi S}{N_\lambda}\right) = \sin^2\left(\frac{\Delta \cdot \tilde{k} \cos\phi}{2}\right) + \sin^2\left(\frac{\Delta \cdot \tilde{k} \sin\phi}{2}\right), \tag{6.4}$$

where ϕ is the propagation direction of the numerical wave with respect to the grid's x-axis. To obtain the numerical dispersion relation for the one-dimensional wave-propagation case, we can assume without loss of generality that $\phi = 0$ in (6.4), yielding

$$\frac{1}{S} \sin\left(\frac{\pi S}{N_\lambda}\right) = \sin\left(\frac{\tilde{k}\Delta}{2}\right) \tag{6.5a}$$

or equivalently

$$\tilde{k} = \frac{2}{\Delta} \sin^{-1}\left[\frac{1}{S} \sin\left(\frac{\pi S}{N_\lambda}\right)\right]. \tag{6.5b}$$

6.3. Extension to three dimensions, Cartesian Yee lattice

The dispersion analysis presented above is now extended to the full three-dimensional case, following the analysis presented by TAFLOVE and BRODWIN [1975]. We consider a normalized, lossless region of space with $\mu = 1$, $\varepsilon = 1$, $\sigma = 0$, $\sigma^* = 0$, and $c = 1$. Letting $j = \sqrt{-1}$, we rewrite Maxwell's equations in compact form as

$$j\nabla \times (\vec{H} + j\vec{E}) = \frac{\partial}{\partial t}(\vec{H} + j\vec{E}) \tag{6.6a}$$

or more simply as

$$j\nabla \times \vec{V} = \frac{\partial \vec{V}}{\partial t}, \tag{6.6b}$$

where $\vec{V} = \vec{H} + j\vec{E}$. Substituting the vector-field traveling-wave expression

$$\vec{V}\big|_{I,J,K}^{n} = \vec{V}_0\, e^{j(\omega n\Delta t - \tilde{k}_x I\Delta x - \tilde{k}_y J\Delta y - \tilde{k}_z K\Delta z)} \tag{6.7}$$

into the Yee space–time central-differencing realization of (6.6b), we obtain

$$\left[\frac{\hat{x}}{\Delta x}\sin\left(\frac{\tilde{k}_x\Delta x}{2}\right) + \frac{\hat{y}}{\Delta y}\sin\left(\frac{\tilde{k}_y\Delta y}{2}\right) + \frac{\hat{z}}{\Delta z}\sin\left(\frac{\tilde{k}_z\Delta z}{2}\right)\right] \times \vec{V}\big|_{I,J,K}^{n}$$
$$= \frac{-j}{\Delta t}\vec{V}\big|_{I,J,K}^{n}\sin\left(\frac{\omega\Delta t}{2}\right), \tag{6.8}$$

where $\hat{x}$, $\hat{y}$, and $\hat{z}$ are unit vectors in the x-, y-, and z-coordinate directions. After performing the vector cross product in (6.8) and writing out the x, y, and z vector component equations, we obtain a homogeneous system (zero right-hand side) of three equations in the unknowns V_x, V_y, and V_z. Setting the determinant of this system equal to zero results in

$$\left[\frac{1}{\Delta t}\sin\left(\frac{\omega\Delta t}{2}\right)\right]^2 = \left[\frac{1}{\Delta x}\sin\left(\frac{\tilde{k}_x\Delta x}{2}\right)\right]^2 + \left[\frac{1}{\Delta y}\sin\left(\frac{\tilde{k}_y\Delta y}{2}\right)\right]^2$$
$$+ \left[\frac{1}{\Delta z}\sin\left(\frac{\tilde{k}_z\Delta z}{2}\right)\right]^2. \tag{6.9}$$

Finally, we denormalize to a nonunity c and obtain the general form of the numerical dispersion relation for the full-vector-field Yee algorithm in three dimensions:

$$\left[\frac{1}{c\Delta t}\sin\left(\frac{\omega\Delta t}{2}\right)\right]^2 = \left[\frac{1}{\Delta x}\sin\left(\frac{\tilde{k}_x\Delta x}{2}\right)\right]^2 + \left[\frac{1}{\Delta y}\sin\left(\frac{\tilde{k}_y\Delta y}{2}\right)\right]^2$$
$$+ \left[\frac{1}{\Delta z}\sin\left(\frac{\tilde{k}_z\Delta z}{2}\right)\right]^2. \tag{6.10}$$

This equation is seen to reduce to (6.3), the numerical dispersion relation for the two-dimensional TM_z mode, simply by letting $\tilde{k}_z = 0$.

6.4. Comparison with the ideal dispersion case

In contrast to (6.10), the analytical (ideal) dispersion relation for a physical plane wave propagating in three dimensions in a homogeneous lossless medium is simply

$$\left(\frac{\omega}{c}\right)^2 = (k_x)^2 + (k_y)^2 + (k_z)^2. \tag{6.11}$$

Although at first glance (6.10) bears little resemblance to the ideal case of (6.11), we can easily show that the two dispersion relations are identical in the limit as Δx, Δy,

Δz, and Δt approach zero. Qualitatively, this suggests that numerical dispersion can be reduced to any degree that is desired if we only use fine enough FDTD gridding.

It can also be shown that (6.10) reduces to (6.11) if the Courant factor S and the wave-propagation direction are suitably chosen. For example, reduction to the ideal dispersion case can be demonstrated for a numerical plane wave propagating along a diagonal of a three-dimensional cubic lattice ($\tilde{k}_x = \tilde{k}_y = \tilde{k}_z = \tilde{k}/\sqrt{3}$) if $S = 1/\sqrt{3}$. Similarly, ideal dispersion results for a numerical plane wave propagating along a diagonal of a two-dimensional square grid ($\tilde{k}_x = \tilde{k}_y = \tilde{k}/\sqrt{2}$) if $S = 1/\sqrt{2}$. Finally, ideal dispersion results for any numerical wave in a one-dimensional grid if $S = 1$. These reductions to the ideal case have little practical value for two- and three-dimensional simulations, occurring only for diagonal propagation. However, the reduction to ideal dispersion in one dimension is very interesting, since it implies that the Yee algorithm (based upon numerical finite-difference approximations) yields an *exact* solution for wave propagation.

6.5. *Anisotropy of the numerical phase velocity*

This section probes a key implication of numerical dispersion relations (6.3) and (6.10). Namely, numerical waves in a two- or three-dimensional Yee space lattice have a propagation velocity that is dependent upon the direction of wave propagation. The space lattice thus represents an anisotropic medium.

Our strategy in developing an understanding of this phenomenon is to first calculate sample values of the numerical phase velocity $\tilde{v}_p$ versus wave-propagation direction ϕ in order to estimate the magnitude of the problem. Then, we will conduct an appropriate analysis to examine the issue more deeply.

6.5.1. *Sample values of numerical phase velocity*

For simplicity, we start with the simplest possible situation where numerical phase-velocity anisotropy arises: two-dimensional $\mathrm{TM_z}$ modes propagating in a square-cell grid. Dispersion relation (6.4) can be solved directly for $\tilde{k}$ for propagation along the major axes of the grid: $\phi = 0°$, $90°$, $180°$, and $270°$. For this case, the solution for $\tilde{k}$ is given by (6.5b), which is repeated here for convenience:

$$\tilde{k} = \frac{2}{\Delta}\sin^{-1}\left[\frac{1}{S}\sin\left(\frac{\pi S}{N_\lambda}\right)\right]. \tag{6.12a}$$

The corresponding numerical phase velocity is given by

$$\tilde{v}_p = \frac{\omega}{\tilde{k}} = \frac{\pi}{N_\lambda \sin^{-1}\left[\frac{1}{S}\sin\left(\frac{\pi S}{N_\lambda}\right)\right]}c. \tag{6.12b}$$

Dispersion relation (6.4) can also be solved directly for $\tilde{k}$ for propagation along the diagonals of the grid $\phi = 45°$, $135°$, $225°$, and $315°$, yielding

$$\tilde{k} = \frac{2\sqrt{2}}{\Delta}\sin^{-1}\left[\frac{1}{S\sqrt{2}}\sin\left(\frac{\pi S}{N_\lambda}\right)\right], \tag{6.13a}$$

$$\tilde{v}_p = \frac{\pi}{N_\lambda\sqrt{2}\sin^{-1}\left[\frac{1}{S\sqrt{2}}\sin\left(\frac{\pi S}{N_\lambda}\right)\right]}c. \tag{6.13b}$$

As an example, assume a grid having $S = 0.5$ and $N_\lambda = 20$. Then (6.12b) and (6.13b) provide unequal $\tilde{v}_p$ values of $0.996892c$ and $0.998968c$, respectively. The implication is that a sinusoidal numerical wave propagating obliquely within this grid has a speed that is $0.998968/0.996892 = 1.00208$ times that of a wave propagating along the major grid axes. This represents a velocity anisotropy of about 0.2% between oblique and along-axis numerical wave propagation.

TAFLOVE and HAGNESS [2000, pp. 115–117] demonstrated that this theoretical anisotropy of the numerical phase velocity appears in FDTD simulations. Fig. 6.1 presents their modeling results for a radially outward-propagating sinusoidal cylindrical wave in a two-dimensional TM_z grid. Their grid was configured with 360×360 square cells with $\Delta x = \Delta y = \Delta = 1.0$. A unity-amplitude sinusoidal excitation was provided

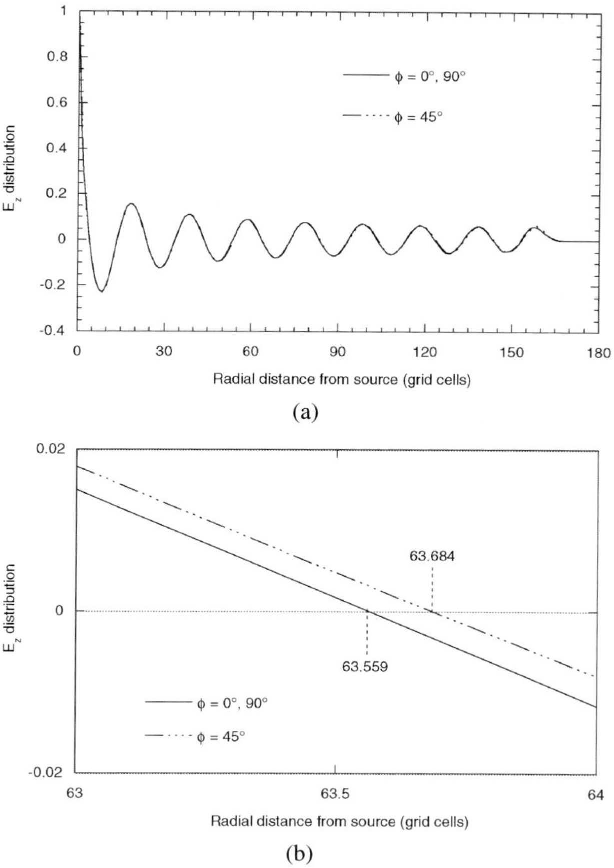

FIG. 6.1. Effect of numerical dispersion upon a radially propagating cylindrical wave in a 2D TM Yee grid. The grid is excited at its center by applying a unity-amplitude sinusoidal time function to a single E_z field component. $S = 0.5$ and the grid sampling density is $N_\lambda = 20$. (a) Comparison of calculated wave propagation along the grid axes and along a grid diagonal. (b) Expanded view of (a) at distances between 63 and 64 grid cells from the source.

to a single E_z component at the center of the grid. Choosing a grid-sampling density of $N_\lambda = 20$ and a Courant factor $S = 0.5$ permitted direct comparison of the FDTD modeling results with the theoretical results for the anisotropy of $\tilde{v}_p$, discussed immediately above.

Fig. 6.1(a) illustrates snapshots of the E_z field distribution vs. radial distance from the source at the center of the grid. Here, field observations are made along cuts through the grid passing through the source point and either parallel to the principal grid axes $\phi = 0°, 90°$ or parallel to the grid diagonal $\phi = 45°$. (Note that, by the 90° rotational symmetry of the Cartesian grid geometry, identical field distributions are obtained along $\phi = 0°$ and $\phi = 90°$.) The snapshots are taken $328\Delta t$ after the beginning of time-stepping. At this time, the wave has not yet reached the outer grid boundary, and the calculated E_z field distribution is free of error due to outer-boundary reflections.

Fig. 6.1(b) is an expanded view of Fig. 6.1(a) at radial distances between 63Δ and 64Δ from the source. This enables evaluation (with three-decimal-place precision) of the locations of the zero-crossings of the E_z distributions along the two observation cuts through the grid. From the data shown in Fig. 6.1(b), the sinusoidal wave along the $\phi = 45°$ cut passes through zero at 63.684 cells, whereas the wave along the $\phi = 0°, 90°$ cut passes through zero at 63.559 cells. Taking the difference, we see that the obliquely propagating wave "leads" the on-axis wave by 0.125 cells. This yields a numerical phase-velocity anisotropy $\Delta\tilde{v}_p/\tilde{v}_p \cong 0.125/63.6 = 0.197\%$. This number is only about 5% less than the 0.208% value obtained using (6.12b) and (6.13b).

To permit determination of $\tilde{k}$ and $\tilde{v}_p$ for any wave-propagation direction ϕ, it would be very useful to derive closed-form equations analogous to (6.12) and (6.13). However, for this general case, the underlying dispersion relation (6.4) is a transcendental equation. TAFLOVE [1995, pp. 97–98] provided a useful alternative approach for obtaining sample values of $\tilde{v}_p$ by applying the following Newton's method iterative procedure to (6.4):

$$\tilde{k}_{\text{icount}+1} = \tilde{k}_{\text{icount}} - \frac{\sin^2(A\tilde{k}_{\text{icount}}) + \sin^2(B\tilde{k}_{\text{icount}}) - C}{A\sin(2A\tilde{k}_{\text{icount}}) + B\sin(2B\tilde{k}_{\text{icount}})}. \tag{6.14a}$$

Here, $\tilde{k}_{\text{icount}+1}$ is the improved estimate of $\tilde{k}$, and $\tilde{k}_{\text{icount}}$ is the previous estimate of $\tilde{k}$. The A, B, and C are coefficients given by

$$A = \frac{\Delta \cdot \cos\phi}{2}, \qquad B = \frac{\Delta \cdot \sin\phi}{2}, \qquad C = \frac{1}{S^2}\sin^2\left(\frac{\pi S}{N_\lambda}\right). \tag{6.14b}$$

Additional simplicity results if Δ is normalized to the free-space wavelength, λ_0. This is equivalent to setting $\lambda_0 = 1$. Then, a very good starting guess for the iterative process is simply 2π. For this case, $\tilde{v}_p$ is given by

$$\frac{\tilde{v}_p}{c} = \frac{2\pi}{\tilde{k}_{\text{final icount}}}. \tag{6.15}$$

Usually, only two or three iterations are required for convergence.

Fig. 6.2 graphs results obtained using this procedure that illustrate the variation of $\tilde{v}_p$ with propagation direction ϕ. Here, for the Courant factor fixed at $S = 0.5$, three

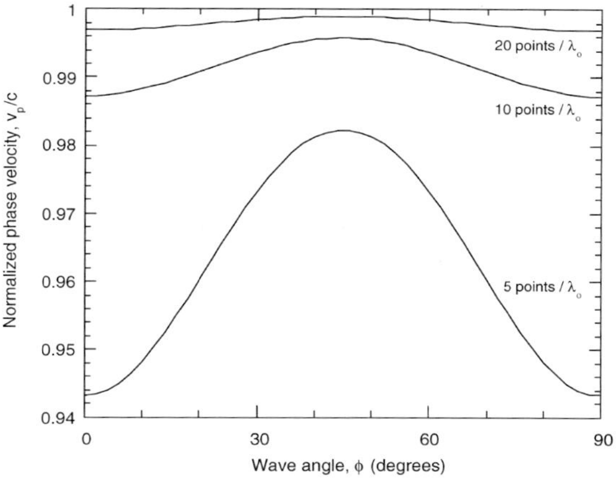

FIG. 6.2. Variation of the numerical phase velocity with wave propagation angle in a 2D FDTD grid for three sampling densities of the square unit cells. $S = c\Delta t/\Delta = 0.5$ for all cases.

different grid sampling densities N_λ are examined: $N_\lambda = 5$ points per λ_0, $N_\lambda = 10$, and $N_\lambda = 20$. We see that $\tilde{v}_p < c$ and is a function of both ϕ and N_λ. $\tilde{v}_p$ is maximum for waves propagating obliquely within the grid ($\phi = 45°$), and is minimum for waves propagating along either grid axis ($\phi = 0°, 90°$).

It is useful to summarize the algorithmic dispersive-error performance by defining two normalized error measures: (1) the physical phase-velocity error $\Delta\tilde{v}_{\text{physical}}$, and (2) the velocity-anisotropy error $\Delta\tilde{v}_{\text{aniso}}$. These are given by

$$\Delta\tilde{v}_{\text{physical}}\big|_{N_\lambda} = \frac{\min[\tilde{v}_p(\phi)] - c}{c} \times 100\%, \tag{6.16}$$

$$\Delta\tilde{v}_{\text{aniso}}\big|_{N_\lambda} = \frac{\max[\tilde{v}_p(\phi)] - \min[\tilde{v}_p(\phi)]}{\min[\tilde{v}_p(\phi)]} \times 100\%. \tag{6.17}$$

$\Delta\tilde{v}_{\text{physical}}$ is useful in quantifying the phase lead or lag that numerical modes suffer relative to physical modes propagating at c. For example, from Fig. 6.2 and (6.12b), $\Delta\tilde{v}_{\text{physical}} = -0.31\%$ for $N_\lambda = 20$. This means that a sinusoidal numerical wave traveling over a $10\lambda_0$ distance in the grid (200 cells) could develop a lagging phase error up to 11°. We note that $\Delta\tilde{v}_{\text{physical}}$ is a function of N_λ. Since the grid cell size Δ is fixed, for an impulsive wave-propagation problem there exists a spread of effective N_λ values for the spectral components comprising the pulse. This causes a spread of $\Delta\tilde{v}_{\text{physical}}$ over the pulse spectrum, which in turn yields a temporal dispersion of the pulse evidenced in the spreading and distortion of its waveform as it propagates.

$\Delta\tilde{v}_{\text{aniso}}$ is useful in quantifying wavefront distortion. For example, a circular cylindrical wave would suffer progressive distortion of its wavefront since the portions propagating along the grid diagonals would travel slightly faster than the portions traveling along the major grid axes. For example, from Fig. 6.2 and (6.12b) and (6.13b), $\Delta\tilde{v}_{\text{aniso}} = 0.208\%$ for $N_\lambda = 20$. The wavefront distortion due to this anisotropy would total about 2.1 cells for each 1000 cells of propagation distance.

It is clear that errors due to inaccurate numerical velocities are cumulative, i.e., they increase linearly with the wave-propagation distance. These errors represent a fundamental limitation of *all* grid-based Maxwell's equations' algorithms, and can be troublesome when modeling electrically large structures. A positive aspect seen in Fig. 6.2 is that both $\Delta\tilde{v}_{\text{physical}}$ and $\Delta\tilde{v}_{\text{aniso}}$ decrease by approximately a 4:1 factor each time the grid-sampling density doubles, indicative of the second-order accuracy of the Yee algorithm. Therefore, finer meshing is one way to control the dispersion error.

As discussed in Section 7, there are proposed means to improve the accuracy of FDTD algorithms to allow much larger structures to be modeled. Specifically, $\Delta\tilde{v}_{\text{aniso}}$ can be reduced to very low levels approaching zero. In this case, residual errors involve primarily the dispersion of $\Delta\tilde{v}_{\text{physical}}$ with N_λ, which can be optimized by the proper choice of Δt. However, the new approaches presently have limitations regarding their ability to model material discontinuities, and require more research.

6.5.2. Intrinsic grid velocity anisotropy

Following TAFLOVE and HAGNESS [2000, pp. 120–123], this section provides a deeper discussion of the numerical phase-velocity errors of the Yee algorithm. We show that the nature of the grid discretization, in a manner virtually independent of the time-stepping scheme, determines the velocity anisotropy $\Delta\tilde{v}_{\text{aniso}}$.

Relation of the time and space discretizations in generating numerical velocity error. In Section 6.5.1, we determined that $\Delta\tilde{v}_{\text{aniso}} = 0.208\%$ for a two-dimensional Yee algorithm having $N_\lambda = 20$ and $S = 0.5$. An important and revealing question is: How is $\Delta\tilde{v}_{\text{aniso}}$ affected by the choice of S, assuming that N_λ is fixed at 20?

To begin to answer this question, we first choose (what will later be shown to be) the largest possible value of S for numerical stability in two dimensions, $S = 1/\sqrt{2}$. Substituting this value of S into (6.12b) and (6.13b) yields

$$\left.\begin{aligned}\tilde{v}_p(\phi = 0^\circ) &= 0.997926c\\ \tilde{v}_p(\phi = 45^\circ) &= c\end{aligned}\right\}\qquad \Delta\tilde{v}_{\text{aniso}} = \frac{c - 0.997926c}{0.997926c} \times 100\% = 0.208\%.$$

To three decimal places, there is no change in $\Delta\tilde{v}_{\text{aniso}}$ from the previous value, $S = 0.5$. We next choose a very small value $S = 0.01$ for substitution into (6.12b) and (6.13b):

$$\left.\begin{aligned}\tilde{v}_p(\phi = 0^\circ) &= 0.995859c\\ \tilde{v}_p(\phi = 45^\circ) &= 0.997937c\end{aligned}\right\}$$

$$\Delta\tilde{v}_{\text{aniso}} = \frac{0.997937c - 0.995859c}{0.995859c} \times 100\% = 0.208\%.$$

Again, there is no change in $\Delta\tilde{v}_{\text{aniso}}$ to three decimal places.

We now suspect that, for a given N_λ, $\Delta\tilde{v}_{\text{aniso}}$ is at most a weak function of S, and therefore is only weakly dependent on Δt. In fact, this is the case. More generally, LIU, Y. [1996] has shown that $\Delta\tilde{v}_{\text{aniso}}$ is only weakly dependent on the specific type of time-marching scheme used, whether leapfrog, Runge–Kutta, etc. Thus, we can say that $\Delta\tilde{v}_{\text{aniso}}$ is virtually an intrinsic characteristic of the space-lattice discretization. Following LIU, Y. [1996], three key points should be made in this regard:

- Numerical-dispersion errors associated with the time discretization are isotropic relative to the propagation direction of the wave.
- The choice of time discretization has little effect upon the phase-velocity anisotropy $\Delta\tilde{v}_{\text{aniso}}$ for $N_\lambda > 10$.
- The choice of time discretization does influence $\Delta\tilde{v}_{\text{physical}}$. However, it is not always true that higher-order time-marching schemes, such as Runge–Kutta, yield less $\Delta\tilde{v}_{\text{physical}}$ than simple Yee leapfrogging. Errors in $\Delta\tilde{v}_{\text{physical}}$ are caused separately by the space and time discretizations, and can either partially reinforce or cancel each other. Thus, the use of fourth-order Runge–Kutta may actually shift the $\tilde{v}_p(\phi)$ profile away from c, representing an increased $\Delta\tilde{v}_{\text{physical}}$ relative to ordinary leapfrogging.

The associated eigenvalue problem. LIU, Y. [1996] has shown that, to determine the relative velocity anisotropy characteristic intrinsic to a space grid, it is useful to set up an eigenvalue problem for the matrix that delineates the spatial derivatives used in the numerical algorithm. Consider as an example the finite-difference system of (3.19) for the case of two-dimensional TM$_z$ electromagnetic wave propagation. The associated eigenvalue problem for the lossless-medium case is written as:

$$-\frac{1}{\mu}\left(\frac{E_z|_{i,j+1/2} - E_z|_{i,j-1/2}}{\Delta y}\right) = \Lambda H_x|_{i,j}, \tag{6.18a}$$

$$\frac{1}{\mu}\left(\frac{E_z|_{i+1/2,j} - E_z|_{i-1/2,j}}{\Delta x}\right) = \Lambda H_y|_{i,j}, \tag{6.18b}$$

$$\frac{1}{\varepsilon}\left(\frac{H_y|_{i+1/2,j} - H_y|_{i-1/2,j}}{\Delta x} - \frac{H_x|_{i,j+1/2} - H_x|_{i,j-1/2}}{\Delta y}\right) = \Lambda E_z|_{i,j}. \tag{6.18c}$$

We note that, at any time step n, the instantaneous values of the E- and H-fields distributed in space across the grid can be Fourier-transformed with respect to the i and j grid coordinates to provide a spectrum of sinusoidal modes. The result is often called the two-dimensional spatial-frequency spectrum, or the plane-wave eigenmodes of the grid. Let the following specify a typical mode of this spectrum having $\tilde{k}_x$ and $\tilde{k}_y$ as, respectively, the x- and y-components of its numerical wavevector:

$$\begin{aligned} E_z|_{I,J} &= E_{z_0}\mathrm{e}^{\mathrm{j}(\tilde{k}_x I\Delta x + \tilde{k}_y J\Delta y)}; \\ H_x|_{I,J} &= H_{x_0}\mathrm{e}^{\mathrm{j}(\tilde{k}_x I\Delta x + \tilde{k}_y J\Delta y)}; \\ H_y|_{I,J} &= H_{y_0}\mathrm{e}^{\mathrm{j}(\tilde{k}_x I\Delta x + \tilde{k}_y J\Delta y)} \end{aligned} \tag{6.19}$$

Upon substituting the eigenmode expressions of (6.19) into (6.18a), we obtain

$$-\frac{1}{\mu}\left(\frac{E_{z_0}\mathrm{e}^{\mathrm{j}[\tilde{k}_x I\Delta x + \tilde{k}_y (J+1/2)\Delta y]} - E_{z_0}\mathrm{e}^{\mathrm{j}[\tilde{k}_x I\Delta x + \tilde{k}_y (J-1/2)\Delta y]}}{\Delta y}\right)$$
$$= \Lambda H_{x_0}\mathrm{e}^{\mathrm{j}(\tilde{k}_x I\Delta x + \tilde{k}_y J\Delta y)}. \tag{6.20}$$

Factoring out the $e^{j(\tilde{k}_x I\Delta x+\tilde{k}_y J\Delta y)}$ term that is common to both sides and then applying Euler's identity yields

$$H_{x_0} = -\frac{2jE_{z0}}{\Lambda\mu\Delta y}\sin\left(\frac{\tilde{k}_y\Delta y}{2}\right). \tag{6.21a}$$

In a similar manner, substituting the eigenmode expressions of (6.19) into (6.18b) and (6.18c) yields

$$H_{y_0} = \frac{2jE_{z0}}{\Lambda\mu\Delta x}\sin\left(\frac{\tilde{k}_x\Delta x}{2}\right), \tag{6.21b}$$

$$E_{z_0} = \frac{2j}{\Lambda\varepsilon}\left[\frac{H_{y0}}{\Delta x}\sin\left(\frac{\tilde{k}_x\Delta x}{2}\right) - \frac{H_{x0}}{\Delta y}\sin\left(\frac{\tilde{k}_y\Delta y}{2}\right)\right]. \tag{6.21c}$$

Substituting H_{x_0} of (6.21a) and H_{y_0} of (6.21b) into (6.21c) yields

$$E_{z_0} = \frac{2j}{\Lambda\varepsilon}\left[\begin{array}{l}\frac{1}{\Delta x}\cdot\frac{2jE_{z0}}{\Lambda\mu\Delta x}\cdot\sin\left(\frac{\tilde{k}_x\Delta x}{2}\right)\cdot\sin\left(\frac{\tilde{k}_x\Delta x}{2}\right)\\ \quad -\frac{1}{\Delta y}\cdot\frac{-2jE_{z0}}{\Lambda\mu\Delta y}\cdot\sin\left(\frac{\tilde{k}_y\Delta y}{2}\right)\cdot\sin\left(\frac{\tilde{k}_y\Delta y}{2}\right)\end{array}\right]. \tag{6.22}$$

Now factoring out the common E_{z_0} term, simplifying, and solving for Λ^2, we obtain

$$\Lambda^2 = -\frac{4}{\mu\varepsilon}\left[\frac{1}{(\Delta x)^2}\sin^2\left(\frac{\tilde{k}_x\Delta x}{2}\right) + \frac{1}{(\Delta y)^2}\sin^2\left(\frac{\tilde{k}_y\Delta y}{2}\right)\right]. \tag{6.23}$$

From the elementary properties of the sine function (assuming that $\tilde{k}_x$ and $\tilde{k}_y$ are real numbers for propagating numerical waves), the right-hand side of (6.23) is negative. Hence, Λ is a pure imaginary number given by

$$\Lambda = j2c\left[\frac{1}{(\Delta x)^2}\sin^2\left(\frac{\tilde{k}_x\Delta x}{2}\right) + \frac{1}{(\Delta y)^2}\sin^2\left(\frac{\tilde{k}_y\Delta y}{2}\right)\right]^{1/2}, \tag{6.24}$$

where $c = 1/\sqrt{\mu\varepsilon}$ is the speed of light in the homogeneous material being modeled. Finally, following the definition provided by LIU, Y. [1996], we obtain the "normalized numerical phase speed" c^*/c intrinsic to the grid discretization, given by

$$\frac{c^*}{c} = \frac{\Lambda_{\text{imag}}}{c\tilde{k}} = \frac{2}{\tilde{k}}\left[\frac{1}{(\Delta x)^2}\sin^2\left(\frac{\tilde{k}_x\Delta x}{2}\right) + \frac{1}{(\Delta y)^2}\sin^2\left(\frac{\tilde{k}_y\Delta y}{2}\right)\right]^{1/2}. \tag{6.25}$$

A convenient closed-form expression for c^*/c can be written by using the approximation $\tilde{k} \cong k$. Then, assuming a uniform square-cell grid, we obtain

$$\frac{c^*}{c} \cong \frac{N_\lambda}{\pi}\left[\sin^2\left(\frac{\pi\cos\phi}{N_\lambda}\right) + \sin^2\left(\frac{\pi\sin\phi}{N_\lambda}\right)\right]^{1/2}, \quad N_\lambda > 10. \tag{6.26}$$

The meaning of c^*/c. The reader is cautioned that c^*/c is *not* the same as $\tilde{v}_p/c$. This is because the derivation of c^*/c utilizes no information regarding the time-stepping process. Thus, c^*/c cannot be used to determine $\Delta\tilde{v}_{\text{physical}}$ defined in (6.16). However, c^*/c does provide information regarding $\Delta\tilde{v}_{\text{aniso}}$ defined in (6.17). Following LIU, Y. [1996], we can expand (6.26) to isolate the leading-order velocity-anisotropy term. This yields a simple expression for $\Delta\tilde{v}_{\text{aniso}}$ that is useful for $N_\lambda > 10$:

$$\Delta\tilde{v}_{\text{aniso}}\big|_{\text{Yee}} \cong \frac{\max\left[\frac{c^*(\phi)}{c}\right] - \min\left[\frac{c^*(\phi)}{c}\right]}{\min\left[\frac{c^*(\phi)}{c}\right]} \times 100\%$$

$$\cong \frac{\pi^2}{12(N_\lambda)^2} \times 100\%. \tag{6.27}$$

For example, (6.27) provides $\Delta\tilde{v}_{\text{aniso}} \cong 0.206\%$ for $N_\lambda = 20$. This is very close to the 0.208% value previously obtained using (6.12b) and (6.13b), the exact solutions of the full numerical dispersion relation for $\phi = 0°$ and $\phi = 45°$, respectively.

In summary, we can use (6.27) to estimate the numerical phase-velocity anisotropy $\Delta\tilde{v}_{\text{aniso}}$ of the Yee algorithm applied to a square-cell grid without having to resort to the Newton's method solution (6.14). This approach provides a convenient means to compare the relative anisotropy of alternative space-gridding techniques, including the higher-order methods and non-Cartesian meshes to be discussed in Section 7.

6.6. *Complex-valued numerical wavenumbers*

SCHNEIDER and WAGNER [1999] found that the Yee algorithm has a low-sampling-density regime that allows complex-valued numerical wavenumbers. In this regime, spatially decaying numerical waves can propagate faster than light, causing a weak, nonphysical signal to appear ahead of the nominal leading edges of sharply defined pulses. This section reviews the theory underlying this phenomenon.

6.6.1. *Case 1: Numerical wave propagation along the principal lattice axes*

Consider again numerical wave propagation along the major axes of a Yee space grid. For convenience, we rewrite (6.12a), the corresponding numerical dispersion relation:

$$\tilde{k} = \frac{2}{\Delta}\sin^{-1}\left[\frac{1}{S}\sin\left(\frac{\pi S}{N_\lambda}\right)\right] \equiv \frac{2}{\Delta}\sin^{-1}(\zeta), \tag{6.28}$$

where

$$\zeta = \frac{1}{S}\sin\left(\frac{\pi S}{N_\lambda}\right). \tag{6.29}$$

SCHNEIDER and WAGNER [1999] realized that, in evaluating numerical dispersion relations such as (6.28), it is possible to choose S and N_λ such that $\tilde{k}$ is complex. In the case of (6.28), it can be shown that the transition between real and complex values of $\tilde{k}$ occurs when $\zeta = 1$. Solving for N_λ at this transition results in

$$N_\lambda\big|_{\text{transition}} = \frac{\pi S}{\sin^{-1}(S)}. \tag{6.30}$$

For a grid sampling density greater than this value, i.e., $N_\lambda > N_\lambda|_{\text{transition}}$, $\tilde{k}$ is a real number and the numerical wave undergoes no attenuation while propagating in the grid. Here, $\tilde{v}_p < c$. For a coarser grid-sampling density $N_\lambda < N_\lambda|_{\text{transition}}$, $\tilde{k}$ is a complex number and the numerical wave undergoes a nonphysical exponential decay while propagating. Further, in this coarse-resolution regime, $\tilde{v}_p$ can exceed c.

Following SCHNEIDER and WAGNER [1999], we now discuss how $\tilde{k}$ and $\tilde{v}_p$ vary with grid sampling N_λ, both above and below the transition between real and complex numerical wavenumbers.

Real-numerical-wavenumber regime. For $N_\lambda > N_\lambda|_{\text{transition}}$ we have from (6.28)

$$\tilde{k}_{\text{real}} = \frac{2}{\Delta}\sin^{-1}\left[\frac{1}{S}\sin\left(\frac{\pi S}{N_\lambda}\right)\right]; \tag{6.31a}$$

$$\tilde{k}_{\text{imag}} = 0. \tag{6.31b}$$

The numerical phase velocity is given by

$$\tilde{v}_p = \frac{\omega}{\tilde{k}_{\text{real}}} = \frac{\pi}{N_\lambda \sin^{-1}\left[\frac{1}{S}\sin\left(\frac{\pi S}{N_\lambda}\right)\right]}c. \tag{6.32}$$

This is exactly expression (6.12b). The wave-amplitude multiplier per grid cell of propagation is given by

$$e^{\tilde{k}_{\text{imag}}\Delta} \equiv e^{-\alpha\Delta} = e^0 = 1. \tag{6.33}$$

Thus, there is a constant wave amplitude with spatial position for this range of N_λ.

Complex-numerical-wavenumber regime. For $N_\lambda < N_\lambda|_{\text{transition}}$, we observe that $\zeta > 1$ in (6.28). Here, the following relation for the complex-valued arc-sine function given by CHURCHILL, BROWN and VERHEY [1976] is useful:

$$\sin^{-1}(\zeta) = -\mathrm{j}\ln\left(\mathrm{j}\zeta + \sqrt{1-\zeta^2}\right). \tag{6.34}$$

Substituting (6.34) into (6.28) yields after some algebraic manipulation

$$\tilde{k}_{\text{real}} = \frac{\pi}{\Delta}; \tag{6.35a}$$

$$\tilde{k}_{\text{imag}} = -\frac{2}{\Delta}\ln\left(\zeta + \sqrt{\zeta^2-1}\right). \tag{6.35b}$$

The numerical phase velocity is then

$$\tilde{v}_p = \frac{\omega}{\tilde{k}_{\text{real}}} = \frac{\omega}{(\pi/\Delta)} = \frac{2\pi f\Delta}{\pi} = \frac{2f\lambda_0}{N_\lambda} = \frac{2}{N_\lambda}c \tag{6.36}$$

and the wave-amplitude multiplier per grid cell of propagation is

$$e^{\tilde{k}_{\text{imag}}\Delta} \equiv e^{-\alpha\Delta} = e^{-2\ln(\zeta+\sqrt{\zeta^2-1})} = \frac{1}{(\zeta+\sqrt{\zeta^2-1})^2}. \tag{6.37}$$

Since $\zeta > 1$, the numerical wave amplitude decays exponentially with spatial position.

We now consider the possibility of $\tilde{v}_p$ exceeding c in this situation. Nyquist theory states that any physical or numerical process that obtains samples of a time waveform every Δt seconds can reproduce the original waveform without aliasing for spectral content up to $f_{\max} = 1/(2\Delta t)$. In the present case, the corresponding minimum free-space wavelength that can be sampled without aliasing is therefore

$$\lambda_{0,\min} = c/f_{\max} = 2c\Delta t. \tag{6.38a}$$

The corresponding minimum spatial-sampling density is

$$N_{\lambda,\min} = \lambda_{0,\min}/\Delta = 2c\Delta t/\Delta = 2S. \tag{6.38b}$$

Then from (6.36), the maximum numerical phase velocity is given by

$$\tilde{v}_{p,\max} = \frac{2}{N_{\lambda,\min}}c = \frac{2}{2S}c = \frac{c}{S}. \tag{6.39a}$$

From the definition of S, this maximum phase velocity can also be expressed as

$$\tilde{v}_{p,\max} = \frac{1}{S}c = \left(\frac{\Delta}{c\Delta t}\right)c = \frac{\Delta}{\Delta t}. \tag{6.39b}$$

This relation tells us that in one Δt, a numerical value can propagate at most one Δ. This is intuitively correct given the local nature of the spatial differences used in the Yee algorithm. That is, a field point more than one Δ away from a source point that undergoes a sudden change cannot possibly "feel" the effect of that change during the next Δt. Note that $\tilde{v}_{p,\max}$ is independent of material parameters and is an inherent property of the grid and its method of obtaining space derivatives.

6.6.2. *Case 2: Numerical wave propagation along a grid diagonal*

We next explore the possibility of complex-valued wavenumbers arising for oblique numerical wave propagation in a square-cell grid. For convenience, we rewrite (6.13a), the corresponding numerical dispersion relation:

$$\tilde{k} = \frac{2\sqrt{2}}{\Delta}\sin^{-1}\left[\frac{1}{S\sqrt{2}}\sin\left(\frac{\pi S}{N_\lambda}\right)\right] \equiv \frac{2\sqrt{2}}{\Delta}\sin^{-1}(\zeta), \tag{6.40}$$

where

$$\zeta = \frac{1}{S\sqrt{2}}\sin\left(\frac{\pi S}{N_\lambda}\right). \tag{6.41}$$

Similar to the previous case of numerical wave propagation along the principal lattice axes, it is possible to choose S and N_λ such that $\tilde{k}$ is complex. In the specific case of (6.40), the transition between real and complex values of $\tilde{k}$ occurs when $\zeta = 1$. Solving for N_λ at this transition results in

$$N_\lambda|_{\text{transition}} = \frac{\pi S}{\sin^{-1}(S\sqrt{2})}. \tag{6.42}$$

We now discuss how $\tilde{k}$ and $\tilde{v}_p$ vary with grid sampling N_λ, both above and below the transition between real and complex numerical wavenumbers.

Real-numerical-wavenumber regime. For $N_\lambda \geqslant N_\lambda|_{\text{transition}}$ we have from (6.40)

$$\tilde{k}_{\text{real}} = \frac{2\sqrt{2}}{\Delta}\sin^{-1}\left[\frac{1}{S\sqrt{2}}\sin\left(\frac{\pi S}{N_\lambda}\right)\right]; \tag{6.43a}$$

$$\tilde{k}_{\text{imag}} = 0. \tag{6.43b}$$

The numerical phase velocity is given by

$$\tilde{v}_p = \frac{\omega}{\tilde{k}_{\text{real}}} = \frac{\pi}{N_\lambda\sqrt{2}\sin^{-1}\left[\frac{1}{S\sqrt{2}}\sin\left(\frac{\pi S}{N_\lambda}\right)\right]}c. \tag{6.44}$$

This is exactly expression (6.13b). The wave-amplitude multiplier per grid cell of propagation is given by

$$\mathrm{e}^{\tilde{k}_{\text{imag}}\Delta} \equiv \mathrm{e}^{-\alpha\Delta} = \mathrm{e}^{0} = 1. \tag{6.45}$$

Thus, there is a constant wave amplitude with spatial position for this range of N_λ.

Complex-numerical-wavenumber regime. For $N_\lambda < N_\lambda|_{\text{transition}}$, we observe that $\zeta > 1$ in (6.40). Substituting the complex-valued arc-sine function of (6.34) into (6.40) yields after some algebraic manipulation

$$\tilde{k}_{\text{real}} = \frac{\pi\sqrt{2}}{\Delta}; \tag{6.46a}$$

$$\tilde{k}_{\text{imag}} = -\frac{2\sqrt{2}}{\Delta}\ln\left(\zeta + \sqrt{\zeta^2 - 1}\right). \tag{6.46b}$$

The numerical phase velocity for this case is

$$\tilde{v}_p = \frac{\omega}{\tilde{k}_{\text{real}}} = \frac{\omega}{(\pi\sqrt{2}/\Delta)} = \frac{\sqrt{2}\,f\lambda_0}{N_\lambda} = \frac{\sqrt{2}}{N_\lambda}c \tag{6.47}$$

and the wave-amplitude multiplier per grid cell of propagation is

$$\mathrm{e}^{\tilde{k}_{\text{imag}}\Delta} \equiv \mathrm{e}^{-\alpha\Delta} = \mathrm{e}^{-2\sqrt{2}\ln(\zeta+\sqrt{\zeta^2-1})} = \frac{1}{(\zeta + \sqrt{\zeta^2 - 1})^{2\sqrt{2}}}. \tag{6.48}$$

Since $\zeta > 1$, the numerical wave amplitude decays exponentially with spatial position.

We again consider the possibility of $\tilde{v}_p$ exceeding c. From our previous discussion of (6.38a) and (6.38b), the minimum free-space wavelength that can be sampled without aliasing is $\lambda_{0,\min} = c/f_{\max} = 2c\Delta t$, and the corresponding minimum spatial-sampling density is $N_{\lambda,\min} = \lambda_{0,\min}/\Delta = 2S$. Then from (6.47), the maximum numerical phase velocity is given by

$$\tilde{v}_{p,\max} = \frac{\sqrt{2}}{N_{\lambda,\min}}c = \frac{\sqrt{2}}{2S}c. \tag{6.49a}$$

From the definition of S, this maximum phase velocity can also be expressed as

$$\tilde{v}_{p,\max} = \frac{\sqrt{2}}{2}\left(\frac{\Delta}{c\Delta t}\right)c = \frac{\sqrt{2}\Delta}{2\Delta t}. \tag{6.49b}$$

This relation tells us that in $2\Delta t$, a numerical value can propagate at most $\sqrt{2}\,\Delta$ along the grid diagonal. We can show that this upper bound on $\tilde{v}_p$ is intuitively correct given the local nature of the spatial differences used in the Yee algorithm. Consider two nearest neighbor field points $P_{i,j}$ and $P_{i+1,j+1}$ along a grid diagonal, and how a sudden change at $P_{i,j}$ could be communicated to $P_{i+1,j+1}$. Now, a basic principle is that the Yee algorithm can communicate field data only along Cartesian (x and y) grid lines, and not along grid diagonals. Thus, at the minimum, $1\Delta t$ would be needed to transfer any part of the field perturbation at $P_{i,j}$ over a distance of 1Δ in the x-direction to $P_{i+1,j}$. Then, a second Δt would be needed, at the minimum, to transfer any part of the resulting field perturbation at $P_{i+1,j}$ over a distance of 1Δ in the y-direction to reach $P_{i+1,j+1}$. Because the distance between $P_{i,j}$ and $P_{i+1,j+1}$ is $\sqrt{2}\,\Delta$, the maximum effective velocity of signal transmission between the two points is $\sqrt{2}\,\Delta/2\Delta t$. By this reasoning, we see that $\tilde{v}_{p,\max}$ is independent of material parameters modeled in the grid. It is an inherent property of the FDTD grid and its method of obtaining space derivatives.

6.6.3. Example of calculation of numerical phase velocity and attenuation

This section provides sample calculations of values of the numerical phase velocity and the exponential attenuation constant for the case of a two-dimensional square-cell Yee grid. These calculations are based upon the numerical dispersion analyses of Sections 6.6.1 and 6.6.2.

Fig. 6.3 graphs the normalized numerical phase velocity and the exponential attenuation constant per grid cell as a function of grid sampling density N_λ. A Courant factor $S = 0.5$ is assumed. From this figure, we note that:

- For propagation along the principal grid axes $\phi = 0°, 90°$, a minimum value of $\tilde{v}_p = (2/3)c$ is reached at $N_\lambda = 3$. This sampling density is also the onset of attenuation. As N_λ is reduced below 3, $\tilde{v}_p$ increases inversely with N_λ. Eventually, $\tilde{v}_p$ exceeds c for $N_\lambda < 2$, and reaches a limiting velocity of $2c$ as $N_\lambda \to 1$. In this limit, as well, the attenuation constant approaches a value of 2.634 nepers/cell.

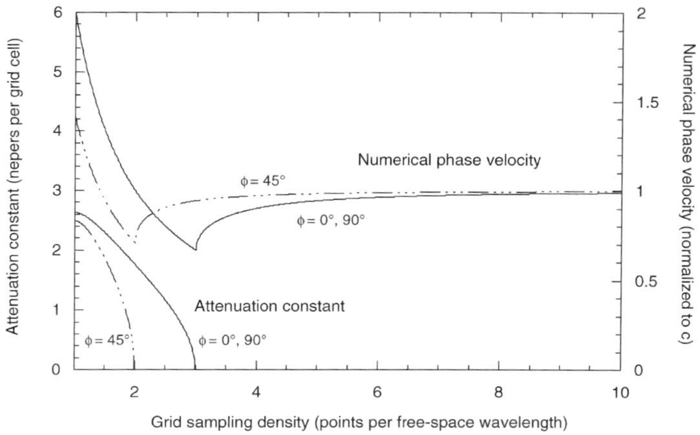

FIG. 6.3. Normalized numerical phase velocity and exponential attenuation constant per grid cell versus grid sampling density for on-axis and oblique wave propagation. $S = 0.5$.

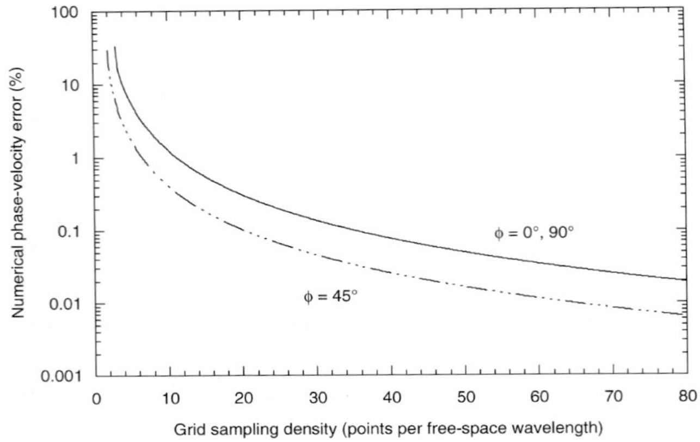

FIG. 6.4. Percent numerical phase-velocity error relative to the free-space speed of light as a function of the grid sampling density for on-axis and oblique wave propagation. $S = 0.5$.

- For propagation along the grid diagonal at $\phi = 45°$, a minimum value of $\tilde{v}_p = (\sqrt{2}/2)c$ is reached at $N_\lambda = 2$. This point is also the onset of exponential attenuation. As N_λ is reduced below 2, $\tilde{v}_p$ increases inversely with N_λ. Eventually, $\tilde{v}_p$ exceeds c for $N_\lambda < \sqrt{2}$, and reaches a limiting velocity of $\sqrt{2}\,c$ as $N_\lambda \to 1$. In this limit, as well, the attenuation constant approaches a value of 2.493 nepers/cell.

Overall, for both the on-axis and oblique cases of numerical wave propagation, we see that very coarsely resolved wave modes in the grid can propagate at superluminal speeds, but are rapidly attenuated.

Fig. 6.4 graphs the percent error in the numerical phase velocity relative to c for lossless wave propagation along the principal grid axes $\phi = 0°, 90°$. In the present example wherein $S = 0.5$, this lossless propagation regime exists for $N_\lambda \geqslant 3$. Fig. 6.4 also graphs the percent velocity error for lossless wave propagation along the grid diagonal $\phi = 45°$. This lossless regime exists for $N_\lambda \geqslant 2$ for $S = 0.5$. As $N_\lambda \gg 10$, we see that the numerical phase-velocity error at each wave-propagation angle diminishes as the inverse square of N_λ. This is indicative of the second-order-accurate nature of the Yee algorithm.

6.6.4. Examples of calculations of pulse propagation in a one-dimensional grid

Fig. 6.5(a) graphs examples of the calculated propagation of a 40-cell-wide rectangular pulse in free space for two cases of the Courant factor: $S = 1$ (i.e., Δt is equal to the value for dispersionless propagation in a one-dimensional grid); and $S = 0.99$. To permit a direct comparison of these results, both "snapshots" are taken at the same absolute time after the onset of time-stepping. There are three key observations:

(1) When $S = 1$, the rectangular shape and spatial width of the pulse are completely preserved. For this case, the abrupt step discontinuities of the propagating pulse are modeled perfectly. In fact, this is expected since $\tilde{v}_p \equiv c$ for all numerical modes in the grid.

(2) When $S = 0.99$, there is appreciable "ringing" located behind the leading and trailing edges of the pulse. This is due to short-wavelength numerical modes in

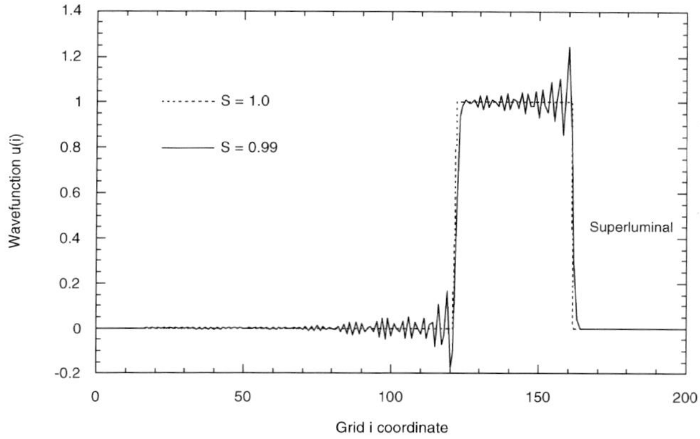

(a)

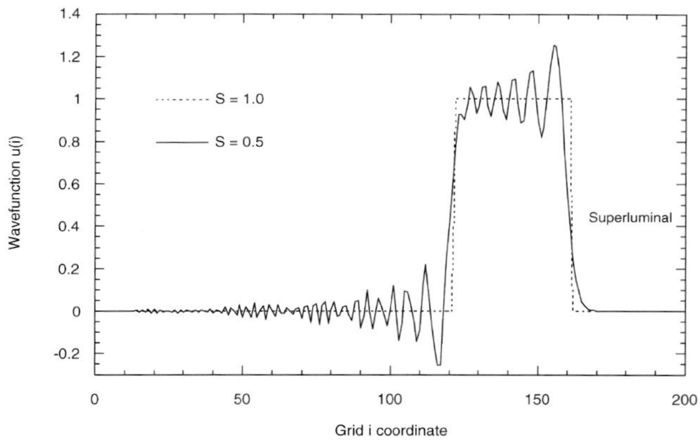

(b)

FIG. 6.5. Effect of numerical dispersion upon a rectangular pulse propagating in free space in a one-dimensional grid for three different Courant factors: $S = 1$, $S = 0.99$, and $S = 0.5$. (a) Comparison of calculated pulse propagation for $S = 1$ and $S = 0.99$. (b) Comparison of calculated pulse propagation for $S = 1$ and $S = 0.5$.

the grid generated at the step discontinuities of the wave. These numerical modes are poorly sampled in space and hence travel slower than c, thereby lagging behind the causative discontinuities.

(3) When $S = 0.99$, a weak superluminal response propagates just ahead of the leading edge of the pulse. This is again due to short-wavelength numerical modes in the grid generated at the step-function wavefront. However, these modes have spatial wavelengths even shorter than those noted in point (2), in fact so short that their grid sampling density drops below the upper bound for complex wavenumbers, and the modes appear in the superluminal, exponentially decaying regime.

Fig. 6.5(b) repeats the examples of Fig. 6.5(a), but for the Courant factors $S = 1$ and $S = 0.5$. We see that the duration and periodicity of the ringing is greater than that

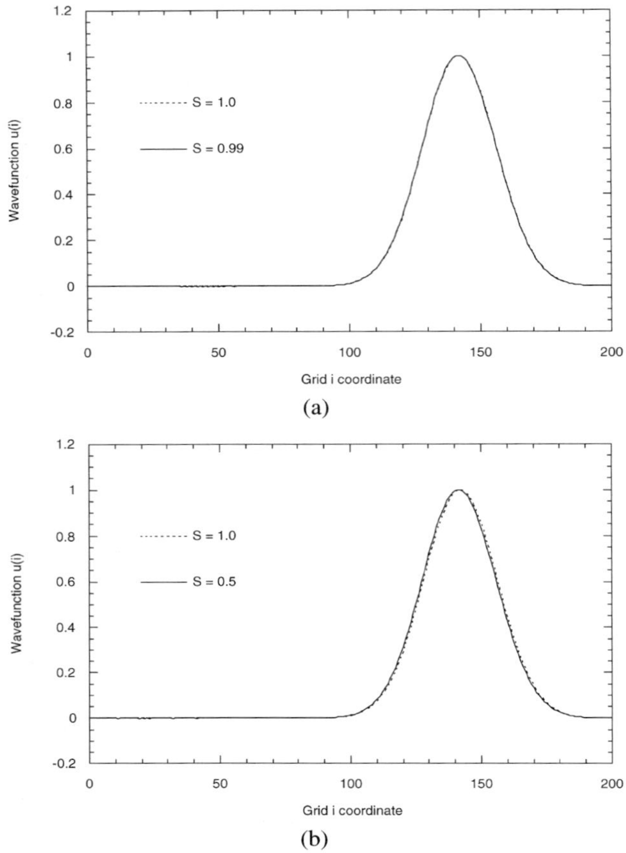

FIG. 6.6. Effect of numerical dispersion upon a Gaussian pulse propagating in free space in a one-dimensional grid for three different Courant factors: $S = 1$, $S = 0.99$, and $S = 0.5$. (a) Comparison of calculated pulse propagation for $S = 1$ and $S = 0.99$. (b) Comparison of calculated pulse propagation for $S = 1$ and $S = 0.5$.

for the $S = 0.99$ case. Further, the superluminal response is more pronounced and less damped.

Figs. 6.6(a) and 6.6(b) repeat the above examples, but for a Gaussian pulse having a 40-grid-cell spatial width between its $1/e$ points. We see that this pulse undergoes much less distortion than the rectangular pulse. The calculated propagation for $S = 0.99$ shows no observable difference (at the scale of Fig. 6.6(a)) relative to the perfect propagation case of $S = 1$. Even for $S = 0.5$, the calculated pulse propagation shows only a slight retardation relative to the exact solution, as expected because $\tilde{v}_p < c$ for virtually all modes in the grid. Further, there is no observable superluminal precursor. All of these phenomena are due to the fact that, for this case, virtually the entire spatial spectrum of propagating wavelengths within the grid is well resolved by the grid's sampling process. As a result, almost all numerical phase-velocity errors relative to c are well below 1%.

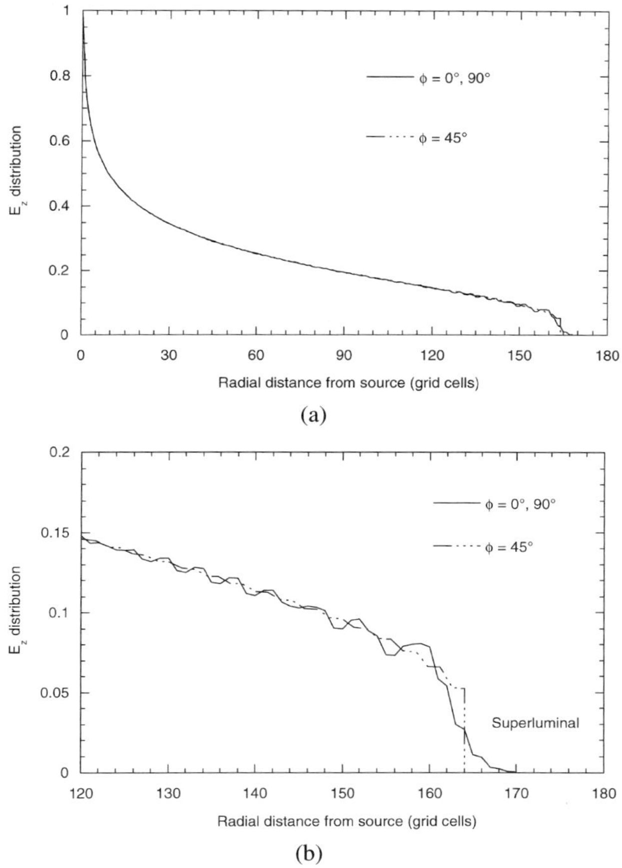

FIG. 6.7. Effect of numerical dispersion upon a radially propagating cylindrical wave in a 2D TM_Z Yee grid. The grid is excited at its center point by applying a unit-step time-function to a single E_z field component. The Courant factor is $S = \sqrt{2}/2$. (a) Comparison of calculated wave propagation along the grid axes and along a grid diagonal. (b) Expanded view of (a) at distances between 120 and 180 grid cells from the source.

This allows the Gaussian pulse to "hold together" while propagating over significant distances within the grid.

6.6.5. *Example of calculation of pulse propagation in a two-dimensional grid*

Fig. 6.7 presents an example of the calculation of a radially outward-propagating cylindrical wave in a two-dimensional TM_Z Yee grid. A 360 × 360-cell square grid with $\Delta x = \Delta y = \Delta = 1.0$ is used in this example. The grid is numerically excited at its center point by applying a unit-step time-function to a single E_z field component. We assume the Courant factor $S = \sqrt{2}/2$, which yields dispersionless propagation for numerical plane-wave modes propagating along the grid diagonals $\phi = 45°$, 135°, 225°, and 315°. In Fig. 6.7(a), we graph snapshots of the E_z distribution vs. radial distance from the source. Here, field observations are made along cuts through the grid passing through the source and either parallel to the principal grid axes $\phi = 0°, 90°$ or parallel to the grid diagonal $\phi = 45°$. The snapshots are taken $232\Delta t$ after the beginning of time-stepping. At this time, the wave has not yet reached the outer grid boundary.

Fig. 6.7(a) illustrates two nonphysical artifacts arising from numerical dispersion. First, for both observation cuts, the leading edge of the wave exhibits an oscillatory spatial jitter superimposed upon the normal field falloff profile. Second, for the observation cuts along the grid axes, the leading edge of the wave exhibits a small, spatially decaying, superluminal component.

To more easily see these artifacts, Fig. 6.7(b) shows an expanded view in the vicinity of the leading edge of the wave. Consider first the oscillatory jitter. Similar to the results shown in Fig. 6.5, this is due to short-wavelength numerical modes in the grid generated at the leading edge of the propagating step-function wave. According to our dispersion theory, these numerical modes are poorly sampled in space and hence travel slower than c, thereby lagging behind the actual wavefront. While the jitter is most pronounced along the grid axes $\phi = 0°, 90°$, it is nonetheless finite along $\phi = 45°$ despite our choice of $S = \sqrt{2}/2$ (which implies dispersionless propagation along grid diagonals). This apparent conflict between theory and numerical experiment is resolved by noting that numerical dispersion introduces a slightly anisotropic propagation characteristic of the background "free space" within the grid versus azimuth angle ϕ. The resulting inhomogeneity of the free-space background scatters part of the radially propagating numerical wave into the ϕ-direction. Thus, no point behind the wavefront can avoid the short-wavelength numerical jitter.

Consider next the superluminal artifact present at the leading edge of the wave shown in Fig. 6.7(b) for $\phi = 0°, 90°$ but not for $\phi = 45°$. This is again due to short-wavelength numerical modes in the grid generated at the leading edge of the outgoing step-function wave. However, these modes have spatial wavelengths so short that their grid sampling density drops below the threshold delineated in (6.30), and the modes appear in the superluminal, exponentially decaying regime. With $S = \sqrt{2}/2$ in the present example, we conclude that the lack of a superluminal artifact along the $\phi = 45°$ cut (and the consequent exact modeling of the step discontinuity at the leading edge of the wave) is due to dispersionless numerical wave propagation along grid diagonals.

7. Algorithms for improved numerical dispersion

7.1. Introduction

The numerical algorithm for Maxwell's equations introduced by YEE [1966] is very robust. Evidence of this claim is provided by the existence of thousands of successful electromagnetic engineering applications and refereed papers derived from the basic Yee algorithm. However, it is clear from Section 6 that Yee's approach is not perfect. For certain modeling problems, numerical dispersion can cause significant errors to arise in the calculated field.

This section reviews a small set of representative strategies aimed at mitigating the effects of numerical dispersion. No attempt is made to provide a comprehensive summary because such an effort would require several sections. The intent here is to provide the flavor of what may be possible in this area.

7.2. *Strategy 1: Center a specific numerical phase-velocity curve about c*

We have seen from Fig. 6.2 that the Yee algorithm yields a family of numerical phase-velocity curves contained in the range $\tilde{v}_p < c$. We also observe that each velocity curve is centered about the value

$$\tilde{v}_{\mathrm{avg}} = \frac{\tilde{v}_p(\phi = 0°) + \tilde{v}_p(\phi = 45°)}{2}, \tag{7.1}$$

where $\tilde{v}_p(\phi = 0°)$ and $\tilde{v}_p(\phi = 45°)$ are given by (6.12b) and (6.13b), respectively. This symmetry can be exploited if a narrowband grid excitation is used such that a specific phase-velocity curve accurately characterizes the propagation of most of the numerical modes in the grid. Then, it is possible to shift the phase-velocity curve of interest so that it is centered about the free-space speed of light c, thereby cutting $\Delta\tilde{v}_{\mathrm{physical}}$ by almost 3:1. Centering is implemented by simply scaling the free-space values of ε_0 and μ_0 used in the finite-difference system of (3.19):

$$\varepsilon_0' = \left(\frac{\tilde{v}_{\mathrm{avg}}}{c}\right)\varepsilon_0; \qquad \mu_0' = \left(\frac{\tilde{v}_{\mathrm{avg}}}{c}\right)\mu_0. \tag{7.2}$$

This scaling increases the baseline value of the model's "free-space" speed of light to compensate for the too-slow value of $\tilde{v}_{\mathrm{avg}}$. By scaling both ε_0 and μ_0, the required shift in $\tilde{v}_{\mathrm{avg}}$ is achieved without introducing any changes in wave impedance.

There are three primary difficulties with this approach: (1) The phase-velocity anisotropy error $\Delta\tilde{v}_{\mathrm{aniso}}$ remains unmitigated. (2) The velocity compensation is only in the average sense over all possible directions in the grid. Hence, important numerical modes can still have phase velocities not equal to c. (3) Propagating wave pulses having broad spectral content cannot be compensated over their entire frequency range. Nevertheless, this approach is so easy to implement that its use can be almost routine.

7.3. *Strategy 2: Use fourth-order-accurate spatial differences*

It is possible to substantially reduce the phase-velocity anisotropy error $\Delta\tilde{v}_{\mathrm{aniso}}$ for the Yee algorithm by incorporating a fourth-order-accurate finite-difference scheme for the spatial first-derivatives needed to implement the curl operator. This section reviews two such approaches. The first, by FANG [1989], is an explicit method wherein a fourth-order-accurate spatial central-difference is calculated at one observation point at a time from two pairs of field values: a pair on each side of the observation point at distances of $\Delta/2$ and $3\Delta/2$. The second approach, by TURKEL [1998], is an implicit method wherein a tridiagonal matrix is solved to obtain fourth-order-accurate spatial derivatives simultaneously at all observation points along a linear cut through the grid.

7.3.1. *Explicit method*

Assuming that Yee leapfrog time-stepping is used, the fourth-order-accurate spatial-difference scheme of FANG [1989] results in the following set of finite-difference expressions for the TM_z mode:

$$\frac{H_x|_{i,j+1/2}^{n+1/2} - H_x|_{i,j+1/2}^{n-1/2}}{\Delta t} = -\frac{1}{\mu_{i,j+1/2}}\left(\frac{-E_z|_{i,j+2}^{n} + 27E_z|_{i,j+1}^{n} - 27E_z|_{i,j}^{n} + E_z|_{i,j-1}^{n}}{24\Delta y}\right), \tag{7.3a}$$

$$\frac{H_y|_{i+1/2,j}^{n+1/2} - H_y|_{i+1/2,j}^{n-1/2}}{\Delta t} = \frac{1}{\mu_{i+1/2,j}}\left(\frac{-E_z|_{i+2,j}^{n} + 27E_z|_{i+1,j}^{n} - 27E_z|_{i,j}^{n} + E_z|_{i-1,j}^{n}}{24\Delta x}\right), \tag{7.3b}$$

$$\frac{E_z|_{i,j}^{n+1} - E_z|_{i,j}^{n}}{\Delta t} = \frac{1}{\varepsilon_{i,j}}\left(\begin{array}{l}\dfrac{-H_y|_{i+3/2,j}^{n+1/2} + 27H_y|_{i+1/2,j}^{n+1/2} - 27H_y|_{i-1/2,j}^{n+1/2} + H_y|_{i-3/2,j}^{n+1/2}}{24\Delta x} \\ \quad - \dfrac{-H_x|_{i,j+3/2}^{n+1/2} + 27H_x|_{i,j+1/2}^{n+1/2} - 27H_x|_{i,j-1/2}^{n+1/2} + H_x|_{i,j-3/2}^{n+1/2}}{24\Delta y}\end{array}\right). \tag{7.3c}$$

The numerical dispersion relation for this algorithm analogous to (6.3) is given by

$$\left[\frac{1}{c\Delta t}\sin\left(\frac{\omega\Delta t}{2}\right)\right]^2 = \frac{1}{(\Delta x)^2}\left[\frac{27}{24}\sin\left(\frac{\tilde{k}_x\Delta x}{2}\right) - \frac{1}{24}\sin\left(\frac{3\tilde{k}_x\Delta x}{2}\right)\right]^2 + \frac{1}{(\Delta y)^2}\left[\frac{27}{24}\sin\left(\frac{\tilde{k}_y\Delta y}{2}\right) - \frac{1}{24}\sin\left(\frac{3\tilde{k}_y\Delta y}{2}\right)\right]^2. \tag{7.4}$$

By analogy with the development in Section 6.5.2 culminating in (6.26), it can be shown that the intrinsic numerical phase-velocity anisotropy for a square-cell grid of this type is given by

$$\frac{c^*}{c} \cong \frac{N_\lambda}{\pi}\sqrt{\begin{array}{l}\left[\frac{27}{24}\sin\left(\frac{\pi\cos\phi}{N_\lambda}\right) - \frac{1}{24}\sin\left(\frac{3\pi\cos\phi}{N_\lambda}\right)\right]^2 \\ \quad + \left[\frac{27}{24}\sin\left(\frac{\pi\sin\phi}{N_\lambda}\right) - \frac{1}{24}\sin\left(\frac{3\pi\sin\phi}{N_\lambda}\right)\right]^2\end{array}} \tag{7.5}$$

and the numerical phase-velocity anisotropy error (by analogy with (6.27)) is given by

$$\Delta\tilde{v}_{\text{aniso}}|_{\text{explicit 4th-order}} \cong \frac{\pi^4}{18(N_\lambda)^4} \times 100\%. \tag{7.6}$$

7.3.2. *Implicit method*

TURKEL [1998] reported the Ty operator, an implicit fourth-order-accurate finite-difference scheme defined on the Yee space lattice for calculating the spatial first-derivatives involved in the curl. To see how the Ty operator is constructed, consider an x-directed cut through the Yee lattice. At every sample point along this cut, we wish to compute with fourth-order accuracy the x-derivatives of the general field component V. By manipulating Taylor's series expansions for V along this line, it was shown in TURKEL [1998, Eq. (2.23)] that

$$\frac{1}{24}\left(\left.\frac{\partial V}{\partial x}\right|_{i+1}+\left.\frac{\partial V}{\partial x}\right|_{i-1}\right)+\left.\frac{11}{12}\frac{\partial V}{\partial x}\right|_{i}=\frac{V_{i+1/2}-V_{i-1/2}}{\Delta x}. \tag{7.7}$$

Here, $\{(\partial V/\partial x)_i\}$ represents the set of initially unknown fourth-order-accurate x-derivatives of V at all grid-points along the observation cut; and $\{V_i\}$ represents the set of known values of V at the same grid-points. Upon writing (7.7) at each grid-point along the cut, a system of simultaneous equations for the unknowns $\{(\partial V/\partial x)_i\}$ is obtained. From the subscripts in (7.7), we see that this linear system has a tridiagonal matrix. This can be efficiently solved to yield $\{(\partial V/\partial x)_i\}$ in one step.

It was shown in TURKEL [1998, Eq. (2.70b)] that the Ty operator results in the following intrinsic grid-velocity anisotropy for a two-dimensional square-cell grid:

$$\frac{c^*}{c}=\frac{12N_\lambda}{\pi}\sqrt{\left[\frac{\sin\left(\frac{\pi\cos\phi}{N_\lambda}\right)}{11+\cos\left(\frac{2\pi\cos\phi}{N_\lambda}\right)}\right]^2+\left[\frac{\sin\left(\frac{\pi\sin\phi}{N_\lambda}\right)}{11+\cos\left(\frac{2\pi\sin\phi}{N_\lambda}\right)}\right]^2}. \tag{7.8}$$

From TURKEL [1998, Eq. (2.71b)], the phase-velocity anisotropy error is

$$\Delta\tilde{v}_{\text{aniso}}\big|_{\text{Ty 4th-order}}\cong\frac{17\cdot 3\cdot 2\cdot\pi^4}{2880(N_\lambda)^4}\times 100\%\cong\frac{\pi^4}{28(N_\lambda)^4}\times 100\%. \tag{7.9}$$

Fig. 7.1 compares the accuracy of the Ty method to the Yee algorithm for a generic two-dimensional wave-propagation problem, a sinusoidal line source radiating in free space after being switched on at $t=0$. Here, Ty(2, 4) and Ty(4, 4) refer to the implicit spatial-differentiation scheme of (7.7) used in conjunction with second-order

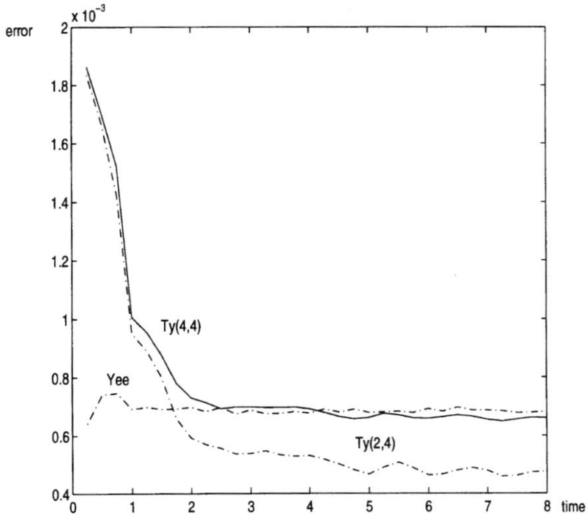

FIG. 7.1. Comparison of high-resolution ($N_\lambda = 40$) Yee and low-resolution ($N_\lambda = 5$) Ty errors in the L_2 norm for a radially propagating sinusoidal wave as a function of the simulated time. *Source*: E. Turkel, Chapter 2 in *Advances in Computational Electrodynamics: The Finite-Difference Time-Domain Method*, A. Taflove, ed., © 1998 Artech House, Inc.

Yee and fourth-order Runge–Kutta time-stepping, respectively. The two Ty approaches are implemented on square-cell grids of sampling-density N_λ with separate, accuracy-optimized Courant factors: $S = 1/18$ for Ty(2, 4) and $S = 1/4$ for Ty(4, 4).

From Fig. 7.1, we see that both Ty schemes run with $N_\lambda = 5$ achieve accuracy comparable to that of the Yee algorithm run with $N_\lambda = 40$ and $S = 2/3$. Under these conditions, both Ty methods are much more efficient than Yee's algorithm, requiring $(40/5)^2{:}1 = 8^2{:}1$ less computer storage and 23:1 less running-time. The permissible coarseness of the Ty grid is decisive in reducing its running-time, more than compensating for the extra operations required by its tridiagonal matrix inversions. These advantages in computer resources scale to the order of $8^3{:}1$ in three dimensions.

7.3.3. *Discussion*

Consider comparing (6.27) with (7.6) and (7.9). This allows us to form approximate ratios of the numerical phase-velocity anisotropy errors of the fourth-order-accurate spatial-differencing schemes discussed above relative to the Yee algorithm:

$$\frac{\Delta\tilde{v}_{\text{aniso}}|_{\text{explicit 4th-order}}}{\Delta\tilde{v}_{\text{aniso}}|_{\text{Yee}}} \cong \frac{2\pi^2}{3(N_\lambda)^2}; \qquad \frac{\Delta\tilde{v}_{\text{aniso}}|_{\text{Ty 4th-order}}}{\Delta\tilde{v}_{\text{aniso}}|_{\text{Yee}}} \cong \frac{3\pi^2}{7(N_\lambda)^2}. \tag{7.10}$$

With the reminder that these ratios were derived based upon assuming $N_\lambda > 10$, we see that greatly reduced $\Delta\tilde{v}_{\text{aniso}}$ error is possible for both fourth-order spatial-differencing schemes. In addition, optimally choosing the Courant number S for each fourth-order spatial technique permits minimizing the overall error, including $\Delta\tilde{v}_{\text{physical}}$. From a growing set of published results similar to those of Fig. 7.1, we conclude that fourth-order-accurate explicit and implicit spatial schemes allow modeling electromagnetic wave-propagation and interaction problems that are at least 8 times the electrical size of those permitted by the Yee algorithm. This is a very worthwhile increase in capability.

However, this improvement is not without cost. Although easy to set up in homogeneous-material regions, the larger stencil needed to calculate fourth-order spatial differences is troublesome when dealing with material interfaces. Metal boundaries are especially challenging since they effectively cause field discontinuities in the grid. Special boundary conditions required for such interfaces significantly complicate the computer software used to render structures in the grid.

7.3.4. *Fourth-order-accurate approximation of jumps in material parameters at interfaces*

As stated above, special boundary conditions must be derived and programmed to deal with material discontinuities when implementing high-order-accuracy finite-difference approximations of spatial derivatives. This is because the nonlocal nature of the numerical space-differentiation process may convey electromagnetic field data across such discontinuities in a nonphysical manner.

TURKEL [1998] reported a means to markedly reduce error due to abrupt dielectric interfaces. This approach replaces the discontinuous permittivity function ε by a fourth-order-accurate smooth implicit approximation. (A similar strategy can be applied to jumps in μ.) Relative to the use of a polynomial approximation to ε, this strategy avoids the overshoot artifact. We note that, with an implicit interpolation, ε varies in the entire domain and not just near the interface. However, far from the interface, the variation is small.

Consider a dielectric interface separating two regions defined along the x-axis of the space lattice. Following TURKEL [1998, Eq. (2.82)], a fourth-order-accurate interpolation of the permittivity distribution with grid position i can be achieved using

$$\frac{1}{8}\begin{bmatrix} 10 & -5 & 4 & -1 & \cdot & \cdot & 0 \\ 1 & 6 & 1 & 0 & \cdot & \cdot & 0 \\ 0 & 1 & 6 & 1 & 0 & \cdot & 0 \\ \cdot & \cdot & \cdot & \cdot & \cdot & \cdot & \cdot \\ 0 & \cdot & \cdot & 0 & 1 & 6 & 1 \\ 0 & \cdot & \cdot & -1 & 4 & -5 & 10 \end{bmatrix} \begin{bmatrix} \varepsilon_1 \\ \varepsilon_2 \\ \cdot \\ \cdot \\ \varepsilon_{p-1} \end{bmatrix} = \frac{1}{2}\left(\begin{bmatrix} \varepsilon_{3/2} \\ \varepsilon_{5/2} \\ \cdot \\ \varepsilon_{p-3/2} \\ \varepsilon_{p-1/2} \end{bmatrix} + \begin{bmatrix} \varepsilon_{1/2} \\ \varepsilon_{3/2} \\ \cdot \\ \varepsilon_{p-5/2} \\ \varepsilon_{p-3/2} \end{bmatrix} \right). \tag{7.11}$$

Here, $[\varepsilon_1, \varepsilon_2, \ldots, \varepsilon_{p-1}]$ is the initially unknown set of values of the smooth approximation to the abrupt dielectric interface; and $[\varepsilon_{1/2}, \varepsilon_{3/2}, \ldots, \varepsilon_{p-1/2}]$ is the known set of permittivities for the original dielectric interface geometry. Inversion of the linear system of (7.11) yields the desired smooth approximation, $[\varepsilon_1, \varepsilon_2, \ldots, \varepsilon_{p-1}]$.

TURKEL [1998] presented a comparative example (shown in Fig. 7.2) of the use of this dielectric interface smoothing technique for both the Yee and Ty algorithms. His example modeled the standing wave within a two-dimensional rectangular cavity comprised of a block of lossless dielectric of $\varepsilon_r = 4$ surrounded by free space. An available exact solution for the sinusoidal space–time variation of the standing-wave mode was used to specify the computational domain's initial conditions and boundary conditions for both the Yee and Ty simulations. It was also used to develop L_2-normed errors of the calculated Yee and Ty fields as a function of the simulated time.

Fig. 7.2(a) compares the error of the Ty method with that of the Yee algorithm for the case where ε at the dielectric interfaces of the cavity is simply set to the arithmetic average of the values on both sides. Here, both the Yee and Ty grids use square unit cells wherein $N_\lambda = 30$ within the dielectric material. The Courant factors are selected as $S_{\text{Yee}} = 2/3$ and $S_{\text{Ty}(2,4)} = 1/18$. We see from this figure that, while the Ty results show less error than the Yee data, the error performance of the Ty scheme is clearly hurt by the treatment of the interfaces, which gives only second-order accuracy.

Fig. 7.2(b) shows the corresponding numerical errors for the case where the permittivity is smoothed as per (7.11). While there is a modest reduction in the error of the Yee data, there is a much greater reduction in the error of the Ty results. In additional studies discussed by TURKEL [1998], it was demonstrated that the Yee error can be reduced to that of Ty by increasing the Yee-grid-sampling density to a level eight times that of Ty, just what was observed for the free-space propagation example discussed previously in the context of Fig. 7.1. Thus, we see that fourth-order-accurate smoothing of abrupt permittivity jumps succeeds in preserving the fourth-order-accuracy advantage of Ty versus Yee observed for the homogeneous-permittivity case.

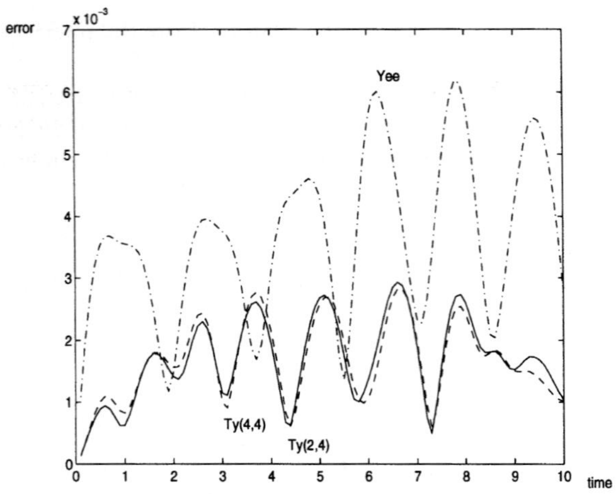

(a)

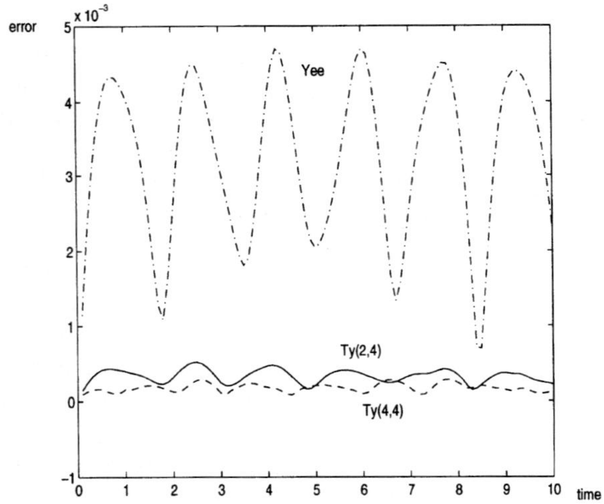

(b)

FIG. 7.2. Comparison of Yee and Ty errors (L_2 norm) for the standing-wave fields within a rectangular dielectric cavity. The same grid-sampling density ($N_\lambda = 30$) is used for both algorithms. (a) Second-order-accurate arithmetic averaging of the permittivity at the dielectric interfaces. (b) Fourth-order-accurate smoothing of the permittivity at the dielectric interfaces as per (4.86). *Source*: E. Turkel, Chapter 2 in *Advances in Computational Electrodynamics: The Finite-Difference Time-Domain Method*, A. Taflove, ed., © 1998 Artech House, Inc.

7.4. Strategy 3: Use hexagonal grids

Regular hexagonal grids in two dimensions have been proposed to reduce the numerical phase-velocity anisotropy well below that of Yee's Cartesian mesh. Here, the primary grid is composed of equilateral hexagons having edge length Δs. Each hexagon can be considered to be the union of six equilateral triangles. Connecting the centroids of these triangles yields a second set of regular hexagons that comprises a dual grid.

Fig. 5.2 illustrated for the TM_z case in two dimensions the two principal ways of arranging E and H vector components about hexagonal grids, as discussed by LIU, Y. [1996] and reviewed in Section 5.2. There, the finite-difference equations for the TM_z mode for the unstaggered, collocated hexagonal grid of Fig. 5.2(a) were given by (5.3), and the finite-difference equations for the TM_z mode for the staggered, uncollocated hexagonal grid of Fig. 5.2(b) were given by (5.4).

Using the analysis method of Section 6.5.2, LIU, Y. [1996] obtained the following expressions for the numerical phase-velocity anisotropy error for the hexagonal grids of Fig. 5.2:

$$\Delta\tilde{v}_{\text{aniso}}\Big|_{\substack{\text{hex.grid,}\\ \text{Fig. 5.2(a)}}} \cong \frac{1 \cdot 2 \cdot \pi^4}{120(N_\lambda)^4} \times 100\% = \frac{\pi^4}{60(N_\lambda)^4} \times 100\%, \tag{7.12}$$

$$\Delta\tilde{v}_{\text{aniso}}\Big|_{\substack{\text{hex.grid,}\\ \text{Fig. 5.2(b)}}} \cong \frac{1 \cdot 2 \cdot \pi^4}{720(N_\lambda)^4} \times 100\% = \frac{\pi^4}{360(N_\lambda)^4} \times 100\%. \tag{7.13}$$

Interestingly, we note that $\Delta\tilde{v}_{\text{aniso}}$ for both hexagonal grids exhibits a *fourth-order* dependence on the grid-sampling density N_λ despite the second-order accuracy of each spatial difference used. As shown by LIU, Y. [1996], this is because the leading second-order error term becomes isotropic for the hexagonal gridding case, with a value exactly equal to the average of its ϕ-dependent Cartesian counterpart.

Comparison of $\Delta\tilde{v}_{\text{aniso}}$ of the Yee algorithm given by (6.27) with $\Delta\tilde{v}_{\text{aniso}}$ of the hexagonal gridding given by (7.12) and (7.13) yields the following error ratios:

$$\frac{\Delta\tilde{v}_{\text{aniso}}|_{\text{hex.grid, Fig. 5.2(a)}}}{\Delta\tilde{v}_{\text{aniso}}|_{\text{Yee}}}(N_\lambda) \cong \frac{\pi^2}{5(N_\lambda)^2} \tag{7.14}$$

and

$$\frac{\Delta\tilde{v}_{\text{aniso}}|_{\text{hex.grid, Fig. 5.2(b)}}}{\Delta\tilde{v}_{\text{aniso}}|_{\text{Yee}}}(N_\lambda) \cong \frac{\pi^2}{30(N_\lambda)^2}. \tag{7.15}$$

Hexagonal gridding is seen to yield velocity-anisotropy errors as little as 1/300th that of the Yee grid at a sampling density of 10 points per wavelength.

We can also compare the velocity-anisotropy errors of hexagonal gridding with those of the fourth-order gridding schemes discussed previously:

$$\frac{\Delta\tilde{v}_{\text{aniso}}|_{\text{hex.grid, Fig. 5.2(a)}}}{\Delta\tilde{v}_{\text{aniso}}|_{\text{explicit 4th-order}}}(N_\lambda) \cong \frac{3}{10}; \qquad \frac{\Delta\tilde{v}_{\text{aniso}}|_{\text{hex.grid, Fig. 5.2(a)}}}{\Delta\tilde{v}_{\text{aniso}}|_{\text{Ty 4th-order}}}(N_\lambda) \cong \frac{7}{15}; \tag{7.16}$$

$$\frac{\Delta\tilde{v}_{\text{aniso}}|_{\text{hex.grid, Fig. 5.2(b)}}}{\Delta\tilde{v}_{\text{aniso}}|_{\text{explicit 4th-order}}}(N_\lambda) \cong \frac{1}{20}; \qquad \frac{\Delta\tilde{v}_{\text{aniso}}|_{\text{hex.grid, Fig. 5.2(b)}}}{\Delta\tilde{v}_{\text{aniso}}|_{\text{Ty 4th-order}}}(N_\lambda) \cong \frac{7}{90}. \tag{7.17}$$

We see that hexagonal gridding yields less velocity-anisotropy error than the two fourth-order-accurate Cartesian spatial-differencing techniques reviewed previously. In the case of the hexagonal grid of Fig. 5.2(b), this dispersion error is lower by more than one order-of-magnitude.

Hexagonal gridding has a second advantage relative to the fourth-order spatial algorithms: it uses only nearest-neighbor field data. Therefore, hexagonal grids can model material discontinuities including metal boundaries as easily as the Yee algorithm. There is no need to develop special boundary conditions.

What, if any, are the limitations in using hexagonal grid algorithms relative to Yee's method? The answer is: none very significant in two dimensions. This is because it is only moderately more complicated to generate (even manually) uniform hexagonal grids than it is to generate uniform Cartesian grids. The difficulty arises in attempting to extend hexagonal gridding to three dimensions. As shown in Fig. 5.3, such an extension involves filling space with tetradecahedron and dual-tetrahedron unit cells. This increases the complexity of the computational mesh to the point where sophisticated computer-based mesh-generation techniques are mandatory.

7.5. Strategy 4: Use discrete Fourier transforms to calculate the spatial derivatives

The fourth and final approach reviewed here for reduction of the numerical dispersion artifact is the *pseudospectral time-domain* (PSTD) method of LIU, Q.H. [1996], LIU, Q.H. [1997]. This technique uses a discrete Fourier transform (DFT) algorithm to represent the spatial derivatives in the Maxwell's equations' computational lattice. The fast Fourier transform (FFT) can also be applied to increase numerical efficiency.

7.5.1. Formulation

The PSTD method works on unstaggered, collocated Cartesian space lattices wherein all field components are located at the same points. An example of such an arrangement is the two-dimensional TM_z grid of Fig. 5.1(a). To see how PSTD works, consider an x-directed cut through this grid. At every sample point along this cut, we wish to compute the x-derivatives of the general field component V. Let $\{V_i\}$ denote the set of initially known values of V at all grid-points along the observation cut, and let $\{(\partial V/\partial x)_i\}$ denote the set of initially unknown x-derivatives of V at the same grid-points. Then, using the differentiation theorem for Fourier transforms, we can write:

$$\left\{ \frac{\partial V}{\partial x}\bigg|_i \right\} = -\mathcal{F}^{-1}\big(j\tilde{k}_x \mathcal{F}\{V_i\}\big), \tag{7.18}$$

where $\mathcal{F}$ and $\mathcal{F}^{-1}$ denote respectively the forward and inverse DFTs, and $\tilde{k}_x$ is the Fourier transform variable representing the x-component of the numerical wavevector. In this manner, the entire set of spatial derivatives of V along the observation cut can be calculated in one step. In multiple dimensions, this process is repeated for each observation cut parallel to the major axes of the space lattice.

According to the Nyquist sampling theorem, the representation in (7.18) is *exact* for $|\tilde{k}_x| \leqslant \pi/\Delta x$, i.e., $\Delta x \leqslant \tilde{\lambda}/2$. Thus, the spatial-differencing process here can be said to be of "infinite order" for grid-sampling densities of two or more points per wavelength. The wraparound effect, a potentially major limitation caused by the periodicity assumed in the FFT, is eliminated by using the perfectly matched layer absorbing boundary condition (to be discussed in Section 10). Finally, the time-differencing for PSTD uses conventional second-order-accurate Yee leapfrogging.

LIU, Q.H. [1996], LIU, Q.H. [1997], derived the following expressions for the wavenumber and phase velocity of a sinusoidal numerical wave of temporal period $T = 2\pi/\omega$ propagating in an arbitrary direction within a three-dimensional PSTD space lattice:

$$|\tilde{\vec{k}}| = \frac{2}{c\Delta t}\sin\left(\frac{\omega\Delta t}{2}\right), \tag{7.19}$$

$$\tilde{v}_p = \frac{\omega}{\tilde{k}} = \frac{\omega}{\frac{2}{c\Delta t}\sin\left(\frac{\omega\Delta t}{2}\right)} = \frac{\omega\Delta t/2}{\sin(\omega\Delta t/2)}c. \tag{7.20}$$

Eq. (7.20) implies that the numerical phase velocity is *independent* of the propagation direction of the wave, unlike any of the methods considered previously. Applying our definitions of numerical phase-velocity error, we therefore have the following figures of merit for the PSTD method:

$$\Delta\tilde{v}_{\text{physical}}|_{\text{PSTD}} = \left[\frac{\omega\Delta t/2}{\sin(\omega\Delta t/2)} - 1\right] \times 100\% = \left[\frac{\pi/N_T}{\sin(\pi/N_T)} - 1\right] \times 100\%, \tag{7.21}$$

$$\Delta\tilde{v}_{\text{aniso}}|_{\text{PSTD}} = 0, \tag{7.22}$$

where we define the temporal sampling density $N_T = T/\Delta t$ time samples per wave-oscillation period.

7.5.2. Discussion

Remarkably, $\Delta\tilde{v}_{\text{aniso}}$ is zero for the PSTD method for *all* propagating sinusoidal waves sampled at $N_\lambda \geqslant 2$. Therefore, to specify the gridding density of the PSTD simulation, we need only a reliable estimate of $\lambda_{\min}$, the fastest oscillating spectral component of significance. This estimate is based upon the wavelength spectrum of the exciting pulse and the size of significant structural details such as material inhomogeneities. Then, the space-cell dimension is set at $\Delta = \lambda_{\min}/2$, regardless of the problem's overall electrical size. This is because our choice of Δ assures zero $\Delta\tilde{v}_{\text{aniso}}$ error, and thus, zero accumulation of this error even if the number of space cells increases without bound. Consequently, we conclude that:

- The density of the PSTD mesh-sampling is independent of the electrical size of the modeling problem.

However, the fact that $\Delta\tilde{v}_{\text{aniso}} = 0$ does *not* mean that PSTD yields perfect results. In fact, (7.21) shows that there remains a numerical phase-velocity error relative to c. This residual velocity error is not a function of the wave-propagation direction ϕ, and is therefore isotropic within the space grid. The residual velocity error arises from the Yee-type leapfrog time-stepping used in the algorithm, and is a function only of N_T. Table 7.1 provides representative values of this residual velocity error.

The key point from Table 7.1 is that N_T limits the accuracy of the PSTD technique when modeling impulsive propagation. This is not an issue for a monochromatic wave where there is only a single value of N_T, and $\Delta\tilde{v}_{\text{physical}}$ can be nulled in the manner of Strategy 1. For a pulse, however, with Δt a fixed algorithm parameter, there exists a spread of equivalent N_T values for the spectral components of the pulse which possess a range of temporal periods T. This causes a spread of $\tilde{v}_p$ over the pulse spectrum, which

TABLE 7.1
Residual numerical phase-velocity error of the PSTD method versus its time-sampling

N_T	$\Delta\tilde{v}_{\text{physical}}$	N_T	$\Delta\tilde{v}_{\text{physical}}$
2	+57%	15	+0.73%
4	+11%	20	+0.41%
8	+2.6%	25	+0.26%
10	+1.7%	30	+0.18%

in turn results in an isotropic progressive broadening and distortion of the pulse waveform as it propagates in the grid. To bound such dispersion, it is important to choose Δt small enough so that it adequately resolves the period $T_{\min}$ of the fastest oscillating spectral component $\lambda_{\min}$. Because this dispersion is cumulative with the wave-propagation distance, we have a second key point:

- The density of the PSTD time sampling must increase with the electrical size of the modeling problem if we apply a fixed upper bound on the maximum total phase error of propagating waves within the mesh.

Despite this need for a small Δt, PSTD can provide a large reduction in computer resources relative to the Yee algorithm for electrically large problems not having spatial details or material inhomogeneities smaller than $\lambda_{\min}/2$. Increased efficiency is expected even relative to the fourth-order-accurate spatial algorithms reviewed previously. LIU, Q.H. [1996], LIU, Q.H. [1997] reported that, within the range of problem sizes from 16–64 wavelengths, the use of PSTD permits an $8^D : 1$ reduction in computer storage and running time relative to the Yee algorithm, where D is the problem dimensionality. While this savings is comparable to that shown for the fourth-order spatial techniques, we expect the PSTD advantage to increase for even larger problems. In fact, the computational benefit of PSTD theoretically increases without limit as the electrical size of the modeling problem expands.

The second topic in our discussion is whether PSTD's global calculation of space derivatives along observation cuts through the lattice (similar to the Ty method) has difficulties at material interfaces. An initial concern is that PSTD might yield nonphysical results for problems having abrupt jumps in ε unless an ε-smoothing technique such as (7.11) is used. However, this is not the case. As reported by LIU, Q.H. [1996], LIU, Q.H. [1997], the PSTD method is successful for dielectric interfaces because, at such discontinuities, the normal derivatives that it calculates via DFT or FFT are implemented on continuous tangential-field data across the interface.

However, structures having metal surfaces comprise a very important set of problems where the required continuity of tangential fields within the PSTD space lattice is effectively violated. Depending upon the orientation and thickness of the metal surfaces, tangential-field discontinuities could appear for two reasons:

(1) A space-cell boundary lies at a metal surface or within a metal layer. The tangential H-field within the space cell drops abruptly to zero at the metal surface, and remains at zero for the remainder of the space cell.

(2) A metal sheet splits the space lattice so there exist distinct lit and shadow regions within the lattice. Here, the tangential H-field on the far (shadowed or shielded)

side of the metal sheet may be physically isolated from the field immediately across the metal sheet on its near (lit) side. Gross error is caused by the global nature of PSTD's spatial-derivative calculation which nonphysically transports field information directly across the shielding metal barrier from the lit to the shadow sides.

Until these problems are solved by the prescription of special boundary conditions, PSTD will likely find its primary applications in Cartesian-mesh modeling of structures comprised entirely of dielectrics. We note that the PSTD usage of collocated field components simplifies rendering such structures within the mesh. It is also ideal for modeling nonlinear optical problems where the local index of refraction is dependent upon a power of the magnitude of the local $\vec{E}$. Here, collocation of the vector components of $\vec{E}$ avoids the need for error-causing spatial interpolations of nearby electric field vector components staggered in space.

8. Numerical stability

8.1. Introduction

In Section 6, we saw that the choice of Δ and Δt can affect the propagation characteristics of numerical waves in the Yee space lattice, and therefore the numerical error. In this section, we show that, in addition, Δt must be bounded to ensure numerical stability. Our approach to determine the upper bound on Δt is based upon the complex-frequency analysis reported by TAFLOVE and HAGNESS [2000, pp. 133–140]. As noted there, the complex-frequency approach is conceptually simple, yet rigorous. It also allows straightforward estimates of the exponential growth rate of unstable numerical solutions.

Subsequently, Section 9 will review a representative approach of a new class of algorithms proposed to eliminate the need to bound Δt. This class replaces Yee's leapfrog time-stepping with an implicit alternating-direction technique.

8.2. Complex-frequency analysis

We first postulate a sinusoidal traveling wave present in the three-dimensional FDTD space lattice and discretely sampled at (x_I, y_J, z_K, t_n), allowing for the possibility of a complex-valued numerical angular frequency, $\tilde{\omega} = \tilde{\omega}_{\text{real}} + \mathrm{j}\tilde{\omega}_{\text{imag}}$. A field vector in this wave can be written as

$$\begin{aligned} \vec{V}\big|_{I,J,K}^{n} &= \vec{V}_0 \mathrm{e}^{\mathrm{j}[(\tilde{\omega}_{\text{real}}+\mathrm{j}\tilde{\omega}_{\text{imag}})n\Delta t-\tilde{k}_x I\Delta x-\tilde{k}_y J\Delta y-\tilde{k}_z K\Delta z]} \\ &= \vec{V}_0 \mathrm{e}^{-\tilde{\omega}_{\text{imag}}n\Delta t}\mathrm{e}^{\mathrm{j}(\tilde{\omega}_{\text{real}}n\Delta t-\tilde{k}_x I\Delta x-\tilde{k}_y J\Delta y-\tilde{k}_z K\Delta z)}, \end{aligned} \tag{8.1}$$

where $\tilde{k}$ is the wavenumber of the numerical sinusoidal traveling wave. We note that (8.1) permits either a constant wave amplitude with time ($\tilde{\omega}_{\text{imag}} = 0$), an exponentially decreasing amplitude with time ($\tilde{\omega}_{\text{imag}} > 0$), or an exponentially increasing amplitude with time ($\tilde{\omega}_{\text{imag}} < 0$).

Given this basis, we proceed to analyze numerical dispersion relation (6.10) allowing for a complex-valued angular frequency:

$$\left[\frac{1}{c\Delta t}\sin\left(\frac{\tilde{\omega}\Delta t}{2}\right)\right]^2 = \left[\frac{1}{\Delta x}\sin\left(\frac{\tilde{k}_x\Delta x}{2}\right)\right]^2 + \left[\frac{1}{\Delta y}\sin\left(\frac{\tilde{k}_y\Delta y}{2}\right)\right]^2 + \left[\frac{1}{\Delta z}\sin\left(\frac{\tilde{k}_z\Delta z}{2}\right)\right]^2. \tag{8.2}$$

We first solve (8.2) for $\tilde{\omega}$. This yields

$$\tilde{\omega} = \frac{2}{\Delta t}\sin^{-1}(\xi), \tag{8.3}$$

where

$$\xi = c\Delta t\sqrt{\frac{1}{(\Delta x)^2}\sin^2\left(\frac{\tilde{k}_x\Delta x}{2}\right) + \frac{1}{(\Delta y)^2}\sin^2\left(\frac{\tilde{k}_y\Delta y}{2}\right) + \frac{1}{(\Delta z)^2}\sin^2\left(\frac{\tilde{k}_z\Delta z}{2}\right)}. \tag{8.4}$$

We observe from (8.4) that

$$0 \leqslant \xi \leqslant c\Delta t\sqrt{\frac{1}{(\Delta x)^2} + \frac{1}{(\Delta y)^2} + \frac{1}{(\Delta z)^2}} \equiv \xi_{\text{upper bound}} \tag{8.5}$$

for all possible real values of $\tilde{k}$, that is, those numerical waves having zero exponential attenuation per grid space cell. $\xi_{\text{upper bound}}$ is obtained when each $\sin^2$ term under the square root of (8.4) simultaneously reaches a value of 1. This occurs for the propagating numerical wave having the wavevector components

$$\tilde{k}_x = \pm\frac{\pi}{\Delta x}; \tag{8.6a}$$

$$\tilde{k}_y = \pm\frac{\pi}{\Delta y}; \tag{8.6b}$$

$$\tilde{k}_z = \pm\frac{\pi}{\Delta z}. \tag{8.6c}$$

It is clear that $\xi_{\text{upper bound}}$ can exceed 1 depending upon the choice of Δt. This yields complex values of $\sin^{-1}(\xi)$ in (8.3), and therefore complex values of $\tilde{\omega}$. To investigate further, we divide the range of ξ given in (8.5) into two subranges.

8.2.1. Stable range: $0 \leqslant \xi \leqslant 1$

Here, $\sin^{-1}(\xi)$ is real-valued and hence, real values of $\tilde{\omega}$ are obtained in (8.3). With $\tilde{\omega}_{\text{imag}} = 0$, (8.1) yields a constant wave amplitude with time.

8.2.2. Unstable range: $1 < \xi < \xi_{\text{upper bound}}$

This subrange exists only if

$$\xi_{\text{upper bound}} = c\Delta t\sqrt{\frac{1}{(\Delta x)^2} + \frac{1}{(\Delta y)^2} + \frac{1}{(\Delta z)^2}} > 1. \tag{8.7}$$

9. Alternating-direction-implicit time-stepping algorithm for operation beyond the Courant limit

9.1. Introduction

Section 8 showed that numerical stability of the Yee algorithm requires placing an upper bound on the time step Δt relative to the space increments Δx, Δy, and Δz. This has allowed the successful application of FDTD methods to a wide variety of electromagnetic wave modeling problems of moderate electrical size and quality factor. Typically, such problems require 10^3–10^4 time-steps to complete a single simulation.

However, there are important potential applications of FDTD modeling where the Courant stability bounds determined in Sections 8.2.3 and 8.2.4 are much too restrictive. Modeling problems that fall into this regime have the following properties:

- The cell size Δ needed to resolve the fine-scale geometric detail of the electromagnetic wave interaction structure is much less than the shortest wavelength λ_{min} of a significant spectral component of the source.
- The simulated time T_{sim} needed to evolve the electromagnetic wave physics to the desired endpoint is related to the cycle time T of λ_{min}.

With Δ fixed by the need to resolve the problem geometry, the requirement for numerical stability in turn specifies the maximum possible time step Δt_{max}. This, in turn, fixes the total number of time steps needed to complete the simulation, $N_{\text{sim}} = T_{\text{sim}}/\Delta t_{\text{max}}$. Table 9.1 lists parameters of two important classes of problems where this decision process results in values of N_{sim} that are so large that standard FDTD modeling in three dimensions is difficult, or even impossible.

If these classes of electromagnetics problems are to be explored using FDTD modeling, we need an advancement of FDTD techniques that permits accurate and numerically stable operation for values of Δt exceeding the Courant limit by much more than 10:1. A candidate, computationally efficient approach for this purpose is to use an *alternating-direction-implicit* (ADI) time-stepping algorithm rather than the usual explicit Yee leapfrogging. In fact, work with ADI FDTD methods in the early 1980s by HOLLAND [1984] and HOLLAND and CHO [1986] achieved promising results for two-dimensional models. However, using these early ADI techniques, it proved difficult to demonstrate numerical stability for the general three-dimensional case, and research in this area was largely discontinued.

Recently, key publications by ZHENG, CHEN and ZHANG [2000] and NAMIKI [2000] have reported the development of unconditionally stable three-dimensional ADI

TABLE 9.1

Two important classes of three-dimensional FDTD modeling problems made difficult or impossible by the Courant limit on Δt

Problem class	Δ	T_{sim}	Δt_{max}	N_{sim}
Propagation of bioelectric signals	~ 1 mm	~ 100 ms	~ 2 ps	$\sim 5 \times 10^{10}$
Propagation of digital logic signals	~ 0.25 μm	~ 1 ns	~ 0.5 fs	$\sim 2 \times 10^{6}$

FDTD algorithms. This section reviews the work by ZHENG, CHEN and ZHANG [2000] wherein for the first time unconditional numerical stability is derived for the full three-dimensional case.

Note that, with any unconditionally stable ADI FDTD algorithm, the upper bound on Δt is relaxed to only that value needed to provide good accuracy in numerically approximating the time derivatives of the electromagnetic field. Thus, in theory, Δt need only be small enough to provide about 20 or more field samples during the cycle time T of the fastest oscillating significant spectral component of the exciting source. For example, in Table 9.1, Δt could be 10 μs rather than 2 ps for studies of signal propagation within human muscles, yielding $N_{\text{sim}} = 10^4$ rather than $N_{\text{sim}} = 5 \times 10^{10}$.

9.2. *Formulation of the Zheng et al. algorithm*

ZHENG, CHEN and ZHANG [2000] reported a new ADI time-stepping algorithm for FDTD that has theoretical unconditional numerical stability for the general three-dimensional case. While this technique uses the same Yee space lattice as conventional FDTD, the six field-vector components are collocated rather than staggered in time. In discussing the formulation of this algorithm, we assume that all of the field components are known everywhere in the lattice at time step n and stored in the computer memory.

9.2.1. *Unsimplified system of time-stepping equations*

The ADI nature of the Zheng et al. algorithm can be best understood by first considering its unsimplified form, and then proceeding to obtain the final simplified system of field update equations. To advance a single time step from n to $n+1$, we perform two subiterations: the first from n to $n+1/2$, and the second from $n+1/2$ to $n+1$. These subiterations are as follows.

Subiteration 1. Advance the 6 field components from time step n to time step $n+1/2$

$$E_x\big|_{i+1/2,j,k}^{n+1/2} = E_x\big|_{i+1/2,j,k}^{n} + \frac{\Delta t}{2\varepsilon\Delta y}\left(H_z\big|_{i+1/2,j+1/2,k}^{n+1/2} - H_z\big|_{i+1/2,j-1/2,k}^{n+1/2}\right) - \frac{\Delta t}{2\varepsilon\Delta z}\left(H_y\big|_{i+1/2,j,k+1/2}^{n} - H_y\big|_{i+1/2,j,k-1/2}^{n}\right), \tag{9.1a}$$

$$E_y\big|_{i,j+1/2,k}^{n+1/2} = E_y\big|_{i,j+1/2,k}^{n} + \frac{\Delta t}{2\varepsilon\Delta z}\left(H_x\big|_{i,j+1/2,k+1/2}^{n+1/2} - H_x\big|_{i,j+1/2,k-1/2}^{n+1/2}\right) - \frac{\Delta t}{2\varepsilon\Delta x}\left(H_z\big|_{i+1/2,j+1/2,k}^{n} - H_z\big|_{i-1/2,j+1/2,k}^{n}\right), \tag{9.1b}$$

$$E_z\big|_{i,j,k+1/2}^{n+1/2} = E_z\big|_{i,j,k+1/2}^{n} + \frac{\Delta t}{2\varepsilon\Delta x}\left(H_y\big|_{i+1/2,j,k+1/2}^{n+1/2} - H_y\big|_{i-1/2,j,k+1/2}^{n+1/2}\right) - \frac{\Delta t}{2\varepsilon\Delta y}\left(H_x\big|_{i,j+1/2,k+1/2}^{n} - H_x\big|_{i,j-1/2,k+1/2}^{n}\right), \tag{9.1c}$$

$$H_x\big|_{i,j+1/2,k+1/2}^{n+1/2} = H_x\big|_{i,j+1/2,k+1/2}^{n} + \frac{\Delta t}{2\mu\Delta z}\left(E_y\big|_{i,j+1/2,k+1}^{n+1/2} - E_y\big|_{i,j+1/2,k}^{n+1/2}\right) - \frac{\Delta t}{2\mu\Delta y}\left(E_z\big|_{i,j+1,k+1/2}^{n} - E_z\big|_{i,j,k+1/2}^{n}\right), \tag{9.2a}$$

$$H_y\big|_{i+1/2,j,k+1/2}^{n+1/2} = H_y\big|_{i+1/2,j,k+1/2}^{n} + \frac{\Delta t}{2\mu\Delta x}\left(E_z\big|_{i+1,j,k+1/2}^{n+1/2} - E_z\big|_{i,j,k+1/2}^{n+1/2}\right) - \frac{\Delta t}{2\mu\Delta z}\left(E_x\big|_{i+1/2,j,k+1}^{n} - E_x\big|_{i+1/2,j,k}^{n}\right), \tag{9.2b}$$

$$H_z\big|_{i+1/2,j+1/2,k}^{n+1/2} = H_z\big|_{i+1/2,j+1/2,k}^{n} + \frac{\Delta t}{2\mu\Delta y}\left(E_x\big|_{i+1/2,j+1,k}^{n+1/2} - E_x\big|_{i+1/2,j,k}^{n+1/2}\right) - \frac{\Delta t}{2\mu\Delta x}\left(E_y\big|_{i+1,j+1/2,k}^{n} - E_y\big|_{i,j+1/2,k}^{n}\right). \tag{9.2c}$$

In each of the above equations, the first finite-difference on the right-hand side is set up to be evaluated implicitly from as-yet unknown field data at time step $n + 1/2$, while the second finite-difference on the right-hand side is evaluated explicitly from known field data at time step n.

Subiteration 2. Advance the 6 field components from time step $n + 1/2$ to $n + 1$

$$E_x\big|_{i+1/2,j,k}^{n+1} = E_x\big|_{i+1/2,j,k}^{n+1/2} + \frac{\Delta t}{2\varepsilon\Delta y}\left(H_z\big|_{i+1/2,j+1/2,k}^{n+1/2} - H_z\big|_{i+1/2,j-1/2,k}^{n+1/2}\right) - \frac{\Delta t}{2\varepsilon\Delta z}\left(H_y\big|_{i+1/2,j,k+1/2}^{n+1} - H_y\big|_{i+1/2,j,k-1/2}^{n+1}\right), \tag{9.3a}$$

$$E_y\big|_{i,j+1/2,k}^{n+1} = E_y\big|_{i,j+1/2,k}^{n+1/2} + \frac{\Delta t}{2\varepsilon\Delta z}\left(H_x\big|_{i,j+1/2,k+1/2}^{n+1/2} - H_x\big|_{i,j+1/2,k-1/2}^{n+1/2}\right) - \frac{\Delta t}{2\varepsilon\Delta x}\left(H_z\big|_{i+1/2,j+1/2,k}^{n+1} - H_z\big|_{i-1/2,j+1/2,k}^{n+1}\right), \tag{9.3b}$$

$$E_z\big|_{i,j,k+1/2}^{n+1} = E_z\big|_{i,j,k+1/2}^{n+1/2} + \frac{\Delta t}{2\varepsilon\Delta x}\left(H_y\big|_{i+1/2,j,k+1/2}^{n+1/2} - H_y\big|_{i-1/2,j,k+1/2}^{n+1/2}\right) - \frac{\Delta t}{2\varepsilon\Delta y}\left(H_x\big|_{i,j+1/2,k+1/2}^{n+1} - H_x\big|_{i,j-1/2,k+1/2}^{n+1}\right), \tag{9.3c}$$

$$H_x\big|_{i,j+1/2,k+1/2}^{n+1} = H_x\big|_{i,j+1/2,k+1/2}^{n+1/2} + \frac{\Delta t}{2\mu\Delta z}\left(E_y\big|_{i,j+1/2,k+1}^{n+1/2} - E_y\big|_{i,j+1/2,k}^{n+1/2}\right) - \frac{\Delta t}{2\mu\Delta y}\left(E_z\big|_{i,j+1,k+1/2}^{n+1} - E_z\big|_{i,j,k+1/2}^{n+1}\right), \tag{9.4a}$$

$$H_y\big|_{i+1/2,j,k+1/2}^{n+1} = H_y\big|_{i+1/2,j,k+1/2}^{n+1/2} + \frac{\Delta t}{2\mu\Delta x}\left(E_z\big|_{i+1,j,k+1/2}^{n+1/2} - E_z\big|_{i,j,k+1/2}^{n+1/2}\right) - \frac{\Delta t}{2\mu\Delta z}\left(E_x\big|_{i+1/2,j,k+1}^{n+1} - E_x\big|_{i+1/2,j,k}^{n+1}\right), \tag{9.4b}$$

$$H_z\big|_{i+1/2,j+1/2,k}^{n+1} = H_z\big|_{i+1/2,j+1/2,k}^{n+1/2} + \frac{\Delta t}{2\mu\Delta y}\left(E_x\big|_{i+1/2,j+1,k}^{n+1/2} - E_x\big|_{i+1/2,j,k}^{n+1/2}\right) - \frac{\Delta t}{2\mu\Delta x}\left(E_y\big|_{i+1,j+1/2,k}^{n+1} - E_y\big|_{i,j+1/2,k}^{n+1}\right). \tag{9.4c}$$

In each of the above equations, the second finite-difference on the right-hand side is set up to be evaluated implicitly from as-yet unknown field data at time step $n + 1$, while the first finite-difference on the right-hand side is evaluated explicitly from known field data at time step $n + 1/2$ previously computed using (9.1) and (9.2).

9.2.2. Simplified system of time-stepping equations

The system of equations summarized above for each subiteration can be greatly simplified. For Subiteration 1, this is done by substituting the expressions of (9.2) for the H-field components evaluated at time step $n + 1/2$ into the E-field updates of (9.1). Similarly, for Subiteration 2, this is done by substituting the expressions of (9.4) for the H-field components evaluated at time step $n + 1$ into the E-field updates of (9.3). This yields the following simplified system of time-stepping equations for the algorithm of ZHENG, CHEN and ZHANG [2000]:

Subiteration 1. Advance the 6 field components from time step n to time step $n+1/2$

$$\left[1+\frac{(\Delta t)^2}{2\mu\varepsilon(\Delta y)^2}\right]E_x\Big|_{i+1/2,j,k}^{n+1/2}-\left[\frac{(\Delta t)^2}{4\mu\varepsilon(\Delta y)^2}\right]\left(E_x\Big|_{i+1/2,j-1,k}^{n+1/2}+E_x\Big|_{i+1/2,j+1,k}^{n+1/2}\right)$$
$$=E_x\Big|_{i+1/2,j,k}^{n}+\frac{\Delta t}{2\varepsilon\Delta y}\left(H_z\Big|_{i+1/2,j+1/2,k}^{n}-H_z\Big|_{i+1/2,j-1/2,k}^{n}\right)$$
$$-\frac{\Delta t}{2\varepsilon\Delta z}\left(H_y\Big|_{i+1/2,j,k+1/2}^{n}-H_y\Big|_{i+1/2,j,k-1/2}^{n}\right)-\left[\frac{(\Delta t)^2}{4\mu\varepsilon\Delta x\Delta y}\right]$$
$$\times\left(E_y\Big|_{i+1,j+1/2,k}^{n}-E_y\Big|_{i,j+1/2,k}^{n}-E_y\Big|_{i+1,j-1/2,k}^{n}+E_y\Big|_{i,j-1/2,k}^{n}\right),\tag{9.5a}$$

$$\left[1+\frac{(\Delta t)^2}{2\mu\varepsilon(\Delta z)^2}\right]E_y\Big|_{i,j+1/2,k}^{n+1/2}-\left[\frac{(\Delta t)^2}{4\mu\varepsilon(\Delta z)^2}\right]\left(E_y\Big|_{i,j+1/2,k-1}^{n+1/2}+E_y\Big|_{i,j+1/2,k+1}^{n+1/2}\right)$$
$$=E_y\Big|_{i,j+1/2,k}^{n}+\frac{\Delta t}{2\varepsilon\Delta z}\left(H_x\Big|_{i,j+1/2,k+1/2}^{n}-H_x\Big|_{i,j+1/2,k-1/2}^{n}\right)$$
$$-\frac{\Delta t}{2\varepsilon\Delta x}\left(H_z\Big|_{i+1/2,j+1/2,k}^{n}-H_z\Big|_{i-1/2,j+1/2,k}^{n}\right)-\left[\frac{(\Delta t)^2}{4\mu\varepsilon\Delta y\Delta z}\right]$$
$$\times\left(E_z\Big|_{i,j+1,k+1/2}^{n}-E_z\Big|_{i,j,k+1/2}^{n}-E_z\Big|_{i,j+1,k-1/2}^{n}+E_z\Big|_{i,j,k-1/2}^{n}\right),\tag{9.5b}$$

$$\left[1+\frac{(\Delta t)^2}{2\mu\varepsilon(\Delta x)^2}\right]E_z\Big|_{i,j,k+1/2}^{n+1/2}-\left[\frac{(\Delta t)^2}{4\mu\varepsilon(\Delta x)^2}\right]\left(E_z\Big|_{i-1,j,k+1/2}^{n+1/2}+E_z\Big|_{i+1,j,k+1/2}^{n+1/2}\right)$$
$$=E_z\Big|_{i,j,k+1/2}^{n}+\frac{\Delta t}{2\varepsilon\Delta x}\left(H_y\Big|_{i+1/2,j,k+1/2}^{n}-H_y\Big|_{i-1/2,j,k+1/2}^{n}\right)$$
$$-\frac{\Delta t}{2\varepsilon\Delta y}\left(H_x\Big|_{i,j+1/2,k+1/2}^{n}-H_x\Big|_{i,j-1/2,k+1/2}^{n}\right)-\left[\frac{(\Delta t)^2}{4\mu\varepsilon\Delta x\Delta z}\right]$$
$$\times\left(E_x\Big|_{i+1/2,j,k+1}^{n}-E_x\Big|_{i+1/2,j,k}^{n}-E_x\Big|_{i-1/2,j,k+1}^{n}+E_x\Big|_{i-1/2,j,k}^{n}\right).\tag{9.5c}$$

We see that (9.5a) yields a set of simultaneous equations for $E_x^{n+1/2}$ when written for each j coordinate along a y-directed line through the space lattice. The matrix associated with this system is tridiagonal, and hence, easily solved. This process is repeated for each y-cut through the lattice where E_x components are located. Similarly, (9.5b) yields a tridiagonal matrix system for each z-cut through the lattice to obtain $E_y^{n+1/2}$, and (9.5c) yields a tridiagonal matrix system for each x-cut through the lattice to obtain $E_z^{n+1/2}$.

To complete Subiteration 1, we next apply (9.2a)–(9.2c). These H-field updating equations are now fully explicit because all of their required E-field component data at time step $n+1/2$ are available upon solving (9.5a)–(9.5c) in the manner described above.

Subiteration 2. Advance the 6 field components from time step $n+1/2$ to $n+1$

$$\left[1+\frac{(\Delta t)^2}{2\mu\varepsilon(\Delta z)^2}\right]E_x\big|_{i+1/2,j,k}^{n+1}-\left[\frac{(\Delta t)^2}{4\mu\varepsilon(\Delta z)^2}\right]\left(E_x\big|_{i+1/2,j,k-1}^{n+1}+E_x\big|_{i+1/2,j,k+1}^{n+1}\right)$$
$$=E_x\big|_{i+1/2,j,k}^{n+1/2}+\frac{\Delta t}{2\varepsilon\Delta y}\left(H_z\big|_{i+1/2,j+1/2,k}^{n+1/2}-H_z\big|_{i+1/2,j-1/2,k}^{n+1/2}\right)$$
$$-\frac{\Delta t}{2\varepsilon\Delta z}\left(H_y\big|_{i+1/2,j,k+1/2}^{n+1/2}-H_y\big|_{i+1/2,j,k-1/2}^{n+1/2}\right)-\left[\frac{(\Delta t)^2}{4\mu\varepsilon\Delta x\Delta z}\right]$$
$$\times\left(E_z\big|_{i+1,j,k+1/2}^{n+1/2}-E_z\big|_{i,j,k+1/2}^{n+1/2}-E_z\big|_{i+1,j,k-1/2}^{n+1/2}+E_z\big|_{i,j,k-1/2}^{n+1/2}\right),\tag{9.6a}$$

$$\left[1+\frac{(\Delta t)^2}{2\mu\varepsilon(\Delta x)^2}\right]E_y\big|_{i,j+1/2,k}^{n+1}-\left[\frac{(\Delta t)^2}{4\mu\varepsilon(\Delta x)^2}\right]\left(E_y\big|_{i-1,j+1/2,k}^{n+1}+E_y\big|_{i+1,j+1/2,k}^{n+1}\right)$$
$$=E_y\big|_{i,j+1/2,k}^{n+1/2}+\frac{\Delta t}{2\varepsilon\Delta z}\left(H_x\big|_{i,j+1/2,k+1/2}^{n+1/2}-H_x\big|_{i,j+1/2,k-1/2}^{n+1/2}\right)$$
$$-\frac{\Delta t}{2\varepsilon\Delta x}\left(H_z\big|_{i+1/2,j+1/2,k}^{n+1/2}-H_z\big|_{i-1/2,j+1/2,k}^{n+1/2}\right)-\left[\frac{(\Delta t)^2}{4\mu\varepsilon\Delta x\Delta y}\right]$$
$$\times\left(E_x\big|_{i+1/2,j+1,k}^{n+1/2}-E_x\big|_{i+1/2,j,k}^{n+1/2}-E_x\big|_{i-1/2,j+1,k}^{n+1/2}+E_x\big|_{i-1/2,j,k}^{n+1/2}\right),\tag{9.6b}$$

$$\left[1+\frac{(\Delta t)^2}{2\mu\varepsilon(\Delta y)^2}\right]E_z\big|_{i,j,k+1/2}^{n+1}-\left[\frac{(\Delta t)^2}{4\mu\varepsilon(\Delta y)^2}\right]\left(E_z\big|_{i,j-1,k+1/2}^{n+1}+E_z\big|_{i,j+1,k+1/2}^{n+1}\right)$$
$$=E_z\big|_{i,j,k+1/2}^{n+1/2}+\frac{\Delta t}{2\varepsilon\Delta x}\left(H_y\big|_{i+1/2,j,k+1/2}^{n+1/2}-H_y\big|_{i-1/2,j,k+1/2}^{n+1/2}\right)$$
$$-\frac{\Delta t}{2\varepsilon\Delta y}\left(H_x\big|_{i,j+1/2,k+1/2}^{n+1/2}-H_x\big|_{i,j-1/2,k+1/2}^{n+1/2}\right)-\left[\frac{(\Delta t)^2}{4\mu\varepsilon\Delta y\Delta z}\right]$$
$$\times\left(E_y\big|_{i,j+1/2,k+1}^{n+1/2}-E_y\big|_{i,j+1/2,k}^{n+1/2}-E_y\big|_{i,j-1/2,k+1}^{n+1/2}+E_y\big|_{i,j-1/2,k}^{n+1/2}\right).\tag{9.6c}$$

We see that (9.6a) yields a set of simultaneous equations for E_x^{n+1} when written for each k coordinate along a z-directed line through the space lattice. The matrix associated with this system is tridiagonal, and hence, easily solved. This process is repeated for each z-cut through the lattice where E_x components are located. Similarly, (9.6b) yields a tridiagonal matrix system for each x-cut through the lattice to obtain E_y^{n+1}, and (9.6c) yields a tridiagonal matrix system for each y-cut through the lattice to obtain E_z^{n+1}.

To complete Subiteration 2, we next apply (9.4a)–(9.4c). These H-field updating equations are now fully explicit because all of their required E-component data at time step $n+1$ are available upon solving (9.6a)–(9.6c) in the manner described above. This completes the ADI algorithm.

9.3. *Proof of numerical stability*

ZHENG, CHEN and ZHANG [2000] provided the following proof of the numerical stability of their ADI algorithm. Assume that for each time step n, the instantaneous values of the E- and H-fields are Fourier-transformed into the spatial spectral domain with wavenumbers $\tilde{k}_x$, $\tilde{k}_y$, and $\tilde{k}_z$ along the x-, y-, and z-directions, respectively. Denoting the composite field vector in the spatial spectral domain at time step n as

$$\mathbf{F}^n = \begin{bmatrix} E_x^n \\ E_y^n \\ E_z^n \\ H_x^n \\ H_y^n \\ H_z^n \end{bmatrix} \tag{9.7}$$

then Subiteration 1 (consisting of the systems (9.5) and (9.2) can be written as

$$\mathbf{F}^{n+1/2} = \overline{\overline{\mathbf{M}}}_1 \mathbf{F}^n, \tag{9.8}$$

where

$$\overline{\overline{\mathbf{M}}}_1 = \begin{bmatrix} \frac{1}{Q_y} & \frac{W_x W_y}{\mu\varepsilon Q_y} & 0 & 0 & \frac{jW_z}{\varepsilon Q_y} & \frac{-jW_y}{\varepsilon Q_y} \\ 0 & \frac{1}{Q_z} & \frac{W_z W_y}{\mu\varepsilon Q_z} & \frac{-jW_z}{\varepsilon Q_z} & 0 & \frac{jW_x}{\varepsilon Q_z} \\ \frac{W_x W_z}{\mu\varepsilon Q_x} & 0 & \frac{1}{Q_x} & \frac{jW_y}{\varepsilon Q_x} & \frac{-jW_x}{\varepsilon Q_x} & 0 \\ 0 & \frac{-jW_z}{\mu Q_z} & \frac{jW_z}{\mu Q_z} & \frac{1}{Q_z} & 0 & \frac{W_x W_z}{\mu\varepsilon Q_z} \\ \frac{jW_z}{\mu Q_x} & 0 & \frac{-jW_x}{\mu Q_x} & \frac{W_x W_y}{\mu\varepsilon Q_x} & \frac{1}{Q_x} & 0 \\ \frac{-jW_y}{\mu Q_y} & \frac{jW_x}{\mu Q_y} & 0 & 0 & \frac{W_z W_y}{\mu\varepsilon Q_y} & \frac{1}{Q_y} \end{bmatrix} \tag{9.9}$$

and

$$W_x = \frac{\Delta t}{\Delta x}\sin\left(\frac{\tilde{k}_x \Delta x}{2}\right); \qquad W_y = \frac{\Delta t}{\Delta y}\sin\left(\frac{\tilde{k}_y \Delta y}{2}\right); \qquad W_z = \frac{\Delta t}{\Delta z}\sin\left(\frac{\tilde{k}_z \Delta z}{2}\right), \tag{9.10}$$

$$Q_x = 1 + \frac{(W_x)^2}{\mu\varepsilon}; \qquad Q_y = 1 + \frac{(W_y)^2}{\mu\varepsilon}; \qquad Q_z = 1 + \frac{(W_z)^2}{\mu\varepsilon}. \tag{9.11}$$

Similarly, it can be shown that Subiteration 2 (consisting of the systems (9.6) and (9.4)) can be written as

$$\mathbf{F}^{n+1} = \overline{\overline{\mathbf{M}}}_2 \mathbf{F}^{n+1/2}, \tag{9.12}$$

where

$$\overline{\overline{\mathbf{M}}}_2 = \begin{bmatrix} \frac{1}{Q_z} & 0 & \frac{W_z W_x}{\mu\varepsilon Q_z} & 0 & \frac{jW_z}{\varepsilon Q_z} & \frac{-jW_y}{\varepsilon Q_z} \\ \frac{W_x W_y}{\mu\varepsilon Q_x} & \frac{1}{Q_x} & 0 & \frac{-jW_z}{\varepsilon Q_x} & 0 & \frac{jW_x}{\varepsilon Q_x} \\ 0 & \frac{W_y W_z}{\mu\varepsilon Q_y} & \frac{1}{Q_y} & \frac{jW_y}{\varepsilon Q_y} & \frac{-jW_x}{\varepsilon Q_y} & 0 \\ 0 & \frac{-jW_z}{\mu Q_y} & \frac{jW_y}{\mu Q_y} & \frac{1}{Q_y} & \frac{W_z W_y}{\mu\varepsilon Q_y} & 0 \\ \frac{jW_z}{\mu Q_z} & 0 & \frac{-jW_x}{\mu Q_z} & 0 & \frac{1}{Q_z} & \frac{W_z W_y}{\mu\varepsilon Q_z} \\ \frac{-jW_y}{\mu Q_x} & \frac{jW_x}{\mu Q_x} & 0 & 0 & \frac{W_x W_z}{\mu\varepsilon Q_x} & \frac{1}{Q_x} \end{bmatrix}. \tag{9.13}$$

Now, we substitute (9.8) into (9.12) to obtain in matrix form the complete single time-step update expression in the spatial spectral domain:

$$\mathbf{F}^{n+1} = \overline{\overline{\mathbf{M}}}_2 \overline{\overline{\mathbf{M}}}_1 \mathbf{F}^n. \tag{9.14}$$

Using the software package MAPLE™, Zheng, Chen and Zhang found that the magnitudes of all of the eigenvalues of the composite matrix $\overline{\overline{\mathbf{M}}} = \overline{\overline{\mathbf{M}}}_2 \overline{\overline{\mathbf{M}}}_1$ equal unity, regardless of the time-step Δt. Therefore, they concluded that their ADI algorithm is *unconditionally stable* for all Δt, and the Courant stability condition is removed.

9.4. *Numerical dispersion*

ZHENG and CHEN [2001] derived the following numerical dispersion relation for their ADI algorithm:

$$\sin^2(\omega t) = \frac{4\mu\varepsilon \begin{bmatrix} \mu\varepsilon(W_x)^2 + \mu\varepsilon(W_y)^2 + \mu\varepsilon(W_z)^2 \\ + (W_x)^2(W_y)^2 + (W_y)^2(W_z)^2 \\ + (W_z)^2(W_x)^2 \end{bmatrix} \left[(\mu\varepsilon)^3 + (W_x)^2(W_y)^2(W_z)^2\right]}{[\mu\varepsilon + (W_x)^2]^2[\mu\varepsilon + (W_y)^2]^2[\mu\varepsilon + (W_z)^2]^2}. \tag{9.15}$$

For Δt below the usual Courant limit, the numerical dispersion given by (9.15) is quite close to that of Yee's leapfrog time-stepping algorithm. For Δt above the usual Courant limit, the dispersive error given by (9.15) increases steadily.

9.5. *Additional accuracy limitations and their implications*

GONZALEZ GARCIA, LEE and HAGNESS [2002] demonstrated additional accuracy limitations of ADI-FDTD not revealed by previously published numerical dispersion analyses such as that given in ZHENG and CHEN [2001]. They showed that some terms of its truncation error grow with Δt^2 multiplied by the spatial derivatives of the fields. These error terms, which are not present in a fully implicit time-stepping method such as the Crank–Nicolson scheme, give rise to potentially large numerical errors as Δt is increased. Excessive error can occur even if Δt is still small enough to highly resolve key temporal features of the modeled electromagnetic field waveform.

As a result, the primary usage of existing ADI-FDTD techniques appears to be for problems involving a fine mesh needed to model a small geometric feature in an overall much-larger structure that is discretized using a coarse mesh; and where, for computational efficiency, it is desirable to use a large time-step satisfying Courant stability for the coarse mesh. While this limits the impact of the excess error introduced locally within the fine mesh, this also limits the usefulness of ADI-FDTD when considering how to model the key problem areas outlined in Table 9.1.

10. Perfectly matched layer absorbing boundary conditions

10.1. Introduction to absorbing boundary conditions

A basic consideration with the FDTD approach to solving electromagnetic wave interaction problems is that many geometries of interest are defined in "open" regions where the spatial domain of the field is ideally unbounded in one or more directions. Clearly, no computer can store an unlimited amount of data, and therefore, the computational domain must be bounded. However, on this domain's outer boundary, only outward numerical wave motion is desired. That is, all outward-propagating numerical waves should exit the domain with negligible spurious reflections returning to the vicinity of the modeled structure. This would permit the FDTD solution to remain valid for all time steps. Depending upon their theoretical basis, outer-boundary conditions of this type have been called either *radiation boundary conditions* (RBCs) or *absorbing boundary conditions* (ABCs). The notation ABC will be used here.

ABCs cannot be directly obtained from the numerical algorithms for Maxwell's equations reviewed earlier. Principally, this is because these algorithms require field data on both sides of an observation point, and hence cannot be implemented at the outermost planes of the space lattice (since by definition there is no information concerning the fields at points outside of these planes). Although backward finite differences could conceivably be used here, these are generally of lower accuracy for a given space discretization and have not been used in major FDTD software.

Research in this area since 1970 has resulted in two principal categories of ABCs for FDTD simulations:

(1) Special analytical boundary conditions imposed upon the electromagnetic field at the outermost planes of the space lattice. This category was recently reviewed by TAFLOVE and HAGNESS [2000, Chapter 6].

(2) Incorporation of impedance-matched electromagnetic wave absorbing layers adjacent to the outer planes of the space lattice (by analogy with the treatment of the walls of an anechoic chamber). ABCs of this type have excellent capabilities for truncation of FDTD lattices in free space, in lossy or dispersive materials, or in metal or dielectric waveguides. Extremely small numerical-wave reflection coefficients in the order of 10^{-4} to 10^{-6} can be attained with an acceptable computational burden, allowing the possibility of achieving FDTD simulations having a dynamic range of 70 dB or more.

This section reviews modern, perfectly matched electromagnetic wave absorbing layers (Category 2 above). The review is based upon the recent publication by GEDNEY and TAFLOVE [2000].

10.2. Introduction to impedance-matched absorbing layers

Consider implementing an ABC by using an impedance-matched electromagnetic wave absorbing layer adjacent to the outer planes of the FDTD space lattice. Ideally, the absorbing medium is only a few lattice cells thick, reflectionless to all impinging waves over their full frequency spectrum, highly absorbing, and effective in the near field of a source or a scatterer. An early attempt at implementing such an absorbing material boundary condition was reported by HOLLAND and WILLIAMS [1983] who utilized a conventional lossy, dispersionless, absorbing medium. The difficulty with this tactic is that such an absorbing layer is matched only to normally incident plane waves.

BERENGER [1994] provided the seminal insight that a nonphysical absorber can be postulated that is matched independent of the frequency, angle of incidence, and polarization of an impinging plane wave by exploiting additional degrees of freedom arising from a novel split-field formulation of Maxwell's equations. Here, each vector field component is split into two orthogonal components, and the 12 resulting field components are then expressed as satisfying a coupled set of first-order partial differential equations. By choosing loss parameters consistent with a dispersionless medium, a perfectly matched planar interface is derived. This strategy allows the construction of what Berenger called a *perfectly matched layer* (PML) adjacent to the outer boundary of the FDTD space lattice for absorption of all outgoing waves.

Following Berenger's work, many papers appeared validating his technique as well as applying FDTD with the PML medium. An important advance was made by CHEW and WEEDON [1994], who restated the original split-field PML concept in a stretched-coordinate form. Subsequently, this allowed TEIXEIRA and CHEW [1997] to extend PML to cylindrical and spherical coordinate systems. A second important advance was made by SACKS, KINGSLAND, LEE and LEE [1995] and GEDNEY [1995, 1996], who re-posed the split-field PML as a lossy, uniaxial anisotropic medium having both magnetic permeability and electric permittivity tensors. The uniaxial PML, or UPML, is intriguing because it is based on a potentially physically realizable material formulation rather than Berenger's non-physical mathematical model.

10.3. Berenger's perfectly matched layer

10.3.1. Two-dimensional TE_z case

This section reviews the theoretical basis of Berenger's PML for the case of a TE_Z-polarized plane wave incident from Region 1, the lossless material half-space $x < 0$, onto Region 2, the PML half-space $x > 0$.

Field-splitting modification of Maxwell's equations. Within Region 2, Maxwell's curl equations (2.12a)–(2.12c) as modified by Berenger are expressed in their time-dependent form as

$$\varepsilon_2 \frac{\partial E_x}{\partial t} + \sigma_y E_x = \frac{\partial H_z}{\partial y}, \tag{10.1a}$$

$$\varepsilon_2 \frac{\partial E_y}{\partial t} + \sigma_x E_y = -\frac{\partial H_z}{\partial x}, \tag{10.1b}$$

$$\mu_2 \frac{\partial H_{zx}}{\partial t} + \sigma_x^* H_{zx} = -\frac{\partial E_y}{\partial x}, \tag{10.1c}$$

$$\mu_2 \frac{\partial H_{zy}}{\partial t} + \sigma_y^* H_{zy} = \frac{\partial E_x}{\partial y}. \tag{10.1d}$$

Here, H_z is assumed to be split into two additive subcomponents

$$H_z = H_{zx} + H_{zy}. \tag{10.2}$$

Further, the parameters σ_x and σ_y denote electric conductivities, and the parameters σ_x^* and σ_y^* denote magnetic losses.

We see that Berenger's formulation represents a generalization of normally modeled physical media. If $\sigma_x = \sigma_y = 0$ and $\sigma_x^* = \sigma_y^* = 0$, (10.1a)–(10.1d) reduce to Maxwell's equations in a lossless medium. If $\sigma_x = \sigma_y = \sigma$ and $\sigma_x^* = \sigma_y^* = 0$, (10.1a)–(10.1d) describe an electrically conductive medium. And, if $\varepsilon_2 = \varepsilon_1$, $\mu_2 = \mu_1$, $\sigma_x = \sigma_y = \sigma$, $\sigma_x^* = \sigma_y^* = \sigma^*$, and

$$\sigma^*/\mu_1 = \sigma/\varepsilon_1 \quad \rightarrow \quad \sigma^* = \sigma\mu_1/\varepsilon_1 = \sigma(\eta_1)^2 \tag{10.3}$$

then (10.1a)–(10.1d) describe an absorbing medium that is impedance-matched to Region 1 for normally incident plane waves.

Additional possibilities present themselves, however. If $\sigma_y = \sigma_y^* = 0$, the medium can absorb a plane wave having field components (E_y, H_{zx}) propagating along x, but does not absorb a wave having field components (E_x, H_{zy}) propagating along y, since in the first case propagation is governed by (10.1b) and (10.1c), and in the second case by (10.1a) and (10.1d). The converse situation is true for waves (E_y, H_{zx}) and (E_x, H_{zy}) if $\sigma_x = \sigma_x^* = 0$. These properties of particular Berenger media characterized by the pairwise parameter sets $(\sigma_x, \sigma_x^*, 0, 0)$ and $(0, 0, \sigma_y, \sigma_y^*)$ are closely related to the fundamental premise of this novel ABC, proved later. That is, if the pairwise electric and magnetic losses satisfy (10.3), then at interfaces normal to x and y, respectively, the Berenger media have zero reflection of electromagnetic waves.

Now consider (10.1a)–(10.1d) expressed in their time-harmonic form in the Berenger medium. Letting the hat symbol denote a phasor quantity, we write

$$\mathrm{j}\omega\varepsilon_2\left(1 + \frac{\sigma_y}{\mathrm{j}\omega\varepsilon_2}\right)\breve{E}_x = \frac{\partial}{\partial y}(\breve{H}_{zx} + \breve{H}_{zy}), \tag{10.4a}$$

$$\mathrm{j}\omega\varepsilon_2\left(1 + \frac{\sigma_x}{\mathrm{j}\omega\varepsilon_2}\right)\breve{E}_y = -\frac{\partial}{\partial x}(\breve{H}_{zx} + \breve{H}_{zy}), \tag{10.4b}$$

$$\mathrm{j}\omega\mu_2\left(1 + \frac{\sigma_x^*}{\mathrm{j}\omega\mu_2}\right)\breve{H}_{zx} = -\frac{\partial \breve{E}_y}{\partial x}, \tag{10.4c}$$

$$\mathrm{j}\omega\mu_2\left(1 + \frac{\sigma_y^*}{\mathrm{j}\omega\mu_2}\right)\breve{H}_{zy} = \frac{\partial \breve{E}_x}{\partial y}. \tag{10.4d}$$

The notation is simplified by introducing the variables

$$s_w = \left(1 + \frac{\sigma_w}{\mathrm{j}\omega\varepsilon_2}\right); \qquad s_w^* = \left(1 + \frac{\sigma_w^*}{\mathrm{j}\omega\mu_2}\right) : w = x, y. \tag{10.5}$$

Then, (10.4a) and (10.4b) are rewritten as

$$\mathrm{j}\omega\varepsilon_2 s_y \breve{E}_x = \frac{\partial}{\partial y}(\breve{H}_{zx} + \breve{H}_{zy}), \tag{10.6a}$$

$$\mathrm{j}\omega\varepsilon_2 s_x \breve{E}_y = -\frac{\partial}{\partial x}(\breve{H}_{zx} + \breve{H}_{zy}). \tag{10.6b}$$

Plane-wave solution within the Berenger medium. The next step is to derive the plane-wave solution within the Berenger medium. To this end, (10.6a) is differentiated with respect to y and (10.6b) with respect to x. Substituting the expressions for $\partial\breve{E}_y/\partial x$ and $\partial\breve{E}_x/\partial y$ from (10.4c) and (10.4d) leads to

$$-\omega^2\mu_2\varepsilon_2\breve{H}_{zx} = -\frac{1}{s_x^*}\frac{\partial}{\partial x}\frac{1}{s_x}\frac{\partial}{\partial x}(\breve{H}_{zx} + \breve{H}_{zy}), \tag{10.7a}$$

$$-\omega^2\mu_2\varepsilon_2\breve{H}_{zy} = -\frac{1}{s_y^*}\frac{\partial}{\partial y}\frac{1}{s_y}\frac{\partial}{\partial y}(\breve{H}_{zx} + \breve{H}_{zy}). \tag{10.7b}$$

Adding these together and using (10.2) leads to the representative wave equation

$$\frac{1}{s_x^*}\frac{\partial}{\partial x}\frac{1}{s_x}\frac{\partial}{\partial x}\breve{H}_z + \frac{1}{s_y^*}\frac{\partial}{\partial y}\frac{1}{s_y}\frac{\partial}{\partial y}\breve{H}_z + \omega^2\mu_2\varepsilon_2\breve{H}_z = 0. \tag{10.8}$$

This wave equation supports the solutions

$$\breve{H}_z = H_0\tau \mathrm{e}^{-\mathrm{j}\sqrt{s_x s_x^*}\,\beta_{2_x}x - \mathrm{j}\sqrt{s_y s_y^*}\,\beta_{2_y}y} \tag{10.9}$$

with the dispersion relationship

$$(\beta_{2_x})^2 + (\beta_{2_y})^2 = (k_2)^2 \quad \rightarrow \quad \beta_{2_x} = \left[(k_2)^2 - (\beta_{2_y})^2\right]^{1/2}. \tag{10.10}$$

Then, from (10.6a), (10.6b), and (10.2), we have

$$\breve{E}_x = -H_0\tau\frac{\beta_{2_y}}{\omega\varepsilon_2}\sqrt{\frac{s_y^*}{s_y}}\,\mathrm{e}^{-\mathrm{j}\sqrt{s_x s_x^*}\,\beta_{2_x}x - \mathrm{j}\sqrt{s_y s_y^*}\,\beta_{2_y}y}, \tag{10.11}$$

$$\breve{E}_y = H_0\tau\frac{\beta_{2_x}}{\omega\varepsilon_2}\sqrt{\frac{s_x^*}{s_x}}\,\mathrm{e}^{-\mathrm{j}\sqrt{s_x s_x^*}\,\beta_{2_x}x - \mathrm{j}\sqrt{s_y s_y^*}\,\beta_{2_y}y}. \tag{10.12}$$

Despite the field splitting, continuity of the tangential electric and magnetic fields must be preserved across the $x = 0$ interface. To enforce this field continuity, we have $s_y = s_y^* = 1$, or equivalently $\sigma_y = 0 = \sigma_y^*$. This yields the phase-matching condition $\beta_{2_y} = \beta_{1_y} = k_1 \sin\theta$. Further, we derive the H-field reflection and transmission coefficients

$$\Gamma = \left(\frac{\beta_{1_x}}{\omega\varepsilon_1} - \frac{\beta_{2_x}}{\omega\varepsilon_2}\sqrt{\frac{s_x^*}{s_x}}\right)\cdot\left(\frac{\beta_{1_x}}{\omega\varepsilon_1} + \frac{\beta_{2_x}}{\omega\varepsilon_2}\sqrt{\frac{s_x^*}{s_x}}\right)^{-1}; \tag{10.13a}$$

$$\tau = 1 + \Gamma. \tag{10.13b}$$

Reflectionless matching condition. Now, assume $\varepsilon_1 = \varepsilon_2$, $\mu_1 = \mu_2$, and $s_x = s_x^*$. This is equivalent to $k_1 = k_2$, $\eta_1 = \sqrt{\mu_1/\varepsilon_1} = \sqrt{\mu_2/\varepsilon_2}$, and $\sigma_x/\varepsilon_1 = \sigma_x^*/\mu_1$ (i.e., σ_x and σ_x^* satisfying (10.3) in a pairwise manner). With $\beta_{2_y} = \beta_{1_y}$, (10.10) now yields $\beta_{2_x} = \beta_{1_x}$. Substituting into (10.13a) gives the reflectionless condition $\Gamma = 0$ for *all* incident angles regardless of frequency ω. For this case, (10.9), (10.11), (10.12), and (10.13b) specify the following transmitted fields within the Berenger medium:

$$\breve{H}_z = H_0 e^{-js_x\beta_{1_x}x - j\beta_{1_y}y} = H_0 e^{-j\beta_{1_x}x - j\beta_{1_y}y} e^{-\sigma_x x\eta_1 \cos\theta}, \tag{10.14}$$

$$\breve{E}_x = -H_0\eta_1 \sin\theta e^{-j\beta_{1_x}x - j\beta_{1_y}y} e^{-\sigma_x x\eta_1 \cos\theta}, \tag{10.15}$$

$$\breve{E}_y = H_0\eta_1 \cos\theta e^{-j\beta_{1_x}x - j\beta_{1_y}y} e^{-\sigma_x x\eta_1 \cos\theta}. \tag{10.16}$$

Within the matched Berenger medium, the transmitted wave propagates with the same speed and direction as the impinging wave while simultaneously undergoing exponential decay along the x-axis normal to the interface between Regions 1 and 2. Further, the attenuation factor $\sigma_x\eta_1 \cos\theta$ is independent of frequency. These desirable properties apply to all angles of incidence. Hence, Berenger's coining of the term "perfectly matched layer" makes excellent sense.

Structure of an FDTD grid employing Berenger's PML ABC. The above analysis can be repeated for PMLs that are normal to the y-direction. This permitted Berenger to propose the two-dimensional TE_z FDTD grid shown in Fig. 10.1 which uses PMLs to greatly reduce outer-boundary reflections. Here, a free-space computation zone is surrounded by PML backed by perfect electric conductor (PEC) walls. At the left and right sides of the grid (x_1 and x_2), each PML has σ_x and σ_x^* matched according to (10.3) along with $\sigma_y = 0 = \sigma_y^*$ to permit reflectionless transmission across the interface

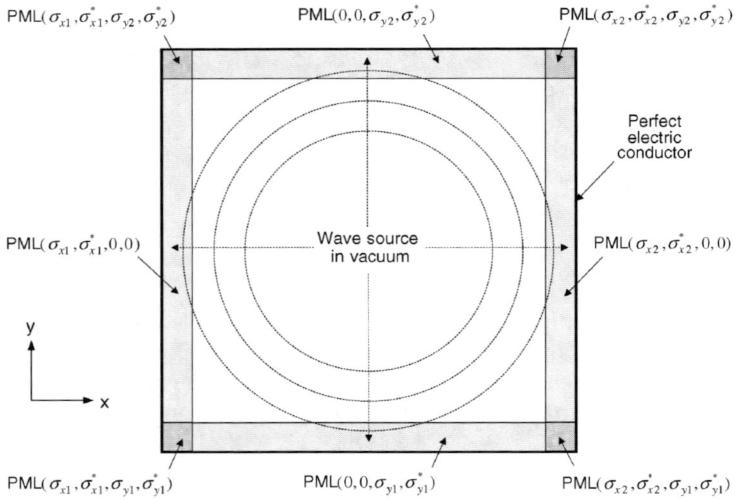

FIG. 10.1. Structure of a two-dimensional TE_z FDTD grid employing the J.P. Berenger PML ABC. *After*: J.P. Berenger, *J. Computational Physics*, 1994, pp. 185–200.

between the free-space and PML regions. At the lower and upper sides of the grid (y_1 and y_2), each PML has σ_y and σ_y^* matched according to (10.3) along with $\sigma_x = 0 = \sigma_x^*$. At the four corners of the grid where there is overlap of two PMLs, all four losses (σ_x, σ_x^*, σ_y, and σ_y^*) are present and set equal to those of the adjacent PMLs.

10.3.2. Two-dimensional TM_z case

The analysis of Section 10.3 can be repeated for the case of a TM_z-polarized incident wave wherein we implement the field splitting $E_z = E_{zx} + E_{zy}$. Analogous to (10.1), Maxwell's curl equations (2.11a)–(2.11c) as modified by Berenger are expressed in their time-dependent form as

$$\mu_2 \frac{\partial H_x}{\partial t} + \sigma_y^* H_x = -\frac{\partial E_z}{\partial y}, \tag{10.17a}$$

$$\mu_2 \frac{\partial H_y}{\partial t} + \sigma_x^* H_y = \frac{\partial E_z}{\partial x}, \tag{10.17b}$$

$$\varepsilon_2 \frac{\partial E_{zx}}{\partial t} + \sigma_x E_{zx} = \frac{\partial H_y}{\partial x}, \tag{10.17c}$$

$$\varepsilon_2 \frac{\partial E_{zy}}{\partial t} + \sigma_y E_{zy} = -\frac{\partial H_x}{\partial y}. \tag{10.17d}$$

A derivation of the PML properties conducted in a manner analogous to that of the TE_z case yields slightly changed results. In most of the equations, the change is only a permutation of ε_2 with μ_2, and of σ with σ^*. However, the PML matching conditions are unchanged. This permits an absorbing reflectionless layer to be constructed adjacent to the outer grid boundary, as in the TE_z case.

10.3.3. Three-dimensional case

KATZ, THIELE and TAFLOVE [1994] showed that Berenger's PML can be realized in three dimensions by splitting all six Cartesian field vector components. For example, the modified Ampere's Law is given by

$$\left(\varepsilon \frac{\partial}{\partial t} + \sigma_y\right) E_{xy} = \frac{\partial}{\partial y}(H_{zx} + H_{zy}), \tag{10.18a}$$

$$\left(\varepsilon \frac{\partial}{\partial t} + \sigma_z\right) E_{xz} = -\frac{\partial}{\partial z}(H_{yx} + H_{yz}), \tag{10.18b}$$

$$\left(\varepsilon \frac{\partial}{\partial t} + \sigma_z\right) E_{yz} = \frac{\partial}{\partial z}(H_{xy} + H_{xz}), \tag{10.18c}$$

$$\left(\varepsilon \frac{\partial}{\partial t} + \sigma_x\right) E_{yx} = -\frac{\partial}{\partial x}(H_{zx} + H_{zy}), \tag{10.18d}$$

$$\left(\varepsilon \frac{\partial}{\partial t} + \sigma_x\right) E_{zx} = \frac{\partial}{\partial x}(H_{yx} + H_{yz}), \tag{10.18e}$$

$$\left(\varepsilon \frac{\partial}{\partial t} + \sigma_y\right) E_{zy} = -\frac{\partial}{\partial y}(H_{xy} + H_{xz}). \tag{10.18f}$$

Similarly, the modified Faraday's Law is given by

$$\left(\mu\frac{\partial}{\partial t}+\sigma_y^*\right)H_{xy}=-\frac{\partial}{\partial y}(E_{zx}+E_{zy}), \tag{10.19a}$$

$$\left(\mu\frac{\partial}{\partial t}+\sigma_z^*\right)H_{xz}=\frac{\partial}{\partial z}(E_{yx}+E_{yz}), \tag{10.19b}$$

$$\left(\mu\frac{\partial}{\partial t}+\sigma_z^*\right)H_{yz}=-\frac{\partial}{\partial z}(E_{xy}+E_{xz}), \tag{10.19c}$$

$$\left(\mu\frac{\partial}{\partial t}+\sigma_x^*\right)H_{yx}=\frac{\partial}{\partial x}(E_{zx}+E_{zy}), \tag{10.19d}$$

$$\left(\mu\frac{\partial}{\partial t}+\sigma_x^*\right)H_{zx}=-\frac{\partial}{\partial x}(E_{yx}+E_{yz}), \tag{10.19e}$$

$$\left(\mu\frac{\partial}{\partial t}+\sigma_y^*\right)H_{zy}=\frac{\partial}{\partial y}(E_{xy}+E_{xz}). \tag{10.19f}$$

PML matching conditions analogous to the two-dimensional cases discussed previously are used. Specifically, if we denote $w = x, y, z$, the matching condition at a normal-to-w PML interface has the parameter pair (σ_w, σ_w^*) satisfy (10.3). This causes the transmitted wave within the PML to undergo exponential decay in the $\pm w$-directions. All other (σ_w, σ_w^*) pairs within this PML are zero. In a corner region, the PML is provided with each matched (σ_w, σ_w^*) pair that is assigned to the overlapping PMLs forming the corner. Thus, PML media located in dihedral-corner overlapping regions have two nonzero and one zero (σ_w, σ_w^*) pairs. PML media located in trihedral-corner overlapping regions have three nonzero (σ_w, σ_w^*) pairs.

10.4. Stretched-coordinate formulation of Berenger's PML

A more compact form of the split-field equations of (10.18) and (10.19) was introduced by CHEW and WEEDON [1994]. Here, the split-field equations are re-posed in a non-split form that maps Maxwell's equations into a complex coordinate space. To this end, the following coordinate mapping is introduced:

$$\tilde{x}\to\int 0^x s_x(x')\,\mathrm{d}x';\qquad \tilde{y}\to\int 0^y s_y(y')\,\mathrm{d}y';\qquad \tilde{z}\to\int 0^z s_z(z')\,\mathrm{d}z'. \tag{10.20}$$

In (10.20), we assume that the PML parameters s_w are continuous functions along the axial directions. The partial derivatives in the stretched coordinate space are then

$$\frac{\partial}{\partial\tilde{x}}=\frac{1}{s_x}\frac{\partial}{\partial x};\qquad \frac{\partial}{\partial\tilde{y}}=\frac{1}{s_y}\frac{\partial}{\partial y};\qquad \frac{\partial}{\partial\tilde{z}}=\frac{1}{s_z}\frac{\partial}{\partial z}. \tag{10.21}$$

Thus, the ∇ operator in the mapped space is defined as

$$\tilde{\nabla}=\hat{x}\frac{\partial}{\partial\tilde{x}}+\hat{y}\frac{\partial}{\partial\tilde{y}}+\hat{z}\frac{\partial}{\partial\tilde{z}}=\hat{x}\frac{1}{s_x}\frac{\partial}{\partial x}+\hat{y}\frac{1}{s_y}\frac{\partial}{\partial y}+\hat{z}\frac{1}{s_z}\frac{\partial}{\partial z}. \tag{10.22}$$

The time-harmonic Maxwell's equations in the complex-coordinate stretched space are then expressed as

$$
\begin{aligned}
\mathrm{j}\omega\varepsilon\breve{\vec{E}} &= \tilde{\nabla}\times\breve{\vec{H}} \\
&= \hat{x}\left(\frac{1}{s_y}\frac{\partial}{\partial y}\breve{H}_z - \frac{1}{s_z}\frac{\partial}{\partial z}\breve{H}_y\right) + \hat{y}\left(\frac{1}{s_z}\frac{\partial}{\partial z}\breve{H}_x - \frac{1}{s_x}\frac{\partial}{\partial x}\breve{H}_z\right) \\
&\quad + \hat{z}\left(\frac{1}{s_x}\frac{\partial}{\partial x}\breve{H}_y - \frac{1}{s_y}\frac{\partial}{\partial y}\breve{H}_x\right),
\end{aligned}
\tag{10.23}
$$

$$
\begin{aligned}
-\mathrm{j}\omega\mu\breve{\vec{H}} &= \tilde{\nabla}\times\breve{\vec{E}} \\
&= \hat{x}\left(\frac{1}{s_y}\frac{\partial}{\partial y}\breve{E}_z - \frac{1}{s_z}\frac{\partial}{\partial z}\breve{E}_y\right) + \hat{y}\left(\frac{1}{s_z}\frac{\partial}{\partial z}\breve{E}_x - \frac{1}{s_x}\frac{\partial}{\partial x}\breve{E}_z\right) \\
&\quad + \hat{z}\left(\frac{1}{s_x}\frac{\partial}{\partial x}\breve{E}_y - \frac{1}{s_y}\frac{\partial}{\partial y}\breve{E}_x\right).
\end{aligned}
\tag{10.24}
$$

A direct relationship can now be shown between the stretched-coordinate form of Maxwell's equations and Berenger's split-field PML. To demonstrate this, we first rewrite the split-field equations of (10.18) for the time-harmonic case:

$$\mathrm{j}\omega\varepsilon s_y\breve{E}_{xy} = \frac{\partial}{\partial y}(\breve{H}_{zx}+\breve{H}_{zy}), \tag{10.25a}$$

$$\mathrm{j}\omega\varepsilon s_z\breve{E}_{xz} = -\frac{\partial}{\partial z}(\breve{H}_{yx}+\breve{H}_{yz}), \tag{10.25b}$$

$$\mathrm{j}\omega\varepsilon s_z\breve{E}_{yz} = \frac{\partial}{\partial z}(\breve{H}_{xy}+\breve{H}_{xz}), \tag{10.25c}$$

$$\mathrm{j}\omega\varepsilon s_x\breve{E}_{yx} = -\frac{\partial}{\partial x}(\breve{H}_{zx}+\breve{H}_{zy}), \tag{10.25d}$$

$$\mathrm{j}\omega\varepsilon s_x\breve{E}_{zx} = \frac{\partial}{\partial x}(\breve{H}_{yx}+\breve{H}_{yz}), \tag{10.25e}$$

$$\mathrm{j}\omega\varepsilon s_y\breve{E}_{zy} = -\frac{\partial}{\partial y}(\breve{H}_{xy}+\breve{H}_{xz}). \tag{10.25f}$$

Then, we add (10.25a) + (10.25b); (10.25c) + (10.25d); and (10.25e) + (10.25f); and use the relationships $E_x = E_{xy} + E_{xz}$, $E_y = E_{yx} + E_{yz}$, and $E_z = E_{zx} + E_{zy}$. This yields

$$\mathrm{j}\omega\varepsilon\breve{E}_x = \frac{1}{s_y}\frac{\partial}{\partial y}\breve{H}_z - \frac{1}{s_z}\frac{\partial}{\partial z}\breve{H}_y, \tag{10.26a}$$

$$\mathrm{j}\omega\varepsilon\breve{E}_y = \frac{1}{s_z}\frac{\partial}{\partial z}\breve{H}_x - \frac{1}{s_x}\frac{\partial}{\partial x}\breve{H}_z, \tag{10.26b}$$

$$\mathrm{j}\omega\varepsilon\breve{E}_z = \frac{1}{s_x}\frac{\partial}{\partial x}\breve{H}_y - \frac{1}{s_y}\frac{\partial}{\partial y}\breve{H}_x \tag{10.26c}$$

which is identical to (10.23). This procedure is repeated for the split-field equations of (10.19) rewritten for the time-harmonic case, leading exactly to (10.24). Specifically, we see that the stretched-coordinate form of the PML is equivalent to the split-field PML; however, it re-poses it in a non-split form.

The principal advantage the complex stretched-coordinate formulation offers is the ease of mathematically manipulating the PML equations, thereby simplifying the understanding of the behavior of the PML. It also provides a pathway to mapping the PML into other coordinate systems such as cylindrical and spherical coordinates, as shown by TEIXEIRA and CHEW [1997], as well as utilizing the split-field PML in frequency-domain finite-element methods based on unstructured discretizations, as shown by RAPPAPORT [1995] and CHEW and JIN [1996].

10.5. An anisotropic PML absorbing medium

The split-field PML introduced by Berenger is a hypothetical, non-physical medium based on a mathematical model. Due to the coordinate-dependence of the loss terms, if such a physical medium exists, it must be anisotropic.

Indeed, a physical model based on an anisotropic, perfectly matched medium can be formulated. This was first discussed by SACKS, KINGSLAND, LEE and LEE [1995]. For a single interface, the anisotropic medium is uniaxial and is composed of both electric and magnetic constitutive tensors. The uniaxial material performs as well as Berenger's PML while avoiding the non-physical field splitting. This section introduces the theoretical basis of the uniaxial PML and compares its formulation with Berenger's PML and the stretched-coordinate PML.

10.5.1. Perfectly matched uniaxial medium

We consider an arbitrarily polarized time-harmonic plane wave propagating in Region 1, the isotropic half-space $x < 0$. This wave is assumed to be incident on Region 2, the half-space $x > 0$ comprised of a uniaxial anisotropic medium having the permittivity and permeability tensors

$$\bar{\bar{\varepsilon}}_2 = \varepsilon_2 \begin{bmatrix} a & 0 & 0 \\ 0 & b & 0 \\ 0 & 0 & b \end{bmatrix}, \tag{10.27a}$$

$$\bar{\bar{\mu}}_2 = \mu_2 \begin{bmatrix} c & 0 & 0 \\ 0 & d & 0 \\ 0 & 0 & d \end{bmatrix}. \tag{10.27b}$$

Here, $\varepsilon_{yy} = \varepsilon_{zz}$ and $\mu_{yy} = \mu_{zz}$ since the medium is assumed to be rotationally symmetric about the x-axis.

The fields excited within Region 2 are also plane-wave in nature and satisfy Maxwell's curl equations. We obtain

$$\vec{\beta}_2 \times \breve{\vec{E}} = \omega\bar{\bar{\mu}}_2\breve{\vec{H}}; \tag{10.28a}$$

$$\vec{\beta}_2 \times \breve{\vec{H}} = -\omega\bar{\bar{\varepsilon}}_2\breve{\vec{E}} \tag{10.28b}$$

where $\vec{\beta}_2 = \hat{x}\beta_{2_x} + \hat{y}\beta_{2_y}$ is the wavevector in anisotropic Region 2. In turn, this permits derivation of the wave equation

$$\vec{\beta}_2 \times \left(\bar{\bar{\varepsilon}}_2^{-1}\vec{\beta}_2\right) \times \breve{\vec{H}} + \omega^2\bar{\bar{\mu}}_2\breve{\vec{H}} = 0. \tag{10.29}$$

Expressing the cross products as matrix operators, this wave equation can be expressed in matrix form as

$$\begin{bmatrix} k_2^2c - (\beta_{2_y})^2b^{-1} & \beta_{2_x}\beta_{2_y}b^{-1} & 0 \\ \beta_{2_x}\beta_{2_y}b^{-1} & k_2^2d - (\beta_{2_x})^2b^{-1} & 0 \\ 0 & 0 & k_2^2d - (\beta_{2_x})^2b^{-1} - (\beta_{2_y})^2a^{-1} \end{bmatrix} \times \begin{bmatrix} \breve{H}_x \\ \breve{H}_y \\ \breve{H}_z \end{bmatrix} = 0, \tag{10.30}$$

where $k_2^2 = \omega^2\mu_2\varepsilon_2$. The dispersion relation for the uniaxial medium in Region 2 is derived from the determinant of the matrix operator. Solving for β_{2_x}, we find that there are four eigenmode solutions. Conveniently, these solutions can be decoupled into forward and backward $\mathrm{TE_z}$ and $\mathrm{TM_z}$ modes, which satisfy the dispersion relations

$$k_2^2 - (\beta_{2_x})^2b^{-1}d^{-1} - (\beta_{2_y})^2a^{-1}d^{-1} = 0: \quad \mathrm{TE_z}(\breve{H}_x, \breve{H}_y = 0), \tag{10.31}$$

$$k_2^2 - (\beta_{2_x})^2b^{-1}d^{-1} - (\beta_{2_y})^2b^{-1}c^{-1} = 0: \quad \mathrm{TM_z}(\breve{H}_z = 0). \tag{10.32}$$

The reflection coefficient at the interface $x = 0$ of Regions 1 and 2 can now be derived. Let us assume a $\mathrm{TE_z}$ incident wave in Region 1. Then, in isotropic Region 1, the fields are expressed as a superposition of the incident and reflected fields as

$$\begin{aligned} \breve{\vec{H}}_1 &= \hat{z}H_0\left(1 + \Gamma e^{2j\beta_{1_x}x}\right)e^{-j\beta_{1_x}x - j\beta_{1_y}y}, \\ \breve{\vec{E}}_1 &= \left[-\hat{x}\frac{\beta_{1_y}}{\omega\varepsilon_1}\left(1 + \Gamma e^{2j\beta_{1_x}x}\right) + \hat{y}\frac{\beta_{1_x}}{\omega\varepsilon_1}\left(1 - \Gamma e^{2j\beta_{1_x}x}\right)\right]H_0 e^{-j\beta_{1_x}x - j\beta_{1_y}y}. \end{aligned} \tag{10.33}$$

The wave transmitted into Region 2 is also $\mathrm{TE_z}$ with propagation characteristics governed by (10.31). These fields are expressed as

$$\begin{aligned} \breve{\vec{H}}_2 &= \hat{z}H_0\tau e^{-j\beta_{2_x}x - j\beta_{2_y}y}, \\ \breve{\vec{E}}_2 &= \left(-\hat{x}\frac{\beta_{2_y}}{\omega\varepsilon_2 a} + \hat{y}\frac{\beta_{2_x}}{\omega\varepsilon_2 b}\right)H_0\tau e^{-j\beta_{2_x}x - j\beta_{2_y}y}, \end{aligned} \tag{10.34}$$

where Γ and τ are the H-field reflection and transmission coefficients, respectively. These are derived by enforcing continuity of the tangential E and H fields across $x = 0$, and are given by

$$\Gamma = \frac{\beta_{1_x} - \beta_{2_x}b^{-1}}{\beta_{1_x} + \beta_{2_x}b^{-1}}; \tag{10.35a}$$

$$\tau = 1 + \Gamma = \frac{2\beta_{1_x}}{\beta_{1_x} + \beta_{2_x}b^{-1}}. \tag{10.35b}$$

Further, for all angles of wave incidence we have

$$\beta_{2_y} = \beta_{1_y} \tag{10.36}$$

due to phase-matching across the $x = 0$ interface. Substituting (10.36) into (10.31) and solving for β_{2_x} yields

$$\beta_{2_x} = \sqrt{k_2^2 bd - (\beta_{1_y})^2 a^{-1} b}. \tag{10.37}$$

Then, if we set $\varepsilon_1 = \varepsilon_2$, $\mu_1 = \mu_2$, $d = b$, and $a^{-1} = b$, we have $k_2 = k_1$ and

$$\beta_{2_x} = \sqrt{k_1^2 b^2 - (\beta_{1_y})^2 b^2} = b\sqrt{k_1^2 - (\beta_{1_y})^2} \equiv b\beta_{1_x}. \tag{10.38}$$

Substituting (10.38) into (10.35a) yields $\Gamma = 0$ for all β_{1_x}. Thus, the interface between Regions 1 and 2 is reflectionless for angles of wave incidence.

The above exercise can be repeated for TM_z polarization. Here, the E-field reflection coefficient is the dual of (10.35a) and is found by replacing b with d (and vice versa), and a with c. For this case, the reflectionless condition holds if $b = d$ and $c^{-1} = d$.

Combining the results for the TE_z and TM_z cases, we see that reflectionless wave transmission into Region 2 occurs when it is composed of a uniaxial medium having the ε and μ tensors

$$\bar{\bar{\varepsilon}}_2 = \varepsilon_1 \bar{\bar{s}}; \tag{10.39a}$$

$$\bar{\bar{\mu}}_2 = \mu_1 \bar{\bar{s}}; \tag{10.39b}$$

$$\bar{\bar{s}} = \begin{bmatrix} s_x^{-1} & 0 & 0 \\ 0 & s_x & 0 \\ 0 & 0 & s_x \end{bmatrix}. \tag{10.39c}$$

This reflectionless property is completely independent of the angle of incidence, polarization, and frequency of the incident wave. Further, from (10.31) and (10.32), the propagation characteristics of the TE- and TM-polarized waves are identical. We call this medium a *uniaxial PML* (UPML) in recognition of its uniaxial anisotropy and perfect matching.

Similar to Berenger's PML, the reflectionless property of the UPML in Region 2 is valid for any s_x. For example, choose $s_x = 1 + \sigma_x / \mathrm{j}\omega\varepsilon_1 = 1 - \mathrm{j}\sigma_x/\omega\varepsilon_1$. Then, from (10.38) we have

$$\beta_{2_x} = (1 - \mathrm{j}\sigma_x/\omega\varepsilon_1)\beta_{1_x}. \tag{10.40}$$

We note that the real part of β_{2_x} is identical to β_{1_x}. Combined with (10.36), this implies that the phase velocities of the impinging and transmitted waves are identical for all incident angles. The characteristic wave impedance in Region 2 is also identical to that in Region 1, a consequence of the fact that the media are perfectly matched.

Finally, substituting (10.36) and (10.40) into (10.34) and (10.35b) yields the fields transmitted into the Region-2 UPML for a TE_z incident wave:

$$\begin{aligned} \breve{\vec{H}}_2 &= \hat{z} H_0 \mathrm{e}^{-\mathrm{j}\beta_{1_x} x - \mathrm{j}\beta_{1_y} y} \mathrm{e}^{-\sigma_x x \eta_1 \cos\theta}, \\ \breve{\vec{E}}_2 &= (-\hat{x} s_x \eta_1 \sin\theta + \hat{y}\eta_1 \cos\theta) H_0 \mathrm{e}^{-\mathrm{j}\beta_{1_x} x - \mathrm{j}\beta_{1_y} y} \mathrm{e}^{-\sigma_x x \eta_1 \cos\theta}. \end{aligned} \tag{10.41}$$

Here, $\eta_1 = \sqrt{\mu_1/\varepsilon_1}$ and θ is the angle of incidence relative to the x-axis. Thus, the transmitted wave in the UPML propagates with the same phase velocity as the incident wave, while simultaneously undergoing exponential decay along the x-axis normal to the interface between Regions 1 and 2. The attenuation factor is independent of frequency, although it is dependent on θ and the UPML conductivity σ_x.

10.5.2. Relationship to Berenger's split-field PML

Comparing the E- and H-fields transmitted into the UPML in (10.41) with the corresponding fields for Berenger's split-field PML in (10.14)–(10.16), we observe identical fields and identical propagation characteristics. Further examination of (10.8) and (10.29) reveals that the two methods result in the same wave equation. Consequently, the plane waves satisfy the same dispersion relation.

However, in the split-field formulation, E_x is continuous across the $x = 0$ boundary, whereas for UPML, E_x is discontinuous and $D_x = s_x^{-1} E_x$ is continuous. This implies that the two methods host different divergence theorems. Within the UPML, Gauss' Law for the E-field is explicitly written as

$$\nabla \cdot \vec{D} = \nabla \cdot (\varepsilon \bar{\bar{s}} \vec{E}) = \frac{\partial}{\partial x}\left(\varepsilon s_x^{-1} E_x\right) + \frac{\partial}{\partial y}(\varepsilon s_x E_y) + \frac{\partial}{\partial z}(\varepsilon s_x E_z) = 0. \tag{10.42}$$

This implies that $D_x = \varepsilon s_x^{-1} E_x$ must be continuous across the $x = 0$ interface since no sources are assumed here. It was shown that ε must be continuous across the interface for a perfectly matched condition. Thus, D_x and $s_x^{-1} E_x$ must be continuous across $x = 0$. Comparing (10.41) with (10.33), this is indeed true for a TE_z-polarized wave.

Next, consider Gauss' Law for Berenger's split-field PML formulation. The ∇ operator in the stretched-coordinate space of interest is defined as

$$\nabla = \hat{x}\frac{\partial}{s_x \partial x} + \hat{y}\frac{\partial}{\partial y} + \hat{z}\frac{\partial}{\partial z}. \tag{10.43}$$

Therefore, we can express the divergence of the electric flux density as

$$\frac{1}{s_x}\frac{\partial}{\partial x}(\varepsilon E_x) + \frac{\partial}{\partial y}(\varepsilon E_y) + \frac{\partial}{\partial z}(\varepsilon E_z) = 0. \tag{10.44}$$

Since ε is continuous across the boundary and s_x^{-1} occurs outside the derivative, both E_x and D_x are continuous.

In summary, Berenger's split-field PML and the UPML have the same propagation characteristics since they both result in the same wave equation. However, the two formulations have different Gauss' Laws. Hence, the E- and H-field components that are normal to the PML interface are different.

10.5.3. A generalized three-dimensional formulation

We now show that properly defining a general constitutive tensor $\bar{\bar{s}}$ allows the UPML medium to be used throughout the entire FDTD space lattice. This tensor provides for both a lossless, isotropic medium in the primary computation zone, *and* individual UPML absorbers adjacent to the outer lattice boundary planes for mitigation of spurious wave reflections.

For a matched condition, the time-harmonic Maxwell's curl equations in the UPML can be written in their most general form as

$$\nabla \times \breve{\vec{H}} = \mathrm{j}\omega\varepsilon\bar{\bar{s}}\breve{\vec{E}}; \tag{10.45a}$$

$$\nabla \times \breve{\vec{E}} = -\mathrm{j}\omega\mu\bar{\bar{s}}\breve{\vec{H}} \tag{10.45b}$$

where $\bar{\bar{s}}$ is the diagonal tensor defined by

$$\bar{\bar{s}} = \begin{bmatrix} s_x^{-1} & 0 & 0 \\ 0 & s_x & 0 \\ 0 & 0 & s_x \end{bmatrix} \begin{bmatrix} s_y & 0 & 0 \\ 0 & s_y^{-1} & 0 \\ 0 & 0 & s_y \end{bmatrix} \begin{bmatrix} s_z & 0 & 0 \\ 0 & s_z & 0 \\ 0 & 0 & s_z^{-1} \end{bmatrix}$$
$$= \begin{bmatrix} s_y s_z s_x^{-1} & 0 & 0 \\ 0 & s_x s_z s_y^{-1} & 0 \\ 0 & 0 & s_x s_y s_z^{-1} \end{bmatrix}. \tag{10.46}$$

Allowing for a nonunity real part κ, the multiplicative components of the diagonal elements of $\bar{\bar{s}}$ are given by

$$s_x = \kappa_x + \frac{\sigma_x}{\mathrm{j}\omega\varepsilon}; \tag{10.47a}$$

$$s_y = \kappa_y + \frac{\sigma_y}{\mathrm{j}\omega\varepsilon}; \tag{10.47b}$$

$$s_z = \kappa_z + \frac{\sigma_z}{\mathrm{j}\omega\varepsilon}. \tag{10.47c}$$

Now, given the above definitions, the following lists all of the special cases involved in implementing the strategy of using $\bar{\bar{s}}$ throughout the entire FDTD lattice.

Lossless, isotropic interior zone
$\bar{\bar{s}}$ is the identity tensor realized by setting $s_x = s_y = s_z = 1$ in (10.46). This requires $\sigma_x = \sigma_y = \sigma_z = 0$ and $\kappa_x = \kappa_y = \kappa_z = 1$ in (10.47).

UPML absorbers at $x_{\min}$ and $x_{\max}$ outer-boundary planes
$\bar{\bar{s}}$ is the tensor given in (10.39), which is realized by setting $s_y = s_z = 1$ in (10.46). This requires $\sigma_y = \sigma_z = 0$ and $\kappa_y = \kappa_z = 1$ in (10.47).

UPML absorbers at $y_{\min}$ and $y_{\max}$ outer-boundary planes
We set $s_x = s_z = 1$ in (10.46). This requires $\sigma_x = \sigma_z = 0$ and $\kappa_x = \kappa_z = 1$ in (10.47).

UPML Absorbers at $z_{\min}$ and $z_{\max}$ outer-boundary planes
We set $s_x = s_y = 1$ in (10.46). This requires $\sigma_x = \sigma_y = 0$ and $\kappa_x = \kappa_y = 1$ in (10.47).

Overlapping UPML absorbers at $x_{\min}$, $x_{\max}$ and $y_{\min}$, $y_{\max}$ dihedral corners
We set $s_z = 1$ in (10.46). This requires $\sigma_z = 0$ and $\kappa_z = 1$ in (10.47).

Overlapping UPML absorbers at $x_{\min}$, $x_{\max}$ and $z_{\min}$, $z_{\max}$ dihedral corners
We set $s_y = 1$ in (10.46). This requires $\sigma_y = 0$ and $\kappa_y = 1$ in (10.47).

Overlapping UPML absorbers at $y_{\min}$, $y_{\max}$ and $z_{\min}$, $z_{\max}$ dihedral corners
We set $s_x = 1$ in (10.46). This requires $\sigma_x = 0$ and $\kappa_x = 1$ in (10.47).

Overlapping UPML absorbers at all trihedral corners
We use the complete general tensor in (10.46).

The generalized constitutive tensor defined in (10.46) is no longer uniaxial by strict definition, but rather is anisotropic. However, the anisotropic PML is still referenced as uniaxial since it is uniaxial in the nonoverlapping PML regions.

10.5.4. Inhomogeneous media

At times, we need to use the PML to terminate an inhomogeneous material region in the FDTD space lattice. An example is a printed circuit constructed on a dielectric substrate backed by a metal ground plane. A second example is a long optical fiber. In such cases, the inhomogeneous material region extends through the PML to the outer boundary of the FDTD lattice. GEDNEY [1998] has shown that the PML can be perfectly matched to such a medium. However, care must be taken to properly choose the PML parameters to maintain a stable and accurate formulation.

Consider an x-normal UPML boundary. Let an inhomogeneous dielectric $\varepsilon(y, z)$, assumed to be piecewise constant in the transverse y- and z-directions, extend into the UPML. From fundamental electromagnetic theory, $D_y = \varepsilon E_y$ must be continuous across any y-normal boundary, and $D_z = \varepsilon E_z$ must be continuous across any z-normal boundary. Then, from Gauss' Law for the UPML in (10.42), we see that s_x must be independent of y and z to avoid surface charge at the boundaries of the discontinuity.

In the previous discussions, the dielectric in the UPML was assumed to be homogeneous. For this case in (10.47a), $s_x = \kappa_x + \sigma_x/\mathrm{j}\omega\varepsilon$ was chosen. However, if $\varepsilon = \varepsilon(y, z)$ and is piecewise constant, then s_x is also piecewise constant in the transverse directions. Thus, surface charge densities result at the material boundaries as predicted by Gauss' Law in (10.42) due to the derivative of a discontinuous function. This nonphysical charge leads to an ill-posed formulation. To avoid this, s_x must be independent of y and z. This holds only if σ_x/ε is maintained constant. This can be done in a brute-force manner by modifying σ_x in the transverse direction such that $\sigma_x(y, z)/\varepsilon(y, z)$ is a constant. A much simpler approach is to normalize σ_x by the relative permittivity, rewriting (10.47a) as

$$s_x = \kappa_x + \sigma'_x/\mathrm{j}\omega\varepsilon_0, \tag{10.48}$$

where ε_0 is the free-space permittivity. In this case, σ'_x is simply a constant in the transverse y- and z-directions, although it is still scaled along the normal x-direction. Now, Gauss' Law is satisfied within the UPML, leading to a well-posed formulation. This also leads to a materially independent formulation of the UPML.

Next, consider Berenger's split-field PML. To understand the constraints of this technique in an inhomogeneous medium, it is simpler to work with its stretched-coordinate representation. In stretched coordinates, Gauss' Law is represented in (10.44). Here, it appears that there are no further constraints on s_x. However, conservation laws require that the charge continuity equation be derived from Ampere's Law and Gauss' Law. To this end, the divergence of Ampere's Law in (10.23) is performed in the stretched coordinates using (10.44). We see that the divergence of the curl of $\vec{H}$ is zero only if s_x is independent of the transverse coordinates y and z. This holds only if σ_x/ε is independent of y and z. Again, this can be easily managed by representing s_x by (10.48), thus leading to a material-independent PML.

In summary, an inhomogeneous medium that is infinite in extent can be terminated by either a split-field PML or UPML medium. Both are perfectly matched to arbitrary electromagnetic waves impinging upon the PML boundary. The method is accurate and stable provided that the PML parameters s_w (s_x, s_y, or s_z) are posed to be independent of the transverse directions. This can be readily accomplished by normalizing σ_w by the relative permittivity, and hence posing $s_w = \kappa_w + \sigma'_w/\mathrm{j}\omega\varepsilon_0$, where σ'_w is constant in the transverse direction.

10.6. Theoretical performance of the PML

10.6.1. The continuous space

When used to truncate an FDTD lattice, the PML has a thickness d and is terminated by the outer boundary of the lattice. If the outer boundary is assumed to be a PEC wall, finite power reflects back into the primary computation zone. For a wave impinging upon the PML at angle θ relative to the w-directed surface normal, this reflection can be computed using transmission line analysis, yielding

$$R(\theta) = \mathrm{e}^{-2\sigma_w \eta d \cos\theta}. \tag{10.49}$$

Here, η and σ_w are, respectively, the PML's characteristic wave impedance and its conductivity, referred to propagation in the w-direction. In the context of an FDTD simulation, $R(\theta)$ is referred to as the "reflection error" since it is a nonphysical reflection due to the PEC wall that backs the PML. We note that the reflection error is the same for both the split-field PML and the UPML, since both support the same wave equation. This error decreases exponentially with σ_w and d. However, the reflection error increases as $\exp(\cos\theta)$, reaching the worst case for $\theta = 90°$. At this grazing angle of incidence, $R = 1$ and the PML is completely ineffective. To be useful within an FDTD simulation, we want $R(\theta)$ to be as small as possible. Clearly, for a thin PML, we must have σ_w as large as possible to reduce $R(\theta)$ to acceptably small levels, especially for θ approaching 90°.

10.6.2. The discrete space

Grading of the PML loss parameters. Theoretically, reflectionless wave transmission can take place across a PML interface regardless of the local step-discontinuity in σ and σ^* presented to the continuous impinging electromagnetic field. However, in FDTD or any discrete representation of Maxwell's equations, numerical artifacts arise due to the finite spatial sampling. Consequently, implementing PML as a single step-discontinuity of σ and σ^* in the FDTD lattice leads to significant spurious wave reflection at the PML surface.

To reduce this reflection error, BERENGER [1994] proposed that the PML losses gradually rise from zero along the direction normal to the interface. Assuming such a grading, the PML remains matched, as seen from the stretched-coordinate theory in Section 10.4. Pursuing this idea, we consider as an example an x-directed plane wave impinging at angle θ upon a PEC-backed PML slab of thickness d, with the front planar interface located in the $x = 0$ plane. Assuming the graded PML conductivity profile $\sigma_x(x)$, we have from (10.20) and (10.14)–(10.16) or (10.41)

$$R(\theta) = \mathrm{e}^{-2\eta \cos\theta \int 0^d \sigma_x(x)\,\mathrm{d}x}. \tag{10.50}$$

Polynomial grading. Several profiles have been suggested for grading $\sigma_x(x)$ (and $\kappa_x(x)$ in the context of the UPML). The most successful use a polynomial or geometric variation of the PML loss with depth x. Polynomial grading is simply

$$\sigma_x(x) = (x/d)^m \sigma_{x,\max}; \tag{10.51a}$$

$$\kappa_x(x) = 1 + (\kappa_{x,\max} - 1) \cdot (x/d)^m. \tag{10.51b}$$

This increases the value of the PML σ_x from zero at $x = 0$, the surface of the PML, to $\sigma_{x,\max}$ at $x = d$, the PEC outer boundary. Similarly, for the UPML, κ_x increases from one at $x = 0$ to $\kappa_{x,\max}$ at $x = d$. Substituting (10.51a) into (10.50) yields

$$R(\theta) = \mathrm{e}^{-2\eta\sigma_{x,\max} d \cos\theta/(m+1)}. \tag{10.52}$$

For a fixed d, polynomial grading provides two parameters: $\sigma_{x,\max}$ and m. A large m yields a $\sigma_x(x)$ distribution that is relatively flat near the PML surface. However, deeper within the PML, σ_x increases more rapidly than for small m. In this region, the field amplitudes are substantially decayed and reflections due to the discretization error contribute less. Typically, $3 \leqslant m \leqslant 4$ has been found to be nearly optimal for many FDTD simulations (see, for example, BERENGER [1996]).

For polynomial grading, the PML parameters can be readily determined for a given error estimate. For example, let m, d, and the desired reflection error $R(0)$ be known. Then, from (10.52), $\sigma_{x,\max}$ is computed as

$$\sigma_{x,\max} = -\frac{(m+1)\ln[R(0)]}{2\eta d}. \tag{10.53}$$

Geometric grading. The PML loss profile for this case was defined by BERENGER [1997] as

$$\sigma_x(x) = \left(g^{1/\Delta}\right)^x \sigma_{x,0}; \tag{10.54a}$$

$$\kappa_x(x) = \left(g^{1/\Delta}\right)^x, \tag{10.54b}$$

where $\sigma_{x,0}$ is the PML conductivity at its surface, g is the scaling factor, and Δ is the FDTD space increment. Here, the PML conductivity increases from $\sigma_{x,0}$ at its surface to $g^{d/\Delta}\sigma_{x,0}$ at the PEC outer boundary. Substituting (10.54a) into (10.50) results in

$$R(\theta) = \mathrm{e}^{-2\eta\sigma_{x,0}\Delta(g^{d/\Delta}-1)\cos\theta/\ln g}. \tag{10.55}$$

For a fixed d, geometric grading provides two parameters: g and $\sigma_{x,0}$. $\sigma_{x,0}$ must be small to minimize the initial discretization error. Large values of g flatten the conductivity profile near $x = 0$, and steepen it deeper into the PML. Usually, g, d, and $R(0)$ are predetermined. This yields

$$\sigma_{x,0} = -\frac{\ln[R(0)]\ln(g)}{2\eta\Delta(g^{d/\Delta}-1)}. \tag{10.56}$$

Typically, $2 \leqslant g \leqslant 3$ has been found to be nearly optimal for many FDTD simulations.

Discretization error. The design of an effective PML requires balancing the theoretical reflection error $R(\theta)$ and the numerical discretization error. For example, (10.53) provides $\sigma_{x,\max}$ for a polynomial-graded conductivity given a predetermined $R(0)$ and m. If $\sigma_{x,\max}$ is small, the primary reflection from the PML is due to its PEC backing, and (10.50) provides a fairly accurate approximation of the reflection error. Now, we normally choose $\sigma_{x,\max}$ to be as large as possible to minimize $R(\theta)$. However, if $\sigma_{x,\max}$ is too large, the discretization error due to the FDTD approximation dominates, and the actual reflection error is potentially orders of magnitude higher than what (10.50) predicts. Consequently, there is an optimal choice for $\sigma_{x,\max}$ that balances reflection from the PEC outer boundary and discretization error.

BERENGER [1996], BERENGER [1997] postulated that the largest reflection error due to discretization occurs at $x = 0$, the PML surface. Any wave energy that penetrates further into the PML and then is reflected undergoes attenuation both before and after its point of reflection, and typically is not as large a contribution. Thus, it is desirable to minimize the discontinuity at $x = 0$. As discussed earlier, one way to achieve this is by flattening the PML loss profile near $x = 0$. However, if the subsequent rise of loss with depth is too rapid, reflections from deeper within the PML can dominate.

Through extensive numerical experimentation, GEDNEY [1996] and HE [1997] found that, for a broad range of applications, an optimal choice for a 10-cell-thick, polynomial-graded PML is $R(0) \approx \mathrm{e}^{-16}$. For a 5-cell-thick PML, $R(0) \approx \mathrm{e}^{-8}$ is optimal. From (10.53), this leads to an optimal $\sigma_{x,\max}$ for polynomial grading:

$$\sigma_{x,\mathrm{opt}} \approx -\frac{(m+1)\cdot(-16)}{(2\eta)\cdot(10\Delta)} = \frac{0.8(m+1)}{\eta\Delta}. \tag{10.57}$$

This expression has proven to be quite robust for many applications. However, its value may be too large when the PML terminates highly elongated resonant structures or sources with a very long time duration, such as a unit step. For a detailed discussion, the reader is referred to GEDNEY [1998].

10.7. Efficient implementation of UPML in FDTD

This section discusses mapping the UPML presented in Section 10.5 into the discrete FDTD space. The FDTD approximation is derived from the time-harmonic Maxwell's curl equations within the generalized uniaxial medium as defined in (10.45)–(10.47).

10.7.1. Derivation of the finite-difference expressions

Starting with (10.45a) and (10.46), Ampere's Law in a matched UPML is expressed as

$$\begin{bmatrix} \frac{\partial \breve{H}_z}{\partial y} - \frac{\partial \breve{H}_y}{\partial z} \\ \frac{\partial \breve{H}_x}{\partial z} - \frac{\partial \breve{H}_z}{\partial x} \\ \frac{\partial \breve{H}_y}{\partial x} - \frac{\partial \breve{H}_x}{\partial y} \end{bmatrix} = \mathrm{j}\omega\varepsilon \begin{bmatrix} \frac{s_y s_z}{s_x} & 0 & 0 \\ 0 & \frac{s_x s_z}{s_y} & 0 \\ 0 & 0 & \frac{s_x s_y}{s_z} \end{bmatrix} \begin{bmatrix} \breve{E}_x \\ \breve{E}_y \\ \breve{E}_z \end{bmatrix}, \tag{10.58}$$

where s_x, s_y, and s_z are defined in (10.47). Directly inserting (10.47) into (10.58) and then transforming into the time domain would lead to a convolution between the tensor coefficients and the E-field. This is not advisable because implementing this convolution would be computationally intensive. As shown by GEDNEY [1995], GEDNEY [1996], a much more efficient approach is to define the proper constitutive relationship to decouple the frequency-dependent terms. Specifically, let

$$\breve{D}_x = \varepsilon \frac{s_z}{s_x} \breve{E}_x; \tag{10.59a}$$

$$\breve{D}_y = \varepsilon \frac{s_x}{s_y} \breve{E}_y; \tag{10.59b}$$

$$\breve{D}_z = \varepsilon \frac{s_y}{s_z} \breve{E}_z. \tag{10.59c}$$

Then, (10.58) is rewritten as

$$\begin{bmatrix} \frac{\partial \breve{H}_z}{\partial y} - \frac{\partial \breve{H}_y}{\partial z} \\ \frac{\partial \breve{H}_x}{\partial z} - \frac{\partial \breve{H}_z}{\partial x} \\ \frac{\partial \breve{H}_y}{\partial x} - \frac{\partial \breve{H}_x}{\partial y} \end{bmatrix} = \mathrm{j}\omega \begin{bmatrix} s_y & 0 & 0 \\ 0 & s_z & 0 \\ 0 & 0 & s_x \end{bmatrix} \begin{bmatrix} \breve{D}_x \\ \breve{D}_y \\ \breve{D}_z \end{bmatrix}. \tag{10.60}$$

Now, we substitute s_x, s_y, and s_z from (10.47) into (10.60), and then apply the inverse Fourier transform using the identity $\mathrm{j}\omega f(\omega) \to (\partial/\partial t) f(t)$. This yields an equivalent system of time-domain differential equations for (10.60):

$$\begin{bmatrix} \frac{\partial H_z}{\partial y} - \frac{\partial H_y}{\partial z} \\ \frac{\partial H_x}{\partial z} - \frac{\partial H_z}{\partial x} \\ \frac{\partial H_y}{\partial x} - \frac{\partial H_x}{\partial y} \end{bmatrix} = \frac{\partial}{\partial t} \begin{bmatrix} \kappa_y & 0 & 0 \\ 0 & \kappa_z & 0 \\ 0 & 0 & \kappa_x \end{bmatrix} \begin{bmatrix} D_x \\ D_y \\ D_z \end{bmatrix} + \frac{1}{\varepsilon} \begin{bmatrix} \sigma_y & 0 & 0 \\ 0 & \sigma_z & 0 \\ 0 & 0 & \sigma_x \end{bmatrix} \begin{bmatrix} D_x \\ D_y \\ D_z \end{bmatrix}. \tag{10.61}$$

The system of equations in (10.61) can be discretized on the standard Yee lattice. It is suitable to use normal leapfrogging in time wherein the loss terms are time-averaged according to the semi-implicit scheme. This leads to explicit time-stepping expressions for D_x, D_y, and D_z. For example, the D_x update is given by

$$D_x\big|_{i+1/2,j,k}^{n+1} = \left(\frac{2\varepsilon\kappa_y - \sigma_y \Delta t}{2\varepsilon\kappa_y + \sigma_y \Delta t}\right) D_x\big|_{i+1/2,j,k}^{n} + \left(\frac{2\varepsilon \Delta t}{2\varepsilon\kappa_y + \sigma_y \Delta t}\right)$$

$$\times\left(\frac{H_z|_{i+1/2,j+1/2,k}^{n+1/2} - H_z|_{i+1/2,j-1/2,k}^{n+1/2}}{\Delta y} - \frac{H_y|_{i+1/2,j,k+1/2}^{n+1/2} - H_y|_{i+1/2,j,k-1/2}^{n+1/2}}{\Delta z}\right). \tag{10.62}$$

Next, we focus on (10.59a)–(10.59c). For example, we consider (10.59a). After multiplying both sides by s_x and substituting for s_x and s_z from (10.47a), (10.47c), we have

$$\left(\kappa_x + \frac{\sigma_x}{\mathrm{j}\omega\varepsilon}\right)\breve{D}_x = \varepsilon\left(\kappa_z + \frac{\sigma_z}{\mathrm{j}\omega\varepsilon}\right)\breve{E}_x. \tag{10.63}$$

Multiplying both sides by $\mathrm{j}\omega$ and transforming into the time domain leads to

$$\frac{\partial}{\partial t}(\kappa_x D_x) + \frac{\sigma_x}{\varepsilon} D_x = \varepsilon\left[\frac{\partial}{\partial t}(\kappa_z E_x) + \frac{\sigma_z}{\varepsilon} E_x\right]. \tag{10.64a}$$

Similarly, from (10.59b) and (10.59c), we obtain

$$\frac{\partial}{\partial t}(\kappa_y D_y) + \frac{\sigma_y}{\varepsilon} D_y = \varepsilon\left[\frac{\partial}{\partial t}(\kappa_x E_y) + \frac{\sigma_x}{\varepsilon} E_y\right], \tag{10.64b}$$

$$\frac{\partial}{\partial t}(\kappa_z D_z) + \frac{\sigma_z}{\varepsilon} D_z = \varepsilon\left[\frac{\partial}{\partial t}(\kappa_y E_z) + \frac{\sigma_y}{\varepsilon} E_z\right]. \tag{10.64c}$$

The time derivatives in (10.64) are discretized using standard Yee leapfrogging and time-averaging the loss terms. This yields explicit time-stepping expressions for E_x, E_y, and E_z. For example, the E_x update is given by

$$E_x|_{i+1/2,j,k}^{n+1} = \left(\frac{2\varepsilon\kappa_z - \sigma_z\Delta t}{2\varepsilon\kappa_z + \sigma_z\Delta t}\right)E_x|_{i+1/2,j,k}^{n} + \left[\frac{1}{(2\varepsilon\kappa_z + \sigma_z\Delta t)\varepsilon}\right] \times \left[(2\varepsilon\kappa_x + \sigma_x\Delta t)D_x|_{i+1/2,j,k}^{n+1} - (2\varepsilon\kappa_x - \sigma_x\Delta t)D_x|_{i+1/2,j,k}^{n}\right]. \tag{10.65}$$

Overall, updating the components of $\vec{E}$ in the UPML requires two steps in sequence: (1) obtaining the new values of the components of $\vec{D}$ according to (10.62), and (2) using these new $\vec{D}$ components to obtain new values of the $\vec{E}$-components according to (10.65).

A similar two-step procedure is required to update the components of $\vec{H}$ in the UPML. Starting with Faraday's Law in (10.45b) and (10.46), the first step involves developing the updates for the components of $\vec{B}$. A procedure analogous to that followed in obtaining (10.62) yields, for example, the following update for B_x:

$$B_x|_{i,j+1/2,k+1/2}^{n+3/2} = \left(\frac{2\varepsilon\kappa_y - \sigma_y\Delta t}{2\varepsilon\kappa_y + \sigma_y\Delta t}\right)B_x|_{i,j+1/2,k+1/2}^{n+1/2} - \left(\frac{2\varepsilon\Delta t}{2\varepsilon\kappa_y + \sigma_y\Delta t}\right) \times\left(\frac{E_z|_{i,j+1,k+1/2}^{n+1} - E_z|_{i,j,k+1/2}^{n+1}}{\Delta y} - \frac{E_y|_{i,j+1/2,k+1}^{n+1} - E_y|_{i,j+1/2,k}^{n+1}}{\Delta z}\right), \tag{10.66}$$

The second step involves updating the $\vec{H}$ components in the UPML using the values of the $\vec{B}$ components just obtained with (10.66) and similar expressions for B_y and B_z. For example, employing the dual constitutive relation $\breve{B}_x = \mu(s_z/s_x)\breve{H}_x$, a procedure analogous to that followed in obtaining (10.65) yields the following update for H_x:

$$H_x\big|_{i,j+1/2,k+1/2}^{n+3/2} = \left(\frac{2\varepsilon\kappa_z - \sigma_z\Delta t}{2\varepsilon\kappa_z + \sigma_z\Delta t}\right) H_x\big|_{i,j+1/2,k+1/2}^{n+1/2} + \left[\frac{1}{(2\varepsilon\kappa_z + \sigma_z\Delta t)\mu}\right] \times \left[(2\varepsilon\kappa_x + \sigma_x\Delta t) B_x\big|_{i,j+1/2,k+1/2}^{n+3/2} - (2\varepsilon\kappa_x - \sigma_x\Delta t) B_x\big|_{i,j+1/2,k+1/2}^{n+1/2}\right]. \tag{10.67}$$

Similar expressions can be derived for H_y and H_z.

NEHRBASS, LEE and LEE [1996] showed that such an algorithm is numerically stable within the Courant limit. Further, ABARBANEL and GOTTLIEB [1997] showed that the resulting discrete fields satisfy Gauss' Law, and the UPML is well posed.

10.7.2. Computer implementation of the UPML

Each $\vec{E}$ and $\vec{H}$ component within the UPML is computed using an explicit two-step time-marching scheme as illustrated in (10.62) and (10.65) for E_x, and in (10.66) and (10.67) for H_x. Based on these updates, the UPML is easily and efficiently implemented within the framework of existing FDTD codes. We now illustrate this in FORTRAN using the time-stepping expressions for E_x given in (10.62) and (10.65). First, we precompute six coefficient arrays to be used in the field updates:

$$C1(j) = \frac{2\varepsilon\kappa_y(j) - \sigma_y(j)\Delta t}{2\varepsilon\kappa_y(j) + \sigma_y(j)\Delta t}, \tag{10.68a}$$

$$C2(j) = \frac{2\varepsilon\Delta t}{2\varepsilon\kappa_y(j) + \sigma_y(j)\Delta t}, \tag{10.68b}$$

$$C3(k) = \frac{2\varepsilon\kappa_z(k) - \sigma_z(k)\Delta t}{2\varepsilon\kappa_z(k) + \sigma_z(k)\Delta t}, \tag{10.68c}$$

$$C4(k) = \frac{1}{[2\varepsilon\kappa_z(k) + \sigma_z(k)\Delta t]\varepsilon}, \tag{10.68d}$$

$$C5(i) = 2\varepsilon\kappa_x(i) + \sigma_x(i)\Delta t, \tag{10.68e}$$

$$C6(i) = 2\varepsilon\kappa_x(i) - \sigma_x(i)\Delta t. \tag{10.68f}$$

Defining the field-updating coefficients in this manner permits a unified treatment of both the lossless interior working volume and the UPML slabs. In effect, UPML is assumed to fill the entire FDTD space lattice. We set $\sigma_w = 0$ and $\kappa_w = 1$ in the working volume to model free space. However, in the UPML slabs, σ_w and κ_w are assumed to have the polynomial-graded profile given in (10.51), or the geometric-graded profile given in (10.54), along the normal axes of the UPML slabs. As a result, the coefficients in (10.68) vary in only one dimension.

When defining the coefficient arrays specified in (10.68), it is critical to assign the proper value to the UPML loss parameters. To this end, σ_w and κ_w are computed at a physical coordinate using (10.51) or (10.54). The appropriate choice of the physical coordinate is at the edge center of the discrete field $E_x(i, j, k)$, which is $[(i+1/2)\Delta x, j\Delta y, k\Delta z]$. Thus, in (10.68e) and (10.68f), $\sigma_x(i)$ and $\kappa_x(i)$ are computed at the physical coordinate $(i + 1/2)\Delta x$. Similarly, in (10.68a) and (10.68b), $\sigma_y(j)$ and $\kappa_y(j)$ are computed at the physical coordinate $j\Delta y$; and in (10.68c) and (10.68d), $\sigma_z(k)$ and $\kappa_z(k)$ are computed at the physical coordinate $k\Delta z$. This is similarly done for the updates of E_y and E_z.

The UPML loss parameters for the H-fields are chosen at the lattice face centers. For example, the physical coordinate of the discrete field $H_x(i, j, k)$ is $[i\Delta x, (j + 1/2)\Delta y, (k + 1/2)\Delta z]$. Thus, for the update of $H_x(i, j, k)$, $\sigma_x(i)$ and $\kappa_x(i)$ are computed at the physical coordinate $i\Delta x$; $\sigma_y(j)$ and $\kappa_y(j)$ are computed at the physical coordinate $(j + 1/2)\Delta y$; and $\sigma_z(k)$ and $\kappa_z(k)$ are computed at the physical coordinate $(k + 1/2)\Delta z$. This is similarly done for the updates of H_y and H_z.

Given the above "all-UPML" strategy, and assuming that the infinite region extending out of the space lattice has homogeneous material properties, then the FORTRAN program segment that executes the time-stepping of E_x *everywhere* in the FDTD space lattice can be written as a simple triply-nested loop:

```
   do 10 k=2,nz-1
     do 10 j=2,ny-1
       do 10 i=1,nx-1
         dstore = dx(i,j,k)
         dx(i,j,k) = C1(j)*dx(i,j,k)
           + C2(j)*( (hz(i,j,k) - hz(i,j-1,k)) / deltay -
             (hy(i,j,k) - hy(i,j,k-1)) / deltaz )
         ex(i,j,k) = C3(k)*ex(i,j,k)
           + C4(k)*( C5(i)*dx(i,j,k) - C6(i)*dstore )
 10 continue
```
(10.69)

Assuming UPML throughout the entire FDTD lattice in this manner has the limitation that the flux densities D_x and B_x must be stored everywhere in the lattice. However, this approach offers the significant advantage of simplifying the modification of existing FDTD codes. An alternative is to write a triply-nested loop for the interior fields and separate loops for the various UPML slabs (segregating the corner regions). In this case, the auxiliary variables need to be stored only in the UPML region, leading to memory savings. Further, in this circumstance, the UPML requires considerably less storage than Berenger's split-field PML since only the normal fields require dual storage, as opposed to the two tangential fields. The memory requirement (real numbers) for the UPML truncation on all six outer lattice boundaries totals

$$\begin{aligned} &6N_xN_yN_z + 8N_{\text{UPML}}(N_xN_y + N_yN_z + N_zN_x) \\ &\quad - 16N_{\text{UPML}}(N_x + N_y + N_z) + (24N_{\text{UPML}})^2, \end{aligned} \tag{10.70}$$

where N_{UPML} is the thickness (in space cells) of the UPML. In contrast, approximately $6N_xN_yN_z$ real numbers must be stored when using FDTD with a local ABC. With these measures, it is straightforward to calculate the percentage of additional memory required to implement the UPML ABC in a cubic lattice ($N_x = N_y = N_z$) relative to the primary field storage. For 4-cell UPML, the storage burden drops below 10% for

$N_x > 90$; and for 10-cell UPML, this burden drops below 10% for $N_x > 240$. Note also that, without the use of the reflection-cancellation techniques of RAMAHI [1998], local ABCs must be placed *much* further out than the UPML, requiring even larger lattices. Consequently, the UPML ABC can lead to an overall decrease in the required memory to achieve a given desired outer-boundary reflectivity.

If we assume UPML throughout the entire FDTD space lattice as in (10.68) and (10.69), the computer memory requirement is $12N_xN_yN_z$ real numbers. This option is suitable when coding simplicity is desired and memory is not a constraint.

10.8. Numerical experiments with Berenger's split-field PML

10.8.1. Outgoing cylindrical wave in a two-dimensional open-region grid

We first review the numerical experiment of KATZ, THIELE and TAFLOVE [1994] which used Berenger's split-field PML in a two-dimensional square-cell FDTD grid to absorb an outgoing cylindrical wave generated by a hard source centered in the grid. Using the methodology of MOORE, BLASCHAK, TAFLOVE and KRIEGSMANN [1988], the accuracy of the Berenger's PML ABC was compared with that of the previously standard, second-order accurate, analytical ABC of MUR [1981]. The PML loss was assumed to be quadratically graded with depth from the interface of the interior free-space computation region. This allowed a direct comparison with the computed results reported by BERENGER [1994].

Fig. 10.2 graphs the global error power within a 100 ×50-cell TE_z test grid for both the Mur ABC and a 16-cell Berenger PML ABC. At $n = 100$ time steps, the global reflection error in the PML grid is about 10^{-7} times the error in the Mur grid, dropping to a microscopic 10^{-12} times the global error in the Mur grid at $n = 500$ time steps.

We next consider the performance of Berenger's PML ABC for this open-region radiation problem as a function of frequency. Here, the local PML reflection coefficient versus frequency is obtained by using the discrete Fourier transform to calculate the

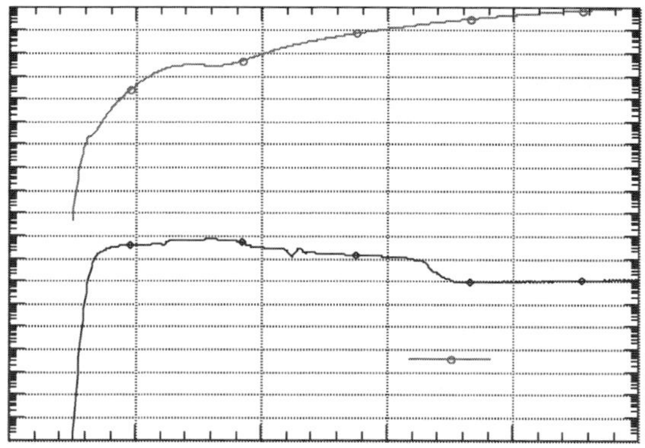

FIG. 10.2. Global error power within a 100 × 50-cell 2D TE_z test grid for both the second-order Mur ABC and a 16-cell quadratically graded Berenger PML ABC, plotted as a function of time-step number on a logarithmic vertical scale. *Source*: D.S. Katz et al., *IEEE Microwave and Guided Wave Letters*, 1994, pp. 268–270, ©1994 IEEE.

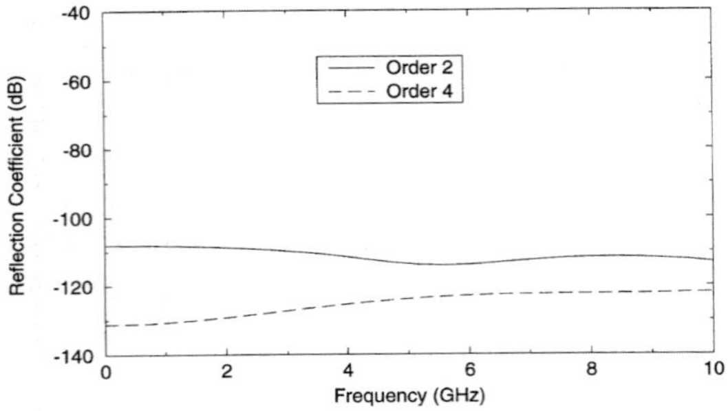

FIG. 10.3. PML reflection coefficient versus frequency for the order-2 (baseline quadratic grading) and order-4 grading cases. Two-dimensional grid with 16-cell-thick PML having $R(0) = 10^{-6}$ used for these calculations.

incident and reflected pulse spectra observed at the midpoint of the 100-cell air/PML interface, and dividing the reflected spectrum by the incident spectrum. The numerical procedure is otherwise similar to that used above, with the exception that the grading of the PML loss is either order 2 or order 4.

Fig. 10.3 graphs the results of this study, comparing the local PML reflection from 0–10 GHz. Here, the PML thickness is 16 cells with $R(0) = 10^{-6}$, and a uniform grid space increment of 1.5 mm (equivalent to $\lambda_0/20$ at 10 GHz) is used.

Fig. 10.3 shows that the local PML reflection coefficient is *virtually flat* from 0–10 GHz. Therefore, Berenger's PML is effective for absorbing ultrawideband pulses. We also observe that the order-4 PML loss grading has 10–24 dB less reflectivity than the baseline quadratic case. Additional studies of this type have shown similar results for a variety of FDTD models. These indicate that the optimum grading of the PML loss is generally not quadratic. It is apparent that a simple grading optimization provides a no-cost means of achieving the widest possible dynamic range of the PML ABC.

The reader is cautioned that double-precision computer arithmetic may be required to achieve the full benefit of PML grading. Simply shifting the test code from a Unix workstation to the Cray C-90 permitted the grading improvement of Fig. 10.3 to be observed. The improvement was not observed on the workstation.

10.8.2. Outgoing spherical wave in a three-dimensional open-region lattice

In this numerical experiment, KATZ, THIELE and TAFLOVE [1994] used quadratically graded Berenger PML in a three-dimensional cubic-cell FDTD lattice to absorb an impulsive, outgoing spherical wave generated by a Hertzian dipole. The Hertzian dipole was simply a single, hard-sourced E_z field component centered in the lattice. Otherwise, the experimental procedure was the same as in Section 10.8.1.

Fig. 10.4 compares the local E-field error due to the second-order Mur and 16-cell PML ABCs for a $100 \times 100 \times 50$-cell three-dimensional test lattice. The observation was made along the x-axis at the outer boundary of the test lattice at time step $n = 100$, the time of maximum excitation of the PML by the outgoing wave. We see that the error due to the PML is on the order of 10^{-3} times that of the Mur ABC.

KATZ, THIELE and TAFLOVE [1994] determined that, if one fixes the PML thickness, increasing the PML loss can reduce both the local and global reflection errors.

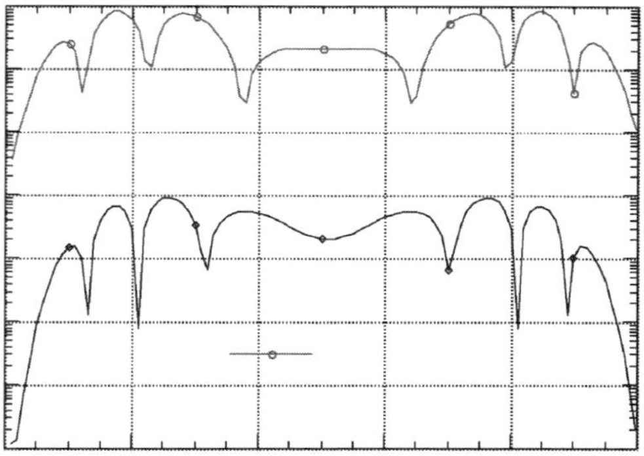

FIG. 10.4. Local E-field error at time-step $n = 100$ along the x-axis at the outer boundary of a $100 \times 100 \times 50$-cell three-dimensional test lattice for Mur's second-order ABC and 16-cell quadratically graded PML, plotted on a logarithmic vertical scale. *Source*: D.S. Katz et al., *IEEE Microwave and Guided Wave Letters*, 1994, pp. 268–270, © 1994 IEEE.

TABLE 10.1

Tradeoff of error reduction for quadratically graded PML relative to Mur's second-order abc versus computer resources for a three-dimensional test lattice of $100 \times 100 \times 50$ cells. *Source*: D.S. Katz et al., *IEEE Microwave and Guided Wave Letters*, 1994, pp. 268–270, © 1994 IEEE

ABC	Avg. local field error reduction relative to second-order Mur	Computer resources one CPU, Cray C-90	If free-space buffer is reduced by 10 cells
Mur	1 (0 dB)	10 Mwords, 6.5 s	–
4-cell PML	22 (27 dB)	16 Mwords, 12 s	7 Mwords, 10 s
8-cell PML	580 (55 dB)	23 Mwords, 37 s	12 Mwords, 27 s
16-cell PML	5800 (75 dB)	43 Mwords, 87 s	25 Mwords, 60 s

However, this benefit levels off when $R(0)$ drops to less than 10^{-5}. Similarly, the local and global error can drop as the PML thickness increases. Here, however, a tradeoff with the computer burden must be factored.

Table 10.1 compares for the test case of Fig. 10.4 the error reduction and computer resources of Mur's second-order ABC with a quadratically graded Berenger PML of varying thickness. Here, the arithmetic average of the absolute values of the E-field errors over a complete planar cut through the $100 \times 100 \times 50$-cell lattice at $y = 0$ and $n = 100$ is compared for the Mur and PML ABCs. The last column indicates the effect of reducing the free-space buffer between the interior working zone and the PML interface by 10 cells relative to that needed for Mur, taking advantage of the transparency of the PML ABC. From these results, a PML that is 4–8 cells thick appears to present a good balance between error reduction and computer burden. Relative to the outer-boundary reflection noise caused by Mur's ABC, PMLs in this thickness range improve the FDTD computational dynamic range by 27–55 dB.

In summary, these results show that Berenger's split-field PML achieves orders-of-magnitude less outer-boundary reflection than previous ABCs when used to model radiating sources in open regions. Depending upon the grading order of the PML loss, 16-cell PML is 60–80 dB less reflective than the second-order Mur ABC. Berenger's PML is also effective over ultrawideband frequency ranges. Unlike previous analytical ABCs (see TAFLOVE and HAGNESS [2000, Chapter 6]) used without the reflection cancellation techniques of RAMAHI [1998], the PML ABC can realize close to its theoretical potential.

10.8.3. Dispersive wave propagation in metal waveguides

FDTD is being used increasingly to model the electromagnetic behavior of not only open-region scattering problems, but also propagation of waves in microwave and optical circuits. An outstanding problem here is the accurate termination of guided-wave structures extending beyond the FDTD lattice boundaries. The key difficulty is that the propagation in a waveguide can be multimodal and dispersive, and the ABC used to terminate the waveguide must be able to absorb energy having widely varying transverse distributions and group velocities v_g.

REUTER, JOSEPH, THIELE, KATZ and TAFLOVE [1994] used Berenger's split-field PML to obtain an ABC for an FDTD model of dispersive wave propagation in a two-dimensional, parallel-plate metal waveguide. This paper assumed a waveguide filled with air and having perfectly conducting walls separated by 40 mm ($f_{\text{cutoff}} =$ 3.75 GHz). The waveguide was assumed to be excited by a Gaussian pulse of temporal width 83.3 ps (full width at half maximum – FWHM) modulating a 7.5-GHz carrier. This launched a $+x$-directed TM_1 mode having the field components E_x, E_y, and H_z towards an 8-cell or 32-cell PML absorber. In effect, the waveguide plunged into the PML, which provided an absorbing "plug". The two waveguide plates continued to the outer boundary of the FDTD grid, where they electrically contacted the perfectly conducting wall backing the PML medium. For the 8-cell PML trial, a quadratic loss grading was assumed with $R(0) = 10^{-6}$; cubic loss grading with $R(0) = 10^{-7}$ was used for the 32-cell PML trial.

Fig. 10.5(a) shows the spectrum of the input pulse used by REUTER, JOSEPH, THIELE, KATZ and TAFLOVE [1994] superimposed upon the normalized group velocity for the TM_1 mode. The incident pulse contained significant spectral energy below cutoff, and the group velocity of the pulse's spectral components varied over an enormous range from zero at f_{cutoff} to about 0.98c well above f_{cutoff}. Because of this huge range, REUTER, JOSEPH, THIELE, KATZ and TAFLOVE [1994] allowed the wave reflected from the PML to fully evolve over many thousands of time steps before completing the simulation. This properly modeled the very slowly propagating spectral components near f_{cutoff}, which generated an equally slowly decaying impulse response for the PML termination. Using a discrete Fourier transformation run concurrently with the FDTD time-stepping, this allowed calculation of the PML reflection coefficient versus frequency by dividing the reflected spectrum by the incident spectrum as observed at the air/PML interface.

Fig. 10.5(b) graphs the resulting reflection coefficient of the waveguide PML ABC versus frequency. For the 8-cell PML, reflections were between -60 dB and -100 dB in the frequency range 4–20 GHz. For the 32-cell PML, the reflection coefficient was below -100 dB in the frequency range 6–18 GHz. (Note that Cray word precision

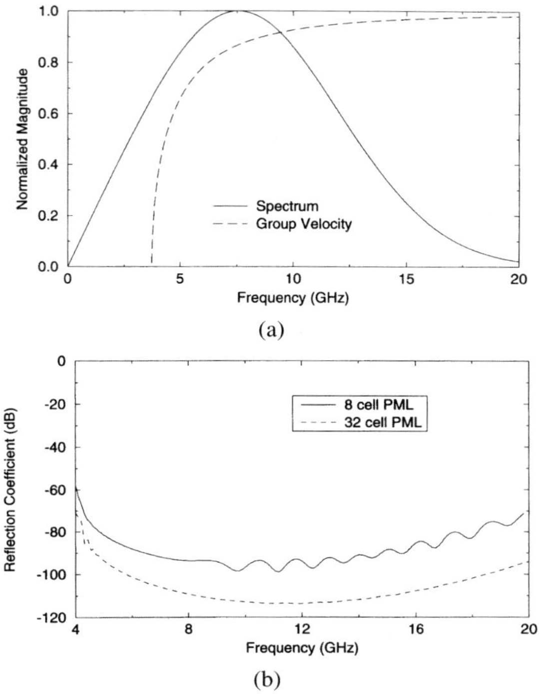

FIG. 10.5. Test of PML ABC for two-dimensional PEC parallel-plate waveguide propagating an ultrawideband pulsed TM_1 mode. (a) Excitation spectrum superimposed upon the group velocity versus frequency (cutoff = 3.75 GHz). (b) PML reflection coefficient versus frequency. *Adapted from*: C.E. Reuter et al., *IEEE Microwave and Guided Wave Letters*, 1994, pp. 344–346, © 1994 IEEE.

was used for these studies.) This example demonstrates the ability of the PML ABC to absorb ultrawideband energy propagating in a waveguide having strong dispersion.

10.8.4. Dispersive and multimode wave propagation in dielectric waveguides

REUTER, JOSEPH, THIELE, KATZ and TAFLOVE [1994] also reported numerical experiments using Berenger's split-field PML to terminate the FDTD model of a two-dimensional, asymmetric, dielectric-slab optical waveguide. This consisted of a 1.5-μm film of permittivity $\varepsilon_r = 10.63$ sandwiched between an infinite substrate of $\varepsilon_r = 9.61$ and an infinite region of air. The excitation introduced at the left edge of the three-layer system was a 17-fs FWHM Gaussian pulse modulating a 200-THz carrier. Fig. 10.6(a) shows the spectrum of this excitation superimposed upon the normalized propagation factors of the three modes supported by the optical waveguide in the frequency range 100–300 THz. We see that the incident pulse contained significant spectral energy over this entire range. Therefore, all three of the waveguide modes were launched. The model of the optical waveguiding system was terminated by extending each of its three dielectric layers into matching PML regions at the right side of the grid. The PML was 16 cells thick with (σ_x, σ_x^*) varying quadratically in the x-direction. Further, the PML loss

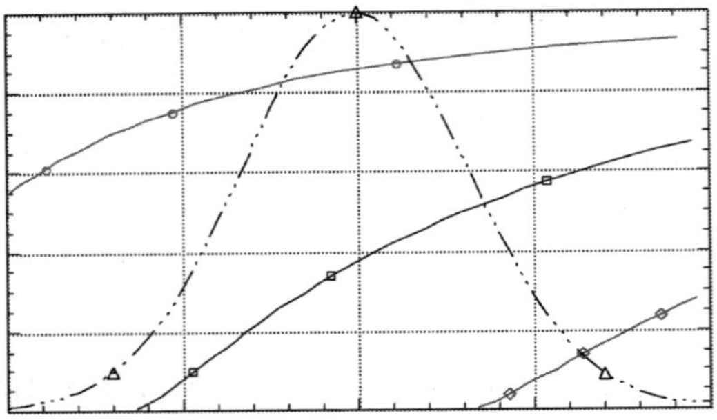

(a)

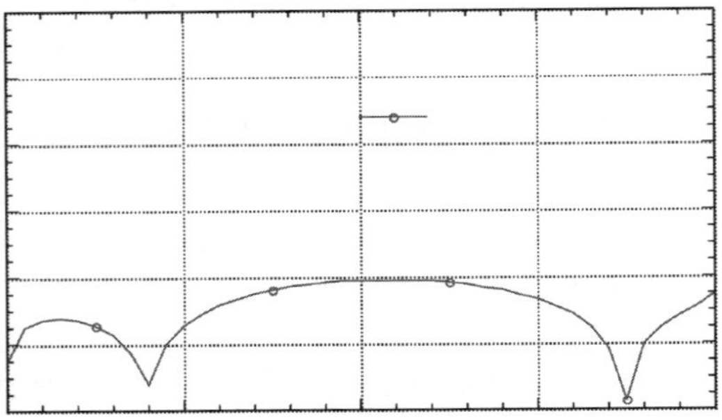

(b)

FIG. 10.6. Test of PML ABC for 2D asymmetric three-layer dielectric optical waveguide propagating a pulsed tri-modal wave: (a) excitation spectrum superimposed upon propagation factors for the three modes; (b) PML reflection coefficient versus frequency. *Source*: C.E. Reuter et al., *IEEE Microwave and Guided Wave Letters*, 1994, pp. 344–346, © 1994 IEEE.

parameter was chosen such that s_x was constant in the transverse direction, as described in Section 10.5.4.

Fig. 10.6(b) graphs the composite reflection coefficient representing the total retrodirected energy in all three regions, as computed at the PML interface. The PML ABC exhibited reflections below -80 dB across the entire spectrum of the incident field. This demonstrates the absorptive capability of the PML ABC for dispersive multimodal propagation.

In summary, these results show that Berenger's split-field PML can achieve highly accurate, ultrawideband terminations of PEC and dielectric waveguides in FDTD space lattices. The PML ABC can provide broadband reflection coefficients better than -80 dB, absorbing dispersive and multimodal energy. Relative to previous approaches for this purpose, the PML ABC has the advantages of being local in time and space, extremely accurate over a wide range of group velocities, and requiring no a priori knowledge of the modal distribution or dispersive nature of the propagating field. PML provides a combination of broadband effectiveness, robustness, and computational efficiency that is unmatched by previous ABCs for FDTD models.

10.9. Numerical experiments with UPML

In this section, the UPML termination of FDTD grids is presented for a number of sample applications. The goal is to provide an understanding of how the UPML parameters and grading functions impact its effectiveness as an ABC. In this manner, the reader can better understand how to properly choose these parameters.

10.9.1. Current source radiating in an unbounded two-dimensional region

Fig. 10.7 illustrates the first example, as reported by GEDNEY and TAFLOVE [2000]. This involves an electric current source $\vec{J}$ centered in a 40 × 40-cell FDTD grid. The source is vertically directed and invariant along the axial direction. Hence, it radiates two-dimensional TE waves. It has the time signature of a differentiated Gaussian pulse

$$J_y(x_0, y_0, t) = -2\big[(t - t_0)/t_w\big]\exp\big\{-\big[(t - t_0)/t_w\big]^2\big\}, \tag{10.71}$$

where $t_w = 26.53$ ps and $t_0 = 4t_w$.

The grid has 1-mm square cells and a time step of 0.98 times the Courant limit. The E-field is probed at two points, A and B, as shown in the figure. Point A is in the same plane as the source and two cells from the UPML, and point B is two cells from the bottom and side UPMLs. Time-stepping runs over 1000 iterations, well past the steady-state response. Both 5-cell and 10-cell UPML ABCs are used with polynomial grading $m = 4$.

For this case, the reference solution $E_{\text{ref}}|_{i,j}^n$ is obtained using a 1240 × 1240-cell grid. An identical current source is centered within this grid, and the field-observation point (i, j) is at the same position relative to the source as in the test grid. The reference grid is sufficiently large such that there are no reflections from its outer boundaries during the time stepping. This allows a relative error to be defined as

$$\text{Rel.error}|_{i,j}^n = \big|E|_{i,j}^n - E_{\text{ref}}|_{i,j}^n\big| / \big|E_{\text{ref max}}|_{i,j}\big|, \tag{10.72}$$

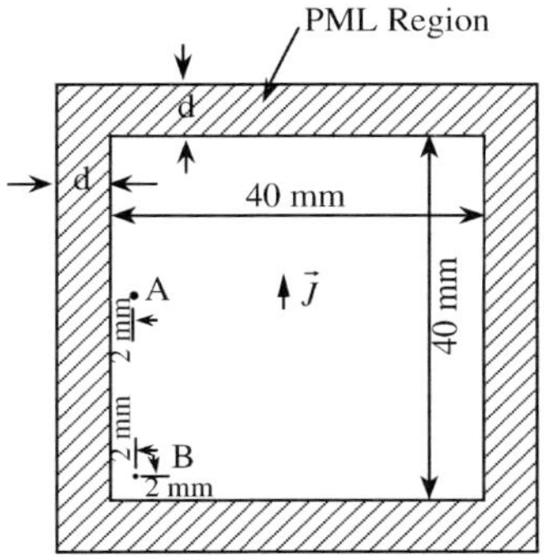

FIG. 10.7. Vertically directed electric current source centered in a 2D FDTD grid. The working volume of 40 × 40 mm is surrounded by UPML of thickness d. E-fields are probed at points A and B.

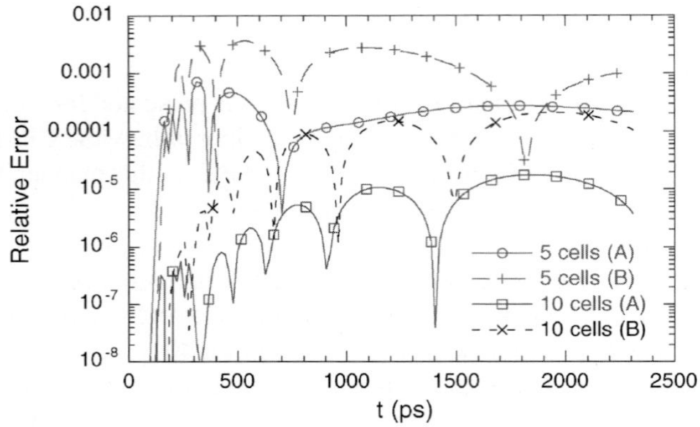

FIG. 10.8. Relative error at points A and B of Fig. 10.7 over 1000 time steps for 5-cell and 10-cell UPMLs with $\sigma_{\max} = \sigma_{\text{opt}}$ and $\kappa = 1$.

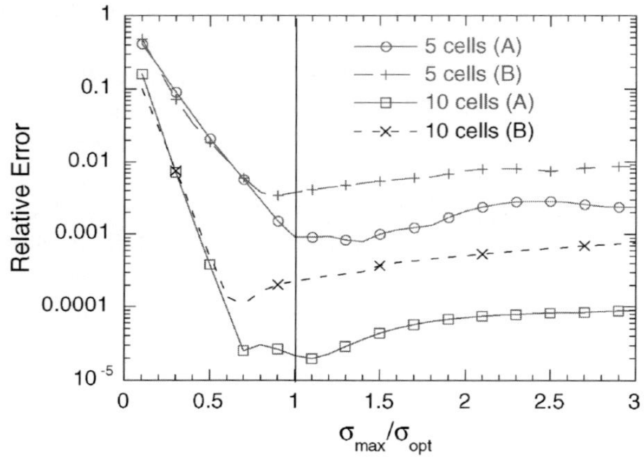

FIG. 10.9. Maximum relative error due to 5-cell and 10-cell UPML ABCs in the grid of Fig. 10.7 versus $\sigma_{\max}/\sigma_{\text{opt}}$ over a 1000-time-step observation period.

where $E|_{i,j}^{n}$ is the field value at grid location (i, j) and time step n in the test grid, and $E_{\text{ref max}}|_{i,j}$ is the maximum amplitude of the reference field at grid location (i, j), as observed during the time-stepping span of interest.

Fig. 10.8 graphs the relative error calculated using (10.72) at points A and B of Fig. 10.7 over the first 1000 time steps of the FDTD run for 5-cell and 10-cell UPMLs. Here, the key UPML parameters are $\sigma_{\max} = \sigma_{\text{opt}}$, where σ_{opt} is given by (10.57), and $\kappa_{\max} = 1$. We note that the error at A is always less than that at B. This is because the wave impinging on the UPML near A is nearly normally incident and undergoes maximum absorption. At B, while the amplitude of the outgoing wave is smaller due to the radiation pattern of the source, the wave impinges on the UPML obliquely at 45°.

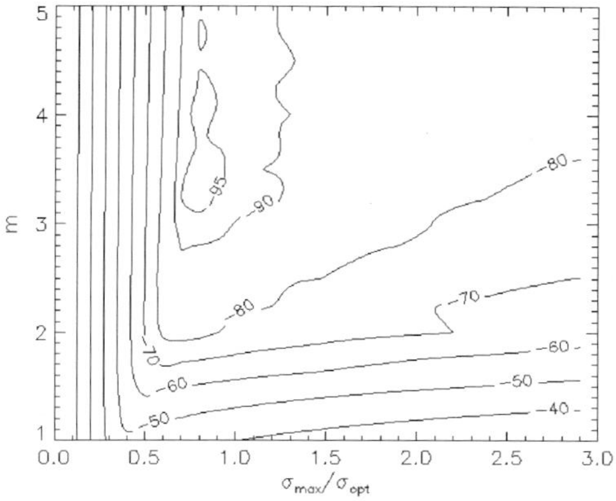

(a)

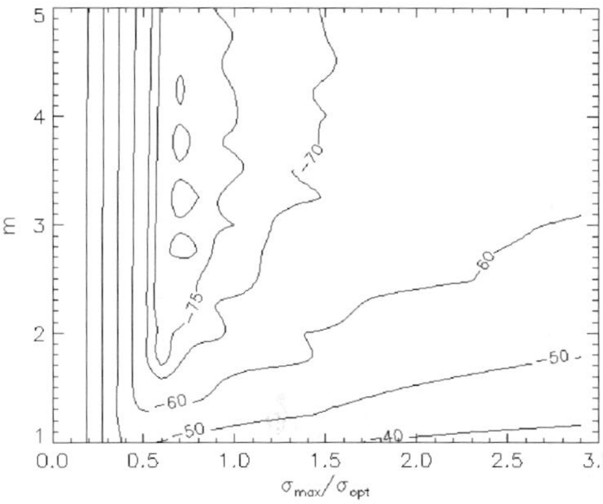

(b)

FIG. 10.10. Contour plots of the maximum relative error in dB in the grid of Fig. 10.7 versus $\sigma_{\max}/\sigma_{\text{opt}}$ and polynomial order m for a 10-cell UPML: (a) at point A; (b) at point B.

Fig. 10.9 provides additional information by graphing as a function of $\sigma_{\max}/\sigma_{\text{opt}}$ the maximum relative error at points A and B during the 1000-time-step simulation. As before, polynomial grading is used with $m = 4$, and the same $\sigma_{\max}$ is used for each of the four UPML absorbers at the outer boundary planes of the grid. We see that the optimal choice for $\sigma_{\max}$ is indeed close to σ_{opt}. Again, the maximum error at B is about an order of magnitude larger than that at A.

Figs. 10.10(a) and 10.10(b) are contour plots resulting from a comprehensive parametric study of the 10-cell UPML. These figures map the maximum relative error at A and B, respectively, during the 1000-time-step simulation. The horizontal axis of each plot provides a scale for $\sigma_{\max}/\sigma_{\text{opt}}$, and the vertical axis provides a scale for the polynomial order m, not necessarily an integer. The minimum error is found for $3 < m < 4$ and $\sigma_{\max} \cong 0.75\sigma_{\text{opt}}$, and is approximately -95 dB at A and -80 dB at B.

GEDNEY and TAFLOVE [2000] reported a similar study involving geometric grading of the UPML parameters. For $g = 2.2$, about -85 dB of reflection error was realized at both A and B for $\ln[R(0)]$ between -12 and -16. It was observed that the effectiveness of the UPML is quite sensitive to the choice of g.

10.9.2. Highly elongated domains

In the previous example, the incident angle of the wave impinging on the UPML never exceeded 45°. However, for highly oblique incidence angles approaching 90°, the reflectivity of the UPML increases markedly and its performance as an ABC degrades.

Fig. 10.11 illustrates an example reported by GEDNEY and TAFLOVE [2000] of just such a situation, a highly elongated 435 × 15-cell (working volume) TE_z grid. Here, the relative error in the E-field was computed at points A (10 mm from the source), B (50 mm from the source), C (100 mm from the source), D (200 mm from the source), and E (400 mm from the source). The current source was polarized such that the radiated signal impinging on the long grid boundary was maximum in amplitude. From the problem geometry, we see that the specularly reflected wave from the long-grid-

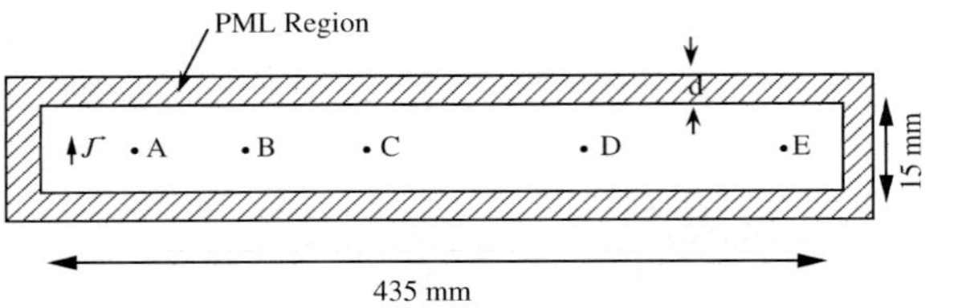

FIG. 10.11. Current element $\vec{J}$ radiating in an elongated FDTD grid (not to scale) terminated by UPML. Distance of each observation point from the source: A, 10 mm; B, 50 mm; C, 100 mm; D, 200 mm; and E, 400 mm.

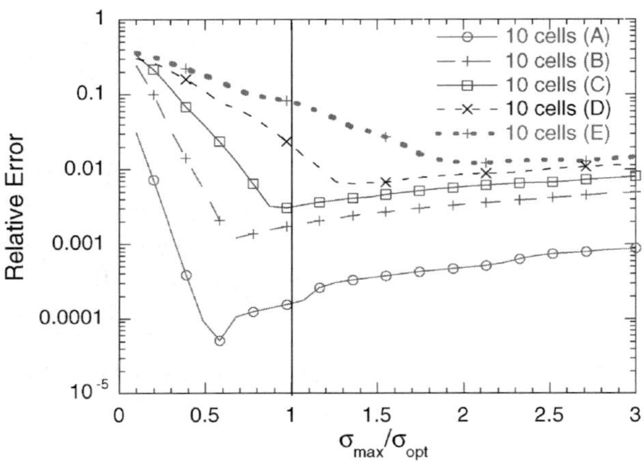

FIG. 10.12. Maximum relative error at points A, B, C, D, and E in the grid of Fig. 10.11 due to a polynomial-scaled 10-cell PML vs. $\sigma_{\max}/\sigma_{\text{opt}}$. Parameters $m = 4$ and $\kappa = 1$ are fixed.

boundary UPML arriving at E was incident on the UPML at 89°, implying that the reflection error at E should be degraded from that observed at A.

The model of Fig. 10.11 used 1-mm square grid cells and Δt set at 0.98 times the Courant limit. The source had the same differentiated Gaussian-pulse waveform used previously. A 10-cell UPML absorber with polynomial spatial scaling ($m = 4$ and $\kappa_{\max} = 1$) was used on all sides.

Fig. 10.12 graphs the maximum relative error recorded at each of the observation points over the initial 1000 time-steps as a function of $\sigma_{\max}/\sigma_{\text{opt}}$. While the error at

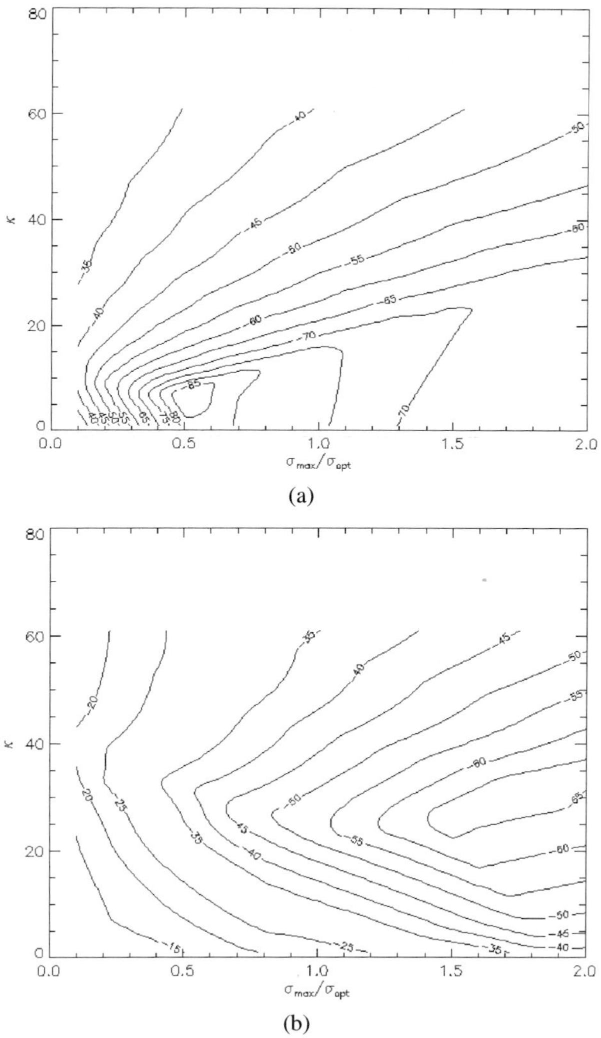

FIG. 10.13. Contour plots of the maximum relative error in dB in the grid of Fig. 10.11 vs. $\kappa_{y\text{-}\max}$ and $\sigma_{y\text{-}\max}/\sigma_{\text{opt}}$ for a polynomial-scaled 10-cell UPML: (a) at point A; (b) at point E.

A can be less than 0.0001, or -80 dB, the error progressively worsens at B–E. We expect that larger values of σ_{max} could reduce the reflection error. However, larger values of σ_{max} lead to larger step discontinuities in the UPML profile, and hence, a larger discretization error.

We note that the radiation due to the current source is characterized by a spectrum of waves containing evanescent as well as propagating modes. However, the evanescent modes are not absorbed by the UPML when $\kappa = 1$. Increasing κ should help this situation. To investigate this possibility, Figs. 10.13(a) and 10.13(b) plot contours of the maximum relative error at A and E, respectively, as a function of $\sigma_{max}/\sigma_{opt}$ and κ_{max}. We see that increasing κ_{max} causes the reflection error at E to decrease by two orders of magnitude to less than -65 dB in the vicinity of $\sigma_{max} \approx 1.6\sigma_{opt}$ and $\kappa_{max} \approx 25$. While this strategy degrades the reflection error at A, the error at A is still less than -65 dB.

10.9.3. Microstrip transmission line

Our final numerical example, reported by GEDNEY and TAFLOVE [2000], involves the use of UPML to terminate a three-dimensional FDTD model of a 50 Ω microstrip transmission line. This is a case wherein an inhomogeneous dielectric medium penetrates into the UPML, and ultralow levels of wave reflection are required.

Fig. 10.14 illustrates the cross-section of the microstrip line. The metal trace was assumed to be 254 μm wide with a negligible thickness compared to its width. This trace was assumed printed on a 254 μm thick alumina substrate having $\varepsilon_r = 9.8$, with the region above the substrate being air. The line was assumed to be excited at one end by a voltage source with a Gaussian profile and a 40-GHz bandwidth.

A uniform lattice discretization $\Delta x = \Delta y = 42.333$ μm was used in the transverse (x, y)-plane of the microstrip line, while $\Delta z = 120$ μm was used along the direction of wave propagation z. Polynomial-graded UPML backed by perfectly conducting walls terminated all lattice outer boundary planes. The metal trace extended completely through the z-normal UPML and electrically contacted the backing wall. In addition, the substrate, ground plane, and air media each continued into their respective adjacent UPMLs, maintaining their permittivities and geometries.

The wave impinging on the z-normal boundary was a quasi-TEM mode supported by the microstrip line. Even though the media were inhomogeneous, the UPML was

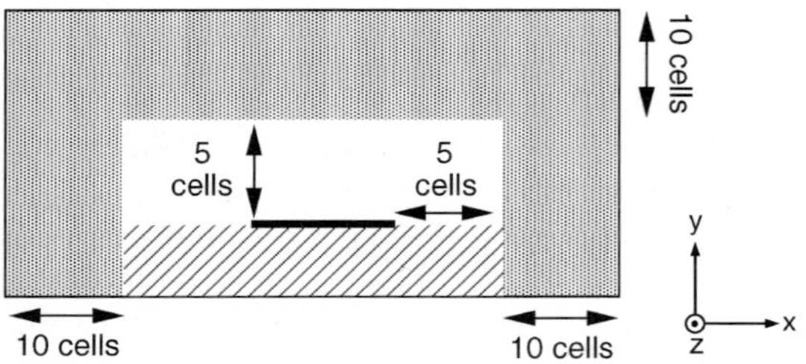

FIG. 10.14. Cross section of a 50 Ω microstrip line printed on a 254 μm thick alumina substrate terminating in UPML. The FDTD lattice sidewalls are also terminated by UPML.

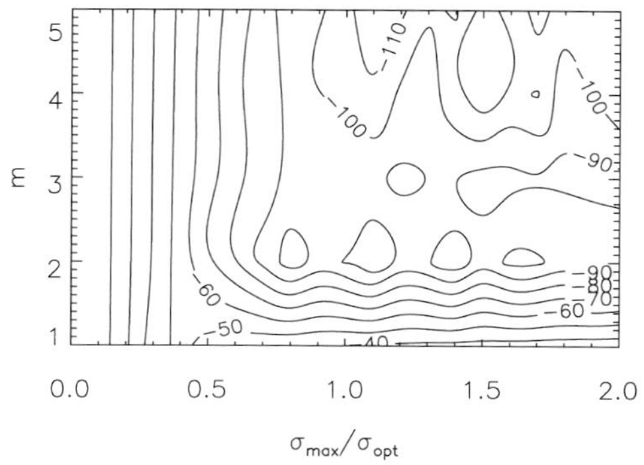

FIG. 10.15. Reflection error (in dB) of a 10-cell z-normal UPML in the microstrip line of Fig. 10.14 as a function of the polynomial grading order m and the normalized conductivity.

perfectly matched to this wave. The goal was to study the reflection behavior of the UPML as a function of its loss profile when used to terminate this line.

Because the conductivity within the UPML was polynomial graded, the optimal value of $\sigma_{z\max}$ could be predicted by (10.57). However, an ambiguity existed because the dielectric penetrating into the z-normal UPML was inhomogeneous. Thus, as recommended in Section 10.5.4, the UPML conductivity was scaled by the relative permittivity as defined in (10.48). This yielded

$$\sigma_{z,\mathrm{opt}} = \frac{0.8(m+1)}{\eta_0 \Delta \sqrt{\varepsilon_{\mathrm{eff}}}}, \tag{10.73}$$

where $\varepsilon_{\mathrm{eff}}$ is the effective relative permittivity for the inhomogeneous media extending into the UPML. For the microstrip line, $\varepsilon_{\mathrm{eff}}$ could be estimated via quasistatic theory (see, for example, POZAR [1998]), and in the case of Fig. 10.14 equals 6.62.

Fig. 10.15 illustrates the results of a parametric study of the reflection-error performance of a 10-cell UPML at the z-normal lattice outer boundary. The parameters investigated were $\sigma_{z\max}$ and the polynomial grading order m. We see that the optimal value of $\sigma_{z\max}$ was well predicted by (10.73), and further choosing m in the range of 3–5 was sufficient for minimizing the reflection error. Additional studies showed that the UPML reflection was better than -100 dB from 0–50 GHz for $m = 4$ and $\sigma_{z,\mathrm{opt}}$.

10.10. UPML terminations for conductive and dispersive media

For certain applications, it is necessary to simulate electromagnetic wave interactions within conductive or frequency-dispersive media of significant spatial extent. Examples include wave propagation within microwave circuits printed on lossy dielectric substrates, and impulsive scattering by objects buried in the earth or embedded within biological tissues. In such cases, it is desirable to simulate the lossy or dispersive material extending to infinity through the use of a PML absorbing boundary. The reader is referred to the work of GEDNEY and TAFLOVE [2000], which discusses in detail the extension of the UPML formulation presented previously in this section for purposes of terminating conductive and dispersive media.

11. Summary and conclusions

This chapter reviewed key elements of the theoretical foundation and numerical implementation of finite-difference time-domain (FDTD) solutions of Maxwell's equations. The chapter included:

- Introduction and background
- Review of Maxwell's equations
- The Yee algorithm
- The nonuniform Yee grid
- Alternative finite-difference grids
- Theory of numerical dispersion
- Algorithms for improved numerical dispersion properties
- Theory of numerical stability
- Alternating-direction implicit time-stepping algorithm for operation beyond the Courant limit
- Perfectly matched layer (PML) absorbing boundary conditions, including Berenger's split-field PML, the stretched-coordinate PML formulation, and the uniaxial anisotropic PML (UPML).

With literally hundreds of papers on FDTD methods and applications published each year, it is clear that FDTD is one of the most powerful and widely used numerical modeling approaches for electromagnetic wave interaction problems. With expanding developer and user communities within an increasing number of disciplines in science and engineering, FDTD technology is continually evolving in terms of its theoretical basis, numerical implementation, and technological applications. The latter now literally approach the proverbial spectral range from dc to light.

References

ABARBANEL, S., GOTTLIEB, D. (1997). A mathematical analysis of the PML method. *J. Comput. Phys.* **134**, 357–363.

BERENGER, J.P. (1994). A perfectly matched layer for the absorption of electromagnetic waves. *J. Comput. Phys.* **114**, 185–200.

BERENGER, J.P. (1996). Perfectly matched layer for the FDTD solution of wave-structure interaction problems. *IEEE Trans. Antennas Propagat.* **44**, 110–117.

BERENGER, J.P. (1997). Improved PML for the FDTD solution of wave-structure interaction problems. *IEEE Trans. Antennas Propagat.* **45**, 466–473.

CHEW, W.C., JIN, J.M. (1996). Perfectly matched layers in the discretized space: an analysis and optimization. *Electromagnetics* **16**, 325–340.

CHEW, W.C., WEEDON, W.H. (1994). A 3D perfectly matched medium from modified Maxwell's equations with stretched coordinates. *IEEE Microwave Guided Wave Lett.* **4**, 599–604.

CHURCHILL, R.V., BROWN, J.W., VERHEY, R.F. (1976). *Complex Variables and Applications* (McGraw-Hill, New York).

DEY, S., MITTRA, R. (1997). A locally conformal finite-difference time-domain algorithm for modeling three-dimensional perfectly conducting objects. *IEEE Microwave Guided Wave Lett.* **7**, 273–275.

ENGQUIST, B., MAJDA, A. (1977). Absorbing boundary conditions for the numerical simulation of waves. *Math. Comput.* **31**, 629–651.

FANG, J. (1989). Time-domain finite difference computations for Maxwell's equations, Ph.D. dissertation, Univ. of California, Berkeley, CA.

GEDNEY, S.D. (1995). An anisotropic perfectly matched layer absorbing medium for the truncation of FDTD lattices, Report EMG-95-006, University of Kentucky, Lexington, KY.

GEDNEY, S.D. (1996). An anisotropic perfectly matched layer absorbing medium for the truncation of FDTD lattices. *IEEE Trans. Antennas Propagat.* **44**, 1630–1639.

GEDNEY, S.D. (1998). The perfectly matched layer absorbing medium. In: Taflove, A. (ed.), *Advances in Computational Electrodynamics: The Finite-Difference Time-Domain Method* (Artech House, Norwood, MA), pp. 263–343.

GEDNEY, S.D., LANSING, F. (1995). Nonuniform orthogonal grids. In: Taflove, A. (ed.), *Computational Electrodynamics: The Finite-Difference Time-Domain Method* (Artech House, Norwood, MA), pp. 344–353.

GEDNEY, S.D., TAFLOVE, A. (2000). Perfectly matched layer absorbing boundary conditions. In: Taflove, A., Hagness, S.C. (eds.), *Computational Electrodynamics: The Finite-Difference Time-Domain Method*, second ed. (Artech House, Norwood, MA), pp. 285–348.

GONZALEZ GARCIA, S., LEE, T.W., HAGNESS, S.C. (2002). On the accuracy of the ADI-FDTD method. *IEEE Antennas Wireless Propagation Lett.* **1**, 31–34.

HARRINGTON, R.F. (1968). *Field Computation by Moment Methods* (MacMillan, New York).

HE, L. (1997). FDTD – Advances in sub-sampling methods, UPML, and higher-order boundary conditions, M.S. thesis, University of Kentucky, Lexington, KY.

HIGDON, R.L. (1986). Absorbing boundary conditions for difference approximations to the multi-dimensional wave equation. *Math. Comput.* **47**, 437–459.

HOLLAND, R. (1984). Implicit three-dimensional finite-differencing of Maxwell's equations. *IEEE Trans. Nucl. Sci.* **31**, 1322–1326.

HOLLAND, R., CHO, K.S. (1986). Alternating-direction implicit differencing of Maxwell's equations: 3D results. Computer Sciences Corp., Albuquerque, NM, Technical report to Harry Diamond Labs., Adelphi, MD, Contract DAAL02-85-C-0200.

HOLLAND, R., WILLIAMS, J. (1983). Total-field versus scattered-field finite-difference. *IEEE Trans. Nucl. Sci.* **30**, 4583–4587.

JURGENS, T.G., TAFLOVE, A., UMASHANKAR, K.R., MOORE, T.G. (1992). Finite-difference time-domain modeling of curved surfaces. *IEEE Trans. Antennas Propagat.* **40**, 357–366.

KATZ, D.S., THIELE, E.T., TAFLOVE, A. (1994). Validation and extension to three-dimensions of the Berenger PML absorbing boundary condition for FDTD meshes. *IEEE Microwave Guided Wave Lett.* **4**, 268–270.

KELLER, J.B. (1962). Geometrical theory of diffraction. *J. Opt. Soc. Amer.* **52**, 116–130.

KOUYOUMJIAN, R.G., PATHAK, P.H. (1974). A uniform geometrical theory of diffraction for an edge in a perfectly conducting surface. *Proc. IEEE* **62**, 1448–1461.

KREISS, H., MANTEUFFEL, T., SCHWARTZ, B., WENDROFF, B., WHITE, J.A.B. (1986). Supraconvergent schemes on irregular meshes. *Math. Comput.* **47**, 537–554.

KRUMPHOLZ, M., KATEHI, L.P.B. (1996). MRTD: new time-domain schemes based on multiresolution analysis. *IEEE Trans. Microwave Theory Tech.* **44**, 555–572.

LIAO, Z.P., WONG, H.L., YANG, B.P., YUAN, Y.F. (1984). A transmitting boundary for transient wave analyses. *Scientia Sinica (series A)* **XXVII**, 1063–1076.

LIU, Q.H. (1996). The PSTD algorithm: a time-domain method requiring only two grids per wavelength, Report NMSU-ECE96–013, New Mexico State Univ., Las Cruces, NM.

LIU, Q.H. (1997). The pseudospectral time-domain (PSTD) method: a new algorithm for solutions of Maxwell's equations. In: Proc. 1997 IEEE Antennas & Propagation Soc. Internat. Symp. **I** (IEEE, Piscataway, NJ), pp. 122–125 (catalog no. 97CH36122).

LIU, Y. (1996). Fourier analysis of numerical algorithms for the Maxwell's equations. *J. Comput. Phys.* **124**, 396–416.

MADSEN, N.K., ZIOLKOWSKI, R.W. (1990). A three-dimensional modified finite volume technique for Maxwell's equations. *Electromagnetics* **10**, 147–161.

MANTEUFFEL, T.A., WHITE, J.A. (1986). The numerical solution of second-order boundary value problems on nonuniform meshes. *Math. Comput.* **47**, 511–535.

MIN, M.S., TENG, C.H. (2001). The instability of the Yee scheme for the "magic time step". *J. Comput. Phys.* **166**, 418–424.

MONK, P. (1994). Error estimates for Yee's method on non-uniform grids. *IEEE Trans. Magnetics* **30**, 3200–3203.

MONK, P., SULI, E. (1994). A convergence analysis of Yee's scheme on non-uniform grids. *SIAM J. Numer. Anal.* **31**, 393–412.

MOORE, T.G., BLASCHAK, J.G., TAFLOVE, A., KRIEGSMANN, G.A. (1988). Theory and application of radiation boundary operators. *IEEE Trans. Antennas Propagation* **36**, 1797–1812.

MUR, G. (1981). Absorbing boundary conditions for the finite-difference approximation of the time-domain electromagnetic field equations. *IEEE Trans. Electromagnetic Compatibility* **23**, 377–382.

NAMIKI, T. (2000). 3-D ADI-FDTD method – unconditionally stable time-domain algorithm for solving full vector Maxwell's equations. *IEEE Trans. Microwave Theory and Techniques* **48**, 1743–1748.

NEHRBASS, J.W., LEE, J.F., LEE, R. (1996). Stability analysis for perfectly matched layered absorbers. *Electromagnetics* **16**, 385–389.

POZAR, D.M. (1998). *Microwave Engineering*, second ed. (Wiley, New York).

RAMAHI, O.M. (1997). The complementary operators method in FDTD simulations. *IEEE Antennas Propagat. Magazine* **39/6**, 33–45.

RAMAHI, O.M. (1998). The concurrent complementary operators method for FDTD mesh truncation. *IEEE Trans. Antennas Propagat.* **46**, 1475–1482.

RAPPAPORT, C.M. (1995). Perfectly matched absorbing boundary conditions based on anisotropic lossy mapping of space. *IEEE Microwave Guided Wave Lett.* **5**, 90–92.

REUTER, C.E., JOSEPH, R.M., THIELE, E.T., KATZ, D.S., TAFLOVE, A. (1994). Ultrawideband absorbing boundary condition for termination of waveguiding structures in FDTD simulations. *IEEE Microwave Guided Wave Lett.* **4**, 344–346.

SACKS, Z.S., KINGSLAND, D.M., LEE, R., LEE, J.F. (1995). A perfectly matched anisotropic absorber for use as an absorbing boundary condition. *IEEE Trans. Antennas Propagat.* **43**, 1460–1463.

SCHNEIDER, J.B., WAGNER, C.L. (1999). FDTD dispersion revisited: Faster-than-light propagation. *IEEE Microwave Guided Wave Lett.* **9**, 54–56.

SHANKAR, V., MOHAMMADIAN, A.H., HALL, W.F. (1990). A time-domain finite-volume treatment for the Maxwell equations. *Electromagnetics* **10**, 127–145.

SHEEN, D. (1991). Numerical modeling of microstrip circuits and antennas, Ph.D. thesis, Massachusetts Institute of Technology, Cambridge, MA.

SHLAGER, K.L., SCHNEIDER, J.B. (1998). A survey of the finite-difference time-domain literature. In: Taflove, A. (ed.), *Advances in Computational Electrodynamics: The Finite-Difference Time-Domain Method* (Artech House, Norwood, MA), pp. 1–62.

SONG, J., CHEW, W.C. (1998). The fast Illinois solver code: requirements and scaling properties. *IEEE Comp. Sci. Engrg.* **5**.

TAFLOVE, A. (1995). *Computational Electrodynamics: The Finite-Difference Time-Domain Method*, first ed. (Artech House, Norwood, MA).

TAFLOVE, A., BRODWIN, M.E. (1975). Numerical solution of steady-state electromagnetic scattering problems using the time-dependent Maxwell's equations. *IEEE Trans. Microwave Theory and Techniques* **23**, 623–630.

TAFLOVE, A., HAGNESS, S.C. (2000). *Computational Electrodynamics: The Finite-Difference Time-Domain Method*, second ed. (Artech House, Norwood, MA).

TAFLOVE, A., UMASHANKAR, K.R., BEKER, B., HARFOUSH, F.A., YEE, K.S. (1988). Detailed FDTD analysis of electromagnetic fields penetrating narrow slots and lapped joints in thick conducting screens. *IEEE Trans. Antennas Propagation* **36**, 247–257.

TEIXEIRA, F.L., CHEW, W.C. (1997). PML-FDTD in cylindrical and spherical coordinates. *IEEE Microwave Guided Wave Lett.* **7**, 285–287.

TULINTSEFF, A. (1992). The finite-difference time-domain method and computer program description applied to multilayered microstrip antenna and circuit configurations, Technical Report AMT: 336.5-92-041, Jet Propulsion Laboratory, Pasadena, CA.

TURKEL, E. (1998). In: Taflove, A. (ed.), *Advances in Computational Electrodynamics: The Finite-Difference Time-Domain Method* (Artech House, Norwood, MA). Chapter 2.

YEE, K.S. (1966). Numerical solution of initial boundary value problems involving Maxwell's equations in isotropic media. *IEEE Trans. Antennas Propagat.* **14**, 302–307.

ZHENG, F., CHEN, Z., ZHANG, J. (2000). Toward the development of a three-dimensional unconditionally stable finite-difference time-domain method. *IEEE Trans. Microwave Theory and Techniques* **48**, 1550–1558.

ZHENG, F., CHEN, Z. (2001). Numerical dispersion analysis of the unconditionally stable 3-D ADI-FDTD method. *IEEE Trans. Microwave Theory and Techniques* **49**, 1006–1009.

Discretization of Semiconductor Device Problems (I)

F. Brezzi [a,b], L.D. Marini [a,b], S. Micheletti [c], P. Pietra [b], R. Sacco [c], S. Wang [d]

[a] *Dipartimento di Matematica "F. Casorati", Università di Pavia, Via Ferrata 1, I-27100 Pavia, Italy*

[b] *Istituto di Matematica Applicata e Tecnologie Informatiche – C.N.R., Via Ferrata 1, I-27100 Pavia, Italy*

[c] *Dipartimento di Matematica "F. Brioschi", Politecnico di Milano, Via Bonardi 9, 20133 Milano, Italy*

[d] *University of Western Australia, Department of Mathematics and Statistics, Nedlands, Western Australia 6907, Australia*

1. Fluid models for transport in semiconductors

In this section we recall the Drift-Diffusion model, already but quickly addressed in Chapter 1, and we describe the initial/boundary conditions which complete the model. We also consider in more detail the Energy-Balance transport model. For a more complete discussion of these issues we suggest the books by MARKOWICH [1986], MARKOWICH, RINGHOFER and SCHMEISER [1990], JEROME [1996], JÜNGEL [2001] and SELBERHERR [1984].

1.1. The Drift-Diffusion model

It is possible to view classical or semiclassical modelling of transport in semiconductors as a hierarchical structure, sweeping from the Boltzmann Transport Equation (BTE) down to the Drift-Diffusion (DD) model, passing through all the systems derived from

Numerical Methods in Electromagnetics
Special Volume (W.H.A. Schilders and E.J.W. ter Maten, Guest Editors) of
HANDBOOK OF NUMERICAL ANALYSIS, VOL. XIII
P.G. Ciarlet (Editor)

ISSN 1570-8659
DOI 10.1016/S1570-8659(04)13004-4

the moments of the BTE with respect to increasing powers of the carrier group velocity (cf. Section 4.6 of Chapter 1). The Hydrodynamic (HD) model and its inertial limit, that is a type of Energy-Transport (ET) model, are examples, among others, of these intermediate steps. The ET model will be briefly presented in Section 1.4. For the HD model we refer to Chapter 5 and to the references therein.

The DD model is by far the best understood of the above models. It was introduced by VAN ROOSBROECK [1950] back in 1950 as a conservation system for electron and hole carriers, along with the electric field determined from the Poisson equation in a self-consistent fashion. Other derivations of the model are also possible, starting from the BTE and making use of the diffusion approximation, see, e.g., (COWELL [1967], RODE [1995], DEGOND, GUYOT-DELAURENS, MUSTIELES and NIER [1990]). For a mathematical theory on the subject we refer to (POUPAUD [1991, 1992], POUPAUD and SCHMEISER [1991], GOLSE and POUPAUD [1992], MARKOWICH and SCHMEISER [1997], MARKOWICH, POUPAUD and SCHMEISER [1995]). As a general comment, the DD model is a good approximation of the underlying physical phenomena when

- All of the scattering processes are elastic. The spatial variation of the relaxation times and of the band structure of the semiconductor are neglected, i.e., slow variations of the doping profile are assumed.
- Degeneration effects in the approximation of the collision integral are neglected.
- The electric field depends mildly on the spatial position and the magnetic induction vanishes.
- The carrier temperature coincides with the lattice temperature, which is constant. This prevents simulating hot carrier and velocity overshoot phenomena.
- The energy bands are parabolic.

From a mathematical viewpoint, when appropriate scalings are employed for the quantities appearing in the DD system, one finds that the conservation part of the system is convection dominated. In particular, there is a scaling which makes the order of magnitude of the electron and hole concentrations equal to 1. The convection domination arises from the potential, solving Poisson equation, which can be singularly perturbed. This issue will be addressed in Section 1.3. Major breakthroughs in the solution of the model occurred during the 1960s thanks to the pioneering work by H.K. Gummel and D.L. Scharfetter. Gummel introduced in [1964] the system decoupling through a nonlinear Gauss–Seidel iteration, and Scharfetter and Gummel provided an exponential fitting approximation for the continuity equations in 1-d (SCHARFETTER and GUMMEL [1969]) (this scheme had been previously introduced in fluidynamics applications by ALLEN and SOUTHWELL [1955]). The nonlinear map (or family of maps) that arises from the Gauss–Seidel iteration is a compact continuous map whose fixed point characterizes the solution of the DD model, and is since then called Gummel's map. It will be the object of Section 2. The exponentially fitted difference scheme proposed in the one-dimensional case in SCHARFETTER and GUMMEL [1969] will be the object of multidimensional extension in Sections 4.1, 4.2, 5.1.1 and 6.2.

1.2. The DD system

The geometrical model of a semiconductor device consists of a bounded and simply connected portion Ω of $\mathbb{R}^d$ $(d = 1, 2, 3)$, which is constituted by a semiconductor part, denoted by Ω_S, and, in the case of Metal-Oxide-Semiconductor (MOS) devices, also by one or more subdomains of thin oxide adjacent to Ω_S and whose union we denote by Ω_O. We also denote by $x \in \mathbb{R}^d$ the independent spatial variable and by $t \in (0, t_f)$ the time variable.

The DD model is given by the following set of equations for the electric field $\underline{E}$ [V cm^{-1}], and the electron and hole current densities $\underline{J}_n$, $\underline{J}_p$ [A cm^{-2}], respectively:

$$\begin{cases} \operatorname{div}(\varepsilon \underline{E}) = \rho & \text{in } Q := \Omega \times (0, t_f), \\ q\frac{\partial n}{\partial t} - \operatorname{div} \underline{J}_n = -qR & \text{in } Q_S := \Omega_S \times (0, t_f), \\ q\frac{\partial p}{\partial t} + \operatorname{div} \underline{J}_p = -qR & \text{in } Q_S := \Omega_S \times (0, t_f), \end{cases} \tag{1.1}$$

where q [A s] is the (positive) electron charge, ε [A s V^{-1} cm^{-1}] is the dielectric constant of the materials, and the following constitutive relations hold:

$$\underline{E} = -\underline{\nabla} \psi \quad \text{in } Q,$$
$$\rho = \begin{cases} q(p - n + C(x)) = q(p - n + N_D^+ - N_A^-) & \text{in } Q_S, \\ 0 & \text{in } Q_O. \end{cases} \tag{1.2}$$

In (1.1)–(1.2) ψ [V] is the electrostatic potential, n, p [cm^{-3}] are the electron and hole concentrations inside the semiconductor ($n|_{\Omega_O} \equiv p|_{\Omega_O} \equiv 0$), ρ [A s cm^{-3}] is the net charge density in the device and is zero inside the oxide, which is assumed to be neutral, $C(x) = N_D^+ - N_A^-$ [cm^{-3}] is the so-called doping profile, which is assumed to be a given datum of the problem in terms of the ionized donors and acceptors concentrations N_D^+ and N_A^-, respectively. Moreover, the oxide is assumed as a perfect insulator; thus we have $\underline{J}_n|_{\Omega_O} \equiv \underline{J}_p|_{\Omega_O} \equiv 0$.

The next step towards the completion of the DD system is to provide constitutive relations for $\underline{J}_n$, $\underline{J}_p$ in the semiconductor and for the source term R [cm^{-3} s^{-1}]. This last one can be interpreted as the net generation/recombination rate of carriers in unit time and volume: $R > 0$ means net recombination, $R < 0$ means net generation. More details about the modeling of R will be given in Section 1.2.1.

Physically, the current density is the product of the elementary charge q, the carrier density, and the mean velocity (drift velocity), i.e.,

$$\underline{J}_n = -qn\underline{v}_n, \qquad \underline{J}_p = qp\underline{v}_p,$$

where $\underline{v}_n$, $\underline{v}_p$ [cm s^{-1}] are the drift velocities of the carriers. We have then to determine some relations which link these velocities to the electric field and to the carrier concentrations. This can be done from the BTE. Another approach consists in carrying out a singular perturbation expansion of the current densities with respect to relaxation times. Actually, introducing the mobilities μ_n, μ_p [cm^2 V^{-1} s^{-1}]

$$\mu_n = q\tau_n/m_n^*, \qquad \mu_p = q\tau_p/m_p^*,$$

in terms of the effective masses m_n^*, m_p^* [kg] and of the relaxation times τ_n, τ_p [s], we obtain (see SELBERHERR [1984])

$$\tau_n \frac{\partial \underline{J}_n}{\partial t} + \underline{J}_n = q\mu_n n\left(\underline{E} + \frac{1}{n}\underline{\nabla}(nK_BT/q)\right),$$
$$\tau_p \frac{\partial \underline{J}_p}{\partial t} + \underline{J}_p = q\mu_p p\left(\underline{E} - \frac{1}{p}\underline{\nabla}(pK_BT/q)\right),$$

where K_B [$\mathrm{V\,A\,s\,K^{-1}}$] is the Boltzmann constant and T [K] denotes the lattice temperature. Given that the relaxation times are "very small" we can proceed with an expansion where the singular perturbation parameters are the relaxation times

$$\underline{J}_n(\tau_n) = \sum_{j=0}^{\infty} \underline{J}_{nj}\tau_n^j, \qquad \underline{J}_p(\tau_p) = \sum_{j=0}^{\infty} \underline{J}_{pj}\tau_p^j.$$

Truncating at the first term ($j = 0$), assuming that the temperature T of the crystal is constant, and assuming Einstein's relations

$$D_n = \mu_n \frac{K_BT}{q}, \qquad D_p = \mu_p \frac{K_BT}{q}, \tag{1.3}$$

for the diffusion coefficients of the carriers D_n, D_p [$\mathrm{cm^2\,s^{-1}}$], we obtain eventually the classical DD relation for $\underline{J}_n$, $\underline{J}_p$

$$\begin{aligned} \underline{J}_n &= q\mu_n n\underline{E} + qD_n\underline{\nabla}n = -q\mu_n n\underline{\nabla}\psi + qD_n\underline{\nabla}n, \\ \underline{J}_p &= q\mu_p p\underline{E} - qD_p\underline{\nabla}p = -(q\mu_p p\underline{\nabla}\psi + qD_p\underline{\nabla}p), \end{aligned} \tag{1.4}$$

in which we can recognize two different contributions to the current densities: a drift term proportional to $n\underline{E}$ ($p\underline{E}$), and a diffusion term proportional to $\underline{\nabla}n$ ($\underline{\nabla}p$).

Let us recall the Maxwell–Boltzmann statistics relating the carrier concentrations to the electrostatic potential and the quasi-Fermi levels φ_n, φ_p

$$n = n_i \exp\left(\frac{\psi - \varphi_n}{V_{th}}\right), \qquad p = n_i \exp\left(\frac{\varphi_p - \psi}{V_{th}}\right), \tag{1.5}$$

where $V_{th} = K_BT/q$ is the thermal voltage and n_i is the intrinsic concentration of the semiconductor (see, e.g., SZE [1981]). Substituting relations (1.5) into (1.4) we obtain the following alternative expression of the current densities:

$$\underline{J}_n = -q\mu_n n\underline{\nabla}\varphi_n, \qquad \underline{J}_p = -q\mu_p p\underline{\nabla}\varphi_p. \tag{1.6}$$

These can be interpreted as two drift currents where electron and holes are driven by the effective fields

$$\underline{E}_n = -\underline{\nabla}\varphi_n, \qquad \underline{E}_p = -\underline{\nabla}\varphi_p.$$

Looking back at (1.5), a new set of independent variables can be derived by setting

$$\rho_n = \exp\left(-\frac{\varphi_n}{V_{th}}\right), \qquad \rho_p = \exp\left(\frac{\varphi_p}{V_{th}}\right). \tag{1.7}$$

The new unknowns ρ_n and ρ_p have been first introduced by SLOTBOOM [1973], and therefore are usually denoted as the *Slotboom* variables associated with electron and hole densities n and p. Substituting (1.7) into (1.5) we get

$$n = n_i \rho_n \exp\left(\frac{\psi}{V_{th}}\right), \qquad p = n_i \rho_p \exp\left(-\frac{\psi}{V_{th}}\right), \tag{1.8}$$

and the current densities become

$$\underline{J}_n = q D_n n_i \exp\left(\frac{\psi}{V_{th}}\right) \underline{\nabla} \rho_n, \qquad \underline{J}_p = -q D_p n_i \exp\left(-\frac{\psi}{V_{th}}\right) \underline{\nabla} \rho_p. \tag{1.9}$$

These can be interpreted as two diffusion currents of the equivalent concentrations $n_i \rho_n$ and $n_i \rho_p$ with diffusion coefficients $D_n \exp(\frac{\psi}{V_{th}})$ and $D_p \exp(-\frac{\psi}{V_{th}})$, respectively.

Three sets of dependent variables have been introduced so far: the primitive variables (ψ, n, p), the set including the Slotboom variables (ψ, ρ_n, ρ_p) and the set comprising the potentials $(\psi, \varphi_n, \varphi_p)$. The first set is the most widely used in computations, although the three unknowns have different physical meaning and attain strongly varying numerical ranges. The second one is useful for analytical purposes, as will be seen in Section 2, since the current continuity equations $(1.1)_2$–$(1.1)_3$ become self-adjoint. However, the set (ψ, ρ_n, ρ_p) can be used in numerical simulation only for low-bias applications due to the enormous dynamic range required by the evaluation on the computer of the Slotboom variables. Compared with the two previous sets of variables, the set $(\psi, \varphi_n, \varphi_p)$ has the advantage of collecting physically homogeneous quantities which have the same order of magnitude, at the price of introducing an exponentially nonlinear diffusion coefficient in the current continuity equations. Moreover, strictly positive concentrations are a priori guaranteed due to (1.5). For further details on a comparison between the various sets of dependent variables (including other possible choices that have not been addressed here) we refer to SELBERHERR [1984], Section 5.2 and to POLAK, DEN HEIJER, SCHILDERS and MARKOWICH [1987].

1.2.1. Boundary conditions and physical modeling

In this section we provide the DD model with appropriate boundary conditions in the case of steady state problems. Let us assume that $\Omega \equiv \Omega_S \cup \Omega_O \subset \mathbb{R}^d$ $(d \leqslant 2)$ is an open bounded set. A two-dimensional example of domain for semiconductor simulation is shown in Fig. 1.1 where the cross-section of a MOS transistor is schematically represented.

The boundary of a semiconductor device can be subdivided in two disjoint parts $\partial\Omega_P$ and $\partial\Omega_A$, respectively. The first one is a physical boundary, e.g., the metal contacts where the external potentials are applied. The second one is an artificial boundary which separates several devices on the same chip by symmetry axes or internal interfaces, or adjacent parts of a same device but with very different properties, e.g., the oxide-semiconductor interfaces. Both parts of the device boundary can be subdivided into disjoint portions pertaining to the semiconductor and to the oxide respectively, so that we have $\partial\Omega_P \equiv \partial\Omega_{P,S} \cup \partial\Omega_{P,O}$ and $\partial\Omega_A \equiv \partial\Omega_{A,S} \cup \partial\Omega_{A,O} \cup \partial\Omega_I$, where $\partial\Omega_I$ denotes the oxide-semiconductor interface. In the example of Fig. 1.1 we have $\partial\Omega_{P,S} \equiv \overline{AB} \cup \overline{EF} \cup \overline{GH}$, $\partial\Omega_{P,O} \equiv \overline{CD}$, $\partial\Omega_{A,S} \equiv \overline{AH} \cup \overline{FG}$, $\partial\Omega_{A,O} \equiv \overline{BC} \cup \overline{DE}$

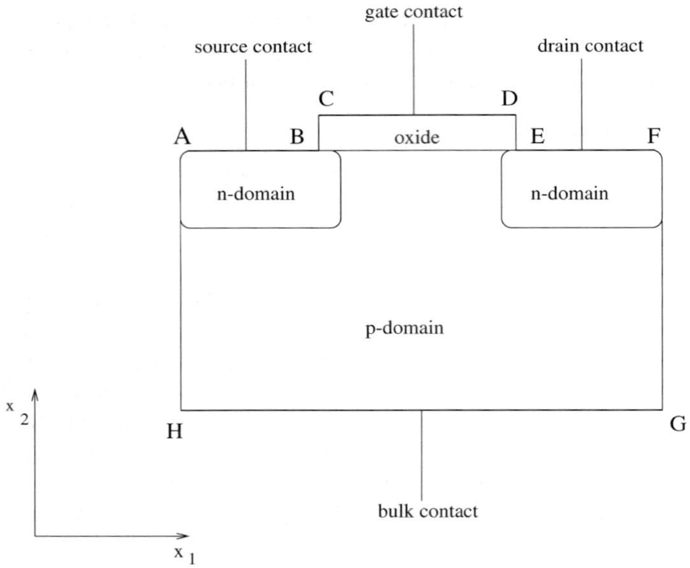

FIG. 1.1. Cross-section of a MOS transistor.

and $\partial\Omega_I \equiv \overline{BE}$. We shall consider Dirichlet–Neumann boundary conditions, typically nonhomogeneous Dirichlet conditions for (ψ, n, p) on the segments $\partial\Omega_P$ (ideal ohmic contacts) while on the remaining parts

we shall let the fluxes of the electric field and of the current densities vanish, i.e., homogeneous Neumann conditions for (ψ, n, p). For a more exhaustive survey of boundary conditions, see MARKOWICH [1986], SELBERHERR [1984], MOCK [1983a].

We start by describing the boundary conditions on $\partial\Omega_P$ and consider first the part $\partial\Omega_{P,S}$. This is made by ohmic contacts where external voltages are applied to electrically drive the semiconductor device. From a mathematical viewpoint, an ideal ohmic contact is a Dirichlet segment where thermodynamic equilibrium is assumed, i.e., the mass-action law

$$np_{|\partial\Omega_{P,S}} = n_i^2, \tag{1.10}$$

and the charge neutrality

$$\rho_{|\partial\Omega_{P,S}} = q(p - n + C)_{|\partial\Omega_{P,S}} = 0 \tag{1.11}$$

hold. By combining (1.10) and (1.11) and denoting by C_{oh} the restriction of the doping profile to the ohmic contact we obtain the following values n_D and p_D for the concentrations of the carriers

$$\begin{aligned} n_{|\partial\Omega_{P,S}} &= n_D := \frac{\sqrt{C_{oh}^2 + 4n_i^2} + C_{oh}}{2}, \\ p_{|\partial\Omega_{P,S}} &= p_D := \frac{\sqrt{C_{oh}^2 + 4n_i^2} - C_{oh}}{2}. \end{aligned} \tag{1.12}$$

Concerning the boundary conditions for the electrostatic potential ψ, the value of ψ at the contacts is obtained by summing the external applied voltage V_{ext} with the so-called built-in potential between the metal and the semiconductor

$$\psi_{|\partial\Omega_{P,S}} = \psi_{D|\partial\Omega_{P,S}} := V_{ext|\partial\Omega_{P,S}} + \psi_{bi|\partial\Omega_{P,S}}. \tag{1.13}$$

The built-in potential ψ_{bi} is computed in order that the semiconductor is in thermal equilibrium when all the applied external voltages are zero. This amounts to assuming that the energy levels of the semiconductor in the neighborhood of the contact are horizontal (flat band approximation).

From the Maxwell–Boltzmann statistics (1.5) we obtain the following relations for the built-in potential (see (1.11))

$$n_i \exp\left(-\frac{\psi_{bi}}{V_{th}}\right) - n_i \exp\left(\frac{\psi_{bi}}{V_{th}}\right) + C_{oh} = 0,$$

from which it follows

$$\psi_{bi|\partial\Omega_{P,S}} = V_{th} \sinh^{-1}\left(\frac{C_{oh}}{2n_i}\right) = V_{th} \sinh^{-1}\left(\left.\frac{N_D^+ - N_A^-}{2n_i}\right|_{\partial\Omega_{P,S}}\right). \tag{1.14}$$

In the case when one of the two dopant species is dominant over the other, (1.14) simplifies into

$$\psi_{bi|\partial\Omega_{P,S}} \simeq V_{th} \ln\left(\frac{N_D^+|_{\partial\Omega_{P,S}}}{n_i}\right), \qquad N_D^+ \gg N_A^-,$$

$$\psi_{bi|\partial\Omega_{P,S}} \simeq -V_{th} \ln\left(\frac{N_A^-|_{\partial\Omega_{P,S}}}{n_i}\right), \qquad N_A^- \gg N_D^+.$$

Let us now consider the boundary conditions on the part $\partial\Omega_{P,O}$. This is made by gate contacts which are located over the oxide region Ω_O and where external voltages are applied to control the current flow between the input-output contacts of the MOS transistor. Since $n = p = 0$ within the oxide, only a boundary condition for the electrostatic potential ψ can be prescribed on $\partial\Omega_{P,O}$ and reads

$$\psi_{|\partial\Omega_{P,O}} = \psi_{D|\partial\Omega_{P,O}} := V_{ext|\partial\Omega_{P,O}} - \Phi_{ms}, \tag{1.15}$$

$\Phi_{ms} = \Phi_m - \Phi_s$ being the metal-semiconductor work function difference, referred to an intrinsic semiconductor, i.e., such that

$$\Phi_s = \chi + \frac{E_c - E_i}{q},$$

χ being the semiconductor affinity, E_c the conduction-band edge, and E_i the intrinsic Fermi level within the semiconductor (see BACCARANI, RUDAN, GUERRIERI and CIAMPOLINI [1986]).

As for the Neumann data, we assume homogeneous conditions on $\partial\Omega_{A,S} \cup \partial\Omega_{A,O}$, i.e., vanishing fluxes of the electric field and of the current densities

$$\left.\frac{\partial\psi}{\partial n}\right|_{\partial\Omega_{A,S}} \equiv \left.\frac{\partial\psi}{\partial n}\right|_{\partial\Omega_{A,O}} \equiv \underline{J}_n \cdot \underline{n}|_{\partial\Omega_{A,S}} \equiv \underline{J}_p \cdot \underline{n}|_{\partial\Omega_{A,S}} \equiv 0, \tag{1.16}$$

where $\underline{n}$ is the unit outward normal vector to the boundaries.

The so-called material interfaces deserve a special treatment. We denote them by $\partial\Omega_I$, e.g., the interface between the oxide and the semiconductor, which in the example of Fig. 1.1 is $\partial\Omega_I \equiv \overline{BE}$. In this case the Neumann condition for the electrical potential is a consequence of the Gauss law. Assuming for simplicity that the surface charge density is zero and letting $[\cdot]$ denote the jump function, and ε_{ox}, ε_{sem} the dielectric constants of the oxide and of the semiconductor, respectively, we have

$$[\psi]_{\partial\Omega_I} = 0, \quad \left[\varepsilon \frac{\partial \psi}{\partial n}\right]_{\partial\Omega_I} = 0, \quad \varepsilon(x) \equiv \begin{cases} \varepsilon_{sem}, & x \in \Omega_S, \\ \varepsilon_{ox}, & x \in \Omega_O. \end{cases} \tag{1.17}$$

In order to complete the description of system (1.1), we provide some constitutive relations for the function R and the mobilities μ_n, μ_p. We recall that in thermal equilibrium the mass-action law (1.10) holds. When the device works under nonequilibrium conditions, e.g., when some external voltages are applied, the carrier concentrations move away from their equilibrium values. In this case some recombination/generation mechanisms arise in order to bring the system back to equilibrium. The mechanisms typically considered in the simulations are the Shockley-Hall-Read, Auger, and Impact Ionization phenomena (for a physical description, see, e.g., SZE [1981]). The mathematical expressions of these mechanisms are

$$R_{SHR} = \frac{pn - n_i^2}{\tau_n^*(p + n_i) + \tau_p^*(n + n_i)}, \tag{1.18}$$

$$R_{AU} = (pn - n_i^2)(C_n n + C_p p), \tag{1.19}$$

$$R_{II} = -\big(\alpha_n(|\underline{E}|)|\underline{J}_n|/q + \alpha_p(|\underline{E}|)|\underline{J}_p|/q\big), \tag{1.20}$$

where τ_n^*, τ_p^* are the carrier lifetimes, C_n, C_p are the Auger coefficients, and α_n, α_p are the ionization rates for electrons and holes, respectively. The recombination/generation rate is eventually computed as

$$R = R_{SHR} + R_{AU} + R_{II}.$$

As for the carrier mobilities, the modeling has to take into account that, for high electric field, the drift velocities ($|v_{n,p}| = \mu_{n,p}|\underline{\nabla}\psi|$) saturate. A commonly employed model is the one suggested by CAUGHEY and THOMAS [1967]

$$\mu_{n,p} = \left(\frac{\mu_{n,p}^0}{1 + \left(\frac{\mu_{n,p}^0|\underline{\nabla}\psi|}{v_{n,p}^{sat}}\right)^{\beta_{n,p}}}\right)^{1/\beta_{n,p}}, \tag{1.21}$$

where $v_{n,p}^{sat}$ are the carrier saturation velocities, $\beta_n = 2$, $\beta_p = 1$ and $\mu_{n,p}^0$ are the low-field mobilities. For the details about the modeling of all the above coefficients we refer to SELBERHERR [1984], Chapter 41.

The analysis of the multi-dimensional boundary-value problem in steady-state is not trivial because of the presence of mixed boundary conditions. Actually, this is a situation in which gradient singularities may occur at the boundary transition points. This issue has been addressed in the literature for the case of linear elliptic equations (see

WIGLEY [1970], KELLOGG [1972], AZZAM and KREYSZIG [1982], GRISVARD [1985] for asymptotic expansions and MURTHY and STAMPACCHIA [1972] for gradient integrability). A study of asymptotic behavior in the two-dimensional case has been carried out in GAMBA [1993], where the case of a singularity at a corner formed by the oxide region of a MOS device is also considered. As for the transient model, some developments can be found in COUGHRAN and JEROME [1990], JEROME [1987].

The first mathematical study of the DD model is due to MOCK [1972], who dealt with the steady-state case. He used a decoupling map acting on the potential (the quasi-Fermi levels are computed as an intermediate step). Since then, many variants of such an approach have been introduced in the literature, all defined by an appropriate decoupling procedure, and they are referred to as Gummel fixed point maps. Complicate geometries or parameter models affect the structure of the fixed point map. An example of such map will be discussed in Section 2. We refer the reader to the literature for existence results (SEIDMAN [1980], MOCK [1983a], MARKOWICH [1984], GAJEWSKI [1985], JEROME [1985], MARKOWICH [1986]). Global uniqueness of the solution of the DD system cannot be expected in the general case, since there are devices, such as thyristors, whose performance is based explicitly on the existence of multiple steady-state solutions (SZE [1981]). However, there are uniqueness results close to thermal equilibrium (MOCK [1983a], MARKOWICH [1986]). For the analysis of the transient problem we refer to MOCK [1983a].

1.3. Scaling

The physical quantities in system (1.1) have different physical dimensions and, in order to compare their orders of magnitude, these quantities have to be made dimensionless first by appropriate scalings. We introduce for system (1.1) two closely related scalings, and we shall refer to these scalings as the De Mari and the Unit scalings.

1. De Mari scaling (see DEMARI [1968]):
   - Potentials scaled by V_{th};
   - Concentrations scaled by the intrinsic concentration n_i;
   - Length scaled by a characteristic Debye length $L_D = \sqrt{\varepsilon V_{th}/(qn_i)}$.
2. Unit scaling (see SELBERHERR [1984] and MARKOWICH [1986]):
   - Potentials scaled by V_{th};
   - Concentrations scaled by $\overline{C} = \sup_{x\in\Omega} |C(x)|$;
   - Length scaled by a characteristic device dimension l.

After any of the above scalings, the scaled dimensionless DD system reads

$$\begin{cases} \operatorname{div}(\lambda^2 \underline{E}) = p - n + C(x), \\ \frac{\partial n}{\partial t} - \operatorname{div} \underline{J}_n = -R, \\ \frac{\partial p}{\partial t} + \operatorname{div} \underline{J}_p = -R, \\ \underline{E} = -\underline{\nabla}\psi, \\ \underline{J}_n = \mu_n(\underline{\nabla} n - n\underline{\nabla}\psi), \\ \underline{J}_p = -\mu_p(\underline{\nabla} p + p\underline{\nabla}\psi), \end{cases} \tag{1.22}$$

TABLE 1.1
De Mari and Unit scaling factors

Quantity	De Mari factor	De Mari factor value	Unit factor	Unit factor value
ψ	V_{th}	0.0258 V	V_{th}	0.0258 V
n, p	n_i	1.482×10^{10} cm^{-3}	$\overline{C}$	10^{18} cm^{-3}
x	L_D	3.357×10^{-3} cm	l	10^{-4} cm
μ_n, μ_p	$D_0 V_{th}^{-1}$	38.68 cm^2 V^{-1} s^{-1}	$\overline{\mu}$	1000 cm^2 V^{-1} s^{-1}
D_n, D_p	D_0	1 cm^2 s^{-1}	$\overline{\mu} V_{th}$	25.8 cm^2 s^{-1}
$\underline{J}_n, \underline{J}_p$	$q D_0 n_i / L_D$	0.71×10^{-6} A cm^{-2}	$q V_{th} \overline{\mu} \overline{C} / l$	4.13×10^6 A cm^{-2}
R	$D_0 n_i / L_D^2$	1.314×10^{15} cm^{-3} s^{-1}	$V_{th} \overline{\mu} \overline{C} / l^2$	2.58×10^{29} cm^{-3} s^{-1}
t	L_D^2 / D_0	1.127×10^{-5} s	$l^2 / (V_{th} \overline{\mu})$	4×10^{-10} s

where for simplicity we used the same unscaled symbols for the variables, the mobility coefficients and the recombination/generation term. For either scalings we have

$$\lambda^2 = \frac{\varepsilon V_{th}}{l_{scal}^2 q C_{scal}}, \qquad l_{scal} = L_D \text{ or } l, \qquad C_{scal} = n_i \text{ or } \overline{C}. \tag{1.23}$$

The entire list of scaling factors used to deduce (1.22) is reported in Table 1.1, where $\overline{\mu}$ denotes the maximum mobility. Moreover, numerical values of the physical quantities in the case of Silicon at 300 K are given.

In the case of the De Mari scaling $\lambda^2 = 1$, whereas in the case of the Unit scaling, $\lambda^2 \simeq 10^{-1} \div 10^{-7}$ (for instance, with the choices of Table 1.1 we have $\lambda^2 \simeq 10^{-6}$). With the De Mari scaling the doping is of the order $10^7 \div 10^{10}$ and the carrier concentrations are expected to have these values too. In the Unit scaling all the concentrations are expected to be maximally of order 1. The scaled Debye length λ acts as a singular perturbation parameter and the behavior of the solution of (1.22) as $\lambda \to 0^+$ (quasi-neutral limit) can be analyzed. We refer to MARKOWICH [1984], ALABEAU [1984], MARKOWICH and RINGHOFER [1984], HENRI and LOURO [1989], MARKOWICH and SCHMEISER [1986] for studies in the stationary case and to MARKOWICH [1986], MARKOWICH, RINGHOFER and SCHMEISER [1990] for an overview. The study of the quasi-neutral limit in the transient case is more intricate and still a subject of active research (see Ringhofer [1987a, 1987b], GASSER, HSIAO, MARKOWICH and WANG [2002], GASSER, LEVERMORE, MARKOWICH and SCHMEISER [2001], GASSER [2001], e.g.).

A different form of scaling is more appropriate for pn-junctions (which are the boundaries between p-regions and n-regions) under extreme reverse biasing conditions. In that case a so-called depletion region forms about the junction, where very few carriers exist and $n, p \simeq 0$ holds. By changing the scaling factor of the potential to $ql^2\overline{C}/\varepsilon$, the singular perturbation parameter (still the scaled Debye length λ) appears in the current continuity equations, rather than in the Poisson equation, and in the depletion region $\Delta\psi \simeq -C$ holds. Consequently, the depletion region does not disappear in the limit when $\lambda \to 0$. Actually, the limiting problem is a free boundary problem where the free boundaries coincide with the edges of the depletion region. We refer to

BREZZI and GASTALDI [1986], BREZZI, CAPELO and GASTALDI [1989], CAFFARELLI and FRIEDMAN [1987], SCHMEISER [1989, 1990], MONTARNAL and PERTHAME [1997] and to MARKOWICH, RINGHOFER and SCHMEISER [1990] for an overview.

1.4. The Energy-Transport model

As already pointed out in Section 1.1, the DD model provides a good description of the electrical behavior of the semiconductor devices only close to thermal equilibrium, but it is not accurate enough for sub-micron device modeling, owing to the rapidly changing fields and temperature effects. In the last years, various models, either of kinetic or of macroscopic type, have been derived in order to improve the physical description of the transport in semiconductor devices. The semiconductor Boltzmann equation gives quite accurate simulation results, but the numerical methods to solve this equation (for instance, with Monte-Carlo methods) are too costly and time consuming to model real problems in semiconductor production mode where simulation results are needed in hours or minutes. Extended drift-diffusion models able to describe temperature effects in sub-micron devices are the so-called Energy-Transport (or Energy-Balance) models. They consist of the conservation laws of mass and energy, together with constitutive relations for the particle and energy currents.

The first Energy-Transport model has been presented by STRATTON [1962]. In the physical literature, Energy-Transport equations have been derived from Hydrodynamic (HD) models usually by neglecting the inertia terms in the momentum transport equation (see, e.g., RUDAN, GNUDI and QUADE [1993], SOUISSI, ODEH, TANG and GNUDI [1994] and references therein). This approach can be made mathematically rigorous by considering a diffusion time scaling (GASSER and NATALINI [1999]). For HD models, which can be regarded as the Euler equations of gas and fluid dynamics for a gas of charged and colliding particles in an electric field, we refer to Chapter 5 where they are extensively discussed.

Another approach is to derive the Energy-Transport model from the semiconductor Boltzmann equation in the diffusive limit, by means of the Hilbert expansion method (BEN ABDALLAH and DEGOND [1996], BEN ABDALLAH, DEGOND and GÉNIEYS [1996]). In this derivation, the dominant scattering mechanisms are assumed to be electron-electron and elastic electron-phonon scattering. The Energy-Transport model for electrons reads as follows:

$$\begin{cases} \operatorname{div}(\varepsilon \underline{E}) = q(C-n), \\ q\frac{\partial n}{\partial t} - \operatorname{div} \underline{J}_n = 0, \\ \frac{\partial U}{\partial t} - \operatorname{div} \underline{S}_n = \underline{E}\cdot \underline{J}_n + W(n, T_n), \\ \underline{E} = -\underline{\nabla}\psi, \\ \underline{J}_n = L_{11}\left(\frac{\underline{\nabla} n}{n} - \frac{q\underline{\nabla}\psi}{K_B T_n}\right) + \left(\frac{L_{12}}{K_B T_n} - \frac{3}{2}L_{11}\right)\frac{\underline{\nabla} T_n}{T_n}, \\ q\underline{S}_n = L_{21}\left(\frac{\underline{\nabla} n}{n} - \frac{q\underline{\nabla}\psi}{K_B T_n}\right) + \left(\frac{L_{22}}{K_B T_n} - \frac{3}{2}L_{21}\right)\frac{\underline{\nabla} T_n}{T_n}. \end{cases} \tag{1.24}$$

The variables are the electron density n, the electron temperature T_n, and the electrostatic potential ψ. Furthermore, $U = \frac{3}{2} n K_B T_n$ is the internal energy, $\underline{J}_n$, $\underline{S}_n$ are the particle current and energy flux densities, respectively. The diffusion coefficients $L_{ij} = L_{ij}(n, T_n)$ and the energy relaxation term $W(n, T_n)$ are non-linear functions of n and T_n, depending on the physical assumptions (distribution function *ansatz*, energy band diagram and so on). For instance, if we assume the parabolic band approximation, non-degenerate Boltzmann statistics (the distribution function being a Maxwellian) and a special form of momentum relaxation time, the diffusion matrix takes the form

$$L = (L_{ij}) = \mu_n K_B T^{eq} n \begin{pmatrix} 1 & \frac{3}{2} K_B T_n \\ \frac{3}{2} K_B T_n & \frac{15}{4} (K_B T_n)^2 \end{pmatrix}, \tag{1.25}$$

and the energy relaxation term is given by

$$W(n, T_n) = -\frac{3}{2} n K_B (T_n - T^{eq}) / \tau_{w_n}, \tag{1.26}$$

where T^{eq} denotes the lattice temperature (assumed to be constant) and τ_{w_n} denotes the energy relaxation time. This corresponds to the model introduced in CHEN, KAN, RAVAIOLI, SHU and DUTTON [1992]. More general diffusion coefficients, including non-parabolic energy band cases, are computed in BEN ABDALLAH and DEGOND [1996], DEGOND, JÜNGEL and PIETRA [2000] and corresponding numerical simulations are shown in Section 7.4.

When the diffusion matrix is taken as

$$L = (L_{ij}) = \mu_n K_B T_n n \begin{pmatrix} 1 & \frac{5}{2} K_B T_n \\ \frac{5}{2} K_B T_n & (\frac{25}{4} + \kappa_0)(K_B T_n)^2 \end{pmatrix}, \tag{1.27}$$

with $\kappa_0 > 0$, system (1.24) corresponds to the model known in the literature as Energy-Balance model. It reads

$$\begin{cases} \operatorname{div}(\varepsilon \underline{E}) = q(C - n), \\ q \frac{\partial n}{\partial t} - \operatorname{div} \underline{J}_n = 0, \\ \frac{\partial U}{\partial t} - \operatorname{div} \underline{S}_n = \underline{E} \cdot \underline{J}_n - \frac{3}{2} n K_B \frac{(T_n - T^{eq})}{\tau_{w_n}}, \\ \underline{E} = -\underline{\nabla} \psi, \\ \underline{J}_n = q D_n \underline{\nabla} n - q \mu_n n \underline{\nabla} (\psi - \frac{K_B T_n}{q}), \\ \underline{S}_n = \kappa_n \underline{\nabla} T_n + \frac{5}{2} \frac{K_B T_n}{q} \underline{J}_n, \end{cases} \tag{1.28}$$

where D_n is the non-constant diffusion coefficient given by $D_n = \mu_n K_B T_n q^{-1}$, and κ_n is the non-constant heat conductivity coefficient corresponding to the Wiedemann–Franz law $\kappa_n = \kappa_0 K_B^2 \mu_n q^{-1} n T_n$.

We complement the equations with physically motivated mixed Dirichlet–Neumann boundary conditions

$$n = n_D, \ T_n = T_D, \ \psi = \psi_D \quad \text{on } \Gamma_D, \tag{1.29}$$

$$\underline{J}_n \cdot \underline{n} = \underline{S}_n \cdot \underline{n} = \underline{\nabla} \psi \cdot \underline{n} = 0 \quad \text{on } \Gamma_N, \tag{1.30}$$

modeling the contacts Γ_D and the insulating boundary parts Γ_N. We have assumed that $\partial\Omega = \Gamma_D \cup \Gamma_N$ and $\Gamma_D \cap \Gamma_N = \emptyset$. The boundary conditions are usually taken as for the DD model: at ohmic contacts, n_D is defined by (1.12), ψ_D is given by (1.13) and $T_D = T^{eq}$. Moreover, initial conditions for n and T_n are prescribed.

We point out that these models reduce to the classical DD model under the simplifying assumption of vanishing relaxation time (i.e., $\tau_{w_n} \to 0$) which amounts to considering the carriers in thermal equilibrium with the crystal lattice, $T_n = T^{eq}$, yielding

$$\underline{J}_n = q D_n \underline{\nabla} n - q\mu_n n \underline{\nabla}\psi. \tag{1.31}$$

The mathematical analysis of Eqs. (1.24) has been recently carried out in DEGOND, GÉNIEYS and JÜNGEL [1997], DEGOND, GÉNIEYS and JÜNGEL [1998], under the assumption of uniformly bounded diffusion coefficients. The existence and uniqueness of weak solutions to both the stationary and the time-dependent (initial) boundary-value problems have been proved. For an overview, we refer also to JÜNGEL [2001]. Existence results under different assumptions (for instance, near-equilibrium situations) have been shown in ALLEGRETTO and XIE [1994], GRIEPENTROG [1999], JEROME and SHU [1996].

2. A nonlinear block iterative solution of the semiconductor device equations: the Gummel map

In this section we introduce the so-called *Gummel map*, a nonlinear block iterative algorithm that splits the DD semiconductor device equations into the successive solution of a nonlinear Poisson equation for the electric potential ψ and two linearized continuity equations for the electron and hole densities n and p.

The basic form of the iterative map was first proposed by GUMMEL [1964]. Since then, the map has become the most established approach in computer simulations due to its good convergence properties (even in case of a badly chosen initial guess), ease of implementation, and reduced computational effort (compared, for instance, to a fully coupled solution of the DD system using Newton's method).

Due to its successful use in semiconductor device simulation, a considerable amount of mathematical work has been carried out to analyze the convergence of Gummel's iteration (see, e.g., the papers KERKHOVEN [1986, 1988] and the books by MARKOWICH [1986], MARKOWICH, RINGHOFER and SCHMEISER [1990] and JEROME [1996]). In the following, we review the basic formalism and notation needed for a mathematically consistent presentation of the algorithm.

With this aim, the scaled DD system is formulated in Section 2.1 in terms of the Slotboom variables introduced in Section 1.1. Using these new unknowns, the linearized carrier continuity equations assume a self-adjoint form which is convenient in view of the analysis.

Sections 2.2 and 2.3 are devoted to the presentation and discussion of the Gummel map as a fixed-point iteration. In Section 2.2 the Gummel map is used as an abstract tool for constructively proving the existence of a (weak) solution to the DD system in the stationary case. In Section 2.3 it is shown how to utilize the Gummel map as an iteration scheme for the approximate computation of the solution of the semiconductor device problem. In particular, we summarize the main theoretical convergence results for the iteration and briefly address the delicate matter of its acceleration, referring for more details on this subject to Chapter 7 of this book.

Finally, in Section 2.4 the differential subproblems involved in the decoupled iterative solution of the DD equations are cast into the unified framework of a reaction-diffusion model problem. This will be the object of the discretization techniques discussed in the forthcoming sections of this chapter.

2.1. *The drift-diffusion equations in self-adjoint form*

We will use henceforth the Unit scaling introduced in Section 1.3. Under this assumption the scaled Slotboom variables are

$$\rho_n = e^{-\varphi_n}, \qquad \rho_p = e^{\varphi_p}, \tag{2.1}$$

while the scaled Maxwell–Boltzmann statistics and the current densities become

$$n = \delta^2 \rho_n e^{\psi}, \qquad p = \delta^2 \rho_p e^{-\psi}, \tag{2.2}$$

and

$$\underline{J}_n = \delta^2 \mu_n e^{\psi} \underline{\nabla} \rho_n, \qquad \underline{J}_p = -\delta^2 \mu_p e^{-\psi} \underline{\nabla} \rho_p, \tag{2.3}$$

where $\delta^2 = n_i/\overline{C}$. Notice that in thermal equilibrium conditions (i.e., zero external applied biases) $\varphi_n = \varphi_p = 0$ and correspondingly $\rho_n = \rho_p = 1$ which clearly implies $\underline{J}_n = \underline{J}_p = \underline{0}$. Moreover, by definition ρ_n and ρ_p are strictly positive quantities.

Using the triplet (ψ, ρ_n, ρ_p), and assuming for simplicity that $\Omega_O = \emptyset$, the DD system (1.22), can be written in $Q_S \equiv Q$ as

$$\begin{cases} -\operatorname{div}(\lambda^2 \underline{E}) = \delta^2 \rho_p e^{-\psi} - \delta^2 \rho_n e^{\psi} + C(x), \\ \delta^2 \frac{\partial(\rho_n e^{\psi})}{\partial t} - \operatorname{div} \underline{J}_n = -R, \\ \delta^2 \frac{\partial(\rho_p e^{-\psi})}{\partial t} + \operatorname{div} \underline{J}_p = -R, \\ \underline{E} = -\underline{\nabla}\, \psi, \\ \underline{J}_n = \delta^2 \mu_n e^{\psi} \underline{\nabla} \rho_n, \\ \underline{J}_p = -\delta^2 \mu_p e^{-\psi} \underline{\nabla} \rho_p, \\ \psi = \psi_D,\ \rho_n = \rho_{nD},\ \rho_p = \rho_{pD} \quad \text{on } \Gamma_D, \\ \underline{E} \cdot \underline{n} = \underline{\nabla} \rho_n \cdot \underline{n} = \underline{\nabla} \rho_p \cdot \underline{n} = 0 \quad \text{on } \Gamma_N, \end{cases} \tag{2.4}$$

where Γ_D and Γ_N denote the Dirichlet and Neumann portions of the boundary, respectively, with $\partial\Omega =: \Gamma = \Gamma_D \cup \Gamma_N$, and the strictly positive Dirichlet boundary data ρ_{nD} and ρ_{pD} are related to the corresponding data for n and p through (1.8) as

$$n_D = \delta^2 \rho_{nD} e^{\psi_D}, \qquad p_D = \delta^2 \rho_{pD} e^{-\psi_D}, \tag{2.5}$$

where n_D, p_D are computed according to (1.12). The advantage of working with the Slotboom variables is that the spatial part of the continuity equations is, for given R, in self-adjoint form, which greatly facilitates the mathematical analysis. On the other hand, for given ρ_n and ρ_p the Poisson equation becomes nonlinear in the unknown ψ, so that a suitable linearization technique (for example, Newton's method) is required to carry out its abstract analysis as well as its discretization.

2.2. *The Gummel map in the stationary case*

In this section we introduce the Gummel map for the solution of the semiconductor device equations. Without loss of generality, only the stationary case will be considered since a semidiscretization in time (using for instance the Backward Euler method) modifies the continuity equations by the addition of a zero-order term proportional to the reciprocal of the time-step Δt in the right and left-hand sides. The presentation follows the guidelines of MARKOWICH [1986], Chapter 3, and JEROME [1996], Chapter 4, to which we refer the interested reader for more details and for the proofs of the results. We also refer to Section 3.1 for the definition of the functional spaces that will be used in the sequel.

Assume that there exist two positive constants $\rho_{\min}$, $\rho_{\max}$ such that

$$0 < \rho_{\min} \leqslant \rho_n(x)_{|\Gamma_D}, \quad \rho_p(x)_{|\Gamma_D} \leqslant \rho_{\max}, \quad \forall x \in \Gamma_D,$$

and let

$$\mathcal{K} = \max\{|\log(\rho_{\min})|, \ |\log(\rho_{\max})|\}.$$

Let $(\rho_n^{(0)}, \rho_p^{(0)}) \in (L^\infty(\Omega))^2$ be a given pair of positive essentially bounded functions such that

$$e^{-\mathcal{K}} \leqslant \rho_n^{(0)}(x), \quad \rho_p^{(0)}(x) \leqslant e^{\mathcal{K}} \quad \text{a.e. in } \Omega. \tag{2.6}$$

The Gummel abstract fixed point iteration consists of:

(1) solving the semilinear Poisson equation

$$\begin{cases} -\operatorname{div}(\lambda^2 \underline{\nabla}\, \psi) = \delta^2 \rho_p^{(0)} e^{-\psi} - \delta^2 \rho_n^{(0)} e^{\psi} + C(x) & \text{in } \Omega, \\ \psi = \psi_D \text{ on } \Gamma_D, \quad \underline{\nabla}\, \psi \cdot \underline{n} = 0 \text{ on } \Gamma_N, \end{cases} \tag{2.7}$$

for $\psi = \psi^{(1)}$;

(2) solving the two decoupled continuity equations

$$\begin{cases} -\operatorname{div}(\mu_n e^{\psi^{(1)}} \underline{\nabla}\, \rho_n) = -R(x, \psi^{(1)}, \rho_n, \rho_p^{(0)}) & \text{in } \Omega, \\ \rho_n = \rho_{n\,D} \text{ on } \Gamma_D, \quad \underline{\nabla}\, \rho_n \cdot \underline{n} = 0 \text{ on } \Gamma_N, \end{cases} \tag{2.8}$$

for $\rho_n = \rho_n^{(1)}$, and

$$\begin{cases} -\operatorname{div}(\mu_p e^{-\psi^{(1)}} \underline{\nabla}\, \rho_p) = -R(x, \psi^{(1)}, \rho_n^{(1)}, \rho_p) & \text{in } \Omega, \\ \rho_p = \rho_{p\,D} \text{ on } \Gamma_D, \quad \underline{\nabla}\, \rho_p \cdot \underline{n} = 0 \text{ on } \Gamma_N, \end{cases} \tag{2.9}$$

for $\rho_p = \rho_p^{(1)}$.

Let

$$N = \{(u, v) \in (L^2(\Omega))^2 \mid e^{-\mathcal{K}} \leqslant u(x), \ v(x) \leqslant e^{\mathcal{K}} \text{ a.e. in } \Omega\}.$$

It can be checked that $\rho_n^{(1)}, \rho_p^{(1)} \in N$. Steps (1)–(2) implicitly define an operator

$$T : N \to N$$

such that

$$T\big(\rho_n^{(0)}, \rho_p^{(0)}\big) = \big(\rho_n^{(1)}, \rho_p^{(1)}\big).$$

Under suitable assumptions on the doping profile C, on the mobilities μ_n and μ_p, and on the net recombination rate R (see MARKOWICH [1986], p. 34), it can be shown that T admits a fixed point $(\rho_n^*, \rho_p^*) \in (H^1(\Omega) \cap L^\infty(\Omega))^2$ with $\rho_n^*, \rho_p^* \in N$. This in turn allows us to prove that there exists a solution $w^* = (\psi^*, \rho_n^*, \rho_p^*) \in (H^1(\Omega) \cap L^\infty(\Omega))^3$ of the DD system (2.4). Explicit bounds for ψ^* can be found in MARKOWICH [1986]. The decoupling abstract procedure described above is the basic Gummel iteration that is commonly employed in semiconductor device simulation. Its actual implementation is the object of the next section.

2.3. *A fixed-point iteration for the approximate solution of the drift-diffusion system*

In the following we use the Gummel fixed-point iteration introduced in the previous section to construct a sequence of approximate solutions of the semiconductor DD system (2.4) in the stationary case. For ease of presentation the algorithm will be described "on the continuous level", although it can be applied to discrete schemes or, conversely, the "continuous" iterative scheme can be discretized appropriately (see, e.g., BANK, ROSE and FICHTNER [1983] or JEROME [1996], Chapter 5).

Let $k \geqslant 0$ be a fixed integer; for given $(\psi^{(k)}, \rho_n^{(k)}, \rho_p^{(k)})$ the Gummel map consists of solving successively the following boundary-value subproblems:

$$\begin{cases} -\operatorname{div}(\lambda^2 \underline{\nabla}\, \psi) = \delta^2 \rho_p^{(k)} e^{-\psi} - \delta^2 \rho_n^{(k)} e^{\psi} + C(x) & \text{in } \Omega, \\ \psi = \psi_D \text{ on } \Gamma_D, \quad \underline{\nabla}\, \psi \cdot \underline{n} = 0 \text{ on } \Gamma_N, \end{cases} \tag{2.10}$$

for $\psi = \psi^{(k+1)}$;

$$\begin{cases} -\operatorname{div}(\mu_n e^{\psi^{(k+1)}} \underline{\nabla}\, \rho_n) = -R(x, \psi^{(k+1)}, \rho_n, \rho_p^{(k)}) & \text{in } \Omega, \\ \rho_n = \rho_{n\,D} \text{ on } \Gamma_D, \quad \underline{\nabla}\, \rho_n \cdot \underline{n} = 0 \text{ on } \Gamma_N, \end{cases} \tag{2.11}$$

for $\rho_n = \rho_n^{(k+1)}$, and

$$\begin{cases} -\operatorname{div}(\mu_p e^{-\psi^{(k+1)}} \underline{\nabla}\, \rho_p) = -R(x, \psi^{(k+1)}, \rho_n^{(k+1)}, \rho_p) & \text{in } \Omega, \\ \rho_p = \rho_{p\,D} \text{ on } \Gamma_D, \quad \underline{\nabla}\, \rho_p \cdot \underline{n} = 0 \text{ on } \Gamma_N, \end{cases} \tag{2.12}$$

for $\rho_p = \rho_p^{(k+1)}$. Steps (2.10)–(2.12) can be simply written as $T(\rho_n^{(k)}, \rho_p^{(k)}) = (\rho_n^{(k+1)}, \rho_p^{(k+1)})$. We prefer to use the triplet $(\psi^{(k)}, \rho_n^{(k)}, \rho_p^{(k)})$, instead of the pair $(\rho_n^{(k)}, \rho_p^{(k)})$, in order to emphasize the fact that $\psi^{(k)}$ is used as an initial guess for solving the nonlinear problem (2.10).

Looking at the structure of the Gummel map, we recognize a nonlinear block Gauss–Seidel iteration which can be subdivided into two main loops: (i) a nonlinear inner iteration for solving the semilinear Poisson equation (2.10); (ii) two iterations for solving the continuity equations (2.11) and (2.12), which will be suitably linearized in the following. Typically, a damped Newton method is used for dealing with (i) (see BANK and ROSE [1981]); this leads to the following inner nonlinear iteration:

(A) set $u^{(0)} = \psi^{(k)}$;

then, for $j = 0, 1, \ldots$ until convergence execute the following steps (B)–(D):

(B) set

$$n^{(j)} = \delta^2 \rho_n^{(k)} e^{u^{(j)}}, \qquad p^{(j)} = \delta^2 \rho_p^{(k)} e^{-u^{(j)}};$$

(C) solve

$$\begin{cases} -\operatorname{div}(\lambda^2 \underline{\nabla} \phi) + (n^{(j)} + p^{(j)})\phi = -\mathcal{R}_\psi(u^{(j)}) \quad \text{in } \Omega, \\ \phi = 0 \text{ on } \Gamma_D, \quad \underline{\nabla}\phi \cdot \underline{n} = 0 \text{ on } \Gamma_N, \end{cases} \tag{2.13}$$

(D) and set

$$u^{(j+1)} = u^{(j)} + t_j \phi, \quad t_j \in (0, 1].$$

The right-hand side in (2.13) is the residual of the Poisson equation at the jth step of Newton's iteration and is defined as

$$\mathcal{R}_\psi(V) = -\operatorname{div}(\lambda^2 \underline{\nabla} V) + \delta^2 \rho_n^{(k)} e^V - \delta^2 \rho_p^{(k)} e^{-V} - C(x).$$

The damping coefficients t_j can be chosen as suggested in BANK and ROSE [1981] in such a way to ensure a monotonical reduction of the norm of the residual. As for the solution of step (ii) of Gummel's iteration, suitable exponentially-fitted box-like schemes are usually employed for a stable and accurate discretization (see, e.g., SELBERHERR [1984], Chapter 6).

A convergence proof of iteration (2.10)–(2.12) has been first given in KERKHOVEN [1986] under the assumption of vanishing recombination/generation rate (i.e., $R = 0$). Using contraction-type arguments, and always assuming $R = 0$, it is possible to prove the convergence of the Gummel map only close to thermal equilibrium, i.e., for small values of the applied external biases ψ_D (see MARKOWICH [1986], Theorem 3.6.5 and JEROME [1996], Theorem 4.1.1). Convergence results under less restrictive conditions do not exist, although numerical experiments show that the iteration is always rapidly converging (even if a bad initial guess is chosen) whenever the magnitude of the current densities and of the recombination-generation rate is not too large (see MARKOWICH [1986], formula (3.6.56)). On the contrary, the convergence of the decoupled iteration becomes very slow in the case of high injection, or in the presence of impact ionization phenonema. This prompts for devising suitable acceleration techniques that improve the performance of Gummel's map.

This latter issue will be more extensively addressed in Chapter 7. Here we just mention the vectorial acceleration methods proposed in GUERRIERI, RUDAN, CIAMPOLINI and BACCARANI [1985] that produce a superlinear asymptotic rate of convergence without excessively increasing the CPU time and the memory resources. A different approach has been pursued in MICHELETTI, QUARTERONI and SACCO [1995] where a variant of Gummel's map based on the use of BI-CGSTAB method has been developed.

Acceleration methods based on Newton–Krylov subspace iterations have been recently proposed in KERKHOVEN and SAAD [1992], and an extension of their techniques to the simulation of optoelectronic semiconductor devices under high voltage operation conditions has been carried out in BOSISIO, MICHELETTI and SACCO [2000].

2.4. The decoupled drift-diffusion system

Each differential subproblem that has to be solved at each step of the Gummel iterative procedure (2.10)–(2.12) can be cast under the following general form of a reaction-diffusion problem

$$\begin{cases} -\operatorname{div}(a\underline{\nabla} u) + \gamma u = f & \text{in } \Omega, \\ u = g & \text{on } \Gamma_D, \\ a\underline{\nabla} u \cdot \underline{n} = 0 & \text{on } \Gamma_N. \end{cases} \tag{2.14}$$

Actually, the linearized Poisson equation that has to be solved within the Newton inner iteration (2.13) can be recovered from (2.14) (after dividing by λ^2) by letting $u = \phi$ and taking

$$a = 1, \qquad \gamma = (n^{(j)} + p^{(j)})/\lambda^2, \qquad f = -\mathcal{R}_\psi(u^{(j)})/\lambda^2, \qquad g = 0.$$

As for the solution of the linearized continuity equations, assume, as in Section 1.2.1, that the recombination-generation term R can be written as

$$R(x, \psi, \rho_n, \rho_p) = F(x, \psi, \rho_n, \rho_p)(\rho_n \rho_p - 1) - G\big(x, |\nabla \psi|, |\underline{J}_n|, |\underline{J}_p|\big)$$

for a positive function $F(\cdot,\cdot,\cdot,\cdot)$ in $(\Omega \times \mathbb{R} \times (0, +\infty)^2)$ and a nonnegative function $G(\cdot,\cdot,\cdot,\cdot)$ in $(\Omega \times (0, +\infty)^3)$. The first term at right-hand side models the Shockley-Read-Hall and Auger recombination rates while the second one is a net generation rate modelling impact ionization phenomena (see (1.18)–(1.20)). Namely, $-G$ is the (scaled) term defined in (1.20), and

$$F_{SHR} = \frac{\delta^2}{(\tau_n^*(\rho_p e^{-\psi} + 1) + \tau_p^*(\rho_n e^{\psi} + 1))}, \tag{2.15}$$

$$F_{AU} = \delta^2 (C_n \rho_n e^{\psi} + C_p \rho_p e^{-\psi}). \tag{2.16}$$

Finally, we have

$$F(x, \psi, \rho_n, \rho_p) = F_{SHR} + F_{AU}.$$

Then we can recover problems (2.11) and (2.12) from (2.14) by setting

$$u = \rho_n, \qquad a = \mu_n e^{\psi^{(k+1)}}, \qquad \gamma = F\big(x, \psi^{(k+1)}, \rho_n^{(k)}, \rho_p^{(k)}\big)\rho_p^{(k)}$$
$$f = F\big(x, \psi^{(k+1)}, \rho_n^{(k)}, \rho_p^{(k)}\big) + G\big(x, |\nabla \psi^{(k+1)}|, |\underline{J}_n^{(k)}|, |\underline{J}_p^{(k)}|\big), \qquad g = \rho_{nD},$$

and

$$u = \rho_p, \qquad a = \mu_p e^{-\psi^{(k+1)}}, \qquad \gamma = F\big(x, \psi^{(k+1)}, \rho_n^{(k+1)}, \rho_p^{(k)}\big)\rho_n^{(k+1)},$$
$$f = F\big(x, \psi^{(k+1)}, \rho_n^{(k+1)}, \rho_p^{(k)}\big) + G\big(x, |\nabla \psi^{(k+1)}|, |\underline{J}_n^{(k+1)}|, |\underline{J}_p^{(k)}|\big), \qquad g = \rho_{pD}.$$

As anticipated in Section 1.2, the use of the Slotboom variables in numerical computation is typically restricted to low-bias applications due to possible overflow/underflow problems. For this reason it is convenient to formulate the Gummel map in terms of the variables (ψ, n, p). By doing so, the linearized Poisson equation remains in the form of the model problem (2.14), while the linearized current continuity equations can be cast in the form of linear advection-diffusion-reaction problems, where the advective term is represented by the electric field (see for the details MARKOWICH [1986], Section 3.6). The discretization of these latter problems using mixed methods will be addressed in Sections 4.1 and 6.2. Next section will address the mixed finite element discretization of the reaction-diffusion problem (2.14).

3. Mixed formulation of second order elliptic problems

We shall introduce here the mixed formulation of the second order elliptic model problem (2.14). From now on, Ω is assumed to be a convex polygonal domain in $\mathbb{R}^2$, with unitary measure, and f, g are given functions, with $f \in L^2(\Omega)$, and $g \in H^{1/2}(\Gamma_D)$. Moreover, $a = a(x)$ and $\gamma = \gamma(x)$ are given regular functions on $\overline{\Omega}$, bounded from above and below, i.e.,

$$\exists a_0, a_M \quad \text{such that} \quad a_M \geqslant a(x) \geqslant a_0 > 0, \tag{3.1}$$

$$\exists \gamma_0, \gamma_M \quad \text{such that} \quad \gamma_M \geqslant \gamma(x) \geqslant \gamma_0 \geqslant 0. \tag{3.2}$$

The key point for deriving the mixed formulation of (2.14) is to introduce the flux $\underline{\sigma} = a\underline{\nabla} u$ as an independent variable, so that (2.14) becomes

$$\begin{cases} a^{-1}\underline{\sigma} - \underline{\nabla} u = 0 & \text{in } \Omega, \\ -\operatorname{div}\underline{\sigma} + \gamma u = f & \text{in } \Omega, \\ u = g & \text{on } \Gamma_D, \\ \underline{\sigma}\cdot\underline{n} = 0 & \text{on } \Gamma_N. \end{cases} \tag{3.3}$$

When treating the (scaled) Poisson equation (2.13), the flux $\underline{\sigma}$, from the physical point of view, will represent the (scaled) electric displacement. When taking into account the (scaled) continuity equations (2.11) and (2.12), $\underline{\sigma}$ will represent the (scaled) current density for electrons and holes, respectively.

We have now to introduce the proper functional setting for writing the variational formulation of (3.3). For that, let us first set a few notation. Some of the functional spaces have been previously used without precise definitions.

3.1. *Notation*

We shall constantly use Sobolev spaces on $\Omega \subset \mathbb{R}^2$, for which we refer to ADAMS [1975], LIONS and MAGENES [1968], NEČAS [1967]. They are based on

$$L^2(\Omega) := \left\{ v \,\middle|\, \int_\Omega |v|^2\,dx = \|v\|^2_{L^2(\Omega)} < +\infty \right\}, \tag{3.4}$$

the space of square integrable (and measurable) functions on Ω. The scalar product in $L^2(\Omega)$ is denoted by

$$(u, v) = \int_\Omega uv\,dx. \tag{3.5}$$

We then define, for m integer $\geqslant 0$,

$$H^m(\Omega) := \{v \mid D^\alpha v \in L^2(\Omega),\ \forall|\alpha| \leqslant m\}, \tag{3.6}$$

where

$$D^\alpha v = \frac{\partial^{|\alpha|} v}{\partial x_1^{\alpha_1} \partial x_2^{\alpha_2}}, \qquad |\alpha| = \alpha_1 + \alpha_2, \tag{3.7}$$

these derivatives being taken in the sense of distributions. On this space we shall use the seminorm

$$|v|^2_{m,\Omega} = \sum_{|\alpha|=m} \|D^\alpha v\|^2_{L^2(\Omega)}, \tag{3.8}$$

and the norm

$$\|v\|^2_{m,\Omega} = \sum_{k\leqslant m} |v|^2_{k,\Omega}. \tag{3.9}$$

The space $L^2(\Omega)$ is then $H^0(\Omega)$, and we shall usually write $\|v\|_{0,\Omega}$ to denote its norm $\|v\|_{L^2(\Omega)}$. We shall also use the space of traces on $\Gamma = \partial\Omega$ of functions in $H^1(\Omega)$:

$$H^{1/2}(\Gamma) := \big(H^1(\Omega)\big)_{|\Gamma}, \tag{3.10}$$

with the norm

$$\|g\|_{1/2,\Gamma} = \inf_{v\in H^1(\Omega),\, v_{|\Gamma}=g} \|v\|_{1,\Omega}. \tag{3.11}$$

The space $H^1_0(\Omega)$ will denote, as usual, the space of functions in $H^1(\Omega)$ vanishing on the boundary:

$$H^1_0(\Omega) := \{v \mid v \in H^1(\Omega),\ v_{|\Gamma} = 0\}. \tag{3.12}$$

For $v \in H^1_0(\Omega)$, or $v \in H^1(\Omega)$ and vanishing on a part of the boundary, the Poincaré inequality holds

$$\|v\|_{0,\Omega} \leqslant C(\Omega)|v|_{1,\Omega}, \tag{3.13}$$

and the seminorm $|.|_{1,\Omega}$ is therefore a norm in $H^1_0(\Omega)$, equivalent to the $\|.\|_{1,\Omega}$ norm. We denote, for $1 \leqslant p < +\infty$,

$$L^p(\Omega) := \left\{v \,\Big|\, \int_\Omega |v|^p\,dx =: \|v\|^p_{L^p(\Omega)} < +\infty\right\}. \tag{3.14}$$

As usual, $L^\infty(\Omega)$ denotes the space of essentially bounded functions, with norm

$$\|v\|_{\infty,\Omega} = \sup_{x\in\Omega} \operatorname{ess} \big|v(x)\big|. \tag{3.15}$$

For $1 \leqslant p \leqslant +\infty$ let

$$W^{1,p}(\Omega) := \{ v \mid D^\alpha v \in L^p(\Omega), \ \forall |\alpha| \leqslant 1 \}, \tag{3.16}$$

equipped with the norm

$$\|v\|^p_{W^{1,p}(\Omega)} = \sum_{|\alpha|\leqslant 1} \|D^\alpha v\|^p_{L^p(\Omega)}. \tag{3.17}$$

Finally, we denote by $H(\mathrm{div};\, \Omega)$ the space of vector-valued functions

$$H(\mathrm{div};\, \Omega) := \{ \underline{\tau} \in (L^2(\Omega))^2 \mid \mathrm{div}\, \underline{\tau} \in L^2(\Omega) \}, \tag{3.18}$$

equipped with the graph norm

$$\|\underline{\tau}\|^2_{H(\mathrm{div};\Omega)} = \|\underline{\tau}\|^2_{0,\Omega} + \|\,\mathrm{div}\, \underline{\tau}\|^2_{0,\Omega}, \tag{3.19}$$

where the symbol $\|.\|_{0,\Omega}$ denotes as well the L^2-norm for vector valued functions.

3.2. *Mixed formulation*

We are now ready to write the variational formulation of system (3.3). For that, define the spaces

$$\Sigma = \{ \underline{\tau} \in (L^2(\Omega))^2 \mid \mathrm{div}\, \underline{\tau} \in L^2(\Omega), \ \underline{\tau} \cdot \underline{n} = 0 \text{ on } \Gamma_N \} \subset H(\mathrm{div};\, \Omega), \tag{3.20}$$

$$V = L^2(\Omega), \tag{3.21}$$

with norms

$$\|\,\underline{\tau}\,\|^2_\Sigma = \|\underline{\tau}\|^2_{H(\mathrm{div};\Omega)}, \tag{3.22}$$

$$\|v\|_V = \|v\|_{0,\Omega}. \tag{3.23}$$

Next, we introduce the following bilinear forms

$$a(\underline{\sigma}, \underline{\tau}) = \int_\Omega a^{-1} \underline{\sigma} \cdot \underline{\tau}\, dx, \quad \underline{\sigma}, \underline{\tau} \in \Sigma, \tag{3.24}$$

$$b(v, \underline{\tau}) = \int_\Omega v\, \mathrm{div}\, \underline{\tau}\, dx, \quad v \in V,\ \underline{\tau} \in \Sigma, \tag{3.25}$$

$$c(u, v) = \int_\Omega \gamma u v\, dx, \quad u, v \in V. \tag{3.26}$$

Then, the mixed formulation of (2.14) is

$$\begin{cases} \text{Find } (\underline{\sigma}, u) \in \Sigma \times V \text{ such that} \\ a(\underline{\sigma}, \underline{\tau}) + b(u, \underline{\tau}) = \langle g, \underline{\tau} \cdot \underline{n} \rangle, & \forall \underline{\tau} \in \Sigma, \\ b(v, \underline{\sigma}) - c(u, v) = -(f, v), & \forall v \in V, \end{cases} \tag{3.27}$$

where the bracket $\langle .\,, . \rangle$ denotes the duality between $H^{1/2}(\partial\Omega)$ and its dual space $H^{-1/2}(\partial\Omega)$. Following BREZZI [1974] and BREZZI and FORTIN [1991], in order to prove existence, uniqueness and stability for problem (3.27) the following properties are needed

(i) the bilinear form $a(.,.)$ is bounded on $\Sigma \times \Sigma$ and coercive on Ker B, that is,

$$\exists M_a > 0 \text{ such that } \quad \left|a(\underline{\sigma},\underline{\tau})\right| \leqslant M_a \|\underline{\sigma}\|_\Sigma \|\underline{\tau}\|_\Sigma, \quad \forall \underline{\tau}, \underline{\sigma} \in \Sigma, \tag{3.28}$$

$$\exists \alpha > 0 \text{ such that } \quad a(\underline{\tau},\underline{\tau}) \geqslant \alpha \|\underline{\tau}\|_\Sigma^2, \quad \forall \underline{\tau} \in \operatorname{Ker} B, \tag{3.29}$$

where $B : \Sigma \to V'$ is the operator associated with the bilinear form $b(.,.)$, and

$$\operatorname{Ker} B = \left\{\underline{\tau} \in \Sigma \mid b(v,\underline{\tau}) = 0 \ \forall v \in V\right\}; \tag{3.30}$$

(ii) the bilinear form $b(.,.)$ is bounded on $V \times \Sigma$, and satisfies the inf-sup condition, that is,

$$\exists M_b > 0 \text{ such that } \quad \left|b(v,\underline{\tau})\right| \leqslant M_b \|v\|_V \|\underline{\tau}\|_\Sigma, \quad \forall v \in V,\ \underline{\tau} \in \Sigma, \tag{3.31}$$

$$\exists \beta > 0 \text{ such that } \quad \inf_{v \in V} \sup_{\underline{\tau} \in \Sigma} \frac{b(v,\underline{\tau})}{\|v\|_V \|\underline{\tau}\|_\Sigma} \geqslant \beta; \tag{3.32}$$

(here and in the following the inf is obviously taken over functions v not identically zero, and we will omit to specify $\|v\|_V \neq 0$);

(iii) the bilinear form $c(.,.)$ is symmetric, bounded on $V \times V$ and positive semidefinite, that is

$$\exists M_c > 0 \text{ such that } \quad \left|c(u,v)\right| \leqslant M_c \|u\|_V \|v\|_V, \quad \forall u, v \in V, \tag{3.33}$$

$$c(v,v) \geqslant 0, \quad \forall v \in V. \tag{3.34}$$

Let us recall the abstract Theorem 1.2 of Chapter II in BREZZI and FORTIN [1991].

THEOREM 3.1. *Let Σ, V be two Hilbert spaces with norms $\|\cdot\|_\Sigma$, $\|\cdot\|_V$, and let $f \in V'$, $g \in \Sigma'$, (V', Σ' being the dual spaces of V, Σ, respectively). Consider the problem*

$$\begin{cases} \textit{Find } (\underline{\sigma}, u) \in \Sigma \times V \textit{ such that} \\ a(\underline{\sigma},\underline{\tau}) + b(u,\underline{\tau}) = \langle g,\underline{\tau}\rangle_{\Sigma' \times \Sigma}, & \forall \underline{\tau} \in \Sigma, \\ b(v,\underline{\sigma}) - c(u,v) = \langle f, v\rangle_{V' \times V}, & \forall v \in V, \end{cases} \tag{3.35}$$

where $\langle \cdot,\cdot \rangle$ denotes the duality brackets. Under assumptions (3.28)–(3.34), *problem* (3.35) *has a unique solution for all $f \in V'$, $g \in \Sigma'$. Moreover, there exists a positive constant $\mathcal{M}$, depending nonlinearly on M_a, M_b, M_c, α, β, such that*

$$\|\underline{\sigma}\|_\Sigma + \|u\|_V \leqslant \mathcal{M}\left(\|g\|_{\Sigma'} + \|f\|_{V'}\right), \tag{3.36}$$

where

$$\|g\|_{\Sigma'} = \sup_{\underline{\tau} \in \Sigma} \frac{\langle g,\underline{\tau}\rangle}{\|\underline{\tau}\|_\Sigma}, \qquad \|f\|_{V'} = \sup_{v \in V} \frac{\langle f, v\rangle}{\|v\|_V}. \tag{3.37}$$

Let us now check that the abstract hypotheses of Theorem 3.1 are verified for problem (3.27). Boundedness of the bilinear form (3.24) follows trivially from (3.1) with $M_a = a_0^{-1}$. Next, notice that Ker B is characterized by elements of $\Sigma \subset H(\operatorname{div}; \Omega)$ with null divergence, so that $\|\underline{\tau}\|_\Sigma \equiv \|\underline{\tau}\|_{0,\Omega}$ for $\underline{\tau} \in \operatorname{Ker} B$. Then property (3.29) holds with $\alpha = a_M^{-1}$ (see (3.1)).

Symmetry, boundedness, and property (3.34) for the bilinear form (3.26) follow immediately from definition and assumptions (3.2); in particular (3.33) holds with $M_c = \gamma_M$. Boundedness of the bilinear form (3.25) holds trivially, with $M_b = 1$, by definition. In order to prove (3.32), we consider, $\forall v \in L^2(\Omega)$, the following auxiliary problem:

$$\begin{cases} -\Delta\varphi = v & \text{in } \Omega, \\ \varphi = 0 & \text{on } \Gamma_D, \\ \underline{\nabla}\varphi \cdot \underline{n} = 0 & \text{on } \Gamma_N. \end{cases} \tag{3.38}$$

Problem (3.38) has a unique solution $\varphi \in H^1(\Omega)$, and $\|\varphi\|_{1,\Omega} \leqslant C\|v\|_{0,\Omega}$. Then, the vector $\underline{\tau} = -\underline{\nabla}\varphi$ verifies $\underline{\tau} \in \Sigma$, $\operatorname{div}\underline{\tau} = v$ and $\|\underline{\tau}\|_\Sigma \leqslant \sqrt{C^2+1}\,\|v\|_V$. Therefore, the inf-sup condition (3.32) holds with $\beta = (C^2+1)^{-1/2}$.

REMARK 3.1. The solution φ of the auxiliary problem (3.38) actually belongs to $W^{1,p}(\Omega)$ for some $p > 2$, and

$$\|\varphi\|_{W^{1,p}(\Omega)} \leqslant C\|v\|_{0,\Omega}. \tag{3.39}$$

(See, e.g., GRISVARD [1985].) Hence, setting

$$\Sigma^* = \left\{\underline{\tau} \in \left(L^p(\Omega)\right)^2 \mid \operatorname{div}\underline{\tau} \in L^2(\Omega),\ \underline{\tau}\cdot\underline{n} = 0 \text{ on } \Gamma_N\right\}, \tag{3.40}$$

we see that the *inf-sup* condition (3.32) holds with Σ replaced by Σ^*:

$$\exists\beta^* > 0 \text{ such that } \quad \inf_{v\in V}\sup_{\underline{\tau}\in\Sigma^*}\frac{b(v,\underline{\tau})}{\|v\|_V\|\underline{\tau}\|_{\Sigma^*}} \geqslant \beta^*. \tag{3.41}$$

We also notice that the *inf-sup* condition (3.32) implies that

$$\text{the operator } B \text{ is surjective in } V\,(\forall v \in V,\ \exists\underline{\tau}\in\Sigma \text{ such that } \operatorname{div}\underline{\tau} = v). \tag{3.42}$$

Moreover, (3.41) implies that

$$\text{the operator } B \text{ has a continuous lifting from } V \text{ into } \Sigma^*. \tag{3.43}$$

We can then restate Theorem 3.1 applied to problem (3.27).

THEOREM 3.2. *For every $g \in H^{1/2}(\Gamma_D)$, and for every $f \in L^2(\Omega)$, there exists a unique $(\underline{\sigma}, u) \in \Sigma^* \times V$ solution of problem* (3.27). *Moreover, the following bound holds*

$$\|\underline{\sigma}\|_\Sigma + \|u\|_V \leqslant \mathcal{M}\left(\|g\|_{H^{1/2}(\Gamma_D)} + \|f\|_{L^2(\Omega)}\right) \tag{3.44}$$

with $\mathcal{M}$ dependent nonlinearly on a_0, γ_M, α, β.

REMARK 3.2. We explicitly point out that the solution of (3.27) is such that u coincides with the solution of (2.14), and $\underline{\sigma} \equiv a\underline{\nabla}u$. Therefore u, a priori seeked in $L^2(\Omega)$, is actually more regular (at least $u \in H^1(\Omega)$), according to the regularity of the solution of (2.14).

3.3. Discretization schemes: an abstract framework

Let $\{\mathcal{T}_h\}_h$ be a family of regular decompositions of $\overline{\Omega}$ into elements K, with boundary ∂K, made in such a way that there is always a vertex of $\mathcal{T}_h$ on the interface between Γ_D and Γ_N. Following CIARLET [1978], a family of decompositions is regular if there exists a constant $\mathcal{K}^*$ such that

$$\frac{D_K}{\rho_K} \leqslant \mathcal{K}^*, \quad \forall K \in \mathcal{T}_h, \tag{3.45}$$

where D_K is the diameter of the smallest circumscribed circle of K and ρ_K is the diameter of the largest inscribed circle in K. Inequality (3.45) immediately implies that

$$D_K \leqslant \mathcal{K}^* \rho_K \leqslant \mathcal{K} h_K, \quad \forall K \in \mathcal{T}_h, \tag{3.46}$$

where h_K denotes the diameter of K, and $\mathcal{K}$ denotes, here and in the sequel, a generic constant depending only on $\mathcal{K}^*$. Finally, we set $h = \max_K h_K$. For every triangulation $\mathcal{T}_h$ we define

$$\widetilde{\Sigma} = \left\{ \underline{\tau} \in L^2(\Omega) \mid \underline{\tau} \in \prod_K H(\text{div}; K),\ \underline{\tau} \cdot \underline{n} = 0 \text{ on } \partial K \cap \Gamma_N,\ \forall K \in \mathcal{T}_h \right\}, \tag{3.47}$$

with the norm

$$\|\underline{\tau}\|^2_{\widetilde{\Sigma}} = \|\underline{\tau}\|^2_{0,\Omega} + \sum_K \|\operatorname{div} \underline{\tau}\|^2_{0,K}, \tag{3.48}$$

and we notice that

$$\|\underline{\tau}\|_{\widetilde{\Sigma}} \equiv \|\underline{\tau}\|_{\Sigma}, \quad \forall \underline{\tau} \in \Sigma. \tag{3.49}$$

For all $K \in \mathcal{T}_h$, we introduce finite dimensional spaces $Q(K)$ for vector functions and $P(K)$ for scalar functions. For the sake of simplicity, we might assume that there exists an integer $k \geqslant 0$ such that, on each K, $Q(K)$ and $P(K)$ consist of polynomials of degree $\leqslant k$. Let us define the following discrete spaces

$$\widetilde{\Sigma}_h = \left\{ \underline{\tau}_h \in \widetilde{\Sigma} \mid \underline{\tau}_{h|K} \in Q(K),\ \forall K \in \mathcal{T}_h \right\}, \tag{3.50}$$

$$V_h = \left\{ v_h \in V \mid v_{h|K} \in P(K),\ \forall K \in \mathcal{T}_h \right\}. \tag{3.51}$$

As we shall see, the discrete solution will be seeked in a more regular space

$$\Sigma_h \subset \widetilde{\Sigma}_h, \tag{3.52}$$

still equipped with the norm $\|\cdot\|_{\widetilde{\Sigma}}$. Σ_h will include ‘some’ continuity of the normal component across the interelements but, in general, a nonconforming approximation will be allowed, with Σ_h not included in Σ. The abstract formulation which we introduce here includes the well known RAVIART and THOMAS [1977] and BREZZI, DOUGLAS JR and MARINI [1985] elements, and the elements introduced in MARINI and PIETRA [1989]. Actual choices of $Q(K)$ and $P(K)$ will be given in Section 3.5.

Definition (3.47) makes the bilinear form (3.25) meaningless on $\widetilde{\Sigma}$. For this reason we introduce the discrete bilinear form

$$b_h(v, \underline{\tau}) = \sum_K \int_K v \operatorname{div} \underline{\tau} \, dx, \qquad v \in V, \ \underline{\tau} \in \widetilde{\Sigma}, \tag{3.53}$$

and we notice that

$$b_h(v, \underline{\tau}) \equiv b(v, \underline{\tau}), \qquad v \in V, \ \underline{\tau} \in \Sigma. \tag{3.54}$$

Accordingly, instead of (3.30) the following weaker definition will be used from now on

$$\operatorname{Ker} \widetilde{B} = \{\underline{\tau} \in \widetilde{\Sigma} \mid b_h(v, \underline{\tau}) = 0 \ \forall v \in V\}, \tag{3.55}$$

$\widetilde{B}$ being the operator $\widetilde{\Sigma} \to V'$ associated with the bilinear form $b_h(.,.)$. Definition (3.55) implies that $\operatorname{Ker} \widetilde{B}$ is made of vectors which are divergence-free element by element

$$\underline{\tau} \in \operatorname{Ker} \widetilde{B} \quad \Rightarrow \quad \operatorname{div} \underline{\tau} = 0 \quad \text{in } K, \ \forall K \in \mathcal{T}_h. \tag{3.56}$$

Similarly we define

$$\operatorname{Ker} B_h = \{\underline{\tau}_h \in \widetilde{\Sigma}_h \mid b_h(v_h, \underline{\tau}_h) = 0 \ \forall v_h \in V_h\}. \tag{3.57}$$

The discrete formulation of (3.27) is then:

$$\begin{cases} \text{Find } (\underline{\sigma}_h, u_h) \in \Sigma_h \times V_h \text{ such that} \\ a(\underline{\sigma}_h, \underline{\tau}_h) + b_h(u_h, \underline{\tau}_h) = \langle g, \underline{\tau}_h \cdot \underline{n} \rangle_{|\Gamma_D}, \quad \forall \underline{\tau}_h \in \Sigma_h, \\ b_h(v_h, \underline{\sigma}_h) - c(u_h, v_h) = -(f, v_h), \quad \forall v_h \in V_h. \end{cases} \tag{3.58}$$

In order to prove existence and uniqueness of the solution of (3.58), and optimal error bounds, we need to state, at this very abstract level, the assumptions on the spaces Σ_h, V_h that allow to derive optimal error estimates. The following two assumptions are crucial guidelines to introduce proper discretizations:

$$\operatorname{Ker} B_h \subset \operatorname{Ker} \widetilde{B}; \tag{3.59}$$

there exists an operator $\Pi_h : \Sigma^* \to \Sigma_h$ such that

$$b_h(v_h, \underline{\tau} - \Pi_h \underline{\tau}) = 0, \qquad \forall v_h \in V_h, \tag{3.60}$$

and

$$\|\Pi_h \underline{\tau}\|_{\widetilde{\Sigma}} \leqslant C \|\underline{\tau}\|_{\Sigma^*}, \qquad \forall \underline{\tau} \in \Sigma^*, \tag{3.61}$$

with C a constant independent of h.

We note that (3.60)–(3.61) imply that the finite dimensional spaces Σ_h and V_h verify a discrete inf-sup condition (see FORTIN [1977]):

$$\exists \beta > 0: \quad \inf_{v_h \in V_h} \sup_{\underline{\tau}_h \in \Sigma_h} \frac{b_h(v_h, \underline{\tau}_h)}{\|v_h\|_V \|\underline{\tau}_h\|_{\widetilde{\Sigma}}} \geqslant \beta. \tag{3.62}$$

Indeed, as a consequence of the inf-sup condition (3.41), which holds for the continuous problem, for $v_h \in V_h \subset V$ there exists $\underline{\tau} \in \Sigma^*$ such that

$$\frac{b(v_h, \underline{\tau})}{\|v_h\|_V \|\underline{\tau}\|_{\Sigma^*}} \geqslant \beta^*. \tag{3.63}$$

Due to (3.60), (3.61), and (3.54), from (3.63) we obtain

$$\frac{b_h(v_h, \Pi_h \underline{\tau})}{\|v_h\|_V \|\Pi_h \underline{\tau}\|_{\widetilde{\Sigma}}} \geqslant C^{-1} \frac{b_h(v_h, \underline{\tau})}{\|v_h\|_V \|\underline{\tau}\|_{\Sigma^*}} = C^{-1} \frac{b(v_h, \underline{\tau})}{\|v_h\|_V \|\underline{\tau}\|_{\Sigma^*}} \geqslant C^{-1}\beta^*. \tag{3.64}$$

Hence, (3.62) holds with $\beta = C^{-1}\beta^*$.

We have the following result.

THEOREM 3.3. *Under assumptions* (3.59) *and* (3.62), *problem* (3.58) *has a unique solution.*

PROOF. For the proof we could apply the abstract Theorem 3.1. However, the problem being finite dimensional, uniqueness implies existence, and the proof can be simplified. For the convenience of the reader we report it here. Let $(\underline{\sigma}_h^*, u_h^*) \in \Sigma_h \times V_h$ be the solution of the homogeneous discrete problem associated with (3.58). Taking $\underline{\tau}_h = \underline{\sigma}_h^*$ in the first equation of (3.58), and using the second equation with $v_h = u_h^*$ we obtain

$$a(\underline{\sigma}_h^*, \underline{\sigma}_h^*) + c(u_h^*, u_h^*) = 0. \tag{3.65}$$

This implies that $\underline{\sigma}_h^* = 0$, and, if $\gamma_0 > 0$, $u_h^* = 0$. Instead, if $\gamma_0 = 0$, we make use of the first equation of (3.58), which reduces to

$$b_h(u_h^*, \underline{\tau}_h) = 0, \quad \forall \underline{\tau}_h \in \Sigma_h. \tag{3.66}$$

Thanks to (3.62), this implies $u_h^* = 0$. □

REMARK 3.3. A direct consequence of (3.59) and (3.56) is that

$$\int_K v_h \operatorname{div} \underline{\tau}_h \, dx = 0, \quad \forall v_h \in P(K) \quad \Rightarrow \quad \operatorname{div} \underline{\tau}_{h|K} = 0, \quad \forall K \in \mathcal{T}_h, \tag{3.67}$$

so that $\dim(P(K)) \geqslant \dim(\operatorname{div}(Q(K)))$. On the other hand, from (3.62) it follows

$$\int_K v_h \operatorname{div} \underline{\tau}_h \, dx = 0, \quad \forall \underline{\tau}_h \in Q(K) \quad \Rightarrow \quad v_{h|K} = 0, \quad \forall K \in \mathcal{T}_h, \tag{3.68}$$

and consequently $\dim(\operatorname{div}(Q(K))) \geqslant \dim(P(K))$. Therefore, assumptions (3.59) and (3.62) imply

$$\dim\bigl(\operatorname{div}\bigl(Q(K)\bigr)\bigr) \equiv \dim\bigl(P(K)\bigr). \tag{3.69}$$

Moreover, for any choice of basis functions $\{v^1, \ldots, v^r\}$, and $\{d^1, \ldots, d^r\}$ in $P(K)$ and $\operatorname{div}(Q(K))$ respectively, the matrix

$$\int_K d^i v^j \, dx, \quad i, j = 1, r, \tag{3.70}$$

is nonsingular.

3.4. Error estimates in the abstract framework

In order to derive error estimates we introduce an interpolation operator P_h from V to V_h, which can be defined locally by

$$\int_K (v - P_h v)\,\mathrm{div}\,\underline{\tau}_h\,dx = 0, \quad \forall \underline{\tau}_h \in Q(K). \tag{3.71}$$

In order to see that P_h is well defined, thanks to (3.69) it is sufficient to check uniqueness, that is,

$$\int_K P_h v\,\mathrm{div}\,\underline{\tau}_h\,dx = 0, \quad \forall \underline{\tau}_h \in Q(K) \quad \Rightarrow \quad P_h v = 0. \tag{3.72}$$

This is an immediate consequence of the (3.70). Next, for any $v \in H^1(\Omega)$ we introduce

$$N_h(v, \underline{\tau}_h) = \sum_K \langle v, \underline{\tau}_h \cdot \underline{n} \rangle_{|\partial K} - \langle v, \underline{\tau}_h \cdot \underline{n} \rangle_{|\Gamma_D}, \quad \underline{\tau}_h \in \Sigma_h, \tag{3.73}$$

and we notice that for u solution of (3.27), which belongs to $H^1(\Omega)$ (see Remark 3.2), we have

$$N_h(u, \underline{\tau}_h) = \sum_K \langle u, \underline{\tau}_h \cdot \underline{n} \rangle_{|\partial K} - \langle g, \underline{\tau}_h \cdot \underline{n} \rangle_{|\Gamma_D}, \quad \underline{\tau}_h \in \Sigma_h. \tag{3.74}$$

It is easy to see that the solution $(\underline{\sigma}, u)$ of (3.27) verifies

$$\begin{cases} a(\underline{\sigma}, \underline{\tau}_h) + b_h(u, \underline{\tau}_h) - N_h(u, \underline{\tau}_h) = \langle g, \underline{\tau}_h \cdot \underline{n} \rangle_{|\Gamma_D}, & \forall \underline{\tau}_h \in \Sigma_h, \\ b_h(v_h, \underline{\sigma}) - c(u, v_h) = -(f, v_h), & \forall v_h \in V_h. \end{cases} \tag{3.75}$$

The term $N_h(u, \underline{\tau}_h)$ is a measure of the possible nonconformity of the space Σ_h, and it will be zero, due to the regularity of u, for conforming choices of Σ_h. By subtracting (3.58) from (3.75), and using (3.71) and (3.60), we obtain the error equations

$$\begin{cases} a(\underline{\sigma} - \underline{\sigma}_h, \underline{\tau}_h) + b_h(P_h u - u_h, \underline{\tau}_h) - N_h(u, \underline{\tau}_h) = 0, & \forall \underline{\tau}_h \in \Sigma_h, \\ b_h(v_h, \Pi_h \underline{\sigma} - \underline{\sigma}_h) - c(u - u_h, v_h) = 0, & \forall v_h \in V_h. \end{cases} \tag{3.76}$$

We can prove the following result.

THEOREM 3.4. *Let $(\underline{\sigma}, u)$ be the solution of* (3.27) *and $(\underline{\sigma}_h, u_h)$ that of* (3.58). *Under assumptions* (3.28)–(3.34), (3.59) *and* (3.62) *there exists a constant C independent of h, such that the following estimate holds*

$$\begin{aligned} &\|\Pi_h \underline{\sigma} - \underline{\sigma}_h\|_{\widetilde{\Sigma}} + \|P_h u - u_h\|_{0,\Omega} \\ &\quad \leqslant C\left(\|\underline{\sigma} - \Pi_h \underline{\sigma}\|_{0,\Omega} + \|u - P_h u\|_{0,\Omega} + \sup_{\underline{\tau}_h \in \Sigma_h} \frac{N_h(u, \underline{\tau}_h)}{\|\underline{\tau}_h\|_{\widetilde{\Sigma}}} \right). \end{aligned}$$

PROOF. We add and subtract $a(\Pi_h \underline{\sigma}, \underline{\tau}_h)$, with $\Pi_h \underline{\sigma}$ defined in (3.60)–(3.61), in the first equation of (3.76), and $c(P_h u, v_h)$, with $P_h u$ defined in (3.71), in the second equation of (3.76), thus obtaining

$$\begin{cases} a(\Pi_h \underline{\sigma} - \underline{\sigma}_h, \underline{\tau}_h) + b_h(P_h u - u_h, \underline{\tau}_h) \\ \quad = a(\Pi_h \underline{\sigma} - \underline{\sigma}, \underline{\tau}_h) + N_h(u, \underline{\tau}_h), & \forall \underline{\tau}_h \in \Sigma_h, \\ b_h(v_h, \Pi_h \underline{\sigma} - \underline{\sigma}_h) - c(P_h u - u_h, v_h) = c(u - P_h u, v_h), & \forall v_h \in V_h. \end{cases} \tag{3.77}$$

Problem (3.77) has the form (3.35) and verifies the hypotheses of Theorem 3.1. Hence, from (3.36) we get

$$\begin{aligned}&\|\Pi_h\underline{\sigma}-\underline{\sigma}_h\|_{\widetilde{\Sigma}}+\|P_hu-u_h\|_{0,\Omega}\\&\quad\leqslant\mathcal{M}\left(\sup_{\underline{\tau}_h\in\Sigma_h}\frac{a(\Pi_h\underline{\sigma}-\underline{\sigma},\underline{\tau}_h)}{\|\underline{\tau}_h\|_{\widetilde{\Sigma}}}+\sup_{\underline{\tau}_h\in\Sigma_h}\frac{N_h(u,\underline{\tau}_h)}{\|\underline{\tau}_h\|_{\widetilde{\Sigma}}}+\sup_{v_h\in V_h}\frac{c(u-P_hu,v_h)}{\|v_h\|_{0,\Omega}}\right).\end{aligned}$$

Then, the result follows. □

REMARK 3.4. Under the same assumptions of Theorem 3.4, we immediately deduce, via triangle inequality and the obvious observation that $\|\cdot\|_{0,\Omega}\leqslant\|\cdot\|_{\widetilde{\Sigma}}$,

$$\begin{aligned}&\|\underline{\sigma}-\underline{\sigma}_h\|_{0,\Omega}+\|u-u_h\|_{0,\Omega}\\&\quad\leqslant C\left(\|\underline{\sigma}-\Pi_h\underline{\sigma}\|_{0,\Omega}+\|u-P_hu\|_{0,\Omega}+\sup_{\underline{\tau}_h\in\Sigma_h}\frac{N_h(u,\underline{\tau}_h)}{\|\underline{\tau}_h\|_{\widetilde{\Sigma}}}\right).\end{aligned}\tag{3.78}$$

Similarly, whenever an estimate for $\|\underline{\sigma}-\Pi_h\underline{\sigma}\|_{\widetilde{\Sigma}}$ is available, we also deduce

$$\begin{aligned}&\|\underline{\sigma}-\underline{\sigma}_h\|_{\widetilde{\Sigma}}+\|u-u_h\|_{0,\Omega}\\&\quad\leqslant C\left(\|\underline{\sigma}-\Pi_h\underline{\sigma}\|_{\widetilde{\Sigma}}+\|u-P_hu\|_{0,\Omega}+\sup_{\underline{\tau}_h\in\Sigma_h}\frac{N_h(u,\underline{\tau}_h)}{\|\underline{\tau}_h\|_{\widetilde{\Sigma}}}\right).\end{aligned}\tag{3.79}$$

3.5. *Examples of mixed finite elements*

We shall provide here examples of finite elements which fit the abstract framework introduced above. The first four examples refer to well known families of mixed finite elements, namely, the Raviart–Thomas (*RT*) and Brezzi–Douglas–Marini (*BDM*) families, while the last two examples were introduced in MARINI and PIETRA [1989].

EXAMPLE 1. The Raviart–Thomas elements (RAVIART and THOMAS [1977]) over a triangular decomposition of Ω. For any integer $k\geqslant 0$, and for any triangle $K\in\mathcal{T}_h$, define

$$Q(K)=\big(P_k(K)\big)^2+\underline{x}P_k(K),\tag{3.80}$$

$$P(K)=P_k(K),\tag{3.81}$$

where $\underline{x}=(x_1,x_2)$, and $P_k(K)$ denotes the set of polynomials of degree $\leqslant k$ in K.

EXAMPLE 2. The Raviart–Thomas elements (RAVIART and THOMAS [1977]) over a rectangular decomposition of Ω. For any integer $k\geqslant 0$, and for any rectangle $K\in\mathcal{T}_h$, define

$$Q(K)=\big(Q_k(K)+x_1Q_k(K)\big)\times\big(Q_k(K)+x_2Q_k(K)\big),\tag{3.82}$$

$$P(K)=Q_k(K),\tag{3.83}$$

where $Q_k(K)$ denotes the set of polynomials of degree $\leqslant k$ in each variable x_1 and x_2.

EXAMPLE 3. The Brezzi–Douglas–Marini elements (BREZZI, DOUGLAS JR and MARINI [1985]) over a triangular decomposition of Ω. For any integer $k \geqslant 1$, and for any triangle $K \in \mathcal{T}_h$, define

$$Q(K) = \big(P_k(K)\big)^2, \tag{3.84}$$
$$P(K) = P_{k-1}(K). \tag{3.85}$$

EXAMPLE 4. The Brezzi–Douglas–Marini elements (BREZZI, DOUGLAS JR and MARINI [1985]) over a rectangular decomposition of Ω. For any integer $k \geqslant 1$, and for any rectangle $K \in \mathcal{T}_h$ define

$$Q(K) = \big(P_k(K)\big)^2 \oplus \big\{\underline{\mathrm{curl}}\big(x_1 x_2^{k+1}\big)\big\} \oplus \big\{\underline{\mathrm{curl}}\big(x_2 x_1^{k+1}\big)\big\}, \tag{3.86}$$
$$P(K) = P_{k-1}(K), \tag{3.87}$$

where $\underline{\mathrm{curl}}\,\varphi = (\partial_2\varphi, -\partial_1\varphi)$, for $\varphi \in H^1(\Omega)$.

The choice of $P(K)$ identifies the space of scalars V_h defined in (3.51), while the choice of $Q(K)$ identifies only the space of vectors $\widetilde{\Sigma}_h$. In order to define the space $\Sigma_h \subset \widetilde{\Sigma}_h$, where the solution of (3.75) is seeked, the regularity assumption must be specified. We shall show that, with the above choices of polynomial spaces, it is possible to define degrees of freedom, and consequently to construct local bases, that guarantee continuity of the normal component of vectors in Σ_h across interelement boundaries, thus giving rise to a conforming approximation:

$$\Sigma_h = \big\{\underline{\tau}_h \in H(\mathrm{div};\Omega) \mid \underline{\tau}_{h|K} \in Q(K),\ \forall K \in \mathcal{T}_h,\ \underline{\tau}_h \cdot \underline{n} = 0 \text{ on } \Gamma_N\big\}. \tag{3.88}$$

Indeed, for each k that identifies the pair $(Q(K), P(K))$ in the above families, let us introduce the spaces of polynomials on the edges of K

$$R(e) = P_k(e) \quad \text{for each edge } e \text{ of } K, \qquad R(\partial K) = \prod_{e \in \partial K} R(e). \tag{3.89}$$

Notice that, in all the previous examples,

$$\underline{\tau} \in Q(K) \quad \Rightarrow \quad \underline{\tau} \cdot \underline{n}_{|\partial K} \in R(\partial K). \tag{3.90}$$

Moreover,

$$\mathrm{div}\, Q(K) = P(K). \tag{3.91}$$

Next, let D be the space

$$D = \big\{\underline{q} \in Q(K) \mid \mathrm{div}\,\underline{q}_{|K} = 0,\ \underline{q} \cdot \underline{n}_{|\partial K} = 0\big\}, \tag{3.92}$$

and notice that $D = \{0\}$ for the low elements of the four families.

PROPOSITION 3.1. *For any $\underline{q} \in Q(K)$, the following relations imply $\underline{q} = 0$:*

$$\int_e \mu \underline{q} \cdot \underline{n}\, ds = 0, \quad \forall \text{ edge } e \text{ of } K,\ \forall \mu \in R(e), \tag{3.93}$$

$$\int_K \underline{q} \cdot \underline{\nabla} v \, dx = 0, \quad \forall v \in P(K), \tag{3.94}$$

$$\int_K \underline{q} \cdot \underline{\varphi} \, dx = 0, \quad \forall \underline{\varphi} \in D. \tag{3.95}$$

PROOF. For any fixed k, we have that $\underline{q} \cdot \underline{n}_{|e} \in R(e)$. Hence, (3.93) implies $\underline{q} \cdot \underline{n}_{|\partial K} = 0$. Moreover, the trace of a function $v \in P(K)$ belongs to $R(\partial K)$ in all the examples. Hence, integration by parts and (3.93)–(3.94) give

$$\int_K \operatorname{div} \underline{q} v \, dx = -\int_K \underline{q} \cdot \underline{\nabla} v \, dx + \int_{\partial K} v \underline{q} \cdot \underline{n} \, ds = 0, \quad \forall v \in P(K), \tag{3.96}$$

which implies $\operatorname{div} \underline{q} = 0$, due to (3.91). Then, (3.93)–(3.94) imply $\underline{q} \in D$. Reciprocally, it is obvious that (3.93)–(3.94) hold for $\underline{\varphi} \in D$. Hence, (3.93)–(3.94) are equivalent to $\underline{q} \in D$, and (3.95) gives $\underline{q} = 0$. □

To see that (3.93)–(3.95) can be used as degrees of freedom, after choosing local bases, it remains to check that (3.93)–(3.94) are linearly independent. For this we refer, e.g., to BREZZI and FORTIN [1991] Section III.3, Lemma 3.1. It is then possible to use $\underline{\tau} \cdot \underline{n}$ among the degrees of freedom, and then to impose continuity at the interelements, thus constructing a conforming approximation. In particular, for the lowest order Raviart–Thomas triangular elements (RT_0) of Example 1 a local basis for $Q(K)$ is uniquely defined by the following degrees of freedom

$$\int_{e^j} \underline{\tau}^i \cdot \underline{n}^j \, ds = \delta_{ij}, \quad i, j = 1, 3. \tag{3.97}$$

Conformity, together with relation (3.91) (which is a stronger property than (3.69)), will allow to simplify (and somewhat improve) the abstract results. In particular, conformity implies that the discrete bilinear form $b_h(.,.)$ defined in (3.53) is not needed here, since it will always be applied to elements of Σ. Hence, it can be replaced by the bilinear form $b(.,.)$ defined in (3.25):

$$b_h(v_h, \underline{\tau}_h) \equiv b(v_h, \underline{\tau}_h), \quad \forall v_h \in V_h, \ \underline{\tau}_h \in \Sigma_h \subset \Sigma. \tag{3.98}$$

From property (3.91) it is immediate to check that the first abstract hypothesis (3.59) is fulfilled. An important consequence of (3.91) is that the operator P_h defined from V to V_h in (3.71) coincides with as the usual L^2-projection

$$\int_K (v - P_h v) w \, dx = 0, \quad \forall w \in P(K). \tag{3.99}$$

It remains to define uniquely the element $\Pi_h \underline{\tau} \in Q(K)$ verifying (3.60)–(3.61). Proposition 3.1 implies that, $\forall \underline{\tau} \in \Sigma^*$, $\Pi_h \underline{\tau} \in \Sigma_h$ can be uniquely defined locally through (3.93)–(3.94)–(3.95) as:

$$\int_e \mu(\underline{\tau} - \Pi_h \underline{\tau}) \cdot \underline{n} \, ds = 0, \quad \forall \text{ edge } e \text{ of } K, \ \forall \mu \in R(e), \tag{3.100}$$

$$\int_K (\underline{\tau} - \Pi_h \underline{\tau}) \cdot \underline{\nabla} v \, dx = 0, \quad \forall v \in P(K), \tag{3.101}$$

$$\int_K (\underline{\tau} - \Pi_h \underline{\tau}) \cdot \underline{q}\, dx = 0, \quad \forall \underline{q} \in D. \tag{3.102}$$

Conditions (3.100)–(3.102) define an interpolation operator Π_h from Σ^* to Σ_h which is uniformly bounded from Σ^* to $\Sigma_h \subset \Sigma$, that is,

$$\|\Pi_h \underline{\tau}\|_\Sigma \leqslant C \|\underline{\tau}\|_{\Sigma^*}, \tag{3.103}$$

with C independent of h and of $\underline{\tau}$, as shown, e.g., in BREZZI and FORTIN [1991]. Moreover, since the trace of a function $v \in P(K)$ belongs to $R(\partial K)$, integration by parts, and (3.100)–(3.101) imply, for all $v \in P(K)$,

$$\int_K \operatorname{div}(\underline{\tau} - \Pi_h \underline{\tau}) v\, dx = -\int_K (\underline{\tau} - \Pi_h \underline{\tau}) \cdot \underline{\nabla} v\, dx + \int_{\partial K} (\underline{\tau} - \Pi_h \underline{\tau}) \cdot \underline{n} v\, ds = 0. \tag{3.104}$$

Collecting the various properties (3.104), (3.98), and (3.99) we deduce that the commuting diagram property (DOUGLAS JR and ROBERTS [1985]) holds

$$\begin{array}{ccccc} \Sigma^* & \xrightarrow{\operatorname{div}} & V & \longrightarrow & 0 \\ \Pi_h \big\downarrow & & \big\downarrow P_h & & \\ \Sigma_h & \xrightarrow{\operatorname{div}} & V_h & \longrightarrow & 0 \end{array} \tag{3.105}$$

This means, in other words, that for every $\underline{\tau} \in \Sigma^*$ one has

$$\operatorname{div}(\Pi_h \underline{\tau}) = P_h \operatorname{div} \underline{\tau}. \tag{3.106}$$

Therefore, the discrete inf-sup condition (3.62) holds. This, together with (3.59) allows to apply the abstract theory of Sections 3.3 and 3.4. In particular, error estimates (3.78)–(3.79) hold with $N_h \equiv 0$, due to the regularity of u and to the continuity of the normal component of elements in Σ_h.

Since in the next sections the lowest order Raviart–Thomas element RT_0 will be extensively used, we report here, for the reader convenience, the definition of the discrete spaces

$$V_h = \{ v_h \in L^2(\Omega) \mid v_{h|K} \in P_0(K),\ \forall K \in \mathcal{T}_h \}, \tag{3.107}$$

$$\Sigma_h = \{ \underline{\tau}_h \in H(\operatorname{div}; \Omega) \mid \underline{\tau}_{h|K} \in Q(K),\ \forall K \in \mathcal{T}_h,\ \underline{\tau}_h \cdot \underline{n} = 0 \text{ on } \Gamma_N \}, \tag{3.108}$$

with $Q(K) = (P_0(K))^2 + \underline{x} P_0(K)$. Notice that $\dim(V_h) = \#$ of elements in $\mathcal{T}_h$, and $\dim(\Sigma_h) = \#$ of edges not belonging to Γ_N. According to (3.97), a basis function $\underline{\tau}^r \in \Sigma_h$ is defined requiring that, for each edge e^s not belonging to Γ_N,

$$\int_{e^s} \underline{\tau}^r \cdot \underline{n}^s\, ds = \delta_{rs}, \tag{3.109}$$

where $\underline{n}^s$ is the normal unit vector to e^s, whose orientation is chosen once and for all. Consequently, considering the two elements having e^s in common, $\underline{n}^s$ is outward for one triangle and inward for the other one. Error estimate (3.79) gives

$$\|\underline{\sigma} - \underline{\sigma}_h\|_\Sigma + \|u - u_h\|_{0,\Omega} \leqslant Ch\big(|\underline{\sigma}|_{1,\Omega} + |\operatorname{div} \underline{\sigma}|_{1,\Omega} + |u|_{1,\Omega}\big). \tag{3.110}$$

Indeed, the interpolation errors are (see CIARLET [1978], BREZZI and FORTIN [1991])

$$\|\underline{\sigma} - \Pi_h\underline{\sigma}\|_\Sigma \leqslant Ch\big(|\underline{\sigma}|_{1,\Omega} + |\operatorname{div}\underline{\sigma}|_{1,\Omega}\big), \qquad \|u - P_h u\|_{0,\Omega} \leqslant Ch|u|_{1,\Omega}. \tag{3.111}$$

The key properties to obtain the commuting diagram property are conformity and (3.91). We provide now an example, taken from MARINI and PIETRA [1989], of conforming approximation which violates (3.91), though remaining in the abstract framework.

EXAMPLE 5. Consider a triangular decomposition of Ω. For each triangle $K \in \mathcal{T}_h$, we choose the finite-dimensional polynomial sets as follows

$$Q(K) = \operatorname{span}\{\underline{\tau}^1, \underline{\tau}^2, \underline{\tau}^3\}, \tag{3.112}$$

$$P(K) = P_0(K), \tag{3.113}$$

where

$$\underline{\tau}^1 = (1,0), \qquad \underline{\tau}^2 = (0,1), \qquad \underline{\tau}^3 = (\psi_1, \psi_2). \tag{3.114}$$

The choice for ψ_1 and ψ_2 is

$$\psi_1, \psi_2 \in P_2(K), \tag{3.115}$$

and, having chosen an edge $\tilde{e}$ of K, $\underline{\tau}^3 = (\psi_1, \psi_2)$ is defined through the following degrees of freedom

$$\begin{cases} \underline{\tau}^3 \cdot \underline{n}_{|\tilde{e}} = 1, \\ \underline{\tau}^3 \cdot \underline{n}_{|e} = 0, \quad e \neq \tilde{e}, \\ \int_K \psi_1\,dx = \int_K \psi_2\,dx = 0, \\ \underline{\tau}^3 \cdot \underline{t}(\tilde{m}) = 0, \end{cases} \tag{3.116}$$

where $\tilde{m}$ is the midpoint of $\tilde{e}$, and $\underline{t}$ denotes the unit tangent vector.

PROPOSITION 3.2. *The degrees of freedom* (3.116) *uniquely define* $\underline{\tau}^3$.

PROOF. Since $\underline{\tau}^3$ is seeked in a space of dimension 12 (see (3.115)), and the number of degrees of freedom (3.116) is precisely 12, it is sufficient to check uniqueness. Let then $\underline{\tau} = (\tau_1, \tau_2)$ be a vector verifying

(i) $\underline{\tau} \cdot \underline{n}_{|e} = 0, \quad \forall e,$

(ii) $\displaystyle\int_K \tau_1\,dx = \int_K \tau_2\,dx = 0,$

(iii) $\underline{\tau} \cdot \underline{t}(\tilde{m}) = 0.$

From (i) we deduce $\underline{\tau} \cdot \underline{n}_{|\partial K} = 0$. From (ii), after integration by parts, we have

$$0 = \int_K \underline{\tau} \cdot \underline{\nabla}\, p_1\,dx = -\int_K \operatorname{div}\underline{\tau}\, p_1\,dx, \quad \forall p_1 \in P_1(K), \tag{3.117}$$

which gives $\operatorname{div} \underline{\tau} = 0$. Hence,

$$\underline{\tau} = \underline{\operatorname{curl}}\, \varphi, \quad \text{with } \varphi \in P_3(K) \text{ and } \frac{\partial \varphi}{\partial s} = 0 \text{ on } \partial K, \tag{3.118}$$

where $\partial\varphi/\partial s$ denotes the tangential derivative of φ along ∂K. Consequently, $\varphi =$ constant on ∂K, i.e., $\varphi = a(\lambda_1\lambda_2\lambda_3 + c)$, and $\underline{\tau} = a\,\underline{\operatorname{curl}}(\lambda_1\lambda_2\lambda_3)$, where λ_i, $i = 1, 3$, being the barycentric coordinates. Condition (iii) becomes $a(\partial(\lambda_1\lambda_2\lambda_3)/\partial n)(\tilde{m}) = 0$, implying $a = 0$, since $(\partial(\lambda_1\lambda_2\lambda_3)/\partial n)(\tilde{m}) \neq 0$. □

With this choice, condition (3.91) is violated, since $\operatorname{div} \underline{\tau}^3$ is not constant on K. However, we have

$$\dim\bigl(\operatorname{div}\bigl(Q(K)\bigr)\bigr) = 1 = \dim\bigl(P(K)\bigr), \tag{3.119}$$

and

$$\int_K \operatorname{div} \underline{\tau}^3 \, dx = \int_{\partial K} \underline{\tau}^3 \cdot \underline{n} \, ds = 1. \tag{3.120}$$

Hence $\int_K \operatorname{div} \underline{\tau}_h \, dx = 0$, with $\underline{\tau}_h \in Q(K)$, implies $\operatorname{div} \underline{\tau}_h = 0$, and the first abstract hypothesis (3.59) is fulfilled.

The space Σ_h can be taken as in (3.88), and an interpolation operator Π_h from Σ^* to Σ_h verifying (3.60)–(3.61) can be defined locally by

$$\int_e (\underline{\tau} - \Pi_h \underline{\tau}) \cdot \underline{n} \, ds = 0, \quad \forall e \text{ edge of } K. \tag{3.121}$$

Π_h is well defined, since the matrix $\int_{e^i} \underline{\tau}^j \cdot \underline{n}^i \, ds$, with $i, j = 1, 2, 3$, is nonsingular, as can be easily seen by construction. Moreover, it can be proved that

$$\|\Pi_h \underline{\tau}\|_{\tilde{\Sigma}} \leqslant C \|\underline{\tau}\|_{\Sigma^*}. \tag{3.122}$$

The second abstract hypothesis is then fulfilled and the abstract theory applies. In particular, since we are using a conforming approximation, error estimate (3.78) holds true with $N_h = 0$. In this case, (3.79) is not applicable since only an interpolation estimate in $L^2(\Omega)$ can be proved

$$\|\underline{\tau} - \Pi_h \underline{\tau}\|_{0,\Omega} \leqslant Ch \|\underline{\tau}\|_{1,\Omega}. \tag{3.123}$$

We explicitly point out that, in contrast with the previous examples, in this case the operator P_h, as defined in (3.71), is not the L^2-projection. Consequently, the commuting diagram property (3.105) does not hold true for this element, but the interpolation estimate for u in (3.111) still applies. Hence, for this element (3.78) gives

$$\|\underline{\sigma} - \underline{\sigma}_h\|_{0,\Omega} + \|u - u_h\|_{0,\Omega} \leqslant Ch\bigl(|\underline{\sigma}|_{1,\Omega} + |u|_{1,\Omega}\bigr). \tag{3.124}$$

The last example we present here has also been introduced in MARINI and PIETRA [1989] and it is an example of nonconforming mixed finite element: the continuity of the normal component of the vector variable is imposed only in weak form and the inclusion of Σ_h in Σ is violated. In contrast with the previous case, for this example property (3.91) will be satisfied.

EXAMPLE 6. Consider a triangular decomposition of Ω. For each triangle $K \in \mathcal{T}_h$ we choose the finite-dimensional polynomial sets as follows

$$Q(K) = \operatorname{span}\{\underline{\tau}^1, \underline{\tau}^2, \underline{\tau}^3\}, \tag{3.125}$$

$$P(K) = P_0(K), \tag{3.126}$$

where

$$\underline{\tau}^1 = (1, 0), \qquad \underline{\tau}^2 = (0, 1), \qquad \underline{\tau}^3 = (\psi_1, \psi_2). \tag{3.127}$$

The choice for ψ_1 and ψ_2 is

$$\psi_1, \psi_2 \in P_1(K), \tag{3.128}$$

and, for a chosen edge $\tilde{e}$ of K, $\underline{\tau}^3 = (\psi_1, \psi_2)$ is defined through the following degrees of freedom

$$\begin{cases} \underline{\tau}^3 \cdot \underline{n}_{|\tilde{e}} = 1/|\tilde{e}|, \\ \int_e \underline{\tau}^3 \cdot \underline{n}\, ds = 0, \ e \neq \tilde{e}, \\ \int_K \psi_1\, dx = \int_K \psi_2\, dx = 0. \end{cases} \tag{3.129}$$

PROPOSITION 3.3. *The degrees of freedom* (3.129) *uniquely define* $\underline{\tau}^3$.

PROOF. Since $\underline{\tau}^3$ is seeked in a space of dimension 6 (see (3.128)), and the number of degrees of freedom (3.129) is precisely 6, it is sufficient to check uniqueness. To fix ideas, let e^1 be the special edge $\tilde{e}$, and let e^2, e^3 be the other two edges (see Fig. 3.1). Let then $\underline{\tau} = (\tau_1, \tau_2)$ be a vector verifying

(i) $\quad \underline{\tau} \cdot \underline{n}^1_{|e^1} = 0,$

(ii) $\quad \displaystyle\int_{e^j} \underline{\tau} \cdot \underline{n}^j\, ds = 0, \quad j \neq 1,$

(iii) $\quad \displaystyle\int_K \tau_1\, dx = \int_K \tau_2\, dx = 0.$

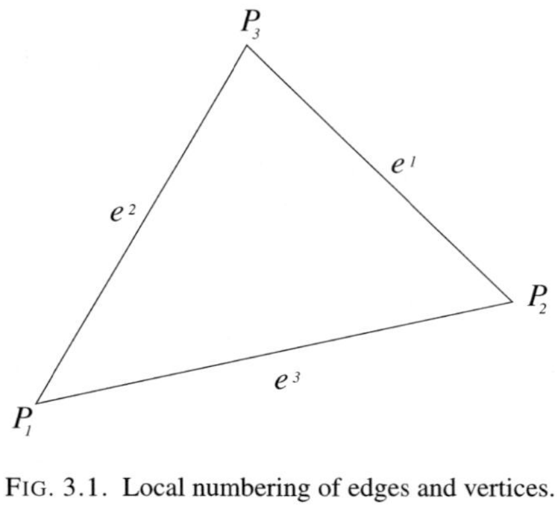

FIG. 3.1. Local numbering of edges and vertices.

Since $\operatorname{div} \underline{\tau}$ is constant, from (i)–(ii) we deduce $\operatorname{div} \underline{\tau} = 0$. From this and (iii), after integration by parts, we have, for all $p_1 \in P_1(K)$,

$$0 = \int_K \underline{\tau} \cdot \underline{\nabla} p_1 \, dx = \int_{\partial K} \underline{\tau} \cdot \underline{n} p_1 \, ds = \int_{e^2} \underline{\tau} \cdot \underline{n}^2 p_1 \, ds + \int_{e^3} \underline{\tau} \cdot \underline{n}^3 p_1 \, ds, \tag{3.130}$$

where in the last step we used (i). Taking $p_1 = \lambda_2$ in (3.130) (λ_2 being the barycentric coordinates verifying $\lambda_{2|e^2} \equiv 0$, $\lambda_2(P_2) = 1$) we obtain

$$\int_{e^3} \underline{\tau} \cdot \underline{n}^3 \lambda_2 \, ds = 0. \tag{3.131}$$

Using Simpson's rule and (ii) we deduce $\underline{\tau} \cdot \underline{n}^3(P_2) = 0$. This, together with (ii), implies $\underline{\tau} \cdot \underline{n}^3 = 0$ on e^3. With the same argument we deduce $\underline{\tau} \cdot \underline{n}^2 = 0$ on e^2. Hence, $\underline{\tau} \cdot \underline{n} = 0$ on ∂K, together with (3.128), ends the proof. □

Due to the choice of $Q(K)$, $\dim(Q(K)) = 3$, but $\underline{\tau} \cdot \underline{n}_{|e} \in P_1(e)$; hence, for $\underline{\tau} \in Q(K)$ continuity of $\underline{\tau} \cdot \underline{n}$ on the interelement edges cannot be imposed, as for the previous examples. However, a weak continuity can be required. More precisely, denoting by $\mathcal{E}_h$ the set of edges of $\mathcal{T}_h$, we define

$$\Sigma_h = \left\{ \underline{\tau}_h \in \left(L^2(\Omega)\right)^2 \,|\, \underline{\tau}_{h|K} \in Q(K),\ \forall K \in \mathcal{T}_h, \right.$$
$$\left. \int_e [\underline{\tau}_h \cdot \underline{n}] \, ds = 0,\ \forall e \in \mathcal{E}_h \setminus \Gamma_D \right\}, \tag{3.132}$$

where $[\underline{\tau}_h \cdot \underline{n}]$ denotes the jump of $\underline{\tau}_h \cdot \underline{n}$ across the edge e when e is an internal edge, and, when e belongs to Γ_N, $[\underline{\tau}_h \cdot \underline{n}]$ simply denotes $\underline{\tau}_h \cdot \underline{n}$.

Note that (3.129) implies that

$$\int_K \operatorname{div} \underline{\tau}^3 \, dx = \int_{\tilde{e}} \underline{\tau}^3 \cdot \underline{n} \, ds = 1 \quad \Rightarrow \quad \operatorname{div} \underline{\tau}^3|_K = 1/|K|, \tag{3.133}$$

where $|K|$ denotes the area of K. Therefore, (3.91) is satisfied, and the first abstract hypothesis (3.59) is fulfilled; moreover, the operator P_h defined from V to V_h in (3.71) is the usual orthogonal L^2-projection.

An interpolation operator Π_h from Σ^* to Σ_h, verifying (3.60)–(3.61), can be defined locally by

$$\int_e (\underline{\tau} - \Pi_h \underline{\tau}) \cdot \underline{n} \, ds = 0, \quad \forall e \text{ edge of } K. \tag{3.134}$$

Π_h is well defined, since the matrix $\int_{e^i} \underline{\tau}^j \cdot \underline{n}^i \, ds$, with $i, j = 1, 2, 3$, is nonsingular, as can be easily seen by construction. We notice that (3.134) implies

$$\operatorname{div} \Pi_h \underline{\tau}_{|K} = P_h \operatorname{div} \underline{\tau}_{|K}, \quad \forall \underline{\tau} \in \Sigma^*. \tag{3.135}$$

Therefore,

$$\|\Pi_h \underline{\tau}\|_{\tilde{\Sigma}} \leqslant C \|\underline{\tau}\|_{\Sigma^*}, \tag{3.136}$$

and the second abstract hypothesis is satisfied. Moreover, the following interpolation estimates hold

$$\|\underline{\tau} - \Pi_h \underline{\tau}\|_{0,\Omega} \leqslant Ch|\underline{\tau}|_{1,\Omega}, \tag{3.137}$$

$$\sum_K \|\operatorname{div}(\underline{\tau} - \Pi_h \underline{\tau})\|^2_{0,K} \leqslant Ch^2 |\operatorname{div} \underline{\tau}|^2_{1,\Omega}. \tag{3.138}$$

In contrast with the previous examples, the term N_h is no longer zero, due to the non-conformity of the discrete space Σ_h, and has to be bounded properly.

PROPOSITION 3.4. *Let Σ_h be the space defined in* (3.132), *and let $w \in H^1(\Omega)$. Then*

$$\sup_{\underline{\tau}_h \in \Sigma_h} \frac{N_h(w, \underline{\tau}_h)}{\|\underline{\tau}_h\|_{\widetilde{\Sigma}}} \leqslant Ch|w|_{1,\Omega}, \tag{3.139}$$

where N_h is defined in (3.73).

PROOF. Let $\mathcal{E}'$ be the set of edges of $\mathcal{T}_h$ not belonging to Γ_D. We can rewrite (3.73) as

$$N_h(w, \underline{\tau}_h) = \sum_{e \in \mathcal{E}'} \int_e w[\underline{\tau}_h \cdot \underline{n}]\, ds. \tag{3.140}$$

Let P_h^e be the piecewise constant interpolant of w defined as

$$\int_e (w - P_h^e w)\, ds = 0, \quad \forall e \in \mathcal{E}'. \tag{3.141}$$

According to (3.132), we have

$$N_h(w, \underline{\tau}_h) = \sum_{e \in \mathcal{E}'} \int_e (w - P_h^e w)[\underline{\tau}_h \cdot \underline{n}]\, ds. \tag{3.142}$$

Any $\underline{\tau}_h \in \Sigma_h$ can be split as $\underline{\tau}_h = \underline{\tau}' + \underline{\tau}''$, with $\underline{\tau}' \in \operatorname{span}\{\underline{\tau}^1, \underline{\tau}^2\}$, and $\underline{\tau}'' \in \operatorname{span}\{\underline{\tau}^3\}$ in K, $\forall K \in \mathcal{T}_h$. Therefore, using (3.141), (3.142) reads

$$N_h(w, \underline{\tau}_h) = \sum_{e \in \mathcal{E}'} \int_e (w - P_h^e w)[\underline{\tau}'' \cdot \underline{n}]\, ds. \tag{3.143}$$

Since $\operatorname{div} \underline{\tau}'' = 0$ in K if and only if $\underline{\tau}'' = 0$ in K due to (3.133), then $\|\operatorname{div} \underline{\tau}''\|_{0,K}$ is a norm. A simple scaling argument in (3.143), classical interpolation estimates, and Cauchy–Schwarz inequality give

$$N_h(w, \underline{\tau}_h) \leqslant C \sum_K \|w - P_h^e w\|_{0,\partial K} \|\operatorname{div} \underline{\tau}''\|_{0,K} h_K^{1/2} \tag{3.144}$$

$$\leqslant C \sum_K h_K |w|_{1,K} \|\operatorname{div} \underline{\tau}''\|_{0,K} \tag{3.145}$$

$$\leqslant Ch|w|_{1,\Omega} \|\underline{\tau}_h\|_{\widetilde{\Sigma}}, \tag{3.146}$$

and the proof is concluded. □

Using (3.137), (3.138), the interpolation estimate for u in (3.111), and (3.139) in (3.79) we deduce

$$\|\underline{\sigma} - \underline{\sigma}_h\|_{\widetilde{\Sigma}} + \|u - u_h\|_{0,\Omega} \leqslant Ch\big(|\underline{\sigma}|_{1,\Omega} + |\operatorname{div}\underline{\sigma}|_{1,\Omega} + |u|_{1,\Omega}\big). \tag{3.147}$$

3.6. *Hybridization of the mixed formulation*

It is well known that problem (3.58) leads to a final linear system whose matrix can be indefinite (this is the case, for instance, when the zeroth-order term is not present in (2.14)). A way to circumvent this problem is to relax the continuity of the normal component of the vectors at the interelement boundaries (even the weak one of the nonconforming case) and to enforce it back through the use of Lagrange multipliers. This technique, introduced in FRAEIJS DE VEUBEKE [1965] in a different context, was used successfully as a trick to deal with the algebraic system. In ARNOLD and BREZZI [1985] it was also studied from the theoretical point of view in the context of mixed formulations for the Raviart–Thomas elements. Error estimates were derived for the Lagrange multipliers, which proved to give an approximation of the scalar variable at the interelements better than that given directly by u_h. In BREZZI, DOUGLAS JR and MARINI [1985] error estimates were proved for the *BDM*-family, while in MARINI and PIETRA [1989] the elements of Examples 5–6 of Section 3.5 were analyzed. In order to describe this procedure, we then introduce a new space of vectors, made of discontinuous piecewise polynomial functions (without boundary conditions) defined as

$$\widehat{\Sigma}_h = \big\{\underline{\tau}_h \in \big(L^2(\Omega)\big)^2 \mid \underline{\tau}_{h|K} \in Q(K),\ \forall K \in \mathcal{T}_h\big\}, \tag{3.148}$$

equipped with the norm

$$\|\underline{\tau}\|_h^2 = \|\underline{\tau}\|_{0,\Omega}^2 + \sum_K h_K^2 \|\operatorname{div}\underline{\tau}\|_{0,K}^2, \quad \underline{\tau} \in \widehat{\Sigma}_h. \tag{3.149}$$

To enforce continuity across interelement boundaries of the normal component of vectors in $\widehat{\Sigma}_h$ we need to define a space for the multipliers. Let $\mathcal{E}_h$ be the set of all edges of the decomposition $\mathcal{T}_h$, and let $\mathcal{E}_h'$ be the set of edges not belonging to Γ_D. For every edge $e \in \mathcal{E}_h$ we introduce a finite dimensional space of scalars $R(e)$. For the sake of simplicity, we assume that, on each edge e, $R(e)$ consists of polynomials of degree $\leqslant k$. We set

$$\Lambda_h = \big\{\mu_h \in L^2(\mathcal{E}_h) \mid \mu_{h|e} \in R(e),\ \forall e \in \mathcal{E}_h\big\}, \tag{3.150}$$

and

$$\Lambda_{h,0} = \big\{\mu_h \in \Lambda_h \mid \mu_{h|e} = 0,\ \forall e \in \mathcal{E}_h \cap \Gamma_D\big\}, \tag{3.151}$$

equipped with the norm

$$\|\mu_h\|_{1/2,h}^2 = \sum_{e\in\mathcal{E}_h} h_e^{-1} \|\mu_h\|_{0,e}^2, \tag{3.152}$$

where h_e denotes the length of the edge e. The norm (3.152) is a sort of $H^{1/2}$-norm, which is natural for a space "of traces". We also have to introduce a bilinear form on $\widehat{\Sigma}_h \times \Lambda_h$:

$$d_h(\mu_h, \underline{\tau}_h) = \sum_K \int_{\partial K} \mu_h \underline{\tau}_h \cdot \underline{n}\, ds, \qquad \mu_h \in \Lambda_h,\ \underline{\tau}_h \in \widehat{\Sigma}_h. \tag{3.153}$$

We denote by $[\underline{\tau} \cdot \underline{n}]_{|e}$ the jump of $\underline{\tau} \cdot \underline{n}$ across the edge e when e is internal; for boundary edges, $[\underline{\tau} \cdot \underline{n}]_{|e}$ simply denotes $\underline{\tau} \cdot \underline{n}$. Then, (3.153) can be equivalently written as

$$d_h(\mu_h, \underline{\tau}_h) = \sum_e \int_e \mu_h [\underline{\tau} \cdot \underline{n}]\, ds, \qquad \mu_h \in \Lambda_h,\ \underline{\tau}_h \in \widehat{\Sigma}_h. \tag{3.154}$$

We define

$$\operatorname{Ker} D_h = \left\{\underline{\tau}_h \in \widehat{\Sigma}_h \mid d_h(\mu_h, \underline{\tau}_h) = 0,\ \forall \mu_h \in \Lambda_{h,0}\right\}, \tag{3.155}$$

having denoted by D_h the operator from $\widehat{\Sigma}_h \to (\Lambda_{h,0})'$ associated with the bilinear form $d_h(.,.)$. Using (3.154) and the definition of $\Lambda_{h,0}$ it is easy to see that

$$\underline{\tau} \in \operatorname{Ker} D_h \quad \Rightarrow \quad \int_e \mu_h [\underline{\tau} \cdot \underline{n}]\, ds = 0, \quad \forall \mu_h \in R(e)\ \forall e \in \mathcal{E}'_h. \tag{3.156}$$

Together with the abstract assumptions (3.59) and (3.62), we need two additional hypotheses:

$$\Sigma_h = \operatorname{Ker} D_h, \tag{3.157}$$

and a discrete inf-sup condition for the bilinear form $d_h(.,.)$

$$\exists \delta > 0: \quad \inf_{\mu_h \in \Lambda_h} \sup_{\underline{\tau}_h \in \widehat{\Sigma}_h} \frac{d_h(\mu_h, \underline{\tau}_h)}{\|\mu_h\|_{1/2,h} \|\underline{\tau}_h\|_h} \geqslant \delta. \tag{3.158}$$

A direct consequence of (3.158) is that, for all $e \in \mathcal{E}_h$,

$$\int_e \mu_h \underline{\tau}_h \cdot \underline{n}\, ds = 0, \quad \forall \underline{\tau}_h \in Q(K) \text{ with } e \subset \partial K \quad \Rightarrow \quad \mu_{h|e} = 0. \tag{3.159}$$

For every function $\xi \in L^2(\Gamma_D)$, define now

$$\Lambda_{h,\xi} = \left\{\mu_h \in \Lambda_h \;\Big|\; \int_e (\mu_h - \xi) \underline{\tau}_h \cdot \underline{n}\, ds = 0,\ \forall \underline{\tau}_h \in \widehat{\Sigma}_h,\ \forall e \in \mathcal{E}_h \cap \Gamma_D\right\}. \tag{3.160}$$

Note that, thanks to (3.159), condition $\int_e (\mu_h - \xi) \underline{\tau}_h \cdot \underline{n}\, ds = 0\ \forall \underline{\tau}_h \in \widehat{\Sigma}_h$ determines uniquely μ_h on e. The affine manifold $\Lambda_{h,\xi}$ can be seen as the subset of Λ_h made of functions μ_h which satisfy $\mu_h = \xi$ on Γ_D in a weak sense.

The discrete formulation of (3.27) is then:

$$\begin{cases} \text{Find } (\hat{\underline{\sigma}}_h, \hat{u}_h, \lambda_h) \in \widehat{\Sigma}_h \times V_h \times \Lambda_{h,g} \text{ such that} \\ a(\hat{\underline{\sigma}}_h, \underline{\tau}_h) + b_h(\hat{u}_h, \underline{\tau}_h) - d_h(\lambda_h, \underline{\tau}_h) = 0, & \forall \underline{\tau}_h \in \widehat{\Sigma}_h, \\ b_h(v_h, \hat{\underline{\sigma}}_h) - c(\hat{u}_h, v_h) = -(f, v_h), & \forall v_h \in V_h, \\ d_h(\mu_h, \hat{\underline{\sigma}}_h) = 0, & \forall \mu_h \in \Lambda_{h,0}. \end{cases} \tag{3.161}$$

THEOREM 3.5. *Problem* (3.161) *has a unique solution* $(\hat{\underline{\sigma}}_h, \hat{u}_h, \lambda_h)$, *and* $(\hat{\underline{\sigma}}_h, \hat{u}_h)$ *coincides with* $(\underline{\sigma}_h, u_h)$, *solution of Problem* (3.58).

PROOF. Let $(\underline{\sigma}^*_h, u^*_h, \lambda^*_h) \in \widehat{\Sigma}_h \times V_h \times \Lambda_{h,0}$ be the solution of the homogeneous discrete problem associated with (3.161). Taking $\underline{\tau}_h = \underline{\sigma}^*_h$ in the first equation of (3.161), and using the second equation with $v_h = u^*_h$ and the third equation with $\mu_h = \lambda^*_h$ we obtain

$$a(\underline{\sigma}^*_h, \underline{\sigma}^*_h) + c(u^*_h, u^*_h) = 0. \tag{3.162}$$

This implies that $\underline{\sigma}^*_h = 0$, and, if $\gamma_0 > 0$, $u^*_h = 0$. Instead, if $\gamma_0 = 0$, the first equation of (3.161) with $\underline{\tau}_h \in \operatorname{Ker} D_h$ reduces to

$$b_h(u^*_h, \underline{\tau}_h) = 0, \quad \forall \underline{\tau}_h \in \operatorname{Ker} D_h. \tag{3.163}$$

Thanks to (3.62) and (3.157), this implies $u^*_h = 0$. Hence, we are left with the equation

$$d_h(\lambda^*_h, \underline{\tau}_h) = 0, \quad \forall \underline{\tau}_h \in \widehat{\Sigma}_h, \tag{3.164}$$

which implies $\lambda^*_h = 0$, due to (3.158). Therefore, uniqueness is proved.

The third equation of problem (3.161) gives $\hat{\underline{\sigma}}_h \in \operatorname{Ker} D_h$, and hence, due to (3.157),

$$\hat{\underline{\sigma}}_h \in \Sigma_h. \tag{3.165}$$

Notice that for $\underline{\tau}_h \in \Sigma_h$ the bilinear form $d_h(\lambda_h, \underline{\tau}_h)$ simplifies to

$$d_h(\lambda_h, \underline{\tau}_h) = \sum_{e \subset \Gamma_D} \int_e \lambda_h \underline{\tau}_h \cdot \underline{n}\, ds = \langle g, \underline{\tau}_h \cdot \underline{n} \rangle_{|\Gamma_D}, \tag{3.166}$$

since $\lambda_h \in \Lambda_{h,g}$. Hence, the first equation of (3.161) with $\underline{\tau}_h \in \Sigma_h$ reads

$$a(\hat{\underline{\sigma}}_h, \underline{\tau}_h) + b_h(\hat{u}_h, \underline{\tau}_h) = \langle g, \underline{\tau}_h \cdot \underline{n} \rangle_{|\Gamma_D}, \quad \forall \underline{\tau}_h \in \Sigma_h. \tag{3.167}$$

Collecting (3.165), (3.167) and the second equation of problem (3.161) we see that $(\hat{\underline{\sigma}}_h, \hat{u}_h)$ is solution of problem (3.58). □

From now on the superscript ^ will be dropped from the variables.

A projection P^e_h from $L^2(\mathcal{E}_h)$ to Λ_h can be defined locally as the usual orthogonal L^2-projection

$$\int_e \mu_h \big(u - P^e_h u\big)\, ds = 0, \quad \forall e \subset \partial K,\ \forall \mu_h \in R(e). \tag{3.168}$$

Let us give now the abstract error estimate for the Lagrange multipliers.

THEOREM 3.6. *Let* $(\underline{\sigma}, u)$ *be the solution of* (3.27), *and* $(\underline{\sigma}_h, u_h, \lambda_h)$ *that of* (3.161). *The following estimate holds*

$$\begin{aligned} &\big\|\lambda_h - P^e_h u\big\|_{1/2,h} \\ &\quad \leqslant C\bigg(\|\underline{\sigma} - \underline{\sigma}_h\|_{0,\Omega} + \|P_h u - u_h\|_{1,h} + \sup_{\underline{\tau} \in \widehat{\Sigma}_h} \frac{d_h(u - P^e_h u, \underline{\tau})}{\|\underline{\tau}\|_h}\bigg), \end{aligned} \tag{3.169}$$

where $\|\cdot\|_{1,h}$ *is defined as*

$$\|v_h\|_{1,h}^2 = \sum_K h_K^{-2}\|v_h\|_{0,K}^2, \quad \forall v_h \in V_h, \tag{3.170}$$

and P_h *is defined in* (3.71).

PROOF. Taking $\mu = \lambda_h - P_h^e u$, from (3.158) we deduce

$$\begin{aligned}\delta\left\|\lambda_h - P_h^e u\right\|_{1/2,h} &\leqslant \sup_{\underline{\tau}\in\widehat{\Sigma}_h} \frac{d_h(\lambda_h - P_h^e u, \underline{\tau})}{\|\underline{\tau}\|_h} \\ &\leqslant \sup_{\underline{\tau}\in\widehat{\Sigma}_h} \frac{d_h(\lambda_h - u, \underline{\tau})}{\|\underline{\tau}\|_h} + \sup_{\underline{\tau}\in\widehat{\Sigma}_h} \frac{d_h(u - P_h^e u, \underline{\tau})}{\|\underline{\tau}\|_h}.\end{aligned} \tag{3.171}$$

Next, let us observe that $(\underline{\sigma}, u)$ solution of (3.27) verifies

$$a(\underline{\sigma}, \underline{\tau}) + b_h(u, \underline{\tau}) - d_h(u, \underline{\tau}) = 0, \quad \forall \underline{\tau} \in \widehat{\Sigma}_h. \tag{3.172}$$

Then, subtracting the first equation of (3.161) from (3.172) and using definition (3.71) we obtain

$$d_h(u - \lambda_h, \underline{\tau}) = a(\underline{\sigma} - \underline{\sigma}_h, \underline{\tau}) + b_h(P_h u - u_h, \underline{\tau}) \quad \forall \underline{\tau} \in \widehat{\Sigma}_h. \tag{3.173}$$

Using (3.53) and the Cauchy–Schwarz inequality

$$\begin{aligned}b_h(P_h u - u_h, \underline{\tau}) &\leqslant \sum_K h_K \|\operatorname{div}\underline{\tau}\|_{0,K} h_K^{-1}\|P_h u - u_h\|_{0,K} \\ &\leqslant \|\underline{\tau}\|_h \left(\sum_K h_K^{-2}\|P_h u - u_h\|_{0,K}^2\right)^{1/2} \\ &= \|\underline{\tau}\|_h \|P_h u - u_h\|_{1,h}.\end{aligned} \tag{3.174}$$

Hence, from (3.173), (3.174) we have

$$\sup_{\underline{\tau}\in\widehat{\Sigma}_h} \frac{d_h(\lambda_h - u, \underline{\tau})}{\|\underline{\tau}\|_h} \leqslant C\left(\|\underline{\sigma} - \underline{\sigma}_h\|_{0,\Omega} + \|P_h u - u_h\|_{1,h}\right). \tag{3.175}$$

Using (3.175) in (3.171) gives (3.169). □

COROLLARY 3.1. *Under the same assumptions as in Theorem* 3.6, *if the decomposition* $\mathcal{T}_h$ *is quasi-uniform, we have*

$$\begin{aligned}\left\|\lambda_h - P_h^e u\right\|_{0,\mathcal{E}_h} \leqslant C\Bigg(& h^{1/2}\|\underline{\sigma} - \underline{\sigma}_h\|_{0,\Omega} + h^{-1/2}\|P_h u - u_h\|_{0,\Omega} \\ &+ h^{1/2} \sup_{\underline{\tau}\in\widehat{\Sigma}_h} \frac{d_h(u - P_h^e u, \underline{\tau})}{\|\underline{\tau}\|_h}\Bigg).\end{aligned} \tag{3.176}$$

PROOF. The result follows multiplying (3.169) by $h^{1/2}$ and using the quasi-uniformity assumption. □

REMARK 3.5. In all practical cases, the space $\widehat{\Sigma}_h$ will be made of piecewise polynomials of some given degree. This implies that the following inequality will easily hold

$$\|\underline{\tau} \cdot \underline{n}\|_{0,e}^2 \leqslant C h_K^{-1} \|\underline{\tau}\|_{0,K}^2, \quad \forall \underline{\tau} \in \widehat{\Sigma}_h,\ \forall K,\ \forall e \subset \partial K, \tag{3.177}$$

where C denotes a constant depending on the degree of polynomials. In these cases we have, using (3.177) and Cauchy–Schwarz inequality,

$$\frac{d_h(u - P_h^e u, \underline{\tau})}{\|\underline{\tau}\|_h} \leqslant C \|u - P_h^e u\|_{1/2,h}, \tag{3.178}$$

so that (3.169) becomes

$$\begin{aligned} \|\lambda_h - P_h^e u\|_{1/2,h} \leqslant C\big(&\|\underline{\sigma} - \underline{\sigma}_h\|_{0,\Omega} + \|P_h u - u_h\|_{1,h} \\ &+ \|u - P_h^e u\|_{1/2,h}\big). \end{aligned} \tag{3.179}$$

Moreover, if the decomposition is quasi-uniform, (3.176) becomes

$$\begin{aligned} \|\lambda_h - P_h^e u\|_{0,\mathcal{E}_h} \leqslant C\big(&h^{1/2}\|\underline{\sigma} - \underline{\sigma}_h\|_{0,\Omega} + h^{-1/2}\|P_h u - u_h\|_{0,\Omega} \\ &+ h^{1/2}\|u - P_h^e u\|_{1/2,h}\big). \end{aligned} \tag{3.180}$$

REMARK 3.6. The above estimates on the multipliers, as (3.169) and (3.176), or (3.179) and (3.180), could be used in order to estimate the error for suitable lifting of λ_h. For instance, if λ_h is piecewise constant on the edges, one can introduce a new approximation u_h^* to u defined as the P_1-nonconforming function that, on each edge e, verifies

$$\int_e (u_h^* - \lambda_h)\, ds = 0. \tag{3.181}$$

Similarly, one can introduce the P_1-nonconforming interpolant of u, that we denote by u_I^*, defined as

$$\int_e (u_I^* - u)\, ds = 0. \tag{3.182}$$

It is immediate to check that, for each $K \in \mathcal{T}_h$

$$\|u_h^* - u_I^*\|_{0,K} \leqslant C \sum_{e \subset \partial K} h_e^{1/2} \|\lambda_h - P_h^e u\|_{0,e}, \tag{3.183}$$

that allows to use the above estimates in order to obtain suitable bounds for $\|u_h^* - u_I^*\|_{0,\Omega}$ (and then for $\|u - u_h^*\|_{0,\Omega}$ by the triangle inequality).

We shall discuss now how the hybridization of the mixed formulation, presented above in the abstract framework, can be applied to the examples of Section 3.5. For this, we associate with each pair $(Q(K), P(K))$ suitable spaces of polynomials defined on the edges $\mathcal{E}_h$.

EXAMPLE 1. For any integer $k \geqslant 0$, and for any triangle $K \in \mathcal{T}_h$, define

$$k \geqslant 0 \quad \begin{cases} Q(K) = (P_k(K))^2 + \underline{x} P_k(K), \\ P(K) = P_k(K), \\ R(e) = P_k(e), \quad \forall e \text{ edge of } K. \end{cases} \tag{3.184}$$

EXAMPLE 2. For any integer $k \geqslant 0$, and for any rectangle $K \in \mathcal{T}_h$, define

$$k \geqslant 0 \quad \begin{cases} Q(K) = (Q_k(K) + x_1 Q_k(K)) \times (Q_k(K) + x_2 Q_k(K)), \\ P(K) = Q_k(K), \\ R(e) = P_k(e), \quad \forall e \text{ edge of } K. \end{cases} \tag{3.185}$$

EXAMPLE 3. For any integer $k \geqslant 1$, and for any triangle $K \in \mathcal{T}_h$, define

$$k \geqslant 1 \quad \begin{cases} Q(K) = (P_k(K))^2, \\ P(K) = P_{k-1}(K), \\ R(e) = P_k(e), \quad \forall e \text{ edge of } K. \end{cases} \tag{3.186}$$

EXAMPLE 4. For any integer $k \geqslant 1$, and for any rectangle $K \in \mathcal{T}_h$, define

$$k \geqslant 1 \quad \begin{cases} Q(K) = (P_k(K))^2 \oplus \{\underline{\text{curl}}(x_1 x_2^{k+1})\} \oplus \{\underline{\text{curl}}(x_2 x_1^{k+1})\}, \\ P(K) = P_{k-1}(K), \\ R(e) = P_k(e), \quad \forall e \text{ edge of } K. \end{cases} \tag{3.187}$$

EXAMPLE 5. For each triangle $K \in \mathcal{T}_h$, define

$$\begin{cases} Q(K) = \text{span}\{\underline{\tau}^1, \underline{\tau}^2, \underline{\tau}^3\} \quad \text{(see (3.114)–(3.116))}, \\ P(K) = P_0(K), \\ R(e) = P_0(e), \quad \forall e \text{ edge of } K. \end{cases} \tag{3.188}$$

EXAMPLE 6. For each triangle $K \in \mathcal{T}_h$, define

$$\begin{cases} Q(K) = \text{span}\{\underline{\tau}^1, \underline{\tau}^2, \underline{\tau}^3\} \quad \text{(see (3.127)–(3.129))}, \\ P(K) = P_0(K), \\ R(e) = P_0(e), \quad \forall e \text{ edge of } K. \end{cases} \tag{3.189}$$

Denoting by $R(\partial K) = \prod_{e \in \partial K} R(e)$, we see that all the triplets $(Q(K), P(K), R(\partial K))$ of the examples above are such that $\operatorname{Ker} D_h$ coincides with the space Σ_h used in Problem (3.58), so that assumption (3.157) is fulfilled. Indeed, for Example 6 definition (3.132) of Σ_h clearly coincides with that of (3.155), as it can be easily seen using (3.156). For Examples 1–5 the space Σ_h is defined in (3.88), and $\underline{\tau}_h \in \Sigma_h$ *iff* $\underline{\tau}_h \in \widehat{\Sigma}_h$ and $[\underline{\tau} \cdot \underline{n}]_{|e} = 0$, $\forall e \in \mathcal{E}_h'$, that is, $\underline{\tau} \cdot \underline{n}$ is continuous across the internal edges and $\underline{\tau} \cdot \underline{n} = 0$ on Γ_N. Moreover, we have that $\underline{\tau} \in Q(K) \Rightarrow \underline{\tau} \cdot \underline{n}_{|\partial K} \in R(\partial K)$, and in particular, $[\underline{\tau} \cdot \underline{n}]_{|e} \in R(e)$. Then, according to (3.156), $\underline{\tau} \in \operatorname{Ker} D_h$ implies that $[\underline{\tau} \cdot \underline{n}]_{|e} = 0$, $\forall e \in \mathcal{E}_h'$. Therefore, $\operatorname{Ker} D_h = \Sigma_h$.

We check now that the inf-sup condition (3.158) holds for the bilinear form $d_h(.,.)$ in all the examples.

PROPOSITION 3.5. *For the mixed finite elements of Examples* 1–6 *it holds*

$$\exists\delta > 0: \quad \inf_{\mu_h\in\Lambda_h} \sup_{\underline{\tau}_h\in\widehat{\Sigma}_h} \frac{d_h(\mu_h,\underline{\tau}_h)}{\|\mu_h\|_{1/2,h}\|\underline{\tau}_h\|_h} \geqslant \delta. \tag{3.190}$$

PROOF. Let us consider $\mu^* \in \Lambda_h$. Let e be an edge in $\mathcal{E}_h$ and let $K = K(e)$ be an element of $\mathcal{T}_h$ having e as an edge. Take $\underline{\tau}^e \in \widehat{\Sigma}_h$ such that $\underline{\tau}^e = 0$ in $\Omega \setminus K$, and in K is the vector $\underline{\tau}^e \in Q(K)$ which satisfies

$$\int_e \mu\underline{\tau}^e\cdot\underline{n}\,ds = h_e^{-1}\int_e \mu^*\mu\,ds, \quad \forall\mu\in R(e), \tag{3.191}$$

$$\int_{\partial K\setminus e} \mu\underline{\tau}^e\cdot\underline{n}\,ds = 0, \quad \forall\mu\in R(\partial K), \tag{3.192}$$

$$\int_K \underline{\tau}^e\cdot\underline{\nabla}v\,dx = 0, \quad \forall v\in P(K), \tag{3.193}$$

$$\int_K \underline{\tau}^e\cdot\underline{\varphi}\,dx = 0, \quad \forall\underline{\varphi}\in D, \tag{3.194}$$

where $D = \{\underline{q}\in Q(K) \mid \operatorname{div}\underline{q} = 0,\ \underline{q}\cdot\underline{n}_{|\partial K} = 0\}$ as defined in (3.92). Such a vector $\underline{\tau}^e$ exists and is unique in $Q(K)$ since (3.191)–(3.194) are a set of degrees of freedom for all $Q(K)$ considered here, as discussed in Section 3.5. Notice that the degrees of freedom (3.191), (3.192) are always present and allow us to impose conditions on the normal component of vectors in $Q(K)$; (3.193) and (3.194) are needed for higher order elements.

A usual scaling argument gives

$$\|\underline{\tau}^e\|_{0,K}^2 + h_K^2\|\operatorname{div}\underline{\tau}^e\|_{0,K}^2 \leqslant Ch_e^{-1}\|\mu^*\|_{0,e}^2. \tag{3.195}$$

Define now $\underline{\tau}^* = \sum_e \underline{\tau}^e$. Since $\underline{\tau}^*$ on a single element K is the sum of at most three $\underline{\tau}^e$, summing (3.195) over all $K\in\mathcal{T}_h$ gives

$$\|\underline{\tau}^*\|_h \leqslant C\left(\sum_e h_e^{-1}\|\mu^*\|_{0,e}^2\right)^{1/2} = C\|\mu^*\|_{1/2,h}. \tag{3.196}$$

Using the definition of $\underline{\tau}^*$ and (3.191) we obtain

$$\int_e \mu[\underline{\tau}^*\cdot\underline{n}]\,ds = \int_e \mu[\underline{\tau}^e\cdot\underline{n}]\,ds = \int_e \mu\underline{\tau}^e\cdot\underline{n}\,ds = h_e^{-1}\int_e \mu^*\mu\,ds, \tag{3.197}$$

and consequently we have

$$d_h(\underline{\tau}^*,\mu) = \sum_e\int_e \mu[\underline{\tau}^*\cdot\underline{n}]\,ds = \sum_e h_e^{-1}\int_e \mu^*\mu\,ds. \tag{3.198}$$

Hence, taking $\mu = \mu^*$ in (3.198) and using (3.196) we have

$$\frac{d_h(\underline{\tau}^*,\mu^*)}{\|\underline{\tau}^*\|_h} \geqslant C^{-1}\frac{\sum_e h_e^{-1}\|\mu^*\|_{0,e}^2}{(\sum_e h_e^{-1}\|\mu^*\|_{0,e}^2)^{1/2}} = C^{-1}\|\mu^*\|_{1/2,h}, \tag{3.199}$$

and the proof is concluded with $\delta = C^{-1}$. □

Due to Proposition 3.5 the abstract error estimate (3.169) holds, (together with (3.176), if the decomposition is quasi-uniform). Notice that, when conforming approximations are considered, the term $d_h(u - P_h^e u, \underline{\tau})$, with $\underline{\tau} \in \widehat{\Sigma}_h$, vanishes. This is the case for all the Examples 1–5. In particular, for the RT_0 element, using (3.110) in (3.176) we obtain

$$\|\lambda_h - P_h^e u\|_{0,\mathcal{E}_h} \leqslant C\big(h^{1/2}\|\underline{\sigma} - \underline{\sigma}_h\|_{0,\Omega} + h^{-1/2}\|P_h u - u_h\|_{0,\Omega}\big) \leqslant Ch^{1/2}. \tag{3.200}$$

The same estimate applies to the element of Example 5, using (3.124) in (3.176). For the nonconforming element of Example 6 we give the following

PROPOSITION 3.6. *Let* $\widehat{\Sigma}_h$ *and* Λ_h *be the spaces defined in* (3.148) *and* (3.150) *associated with the element* (3.189). *For* $w \in H^1(\Omega)$ *it holds*

$$\sup_{\underline{\tau} \in \widehat{\Sigma}_h} \frac{d_h(w - P_h^e w, \underline{\tau})}{\|\underline{\tau}\|_h} \leqslant C|w|_{1,\Omega}, \tag{3.201}$$

where P_h^e *is defined in* (3.168).

PROOF. We can use a scaling argument on each edge $e \in \mathcal{E}_h$ and obtain for $\underline{\tau} \in \widehat{\Sigma}_h$

$$\int_e (w - P_h^e w)\underline{\tau} \cdot \underline{n}\, ds \leqslant Ch_K^{1/2}|w|_{1,K}\|\underline{\tau} \cdot \underline{n}\|_{0,e} \leqslant C|w|_{1,K}\|\underline{\tau}\|_{0,K}. \tag{3.202}$$

Then, (3.201) easily follows. □

Consequently, using (3.147) and (3.201) in (3.176) we deduce

$$\begin{aligned}\|\lambda_h - P_h^e u\|_{0,\mathcal{E}_h} &\leqslant C\big(h^{1/2}\|\underline{\sigma} - \underline{\sigma}_h\|_{0,\Omega} + h^{-1/2}\|P_h u - u_h\|_{0,\Omega} + h^{1/2}|u|_{1,\Omega}\big)\\ &\leqslant Ch^{1/2}.\end{aligned} \tag{3.203}$$

REMARK 3.7. The order $\mathcal{O}(h^{1/2})$ in estimates (3.200), (3.203) is not in disagreement with the estimates for the scalar variable u. Indeed, (3.200) (or (3.203)) in (3.183) gives back the $\mathcal{O}(h)$ order of convergence proved for these elements. Moreover, for the RT_0 element we can use the superconvergence result (see DOUGLAS JR and ROBERTS [1985])

$$\|P_h u - u_h\|_{0,\Omega} \leqslant Ch^2, \tag{3.204}$$

thus obtaining in (3.200)

$$\|\lambda_h - P_h^e u\|_{0,\mathcal{E}_h} \leqslant Ch^{3/2}. \tag{3.205}$$

This, in turns, using (3.183), gives

$$\|u_h^* - u_I^*\|_{0,\Omega} \leqslant Ch^2, \tag{3.206}$$

which is an optimal error bound. Estimate (3.204) (and then, in the end, (3.206)) can also be proved for the lowest order elements of the families of the first four examples presented here. Actually, one can consider other values of k as well, obtaining better convergence estimates, analogous to (3.204), (3.205), and finally (3.206), for suitably defined higher order interpolants (instead of u_I^*) and liftings (instead of u_h^*). We refer for that, e.g., to ARNOLD and BREZZI [1985], BREZZI, DOUGLAS JR and MARINI [1985]. Finally, we recall that for all the examples presented here estimates for $\|\lambda_h - P_h^e u\|_{0,\mathcal{E}_h}$ can be obtained directly, case by case, dropping the quasi-uniformity assumption on the mesh. We refer to ARNOLD and BREZZI [1985], BREZZI, DOUGLAS JR and MARINI [1985], MARINI and PIETRA [1989] for the proofs.

3.7. *Algebraic treatment of problem (3.161)*

Problem (3.161) is easier to deal with than (3.75). Indeed, the linear system associated with (3.161) takes the matrix form

$$\begin{pmatrix} A & B & -D \\ B^t & -C & 0 \\ -D^t & 0 & 0 \end{pmatrix} \begin{pmatrix} \underline{\sigma}_h \\ u_h \\ \lambda_h \end{pmatrix} = \begin{pmatrix} 0 \\ -F \\ 0 \end{pmatrix}, \tag{3.207}$$

with obvious meaning of the notation. The matrix A is now a block-diagonal matrix, each block being a matrix of dimension equal to the dimension of $Q(K)$, easy to invert. Hence, the variable $\underline{\sigma}_h$ can be eliminated by static condensation,

$$\underline{\sigma}_h = A^{-1}(D\lambda_h - Bu_h), \tag{3.208}$$

leading to the new system

$$\begin{pmatrix} B^t A^{-1} B + C & -B^t A^{-1} D \\ -D^t A^{-1} B & D^t A^{-1} D \end{pmatrix} \begin{pmatrix} u_h \\ \lambda_h \end{pmatrix} = \begin{pmatrix} F \\ 0 \end{pmatrix}. \tag{3.209}$$

Since no continuity assumption is made on V_h, the matrix $B^t A^{-1} B + C$ is also block-diagonal, so that the variable u_h can be eliminated by static condensation,

$$u_h = (B^t A^{-1} B + C)^{-1}(F + B^t A^{-1} D\lambda_h). \tag{3.210}$$

This leads to a final system, acting on the unknown λ_h only, of the form

$$\mathcal{M}\lambda_h = \mathcal{G}, \tag{3.211}$$

where $\mathcal{M}$ and $\mathcal{G}$ are given by

$$\mathcal{M} = D^t A^{-1} D - D^t A^{-1} B(B^t A^{-1} B + C)^{-1} B^t A^{-1} D, \tag{3.212}$$

$$\mathcal{G} = D^t A^{-1} B(B^t A^{-1} B + C)^{-1} F. \tag{3.213}$$

We present in detail the structure of the matrix $\mathcal{M}$ and of the right-hand side $\mathcal{G}$ defined in (3.212) and (3.213). Since for applications to semiconductor device simulation the low regularity of the solution makes the use of high order elements unsuitable, we discuss here only the case of lowest order elements, namely, the lowest order Raviart–Thomas element on triangles (RT_0) and the two elements of Examples 5–6. We construct the element matrix $\mathcal{M}^K$ and the element right-hand side $\mathcal{G}^K$ associated with the current element K. In order to give a compact presentation, we set some common notation and recall the definition of the triplet $(Q(K), P(K), R(\partial K))$ for the three cases under consideration.

$$\begin{aligned} Q(K) &= \operatorname{span}\{\underline{\tau}^1, \underline{\tau}^2, \underline{\tau}^3\}, \\ P(K) &= P_0(K), \\ R(\partial K) &= \prod_{e \in \partial K} P_0(e), \end{aligned} \tag{3.214}$$

where

$$\underline{\tau}^1 = (1, 0), \qquad \underline{\tau}^2 = (0, 1), \qquad \underline{\tau}^3 = (\psi_1, \psi_2). \tag{3.215}$$

The choice for ψ_1 and ψ_2 makes the difference of the three cases. For RT_0 we take

$$\underline{\tau}^3 = \underline{x} - \underline{x}_B, \quad \underline{x}_B = \text{ coordinates of the centroid of } K. \tag{3.216}$$

The reason for this choice is that $\int_K \psi_1 \, dx = \int_K \psi_2 \, dx = 0$, and therefore each block of A becomes diagonal. We refer to the definition (3.115)–(3.116) for the element of Example 5 and to (3.128)–(3.129) for the element of Example 6.

As basis function in $P_0(K)$ we make the natural choice $v = 1$ in K, and as basis function in $R(\partial K)$ we take $\mu = 1$ on one edge e and $\mu = 0$ on the others.

The functions $a^{-1}(x)$ and $\gamma(x)$ appearing in the bilinear forms (3.24) and (3.26) are approximated by piecewise constant functions defined in each element K by

$$\overline{\alpha} := \frac{1}{|K|}\left(\int_K a^{-1}(x) \, dx\right), \tag{3.217}$$

$$\overline{\gamma} := \frac{1}{|K|}\left(\int_K \gamma(x) \, dx\right). \tag{3.218}$$

We introduce the following notation

$$\underline{\nu}^i = \underline{n}^i |e^i|, \quad i = 1, 3, \tag{3.219}$$

$$\delta = \int_K (\psi_1^2 + \psi_2^2) \, dx, \tag{3.220}$$

$$\beta = \int_K \operatorname{div} \underline{\tau}^3 \, dx, \tag{3.221}$$

$$\eta_i = \int_{e^i} \underline{\tau}^3 \cdot \underline{n} \, ds, \quad i = 1, 3. \tag{3.222}$$

Then the element matrices are

$$A^K = \overline{\alpha} \begin{pmatrix} |K| & 0 & 0 \\ 0 & |K| & 0 \\ 0 & 0 & \delta \end{pmatrix}, \qquad B^K = \begin{pmatrix} 0 \\ 0 \\ \beta \end{pmatrix}, \tag{3.223}$$

$$D^K = \begin{pmatrix} \nu_1^1 & \nu_1^2 & \nu_1^3 \\ \nu_2^1 & \nu_2^2 & \nu_2^3 \\ \eta_1 & \eta_2 & \eta_3 \end{pmatrix}, \qquad C^K = \overline{\gamma}|K|, \qquad F^K = \int_K f \, dx. \tag{3.224}$$

The coefficients of the matrix $\mathcal{M}^K$ are then given by

$$m_{ij}^K = (\overline{\alpha})^{-1} \frac{\underline{\nu}^i \cdot \underline{\nu}^j}{|K|} + \frac{\overline{\gamma}|K|}{\beta^2 + \overline{\alpha}\delta\overline{\gamma}|K|} \eta_i \eta_j, \quad i, j = 1, 3, \tag{3.225}$$

and the right-hand side by

$$g_i^K = \frac{\beta \eta_i}{\beta^2 + \overline{\alpha}\delta\overline{\gamma}|K|} \int_K f \, dx, \quad i = 1, 3. \tag{3.226}$$

If $\gamma(x) = 0$ in (2.14), the coefficients of $\mathcal{M}^K$ reduce to

$$m_{ij}^K = (\overline{\alpha})^{-1} \frac{\underline{\nu}^i \cdot \underline{\nu}^j}{|K|}. \tag{3.227}$$

The matrix $\mathcal{M}$ thus corresponds to a nonconforming piecewise linear approximation of (2.14), where the function $a(x)$ is approximated by its harmonic average:

$$a(x)_{|K} \simeq \frac{|K|}{\int_K a^{-1}(x) \, dx}. \tag{3.228}$$

The final matrix $\mathcal{M}$ is then symmetric and positive definite. Moreover, if the decomposition is of weakly acute type (i.e., every angle θ of every triangle is $\theta \leqslant \pi/2$), then

$$m_{ii}^K > 0, \quad m_{ij}^K \leqslant 0, \quad i \neq j, \ i, j = 1, 3. \tag{3.229}$$

Hence, in this case, $\mathcal{M}$ is an M-matrix. In the general case $\gamma(x) \geqslant 0$, we see that the quantity $\overline{\gamma}|K|/(\beta^2 + \overline{\alpha}\delta\overline{\gamma}|K|)$ is always nonnegative. As far as the η_i are concerned, we have to discuss separately what happens in the three cases considered above. For the RT_0 element, a simple computation shows that

$$\eta_i = \frac{2|K|}{3} > 0, \quad i = 1, 3. \tag{3.230}$$

Therefore, the off-diagonal coefficients of $\mathcal{M}^K$ might be positive even if the triangulation is weakly acute, and the final matrix is not, in general, an M-matrix.

Instead, the other two elements are designed to guarantee a final M-matrix. Indeed, to fix ideas, let e^1 be the special edge ($\tilde{e}$ in the definitions (3.116) or (3.129)). Then,

$$\tilde{e} = e^1 \quad \Rightarrow \quad \eta_1 > 0, \quad \eta_2 = \eta_3 = 0, \tag{3.231}$$

with $\eta_1 = |e^1|$ for Example 5, and $\eta_1 = 1$ for Example 6. Incidentally we also notice that, for both elements, definition (3.221) gives $\beta = \eta_1$. Hence, the coefficients of $\mathcal{M}^K$ are given by

$$m_{ij}^K = \begin{cases} (\overline{\alpha})^{-1} \frac{\underline{\nu}^1 \cdot \underline{\nu}^1}{|K|} + \frac{\overline{\gamma}|K|}{\beta^2 + \overline{\alpha}\delta\overline{\gamma}|K|} \eta_1^2, & \text{for } i = j = 1, \\ (\overline{\alpha})^{-1} \frac{\underline{\nu}^i \cdot \underline{\nu}^j}{|K|}, & \text{otherwise,} \end{cases} \tag{3.232}$$

and the right-hand side is given by

$$g_i^K = \begin{cases} \frac{\eta_1^2}{\beta^2 + \overline{\alpha}\delta\overline{\gamma}|K|} \int_K f \, dx, & \text{for } i = 1, \\ 0, & \text{otherwise}. \end{cases} \tag{3.233}$$

It is then clear that the final matrix is, for both elements, always an M-matrix, if the decomposition is of weakly acute type, since the zeroth-order term gives contribution, with the positive sign, only to the coefficient m_{11}. For an example of a mixed finite element over rectangles satisfying the M-matrix property we refer to MARINI and PIETRA [1991].

Another approach which yields the M-matrix property for the RT_0 element, even in the case $\gamma(x) \geqslant 0$, will be examined in the following Section 3.8.

3.8. *Numerical quadrature: towards finite volumes*

In this section we analyze the family of mixed finite volume methods proposed in MICHELETTI, SACCO and SALERI [2001] for the approximation of the reaction-diffusion problem (2.14). All of the methods are a suitable discretization of the dual mixed formulation (3.27) and employ the lowest-order Raviart–Thomas (RT_0) finite element spaces (3.107)–(3.108) plus a suitable quadrature formula for the matrix corresponding to $a(\underline{\sigma}_h, \underline{\tau}_h)$. This allows the use of different averages of the inverse diffusion coefficient a^{-1} to enforce the constitutive law $(3.3)_1$ for the fluxes at the interelement boundaries in a finite volume fashion.

The mixed finite volume formulation (MFV) addressed in this section is to be viewed as an alternative approach to the hybridization procedure discussed in Section 3.6, and has been the object of several researches in the recent literature. The central issue of the MFV formulation is to perform a *lumping* of the matrix A associated with $a(\underline{\sigma}_h, \underline{\tau}_h)$ in (3.58) through some suitable quadrature formula. In the case of rectangular grids, this strategy has been first proposed in POLAK, SCHILDERS and COUPERUS [1988] in the case $a(x) = 1$. Theoretical and implementational issues that link "nodal" finite elements, mesh-centered finite differences and mixed-hybrid finite elements have been addressed in HENNART and DEL VALLE [1993, 1996]. In MOLENAAR [1995] a theoretical analysis shows that, under appropriate smoothness assumptions, the quadrature error (in the evaluation of A and of the right-hand side in (3.58)) does not spoil the accuracy of the mixed method with exact integration.

In the case of triangular elements, similar conclusions have been drawn in BARANGER, MAITRE and OUDIN [1994], BARANGER, MAITRE and OUDIN [1996], where, in the case of the Laplace operator, a quadrature formula to diagonalize the matrix A is proposed and analyzed following HAUGAZEAU and LACOSTE [1993]. An extension of this latter lumping procedure has been carried out in AGOUZAL, BARANGER, MAITRE and OUDIN [1995], SACCO and SALERI [1997a], SACCO and SALERI [1997b], MICHELETTI and SACCO [1999] for diffusion and convection-diffusion problems. In particular, in AGOUZAL, BARANGER, MAITRE and OUDIN [1995] $a(x)$ is approximated by its harmonic average over each triangle. In this section we analyze a family of averages, all characterized by being piecewise constant over the dual tessellation of the domain. Moreover we also include in the analysis the zeroth-order term.

The case of an elliptic problem where $a(x)$ is a symmetric positive definite tensor has been studied in ARBOGAST, WHEELER and YOTOV [1997], CAI, JONES, MCCORMICK and RUSSELL [1996], where RT_0 mixed finite elements with numerical integration are considered on both rectangular and logically rectangular grids. Another numerical scheme on this subject has been proposed in ARBOGAST, DAWSON, KEENAN, WHEELER and YOTOV [1998], where an expanded mixed finite element method capable to handle the case of a discontinuous tensor $a(x)$ and general shape elements and geometry is derived. It can be checked that in the special case of a reference equilateral triangle and RT_0 finite elements, the quadrature rule proposed in ARBOGAST, DAWSON, KEENAN, WHEELER and YOTOV [1998] to diagonalize the mass matrix yields the same result as the lumping formula of BARANGER, MAITRE and OUDIN [1994]. However, the resulting method in ARBOGAST, DAWSON, KEENAN, WHEELER and YOTOV [1998] gives a ten-point finite difference stencil while the method we are going to analyze in this section gives a four-point finite difference stencil.

The common feature of the approaches based on a lumping procedure is that the mixed formulation using RT_0 finite elements can be interpreted as a finite volume method acting on the scalar unknown u_h. This connection has been also recently investigated in EWING, SAEVAREID and SHEN [1998], where cell-centered finite difference schemes are constructed on triangular grids of regular shape (equilateral, and isosceles right triangles). In this respect we also mention VASSILEVSKI, PETROVA and LAZAROV [1992] where finite difference schemes on triangular cell-centered grids are derived under the assumption of weakly acute triangulation. In this case, and using the harmonic average of $a(x)$ along the Voronoi edge, the methods of VASSILEVSKI, PETROVA and LAZAROV [1992] and of the present section coincide, although the error analysis in VASSILEVSKI, PETROVA and LAZAROV [1992] is carried out only in the case $a(x) = 1$.

Let us now give a brief outline of the contents of this section. The dual tessellation $\mathcal{L}_h$ associated with $\mathcal{T}_h$ is first introduced in Section 3.8.1. Notice that $\mathcal{T}_h$ is only required to be a Delaunay triangulation, while previous schemes in the literature assume that $\mathcal{T}_h$ must be a weakly acute triangulation (see BREZZI, MARINI and PIETRA [1989a], MARINI and PIETRA [1989], BANK, BÜRGLER, FICHTNER and SMITH [1990], VASSILEVSKI, PETROVA and LAZAROV [1992]). The Delaunay property allows for the presence of obtuse triangles in the mesh while in a weakly acute triangulation all the angles are required to be $\leqslant \pi/2$. Next, the mixed finite volume schemes are derived in Section 3.8.2. The methods are based on the combined use of a piecewise constant approximation $\overline{\alpha}$ of $\alpha \equiv a^{-1}$ over $\mathcal{L}_h$ and of the quadrature formula proposed in BARANGER, MAITRE and OUDIN [1994]. The resulting discretization is a cell-centered finite volume scheme where the degrees of freedom for u_h are taken at the circumcenters of each element of $\mathcal{T}_h$, while the interelement fluxes are computed using the values of u_h at the circumcenters of two neighboring triangles of $\mathcal{T}_h$.

The error analysis is carried out in Section 3.8.3, where we proved the (optimal) $\mathcal{O}(h)$ convergence of this family of methods with respect to the $H(\text{div}; \Omega) \times L^2(\Omega)$-norm, which is the standard norm for the analysis of dual mixed methods. It is worth noting that the derivation of the methods, as well as their convergence analysis, allows us in every respect to call them "mixed finite volume schemes" or, equivalently, "mixed finite element schemes with numerical integration". In Section 3.8.4 we consider three choices of $\overline{\alpha}$: two averages are suitable approximations of the *harmonic average* of a (see BABUŠKA and OSBORN [1983]) while the third one is the trapezoidal rule.

3.8.1. Geometry and notation

In view of the mixed finite volume discretization of problem (3.27) we introduce the following

DEFINITION 3.1. $\mathcal{T}_h$ is a Delaunay triangulation if, for every $K \in \mathcal{T}_h$, the closed circumcircle of K contains no other vertices than those belonging to K. Moreover, $\mathcal{T}_h$ is a degenerate Delaunay mesh when the above property only holds for the open circumcircle (cf. DELAUNAY [1934]).

We assume henceforth that $\mathcal{T}_h$ is a Delaunay triangulation. Moreover, we shall indicate by $\mathrm{N_E}$ and $\mathrm{N_T}$ the total number of edges and triangles of $\mathcal{T}_h$, respectively. Throughout this section, any geometrical entity will be always understood as being an open bounded subset of $\mathbb{R}^2$ or $\mathbb{R}$.

For any pair of neighboring triangles K_i and K_j of $\mathcal{T}_h$, $i, j = 1, \dots, \mathrm{N_T}$, let l be the common edge. We also consider the dual tessellation $\mathcal{L}_h$ of $\mathcal{T}_h$ and denote its elements by "lumping regions" (see Fig. 3.2, left). The lumping region $\mathcal{L}^l$ corresponding to edge l is obtained by joining the common vertices and the two circumcenters C_i and C_j (see Fig. 3.3). We set also $\mathcal{L}^l_k = \mathcal{L}^l \cap K_k$, for $k = i, j$.

Throughout this section it is understood that any geometrical entity referred to by a superscript, say r, and/or a subscript, say k, has the following meaning: the superscript and the subscript refer to an edge and a triangle, respectively. When both the superscript and the subscript appear, we are referring to an entity depending on the edge e^r that shares the triangle K_k. Moreover, $T(r)$ is the index set of the triangles shared by the edge e^r, and it is understood that the cardinality of $T(r)$ is one or two depending on whether or not the edge $e^r \in \partial\Omega$. By $E(k)$ we denote the index set of the edges of the triangle K_k. For any triangle K_k we have: three indices of the edges of K_k (say r, i, j), three corresponding edges e^r, e^i, e^j and the vectors $\underline{e}^r_k, \underline{e}^i_k, \underline{e}^j_k$ obtained by orienting the boundary of K_k counterclockwise. Observe that $\underline{e}^r_{k'} = -\underline{e}^r_k$ for $k', k \in T(r)$.

We assume the regularity assumptions (3.45)–(3.46) on $\mathcal{T}_h$, namely (see Fig. 3.2, right):

$$\frac{D_K}{\rho_K} \leqslant \mathcal{K}^*, \quad \forall K \in \mathcal{T}_h, \qquad D_K \leqslant \mathcal{K}^* \rho_K \leqslant \mathcal{K} h_K, \quad \forall K \in \mathcal{T}_h, \tag{3.234}$$

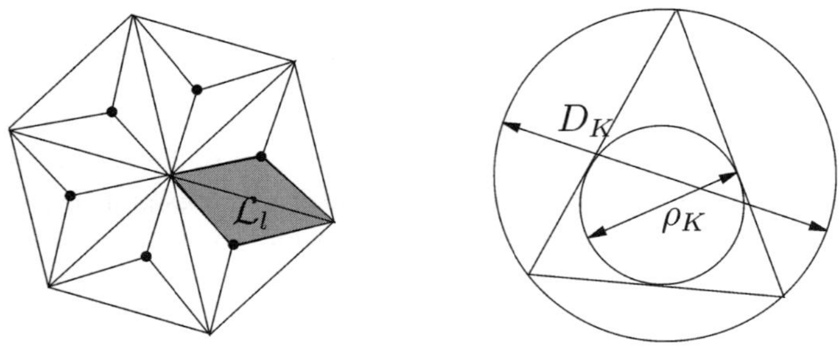

FIG. 3.2. Primal triangulation $\mathcal{T}_h$ with the corresponding lumping regions $\mathcal{L}^l$ (left), mesh parameters (right).

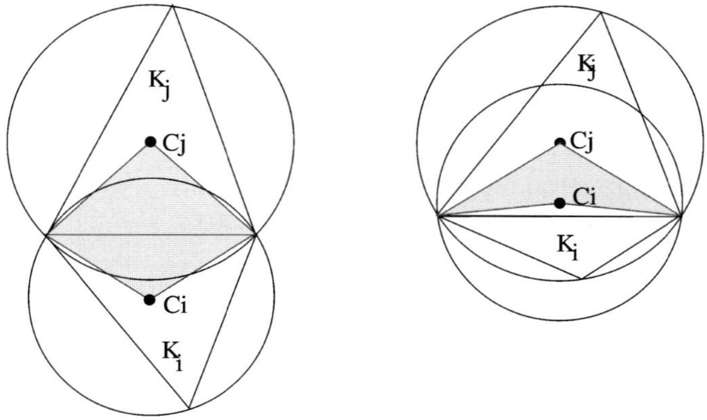

FIG. 3.3. Examples of lumping regions for acute (left) and obtuse (right) triangles.

where we recall that $\mathcal{K}$ denotes, here and in the sequel, a generic constant depending only on $\mathcal{K}^*$. It follows that the triangulation $\mathcal{T}_h$ satisfies

$$h_{\mathcal{L}^l} \leqslant D \leqslant \mathcal{K}h, \quad \forall \mathcal{L}^l \in \mathcal{L}_h,$$

where $D = \max_K D_K$, $h = \max_K h_K$ and $h_{\mathcal{L}^l}$ is the diameter of $\mathcal{L}^l$ (see Fig. 3.2, left).

3.8.2. A family of finite volume methods

Let us now deal with the saddle-point problem (3.58) using the lowest-order Raviart–Thomas approximation spaces (see (3.107)–(3.109)). For any $K_k \in \mathcal{T}_h$ and for any integrable function φ, define its mean value as

$$\varphi_k = \frac{1}{|K_k|}\int_{K_k} \varphi \, dx, \quad \forall K_k \in \mathcal{T}_h. \tag{3.235}$$

Taking v_h equal to the characteristic function χ_k of the triangle K_k in the second equation of (3.58) we obtain

$$\sum_{r\in E(k)} \int_{e^r} \underline{\sigma}_h \cdot \underline{n}_k^r \, ds - \int_{K_k} \gamma u_h \, dx = -\int_{K_k} f \, dx, \quad \forall K_k \in \mathcal{T}_h, \tag{3.236}$$

where $\underline{n}_k^r$ is the outward unit normal vector to edge e^r. Eq. (3.236) can be rewritten as

$$\sum_{r\in E(k)} \Phi_k^r - u_k \gamma_k |K_k| = -f_k |K_k|, \quad \forall K_k \in \mathcal{T}_h, \tag{3.237}$$

where Φ_k^r is the outward flux of $\underline{\sigma}_h$ from the triangle K_k through the edge e^r:

$$\Phi_k^r := \int_{e^r} \underline{\sigma}_h \cdot \underline{n}_k^r \, ds, \quad \text{with } \Phi_k^r = 0 \text{ on } e^r \in \Gamma_N. \tag{3.238}$$

Clearly, (3.237) is a genuine finite volume discretization for (2.14), provided that we are able to express each flux Φ_k^r in terms of the values u_j, $j \in T(r)$, only. With this aim, let us consider the first equation in (3.58), and denote by e^r an internal edge of K_k and by K_j the element of $\mathcal{T}_h$ such that $k, j \in T(r)$, i.e., $\partial K_k \cap \partial K_j = e^r$. Let $\underline{\tau}_h^r$ be the basis function defined in (3.109), where the normal is taken outward to K_k, so that $\operatorname{div} \underline{\tau}^r_{h|K_k} = 1/|K_k|$ and $\operatorname{div} \underline{\tau}^r_{h|K_j} = -1/|K_j|$. Taking $\underline{\tau}_h = \underline{\tau}_h^r$ in the first equation of problem (3.58), we get

$$\int_{K_k} \alpha \underline{\sigma}_h \cdot \underline{\tau}_h^r \, dx + \int_{K_j} \alpha \underline{\sigma}_h \cdot \underline{\tau}_h^r \, dx + u_k - u_j = 0, \tag{3.239}$$

where we have set $\alpha := a^{-1}$. For any $K_k \in \mathcal{T}_h$ let now $r, r' \in E(k)$; we introduce the following exact and approximate bilinear forms restricted over K_k

$$\begin{aligned} a^{K_k}(\underline{\tau}_h^{r'}, \underline{\tau}_h^r) &= (\alpha \underline{\tau}_h^{r'}, \underline{\tau}_h^r)_{0,K_k} = \int_{K_k} \alpha \underline{\tau}_h^{r'} \cdot \underline{\tau}_h^r \, dx, \\ a_h^{K_k}(\underline{\tau}_h^{r'}, \underline{\tau}_h^r) &= \overline{\alpha}^r (\underline{\tau}_h^{r'}, \underline{\tau}_h^r)_{h,0,K_k} = \overline{\alpha}^r \delta_{rr'} \omega_k^r, \end{aligned} \tag{3.240}$$

where the expression for ω_k^r is provided in the next proposition, and $\overline{\alpha}^r$ is a suitable average of α over $\mathcal{L}^r$.

PROPOSITION 3.7. *Let $\underline{\tau}_h^r$, $\underline{\tau}_h^{r'}$ be two basis functions in Σ_h associated with two edges of $K_k \in \mathcal{T}_h$. The quadrature formula*

$$\int_{K_k} \underline{\tau}_h^r \cdot \underline{\tau}_h^{r'} \, dx \sim \delta_{rr'} \omega_k^r \tag{3.241}$$

is exact on constant vectors iff *the ω_k^r's are chosen as*

$$\omega_k^r = -\frac{\underline{e}_k^i \cdot \underline{e}_k^j}{4|K_k|}, \quad i, j \in E(k),\ i \neq j \neq r. \tag{3.242}$$

PROOF. We provide in the following an alternative proof to the original one given in BARANGER, MAITRE and OUDIN [1994]. For ease of exposition let us switch to a local numbering of the indices of the geometrical quantities. In particular, we shall use the local indices 1, 2, 3 in place of the global ones. Then let K be a generic triangle with edges e^1, e^2, e^3, and let $\underline{t}^i$ and $\underline{n}^i$ ($i = 1, 2, 3$) be the unit tangent and normal vectors to edge e^i (see Fig. 3.4 for the local numbering and orientation). Finally, let $\underline{e}^i = \underline{t}^i |e^i|$, $\underline{\nu}^i = \underline{n}^i |e^i|$ ($i = 1, 2, 3$), where $|e^i|$ denotes the length of edge e^i. We require the numerical quadrature to be exact for constant vectors, i.e., we require the formula to integrate exactly

$$\int_K \underline{\tau} \cdot \underline{\sigma} \, dx, \quad \forall \text{ constant vectors } \underline{\tau}, \underline{\sigma}. \tag{3.243}$$

For this, let us first consider $\underline{\sigma} = \underline{\tau} = \underline{c}$, $\underline{c}$ being any constant vector. Let $\underline{\tau}^i$, $i = 1, 3$, be the local basis for RT_0 elements, defined by the following choice of degrees of freedom (see (3.97)):

$$\int_{e^j} \underline{\tau}^i \cdot \underline{n}^j \, ds = \delta_{ij}, \quad i, j = 1, 3. \tag{3.244}$$

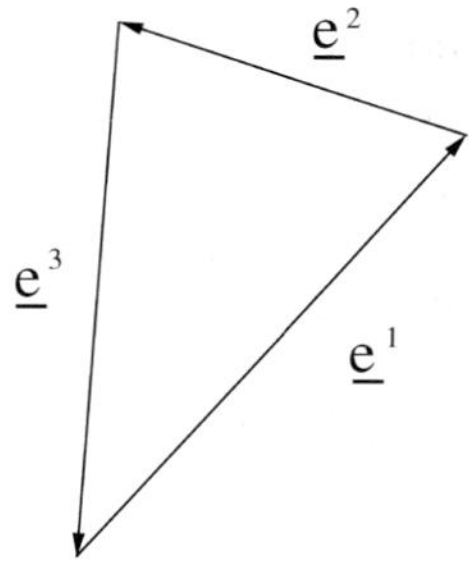

FIG. 3.4. Local numbering and orientation of the edges of a triangle.

Then $\underline{c}$ can be expressed as

$$\underline{c} = \sum_{i=1}^{3} c_i \underline{\tau}^i,$$

where the coefficients c_i can be easily obtained using (3.244):

$$(\underline{c}, \underline{\nu}^j) = (\underline{c}, \underline{n}^j)|e^j| = \int_{e^j} \underline{c} \cdot \underline{n}^j \, ds = \int_{e^j} \left(\sum_{i=1}^{3} c_i \underline{\tau}^i \right) \cdot \underline{n}^j \, ds = c_j.$$

By imposing the quadrature formula to be exact on constant vectors $\underline{\tau} = \underline{c}$ we obtain

$$\begin{aligned} |\underline{c}|^2|K| &\equiv \int_K \underline{c} \cdot \underline{c} \, dx = \int_K \left(\sum_{i=1}^{3} c_i \underline{\tau}^i \right) \cdot \left(\sum_{j=1}^{3} c_j \underline{\tau}^j \right) dx \\ &= \sum_{j=1}^{3} c_j^2 \omega^j = \sum_{j=1}^{3} (\underline{c}, \underline{\nu}^j)^2 \omega^j. \end{aligned} \tag{3.245}$$

Choosing now $\underline{c} = \underline{e}^i$, $i = 1, 2, 3$, and observing that $|(\underline{e}^i, \underline{\nu}^j)| = 2(1 - \delta_{ij})|K|$, from (3.245) we obtain

$$|e^i|^2|K| = 4|K|^2 \sum_{j=1}^{3} (1 - \delta_{ij})^2 \omega^j, \quad i = 1, 2, 3,$$

that is, the linear system

$$\sum_{j=1}^{3} (1 - \delta_{ij}) \omega^j = \frac{|e^i|^2}{4|K|}, \quad i = 1, 2, 3,$$

using the obvious fact that $(1 - \delta_{ij})^2 \equiv 1 - \delta_{ij}$. The solution to this system, given by

$$\omega^i = \frac{\sum_{j=1}^{3} (1 - 2\delta_{ij})|e^j|^2}{8|K|}, \tag{3.246}$$

can be further simplified by noting that, with the orientation of Fig. 3.4, $\underline{e}^1 + \underline{e}^2 + \underline{e}^3 = \underline{0}$, so that, for instance, $|e^1|^2 = |e^2|^2 + |e^3|^2 + 2(\underline{e}^2, \underline{e}^3)$, and analogous relations hold cyclically. Therefore, from (3.246) we have

$$\omega^1 = -\frac{(\underline{e}^2, \underline{e}^3)}{4|K|}, \qquad \omega^2 = -\frac{(\underline{e}^3, \underline{e}^1)}{4|K|}, \qquad \omega^3 = -\frac{(\underline{e}^1, \underline{e}^2)}{4|K|}. \tag{3.247}$$

Notice that these quantities are not necessarily all positive, as their sign depends on the angles of the triangle. It remains to prove that (3.247) gives a solution to (3.245) also for all possible choices of the constant vector $\underline{c}$. First, since (3.245) is homogeneous in $\underline{c}$, this can be taken of length 1. Second, letting $\underline{c} = (\cos\theta, \sin\theta)$, $\theta \in \mathbb{R}$ and substituting in (3.245) leads to a homogeneous polynomial of degree 2 in $\cos\theta$, $\sin\theta$ that must vanish identically in θ. Thus, it suffices to require that this occurs only for three different values of θ. Since the triangle is not degenerate, $\underline{e}^i$ $(i = 1, 2, 3)$ are three independent vectors which correspond to three different values for θ.

To conclude, we have to check that the formula is exact for any pair of constant vectors $\underline{\tau} = \underline{c}^1 \neq \underline{\sigma} = \underline{c}^2$. To this end, we observe that the bilinear form

$$\Phi(\underline{c}^1, \underline{c}^2) := (\underline{c}^1, \underline{c}^2)|K| - \sum_{j=1}^{3} (\underline{c}^1, \underline{\nu}^j)(\underline{c}^2, \underline{\nu}^j)\omega^j \tag{3.248}$$

is symmetric and verifies, for ω^j as in (3.247), $\Phi(\underline{c}, \underline{c}) = 0$. Using this, for every $\underline{c}^1$ and $\underline{c}^2$ we can write $\Phi(\underline{c}^1 + \underline{c}^2, \underline{c}^1 + \underline{c}^2) = 0$, and derive $\Phi(\underline{c}^1, \underline{c}^2) \equiv 0$. □

REMARK 3.8. For each triangle K_k and for each edge e^r, the quantities ω_k^r can also be computed using the formula

$$\omega_k^r = \frac{d_k^r}{|e^r|}, \tag{3.249}$$

where d_k^r is the "distance" between C_k and the edge e^r, in a sense that will be made clear in a while. For the moment we notice that these quantities are positive when the angle opposite to the edge e^r is acute, and negative when it is obtuse. This expression is very important in view of the finite volume interpretation of the numerical method obtained with the quadrature formula $(3.240)_2$. However, we point out that expression (3.242) is easier to compute, and it is actually used for the implementation of the method.

To prove (and explain) (3.249) note that, referring to Fig. 3.5, the scalar product in the numerator of (3.242) can be written as $|e^i||e^j|\cos\theta^r$, where $0 < \theta^r < \pi$ is the

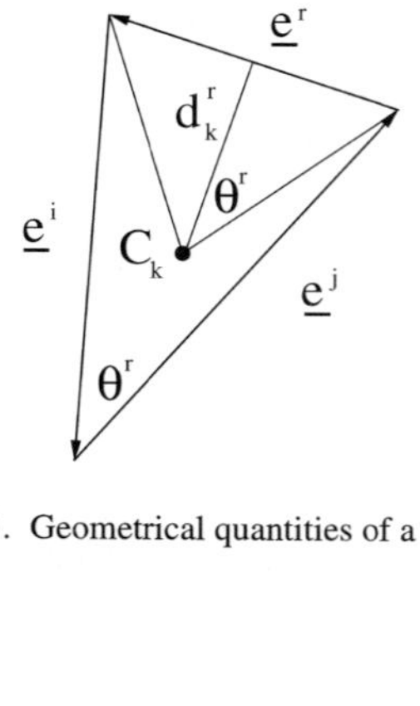

FIG. 3.5. Geometrical quantities of a triangle.

angle (opposite to the edge e^r) between edges e^i and e^j; on the other hand, $2|K_k| = |e^i||e^j|\sin\theta^r$. This gives $2\omega_k^r = \cot\theta^r$ and, since C_k is the circumcenter of the triangle, θ^r is also half the angle opposite to the edge e^r seen by C_k. This finally yields $\cot\theta^r = 2d_k^r/|e^r|$ and thus relation (3.249), where it is understood that d_k^r is positive when θ^r is acute (C_k inside K_k) and it is negative when θ^r is obtuse (C_k outside of K_k).

The construction of the piecewise constant function $\overline{\alpha}^r$ will be fully discussed in Sections 3.8.3 and 3.8.4; here we just emphasize the two basic properties of the average:
(i) $\overline{\alpha}^r$ is constant on each lumping region $\mathcal{L}^r$;
(ii) $\overline{\alpha}^r$ is some average of α on a suitable subset of each lumping region $\mathcal{L}^r$.
We notice that in the case $\alpha = 1$, the approximate bilinear form a_h^K coincides with the quadrature formula proposed in BARANGER, MAITRE and OUDIN [1994], where it is shown that the quadrature error is $\mathcal{O}(h_K)$. Let then $(\underline{\sigma}_h^*, u_h^*) \in \Sigma_h \times V_h$ be the solution of the dual mixed system (3.58) in the presence of some quadrature error.

Using $(3.240)_2$ in (3.239) and recalling that $\overline{\alpha}^r$ is constant over the *whole* lumping region $\mathcal{L}^r$, we end up with the following equation for the approximate interelement flux through the edge $e^r \notin \Gamma_N$,

$$\Phi_k^{r,*} = (\overline{\alpha}^r)^{-1}\left(\frac{u_j^* - u_k^*}{d^r}\right)|e^r|, \quad d^r = d_k^r + d_j^r, \quad k, j \in T(r). \tag{3.250}$$

Eq. (3.250 holds for any edge e^r in the interior of Ω; if e^r lies on Γ_D, it is understood that the value u_j^* is set equal to the average g^r of the Dirichlet datum g over e^r and that $d^r = d_k^r$ is the "distance" between the circumcenter of K_k and e^r (see BOSISIO, MICHELETTI and SACCO [2000]).

Substituting the exact fluxes Φ_k^r in (3.237) with the corresponding approximations (3.250), we finally obtain the family of cell-centered finite volume schemes in the new unknown u_h^*

$$\sum_{\substack{r\in E(k)\\ e^r\notin\Gamma_N}} (\overline{\alpha}^r)^{-1}\left(\frac{u_k^* - u_{j(r)}^*}{d^r}\right)|e^r| + u_k^*\gamma_k|K_k| = f_k|K_k|, \quad \forall K_k \in \mathcal{T}_h, \tag{3.251}$$

where, for an internal e^r, $j(r) \in T(r)$ is the index of the triangle shared by the edge e^r opposite to the triangle K_k, and $u_{j(r)}^*$ is the unknown on such a triangle. For an edge $e^r \in \Gamma_D$, we set $u_{j(r)}^* = g^r$. In this last case, g^r is the L^2-projection of g on the space of the constant functions over e^r, as in (3.160). Finally, we recall that $\Phi_k^{r,*} = 0$ on $e^r \in \Gamma_N$.

The set of linear algebraic equations (3.251) can be written in matrix form as

$$W^*\mathbf{u}^* = \mathbf{f}^*, \tag{3.252}$$

where the ith component of $\mathbf{f}^*$ is $f_i|K_i|$ and the ijth nonzero entries of the $\mathrm{N_T} \times \mathrm{N_T}$ matrix W^* are

$$W_{ij}^* = \begin{cases} \sum_{\substack{r\in E(i)\\ e^r\notin\Gamma_N}} (\overline{\alpha}^r)^{-1}\frac{|e^r|}{d^r} + \gamma_i|K_i|, & \text{if } i = j, \\ -(\overline{\alpha}^{r(j)})^{-1}\frac{|e^{r(j)}|}{d^{r(j)}}, & \text{if } j \in T(E(i)),\ j \neq i, \end{cases} \tag{3.253}$$

where $r(j) = E(j) \cap E(i)$ refers to the edge shared by the triangles K_i and K_j.

LEMMA 3.1. *Assume that $\mathcal{T}_h$ is a Delaunay triangulation. Then, the matrix W^* in* (3.252) *is a symmetric, positive definite and irreducibly diagonally dominant M-*matrix.

PROOF. We first notice that the Delaunay property for $\mathcal{T}_h$ implies that the quantities d^r's in (3.253) are positive. As a consequence, $W^*_{ii} > 0$ and $W^*_{ij} \leqslant 0$ in (3.253), for $i = 1, \ldots, \mathrm{N_T}$ and $j \in T(E(i))$, $j \neq i$. The expressions (3.253) show that W^* is a symmetric matrix with at most four nonzero entries on each row. Let s_i denote the sum of the entries of each row i of W^*; then

$$s_i = \begin{cases} \gamma_i |K_i|, & \text{if } \partial K_i \cap \Gamma_D = \emptyset, \\ \sum_{\substack{r \in E(i) \\ e^r \in \Gamma_D}} (\overline{\alpha}^r)^{-1} \frac{|e^r|}{d^r} + \gamma_i |K_i|, & \text{if } \partial K_i \cap \Gamma_D \neq \emptyset. \end{cases}$$

We see that the sum of the entries of the rows corresponding to triangles intersecting Γ_D is strictly positive while it is nonnegative for the rows corresponding to internal triangles. Thus, W^* is a symmetric, positive definite irreducibly diagonally dominant M-matrix (see VARGA [1962], Corollary 1, p. 85). □

REMARK 3.9. The M-matrix property ensures that system (3.252) is uniquely solvable. This property along with the fact that W^* is also positive definite and irreducibly diagonally dominant ensures that the family of finite volume schemes (3.251) verifies a discrete maximum principle (see VARGA [1962], CIARLET and RAVIART [1973], ROOS, STYNES and TOBISKA [1996]). In particular, provided $\mathbf{f}^* \geqslant 0$, the solution $\mathbf{u}^*$ of system (3.252) turns out to be nonnegative.

The M-matrix property is very desirable in applications. As an example, consider the case where (2.14) is the linearized current continuity equation using the drift-diffusion or energy-balance transport models. In such a case, the unknown u is the scaled concentration and therefore it must be nonnegative.

REMARK 3.10 (*Extension to the* 3D *case*). Let us briefly comment about the extension of the finite volume formulation at hand to the case where Ω is a polygonal domain in $\mathbb{R}^3$.

The extension is straightforward if Ω is of the form $\Omega = \Omega_{xy} \times (0, H)$ with $\Omega_{xy} \subset \mathbb{R}^2$ and $H > 0$, and $\mathcal{T}_h$ is a partition of Ω into prisms obtained as "tensor product" of a decomposition of Ω_{xy} and a decomposition of $[0, H]$. We refer to ARBOGAST and CHEN [1995] for the definition and to MIGLIO, QUARTERONI and SALERI [1999] for the use of mixed approximations with Raviart–Thomas finite elements on prismatic triangulations.

The extension of the mixed finite volume method analyzed in this section to the case where $\mathcal{T}_h$ is a decomposition of Ω into simplices is less straightforward. In particular, if the finite element mesh is made of tetrahedra a consistent diagonalization of the mass matrix has been provided in BARANGER, MAITRE and OUDIN [1996], in the case of the Laplace operator, under restrictive conditions on the mesh. Under the same assumptions, the present mixed finite volume scheme can be formulated also in three dimensions.

In the case of a general 3D triangulation the method of this section as well as the related approaches proposed in BARANGER, MAITRE and OUDIN [1994, 1996], AGOUZAL, BARANGER, MAITRE and OUDIN [1995] cannot be extended, as pointed out in THOMAS and TRUJILLO [1997].

3.8.3. *Analysis of the finite volume scheme*

In this section we analyze the finite volume scheme defined in (3.251) by setting it into an abstract framework of Galerkin methods with numerical quadratures for the approximation of saddle point problems of the form (3.35). Concerning this issue, error estimates and analysis can be found in BREZZI and FORTIN [1991], Section II.2.4 and ROBERTS and THOMAS [1991], Chapter 3, Section 11. For the case under consideration, the choice of RT_0 finite element spaces for Σ_h, V_h is such that the family of finite volume schemes (3.251) can be written in the form:

$$\begin{cases} \text{Find } \underline{\sigma}_h \in \Sigma_h \text{ and } u_h \in V_h \text{ such that} \\ a_h(\underline{\sigma}_h, \underline{\tau}_h) + b_h(u_h, \underline{\tau}_h) = \langle g, \underline{\tau}_h \rangle_h, \quad \forall \underline{\tau}_h \in \Sigma_h, \\ b_h(v_h, \underline{\sigma}_h) - c_h(u_h, v_h) = -\langle f, v_h \rangle_h, \quad \forall v_h \in V_h, \end{cases} \tag{3.254}$$

where $a_h(\cdot,\cdot)$, $b_h(\cdot,\cdot)$, $c_h(\cdot,\cdot)$ and $\langle\cdot,\cdot\rangle_h$ are suitable approximations of the corresponding continuous counterparts through the use of numerical quadratures. In this case we have

$$\begin{aligned} &a_h(\underline{\sigma}_h, \underline{\tau}_h) = \sum_{K_k \in \mathcal{T}_h} a_h^{K_k}(\underline{\sigma}_h, \underline{\tau}_h), \qquad b_h(v_h, \underline{\tau}_h) = b(v_h, \underline{\tau}_h), \\ &c_h(u_h, v_h) = \sum_{K_k \in \mathcal{T}_h} \gamma_k\, (u_h, v_h)_{0,K_k} \equiv c(u_h, v_h), \\ &\langle f, v_h \rangle_h = \sum_{K_k \in \mathcal{T}_h} f_k (v_h, 1)_{0,K_k} \equiv (f, v_h), \qquad \langle g, \underline{\tau}_h \rangle_h = \langle g, \underline{\tau}_h \cdot \underline{n} \rangle_{|\Gamma_D}. \end{aligned} \tag{3.255}$$

(See (3.25)–(3.27) and Section 3.8.2 for notation and definitions.) Hypotheses (3.31)–(3.34) are all verified, while (3.28) and (3.29) for $a_h(\cdot,\cdot)$ need to be checked in order to apply the abstract Theorem 3.1. For that we recall that definition (3.57) of $\operatorname{Ker} B_h$ reads in this case

$$\operatorname{Ker} B_h = \{ \underline{\tau}_h \in \Sigma_h \mid b(v_h, \underline{\tau}_h) = 0, \ \forall v_h \in V_h \}. \tag{3.256}$$

LEMMA 3.2. *There exist positive constants $\overline{\mathcal{A}}$ and $\underline{\mathcal{A}}$, independent of h, such that*

$$\begin{aligned} &|a_h(\underline{\sigma}_h, \underline{\tau}_h)| \leqslant \overline{\mathcal{A}} \|\underline{\sigma}_h\|_\Sigma \|\underline{\tau}_h\|_\Sigma, \quad \forall \underline{\sigma}_h, \underline{\tau}_h \in \Sigma_h, \\ &a_h(\underline{\tau}_h, \underline{\tau}_h) \geqslant \underline{\mathcal{A}} \|\underline{\tau}_h\|_\Sigma^2, \qquad \forall \underline{\tau}_h \in \operatorname{Ker} B_h. \end{aligned} \tag{3.257}$$

PROOF. To prove $(3.257)_1$, let us first observe that, $a_h(\underline{\sigma}_h, \underline{\tau}_h)$ being symmetric and positive semidefinite, we have

$$a_h(\underline{\sigma}_h, \underline{\tau}_h) \leqslant \big(a_h(\underline{\sigma}_h, \underline{\sigma}_h) a_h(\underline{\tau}_h, \underline{\tau}_h)\big)^{1/2}, \quad \forall \underline{\sigma}_h, \underline{\tau}_h \in \Sigma_h. \tag{3.258}$$

Then it is sufficient to prove that

$$a_h(\underline{\tau}_h, \underline{\tau}_h) \leqslant \overline{\mathcal{A}} \| \underline{\tau}_h \|_{\Sigma}^2, \quad \underline{\tau}_h \in \Sigma_h. \tag{3.259}$$

Consider a triangle $K_k \in \mathcal{T}_h$ and set $\underline{\tau}_{h|K_k} = \sum_{i \in E(k)} \Phi^i_{\underline{\tau}_h} \underline{\tau}^i_h$ and $\omega^i_k = d^i_k / |e^i|$, where $\Phi^i_{\underline{\tau}_h}$ is the flux of $\underline{\tau}_h$ across the edge e^i, and d^i_k is the "distance" between the circumcenter C_k of K_k and the edge e^i (see Section 3.8.2). We also set $\mathcal{R}_{K_k} = \bigcup_{i \in E(k)} \mathcal{L}^i$. For brevity, the subscript k will be dropped. On the element K the approximate bilinear form can be written as

$$a_h^K(\underline{\tau}_h, \underline{\tau}_h) = \sum_{i \in E(k)} \overline{\alpha}^i \big(\Phi^i_{\underline{\tau}_h}\big)^2 \omega^i_k.$$

Notice that, using (3.234), we have

$$\max_{i \in E(k)} |\omega^i_k| \leqslant \frac{D_K}{\rho_K} \leqslant \mathcal{K}^*, \tag{3.260}$$

which holds both for acute and obtuse triangles. Therefore, since $\overline{\alpha}$ is an average of α

$$\big|a_h^K(\underline{\tau}_h, \underline{\tau}_h)\big| \leqslant \|\alpha\|_{\infty, \mathcal{R}_K} \sum_{i \in E(k)} \big|\Phi^i_{\underline{\tau}_h}\big|^2 |\omega^i_k| \leqslant \mathcal{K}^* \|\alpha\|_{\infty, \Omega} \sum_{i \in E(k)} \big|\Phi^i_{\underline{\tau}_h}\big|^2. \tag{3.261}$$

Recalling the definition of flux through an edge e^i of $\underline{\tau}_h \in \Sigma_h$

$$\Phi^i_{\underline{\tau}_h} = \int_{e^i} \underline{\tau}_h \cdot \underline{n}^i \, ds,$$

Cauchy–Schwarz inequality gives

$$\big|\Phi^i_{\underline{\tau}_h}\big|^2 \leqslant |e^i| \|\underline{\tau}_h \cdot \underline{n}^i\|^2_{0, e^i}. \tag{3.262}$$

Using (3.177), i.e., $\|\underline{\tau}_h \cdot \underline{n}^i\|^2_{0, e^i} \leqslant C h_K^{-1} \|\underline{\tau}_h\|^2_{0, K}$, we deduce

$$\big|\Phi^i_{\underline{\tau}_h}\big|^2 \leqslant C \|\underline{\tau}_h\|^2_{0, K}, \quad \forall \underline{\tau}_h \in \Sigma_h. \tag{3.263}$$

It can be easily checked that, in the present case of RT_0 elements, $C \leqslant 6$. Substituting in (3.261) gives

$$\big|a_h^K(\underline{\tau}_h, \underline{\tau}_h)\big| \leqslant 18 \mathcal{K}^* \|\alpha\|_{\infty, \Omega} \|\underline{\tau}_h\|^2_{0, K}, \tag{3.264}$$

from which, summing over all the triangles of $\mathcal{T}_h$, the estimate $(3.257)_1$ follows with $\overline{\mathcal{A}} = 18 \mathcal{K}^* \|\alpha\|_{\infty, \Omega}$.

Let us now prove $(3.257)_2$. We remark that, in this case, it is not possible to work at the element level since the geometric quantities ω^r_k's associated with an element K_k are not, a priori, both positive. Actually, we are dealing with a Delaunay mesh and this guarantees only that $\sum_{k \in T(r)} \omega^r_k \geqslant 0$, the equal sign holding in the case of a degenerate Delaunay mesh. Thus, we are led to working with the full bilinear form. Since any function $\underline{\tau}_h \in \operatorname{Ker} B_h$ is such that $\underline{\tau}_{h|K} \in (P_0(K))^2$, $\forall K \in \mathcal{T}_h$, we have $\| \underline{\tau}_h \|_{\Sigma} = \| \underline{\tau}_h \|_{0, \Omega}$. Moreover, Proposition 3.7 ensures that the quadrature formula is exact on piecewise constant vectors, that is,

$$\sum_{i\in E(k)} \left(\Phi^i_{\underline{\tau}_h}\right)^2 \omega^i_k = \|\underline{\tau}_h\|^2_{0,K}, \qquad \underline{\tau}_h \in \operatorname{Ker} B_h.$$

Thus

$$\begin{aligned} a_h(\underline{\tau}_h, \underline{\tau}_h) &= \sum_{r=1}^{\mathrm{N_E}} \sum_{k\in T(r)} \overline{\alpha}^r \left(\Phi^r_{\underline{\tau}_h}\right)^2 \omega^r_k \geqslant \alpha_0 \sum_{r=1}^{\mathrm{N_E}} \sum_{k\in T(r)} \left(\Phi^r_{\underline{\tau}_h}\right)^2 \omega^r_k \\ &= \alpha_0 \sum_{k=1}^{\mathrm{N_T}} \sum_{r\in E(k)} \left(\Phi^r_{\underline{\tau}_h}\right)^2 \omega^r_k = \alpha_0 \sum_{k=1}^{\mathrm{N_T}} \|\underline{\tau}_h\|^2_{0,K} = \alpha_0 \|\underline{\tau}_h\|^2_{0,\Omega}, \end{aligned}$$

where $\alpha_0 = \inf_{x\in\Omega} \alpha(x)$. We finally get $(3.257)_2$ with $\underline{A} = \alpha_0$. □

We are in position to state the main result of this section.

THEOREM 3.7. *For every $g \in H^{1/2}(\Gamma_D)$, and for every $f \in L^2(\Omega)$ problem* (3.254) *has a unique solution $(\underline{\sigma}_h, u_h)$, and the following a priori error estimate holds*

$$\begin{aligned} \|\underline{\sigma} - \underline{\sigma}_h\|_\Sigma + \|u - u_h\|_V \leqslant \mathcal{M} \inf_{\substack{\underline{w}_h\in\Sigma_h \\ v_h\in V_h}} \Bigg\{ &\|\underline{\sigma} - \underline{w}_h\|_\Sigma + \|u - v_h\|_V \\ &+ \sup_{\underline{\tau}_h\in\Sigma_h} \frac{|a(\underline{w}_h, \underline{\tau}_h) - a_h(\underline{w}_h, \underline{\tau}_h)|}{\|\underline{\tau}_h\|_\Sigma} \Bigg\}, \end{aligned} \tag{3.265}$$

with $(\underline{\sigma}, u)$ solution of problem (3.27), *and $\mathcal{M}$ a constant independent of h.*

PROOF. It is an extension of the proof of Proposition 2.11, BREZZI and FORTIN [1991], which holds in the case of exact integration and it is here omitted. The interested reader can consult MICHELETTI, SACCO and SALERI [2001]. □

Let us deal now with the quadrature errors occurring in the approximate bilinear form $a_h(\cdot,\cdot)$. We recall that $\mathcal{K}$ denotes a generic constant depending only on the local geometrical properties of the mesh, namely on $\mathcal{K}^*$ (see (3.45)–(3.46)). With a slight change of notation, we let K be a generic triangle and C_K its circumcenter. We shall use the following result:

PROPOSITION 3.8. *Let $\mathcal{R}_K$ be the union of K and of the three lumping regions associated with its edges. Then under assumptions* (3.45)–(3.46) *on the mesh $\mathcal{T}_h$, for every $\overline{x} \in \overline{\mathcal{R}}_K$ we have*

$$\sup_{x\in\mathcal{R}_K} \left|v(x) - v(\bar{x})\right| \leqslant \mathcal{K} h_K |v|_{1,\infty,\mathcal{R}_K}, \quad \forall v \in C^1(\overline{\mathcal{R}}_K). \tag{3.266}$$

Inequality (3.266) follows immediately from the mean-value theorem. For more general inequalities of this type see, e.g., CIARLET [1991], Section 15, e.g., Theorem 15.3.

Let us now give a bound for the quadrature error associated with $a_h(\cdot,\cdot)$. With this aim, let $\underline{w}_h$ and $\underline{\tau}_h$ be any two functions in Σ_h. For any $K_k \in \mathcal{T}_h$ we set $\underline{w}_{h|K_k} = \sum_{r\in E(k)} \Phi^r_{\underline{w}_h} \underline{\tau}^r_h$ and $\underline{\tau}_{h|K_k} = \sum_{r\in E(k)} \Phi^r_{\underline{\tau}_h} \underline{\tau}^r_h$. Dropping subscript k, we recall the result proved in BARANGER, MAITRE and OUDIN [1994]

$$\left|(\underline{w}_h,\underline{\tau}_h)_{0,K} - (\underline{w}_h,\underline{\tau}_h)_{h,0,K}\right| \leqslant \mathcal{K} h_K \|\underline{w}_h\|_{H(\mathrm{div};K)} \|\underline{\tau}_h\|_{H(\mathrm{div};K)}. \tag{3.267}$$

Let $\tilde{\alpha}_K$ be a suitable average of the function α over K, for instance equal to the value of α at the centroid of K. This choice clearly satisfies (3.266) where $v = \alpha$. The quadrature error $a^K(\underline{w}_h,\underline{\tau}_h) - a_h^K(\underline{w}_h,\underline{\tau}_h)$ can then be split as

$$a^K(\underline{w}_h,\underline{\tau}_h) - a_h^K(\underline{w}_h,\underline{\tau}_h) = \mathcal{E}_1(\underline{w}_h,\underline{\tau}_h) + \mathcal{E}_2(\underline{w}_h,\underline{\tau}_h) + \mathcal{E}_3(\underline{w}_h,\underline{\tau}_h)$$

where

$$\begin{aligned}
\mathcal{E}_1(\underline{w}_h,\underline{\tau}_h) &= (\alpha\underline{w}_h,\underline{\tau}_h)_{0,K} - (\tilde{\alpha}_K\underline{w}_h,\underline{\tau}_h)_{0,K} = \int_K (\alpha - \tilde{\alpha}_K)\underline{w}_h\cdot\underline{\tau}_h\,dx,\\
\mathcal{E}_2(\underline{w}_h,\underline{\tau}_h) &= (\tilde{\alpha}_K\underline{w}_h,\underline{\tau}_h)_{0,K} - (\tilde{\alpha}_K\underline{w}_h,\underline{\tau}_h)_{h,0,K}\\
&= \int_K \tilde{\alpha}_K\underline{w}_h\cdot\underline{\tau}_h\,dx - \sum_{r\in E(k)} \tilde{\alpha}_K \Phi^r_{\underline{w}_h}\Phi^r_{\underline{\tau}_h}\omega^r_k,\\
\mathcal{E}_3(\underline{w}_h,\underline{\tau}_h) &= (\tilde{\alpha}_K\underline{w}_h,\underline{\tau}_h)_{h,0,K} - a_h^K(\underline{w}_h,\underline{\tau}_h) = \sum_{r\in E(k)} (\tilde{\alpha}_K - \overline{\alpha}^r)\Phi^r_{\underline{w}_h}\Phi^r_{\underline{\tau}_h}\omega^r_k.
\end{aligned}$$

Using the Cauchy–Schwarz inequality and bounding the term $\alpha - \tilde{\alpha}_K$ over $K \subset \mathcal{R}_K$ using (3.266) we get

$$\left|\mathcal{E}_1(\underline{w}_h,\underline{\tau}_h)\right| \leqslant C h_K |\alpha|_{1,\infty,\mathcal{R}_K} \|\underline{w}_h\|_{H(\mathrm{div};K)} \|\underline{\tau}_h\|_{H(\mathrm{div};K)}$$

while (3.267) yields

$$\left|\mathcal{E}_2(\underline{w}_h,\underline{\tau}_h)\right| \leqslant \tilde{\alpha}_K \mathcal{K} h_K \|\underline{w}_h\|_{H(\mathrm{div};K)} \|\underline{\tau}_h\|_{H(\mathrm{div};K)}.$$

Recalling that $\tilde{\alpha}_K$ and $\overline{\alpha}^r$ are averages of α over $K \subset \mathcal{R}_K$ and $\mathcal{L}^r \subset \mathcal{R}_K$, respectively, we can immediately conclude that

$$\begin{aligned}
\left|\mathcal{E}_3(\underline{w}_h,\underline{\tau}_h)\right| &\leqslant \Big(\max_{x\in\mathcal{R}_K}\alpha(x) - \min_{x\in\mathcal{R}_K}\alpha(x)\Big) \sum_{r\in E(k)} |\Phi^r_{\underline{w}_h}||\Phi^r_{\underline{\tau}_h}||\omega^r_k|\\
&\leqslant |\alpha|_{1,\infty,\mathcal{R}_K} \mathcal{K} h_K \sum_{r\in E(k)} |\Phi^r_{\underline{w}_h}||\Phi^r_{\underline{\tau}_h}||\omega^r_k|\\
&\leqslant \mathcal{K}|\alpha|_{1,\infty,\mathcal{R}_K} h_K \|\underline{w}_h\|_{H(\mathrm{div};K)} \|\underline{\tau}_h\|_{H(\mathrm{div};K)},
\end{aligned}$$

where we have used (3.266), (3.260), Cauchy–Schwarz inequality and (3.263). Gathering the previous estimates yields, $\forall K \in \mathcal{T}_h$

$$\begin{aligned}
&\left|a^K(\underline{w}_h,\underline{\tau}_h) - a_h^K(\underline{w}_h,\underline{\tau}_h)\right|\\
&\quad \leqslant \mathcal{K}\big(|\alpha|_{1,\infty,\Omega} + \|\alpha\|_{\infty,\Omega}\big) h_K \|\underline{w}_h\|_{H(\mathrm{div};K)} \|\underline{\tau}_h\|_{H(\mathrm{div};K)}.
\end{aligned}$$

Summing over all triangles and using the Cauchy–Schwarz inequality finally gives

$$\sup_{\underline{\tau}_h \in \Sigma_h} \frac{|a(\underline{w}_h, \underline{\tau}_h) - a_h(\underline{w}_h, \underline{\tau}_h)|}{\| \underline{\tau}_h \|_\Sigma} \leqslant Ch \| \underline{w}_h \|_{H(\mathrm{div};\Omega)}, \quad \forall \underline{w}_h \in \Sigma_h, \tag{3.268}$$

with $C = \mathcal{K}(|\alpha|_{1,\infty,\Omega} + \|\alpha\|_{\infty,\Omega})$. In particular, taking in (3.268) $\underline{w}_h = \Pi_h \underline{\sigma}$, as defined in (3.100), and recalling that (see, ROBERTS and THOMAS [1991], p. 583)

$$\| \Pi_h \underline{\sigma} \|_\Sigma \leqslant C\big(h |\underline{\sigma}|_{1,\Omega} + \| \underline{\sigma} \|_\Sigma\big),$$

we get

$$\sup_{\underline{\tau}_h \in \Sigma_h} \frac{|a(\Pi_h \underline{\sigma}, \underline{\tau}_h) - a_h(\Pi_h \underline{\sigma}, \underline{\tau}_h)|}{\| \underline{\tau}_h \|_\Sigma} \leqslant Ch\big(h |\underline{\sigma}|_{1,\Omega} + \| \underline{\sigma} \|_\Sigma\big). \tag{3.269}$$

Taking in (3.265) $v_h = P_h u$, as defined in (3.99), $\underline{w}_h = \Pi_h \underline{\sigma}$, and using the interpolation estimates (3.110) and (3.269), we finally obtain the convergence result

$$\| \underline{\sigma} - \underline{\sigma}_h \|_\Sigma + \|u - u_h\|_V \leqslant Ch\big(|u|_{1,\Omega} + |\underline{\sigma}|_{1,\Omega} + |\operatorname{div} \underline{\sigma}|_{1,\Omega}\big). \tag{3.270}$$

We have proved the following

THEOREM 3.8. *Assume that the solution* $(\underline{\sigma}, u)$ *of problem* (3.27) *is such that* $(\underline{\sigma}, u) \in (H^1(\Omega))^2 \times H^1(\Omega)$ *and* $\operatorname{div} \underline{\sigma} \in H^1(\Omega)$, *and that* $a \in W^{1,\infty}(\Omega)$. *Then, there exists a positive constant* $\mathcal{C}$, *independent of* h, *such that the solution* $(\underline{\sigma}_h, u_h)$ *of problem* (3.254), *with the choices* (3.255), *satisfies*

$$\| \underline{\sigma} - \underline{\sigma}_h \|_\Sigma + \|u - u_h\|_V \leqslant \mathcal{C}h. \tag{3.271}$$

We emphasize that Theorem 3.8 holds irrespectively of the choice of the average $\overline{\alpha}^r$ in (3.251), the only requirement being that

$$\min_{x \in \mathcal{L}^r} \alpha(x) \leqslant \overline{\alpha}^r \leqslant \max_{x \in \mathcal{L}^r} \alpha(x), \quad \forall \mathcal{L}^r \in \mathcal{L}_h. \tag{3.272}$$

REMARK 3.11. We remark that the $\mathcal{O}(h)$ convergence proved above for the mixed finite volume methods (3.251) is optimal and that it has been obtained without giving up the *M-matrix* property of the schemes. This feature, together with the reduced computational cost, makes these methods quite attractive and competitive with respect to the standard dual mixed approaches with exact integration analyzed in Section 3.5 and to dual mixed schemes with numerical integration recently proposed for the approximation of problem (3.27) in the case $\gamma = 0$ (BARANGER, MAITRE and OUDIN [1994, 1996]). Finally, the genuine finite volume flavor of these methods allows for the basic conservation properties (mass and interelement fluxes) to be satisfied, even in the presence of jumps in the coefficient of the problem (as happens for instance in porous media flows governed by Darcy's law (EWING, SAEVAREID and SHEN [1998])) or steep internal/boundary layers in the scalar unknown u (as happens in semiconductor device simulation using the drift-diffusion model (JEROME [1996], MARKOWICH [1986], MOLENAAR [1995], SACCO and SALERI [1997a])).

3.8.4. Several choices of the averages

In this section we characterize the choice of the average $\overline{\alpha}$ of the inverse diffusion coefficient α. Being this matter one-dimensional, we restrict our attention on the affine-equivalent interval $\mathcal{I} = [0, L]$, for any $L > 0$.

The obvious candidate for $\overline{\alpha}$ is the mean value of α over $\mathcal{I}$. Since we are interested in the inverse of $\overline{\alpha}$ (see (3.251)) and $\alpha = a^{-1}$, it follows that

$$\overline{\alpha}^{-1} = \left(\frac{\int_0^L a^{-1}(x)\,dx}{L}\right)^{-1} =: \mathcal{H}_{\mathcal{I}}(a) \tag{3.273}$$

where $\mathcal{H}_{\mathcal{I}}(a)$ denotes the *harmonic average* of a over the interval $\mathcal{I}$.

Use of harmonic averaging for the diffusion coefficient a is quite natural in mixed methods (see BABUŠKA and OSBORN [1983], BREZZI and FORTIN [1991]) and has been proved in one dimension to provide better results than the mean value, in particular when a exhibits sharp variations on $\mathcal{I}$ or is even discontinuous. Typical instances of such a behavior are flows in porous media (see EWING, SAEVAREID and SHEN [1998] and the references cited therein) or electron and hole carrier flow in a semiconductor device.

It is clear that, except for special cases, the evaluation of (3.273) cannot be carried out. This, of course, demands for the use of a suitable quadrature formula. With this aim, define for any function $\phi : \mathcal{I} \to \mathbb{R}^+$ the *exponential interpolant* to ϕ as

$$\mathcal{E}_{\mathcal{I}}\phi(x) := \exp\bigl\{\bigl[\ln\bigl(\phi(x)\bigr)\bigr]_I\bigr\} = \phi_0\left(\frac{\phi_L}{\phi_0}\right)^{x/L}, \quad x \in \mathcal{I}, \tag{3.274}$$

where $\phi_0 = \phi(0)$, $\phi_L = \phi(L)$ and v_I is the P_1-interpolant to v. The quadrature formula approximating (3.273) can then be defined as

$$\begin{aligned}\overline{\alpha}^{-1} \simeq \left(\frac{\int_0^L \mathcal{E}_{\mathcal{I}} a^{-1}(x)\,dx}{L}\right)^{-1} &= \frac{\ln(1/a_L) - \ln(1/a_0)}{1/a_L - 1/a_0} \\ &= \left(\frac{\ln(a_0) - \ln(a_L)}{a_0 - a_L}\right) a_0 a_L\end{aligned} \tag{3.275}$$

which clearly satisfies (3.272). Assuming $\alpha \in W^{1,\infty}(\mathcal{I})$, it is easy to prove the following bound for the interpolation error

$$\|\alpha - \mathcal{E}_{\mathcal{I}}\alpha\|_{\infty,\mathcal{I}} \leqslant Ch\frac{\alpha_M}{\alpha_m}|\alpha|_{1,\infty,\mathcal{I}}$$

where α_M and α_m are the maximum and the minimum values of α over $\mathcal{I}$, respectively. We remark that (3.275) is *exact* if $a(x) = e^{lx+m}$, $l, m \in \mathbb{R}$, $x \in \mathcal{I}$, which is the case of the numerical approximation of the drift-diffusion semiconductor device equations when the self-adjoint form (1.9) is used for the current densities and when a linear variation of the (scaled) electric potential ψ is assumed over $\mathcal{I}$. The resulting discretization scheme will be addressed in Section 4.2.

Our second choice for $\overline{\alpha}^{-1}$ is the harmonic average of the piecewise constant extension of a over $\mathcal{I}$

$$\overline{\alpha}^{-1} \simeq \left(\frac{\int_0^{\bar{x}} a_0^{-1}\,dx + \int_{\bar{x}}^L a_L^{-1}\,dx}{L}\right)^{-1} = \frac{a_0 a_L}{a_L\bar{x}/L + a_0(1 - \bar{x}/L)}, \tag{3.276}$$

where $\bar{x} \in \mathcal{I}$. In our application, $\mathcal{I}$ will be the segment joining the circumcenters of two adjacent triangles K and K' and $\bar{x}$ will be the intersection between $\mathcal{I}$ and the edge common to K and K'. The average (3.276) seems to be quite promising in presence of discontinuities of a.

The last choice that we consider for $\overline{\alpha}^{-1}$ employs linear interpolation for a. This leads to the trapezoidal quadrature formula

$$\overline{\alpha}^{-1} \simeq \frac{\int_0^L a_I(x)\,dx}{L} = \frac{a_0 + a_L}{2}. \tag{3.277}$$

Numerical experiments using the three averages introduced above will be presented in Section 7.

4. Application to continuity equations

We shall describe in this section mixed discretizations of the scaled current continuity equations $(1.22)_2$, $(1.22)_3$. For simplicity, we shall deal only with the scaled continuity equation $(1.22)_3$ for the positive charge density p. Moreover, we shall consider the stationary case and a constant mobility $\mu_p \equiv 1$. The problem under investigation has the form

$$\begin{cases} \text{Find } p \in H^1(\Omega) \text{ such that} & \\ -\operatorname{div}(\underline{\nabla}\, p + p\underline{\nabla}\,\psi) = -R(p,n) & \text{in } \Omega \subset \mathbb{R}^2, \\ p = g := p_D & \text{on } \Gamma_D \subset \partial\Omega, \\ (\underline{\nabla}\, p + p\underline{\nabla}\,\psi)\cdot \underline{n} = 0 & \text{on } \Gamma_N \subset \partial\Omega, \end{cases} \tag{4.1}$$

and the current is given by

$$\underline{J} = -(\underline{\nabla}\, p + p\underline{\nabla}\,\psi). \tag{4.2}$$

In the simulation of the DD model the solution of (4.1) is an intermediate step inside an iterative process, as presented in Section 2, and we shall assume that ψ and n are known. Moreover, during the iterative solution procedure, Eq. $(4.1)_1$ is usually linearized (see Section 2.4) so that problem (4.1) becomes

$$\begin{cases} \text{Find } p \in H^1(\Omega) \text{ such that} & \\ -\operatorname{div}(\underline{\nabla}\, p + p\underline{\nabla}\,\psi) + cp = f & \text{in } \Omega, \\ p = g & \text{on } \Gamma_D, \\ (\underline{\nabla}\, p + p\underline{\nabla}\,\psi)\cdot \underline{n} = 0 & \text{on } \Gamma_N, \end{cases} \tag{4.3}$$

where $f = f(x)$ is a function independent of p, and $c = c(x)$ is a nonnegative function independent of p, which can be assumed piecewise constant. Finally, ψ is assumed to be piecewise linear (resulting from a discretization of the Poisson equation). Problem (4.3) is not directly suited to the application of the schemes described in Section 3 because it is not self-adjoint. Using the Slotboom variable ρ (see (1.8))

$$p = \rho e^{-\psi}, \tag{4.4}$$

(4.3) can be written in the symmetric form as

$$\begin{cases} \text{Find } \rho \in H^1(\Omega) \text{ such that} \\ -\operatorname{div}(e^{-\psi}\underline{\nabla}\rho) + ce^{-\psi}\rho = f & \text{in } \Omega, \\ \rho = \chi := e^{\psi} g & \text{on } \Gamma_D, \\ \underline{\nabla}\rho \cdot \underline{n} = 0 & \text{on } \Gamma_N, \end{cases} \tag{4.5}$$

and the current is now given by

$$\underline{J} = -e^{-\psi}\underline{\nabla}\rho. \tag{4.6}$$

The symmetric problem (4.5) is now of the form (2.14) with $a = e^{-\psi}$ and $\gamma = ce^{-\psi}$. We point out that other approaches, based on the use of the quasi-Fermi potentials defined in (1.5), also lead to self-adjoint problems. However, in this case the equations are exponentially nonlinear. For mixed approximations of the quasi-Fermi potential formulation we refer to HECHT, MARROCCO, CAQUOT and FILOCHE [1991], HECHT and MARROCCO [1994].

Problem (4.5) will not be discretized directly because the presence of the exponential functions can be a source of numerical troubles in many relevant situations. The idea of the schemes presented here is to discretize the symmetric equation (4.5) with a mixed finite element scheme in the hybridization form presented in Section 3.6 or with a mixed finite volume schemes presented in Section 3.8, to write the system in matrix form as in (3.211)–(3.213) or in (3.251)–(3.253), to go back to the original variable p by using a suitable discrete version of transformation (4.4), and then to solve for p.

4.1. Exponential fitting mixed finite elements

We shall describe in this section a mixed approximation to the scaled current continuity equations (4.3). For the case $c = 0$ a mixed scheme based on the lowest order Raviart–Thomas element RT_0 has been introduced in BREZZI, MARINI and PIETRA [1987] and extensively discussed in BREZZI, MARINI and PIETRA [1989a, 1989b]. For the case $c \neq 0$ mixed schemes based on the elements of Examples 5–6, introduced and analyzed in MARINI and PIETRA [1989], have been applied and discussed in MARINI and PIETRA [1990]. In the following we shall give a compact presentation which includes the three different elements.

The presence of the exponentials in the definition of a and γ requires special care. For the approximation of $a^{-1} = e^{\psi}$, following (3.217), we define $\overline{\psi}$ as the piecewise constant function given in each element K by

$$e^{\overline{\psi}}_{|K} := \frac{1}{|K|}\int_K e^{\psi}\,dx, \tag{4.7}$$

and we take

$$\overline{\alpha} = e^{\overline{\psi}}. \tag{4.8}$$

The approximation of $\gamma = ce^{-\psi}$ suggested in (3.218) is not suited here for reasons which will be made clear at the end of this section. We proceed as follows. Let $\widetilde{\psi}$ denote the piecewise constant function defined in each element K via a harmonic average on a special edge $\tilde{e}$. In order to define $\tilde{e}$, let $V_{\max}$, and $V_{\min}$ be the vertices of K where ψ assumes maximum and minimum value respectively, and let V_{med} be the third vertex. The edge $\tilde{e}$ is taken as the edge connecting $V_{\max}$ and V_{med}, and $\widetilde{\psi}$ is computed by

$$e_{|K}^{-\widetilde{\psi}} := |\tilde{e}| \Big/ \left(\int_{\tilde{e}} e^{\psi}\, ds \right), \qquad \tilde{e} = V_{\max} V_{\text{med}}, \tag{4.9}$$

or, with the notation of (3.273), by $e_{|K}^{-\widetilde{\psi}} := \mathcal{H}_{\tilde{e}}(e^{\psi})$. Therefore, instead of (3.218), we have

$$\overline{\gamma} = c e^{-\widetilde{\psi}}. \tag{4.10}$$

For the reader's convenience, we recall the definition of the finite dimensional spaces already introduced in (3.148), (3.51), (3.160)

$$\begin{aligned}
&\widehat{\Sigma}_h = \left\{ \underline{\tau}_h \in \left(L^2(\Omega)\right)^2 \mid \underline{\tau}_{h|K} \in Q(K),\ \forall K \in \mathcal{T}_h \right\},\\
&V_h = \left\{ v_h \in V \mid v_{v|K} \in P_0(K),\ \forall K \in \mathcal{T}_h \right\},\\
&\Lambda_{h,\xi} = \left\{ \mu_h \in L^2(\mathcal{E}_h) \mid \mu_{h|e} \in P_0(e),\ \forall e \in \mathcal{E}_h,\ \int_e (\mu_h - \xi)\, ds = 0,\ \forall e \in \mathcal{E}_h \cap \Gamma_D \right\},
\end{aligned}$$

where, as usual, $P_0(D)$ denotes the set of constant functions in the domain D. Moreover, $Q(K)$ denotes here a set of polynomial vectors with $Q(K) = \text{span}\{\underline{\tau}^1, \underline{\tau}^2, \underline{\tau}^3\}$. In the three cases $\underline{\tau}^1 = (1, 0)$, $\underline{\tau}^2 = (0, 1)$; instead $\underline{\tau}^3$ is defined by (3.216) for the RT_0 element, by (3.115)–(3.116) for the element of Example 5 and by (3.128)–(3.129) for the element of Example 6.

Specializing (3.161) to this case, the discrete formulation of (4.5), with $\underline{J}$ defined as in (4.6), becomes

$$\begin{cases}
\text{Find } (\underline{J}_h, \rho_h, \lambda_h) \in \widehat{\Sigma}_h \times V_h \times \Lambda_{h,\chi} \text{ such that} & \\
\int_\Omega e^{\overline{\psi}} \underline{J}_h \cdot \underline{\tau}_h\, dx - \sum_K \int_K \rho_h \operatorname{div} \underline{\tau}_h\, dx & \\
\quad + \sum_K \int_{\partial K} \lambda_h \underline{\tau}_h \cdot \underline{n}\, ds = 0, & \forall \underline{\tau}_h \in \widehat{\Sigma}_h,\\
- \sum_K \int_K v_h \operatorname{div} \underline{J}_h\, dx - \sum_K \int_K c e^{-\widetilde{\psi}} \rho_h v_h\, dx & \\
\quad = - \int_\Omega f v_h\, dx, & \forall v_h \in V_h,\\
\sum_K \int_{\partial K} \mu_h \underline{J}_h \cdot \underline{n}\, ds = 0, & \forall \mu_h \in \Lambda_{h,0}.
\end{cases} \tag{4.11}$$

Performing the elimination of $\underline{J}_h$ and ρ_h by static condensation as in Section 3.7, we obtain a final matrix $\mathcal{M}$ acting only on the Lagrange multipliers λ_h, which we recall to be an approximation of ρ on the edges. In the present case, the coefficients of the element matrix $\mathcal{M}^K$ defined in (3.225) take the form

$$m_{ij}^K = e^{-\overline{\psi}} \frac{\underline{\nu}^i \cdot \underline{\nu}^j}{|K|} + \frac{c e^{-\widetilde{\psi}} |K|}{\beta^2 + \delta c e^{\overline{\psi} - \widetilde{\psi}} |K|} \eta_i \eta_j, \quad i, j = 1, 3, \tag{4.12}$$

and the right-hand side takes the form

$$g_i^K = \frac{\beta\eta_i}{\beta^2 + \delta c e^{\overline{\psi}-\widetilde{\psi}}|K|} \int_K f\,dx, \quad i = 1, 3, \tag{4.13}$$

where δ, β, η_i are defined in (3.220)–(3.222). We are now ready to introduce an approximation of the original variable p. Since λ_h lives on the edges, a discrete version of the inverse transformation of (4.4) must be defined edge by edge. For this, for every $\zeta \in L^2(\mathcal{E}_h)$, we define ζ^I to be the L^2-projection of ζ onto $\Lambda_{h,\zeta}$, that is,

$$\zeta^I{}_{|e} = \frac{1}{|e|} \int_e \zeta\,ds, \quad \forall e \in \mathcal{E}_h. \tag{4.14}$$

The discrete change of variable is then

$$\lambda_h = (e^{\psi})^I p_h, \tag{4.15}$$

and (4.11) becomes

$$\begin{cases} \text{Find } (\underline{J}_h, \rho_h, p_h) \in \widehat{\Sigma}_h \times V_h \times \Lambda_{h,g} \text{ such that} \\ \int_\Omega e^{\overline{\psi}} \underline{J}_h \cdot \underline{\tau}_h\,dx - \sum_K \int_K \rho_h \operatorname{div} \underline{\tau}_h\,dx \\ \quad + \sum_K \int_{\partial K} (e^{\psi})^I p_h \underline{\tau}_h \cdot \underline{n}\,ds = 0, & \forall \underline{\tau}_h \in \widehat{\Sigma}_h, \\ -\sum_K \int_K v_h \operatorname{div} \underline{J}_h\,dx - \sum_K \int_K c e^{-\widetilde{\psi}} \rho_h v_h\,dx \\ \quad = -\int_\Omega f v_h\,dx, & \forall v_h \in V_h, \\ \sum_K \int_{\partial K} \mu_h \underline{J}_h \cdot \underline{n}\,ds = 0, & \forall \mu_h \in \Lambda_{h,0}, \end{cases} \tag{4.16}$$

where p_h is an approximation of p on the edges. We point out that (4.15) is intended on $\mathcal{E}_h \backslash \Gamma_D$ only, unless g is piecewise constant on Γ_D (which is the case in most applications). In that case, $\lambda_h \in \Lambda_{h,\chi}$ and $p_h \in \Lambda_{h,g}$ are equivalent.

Performing transformation (4.15) at the matrix level (for the nodes not belonging to Γ_D) amounts to multiplying the matrix $\mathcal{M}$ columnwise by the value of $(e^{\psi})^I$ on the corresponding edge, thus giving rise to the matrix $\widetilde{\mathcal{M}}$ acting on the variable p_h. The algebraic system to be solved is then

$$\widetilde{\mathcal{M}} p_h = \mathcal{G}. \tag{4.17}$$

The matrix $\widetilde{\mathcal{M}}$ is not symmetric anymore (as expected, since it corresponds to the non symmetric problem (4.1)), but if $\mathcal{M}$ is an M-matrix this property is preserved in $\widetilde{\mathcal{M}}$, since $(e^{\psi})^I$ is always positive. The coefficients of the stiffness matrix $\widetilde{\mathcal{M}}$ take the form

$$\begin{aligned} \widetilde{m}_{ij}^K &= (e^{\psi})^I{}_{|e^j} e^{-\overline{\psi}} \frac{\underline{\nu}^i \cdot \underline{\nu}^j}{|K|} \\ &\quad + (e^{\psi})^I{}_{|e^j} e^{-\widetilde{\psi}} \frac{c|K|}{\beta^2 + \delta c e^{\overline{\psi}-\widetilde{\psi}}|K|} \eta_i \eta_j, \quad i, j = 1, 3. \end{aligned} \tag{4.18}$$

In the case of the elements of Examples 5–6, formula (4.18) can be simplified. Indeed, taking the special edge $\tilde{e}$ (denoted also by e^1 in (3.231)) appearing in the definition of these elements ((3.116), (3.129)) as the edge $V_{\max} V_{\text{med}}$ (see (4.9)), one can see that

$$(e^{\psi})^{I}{}_{|e^1} e^{-\widetilde{\psi}} = 1, \tag{4.19}$$

and (4.18) reduces to (see also (3.232))

$$\widetilde{m}_{ij}^{K} = \begin{cases} (e^{\psi})^{I}{}_{|e^1} e^{-\overline{\psi}} \frac{\underline{\nu}^1 \cdot \underline{\nu}^1}{|K|} + \frac{c|K|}{\beta^2 + \delta c e^{\overline{\psi} - \widetilde{\psi}} |K|} \eta_1^2, & \text{for } i = j = 1, \\ (e^{\psi})^{I}{}_{|e^j} e^{-\overline{\psi}} \frac{\underline{\nu}^i \cdot \underline{\nu}^j}{|K|}, & \text{otherwise}, \end{cases} \tag{4.20}$$

with $\eta_1 = |e^1|$ for Example 5, and $\eta_1 = 1$ for Example 6. Using the results presented at the end of Section 3.7, if the triangulation is of weakly acute type, the two elements of Examples 5–6 provide a final matrix $\widetilde{\mathcal{M}}$ which is an M-matrix for all $c \geqslant 0$. Instead, the element RT_0 guarantees the M-matrix property only if $c = 0$.

REMARK 4.1. Few remarks on numerical "tricks". For the computation of $\int_K f\,dx$ in (4.13) a quadrature formula which is exact for constant f can be used. Exact integration can be used for computing $e^{\overline{\psi}}$, $e^{\widetilde{\psi}}$ defined in (4.7), (4.9), since ψ is piecewise linear. The choice of $\tilde{e}$ as the edge which connects the vertex with the largest potential value and the vertex with the second largest potential value takes care of all possible cases for the potential: $\psi \equiv$ constant on K, $\psi =$ constant $= \psi^M$ on one edge, and ψ having the maximum in one vertex only.

As already pointed out, the electric field $\underline{E}$ $(= -\underline{\nabla}\,\psi)$ can be quite large in a portion of the domain, so that the presence of exponentials in the coefficients might be a source of numerical problems. A (rough) analysis of the behaviour of the coefficients when the electric field becomes larger and larger will be performed. It is more convenient to set

$$\psi = \frac{\psi_0}{l}, \tag{4.21}$$

and assume that $\underline{\nabla}\,\psi_0$ is smooth everywhere and l is a small number. Accordingly, Eq. (4.3) becomes

$$-\operatorname{div}\left(\underline{\nabla}\,p + p\frac{\underline{\nabla}\,\psi_0}{l}\right) + cp = f. \tag{4.22}$$

The nature of Eq. (4.22) is such that, as $l \to 0$, the differential term behaves like l^{-1}, while the zeroth-order term is of order 1. Hence, for very small l (say $l \ll |\underline{\nabla}\,\psi_0| h_K$), our discrete scheme must reproduce the behavior of the continuous equation (4.22). To check that, recall that ψ (and then ψ_0) is assumed piecewise linear, and denote by ψ^M the maximum of ψ on K and by ψ^{M_j} the maximum of ψ on the edge e^j. We only consider the generic case where the maximum is reached at one vertex. When $l \ll |\underline{\nabla}\,\psi_0| h_K$, a simple computation shows that

$$e^{\overline{\psi}} = \frac{1}{|K|} \int_K e^{\psi_0/l}\,dx \simeq l^2 e^{\psi_0^M/l} = l^2 e^{\psi^M}, \tag{4.23}$$

$$(e^{\psi})^{I}{}_{|e^j} = \frac{1}{|e^j|} \int_{e^j} e^{\psi_0/l}\,ds \simeq l e^{\psi_0^{M_j}/l} = l e^{\psi^{M_j}}. \tag{4.24}$$

Moreover, recalling that $\tilde{e} = V_{\max} V_{\mathrm{med}}$, we have

$$e^{\widetilde{\psi}} \simeq l e^{\psi^M}. \tag{4.25}$$

Then, we obtain

$$(e^{\psi})^I{}_{|e^j} e^{-\overline{\psi}} \simeq \begin{cases} l^{-1}, & \text{if } \psi^{M_j} = \psi^M, \\ 0, & \text{otherwise,} \end{cases} \tag{4.26}$$

$$(e^{\psi})^I{}_{|e^j} e^{-\widetilde{\psi}} \simeq \begin{cases} 1, & \text{if } \psi^{M_j} = \psi^M, \\ 0, & \text{otherwise,} \end{cases} \tag{4.27}$$

$$e^{\overline{\psi}} e^{-\widetilde{\psi}} \simeq l. \tag{4.28}$$

Hence, coefficients (4.18) behave as

$$\widetilde{m}_{ij}^K \simeq \begin{cases} l^{-1} \frac{\nu^i \cdot \nu^j}{|K|} + \frac{c|K|}{\beta^2 + \delta c l |K|} \eta_i \eta_j, & \text{if } \psi^{M_j} = \psi^M, \\ 0, & \text{otherwise.} \end{cases} \tag{4.29}$$

The reason for the choice (4.9) is now clear. The expected behaviour in terms of the order of magnitude with respect to l is preserved and, moreover, no bad blow-up occurs. Different (and maybe more natural) choices for $e^{-\widetilde{\psi}}$ could lead to a coefficient for the zeroth-order term in which $(e^{\psi})^I{}_{|e^j} e^{-\widetilde{\psi}}$ is not of order 1 when $\psi^{M_j} = \psi^M$. Then, the presence of this factor could give rise to schemes whose structure does not fit the structure of the continuous problem and which produce poor results, unless the mesh size is very small.

To conclude this section, we summarize the main features of the discretization schemes presented here.

- Current conservation.

  The discretization schemes presented here are based on mixed finite elements and enforce some continuity of the normal component of the current at the interelements. As extensively discussed in Section 3, strong continuity is imposed when conforming approximations of $H(\mathrm{div}; \Omega)$ are considered (RT_0 or the element of Example 5), weak continuity (in the sense that the jump of the normal component of the current has zero mean value at the interelements) is imposed for the element of Example 6. This property is particularly relevant here, since the current is possibly the most important output of device simulation.
- Automatic upwinding effects.

  The expression (4.29) tells us that whenever $|\underline{\nabla}\, \psi|$ is large, the coefficient corresponding to the node on the edge where ψ does not reach its maximum is zero (with respect to the machine precision). Such a node can be regarded as downwind node (wind $= -\underline{\nabla}\, \psi$) and the scheme as an upwind scheme. In a sense, the scheme automatically adapts to the changed nature of the problem when advection becomes bigger than diffusion, and chooses the upwind nodes with no extra computational cost. The numerical tests presented in Section 7.3 will show that the numerical diffusion introduced by the schemes is very small in applications to semiconductor device problems.

- M-matrix property.

  The two elements of Examples 5–6 provide a final matrix $\widetilde{\mathcal{M}}$ which is an M-matrix for all $c \geqslant 0$. Instead, the element RT_0 guarantees the M-matrix property only if $c = 0$. The M-matrix property implies a sort of discrete maximum principle, and this entails, in particular, that the discrete solution is nonnegative if the boundary data are nonnegative. This is highly desirable in this case since, on the one hand, the variables we are dealing with (charge densities) are intrinsically nonnegative. On the other hand, the possible development of spurious oscillations might heavily pollute the nonlinear Gauss–Seidel iterations, thus seriously compromising the results.

4.2. *MFV approximation of the continuity equation*

In this section we address the discretization of the current continuity equation (4.3) using the MFV method discussed in Section 3.8.2. Using the symmetric form (4.5) of (4.3), the MFV discrete formulation (3.251) reads in the present case

$$\begin{cases} \sum_{r\in E(k)} e^{-\overline{\psi}_r}\left(\frac{\rho_k^*-\rho_{j(r)}^*}{d^r}\right)|e^r| + \rho_k^* c_k e^{-\psi_k}|K_k| = f_k|K_k|, & \forall K_k \in \mathcal{T}_h, \\ \rho_{j(r)}^* = \chi^r = e^{\psi_r} g^r, & \forall e^r \in \Gamma_D, \end{cases} \tag{4.30}$$

where $j(r) = T(r)\backslash k$. We note that in the formulation (3.251) v_k denotes the mean value of the function v over the triangle K_k. However, in practical implementation it is easier to use, instead of the mean value, the value of v at the circumcenter C_k of every $K_k \in \mathcal{T}_h$. Similarly, $\chi^r = e^{\psi_r} g^r$ is taken as the value of the function $e^{\psi} g$ at the midpoint of each edge $e^r \in \Gamma_D$. The quantity $e^{-\overline{\psi}_r}$ could be set equal to one of the three averages discussed in Section 3.8.4. In the context of semiconductor device simulation, the more appropriate choice is the harmonic average (3.275) of e^{ψ} along the segment d_r joining the circumcenters C_k and $C_{j(r)}$.

With this choice we obtain

$$e^{-\overline{\psi}_r} = e^{-\psi_k}\frac{\psi_{j(r)} - \psi_k}{e^{\psi_{j(r)}-\psi_k} - 1} \equiv e^{-\psi_k}\mathrm{B}(\psi_{j(r)} - \psi_k), \quad \forall r \in E(k), \tag{4.31}$$

where

$$\mathrm{B}(t) = \begin{cases} \frac{t}{e^t - 1}, & t \neq 0, \\ 1, & t = 0, \end{cases} \tag{4.32}$$

is the Bernoulli function. Substituting (4.31) into (4.30) leaves us with solving the following symmetric and positive definite linear system acting on the variable $\boldsymbol{\rho}^*$

$$W^*\boldsymbol{\rho}^* = \mathbf{f}^*, \tag{4.33}$$

whose nonzero matrix entries are

$$W_{ij}^* = \begin{cases} \sum_{\substack{k\in T(E(i))\\ k\neq i}} e^{-\psi_i}\mathrm{B}(\psi_k - \psi_i)\frac{|e^{r(k)}|}{d^{r(k)}} & \\ \quad + c_i e^{-\psi_i}|K_i|, & \text{if } i = j, \\ -e^{-\psi_i}\mathrm{B}(\psi_{r(j)} - \psi_i)\frac{|e^{r(j)}|}{d^{r(j)}}, & \text{if } j \in T(E(i)),\ j \neq i, \end{cases} \tag{4.34}$$

where $r(k) = E(k) \cap E(i)$ and $r(j) = E(j) \cap E(i)$. As already pointed out in the previous section, the presence of the exponential terms in (4.34) is a source of trouble in numerical computations due to potential occurrence of overflow/underflow problems. Therefore, we go back to the primitive variable p by applying (4.4) at the discrete level, i.e.,

$$\rho_m^* = p_m e^{\psi_m}, \quad \forall K_m \in \mathcal{T}_h. \tag{4.35}$$

Using (4.35) into system (4.33)–(4.34) amounts to multiplying the matrix W^* columnwise by the value of e^{ψ_m} on the corresponding triangle K_m and leads to solving the following linear system acting on the variable $\mathbf{p}^*$

$$\widetilde{W}^* \mathbf{p}^* = \mathbf{f}^*. \tag{4.36}$$

Noting that

$$e^t \mathrm{B}(t) = \mathrm{B}(-t), \quad \forall t \in \mathbb{R}, \tag{4.37}$$

the nonzero entries of matrix $\widetilde{W}^*$ are given by

$$\widetilde{W}_{ij}^* = \begin{cases} \sum_{\substack{k \in T(E(i)) \\ k \neq i}} \mathrm{B}(\psi_k - \psi_i) \frac{|e^{r(k)}|}{d^{r(k)}} + c_i |K_i|, & \text{if } i = j, \\ -\mathrm{B}(-(\psi_{r(j)} - \psi_i)) \frac{|e^{r(j)}|}{d^{r(j)}}, & \text{if } j \in T(E(i)),\ j \neq i. \end{cases} \tag{4.38}$$

The matrix $\widetilde{W}^*$ is no longer symmetric and positive definite; however, under the assumption that $\mathcal{T}_h$ is a Delaunay triangulation, $\widetilde{W}^*$ turns out to be an M-matrix with strictly positive inverse. As a consequence, a sort of discrete maximum principle holds for the discrete formulation and $\mathbf{p}^* > 0$ provided that $\mathbf{f}^* > 0$, as it is desirable since the variable p has the physical meaning of a (scaled) hole density.

REMARK 4.2. The mixed finite volume (4.36)–(4.38) can be regarded as a two-dimensional generalization of the exponentially fitted Scharfetter–Gummel method SCHARFETTER and GUMMEL [1969] for the discretization of the current continuity equation (4.3), which will be object of investigation of Section 5. To check this assertion, let us integrate (4.3) over a single element $K_k \in \mathcal{T}_h$; using the divergence theorem, we obtain

$$\int_{\partial K_k} \underline{J} \cdot \underline{n}\, ds + \int_{K_k} cp\, dx = \int_{K_k} f\, dx,$$

where the current $\underline{J}$ is defined in (4.2). The approximate evaluation of the net flux throughout the control volume K_k can be performed as

$$\sum_{r \in E(k)} \mathcal{J}_{k,j(r)} |e^r| + c_k p_k |K_k| = f_k |K_k|, \quad \forall K_k \in \mathcal{T}_h, \tag{4.39}$$

where, for any couple of adjoining triangles K_k, $K_{j(r)}$, $\mathcal{J}_{k,j(r)}$ is the *constant* hole current density flowing along the "pipeline" linking the circumcenters C_k, $C_{j(r)}$ of the two elements and evaluated according to the classical Scharfetter–Gummel formula

(SCHARFETTER and GUMMEL [1969])

$$\mathcal{J}_{k,j(r)} = \frac{p_k \mathrm{B}(\psi_{j(r)} - \psi_k) - p_{j(r)} \mathrm{B}(-(\psi_{j(r)} - \psi_k))}{d_r}. \tag{4.40}$$

Substituting (4.40) into (4.39) we exactly get the mixed-finite volume discretization (4.36).

REMARK 4.3. The mixed-finite volume scheme (4.36) can be proved to recover the *exact* solution $(p, \underline{J})$ at the circumcenters of $\mathcal{T}_h$ when $c = 0$, $f = 0$, ψ is linear in Ω, and suitable Dirichlet–Neumann boundary conditions are assumed in problem (4.3) in such a way that $\underline{J} = \underline{\text{const}}$ (see for the proof SACCO and SALERI [1997b] and VAN NOOYEN [1995] in the case of triangles and rectangles, respectively). This nice property is a special instance of the "patch-test" (see ROBERTS and THOMAS [1991], Chapter V, Section 34 and HENNART and DEL VALLE [1996]) and turns out to be a sound indication for good behavior of a numerical scheme to deal with advection-dominated flow problems, as previously remarked in VAN NOOYEN [1995], SACCO, GATTI and GOTUSSO [1995].

REMARK 4.4. The expression (4.31) used to compute the average of e^{ψ} has been chosen by observing that its use in the one-dimensional case leads to a proper exponential fitting interpolation for the carrier concentration p. This gives rise to difference schemes of optimal order (see ROOS, STYNES and TOBISKA [1996] for a complete survey of this subject). At the same time, the Bernoulli weights in (4.38) ensure robustness and stability of the approximation irrespectively of the potential drop $\psi_{r(j)} - \psi_k$ across e^r and provide automatically the suitable upwinding effect, exactly as happens for the standard mixed formulation discussed in Section 4.1.

5. Other approaches

In this section we shall discuss a number of other finite element methods for the semiconductor device equations. We shall start with the Scharfetter–Gummel box scheme which is very popular in the community of semiconductor device modelling. We shall then discuss some extensions of the box scheme and other stable finite element methods.

5.1. The Scharfetter–Gummel box/finite volume scheme

The key idea of the Scharfetter–Gummel box or finite volume method is to approximate the flux of a boundary value problem along each edge in a mesh by a constant, which yields an exponential approximation to the potential function of the problem. Therefore, it is also called an exponentially fitted method. This method was proposed by Scharfetter and Gummel for a one-dimensional problem (SCHARFETTER and GUMMEL [1969]). The same idea was earlier introduced by ALLEN and SOUTHWELL [1955] in a different context. The one-dimensional Scharfetter–Gummel scheme approximation has been combined with higher dimensional box scheme by several authors to form the so-called Scharfetter–Gummel box method. (cf., for example, BUTURLA, COTTRELL, GROSSMAN and SALSBURG [1981], MCCARTIN [1985], BANK, BÜRGLER, FICHTNER and SMITH [1990], MILLER and WANG [1994b]). In what follows we discuss the method in two dimensions. The extension to three dimensions is straightforward.

Let us consider the decoupled and linearized equations in the Slotboom variables given in Section 2. These equations are of the form (2.14) in which $a = \lambda^2$ for the Poisson equation, $a = \mu_n e^{\psi}$ for the continuity equation for ρ_n, and $a = \mu_p e^{-\psi}$ for the continuity equation for ρ_p. Without loss of generality, we assume that $g = 0$ on Γ_D. The case of non-homogeneous Dirichlet boundary conditions can be transformed into this one by subtracting from both sides of (2.14) a known function satisfying a given non-homogeneous boundary condition. We now discuss the discretization of this boundary value problem by the box method. We shall first formulate it as a finite element method and then we present a stability and error analysis for the method. For brevity, we only consider the case that $a = e^{\psi}$, and we let $\underline{\sigma} = e^{\psi}\underline{\nabla} u$. The discretization for the other two cases and for the case that μ_n is not constant are similar to this one.

5.1.1. *Formulation of the method*

Classically, the Scharfetter–Gummel box method can be formulated as a finite volume method with a constant approximation to the flux projected on each of the mesh edges. This formulation can be found, for example, in BUTURLA, COTTRELL, GROSSMAN and SALSBURG [1981], MCCARTIN [1985], BANK, BÜRGLER, FICHTNER and SMITH [1990]. Though the formulation is easy to understand, it is not very suitable for stability and error analysis. In what follows, we shall formulate it as a non-conforming Petrov–Galerkin finite element method as described in MILLER and WANG [1994a]. This can be viewed as a generalized finite element method and is closely related to a mixed finite element formulation (cf. BABUŠKA and OSBORN [1983]). It can also be viewed as a lumped form of a primal mixed finite element method proposed in MILLER and WANG [1991] and MILLER and WANG [1994c]. Alternative analysis can be found in MOCK [1983b].

To discuss the method we first define primal and dual decompositions of Ω. Let $\mathcal{T}_h$ be a Delaunay triangulation of Ω (see Definition 3.1). We also let $\mathcal{X}_h = \{x_i\}_1^N$ denote the vertices of $\mathcal{T}_h$ and $\mathcal{E}_h = \{e_i\}_1^M$ the edges of $\mathcal{T}_h$. We assume that the nodes in $\mathcal{X}_h$ and the edges in $\mathcal{E}_h$ are numbered in such a way that $\mathcal{X}'_h = \{x_i\}_1^{N'}$ and $\mathcal{E}'_h = \{e_i\}_1^{M'}$ are the set of nodes in $\mathcal{X}_h$ not on Γ_D and the set of edges in $\mathcal{E}_h$ not on Γ_D, respectively. The first nontriangular mesh, dual to $\mathcal{T}_h$, is the tessellation denoted by $\mathcal{L}_h$ in Section 3.8.1 (see Fig. 3.2, left). For convenience, we repeat here the definition with a slight change of notation more suited for our purposes. With each edge $e_{i,j} \in \mathcal{E}_h$ connecting two vertices x_i, x_j we associate an open box $B_{i,j}$ which is the interior of the polygon having as its vertices x_i, x_j, and the circumcenters of the triangles having $e_{i,j}$ as a common edge. If $e_{i,j}$ is not on $\partial\Omega$ the region $B_{i,j}$ consists of two triangles, otherwise of only one triangle. A second nontriangular mesh, also dual to $\mathcal{T}_h$, is defined as follows.

DEFINITION 5.1. The Dirichlet tessellation $\mathcal{D}_h$, corresponding to the triangulation $\mathcal{T}_h$, is defined by $\mathcal{D}_h = \bigcup D_i$, $i = 1, N$, where, for each $x_i \in \mathcal{X}_h$, the tile D_i is given by

$$D_i = \{x \in \Omega \colon |x - x_i| < |x - x_j|, \ \forall x_j \in \mathcal{X}_h, \ j \neq i\}. \tag{5.1}$$

We remark that for each $x_i \in \mathcal{X}_h$, the boundary ∂D_i of the tile D_i is the polygon having as its vertices the circumcenters of all triangles with the common vertex x_i. Each segment of ∂D_i is perpendicular to one of the edges sharing the vertex x_i. The subset of $\mathcal{D}_h$ corresponding to X'_h is denoted by $\mathcal{D}'_h = \bigcup D_i$, for $i = 1, N'$.

With the two meshes $\mathcal{T}_h$ and $\mathcal{D}_h$ we associate a trial space $V_h \subset L^2(\Omega)$ and a test space $Q_h \subset L^2(\Omega)$, respectively, each of dimension N'.

To construct Q_h we define a set of piecewise constant basis functions ξ_i ($i = 1, 2, \ldots, N$) corresponding to the mesh $\mathcal{D}_h$ as follows

$$\xi_i = \begin{cases} 1, & \text{on } D_i, \\ 0, & \text{otherwise.} \end{cases}$$

We then define $Q_h = \operatorname{span}\{\xi_i\}_1^{N'}$. To construct V_h we proceed as follows.

DEFINITION 5.2. For each edge $e_{i,j} \in \mathcal{E}'_h$, let v be a regular function on $e_{i,j}$. We define $\widehat{v}$ the extension of v to the box $B_{i,j}$ taken constant along all the perpendiculars to $e_{i,j}$.

For each edge $e_{i,j} \in \mathcal{E}'_h$ connecting the vertices x_i and x_j we define an exponential function $\phi_{i,j}$ on $e_{i,j}$ solution of

$$\begin{cases} \frac{d}{ds}\left(e^{\psi} \frac{d\phi_{i,j}}{ds}\right) = 0, & \text{on } e_{i,j}, \\ \phi_{i,j}(x_i) = 1, & \phi_{i,j}(x_j) = 0, \end{cases}$$

where d/ds denotes the derivative along the edge $e_{i,j}$ from x_i to x_j. We then extend $\phi_{i,j}$ to $B_{i,j}$ as in Definition 5.2. Finally, the basis functions $\phi_i(x)$'s in V_h are taken as

$$\phi_i(x) = \begin{cases} \widehat{\phi_{i,j}}, & \text{on } B_{i,j}, \ \forall j \in I_i, \\ 0, & \text{elsewhere,} \end{cases}$$

where

$$I_i = \{j \neq i \colon \exists e_{i,j} \in \mathcal{E}'_h \text{ connecting } x_i \text{ and } x_j\} \tag{5.2}$$

denotes the index set of all neighbouring vertices of x_i. The support of ϕ_i is star-shaped (see Fig. 3.2, left). Setting $V_h = \operatorname{span}\{\phi_i\}_1^{N'}$, we obviously have $V_h \subset L^2(\Omega)$.

For simplicity we make the following assumption.

ASSUMPTION 5.1. For every $K \in \mathcal{T}_h$ the function ψ is linear on K.

Assumption 5.1 is true in practice. Indeed, the system of equations is normally solved iteratively using the Gummel method discussed in Section 2 of this chapter, and thus ψ is the numerical solution of the decoupled Poisson equation. Therefore, we can always use the piecewise linear interpolant of this numerical solution.

We notice that any $v_h \in V_h$ has the form $v_h(x) = \sum \phi_i(x) v_i$, with $v_i = v_h(x_i)$. Consequently v_h satisfies

$$\begin{cases} \frac{d}{ds}(e^{\psi} \frac{dv_h}{ds}) = 0, & \text{on } e_{i,j}, \\ v_h(x_i) = v_i, \quad v_h(x_j) = v_j. \end{cases}$$

Solving this two-point boundary value problem, and thanks to Assumption 5.1, we have the following flux representation on $e_{i,j}$

$$\sigma_{i,j}(v_h) := e^{\psi} \frac{dv_h}{ds} = e^{\psi_i} \mathrm{B}(\psi_i - \psi_j) \frac{v_j - v_i}{|e_{i,j}|}, \tag{5.3}$$

where $\psi_i = \psi(x_i)$, and $\mathrm{B}(x)$ denotes the Bernoulli function defined in (4.32). The constant value (5.3) is then extended to the whole box $B_{i,j}$. For any sufficiently smooth function w we can define the V_h-interpolant w_I of w, given by $w_I(x) = \sum \phi_i(x) w(x_i)$. Let $\underline{\sigma}(w) := e^{\psi} \underline{\nabla} w$ be the flux associated with w, and let $\underline{\sigma}_{i,j}(w_I)$ the flux associated with w_I as in (5.3). Denoting by $\underline{e}_{i,j}$ the unit tangent vector on $e_{i,j}$, oriented from x_i to x_j, it is easy to see that $\sigma_{i,j}(w_I)$ is the projection of $\underline{\sigma}(w) \cdot \underline{e}_{i,j} \in L^2(e_{i,j})$ onto the space of constant polynomials on $e_{i,j}$ with respect to the weighted inner product $\int_{e_{i,j}} e^{-\psi} fg \, ds$, $f, g \in L^2(e_{i,j})$. Thus, using Taylor expansion we easily obtain

$$\left\| \underline{\sigma}(w) \cdot \underline{e}_{i,j} - \sigma_{i,j}(w_I) \right\|_{\infty, e_{i,j}} \leqslant C |e_{i,j}| \left| \underline{\sigma}(w) \cdot \underline{e}_{i,j} \right|_{1,\infty,e_{i,j}}.$$

Moreover, since $e_{i,j} \subset B_{i,j}$, a continuity argument gives

$$\left\| \underline{\sigma}(w) \cdot \underline{e}_{i,j} - \sigma_{i,j}(w_I) \right\|_{\infty, B_{i,j}} \leqslant Ch \left| \underline{\sigma}(w) \right|_{1,\infty,B_{i,j}}, \tag{5.4}$$

where C is a positive constant, independent of h and w.

We introduce the mass lumping operator $P : C^0(\bar{\Omega}) \mapsto Q_h$ such that

$$P(v)(x) = \sum_{x_i \in \mathcal{X}_h} v(x_i) \xi_i(x), \quad \text{for all } x \in \overline{\Omega}. \tag{5.5}$$

Using the trial and test spaces V_h and Q_h, we now define the following Petrov–Galerkin problem corresponding to (2.14)

PROBLEM 5.1. Find $u_h \in V_h$ such that for all $q_h \in Q_h$

$$a(u_h, q_h) + \big(P(\gamma u_h), q_h\big) = (\hat{f}, q_h) \tag{5.6}$$

where $\hat{f}$ is an approximation to f and $a(\cdot,\cdot)$ denotes the bilinear form on $V_h \times Q_h$ defined by

$$a(u_h, q_h) = -\sum_{i=1}^{N'} \int_{\partial D_i \setminus \partial \Omega} e^{\hat{\psi}} \underline{\nabla} u_h \cdot \underline{n} q_h \, ds, \tag{5.7}$$

where $\hat{\psi}$ is taken as in Definition 5.2, and $\underline{n}$ is the unit outward normal vector on ∂D_i.

We shall prove that Problem 5.1 has a unique solution by showing that the associated linear system is nonsingular. The solution we are looking for will have the form $u_h(x) = \sum_{i=1}^{N'} u_i \phi_i(x)$. Taking $q_h = \xi_j$ in (5.6), we get

$$-\int_{\partial D_j \setminus \partial\Omega} e^{\hat{\psi}} \underline{\nabla} u_h \cdot \underline{n}\, ds + \gamma_j u_j |D_j| = \int_{D_j} \hat{f}\, dx, \quad j = 1, 2, \ldots, N',$$

where $\gamma_j = \gamma(x_j)$. Let the line segment $l_{j,k} = \partial D_j \cap \partial D_k$, so that $\partial D_j = \bigcup_{k \in I_j} l_{j,k}$, where I_j is the index set defined in (5.2). It is easy to check that

$$|l_{j,k}| = \frac{2|B_{j,k}|}{|e_{j,k}|}, \quad j, k = 1, \ldots N', \ j \neq k. \tag{5.8}$$

Therefore, we have from the above equality

$$-\sum_{k \in I_j} \int_{l_{j,k}} e^{\hat{\psi}} \underline{\nabla} u_h \cdot \underline{n}\, ds + \gamma_j u_j |D_j| = \int_{D_j} \hat{f}\, dx, \quad j = 1, 2, \ldots, N'.$$

Noticing that $\underline{n} = \underline{e}_{j,k}$, and $e^{\hat{\psi}}_{|l_{j,k}} =$ constant $=$ value of e^{ψ} in the midpoint of $e_{j,k}$, we can use (5.3) to obtain

$$\sum_{k \in I_j} e^{\psi_j} \mathrm{B}(\psi_j - \psi_k) \frac{u_j - u_k}{|e_{j,k}|} |l_{j,k}| + \gamma_j u_j |D_j| = \int_{D_j} \hat{f}\, dx, \quad j = 1, 2, \ldots, N'. \tag{5.9}$$

The coefficient matrix of this linear system is a symmetric and positive-definite M-matrix, since it is diagonally dominant with positive diagonal elements and negative off-diagonal elements (cf., for example, VARGA [1962], p. 85). Each element of this coefficient matrix depends exponentially on ψ_i for some i. Therefore, the matrix may be computationally very stiff as the values may vary by several orders of magnitude across an element. This drawback can be overcome by transforming the Slotboom variables back to the original ones (electron density in this case) at the discrete level, as done in Section 4. Setting

$$u_i = e^{-\psi_i} w_i, \quad i = 1, 2, \ldots, N,$$

and substituting into (5.9) we have

$$\left(\sum_{k \in I_j} \mathrm{B}(\psi_j - \psi_k) \frac{|l_{j,k}|}{|e_{j,k}|} + \gamma_j e^{-\psi_j} |D_j|\right) w_j$$
$$- \sum_{k \in I_j} e^{(\psi_j - \psi_k)} \mathrm{B}(\psi_j - \psi_k) \frac{|l_{j,k}|}{|e_{j,k}|} w_k = \int_{D_j} \hat{f}\, dx.$$

Thanks to (4.37) the above reduces to

$$\left(\sum_{k \in I_j} \mathrm{B}(\psi_j - \psi_k) \frac{|l_{j,k}|}{|e_{j,k}|} + \gamma_j e^{-\psi_j} |D_j|\right) w_j$$
$$- \sum_{k \in I_j} \mathrm{B}(\psi_k - \psi_j) \frac{|l_{j,k}|}{|e_{j,k}|} w_k = \int_{D_j} \hat{f}\, dx,$$

for $j = 1, 2, \ldots, N'$. Obviously the entries of the coefficient matrix of the above are more balanced than those of (5.9), although the matrix is not symmetric anymore, unless ψ is constant. However, it is diagonally dominant with respect to its columns. Furthermore, it can be shown that the system matrix is an M-matrix (cf. MILLER and WANG [1994a, 1994b]), and thus can be solved by a preconditioned conjugate gradient method, for example the CGS method or the Bi-CGSTAB as discussed in Chapter 9 of this handbook.

5.1.2. Convergence of the approximate solution

We now show stability and convergence of the method just described with respect to a suitable discrete norm defined on V_h. For this it is convenient to rewrite Problem 5.1 in an equivalent form where trial and test spaces coincide. Let $b(\cdot,\cdot)$ be a bilinear form on $V_h \times V_h$ defined by

$$b(u_h, v_h) = a\big(u_h, P(v_h)\big) + \big(P(\gamma u_h), P(v_h)\big). \tag{5.10}$$

Consider the following Bubnov–Galerkin problem:

PROBLEM 5.2. Find $u_h \in V_h$ such that for all $v_h \in V_h$

$$b(u_h, v_h) = \big(\hat{f}, P(v_h)\big). \tag{5.11}$$

It is easy to prove the following lemma:

LEMMA 5.1. *The mass lumping operator defined in* (5.5) *is surjective from* V_h *to* Q_h.

It is obvious from this lemma that Problem 5.2 is equivalent to Problem 5.1, hence we shall concentrate on Problem 5.2. On V_h we define the discrete energy norm

$$\|v_h\|^2 = \|v_h\|_h^2 + \sum_{i=1}^{N'} \gamma_i v_i^2 |D_i|, \quad \forall v_h = \sum_{i=1}^{N'} v_i \phi_i \in V_h, \tag{5.12}$$

where

$$\|v_h\|_h^2 = \sum_{e_{i,j} \in \mathcal{E}'_h} \left(\frac{v_j - v_i}{|e_{i,j}|} \right)^2 |B_{i,j}|, \quad \forall v_h \in V_h. \tag{5.13}$$

It is trivial to check that $\|\cdot\|_h$ is a norm on V_h, and so is $\|\cdot\|$. Let

$$\beta = \min_{e_{i,j} \in \mathcal{E}'_h} \frac{|e_{i,j}|}{\int_{e_{i,j}} e^{-\psi}\, ds}. \tag{5.14}$$

Since $|\psi|$ is bounded, there exists a $\beta_0 > 0$ such that $\beta \geqslant \beta_0 > 0$. It is also easy to verify that $\beta = \min_{e_{i,j} \in \mathcal{E}'_h} e^{\psi_i} \mathrm{B}(\psi_i - \psi_j)$ because of Assumption 5.1. The following theorem shows the coercivity of the bilinear form $b(\cdot,\cdot)$ with respect to the norm (5.12).

THEOREM 5.1. *For all* $v_h \in V_h$ *we have*

$$b(v_h, v_h) \geqslant \min\{1, 2\beta\} \|v_h\|^2. \tag{5.15}$$

PROOF. If $v_h = 0$, then (5.15) holds. Let $v_h \neq 0$. Using the method for the derivation of (5.9) we have

$$\begin{aligned} a(v_h, P(v_h)) &= -\sum_{i=1}^{N'} \int_{\partial D_i \setminus \partial\Omega} e^{\hat{\psi}} \underline{\nabla} v_h \cdot \underline{n} P(v_h)\, ds \\ &= \sum_{e_{i,j} \in \mathcal{E}'_h} (v_j - v_i) \int_{l_{i,j}} e^{\hat{\psi}} \underline{\nabla} v_h \cdot \underline{e}_{i,j}\, ds \\ &= \sum_{e_{i,j} \in \mathcal{E}'_h} (v_j - v_i) e^{\psi_i} \mathrm{B}(\psi_i - \psi_j) \frac{v_j - v_i}{|e_{i,j}|} |l_{i,j}| \geqslant 2\beta \|v_h\|_h^2, \end{aligned}$$

where in the last step we used the relation (5.8). From (5.10) and (5.12) we finally have

$$\begin{aligned} b(v_h, v_h) &= a(v_h, P(v_h)) + (P(\gamma v_h), P(v_h)) \\ &\geqslant 2\beta \|v_h\|_h^2 + \sum_{i=1}^{N'} \gamma_i v_i^2 |D_i| \geqslant \min\{1, 2\beta\} \|v_h\|^2. \end{aligned}$$ □

Theorem 5.1 implies that the solution to Problem 5.2 is stable with respect to the norm $\|\cdot\|$.

LEMMA 5.2. *For any $v_h \in V_h$, there is a constant $C > 0$, independent of h and v_h, such that*

$$\|P(v_h)\|_{0,\Omega} \leqslant C \|v_h\|_h. \tag{5.16}$$

PROOF. The proof of this can be found in MILLER and WANG [1991], Lemma 3.4. □

For any $\underline{p} \in (W^{1,\infty}(\Omega))^2$ we define

$$|\underline{p}|_{1,\infty,h} = \left(\sum_{e_{i,j} \in \mathcal{E}'_h} |\underline{p}|^2_{1,\infty,B_{i,j}} |B_{i,j}| \right)^{1/2}. \tag{5.17}$$

Obviously $|\cdot|_{1,\infty,h}$ is only a seminorm on $(W^{1,\infty}(\Omega))^2$. The following theorem establishes the convergence of the approximate solution u_h to the V_h-interpolant of u.

THEOREM 5.2. *Let u_h be the solution of Problem 5.2 and let u_I be the V_h-interpolant of the solution of problem* (2.14). *Then there is a constant $C > 0$, independent of h, u and ψ, such that*

$$\|u_h - u_I\| \leqslant \frac{C}{\min\{1, 2\beta\}} \left(h |\underline{\sigma}|_{1,\infty,h} + \|\gamma u - P(\gamma u)\|_0 + \|f - \hat{f}\|_0 \right). \tag{5.18}$$

PROOF. For the proof we refer the interested reader to MILLER and WANG [1994a]. □

We remark that depending on the decoupling technique used for the two continuity equations, we may have $\gamma = 0$ in the above (as the case in Gummel's original work). In such cases the error estimate (5.18) depends only on the seminorm of the flux $\underline{\sigma}$ and the approximation error of the source term, while error bounds in the energy norm for classical linear finite element methods depend on $\|u\|_{2,\Omega}$.

Finally we remark that the approximate flux $\underline{\sigma}_h := e^{\hat{\psi}}\underline{\nabla} u_h$ does not converge to the exact flux $\underline{\sigma} = e^{\psi}\underline{\nabla} u$. This is because in each box $B_{i,j}$, $\underline{\sigma}_h \equiv \sigma_{i,j}(u_h)\underline{e}_{i,j}$ which converges locally only to $\underline{\sigma} \cdot \underline{e}_{i,j}$ (see (5.4)). However, by post processing it is easy to define an approximate flux which converges to the exact one. For example, we can define

$$\underline{\sigma}_{h|B_{i,j}} = \sigma_{i,j}(u_h)\underline{e}_{i,j} + \frac{\int_{l_{i,j}} \underline{\nabla} u_h \cdot \underline{l}_{i,j}\, ds}{\int_{l_{i,j}} e^{-\psi}\, ds}\underline{l}_{i,j}$$

for all $e_{i,j} \in \mathcal{E}'_h$, where $\underline{l}_{i,j}$ denotes the unit tangential vector along $l_{i,j}$. Moreover the computed ohmic contact currents are convergent, as is shown in the next subsection.

5.1.3. The evaluation of the ohmic contact currents

The ultimate goal of device simulation is to find the terminal currents. We now consider the evaluation of the ohmic contact currents. For simplicity, we restrict our attention to a device with a finite number of ohmic contacts, and so Γ_D is a finite set of separated contacts Γ_c's. For any $\Gamma_c \subset \Gamma_D$, let $\{x_i^c\}_1^{N_c}$ denote the mesh nodes on Γ_c.

Let ξ^c be a piecewise constant function satisfying

$$\xi^c(x) = \begin{cases} 1, & x \in \bigcup_{i=1}^{N_c} D_i^c, \\ 0, & \text{otherwise}, \end{cases} \tag{5.19}$$

where D_i^c denotes the element in $\mathcal{D}_h$ containing x_i^c. Taking, for simplicity, $\gamma = 0$ in (2.14), multiplying by ξ^c and integrating by parts we have

$$-\int_{\Gamma_c} \underline{\sigma} \cdot \underline{n}\, ds - \sum_{i=1}^{N_c} \int_{\partial D_i^c \setminus \Gamma_c} \underline{\sigma} \cdot \underline{n}\, ds = (f, \xi^c).$$

Thus, the (scalar) outflow current through Γ_c is

$$J_c := \int_{\Gamma_c} \underline{\sigma} \cdot \underline{n}\, ds = -\sum_{i=1}^{N_c} \int_{\partial D_i^c \setminus \Gamma_c} \underline{\sigma} \cdot \underline{n}\, ds - (f, \xi^c). \tag{5.20}$$

Replacing $\underline{\sigma}$ by the approximate flux $\underline{\sigma}_h = e^{\hat{\psi}}\underline{\nabla} u_h$ and f by the approximation $\hat{f}$, we define the following approximate outflow current through Γ_c

$$J_{c,h} := -\sum_{i=1}^{N_c} \int_{\partial D_i^c \setminus \Gamma_c} \underline{\sigma}_h \cdot \underline{n}\, ds - (\hat{f}, \xi^c). \tag{5.21}$$

From (5.19), (5.3) and the argument used in the derivation of (5.9), we obtain

$$J_{c,h} = -\sum_{j=1}^{N_c}\left[\int_{\partial D_j^c \setminus \Gamma_c} \underline{\sigma}_h \cdot \underline{n}\, ds + \int_{D_j^c} \hat{f}\, dx\right]$$

$$= \sum_{j=1}^{N_c}\left[\sum_{k \in I_j,\, x_k \notin \Gamma_c} e^{\psi_j} \mathrm{B}(\psi_j - \psi_k)\frac{2|B_{j,k}|}{|e_{j,k}|}\frac{u_j - u_k}{|e_{j,k}|} - \int_{D_j^c} \hat{f}\, dx\right]$$

where I_j is the index set of neighbouring nodes of x_j as defined in (5.2).

The convergence and the conservation of the computed ohmic contact currents are established in the following theorem.

THEOREM 5.3. *Let J_c and $J_{c,h}$ be the exact and the computed outflow currents through $\Gamma_c \subset \Gamma_D$, respectively. Then, there exists a constant $C > 0$, independent of h, ψ and u, such that*

$$|J_c - J_{c,h}| \leqslant C\big(h|\underline{\sigma}|_{1,\infty,\mathcal{B}_h} + \|f - \hat{f}\|_0\big).$$

Furthermore

$$\sum_{\Gamma_c \subset \Gamma_D} J_{c,h} = -\int_{\Omega} \hat{f}\, dx.$$

PROOF. The proof is omitted and we refer the interested reader to MILLER and WANG [1994a]. □

5.2. *Finite element methods based on ad hoc chosen basis functions*

The key idea of Scharfetter–Gummel box method is to approximate the flux by a constant on each mesh line. This is in contrast to classic approaches in which a potential function is approximated by a piecewise polynomial. Physically, a flux is better behaved than the corresponding potential function. It is often bounded with respect to the singular perturbation parameter, though this is not proved. Thus, approximation of a flux by piecewise polynomials such as piecewise constant provides a more stable and accurate numerical scheme for the semiconductor device equations. From the construction of the basis functions in the previous section we see that, in each box element in $\mathcal{L}_h$, the Scharfetter–Gummel method is based on a 'divergence-free' approximation of the flux along the mesh edge and its perpendiculars. A more general choice is to seek divergence-free basis functions on the elements, i.e., seek solutions to $\operatorname{div} \underline{\sigma} = 0$ in each element with appropriate boundary condition. However, solving this problem is equivalent to solving the original problem (2.14). Note that if we only require that such a basis function is unity at one node and zero at all other nodes of the element, it is rather arbitrary. So, we may restrict ourselves to the case that

$$\frac{\partial \sigma_x}{\partial x} = 0 = \frac{\partial \sigma_y}{\partial y},$$

where σ_x and σ_y denote the two components of $\underline{\sigma}$. This also courses the problem that there are only two degrees of freedom so that only two of the three vertices in each triangle are needed to determine this basis function. Therefore, on each triangle one can define three hat functions which are unity at one vertex and zero at the other two vertices of the triangle, though the difference between any two of them is small. Of course we may choose only one hat function or the average of the three as our basis. In BANK, BÜRGLER, FICHTNER and SMITH [1990] the authors proposed such a divergence-free flux approximation on triangular elements. An improvement of this divergence-free basis function on triangles is given in SACCO and STYNES [1998]. All these basis functions may be non-conforming, i.e., they are not continuous across element boundaries. A similar approach can also be found in MARKOWICH and ZLÁMAL [1988] in which a rigorous analysis of the method is also given. A novel flux approximation technique based on that of the Scharfetter–Gummel was proposed in SEVER [1988], which makes use of three degrees of freedom on each triangle. This technique has the property that, at each point in a triangle, the approximate flux is constant along each of the lines connecting the vertices of the triangle and the point. The same idea was also proposed independently in SACCO, GATTI and GOTUSSO [1995]. Based on this idea, a set of conforming basis functions is proposed in WANG [1997] and a rigorous analysis for the method applied to semiconductor device equations is given in WANG [1999]. This analysis is based on a weighted $L^2(\Omega)$ inner product weak formulation which coincides with that in GATTI, MICHELETTI and SACCO [1998], while the finite element basis functions used in WANG [1999] coincide with those in SACCO, GATTI and GOTUSSO [1995]. It is interesting to note that the method can also be formulated as a mixed finite element method. In what follows we shall discuss the formulation and analysis of the method. For brevity, we shall omit most of the proofs and refer the reader to the relevant references.

5.2.1. *The weak formulation*

For brevity we only consider the discretization of the scaled current continuity equation for the electron density. The corresponding convection-diffusion problem is of the form

$$\begin{cases} -\operatorname{div}\underline{\sigma} + \gamma u = f & \text{in } \Omega, \\ u = g & \text{on } \Gamma_D, \\ \underline{\sigma}\cdot\underline{n} = 0 & \text{on } \Gamma_N. \end{cases} \tag{5.22}$$

with the flux defined by

$$\underline{\sigma} = \underline{\nabla} u - u\underline{\nabla}\psi.$$

Without loss of generality, we assume that $g = 0$ as in the previous section. Now, we define a weighted inner product $(\cdot,\cdot)_\psi$ on $L^2(\Omega)$ and on $(L^2(\Omega))^2$ by

$$(v, w)_\psi = (e^{-\psi} v, w). \tag{5.23}$$

The $L^2(\Omega)$-norm corresponding to this weighted inner product is denoted by $\|\cdot\|_{0,\psi}$. Using this inner product we define the following variational problem corresponding to (5.22):

PROBLEM 5.3. Find $u \in H^1_D(\Omega) := \{v \in H^1(\Omega) \mid v_{|\Gamma_D} = 0\}$ such that for all $v \in H^1_D(\Omega)$

$$A_\psi(u, v) = (f, v)_\psi, \tag{5.24}$$

where $A_\psi(\cdot,\cdot)$ is the bilinear form on $(H^1(\Omega))^2$ defined by

$$A_\psi(u, v) = (\underline{\nabla} u - \underline{\nabla}\,\psi u, \underline{\nabla} v - \underline{\nabla}\,\psi v)_\psi + (\gamma u, v)_\psi. \tag{5.25}$$

Since $A_\psi(\cdot,\cdot)$ is symmetric and positive definite on $H^1_D(\Omega)$, we can associate with A_ψ the norm

$$\|v\|^2_{1,\psi} = A_\psi(v, v), \quad v \in H^1_D(\Omega). \tag{5.26}$$

Existence and uniqueness of the solution of (5.24) follow immediately from (5.26).

5.2.2. *The finite element method*

Let $\{\mathcal{T}_h\}_h$ be a regular sequence of decompositions of $\overline{\Omega}$ into triangles (see (3.45)). As in the previous section we denote by $\mathcal{X}_h = \{x_i\}_1^N$ the set of vertices of $\mathcal{T}_h$, numbered in such a way that $\{x_i\}_1^{N'}$ is the set of vertices not on Γ_D. As before, we make Assumption 5.1, so that the vector $\underline{a} = \underline{\nabla}\,\psi$ is constant on each triangle $K \in \mathcal{T}_h$.

Corresponding to the mesh $\mathcal{T}_h$, we now construct a space $S_h \subset H^1_D(\Omega)$ of dimension N' using the basis functions $\{\phi_i\}_1^{N'}$ defined below. Let $K \in \mathcal{T}_h$ be a triangle with vertices x_i, x_j and x_k. We define a local function $\phi_i(x)$ on K associated with x_i as follows. For any point $x \in K$ we denote by $\underline{l}_m$ $(m = i, j, k)$ the vector of length $|l_m|$ connecting x_m to x, and by $\underline{e}_m := \underline{l}_m/|l_m|$ $(m = i, j, k)$ the unit vector from x_m to x (cf. Fig. 5.1). We now consider the following two-point boundary value problem on the segment l_m: find $g_i(s)$ such that

$$\begin{cases} \frac{d}{ds}(\underline{p}_i \cdot \underline{e}_m) := \frac{d}{ds}\left(\frac{dg_i(s)}{ds} - \underline{a}\cdot\underline{e}_m g_i(s)\right) = 0, & s \in (0, |l_m|), \\ g_i(0) = \delta_{im}, \quad g_i(|l_m|) = \phi_i(x) \end{cases}$$

for $m = i, j, k$ (i.e., $\underline{p}_i \cdot \underline{e}_m$ is constant on l_m), where δ_{im} denotes the Kronecker delta and $\phi_i(x)$ is yet to be determined. Solving the above boundary value problem yields

$$\underline{p}_i \cdot \underline{e}_m = \frac{1}{|l_m|}\left[\mathrm{B}(a_m)\phi_i(x) - \mathrm{B}(-a_m)\delta_{im}\right], \quad \forall x \in l_m,\ m = i, j, k,$$

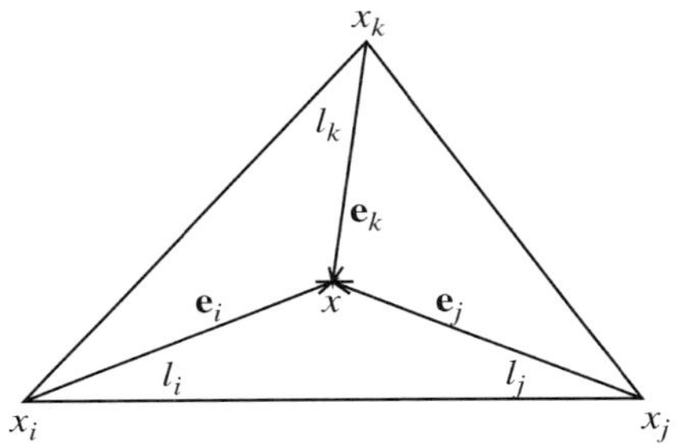

FIG. 5.1. Notation associated with the triangle K.

where $a_m = \underline{a} \cdot \underline{l}_m$ and $\mathrm{B}(z)$ denotes the Bernoulli function defined in (4.32). The above equation motivates us to define the following problem.

PROBLEM 5.4. Find, for all $x \in \overline{K}$, $\phi_i(x)$ and $\underline{p}_i = (p_{i,1}, p_{i,2})$ such that

$$D(x) \begin{pmatrix} p_{i,1} \\ p_{i,2} \\ \phi_i(x) \end{pmatrix} = \begin{pmatrix} -\mathrm{B}(-a_i) \\ 0 \\ 0 \end{pmatrix} \tag{5.27}$$

where $D(x)$ is a 3×3 matrix defined by

$$D(x) = \begin{pmatrix} l_{i,1} & l_{i,2} & -\mathrm{B}(a_i) \\ l_{j,1} & l_{j,2} & -\mathrm{B}(a_j) \\ l_{k,1} & l_{k,2} & -\mathrm{B}(a_k) \end{pmatrix}. \tag{5.28}$$

Solving Problem 5.4 for all $x \in \overline{K}$ defines the point values of the function ϕ_i and an auxiliary flux $\underline{p}_i$. Similarly we can define functions ϕ_j and ϕ_k associated with x_j and x_k respectively. The following theorem shows that Problem 5.4 is uniquely solvable for all $x \in \overline{K}$, and that ϕ_i, ϕ_j and ϕ_k form a system of local basis functions.

THEOREM 5.4. *Let $K \in \mathcal{T}_h$. Then, for any $x \in \overline{K}$, there exists a unique solution to Problem 5.4. Furthermore, we have*

$$\phi_i(x_i) = 1, \quad \phi_i(x) = 0, \quad \forall x \in \overline{x_j x_k}, \tag{5.29}$$

$$\phi_i + \phi_j + \phi_k = 1, \quad \underline{p}_i + \underline{p}_j + \underline{p}_k = -\underline{a} \quad \text{in } \overline{K}, \tag{5.30}$$

where $\overline{x_j x_k}$ denotes the edge of K connecting x_j and x_k.

PROOF. To prove that Problem 5.4 is uniquely solvable we need only to show that for any $x \in \overline{K}$ the system matrix $D(x)$ is non-singular, or $\det D(x) \neq 0$. From (5.28) we have, by direct computation,

$$\begin{aligned} \det D(x) &= \mathrm{B}(a_k)(l_{j,1}l_{i,2} - l_{j,2}l_{i,1}) + \mathrm{B}(a_i)(l_{k,1}l_{j,2} - l_{k,2}l_{j,1}) \\ &\quad + \mathrm{B}(a_j)(l_{i,1}l_{k,2} - l_{i,2}l_{k,1}) \\ &= -\big[\mathrm{B}(a_k)\underline{e}_z \cdot (\underline{l}_i \times \underline{l}_j) + \mathrm{B}(a_i)\underline{e}_z \cdot (\underline{l}_j \times \underline{l}_k) + \mathrm{B}(a_j)\underline{e}_z \cdot (\underline{l}_k \times \underline{l}_i)\big] \end{aligned} \tag{5.31}$$

with $\underline{e}_z = (0, 0, 1)$ the unit vector perpendicular to K. From the orientations of $\underline{l}_i$, $\underline{l}_j$, $\underline{l}_k$ and $\underline{e}_z$ (cf. Fig. 5.1) we see that $\underline{e}_z \cdot (\underline{l}_i \times \underline{l}_j)$, $\underline{e}_z \cdot (\underline{l}_j \times \underline{l}_k)$ and $\underline{e}_z \cdot (\underline{l}_k \times \underline{l}_i)$ are all nonnegative, and at least two of them are positive. Furthermore, since $\mathrm{B}(\cdot)$ is always positive and at least two of $|l_i|$, $|l_j|$ and $|l_k|$ are not zero, we have $\det D(x) \neq 0$.

Solving (5.27) we obtain, in particular, the explicit expression for $\phi_i(x)$:

$$\phi_i(x) = \frac{-\mathrm{B}(-a_i)\underline{e}_z \cdot (\underline{l}_j \times \underline{l}_k)}{\det D(x)}. \tag{5.32}$$

When $x = x_i$ we have $|l_i| = 0$, $a_i = 0$, and thus $\det D(x_i) = -\mathrm{B}(0)\underline{e}_z \cdot (\underline{l}_j \times \underline{l}_k)$. Hence, from (5.32) we deduce $\phi_i(x_i) = 1$. When $x \in \overline{x_j x_k}$, we have $\underline{l}_j \times \underline{l}_k = 0$ and (5.32) gives $\phi_i(x) = 0$. Thus, (5.29) is proved. In order to show that (5.30) is verified, let $\phi = \phi_i + \phi_j + \phi_k$ and $(p_1, p_2) := \underline{p} = \underline{p}_i + \underline{p}_j + \underline{p}_k$. Since ϕ_m and $\underline{p}_m$ satisfy (5.27) for $m = i, j, k$, summing the three linear systems yields

$$D(x)\begin{pmatrix} p_1 \\ p_2 \\ \phi \end{pmatrix} = \begin{pmatrix} -\mathrm{B}(-a_i) \\ -\mathrm{B}(-a_j) \\ -\mathrm{B}(-a_k) \end{pmatrix}.$$

We now show that $\phi = 1$ and $\underline{p} = -\underline{a}$ satisfy the above linear system. Using (4.37) and substituting $\phi = 1$ into the above linear system we have, for $m = i, j, k$,

$$\underline{l}_m \cdot \underline{p} = \mathrm{B}(a_m) - \mathrm{B}(-a_m) = \mathrm{B}(a_m)(1 - e^{a_m}) = -\underline{l}_m \cdot \underline{a}.$$

Therefore we have $\underline{p} = -\underline{a}$, proving (5.30). □

For each triangle having x_i as a vertex we have defined a local function ϕ_i and an auxiliary flux $\underline{p}_i$ associated with ϕ_i as above. Combining all the local functions associated with x_i we obtain a hat function ϕ_i defined on the union of all the triangles sharing x_i, denoted by Ω_i. From Theorem 5.4 we see that this ϕ_i is unity at x_i and 0 on $\partial\Omega_i$. This hat function ϕ_i can then be extended to Ω by defining $\phi_i(x) = 0$ for all $x \in \Omega \setminus \Omega_i$. If we can show that ϕ_i is continuous across inter-element boundaries in Ω_i, then $\phi_i \in C^0(\overline{\Omega}) \cap H^1_D(\Omega)$. On the edge $\overline{x_i x_j}$ we have $\underline{l}_i \times \underline{l}_j = 0$ and $\underline{l}_i = -|l_i|\underline{e}_j$. Hence, for $x \in \overline{x_i x_j}$ (5.31) becomes

$$\det D(x) = -\big(\mathrm{B}(a_i)|l_j| + \mathrm{B}(a_j)|l_i|\big)\underline{e}_z \cdot (\underline{e}_j \times \underline{l}_k), \tag{5.33}$$

which, substituted in (5.32), gives

$$\phi_i(x) = \frac{\mathrm{B}(-a_i)|l_j|}{\mathrm{B}(a_i)|l_j| + \mathrm{B}(a_j)|l_i|}. \tag{5.34}$$

Hence the function $\phi_i(x)$ on the edge $\overline{x_i x_j}$ depends only on the edge and not on the triangle. We comment that the continuity of the basis function ϕ_i does not depend on the continuity of $\underline{a}$, but depends on the continuity of its tangent component along each edge having x_i as a vertex. Although $\underline{a} = \underline{\nabla}\,\psi$ is not continuous across element edges, its tangent component along each edge is continuous, since ψ is assumed piecewise linear and continuous. To visualize this kind of hat functions, we divide $[0, 1] \times [0, 1]$ into four triangles by the two diagonals of this square and solve (5.27) on these triangles. The computed hat functions associated with the mid-point of the square corresponding to $\underline{a} = (5, 1)$ and $\underline{a} = (1, 10)$ are shown in Fig. 5.2. From this we see that the hat functions are 1 at the mid-point and zero along the boundary. They are also continuous across the inter-element boundaries.

We remark that when $\underline{a} \equiv 0$, the basis function ϕ_i reduces to the standard piecewise linear basis function. We comment that although (5.27) defines an auxiliary flux $\underline{p}_i$, in general, $\underline{p}_i \neq \underline{\sigma}_{\phi_i} := \underline{\nabla}\,\phi_i - \underline{a}\phi_i$ on a triangle K having x_i as one vertex. Nevertheless, it can be shown that $\underline{p}_i = \underline{\sigma}_{\phi_i}$ at the three vertices of K (WANG [1999]).

We now set $S_h = \mathrm{span}\{\phi_i\}_1^{N'}$; from the above discussion we see that $S_h \subset C^0(\overline{\Omega}) \cap H^1_D(\Omega)$. Using the finite element space S_h we define the following discrete problem.

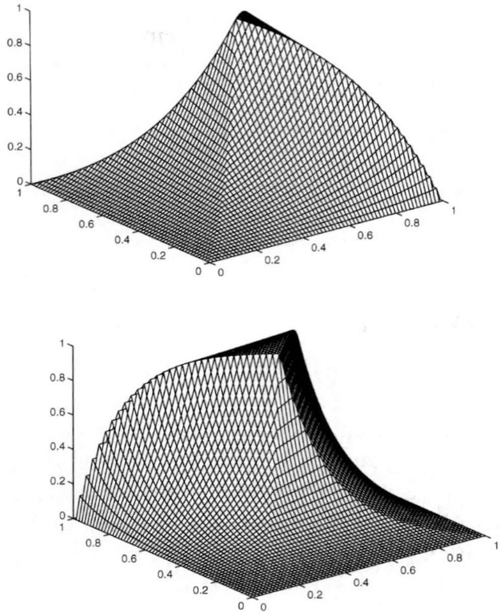

FIG. 5.2. Hat functions for different values of $\underline{a}$.

PROBLEM 5.5. Find $u_h \in S_h$ such that for all $v_h \in S_h$

$$A_\psi(u_h, v_h) = (f, v_h)_\psi, \tag{5.35}$$

where $A_\psi(\cdot,\cdot)$ is the bilinear form defined by (5.25).

Problem 5.5 is a discrete problem corresponding to Problem 5.3. Since $S_h \subset H^1_D(\Omega)$, existence and uniqueness of the solution of this problem follows immediately.

We define a seminorm $|\cdot|_{1,\infty,\psi,h}$ on $(W^{1,\infty}(\Omega))^2$ by

$$|\underline{p}|_{1,\infty,\psi,h} = \left(\sum_{K\in\mathcal{T}_h}\int_K e^{-\psi}\,dx\,|\underline{p}|^2_{1,\infty,K}\right)^{1/2}.$$

The convergence of the solution to Problem 5.5 to that of Problem 5.3 is established in the following theorem.

THEOREM 5.5. *Let u and u_h be the solutions to Problems 5.5 and 5.3 respectively, and let $\underline{\sigma}$ and $\underline{\sigma}_{u_h}$ be the respective associated fluxes ($\underline{\sigma} = \underline{\nabla} u - \underline{\nabla}\,\psi u$ and $\underline{\sigma}_{u_h|K} = (\underline{\nabla} u_h - \underline{\nabla}\,\psi u_h)_{|K}$). Then there exists a constant $C > 0$, independent of h and u, such that*

$$\|u - u_h\|_{1,\psi} \leqslant Ch|\underline{\sigma}|_{1,\infty,\psi,h},$$

$$\|\underline{\sigma} - \underline{\sigma}_{u_h}\|_{0,\psi} \leqslant Ch|\underline{\sigma}|_{1,\infty,\psi,h}.$$

PROOF. The proof of this theorem can be found in WANG [1999]. □

This theorem shows that the solution of Problem 5.3 and its associated flux converge to the exact ones with the convergence rate of order $\mathcal{O}(h)$. The error estimates depend only on the weighted first order seminorms of the exact flux, in contrast with the standard piecewise linear finite element method in which the error bound depends on $\|u\|_{2,\Omega}$. Also, the variable used in this analysis is the electron or hole concentration rather than one of the Slotboom variables. The latter is physically less interesting than the former.

5.2.3. Evaluation of terminal currents

We now consider the evaluation of the ohmic contact currents, which is often the final goal of device simulation. For simplicity, we restrict again our attention to a device with a finite number of ohmic contacts, and so Γ_D is a finite set of separated contacts. We assume that the mesh $\mathcal{T}_h$ is such that the end-points of any contact are mesh nodes of $\mathcal{T}_h$. From the definition of ohmic contacts we know that the potential drop within a contact is negligible (cf. SZE [1981], p. 304). Thus ψ is constant on each ohmic contact. Let $V_h := \operatorname{span}\{\phi_i\}_1^N \subset C^0(\overline{\Omega}) \cap H^1(\Omega)$. Obviously, if $v \in V_h$ and $v_{|\Gamma_D} = 0$, then $v \in S_h$. For any $\Gamma_c \subset \Gamma_D$, we choose $\phi_c \in V_h$ satisfying

$$\phi_c(x) = \begin{cases} e^{\psi_c}, & x \in \Gamma_c, \\ 0, & x \in \Gamma_D \setminus \Gamma_c, \end{cases}$$

where ψ_c denotes the (constant) value of ψ on Γ_c. Taking $\gamma = 0$ in (5.22) (with $\underline{\sigma} = \underline{\nabla} u - \underline{\nabla}\,\psi u$), multiplying by $e^{-\psi}\phi_c$ and integrating by parts we have

$$-\int_{\Gamma_c} \underline{\sigma} \cdot \underline{n}\, ds + (\underline{\sigma}, \underline{\sigma}_{\phi_c})_\psi = (f, \phi_c)_\psi,$$

where $\underline{\sigma}_{\phi_c} = \underline{\nabla}\,\phi_c - \underline{\nabla}\,\psi\phi_c$. Thus, the outflow current through Γ_c is

$$J_c := \int_{\Gamma_c} \underline{\sigma} \cdot \underline{n}\, ds = (\underline{\sigma}, \underline{\sigma}_{\phi_c})_\psi - (f, \phi_c)_\psi.$$

Replacing $\underline{\sigma}$ by the approximate flux $\underline{\sigma}_{u_h}$ we obtain the following approximate outflow current through Γ_c

$$J_{c,h} := (\underline{\sigma}_{u_h}, \underline{\sigma}_{\phi_c})_\psi - (f, \phi_c)_\psi.$$

The convergence and the conservation of the computed ohmic contact currents are established in the following theorem.

THEOREM 5.6. *Let J_c and $J_{c,h}$ be respectively the exact and the computed outflow currents through $\Gamma_c \subset \Gamma_D$ defined above. Then, there exists a constant $C > 0$, independent of h and u, such that*

$$|J_c - J_{c,h}| \leqslant Ch|\underline{\sigma}|_{1,\infty,\psi,h}\|\phi_c\|_{1,\psi}.$$

Furthermore,

$$\sum_{\Gamma_c \subset \Gamma_D} J_{c,h} = -\int_\Omega f\, dx.$$

PROOF. For the proof we refer to WANG [1999]. □

5.3. Stabilized finite element methods

Solutions to the boundary value problem (2.14) normally show sharp interior layers so that the application of a conventional finite element method to the problem often yields numerical solutions with non-physical spurious oscillations. This is because classical methods are numerically not stable, or the discretized diffusion term is too small so that the corresponding system matrix is almost singular. To overcome this difficulty, stabilized finite elements have been developed. The simplest case is to add to the system an artificial diffusion term with a coefficient of the order $\mathcal{O}(h)$. However, although this technique stabilizes the discretization, it introduces much artificial/computational diffusion, especially in the direction perpendicular to the characteristic or streamline direction (called cross-wind direction), and thus yields a substantial cross-wind dissipation so that a sharp interior layer is smeared out. Furthermore, the resulting scheme is of first order accuracy at most. A better approach is to add a diffusion term with a positive coefficient δ to the streamline direction only. This forms the base of the so called streamline diffusion method. This approach, of course, introduces less cross-wind diffusion. The artificial streamline diffusion term is rather arbitrary and can be conforming or non-conforming. It can also be used in a Galerkin or Petrov–Galerkin formulation. Several methods have been developed for semiconductor device equations. For example, BANK, BÜRGLER, FICHTNER and SMITH [1990], SHARMA and CAREY [1989a, 1989b], MICHELETTI [2001]. In particular, the streamline upwind Petrov–Galerkin (SUPG) method used in SHARMA and CAREY [1989a], originally proposed in HUGHES and BROOKS [1982], BROOKS and HUGHES [1982] has been very successful for solving fluid flow problems with layers. We now give a brief account of the method for problem (5.22). For discussion brevity we assume $\gamma = 0$, and, as before, $g = 0$.

Multiplying Eq. (5.22) by $v \in H_D^1(\Omega)$, integrating by parts, and using $\underline{\sigma} = \underline{\nabla} u - \underline{\nabla}\,\psi u$ gives the usual variational formulation: find $u \in H_D^1(\Omega)$ such that

$$\sum_{K\in\mathcal{T}_h} \int_K (\underline{\nabla} u - \underline{\nabla}\,\psi u)\cdot\underline{\nabla} v\,dx = \sum_{K\in\mathcal{T}_h} \int_K f v\,dx, \quad \forall v \in H_D^1(\Omega). \tag{5.36}$$

We then add a stabilizing term, which does not alter consistency, and write

$$\sum_{K\in\mathcal{T}_h} \int_K (\underline{\nabla} u - \underline{\nabla}\,\psi u)\cdot\underline{\nabla} v\,dx + \sum_{K\in\mathcal{T}_h} \delta \int_K \big(f + \operatorname{div}(\underline{\nabla} u - \underline{\nabla}\,\psi u)\big)\underline{\nabla}\,\psi\cdot\underline{\nabla} v\,dx$$
$$= \sum_K \int_K f v\,dx,$$

for all $v \in H_D^1(\Omega)$, where δ is a positive stabilizing parameter to be chosen. At the continuous level we are not modifying the formulation, as we are adding a term containing the residual on the exact solution, which is zero. Let $V_h \subset H_D^1(\Omega)$ be the finite element space with the conventional piecewise linear basis functions constructed on $\mathcal{T}_h$. The approximate problem reads find $u_h \in V_h$ such that

$$\sum_{K \in \mathcal{T}_h} \int_K (\underline{\nabla} u_h - \underline{\nabla} \psi u_h) \cdot \underline{\nabla} v_h \, dx - \sum_{K \in \mathcal{T}_h} \delta \int_K (\underline{\nabla} \psi \cdot \underline{\nabla} u_h)(\underline{\nabla} \psi \cdot \underline{\nabla} v_h) \, dx$$

$$= \sum_{K \in \mathcal{T}_h} \int_K f (v_h - \delta \underline{\nabla} \psi \cdot \underline{\nabla} v_h) \, dx$$

for all $v_h \in V_h$. Indeed, since u_h and ψ are piecewise linear, $\Delta u_h = 0$, and $\text{div}(\underline{\nabla} \psi u_h) = \underline{\nabla} \psi \underline{\nabla} u_h$ in K, $\forall K \in \mathcal{T}_h$.

This approach has the merit that it can reduce numerical cross-wind dissipation, and thus it is expected that it can capture sharp interior layers. However, the discretized system does not satisfy the maximum principle, or the system matrix is no longer an M-matrix. So, non-physical oscillations may occur near the layers. One way to improve the numerical stability of this formulation is to add a so-called shock-capturing term which may be non-linear. Analysis of this type methods can be found in HUGHES and BROOKS [1982], BROOKS and HUGHES [1982], JOHNSON, NAVERT and PITKARANA [1984], just to name a few. An overview of streamline diffusion type methods for convection-diffusion equations is given in JOHNSON [1987] in which a list of further readings can also be found.

5.4. *Combinations of various numerical methods*

Several combinations of finite element and other methods have been developed by various authors for the semiconductor device equations. In what follows we give a brief review of these methods.

In COCKBURN and TRIANDAF [1994] a numerical method for one-dimensional time-dependent semiconductor device equations with zero diffusion coefficient was proposed. This method combines a mixed finite element method for the electric field with an explicit upwind finite element method for the electron continuity equation. A rigorous analysis of the method is given. The same method is analyzed in CHEN and COCKBURN [1994] for the case that the electron concentration does not have discontinuity, and an improved error bound was obtained. The relevance of these methods is mostly theoretical, as the system under consideration is over-simplified. In CHEN and COCKBURN [1995] the author proposed a combined numerical method for a two-dimensional system containing the Poisson equation for ψ and the continuity equation for n. The method is a combination of a mixed finite element discretization for the Poisson equation and a discontinuous upwind finite element discretization for the continuity equation. A mathematical analysis was given and some numerical results were presented. Similar combinations of methods, using the finite element method only for the discretization of Poisson's equation to obtain a global representation of the electric field, have also been described for some particular types of semiconductor device simulations. For example, in HECHT, MARROCCO, CAQUOT and FILOCHE [1991], a combination of a mixed finite element method with an alternating direction implicit (ADI) method is presented for the modelling of semiconductor heterojunction structures.

6. Discretization schemes for Energy-Transport and Energy-Balance models

In the physical literature, the Energy-Transport equations have been investigated numerically for several years (APANOVICH, BLAKEY, COTTLE, LYUMKIS, POLSKY, SHUR and TCHERNIAEV [1995], CHEN, KAN, RAVAIOLI, SHU and DUTTON [1992], CHEN, SANGIORGI, PINTO, KAN, RAVAIOLI and DUTTON [1992], SOUISSI, ODEH, TANG and GNUDI [1994], VISOCKY [1994]), usually using Scharfetter–Gummel-type discretizations. Entropy-based finite difference schemes have been proposed in RINGHOFER [2001]; JEROME and SHU [1994] solved the equations employing ENO (essentially non-oscillatory) methods, while mixed finite element discretization for the dual entropy formulation has been used in MARROCCO and MONTARNAL [1996], MARROCCO, MONTARNAL and PERTHAME [1996], LAB and CAUSSIGNAC [1999]. In Section 6.1 we briefly present a mixed exponential fitting discretization for Energy-Transport models. These schemes are an extension of the schemes presented in Sections 3 and 4.1 for the Drift-Diffusion continuity equation, referring to DEGOND, JÜNGEL and PIETRA [2000], HOLST, JÜNGEL and PIETRA [2003] for a complete discussion. In Section 6.2 we address the extension of the MFV scheme discussed in Section 4.2 to the case of the Energy-Balance transport model (1.28).

6.1. *Mixed finite element discretization for Energy-Transport models*

We consider here the stationary Energy-Transport equations in a scaled and dimensionless form. Moreover, in order to simplify the subsequent presentation, we drop the subscript n appearing in equations (1.24) and we redefine the current and energy densities as

$$\underline{J}^1 := \underline{J}_n; \qquad \underline{J}^2 := \underline{S}_n. \tag{6.1}$$

We use the Unit scaling of Table 1.1 of Section 1.3 and the lattice temperature T^{eq} as scaling factor for the electron temperature

$$\begin{aligned}
&n \to \overline{C}n, \qquad C \to \overline{C}C, \qquad T \to T^{eq}T, \qquad \psi \to V_{th}\psi, \qquad \mu \to \overline{\mu}\mu,\\
&x \to lx, \qquad \underline{J}^1 \to (q\overline{\mu}V_{th}\overline{C}/l)\underline{J}^1, \qquad \underline{J}^2 \to (q\overline{\mu}V_{th}^2\overline{C}/l)\underline{J}^2,\\
&L_{ij} \to \big((qV_{th})^{i+j-1}\overline{\mu}\overline{C}\big)L_{ij}, \qquad W \to (q\overline{\mu}V_{th}^2\overline{C}/l^2)W.
\end{aligned}$$

System (1.24) in the stationary case takes now the form

$$\begin{cases}
\lambda^2\Delta\psi = n - C,\\
-\operatorname{div}\underline{J}^1 = 0,\\
-\operatorname{div}\underline{J}^2 = -\underline{\nabla}\psi\cdot\underline{J}^1 + W(n,T),\\
\underline{J}^1 = L_{11}\big(\frac{\underline{\nabla}n}{n} - \frac{\underline{\nabla}\psi}{T}\big) + \big(\frac{L_{12}}{T} - \frac{3}{2}L_{11}\big)\frac{\underline{\nabla}T}{T},\\
\underline{J}^2 = L_{21}\big(\frac{\underline{\nabla}n}{n} - \frac{\underline{\nabla}\psi}{T}\big) + \big(\frac{L_{22}}{T} - \frac{3}{2}L_{21}\big)\frac{\underline{\nabla}T}{T},
\end{cases} \tag{6.2}$$

where $\lambda^2 = \varepsilon V_{th}/(q\overline{C}l^2)$ denotes, as usual, the square of the scaled Debye length.

The starting point for the numerical discretization of the Energy-Transport system (6.2) is the observation that a suitable choice of variables allows to write the current and the energy densities $(6.2)_4$–$(6.2)_5$ in a Drift-Diffusion form in a very general context. More precisely, the following physical assumptions are imposed

- The energy band diagram is spherically symmetric and monotone with respect to the modulus of the wave vector $\vec{k}$.
- The electron density is given by non-degenerate Boltzmann statistics.
- The energy relaxation term is given by a Fokker–Planck approximation (see DEGOND, JÜNGEL and PIETRA [2000], Section 2.2).

Under these assumptions, explicit expressions for the diffusion matrix $L = (L_{ij})$ and the energy relaxation term W in terms of n, T can be given. We refer to DEGOND, JÜNGEL and PIETRA [2000] for details of the computation, which goes through the derivation of a so-called *spherical harmonic expansion* (SHE) model (derived from the Boltzmann equation in the diffusion limit, under the assumption of dominant elastic scattering) and, afterwards, through a diffusion approximation, making electron-electron or electron-phonon scattering large. *Any* Energy-Transport model derived in this way allows for a Drift-Diffusion formulation of the form

$$\underline{J}^i = \underline{\nabla}\, g^i(n,T) - g^i(n,T)\frac{\underline{\nabla}\,\psi}{T}, \quad i = 1, 2, \tag{6.3}$$

where g^1 and g^2 are nonlinear functions of n and T. (In fact, $g^1 = L_{11}$ and $g^2 = L_{21}$.) For constant temperature, this expression reduces to the standard drift-diffusion current definition.

Moreover, the energy relaxation term can be written in the form

$$W = c^1 g^1 - c^2 g^2, \quad \text{with } c^i = c^i(g^1, g^2) \geqslant 0, \tag{6.4}$$

and the continuity equation $(6.2)_3$, (6.3) in the variables g^1 and g^2 reads as follows

$$-\operatorname{div} \underline{J}^2 + c^2 g^2 = f, \qquad \underline{J}^2 = \underline{\nabla}\, g^2 - g^2 \frac{\underline{\nabla}\,\psi}{T(g^1, g^2)},$$

where f contains the Joule heating term $\underline{J}^1 \cdot \underline{\nabla}\,\psi$ and the term $c^1 g^1$. Finally, T can be defined in terms of g^1, g^2, whenever the diffusion matrix $L = (L_{ij})$ is positive definite (see DEGOND, JÜNGEL and PIETRA [2000], Lemma 2.1). Notice that this property has to be satisfied in order to get a well-posed mathematical problem.

The particular expression of g^1, g^2, c^1 and c^2 clearly depends on the actual choice of the energy band diagram and of the time relaxation model. For the parabolic band case and a special choice of the relaxation term (corresponding to the model studied in CHEN, KAN, RAVAIOLI, SHU and DUTTON [1992]), the diffusion matrix $L = (L_{ij})$ is given by the scaled version of (1.25) and we have

$$g^1(n,T) = n, \qquad g^2(n,T) = \frac{3}{2} nT. \tag{6.5}$$

The energy relaxation term is of the form (see (1.26))

$$W = -\frac{3}{2}\frac{n(T - T^{eq})}{\tau_0} = \frac{1}{\tau_0}\left(\frac{3}{2}g^1 - g^2\right). \tag{6.6}$$

Here, τ_0 is the (scaled) energy relaxation time. In the general case, τ_0 depends on the temperature T (see DEGOND, JÜNGEL and PIETRA [2000]).

In the following, we shall consider the discretization of the Energy-Transport model in the (g^1, g^2, ψ) variables:

$$\begin{cases} \lambda^2 \Delta \psi = n(g^1, g^2) - C(x), & \\ -\mathrm{div} \underline{J}^1 = 0, & \\ -\mathrm{div} \underline{J}^2 + c^2(g^1, g^2) g^2 = c^1(g^1, g^2) g^1 - \underline{J}^1 \cdot \underline{\nabla} \psi & \text{in } \Omega, \\ \underline{J}^i = \underline{\nabla} g^i - g^i \frac{\underline{\nabla} \psi}{T(g^1, g^2)}, \quad i = 1, 2, & \\ g^1 = g^1_D, \quad g^2 = g^2_D, \quad \psi = \psi_D & \text{on } \Gamma_D, \\ \underline{J}^1 \cdot \underline{n} = \underline{J}^2 \cdot \underline{n} = \underline{\nabla} \psi \cdot \underline{n} = 0 & \text{on } \Gamma_N, \end{cases} \tag{6.7}$$

where $g^i_D = g^i(n_D, T_D)$, $i = 1, 2$, and

$$c^1(g^1, g^2) = \frac{3}{2\tau_0}, \qquad c^2(g^1, g^2) = \frac{1}{\tau_0}, \tag{6.8}$$

$$T(g^1, g^2) = \frac{2g^2}{3g^1}, \qquad n(g^1, g^2) = g^1, \tag{6.9}$$

when (6.5) and (6.6) are chosen. Although in the example given here g^1 coincides with n and the relaxation time τ_0 does not depend on T, we prefer to write (6.7) in the general setting, which includes other choices of energy band diagrams and relaxation time models. In particular, when a non-parabolic band in the sense of KANE [1957] is chosen, the dependence on T is non-local, but the discretization scheme presented here can be used without changes (the computation of $c^1(g^1, g^2)$, $c^2(g^1, g^2)$, $T(g^1, g^2)$, and $n(g^1, g^2)$ requires, of course, more effort than in the case (6.8) and (6.9)).

The Drift-Diffusion form (6.3) of the fluxes in the Energy-Transport system suggests that numerical schemes developed for the linear (in the charge variable) Drift-Diffusion continuity equation might be employed here. In DEGOND, JÜNGEL and PIETRA [2000], HOLST, JÜNGEL and PIETRA [2003] extensions of the exponential fitting mixed finite element methods presented in Section 4.1 have been developed (in the one-dimensional, and the two-dimensional case, respectively). We refer also to JÜNGEL and PIETRA [1997] for a variant of these schemes in the case of a nonlinear (in the charge variable) Drift-Diffusion model (JÜNGEL [2001]). In the following, we shall describe in detail the discretization of the energy flux continuity equation $(6.7)_3$. The discretization of equations $(6.7)_2$ is similar but simpler (since the zeroth-order term and the right-hand side of $(6.7)_2$ are zero). For the Poisson equation $(6.7)_1$, P_1-nonconforming elements are used (see CROUZEIX and RAVIART [1973]). Therefore, in the forthcoming, we assume ψ to be linear on each element.

We recall that the numerical scheme of Section 4.1 is based on the following ingredients: transform the problem by means of the Slotboom change of variable to a symmetric form, then discretize the symmetric form with mixed finite elements, and, finally, use a suitable discrete change of variable to return to the original density variable. Due to the non-constant electron temperature, a *global* Slotboom variable does not exist in the present case. However, we can define a *local* "Slotboom variable" ρ, assuming that the temperature $T = T(g^1, g^2)$ is a prescribed piecewise constant function, called $\overline{T}$, and defined in the global iteration process which solves the entire system (6.7). Therefore, $\underline{\nabla}\,\psi/\overline{T}$ is constant on each element $K \in \mathcal{T}_h$, and we can define (see (4.4))

$$\rho = e^{-\psi/\overline{T}} g, \qquad \text{in } K, \tag{6.10}$$

under the assumption $\overline{T} > 0$. For simplicity of notation, here and in the following the superscript$_2$ is dropped. Eqs. (6.7)$_4$ and (6.7)$_3$ can be rewritten on each triangle K as follows:

$$e^{-\psi/\overline{T}} \underline{J} - \underline{\nabla}\,\rho = 0, \tag{6.11}$$

$$-\operatorname{div} \underline{J} + \overline{c} e^{\psi/\overline{T}} \rho = -\underline{J}^1 \cdot \underline{\nabla}\,\psi + \overline{c^1} g^1, \tag{6.12}$$

where $\overline{c^1} = c^1(\overline{T})$, and $\overline{c} = c^2(\overline{T})$. In order to approximate the exponential functions in (6.11), (6.12), following (4.7) and (4.9), we define $\overline{\psi}$ as the piecewise constant function given in each element K by

$$e^{-\overline{\psi}/\overline{T}}{}_{|K} := \frac{1}{|K|} \int_K e^{-\psi/\overline{T}}\, dx, \tag{6.13}$$

and we define $\widetilde{\psi}$ as the piecewise constant function defined in each element K by

$$e^{\widetilde{\psi}/\overline{T}}{}_{|K} := |\tilde{e}| \Big/ \left(\int_{\tilde{e}} e^{-\psi/\overline{T}_{|K}}\, ds \right), \qquad \tilde{e} = V_{\min} V_{\mathrm{med}}, \tag{6.14}$$

where the special edge $\tilde{e}$ is the edge connecting the vertices with the smallest values of ψ in K.

Since $\overline{T}$ is piecewise constant and it might take different values in two triangles having an edge e in common, $\int_e e^{-\psi/\overline{T}}\, ds$ is not uniquely defined on e. Therefore, formula (4.14) cannot be used in the present case. Here, for each element K, we introduce on each edge $e \subset \partial K$ the following constant function

$$\left(e^{-\psi/\overline{T}}\right)^{I_K}_{|e} := \frac{1}{|e|} \int_e e^{-\psi/\overline{T}_{|K}}\, ds. \tag{6.15}$$

Notice that the change of sign of ψ with respect to the analogous formulas of Section 4.1 is due to the different sign of ψ in Eq. (6.7)$_4$ (given for negative charges) and in Eq. (4.2) (given for positive charges).

Due to the presence of the zeroth-order term in (6.7)$_3$ we shall use, instead of the well known RT_0 element, the mixed finite element of Example 5 (Section 3.5), which guarantees positivity of the solution, and, in contrast with the element of Example 6, gives a conforming approximation of $H(\operatorname{div}; \Omega)$, i.e., the discrete current vector has continuous normal component across the interelement boundaries.

For the reader's convenience, we recall the definition of the finite dimensional spaces already introduced in (3.148), (3.51), (3.160),

$$\widehat{\Sigma}_h = \left\{ \underline{\tau}_h \in \left(L^2(\Omega)\right)^2 \mid \underline{\tau}_{h|K} \in Q(K),\ \forall K \in \mathcal{T}_h \right\},$$

$$V_h = \left\{ v_h \in V \mid v_{v|K} \in P_0(K),\ \forall K \in \mathcal{T}_h \right\},$$

$$\Lambda_{h,\xi} = \left\{ \mu_h \in L^2(\mathcal{E}_h) \mid \mu_{h|e} \in P_0(e),\ \forall e \in \mathcal{E}_h,\ \int_e (\mu_h - \xi)\, ds = 0,\ \forall e \in \mathcal{E}_h \cap \Gamma_D \right\},$$

where, as usual, $P_0(D)$ denotes the set of constant functions in the domain D. Moreover, $Q(K)$ denotes here the set of polynomial vectors with $\dim(Q(K)) = 3$ defined by (3.112), (3.114)–(3.116), where the special edge $\tilde{e}$ is the edge connecting the vertices with the smallest values of ψ in K (see also (6.14)).

Specializing (4.16) to this case, the discrete formulation of $(6.7)_3$, $(6.7)_4$, becomes

$$\begin{cases} \text{Find } (\underline{J}_h, \rho_h, g_h) \in \widehat{\Sigma}_h \times V_h \times \Lambda_{h,g_D} \text{ such that} \\ \sum_K \big(\int_K (e^{-\overline{\psi}/\overline{T}} \underline{J}_h \cdot \underline{\tau}_h + \rho_h \operatorname{div} \underline{\tau}_h)\, dx \\ \quad - \int_{\partial K} (e^{-\psi/T})^{I_K} g_h \underline{\tau}_h \cdot \underline{n}\, ds \big) = 0, \\ -\sum_K \int_K (v_h \operatorname{div} \underline{J}_h\, dx + \overline{c} e^{\widetilde{\psi}/\overline{T}} \rho_h v_h)\, dx \\ \quad = \sum_K \int_K (-\underline{J}_h^1 \cdot \underline{\nabla} \psi + \overline{c^1 \overline{g}_h^1}) v_h\, dx, \\ \sum_K \int_{\partial K} \mu_h \underline{J}_h \cdot \underline{n}\, ds = 0, \end{cases} \tag{6.16}$$

for all $\underline{\tau}_h \in \widehat{\Sigma}_h$, $v_h \in V_h$, and $\mu_h \in \Lambda_{h,0}$, respectively. The vector $\underline{J}_h^1 \in \Sigma \cap \widehat{\Sigma}_h$ is the approximation of the current density $\underline{J}^1$ (Σ being defined in (3.20)) and $\overline{g}_h^1 \in V_h$ is a piecewise constant approximation of g^1 provided by an analogous discretization of $(6.7)_2$, $(6.7)_4$. The first equation in (6.16) is a weak discrete version of (6.11), the second equation corresponds to a discrete version of (6.12), and the third equation imposes a continuity requirement of the normal component of $\underline{J}_h$ at the interelement boundaries. Notice that ρ has been introduced as a "trick" and its computation is of no interest (actually it will be eliminated by static condensation). The "density" variable g is approximated by g_h on the edges.

Performing the elimination of $\underline{J}_h$ and ρ_h by static condensation as in Section 3.7 and in Section 4.1, we obtain an algebraic system in the variable g_h of the form

$$\widetilde{\mathcal{M}} g_h = \mathcal{G}. \tag{6.17}$$

When computing the element matrix $\widetilde{\mathcal{M}}^K$, the edges of K are numbered counter–clockwise starting from e^1, chosen as the special edge $\tilde{e} = V_{\min} V_{\text{med}}$, used in (6.14) and in the definition of $Q(K)$ (see (3.116)). As in (4.19) one can see that

$$\left(e^{-\psi/T}\right)^{I_K}_{|e^1} e^{\widetilde{\psi}/\overline{T}} = 1. \tag{6.18}$$

The coefficients of $\widetilde{\mathcal{M}}^K$ have the form (see also (3.232) and (4.18))

$$\widetilde{m}_{ij}^K = \begin{cases} \left(e^{-\psi/\overline{T}}\right)^{I_K}_{|e^1} e^{\overline{\psi}/\overline{T}} \frac{\underline{\nu}^1 \cdot \underline{\nu}^1}{|K|} + \frac{\overline{c}|K|}{\beta^2 + \delta \overline{c} e^{(\widetilde{\psi} - \overline{\psi})/\overline{T}} |K|} \gamma_1^2, & \text{for } i = j = 1, \\ \left(e^{-\psi/\overline{T}}\right)^{I_K}_{|e^j} e^{\overline{\psi}/\overline{T}} \frac{\underline{\nu}^i \cdot \underline{\nu}^j}{|K|}, & \text{otherwise}, \end{cases} \tag{6.19}$$

for $i, j = 1, 3$, and the coefficients of the element right-hand side $\mathcal{G}^K$ are given by (see also (3.233))

$$g_i^K = \begin{cases} \frac{\beta \gamma_1}{\beta^2 + \delta \overline{c} e^{(\widetilde{\psi} - \overline{\psi})/\overline{T}} |K|} \int_K f\, dx, & \text{for } i = 1, \\ 0, & \text{otherwise}, \end{cases} \tag{6.20}$$

with $f = -\underline{J}_h^1 \cdot \underline{\nabla}\psi + \overline{c^1}\overline{g}_h^1$. The definition of δ, β, γ_1 is given in (3.220)–(3.222), and it is related to $Q(K)$. As for (4.17) with (4.20), the matrix $\widetilde{\mathcal{M}}$ is an M-matrix, if the decomposition is of weakly acute type.

For the definition of $\overline{T}$ and a description of the global iteration process we refer to HOLST, JÜNGEL and PIETRA [2003], where a generalization of the Gummel map presented in Section 2 has been proposed, together with a Newton-type algorithm.

6.2. MFV discretization of the Energy-Balance model

In this section we shall consider the discretization of the Energy-Balance (EB) model (1.28) introduced in Section 1.4.

The fundamental constraints for the discretization are the conservation of electric field, current and energy fluxes, and the nonnegativity of the concentrations and of the temperatures of the carriers. With this aim, we propose an efficient and accurate solver for the EB equations in steady-state conditions. The Gummel decoupled algorithm introduced in Section 2 is employed to solve iteratively the full system, which consists of a linearized Poisson equation and of a linearized current continuity and energy-balance equations. The discretization of these linearized equations is based on the use of the cell-centered MFV method discussed in Section 3.8.

In the case of the linearized Poisson equation, this leads to solving a linear system whose coefficient matrix is symmetric, positive definite and diagonally dominant, while in the case of the linearized current continuity and energy-balance equations a stabilization procedure is developed by adding a suitable artificial diffusion term to the discretization of the constitutive laws $(1.28)_5$ and $(1.28)_6$. The artificial diffusion is associated with each edge of the finite element triangulation and can be written in terms of the jumps of the approximate scalar unknown across the edge and of the convective flux.

A special choice of the artificial diffusion term is considered which yields the exponentially-fitted upwinded Scharfetter–Gummel (SG) scheme (SCHARFETTER and GUMMEL [1969]). This method provides an *optimal* upwinding approximation of the interelement fluxes and is employed in numerical computations. The resulting matrices in the discretization of $(1.28)_2$ and $(1.28)_3$ turn out to be diagonally dominant M-matrices with respect to the columns.

The outline of the section is as follows. The dual mixed formulation of a convection-diffusion model problem is considered in Section 6.2.1. The approximation of the model problem using RT_0 finite elements of lowest degree is then carried out in Section 6.2.2, while a detailed description of the stabilization procedure and of the stability analysis of the method are addressed in Sections 6.2.3 and 6.2.4, respectively. In this latter section, sufficient conditions for the stiffness matrix of the scheme to be a nonsingular M-matrix and a coercivity result in a discrete energy norm are provided. In Section 6.2.5 we specialize the general stabilization approach to both the DD and EB transport models, providing the expressions of the interelement fluxes which extend the SG discretization to the two-dimensional case. The section is concluded with a summary of the linear algebraic systems that must be solved in the case of a bipolar model (comprising electrons and holes) at each step of the Gummel algorithm.

6.2.1. The mixed formulation

In this section we address the dual mixed formulation of the differential subproblems obtained applying the iterative Gummel method described in Section 2 to the EB system (1.28) in the stationary case. We shall focus our attention on the discretization of the electron current continuity and energy equations only, since the treatment of the linearized Poisson equation has already been dealt with in Section 3.8. Moreover, since similar equations hold for the holes, we shall supplement the current equation $(1.28)_2$ and the energy-balance equation $(1.28)_3$ with nonvanishing right-hand sides.

At each step of the Gummel loop the following problems must be solved for $\hat{n} = qn$ and $\phi_n = K_B T_n / q$:

$$-\operatorname{div}\big(D_n \underline{\nabla} \hat{n} - \mu_n \hat{n} \underline{\nabla}(\psi - \phi_n)\big) + \hat{p} \overline{R} \hat{n} = \overline{G} \tag{6.21}$$

and

$$\begin{aligned} &-\operatorname{div}\left(\lambda_n \underline{\nabla} \phi_n + \underline{J}_n \frac{5}{2} \phi_n\right) + \frac{3}{2}\left(\hat{p}\hat{n}\overline{R} + \frac{\hat{n}}{\tau_{w_n}}\right)\phi_n \\ &\quad = \underline{E} \cdot \underline{J}_n + \frac{3}{2}\left(\overline{G}\chi_n + \frac{\hat{n}}{\tau_{w_n}} V_{th}\right), \end{aligned} \tag{6.22}$$

where $\hat{p} = qp$ and $\lambda_n = \frac{q}{K_B}\kappa_n$ is the modified thermal conductivity. In Eq. (6.21) $\hat{n}$ is the unknown whereas ψ, ϕ_n, $\hat{p}$, $\overline{R}$ and $\overline{G}$ are given functions corresponding to the previous Gummel iteration, namely

$$\overline{R} = \frac{1}{\tau_n^*(\hat{p} + \hat{n}_i) + \tau_p^*(\hat{n} + \hat{n}_i)} + \frac{1}{q^2}(C_n \hat{n} + C_p \hat{p}),$$

$$\overline{G} = \hat{n}_i^2 \overline{R} + \alpha_n(|\underline{E}|)|\underline{J}_n| + \alpha_p(|\underline{E}|)|\underline{J}_p|,$$

where $\hat{n}_i = qn_i$. Notice that $\overline{R}$ and $\overline{G}$ are strictly positive functions since they are the recombination and the generation terms, respectively, constructed according to the procedure discussed in Section 2.4, provided that the n, p variables are employed.

In Eq. (6.22) the only unknown is ϕ_n, while χ_n denotes the function ϕ_n at the previous Gummel step.

Eqs. (6.21) and (6.22) can be cast in the form of a stationary convection-diffusion-reaction problem

$$\begin{cases} -\operatorname{div} \underline{\sigma} + \gamma u = f, \\ \underline{\sigma} = a \underline{\nabla} u - \underline{\beta} u, \end{cases} \tag{6.23}$$

where $\underline{\sigma}$ and u are the vector and scalar unknowns, respectively. System (6.23) is supplemented with suitable boundary conditions, depending on the problem we are dealing with, of the type

$$u = g \quad \text{on } \Gamma_D, \qquad \underline{\sigma} \cdot \underline{n} = 0 \quad \text{on } \Gamma_N. \tag{6.24}$$

It is clear that the model problem (2.14) extensively discussed in previous sections can be recovered from (6.23) by simply setting $\underline{\beta} = \underline{0}$. Comparing (6.23) with (6.21) we see that $u = \hat{n}$, $a = D_n$, $\underline{\beta} = \mu_n \underline{\nabla}(\psi - \phi_n)$, $\gamma = \hat{p}\overline{R}$, $f = \overline{G}$, and the boundary condition is $g = qn_D$ (see (1.29)). Comparing (6.23) with (6.22) yields $u = \phi_n$, $a = \lambda_n$, $\underline{\beta} = -(5/2)\underline{J}_n$, $\gamma = (3/2)(\hat{p}\hat{n}\overline{R} + \hat{n}/\tau_{w_n})$, $f = \underline{E} \cdot \underline{J}_n + (3/2)(\overline{G}\chi_n + \hat{n}V_{th}/\tau_{w_n})$, and the boundary condition is $g = K_B T_D / q$ (see (1.29)).

Referring to the model problem (6.23), we remark that the choice of the decoupling splitting in (6.21) and (6.22) aims at ensuring that u is positive through guaranteeing that γ and f are positive. It is worth noticing that the sign of the first term at the right-hand side in (6.22) is not a priori established, although from a physical standpoint it is expected to be nonnegative since it represents the dissipative Joule effect.

In analogy with the presentation of Section 3.2, the mixed formulation of (6.23) reads

$$\begin{cases} \text{Find } (\underline{\sigma}, u) \in \Sigma \times V \text{ such that} \\ (\alpha\underline{\sigma}, \underline{\tau}) + (\alpha u \underline{\beta}, \underline{\tau}) + (\operatorname{div} \underline{\tau}, u) = \langle g, \underline{\tau} \cdot \underline{n} \rangle_{|\Gamma_D}, & \forall \underline{\tau} \in \Sigma, \\ (\operatorname{div} \underline{\sigma}, v) - (\gamma u, v) = -(f, v), & \forall v \in V, \end{cases} \tag{6.25}$$

where the spaces Σ and V are defined in (3.20)–(3.21), and $\alpha := a^{-1}$. Notice that setting $\underline{\beta} = \underline{0}$ we recover the mixed formulation (3.27) that has been thoroughly analyzed in Section 3. A way to show existence and uniqueness of the solution of problem (6.25) is to check that a solution of (6.23) (in the distributional sense) is a solution of (6.25) and viceversa. Uniqueness then follows from uniqueness of the solution of (6.23), which is guaranteed if the usual coercivity conditions hold

$$\gamma + \frac{1}{2} \operatorname{div} \underline{\beta} \geqslant b_0 \geqslant 0 \quad \text{a.e. in } \Omega, \quad \underline{\beta} \cdot \underline{n} \leqslant 0 \quad \text{on } \Gamma_N. \tag{6.26}$$

A sufficient condition (though not always applicable) to obtain uniqueness for (6.25) is $\| \underline{\beta} \|_\infty^2 < 4\gamma_0 a_0 / a_M$ where a_0, a_M, and γ_0 are defined in (3.1) and (3.2).

6.2.2. *Mixed finite element discretization*

The discrete form of (6.25) reads:

$$\begin{cases} \text{Find } (\underline{\sigma}_h, u_h) \in \Sigma_h \times V_h \text{ such that} \\ (\alpha\underline{\sigma}_h, \underline{\tau}_h) + (\alpha u_h \underline{\beta}, \underline{\tau}_h) + (\operatorname{div} \underline{\tau}_h, u_h) = \langle g, \underline{\tau}_h \cdot \underline{n} \rangle_{|\Gamma_D}, & \forall \underline{\tau}_h \in \Sigma_h, \\ (\operatorname{div} \underline{\sigma}_h, v_h) - (\gamma u_h, v_h) = -(f, v_h), & \forall v_h \in V_h, \end{cases} \tag{6.27}$$

where (Σ_h, V_h) is the RT_0 mixed finite element space given in (3.107)–(3.108). For future purposes it is convenient to assume that the convective field $\underline{\beta}$ in (6.27) has continuous normal components across each edge of the triangulation. We therefore assume that $\underline{\beta}$ is itself an RT_0 finite element vector field. The algebraic form of (6.27) reads

$$\begin{pmatrix} A & B \\ C & D \end{pmatrix} \begin{pmatrix} \Phi_h \\ U_h \end{pmatrix} = \begin{pmatrix} G_h \\ F_h \end{pmatrix}, \tag{6.28}$$

where Φ_h is the vector of the unknown fluxes of $\underline{\sigma}_h$ across each edge of $\mathcal{T}_h$, and U_h is the vector of the unknown values of u_h on each $K \in \mathcal{T}_h$. Eliminating Φ_h leads to the following scheme for U_h

$$(D - CA^{-1}B)U_h = F_h - CA^{-1}G_h.$$

The matrix $M \equiv D - CA^{-1}B$ is full and, in general, neither symmetric nor positive definite, so that solving this system can be quite expensive. Moreover, it is well known that M is not an M-matrix for any value of γ, as pointed out in BREZZI and FORTIN [1991], MARINI and PIETRA [1990] in the case of the numerical approximation of the DD current continuity equations.

The stabilization procedure developed in the forthcoming sections will allow us to circumvent the drawbacks of the RT_0 approximation, leading to a family of stable cell-centered finite volume methods that preserve the good approximation properties for $\underline{\sigma}_h$ provided by the mixed approach, though at a reduced computational cost.

6.2.3. *Mixed finite volume stabilization*

In this section we apply the MFV formulation considered in Section 3.8 to the convection-diffusion-reaction problem (6.23). With this aim, we shall extend the stabilization procedure for convection-diffusion problems proposed in SACCO and SALERI [1997b] to the case of a varying diffusion coefficient a. Without loss of generality, from now on we shall assume $g \equiv 0$ in order to simplify the exposition.

Throughout, we shall adopt the same notation as in Section 3.8, with the following extension: for each $K_k \in \mathcal{T}_h$ and for any edge e^r, with $r \in E(k)$, we define $\underline{n}^r$ as the unit normal vector chosen to ensure that

$$\beta^r := \int_{e^r} \underline{\beta} \cdot \underline{n}^r \, ds \geqslant 0. \tag{6.29}$$

Let us consider the second equation in (6.27); taking $v_h = \chi_k$ (the characteristic function of triangle K_k) and proceeding as in Section 3.8.2, we get the discrete conservation law (see (3.237))

$$\sum_{\substack{r \in E(k) \\ e^r \notin \Gamma_N}} \Phi_k^r - \gamma_k u_k |K_k| = -f_k |K_k|, \qquad \forall K_k \in \mathcal{T}_h. \tag{6.30}$$

We recall that $\Phi_k^r = 0$ on $e^r \in \Gamma_N$. To end up with a finite volume scheme we must write the flux Φ_k^r as a function of the values u_j, $j \in T(r)$, only. With this aim, we diagonalize the first two bilinear forms in $(6.27)_1$ using the quadrature formula $(3.240)_2$. In particular, in order to apply to the convective term the same diagonalization procedure used for the diffusive term, a *unique* value for u_h needs to be defined at each edge. For this purpose, for any edge e^r we define

$$u^r = \frac{u_k + u_j}{2}, \quad \text{on } e^r \notin \Gamma, \qquad u^r = \frac{u_k}{2}, \quad \text{on } e^r \in \Gamma_D, \tag{6.31}$$

where the index $j \in T(r) \backslash k$. Taking in the first equation of (6.27) $\underline{\tau}_h = \underline{\tau}_h^r$ as the RT_0 basis function associated with edge e^r and such that $\underline{\tau}_h^r \cdot \underline{n}^r > 0$, we get the following constitutive equation for the edge advective flux

$$\sum_{m \in T(r)} \int_{K_m} \alpha u_h \underline{\beta} \cdot \underline{\tau}_h^r \, dx \simeq \overline{\alpha}^r u^r \sum_{m \in T(r)} \int_{K_m} \underline{\beta} \cdot \underline{\tau}_h^r \, dx \simeq \overline{\alpha}^r u^r \beta^r \frac{d^r}{|e^r|}. \tag{6.32}$$

After diagonalizing the diffusive term as done in Section 3.8.2 and using (6.32) we obtain

$$\Phi_k^r = (\overline{\alpha}^r)^{-1}\frac{u_j - u_k}{d^r}|e^r| - \beta^r u^r S_k^r,\ \forall e^r \notin \Gamma_N, \qquad \Phi_k^r = 0 \text{ on } e^r \in \Gamma_N, \tag{6.33}$$

where we set

$$S_k^r = \underline{n}_k^r \cdot \underline{n}^r.$$

Substituting the above expression into (6.30) yields a finite volume scheme for u_h of cell-centered type. This is easily checked to become unstable in the advection-dominated case, so that a stabilization procedure is required.

In order to stabilize (6.27) or, equivalently, to introduce flux upwinding, we consider a general *artificial diffusion* function $\rho_h : \mathcal{L}_h \to \mathbb{R}$, such that ρ_h is piecewise constant over $\mathcal{L}_h$ and for every lumping region $\mathcal{L}^r \in \mathcal{L}_h$, $\rho_h^r \geqslant 0$ and $\lim_{h\to 0} \rho_h^r = 0$, ($\rho_h^r := \rho_{h|\mathcal{L}^r}$). Next, we introduce the stabilized dual mixed discretization

$$\begin{cases} \text{Find } \underline{\sigma}_h^* \in \Sigma_h,\ u_h^* \in V_h \text{ such that } \forall(\underline{\tau}_h, v_h) \in \Sigma_h \times V_h \\ (\alpha \underline{\sigma}_h^*, \underline{\tau}_h) + (\alpha u_h^* \underline{\beta}, \underline{\tau}_h) + (\operatorname{div} \underline{\tau}_h, u_h^*) \\ \quad + \sum_{K_k \in \mathcal{T}_h} \int_{\partial K_k} \rho_h u_h^* \underline{\tau}_h \cdot \underline{n}_k \, ds = 0, \\ (\operatorname{div} \underline{\sigma}_h^*, v_h) - (\gamma u_h^*, v_h) = -(f, v_h), \end{cases} \tag{6.34}$$

where $\underline{n}_k$ is the unit outward normal vector along ∂K_k.

Our goal is to choose ρ_h in such a way that problem (6.34) becomes stable irrespectively of the strength of the *local Péclet number*

$$\mathbb{P}\mathsf{e}^r = \frac{1}{2}\overline{\alpha}^r \hat{\beta}^r d^r, \tag{6.35}$$

where $\hat{\beta}^r := \beta^r / |e^r| \geqslant 0$. Proceeding analogously as in the nonstabilized case, we get the following expression for the fluxes

$$\begin{aligned} &\Phi_k^r = (\overline{\alpha}^r)^{-1}(1 + \rho_h^r)\frac{u_j - u_k}{d^r}|e^r| - \beta^r u^r S_k^r, \quad \forall e^r \notin \Gamma_N, \\ &\Phi_k^r = 0, \quad \text{on } e^r \in \Gamma_N. \end{aligned} \tag{6.36}$$

Substituting the above expression into (6.30) yields the following cell-centered finite volume scheme for u_h^*

$$\begin{cases} \sum_{\substack{r\in E(k) \\ e^r \notin \Gamma_N}} (\overline{\alpha}^r)^{-1}(1 + \rho_h^r)\left(\frac{u_k^* - u_{j(r)}^*}{d^r}\right)|e^r| \\ \quad + \beta^r u^r S_k^r + u_k^* \gamma_k |K_k| = f_k |K_k|, \qquad \forall K_k \in \mathcal{T}_h, \\ u_{j(r)}^* = 0, \qquad \forall e^r \in \Gamma_D, \end{cases} \tag{6.37}$$

where $j(r) = T(r)\backslash k$. The set of linear algebraic equations (6.37) can be written in matrix form as

$$W^* \mathbf{u}^* = \mathbf{f}^*, \tag{6.38}$$

where the ith component of $\mathbf{f}^*$ is $f_i|K_i|$ and the ijth nonzero entries of the $\mathrm{N_T} \times \mathrm{N_T}$ matrix W^* are

$$W_{ij}^* = \begin{cases} \sum_{\substack{r \in E(i) \\ e^r \notin \Gamma_N}} (\overline{\alpha}^r)^{-1}(1+\rho_h^r)\frac{|e^r|}{d^r} + \frac{1}{2}\beta^r S_i^r + \gamma_i |K_i|, & \text{if } i = j, \\ -(\overline{\alpha}^{r(j)})^{-1}(1+\rho_h^{r(j)})\frac{|e^{r(j)}|}{d^{r(j)}} + \frac{1}{2}\beta^r S_i^r, & \text{if } j \in T(E(i)),\ j \neq i, \end{cases} \tag{6.39}$$

where $r(j) = E(j) \cap E(i)$ refers to the edge shared between triangles K_i and K_j.

6.2.4. *Stability analysis*

In this section we prove stability of the MFV method (6.38)–(6.39). With this aim, let us write system (6.38) in the abstract form

$$\begin{cases} \text{Find } u_h \in V_h \text{ such that} \\ a_h^S(u_h, v_h) = (f, v_h), \quad \forall v_h \in V_h, \end{cases}$$

where the discrete stabilized bilinear form $a_h^S(u_h, v_h) : V_h \times V_h \to \mathbb{R}$ is defined as

$$\begin{cases} a_h^S(u_h, v_h) = \sum_{K_k \in \mathcal{T}_h} \int_{K_k} (-\operatorname{div} \underline{\sigma}_h^S(u_h) + \gamma u_h) v_h \, dx, \\ \underline{\sigma}_h^S(u_h)_{|K_k} = \sum_{r \in E(k)} \Phi_k^r \underline{\tau}_k^r, \end{cases} \tag{6.40}$$

where the fluxes Φ_k^r are given in (6.36), and $\underline{\tau}_k^r = S_k^r \underline{\tau}_{h|K_k}^r$. Moreover, we denote by $\mathcal{E}_h''$ the set of all edges not belonging to Γ_N.

THEOREM 6.1. *For each $\rho_h^r \geqslant 0$, we have*

$$a_h^S(v_h, v_h) \geqslant \sum_{e^r \in \mathcal{E}_h''} \kappa^r |\mathcal{L}^r| \left(\frac{v_k - v_j}{d^r}\right)^2 + b_0 \|v_h\|_{0,\Omega}^2, \tag{6.41}$$

with

$$\kappa^r = 2(\overline{\alpha}^r)^{-1}(1+\rho_h^r). \tag{6.42}$$

PROOF. Let v_h be a function in V_h. With a (minor) abuse of notation we denote, for every edge $e^r \in \mathcal{E}_h''$, by v_k and v_j (instead of $v_{k(r)}$, $v_{j(r)}$) the values of v_h in the triangles sharing edge e^r, with the convention that $v_{j(r)} = 0$ when $e^r \in \Gamma_D$ (as in (6.37)). Multiplying the left-hand-side of (6.37) by v_k and summing over k, we have:

$$\begin{aligned} & a_h^S(v_h, v_h) \\ & \quad = \sum_{K_k \in \mathcal{T}_h} \left(\sum_{\substack{r \in E(k) \\ e^r \notin \Gamma_N}} \left[(\overline{\alpha}^r)^{-1}(1+\rho_h^r)\left(\frac{v_k - v_j}{d^r}\right)|e^r| + \beta^r v^r S_k^r \right] v_k \right) + \sum_{K_k \in \mathcal{T}_h} \gamma_k v_k^2 |K_k| \\ & \quad = \sum_{e^r \in \mathcal{E}_h''} (1+\rho_h^r) \frac{(\overline{\alpha}^r)^{-1}|e^r|}{d^r} \left[(v_k - v_j)v_k + (v_j - v_k)v_j \right] \end{aligned}$$

$$+\sum_{K_k\in\mathcal{T}_h}\sum_{\substack{r\in E(k)\\ e^r\notin\Gamma_N}}\beta^r v^r S_k^r v_k+\sum_{K_k\in\mathcal{T}_h}\gamma_k v_k^2|K_k| \tag{6.43}$$

$$=\underbrace{\sum_{e^r\in\mathcal{E}_h''}(1+\rho_h^r)\frac{(\overline{\alpha}^r)^{-1}|e^r|}{d^r}(v_k-v_j)^2}_{I}+\underbrace{\sum_{K_k\in\mathcal{T}_h}\sum_{\substack{r\in E(k)\\ e^r\notin\Gamma_N}}\beta^r v^r S_k^r v_k}_{II}+\underbrace{\sum_{K_k\in\mathcal{T}_h}\gamma_k v_k^2|K_k|}_{III}.$$

Recalling definition (6.31) of v^r, and observing that $\beta^r S_k^r=\int_{e^r}\underline{\beta}\cdot\underline{n}_k^r$, we deduce

$$\begin{aligned}
II&=\frac{1}{2}\sum_{K_k\in\mathcal{T}_h}v_k^2\sum_{\substack{r\in E(k)\\ e^r\notin\Gamma_N}}\int_{e^r}\underline{\beta}\cdot\underline{n}_k^r\,ds+\frac{1}{2}\sum_{K_k\in\mathcal{T}_h}v_k\sum_{\substack{r\in E(k)\\ e^r\notin\Gamma}}v_j\int_{e^r}\underline{\beta}\cdot\underline{n}_k^r\,ds\\
&=\frac{1}{2}\sum_{K_k\in\mathcal{T}_h}v_k^2\int_{\partial K_k}\underline{\beta}\cdot\underline{n}_k\,ds-\frac{1}{2}\sum_{K_k\in\mathcal{T}_h}v_k^2\sum_{\substack{r\in E(k)\\ e^r\in\Gamma_N}}\int_{e^r}\underline{\beta}\cdot\underline{n}_k^r\,ds \qquad (6.44)\\
&\geqslant\frac{1}{2}\sum_{K_k\in\mathcal{T}_h}\int_{K_k}v_k^2\operatorname{div}\underline{\beta}\,dx,
\end{aligned}$$

where in the last step we used the fact that $\underline{\beta}\cdot\underline{n}\leqslant 0$ on Γ_N (see (6.26)). Then, using again (6.26) we obtain

$$II+III=\sum_{K_k\in\mathcal{T}_h}\int_{K_k}\left(\gamma+\frac{1}{2}\operatorname{div}\underline{\beta}\right)v_k^2\,dx\geqslant b_0\|v_h\|_{0,\Omega}^2. \tag{6.45}$$

By multiplying and dividing by d^r each term in I, and using $d^r|e^r|=2|\mathcal{L}^r|$ we obtain

$$I=2\sum_{e^r\in\mathcal{E}_h''}(1+\rho_h^r)(\overline{\alpha}^r)^{-1}|\mathcal{L}^r|\left(\frac{v_k-v_j}{d^r}\right)^2. \tag{6.46}$$

Finally, using (6.46), and (6.45) in (6.43) we deduce

$$a_h^S(v_h,v_h)\geqslant 2\sum_{e^r\in\mathcal{E}_h''}(1+\rho_h^r)(\overline{\alpha}^r)^{-1}|\mathcal{L}^r|\left(\frac{v_k-v_j}{d^r}\right)^2+b_0\|v_h\|_{0,\Omega}^2. \tag{6.47}$$

□

As a consequence of Theorem 6.1 it is easily seen that the diagonal entries of matrix (6.39) are strictly positive. Therefore, by requiring the off-diagonal entries to be nonpositive, an M-matrix is obtained. The conditions on ρ_h which enforce such a property are stated in the following proposition, which extends the analogous result given in SACCO and SALERI [1997b].

PROPOSITION 6.1. *Let the edge artificial viscosity ρ_h^r be chosen in such a way that for each $\mathcal{L}^r\in\mathcal{L}_h$ we have*

$$\rho_h^r\geqslant\max\{0,\mathbb{P}\mathrm{e}^r-1\}. \tag{6.48}$$

Then the stiffness matrix W^* of the stabilized dual mixed finite volume scheme turns out to be an irreducible diagonally dominant M-matrix with respect to its columns.

A consequence of Proposition 6.1 is that the elemental values U_k are nonnegative, provided $\mathbf{f}^* \geqslant 0$, irrespectively of the strength of the local Péclet number.

We notice that, by taking

$$\rho_h^r = \max\{0, \mathbb{P}\mathrm{e}^r - 1\}, \tag{6.49}$$

(6.42) becomes

$$\kappa^r = \max\{2(\overline{\alpha}^r)^{-1}, \hat{\beta}^r d^r\}. \tag{6.50}$$

On the other hand, by taking

$$\rho_h^r = \mathbb{P}\mathrm{e}^r = \frac{1}{2}\overline{\alpha}^r \hat{\beta}^r d^r, \tag{6.51}$$

we have

$$\kappa^r = 2(\overline{\alpha}^r)^{-1} + \hat{\beta}^r d^r, \tag{6.52}$$

which, inserted into (6.41), produces an estimate similar to those that are typically obtained for stabilized formulations of advection-diffusion problems (see, for instance ROOS, STYNES and TOBISKA [1996]). For a detailed analysis we refer to BREZZI, MARINI, MICHELETTI, PIETRA and SACCO [submitted for publication].

REMARK 6.1. We note that the choice (6.51) gives the same scheme as the one that would be obtained, starting from the non stabilized MFV scheme (i.e., (6.34) with $\rho_h = 0$), using classical upwind for the convective term. Indeed, let us rewrite the non stabilized fluxes (6.33)

$$\Phi_k^r = (\overline{\alpha}^r)^{-1} \frac{u_j - u_k}{d^r} |e^r| - \beta^r u^r S_k^r, \quad \forall e^r \notin \Gamma_N. \tag{6.53}$$

Taking, instead of (6.31), the upwind choice

$$u_{uw}^r = \begin{cases} u_k, & \text{if } S_k^r = 1, \\ u_j, & \text{if } S_k^r = -1, \end{cases} \tag{6.54}$$

it is easy to see that

$$\beta^r u_{uw}^r S_k^r = \beta^r u^r S_k^r + \beta^r \frac{u_k - u_j}{2}. \tag{6.55}$$

Hence, taking u_{uw}^r instead of u^r in (6.53) amounts to adding the term

$$\beta^r \frac{u_j - u_k}{2}.$$

On the other hand, taking instead of (6.53):

$$\Phi_k^r = (\overline{\alpha}^r)^{-1}(1 + \mathbb{P}\mathrm{e}^r) \frac{u_j - u_k}{d^r} |e^r| - \beta^r u^r S_k^r, \quad \forall e^r \notin \Gamma_N, \tag{6.56}$$

corresponds to adding to (6.53) the term

$$(\overline{\alpha}^r)^{-1}\mathbb{P}e^r\frac{u_j-u_k}{d^r}|e^r| = (\overline{\alpha}^r)^{-1}\frac{1}{2}\overline{\alpha}^r\hat{\beta}^r d^r\frac{u_j-u_k}{d^r}|e^r| = \beta^r\frac{u_j-u_k}{2}.$$

REMARK 6.2. The presence of the factor 2 multiplying $|\mathcal{L}^r|$ in (6.46) should not be surprising. Indeed, it can be checked that, taking for the sake of simplicity a uniform mesh made of equilateral triangles of edge h, we have, for a smooth enough function v,

$$\sum_{e^r\in\mathcal{E}_h}\left(\frac{v_k-v_j}{d^r}\right)^2 2|\mathcal{L}^r| \simeq \int_\Omega |\underline{\nabla} v|^2\,dx \quad \text{for } h\to 0.$$

6.2.5. The Scharfetter–Gummel stabilization

We provide in this section a choice of ρ_h that fulfils the stability requirement (6.48) and extends to the two-dimensional case the Scharfetter–Gummel (SG) exponentially fitted difference scheme (SCHARFETTER and GUMMEL [1969]) that is widely employed in contemporary semiconductor device simulation. We shall consider in the following both the DD and EB cases (in unscaled forms).

The DD case. The classical SG method, hereafter denoted by SG-MFV, can be recovered by setting in the flux expression in (6.40)

$$\rho_h^r = \mathbb{P}e^r - 1 + \mathrm{B}(2\mathbb{P}e^r), \qquad \forall \mathcal{L}^r \in \mathcal{L}_h, \tag{6.57}$$

where for any $z\in\mathbb{R}$, $\mathrm{B}(z)$ is the Bernoulli function defined in (4.32). It is worth noting that for high Péclet numbers the SG mixed finite volume scheme degenerates into the standard Engquist–Osher (EO) upwinding procedure, which is well-known to be first-order accurate. However, as $\mathbb{P}e^r\to 0$, the amount of extra viscosity introduced by the SG flux approximation is $\mathcal{O}(h^2)$, whilst the EO one is $\mathcal{O}(h)$. After some algebra the flux across edge e^r reads

$$\Phi_k^r = qD_n\frac{u_j\mathrm{B}(\Delta\hat{\psi}) - u_k\mathrm{B}(-\Delta\hat{\psi})}{d^r}|e^r|, \tag{6.58}$$

where $\Delta z = z_j - z_k$ and $\hat{\psi} = \psi/V_{th}$. As for the analysis of the SG-MFV method, the convergence estimate (3.271) has been proved in MICHELETTI, SACCO and SALERI [2001] under the assumption $\mathrm{curl}\,\underline{\beta} = \underline{0}$, which is actually the case here, since $\underline{\beta}$ is the gradient of a potential. We finally mention that the SG-MFV scheme gives the *exact solution* at the circumcenters when $a = \text{const}$, $\gamma = f = 0$, ψ is linear in Ω, and suitable boundary conditions are assumed for u and $\underline{\sigma}$. In this case the method passes the Constant-Current Patch-Test (SACCO and SALERI [1997b]), which is a sound indication for a good behaviour of the numerical method in presence of steep layers arising in advection-dominated flows (see also VAN NOOYEN [1995]).

The EB case. Here we address the generalization of the SG method to the EB model. In particular we introduce two different expressions for the edge fluxes of the electron current (see also FORGHIERI, GUERRIERI, CIAMPOLINI, GNUDI, RUDAN and BACCARANI [1988]).

As usual for this approach, we consider a one-dimensional reference interval $[0, L]$ (which can be mapped into the segment joining the circumcenters K_k and K_j) and set $\overline{R} = \overline{G} = 0$, so that the continuity equation reads

$$\frac{d}{dx}\left[q\mu_n\phi_n\left(\frac{dn}{dx} - \frac{n}{\phi_n}\frac{d(\psi - \phi_n)}{dx}\right)\right] = 0.$$

Supposing μ_n constant and both ψ and ϕ_n linearly varying, after some algebra we obtain

$$J_n = q\mu_n\left(\frac{\Delta\phi_n - \Delta\psi}{L}\right)\left[\frac{n_0\left(\frac{\phi_n(L)}{\phi_n(0)}\right)^{(\Delta\psi/\Delta\phi_n - 1)} - n_L}{\left(\frac{\phi_n(L)}{\phi_n(0)}\right)^{(\Delta\psi/\Delta\phi_n - 1)} - 1}\right], \tag{6.59}$$

where J_n is the one-dimensional electron current density and $n_L = n(L)$, $n_0 = n(0)$. Similarly as what is done in the heat exchanger theory, we introduce the *average logarithmic temperature*

$$T_n^{\ell m} = \frac{T_n(L) - T_n(0)}{\ln\left(\frac{T_n(L)}{T_n(0)}\right)} = \frac{\Delta T_n}{\ln\left(\frac{T_n(L)}{T_n(0)}\right)},$$

and the *average logarithmic thermal potential* $\phi_n^{\ell m} = K_b T_n^{\ell m}/q$. Using these definitions we can rewrite the expression of the current as

$$J_n = \frac{q\mu_n\phi_n^{\ell m}}{L}\left[n_L\,\mathrm{B}\left(\frac{\Delta\psi - \Delta\phi_n}{\phi_n^{\ell m}}\right) - n_0\,\mathrm{B}\left(-\frac{\Delta\psi - \Delta\phi_n}{\phi_n^{\ell m}}\right)\right]. \tag{6.60}$$

Comparing this expression with (6.58) we note that:

- the thermal potential V_{th} has been replaced by the average logarithmic thermal potential $\phi_n^{\ell m}$;
- the argument of the Bernoulli functions is the difference between the electric and thermal potentials instead of being simply the potential drop across the interval.

This guarantees that if the electron temperature is constant then we exactly recover the DD expression.

An alternative form of the EB current can be provided by assuming the validity of the mobility model as in RUDAN, GNUDI and QUADE [1993]. It turns out that the diffusion coefficient is independent of T_n since

$$D_n = \mu_n(T_n)\phi_n = \left(\mu_{n0}\frac{T_0}{T_n}\right)\phi_n = \mu_{n0}V_{th} = D_{n0}. \tag{6.61}$$

The electron current can thus be written as

$$J_n = qD_{n0}\left(\frac{dn}{dx} - \frac{n}{\phi_n}\frac{d(\psi - \phi_n)}{dx}\right). \tag{6.62}$$

As above, assuming that J_n is constant on $[0, L]$ and that both potentials are linearly varying we eventually get

$$J_n = \frac{qD_{n0}}{L}\phi_n^{\ell m}\left[\frac{n_L}{\phi_n(L)}\mathrm{B}\left(\frac{\Delta\psi - 2\Delta\phi_n}{\phi_n^{\ell m}}\right) - \frac{n_0}{\phi_n(0)}\mathrm{B}\left(-\frac{\Delta\psi - 2\Delta\phi_n}{\phi_n^{\ell m}}\right)\right]. \tag{6.63}$$

Comparing this expression with (6.60) we note three main differences:

- the diffusion coefficient has replaced the product between the mobility and the average logarithmic thermal potential. This in turn implies that with the same electric potential, the current is smaller when the temperature raises due to the decrease of μ_n with respect to an increase of T_n;
- in the argument of the Bernoulli functions in (6.63) the variation of the thermal potential is multiplied by a factor 2 and guarantees the correct behaviour of (6.63) in all the limiting cases;
- each of the two concentrations n_i $(i = 0, L)$ is also multiplied by the ratio $\phi_n^{\ell m}/\phi_n(i)$, that is, we have another weight which takes into account the variation of the temperature across the interval.

Also in this case if the electron temperature is constant equation (6.63) recovers the DD limit since $\phi_n^{\ell m}$ tends to V_{th} as $\Delta T_n \to 0$.

We conclude this section by casting the flux expressions corresponding to (6.60) and (6.63) within the general stabilization framework summarized in (6.40). Indeed, with regard to the former we have

$$\begin{cases} (\overline{\alpha}^r)^{-1} = q\mu_n\phi_n^{\ell m}, \\ \beta^r = \frac{\Delta(\psi-\phi_n)}{d^r}|e^r|, \\ \rho_h^r = \mathbb{P}\mathrm{e}^r - 1 + \mathrm{B}(2\mathbb{P}\mathrm{e}^r), \end{cases} \tag{6.64}$$

where the local Péclet number is evaluated according to (6.35). About expression (6.63) we note that it can be obtained in a straightforward manner from the following equivalent form of the current (6.62)

$$J_y = qD_{n0}\left(\frac{d(y\phi_n)}{dx} - y\frac{d(\varphi_n)}{dx}\right),$$

where $y = n/\phi_n$, $\varphi_n = \psi - \phi_n$ and applying (6.60) to the new variables. Therefore we get, with respect to y

$$\begin{cases} (\overline{\alpha}^r)^{-1} = qD_{n0}\phi_n^{\ell m}, \\ \beta^r = \frac{\Delta(\psi-\phi_n)}{d^r}|e^r|, \\ \rho_h^r = \mathbb{P}\mathrm{e}^r - 1 + \mathrm{B}(2\mathbb{P}\mathrm{e}^r). \end{cases} \tag{6.65}$$

Scharfetter–Gummel approximation to the energy fluxes. Recalling $(1.28)_6$ and the definition of ϕ_n, the electron energy flux can be written as

$$\underline{S}_n = \lambda_n \underline{\nabla}\phi_n + \frac{5}{2}\phi_n \underline{J}_n, \tag{6.66}$$

where $\lambda_n = q\kappa_n/K_B$. As before, we consider a one-dimensional reference interval $[0, L]$; assuming λ_n to be constant and replacing n with a suitable average (see FORGHIERI, GUERRIERI, CIAMPOLINI, GNUDI, RUDAN and BACCARANI [1988]), we get

$$S_n = \lambda_n\left(\frac{d\phi_n}{dx} + \frac{5J_n}{2\lambda_n}\phi_n\right). \tag{6.67}$$

This flux expression is the analogue of the hole current in the DD case where the electric field is replaced by $\frac{5J_n}{2\lambda_n}$. Assuming as usual that S_n is constant yields

$$S_n = \frac{\lambda_n}{L}\left[\phi_n(L)\,\mathrm{B}\left(-\frac{5J_n}{2\lambda_n}L\right) - \phi_n(0)\,\mathrm{B}\left(\frac{5J_n}{2\lambda_n}L\right)\right]. \tag{6.68}$$

Exactly as happens in the DD case for the hole current, if $|(5J_nL)/(2\lambda_n)| \gg 1$ then an upwinding effect is obtained otherwise the finite-difference approximation of the temperature gradient is recovered.

Discretization of the EB equations. We summarize below all the systems of equations obtained applying the MFV method to discretize each of the EB equations in the Gummel procedure. For any $K_k \in \mathcal{T}_h$, we let as usual $r \in E(k)$, $j \in \{T(E(k))\backslash k\}$, $\gamma^r = |e^r|/d^r$, $\Delta\psi = \psi_j - \psi_k$, $\Delta\phi_\nu = \phi_{\nu j} - \phi_{\nu k}$ ($\nu = n, p$) and define

$$\theta_n = \frac{\Delta\psi - 2\Delta\phi_n}{\phi_n^{\ell m}}, \qquad \theta_p = \frac{\Delta\psi + 2\Delta\phi_p}{\phi_p^{\ell m}}, \tag{6.69}$$

where the average logarithmic thermal potentials $\phi_\nu^{\ell m}$ are evaluated between K_k and K_j. The discrete momentum and energy conservation equations read

$$\sum_{j\in\{T(E(k))\backslash k\}} D_{n0}\gamma^r\phi_n^{\ell m}\left[\frac{\hat{n}_k}{\phi_{nk}}\mathrm{B}(-\theta_n) - \frac{\hat{n}_j}{\phi_{nj}}\mathrm{B}(\theta_n)\right] + \hat{n}_k\hat{p}_k R_k^{(1)}|K_k| = G_k^{(1)}|K_k|, \tag{6.70}$$

$$\sum_{j\in\{T(E(k))\backslash k\}} D_{p0}\gamma^r\phi_p^{\ell m}\left[\frac{\hat{p}_k}{\phi_{pk}}\mathrm{B}(\theta_p) - \frac{\hat{p}_j}{\phi_{pj}}\mathrm{B}(-\theta_p)\right] + \hat{n}_k\hat{p}_k R_k^{(2)}|K_k| = G_k^{(2)}|K_k|, \tag{6.71}$$

$$\begin{aligned}&\sum_{j\in\{T(E(k))\backslash k\}} \lambda_n\gamma^r\left[\phi_{n_k}\mathrm{B}\left(\frac{5j_n^r}{2\lambda_n}|d^r|\right) - \phi_{n_j}\mathrm{B}\left(-\frac{5j_n^r}{2\lambda_n}|d^r|\right)\right]\\ &\quad + \frac{3}{2}\left(R_k^{(1)}\hat{p}_k + \frac{1}{\tau_{w_n}}\right)\hat{n}_k|K_k|\phi_{n_k}\\ &= \underline{E}\cdot\underline{J}_n|K_k| + \frac{3}{2}\left(G_k^{(1)}\phi_{n_k} + \frac{n}{\tau_{w_n}}V_{th}\right)|K_k|,\end{aligned} \tag{6.72}$$

$$\begin{aligned}&\sum_{j\in\{T(E(k))\backslash k\}} \lambda_p\gamma^r\left[\phi_{p_k}\mathrm{B}\left(-\frac{5j_p^r}{2\lambda_p}|d^r|\right) - \phi_{p_j}\mathrm{B}\left(\frac{5j_p^r}{2\lambda_p}|d^r|\right)\right]\\ &\quad + \frac{3}{2}\left(R_k^{(2)}\hat{n}_k + \frac{1}{\tau_{w_p}}\right)\hat{p}_k|K_k|\phi_{p_k}\\ &= \underline{E}\cdot\underline{J}_p|K_k| + \frac{3}{2}\left(G_k^{(2)}\phi_{p_k} + \frac{p}{\tau_{w_p}}V_{th}\right)|K_k|,\end{aligned} \tag{6.73}$$

where $R_k^{(l)}$ and $G_k^{(l)}$ ($l = 1, 2$) are the recombination and generation terms computed using the values of ψ, $\hat{n}$, $\hat{p}$, T_n and T_p of the previous Gummel step ($l = 1$) and the values of $\hat{n}$ computed by solving (6.70) ($l = 2$), respectively. The electron current flowing between K_k and K_j given by (6.63) is denoted by j_n^r and the hole current is

$$j_p^r = \frac{q D_{p0}}{d^r} \phi_p^{\ell m} \left[\frac{p_k}{\phi_{p_k}} \mathrm{B}(\theta_p) - \frac{p_j}{\phi_{p_j}} \mathrm{B}(-\theta_p) \right]. \tag{6.74}$$

Each of the above systems shares the feature that the corresponding coefficient matrix is a diagonally dominant M-matrix with respect to its columns.

7. Numerical results

The first set of numerical experiments, collected in Section 7.1, is concerned with the discretization of the symmetric problem (2.14) with the methods discussed in Section 3.8. Next, Section 7.2 aims at showing the performances of the discretization schemes presented in Sections 4.1–4.2 for the linear continuity equation (4.3). Section 7.3 presents simulations of a *pn*-diode with the DD model. Finally, Section 7.4 collects results obtained with the ET and the EB models, both for one and two dimensional geometry.

7.1. Mixed finite volumes

We assume that, in (2.14), Ω is the unit square; five unstructured grids of triangular elements have been employed, with average mesh sizes of 2^{-k}, for $k = 3, \ldots, 7$. Suitably accurate integration formulae have been used to evaluate the $L^2(\Omega)$ and $H(\mathrm{div}; \Omega)$ norms of the discretization errors $u - u_h$ and $\underline{\sigma} - \underline{\sigma}_h$. Precisely, a symmetric formula using 10 nodes on each element of the triangular grid has been employed. We emphasize that all the numerical results presented below have been obtained using the midpoint quadrature formula to approximate the right-hand side in (3.237). In all the reported figures the symbols have the following meaning: '*' refers to the scheme using the average (3.277), 'o' refers to the use of the average (3.275) and '□' refers to use of the average (3.276). We aim at checking the theoretical convergence properties of the method of Section 3.8. Therefore, we take as exact solution the smooth function

$$u(x_1, x_2) = 16 x_1 x_2 (x_1 - 1)(x_2 - 1),$$

with $\|u\|_{\infty,\Omega} = 1$. The diffusion coefficient a is chosen as

$$a(x_1, x_2) = \left(\frac{\nu_a + \nu_b}{2} \right) \left(1 + \frac{\nu_b - \nu_a}{\nu_a + \nu_b} \tanh\big(K(x_2 - x_1)\big) \right),$$

where $\nu_b > \nu_a > 0$ and K is a suitable constant. Notice that for $\nu_b/\nu_a \gg 1$ and $K \gg 1$, a tends to a step function with jump equal to $\nu_b - \nu_a$ across the main diagonal of the unit square.

The first set of numerical results refers to the case $\nu_a = 1$, $\nu_b = 10$ and $K = 2$. The maximum values of $|\underline{\sigma}|$ and $\mathrm{div}\,\underline{\sigma}$ are 36.05 and 129.6, respectively. Fig. 7.1 shows the L^2-norm (left) and the discrete L^∞-norm (right) of the error $u - u_h$, this latter norm being evaluated as

$$\|u - u_h\|_{\infty,C} = \max_{K_k \in \mathcal{T}_h} \big| u(\underline{x}_{C_k}) - u_k \big|.$$

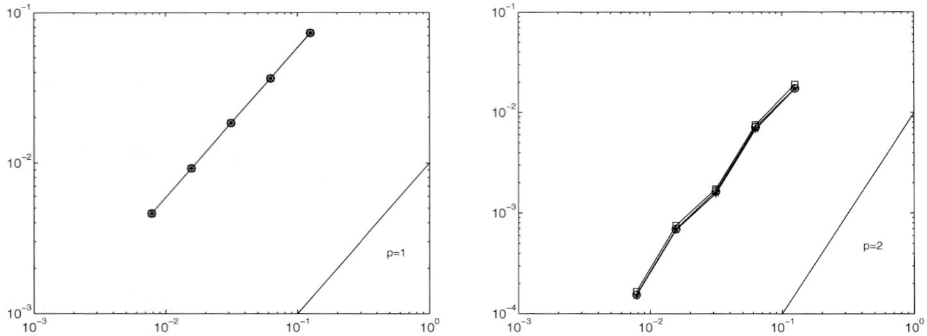

FIG. 7.1. Discretization error $u - u_h$: L^2-norm (left) and discrete L^∞-norm (right) ($\nu_a = 1$, $\nu_b = 10$, $K = 2$).

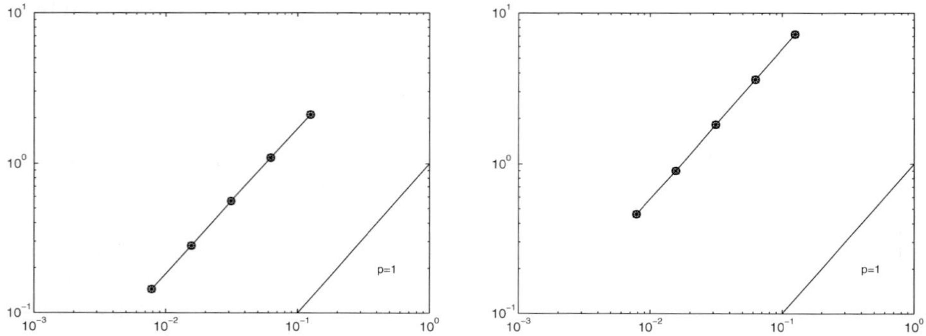

FIG. 7.2. L^2-norm of the discretization error $\underline{\sigma} - \underline{\sigma}_h$ (left) and graph norm of the discretization error $U - U_h = (u - u_h, \underline{\sigma} - \underline{\sigma}_h)$ (right) ($\nu_a = 1$, $\nu_b = 10$, $K = 2$).

Linear convergence with respect to the mesh size h is clearly observed for all of the considered averages in the case of the L^2-norm while second-order convergence is exhibited by the method when the discrete L^∞-norm is used to measure convergence. This appears to be a (nodal) superconvergence of the mixed finite volume schemes and is to be ascribed to the special choice of the point of measurement of the convergence (i.e., the circumcenter of the triangle) which coincides with the center of the stencil of the method. Fig. 7.2 shows the L^2-norm of the discretization error $\underline{\sigma} - \underline{\sigma}_h$ (left) and the graph norm of the error $U - U_h \equiv (u - u_h, \underline{\sigma} - \underline{\sigma}_h)$ (right) defined as

$$\|U - U_h\|_{\Sigma \times V} = \left(\|u - u_h\|_{0,\Omega}^2 + \|\underline{\sigma} - \underline{\sigma}_h\|_{H(\mathrm{div};\Omega)}^2\right)^{1/2}.$$

Linear convergence is clearly observed for all of the three averages.

The second set of numerical results refers to the case $\nu_a = 1$, $\nu_b = 1000$ and $K = 100$. The maximum values of $|\underline{\sigma}|$ and $\operatorname{div} \underline{\sigma}$ are 4000 and 16308.8, respectively (clearly much bigger than in the previous case). Fig. 7.3 shows that linear convergence in the L^2-norm is exhibited by all the considered averages, while the schemes appear to be second-

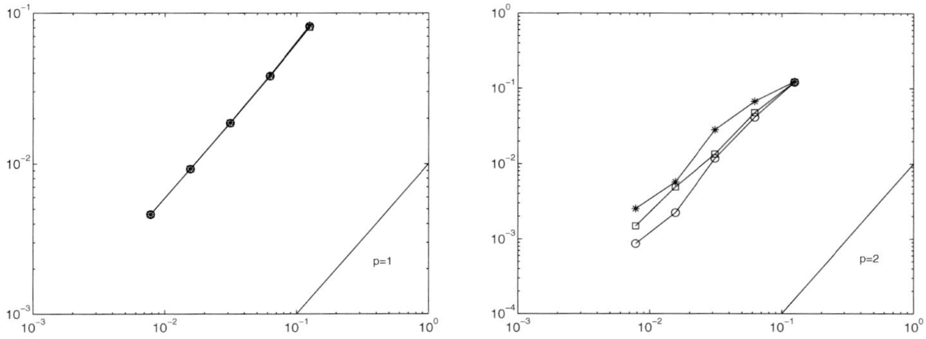

FIG. 7.3. Discretization error $u - u_h$: L^2-norm (left) and discrete L^∞-norm (right) ($\nu_a = 1$, $\nu_b = 1000$, $K = 100$).

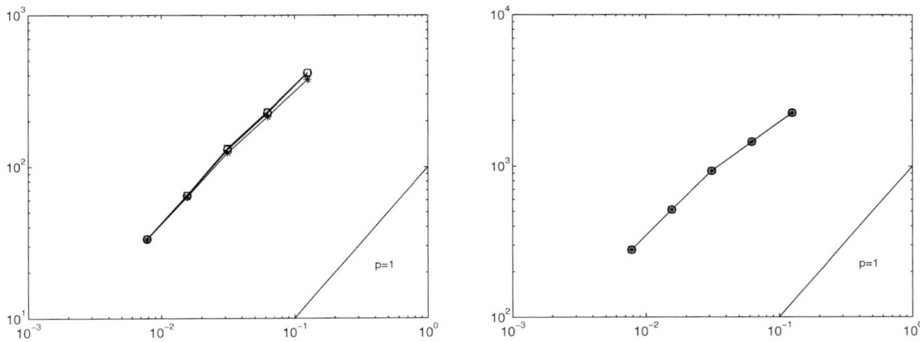

FIG. 7.4. L^2-norm of the discretization error $\underline{\sigma} - \underline{\sigma}_h$ (left) and graph norm of the discretization error $U - U_h = (u - u_h, \underline{\sigma} - \underline{\sigma}_h)$ (right) ($\nu_a = 1$, $\nu_b = 1000$, $K = 100$).

order accurate in the discrete L^∞-norm. Linear convergence is also observed for the discretization errors $\underline{\sigma} - \underline{\sigma}_h$ and $U - U_h$, as shown in Fig. 7.4.

7.2. *Linear continuity equation*

The examples presented here intend to compare the mixed finite element/finite volume methods described in Section 4 for the linear continuity equation (4.3), with ψ, c and f prescribed. In particular, the case $c \neq 0$ will be explored, in situations where the solution exhibits boundary and/or internal layers. The domain Ω is the unit square.

TEST 1. The first example deals with a case of a boundary layer produced by a reaction term dominant with respect to diffusion and convection: ψ has a constant gradient $\underline{\nabla}\,\psi = -(1, 0.5)$, $f \equiv c \equiv 10^6$, and homogeneous Dirichlet boundary conditions for p are taken on the entire border $\partial\Omega$. In Figs. 7.5 and 7.6 the solution has been computed with the lowest order Raviart–Thomas elements (RT_0) and with the two elements of

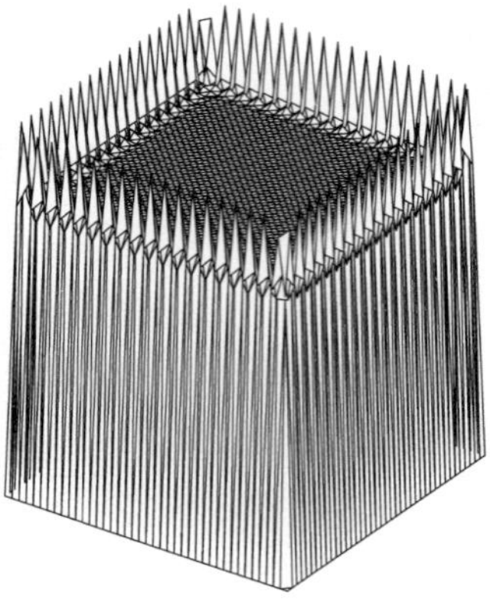

FIG. 7.5. Test 1 with RT_0 elements.

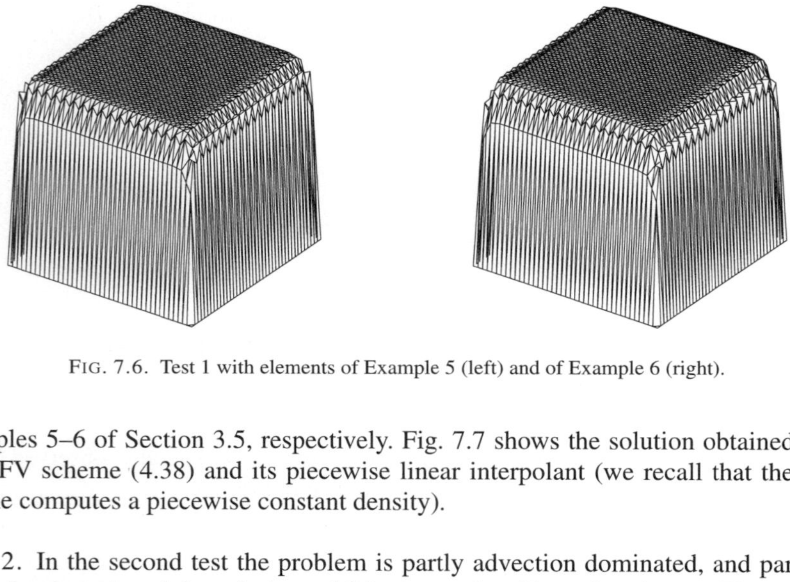

FIG. 7.6. Test 1 with elements of Example 5 (left) and of Example 6 (right).

Examples 5–6 of Section 3.5, respectively. Fig. 7.7 shows the solution obtained using the MFV scheme (4.38) and its piecewise linear interpolant (we recall that the MFV scheme computes a piecewise constant density).

TEST 2. In the second test the problem is partly advection dominated, and partly reaction dominated, and the solution exhibits internal and boundary layers. The potential ψ has a large constant gradient $\underline{\nabla}\,\psi = -10^4 \times (1, 0.5)$, $f \equiv 0$ and c changes abruptly: $c = 10^{10}$ in the quarter of the circle centered in $B = (1, 0)$ and with radius 0.5, $c = 0$ elsewhere. The boundary conditions for p are: $p = 0$ on $EB \cup BC$ and $p = 1$ on the rest of $\partial\Omega$ (see Fig. 7.8 (left)). Figs. 7.8 (right) and 7.9 report the results obtained by applying the RT_0 elements and the two elements of Examples 5–6, respectively. Fig. 7.10 shows the solution of the MFV scheme (4.38) and its piecewise linear interpolant.

As expected, in both test problems the Raviart–Thomas element (with exact integration) provides sharp fronts, but the lack of monotonicity properties shows up in heavy overshooting and undershooting. A remedy to this is the use of a proper numerical quadrature, as done with the MFV schemes. The elements of Examples 5–6 are both

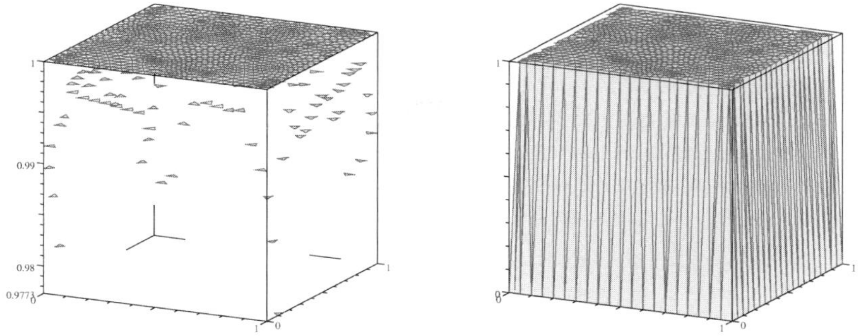

FIG. 7.7. Test 1 with the MFV scheme (left) and after interpolation (right).

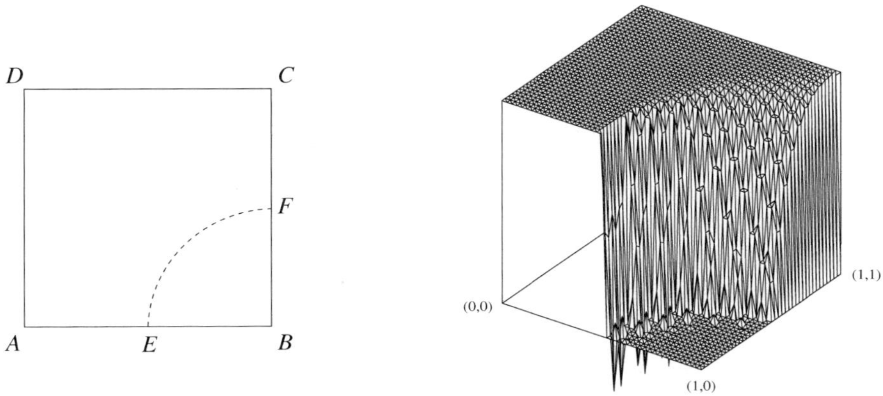

FIG. 7.8. Test 2: computational domain (left) and RT_0-solution (right).

monotonic, and show little differences, though the element of Example 5 seems to be slightly less diffusive.

TEST 3. The third example deals with the solution of the electron continuity equation (see (4.3) with ψ changed in $-\psi$, and p changed in n). The domain Ω is the unit square. The Dirichlet boundary conditions for n are:

- $n = 10^3$ for $0 \leqslant x_1 \leqslant 0.2$, $x_2 = 0$ and $x_1 = 0$, $0 \leqslant x_2 \leqslant 0.2$,
- $n = 10^{17}$ for $x_1 = 1$, $0.8 \leqslant x_2 \leqslant 1$ and $0.8 \leqslant x_1 \leqslant 1$, $x_2 = 1$.

On the rest of the boundary homogeneous Neumann conditions are given. The reaction coefficient is $c = 0$ and the given electrostatic potential is $\psi(x_1, x_2) = \psi_0(x_1, x_2)/\ell$, with

$$\psi_0(\rho) = \begin{cases} 0, & 0 \leqslant \rho \leqslant 0.8, \\ 2(\rho - 0.8), & 0.8 \leqslant \rho \leqslant 0.9, \\ 0.2, & \rho \geqslant 0.9, \end{cases} \tag{7.1}$$

where $\rho = (x_1^2 + x_2^2)^{1/2}$, and $\ell = 10^{-4}$. The right hand side is $f = 0.5 \times 10^{16}$ in the annular region $0.82 \leqslant \rho \leqslant 0.88$, and $f = 0$ elsewhere. This choice of ψ, f, as well as

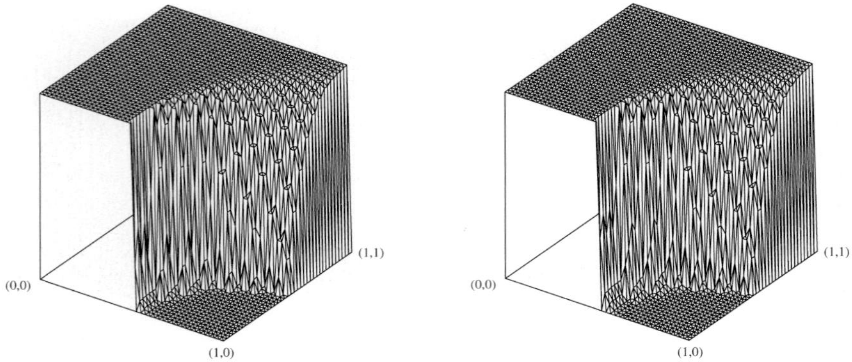

FIG. 7.9. Test 2 with elements of Example 5 (left) and of Example 6 (right).

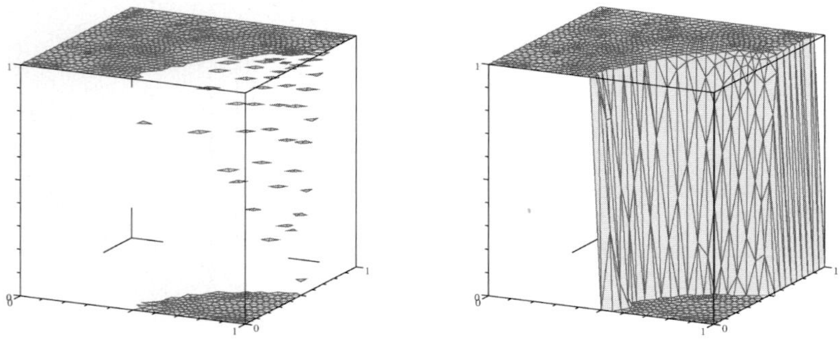

FIG. 7.10. Test 2 with the MFV scheme (left) and after interpolation (right).

that of the Dirichlet boundary values for n, aims at reproducing the physical behavior of a curved pn-junction diode subject to a high reverse bias.

Fig. 7.11 shows (in semilogarithmic scale) the solution computed with the MFV scheme on an unstructured mesh and its piecewise linear continuous interpolant. The computed current field $\underline{J}_h$ is displayed in Fig. 7.12. Fig. 7.13 shows the result obtained with the mixed exponential fitting scheme of Section 4.1 (based on the RT_0 elements).

The net generation effect on the electron concentration is very well reproduced, as confirmed by the constant behavior within the anular region and by the net generation current lines collected at the upper right corner of the domain. Notice also the effectiveness of the schemes in capturing the internal layers without any spurious oscillations, as well as in tracking the current lines throughout the semiconductor, especially along the Neumann boundary where they turn out to be rigorously tangential. We point out that the undershooting about the arc $\rho = 0.8$ is an actual feature of the solution and it is not due to numerical pollution.

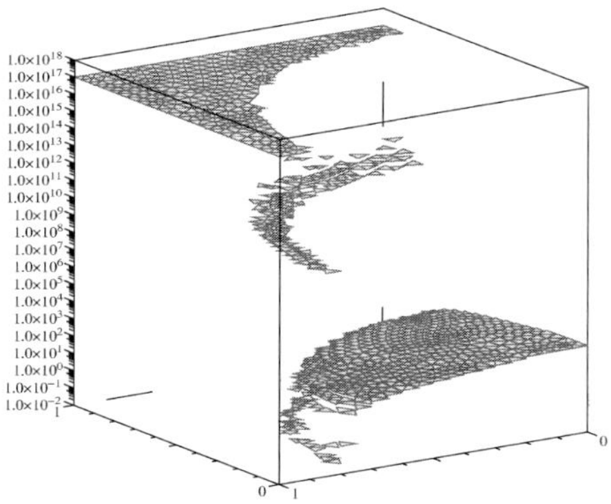

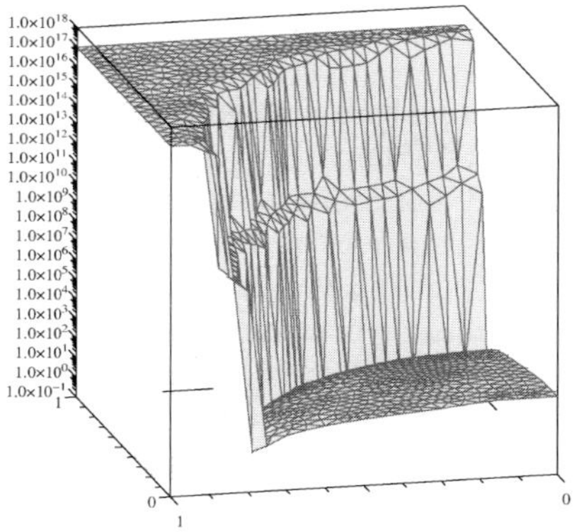

FIG. 7.11. Electron concentration as computed with the MFV scheme (left) and after interpolation (right).

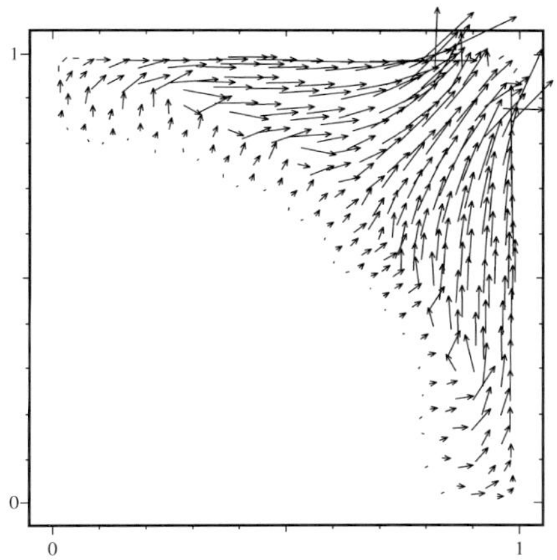

FIG. 7.12. Electron current field computed with the MFV scheme.

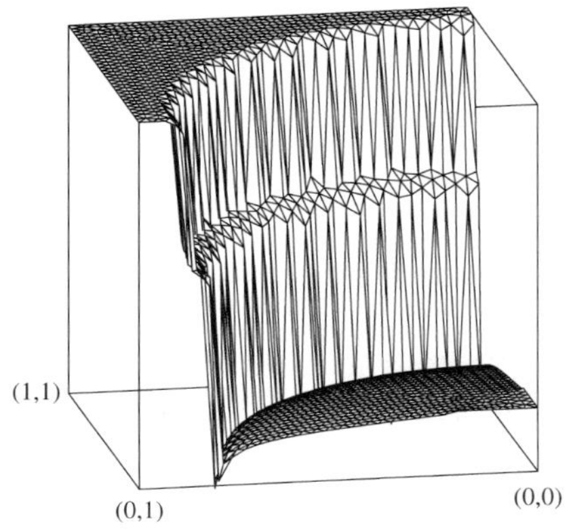

FIG. 7.13. Electron concentration computed with RT_0 elements.

7.3. *Drift-Diffusion model: a pn-diode*

We present here simulations of a *pn*-diode, neglecting recombination-generation effects. The DD system is solved in the (ψ, n, p) variables, the Poisson equation is discretized with P_1-nonconforming elements (see CROUZEIX and RAVIART [1973]) and the continuity equations are approximated with the mixed exponential fitting scheme based on the RT_0 elements of Section 4.1. The semiconductor material is silicon and the device is assumed to be a square of size 10^{-3} cm $\times$ 10^{-3} cm. The numerical values of physical parameters are given as in Table 7.1. The electron and hole mobilities are taken constant, with $\mu_n = \mu_p = 10^3\ \mathrm{cm}^2\,\mathrm{V}^{-1}\,\mathrm{s}^{-1}$. The geometry of the device and the triangular mesh are shown in Fig. 7.14. The choice of a rather coarse quasi-uniform decomposition will emphasize the good performances of the numerical method which produces little numerical diffusion and does not develop any spurious oscillation.

First, we consider a strongly reverse biased diode with an applied bias of $U = -2.0$ V, defined as the difference of the applied potential at the contacts of the n-region, p-region, respectively. A moderate doping profile $C(x) = -10^{15}\ \mathrm{cm}^{-3}$ in the p-region and $C(x) = 10^{15}\ \mathrm{cm}^{-3}$ in the n-region (abrupt junction) is chosen. Then, the squared scaled Debye length defined in (1.23) is $\lambda^2 = 2.56 \times 10^{-4}$. The corresponding electron and hole densities are depicted in Fig. 7.15. Fig. 7.16 represents their level curves and

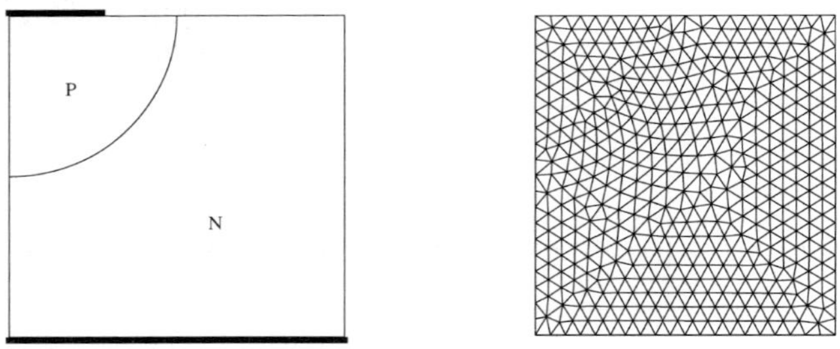

FIG. 7.14. *pn*-junction diode: geometry (left) and mesh (right).

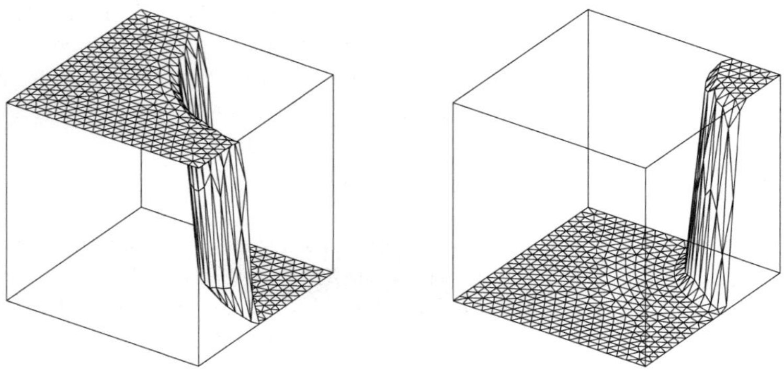

FIG. 7.15. Reverse biased diode ($U = -2.0$ V, $|C(x)| = 10^{15}$): electron density (left) and hole density (right).

FIG. 7.16. Level curves of the electron and hole densities of Fig. 7.15.

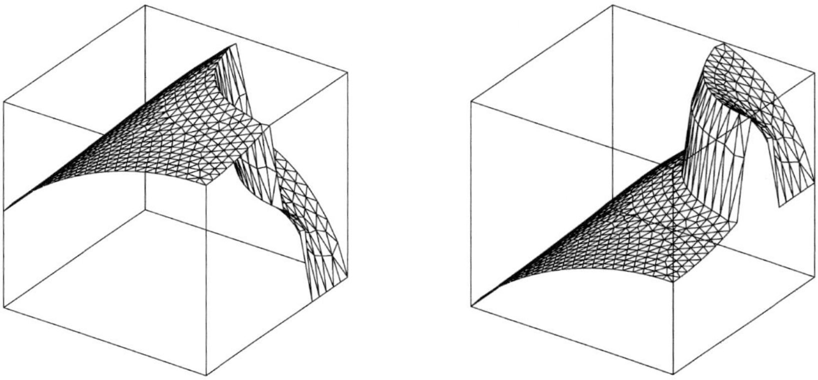

FIG. 7.17. Forward biased diode ($U = 0.8$ V, $|C(x)| = 10^{16}$): electron density (left) and hole density (right).

it shows, about the junction, the so-called depletion region where both the electron and hole densities are negligible.

In the case of a forward biased diode, the computations are carried out with a doping profile $C(x) = -10^{16}\ \mathrm{cm}^{-3}$ in the p-region and $C(x) = 10^{16}\ \mathrm{cm}^{-3}$ in the n-region, and with an applied voltage $U = 0.8$ V. The corresponding squared scaled Debye length is $\lambda^2 = 2.56 \times 10^{-5}$. The carrier densities are shown in Fig. 7.17. The total current density flow from the anode to the cathode is reported in Fig. 7.18 ($\underline{J}_{tot} = \underline{J}_n + \underline{J}_p$).

7.4. Energy-Transport and Energy-Balance models

We present first the simulation of a one-dimensional n^+nn^+ ballistic silicon diode which is a simple model for the channel of a MOS transistor. The spherically symmetric energy-band diagram ε of the semiconductor crystal, function of the modulus $k = |\vec{k}|$ of the wave vector $\vec{k}$, is given by the Kane model (KANE [1957]):

$$\varepsilon(1 + \alpha\varepsilon) = \frac{\hbar^2}{2m_0}|\vec{k}|^2,$$

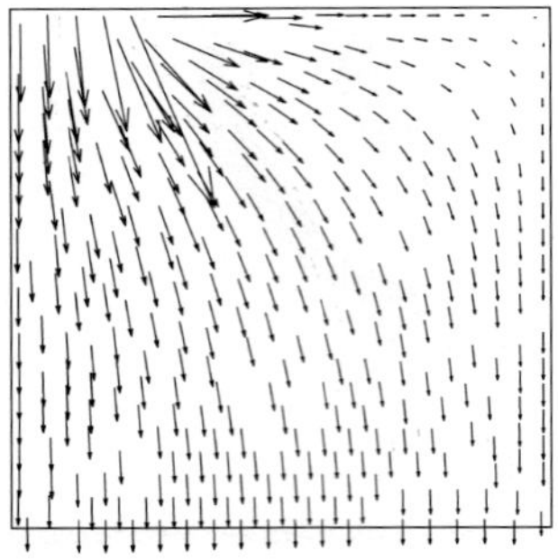

FIG. 7.18. Forward biased diode ($U = 0.8$ V, $|C(x)| = 10^{16}$): total current density.

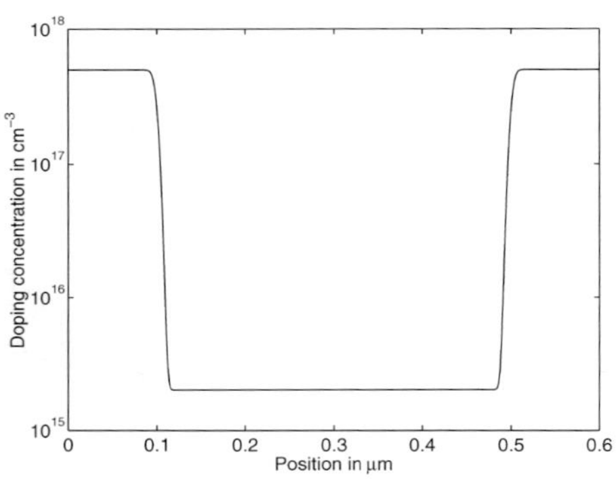

FIG. 7.19. Doping concentration in the n^+nn^+ diode.

where $\hbar$ is the reduced Planck constant, m_0 the effective electron mass, and $\alpha > 0$ the non-parabolicity parameter. When $\alpha = 0$ the energy-band is parabolic and we take the diffusion matrix L as in (1.25). For $\alpha > 0$ the coefficients L_{ij} of the diffusion matrix depend nonlocally on the temperature T and we refer to DEGOND, JÜNGEL and PIETRA [2000] for their explicit expression. The semiconductor domain is given by the interval $\Omega = (0, \ell)$ with $\ell > 0$. In the n^+-regions the maximal doping concentration is $C_1 = 5 \times 10^{17}$ cm^{-3}; in the n-channel the minimal doping profile is $C_0 = 2 \times 10^{15}$ cm^{-3}. The doping profile is shown in Fig. 7.19. The length of each n^+-region is 0.1 μm, whereas the length of the channel region equals 0.4 μm.

The numerical values of the physical parameters (for a silicon diode) are given in Table 7.1. On the boundary points $x = 0$ and $x = \ell$ we assume that the (unscaled) total space charge $C - n$ vanishes and that the (unscaled) temperature has the ambient temperature:

$$n(0) = n(\ell) = C_1, \qquad T(0) = T(\ell) = T^{eq}, \qquad \psi(0) = 0, \qquad \psi(\ell) = V_{ext},$$

where $V_{ext} > 0$ is the applied voltage. We take the value $V_{ext} = 1.5$ V.

We perform numerical tests on a uniform mesh of 100 nodes with the mixed exponential fitting scheme of Section 6.1. In Fig. 7.20 we present the electron temperature

TABLE 7.1
Physical parameters

Parameter	Physical meaning	Numerical value
q	elementary charge	1.6×10^{-19} A s
ε_s	permittivity constant	10^{-12} A s V^{-1} cm^{-1}
μ_0	(low field) mobility constant	1.5×10^3 $\mathrm{cm}^2\,\mathrm{V}^{-1}\,\mathrm{s}^{-1}$
V_{th}	thermal voltage at $T^{eq} = 300$ K	0.026 V
ℓ	length of the device	0.6 μm
ℓ_0	length of the n^+ region	0.1 μm
C_0	doping concentration in the n region	2×10^{15} cm^{-3}
C_1	doping concentration in the n^+ region	5×10^{17} cm^{-3}
τ_0	energy relaxation time	0.4×10^{-12} s
α	non-parabolicity parameter	0.5 $(\mathrm{eV})^{-1}$

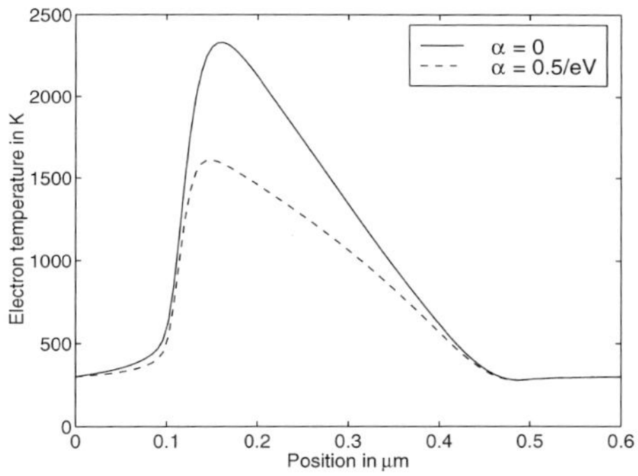

FIG. 7.20. Electron temperature in a ballistic diode.

for vanishing and non-vanishing non-parabolicity parameter α. As expected, the temperature in the n-channel is high, i.e., the electrons are *hot*. The maximal temperature values are $T = 2330$ K for $\alpha = 0$ and $T = 1610$ K for $\alpha = 0.5$ $(\mathrm{eV})^{-1}$. Therefore, the temperature is reduced due to the non-parabolicity effects. In Fig. 7.21 the electron mean velocity is shown. The mean velocity u is defined by $u = J^1/(qn)$. The spurious velocity overshoot spike at the left junction almost vanishes for $\alpha = 0.5$ $(\mathrm{eV})^{-1}$. The maximal velocities are $u = 1.44 \times 10^7$ $\mathrm{cm\,s}^{-1}$ for $\alpha = 0$ and $u = 1.25 \times 10^7$ $\mathrm{cm\,s}^{-1}$ for $\alpha = 0.5$ $(\mathrm{eV})^{-1}$.

Finally, we present the simulation of two realistic semiconductor devices carried out by applying the SG-MFV method of Section 6.2 for the discretization of the EB equations for electrons and holes.

The first device is a pn-diode similar to that studied in QUADE, RUDAN and SCHÖLL [1991]. The structure of the device is one-dimensional, corresponding to the interval $0 \leqslant x \leqslant \ell$ with $\ell = 1.23$ μm. The doping profile C is piecewise constant with $C(x) = 10^{17}$ cm^{-3} for $0 \leqslant x < \ell/2$, and $C(x) = -10^{17}$ cm^{-3} for $\ell/2 < x \leqslant \ell$. Constant carrier mobilities have been assumed, namely $\mu_n = 800$ $\mathrm{cm}^2\,\mathrm{V}^{-1}\,\mathrm{s}^{-1}$ and

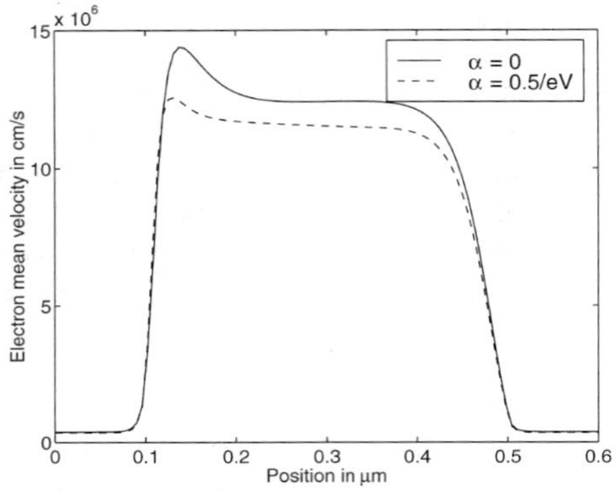

FIG. 7.21. Electron mean velocity in a ballistic diode.

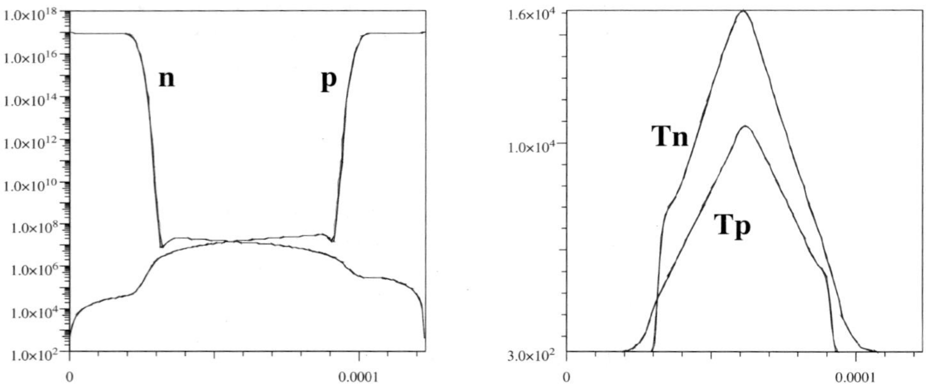

FIG. 7.22. Hole and electron concentrations (left) and hole and electron temperatures (right).

$\mu_p = 400\ \mathrm{cm}^2\,\mathrm{V}^{-1}\,\mathrm{s}^{-1}$. The SHR recombination/generation model has been considered with $\tau_n^* = \tau_p^* = 1 \times 10^{-6}$ s and the Wiedemann–Franz law described in Section 1.4 has been used to model the thermal conductivity with $\kappa_0 = 1/2$, while the carrier energy relaxation times have been set equal to $\tau_{w_\nu} = 0.5 \times 10^{-12}$ s. The diode has been simulated under reverse-bias conditions, with the applied voltage varying from 0 to 50 V. Fig. 7.22 shows the carrier concentrations n (electrons) and p (holes) and the corresponding temperatures T_n and T_p at an applied bias of 28 V. The comparison with the results of QUADE, RUDAN and SCHÖLL [1991] is quite satisfactory. In particular, we point out the strong carrier heating in the middle of the device due to the presence of a high electric field.

The next example refers to a realistic 1 μm channel-length nMOS transistor operating at room temperature $T^{eq} = 300$ K. The MOS structure is composed by a silicon semiconductor wafer Ω_S coated with a thin layer of insulating oxide Ω_O, as shown in Fig. 7.23 (left). The ohmic contacts (marked in bold) are respectively denoted by Source (S), Gate (G), Drain (D) and Back (B). The silicon-oxide interface is $\partial\Omega_I$; Ω_S is the

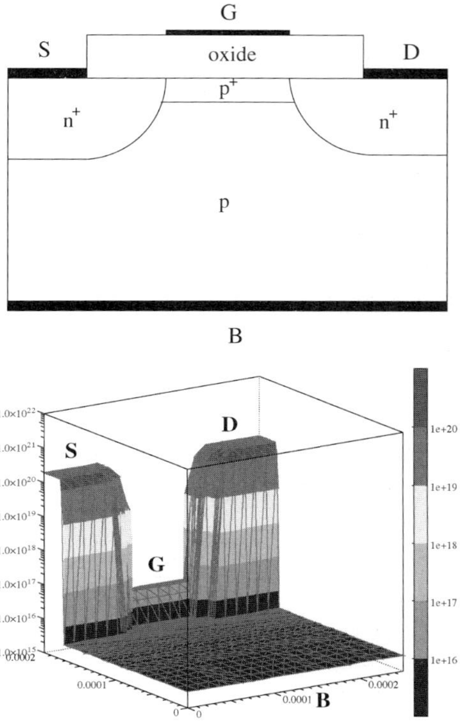

FIG. 7.23. MOS structure (top) and doping profile (bottom).

rectangle $0 \leqslant x_1 \leqslant 2.5$ μm, $0 \leqslant x_2 \leqslant 2$ μm and Ω_O is the rectangle $0.5 \leqslant x_1 \leqslant 2$ μm, $2 \leqslant x_2 \leqslant 2.025$ μm (oxide thickness $t_{ox} = 25$ nm).

The absolute value of the doping profile is plotted in logarithmic scale in Fig. 7.23 (right), with a p-type substrate of 3×10^{15} cm^{-3} (the *semiconductor bulk*), two n^+-type regions (source and drain) of 2×10^{20} cm^{-3} and a p^+ implantation of 3×10^{16} cm^{-3} in the bulk region close to the silicon-oxide interface. The following model has been used for carrier mobilities

$$\mu_\nu = \frac{\mu_{\nu,0}}{\left(1 + \left(\frac{\mu_{\nu,0} E_\nu^{\|}}{v_\nu^{sat}}\right)^{\beta_\nu}\right)^{1/\beta_\nu}} \frac{1}{\sqrt{1 + \frac{E_\nu^{\perp}}{E_\nu^{crit}}}}, \qquad \nu = n, p,$$

where $\mu_{n,0} = 800$ cm^2 V^{-1} s^{-1}, $\mu_{p,0} = 400$ cm^2 V^{-1} s^{-1}, $\beta_n = 2$, $\beta_p = 1$, $v_n^{sat} = 1.1 \times 10^7$ cm s^{-1}, $v_p^{sat} = 9.5 \times 10^6$ cm s^{-1}, $E_n^{crit} = 6.49 \times 10^4$ V cm^{-1}, $E_p^{crit} = 1.87 \times 10^4$ V cm^{-1}, and

$$E_\nu^{\|} = \frac{|\underline{E} \cdot \underline{J}_\nu|}{|\underline{J}_\nu|}, \qquad E_\nu^{\perp} = \frac{|\underline{E} \times \underline{J}_\nu|}{|\underline{J}_\nu|}.$$

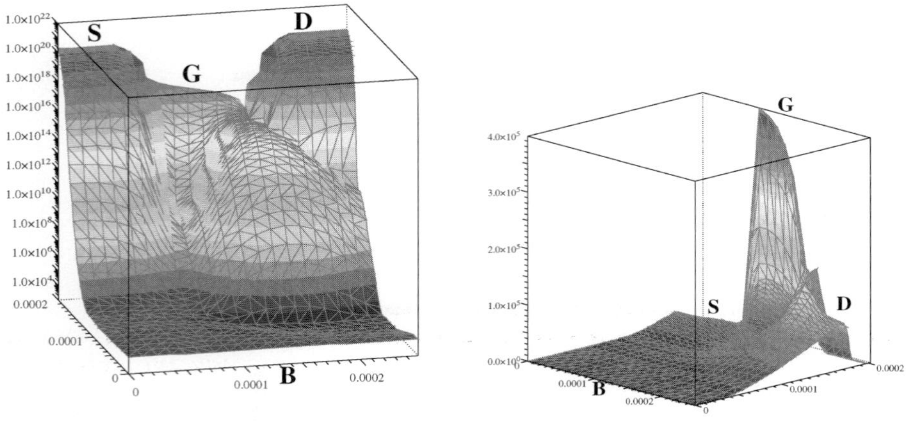

FIG. 7.24. Electron concentration (left) and electric field (right).

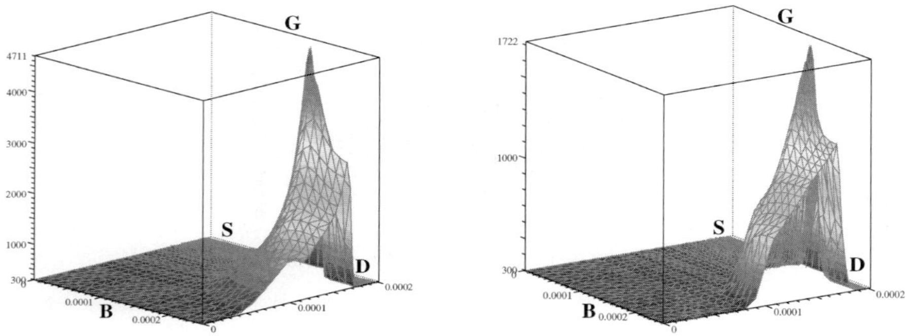

FIG. 7.25. Electron (left) and hole (right) temperatures.

As for the recombination/generation mechanisms, only the SHR model has been applied using the following doping-dependent model for the carrier lifetimes

$$\tau_\nu^* = \frac{\tau_{\nu,0}}{1 + \frac{N_D^+ + N_A^-}{N_\nu^{ref}}},$$

with $\tau_{\nu,0} = 5 \times 10^{-5}$ s and $N_\nu^{ref} = 5 \times 10^{16}\ \mathrm{cm}^{-3}$. The Wiedemann–Franz law described in Section 1.4 has been used to model the thermal conductivity with $\kappa_0 = 0.5$, while the carrier energy relaxation times have been set equal to $\tau_{w_\nu} = 0.5 \times 10^{-12}$ s.

The device has been simulated with 5 V applied between the gate and the source and with a drain-source voltage varying from 0 to 5 V. In Figs. 7.24–7.25 the electron concentration, electric field and carrier temperatures are shown at a drain-source voltage of 4.8 V. The carrier heating around the drain region, where a high peak of electric field exists, is clearly visible. This, in turn, is the reason for the large flooding of electrons near the drain end of the channel, due to their increased thermal diffusivity.

References

ADAMS, D.A. (1975). *Sobolev Spaces* (Academic Press, New York).

AGOUZAL, A., BARANGER, J., MAITRE, J.F., OUDIN, F. (1995). Connection between finite volume and mixed finite element methods for a diffusion problem with nonconstant coefficients Application to a convection-diffusion problem. *East-West J. Numer. Math.* **34**, 237–254.

ALABEAU, F. (1984). A singular perturbation analysis of the semiconductor device and the electrochemistry equations.

ALLEGRETTO, W., XIE, H. (1994). Nonisothermal semiconductor systems. In: Liu, X., Siegel, D. (eds.), *Comparison Methods and Stability Theory*. In: Lecture Notes in Pure and Applied Mathematics **162** (Marcel Dekker, New York).

ALLEN, D.N. DE G., SOUTHWELL, R.V. (1955). Relaxation methods applied to determine the motion, in two dimensions, of a viscous fluid past a fixed cylinder. *Quart. J. Mech. Appl. Math.* **8**, 129–145.

APANOVICH, Y., BLAKEY, P., COTTLE, R., LYUMKIS, E., POLSKY, B., SHUR, A., TCHERNIAEV, A. (1995). Numerical simulations of submicrometer devices including coupled nonlocal transport and nonisothermal effects. *IEEE Trans. Electr. Dev.* **42**, 890–897.

ARBOGAST, T., CHEN, Z. (1995). On the implementation of mixed methods as nonconforming methods for second-order elliptic problems. *Math. Comp.* **64** (211), 943–972.

ARBOGAST, T., DAWSON, C.N., KEENAN, P.T., WHEELER, M.F., YOTOV, I. (1998). Enhanced cell-centered finite differences for elliptic equations on general geometry. *SIAM J. Sci. Comput.* **19** (2), 404–425.

ARBOGAST, T., WHEELER, M.F., YOTOV, I. (1997). Mixed finite elements for elliptic problems with tensor coefficients as cell-centered finite differences. *SIAM J. Numer. Anal.* **34** (2), 826–852.

ARNOLD, D.N., BREZZI, F. (1985). Mixed and nonconforming finite element methods: implementation, postprocessing and error estimates. *RAIRO Modél. Math. Anal. Numér.* **19**, 7–32.

AZZAM, A., KREYSZIG, E. (1982). On solutions of elliptic equations satisfying mixed boundary conditions. *SIAM J. Math. Anal.* **13**, 254–262.

BABUŠKA, I., OSBORN, J.E. (1983). Generalized finite element methods: their performance and their relation to mixed methods. *SIAM J. Numer. Anal.* **203**, 510–536.

BACCARANI, G., RUDAN, M., GUERRIERI, R., CIAMPOLINI, P. (1986). Physical models for numerical device simulation. In: Engl, W.H. (ed.), *Process and Device Modeling* (Elsevier/North-Holland, Amsterdam), pp. 107–158.

BANK, R.E., ROSE, D.J. (1981). Global approximate newton methods. *Numer. Math.* **37**, 279–295.

BANK, R.E., ROSE, D.J., FICHTNER, W. (1983). Numerical methods for semiconductor device simulation. *IEEE Trans. Electr. Dev.* **ED-30**, 1031–1041.

BANK, R.E., BÜRGLER, J.F., FICHTNER, W., SMITH, R.K. (1990). Some upwinding techniques for finite element approximations of convection-diffusion equations. *Numer. Math.* **58**, 185–202.

BARANGER, J., MAITRE, J.F., OUDIN, F. (1994). Application de la théorie des éléments finis mixtes à l'étude d'une classe de schémas aux volumes différences finis pour les problèmes elliptiques. *C. R. Acad. Sci. Paris, Sér. I* **319**, 401–404.

BARANGER, J., MAITRE, J.F., OUDIN, F. (1996). Connection between finite volume and mixed finite element methods. *M^2AN* **30** (4), 445–465.

BEN ABDALLAH, N., DEGOND, P. (1996). On a hierarchy of macroscopic models for semiconductors. *J. Math. Phys.* **37**, 3308–3333.

BEN ABDALLAH, N., DEGOND, P., GÉNIEYS, S. (1996). An energy-transport model for semiconductors derived from the Boltzmann equation. *J. Stat. Phys.* **84**, 205–231.

BOSISIO, F., MICHELETTI, S., SACCO, R. (2000). A discretization scheme for an extended drift-diffusion model including trap-assisted phenomena. *J. Comp. Phys.* **159**, 197–212.

BREZZI, F. (1974). On the existence, uniqueness and approximation of saddle-point problems arising from lagrangian multipliers. *RAIRO Anal. Numer.* **8**, 129–151.

BREZZI, F., CAPELO, A.C.S., GASTALDI, L. (1989). A singular perturbation analysis of reversed-biased semiconductor diodes. *SIAM J. Math. Anal.* **20**, 372–387.

BREZZI, F., DOUGLAS JR, J., MARINI, L.D. (1985). Two families of mixed finite elements for second order elliptic problems. *Numer. Math.* **47**, 217–235.

BREZZI, F., FORTIN, M. (1991). *Mixed and Hybrid Finite Element Methods* (Springer, New York).

BREZZI, F., GASTALDI, L. (1986). Mathematical properties of one-dimensional semiconductors. *Matem. Apl. e Comput.* **5**, 123–137.

BREZZI, F., MARINI, L.D., PIETRA, P. (1987). Méthodes d'éléments finis mixtes et schéma de Scharfetter–Gummel. *C. R. Acad. Sci. Paris, Sér. I* **305**, 599–604.

BREZZI, F., MARINI, L.D., PIETRA, P. (1989a). Two-dimensional exponential fitting and applications to semiconductor device equations. *SIAM J. Numer. Anal.* **26**, 1342–1355.

BREZZI, F., MARINI, L.D., PIETRA, P. (1989b). Numerical simulation of semiconductor devices. *Comp. Meths. Appl. Mech. Engrg.* **75**, 493–514.

BREZZI, F., MARINI, L.D., MICHELETTI, S., PIETRA, P., SACCO, R. (submitted for publication). Stability and error analysis of mixed finite volume methods for advection-dominated problems.

BROOKS, A., HUGHES, T.J.R. (1982). Streamline upwind Petrov–Galerkin formulation for convection-dominated flows with particular emphasis on incompressible Navier–Stokes equations. *Comp. Meth. Appl. Mech. Eng.* **32**, 199–259.

BUTURLA, E., COTTRELL, P., GROSSMAN, B.M., SALSBURG, K.A. (1981). Finite element analysis of semiconductor devices: the FIELDAY program. *IBM J. Res. Develop.* **25** (4), 218–231.

CAFFARELLI, L., FRIEDMAN, A. (1987). A singular perturbation problem for semiconductors. *Boll. UMI* **7**, 409–421.

CAI, Z., JONES, J.E., MCCORMICK, S.F., RUSSELL, T.F. (1996). Control-volume mixed finite element methods. UCD/CCM Report No. 89 (University of Colorado, Denver).

CAUGHEY, D.M., THOMAS, R.E. (1967). Carrier mobilities in silicon empirically related to doping and field. *Proc. IEEE* **52**, 2192–2193.

CHEN, D., KAN, E., RAVAIOLI, U., SHU, C., DUTTON, R. (1992). An improved energy transport model including non-parabolic and non-Maxwellian distribution effects. *IEEE Electr. Dev. Lett.* **13**, 26–28.

CHEN, D., SANGIORGI, E., PINTO, M., KAN, E., RAVAIOLI, U., DUTTON, R. (1992). Analysis of spurious velocity overshoot in hydrodynamic simulations. *NUPAD IV*, 109–114.

CHEN, Z., COCKBURN, B. (1994). Error estimates for a finite element method for the drift-diffusion semiconductor device equations. *SIAM J. Numer. Anal.* **31**, 1062–1089.

CHEN, Z., COCKBURN, B. (1995). Analysis of a finite element method for the drift-diffusion semiconductor device equations: the multidimensional case. *Numer. Math.* **71**, 1–28.

CIARLET, PH.G. (1978). *The Finite Element Method for Elliptic Problems* (North-Holland, Amsterdam).

CIARLET, PH.G. (1991). Basic error estimates. In: Ciarlet, Ph.G., Lions, J.L. (eds.), *Handbook of Numerical Analysis, vol. II, Finite Element Methods (Part I)* (North-Holland, Amsterdam).

CIARLET, PH.G., RAVIART, P.A. (1973). Maximum principle and uniform convergence for the finite element method. *Comput. Methods Appl. Mech. Engrg.* **2**, 17–31.

COCKBURN, B., TRIANDAF, I. (1994). Error estimates for a finite element method for the drift-diffusion semiconductor device equations: The zero diffusion case. *Math. Comp.* **63**, 1–28.

COUGHRAN, W.M., JEROME, J.W. (1990). Modular algorithms for transient semiconductor device simulation, Part I: Analysis of the outer iteration. In: Bank, R.E. (ed.), *Computational Aspects of VLSI Design with an Emphasis on Semiconductor Device Simulation*. In: Lecture in Applied Mathematics **25** (American Mathematical Society, Providence, RI), pp. 107–149.

CROUZEIX, M., RAVIART, P.A. (1973). Conforming and nonconforming finite element methods for solving the stationary Stokes equation. *RAIRO* **7**, 33–76.

COWELL, E.M. (1967). *High-Field Transport in Semiconductor*, Solid State Physics **g** (Academic Press, New York).

DEGOND, P., GÉNIEYS, S., JÜNGEL, A. (1997). A system of parabolic equations in nonequilibrium thermodynamics including thermal and electrical effects. *J. Math. Pures Appl.* **76**, 991–1015.

DEGOND, P., GÉNIEYS, S., JÜNGEL, A. (1998). A steady-state system in nonequilibrium thermodynamics including thermal and electrical effects. *Math. Meth. Appl. Sci.* **21**, 1399–1413.

DEGOND, P., GUYOT-DELAURENS, F., MUSTIELES, F.J., NIER, F. (1990). Semiconductor modelling via the Boltzmann equation. In: Bank, R.E., Bulirsch, R., Merten, K. (eds.), *Mathematical Modelling and Simulation of Electrical Circuits and Semiconductor Devices* (Birkäuser, Basel), pp. 153–167.

DEGOND, P., JÜNGEL, A., PIETRA, P. (2000). Numerical discretization of energy-transport model for semiconductors with non-parabolic band structure. *SIAM J. Sci. Comp.* **22**, 986–1007.

DELAUNAY, B. (1934). Sur la sphère vide. *Izv. Akad. Nauk. SSSR., Math. and Nat. Sci. Div.* **6**, 793–800.

DEMARI, A. (1968). An accurate numerical steady state one-dimensional solution of the p-n junction. *Solid-State Electron.* **11**, 33–58.

DOUGLAS JR, J., ROBERTS, J.E. (1985). Global estimates for mixed methods for second order elliptic equations. *Math. Comp.* **44** (169), 39–52.

EWING, R.E., SAEVAREID, O., SHEN, J. (1998). Discretization schemes on triangular grids. *Comput. Methods Appl. Mech. Engrg.* **152**, 219–238.

FORGHIERI, A., GUERRIERI, R., CIAMPOLINI, P., GNUDI, A., RUDAN, M., BACCARANI, G. (1988). A new discretization strategy of the semiconductor equations comprising momentum and energy balance. *IEEE Trans. CAD* **7**, 231–241.

FORTIN, M. (1977). An analysis of the convergence of mixed finite element methods. *RAIRO Anal. Numer.* **11**, 341–354.

FRAEIJS DE VEUBEKE, B.X. (1965). Displacement and equilibrium models in the finite element method. In: Zienkiewicz, O.C., Hollister, G. (eds.), *Stress Analysis*.

GAJEWSKI, H. (1985). On existence, uniqueness and asymptotic behavior of solutions of the basic equations for carrier transport in semiconductors. *ZAMM* **65**, 101–108.

GAMBA, I.M. (1993). Asymptotic behavior at the boundary of a semiconductor device in two space dimensions. *Annali Mat. Pura Appl.* **163**, 43–91.

GASSER, I. (2001). The initial time layer problem and the quasineutral limit in a nonlinear drift diffusion model for semiconductors. *NoDEA*, 237–249.

GASSER, I., HSIAO, L., MARKOWICH, P.A., WANG, S. (2002). Quasi-neutral limit of a nonlinear drift-diffusion model for semiconductors. *J. Math. Anal. Appl.*

GASSER, I., LEVERMORE, D., MARKOWICH, P.A., SCHMEISER, C. (2001). The initial time layer problem and the quasi-neutral limit in the semiconductor drift-diffusion model. *Euro. J. Appl. Math.* **12**, 497–512.

GASSER, I., NATALINI, R. (1999). The energy-transport and the drift-diffusion equations as relaxation limits of the hydrodynamic model for semiconductors. *Quart. Appl. Math.* **57**, 269–282.

GATTI, E., MICHELETTI, S., SACCO, R. (1998). A new Galerkin framework for the drift-diffusion equation in semiconductors. *East-West J. Numer. Math.* **6** (2).

GOLSE, F., POUPAUD, F. (1992). Limite fluide des equations de Boltzmann des semiconducteuts pour une statistique de Fermi–Dirac. *Asympt. Anal.* **6**, 135–160.

GRIEPENTROG, J. (1999). An application of the implicit function theorem to an energy model of the semiconductor theory. *Z. Angew. Math. Mech.* **79**, 43–51.

GRISVARD, P. (1985). *Elliptic Problems in Nonsmooth Domains*, Monographs and Studies in Mathematics **24** (Pitman, London).

GUERRIERI, R., RUDAN, M., CIAMPOLINI, P., BACCARANI, G. (1985). Vectorial convergence acceleration techniques in semiconductor device analysis. In: Miller, J.J.H. (ed.), *Proceedings of the NASECODE IV Conference* (Boole Press, Dublin), pp. 293–298.

GUMMEL, H.K. (1964). A self-consistent iterative scheme for one-dimensional steady-state transistor calculations. *IEEE Trans. Electr. Dev.* **ED-11**, 455–465.

LAB, C., CAUSSIGNAC, P. (1999). An energy-transport model for semiconductor heterostructure devices: Application to AlGaAs/GaAs MODFETs. *COMPEL* **18**, 61–76.

HAUGAZEAU, Y., LACOSTE, P. (1993). Condensation de la matrice masse pour les éléments finis mixtes de $H(rot)$. *C. R. Acad. Sci. Paris, Sér. I* **316**, 509–512.

HECHT, F., MARROCCO, A. (1994). Mixed finite element simulation of heterojunction structures including a boundary layer model for the quasi-Fermi levels. *COMPEL* **13**, 757–770.

HECHT, F., MARROCCO, A., CAQUOT, E., FILOCHE, M. (1991). Semiconductor device modelling for heterojunctions structures with mixed finite elements. *COMPEL* **10**, 425–438.

HENNART, J.P., DEL VALLE, E. (1993). On the relationship between nodal schemes and mixed-hybrid finite elements. *Numer. Meth. Part. Diff. Eq.* **9**, 411–430.

HENNART, J.P., DEL VALLE, E. (1996). Mesh-centered finite differences from nodal finite elements. Rapport de recherche n. 2979 (INRIA).

HENRI, J., LOURO, B. (1989). Singular perturbation theory applied to electrochemistry equations in the case of electroneutrality. *Nonlinear Anal. TMA* **13**, 787–801.

HOLST, S., JÜNGEL, A., PIETRA, P. (2003). A mixed finite element discretization of the Energy-Transport model for semiconductors. *SIAM J. Sci. Comp.* **24**, 2058–2075.

HUGHES, T.J.R., BROOKS, A. (1982). A theoretical framework for Petrov–Galerkin methods with discontinuous weighting functions. Application to the streamline upwind procedure. In: Gallagher, et al. (eds.), Finite Elements in Fluids **4**, pp. 47–65.

JEROME, J.W. (1985). Consistency of semiconductor modeling: an existence/stability analysis for the stationary Van Roosbroeck system. *SIAM J. Appl. Math.* **45**, 565–590.

JEROME, J.W. (1987). Evolution systems in semiconductor device modeling: A cyclic uncoupled line analysis for the Gummel map. *Math. Methods Appl. Sci.* **9**, 455–492.

JEROME, J.W. (1996). *Analysis of Charge Transport* (Springer, Berlin).

JEROME, J.W., SHU, C.-W. (1994). Energy models for one-carrier transport in semiconductor devices. In: Coughran, W., Colde, J., Lloyd, P., White, J. (eds.), *Semiconductors, Part II*. In: IMA Volumes in Mathematics and its Applications **59** (Springer, New York), pp. 185–207.

JEROME, J.W., SHU, C.-W. (1996). Energy transport systems for semiconductors: Analysis and simulation. In: Lakshmikantham, V. (ed.), *Proceedings of the First World Congress of Nonlinear Analysts* (Walter de Gruyter, Berlin), pp. 3835–3846.

JOHNSON, C., NAVERT, U., PITKARANA, J. (1984). Finite element methods for linear hyperbolic problems. *Comput. Meth. Appl. Mech.* **45**, 285–312.

JOHNSON, C. (1987). *Numerical Solutions of Partial Differential Equations by the Finite Element Method* (Cambridge University Press, Cambridge).

JÜNGEL, A. (2001). *Quasi-Hydrodynamic Semiconductor Equations* (Birkhäuser, Basel).

JÜNGEL, A., PIETRA, P. (1997). A discretization scheme of a quasi-hydrodynamic semiconductor model. *Math. Models Meth. Appl. Sci.* **7**, 935–955.

KANE, E. (1957). *J. Phys. Chem. Solids*, 1–249.

KELLOGG, R.B. (1972). Higher order singularities for interface problems. In: Aziz, A.K. (ed.), *The Mathematical Foundations of the Finite Element Method with Applications to Partial Differential Equations* (Academic Press, New York), pp. 589–602.

KERKHOVEN, T. (1986). A proof of convergence of Gummel's algorithm for realistic device geometries. *SIAM J. Numer. Anal.* **23**, 1121–1137.

KERKHOVEN, T. (1988). A spectral analysis of the decoupling algorithm for semiconductor simulation. *SIAM J. Numer. Anal.* **25**, 1299–1312.

KERKHOVEN, T., SAAD, Y. (1992). On acceleration methods for coupled nonlinear elliptic systems. *Numer. Math.* **60**, 525–548.

LIONS, J.L., MAGENES, E. (1968). *Problèmes aux limites non-homogènes et applications* (Dunod, Paris).

MARINI, L.D., PIETRA, P. (1989). An abstract theory for mixed approximations of second order elliptic problems. *Mat. Aplic. Comp.* **8**, 219–239.

MARINI, L.D., PIETRA, P. (1990). New mixed finite element schemes for current continuity equations. *Compel* **9**, 257–268.

MARINI, L.D., PIETRA, P. (1991). A monotonic mixed finite element on rectangles for second order elliptic problems. *Mat. Aplic. Comp.* **10**, 263–277.

MARKOWICH, P.A. (1984). A singular perturbation analysis for the fundamental semiconductor device equations. *SIAM J. Appl. Math.* **44**, 896–928.

MARKOWICH, P.A. (1986). *The Stationary Semiconductor Device Equations* (Springer, Wien).

MARKOWICH, P.A., POUPAUD, F., SCHMEISER, C. (1995). Diffusion approximation of nonlinear electron phonon collision mechanisms. *RAIRO M^2AN* **29**, 857–869.

MARKOWICH, P.A., RINGHOFER, C.A. (1984). A singularly perturbed boundary value problem modelling a semiconductor device. *SIAM J. Appl. Math.* **44**, 231–256.

MARKOWICH, P.A., RINGHOFER, C.A., SCHMEISER, C. (1990). *Semiconductor Equations* (Springer, Wien).

MARKOWICH, P.A., SCHMEISER, C. (1986). Uniform asymptotic representations of the basic semiconductor device equations. *IMA J. Appl. Math.* **36**, 43–57.

MARKOWICH, P.A., SCHMEISER, C. (1997). The drift-diffusion limit for electron-phonon interaction in semiconductors. *M^3AS* **7**, 707–729.

MARKOWICH, P.A., ZLÁMAL, M. (1988). Inverse-average-type finite element discretisations of selfadjoint second-order elliptic problems. *Math. Comp.* **51** (184), 431–449.

MARROCCO, A., MONTARNAL, P. (1996). Simulation des modèles energy-transport à l'aide des éléments finis mixtes. *C. R. Acad. Sci. Paris, Sér. I* **323**, 535–541.

MARROCCO, A., MONTARNAL, P., PERTHAME, B. (1996). Simulation of the energy-transport and simplified hydrodynamic models for semiconductor devices using mixed finite elements. In: *Proceedings ECCOMAS 96* (Wiley, London).

MCCARTIN, B.J. (1985). Discretization of the semiconductor device equations. In: Miller, J.J.H. (ed.), *New Problems and New Solutions for Device and Process Modelling* (Boole Press, Dublin).

MICHELETTI, S. (2001). Stabilized finite elements for semiconductor device simulation. *Comput. Visual. Sci.* **3** (4), 177–183.

MICHELETTI, S., QUARTERONI, A., SACCO, R. (1995). Current-voltage characteristics simulation of semiconductor devices using domain decomposition. *J. Comp. Phys.* **119**, 46–61.

MICHELETTI, S., SACCO, R. (1999). Stabilized mixed finite elements for fluid models in semiconductors. *Comput. Visual. Sci.* **2**, 139–147.

MICHELETTI, S., SACCO, R., SALERI, F. (2001). On some mixed finite element methods with numerical integration. *SIAM J. Sci. Comput.* **231**, 245–270.

MIGLIO, E., QUARTERONI, A., SALERI, F. (1999). Finite element approximation of quasi-3D shallow water equations. *Comp. Meth. Appl. Mech. Engrg.* **174**, 355–369.

MILLER, J.J.H., WANG, S. (1991). A triangular mixed finite element method for the stationary semiconductor device equations. *RAIRO Modél. Math. Anal. Numér.* **25** (2), 441–463.

MILLER, J.J.H., WANG, S. (1994a). An analysis of the Scharfetter–Gummel box method for the stationary semiconductor device equations. *RAIRO Modél. Math. Anal. Numér.* **28** (2), 123–140.

MILLER, J.J.H., WANG, S. (1994b). A new non-conforming Petrov–Galerkin finite-element method with triangular elements for a singularly perturbed advection-diffusion problem. *IMA J. Numer. Anal.* **14**, 257–276.

MILLER, J.J.H., WANG, S. (1994c). A tetrahedral mixed finite element method for the stationary semiconductor continuity equations. *SIAM J. Numer. Anal.* **311**, 196–216.

MOCK, M.S. (1972). On equations describing steady-state carrier distribution in a semiconductor device. *Comm. Pure Appl. Math.* **25**, 781–792.

MOCK, M.S. (1983a). *Analysis of Mathematical Models of Semiconductor Devices* (Bool Press, Dublin).

MOCK, M.S. (1983b). Analysis of a discretization algorithm for stationary continuity equations in semiconductor device models. *COMPEL* **2** (4), 117–139.

MOLENAAR, J. (1995). Adaptive multigrid applied to a bipolar transistor problem. *Appl. Numer. Math.* **17**, 61–83.

MONTARNAL, P., PERTHAME, B. (1997). Asymptotic analysis of the drift-diffusion equations and Hamilton-Jacobi equations. *M^3AS* **7**, 61–80.

MURTHY, M.K.V., STAMPACCHIA, G. (1972). A variational inequality with mixed boundary conditions. *Israel J. Math.* **13**, 188–224.

NEČAS, J. (1967). *Les méthodes directes en théorie des équations elliptiques* (Masson, Paris).

POLAK, S.J., DEN HEIJER, C., SCHILDERS, W.H.A., MARKOWICH, P.A. (1987). Semiconductor device modelling from the numerical point of view. *Int. J. Numer. Meth. Engrg.* **24**, 763–838.

POLAK, S.J., SCHILDERS, W.H.A., COUPERUS, H.D. (1988). A finite element method with current conservation. In: Baccarani, G., Rudan, M. (eds.), *Proc. SISDEP-88, Bologna*, pp. 453–462.

POUPAUD, F. (1991). Diffusion approximation of the linear semiconductor equation: analysis of boundary layers. *Asympt. Anal.* **4**, 293–317.
POUPAUD, F. (1992). Runaway phenomena and fluid approximation under high fields in semiconductor kinetic theory. *ZAMM* **72**, 359–372.
POUPAUD, F., SCHMEISER, C. (1991). Charge transport in semiconductors with degeneracy effects. *Math. Methods Appl. Sci.* **14**, 301–318.
QUADE, W., RUDAN, M., SCHÖLL, E. (1991). Hydrodynamic simulation of impact ionization effects in P-N junctions. *IEEE Trans. on CAD* **10**, 1287–1294.
RAVIART, P.A., THOMAS, J.M. (1977). A mixed finite element method for second order elliptic problems. In: Galligani, I., Magenes, E. (eds.), *Mathematical Aspects of the Finite Element Method*. In: Lecture Notes in Math. **606** (Springer, New York), pp. 292–315.
RINGHOFER, C.A. (1987a). An asymptotic analysis of a transient P-N junction model. *SIAM J. Appl. Math.* **47**, 624–642.
RINGHOFER, C.A. (1987b). A singular perturbation analysis for the transient semiconductor device equations in one space dimension. *IMA J. Appl. Math.* **39**, 17–32.
RINGHOFER, C.A. (2001). An entropy-based finite difference method for the energy-transport system. *Math. Models Meth. Appl. Sci.* **11**, 769–795.
ROBERTS, J.E., THOMAS, J.M. (1991). Mixed and hybrid methods. In: Ciarlet, Ph.G., Lions, J.L. (eds.), *Handbook of Numerical Analysis, vol. II, Finite Element Methods (Part I)* (North-Holland, Amsterdam).
RODE, D.L. (1995). Low-field electron transport. In: Semiconductors and Semimetals **10** (Academic Press, New York), pp. 1–52.
ROOS, H.G., STYNES, M., TOBISKA, L. (1996). *Numerical Methods for Singularly Perturbed Differential Equations* (Springer, Berlin).
RUDAN, M., GNUDI, A., QUADE, W. (1993). A generalized approach to the hydrodynamic model of semiconductor equations. In: Baccarani, G. (ed.), *Precess and Device Modeling for Microelectronics* (North-Holland/Elsevier, Amsterdam), pp. 109–154.
SACCO, R., GATTI, E., GOTUSSO, L. (1995). The patch-test as a validation of a new finite element for the solution of convection-diffusion equations. *Comput. Methods Appl. Mech. Engrg.* **124**, 113–124.
SACCO, R., SALERI, F. (1997a). Mixed finite volume methods for semiconductor device simulation. *Numer. Meth. Part. Diff. Eq.* **13**, 215–236.
SACCO, R., SALERI, F. (1997b). Stabilized mixed finite volume methods for convection-diffusion problems. *East-West J. Numer. Math.* **4** (5), 291–311.
SACCO, R., STYNES, M. (1998). Finite element methods for convection-diffusion problems using exponential splines on triangles. *Comp. Math. Appl.* **353**, 35–45.
SCHARFETTER, D.L., GUMMEL, H.K. (1969). Large signal analysis of a silicon Read diode oscillator. *IEEE Trans. Electr. Dev.* **ED-16**, 64–77.
SCHMEISER, C. (1989). On strongly reverse biased semiconductor diodes. *SIAM J. Appl. Math.* **49**, 1734–1748.
SCHMEISER, C. (1990). A singular perturbation analysis of reverse biased pn-junctions. *SIAM J. Math. Anal.* **21**, 313–326.
SEIDMAN, T.I. (1980). Steady state solutions of reaction-diffusion systems with electrostatic convection. *Nonlinear Anal.* **4**, 632–637.
SELBERHERR, S. (1984). *Analysis and Simulation of Semiconductor Devices* (Springer, Wien).
SEVER, M. (1988). Discretization of time-dependent continuity equations. In: Miller, J.J.H. (ed.), *Proceedings of the 6th International NASECODE Conference* (Boole Press, Dublin), pp. 71–83.
SHARMA, M., CAREY, G. (1989a). Semiconductor device simulation using adaptive refinement and flux upwinding. *IEEE Trans. on CAD* **8**, 590–598.
SHARMA, M., CAREY, G. (1989b). Semiconductor device modeling using flux upwinding finite elements. *COMPEL* **84**, 219–224.
SLOTBOOM, J.W. (1973). Computer-aided two-dimensional analysis of bipolar transistors. *IEEE Trans. Electr. Dev.* **ED-20**, 669–679.
SOUISSI, K., ODEH, F., TANG, H., GNUDI, A. (1994). Comparative studies of hydrodynamic and energy transport models. *COMPEL* **13**, 439–453.

STRATTON, R. (1962). Diffusion of hot and cold electrons in semiconductor barriers. *Phys. Rev.* **126**, 2002–2014.

SZE, S.M. (1981). *The Physics of Semiconductor Devices*, second ed. (Wiley, New York).

THOMAS, J.M., TRUJILLO, D. (1997). Finite volume methods for elliptic problems: convergence on unstructured meshes. In: Conca, C., Gatica, G. (eds.), *Numerical Methods in Mechanics* (Addison Wesley, Reading, MA), pp. 163–174.

VAN ROOSBROECK, W. (1950). Theory of flow of electrons and holes in germanium and other semiconductors. *Bell System Tech. J.* **29**, 560–607.

VAN NOOYEN, R.R.P. (1995). A Petrov–Galerkin mixed finite element method with exponential fitting. *Numer. Meth. Part. Diff. Eq.* **11**, 501–524.

VARGA, R.S. (1962). *Matrix Iterative Analysis* (Prentice-Hall, Englewood Cliffs, NJ).

VASSILEVSKI, P.S., PETROVA, S.I., LAZAROV, R.D. (1992). Finite difference schemes on triangular cell-centered grids with local refinement. *SIAM J. Sci. Stat. Comput.* **13** (6), 1287–1313.

VISOCKY, P. (1994). A method for transient semiconductor device simulation using hot-electron transport equations. In: Miller, J. (ed.), *Proc. of the Nasecode X Conf.* (Boole Press, Dublin).

WANG, S. (1997). A novel exponentially fitted triangular finite element method for an advection-diffusion problem with boundary layers. *J. Comp. Phys.* **134**, 253–260.

WANG, S. (1999). A new exponentially fitted triangular finite element method for the continuity equations in the drift-diffusion model of semiconductor devices. *RAIRO Modél. Math. Anal. Numér.* **33** (1), 99–112.

WIGLEY, N.M. (1970). Mixed boundary value problems in plane domains with corners. *Math. Z.* **115**, 33–52.

Discretization of Semiconductor Device Problems (II)

A.M. Anile[a], N. Nikiforakis[b], V. Romano[a], G. Russo[a]

[a]*Dipartimento di Matematica, Universitá di Catania, viale A. Doria 6-95125, Catania, Italy*

[b]*Department of Applied Mathematics and Theoretical Physics, University of Cambridge, Silver st., Cambridge CB3 9EW, UK*

E-mail addresses: anile@dmi.unict.it (A.M. Anile); N.Nikiforakis@damtp.cam.ac.uk (N. Nikiforakis); romano@dmi.unict.it (V. Romano); russo@dmi.unict.it (G. Russo)
URLs: http://www.dipmat.unict.it/~anile (A.M. Anile); http://www.damtp.cam.ac.uk/user/nn10005/ (N. Nikiforakis); http://www.dmi.unict.it/~romano (V. Romano); http://www.dmi.unict.it/~russo (G. Russo)

Abstract

Enhanced functional integration in modern electron devices and the ensuing device miniaturization requires an accurate modeling of transient energy transport in semiconductors in order to describe high-field phenomena such as hot electrons, impact ionization and high frequency oscillations. For these reasons macroscopic models like the drift-diffusion equations (and the augmented ones) are no longer adequate and it has became almost mandatory to resort to a set of moment equations obtained from the semiconductor Boltzmann transport equation, which form a system of hyperbolic equations. From a computational point of view this has prompted the use of suitable numerical schemes which are able to cope with the dominantly hyperbolic nature of the problem, the coupling with the Poisson equation and the stiffness of the source term.

Introduction

Enhanced functional integration in modern electron devices requires an accurate modeling of energy transport in semiconductors in order to describe high-field phenom-

Numerical Methods in Electromagnetics
Special Volume (W.H.A. Schilders and E.J.W. ter Maten, Guest Editors) of
HANDBOOK OF NUMERICAL ANALYSIS, VOL. XIII
P.G. Ciarlet (Editor)

ISSN 1570-8659
DOI 10.1016/S1570-8659(04)13005-6

ena such as hot electron propagation, impact ionization and heat generation in the bulk material. The standard drift-diffusion models cannot cope with high-field phenomena because they do not comprise energy as a dynamical variable.

Furthermore for many applications in optoelectronics one needs to describe the transient interaction of electromagnetic radiation with carriers in complex semiconductor materials and since the characteristic times are of order of the electron momentum or energy flux relaxation times, some higher moments of the distribution function are necessarily involved. Therefore these phenomena cannot be described within the framework of the drift-diffusion equations (which are valid only in the quasi-stationary limit). Generalizations of the drift-diffusion equations have been sought, which would incorporate energy as a dynamical variable and also would not be restricted to quasi-stationary situations. These models are loosely speaking called hydrodynamical models. They are obtained from the infinite hierarchy of the moment equations of the Boltzmann transport equation by a suitable truncation procedure. This requires making suitable assumptions on: (i) closing the hierarchy by finding appropriate expressions for the $N+1$ order moment in terms of the previous ones; (ii) modeling the production terms on the right-hand side of the moment equations which arise from the moments of the collision terms in the Boltzmann transport equation.

One of the earliest hydrodynamical models currently used in applications was originally put forward by BLOTEKJAER [1970] and subsequently investigated by BACCARANI and WORDEMAN [1982] and by other authors (see references in RUDAN and BACCARANI [2001]). This model is implemented in simulation codes currently used in the microelectronic industry. It consists of a set of balance equations for carrier density, momentum and energy obtained from the Boltzmann transport equation in the parabolic band approximation, closed by a postulated Fourier law as constitutive equation for the heat flux. The production terms for momentum and energy are assumed to be of the relaxation type and the relaxation times are obtained by phenomenological arguments. Other models have also been investigated, some including also nonparabolic band approximation effects (HÄNSCH [1991], THOMA, EDMUNDS, MEINERZHAGEN, PEIFER and ENGL [1991], STETTLER, ALAM and LUNDSTROM [1993], BORDOLON, WANG, MAZIAR and TASCH [1991], WOOLARD, TIAN, TREW, LITTLEJOHN and KIM [1991]).

Most implemented hydrodynamical models suffer from serious theoretical drawbacks due to the *ad hoc* treatment of the closure problem (lacking a physically convincing motivation) and the modeling of the production terms (usually assumed to be of the relaxation type and this, as we shall see, leads to serious inconsistencies with the Onsager reciprocity relations). These difficulties are overcome by the Extended Hydrodynamical Models, obtained by applying the Maximum Entropy Principle to the closure of the moment equations (ANILE and PENNISI [1992], ANILE, MUSCATO, MACCORA and PIDATELLA [1996], ANILE, ROMANO and RUSSO [1998], ANILE and ROMANO [1999], ANILE, JUNK, ROMANO and RUSSO [2000], ROMANO [2000]).

From the mathematical viewpoint the equations of the Extended Models have the structure of a quasi-linear hyperbolic system with source terms. The latter contains a relaxation term and a nonlocal drift term, which is due to a self consistent electric field, coupled to a Poisson equation. Most numerical methods which have been used to solve the equations of hydrodynamical models are simply an adaptation of methods used for the drift-diffusion equations, e.g., methods appropriate for parabolic systems. However, unless some special (and usually unwarranted) approximations are made, the evolution equations of hydrodynamical models form a hyperbolic system. This is for instance the case when one is interested in the oscillations induced by interaction with electromagnetic waves of sufficiently high frequency. Therefore, in general, the methods which should be employed for the numerical solution of the governing equations must have certain features to ensure that the *correct* weak solution is captured at the correct point in space and time. Moreover the numerical scheme should not introduce spurious features, like unphysical oscillations in the vicinity of strong gradients.

The final goal of the Extended Hydrodynamical Models is the accurate multidimensional calculations of the full time dependent Extended Models equations applied to simulate the behaviour of realistic devices (BJT, MOSFET, resonant diode, etc.) in both transient and steady state regimes. In this context accuracy means being able to capture small scale wave features (e.g., related to Gunn type oscillations), as well as the bulk behaviour. This implies that it is mandatory to use methods which do not suffer from excess numerical diffusion or spurious oscillations in the vicinity of steep gradients (otherwise the viscosity inherent in a lower-order method would corrupt the solution at late times). Furthermore, a high order of accuracy is required if the numerical simulations are to be extended to the two-dimensional case (otherwise the required number of grid points would be prohibitively large).

Also although there are "source terms" the conservation properties of the hyperbolic left-hand side must be maintained. These requirements point us to the high resolution methods for hyperbolic systems. One of the aims of the present work is to report on the application for the numerical simulation of hydrodynamical models of a technique that has shown itself to be very useful in the area of computational fluid dynamics: the adaptive mesh refinement (AMR) approach originally introduced by BERGER and OLIGER [1984]. The AMR method allows the local spatial resolution to increase or decrease dynamically according to the requirements of the evolving solution; therefore computational resources are not wasted in maintaining uninteresting parts of the solution at unnecessarily high resolutions and this is a crucial features for 2D or 3D simulations. The AMR method has usually been applied to systems of partial differential equations which are purely hyperbolic in character. Its application to hydrodynamical semiconductor models is straightforward because these models involve also elliptic (arising from Poisson's equation), sometimes also parabolic modes (if heat conduction is approximately described by appropriate generalizations of Fourier's law) as well as hyperbolic modes.

The plan of the article is the following. Section 1 (written by A.M. Anile and V. Romano) consists of a general overview of the theory underlying hydrodynamical models for carrier transport in semiconductors with a special emphasis on the Extended Hydrodynamical Models. Section 2 (written by V. Romano and G. Russo) is a self-contained exposition of some recently introduced numerical methods for solving hyperbolic systems of conservation laws with particular regard to the hyperbolic models of semiconductors. These methods are then employed in Section 3 (written also by V. Romano and G. Russo) for solving numerically the extended hydrodynamical models previously introduced for the benchmark problems of the ballistic diode and the MESFET. Section 4 (written by A.M. Anile and N. Nikiforakis) consists of a full report of very recent works on the applications of adaptative mesh refinement to the solution of hydrodynamical equations.

1. Introduction to the hydrodynamical models of silicon semiconductors

1.1. Semiclassical kinetic model

Semiconductors are characterized by a sizable energy gap between the valence and the conduction bands, which are almost fully filled at thermal equilibrium. Upon thermal excitation electrons from the valence band can jump to the conduction band leaving behind holes (in the language of quasi-particles). Therefore the transport of charge is achieved through both negatively charged (electrons) and positively charged (holes) carriers.

The energy band structure of crystals can be obtained by the quantum theory of solids (ASHCROFT and MERMIN [1976]) at the cost of intensive numerical calculations. However, in order to describe electron transport, for most applications, a simplified description is adopted which is based on a simple analytical model. This is the so-called parabolic band and effective mass approximation, where the energy curve corresponding to a given energy band is approximated by a parabola near its minimum. In the sequel, for the sake of simplicity, only one conduction band will be considered.

In the parabolic band approximation, if we denote by $\mathcal{E}$ the energy of the considered conduction band measured from the band minimum, we have that the first Brillouin zone, $\mathcal{B}$, coincides with $\mathbb{R}^3$ and

$$\mathcal{E} = \frac{\hbar^2 |\mathbf{k}|^2}{2m^*}, \tag{1.1}$$

with m^* the effective electron mass (for silicon $m^* = 0.32 m_e$, with m_e the electron mass in vacuum), $\hbar\mathbf{k}$ the *crystal momentum* and $\hbar$ the Planck constant h divided by 2π.

In the approximation of the Kane dispersion relation (JACOBONI and REGGIANI [1983], JACOBONI and LUGLI [1989]), which takes into account the nonparabolicity at high energy, $\mathcal{E}$ still depends only on k, the modulus of $\mathbf{k}$, $\mathcal{B} = \mathbb{R}^3$, but

$$\mathcal{E}(k)\big[1 + \alpha\mathcal{E}(k)\big] = \frac{\hbar^2 k^2}{2m^*}, \quad \mathbf{k} \in \mathcal{B}, \tag{1.2}$$

where α is the nonparabolicity parameter (for silicon $\alpha = 0.5\ \mathrm{eV}^{-1}$).

The electron velocity $v(\mathbf{k})$ in a generic band depends on the energy $\mathcal{E}$ measured from the conduction band minimum by the relation

$$v(\mathbf{k}) = \frac{1}{\hbar}\nabla_{\mathbf{k}}\mathcal{E}.$$

Explicitly we get for parabolic band

$$v_i = \frac{\hbar k_i}{m^*}, \tag{1.3}$$

while in the approximation of the Kane dispersion relation

$$v_i = \frac{\hbar k_i}{m^*[1 + 2\alpha\mathcal{E}(k)]}. \tag{1.4}$$

In silicon the conduction band with lower energy has six equivalent valleys located along the main crystallographic directions Δ at about 85% from the center of the first Bruillouin zone near the X point (see ASHCROFT and MERMIN [1976] for details about crystal theory).

The above description of electron motion is valid for an ideal perfectly periodic crystal. Real semiconductors cannot be considered as ideal periodic crystals for several reasons. In fact strict periodicity is destroyed by:

- doping with impurities (which is done in order to control the electrical conductivity);
- thermal vibrations of the ions off their equilibrium positions in the lattice;
- electron–electron interactions.

These effects can be taken into account in a perturbative way by describing the interaction of the electrons with the lattice of ions as being only approximately periodic. The weak deviations from periodicity are treated as small perturbations of the background periodic ion potential. In particular the effect of the thermal vibrations of the ions on the electron dynamics can be described quantum mechanically as *scattering with quasi-particles (phonons) representing the thermal lattice vibrations.*

In a semiclassical kinetic description the electron wave packets are considered highly localized and the effects destroying the perfect periodicity are taken into account *by introducing a nonzero right-hand side in the semiclassical Vlasov equation.* In this way one obtains the *semiclassical Boltzmann equation for electrons in the conduction band in semiconductors*

$$\frac{\partial f}{\partial t} + v_i(\mathbf{k})\frac{\partial f}{\partial x_i} - \frac{eE_i}{\hbar}\frac{\partial f}{\partial k_i} = \mathcal{C}[f], \tag{1.5}$$

where $\mathcal{C}[f]$ represents the effects due to scattering with phonons, impurities and with other electrons. Hereafter we assume summation over repeated indices. A similar equation can be deduced for electrons in the valance bands (*holes*).

The electric field is calculated by solving the Poisson equation for the electric potential ϕ

$$E_i = -\frac{\partial \phi}{\partial x_i}, \tag{1.6}$$

$$\nabla \cdot (\varepsilon \nabla \phi) = -e(N_D - N_A - n), \tag{1.7}$$

N_D and N_A being the donor and acceptor density, respectively (which are fixed ions implanted in the semiconductors and depending only on the position) and n the electron number density

$$n = \int_B f \, \mathrm{d}^3\mathbf{k}.$$

The expression of the collision term has been investigated in the quantum theory of scattering.

The main scattering mechanisms in a semiconductor are the electron–phonon interaction, the interaction with impurities, electron–electron scatterings and interaction with stationary imperfections of the crystal as vacancies, external and internal crystal boundaries. In many situations the electron–electron collision term can be neglected since the electron density is not too high and one can linearize the collision operator by neglecting the degeneracy terms. Under the above approximation, for each type of interaction mechanism, the collision operator can be schematically written as

$$\mathcal{C}[f] = \int_{\mathcal{B}} \left[P(\mathbf{k}', \mathbf{k}) f(\mathbf{k}') - P(\mathbf{k}, \mathbf{k}') f(\mathbf{k}) \right] \mathrm{d}^3\mathbf{k}' \tag{1.8}$$

with $P(\mathbf{k}, \mathbf{k}')$ the scattering probability per unit time from a state $\mathbf{k}$ to a state $\mathbf{k}'$. The first term in (1.8) represents the gain and the second one the loss.

From the principle of detailed balance (see MARKOWICH, RINGHOFER and SCHMEISER [1990]) it follows that

$$P(\mathbf{k}', \mathbf{k}) = P(\mathbf{k}, \mathbf{k}') \exp\left(-\frac{\mathcal{E} - \mathcal{E}'}{k_B T_L} \right), \tag{1.9}$$

where $\mathcal{E} = \mathcal{E}(\mathbf{k})$ and $\mathcal{E}' = \mathcal{E}(\mathbf{k}')$, k_B being the Boltzmann constant and T_L being the lattice temperature which will be taken as constant.

In the case of acoustic phonon scattering in the elastic approximation we have (JACOBONI and REGGIANI [1983], JACOBONI and LUGLI [1989])

$$P(\mathbf{k}, \mathbf{k}') = \frac{k_B T_L \Xi_d^2}{4\pi^2 \hbar \rho v_s^2} \delta(\mathcal{E} - \mathcal{E}'), \tag{1.10}$$

where δ is the Dirac delta function, Ξ_d is the deformation potential of acoustic phonons, ρ the mass density of the material and v_s the sound speed of the longitudinal acoustic mode.

In the case of nonpolar optical phonon interaction (which is very important in Silicon), the scattering rate reads (JACOBONI and REGGIANI [1983], JACOBONI and LUGLI [1989])

$$P_{\pm}(\mathbf{k}, \mathbf{k}') = \frac{(D_t K)^2}{8\pi^2 \rho \omega} \left(n_B + \tfrac{1}{2} \mp \tfrac{1}{2} \right) \delta(\mathcal{E}' - \mathcal{E} \mp \hbar\omega), \tag{1.11}$$

where $D_t K$ is the deformation potential for nonpolar optical phonons, $\hbar\omega$ is the longitudinal optical phonon energy and n_B is the phonon equilibrium distribution according to the Bose–Einstein statistics

$$n_B = \frac{1}{\exp(\hbar\omega / k_B T_L) - 1}.$$

The upper sign refers to absorption processes and the lower sign refers to emission processes. The total scattering rate is given by the sum of these two terms

$$P(\mathbf{k}, \mathbf{k}') = \frac{(D_t K)^2}{8\pi^2 \rho \omega} \left[n_B \delta(\mathcal{E}' - \mathcal{E} - \hbar\omega) + (n_B + 1) \delta(\mathcal{E}' - \mathcal{E} + \hbar\omega) \right]. \tag{1.12}$$

In this article the contribution due to the impurities will be neglect.

1.2. Macroscopic models

Macroscopic models are obtained from the moment equations of the Boltzmann transport equation suitably truncated at a certain order N. The truncation procedure requires solving the following two important problems:

(i) the closure for higher order fluxes;

(ii) the closure for the production terms.

If a N-moment model is considered, the closure problem consists in finding an appropriate expression for the higher order moments and the production terms as suitable functions (constitutive relations) of the first N moments.

The macroscopic balance equations are deduced as moment equations of the Boltzmann transport equation as in gasdynamics. By multiplying Eq. (1.5) by a function $\psi(\mathbf{k})$ and integrating over $\mathcal{B}$, one finds the *moment equation*

$$\frac{\partial M_\psi}{\partial t} + \int_{\mathcal{B}} \psi(\mathbf{k}) v_i(\mathbf{k}) \frac{\partial f}{\partial x_i} \, \mathrm{d}^3\mathbf{k} - eE_j \int_{\mathcal{B}} \psi(\mathbf{k}) \frac{\partial}{\partial k_j} f \, \mathrm{d}^3\mathbf{k}$$
$$= \int_{\mathcal{B}} \psi(\mathbf{k}) \mathcal{C}[f] \, \mathrm{d}^3\mathbf{k}, \tag{1.13}$$

with

$$M_\psi = \int_{\mathcal{B}} \psi(\mathbf{k}) f \, \mathrm{d}^3\mathbf{k},$$

the moment relative to the weight function ψ.

Since

$$\int_{\mathcal{B}} \psi(\mathbf{k}) \frac{\partial f}{\partial k_j} \, \mathrm{d}^3\mathbf{k} = \int_{\partial\mathcal{B}} \psi(\mathbf{k}) f n_j \, \mathrm{d}\sigma - \int_{\mathcal{B}} f \frac{\partial \psi(\mathbf{k})}{\partial k_j} \, \mathrm{d}^3\mathbf{k},$$

with $\mathbf{n}$ outward unit normal field on the boundary $\partial\mathcal{B}$ of the domain $\mathcal{B}$ and $\mathrm{d}\sigma$ surface element of $\partial\mathcal{B}$, Eq. (1.13) becomes

$$\frac{\partial M_\psi}{\partial t} + \frac{\partial}{\partial x_i} \int_{\mathcal{B}} f \psi(\mathbf{k}) v_i(\mathbf{k}) \, \mathrm{d}^3\mathbf{k} + eE_j \left[\int_{\mathcal{B}} f \frac{\partial \psi(\mathbf{k})}{\partial k_j} \, \mathrm{d}^3\mathbf{k} - \int_{\partial\mathcal{B}} \psi(\mathbf{k}) f n_j \, \mathrm{d}\sigma \right]$$
$$= \int_{\mathcal{B}} \psi(\mathbf{k}) \mathcal{C}(f) \, \mathrm{d}^3\mathbf{k}. \tag{1.14}$$

The term

$$\int_{\partial\mathcal{B}} \psi(\mathbf{k}) f \mathbf{n} \, \mathrm{d}\sigma$$

vanishes either when $\mathcal{B}$ is expanded to $\mathbb{R}^3$ (because in order to guarantee the integrability condition f must tend to zero sufficiently fast as $k \mapsto \infty$) or when $\mathcal{B}$ is compact and $\psi(\mathbf{k})$ is periodic and continuous on ∂B. This latter condition is a consequence of the periodicity of f on $\mathcal{B}$ and the symmetry of $\mathcal{B}$ with respect to the origin.

Various models employ different expression of $\psi(\mathbf{k})$ and number of moments. Moreover a unipolar or bipolar version can be formulated. In the sequel only the motion of electrons in the valence bands will be considered and the motion of the holes neglected.

1.3. Hydrodynamical models: the BBW model

The energy bands are assumed to be of parabolic type and the classical moment equations are considered: continuity equation and the balance equations of linear momentum (particle flux) and energy[1]

$$\frac{\partial n}{\partial t} + \frac{\partial (nV_i)}{\partial x_j} = 0, \tag{1.15}$$

$$\frac{\partial (nV_i)}{\partial t} + \frac{\partial (nP_{ij})}{\partial x_j} + \frac{neE_i}{m^*} = nC_{Pi}, \tag{1.16}$$

$$\frac{\partial (nW)}{\partial t} + \frac{\partial (nS_j)}{\partial x_j} + neV_k E^k = nC_W, \tag{1.17}$$

where

$$n = \int_{\mathcal{B}} f \, \mathrm{d}^3\mathbf{k} \quad \text{is the electron density}, \tag{1.18}$$

$$V_i = \frac{1}{n} \int_{\mathcal{B}} v_i f \, \mathrm{d}^3\mathbf{k} \quad \text{is the average electron velocity}, \tag{1.19}$$

$$W = \frac{1}{n} \int_{\mathcal{B}} \mathcal{E}(k) f \, \mathrm{d}^3\mathbf{k} \quad \text{is the average electron energy}, \tag{1.20}$$

$$S^i = \frac{1}{n} \int_{\mathcal{B}} f v_i \mathcal{E}(k) \, \mathrm{d}^3\mathbf{k} \quad \text{is the energy flux}, \tag{1.21}$$

$$P_{ij} = \frac{1}{n} \int_{\mathcal{B}} f v_i v_j \, \mathrm{d}^3\mathbf{k} \quad \text{is the pressure tensor}, \tag{1.22}$$

$$C_{Pi} = \frac{1}{n} \int_{\mathcal{B}} \mathcal{C}[f] v_i \, \mathrm{d}^3\mathbf{k} \quad \text{is the linear momentum production}, \tag{1.23}$$

$$C_W = \frac{1}{n} \int_{\mathcal{B}} \mathcal{C}[f] \mathcal{E}(k) \, \mathrm{d}^3\mathbf{k} \quad \text{is the energy production}. \tag{1.24}$$

This approach dates back to the pioneering work of BLOTEKJAER [1970] and then BACCARANI and WORDEMAN [1982]. Because of its widespread popularity we denote this model by BBW (Blotekjaer–Baccarani–Wordeman).

Let us introduce the random component $m^*\mathbf{c}$ of $\hbar\mathbf{k}$, then

$$\hbar\mathbf{k} = m^*(\mathbf{V} + \mathbf{c}),$$

and one can decompose the tensor P_{ij} as

$$nP_{ij} = nV_i V_j + \int \mathrm{d}^3\mathbf{k} \, f c_i c_j.$$

[1] Unless otherwise stated, summation over repeated indices is assumed.

If one splits the tensor $\hat{\theta}_{ij} = \int d^3\mathbf{k}\, c_i c_j f$ into an isotropic and traceless part[2]

$$\hat{\theta}_{ij} = \tfrac{1}{3}\hat{\theta}_k^k \delta_{ij} + \hat{\theta}_{\langle ij\rangle} \quad \text{where } \hat{\theta}_k^k = \int d^3\mathbf{k}\, f\mathbf{c}^2,$$

then

$$P_{ij} = nV_i V_j + \tfrac{1}{3}\hat{\theta}_k^k \delta_{ij} + \hat{\theta}_{\langle ij\rangle}. \tag{1.25}$$

In the model of Baccarani and Wordeman the anisotropic tensor $\hat{\theta}_{\langle ij\rangle}$ is neglected, $\hat{\theta}_{\langle ij\rangle} = 0$. For the energy density one finds

$$nW = \frac{m^*}{2}\left(nV^2 + \hat{\theta}_k^k\right).$$

Now we define the electron temperature T (distinct from the lattice temperature T_L) by assuming, in analogy with kinetic energy of monatomic gas, the equation of state for ideal gas

$$nW - \frac{nm^*\mathbf{V}^2}{2} = \frac{3nk_BT}{2} \tag{1.26}$$

whence

$$\hat{\theta}^k{}_k = \frac{3nk_BT}{m^*} \quad \text{and} \quad \hat{\theta}_{ij} = \frac{nk_BT}{m^*}\delta_{ij}.$$

Concerning the productions, the momentum rate of change is assumed to be of the relaxation time type

$$C_{Pi} = -\frac{V_i}{\tau_p} \tag{1.27}$$

hence the momentum equation (1.16) rewrites

$$\frac{\partial(nV_i)}{\partial t} + \frac{\partial}{\partial x_j}\left[nV_iV_j + \frac{nk_BT}{m^*}\delta_{ij}\right] + \frac{nqE_i}{m^*} = -\frac{nV_i}{\tau_p}. \tag{1.28}$$

Furthermore we can decompose the energy flow $\mathbf{S}$ as

$$\mathbf{S} = W\mathbf{V} + k_BT\mathbf{V} + \mathbf{q}, \tag{1.29}$$

where $\mathbf{q}$ is the heat flow vector

$$n\mathbf{q} = \frac{m^*}{2}\int d^3\mathbf{k}\, f\mathbf{c}^2\mathbf{c}. \tag{1.30}$$

In the BBW model it is assumed that *the heat flow vector is given by the Fourier law* (closure assumption)

$$\mathbf{q} = -\kappa\nabla T. \tag{1.31}$$

[2]For a second order symmetric tensor of components A_{ij} one has $A_{\langle ij\rangle} = A_{ij} - \frac{1}{3}A_k^k\delta_{ij}$, A_k^k being the trace of A.

In the original Baccarani and Wordeman model the Fourier law is simply assumed as a phenomenological law. A better justification (which however leads to a more complicated expression) has been given by several authors (HÄNSCH [1991], ANILE and PENNISI [1992]). The closure assumptions of the BBW model for fluxes are open to criticism. In particular the assumption that the heat flow is described by the Fourier law is rather questionable (STETTLER, ALAM and LUNDSTROM [1993], CHENG, LIANGYING, FITHEN and YANSHENG [1997]).

Modeling the relaxation times is also a rather delicate question. In the original Baccarani and Wordeman formulation τ_p is determined by the following consideration. One introduce the electron mobility μ_n, related to the momentum relaxation time by

$$\mu_n = \frac{e\tau_p}{m^*}$$

and assumes that the Einstein relation relating mobility to diffusivity

$$D_n = k_B T \mu_n$$

holds also outside thermal equilibrium and that D_n is constant and equal to the low field diffusivity D_0. Hence

$$D_0 = k_B T_L \mu_{n0} = k_B T \mu_n,$$

where μ_{n0} is the low field mobility, whence

$$\tau_p = \frac{m^* \mu_{n0} T_L}{eT}.$$

The energy production term is also assumed to be of relaxation type

$$C_W = -n \frac{W - W_0}{\tau_w},$$

where τ_w is the energy relaxation time and $W_0 = \frac{3k_B T_L}{2}$ is the equilibrium energy.

The relaxation time τ_w is obtained by approximating it with the corresponding expression of the stationary and homogeneous case

$$\tau_w = \frac{W - W_0}{e|\mathbf{E}|V}$$

and by expressing the electric field $|\mathbf{E}|$ as a function of temperature by using the Caughey–Thomas formula for the high-field mobility (see SELBERHERR [1984])

$$\mu_n = \mu_{n0}\left[1 + \left(\frac{\mu_{n0}|\mathbf{E}|}{v_s}\right)^2\right]^{-1/2},$$

where $v_s = 1.0 \times 10^7$ cm/s is the saturation velocity. Since in the homogeneous case

$$V_i = \frac{\tau_p e E_i}{m^*} = \mu_n E_i,$$

then

$$\tau_w = \frac{m^* \mu_{n0} T_L}{2eT} + \frac{3k_B \mu_{n0} T T_L}{2ev_s^2(T + T_L)}.$$

The modeling of the thermal conductivity coefficient is obtained from the Wiedemann–Franz law (which is justified only near thermal equilibrium)

$$\kappa = \left(\frac{5}{2} + c\right)\left(\frac{k_B}{e}\right)^2 nq\mu_n T, \tag{1.32}$$

where the constant c is related to the exponent in the expression for the relaxation time

$$\tau(\mathcal{E}) = \tau_0\left(\frac{\mathcal{E}}{\mathcal{E}_0}\right)^c.$$

In the original model of Baccarani and Wordeman the choice $c = -1$ is made and therefore

$$\kappa = \frac{3}{2}\left(\frac{k_B}{e}\right)^2 ne\mu_n T.$$

The BBW hydrodynamical model for electrons consists then of the following equations.

– continuity equation

$$\frac{\partial n}{\partial t} + \nabla \cdot (n\mathbf{V}) = 0; \tag{1.33}$$

– momentum equation

$$\frac{\partial (nV_i)}{\partial t} + \frac{\partial}{\partial x_j}\left(nV_iV_j + \frac{nk_BT}{m^*}\delta_{ij}\right) + \frac{nqE_i}{m^*} = -\frac{nV_i}{\tau_p}; \tag{1.34}$$

– energy equation

$$\frac{\partial}{\partial t}\left(\frac{1}{2}nm^*\mathbf{V}^2 + \frac{3}{2}nk_BT\right) + \nabla \cdot \left[\left(\frac{1}{2}nm^*V^2 + \frac{5}{2}nk_BT\right)\mathbf{V} - \kappa\nabla T\right] + ne\mathbf{E}\cdot\mathbf{V} = -\frac{W - W_0}{\tau_w}; \tag{1.35}$$

– Poisson's equation

$$\nabla \cdot (\varepsilon\nabla\phi) = e(N_A - N_D + n - p). \tag{1.36}$$

These equations, were not for the collision terms, would be the same as the balance equations for a charged heat conducting fluid coupled to Poisson's equation.

GARDNER, JEROME and ROSE [1989] and GARDNER [1991], GARDNER [1993] numerically integrated the BBW model for the ballistic diode in the stationary case. In GARDNER [1991] the system of equations was discretized by using central differences (if the flow is everywhere subsonic) or second order upwind method (for transonic flow). The discretized system is then linearized by using Newton's method with a damping factor. In this way Gardner was able to show evidence for an electron shock wave in the diode. In ANILE, MACCORA and PIDATELLA [1995] Gardner's results have been recovered by using a viscosity method.

Numerical solutions in the nonstationary case have been obtained in FATEMI, JEROME and OSHER [1991] by using an ENO scheme. The same results have been obtained in ROMANO and RUSSO [2000] by using a central scheme.

We remark that, as proved in ANILE and MUSCATO [1995], ANILE and MUSCATO [1996] the Onsager conditions do not hold for the BBW model and this highlights the poor physical consistency of such a closure.

1.4. The extended hydrodynamical model

In a series of articles (ANILE and PENNISI [1992], ANILE and MUSCATO [1995], ANILE and MUSCATO [1996], ANILE, ROMANO and RUSSO [1998], ANILE, JUNK, ROMANO and RUSSO [2000], ANILE and ROMANO [1999], ANILE and ROMANO [2000], ROMANO [2000]) a general framework for getting closure relation is proposed. At variance with previous treatments, it is not an *ad hoc* procedure but it is based on the application of the entropy principle within the framework of Extended Thermodynamics (MÜLLER and RUGGERI [1998], JOU, CASAS-VAZQUEZ and LEBON [1993]) or equivalently the Maximum Entropy Principle or the moment theory of LEVERMORE [1995], LEVERMORE [1996]. Apart from the usual balance equations for carriers density, momentum and energy, this class of models comprises evolution equations for the heat flux and shear stress.

The resulting system is hyperbolic in a suitable domain of the space of variables. In the stationary case, by linearizing the heat flux equation for small temperature gradients (*Maxwellian iteration*) one obtains an extension of the Fourier law which includes also a convective term. With the addition of this term, the Onsager relations for small deviations from thermodynamical equilibrium are verified (at variance with the BBW model). Furthermore the heat conductivity turns out to be directly related to the energy-flux relaxation time and does not contain any undetermined free parameters (at variance with the BBW model).

We illustrate the main guidelines upon which to construct the model. Then specific results will be presented in the case of the Kane dispersion relation and, as limiting case, in the parabolic band approximation.

Concerning the moment equations, several choices of the weight function ψ can be made and they lead to different balance equations for macroscopic quantities. We take the following set of weight functions: 1, $\hbar\mathbf{k}$, $\mathcal{E}$ and $\mathcal{E}\mathbf{v}$.

By considering such expressions for ψ one obtains the continuity equation (indeed a term due to the generation-recombination mechanism should appear in the right-hand side, but this effect is relevant for times of order 10^{-9} s and in most applications can be neglected because the characteristic times are of order of a fraction of picosecond), the balance equation for the crystal momentum, the balance equation for the electron energy, and the balance equation for the electron energy flux. Since only the Kane dispersion relation or the parabolic case will be considered in the sequel, we neglect the boundary integral terms in the moment equations. Then the explicit form of the macroscopic balance equations reads

$$\frac{\partial n}{\partial t} + \frac{\partial (n V_i)}{\partial x_i} = 0, \tag{1.37}$$

$$\frac{\partial(nP_i)}{\partial t} + \frac{\partial(nU_{ij})}{\partial x_j} + neE_i = nC_{Pi}, \tag{1.38}$$

$$\frac{\partial(nW)}{\partial t} + \frac{\partial(nS_j)}{\partial x_j} + neV_kE^k = nC_W, \tag{1.39}$$

$$\frac{\partial(nS_i)}{\partial t} + \frac{\partial(nF_{ij})}{\partial x_j} + neE_jG_{ij} = nC_{Wi}, \tag{1.40}$$

where

$$n = \int_{\mathcal{B}} f \, \mathrm{d}^3\mathbf{k} \quad \text{is the electron density,}$$

$$V_i = \frac{1}{n}\int_{\mathcal{B}} f v_i \, \mathrm{d}^3\mathbf{k} \quad \text{is the average electron velocity,}$$

$$W = \frac{1}{n}\int_{\mathcal{B}} \mathcal{E}(k) f \, \mathrm{d}^3\mathbf{k} \quad \text{is the average electron energy,}$$

$$S_i = \frac{1}{n}\int_{\mathcal{B}} f v_i \mathcal{E}(k) \, \mathrm{d}^3\mathbf{k} \quad \text{is the energy flux,}$$

$$P_i = \frac{1}{n}\int_{\mathcal{B}} f \hbar k_i \, \mathrm{d}^3\mathbf{k} \quad \text{is the average crystal momentum,}$$

$$U_{ij} = \frac{1}{n}\int_{\mathcal{B}} f v_i k_j \, \mathrm{d}^3\mathbf{k} \quad \text{is the flow of crystal momentum,}$$

$$G_{ij} = \frac{1}{n}\int_{\mathcal{B}} \frac{1}{\hbar} f \frac{\partial}{\partial k_j}(\mathcal{E} v_i) \, \mathrm{d}^3\mathbf{k},$$

$$F_{ij} = \frac{1}{n}\int_{\mathcal{B}} f v_i v_j \mathcal{E}(k) \, \mathrm{d}^3\mathbf{k} \quad \text{is the flux of energy flux,}$$

$$C_{Pi} = \frac{1}{n}\int_{\mathcal{B}} \mathcal{C}[f]\hbar k_i \, \mathrm{d}^3\mathbf{k} \quad \text{is the production of the crystal momentum balance equation,}$$

$$C_W = \frac{1}{n}\int_{\mathcal{B}} \mathcal{C}[f]\mathcal{E}(k) \, \mathrm{d}^3\mathbf{k} \quad \text{is the production of the energy balance equation,}$$

$$C_{Wi} = \frac{1}{n}\int_{\mathcal{B}} \mathcal{C}[f] v_i \mathcal{E}(k) \, \mathrm{d}^3\mathbf{k} \quad \text{is the production of the energy flux balance equation.}$$

Analogous equations can be written for holes if a two component charge carrier model is employed.

We remark that in general the average crystal momentum P_i does not coincide with the average electron momentum, but it is related to the latter by

$$P_i = m^*(V_i + 2\alpha S_i). \tag{1.41}$$

As remarked several times, the moment equations do not constitute a set of closed relations because of the fluxes and production terms. Now we will present a physically sound procedure for getting the required closure relations.

If we assume as fundamental variables n, V_i, W and S_i, which have a direct physical interpretation, the closure problem consists in expressing P_i, U_{ij}, F_{ij} and G_{ij} and the moments of the collision term C_{Pi}, C_W and C_{Wi} as functions of n, V_i, W and S_i.

We stress that the role of the mean velocity V_i here is radically different from that played in gas dynamics. In fact, for a simple gas the explicit dependence of fluxes on the velocity can be predicted by requiring Galilean invariance of the constitutive functions. Instead Eqs. (1.37)–(1.40) are not valid in an arbitrary Galilean reference frame, but they hold only in a frame where the crystal is at rest (in the applications it can be considered as inertial and it is possible to neglect the inertial forces). Therefore V_i is the velocity relative to the crystal and the dependence on it in the constitutive functions cannot be removed by a Galilean transformation.

The Maximum Entropy Principle (hereafter MEP) leads to a systematic way for obtaining constitutive relations on the basis of information theory (see MÜLLER and RUGGERI [1998], JOU, CASAS-VAZQUEZ and LEBON [1993], LEVERMORE [1995], LEVERMORE [1996], DREYER [1987] for a review).

According to the MEP, if a given number of moments M_A are known, the distribution function f_{ME} which can be used to evaluate the unknown moments of f corresponds to the extremal of the entropy functional under the constraints that it yields exactly the known moments M_A

$$\int_{\mathcal{B}} \psi_A f_{ME}\, \mathrm{d}^3\mathbf{k} = M_A. \tag{1.42}$$

Since the electrons interact with the phonons describing the thermal vibrations of the ions placed at the points of the crystal lattice, in principle we should deal with a two component system (electrons and phonons). However, if one considers the phonon gas as a thermal bath at constant temperature T_L, only the electron component of the entropy must be maximized. Moreover, by considering the electron gas as sufficiently dilute, one can take for the electron gas the expression of the entropy obtained as limiting case of that arising in the Fermi statistics

$$s = -k_B \int_{\mathcal{B}} (f \log f - f)\, \mathrm{d}^3\mathbf{k}. \tag{1.43}$$

If we introduce the Lagrange multipliers Λ_A, the problem to maximize s under the constraints (1.42) is equivalent to maximize the Legendre transform of s,

$$s' = \Lambda_A M_A - s,$$

without constraints,

$$\delta s' = 0.$$

This gives

$$\left[\log f + \frac{\Lambda_A \psi_A}{k_B}\right]\delta f = 0.$$

Since the latter relation must hold for arbitrary δf, it follows

$$f_{ME} = \exp\left[-\frac{1}{k_B}\Lambda_A \psi_A\right]. \tag{1.44}$$

If n, V_i, W and S_i are assumed as fundamental variables, then

$$\psi_A = (1, \mathbf{v}, \mathcal{E}, \mathcal{E}\mathbf{v})$$

and

$$\Lambda_A = \left(\lambda, k_B\lambda_i, k_B\lambda^W, k_B\lambda_i^W\right)$$

with λ Lagrange multiplier relative to the density n, λ^W Lagrange multiplier relative to the energy W, λ_i Lagrange multiplier relative to the velocity v_j and λ_i^W Lagrange multiplier relative to the energy flux S_j. Therefore the maximum entropy distribution function reads

$$f_{ME} = \exp\left[-\left(\frac{1}{k_B}\lambda + \lambda^W\mathcal{E} + \lambda_i v_i + \lambda_i^W v_i\mathcal{E}\right)\right], \tag{1.45}$$

with Λ_A functions of the moments M_A.

In order to get the dependence of the Λ_A's from the M_A, one has to invert the constraints (1.42). Then by taking the moments of f_{ME} and $C[f_{ME}]$ one finds the closure relations for the fluxes and the production terms of the system (1.37)–(1.40). On account of the analytical difficulties this can be achieved only with a numerical procedure. For example, in the case of gas dynamics, in LE TALLEC and PERLAT [1997] the multipliers have been used as independent unknowns, and the conserved field and the flux have been computed by performing a numerical integration in velocity space. Here we overcome the problem looking for an asymptotic form of f_{ME} as follows.

At equilibrium the distribution function is isotropic

$$f_{EQ} = \exp\left[-\left(\frac{1}{k_B}\lambda_E + \frac{\mathcal{E}}{k_B T_0}\right)\right], \tag{1.46}$$

that is at equilibrium

$$\lambda_E^W = \frac{1}{k_B T_0}, \quad \lambda_{iE} = 0, \quad \lambda_{iE}^W = 0.$$

Monte Carlo simulations for electron transport in Si show that the anisotropy of f is small even far from equilibrium.

Upon such a consideration we make the *ansatz* of small anisotropy for f_{ME}, as measured by the small anisotropy parameter δ.

For details about the closure relations see ANILE and ROMANO [1999], ROMANO [2000]. Here we summarize the results up to first order in δ. Concerning the tensors U_{ij}, F_{ij} and G_{ij}, one has

$$U_{ij} = U\delta_{ij}, \quad F_{ij} = F\delta_{ij}, \quad G_{ij} = G\delta_{ij}, \tag{1.47}$$

with

$$U = \frac{2}{3d_0}\int_0^\infty \left[\mathcal{E}(1+\alpha\mathcal{E})\right]^{3/2}\exp\left(-\lambda^{W(0)}\mathcal{E}\right)\mathrm{d}\mathcal{E}, \tag{1.48}$$

$$F = \frac{2}{3m^* d_0}\int_0^\infty \exp\left(-\lambda^{W(0)}\mathcal{E}\right)\frac{\mathcal{E}[\mathcal{E}(1+\alpha\mathcal{E})]^{3/2}}{1+2\alpha\mathcal{E}}\,\mathrm{d}\mathcal{E}, \tag{1.49}$$

$$G = \frac{1}{m^* d_0}\int_0^\infty \exp\left(-\lambda^{W(0)}\mathcal{E}\right)\left[1 + \frac{2(1+\alpha\mathcal{E})}{3(1+2\alpha\mathcal{E})^2}\right]\mathcal{E}^{3/2}\sqrt{1+\alpha\mathcal{E}}\,\mathrm{d}\mathcal{E}. \tag{1.50}$$

$\lambda^{W(0)}$ is the expression of the Lagrange multipliers relative to the energy up to first order in δ (see ANILE and ROMANO [1999]).

C_{Pi} and C_{Wi} have the form

$$\begin{pmatrix} C_{Pi} \\ C_{Wi} \end{pmatrix} = \begin{pmatrix} c_{11} & c_{12} \\ c_{21} & c_{22} \end{pmatrix} \begin{pmatrix} V_i \\ S_i \end{pmatrix}.$$

The production terms are the sum of the term due to the elastic scatterings (acoustical phonon scattering) and that due to inelastic phonon scatterings. Therefore the production matrix $C = (c_{ij})$ is given by the sum $C = C^{(ac)} + C^{(np)}$.

Concerning the acoustic phonon scattering, the contribution to the energy balance equation is zero while the production matrix $C^{(ac)} = (c_{ij}^{(ac)})$ can be written as $C^{(ac)} = A^{(ac)}B$. The coefficients b_{ij} of the matrix B are given by ANILE and ROMANO [1999]

$$b_{11} = \frac{a_{22}}{\Delta}, \quad b_{12} = -\frac{a_{12}}{\Delta}, \quad b_{22} = \frac{a_{11}}{\Delta}$$

with

$$a_{11} = -\frac{2p_0}{3m^* d_0}, \quad a_{12} = -\frac{2p_1}{3m^* d_0}, \quad a_{22} = -\frac{2p_2}{3m^* d_0}, \quad \Delta = a_{11}a_{22} - a_{12}^2,$$

$$d_0 = \int_0^\infty \sqrt{\mathcal{E}(1+\alpha\mathcal{E})}\,(1+2\alpha\mathcal{E})\exp\bigl(-\lambda^{W(0)}\mathcal{E}\bigr)\,\mathrm{d}\mathcal{E},$$

$$p_k = \int_0^\infty \frac{[\mathcal{E}(1+\alpha\mathcal{E})]^{3/2}\mathcal{E}^k}{1+2\alpha\mathcal{E}}\exp\bigl(-\lambda^{W(0)}\mathcal{E}\bigr)\,\mathrm{d}\mathcal{E}, \quad k = 0, 1, \ldots.$$

The coefficients of the matrix $A^{(ac)}$ read

$$a_{11}^{(ac)} = \frac{\overline{K}_{ac}}{d_0}\int_0^\infty \mathcal{E}^2(1+\alpha\mathcal{E})^2(1+2\alpha\mathcal{E})\exp\bigl(-\lambda^{W(0)}\mathcal{E}\bigr)\,\mathrm{d}\mathcal{E},$$

$$a_{12}^{(ac)} = \frac{\overline{K}_{ac}}{d_0}\int_0^\infty \mathcal{E}^3(1+\alpha\mathcal{E})^2(1+2\alpha\mathcal{E})\exp\bigl(-\lambda^{W(0)}\mathcal{E}\bigr)\,\mathrm{d}\mathcal{E},$$

$$a_{21}^{(ac)} = \frac{\overline{K}_{ac}}{m^* d_0}\int_0^\infty \mathcal{E}^3(1+\alpha\mathcal{E})^2\exp\bigl(-\lambda^{W(0)}\mathcal{E}\bigr)\,\mathrm{d}\mathcal{E},$$

$$a_{22}^{(ac)} = \frac{\overline{K}_{ac}}{m^* d_0}\int_0^\infty \mathcal{E}^4(1+\alpha\mathcal{E})^2\exp\bigl(-\lambda^{W(0)}\mathcal{E}\bigr)\,\mathrm{d}\mathcal{E},$$

where

$$\overline{K}_{ac} = \frac{8\pi\sqrt{2}(m^*)^{3/2}K_{ac}}{3\hbar^3}, \qquad K_{ac} = \frac{k_B T_L \Xi_d^2}{4\pi^2\hbar\rho v_s^2}.$$

Concerning the nonpolar phonon scattering the production term of the energy balance equation is given by $C_W = \sum_{A=1}^{6} C_{W_A}$, where for each valley (ROMANO [2001])

$$C_{W_A} = \frac{3}{2}\frac{\overline{K}_{np}}{d_0}\sum_{\pm}\left(n_B + \frac{1}{2} \mp \frac{1}{2}\right)\left[\exp\left(\pm\frac{\hbar\omega_{np}}{k_B T_L} \mp \lambda^{W(0)}\hbar\omega_{np}\right) - 1\right]\eta^{\pm},$$

with

$$\eta^{\pm} = \int_{\hbar\omega_{np} H(1\mp 1)}^{\infty} \sqrt{\mathcal{E}(1+\alpha\mathcal{E})}(1+2\alpha\mathcal{E}) \exp(-\lambda^{W(0)\mathcal{E}})\, \mathrm{d}\mathcal{E},$$

$$\mathcal{N}_{\pm} = \sqrt{(\mathcal{E} \pm \hbar\omega_{np})[1+\alpha(\mathcal{E} \pm \hbar\omega_{np})]}\big[1+2\alpha(\mathcal{E} \pm \hbar\omega_{np})\big],$$

and

$$\overline{K}_{np} = \frac{8\pi\sqrt{2}(m^*)^{3/2} K_{np}}{3\hbar^3}, \qquad K_{np} = Z_f \frac{(D_t K)^2}{8\pi^2 \rho\omega_{np}}.$$

H is the Heaviside function

$$H(x) = \begin{cases} 1 & \text{if } x > 0, \\ 0 & \text{otherwise.} \end{cases}$$

The coefficients of the production matrix $C^{(np)} = (c_{ij}^{(np)})$ are given by $c_{ij}^{(np)} = \sum_{A=1}^{6} c_{A_{ij}}^{(np)}$. For each valley one has $C^{(np)} = A^{(np)} B$, where the matrix $A^{(np)}$ has components (see ROMANO [2001])

$$a_{11}^{(np)} = \frac{\overline{K}_{np}}{d_0} \sum_{\pm} \left(n_B + \frac{1}{2} \mp \frac{1}{2}\right) \int_{\hbar\omega_{np} H(1\mp 1)}^{\infty} \mathcal{N}_{\pm} \mathcal{E}^{3/2} (1+\alpha\mathcal{E})^{3/2} \times \exp(-\lambda^{W(0)} \mathcal{E})\, \mathrm{d}\mathcal{E},$$

$$a_{12}^{(np)} = \frac{\overline{K}_{np}}{d_0} \sum_{\pm} \left(n_B + \frac{1}{2} \mp \frac{1}{2}\right) \int_{\hbar\omega_{np} H(1\mp 1)}^{\infty} \mathcal{N}_{\pm} \mathcal{E}^{5/2} (1+\alpha\mathcal{E})^{3/2} \times \exp(-\lambda^{W(0)} \mathcal{E})\, \mathrm{d}\mathcal{E},$$

$$a_{21}^{(np)} = \frac{\overline{K}_{np}}{m^* d_0} \sum_{\pm} \left(n_B + \frac{1}{2} \mp \frac{1}{2}\right) \int_{\hbar\omega_{np} H(1\mp 1)}^{\infty} \mathcal{N}_{\pm} \frac{\mathcal{E}^{5/2} (1+\alpha\mathcal{E})^{3/2}}{1+2\alpha\mathcal{E}} \times \exp(-\lambda^{W(0)} \mathcal{E})\, \mathrm{d}\mathcal{E},$$

$$a_{22}^{(np)} = \frac{\overline{K}_{np}}{m^* d_0} \sum_{\pm} \left(n_B + \frac{1}{2} \mp \frac{1}{2}\right) \int_{\hbar\omega_{np} H(1\mp 1)}^{\infty} \mathcal{N}_{\pm} \frac{\mathcal{E}^{7/2} (1+\alpha\mathcal{E})^{3/2}}{1+2\alpha\mathcal{E}} \times \exp(-\lambda^{W(0)} \mathcal{E})\, \mathrm{d}\mathcal{E}.$$

In order to speed up the computation, in the numerical code we do not evaluate τ_W and the coefficients c_{ij} at each time step by using the above formulas. Instead we calculate in advance a numerical table of the variables as functions of the energy W and during the simulation we determine particular values by interpolation. In Table 1.1 we report the values of the physical parameters used in the simulations. The coupling constants and the values of the energy phonons for each valley are reported in Table 1.2 (JACOBONI and REGGIANI [1983]).

The parabolic band limit of the closures for the production terms is recovered from the results obtained in the case of Kane dispersion relation as $\mathcal{E} \mapsto 0$ (see ANILE and

TABLE 1.1
Values of the physical parameters used for silicon

m_e	electron rest mass	9.1095×10^{-28} g
m^*	effective electron mass	$0.32 m_e$
T_L	lattice temperature	300 K
ρ	density	2.33 g/cm^3
v_s	longitudinal sound speed	9.18×10^5 cm/sec
Ξ_d	acoustic-phonon deformation potential	9 eV
α	nonparabolicity factor	0.5 eV^{-1}
ε_r	relative dielectric constant	11.7
ε_0	vacuum dieletric constant	8.85×10^{-18} C/V µm

TABLE 1.2
Coupling constants and phonon energies for the inelastic scatterings in silicon

A	Z_f	$\hbar\omega$ (meV)	$D_t K$ (10^8 eV/cm)
1	1	12	0.5
2	1	18.5	0.8
3	4	19.0	0.3
4	4	47.4	2.0
5	1	61.2	11
6	4	59.0	2.0

ROMANO [1999], ROMANO [2000] for more details). Concerning the fluxes one has

$$U_{ij}^P = \frac{2}{3} W \delta_{ij}, \qquad m^* F_{ij}^P = \frac{10}{9} W^2 \delta_{ij}, \qquad G_{ij} = \frac{1}{m^*}(U_{ij} + W\delta_{ij}).$$

Concerning the production terms one finds what follows.

For the acoustic phonon scattering we have

$$a_{11}^{(ac)} = \frac{32}{3} \frac{\sqrt{2\pi} K_{ac}}{\hbar^3} (m^*)^{3/2} \left(\frac{2}{3} W\right)^{3/2},$$

$$a_{12}^{(ac)} = 32 \frac{\sqrt{2\pi} K_{ac}}{\hbar^3} (m^*)^{3/2} \left(\frac{2}{3} W\right)^{5/2},$$

$$a_{21}^{(ac)} = \frac{a_{12}^{(ac)}}{m^*},$$

$$a_{22}^{(ac)} = 128 \frac{\sqrt{2\pi m^*} K_{ac}}{\hbar^3} \left(\frac{2}{3} W\right)^{7/2}.$$

For the nonpolar optical phonon scattering one obtains

$$C_W = \left(\frac{2}{3} W\right)^{-1/2} \frac{2\sqrt{2\pi} (m^*)^{3/2} (\hbar\omega_{np})^2}{\hbar^3} K_{np} \sum_{\pm} \left(n_B + \frac{1}{2} \mp \frac{1}{2}\right) e^{\pm\zeta}$$
$$\times \left[\exp\left(\pm \frac{\hbar\omega_{np}}{k_B T_L} \mp 2\zeta\right) - 1\right] \left[K_2(\zeta) \mp K_1(\zeta)\right],$$

$$a_{11}^{(np)} = \frac{4}{3}\left(\frac{2}{3}W\right)^{-1/2} \frac{\sqrt{2\pi}(m^*)^{3/2}(\hbar\omega_{np})^2}{\hbar^3} K_{np} \sum_{\pm}\left(n_B + \frac{1}{2} \mp \frac{1}{2}\right) e^{\pm\zeta}$$
$$\times \left[K_2(\zeta) \mp K_1(\zeta)\right],$$
$$a_{12}^{(np)} = \frac{4}{3}\sqrt{\frac{2}{3}W} \frac{\sqrt{2\pi}(m^*)^{3/2}(\hbar\omega_{np})^2}{\hbar^3} K_{np} \sum_{\pm}\left(n_B + \frac{1}{2} \mp \frac{1}{2}\right) e^{\pm\zeta}$$
$$\times \left\{3K_2(\zeta) + 2\zeta\left[K_1(\zeta) \mp K_2(\zeta)\right]\right\},$$
$$a_{21}^{(np)} = \frac{a_{12}^{(np)}}{m^*},$$
$$a_{22}^{(np)} = \frac{4}{3}\left(\frac{2}{3}W\right)^{3/2} \frac{\sqrt{2\pi m^*}(\hbar\omega_{np})^2}{\hbar^3} K_{np} \sum_{\pm}\left(n_B + \frac{1}{2} \mp \frac{1}{2}\right) e^{\pm\zeta}$$
$$\times \left[K_2(\zeta)(12 \mp 9\zeta + 4\zeta^2) + K_1(\zeta)(3\zeta \mp 4\zeta^2)\right],$$

with $\zeta = 3\hbar\omega_{np}/(4W)$ and

$$K_\nu = \frac{\sqrt{\pi}(z/2)^\nu}{\Gamma(\nu + \frac{1}{2})} \int_0^\infty \exp(-z\cosh t)\sinh^{2\nu} t \, \mathrm{d}t, \qquad z, \nu > 0,$$

the modified Bessel functions of second kind. Γ is the Gamma function.

1.5. *The formal properties of the hydrodynamical model*

In this section we will investigate (ROMANO [2000]) the formal properties of the system (1.37)–(1.40). We will prove that it forms a hyperbolic system in the physically relevant region of the space of the dependent variables.

Let us consider the quasilinear system of PDEs

$$\frac{\partial}{\partial t} F^{(0)}(\mathbf{U}) + \sum_{i=1}^{3} \frac{\partial}{\partial x_i} F^{(i)}(\mathbf{U}) = B(\mathbf{U}), \tag{1.51}$$

with

$$F: \Omega \mapsto \mathbb{R}^m$$

sufficiently smooth function and $\Omega \subset \mathbb{R}^m$. If we consider a smooth solution, we can introduce the Jacobian matrices

$$A^{(\beta)} = \nabla_{\mathbf{U}} F^{(\beta)}, \quad \beta = 0, 1, 2, 3.$$

We recall that the system (1.51) is said *hyperbolic in the t-direction* if $\det(A^{(0)}(\mathbf{U})) \neq 0$ and the eigenvalue problem

$$\det\left(\sum_{i=1}^{3} n_i A^{(i)}(\mathbf{U}) - \lambda A^{(0)}(\mathbf{U})\right) = 0 \tag{1.52}$$

has real eigenvalues and the eigenvectors span $\mathbb{R}^m$ for all unit vectors $\mathbf{n} = (n_1, n_2, n_3)$.

In the case of the system (1.37)–(1.40) we have (indeed it is computationally more convenient to substitute the equation for P_i with a linear combination of Eqs. (1.38) and (1.40) in order to have an equation for nV_i)

$$\mathbf{U}=\begin{pmatrix} n \\ \mathbf{V} \\ W \\ \mathbf{S} \end{pmatrix}, \qquad F^{(0)}=n\begin{pmatrix} 1 \\ m^*\mathbf{V} \\ W \\ \mathbf{S} \end{pmatrix}, \qquad F^{(i)}=n\begin{pmatrix} V_i \\ (U-2\alpha m^* F)e_i \\ S_i \\ nFe_i \end{pmatrix},$$
$$i=1,\dots,3,$$

where e_i is the ith column of the 3×3 identity matrix, and the Jacobian matrices are given by

$$A^{(0)}=\begin{pmatrix} 1 & 0 & 0 & 0 \\ m^*\mathbf{V} & m^*nI_3 & 0 & 0 \\ W & 0 & n & 0 \\ \mathbf{S} & 0 & 0 & nI_3 \end{pmatrix},$$
$$A^{(n)}=\sum_{i=1}^{3} n_i A^{(i)}=\begin{pmatrix} \mathbf{n}\cdot\mathbf{V} & n\mathbf{n}^{\mathrm{T}} & 0 & 0 \\ (U-2\alpha m^* F)\mathbf{n} & 0 & n(U'-2\alpha m^* F')\mathbf{n} & 0 \\ \mathbf{n}\cdot\mathbf{S} & 0 & 0 & n\mathbf{n}^{\mathrm{T}} \\ F\mathbf{n} & 0 & nF'\mathbf{n} & 0 \end{pmatrix},$$

where the prime denote partial derivation with respect to W.

Let us introduce the region $\widehat{\Omega}=\{\mathbf{U}\in\mathbb{R}^8\colon\ n>0,\ W>0\}$ and the functions

$$g_1(W)=\bigl(U+m^*F'-WU'+2\alpha m^*(WF'-F)\bigr)^2 - 4m^*(UF'-U'F), \tag{1.53}$$
$$g_2(W)=U+m^*F'-WU'+2\alpha m^*(WF'-F)-\sqrt{g_1(W)}, \tag{1.54}$$
$$g_3(W)=(U-2\alpha m^*F)F'-(U'-2\alpha m^*F')F. \tag{1.55}$$

In ROMANO [2001] the following algebraic lemma has been proved

PROPOSITION 1. *If the inequalities*

$$g_1(W)>0,\quad g_2(W)>0,\quad g_3(W)>0 \tag{1.56}$$

are satisfied, in the region $\widehat{\Omega}$ *the system* (1.37)–(1.40) *is hyperbolic and the eigenvalues are given by*

$$\lambda_{1,2,3,4}=0, \tag{1.57}$$
$$\lambda_{\pm\pm}=\pm\tfrac{\sqrt{2}}{2}\bigl\{U+m^*F'-WU'+2\alpha m^*(WF'-F) \pm\bigl[\bigl(U+m^*F'-WU'+2\alpha m^*(WF'-F)\bigr)^2 - 4m^*(UF'-U'F)\bigr]^{1/2}\bigr\}^{1/2}. \tag{1.58}$$

We remark that in the one-dimensional case the system becomes strictly hyperbolic with eigenvalues $\lambda_{\pm\pm}$.

Now let us check the conditions (1.56). In the parabolic band limit one has

$$g_1(W) = \frac{160}{81}W^2, \quad g_2(W) = \frac{4}{9}(5 - \sqrt{10})W, \quad g_3(W) = \frac{20}{27}W^2$$

and the conditions (1.56) are trivially satisfied in $\widehat{\Omega}$. The eigenvalues are

$$\lambda_{1,2,3,4} = 0 \quad \text{and} \quad \lambda_{\pm\pm} = \pm\sqrt{(10 \pm 2\sqrt{10})W}. \tag{1.59}$$

In the case of the Kane dispersion relation we have numerically evaluated the function $g_1(W)$, $g_2(W)$ and $g_3(W)$ for the range of values of W typically encountered in the electron devices. Fig. 1.1 shows that the relations (1.56) are satisfied also in the nonparabolic case.

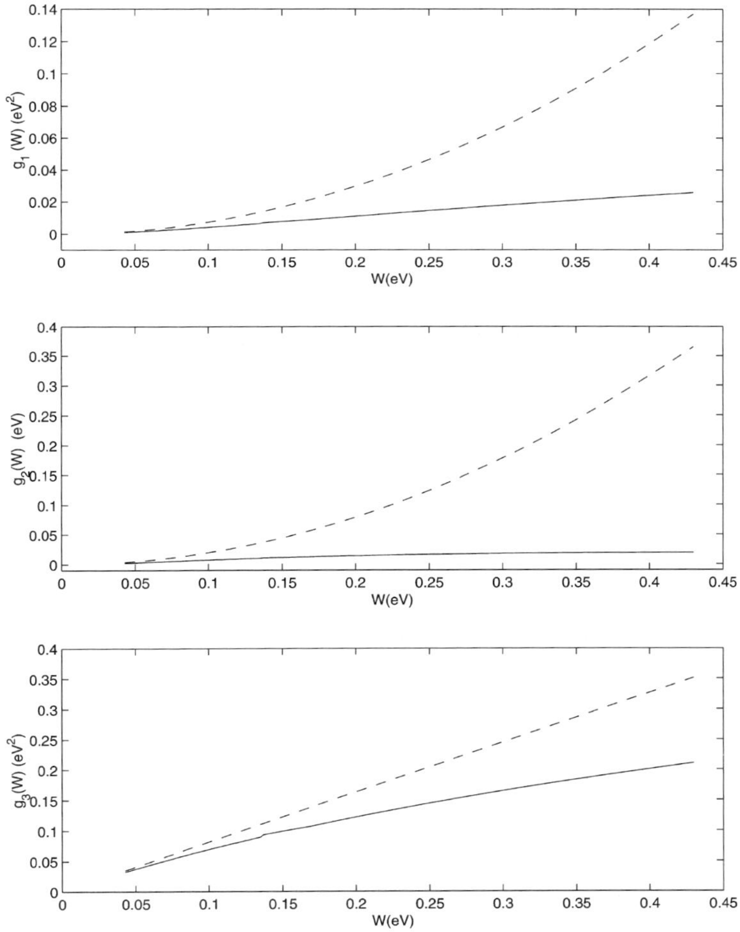

FIG. 1.1. The functions $g_1(W)$, $g_2(W)$ and $g_3(W)$ versus the energy W (eV) in the parabolic case (dashed line) and for the Kane dispersion relation (continuous line).

Therefore we can conclude that *at least for the values of W of practical interest the system* (1.37)–(1.40) *is hyperbolic*.

2. Numerical methods

The numerical solution of hydrodynamical models is a delicate problem, and it requires suitable techniques. Hydrodynamical models have the mathematical structure of hyperbolic systems with a source term of the form given by Eq. (1.51).

The source term $B(\mathbf{U})$ generically represents relaxation terms and drift terms containing the electric field. The latter is related to the charge density through the Poisson equation.

In some reduced hydrodynamical models, the term $B(\mathbf{U})$ may contain second order derivatives, describing diffusion and shear stress effects, and therefore the system has a degenerate parabolic character.

In spite of its mixed character, however, the system is dominantly hyperbolic, and the numerical techniques appropriate for its discretization are derived from those used in the context of hyperbolic systems of conservation laws.

In this section we focus our attention on the description of modern shock capturing methods for conservation laws, in one and several space dimensions. At the end of the section we show how these methods can be combined with Poisson solvers and with a proper treatment of the source, in order to obtain accurate numerical solutions of systems of the form (1.51).

Applications will be presented in the next section.

2.1. General description

Let us consider system (1.51). Let us choose the conservative variables as unknown field vector. Then the system writes

$$\frac{\partial}{\partial t}\mathbf{U} + \sum_{i=1}^{3} \frac{\partial}{\partial x_i} F_i(\mathbf{U}) = B(\mathbf{U}), \tag{2.1}$$

Several strategies can be used to discretize with respect to time.

The simplest one is a fractional step method. The equation is subdivided into simpler steps, each of which can be solved separately. This approach has the advantage of being modular, and therefore each step can be solved with the most appropriate technique. Given the numerical solution at time step n, $\mathbf{U}^n$, the solution after one time step Δt can be obtained by solving the two following steps

$$\frac{\partial \widetilde{\mathbf{U}}}{\partial t} = B(\widetilde{\mathbf{U}}), \qquad \widetilde{\mathbf{U}}(0) = \mathbf{U}^n, \tag{2.2}$$

$$\frac{\partial \widehat{\mathbf{U}}}{\partial t} + \sum_{i=1}^{3} \frac{\partial}{\partial x_i} F_i(\widehat{\mathbf{U}}) = 0, \qquad \widehat{\mathbf{U}}(0) = \widetilde{\mathbf{U}}(\Delta t) \tag{2.3}$$

and then assign $\mathbf{U}^{n+1} = \widehat{\mathbf{U}}(\Delta t)$. This splitting strategy is called simple splitting, and it is first order accurate in time, even if the two steps are solved exactly. A better splitting strategy is the so-called Strang splitting (STRANG [1968]), which guarantees second order accuracy, if both steps are not stiff, and if each of them is solved with at least second order accuracy. Strang splitting consists in solving Eq. (2.2) for half time step. Then Eq. (2.3) is solved for a time step Δt, and finally Eq. (2.2) is solved again for a time step $\Delta t/2$. A different second order splitting scheme will be considered later, when we discuss the applications.

As an alternative to splitting, nonsplitting schemes will be also presented in Section 2.8.

The convection step consists in the solution of a hyperbolic system of conservation laws. The solution of such system may develop discontinuities, such as shocks, in finite time. Suitable methods, known as *shock capturing methods*, have been developed for the numerical approximation of conservation laws.

2.2. *Shock capturing schemes*

Solutions of conservation laws may develop jump discontinuities in finite time. To understand how to obtain numerical approximations that converge to the (discontinuous) solution has been a nontrivial task. The mathematical theory of a quasilinear hyperbolic systems of conservation laws has been used as a guideline in the development of modern numerical schemes for conservation laws (LAX [1973]). Such schemes can be divided into two broad classes: front tracking and shock capturing schemes. In front tracking schemes the surface of discontinuity is computed explicitly, and the evolution of the field on the two sides of the surface is followed by computing the flow in the smooth regions, with additional sets of boundary conditions on the surface. It is not necessary to compute derivatives of the field across the discontinuity surface, and therefore the problem of spurious oscillations across it is overcome. Although such schemes can provide an accurate description of the shock motion, they are not very popular, since in general they require some “a priori” knowledge of the flow, and they are more complicated to treat, especially for complex flows (CHERN, GLIMM, MCBRYAN, PLOHR and YANIV [1986]).

In shock capturing schemes, on the contrary, the location of the discontinuity is *captured* automatically by the scheme as a part of the solution where sharp fronts develop.

Although shock capturing schemes for conservation laws may be based on finite element methods (see, for example, the review papers by C. Johnson and by B. Cockburn in COCKBURN, JOHNSON, SHU and TADMOR [1998] or the recent book COCKBURN, KARNIADAKIS and SHU [2000] and references therein), most shock capturing schemes are based on finite volume or finite difference discretization in space.

A good introductory book which deals with wave propagation and shock capturing schemes (mainly upwind schemes, in one space dimension) is the book by LEVEQUE [1990]. A mathematically oriented reference book on the numerical solutions of conservation laws is the book by GODLEWSKI and RAVIART [1996]. Several schemes, mainly based on Riemann solver, with a lot of numerical examples are considered in the book by TORO [1999]. A good review of modern numerical techniques for the treatment of hyperbolic systems of conservation laws is given in the lecture notes of a CIME course held in 1998 (COCKBURN, JOHNSON, SHU and TADMOR [1998]).

Most schemes for the numerical solution of conservation laws are based on the Godunov scheme, and on the numerical solution of the Riemann problem (LAX [1973]). For numerical purpose it is often more convenient to resort to approximate solvers, such as the one proposed by Roe for gas dynamics (ROE [1981]). Although such solution or its numerical approximation is known in many cases of physical relevance, in the case of hydrodynamical models of semiconductors the eigenvalues and eigenvectors of the matrix of the system are not known analytically for some models, and even if they are known, the solution to the Riemann problem or its numerical approximation is hard to compute. In such cases it is desirable to use schemes that do not require the knowledge of the solution of the Riemann problem.

We call such schemes "central schemes". The prototype central scheme is Lax–Friedrichs scheme. It is well known that Lax–Friedrichs scheme is more dissipative than first order upwind scheme, however it is simpler to use, since it does not require the knowledge of the sign of the flux derivative or the eigenvector decomposition of the system matrix (see, for example, LEVEQUE [1990] for a comparison between Lax–Friedrichs and upwind schemes).

Second order central schemes have been introduced in NESSYAHU and TADMOR [1990] and SANDERS and WEISER [1989]. After that, central schemes have developed in several directions. We mention here the improvement of second order central scheme and the development of semidiscrete central scheme in one space dimension (KURGANOV and TADMOR [2000]), the development of high order central schemes in one space dimension (LIU and TADMOR [1998], BIANCO, PUPPO and RUSSO [1999], LEVY, PUPPO and RUSSO [2001]), central schemes in several space dimensions on rectangular grids (ARMINJON and VIALLON [1995], ARMINJON, VIALLON and MADRANE [1997], JIANG and TADMOR [1998], LEVY, PUPPO and RUSSO [2000], LEVY, PUPPO and RUSSO [2001]), and on unstructured grids (ARMINJON and VIALLON [1999]), the development of central schemes to hyperbolic systems with source term (BEREUX and SAINSAULIEU [1997], LIOTTA, ROMANO and RUSSO [1999], LIOTTA, ROMANO and RUSSO [2000], PARESCHI [2001]). Central schemes have been applied to a variety of different problems. Here we mention the application to the hydrodynamical models of semiconductors (ANILE, JUNK, ROMANO and RUSSO [2000], ANILE, NIKIFORAKIS and PIDATELLA [1999], ANILE, JUNK, ROMANO and RUSSO [2000], ROMANO and RUSSO [2000], TROVATO and FALSAPERLA [1998]).

Closely related to the above mentioned central schemes are finite volume and finite difference schemes which do not make use of the characteristic structure of the system, or which do not require (exact or approximate) Riemann solvers. Such schemes are, for example, conservative schemes which use the local Lax–Friedrichs flux function. High order schemes of this type have been developed mainly by Shu, using Essentially Non-Oscillatory (ENO) and Weighted Essentially Non-Oscillatory space discretization (COCKBURN, JOHNSON, SHU and TADMOR [1998] and references therein, and JIANG and SHU [1996]). The semi-discrete central scheme developed by Kurganov and Tadmor for conservation laws is equivalent to a the second order finite-volume scheme which uses a local Lax–Friedrichs flux function. ENO and WENO type schemes have been used in the context of hydrodynamical models of semiconductors (see, for example, FATEMI, JEROME and OSHER [1991], JEROME and SHU [1994]), however the structure of the hyperbolic part of the model they considered was very similar to that of compressible Euler equations of gas dynamics, and therefore Riemann-based schemes could be used.

Finally we mention that a comparison between modern finite element, finite difference and finite volume schemes can be found in the papers SHU [2001], ZHOU, LI and SHU [2001].

2.3. *Conservative schemes*

We start by considering conservation laws in one space dimension of the form

$$\frac{\partial u}{\partial t} + \frac{\partial f(u)}{\partial x} = 0. \tag{2.4}$$

Let us discretize the equation by dividing space into cells $I_j = [x_{j-1/2}, x_{j+1/2}]$, and time in discrete levels t_n. In finite difference schemes, the unknown is an approximation of the pointwise value of the field u at the center of the cells. In finite volume schemes, the unknown represents an approximation of the cell average of the field:

$$\bar{u}_j^n \approx \frac{1}{\Delta x_j} \int_{x_{j-1/2}}^{x_{j+1/2}} u(x, t_n) \Delta x.$$

Finite difference schemes are more efficient for multidimensional computation, when high order accuracy is required. On the other hand, they require a regular grid with constant (or at least smoothly varying) grid spacing. Finite volume schemes, on the other hand, are very flexible, and they can be implemented on unstructured grids. We shall mainly consider here finite volume schemes.

For simplicity we assume that the cells are all of the same size $\Delta x = h$, so that the center of cell j is $x_j = x_0 + jh$. This assumption is not necessary for finite volume schemes.

Integrating the conservation law over a cell in space–time $I_j \times [t_n, t_{n+1}]$ (see Fig. 2.1) one has

$$\begin{aligned} \int_{x_{j-1/2}}^{x_{j+1/2}} u(x, t_{n+1}) = & \int_{x_{j-1/2}}^{x_{j+1/2}} u(x, t_n) \\ & - \int_{t_n}^{t_{n+1}} \big(f\big(u(x_{j+1/2}, t)\big) - f\big(u(x_{j-1/2}, t)\big)\big)\, \mathrm{d}t. \end{aligned} \tag{2.5}$$

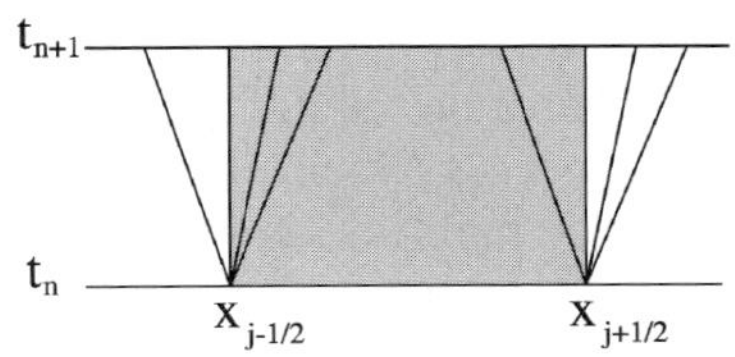

FIG. 2.1. Integration over a cell and Godunov methods.

This (exact) relation suggests the use of numerical scheme of the form

$$\bar{u}_j^{n+1} = \bar{u}_j^n - \frac{\Delta t}{h}(F_{j+1/2} - F_{j-1/2}), \tag{2.6}$$

where $\bar{u}_j^n$ denotes an approximation of the cell average of the solution on cell j at time t_n, and $F_{j+1/2}$, which approximates the integral of the flux on the boundary of the cell, is the so-called *numerical flux*, and depends on the cell average of the cells surrounding point $x_{j+1/2}$. In the simplest case it is

$$F_{j+1/2} = F(\bar{u}_j, \bar{u}_{j+1})$$

with $F(u, u) = f(u)$ for consistency. Such schemes, called conservative schemes, have the properties that they satisfy a conservation property at a discrete level. This is essential in providing the correct propagation speed for discontinuities, which depends uniquely on the conservation properties of the system.

Furthermore, Lax–Wendroff theorem (LAX and WENDROFF [1960]) ensures that if $u(x, t)$ is the limit of a sequence of discrete solutions $\bar{u}_j^n$ of a consistent conservative scheme, obtained as the discretization parameter h vanishes, then $u(x, t)$ is a weak solution of the original equation.

Lax–Wendroff theorem assumes that the sequence of numerical solutions converges strongly to a function $u(x, t)$. Convergence of numerical schemes is studied through the TVD (Total Variation Diminishing) property. A discrete entropy condition is used to guarantee that the numerical solution converges to the unique entropic weak solution of Eq. (2.4) (in the scalar case). For a discussion on these issues see the book by LEVEQUE [1990] and GODLEWSKI and RAVIART [1996].

2.4. *Godunov scheme*

The numerical flux function identifies the conservative scheme. A class of widely developed methods is that of *upwind schemes* (see, for example, LEVEQUE [1990]), in which the numerical flux takes into account the characteristic structure of the system.

The prototype of upwind schemes for conservation laws is the Godunov scheme. It is based on two fundamental ideas. The first is that the solution is reconstructed from cell averages at each time step as a piecewise polynomial in x. In its basic form, the solution is reconstructed as a piecewise constant function

$$u(x, t_n) \approx R(x; \bar{u}^n) = \sum_j \bar{u}_j^n \chi_j(x), \tag{2.7}$$

where $\chi_j(x)$ is the indicator function of interval $I_j = [x_{j-1/2}, x_{j+1/2}]$. The second is that for a piecewise constant function, the solution of the system, for short time, can be computed as the solution of a sequence of Riemann problems.

A Riemann problem is a Cauchy problem for a system of conservation laws, where the initial condition is given by two constant vectors separated by a discontinuity

$$u(x, 0) = \begin{cases} u_- & x < 0, \\ u_+ & x > 0. \end{cases} \tag{2.8}$$

For the scalar equation the solution to the Riemann problem is known analytically. For a system of conservation laws, the solution consists of a similarity solution that depends on the variable x/t. In several cases of interest, such as gas dynamics with polytropic gas, the solution to the Riemann problem is known analytically (LAX [1973]).

Sometimes, for efficiency reason, it is more convenient to use approximate solutions to the Riemann problem (ROE [1981]).

Once the solution to the Riemann problem is known, it can be used for the construction of the Godunov scheme.

Let us denote by $u^*(u_-, u_+)$ the solution of the Riemann problem at $x = 0$. Then the *exact* solution of Eq. (2.4) can be computed from Eq. (2.5):

$$\bar{u}_j^{n+1} = \bar{u}_j^n - \frac{\Delta t}{\Delta x}\big(f\big(u^*(\bar{u}_{j-1}, \bar{u}_j)\big) - f\big(u^*(\bar{u}_j, \bar{u}_{j+1})\big)\big). \tag{2.9}$$

This relation is exact if the Riemann fan does not reach the boundary of the cell (see Fig. 2.1), i.e., if the following CFL condition (COURANT, FRIEDRICHS and LEWY [1967]) is satisfied

$$\Delta t < \frac{\Delta x}{\rho(A)}, \tag{2.10}$$

where $\rho(A) = \max_{1 \leqslant i \leqslant d} |\lambda_i(A)|$ is the spectral radius of the Jacobian matrix $A = \nabla_u f$, λ_i denoting the ith eigenvalue.

If this condition is not satisfied then oscillatory instabilities develop.

Once the cell averages are computed at the new time t_{n+1}, then the solution at this time is again approximated by a piecewise constant solution of the form (2.7).

Godunov scheme is first order accurate, it is Total Variation Diminishing, and it satisfies a discrete entropy inequality. When applied to a linear system, Godunov method is equivalent to first order upwind scheme (see LEVEQUE [1990]).

Higher order version of Godunov scheme can be constructed. They are based on high order nonoscillatory reconstruction and on the solution to the generalized Riemann problem.

High order nonoscillatory reconstruction is a crucial step. It can be obtained by using either ENO or WENO techniques. We shall briefly mention them later.

For the moment, we assume we are able to compute such reconstruction of the form

$$u(x, t_n) \approx R(x; \bar{u}^n) = \sum_j R_j(x)\chi_j(x). \tag{2.11}$$

Then high order Godunov-type schemes are obtained by solving the system

$$\frac{\mathrm{d}\bar{u}_j}{\mathrm{d}t} = -\frac{f(u^*(u_{j+1/2}^-, u_{j+1/2}^+)) - f(u^*(u_{j-1/2}^-, u_{j-1/2}^+))}{h}, \tag{2.12}$$

where $u_{j+1/2}^- = R_j(x_{j+1/2})$, $u_{j+1/2}^+ = R_{j+1}(x_{j+1/2})$. Because the values $u_{j+1/2}^-$ and $u_{j+1/2}^+$ depend on the reconstruction, which depends on the cell averages, it turns out that system (2.13) is a system of ordinary differential equations for the evolution of cell averages.

Such system may be solved by a suitable ODE solver, for example, a Runge–Kutta method, which maintains the accuracy of the spatial discretization (see the paper by Shu in COCKBURN, JOHNSON, SHU and TADMOR [1998] and references therein).

These methods are based on the (exact or approximate) solution to the Riemann problem. Such solution is not always available or inexpensive. As an alternative to these high order extension of the Godunov method, simpler schemes can be constructed, which make use of less expensive numerical flux function.

The general structure of such schemes is given by

$$\frac{\mathrm{d}\bar{u}_j}{\mathrm{d}t} = -\frac{F(u^-_{j+1/2}, u^+_{j+1/2}) - F(u^-_{j-1/2}, u^+_{j-1/2})}{h}, \tag{2.13}$$

where $u^-_{j+1/2}$ and $u^+_{j+1/2}$ are defined as above, and the numerical flux function $F(u^-, u^+)$ defines the scheme. The simplest choice of the numerical flux function is the so-called Local Lax–Friedrichs flux:

$$F(u^-, u^+) = \tfrac{1}{2}\big(f(u^-) + f(u^+) - \alpha(u^+ - u^-)\big), \tag{2.14}$$

where $\alpha = \max(\rho(A(u^-)), \rho(A(u^+)))$, and $\rho(A)$ denotes the spectral radius of matrix A. The advantage of the local Lax–Friedrichs flux is that it does not require the knowledge of the solution to the Riemann problem, nor the exact knowledge of the eigenvalues and eigenvectors of the Jacobian matrix. Only an estimate of the largest eigenvalue is needed.

The disadvantage of this flux with respect to the Riemann solver is that it introduces a larger numerical dissipation.

Other flux functions are available. Common requirements that they have to satisfy are the following: they have to be

(i) locally Lipschitz continuous in both argument;
(ii) nondecreasing in the first argument and nonincreasing in the second argument (symbolically $F(\uparrow, \downarrow)$);
(iii) consistent with the flux function, i.e., $F(u, u) = f(u)$.

Popular flux functions (for scalar equation), besides the local Lax–Friedrichs described above, are the Godunov flux

$$F(a, b) = \begin{cases} \min_{a \leqslant u \leqslant b} f(u) & \text{if } a \leqslant b, \\ \max_{b \leqslant u \leqslant a} f(u) & \text{if } a > b \end{cases}$$

and the Engquist–Osher flux (HARTEN, ENGQUIST, OSHER and CHAKRAVARTHY [1987]):

$$F(a, b) = \int_0^a \max\big(f'(u), 0\big)\,\mathrm{d}u + \int_0^b \min\big(f'(u), 0\big)\,\mathrm{d}u + f(0).$$

Godunov flux is the least dissipative, and the Lax–Friedrichs the most dissipative among the three.

Notice, however, that the numerical dissipation is proportional to the jump $u^+ - u^-$, which is extremely small for high order schemes and smooth solution. For a scheme of order p, in fact, it is $u^+ - u^- = \mathrm{O}(h^p)$. Therefore the numerical dissipation becomes

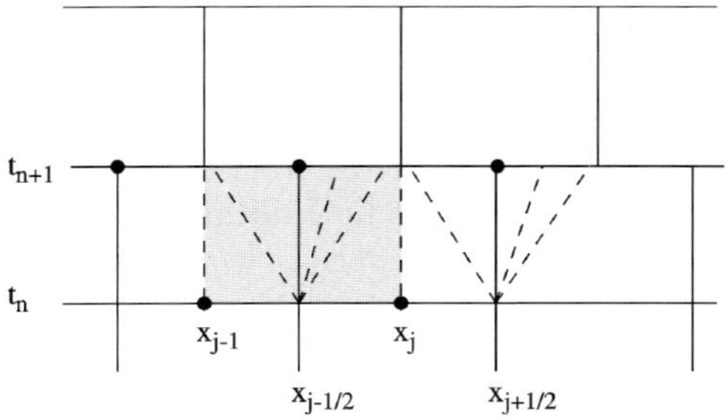

FIG. 2.2. Integration on a staggered grid.

large only near discontinuities, i.e. where it is most needed. As a result, the difference between exact Riemann flux and local Lax–Friedrichs flux is very pronounced for low order schemes, but it is not so dramatic for very high order schemes (COCKBURN, JOHNSON, SHU and TADMOR [1998]).

Another family of schemes, which are intrinsically central, can be derived as follows. Let us integrate Eq. (2.1) on a staggered grid, as shown in Fig. 2.2.

Integrating on the staggered grid one obtains

$$\int_{x_j}^{x_{j+1}} u(x, t_{n+1}) = \int_{x_j}^{x_{j+1}} u(x, t_n) - \int_{t_n}^{t_{n+1}} \big(f\big(u(x_{j+1}, t)\big) - f\big(u(x_j, t)\big)\big)\, \mathrm{d}t. \tag{2.15}$$

Once again, this formula is *exact*. In order to convert it into a numerical scheme one has to approximate the staggered cell average at time t_n, and the time integral of the flux on the border of the cells.

Let us assume that the function $u(x, t_n)$ is reconstructed form cell averages as a piecewise polynomial function. Then the function is smooth at the center of the cell and its discontinuities are at the edge of the cell. If we integrate the equation on a staggered cell, then there will be a fan of characteristics propagating from the center of the staggered cell, while the function on the edge (dashed vertical lines in the figure) will remain smooth, provided the characteristic fan does not intersect the edge of the cell, i.e. provided a suitable CLF condition of the form

$$\Delta t < \frac{\Delta x}{2\rho(A)} \tag{2.16}$$

is satisfied.

The simplest central scheme is obtained by piecewise constant reconstruction of the function, and by using a first order quadrature rule in the evaluation of the integrals. The resulting scheme is

$$u^n_{j+1/2} = \tfrac{1}{2}\big(u^n_j + u^n_{j+1}\big) - \lambda\big(f(u^n_{j+1}) - f(u^n_j)\big), \tag{2.17}$$

where $\lambda = \Delta t/\Delta x$ denotes the mesh ratio. Such scheme is just Lax–Friedrichs scheme on a staggered grid.

2.5. *The Nessyahu–Tadmor central scheme*

A second order scheme is obtained by using a piecewise linear approximation for the reconstruction of the function, and a second order quadrature rule (for example, the midpoint rule) for the computation of the time integral of the flux on the edges of the cell.

Such scheme has been proposed by NESSYAHU and TADMOR [1990], and independently by SANDERS and WEISER [1989].

The staggered cell average is given by

$$u^n_{j+1/2} = \frac{1}{h}\int_{x_j}^{x_{j+1}} R(x; u^n)\,\mathrm{d}x = \frac{1}{h}\left(\int_{x_{j+1/2}}^{x_{j+1}} L_j(x)\,\mathrm{d}x + \int_{x_j}^{x_{j+1/2}} L_{j+1}(x)'\,\mathrm{d}x\right)$$
$$= \frac{1}{2}\left(u^n_j + u^n_{j+1}\right) - \frac{1}{8}\left(u'_{j+1} - u'_j\right),$$

where $L_j(x) = u^n_j + u'_j(x - x_j)/h$ is the linear reconstruction in cell I_j. Here u'_j/h denotes a first order approximation of the derivative of the function in the cell. The value of the field u at the node of the midpoint rule, $u^{n+1/2}_j$, can be computed by first order Taylor expansion, which is equivalent to forward Euler scheme,

$$u^{n+1/2}_j = u^n_j - (\lambda/2) f'_j.$$

Here f'_j/h denotes a first order approximation of the space derivative of the flux.

In order to prevent spurious oscillations in the numerical solution, it is essential that these derivatives are computed by using a suitable *slope limiter*. Several choices are possible for the slope limiter. The simplest one is the MinMod limiter. It is defined according to

$$\mathrm{MM}(a, b) = \begin{cases} \min(a, b) & \text{if } a < 0 \text{ and } b < 0, \\ \max(a, b) & \text{if } a > 0 \text{ and } b > 0, \\ 0 & \text{if } ab < 0. \end{cases}$$

Such simple limiter, however, degrades the accuracy of the scheme near extrema. A better limiter is the so-called UNO (Uniform Non-Oscillatory) limiter, proposed by HARTEN, ENGQUIST, OSHER and CHAKRAVARTHY [1987].

Such limiter can be written as

$$u'_j = \mathrm{MM}\left(d_{j-1/2} + \tfrac{1}{2}\mathrm{MM}(D_{j-1}, D_j), d_{j+1/2} - \tfrac{1}{2}\mathrm{MM}(D_j, D_{j+1})\right), \tag{2.18}$$

where

$$d_{j+1/2} = u_{j+1} - u_j, \qquad D_j = u_{j+1} - 2u_j + u_{j-1}.$$

Other limiters are possible (see NESSYAHU and TADMOR [1990]).

The quantity f'_j can be computed either by applying a slope limiter to $f(\mathbf{u}^n_j)$ or by using the relation

$$f'_j = A\left(u^n_j\right)u'_j.$$

After one time step, one finds the solution $\{u^{n+1}_{j+1/2}\}$ on the staggered cells. Then one repeats a similar step, and after the second step, the solution at time t_{n+2}, $\{u^{n+1}_{j}\}$ is determined on the original grid.

Theoretical properties of the scheme, such as TVD property and the so-called "cell entropy inequality" are discusses in NESSYAHU and TADMOR [1990].

2.6. *Multidimensional central schemes*

Central schemes can be extended to problems in several dimensions. Second order central schemes on rectangular grids have been considered by JIANG and TADMOR [1998], SANDERS and WEISER [1989], and by ARMINJON and VIALLON [1995], ARMINJON, VIALLON and MADRANE [1997]. They have been extended to unstructured grids by ARMINJON and VIALLON [1999]. High order central schemes in two dimensions have been considered in the papers LEVY, PUPPO and RUSSO [1999], LEVY, PUPPO and RUSSO [2000], LEVY, PUPPO and RUSSO [2001].

Consider the two-dimensional system of conservation laws

$$u_t + f(u)_x + g(u)_y = 0, \tag{2.19}$$

subject to the initial values

$$u(x, y, t = 0) = u_0(x, y),$$

and to boundary conditions, which we do not specify at this point. The flux functions f and g are smooth vector valued functions, $f, g : \mathbb{R}^d \to \mathbb{R}^d$. The system (2.19) is assumed to be hyperbolic in the sense that for any unit vector $(n_x, n_y) \in \mathbb{R}^2$, the matrix $n_x \nabla_u f + n_y \nabla_u g$ has real eigenvalues and its eigenvectors form a basis of $\mathbb{R}^d$. In order to integrate numerically (2.19), we introduce a rectangular grid which for simplicity will be assumed to be uniform with mesh sizes $h = \Delta x = \Delta y$ in both directions. We will denote by $I_{i,j}$ the cell centered around the grid point $(x_i, y_j) = (i\Delta x, j\Delta y)$, i.e., $I_{i,j} = [x_i - h/2, x_i + h/2] \times [y_j - h/2, y_j + h/2]$. Let Δt be the time step and denote by $u^n_{i,j}$ the approximated point-value of the solution at the (i, j)th grid point at time $t^n = n\Delta t$. Finally, let $\bar{u}^n_{i,j}$ denote the cell average of a function u evaluated at the point (x_i, y_j),

$$\bar{u}^n_{i,j} = \frac{1}{h^2} \int_{I_{i,j}} u(x, y, t^n)\, \mathrm{d}x\, \mathrm{d}y.$$

Given the cell-averages $\{\bar{u}^n_{i,j}\}$ at time t^n, Godunov-type methods provide the cell-averages at the next time-step, t^{n+1}, in the following way: first, a piecewise-polynomial reconstruction is computed from the data $\{\bar{u}^n_{i,j}\}$ resulting with

$$u^n(x, y) = \sum_{i,j} R_{i,j}(x, y) \chi_{i,j}(x, y). \tag{2.20}$$

Here, $R_{i,j}(x, y)$ is a suitable polynomial (which has to satisfy conservation, accuracy and nonoscillatory requirements), while $\chi_{i,j}(x, y)$ is the characteristic function of the cell $I_{i,j}$. Thus, in general, the function $u^n(x, y)$ will be discontinuous along the boundaries of each cell $I_{i,j}$.

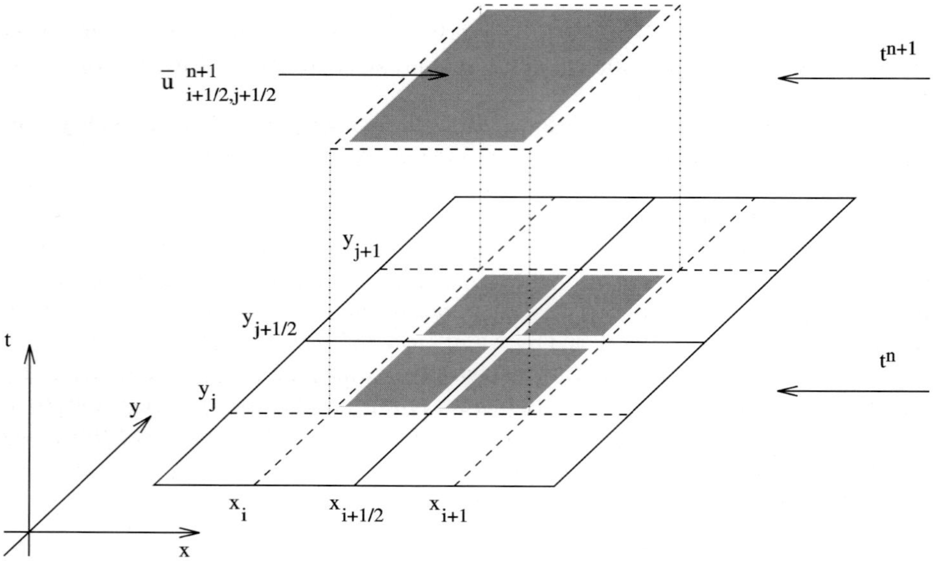

FIG. 2.3. The two-dimensional stencil.

In order to proceed, the reconstruction, $u^n(x, y)$, is evolved according to some approximation of (2.19) for a time step Δt. We will use the fact that the solution remains smooth at the vertical edges of the staggered control volume, $I_{i+1/2,j+1/2} \times [t^n, t^{n+1}]$, provided that the time-step Δt satisfies the CFL condition

$$\Delta t < \frac{h}{2} \frac{1}{\max(|\sigma_x|, |\sigma_y|)}.$$

Here, $I_{i+1/2,j+1/2} = [x_i, x_{i+1}] \times [y_j, y_{j+1}]$ (see Fig. 2.3; the edges at which the solution remains smooth are denoted by dotted vertical lines), and σ_x and σ_y are the largest (in modulus) eigenvalues of the Jacobian of f and g, respectively.

An exact integration of the system (2.19) with data $u^n(x, y)$ over the control volume $I_{i+1/2,j+1/2} \times [t^n, t^{n+1}]$ results with

$$\begin{aligned}
\bar{u}^{n+1}_{i+\frac{1}{2},j+\frac{1}{2}} = {} & \frac{1}{h^2} \iint_{I_{i+\frac{1}{2},j+\frac{1}{2}}} u^n(x, y)\,\mathrm{d}x\,\mathrm{d}y \\
& - \frac{1}{h^2} \int_{\tau=t^n}^{t^{n+1}} \left\{ \int_{y=y_j}^{y_{j+1}} \left[f\big(u(x_{i+1}, y, \tau)\big) - f\big(u(x_i, y, \tau)\big) \right] \mathrm{d}y \right\} \mathrm{d}\tau \\
& - \frac{1}{h^2} \int_{\tau=t^n}^{t^{n+1}} \left\{ \int_{x=x_i}^{x_{i+1}} \left[g\big(u(x, y_{j+1}, \tau)\big) - g\big(u(x, y_j, \tau)\big) \right] \mathrm{d}x \right\} \mathrm{d}\tau.
\end{aligned} \tag{2.21}$$

The first integral on the RHS of (2.21) is the cell-average of the function $u^n(x, y)$ on the staggered cell $I_{i+1/2,j+1/2}$. Given the reconstructed function $u^n(x, y)$, (2.20), this term can be computed exactly: it will consist of a contribution of four terms, resulting from averaging $R_{i+1,j+1}(x,y)$, $R_{i,j+1}(x,y)$, $R_{i+1,j}(x,y)$, and $R_{i,j}(x,y)$, on the corresponding quarter-cells.

The advantage of the central framework appears in the evaluation of the time integrals appearing in (2.21). Since the solution remains smooth on the segments $(x_i, y_j) \times [t^n, t^{n+1}]$, we can evaluate the time integrals with a quadrature rule using only nodes lying in these segments.

A second order scheme is obtained by approximating the integral of the flux f as

$$\int_{\tau=t^n}^{t^{n+1}} \int_{y=y_j}^{y_{j+1}} f(x_i, y, \tau)\,\mathrm{d}y\,\mathrm{d}\tau \approx \frac{h\Delta t}{2}\big(f(u_{i,j}^{n+1/2}) + f(u_{i,j+1}^{n+1/2})\big) \tag{2.22}$$

and likewise for the integral of the flux g. By applying the same discretization used for the Nessyahu–Tadmor scheme, one obtains its two-dimensional counterpart, which has been introduced in ARMINJON and VIALLON [1999] and, independently, in JIANG and TADMOR [1998]. First, in each cell (i, j) the field u is reconstructed by a piecewise linear approximation,

$$L_{i,j}(x,y) = u_{i,j}^n + u'_{i,j}\frac{x - x_i}{h} + u'_{i,j}\frac{y - y_j}{h},$$

where u'/h and u'/h denote, respectively, first order approximations of x- and y-partial derivatives, and can be computed by using a suitable slope limiter, as in the one-dimensional case. The resulting scheme has a compact form similar to the one-dimensional one, and can be written as

$$\begin{aligned} u_{i+1/2,j+1/2}^{n+1} = u_{i+1/2,j+1/2}^{n} &- \frac{\lambda}{2}\big(f(u_{i+1,j}^{n+1/2}) + f(u_{i+1,j+1}^{n+1/2}) - f(u_{i,j}^{n+1/2}) - f(u_{i,j+1}^{n+1/2})\big) \\ &- \frac{\lambda}{2}\big(g(u_{i,j+1}^{n+1/2}) + g(u_{i+1,j+1}^{n+1/2}) - g(u_{i,j}^{n+1/2}) - g(u_{i+1,j}^{n+1/2})\big), \end{aligned} \tag{2.23}$$

where $\lambda = \Delta t/h$, and

$$\begin{aligned} u_{i+1/2,j+1/2}^{n} = \frac{1}{4}\big(u_{i,j}^n + u_{i+1,j}^n + u_{i+1,j+1}^n + u_{i,j+1}^n\big) + \frac{1}{16}\big(u'_{i,j} - u'_{i+1,j} \\ + u'_{i,j+1} - u'_{i+1,j+1} + u'_{i,j} - u'_{i,j+1} + u'_{i+1,j} - u'_{i+1,j+1}\big) \end{aligned} \tag{2.24}$$

and the predictor values are evaluated as

$$u_{i,j}^{n+1/2} = u_{i,j}^n - \frac{\lambda}{2}f'_{i,j} - \frac{\lambda}{2}f'_{i,j}. \tag{2.25}$$

Once again, the first order approximation $f'_{i,j}$ and $g'_{i,j}$ can be computed either by a slope limiter acting on $f(u_{i,j})$ and $g(u_{i,j})$ or by

$$f' = A(u_{i,j})u'_{i,j}, \qquad g' = B(u_{i,j})u'_{i,j},$$

where A and B are the Jacobian matrices $A = \nabla_u f$, $B = \nabla_u g$.

Application of this method to hydrodynamical models of semiconductors will be described later.

2.7. Relaxation step

The integration of the relaxation equations has to be performed by an implicit scheme (fully implicit or linearly implicit). Sometimes the equations can be integrated analytically.

The simple splitting scheme is only first order accurate. A more accurate splitting strategy has been proposed (LIOTTA, ROMANO and RUSSO [1999], LIOTTA, ROMANO and RUSSO [2000]). The resulting scheme is second order accurate for nonstiff source and reduces to first order in the stiff case.

In developing the numerical scheme we keep in mind the following guidelines:

- truncation error analysis is used to obtain second order accuracy in the rarefied regime ($\varepsilon = \mathrm{O}(1)$);
- the collision step is well posed $\forall \varepsilon > 0$ and its solution relaxes to a local Maxwellian as $\varepsilon \to 0$;
- the scheme should be unconditionally stable in the collision step;
- the limiting scheme obtained as $\varepsilon \to 0$ is a consistent numerical scheme for the equilibrium subsystem.

Truncation error analysis is performed on the following linear system

$$\partial_t U + AU + BU = 0, \tag{2.26}$$

where $U \in \mathbf{R}^m$ and A, B are constant matrices which represent the discrete operator of the flux and source, respectively. We assume that A is a second order discretization (in the applications of the next section we will take the Nessyahu–Tadmor scheme).

The convection step solves the equation

$$\partial_t U + AU = 0. \tag{2.27}$$

If we indicate by $\mathcal{T}$ the discrete operator associated to the convection step scheme, after performing the convection step (2.27) starting from U^n one obtains

$$\mathcal{T}U^n = U^n - A\Delta t U^n + \tfrac{1}{2}A^2\Delta t^2 U^n + \mathrm{O}(\Delta t^3).$$

Following the approach used in CAFLISCH, RUSSO and JIN [1997], we write our splitting scheme as a combination of relaxation steps (implicit Euler) and convection steps, which are suitably assembled to give a second order accurate solution. We stress that at variance of what done in CAFLISCH, RUSSO and JIN [1997], we assume that the convection step is discretized by a second order scheme.

Neglecting higher order terms in the expansion, we write the scheme as

$$U_1 = U^n - \alpha\Delta t B U_1, \tag{2.28}$$

$$U_2 = U_1 - \tilde{\alpha}\Delta t A U_1 + \tfrac{1}{2}\tilde{\alpha}^2\Delta t^2 A^2 U_1, \tag{2.29}$$

$$U_3 = U_2 - \beta\Delta t B U_3 - \gamma\Delta t B U_1, \tag{2.30}$$

$$U_4 = U_3 - \tilde{\beta}\Delta t A U_3 + \tfrac{1}{2}\tilde{\beta}^2\Delta t^2 A^2 U_3, \tag{2.31}$$

$$U_5 = \xi U_1 + \eta U_4, \tag{2.32}$$

$$U^{n+1} = U_5 - \mu\Delta t B U^{n+1}, \tag{2.33}$$

where the parameters $\alpha, \beta, \tilde{\alpha}, \tilde{\beta}, \xi, \eta, \mu$ have to be determined.

The exact solution of (2.26) at time $t = \Delta t$ is given by

$$U(\Delta t) = \mathrm{e}^{-(A+B)\Delta t} U(0). \tag{2.34}$$

By applying scheme (2.28)–(2.33) to Eq. (2.26) and writing the difference equation in the compact form

$$U^{n+1} = \mathcal{C} U^n, \tag{2.35}$$

we achieve a second order accuracy if we impose that

$$\mathcal{C} U(0) - \mathrm{e}^{-(A+B)\Delta t} U(0) = \mathrm{O}(\Delta t^3). \tag{2.36}$$

This gives the following constraints on the parameters

$$\eta + \xi = 1, \tag{2.37}$$

$$\eta(\tilde{\alpha} + \tilde{\beta}) = 1, \tag{2.38}$$

$$\eta(\alpha + \beta + \gamma) + \mu(\xi + \eta) + \alpha\xi = 1, \tag{2.39}$$

$$\eta(\tilde{\alpha} + \tilde{\beta})^2 = 1, \tag{2.40}$$

$$2\eta(\alpha\tilde{\alpha} + \alpha\tilde{\beta} + \tilde{\beta}\gamma + \tilde{\beta}\beta) = 1, \tag{2.41}$$

$$2\eta(\tilde{\alpha}\beta + \mu\tilde{\alpha} + \mu\tilde{\beta}) = 1, \tag{2.42}$$

$$2\eta(\alpha^2 + \alpha\gamma + \alpha\beta + \beta\gamma + \beta^2) + 2\mu\eta(\alpha + \beta + \gamma) + 2(\xi + \eta)\mu^2 + 2\alpha\xi(\mu + \alpha) = 1. \tag{2.43}$$

The real solutions of the previous nonlinear algebraic equations can be expressed in terms of the parameter β,

$$\tilde{\alpha} = 0, \quad \tilde{\beta} = 1, \quad \xi = 0, \quad \eta = 1,$$
$$\mu = \frac{1}{2}, \quad \alpha = \frac{\beta}{2\beta - 1}, \quad \gamma = \frac{1}{2} - \frac{\beta}{2\beta - 1} - \beta.$$

Because some coefficients are zero, it is possible to put the splitting scheme in the simpler form

$$U_1 = U^n - \alpha\Delta t B U_1, \tag{2.44}$$

$$U_2 = U_1 - \beta\Delta t B U_2 - \gamma\Delta t B U_1, \tag{2.45}$$

$$U_3 = \mathcal{T}(U_2), \tag{2.46}$$

$$U^{n+1} = U_3 - \tfrac{1}{2}\Delta t B U^{n+1}. \tag{2.47}$$

A set of parameters satisfying (2.37)–(2.43) is given by

$$\beta = 1, \quad \alpha = 1, \quad \gamma = -\tfrac{3}{2}. \tag{2.48}$$

We remark that the Strang splitting is not included as particular case of our scheme because here the relaxation steps are only first order accurate. Moreover the obtained splitting scheme is valid in any spatial dimension.

The scheme is not optimal, because it uses three relaxation steps in order to obtain a second order scheme. However in this case the relaxation step is not the most expensive one, and therefore the cost of the extra step is negligible.

2.8. Nonsplitting schemes

There are several cases where the splitting approach is not very effective. When the relaxation term is stiff, for example, Strang splitting loses second order accuracy, and more sophisticated splitting techniques are necessary to obtain a second order scheme.

In CAFLISCH, RUSSO and JIN [1997], a second order scheme for hyperbolic systems with relaxation is derived, which maintain second order accuracy both in the stiff and nonstiff limit. These methods have been developed in the context of upwind schemes.

The same approach used in CAFLISCH, RUSSO and JIN [1997] cannot be straightforwardly extended to central schemes.

A natural way to treat source term in central scheme is to include the source in the integration over the cell in space–time.

Let us consider a general system of balance laws of the form

$$\frac{\partial u}{\partial t}+\frac{\partial f(u)}{\partial x}=\frac{1}{\varepsilon}g(u). \tag{2.49}$$

The parameter ε represents the relaxation time. If it is very small than we say the relaxation term is *stiff*.

Integrating Eq. (2.49) over the space–time (see Fig. 2.2) one has

$$\int_{x_j}^{x_{j+1}} u(x,t_{n+1}) = \int_{x_j}^{x_{j+1}} u(x,t_n) - \int_{t_n}^{t_{n+1}} \bigl(f\bigl(u(x_{j+1},t)\bigr) - f\bigl(u(x_j,t)\bigr)\bigr)\,\mathrm{d}t + \int_{t_n}^{t_{n+1}}\int_{x_j}^{x_{j+1}} g(x,t)\,\mathrm{d}x\,\mathrm{d}t. \tag{2.50}$$

Numerical schemes are obtained by a suitable discretization of the integrals.

Here we only consider second order schemes.

We use a piecewise linear reconstruction in each cell, as in the Nessyahu–Tadmor scheme, and midpoint rule for the computation of the flux integral.

Different schemes are obtained, according to the discretization of the integral of the source term. If the source is not stiff, then a fully explicit time discretization can be used, resulting in the following scheme (see Fig. 2.4)

$$u_{j+1/2}^{n+1} = u_{j+1/2}^{n} + \frac{\Delta t}{\Delta x}\bigl(f(u_j^{n+1/2}) - f(u_{j+1}^{n+1/2})\bigr) + \frac{\Delta t}{2\varepsilon}\bigl(g(u_j^{n+1/2}) + g(u_{j+1}^{n+1/2})\bigr), \tag{2.51}$$

where

$$u_{j+1/2}^{n} = \frac{1}{2}\bigl(u_j^n + u_{j+1}^n\bigr) + \frac{1}{8}\bigl(u'_j - u'_{j+1}\bigr) \tag{2.52}$$

and the predictor values $u_j^{n+1/2}$ are computed by

$$u_j^{n+1/2} = u_j^n - \frac{\lambda}{2}f'_j + \frac{\Delta t}{2\varepsilon}g\bigl(u_j^n\bigr). \tag{2.53}$$

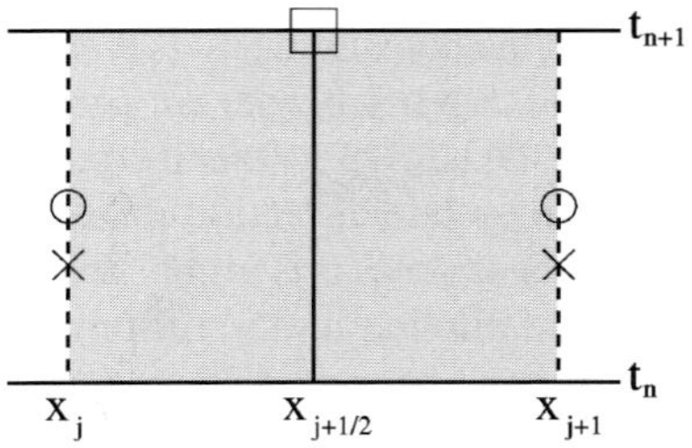

FIG. 2.4. Nodes in space time for the second order Uniform Central Scheme.

Note that this scheme is fully explicit, therefore it is subject to stability restriction due to both the flux and the source term. If the stability restriction of the source term is more severe than the one due to the flux, then not only efficiency, but also accuracy of the calculation will be affected. In this case, in fact, one has to use a Courant number much smaller than the one allowed by the CFL restriction. It is well known that in this case the numerical dissipation will be larger than necessary, and the accuracy of the scheme will be poor. This problem can be partially circumvented, for moderately stiff source, by the use of semidiscrete schemes. For a discussion of this issue see, for example, KURGANOV and TADMOR [2000]. When the stiffness increases it is better to treat the source implicitly. This can be done at the stage of the predictor, as

$$u_j^{n+1/2} = u_j^n - \frac{\lambda}{2} f_j' + \frac{\Delta t}{2} g\big(u_j^{n+1/2}\big). \tag{2.54}$$

The time discretization used for the source is the midpoint implicit scheme, which is a Gauss-collocation scheme with one level. Such scheme is A-stable, but not L-stable, and therefore it is not suitable for very stiff source (see HAIRER and WANNER [1987]). A numerical scheme which is stable and accurate even for very stiff source has been proposed in LIOTTA, ROMANO and RUSSO [2000]. It is obtained by using two predictor stages, one for the flux, and one which is needed for the source. The nodes for the source are chosen according to stability requirements. The scheme can be written in the form

$$\begin{aligned} u_j^{n+1/2} &= u_j^n - \frac{\lambda}{2} f_j' + \frac{\Delta t}{2\varepsilon} g\big(u_j^{n+1/2}\big), \\ u_j^{n+1/3} &= u_j^n - \frac{\lambda}{3} f_j' + \frac{\Delta t}{3\varepsilon} g\big(u_j^{n+1/3}\big), \\ u_{j+1/2}^{n+1} &= \bar{u}_{j+1/2}^n - \lambda\big(f(u_{j+1}^{n+1/2}) - f(u_j^{n+1/2})\big) \\ &\quad + \frac{\Delta t}{8\varepsilon}\big(3g(u_j^{n+1/3}) + 3g(u_{j+1}^{n+1/3}) + 2g(u_{j+1/2}^{n+1})\big), \end{aligned} \tag{2.55}$$

where $u_{j+1/2}^n$ is computed by (2.52). We shall call the above scheme Uniformly accurate Central Scheme of order 2 (UCS2).

Such scheme has the following properties. When applied to hyperbolic systems with relaxation, it is second order accurate in space and time both in the nonstiff case (i.e., $\varepsilon = 1$) and in the stiff limit (i.e., $\varepsilon = 0$). A small degradation of accuracy is observed for intermediate values of the relaxation parameter ε.

The time discretization used in the above scheme is a particular case of Runge–Kutta Implicit–Explicit (IMEX) scheme. Such schemes are particularly important when one has to solve systems that contain the sum of a nonstiff (possibly expensive to compute) and a stiff term. These systems may be convection-diffusion equations, or hyperbolic systems with stiff relaxation. In these cases it is highly desirable to use a scheme which is explicit in the nonstiff term, and implicit in the stiff term.

Runge–Kutta IMEX schemes have been studied in ASCHER, RUUTH and SPITERI [1997] and in PARESCHI and RUSSO [2000].

3. Applications to 1D problems

The numerical method is based on the splitting technique described in the previous section. Here we use the one-dimensional version of the scheme. The 2D extension will be employed in the next section.

Let us consider the system (1.51) in the one-dimensional case

$$\frac{\partial}{\partial t}\mathbf{U} + \frac{\partial}{\partial x}F(\mathbf{U}) = B(\mathbf{U}, \mathbf{E}), \tag{3.1}$$

where $F = F^{(1)}$.

Each convective step has the form of predictor–corrector NT scheme on a staggered grid, derived in Section 2.5.

In order to couple the convection step with the relaxation step, it is convenient to make two convection steps of step size $\Delta t/2$, so that the solution is computed on the same grid. A complete convection step of step size Δt is obtained as a sequence of two intermediate steps of step size $\Delta t/2$.

The values of $\mathbf{U}'_j/\Delta x$ and $F'_j/\Delta x$ used in the two steps of NT scheme are a first order approximation of the space derivatives of the field and of the flux, computed from cell averages by using a Uniform Non-Oscillatory reconstruction (2.18).

The electric potential is calculated by the discretized Poisson equation

$$\varepsilon(\phi_{i+1} - 2\phi_i + \phi_{i-1}) = -e\big(N_D(x_i) - N_A(x_i) - n_i\big)$$

with Dirichlet boundary conditions. The tridiagonal system is solved by a standard procedure. We assume that the dielectric constant is the same in the whole computational domain.

The relaxation step requires to solve the system of ODEs

$$\begin{aligned}
\frac{\mathrm{d}n}{\mathrm{d}t} &= 0,\\
\frac{\mathrm{d}V}{\mathrm{d}t} &= -\frac{eE}{m^*} + 2\alpha eEG + \left(\frac{c_{11}}{m^*} - 2\alpha c_{21}\right)V + \left(\frac{c_{12}}{m^*} - 2\alpha c_{22}\right)S,\\
\frac{\mathrm{d}W}{\mathrm{d}t} &= -eVE - \frac{W - W_0}{\tau_W},\\
\frac{\mathrm{d}S}{\mathrm{d}t} &= -eEG + c_{21}V + c_{22}S.
\end{aligned}$$

Here we dropped the subscript 1 in V, E, and S, and we omitted the grid index j, since all the quantities are computed at the same cell center x_j.

By freezing the energy relaxation time, the coefficients c_{lp} and the electric field at $t = t_n$, we discretize the previous equations for each node j in a semi-implicit form as

$$\begin{aligned}
&\frac{n^{n+1} - n^n}{\Delta t} = 0,\\
&\frac{V^{n+1} - V^n}{\Delta t} = \frac{eE^n}{m^*} + 2\alpha e E^n G^n + \left(\frac{c_{11}^n}{m^*} - 2\alpha c_{21}^n\right) V^{n+1} + \left(\frac{c_{12}^n}{m^*} - 2\alpha c_{22}^n\right) S^{n+1},\\
&\frac{W^{n+1} - W^n}{\Delta t} = -eV^n E^n - \frac{W^{n+1} - W_0}{\tau_W},\\
&\frac{S^{n+1} - S^n}{\Delta t} = -eE^n G^n + c_{21}^n V^{n+1} + c_{22}^n S^{n+1}.
\end{aligned}$$

The equations can be solved for the quantities at the new time step, yielding

$$\begin{aligned}
&n^{n+1} = n^n,\\
&Vn + 1 = \frac{1}{\Delta^n}\left[(1 - c_{22}^n \Delta t) d_1^n + d_2^n \Delta t \left(\frac{c_{12}^n}{m^*} - 2\alpha c_{22}^n\right)\right],\\
&W^{n+1} = \left(1 + \frac{\Delta t}{\tau_W^n}\right)^{-1}\left[W^n + \left(-eE^n V^n + \frac{W_0}{\tau_W^n}\right)\Delta t\right],\\
&S^{n+1} = \frac{1}{\Delta^n}\left\{c_{21}^n d_1^n \Delta t + d_2^n\left[1 - \left(\frac{c_{11}^n}{m^*} - 2\alpha c_{21}^n\right)\Delta t\right]\right\}.
\end{aligned}$$

where

$$\begin{aligned}
&\Delta^n = (1 - c_{22}\Delta t)\left[1 - \left(\frac{c_{11}^n}{m^*} - 2\alpha c_{21}^n\right)\Delta t\right] - c_{21}\left(\frac{c_{12}^n}{m^*} - 2\alpha c_{22}^n\right)(\Delta t)^2,\\
&d_1^n = V^n + \left(-\frac{eE^n}{m^*} + 2\alpha e E^n G^n\right)\Delta t,\\
&d_2^n = S^n - eE^n G^n \Delta t.
\end{aligned}$$

Second order accuracy in time is obtained by using the splitting scheme (2.44)–(2.47) taking into account also the Poisson solver.

Given the fields at time t_n, (U^n, E^n), the fields at time t_{n+1} are obtained by

$$\begin{aligned}
\mathbf{U}_1 &= \mathbf{U}^n - R(\mathbf{U}_1, E^n, \Delta t),\\
\mathbf{U}_2 &= \tfrac{3}{2}\mathbf{U}^n - \tfrac{1}{2}\mathbf{U}_1,\\
\mathbf{U}_3 &= \mathbf{U}_2 - R(\mathbf{U}_3, E^n, \Delta t),\\
\mathbf{U}_4 &= \mathcal{C}_{\Delta t}\mathbf{U}_3,\\
E_1^{n+1} &= \mathcal{P}(\mathbf{U}_4),\\
\mathbf{U}^{n+1} &= \mathbf{U}_4 - R(\mathbf{U}^{n+1}, E^{n+1}, \Delta t/2),
\end{aligned} \tag{3.2}$$

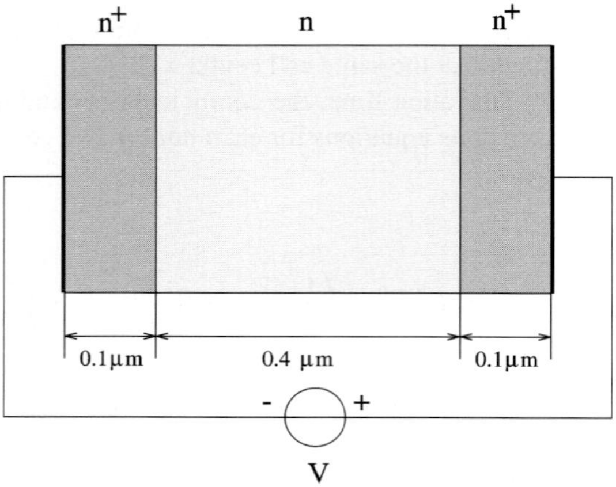

FIG. 3.1. Schematic representation of a n^+-n-n^+ diode.

TABLE 3.1
Length of the channel, doping concentration and applied voltage in the test cases for the diode

Test #	Channel length L_c (μm)	N_D^+ ($\times 10^{17}$ cm^{-3})	N_D ($\times 10^{17}$ cm^{-3})	V_b (V)
1	0.4	5	0.02	2
2	0.3	10	0.1	1
3	0.2	10	0.1	1

where R represents the discrete operator corresponding to the relaxation step, $\mathcal{C}_{\Delta t}$ is the discrete operator corresponding to NT scheme and $\mathcal{P}(U)$ gives the solution to Poisson's equation.

We remark here that the relaxation step does not alter the density n, and therefore the electric field has to be computed only just after the convection step.

As first problem we simulate a ballistic n^+-n-n^+ silicon diode (Fig. 3.1) (see ROMANO [2001] for more details). The n^+ regions are 0.1 μm long while the channel has different length. Moreover several doping profiles will be considered according to Table 3.1.

Initially the electron temperature is equal to the lattice temperature T_L, the charges are at rest and the density is equal to the doping concentration

$$n(x,0) = n_0(x), \quad W(x,0) = \tfrac{3}{2}k_B T_L, \quad V(x,0) = 0, \quad S(x,0) = 0.$$

Regarding the boundary conditions, in principle the number of independent conditions on each boundary should be equal to the number of characteristics entering the domain. However, we impose, in analogy with similar cases (ANILE, JUNK, ROMANO and RUSSO [2000], FATEMI, JEROME and OSHER [1991]) a double number of boundary conditions. More precisely, we give conditions for all the variables in each boundary, located at $x = 0$ and $x = L$,

$$n(0,t) = n(L,t) = N_D^+, \tag{3.3}$$

$$\frac{\partial}{\partial x} W(0,t) = \frac{\partial}{\partial x} W(L,t) = 0, \tag{3.4}$$

$$\frac{\partial}{\partial x} V(0,t) = \frac{\partial}{\partial x} V(L,t) = 0, \tag{3.5}$$

$$\frac{\partial}{\partial x} S(0,t) = \frac{\partial}{\partial x} S(L,t) = 0, \tag{3.6}$$

$$\phi(0) = 0 \quad \text{and} \quad \phi(L) = V_b, \tag{3.7}$$

where V_b is the applied bias voltage. In all the numerical solutions there is no sign of spurious oscillations near the boundary, indicating that the conditions (3.3)–(3.6) are in fact compatible with the solution of the problem.

The doping profile is regularized according to the function

$$n_0(x) = n_0 - d_0\left(\tanh\frac{x - x_1}{s} - \tanh\frac{x - x_2}{s}\right),$$

where $s = 0.01$ μm, $n_0 = n_0(0)$, $d_0 = n_0(1 - N_D/N_D^+)/2$, $x_1 = 0.1$ μm, and $x_2 = x_1 + L_c$ with L_c channel length. The total length of the device is $L = L_c + 0.2$ μm. In Fig. 3.2 the doping profile for the test case 1 is plotted.

A grid with 400 nodes has been used. The stationary solution is reached within a few picoseconds (about five), after a short transient with wide oscillations.

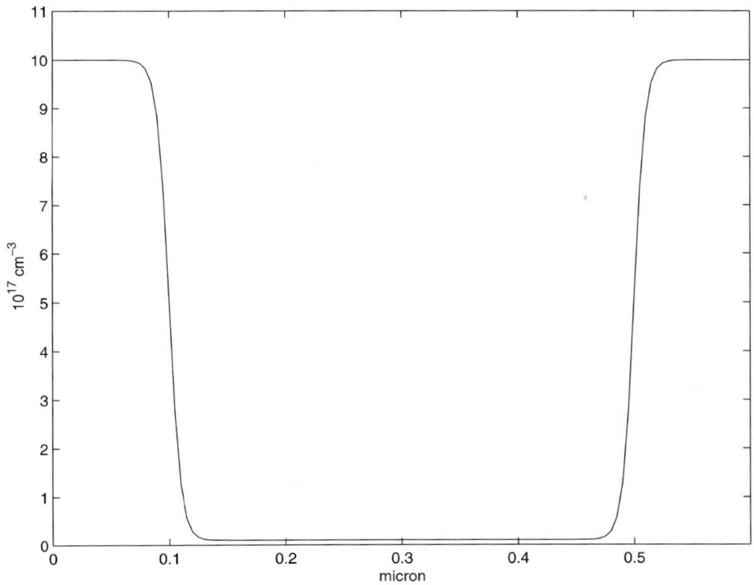

FIG. 3.2. Doping profile for the test case 1.

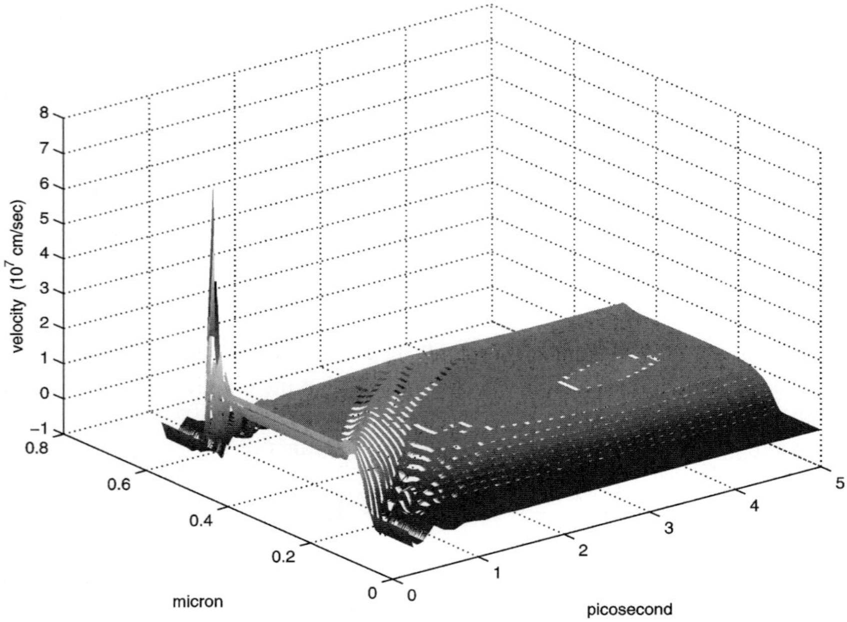

FIG. 3.3. Time dependent numerical result for the velocity of the test case 1 with the Kane dispersion relation.

As first case we consider the test problem 1 (length of the channel 0.4 micron) with $V_b = 2$ V. In Fig. 3.3 the time dependent solution for the velocity is plotted. The stationary solution (after 5 ps) is shown in Fig. 3.4 (continuous line) along with the numerical results for the other variables. At variance with the numerical results obtained in ANILE, JUNK, ROMANO and RUSSO [2000] for the parabolic band case by using a quadratic closure in δ, the new numerical solutions do not present irregularities. This can be probably ascribed to the absence of the nonlinearities in the dissipative variables.

The simulation for the parabolic band approximation is also shown (Fig. 3.4, dashed line), but it is evident, like in the bulk case, that the results are rather poor.

The other test cases have been numerically integrated with $V_b = 1$ V (Figs. 3.5, 3.6). The behaviour of the solution looks again physically reasonable and encouraging: the spurious spike across the second junction is here less apparent than several other hydrodynamical models. The results with the parabolic band are again rough when compared with those obtained in the nonparabolic case.

As further example of application we present the simulation of a *nanoscale* device (see also MUSCATO and ROMANO [2001], ROMANO [2001]): a one-dimensional Si n^+-n-n^+ diode of length 0.25 μm with a channel of 0.05 μm. The donor density N_D is a stepwise function with values 5×10^{18} cm^{-3} in the n^+-region and 10^{15} cm^{-3} in the n-region. Moreover a constant concentration of acceptors $N_A = 5 \times 10^{16}$ cm^{-3} is considered. In Figs. 3.7, 3.8, 3.9 we show the numerical result for the velocity, energy and electric field in the stationary regime (after about five picoseconds) with a $V_b =$ 0.6 V.

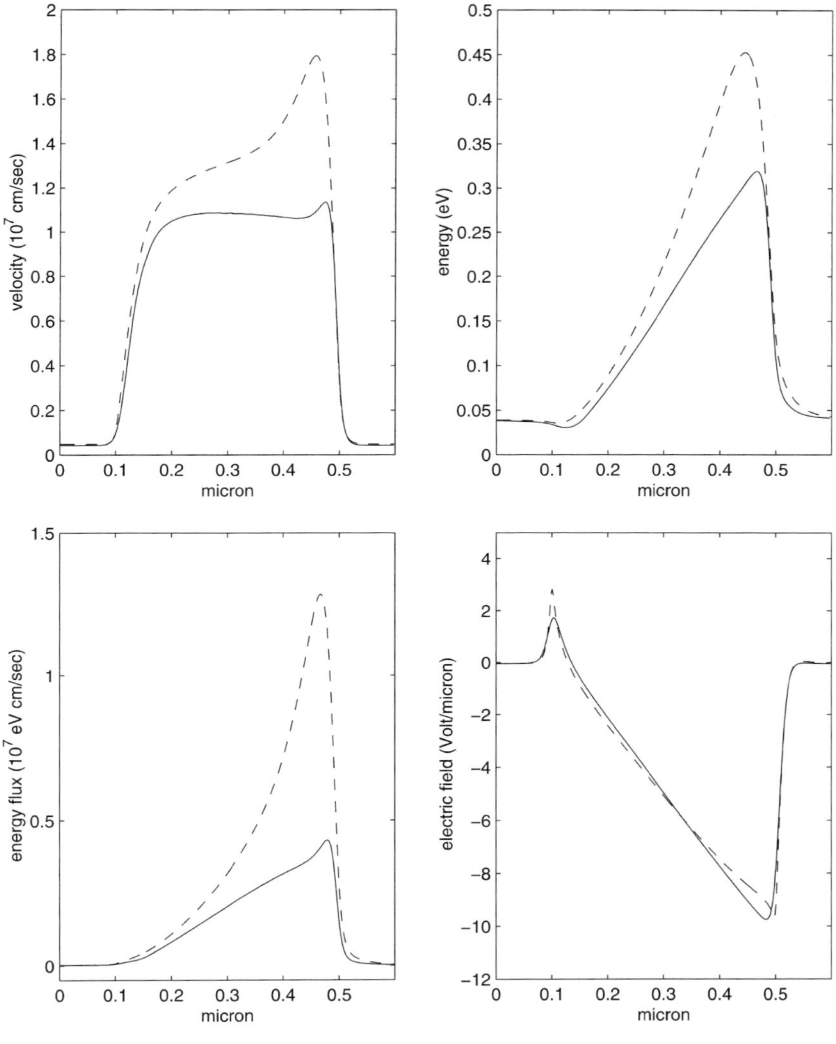

FIG. 3.4. Numerical results of the test case 1 after 5 ps in the parabolic band case (dashed line) and for the Kane dispersion relation (continuous line).

3.1. Simulation of a silicon MESFET

In this section we check the validity of our hydrodynamical model and the efficiency of the numerical method by simulating a two-dimensional Metal Semiconductor Field Effect Transistor (MESFET) (for more details see ROMANO [2002]).

We need to extend the scheme to the two-dimensional case starting from the two-dimensional version of the Nessyahu and Tadmor scheme (JIANG and TADMOR [1998]) and employing the splitting scheme (2.44)–(2.47).

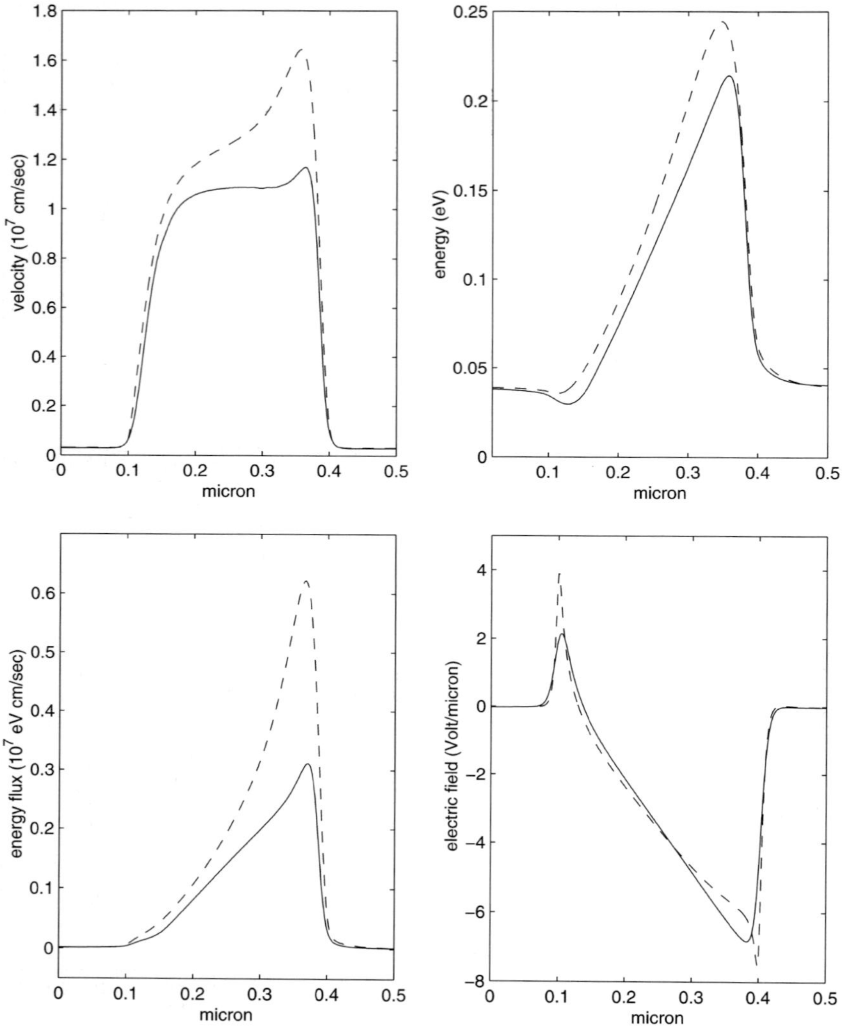

FIG. 3.5. Numerical results of the test case 2 after 5 ps in the parabolic band case (dashed line) and for the Kane dispersion relation (continuous line).

If one introduces an uniform grid (x_i, y_j), with $x_{i+1} - x_i = \Delta x =$ constant and $y_{i+1} - y_i = \Delta y =$ constant, and denote by $\Delta t = t^{n+1} - t^n$ the time step, then the convective part of the scheme is given by Eqs. (2.23)–(2.25).

In order to couple the convection step with the relaxation step, it is convenient, as in the 1D case, to make two convection steps of step size $\Delta t/2$, so that the solution is computed on the same grid. A complete convection step is obtained as a sequence of two intermediate steps of time step size $\Delta t/2$.

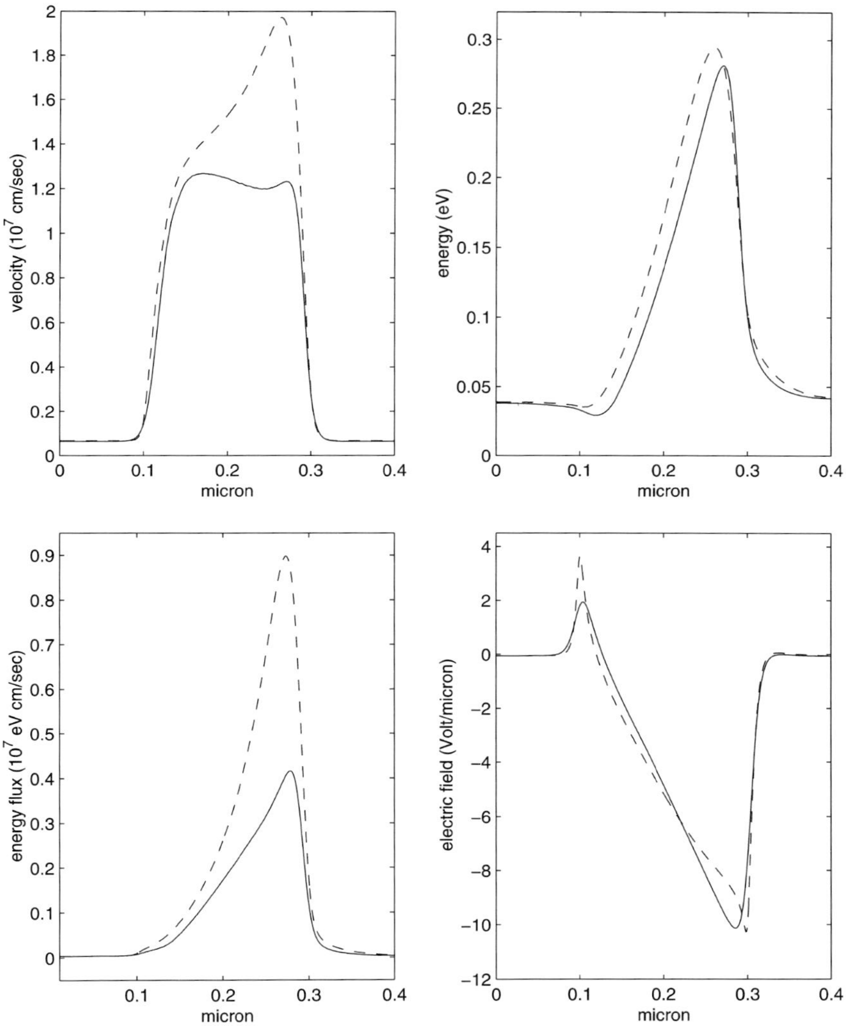

FIG. 3.6. Numerical results of the test case 3 after 5 ps in the parabolic band case (dashed line) and for the Kane dispersion relation (continuous line).

The electric potential is calculated from the Poisson (1.7) equation by central differencing and by resorting to the conjugate gradient method to solve the resulting linear system.

In Fig. 3.10 we show an example of the discretization used for the Poisson equation. All the cells are numbered from 1 to the total number N of cells. The potential is defined at the center of each cell. The computational domain is extended by including a certain number of ghost cells, whose center is denoted by a cross. The value of the potential in the ghost cells is determined by the boundary conditions. For example (see Fig. 3.10),

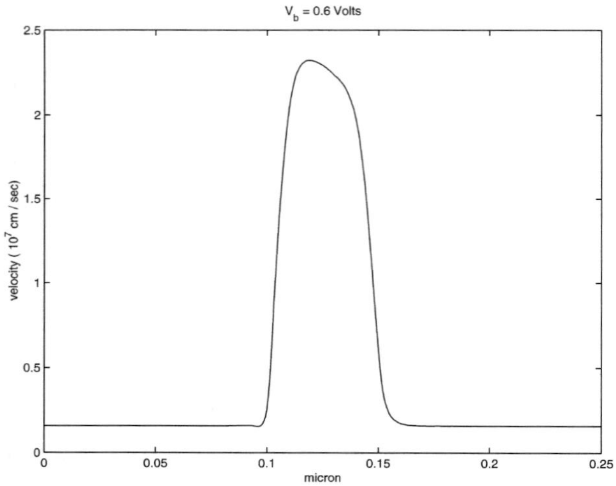

FIG. 3.7. Numerical result of velocity versus position for the for the nanoscale device after 5 ps.

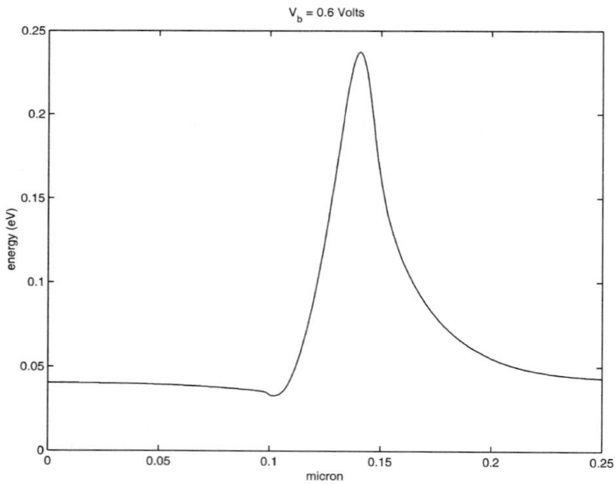

FIG. 3.8. Numerical result of for the energy versus position for the nanoscale device after 5 ps.

Dirichlet boundary condition above cell 1 gives

$$\frac{\phi_{1N} - \phi_1}{2} = \phi_s, \quad \Rightarrow \quad \phi_{1N} = 2\phi_s - \phi_1$$

and Neumann null condition to the left of cell one gives $\phi_{1W} = \phi_1$, therefore the equation for the first cell becomes

$$4\phi_1 - \phi_2 - \phi_5 = h^2 \rho_1 + 2\phi_s,$$

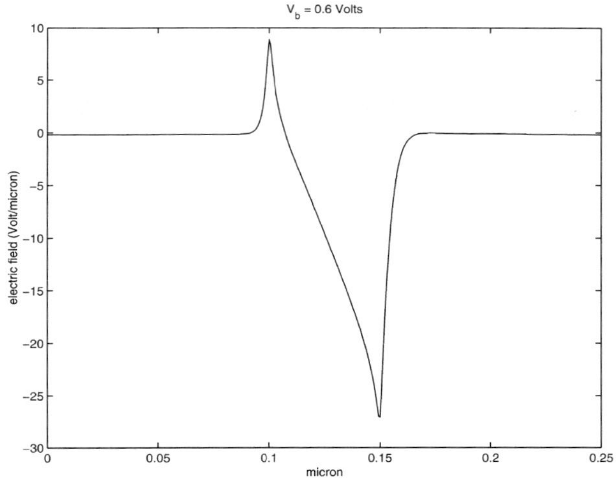

FIG. 3.9. Numerical result of for the electric field versus position for the nanoscale device after 5 ps.

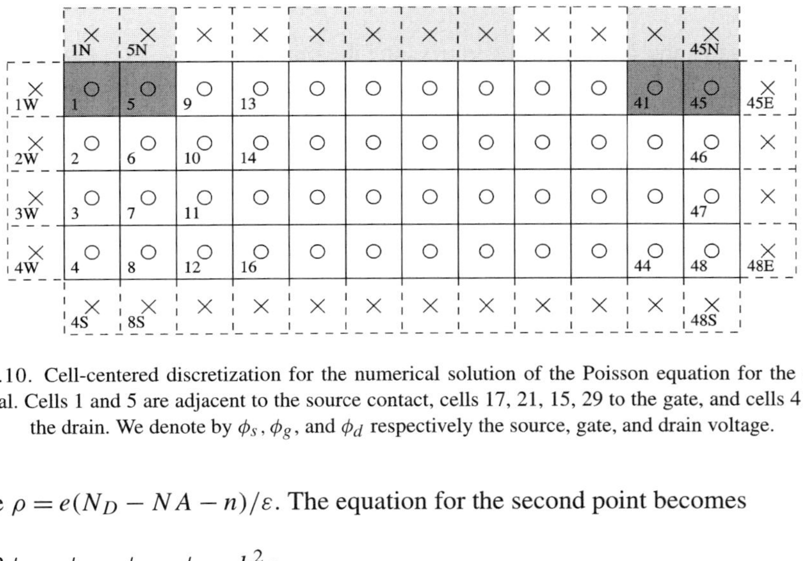

FIG. 3.10. Cell-centered discretization for the numerical solution of the Poisson equation for the electric potential. Cells 1 and 5 are adjacent to the source contact, cells 17, 21, 15, 29 to the gate, and cells 41, 45 to the drain. We denote by ϕ_s, ϕ_g, and ϕ_d respectively the source, gate, and drain voltage.

where $\rho = e(N_D - NA - n)/\varepsilon$. The equation for the second point becomes

$$3\phi_2 - \phi_1 - \phi_3 - \phi_6 = h^2 \rho_2.$$

By this procedure, the linear system for the N-dimensional vector $\Phi = (\phi_1, \ldots, \phi_N)$ is written in the form

$$A\Phi = \mathbf{b},$$

where the A is sparse, symmetric and positive definite. In the example shown in Fig. 3.10 matrix A is given by

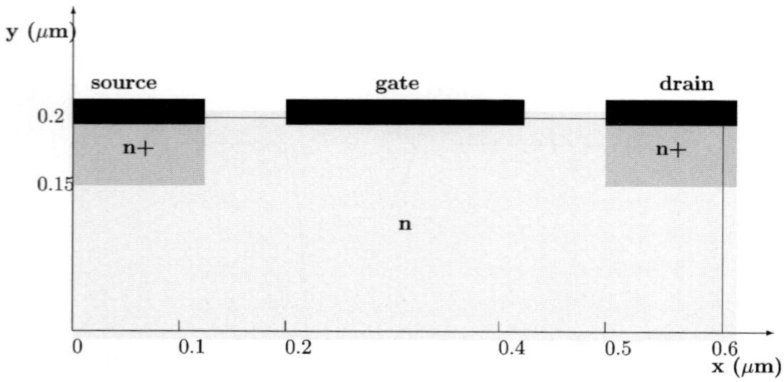

FIG. 3.11. Schematic representation of a two-dimensional MESFET.

$$A = \begin{pmatrix} G_1 & -I & 0 & \cdots & 0 \\ -I & G_2 & -I & \cdots & 0 \\ \vdots & \ddots & \ddots & \ddots & \vdots \\ 0 & \cdots & -I & G_{11} & -I \\ 0 & \cdots & 0 & -I & G_{12} \end{pmatrix},$$

where I denotes the 4×4 identity matrix, and the matrices $G_1, \ldots, G_{12}$ are given by

$$G_1 = G_{12} = \begin{pmatrix} 4 & -1 & 0 & 0 \\ -1 & 3 & -1 & 0 \\ 0 & -1 & 3 & -1 \\ 0 & 0 & -1 & 2 \end{pmatrix},$$

$$G_3 = G_4 = G_9 = G_{10} = \begin{pmatrix} 3 & -1 & 0 & 0 \\ -1 & 4 & -1 & 0 \\ 0 & -1 & 4 & -1 \\ 0 & 0 & -1 & 3 \end{pmatrix},$$

$$G_2 = G_5 = G_6 = G_7 = G_8 = G_{11} = \begin{pmatrix} 5 & -1 & 0 & 0 \\ -1 & 4 & -1 & 0 \\ 0 & -1 & 4 & -1 \\ 0 & 0 & -1 & 3 \end{pmatrix}.$$

The symmetry of the matrix is evident. The matrix is diagonally dominant, with positive elements on the main diagonal. The fact that the matrix is positive definite can be deduced as a consequence of first and second Gershgorin theorems (see, for example, SAAD [1996], pp. 109–111).

The system can be efficiently solved by an iterative scheme such as the conjugate gradient method (see, for example, GOLUB, VAN LOAN and CHARLES [1996]). We remark that because we are solving a time dependent problem, the electric potential at the new time step is only a small perturbation of the electric potential at the previous time step, and therefore one starts the iterative process with a good initial guess.

The scheme we have shown makes use of a fixed grid. A multigrid approach will be present in the next section.

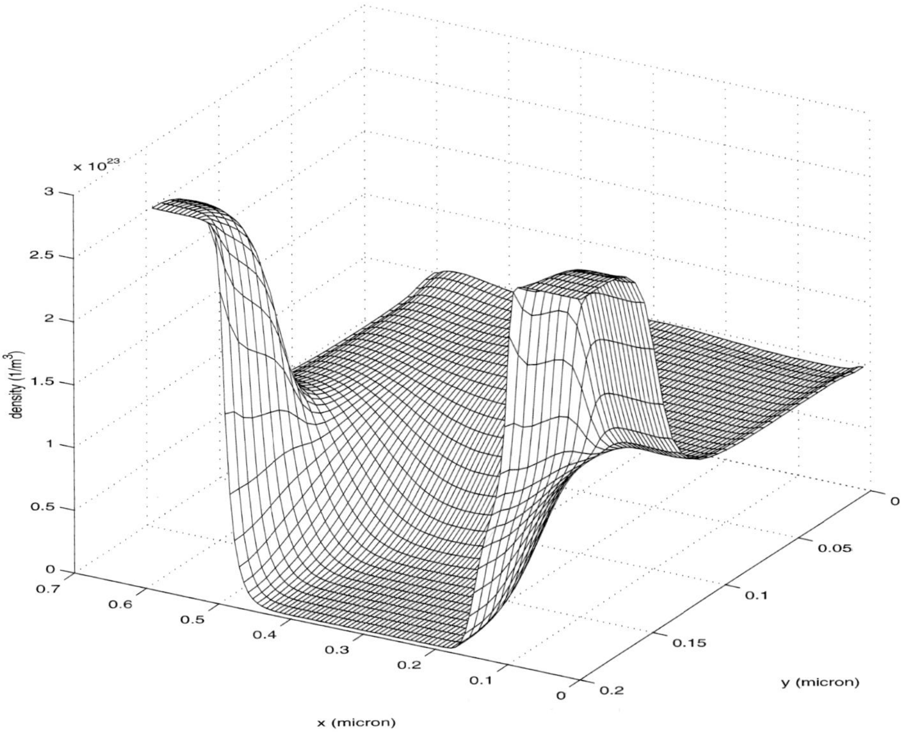

FIG. 3.12. Stationary solution (after 5 ps) for the density for $\phi_b = 1$ V.

In the relaxation step one has to solve the following system of ODEs

$$\frac{\mathrm{d}n}{\mathrm{d}t} = 0,$$

$$\frac{\mathrm{d}V_k}{\mathrm{d}t} = -\frac{eE_k}{m^*} + 2\alpha e E_k G + \left(\frac{c_{11}}{m^*} - 2\alpha c_{21}\right) V_k + \left(\frac{c_{12}}{m^*} - 2\alpha c_{22}\right) S_k, \quad k = 1, 2,$$

$$\frac{\mathrm{d}W}{\mathrm{d}t} = -e \sum_{l=1}^{2} V_l E_l - \frac{W - W_0}{\tau_W},$$

$$\frac{\mathrm{d}S_k}{\mathrm{d}t} = -eE_k G + c_{21} V_k + c_{22} S_k, \quad k = 1, 2.$$

By freezing the energy relaxation time, the coefficients c_{lp} and the electric field at $t = t^n$, we can integrate numerically these equations for each grid point (x_i, y_j) in a semi-implicit way, exactly as in the one-dimensional case

$$n^{n+1} = n^n,$$

$$V_k^{n+1} = \frac{1}{\Delta^n}\left[\left(1 - c_{22}^n \Delta t\right) d_{1_k}^n + d_{2_k}^n \Delta t \left(\frac{c_{12}^n}{m^*} - 2\alpha c_{22}^n\right)\right], \quad k = 1, 2,$$

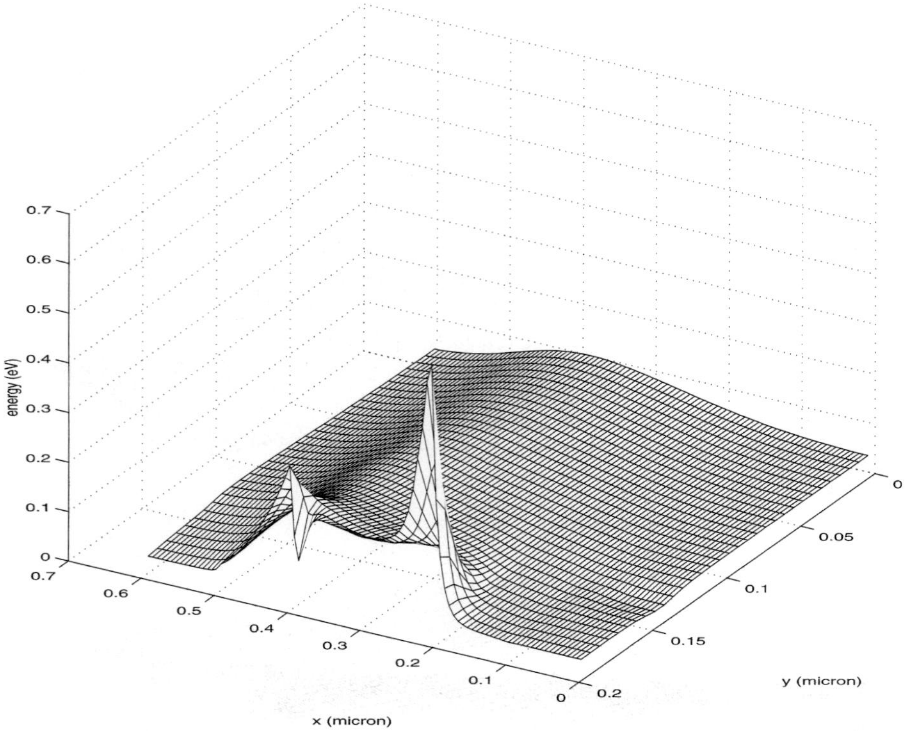

FIG. 3.13. Stationary solution (after 5 ps) for the energy density for $\phi_b = 1$ V.

$$W^{n+1} = \left(1 + \frac{\Delta t}{\tau_W^n}\right)^{-1} \left[W^n + \left(-e\sum_{l=1}^{2} E_l^n V_l^n + \frac{W_0}{\tau_W^n}\right)\Delta t\right],$$

$$S_k^n = \frac{1}{\Delta^n}\left\{c_{21}^n d_{1_k}^n \Delta t + d_{2_k}^n \left[1 - \left(\frac{c_{11}^n}{m^*} - 2\alpha c_{21}^n\right)\Delta t\right]\right\}, \quad k = 1, 2,$$

where

$$\Delta^n = (1 - c_{22}\Delta t)\left[1 - \left(\frac{c_{11}^n}{m^*} - 2\alpha c_{21}^n\right)\Delta t\right] - c_{21}\left(\frac{c_{12}^n}{m^*} - 2\alpha c_{22}^n\right)(\Delta t)^2,$$

$$d_{1_k}^n = V_k^n + \left(-\frac{eE_k^n}{m^*} + 2\alpha e E_k^n G^n\right)\Delta t, \quad k = 1, 2,$$

$$d_{2_k}^n = S_k^n - eE_k^n G^n \Delta t, \quad k = 1, 2.$$

Since the field quantities refer to the same gridpoint i, j, we omit to write it.

In order to get a full second order scheme we combine the relaxation and convective steps in the same way as in the previous subsection because the analysis of the splitting accuracy does not really depend on the dimension of the space. Given the field U^n and E^n at time t^n, the field at time t^{n+1} is obtained by the splitting scheme.

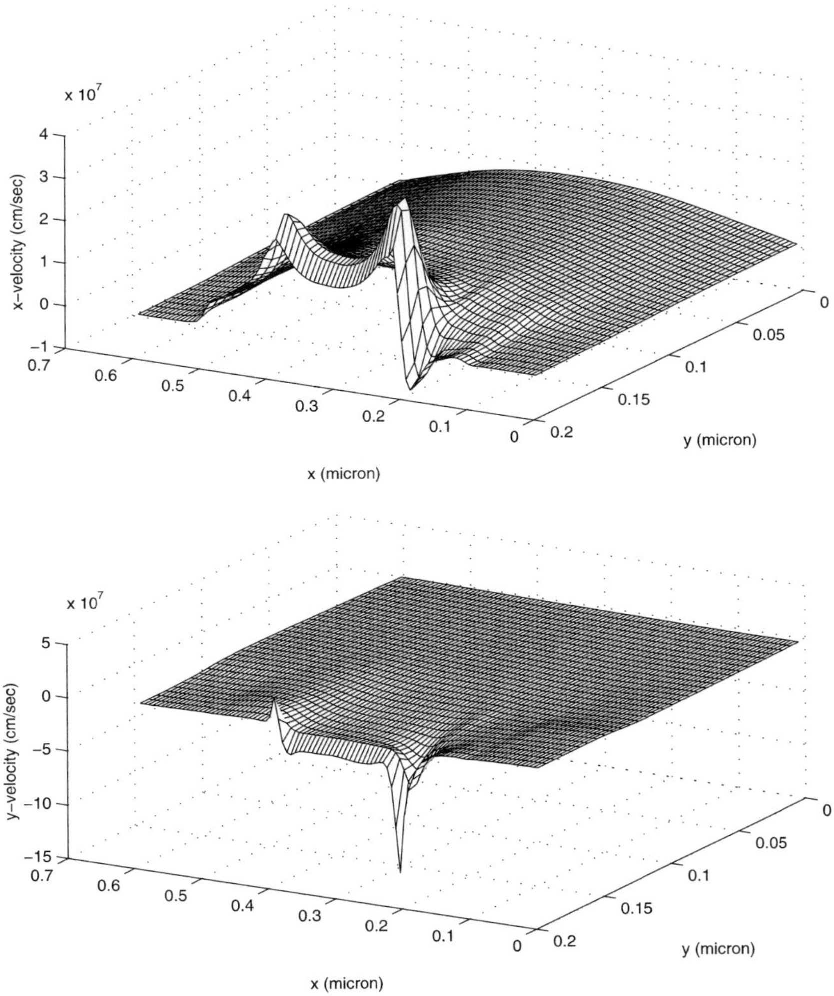

FIG. 3.14. Stationary solution (after 5 ps) for the x-component and y-component of the velocity for $\phi_b = 1$ V.

The shape of the device is taken as rectangular and it is pictured in Fig. 3.11.

The axes of the reference frame are chosen parallel to the edges of the device. We take the dimensions of the MESFET to be such that the numerical domain is

$$\Omega = [0, 0.6] \times [0, 0.2],$$

where the unit length is the micron.

The regions of high doping n^+ are the subset

$$[0, 0.1] \times [0.15, 0.2] \cup [0.5, 0.6] \times [0.15, 0.2].$$

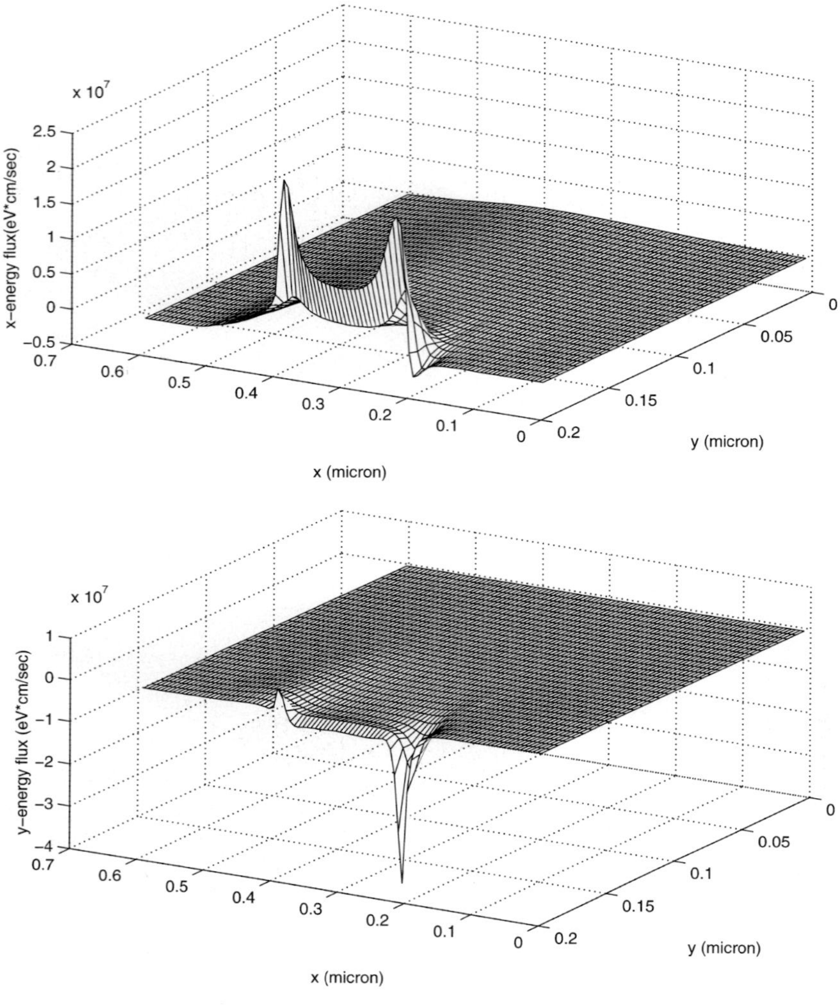

FIG. 3.15. Stationary solution (after 5 ps) for the energy-flux for $\phi_b = 1$ V.

The contacts at the source and drain are 0.1 μm wide and the contact at the gate is 0.2 μm wide. The distance between the gate and the other two contacts is 0.1 μm. A uniform grid of 96 points in the x direction and 32 points in the y direction is used. The same doping concentration as in JEROME and SHU [1994], SHEN, CHENG and LIOU [2000], YIP, SHEN and CHENG [2000] is considered

$$n_D(x) - n_A(x) = \begin{cases} 3 \times 10^{17}\ \text{cm}^{-3} & \text{in the } n^+ \text{ regions,} \\ 10^{17}\ \text{cm}^{-3} & \text{in the } n \text{ region,} \end{cases}$$

with abrupt junctions. n_D and n_A are the number densities of donors and acceptors, respectively.

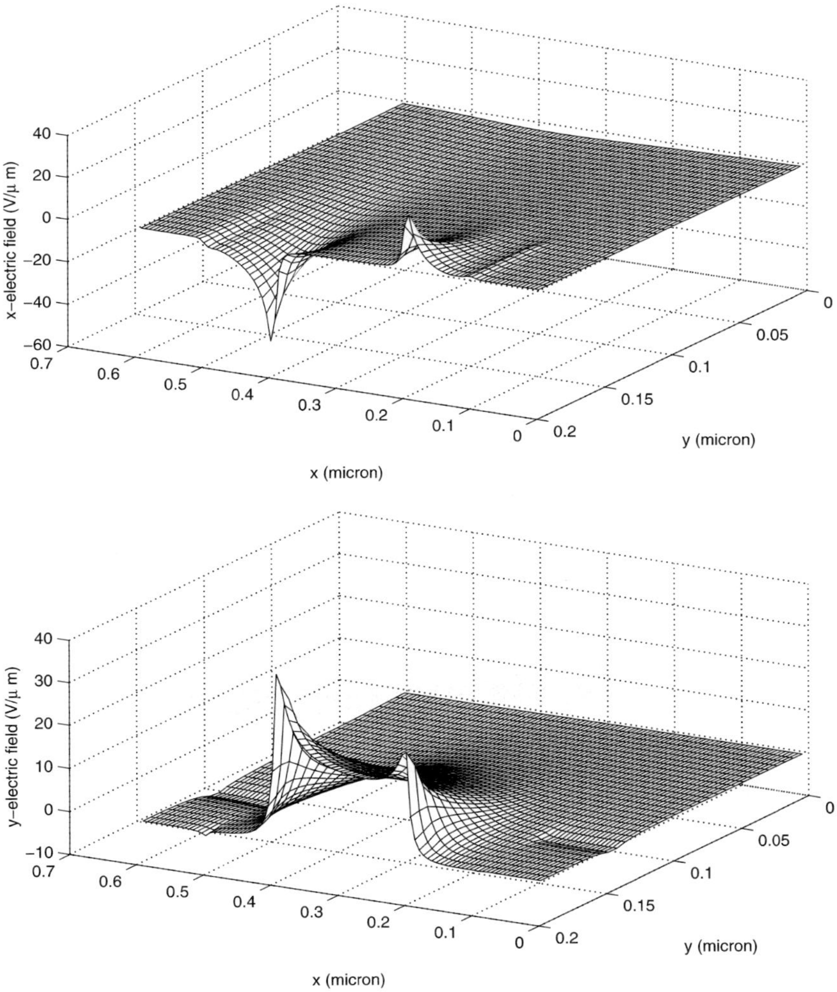

FIG. 3.16. Stationary solution (after 5 ps) for the x-component and y-component of the electric field for $\phi_b = 1$ V.

We denote by Γ_D that part of $\partial\Omega$, the boundary of Ω, which represents the source, gate and drain

$$\Gamma_D = \Big\{(x, y)\colon\ y = 0.2, 0 \leqslant x \leqslant 0.1, 0.2 \leqslant x \leqslant 0.4, 0.5 \leqslant x \leqslant 0.6\Big\}.$$

The other part of $\partial\Omega$ is labelled as Γ_N. The boundary conditions are assigned as follows:

$$n = \begin{cases} n^+ & \text{at source and drain,} \\ n_g & \text{at gate,} \end{cases} \tag{3.8}$$

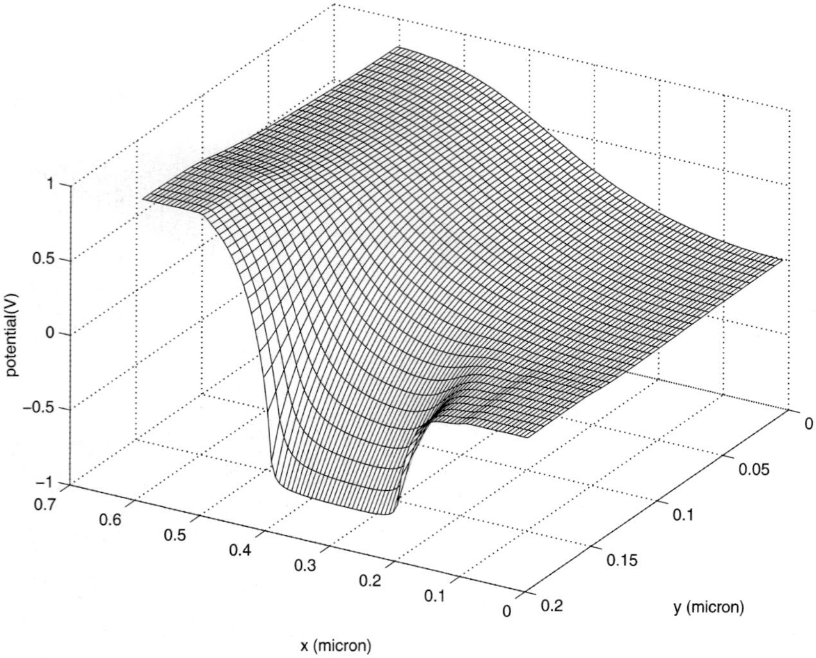

FIG. 3.17. Stationary solution (after 5 ps) for the electric potential for $\phi_b = 1$ V.

$$\phi = \begin{cases} 0 & \text{at the source,} \\ \phi_g & \text{at the gate,} \\ \phi_b & \text{at the drain,} \end{cases} \tag{3.9}$$

$$\begin{cases} W = W_0, & \mathbf{V}\cdot\mathbf{t} = 0, \\ \nabla V_i \cdot \mathbf{n} = 0, & \nabla S_i \cdot \mathbf{t} = 0, \quad \nabla S_i \cdot \mathbf{n} = 0 \end{cases} \quad i = 1, 2 \text{ on } \Gamma_D, \tag{3.10}$$

$$\begin{cases} \nabla n \cdot \mathbf{n} = 0, & \nabla W \cdot \mathbf{n} = 0, \quad \nabla\phi \cdot \mathbf{n} = 0, \\ \nabla V_i \cdot \mathbf{n} = 0, & \nabla S_i \cdot \mathbf{n} = 0 \end{cases} \quad i = 1, 2 \text{ on } \Gamma_N. \tag{3.11}$$

Here ∇ is the two-dimensional gradient operator while $\mathbf{n}$ and $\mathbf{t}$ are the unit outward normal vector and the unit tangent vector to $\partial\Omega$, respectively. n^+ is the doping concentration in the n^+ region and n_g is the density at the gate, which is considered to be a Schottky contact (see SELBERHERR [1984]),

$$n_g = 3.9 \times 10^5 \text{ cm}^{-3}.$$

ϕ_b is the bias voltage and ϕ_g is the gate voltage. In all the simulations we set $\phi_g = -0.8$ while ϕ_b varies.

In the standard hydrodynamical model considered in the literature (e.g., Blotekjaer, Baccarani et al.), the energy flux $\mathbf{S}$ is not a field variable and it is not necessary to prescribe boundary conditions for it. The relations $(3.10)_{5,6}$ and $(3.11)_5$ have no theoretical justification. They are assumed because they seem physically reasonable. Of course a more thorough investigation of this point would be worth-while.

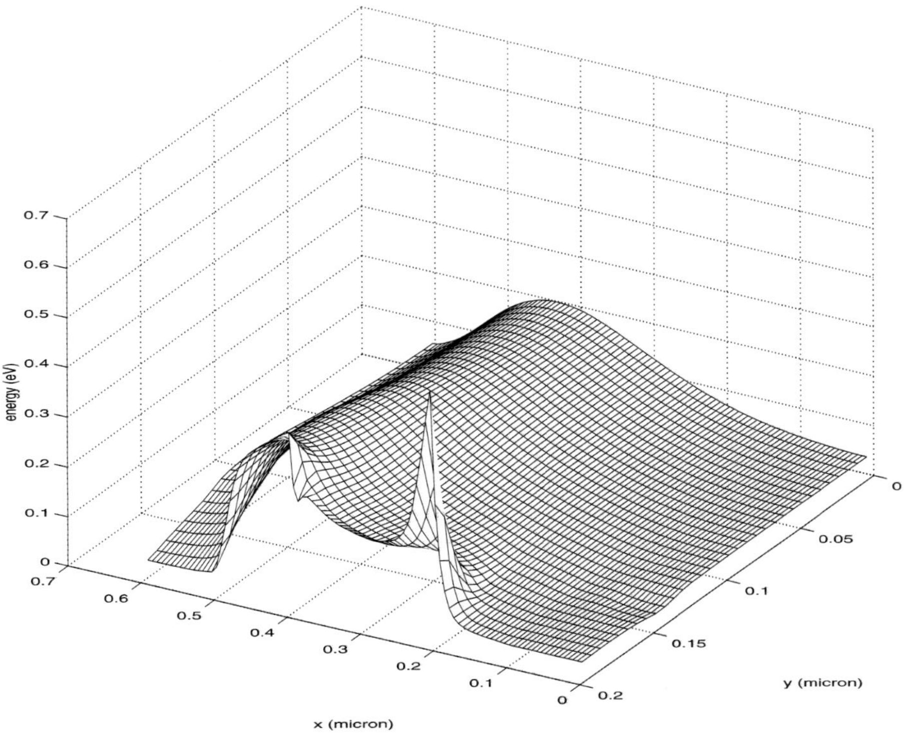

FIG. 3.18. Stationary solution (after 5 ps) for the energy density for $\phi_b = 2$ V.

We start the simulation with the following initial conditions:

$$n(x, y, 0) = n_D(x, y) - n_A(x, y), \qquad W = W_0 = \tfrac{3}{2} k_B T_L,$$
$$V_i = 0, \qquad S_i = 0, \quad i = 1, 2.$$

T_L is the room temperature of 300 K.

The main numerical problems in this work arise from the discontinuous doping and the boundary conditions at the Schottky barrier which gives rise there to sharp changes in the density of several orders of magnitude. The use of a *shock-capturing* scheme is almost mandatory for this problem.

In the first case we take $\phi_b = 1$ V. The stationary solution is reached in a few picoseconds (less than five). After the initial restless behaviour the solution becomes smooth and no signs of spurious oscillations are present. The numerical scheme seems suitably robust and is able to capture the main features of the solution. Only the Kane dispersion relation will be considered here because the results obtained in the parabolic band approximation are rather unsatisfactory when high electric fields are involved as shown in ROMANO [2001] for a silicon n^+-n-n^+ diode.

The density is plotted in Fig. 3.12. As expected there is a depletion region beneath the gate. Moreover one can see that the drain is less populated than the source.

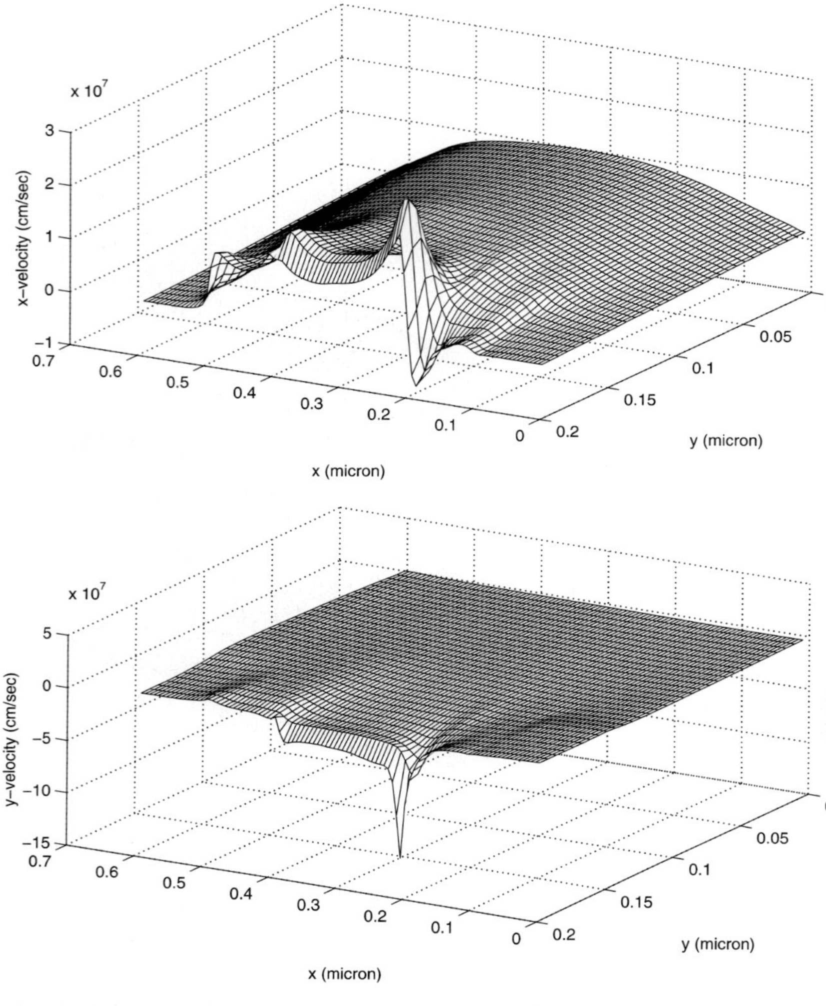

FIG. 3.19. Stationary solution (after 5 ps) for the x-component and y-component of the velocity for $\phi_b = 2$ V.

Concerning the energy (Fig. 3.13) there are sudden variations near the gate edges. The mean energy of the electrons reaches a maximum value of about 0.35 eV in the part of the gate closest to the source.

The results for the velocity are shown in Fig. 3.14. The higher values of the x-component are at the edges of the gate contact. This happens also for the y-component, but with a huge peak at the gate edge closest to the source.

The shape of the energy flux (Fig. 3.15) is qualitatively similar to that of the velocity.

Very large tangential and normal components of the electric field (Fig. 3.16) are present again at the edges of the gate. For completeness the electric potential is also presented in Fig. 3.17.

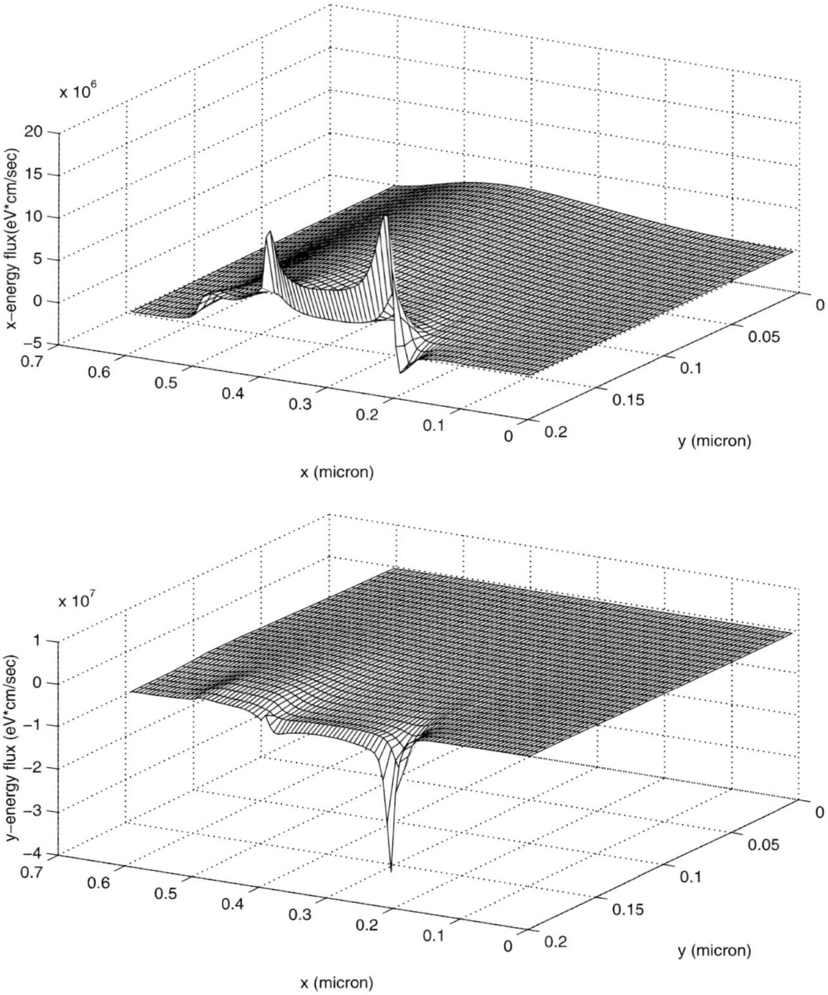

FIG. 3.20. Stationary solution (after 5 ps) for the energy-flux for $\phi_b = 2$ V.

As a second test we take $\phi_b = 2$ V. There are not significant differences to the case $\phi_b = 1$ V for density and electric field. Concerning the other variables, the behaviour of the solution is qualitatively similar to that of the case $\phi_b = 1$ V, but with higher values of the fields. The stationary solution for energy, velocity and energy flux is shown in Figs. 3.18–3.20.

If we compare our results with those obtained in SHEN, CHENG and LIOU [2000], where the standard model BBW with relaxation times extracted from Monte Carlo data has been employed, one notes that the numerical solutions for density and electric field are very similar while the solutions for velocity and energy present some qualitative and quantitative differences.

4. Application of adaptive mesh refinement

4.1. Overview of variable-resolution techniques

Variable resolution, or adaptive computational grids, bypass the problem of reaching the limit of current computer hardware memory by making better use of a smaller number of computational cells. These techniques recognize the fact that there are parts of a solution which need an increased population of cells in order to be captured accurately, while at the same time there are large areas where a coarser grid would be sufficient.

There are a number of underlying considerations to be taken into account when it comes to selecting a type of grid, even before we consider variable resolution. These include viability of generating a good grid, implications for the method of solution, generality of use, treatment of boundaries (especially in complex geometries) and the associated computational expense (regarding overall memory storage requirement and speed of setting-up the grid). Adaptive grids are meant to increase the efficiency of the simulation, so there are some additional issues to be taken into consideration. These include independence of the adaptive software from the numerical method, temporal as well as spacial adaption, efficient implementation on modern computer architectures and the size of the algorithm-related overheads (CPU time for the management and storage of the adaptive grids as compared to the cost of solving on a nonadapted, but finer grid of the same resolution).

The various adaptive meshing techniques can be broadly classified by considering a number of their distinguishing features like time dependence, locality, cell structure and hierarchical logic. Some of the main types are shown in Fig. 4.1 (see also the review by MAVRIPLIS [1996]).

Time dependence of the grid structure is a fundamental attribute of any technique, i.e., whether there is a temporal distribution of the population of the computational cells as well as spatial; the corresponding techniques are referred to as dynamic or static. An obvious application of the static techniques are steady flows, where the grid density may be defined at the beginning of the integration in anticipation of demanding areas of the flow. In unsteady flows, dynamic adaptive methods alter in time the position, size, shape and number of regions of high grid node density in response to evolving flow field structures. Continuous mesh adaption is achieved by automatic refinement and coarsening of the grid according to a set of rules derived from the dynamics or other features of the flow, e.g., a steep gradient of one or more dependent variables.

The *locality* of the refinement distinguishes the various techniques into global or local ones. Global techniques redistribute a given number of cells or grid nodes (Fig. 4.1B, C and D), while local ones add and subtract cells as required (Fig. 4.1E, F). A shortcoming of global techniques when used for simulations where features may appear, disappear or merge, is that it has to be known in advance how many cells will be needed at the most demanding time of the integration. Also, these methods are difficult to extend to three dimensions and the shape of the computational cells dictates the need for specialist numerical schemes. Local techniques on the other hand add and subtract cells to suit the evolution of flow features; no special methods are required for the integration of

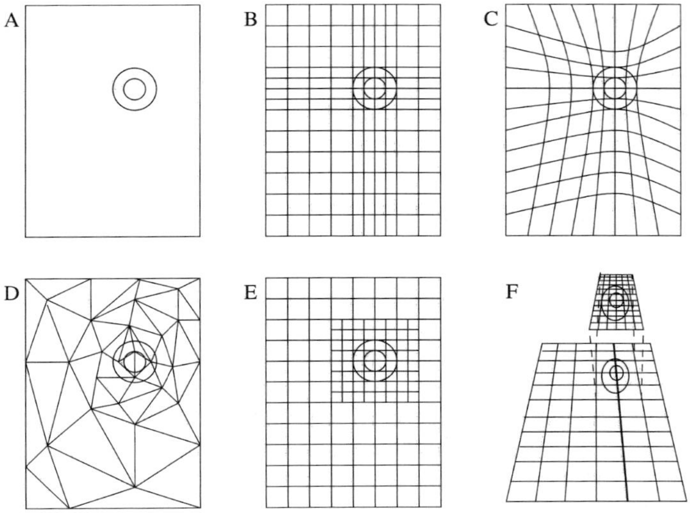

FIG. 4.1. A selection of different approaches to adaptive space discretization. A: the feature that needs high resolution in physical space, B and C: global structured grids, D: global unstructured grid, E: local structured (cell subdivision) grid, F: local structured (AMR) meshes (NIKIFORAKIS [1998]).

the governing equation of flow and extensions to the third dimension are, as a rule, derivatives of their unadapted versions.

The *cell structure* provides a further underlying classification to structured, unstructured and hybrid methods. Care has to be taken in all cases to produce an optimal population of grid cells at the important regions, so that there are no cells with high aspect ratio, which may reduce computational efficiency and/or accuracy.

Another fundamental attribute of any technique is its *hierarchical logic*. To illustrate what we mean by that, let us consider a structured grid consisting of rectangular cells. The obvious way to increase and decrease the resolution locally is to subdivide a number of cells at a certain area(s) of the grid, a process which lends its name to this technique: cell subdivision. An alternative way would be to create separate fine mesh patches which co-exist overlaying a coarse mesh. These overlaying mesh patches are part of a hierarchical system, the latter being a distinguishing feature of a class of methods known as *adaptive mesh refinement* (AMR). A unique feature of AMR is that the meshes can be integrated separately and at a different timestep, thus facilitating adaption *in time* as well as in space.

AMR is our preferred technique for semiconductor applications because time, as well as space refinement has a significant impact on the efficiency of the computational code. Also it is highly desirable to retain the structured character of the computational cells, because of the implications on existing peripheral but vital issues of parametrizations, data structures, graphics, etc. This technique is inherently suitable for running on massively parallel processing computers, by the fact that operations are carried out on standalone discrete meshes.

4.2. *Adaptive Mesh Refinement*

In this chapter the basic principles behind Adaptive Mesh Refinement (AMR) algorithms are outlined; detailed technical descriptions of this technique as it applies to hyperbolic systems can be found in the articles by Berger and co-workers (BERGER and OLIGER [1984], BERGER and COLELLA [1989], BELL, BERGER, SALTZMAN and WELCOME [1994]) and in the articles by various researchers who continued development (e.g., QUIRK and HENEBUTTE [1993], SKAMAROCK and KLEMP [1993] and NIKIFORAKIS, BILLETT, BODEN and PYLE [2001]).

AMR is a technique that dynamically alters the resolution of the computational domain in response to evolving flow-features which are difficult to capture at low resolutions. These features are identified by criteria based on flow properties.

In principle, the computational domain is discretised by a set of embedded meshes (each one forming a rectangular patch), which can be thought of as hierarchical set of grids. At the bottom of the hierarchy lies a coarse, base grid, which completely covers the computational domain in a fundamental way. The resolution of this grid is not altered during the computation, and should be fine enough to capture the bulk of the flow features. Additional, offspring, finer grid patches (noted as G1 and G2 in Fig. 4.2) are nested within the internal boundaries of this underlying grid to increase the resolution locally.

We will reserve the term *grid* to refer to the underlying grid, and use the term *mesh* for the grid patches. A collection of meshes of the same resolution is known as grid level. The size, location and number of these meshes can automatically evolve in time. The area(s) flagged for refinement are determined by using time-dependent information from the solution. The regions which need refinement are the ones where truncation errors are high; various studies (QUIRK [1991], BODEN [1997]) have shown that simple dependent-variable-based criteria (e.g., density jumps or vorticity) will reflect the distribution of truncation error, if chosen appropriately. The hierarchical grid approach allows flow features to transfer smoothly from a coarse domain to a finer one (and vice versa). The grid structure adapts to the evolving flow, resulting in the gliding of fine grids along the underlying coarser ones to cover continuously the evolution of the tracked flow features. More grid patches are automatically added or taken away in response to the changes of the flow topology.

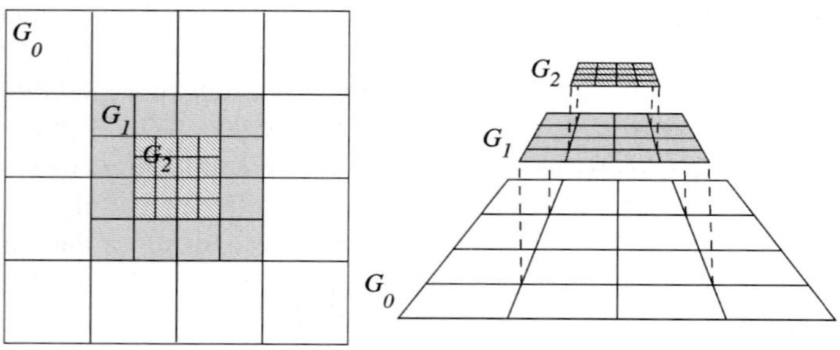

FIG. 4.2. A three-level grid hierarchy as used by Adaptive Mesh Refinement (NIKIFORAKIS [1998]).

The governing equations are integrated from the coarser to the finer grids in succession; a solution exists on all levels at all time, so that the boundary conditions for the finer meshes may be provided by the underlying coarser ones or ultimately the base grid. The solution is interpolated in time from the coarser grids to the finer grid boundaries, which absorb the information via rows of additional cells, known as ghost cells. These are not specific to AMR, but relate to the treatment of computational boundaries by finite volume methods. AMR utilizes them to facilitate internal (mesh connectivity) boundaries. Meshes on a particular grid level are defined in terms of the underlying coarse mesh only as an integer subdivision of the coarse cells, which implies that all meshes are aligned to the x- and y-directions. Early algorithms allowed meshes to be nonaligned (BERGER and OLIGER [1984], SKAMAROCK and KLEMP [1993]), but this is thought to unnecessarily increase computational costs and code complexity. The latter can be further reduced if there is no mesh overlap on a particular level.

For purely hyperbolic systems of equations, the size of the maximum allowable timestep (by the CFL condition required for stability) on the fine grids is smaller than that on the coarse grid. In every other adaptive technique the allowable timestep of the finest mesh would determine the timestep of the complete integration. AMR is unique in the respect that it allows every grid level to advance at its own timestep, which is larger for the coarse ones. For a given large timestep determined by the base grid, which is the timestep of the iteration, the finer meshes match it by a number of smaller ones, a process known as *sub-cycling*. This effectively allows for *refinement in time as well as in space*, i.e., the presence of a few extremely fine computational cells in a small part of the flow domain will not severely restrict the rate at which the rest of the solution is advancing.

Additional complications arise when a mixed elliptic/parabolic/hyperbolic system has to be simultaneously solved on the same computational grid, as the case is with hydrodynamical models of semiconductors. We will elaborate on this and other aspects related to these models in the following sections.

4.3. *Application of AMR to semiconductor simulations*

Our main aim is to present the application of AMR to the hydrodynamical semiconductor models described in the previous chapters of this handbook. Although the technique is fairly mature, there are a number of difficulties introduced by the nature of these models.

To illustrate the technique in practice and highlight the additional difficulties encountered in semiconductor hydrodynamical simulations we consider a representative system of the general form (in the case of the parabolic band approximation):

$$u(t,x,y)=[n,n\mathbf{v},E,\ldots]^{\mathrm{T}},\qquad E=\tfrac{1}{2}m^*nv^2+\tfrac{3}{2}p,\qquad p=k_BnT,\tag{4.1}$$

$$\frac{\partial u}{\partial t}+\frac{\partial f(u)}{\partial x}+\frac{\partial g(u)}{\partial y}=s(u,E_f)+\frac{\partial}{\partial x}\left(\hat{\kappa}\frac{\partial T}{\partial x}\right)+\frac{\partial}{\partial y}\left(\hat{\kappa}\frac{\partial T}{\partial y}\right),\tag{4.2}$$

$$\frac{\partial^2\phi}{\partial x^2}+\frac{\partial^2\phi}{\partial y^2}=-\frac{\partial E_fx}{\partial x}-\frac{\partial E_fy}{\partial y}=\frac{e}{\varepsilon}(n-n_D),\tag{4.3}$$

where the variables comprising the state vector u may vary between models but usually include the number density n of electrons, a momentum analog $n\mathbf{v}$ (where $\mathbf{v}$ is the mean electron velocity), and the total electron energy per unit volume E. The energy density is related to the temperature T and pressure p of the 'electron gas' through Eq. (4.1) (where k_B is the Boltzmann constant and m^* is the effective electron mass).

The evolution for the unknown variables u is represented by the system described by Eqs. (4.2). The flux vector f and the source vector s are functions of the unknown variables (although they do not have any dependence on the spatial or temporal derivatives of the variables). The left-hand side of Eq. (4.2) has the form of a system of conservation laws and describes the transport of the unknown variables. The source part of the system is a mixture of forcing terms, based on the electric field E_f, and terms describing scattering processes. The final terms in Eq. (4.2) account for heat conduction. The vector $\hat{\kappa}$ of conduction coefficients is either identically zero or has only one nonzero component which may depend on the values of the variables u. The evolution system (4.2) is typically either hyperbolic in character (if $\hat{\kappa}$ is identically zero) or mixed hyperbolic-parabolic (if $\hat{\kappa}$ has a nonzero component). The electric field E_f and the electric potential ϕ are related to the electron density n through the Poisson equation (4.3) in which e is the magnitude of the electron charge, ε is the dielectric constant of silicon, and $n_D(x)$ is the number density of donor atoms in the material.

The issues that have to be addressed are the positioning of the grids, the timestep management over the domain, the coupling of the elliptic to the hyperbolic (or hyperbolic/parabolic) mode and the selection of suitable numerical methods for the solution of the hyperbolic or hyperbolic-parabolic evolution system (4.2) and the elliptic equation (4.3).

4.4. *Refinement criteria and time-adaption*

Before we consider suitable schemes to solve this system of equations, the regions of high resolution have to be positioned on the computational domain; this is achieved automatically during every iteration by an internal inference procedure which is based on a number of user-defined standards, known as *refinement criteria*. The selection of appropriate refinement criteria when using AMR (and other variable grid approaches) is an issue which needs particular care in the area of hydrodynamic semiconductor simulations. There is not as yet a reliable theory to suggest in which regions of the computational domain a low density of grid cells can be used without adversely affecting the solution accuracy, or conversely in which regions a high resolution is desirable. This is also true for AMR applications in general; while early work (BERGER and OLIGER [1984]) suggested that an estimate of the local truncation error of the numerical solution could be used to predict an appropriate level of resolution, in practice heuristic criteria (such as having an increased resolution near to shock waves) have proved to be more effective.

For hydrodynamics semiconductor simulations, the velocity variable v can be used as an indicator of where complex behaviour is present, and the AMR algorithm refines any region in which the gradient or the curvature of the velocity variable v is large (that is, where either of the quantities $|v_{n+1} - v_n|$ or $|v_{n+1} - 2v_n + v_{n-1}|$ has a value above a specified threshold). In addition, it is found that a modest degree of refinement is needed in the highly doped regions of the device during the early stages of the simulation in order to produce an acceptable degree of accuracy there. The second derivative of the solution is also a useful indicator of when increased resolution is desirable: large values suggest that additional points in the solution cannot be inferred by linear interpolation. It may also be noted that these refinement criteria tend to increase the resolution in regions where the solution has steep gradients or extrema, and these are also the regions where integration methods are known to lose accuracy.

The fact that the solution exists separately on every grip patch, lends itself to timestep management for individual-meshes, i.e., adaption in time, as mentioned before. To illustrate this concept, consider the case of one coarse grid and one fine grid, as shown in Fig. 4.3. Stability requirements impose a maximum size on the time steps used to advance the numerical solution, this being proportional to the cell size used to discretize the domain. One way to ensure stability would be to advance all of the grids in the AMR grid structure at the same time step, the smallest time step required by the finest grid. This is obviously inefficient, since more work must be done to advance coarse grids than is necessary. Also, because of numerical diffusion (see, for example, LEVEQUE [1990] or TORO [1999]) there is typically a loss of accuracy involved when a numerical solution is advanced at time steps much smaller than the maximum value.

This problem can be bypassed within the AMR framework because different time step sizes can be used for different grids; for example, a sub-grid which is finer by a factor of two than the base grid will use time steps of half the size. This is implemented by sub-cycling the advancement of the grids so that finer grids are advanced multiple times compared to coarse grids, as shown in Fig. 4.3. For the situation shown in Fig. 4.3, if u is the numerical solution, then initially the value of $u(t_0)$ is known both on the coarse grid and the fine grid. The algorithm first advances the solution on the coarse grid to give a value for $u(t_1)$ there, and then advances the solution on the fine grid by one step to get $u(t_{1/2})$. In order for the second step on the fine grid to be taken, boundary data for it must be prescribed; this is achieved by interpolation of the coarse grid values $u(t_0)$

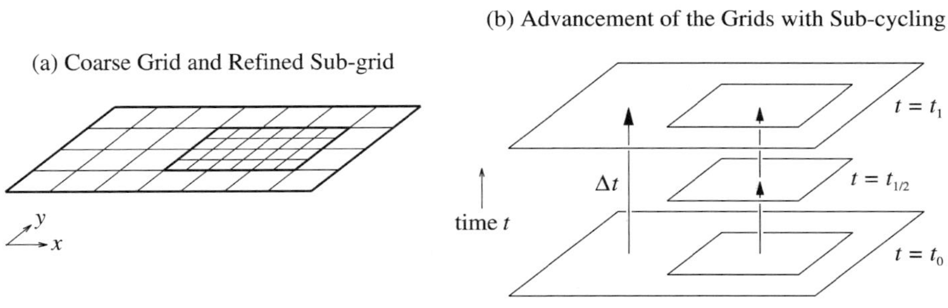

FIG. 4.3. Sub-cycling in time of an AMR grid structure. (a) A simple compound grid consisting of a coarse base grid partially covered by a sub-grid which is finer by a factor of two. In this figure, a two-dimensional spatial domain is pictured since the relationship between the grids is most clear in this case. (b) Advancement of the grid structure in time from $t = t_0$ to $t = t_1$. The coarse grid is advanced in one single step of size Δt, whereas the fine grid is sub-cycled, taking two steps of size $\Delta t/2$ such that it exists at an intermediate time $t = t_{1/2} \equiv (t_0 + t_1)/2$.

and $u(t_1)$, that is,

$$u(t_{1/2}) \approx \tfrac{1}{2}\big(u(t_0) + u(t_1)\big)$$

on the coarse grid.

Following this approach, a numerical solution can be advanced to the same time throughout the AMR grid structure, with recursion being used for situations more complicated than the example of Fig. 4.3; full details are given by BERGER and OLIGER [1984]. In doing so, there is an underlying assumption that the numerical scheme used to advance the solution is expected to use a single, explicit step. This assumption is only valid when the equations solved are purely hyperbolic.

4.5. An operator splitting approach

When the system is mixed hyperbolic/elliptic, errors give rise at the boundaries of the coarse/fine mesh patches. To rectify this problem, a splitting approach is used to time step the equations by separating them into four components which are treated in turn. Full sub-cycling of the AMR algorithm is possible for the hydrodynamical models provided that the electric field is slow changing and the component parts of the evolution system are advanced in a particular order.

Each update of the solution data by one time step is broken into four stages: first, Poisson's equation is solved to determine the electric field; second, the solution is updated taking into account only the scattering and forcing terms; third, the solution is updated based only on the conservation law part of the system; finally, the heat conduction terms in the energy equation (whenever are present) are accounted for. This is a simple first-order splitting method, while higher-order ones can be derived (see, for example, ROMANO and RUSSO [2000]), experimentation suggests that there is only a small loss in accuracy when first-order splitting is used, and furthermore sub-cycling of the AMR algorithm is easier to achieve in this case. For the splitting approach implemented here, a numerical solution u_0 at time t_0 is advanced to a solution u_1 at time $t_1 = t_0 + \Delta t$ according to

$$E_f = \mathrm{P}(u_0), \tag{4.4}$$

$$u^\star = \mathrm{S}(u_0, E_f; \Delta t), \tag{4.5}$$

$$u^{\star\star} = \mathrm{T}(u^\star; \Delta t), \tag{4.6}$$

$$u_1 = \mathrm{H}(u^{\star\star}; \Delta t), \tag{4.7}$$

where E_f is the electric field. The operator P solves the Poisson equation (4.3). The source terms in Eq. (4.2) which represent forcing and scattering processes comprise the operator S, while the transport terms in that equation comprise the operator T. If parabolic heat conduction terms are present in the model, then they are incorporated in the operator H.

One of the benefits of this splitting numerical approach is that any model that fits the general form of Eqs. (4.1), (4.2) and (4.3) can be implemented in this algorithm. Also, reliable numerical schemes can be used on each component of the system (which can be updated as new methods appear), and there is not a strong dependence of the numerical code on the equations being solved. The different stages of this operator-splitting approach are discussed separately in the subsections below.

4.6. The Poisson equation for the electric field

The first stage of the update procedure solves Poisson's equation (4.3) subject to boundary conditions on the electric potential ϕ. This equation must be solved simultaneously over all component meshes in the simulation domain (or sometimes a subset of the domain if full AMR sub-cycling is used). This contrasts with how the AMR algorithm tackles hyperbolic problems by updating each component mesh individually. To solve Poisson's equation a multigrid method can be employed, not least because this class of methods has some similarities with AMR and are very efficient.

For the purposes of this exercise, the approach of MINION [1996] is followed, who had to solve a Poisson's equation as part of an AMR projection scheme for incompressible Euler flows. For a single uniform grid, the scheme constructs a sequence of auxiliary grids, the first a factor of two coarser than the initial one, the second a factor of four coarser, and so on. Given an approximation to the solution on the original grid, Poisson's equation is cast into residual form, and the problem is coarsened (or 'restricted') and solved recursively on the next grid in the sequence. When the coarsest grid is reached, Poisson's equation is solved exactly using a single grid scheme, and error corrections are interpolated (or 'prolongated') back up the sequence of grids. At each level, Gauss–Seidel relaxation steps with red-black ordering are used to 'smooth' the solution. The complete multigrid cycle is iterated until the residual is smaller than some specified tolerance.

Within an AMR framework, multiple grid levels already exist, and (provided that the AMR refinement factors are powers of two) a multigrid arrangement can be constructed straightforwardly by filling in intermediate and coarser levels. The multigrid algorithm becomes more complicated however because fine grids do not in general completely cover coarser grids. Relaxation operations and residual calculations may need to be performed over several levels simultaneously, and special treatment must be given to the internal boundaries between coarse and fine meshes (in a similar way to the use of 'flux corrections' in AMR simulations of hyperbolic systems).

Although full sub-cycling of the AMR code has been found to produce accurate results, in practice the nature of the multigrid solver means that a simulation will be very much faster if Poisson's equation is solved across the entire domain (interpolating source values in time as necessary) on every time step. In fact, since the electric potential varies only very slowly after the early stages of a simulation, typically few iterations of the multigrid cycle are needed to update its value, and so Poisson's equation works out as relatively inexpensive to solve, even without full sub-cycling.

4.7. Scattering and forcing terms

If the simulation variables $u(t,x,y) = [n, n\mathbf{v}, E, \ldots]^{\mathrm{T}}$ are updated only using the terms on the right-hand sides of Eqs. (4.2) (which model collision processes and acceleration due to the electric field), then the evolution equations reduce to a system of ordinary differential equations,

$$\frac{\partial u}{\partial t} = s(u, E_f). \tag{4.8}$$

Solution of this system (with the electric field E_f fixed at the value determined by the solution of Poisson's equation) comprises the second stage of the update procedure for the hydrodynamical model.

The solution in each grid cell is advanced using a second-order explicit Runge–Kutta method (see, for example, PRESS, TEUKOLSKY, VETTERLING and FLANNERY [1992]). To ensure stability, multiple Runge–Kutta steps may be employed: for a time step Δt, the solution is advanced by taking k sub-steps of size $\Delta t/k$. (Typically $k = 4$ in this work.)

4.8. The nonlinear hyperbolic system

For the third stage of the update procedure, the solution u at the end of the second stage is used as initial data for the solution of a system of conservation laws

$$\frac{\partial u}{\partial t} + \frac{\partial f(u)}{\partial x} + \frac{\partial g(u)}{\partial y} = 0, \quad u = [n, n\mathbf{v}, E, \ldots]^{\mathrm{T}} \tag{4.9}$$

formed from the transport part of the hydrodynamical equations (4.2).

The evolution system (4.9) is hyperbolic and it is solved using directional splitting: the solution is advanced by a time step Δt by solving first one then the other of the two directional sub-systems

$$\frac{\partial u}{\partial t} + \frac{\partial f(u)}{\partial x} = 0 \quad \text{and} \quad \frac{\partial u}{\partial t} + \frac{\partial g(u)}{\partial y} = 0, \tag{4.10}$$

with the order of the updates reversing on each step. The use of splitting to solve Eq. (4.9) is not a part of the overall splitting scheme used to advance the hydrodynamical model; an unsplit method could equally be used. However, directional splitting has been found to be reliable in the past, and it has the advantage over multi-dimensional methods that it is very simple to implement.

The one-dimensional transport systems are solved using the SLIC (Slope Limiter Centred – TORO [1999]), which was previously used in ANILE, NIKIFORAKIS and PIDATELLA [1999], ANILE, JUNK, ROMANO and RUSSO [2000], HERN, ANILE and NIKIFORAKIS [2001]. The scheme is conservative, explicit and second-order accurate on smooth regions of the solution, and it accurately resolves discontinuities without introducing unphysical oscillations.

The SLIC method is considered particularly useful in the present work because (in contrast to Riemann problem-based schemes) it does not require characteristic information about the system of equations being solved. This means that only minor changes need to be made to the code to allow it to solve different hydrodynamical models, and furthermore there are no difficulties in evolving solutions to those hydrodynamical models for which characteristic information cannot easily be obtained.

4.9. Heat conduction

The final stage of the update procedure advances the energy variable E by a time step Δt based on the heat conduction term in Eq. (4.2),

$$\frac{\partial E}{\partial t} = \frac{\partial}{\partial x}\left(\kappa \frac{\partial T}{\partial x}\right) + \frac{\partial}{\partial y}\left(\kappa \frac{\partial T}{\partial y}\right). \tag{4.11}$$

Assuming the conduction coefficient κ to be fixed during the time step, and recalling the relationship between energy E and temperature T, Eq. (4.11) is seen to have the form of a linear scalar diffusion equation with space-dependent coefficients. The equation is parabolic, and is most effectively solved using implicit methods. There are many standard methods for solving diffusion equations, and the ADI (alternating direction implicit) scheme is adopted here. This has the advantage that it only requires implicit equations to be solved in one dimension at a time, which makes it very much less computationally expensive than a fully multi-dimensional implicit approach. The ADI method is stable for any size of time step; however (because it does nothing to damp high-frequency modes) it may still cause oscillatory behaviour at discontinuities in the temperature variable. On the occasions when this has proved problematic, an alternative solution scheme has been used: Eq. (4.11) can be solved through directional splitting in the same way as described for Eq. (4.9), with fully implicit steps being used to solve each of the one-dimensional component equations. While the latter method is resilient against oscillations, this is at the expense of the formal order of accuracy of the solution.

Additional boundary data must be specified when using implicit schemes compared to explicit ones. The implications this has for internal mesh boundaries within the AMR framework is discussed in HERN, ANILE and NIKIFORAKIS [2001].

4.10. Numerical simulations

The AMR approach described in the previous sections is designed primarily for solving time-dependent problems, but it can also be used to produce solutions to steady state problems. In this section we consider two case studies, in one and two space dimensions, to demonstrate the use of AMR for hydrodynamic semiconductor simulations.

It should be noted that although the accuracy of these simulations has been checked (for the steady-state) against other numerical studies, the purpose of this exercise is to demonstrate the efficient use of the AMR technique, not to make any claims regarding the details of the solutions.

4.11. A silicon diode

The salient features of the approach are best illustrated by considering a one-dimensional problem. To this end the ballistic n^+-n-n^+ silicon diode is used here. This is a standard test problem for numerical semiconductor simulations in one dimension. The sub-micron device can be considered as a model for a channel in a MOSFET. The numerical results shown here are for the extended hydrodynamical model introduced in Section 1.4 suitably simplified by considering the parabolic band limit and also by expressing the production terms as simple relaxation type terms. This simplified model we call the reduced hyperbolic model. In this case the vectors U, $F(U)$ and $S(U)$ are:

$$U = \begin{bmatrix} n \\ nv \\ 3p/m^* \\ 2q/m^* \end{bmatrix}, \tag{4.12}$$

$$F(U) = \begin{bmatrix} nv \\ p/m^* \\ 2q/m^* \\ \frac{8}{15}q/m^* \\ 5p^2/n(m^*)^2 \end{bmatrix}, \tag{4.13}$$

$$S(U) = \begin{bmatrix} 0 \\ -nv/\tau_p - neE_f/m^* \\ -2(E - E_0)/m^*\tau_W - 2nevE_f/m^* \\ 1/\tau_q(2q/m^*) \\ -eE_f/m^*(5p/m^*) \end{bmatrix}. \tag{4.14}$$

Here n is the electron density,v is the electron velocity, p is the electron fluid pressure, m^* is the effective electron mass q is the energy flux, τ_p is the relaxation time for momentum, τ_w is the relaxation time for energy, τ_q is the relaxation time for the energy flux, e is the absolute value of the electron charge, E_f is the electric field, E is the energy density

$$E = (1/2)m * v^2 + (3/2).$$

E_0 is the thermal equilibrium energy density.

We remark that for this reduced hyperbolic model the interpretation of q is not that of heat flux but of total energy flux.

Still for the sake of simplicity the relaxation times are obtained as functions of energy E from fitting to MC simulation for the same benchmark device (see ANILE, JUNK, ROMANO and RUSSO [2000]).

The sub-micron n^+-n-n^+ device simulated here has a total length of $L = 0.6$ μm divided up into a source region of length 0.1 μm, a channel of length 0.4 μm, and a drain region of length 0.1 μm. The device is doped with donor atoms in a profile

$$n_D(x) = \begin{cases} 10^{18} & \text{for } x < 0.1\ \mu\text{m}, \\ 10^{16} & \text{for } 0.1\ \mu\text{m} \leqslant x \leqslant 0.5\ \mu\text{m}, \ (\text{donors/cm}^3) \\ 10^{18} & \text{for } 0.5\ \mu\text{m} < x. \end{cases} \tag{4.15}$$

The values used for the lattice temperature, the dielectric constant, and the effective electron mass in the silicon device are

$$T_0 = 300\ \text{K}, \quad \varepsilon = 11.7\varepsilon_0, \quad m^* = 0.32m_e, \tag{4.16}$$

where ε_0 is the vacuum dielectric constant, and m_e is the electron mass.

It may be noted that the doping profile of Eq. (4.15) is a discontinuous function, in contrast to the profiles used in much of the literature which are smooth across the device junctions. (The discontinuous profile (4.15) is the same as the doping used by ANILE, JUNK, ROMANO and RUSSO [2000].) A discontinuous doping profile is adopted here to highlight the ability of the numerical code (in particular the SLIC scheme), to accurately evolve solutions which include steep gradients without producing unphysical oscillations. It should be observed however that slight differences in the doping profile for the n^+-n-n^+ device (such as the degree of smoothness at the junctions) can have significant effects on the form of the solution.

As initial data for the simulations that follow, the following choice is made:

$$n = n_D, \quad v = 0, \quad T = T_0, \quad q = 0 \quad \text{at time } t = 0, \tag{4.17}$$

where the energy flux q is only needed in the reduced model, and the values for the energy density E and the pressure p are derived from the temperature T via Eq. (4.1).

The simulations use 'transmissive' boundary conditions for the unknown variables u. (At the right boundary, ghost cells $u_{N+1}, u_{N+2}, \dots$ are given the values of u_N, $u_{N-1}, \dots$, with the left boundary being treated similarly.) The electric field across the device is calculated based on a specified voltage change between the terminals:

$$\phi|_{x=0.6\ \mu\text{m}} - \phi|_{x=0.0\ \mu\text{m}} = V_{\text{bias}}, \tag{4.18}$$

where V_{bias} can vary between simulations. These boundary conditions are found to be stable in practice and have been observed to allow waves to leave the domain of the simulation cleanly without introducing spurious reflections. In particular, if the extent of the computational domain is increased by moving the boundaries outwards by 0.3 μm (while applying the potential difference (4.18) across the same region as before) then the solution at the original edges of the domain is found to be almost unchanged, both for time-dependent and steady state solutions.

Results are shown in Figs. 4.4 and 4.5. For a constant applied voltage V_{bias}, complex transient behaviour is seen in the solution during the first few pico-seconds, after which time the solution gradually settles down to a steady state which is largely independent of the transient behaviour that has gone before. Since the transient features often appear on small spatial scales, the ability of the AMR code to locally increase the resolution of the numerical solution is well tested by this problem, and an improvement in efficiency is seen compared to simulations based on unrefined grids. The simulation is run up to a time of $t = 1$ ps, which is sufficient in which to observe the most complex behaviour of the transient phase of the solution.

A uniform base grid of resolution $\Delta x = L/200$ (where $L = 0.6$ μm is the length of the device) is used by the simulation, and grid refinement is by a factor of 2 at each level. A maximum resolution of $\Delta x = L/1600$ is used during the simulation.

Fig. 4.4 shows how the different levels of refinement used in the AMR simulation adapt in time to follow evolving features of the transient solution. The figure shows the spatial and temporal domain of the simulation shaded according to the degree of refinement used, from white at the coarsest resolution ($\Delta x = L/200$) to black where the resolution is at its maximum ($\Delta x = L/1600$). The figure illustrates the overall behaviour of the transient solution, with complex features originating at the device junctions and then spreading out into the rest of the domain.

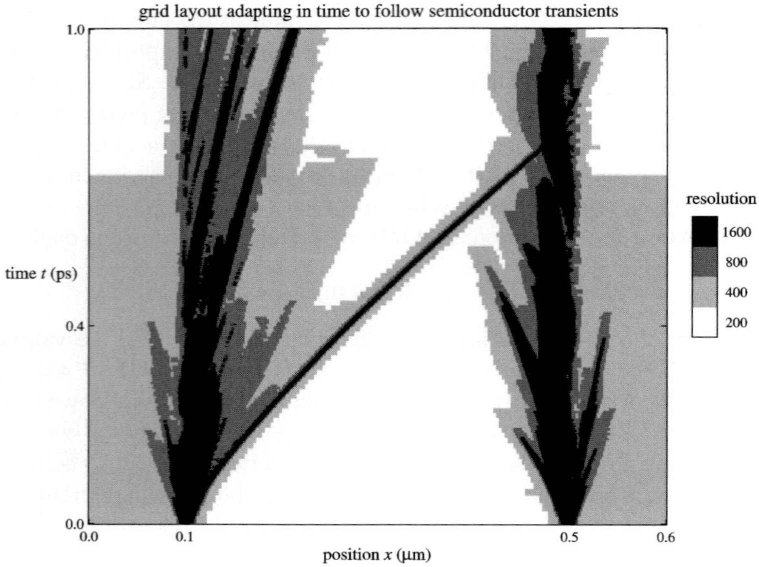

FIG. 4.4. Spatial and temporal variations in resolution for an AMR simulation of transient behaviour in a silicon diode. The space–time domain is shaded to indicate the degree of refinement used, with white at one extreme indicating the base resolution of $\Delta x = L/200$ (where $L = 0.6$ μm is the length of the device) and black at the other indicating the maximum resolution of $\Delta x = L/1600$. Numerical results from this simulation are shown at the time $t = 0.4$ ps in Fig. 4.5.

The value of the AMR approach is that it can produce results which are of a comparable accuracy to those produced using traditional methods, but at a reduced computational cost. It has been found that for simulations in which the mesh refinement capabilities of the code are not used, an accurate solution for the transient behaviour of the n^+-n-n^+ device can be obtained by using a single uniform grid of resolution $\Delta x = L/1600$. (The solution is considered as 'accurate' because there is no discernible difference between the plotted results at this resolution and those at a higher resolution of $\Delta x = L/3200$.) The test of the AMR simulation is then whether its results are comparable in accuracy to those of the high resolution unrefined simulation, and, if so, whether an improvement in computational efficiency is seen.

In Fig. 4.5 the velocity variable v is plotted at a time $t = 0.4$ ps for the AMR simulation and also for two unrefined simulations having resolutions the same as the maximum and the minimum AMR resolutions. The AMR results can be seen to be in very good agreement with the results from the high resolution unrefined simulation, and this level of agreement is present throughout the simulation for all of the evolved variables. The results from the low resolution unrefined simulation provide a contrast with this: while they capture the overall form of the solution adequately, the details are not well resolved.

Returning to the issue of the efficiency of the AMR method, as a measure of the amount of computational work performed during a simulation, the total number of grid cell advancements is used. For the high resolution unrefined simulation shown

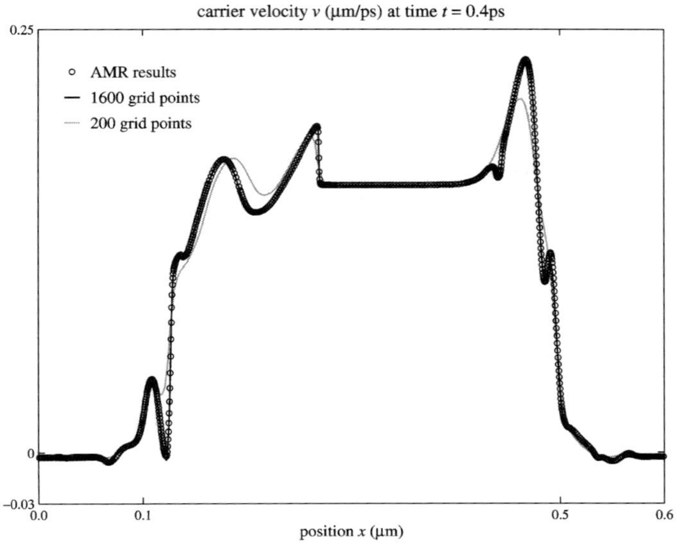

FIG. 4.5. Results at different resolutions for the electron velocity v in simulations of transient semiconductor behaviour. The problem set-up is the same as in Fig. 4.4. Results from an AMR simulation with minimum resolution $\Delta x = L/200$ and maximum resolution $\Delta x = L/1600$ are plotted as circles, with one circle for each computational point. Results from two unrefined simulations with resolutions the same as the maximum and minimum AMR resolutions are plotted as black and grey lines, respectively. The grid arrangement throughout the AMR simulation is plotted in Fig. 4.4.

in Fig. 2.3, 1921 time steps are taken to reach time $t = 1$ ps, and multiplying this by the 1600 cells in the grid gives a total work load of 3 073 600 units. For the AMR simulation which produces similar results the work load is found to be 739 174 units, smaller by a factor 4.2. The work load gives a reasonable estimate of the CPU time used by a simulation, but does not take into account the time taken performing grid management operations (which may usually be assumed to be small compared to the grid integration time) or computer hardware effects such as memory cache behaviour. In terms of actual running time, the improvement of the AMR simulation over the unrefined simulation is estimated to be closer to a factor of 3.

The results of this section demonstrate that, by locally varying the resolution used, the AMR method can produce accurate numerical solutions more efficiently than a traditional single grid approach. The gain in efficiency is most significant if small features of the solution are of particular interest. For example, if the AMR code is used to reproduce Fig. 4.5 but with a resolution of $\Delta x = L/3200$ in the neighbourhood of the shock front, then the work load is found to be around 14 times smaller than for a simulation using a single grid of uniform high resolution.

4.12. A MESFET Device

To demonstrate the technique in two space dimensions, a two-dimensional silicon MESFET (Metal-Semiconductor Field-Effect Transistor) is used here. The device, the specifications of which are taken from the paper by JEROME and SHU [1994], has been used by a number of researchers for testing numerical methods and carrier transport models.

The BBW hydrodynamical model, which has been discussed in Section 1.3 of this volume, is adopted for describing electron transport in the MESFET device. Although it has many shortcomings from the theoretical viewpoint (addressed by alternative hydrodynamical models which have recently been developed), the BBW model is widely used in practice.

The MESFET device is 0.6 μm by 0.2 μm in the (x, y)-plane, with symmetry assumed along the z-direction. Along the top edge of the device ($y = 0.2$ μm) there are three contacts: the source ($0 \leqslant x \leqslant 0.1$ μm), the gate ($0.2\ \mu\text{m} \leqslant x \leqslant 0.4$ μm), and the drain ($0.5\ \mu\text{m} \leqslant x \leqslant 0.6$ μm). The geometry of the device is shown in Fig. 4.6.

The device is doped with donor atoms according to

$$n_{\text{dope}}(x, y) = \begin{cases} n_{\text{high}} & \text{for } (x \leqslant 0.1\ \mu\text{m or } x \geqslant 0.5\ \mu\text{m}) \text{ and } y \geqslant 0.15\ \mu\text{m}, \\ n_{\text{low}} & \text{otherwise}, \end{cases} \tag{4.19}$$

to give highly-doped (n^+) regions next to the source and drain (see Fig. 4.6). Note that the profile is discontinuous between regions of high and low doping. The initial data for the simulation is taken as

$$n = n_{\text{dope}}, \qquad \mathbf{v} = 0, \qquad T = T_0. \tag{4.20}$$

The boundary conditions vary around the perimeter of the device. At the contacts (source, gate and drain) the velocity and temperature satisfy

$$v_x = 0, \qquad \frac{\partial v_y}{\partial y} = 0, \qquad T = T_0, \tag{4.21}$$

while the density and electric potential are set differently for each contact;

$$\text{Source:} \quad n = n_{\text{high}}, \qquad \phi = \frac{k_B T_0}{e} \ln(n/n_i), \tag{4.22}$$

$$\text{Gate:} \quad n = n_{\text{gate}}, \qquad \phi = \frac{k_B T_0}{e} \ln(n/n_i) + \phi_{\text{gate}}, \tag{4.23}$$

$$\text{Drain:} \quad n = n_{\text{high}}, \qquad \phi = \frac{k_B T_0}{e} \ln(n/n_i) + \phi_{\text{bias}}. \tag{4.24}$$

At all other points on the boundary, the variables (including the potential) are set to have zero derivatives in the direction normal to the boundary.

The density values used in the simulation are

$$n_{\text{high}} = 3 \times 10^{23}\ \text{m}^{-3}, \qquad n_{\text{low}} = 1 \times 10^{23}\ \text{m}^{-3}, \qquad n_{\text{gate}} = 3.9 \times 10^{11}\ \text{m}^{-3}, \tag{4.25}$$

where the extremely low density at the gate compared to elsewhere in the device should be noted. The potential differences applied between the three contacts are

$$\phi_{\text{gate}} = -0.8\ \text{V}, \qquad \phi_{\text{bias}} = 2\ \text{V}. \tag{4.26}$$

For fixed applied voltages, this MESFET device reaches a steady state in a time of 5–10 ps.

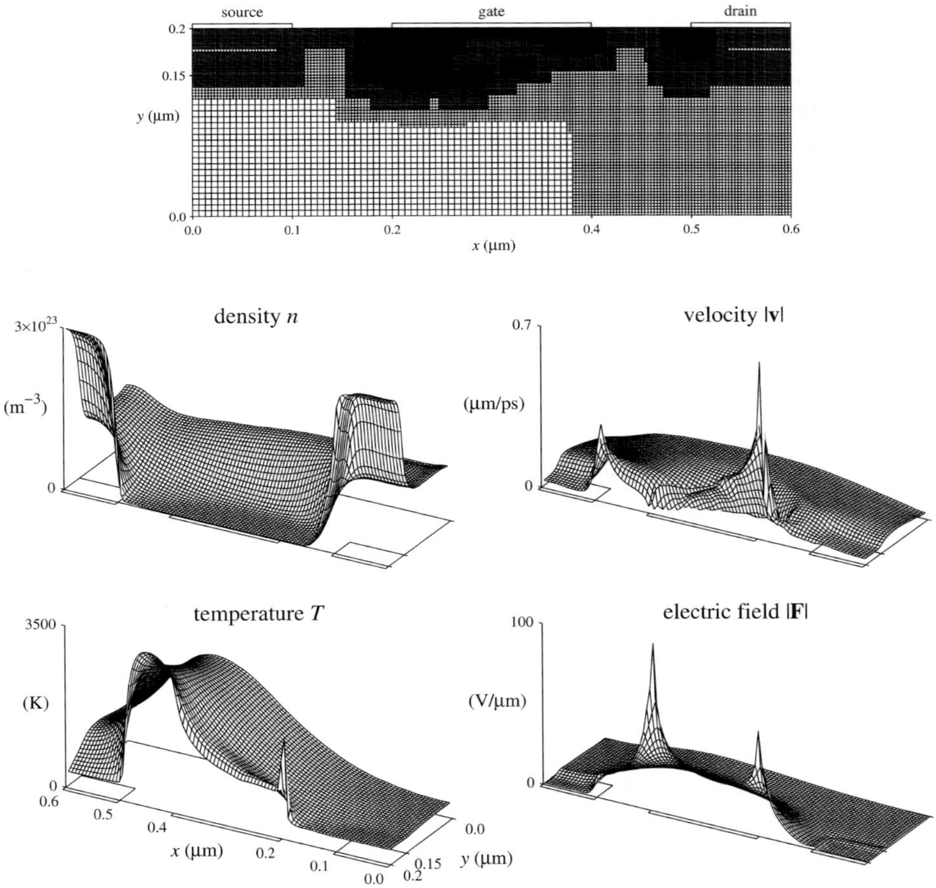

FIG. 4.6. Pattern of refinement and hydrodynamical variables for the MESFET simulation at time $t = 7.5$ ps (steady state). The number density and electron temperature are shown together with the magnitudes of the velocity and electric field vectors. Every grid cell is plotted for the pattern of refinement, but not for the hydrodynamical variables, where data is plotted on a coarse mesh of 96×32 points, and consequently much of the detail in the refined regions of the simulation is not visible here. The edges of the n^+ doping regions are indicated on the lower face of the plotting domain together with the positions of the contacts.

The results presented in Fig. 4.6 are from an AMR simulation using a base grid of 96×32 cells and three levels of refinement, each by a factor $\times 2$. This means that in the most highly refined regions of the simulation, the cell size is the same as for a single uniform grid of 768×256 cells.

Criteria for deciding where regions of refinement should be positioned are based both on the geometry of the device and the instantaneous behaviour of the solution. Two levels of refinement are always used to cover the discontinuity in the device's doping profile, Eq. (4.19). This ensures that the initial data for the simulation, Eq. (4.20), is well resolved, and it also captures the nonsmooth behaviour in the electric field caused by the source term in Poisson's equation. In addition, two levels of refinement are always used along the top ($y = 0.2$ µm) edge of the device where the contacts are located, since this is where the most complex behaviour of the solution is observed to take place.

Two indicators based on the behaviour of the solution are used to position additional regions of refinement. The first indicator, ε_1, is the maximum gradient of $|\mathbf{v}|$, the magnitude of the velocity vector, taken over all directions:

$$\varepsilon_1 = |\nabla v| \quad \text{where } v = |\mathbf{v}|.$$

The second indicator, ε_2, is the curvature of the temperature variable T, defined as the spectral radius (maximum absolute eigenvalue) of the matrix

$$C = \begin{pmatrix} \partial^2 T/\partial x^2 & \partial^2 T/\partial x \partial y \\ \partial^2 T/\partial y \partial x & \partial^2 T/\partial y^2 \end{pmatrix},$$

normalized by the maximum current temperature value:

$$\varepsilon_2 = \text{radius}(C) \times h^2/T_{\max},$$

where h is the grid step size. In both cases, finite differences are used to evaluate the spatial derivatives. The base grid is refined if ε_1 exceeds a value of 0.3×10^{13} or ε_2 exceeds a value of 0.5×10^{14} (where the tolerance values are in standard units). Higher tolerance values are used when refining finer grids: the tolerance for ε_1 is increased by a factor 1.4 for each level of refinement, while the tolerance for ε_2 is increased by a factor 8.

4.13. Conclusions

Various grid-refinement techniques have been outlined in this chapter and one of them, namely adaptive mesh refinement (AMR), has been discussed in more detail. One- and two-dimensional AMR schemes for numerically solving hydrodynamical semiconductor models have been developed and tested. The results demonstrate the potential of this technique for improving the efficiency of numerical simulations of hydrodynamical semiconductor models.

The flexible integration scheme (based on operator splitting) implemented within the AMR codes enable them to be used with a wide range of hydrodynamical models. As evidence of this, results are presented from simulations of two models (the Bløtekjær model, and the reduced hyperbolic model of Anile et al.).

The AMR approach is considered to be more effective than other approaches for varying the resolution of simulations. In particular, the ability to sub-cycle the solution in time leads to an improvement in efficiency of the AMR algorithm and may also reduce the amount of numerical diffusion in the results. However, this sub-cycling causes difficulties when the problem being solved is not purely hyperbolic, and in this work attention is paid to the solution of systems which include elliptic and parabolic modes.

If there is a weak point in the AMR approach, however, it is that currently there is no theoretical basis for predicting what an effective refinement criteria will be for a particular problem. In most work with AMR such criteria are determined through trial and error, and, in fact, one aim of the present one-dimensional study has been to experiment with and calibrate different refinement criteria which could be used when setting up two-dimensional simulations.

The extent to which a simulation benefits from the use of AMR very much depends on the nature of the problem being investigated. For the one-dimensional example problem demonstrated here, the computational work load of an AMR simulation is found to be about four times smaller than for an unrefined simulation of comparable accuracy, but the improvement in efficiency is shown to increase to around an order of magnitude if the requirements of the problem are well suited to a variable grid approach. The massive computational needs of simulations in higher dimensions makes the use of some sort of variable grid approach highly desirable.

References

ANILE, A.M., JUNK, M., ROMANO, V., RUSSO, G. (2000). Cross-validation of numerical schemes for extended hydrodynamical models of semiconductors. *Math. Models Methods Appl. Sci.* **10**, 833–861.

ANILE, A.M., MACCORA, C.R., PIDATELLA, M. (1995). Simulation of n^+-n-n^+ device by a hydrodynamic model: subsonic and supersonic flow. *Compel* **14**, 1–18.

ANILE, A.M., MUSCATO, O. (1995). Improved hydrodynamical model for carrier transport in semiconductors. *Phys. Rev. B* **51**, 16728–16740.

ANILE, A.M., MUSCATO, O. (1996). Extended thermodynamics tested beyond the linear regime: the case of electron transport in silicon semiconductors. *Continuum Mech. Thermodyn.* **8**, 131–142.

ANILE, A.M., MUSCATO, O., MACCORA, C., PIDATELLA, R.M. (1996). Hydrodynamical models for semiconductors. In: Neunzert, H. (ed.), *Progress in Industrial Mathematics: Proceedings of the 1994 ECMI Conference* (Wiley and Teubner, New York).

ANILE, A.M., NIKIFORAKIS, N., PIDATELLA, R.M. (1999). FLIC scheme for the numerical solution of semiconductor transport equations, Preprint.

ANILE, A.M., PENNISI, S. (1992). Thermodynamic derivation of the hydrodynamical model for charge transport in semiconductors. *Phys. Rev. B* **46**, 13 186–13 193.

ANILE, A.M., ROMANO, V. (1999). Nonparabolic band transport in semiconductors: closure of the production terms. *Cont. Mech. Thermodyn.* **11**, 307–325.

ANILE, A.M., ROMANO, V. (2000). Hydrodynamical modeling of charge carrier transport in semiconductors. *Meccanica* **35**, 249–296.

ANILE, A.M., ROMANO, V., RUSSO, G. (1998). Hyperbolic hydrodynamical model of carrier transport in semiconductors. *VLSI Design* **8**, 521–526.

ARMINJON, P., VIALLON, M.-C. (1995). Généralisation du schéma de Nessyahu–Tadmor pour une équation hyperbolique à deux dimensions d'espace [Generalization of the Nessyahu–Tadmor scheme for hyperbolic equations in two space dimensions]. *C. R. Acad. Sci. Paris Sér. I Math.* **320**, 85–88.

ARMINJON, P., VIALLON, M.-C., MADRANE, A. (1997). A finite volume extension of the Lax–Friedrichs and Nessyahu–Tadmor schemes for conservation laws on unstructured grids. *Int. J. Comput. Fluid Dyn.* **9**, 1–22.

ARMINJON, P., VIALLON, M.-C. (1999). Convergence of a finite volume extension of the Nessyahu–Tadmor scheme on unstructured grids for a two-dimensional linear hyperbolic equation. *SIAM J. Numer. Anal.* **36**, 738–771.

ASCHER, U., RUUTH, S., SPITERI, R.J. (1997). Implicit–explicit Runge–Kutta methods for time dependent Partial Differential Equations. *Appl. Numer. Math.* **25**, 151–167.

ASHCROFT, N.C., MERMIN, N.D. (1976). *Solid State Physics* (Holt-Sounders, Philadelphia).

BACCARANI, G., WORDEMAN, M.R. (1982). An investigation on steady-state velocity overshoot in silicon. *Solid-State Electron.* **29**, 970–977.

BELL, J., BERGER, M.J., SALTZMAN, J., WELCOME, M. (1994). Three-dimensional adaptive mesh refinement for hyperbolic conservation laws. *SIAM J. Sci. Comput.* **15**, 127–138.

BERGER, M.J., COLELLA, P. (1989). Local adaptive mesh refinement for shock hydrodynamics. *J. Comput. Phys.* **82**, 64–84.

BERGER, M.J., OLIGER, J. (1984). Adaptive mesh refinement for hyperbolic partial differential equations. *J. Comput. Phys.* **53**, 484–512.

BEREUX, F., SAINSAULIEU, L. (1997). A Roe-type Riemann solver for hyperbolic systems with relaxation based on time-dependent wave decomposition. *Numer. Math.* **77**, 143–185.

BIANCO, F., PUPPO, G., RUSSO, G. (1999). High order central schemes for hyperbolic systems of conservation laws. *SIAM J. Sci. Comput.* **21**, 294–322.

BLOTEKJAER, K. (1970). Transport equations for electron in two-valley semiconductors. *IEEE Trans. Electron Devices* **ED-17**, 38–47.

BODEN, E.P. (1997). An adaptive gridding technique for conservation laws on complex domains. PhD Thesis, College of Aeronautics, Cranfield University.

BORDOLON, T.J., WANG, X.L., MAZIAR, C.M., TASCH, A.F. (1991). *Solid State El.* **34**, 617–624.

CAFLISCH, R.E., RUSSO, G., JIN, S. (1997). Uniformly accurate schemes for hyperbolic system with relaxation. *SIAM J. Numer. Anal.* **34**, 246–281.

CHENG, M.-C., LIANGYING, G., FITHEN, I., YANSHENG, K. (1997). A study of the nonparabolic hydrodynamic modelling of a sub-micrometre n^+-n-n^+ device. *J. Phys. D: Appl. Phys.* **30**, 2343–2353.

CHERN, I.-L., GLIMM, J., MCBRYAN, O., PLOHR, B., YANIV, S. (1986). Front tracking for gas dynamics. *J. Comput. Phys.* **62**, 83–110.

COCKBURN, B., JOHNSON, C., SHU, C.-W., TADMOR, E. (1998). In: Quarteroni, A. (ed.), *Advanced Numerical Approximation of Nonlinear Hyperbolic Equations*. In: Lecture Notes in Mathematics (Springer-Verlag, Berlin).

COCKBURN, B., KARNIADAKIS, G., SHU, C.-W. (2000). In: Cockburn, B., Karniadakis, G., Shu, C.-W. (eds.), *Discontinuous Galerkin Methods: Theory, Computation and Applications*. In: Lecture Notes in Computational Science and Engineering **11** (Springer).

COURANT, R., FRIEDRICHS, K., LEWY, H. (1967). On the partial difference equations of mathematical physics. *IBM J. Res. Develop.* **11**, 215–234.

DREYER, W. (1987). Maximization of the entropy in non-equilibrium. *J. Phys. A: Math. Gen.* **20**, 6505–6517.

FATEMI, E., JEROME, J., OSHER, S. (1991). Solution of hydrodynamic device model using high-order nonoscillatory shock capturing algorithms. *IEEE Trans. Comput.-Aided Design* **10**, 232–244.

GARDNER, C.L. (1991). Numerical simulation of a steady-state electron shock wave in a sub-micrometer semiconductor device. *IEEE Trans. Electron Devices* **38**, 392–398.

GARDNER, C.L. (1993). Hydrodynamic and Monte Carlo simulation of an electron shock wave in a 1-μm n^+-n-n^+ diode. *IEEE Trans. Electron Devices* **ED-40**, 455–457.

GARDNER, C.L., JEROME, J.W., ROSE, D.J. (1989). Numerical methods for the hydrodynamic device model: subsonic flow. *IEEE Trans. CAD* **CAD-8**, 501–507.

GODLEWSKI, E., RAVIART, P.-A. (1996). *Numerical Approximation of Hyperbolic Systems of Conservation Laws*, Applied Mathematical Sciences **118** (Springer-Verlag, New York).

GOLUB, G.H., VAN LOAN, O., CHARLES, F. (1996). *Matrix Computations*, third ed., Johns Hopkins Studies in the Mathematical Sciences (Johns Hopkins University Press, Baltimore).

HÄNSCH, W. (1991). *The Drift-Diffusion Equation and its Application in MOSFET Modeling* (Springer-Verlag, Wien).

HAIRER, E., WANNER, G. (1987). *Solving Ordinary Differential Equations II. Stiff and Differential-Algebraic Problems* (Springer-Verlag, New York).

HARTEN, A., ENGQUIST, B., OSHER, S., CHAKRAVARTHY, S. (1987). Uniformly high order accurate essentially non-oscillatory schemes III. *J. Comput. Phys.* **71**, 231–303.

HERN, S.D., ANILE, A.M., NIKIFORAKIS, N. (2001). A mesh refinement approach for hydrodynamical semiconductor simulations, *J. Comp. Phys.* (also in TMR preprint archive).

JACOBONI, C., LUGLI, P. (1989). *The Monte Carlo Method for Semiconductor Device Simulation* (Springer-Verlag, Wien–New York).

JACOBONI, C., REGGIANI, L. (1983). The Monte Carlo method for the solution of charge transport in semiconductors with applications to covalent materials. *Rev. Mod. Phys.* **55**, 645–705.

JEROME, J.W., SHU, C.-W. (1994). Energy models for one-carrier transport in semiconductor devices. In: Coughran, W.M., Cole, J., Lloyd, P., White, J.K. (eds.), *Semiconductors, Part II*. In: IMA Volumes in Mathematics and its Applications (Springer-Verlag).

JIANG, G.-S., SHU, C.-W. (1996). Efficient implementation of weighted ENO schemes. *J. Comput. Phys.* **126**, 202–228.

JIANG, G.-S., TADMOR, E. (1998). Nonoscillatory central schemes for multidimensional hyperbolic conservation laws. *SIAM J. Sci. Comput.* **19**, 1892–1917.

JOU, D., CASAS-VAZQUEZ, J., LEBON, G. (1993). *Extended Irreversible Thermodynamics* (Springer-Verlag, Berlin).

KURGANOV, A., TADMOR, E. (2000). New high-resolution central schemes for nonlinear conservation laws and convection-diffusion equations. *J. Comput. Phys.* **160**, 214–282.

LAX, P.D. (1973). *Hyperbolic Systems of Conservation Laws and the Mathematical Theory of Shock Waves* (SIAM, Philadelphia).

LAX, P.D., WENDROFF, B. (1960). Systems of conservation laws. *Comm. Pure Appl. Math.* **13**, 217–237.

LE TALLEC, P., PERLAT, J.P. (1997). Numerical analysis of the Levermore's moment system, INRIA Research Report 3124.

LEVEQUE, R. (1990). *Numerical Methods for Conservation Laws* (Birkhäuser, Basel).

LEVERMORE, C.D. (1995). Moment closure hierarchies for the Boltzmann–Poisson equation. *VLSI Design* **8**, 97–101.

LEVERMORE, C.D. (1996). Moment closure hierarchies for kinetic theories. *J. Stat. Phys.* **83**, 331–407.

LEVY, D., PUPPO, G., RUSSO, G. (1999). Central WENO schemes for hyperbolic systems of conservation laws. *Math. Model. Numer. Anal.* **33**, 547–571.

LEVY, D., PUPPO, G., RUSSO, G. (2000). A third order central WENO scheme for 2D conservation laws. *Appl. Numer. Math.* **33**, 407–414.

LEVY, D., PUPPO, G., RUSSO, G. (2001). Central WENO Schemes for Multi-dimensional Hyperbolic Systems of Conservation Laws, submitted.

LIOTTA, F., ROMANO, V., RUSSO, G. (1999). Central schemes for systems of balance laws. *Internat. Ser. Numer. Math.* **30**, 651–660.

LIOTTA, F., ROMANO, V., RUSSO, G. (2000). Central schemes for balance laws of relaxation type. *SIAM J. Numer. Anal.* **38**, 1337–1356.

LIU, X.-D., TADMOR, E. (1998). Third order nonoscillatory central scheme for hyperbolic conservation laws. *Numer. Math.* **79**, 397–425.

MARKOWICH, P., RINGHOFER, C.A., SCHMEISER, C. (1990). *Semiconductor Equations* (Springer-Verlag, Wien).

MAVRIPLIS, D.J. (1996). Mesh generation and adaptivity for complex geometries and flows. In: Peyret, R. (ed.), *Handbook of CFD* (Academic Press).

MINION, M.L. (1996). A projection method for locally refined grids. *J. Comput. Phys.* **127**, 158–178.

MÜLLER, I., RUGGERI, T. (1998). *Rational Extended Thermodynamics* (Springer-Verlag, Berlin).

MUSCATO, O., ROMANO, V. (2001). Simulation of submicron silicon diodes with a non-parabolic hydrodynamical model based on the maximum entropy principle. *VLSI Design* **13**, 273–279.

NESSYAHU, H., TADMOR, E. (1990). Non-oscillatory central differencing for hyperbolic conservation laws. *J. Comput. Phys.* **87**, 408–463.

NIKIFORAKIS, N. (1998). Lecture notes of the Part III (MSc) course: Introduction to Computational Fluid Dynamics, DAMTP, University of Cambridge.

NIKIFORAKIS, N., BILLETT, S.J., BODEN, E., PYLE, J.A., TORO, E.F. (2001). Adaptive mesh refinement for global atmospheric modelling, Preprint.

PARESCHI, L., RUSSO, G. (2000). Implicit–explicit Runge–Kutta schemes for stiff systems of differential equations. In: Trigiante, D. (ed.), *Recent Trends in Numerical Analysis* (Nova Science).

PARESCHI, L. (2001). Central differencing based numerical schemes for hyperbolic conservation laws with relaxation terms. *SIAM J. Numer. Anal.* **39**, 1395–1417.

PRESS, W., TEUKOLSKY, S.A., VETTERLING, W.T., FLANNERY, B.P. (1992). *Numerical Recipes*, second ed. (Cambridge University Press).

QUIRK, J.J. (1991). An adaptive grid algorithm for computational shock hydrodynamics. PhD Thesis, College of Aeronautics, Cranfield Inst. Of Tech.

QUIRK, J.J., HENEBUTTE, U.R. (1993). A parallel adaptive mesh refinement algorithm, ICASE report 93–63.

ROE, P.L. (1981). Approximate Riemann solvers, parameter vectors, and difference schemes. *J. Comput. Phys.* **43**, 357–372.

ROMANO, V. (2000). Nonparabolic band transport in semiconductors: closure of the production terms in the moment equations. *Cont. Mech. Thermodyn.* **12**, 31–51.

ROMANO, V. (2001). Nonparabolic band hydrodynamical model of silicon semiconductors and simulation of electron devices. *Math. Methods Appl. Sci.* **24**, 439–471.

ROMANO, V. (2002). Simulation of a silicon MESFET with a non-parabolic hydrodynamical model based on the maximum entropy principle. *J. Comput. Phys.* **176**, 70–92.

ROMANO, V., RUSSO, G. (2000). Numerical solution for hydrodynamical models of semiconductors. *Math. Models Methods Appl. Sci.* **10**, 833–861.

RUDAN, M., BACCARANI, G. (2001). On the structure and closure condition of the hydrodynamical model, IEEE TVLSI, in press.

SAAD, Y. (1996). *Iterative Methods for Sparse Linear Systems* (PWS Publishing Company, Boston).

SANDERS, R., WEISER, A. (1989). A high order staggered grid method for hyperbolic systems of conservation laws in one space dimension. *Comput. Methods Appl. Mech. Engrg.* **75**, 91–107.

SHEN, M., CHENG, M.-C., LIOU, J.J. (2000). A generalized finite element method for hydrodynamic modeling of short-channel devices, Preprint.

SKAMAROCK, W.C., KLEMP, J.B. (1993). Adaptive grid refinement for two-dimensional and three-dimensional nonhydrostatic atmospheric flow. *Mon. Weakly Rev.* **121**, 788–804.

SELBERHERR, S. (1984). *Analysis and Simulation of Semiconductor Devices* (Springer-Verlag, Wien–New York).

SHU, C.-W. (2001). High order finite difference and finite volume WENO schemes and discontinuous Galerkin methods for CFD, *Internat. J. Comput. Fluid Dynamics*, in press.

STETTLER, M.A., ALAM, M.A., LUNDSTROM, M.S. (1993). A critical examination of the assumptions underlying macroscopic transport equations for silicon device. *IEEE Trans. Electron Devices* **40**, 733–739.

STRANG, G. (1968). On the construction and comparison of difference schemes. *SIAM J. Numer. Anal.* **5**, 506.

THOMA, R., EDMUNDS, A., MEINERZHAGEN, B., PEIFER, H.J., ENGL, W. (1991). Hydrodynamic equations for semiconductors with nonparabolic band structure. *IEEE Trans. Electron Devices* **38**, 1343–1353.

TORO, E.F. (1999). *Riemann Solvers and Numerical Methods for Fluid Dynamics. A Practical Introduction*, second ed. (Springer-Verlag, Berlin).

TROVATO, M., FALSAPERLA, P. (1998). Full nonlinear closure for a hydrodynamical model of transport in silicon. *Phys. Rev. B: Condens. Matt.* **57**, 4456–4471.

WOOLARD, D.L., TIAN, H., TREW, R.J., LITTLEJOHN, M.A., KIM, W. (1991). Hydrodynamic electron-transport model: Nonparabolic corrections to the streaming terms. *Phys. Rev. B* **44**, 11119–11132.

ZHOU, T., LI, Y., SHU, C.-W. (2001). Numerical comparison of WENO finite volume and Runge–Kutta discontinuous Galerkin methods. *Appl. Numer. Math.*, in press.

YIP, W.-K., SHEN, M., CHENG, M.-C. (2000). Hydrodynamic modeling of short-channel devices using an upwind flux vector splitting scheme, Preprint.

Further reading

ANILE, A.M., HERN, S.D. (2001). Two-valley hydrodynamical models for electron transport in gallium arsenide: simulation of Gunn oscillations, *VLSI Design*, in press (also in TMR preprint archive).

ANILE, A.M., MUSCATO, O., ROMANO, V. (2000). Moment equations with maximum entropy closure for carrier transport in semiconductors devices: validation in bulk silicon. *VLSI Design* **10**, 335–354.

ANILE, A.M., NIKIFORAKIS, N., PIDATELLA, R.M. (2000). Assessment of a high resolution centred scheme for the solution of hydrodynamical semiconductor equations. *SIAM J. Sci. Comput.* **22**, 1533–1548.

ANILE, A.M., ROMANO, V., RUSSO, G. (2000). Extended hydrodynamical model of carrier transport in semiconductors. *SIAM J. Appl. Math.* **11**, 74–101.

BEN ABDALLAH, N., DEGOND, P., GENIEYS, S. (1996). An energy-transport model for semiconductors derived from the Boltzmann equation. *J. Stat. Phys.* **84**, 205–225.

BEN ABDALLAH, N., DEGOND, P. (1996). On a hierarchy of macroscopic models for semiconductors. *J. Math. Phys.* **37**, 3306–3333.

ENGQUIST, B., OSHER, S. (1981). One sided difference approximations for nonlinear conservation laws. *Math. Comp.* **36**, 321–351.

ENGQUIST, B., SJÖGREEN, B. (1998). The convergence rate of finite difference schemes in the presence of shocks. *SIAM J. Numer. Anal.* **35**, 2464–2485.

FAWCETT, W., BOARDMANN, D.A., SWAIN, S. (1970). *J. Chem. Solids* **31**.

FISCHETTI, M.V., LAUX, S. (1993). Monte Carlo study of electron transport in silicon inversion layers. *Phys. Rev. B* **48**, 2244–2274.

GNUDI, A., ODEH, F., RUDAN, M. (1990). Investigation of non-local transport phenomena in small semiconductor devices. *European Trans. Telecomm. Related Technol.* **1**, 307–312.

HARTEN, A., OSHER, S. (1987). Uniformly high-order accurate non-oscillatory schemes. *SIAM J. Numer. Anal.* **24**, 279–309.

JEROME, J., SHU, C.-W. (1995). Transport effects and characteristic modes in the modeling and simulation of submicron devices. *IEEE Trans. Comput.-Aided Design Integr. Circuits Syst.* **14**, 917–923.

ODEH, F., RUDAN, M., WHITE, J. (1987). Numerical solution of the hydrodynamic model for a one-dimensional semiconductor device. *Compel* **6**, 151–170.

POUPAUD, F. (1990). On a system of nonlinear Boltzmann equations of semiconductor physics. *SIAM J. Appl. Math.* **50**, 1593–1606.

POUPAUD, F. (1992). Runaway phenomena and fluid approximation under high fields in semiconductor kinetic theory. *Z. Angew. Math. Mech.* **72**, 359–372.

SHU, C.-W., OSHER, S. (1989). Efficient implementation of essentially non-oscillatory shock-capturing schemes II. *J. Comp. Phys.* **83**, 32–78.

STRATTON, R. (1962). Diffusion of hot and cold electrons in semiconductor barriers. *Phys. Rev.* **126**, 2002–2014.

TOMIZAWA, K. (1993). *Numerical Simulation of Submicron Semiconductor Devices* (Artech House, Boston).

VOLGELSANG, T., HÄNSCH, W. (1991). A novel approach for including band structure effects in a Monte Carlo simulation of electron transport in silicon. *J. Appl. Phys.* **70**, 1493–1499.

WACHUTKA, G. (1991). Unified framework for thermal electrical, magnetic and optical semiconductor devices modeling. *Compel* **10**, 311–321.

ZENNARO, M. (1986). Natural continuous extensions of Runge–Kutta methods. *Math. Comp.* **46**, 119–133.

Modelling and Discretization of Circuit Problems

Michael Günther

University of Wuppertal, Department of Mathematics, Gaußstr. 20, D-42119 Wuppertal, Germany

Uwe Feldmann

Infineon Technologies, Balanstr. 73, D-81541 München, Germany

Jan ter Maten

Philips Research Laboratories (NatLab), Electronic Design & Tools, Analogue Simulation, Prof. Holstlaan 4, NL-5656 AA Eindhoven, The Netherlands

Numerical Methods in Electromagnetics
Special Volume (W.H.A. Schilders and E.J.W. ter Maten, Guest Editors) of
HANDBOOK OF NUMERICAL ANALYSIS, VOL. XIII
P.G. Ciarlet (Editor)

ISSN 1570-8659
DOI 10.1016/S1570-8659(04)13006-8

Contents

Preface 527

Chapter I. DAE-Systems – the Modelling Aspect 529

1. The Schmitt trigger – an introductory example 529
2. Principles and basic equations 531
3. Conventional versus charge/flux oriented formulation 534
4. Modified nodal analysis 537
5. Why differential-algebraic equations? 540

Chapter II. DAE-index – the Structural Aspect 545

6. The index concept for linear systems 545
7. Network topology and DAE-index for RLC networks 548
8. Networks with controlled sources 552
9. Effects of refined modelling – a bipolar ringoscillator 557

Chapter III. Numerical Integration Schemes 563

10. The conventional approach: Implicit linear multi-step formulas 563
11. A second approach: One-step methods 574
12. Oscillatory circuits and numerical damping: A comparison 579

Chapter IV. Numerical Treatment of Large Problems 585

13. Numerical properties of an MOS ringoscillator model 586
14. Classification of existing methods 590
15. Parallelization 597
16. Hierarchical simulation 610
17. Multirate integration 612

Chapter V. Periodic Steady-State Problems 617

18. RF simulation 617
19. The basic two-step approach 619
20. The PSS problem 621

21. Perturbation analysis 623
22. Algorithms for the PSS problem 629

REFERENCES 649

Further reading 658

Preface

Microelectronics is the core technology for numerous industrial innovations. Progress in microelectronics is highlighted by milestones in chip technology, i.e., microprocessor and memory chips. This ongoing increase in performance and memory density – accompanied with decreasing prices – would not have been possible without extensive use of computer simulation techniques, especially circuit simulation.

An important analysis type in circuit simulators is time domain analysis, which calculates the time-dependent (transient) behaviour of electrical signals in a circuit responding to time-varying input signals. A network description of the circuit is generated automatically in computer-aided electronics-design systems from designer's drafts or fabrication data files. An input processor translates this network description into a data format reflecting the mathematical model of the system. The mathematical network equations are based on the application of basic physical laws like energy or charge conservation onto network topology and characteristic equations for the network elements. This automatic modeling approach preserves the topological structure of the network and does not aim at systems with a minimal set of unknowns. Hence an initial-value problem of differential-algebraic equations (DAEs) is generated which covers characteristic time constants of several orders of magnitude (stiff equations) and suffers from poor smoothness properties of modern transistor model equations.

In the first part of this article (Chapters I–III) we aim at filtering out the numerical analysis aspects time domain analysis is based on: The numerical integration of the very special differential-algebraic network equations. This task comprises the simulation core of all simulation packages. Although modelling, discretization and numerical integration can be clearly distinguished as different steps, all these levels are strongly interwoven (and therefore also somehow hidden) in commercial packages.

In Chapter I we discuss how these mathematical models are generated on the basis of a network approach with compact (lumped) models. The structural properties of these DAE models can be described by the DAE-index concept. We will learn in Chapter II that these properties are fixed by the topological structure of the network model in most cases. However, if more general models for the network elements are incorporated, or refined models are used to include second order and parasitic effects then special circuit configurations may be built, which render ill-conditioned problems. These investigations form the basis for constructing numerical integration schemes that are tailored to the respective properties of the network equations. In Chapter III we describe the direct integration approach based on multi-step schemes, which is used in the extremely widespread simulator SPICE (NAGEL [1975]) and has become a standard since almost 30 years. We include in our discussion a comparison with one-step methods, since recent developments have revealed an interesting potential for such schemes.

The second part (Chapters IV and V) deals with two challenges circuit simulation is faced actually in industry: The simulation of very large circuits with up to millions of transistors such as memory chips on the one hand, and oscillatory circuits with eventually widely separated time constants, appearing in radio frequency (RF) design on the other hand. For the reason of efficiency and robustness, and to make numerical simulation feasible at all, the time domain approach discussed in the first part has to be adapted in both cases. These fields are very much driven by actual industrial needs, and hence are rapidly evolving. So we can in the second part only describe the state of the art, rather than present an established mathematical theory, as is meanwhile available for numerical integration of DAE systems. Nevertheless we hope that the second part as well as the first one helps to get some feeling about the nature of the underlying problems and the attempts to solve them, and may be useful for both mathematical researchers and the users of the codes.

CHAPTER I

DAE-Systems – the Modelling Aspect

In computational engineering the network modelling approach forms the basis for computer-aided analysis of time-dependent processes in multibody dynamics, process simulation or circuit design. Its principle is to connect compact elements via ideal nodes, and to apply some kind of conservation rules for setting up equations. The mathematical model, a set of so-called network equations, is generated automatically by combining network topology with characteristic equations describing the physical behaviour of network elements under some simplifying assumptions. Usually, this automatic modelling approach tries to preserve the topological structure of the network and does not take care to get systems with a minimal set of unknowns. As a result, coupled systems of implicit differential and nonlinear equations, shortly, differential-algebraic equations (DAEs), of the general type

$$f(x,\dot{x},t)=0 \quad \text{with } \det\left(\frac{\partial f}{\partial \dot{x}}\right)\equiv 0$$

may be generated. From a mathematical point of view, these systems may represent ill-posed problems, and hence are more difficult to solve numerically than systems of ordinary differential equations (ODEs).

In this first chapter we have to answer the questions on *how* and *why*: How does one generate the differential-algebraic network equations that model the circuits? And why using at all a DAE approach with a redundant set of network variables, and not an ODE model?

In the subsequent Chapters II and III answers are given to the remaining questions: What are the structural properties of the arising DAE systems? Do they have an impact on numerical discretization? Which integration schemes are used to solve the systems numerically in a robust and efficient manner?

Let us start our discussion with

1. The Schmitt trigger – an introductory example

The Schmitt trigger (KAMPOWSKY, RENTROP and SCHMIDT [1992]) shown in Fig. 1.1 is used to transform analogue ones into digital signals. The circuit is characterized by two stable states: If the input signal V_{in} exceeds a certain threshold, then the associated stable state is obtained as an output signal at node 5. The circuit consists of five linear resistors with conductances $G_1,\ldots,G_5$ between the input voltage source V_{in} and node 1,

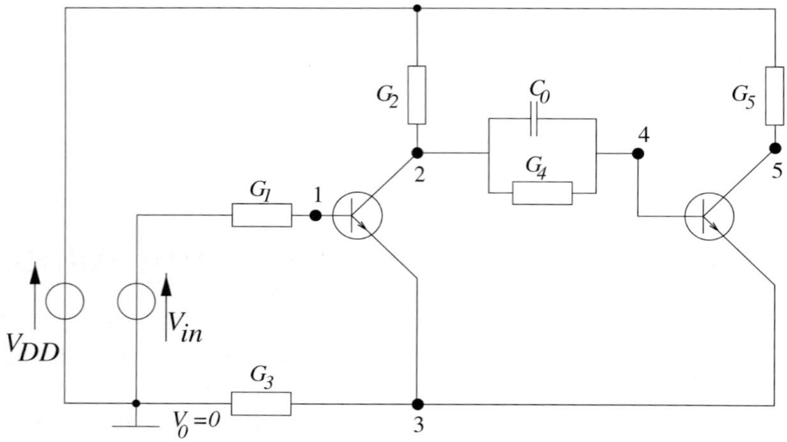

FIG. 1.1. Schmitt trigger circuit.

the power supply voltage source V_{DD} and nodes 2 and 5, between node 3 and ground, and between nodes 2 and 4. The dynamic behaviour of the circuit is caused by the linear capacitor with capacitance C_0 between nodes 2 and 4. The nonlinear characteristic is introduced by two bipolar transistors of npn type at nodes 1, 2, 3 and 4, 5, 3.

To derive a mathematical model for the Schmitt trigger that determines the time-dependent voltage courses of the five node potentials $u_1, \ldots, u_5$ at nodes $1, \ldots, 5$, we may build up the current balances for all nodes except ground. To apply this so-called *nodal analysis*, we first have to replace all branch currents by voltage-depending functions. For the one-port elements *capacitor* and *resistor*, the characteristic equation relating branch current $I(t)$ and branch voltage $U(t)$ is given in admittance form, i.e., $I(t)$ is given explicitly as a function of $U(t)$:

- *Ohm's law for a linear resistor*: $I(t) = GU(t)$ with *conductance* G.
- *Faraday's law for a linear capacitor*: $I(t) = C\dot{U}(t)$ with *capacitance* C.

Using a compact model, the multi-port element *bipolar transistor of npn-type* shown in Fig. 1.2 can be described by three branch currents $I_B(t)$, $I_C(t)$ and $I_E(t)$ entering the base, collector and emitter terminal of the transistor with corresponding node potentials $U_B(t)$, $U_C(t)$ and $U_E(t)$. If $U_C(t) > U_B(t) > U_E(t)$ then the three currents are given

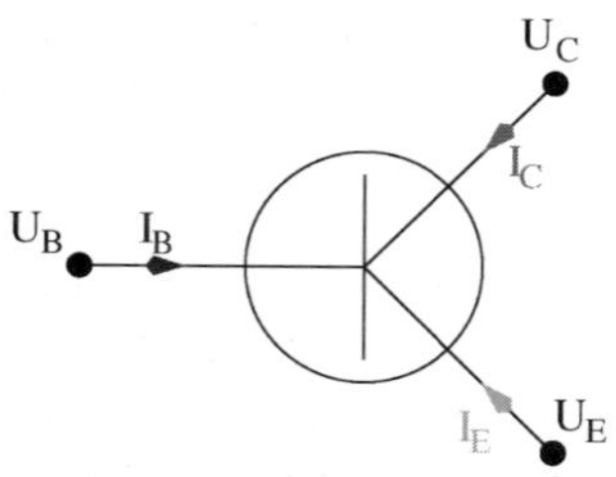

FIG. 1.2. Bipolar transistor of npn type.

in a first order model by

$$\begin{aligned}
I_B(t) &= g\big(U_B(t) - U_E(t)\big),\\
I_C(t) &= \alpha \cdot g\big(U_B(t) - U_E(t)\big),\\
I_E(t) &= -(1+\alpha) \cdot g\big(U_B(t) - U_E(t)\big),
\end{aligned}$$

with the characteristic exponential current $g(U) := \beta \cdot [\exp(U/U_T) - 1]$ of a pn-junction. The parameter α denotes the amplification factor, β the saturation current and U_T the thermal voltage at room temperature. For more details see GÜNTHER and FELDMANN [1999a], KAMPOWSKY, RENTROP and SCHMIDT [1992].

Now we have collected all ingredients to apply nodal analysis (i.e., apply Kirchhoffs' current law) to nodes 1–5. We get:

$$\begin{aligned}
&\boxed{1} \quad 0 = G_1 \cdot (u_1 - V_{in}) + g(u_1 - u_3),\\
&\boxed{2} \quad 0 = G_2 \cdot (u_2 - V_{DD}) + C_0 \cdot (\dot{u}_2 - \dot{u}_4) + G_4 \cdot (u_2 - u_4) + \alpha \cdot g(u_1 - u_3),\\
&\boxed{3} \quad 0 = -(1+\alpha) \cdot g(u_1 - u_3) + G_3 \cdot u_3 - (1+\alpha) \cdot g(u_4 - u_3),\\
&\boxed{4} \quad 0 = G_4 \cdot (u_4 - u_2) + C_0 \cdot (\dot{u}_4 - \dot{u}_2) + g(u_4 - u_3),\\
&\boxed{5} \quad 0 = G_5 \cdot (u_5 - V_{DD}) + \alpha \cdot g(u_4 - u_3).
\end{aligned}$$

Reformulated as a linear implicit system, we have

$$\begin{pmatrix} 0 & 0 & 0 & 0 & 0\\ 0 & C_0 & 0 & -C_0 & 0\\ 0 & 0 & 0 & 0 & 0\\ 0 & -C_0 & 0 & C_0 & 0\\ 0 & 0 & 0 & 0 & 0 \end{pmatrix} \cdot \begin{pmatrix} \dot{u}_1\\ \dot{u}_2\\ \dot{u}_3\\ \dot{u}_4\\ \dot{u}_5 \end{pmatrix} + \begin{pmatrix} G_1 \cdot (u_1 - V_{in}) + g(u_1 - u_3)\\ G_2 \cdot (u_2 - V_{DD}) + G_4 \cdot (u_2 - u_4) + \alpha \cdot g(u_1 - u_3)\\ G_3 \cdot u_3 - (1+\alpha) \cdot g(u_1 - u_3) - (1+\alpha) \cdot g(u_4 - u_3)\\ G_4 \cdot (u_4 - u_2) + g(u_4 - u_3)\\ G_5 \cdot (u_5 - V_{DD}) + \alpha \cdot g(u_4 - u_3) \end{pmatrix} = 0. \tag{1.1}$$

The 5×5 capacitance matrix is not regular and has only rank 1: The *network equations* (1.1) are a mixed system of one differential equation (difference of lines 2 and 4) and four algebraic equations (line 1, sum of lines 2 and 4, line 3, line 5). Hence, it is impossible to transform this system of *differential-algebraic equations* (DAEs) analytically to a system of ordinary differential equations (ODEs) by pure algebraic transformations.

With this example in mind, we can now inspect the mathematical modelling of electrical circuits – the set-up of differential-algebraic network equations – in the general case.

2. Principles and basic equations

In contrast to a field theoretical description based on Maxwell's equations, which is not feasible due to the large complexity of integrated electric circuits, the network approach is based on integral quantities – the three spatial dimensions of the circuit are translated into the network topology. The time behaviour of the system is given by the network quantities *branch currents* $I(t) \in \mathbb{R}^{n_I}$, *branch voltages* $U(t) \in \mathbb{R}^{n_I}$ and *node voltages* $u(t) \in \mathbb{R}^{n_u}$, the voltage drop of the nodes versus the ground node. As will be seen later, it may be convenient to include more physical quantities like *electrical charges* $q(t) \in \mathbb{R}^{n_q}$ and *magnetic fluxes* $\phi(t) \in \mathbb{R}^{n_\phi}$ into the set of variables as well.

Network topology laws. The network model consists of elements and nodes, and the latter are assumed to be electrically ideal. The composition of basic elements is governed by Kirchhoff's laws which can be derived by applying Maxwell's equations in the stationary case to the network topology:

– *Kirchhoff's voltage law* (*KVL*). The algebraic sum of voltages along each loop of the network must be equal to zero at every instant of time. Often this law is used only for getting a relation between branch voltages $U(t)$ and node voltages $u(t)$ in the form:

$$A^\top \cdot u(t) = U(t) \tag{2.1}$$

with an incidence matrix $A \in \{-1, 0, 1\}^{n_u \times n_I}$, which describes the branch-node connections of the network graph.

– *Kirchhoff's current law* (*KCL*). The algebraic sum of currents traversing each cutset of the network must be equal to zero at every instant of time. As a special case we get that the sum of currents leaving any circuit node[1] is zero:

$$A \cdot I(t) = 0. \tag{2.2}$$

When applying KCL to the terminals of an element, one obtains by integration over time the requirement of *charge neutrality*, that is the sum of charges q_{kl} over all terminals k of an element l must be constant:

$$\sum_{k(l)} q_{kl} = \text{const.} \tag{2.3}$$

Hereby the constant can be set to zero without loss of generality.

Basic elements and their constitutive relations. Besides these purely topological relations additional equations are needed for the subsystems to fix the network variables uniquely. These so-called characteristic equations describe the physical behaviour of the network elements.

One-port or *two-terminal* elements given in Fig. 2.1 are described by equations relating their branch current I and branch voltage $U = u_+ - u_-$. Here the arrows in the figure indicate that the branch current is traversing from the "+"-node of higher potential u_+ to the "−"-node of lower potential u_-. The characteristic equations for the basic elements resistor, inductor and capacitor are derived by field theoretical arguments from Maxwell's equations assuming quasistationary behaviour (MEETZ and ENGL [1980]). In doing so, one abstracts on Ohmic losses for a resistor, on generation of magnetic

[1] The sign is only a matter of convention.

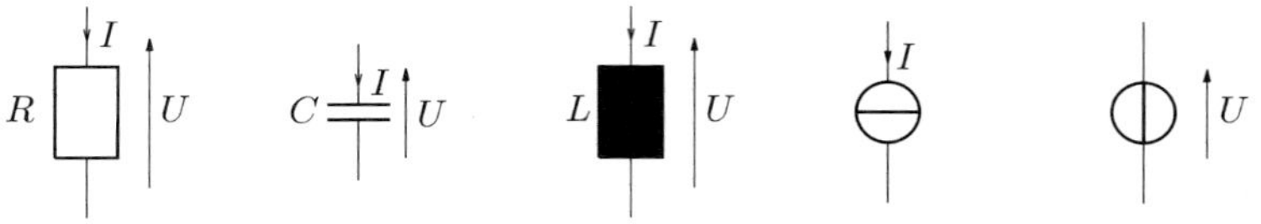

FIG. 2.1. Basic network elements: Linear resistor ($I = U/R = G \cdot U$), capacitor ($I = C \cdot \dot{U}$), inductor ($U = L \cdot \dot{I}$), independent current source ($I = s_1(t)$) and independent voltage source ($U = s_2(t)$).

fluxes for an inductor, and on charge storage for a capacitor, by neglecting all other effects. The set of basic elements is completed by ideal independent, i.e., purely time-dependent current and voltage sources.

Interconnects and semiconductor devices (i.e., transistors) are modelled by multi-terminal elements (*multi-ports*), for which the branch currents entering any terminal and the branch voltages across any pair of terminals are well-defined quantities. One should note that also for these elements Kirchhoff's laws are valid, i.e., the sum of all branch currents flowing into the element is zero, and the sum of branch voltages along an arbitrary closed terminal loop is zero. Hence, n-terminal elements are uniquely determined by $n - 1$ branch currents into $n - 1$ terminals and $n - 1$ branch voltages between these $n - 1$ terminals, and a reference pole. Controlled current sources are used to describe the static branch current; alternatively, controlled voltage sources may be used to describe branch voltages, see Fig. 2.2 for the symbols. Dynamic behaviour is described by inserting capacitive or inductive branches between the terminals.

These constitutive relations describe the terminal characteristic, which in the *classical approach* is a relation between terminal currents I and branch voltages U and/or their time derivatives for each terminal. If the terminal currents are explicitly given, then the element equations are said to be in *admittance form* (independent current source, voltage/current controlled current source, linear resistor and linear capacitor); if the branch voltages are explicitly given, then the equations are said to be in *impedance form* (independent voltage source, voltage/current controlled voltage source and linear inductor).

As a more flexible and universal approach a *charge/flux-oriented formulation* can be taken for energy storing elements which reflects better the underlying physics of the circuit devices, see, e.g., CALAHAN [1968], CHUA and LIN [1975], WARD and DUTTON [1978]. It requires the inclusion of terminal charges q and branch fluxes ϕ into the set of network variables.

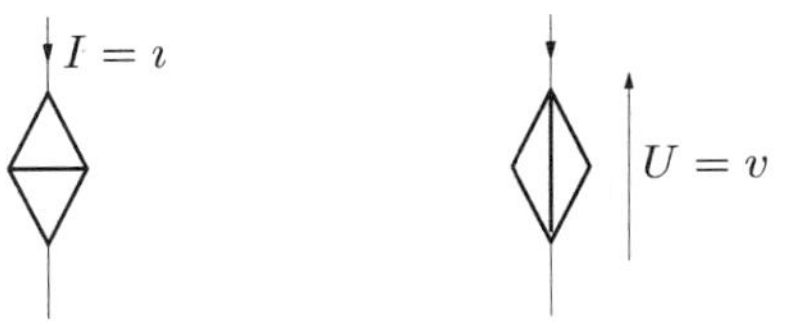

FIG. 2.2. Controlled sources: Voltage/current controlled current source ($I = \imath(U_{control}, I_{control})$), voltage/current controlled voltage source ($U = v(U_{control}, I_{control})$).

3. Conventional versus charge/flux oriented formulation

At this point the reader may pose the question why charges and fluxes are introduced at all to model characteristic equations of energy storing elements in a charge/flux oriented way – and not the classical capacitors and inductors are used. The answer to this question contains modelling, numerical and software engineering aspects. We will now focus on the first two aspects. Arguments for the latter will be addressed in the next section.

Modelling. The use of a charge/flux-oriented formulation can be motivated by inspecting the case of nonlinear capacitors and inductors: $C = C(U)$, $L = L(I)$. The problem here is that there is no generic way to get the capacitor current and inductor voltage in this case. Rather there exist different approaches in the literature. We discuss them for capacitors here, the relations for inductors are similar. See Table 3.1 for an overview.

The most popular is an interpretation of C as *differential capacitance*, with

$$I = C(U) \cdot \dot{U}$$

as capacitor current. However also an interpretation of C as *general nonlinear capacitance* with

$$I = \frac{\mathrm{d}}{\mathrm{d}t}\bigl(C(U) \cdot U\bigr)$$

can be found. This interpretation can be transformed into the first one by using

$$\tilde{C}(U) = \frac{\partial C(U)}{\partial U} \cdot U + C(U)$$

as differential capacitance.

A more natural access for the handling of nonlinear capacitances would be to introduce the terminal charges

$$q = q_C(U)$$

and apply the formula

$$I = \frac{\mathrm{d}q}{\mathrm{d}t}$$

TABLE 3.1
Constitutive relations for energy storing elements

Conventional formulation	Charge/flux-oriented formulation
Linear capacitor $I = C \cdot \dot{U}$	Charge/current defining element $q = q_C(U, I), \quad I = \mu(U) \cdot q + \dot{q} = I_{dc} + \dot{q}$
Linear inductor $U = L \cdot \dot{I}$	Flux/voltage defining elements $\phi = \phi_L(U, I), \quad U = \dot{\phi}$

for getting the capacitor current. Another argument for this approach is, that for the classical capacitance definition the controlling branch voltage U is restricted to be the voltage drop over the capacitor itself, which is too much restrictive to handle large classes of circuits. Hence charge-oriented models are highly desirable, not only for getting more flexibility but also since they are consistent with the physical reality: Both static and dynamic behaviour can be derived from one single set of equations, see the equation for the current in the second column of Table 3.1 (MIURA-MATTAUSCH, FELDMANN, RAHM, BOLLU and SAVIGNAC [1996]). Unfortunately, it is in practice often too difficult to develop such models for real circuit elements with sufficient accuracy. So this ideal principle is often violated in practice, and static and dynamic behaviour are modeled separately.

Charge conservation. A mixture of modelling and numerical aspects is the possibility to correctly model and analyse the charge flow in the circuit. In the following, we will concentrate on the latter item.

The original intent of the charge/flux-oriented formulation was to assure *charge conservation*. This property is crucial for the analysis of many analog circuits like switched capacitor filters, charge pumps, dynamic memories etc., which work on the basis of properly balanced charge flow.

In the following we merely look at charge conservation. The relations for flux conservation are similar. Ideally, charge conservation is assured if

- the principle (2.3) of charge neutrality is observed for each charge storing element, and
- during numerical integration of the network equations no erroneous charges are "created".

In practice, the latter condition can be replaced by the weaker requirement that

- the charge error due to numerical procedures can be made arbitrarily small if the network equations are solved with increasing accuracy.

How can charge conservation be obtained with the different formulations? First we look at the *conventional approach*. Here charges are obtained indirectly via numerical integration of capacitances $C = C(u)$:

$$q(t) = q(t_0) + \int_{u(t_0)}^{u(t)} C(v)\,\mathrm{d}v.$$

Here t_0 is the starting time, and t is the actual time point. The numerically computed voltage u will differ from the exact value u^*:

$$u(t) = u^*(t) + \Delta u(t).$$

So we obtain as a first-order approximation from the capacitance-oriented formulation

$$q(t) \approx q(t_0) + \int_{u(t_0)}^{u(t)} C(v^*)\,\mathrm{d}v + \int_{u(t_0)}^{u(t)} \frac{\partial C}{\partial v} \Delta v\,\mathrm{d}v,$$

while the exact value for the charge is given by

$$q^*(t) = q(t_0) + \int_{u(t_0)}^{u^*(t)} \frac{\partial q(v^*)}{\partial v}\,\mathrm{d}v.$$

Insertion yields

$$\begin{aligned} q(t) \approx q^*(t) &+ \int_{u(t_0)}^{u^*(t)} \left[C(v^*) - \frac{\partial q(v^*)}{\partial v} \right] \mathrm{d}v \\ &+ \left(\int_{u(t_0)}^{u(t)} - \int_{u(t_0)}^{u^*(t)} \right) C(v^*)\,\mathrm{d}v + \int_{u(t_0)}^{u(t)} \frac{\partial C}{\partial v} \Delta v\,\mathrm{d}v. \end{aligned}$$

The latter two integrals can be made arbitrarily small by improving the accuracy of the numerical procedures. However the first integral is independent of numerical accuracy, and hence the charge obtained from the conventional formulation will approximate the exact value only if

$$C(u^*) - \frac{\partial q_C(u^*)}{\partial u} = 0 \tag{3.1}$$

holds. This requirement concerns the capacitance model. It means that with the conventional formulation charge conservation can only be obtained if the *capacitance matrix* is the Jacobian of a charge vector, i.e., *has a generic function*. A sufficient condition is that the capacitance is controlled only by the branch voltage of the capacitor itself. So in these cases there is a chance to get charge conservation even with a capacitance-oriented formulation, provided that the numerical solution is sufficiently accurate. However, in many models developed so far, the requirement (3.1) is violated – because it implies additional restrictions on the capacitor model for real circuit devices, which is difficult to develop anyway. A well known counterexample is the model of Meyer for MOS capacitances, which has been discussed extensively in the literature, since it has been found that it violates the charge conservation requirement (MEYER [1971], SAKALLAH, YEN and GREENBERG [1990], WARD and DUTTON [1978]). See GÜNTHER and FELDMANN [1999b] for more details.

With the *charge/flux-oriented formulation* it is not difficult to obtain charge conservation. The first requirement is automatically met with the construction, that for each charge storing element one terminal charge is just the negative sum of all others. To check the second requirement, we expand the numerical approximation of the charge vector around the exact solution:

$$\begin{aligned} q(t) = q_C\big(u(t)\big) &= q_C\big(u^*(t)\big) + \frac{\partial q_C(u^*(t))}{\partial u} \cdot \Delta u + \mathrm{O}(\Delta u^2) \\ &= q^*(t) + \frac{\partial q_C(u^*)}{\partial u} \cdot \Delta u + \mathrm{O}(\Delta u^2). \end{aligned}$$

Hence $q(t)$ will approximate the exact charge, as Δu becomes smaller with increasing numerical accuracy.

4. Modified nodal analysis

The electrical network is now fully described by both Kirchhoff's laws and the characteristic equations in charge/flux oriented formulation. Based on these relations, most computer programs employ one of three schemes to set up the network equations: *Sparse Tableau Approach* (STA, see HACHTEL, BRAYTON and GUSTAVSON [1971]), *Nodal Analysis* (NA, see CHUA and LIN [1975], or *Modified Nodal Analysis* (MNA, see HO, RUEHLI and BRENNAN [1975]).

STA is rather canonical: All basic equations are set up explicitly in a system which contains all network variables as unknowns, i.e., node voltages u, branch voltages U and branch currents I. However, even for small circuits, a very large number of mainly short equations is generated.

We should mention that for theoretical investigations mostly an even more flexible extension of STA called *Hybrid Analysis* is used, which takes Kirchhoff's equations in their general form for loops of branch voltages and cutsets of branch currents rather than (2.1), (2.2).

NA. Contrary to STA, the aim of NA is to keep the network equations as compact as possible, so the vector of unknowns x contains only node voltages u. Since voltage sources have no admittance representation, they need a special treatment (CHUA and LIN [1975]) by which the number of KCL equations and of components of x is reduced by one for each voltage source. Hence the components u_1 of x are a subset of the node voltages u.

Current-controlled sources are difficult to implement, and inductors may lead to integro-differential network equations. Thus NA is not well suited for modelling circuits which contain these elements.

MNA represents a compromise between STA and NA, combining the advantages of both methods. It shares the universality with STA, but has the advantage of a smaller number of unknowns and equations: In addition to the node voltages, the branch currents $\jmath_V$ and $\jmath_L$ of just those elements are included into the vector x of unknowns which have no simple characteristic equations in admittance form, i.e., voltage sources and inductors/flux sources. Therefore it is most commonly used in industrial applications to generate the network equations.

Charge/flux oriented formulation of MNA. To set up the MNA network equations, KCL (2.2) is applied to each node except ground, and the admittance form representation for the branch current of resistors, current sources, capacitors and charge sources is directly inserted. The impedance form equations of voltage sources, inductors, and flux sources are explicitly added to the system of equations. They implicitly define the branch currents of these elements. Finally, all branch voltages are converted into node voltages with the help of KVL (2.1). Splitting the incidence matrix A into the element related incidence matrices A_C, A_L, A_R, A_V and A_I for charge and flux storing elements, resistors, voltage and current sources, one obtains from MNA the network equations in charge/flux oriented formulation (ESTÉVEZ SCHWARZ and TISCHENDORF [2000]):

$$A_C\dot{q} + A_R r(A_R^\top u, t) + A_L \jmath_L + A_V \jmath_V + A_I \imath(A^\top u, \dot{q}, \jmath_L, \jmath_V, t) = 0, \qquad (4.1a)$$

$$\dot{\phi} - A_L^\top u = 0, \tag{4.1b}$$

$$v(A^\top u, \dot{q}, \jmath_L, \jmath_V, t) - A_V^\top u = 0, \tag{4.1c}$$

$$q - q_C(A_C^\top u) = 0, \tag{4.1d}$$

$$\phi - \phi_L(\jmath_L) = 0 \tag{4.1e}$$

with

node voltages u,

branch currents through voltage and flux controlled elements $\jmath_V$ and $\jmath_L$,

charges and fluxes q and ϕ,

voltage dependent resistors r,

voltage and current dependent charge and flux sources q_C and ϕ_L,

controlled current and voltage sources $\imath$ and v.

For an illustration of A_C, A_L, A_R, A_V and A_I we refer to the following example.

The Schmitt trigger again. Let us now return to the Schmitt trigger introduced in Section 1. To apply Modified Nodal Analysis, we have to introduce additional nodes 6 and 7 as terminals of the voltage sources V_{in} and V_{DD}. This yields additional node potentials u_6, u_7 and branch currents $\jmath_{V_1}, \jmath_{V_2}$ through the voltage sources V_{in} and V_{DD} as new network variables. With $u := (u_1, \ldots, u_7)^\top$ and $\jmath_V := (\jmath_{V_1}, \jmath_{V_2})^\top$, the network equations (1.1) for the Schmitt trigger can be written in the charge/flux-oriented form (4.1a), (4.1c), (4.1d) by defining

$$A_C = \begin{pmatrix} 0 \\ 1 \\ 0 \\ -1 \\ 0 \\ 0 \\ 0 \end{pmatrix}, \qquad A_R = \begin{pmatrix} 1 & 0 & 0 & 0 & 0 \\ 0 & 1 & 0 & 1 & 0 \\ 0 & 0 & 1 & 0 & 0 \\ 0 & 0 & 0 & -1 & 0 \\ 0 & 0 & 0 & 0 & 1 \\ -1 & 0 & 0 & 0 & 0 \\ 0 & -1 & 0 & 0 & -1 \end{pmatrix},$$

$$A_V = \begin{pmatrix} 0 & 0 \\ 0 & 0 \\ 0 & 0 \\ 0 & 0 \\ 0 & 0 \\ 1 & 0 \\ 0 & 1 \end{pmatrix}, \qquad A_I = \begin{pmatrix} 1 & 0 & 0 & 0 \\ 0 & 1 & 0 & 0 \\ -1 & -1 & -1 & -1 \\ 0 & 0 & 1 & 0 \\ 0 & 0 & 0 & 1 \\ 0 & 0 & 0 & 0 \\ 0 & 0 & 0 & 0 \end{pmatrix},$$

$$C = C_0, \qquad q_C(A_C^\top u) = C A_C^\top u,$$

$$G = \operatorname{diag}(G_1, \ldots, G_5), \qquad r(A_R^\top u, t) = G A_R^\top u,$$

$$v(A^\top u, \dot{q}, \jmath_L, \jmath_V, t) = \begin{pmatrix} V_{in}(t) \\ V_{DD}(t) \end{pmatrix},$$

$$\imath(A^\top u, \dot{q}, \jmath_L, \jmath_V, t) = \begin{pmatrix} g(u_1 - u_3) \\ \alpha \cdot g(u_1 - u_3) \\ g(u_4 - u_3) \\ \alpha \cdot g(u_4 - u_3) \end{pmatrix}.$$

Note that the circuit does not contain inductors. Hence $\jmath_L = \{\}$, and the contribution $A_L \jmath_L$ does not appear in (4.1a).

Conventional formulation of MNA. Inserting flux and charge relations (4.1d), (4.1e) into the first equations, one achieves the analytically equivalent *conventional formulation* of MNA

$$\begin{aligned} &A_C C(A_C^\top u) A_C^\top \dot{u} + A_R r(A_R^\top u, t) + A_L \jmath_L \\ &\quad + A_V \jmath_V + A_I \imath(A^\top u, \dot{q}_C(A_C^\top u), \jmath_L, \jmath_V, t) = 0, && (4.2a) \\ &L(\jmath_L, t)\dot{\jmath}_L - A_L^\top u = 0, && (4.2b) \\ &A_V^\top u - v(A^\top u, \dot{q}_C(A_C^\top u), \jmath_L, \jmath_V, t) = 0, && (4.2c) \end{aligned}$$

with generalized capacitance, inductance and conductance matrices

$$C(w) := \frac{\partial q_C(w)}{\partial w}, \quad L(w) := \frac{\partial \phi_L(w)}{\partial w} \quad \text{and} \quad G(w,t) := \frac{\partial r(w,t)}{\partial w}.$$

These matrices are positive-definite, but not necessarily symmetrical, in contrast to the capacitance, inductance and conductance matrices gained from the two-terminal elements capacitor, inductor and resistor used, for example, in the Schmitt trigger example.

Structure of MNA network equations. Generally, the following properties hold: The matrices

$$\widetilde{C}(A_C^\top u) := A_C C(A_C^\top u) A_C^\top, \quad \text{and} \quad \widetilde{G}(A_R^\top u, t) := A_R G(A_R^\top u, t) A_R^\top$$

are usually very sparse and have structural symmetry.

In some respect, the fine structure of the network equations depends on the type of network elements, on the network topology and on the modelling level:

Type of network elements. There are the trivial conclusions, that the system degenerates to a purely algebraic one if the circuit contains neither capacitors nor inductors (energy storing elements), and that the system is homogeneous if there are no time-dependent elements. If there are no controlled sources, then the Jacobian matrix

$$D(A_R^\top u, t) := \begin{pmatrix} \widetilde{G}(A_R^\top u, t) & A_L & A_V \\ -A_L^\top & 0 & 0 \\ -A_V^\top & 0 & 0 \end{pmatrix} \tag{4.3}$$

with respect to u, $\jmath_L$ and $\jmath_V$ has structural symmetry.

Network topology. Due to Kirchhoff's laws, cutsets of current sources and loops of voltage sources are forbidden. This implies that the matrix (A_C, A_R, A_V, A_L) has full row rank and the matrix A_V has full column rank. If there is a loop of independent voltage sources and/or inductors, or a cutset of independent current sources and/or capacitors, the Jacobian matrix $D(A_R^\top u, t)$ is singular. In these cases no steady-state solution can be computed, and so most circuit analysis programs check and refuse these conditions, which are purely topological. But note that in the nonlinear case the Jacobian matrix also may become numerically singular, e.g., due to vanishing partial derivatives or in the case of bifurcation (in the autonomous case this deals with free oscillators).

The matrix $\widetilde{C}(A_C^\top u)$ is singular, if there are nodes which have no path to ground via energy storing elements. If the circuit contains voltage sources, the MNA network equations contain algebraic relations of type (4.1c) and (4.2c), respectively. This is true in most circuits, and so mostly the equations are DAEs, i.e., the Jacobian matrix

$$B(A_C^\top u, t) := \begin{pmatrix} \widetilde{C}(A_C^\top u) & 0 & 0 \\ 0 & L(j_L) & 0 \\ 0 & 0 & 0 \end{pmatrix},$$

with respect to $\dot{u}$, j_L and j_V is singular.

Modelling level. Additionally, the modelling level defines some properties of the systems. $\widetilde{C}(A_C^\top u)$ is symmetrical in case of linear or nonlinear differential capacitances, but symmetry may be lost in case of general nonlinear capacitances or nonlinear charge models, as are used for example in MOS transistor models (GÜNTHER and FELDMANN [1999a], GÜNTHER and FELDMANN [1999b]).

Charge/flux oriented or conventional MNA? On which formulation – charge/flux oriented or conventional – should the numerical discretization be based, if MNA is used for the automatic generation of network equations? From a structural aspect, the conventional MNA formulation yields a standard form of numerical integration problems, while the charge/flux oriented formulation does not. There are however several reasons, not to transform (4.1) into (4.2) before applying numerical discretization schemes, although they are analytically equivalent:

Structure. (4.1) is of linear-implicit nonlinear form, while (4.2) is of nonlinear-implicit nonlinear form. This may have an impact on the choice of a suitable integrator.

Numerics. Information on the charge/flux level is lost in the conventional approach, and charge conservation may only be maintained approximately in numerical integration schemes.

Implementation. Implicit numerical integration schemes for the conventional MNA equations (4.2) require second partial derivatives of q_C and ϕ_L. These derivative informations, however, are not available in circuit simulation packages, may even not exist because of the lack of smoothness in transistor models.

5. Why differential-algebraic equations?

The charge/flux-oriented formulation of energy storing elements *and* MNA network equations supply us with a first argument for using differential-algebraic equations in electrical circuit modelling. In the following we will assemble more arguments why using a DAE approach with a redundant set of network variables, and not an ODE model.

First of all, one has to distinguish between two different ways to obtain ODE models:

- *Generating a state-space model with a minimal set of unknowns.* Drawbacks of this approach include software engineering, modelling, numerical and designer-oriented arguments. The state-space form cannot be generated in an automatic way, and may exist only locally. The use of independent subsystem modelling, which is essential for the performance of todays VLSI circuits, is limited, and the advantage of sparse matrices in the linear algebra part is lost. Finally, the topological information of the system is hidden for the designer, with state variables losing their technical interpretation.
- *Regularizing the DAE to an ODE model by including parasitic effects.* It is commonly believed that the DAE character of the network equations is only caused by a high level of abstraction, based on simplifying modelling assumptions and neglection of parasitic effects. So one proposal is to regularize a DAE into an ODE model by including parasitic effects. However, this will yield singularly perturbed problems, which will not at all be preferable to DAE models in numerical respect. Beyond it, refined models obtained by including parasitics may make things worse and lead to problems which are more ill-posed.

So we have to inspect feasibility of state-space formulation, subcircuit partitioning and regularization based on including parasitic effects.

State-space formulation: State equations. It is well known that for a large class of nonlinear circuits it is possible to write the network equations as an explicit system of ordinary differential equations of first order, the so-called *State Equations* (CHUA and LIN [1975]). For this purpose the vector x_1 of unknowns is constructed only from capacitor voltages and inductor currents[2] – respectively of capacitor charges and inductor fluxes, if a charge/flux-oriented formulation is preferred. In case of special circuit configurations like loops of capacitors or cutsets of inductors, algebraic constraints on the state variables have to be observed (refer also to Section 7), so perhaps not all of them are included into x_1. The resulting system of equations is:

$$\dot{x}_1 = f_1\big(x_1, s(t), \dot{s}(t)\big),$$
$$x_2 = f_2\big(x_1, s(t), \dot{s}(t)\big).$$

Here s describes independent sources, and x_2 contains those network variables which are not included in x_1, e.g., voltages of nodes to which no capacitive element is connected, or branch currents of voltage sources. The second equation serves for computing x_2, once the set of explicit differential equations (first part) has been solved for x_1 (BOWERS and SEDORE [1971]). Note that the equations may contain time derivatives of the input source waveforms.

However, as we can conclude from an extensive literature, the existence of this form is not at all trivial for general classes of nonlinear circuits (CHUA [1984], CHUA and LIN [1975], MATHIS [1987]), and therefore the algorithms for setting up the equations are difficult to program and time consuming. Furthermore, compared to NA or MNA, the number of unknowns is extremely large for actual integrated circuits containing a

[2]Note that in network theory just the latter variables are called *state variables*, which is somewhat different from the use of this notation in numerics.

large number of parasitic capacitors. And most important, the structure of the equations does not reflect the structure of the circuit. Therefore this formulation is no longer used in actual circuit simulation programs.

Subcircuit partitioning. The design of memory chips and advanced digital/analog circuits demands the numerical simulation of networks with several ten thousand transistors. Parallel simulation is then valuable to reduce runtime, which otherwise would be prohibitive for such large applications. For this purpose, domain decomposition methods may be employed, requiring to partition the circuit into subblocks which are decoupled by introducing virtual voltage and/or current sources as coupling units at the boundaries (WEVER and ZHENG [1996], ARNOLD and GÜNTHER [2001]) (see Section 15).

Regard now, for example, two subcircuits only connected by resistive paths, with only linear energy storing elements and resistors, and without any sources. With the partitioned vectors of node voltages $u = (u_1, u_2)^\top$ and branch currents through inductors $j_L = (j_{L1}, j_{L2})^\top$, the network equations read

$$A_C\dot{q} + A_R r(A_R^\top u) + A_L j_L + A_V j_V = 0, \tag{5.1a}$$

$$\dot{\phi} - A_L^\top u = 0, \tag{5.1b}$$

$$A_V^\top u = 0, \tag{5.1c}$$

where

$$\begin{aligned} &A_C := \operatorname{diag}(A_{C_1}, A_{C_2}), && q = q_C(A_C^\top u) := \operatorname{diag}(C_1, C_2)A_C^\top u, \\ &A_L := \operatorname{diag}(A_{L_1}, A_{L_2}), && \phi = \phi_L(j_L) := \operatorname{diag}(L_1, L_2) j_L, \\ &r(A_R^\top u) = G A_R^\top u \end{aligned}$$

and (5.1c) describes the virtual voltage sources and j_V are their branch currents. We will have to deal with DAE models even if the designers assure that all subcircuits are represented by ODE models! This is easily explained by the fact that the network equations (5.1) correspond to Lagrange equations of the first kind: Defining the electric and magnetic energies of both networks by

$$V(u) = \frac{1}{2}\sum_{i=1}^{2} u_i^\top A_{C_i} C_i A_{C_i}^\top u_i, \qquad T(j_L) = \frac{1}{2}\sum_{i=1}^{2} j_{L_i}^\top L_i j_{L_i}$$

yields the Lagrangian

$$\mathcal{L} := T(j_L) - V(u) + \lambda^\top\big(A_V^\top u - 0\big),$$

where the characteristic equations for the virtual voltage sources are added via Lagrangian multipliers λ. $\mathcal{L}$ fulfills the equation

$$\frac{\mathrm{d}}{\mathrm{dt}}\frac{\partial \mathcal{L}}{\partial \dot{x}} - \frac{\partial \mathcal{L}}{\partial x} = \frac{\partial \mathcal{W}}{\partial \dot{x}}$$

for $x = \{q, \phi, \lambda\}$, $\dot{x} = \{j_L, u, j_V\}$, with the dissipative function $\mathcal{W}$ given by

$$\mathcal{W} := \sum_{i=1}^{2} j_{L_i}^\top A_{L_i}^\top u_i + \frac{1}{2} u^\top A_R G A_R^\top u.$$

Here we used the integral quantities charges and fluxes as state variables and Lagrangian multipliers:

$$q(t) := \int_0^t j_L(\tau)\,\mathrm{d}\tau, \quad \phi(t) := \int_0^t u(\tau)\,\mathrm{d}\tau, \quad \lambda(t) := \int_0^t j_V(\tau)\,\mathrm{d}\tau.$$

One notes that the Lagrangian depends only on derivatives of the state variables. This is caused by the fact that the characteristic equations for energy storing elements are differential equations of first order in the state variables u and j_L.

Regularization based on including parasitic effects. In general, regularization is based on the assumption that the differential-algebraic form of the network equations is caused by a too high level of simplification in the modelling process, and therefore an ODE formulation can be reached by adding proper "parasitic" effects or elements to the circuit model (FELDMANN and GÜNTHER [1999]). One rule-of-thumb is to include parasitic capacitors to ground at each node to get a regular capacitance matrix, and thus an ODE model for the circuit. However, this approach fails, if, for example, a cutset of current source and inductor with inductance L is regularized by adding a small capacitor with capacitance C bridging the cutset (GÜNTHER and FELDMANN [1999b]).

The drawback is that we are confronted with a singularly-perturbed ODE system if C is too small: An additional oscillation with frequency $\omega_1 = 1/\sqrt{LC}$, the eigenfrequency of the regularized system, is invoked through regularization, which overlays the principle voltage courses, and the numerical problems even increase. One example for such an inappropriate regularization is given by the ring modulator, whose numerical problems have been discussed extensively in the literature (DENK and RENTROP [1991], HORNEBER [1985], KAMPOWSKY, RENTROP and SCHMIDT [1992]): Parasitic capacitances in the proposed range of some pF yield additional high-frequency oscillations in the GHz-range, which drastically slows down numerical simulation. Numerical regularization effects become visible only for capacitances thousand times larger, which are not realistic (FELDMANN and GÜNTHER [1999]). On the other hand, the DAE model without parasitic capacitors leads to physically correct results, without any numerical problems, if appropriate integration schemes are used.

Besides that, it is not trivial to make sure that a refined modelling based on including parasitic effects will always yield ODE models. Even worse, the numerical problems may increase with the refinement of the model, as will be shown in Section 9 for different levels in the refined modelling of a bipolar ring oscillator. This result can be explained easily by the fact that the DAE index, a measure for the structural properties of DAE systems, increases.

CHAPTER II

DAE-index – the Structural Aspect

So we are faced with network equations of differential-algebraic type when simulating electrical circuits. Before attacking them numerically, we have to reveal the analytical properties of DAEs. In a first step we inspect linear systems and apply, in a second step, the results to nonlinear systems. We will see that for a rather general class of circuits the network topology determines the structural properties of the DAE network equations. However, if more general models for the network elements are incorporated, special circuit configurations apply or refined models are used to include second order and parasitic effects, one may have to cope with ill-conditioned problems.

6. The index concept for linear systems

If a circuit contains only linear elements – or if the system is linearized at an operating point $(x(t_0), \dot{x}(t_0))$, in order to investigate the system behaviour for small signal excitations from that operating point – then the corresponding network equations represent differential-algebraic equations in linear implicit form:

$$B\dot{x} + Dx = s(t), \qquad x(t_0) = x_0. \tag{6.1}$$

If we further assume MNA form then $x = (u, \jmath_L, \jmath_V)^\top$ and

$$B = \begin{pmatrix} A_C C A_C^\top & 0 & 0 \\ 0 & L & 0 \\ 0 & 0 & 0 \end{pmatrix}, \qquad D = \begin{pmatrix} A_R G A_R^\top & A_L & A_V \\ -A_L^\top & 0 & 0 \\ -A_V^\top & 0 & 0 \end{pmatrix},$$

$$s = \begin{pmatrix} -A_I \imath(t) \\ 0 \\ -v(t) \end{pmatrix}.$$

Such linear-implicit systems constitute the starting point to classify differential-algebraic equations by a structural property known as the index. In the linear case, this property depends only on the structure of B and D:

ODE-case: B regular. This holds, iff the circuit contains no voltage sources and there are no nodes which have no path to ground via capacitors. In this case, system (6.1) represents a linear-implicit system of ODEs, and can be transformed into the explicit ODE system

$$\dot{x} = B^{-1}\big(-D \cdot x + s(t)\big).$$

DAE-case: B singular. In the following we will assume D to be regular. This requirement allows for computing equilibria solutions by an operating point analysis to determine initial values, and can be assured by proper demands on the network topology (GÜNTHER and FELDMANN [1999a], GÜNTHER, HOSCHEK and RENTROP [2000]). Thus multiplying (6.1) with D^{-1} from the left-hand side leads to

$$D^{-1}B \cdot \dot{x} + x = D^{-1} \cdot s(t). \tag{6.2}$$

By Jordan decomposition of

$$D^{-1}B = T^{-1}\begin{pmatrix} \widetilde{B} & 0 \\ 0 & N \end{pmatrix} T$$

with a regular, time independent, matrix T, Eq. (6.2) can be written after multiplication by T from the left-hand side as

$$\begin{pmatrix} \widetilde{B} & 0 \\ 0 & N \end{pmatrix} T\dot{x} + Tx = TD^{-1}s(t) \tag{6.3}$$

with a regular matrix $\widetilde{B}$ and a nilpotent matrix N. N belongs to the eigenvalue 0 and is of nilpotency ν, i.e., ν is the smallest number such that $N^\nu = 0$, but $N^{\nu-1} \neq 0$. The transformation

$$\begin{pmatrix} y \\ z \end{pmatrix} := Tx, \quad \text{and} \quad \begin{pmatrix} \eta(t) \\ \delta(t) \end{pmatrix} := TD^{-1}s(t)$$

with differential variables y and algebraic variables z decouples this system into an explicit ODE and a nilpotent part:

$$\dot{y} = \widetilde{B}^{-1}\big(\eta(t) - y\big), \tag{6.4}$$

$$N\dot{z} = \delta(t) - z. \tag{6.5}$$

The nilpotent part has to be investigated further:

- *Index-1 case*: $\nu = 1$, *i.e.*, $N = 0$
  Now the nilpotent part reads

  $$z = \delta(t); \tag{6.6}$$

  the algebraic variables are explicitly given by the input signal. After one differentiation an explicit ODE system for z is obtained.
- *Higher-index case*: $\nu \geqslant 2$
  The algebraic variables are only given explicitly after a differentiation process: Differentiation of (6.5) and multiplication with N from the left-hand side yields

  $$N^2\ddot{z} + N\dot{z} = N\dot{\delta}(t) \quad \Rightarrow \quad z = \delta(t) - N\dot{\delta}(t) + N^2\ddot{z}.$$

  If $\nu = 2$ holds, we cease the process, otherwise it has to be repeated until $N^\nu = 0$:

  $$z = \delta(t) - N\dot{\delta}(t) + N^2\ddot{\delta}(t) - \cdots + (-1)^{\nu-1}N^{\nu-1}\delta^{(\nu-1)}(t). \tag{6.7}$$

  Now the solution depends not only on the input signal, but also on its derivatives! A last differentiation (i.e., the νth one) leads to the desired explicit ODE system

  $$\dot{z} = \dot{\delta}(t) - N\ddot{\delta}(t) + N^2\delta^{(3)}(t) - \cdots + (-1)^{\nu-1}N^{\nu-1}\delta^{(\nu)}(t). \tag{6.8}$$

Here we have assumed that δ is $(\nu-1)$-times differentiable to get a continuous solution z. On the other hand, if we allow discontinuous input signals, solutions may only exist in the sense of distributions (RABIER and RHEINBOLDT [1996]).

Summing up, the solution behaviour of a linear-implicit system of differential equations differs from standard ODE theory in the following sense:

- The solution has to fulfill an algebraic constraint, since $z(t_0)$ is fixed by δ and its higher derivatives at the initial time point t_0. Especially, the solutions do not depend continuously differentiable on the initial values. For $\nu = 1$, this constraint is explicitly given by (6.6). In the higher-index case, however, the constraint is hidden: A differentiation process is necessary to obtain (6.7).
- The system is sensitive to perturbations. Take as example a signal noise, modelled by the input signal δ: Although δ may be very small, its higher derivatives may be arbitrarily large. A severe amplification of perturbations may occur for higher-index problems: We are faced with ill-posed problems.

These analytical results suggest that no severe numerical problems arise in index-1 systems: The algebraic constraint is explicitly given; hence implicit numerical integration schemes for stiff systems such as BDF (GEAR [1971]) or ROW-type methods (RENTROP, ROCHE and STEINEBACH [1989]) (see Chapter III), which contain a nonlinear equation solver, are suitable to treat these problems. Additionally, no amplification of round-off errors is to be expected since the system is not sensitive to perturbations.

However, severe numerical problems may arise for systems with nilpotency $\nu \geqslant 2$: There are hidden algebraic constraints, which can be resolved only by an unstable differentiation process. Regarding perturbations δ entering the right hand side due to inaccurate solutions or due to roundoff errors, terms of order $\delta/h^{\nu-1}$ will enter the solution, where h is the small time discretization parameter.

Since the value of ν defines the behaviour of the system (6.1), both in theoretical and numerical respect, ν is called the *algebraic index* of the linear implicit system (6.1). Additionally, the observations made above motivate three different point of views:

Differential index: To obtain an explicit differential system instead of the linear-implicit system (6.1), we had to differentiate the nilpotent part (6.5). Since numerical differentiation is an unstable procedure, the number of differentiation steps needed to get an explicit ODE system is a measure for the numerical problems to be expected when solving systems of type (6.1). Hence the minimum number of differentiations required is called the *differential index* ν_d of the linear-implicit linear system (6.1).

Perturbation index: We have seen that derivatives of the perturbation enter the solution of (6.1). This observation leads to a new kind of index, which measures the sensitivity of the solutions to perturbations in the equations: The linear-implicit system (6.1) has *perturbation index* ν_p, if derivatives of perturbations up to degree ν_p enter the derivative of the solution.

Tractability index: Finally it was shown that the decomposition of a DAE system into the part governed by regular ODEs, the algebraic part, and the part which can only be solved by performing a differentiation process gives much insight into the nature of the problem. This is especially helpful for analysis and construction of new methods. Griepentrog and März developed a calculus for doing this by using properly constructed chains of projectors, which led to the tractability index concept (GRIEPENTROG and

MÄRZ [1986], MÄRZ [1992]). We restrict here to the definition of index 1 and 2. To this end we introduce

$$N := \ker B, \qquad S := \{z\colon\ Dz \in \operatorname{im} B\}$$

and define: The system (6.1) with B being singular has tractability index 1, if $N \cap S = \{0\}$, i.e., $B_1 := B + DQ$ is nonsingular for a constant projector Q onto N.

If it is not of index 1 then we introduce

$$P := I - Q, \qquad N_1 := \ker B_1, \qquad S_1 := \{z\colon\ DPz \in \operatorname{im} B_1\}$$

and define: The system has tractability index 2, if $N_1 \cap S_1 = \{0\}$, i.e., $B_2 := B_1 + DPQ_1$ is nonsingular for a constant projector Q_1 onto N_1.

In the index-2 case, $N \cap S$ comprises just those components, which can be solved only by a differentiation process. An outcome of this index notation is an exact identification, which part of the DAE system needs which smoothness condition to be solvable.

Although the different index concepts were developed for different purposes, it turns out that in most nonpathological cases all of them yield the same number, or differ at most by one. So we are free to select one of them which suits best to our actual item of interest, or is the easiest to compute.

All definitions can be generalized in a straightforward way to nonlinear DAE systems (GEAR [1988, 1990], HAIRE, LUBICH and ROCHE [1989], GRIEPENTROG and MÄRZ [1986]).

It remains to determine the index of the charge/flux-oriented network equations (4.1). Due to the charge and flux defining equations (4.1d), (4.1e), the index is always $\geqslant 1$ if the circuit contains energy storing elements at all.

7. Network topology and DAE-index for RLC networks

In the linear case, the two-terminal elements capacitor, inductor and resistor are described by linear functions with *positive* capacitance, inductance and resistance. Hence the matrices

$$C := \frac{\partial q_C(w)}{\partial w}, \qquad L := \frac{\partial \phi_L(w)}{\partial w}, \qquad G := \frac{\partial r(w)}{\partial w}$$

of capacitances, inductances and resistances are symmetrical positive-definite. In other words, the elements are strictly passive.

Generalizing this property to the nonlinear case, the local strict passivity of nonlinear capacitors, inductors and resistors corresponds to the positive-definiteness (but not necessarily symmetry) of the so-called generalized capacitance, inductance and conductance matrices

$$C(w) := \frac{\partial q_C(w)}{\partial w}, \quad L(w) := \frac{\partial \phi_L(w)}{\partial w} \quad \text{and} \quad G(w,t) := \frac{\partial r(w,t)}{\partial w}.$$

already introduced in Section 4. If this property of positive-definiteness holds, the network is called an RLC-network.

Topological conditions. Let us first investigate RLC-networks with independent voltage and current sources only. To obtain the perturbation index of (4.1), we perturb the right-hand side of (4.1a)–(4.1c) with a slight perturbation $\delta = (\delta_C, \delta_L, \delta_V)^\top$ on the right-hand side. The corresponding solution of the perturbed system is denoted by $x^\delta := (u^\delta, j_L^\delta, j_V^\delta)^\top$. One can show that the difference $x^\delta - x$ between perturbed and unperturbed solution is bounded by the estimate

$$\begin{aligned}\|x^\delta(t) - x(t)\| \leqslant \text{const} \cdot \Big(\|x^\delta(0) - x(0)\| + \max_{\tau\in[0,t]} \|\delta\| \\ + \max_{\tau\in[0,t]} \|Q_{CRV}^\top \dot{\delta}_C\| + \max_{\tau\in[0,t]} \|\bar{Q}_{V-C}^\top \dot{\delta}_V\|\Big)\end{aligned} \tag{7.1}$$

using orthogonal projectors Q_C, Q_{CRV} and $\bar{Q}_{V-C}$ onto $\ker A_C^\top$, $\ker(A_C A_R A_V)^\top$ and $\ker Q_C^\top A_V$, respectively TISCHENDORF [1999]. Since $Q_{CRV}^\top A_C = 0$ holds, the index does not raise, if also perturbations δ_q and δ_ϕ are allowed in the charge and flux defining Eqs. (4.1d)–(4.1e).

Thus the index of the network equations is one, iff the following two topological conditions hold:

T1: There are no loops of only charge sources (capacitors) and voltage sources (no VC-loops): $\ker Q_C^\top A_V = \{0\}$ and thus $Q_{CRV} = 0$.

T2: There are no cutsets of flux sources (inductors) and/or current sources (no LI-cutsets): $\ker(A_C A_R A_V)^\top = \{0\}$ and thus $\overline{Q}_{V-C} = 0$.

In this case, we are faced with well-posed problems. If however T1 or T2 is violated then we have to cope with ill-posed problems of index 2. The generic index-2 configurations for such a violation are shown in Fig. 7.1: A VC-loop consisting of voltage source and capacitor, and an LI-cutset of current source and inductor[3].

VC loops. Let us investigate one important case of networks with VC-loops. As we have seen in Section 5, one rule-of-thumb is to regularize a circuit by adding at each node parasitic capacitors to ground. This approach yields $\ker A_C^\top = \{0\}$, and condition T2 is fulfilled. However, T1 is violated due to $Q_C = 0$ iff the network contains voltage sources: Every voltage source leads to a loop of capacitors and this voltage source.

We determine now the index for both cases. After differentiating the characteristic equations for charge, flux and voltage sources, we get with $\widetilde{C}(w) := A_C C(w) A_C^\top$ a system of the type

$$\begin{pmatrix} \widetilde{C}(A_C^\top u^\delta) & 0 & A_V \\ 0 & L(j_L^\delta) & 0 \\ A_V^\top & 0 & 0 \end{pmatrix} \begin{pmatrix} \dot{u}^\delta \\ \dot{j_L}^\delta \\ j_V^\delta \end{pmatrix} + f(u, j_L, t) = \begin{pmatrix} \delta_C(t) + A_C \dot{\delta}_q(t) \\ \delta_L(t) + \dot{\delta}_\phi(t) \\ \dot{\delta}_V(t) \end{pmatrix} \tag{7.2}$$

with u^δ, j_L^δ, j_V^δ being the solution of (4.1), perturbed with $\delta = (\delta_C, \delta_L, \delta_V, \delta_q, \delta_\phi)^\top$ on the right-hand side. With Kirchhoff's voltage law we have $\ker A_V = 0$, and thus the system can be resolved for $(\dot{u^\delta}, \dot{j_L^\delta}, j_V^\delta)$. For networks without voltage sources the index is one, otherwise two.

[3]The conductance is only inserted in the figures as a representative to show possible augmentations of the circuit without having an impact on the index.

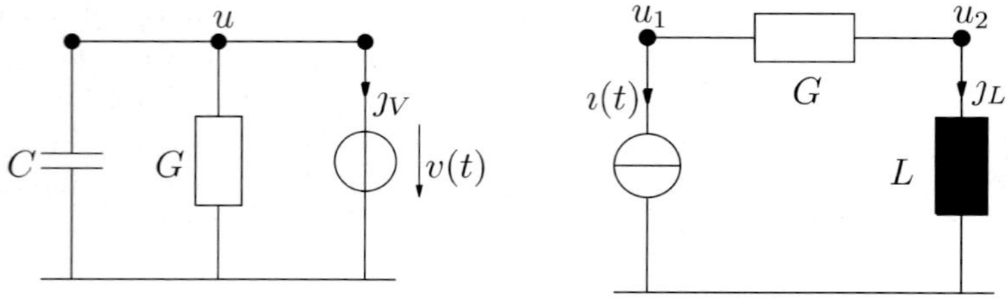

$$\begin{aligned} \dot q + G\cdot u + \jmath_V &= 0 \\ u - v(t) &= 0 \\ q - C\cdot u &= 0 \end{aligned}$$

$$\Rightarrow \jmath_V = -C\cdot \dot v(t) - G\cdot v(t)$$

Loop of voltage source and capacitor

$$\begin{aligned} \imath(t) + G\cdot(u_1 - u_2) &= 0 \\ \jmath_L - G\cdot(u_1 - u_2) &= 0 \\ \dot\phi - u_2 &= 0 \\ \phi - L\cdot \jmath_L &= 0 \end{aligned}$$

$$\Rightarrow \begin{aligned} u_2 &= -L\cdot \dot\imath(t) \\ u_1 &= -L\cdot \dot\imath(t) - 1/G\cdot \imath(t) \end{aligned}$$

Cutset of current source and inductor

FIG. 7.1. Index-2 generic configurations.

One notes that system (5.1) generated by subcircuit partitioning in Section 4 represents a special linear case of (7.2): $C(A_C^\top u) = C$ and $L(\jmath_L) = L$. In this case, we can derive sharper perturbation estimates than (7.1). The difference between the differential part $y := (u, \jmath_L)$ and $y^\delta := (u^\delta, \jmath_L^\delta)$ of unperturbed and perturbed solution is bounded by

$$\left\|y(t) - y^\delta(t)\right\| \leqslant \text{const}\cdot\left(\left\|y(0) - y^\delta(0)\right\| + \max_{\tau\in[0,t]}\|\delta_1\| + \max_{\tau\in[0,t]}\left\|\int_0^\tau \delta_0(\tau)\,\mathrm{d}\tau\right\|\right)$$

with $\delta_0 = (\delta_L, \delta_C)$ and $\delta_1 = (\delta_V, \delta_q, \delta_\phi)$ – no derivatives of perturbations enter the estimate for the differential variables. But for the algebraic components $\jmath_V$ one gets with the sharp estimate

$$\left\|\jmath_V(t) - \jmath_V^\delta(t)\right\| \leqslant \text{const}\cdot\left(\left\|y(0) - y^\delta(0)\right\| + \max_{\tau\in[0,t]}\|\delta\| + \max_{\tau\in[0,t]}\|\dot\delta_1\|\right)$$

an index-2 behaviour, as expected. In general however, derivatives of perturbations cannot be neglected in the bounds of both differential and algebraic components (ARNOLD [1997]).

Generalization. A generalization of these results for RLC-networks with independent voltage and current sources to special linear controlled sources is given in REISSIG [1998], where linear active networks are considered of capacitors, inductors, resistors, ideal transformers and gyrators. The results hold also for RLC-networks with a rather large class of nonlinear voltage and current sources: The index depends only on the topology; in general, the index is one, and two only for special circuit configurations (GÜNTHER and FELDMANN [1999b], ESTÉVEZ SCHWARZ and TISCHENDORF [2000],

TISCHENDORF [1999]). This class of sources contains, for example, controlled current sources not being part of VC-loops that are controlled by $(A_C A_V A_R)^\top u$, $\jmath_V$ and t.

One example for such an RLC-network is given by the Schmitt trigger already introduced in Section 1. Inspecting the charge-/flux-oriented network equations (4.1) derived for the Schmitt trigger in Section 4, we see that the current sources I_B, I_C and I_E describing the bipolar transistors are only controlled by the branch voltages $A_V^\top u$ and $A_R^\top u$. Since $\ker Q_C^\top A_V = \{0\}$ due to

$$Q_C = \begin{pmatrix} 1 & 0 & 0 & 0 & 0 & 0 & 0 \\ 0 & 1/2 & 1/2 & 0 & 0 & 0 & 0 \\ 0 & 0 & 1 & 0 & 0 & 0 & 0 \\ 0 & 1/2 & 1/2 & 0 & 0 & 0 & 0 \\ 0 & 0 & 0 & 0 & 1 & 0 & 0 \\ 0 & 0 & 0 & 0 & 0 & 1 & 0 \\ 0 & 0 & 0 & 0 & 0 & 0 & 1 \end{pmatrix}$$

and $\ker(A_C A_R A_V)^\top = \{0\}$ hold, the Schmitt trigger yields an index-1 problem.

These results obtained for RLC circuits rest on two different types of assumptions: Positive-definiteness of generalized capacitance, inductance and conductance matrices on the one hand, and no arbitrary controlled sources on the other hand. If one of these demands is violated, the index may depend not only on whether the topology conditions T1 and T2 hold or not, but also on circuit and model parameters and – for circuits containing nonlinear elements – on their bias conditions. In addition, the index can be larger than two.

Violation of positive-definiteness. Independent charge and flux sources, which may model α-radiation or external magnetic fields in a somewhat higher level of abstraction,

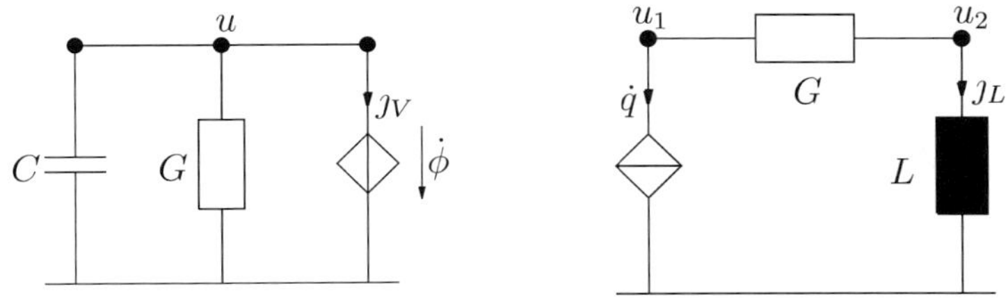

$$\begin{aligned} \dot q + G\cdot u + \jmath_V &= 0 \\ \dot\phi - u &= 0 \\ q - C\cdot u &= 0 \\ \phi - \phi_L(t) &= 0 \end{aligned} \qquad \begin{aligned} \dot q + G\cdot(u_1 - u_2) &= 0 \\ \jmath_L - G\cdot(u_1 - u_2) &= 0 \\ \dot\phi - u_2 &= 0 \\ q - q_C(t) &= 0 \\ \phi - L\cdot \jmath_L &= 0 \end{aligned}$$

$$\Rightarrow \jmath_V = -C\cdot\ddot\phi_L(t) - G\cdot\dot\phi_L(t) \qquad \Rightarrow \begin{aligned} u_2 &= -L\cdot\ddot q_C(t) \\ u_1 &= -L\cdot\ddot q_C(t) - 1/G\cdot\dot q_C(t) \end{aligned}$$

Loop of flux source and capacitor   Cutset of charge source and inductor

FIG. 7.2. Index-3 configurations: ΦC-loop and QL-cutset.

can destroy the positive-definiteness of generalized capacitance and inductance matrices. Generic examples for this case are the circuits of Fig. 7.2: ΦC-loop and QL-cutsets. We see that u_1 and u_2 (respectively j_V) are index-3 variables in the cutset (loop) circuit. It has to be checked whether this mechanism may also lead to index-3 problems in case of MOS circuits with charge models whose derivatives vanish under certain bias conditions.

8. Networks with controlled sources

Higher index can also be generated by controlled sources. One example is the coupling of index-2 problems via controlled sources. Another example is that although both topological conditions T1 and T2 hold, circuit parameters may have an impact on the structural properties of the network equations if a network contains controlled sources, even it is linear. First we discuss the effects of coupling circuits with controlled sources; then we analyze a *differentiator circuit* and a *Miller integrator*, for illustration of the second phenomenon.

Before doing this we should note that controlled sources are indispensable elements in circuit simulation, which are extensively used in semiconductor models as well as in macro models of a somewhat higher level of abstraction, and for modelling signal propagation on and between interconnects. An instructive example for the latter use and its effects on the index is discussed in Section 9.

Coupling of higher-index configurations via controlled sources may raise the index of the driven circuit part by one or two per controlled source, if the controlling network variable itself is of higher index. The question, in which cases the index gets higher, is difficult to answer. Below we will give some simple generic cases. Surprisingly, much sophisticated research in this field was done twenty years ago, although the index notion was not yet introduced at all. The motivation was to develop algorithms for setting up network equations in the *State Variable* approach (GÜNTHER and FELDMANN [1999a]) for general classes of networks including controlled sources, and the methods developed for this purpose aimed just to capture as many circuit configurations as possible, which in our notation are of index $\leqslant 2$. Most of them start from properly constructed *normal trees* spanning the network graph. An overview can be found in CALAHAN [1972], CHUA and LIN [1975].

There are eight possibilities to couple cutsets of current sources/inductors (JL cutsets) and loops of voltage sources/capacitors (VC loops) via either a voltage-controlled element or a current-controlled element. It turns out that only those configurations will have an increased index where the controlling variable itself is of higher index. These configurations are listed in Table 8.1. Here the subscript C (D) denotes network elements and variables of the controlling (driven) circuit. The argument t characterizes the input variable, and a prime $'$ indicates the derivative with respect to the controlling variable.

Extensions are possible by replacing in the VC loops and JL cutsets the inductor or voltage source by a flux source, and the current source or capacitor by a charge source.

TABLE 8.1
Index-3 coupling of index-2 circuits via controlled sources

Case	Controlling circuit	Input variable	Driven circuit	Controlled source	Output variable	Index
1	JL cutset	$J(t)$	JL cutset	$J_D(u_C)$	$u_D = L_C L_D J'_D \ddot{J}(t)$	3
2	JL cutset	$J(t)$	VC loop	$V_D(u_C)$	$I_D = L_C C_D V'_D \ddot{J}(t)$	3
3	VC loop	$V(t)$	VC loop	$V_D(I_C)$	$I_D = C_C C_D V'_D \ddot{V}(t)$	3
4	VC loop	$V(t)$	JL cutset	$J_D(I_C)$	$u_D = C_C L_D J'_D \ddot{V}(t)$	3

These extensions lead to 32 further high-index configurations, and one can get a circuit configuration of index 5 with only 4 elements (GÜNTHER and FELDMANN [1999b]). Further extensions are possible by replacing the sources by norators.

The higher-index configurations described here may be recursively used, thus obtaining circuit configurations of arbitrary high index.

Differentiator circuit. We must expect a higher-index problem if the circuit itself acts as a differentiator. A differentiator circuit with input source $v(t)$ and output voltage u_3 is given in Fig. 8.1 where the operational amplifier (see Fig. 8.2) with amplification factor a is a special case for a voltage-controlled voltage source.

From its MNA equations

$$A_R G A_R^\top u + A_L \jmath_L + A_V \jmath_V = 0,$$

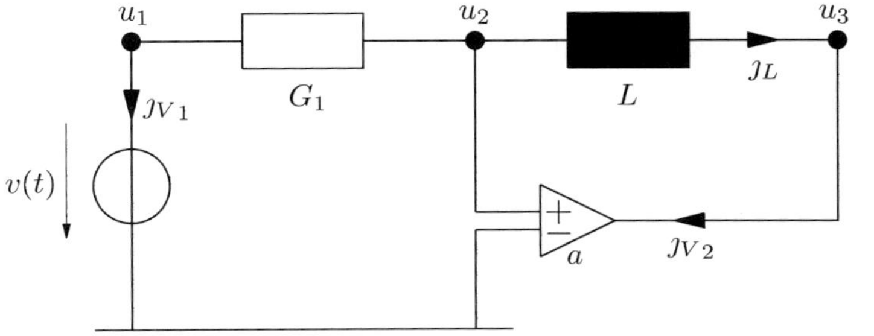

FIG. 8.1. Differentiator circuit.

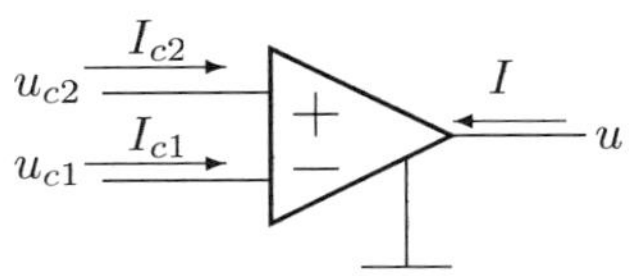

$$u = a \cdot U_{control},$$
$$U_{control} = u_{c2} - u_{c1}$$
$$I_{c1} = -I_{c2} = 0$$

FIG. 8.2. Operational amplifier: Network symbol and characteristic equations.

$$
\begin{aligned}
&\dot{\phi} - A_L^\top u = 0,\\
&A_V^\top u - \begin{pmatrix} v(t) \\ a u_2 \end{pmatrix} = 0,\\
&\phi - L \cdot j_L = 0
\end{aligned}
$$

with $G := G_1$ and

$$
A_R = \begin{pmatrix} 1 \\ -1 \\ 0 \end{pmatrix}, \quad A_L = \begin{pmatrix} 0 \\ 1 \\ -1 \end{pmatrix} \quad \text{and} \quad A_V = \begin{pmatrix} 1 & 0 \\ 0 & 0 \\ 0 & 1 \end{pmatrix}
$$

one obtains index 1.

For the limit case of an ideal operational amplifier, i.e., $a \to \infty$, the element relation has $U_{control} = 0$ ($u_2 = 0$) as a limiting case, and neither the output voltage U (u_3) nor the output current I (j_{V2}) are determined by the characteristic equations. So, its controlling nodes are connected by a *nullator* (i.e., an element with vanishing branch voltage and branch current), and its output nodes are connected by a *norator* (i.e., an element with arbitrary branch voltage and branch current).

In this case, the MNA structure (4.1) is destroyed, since

$$
A_V^\top u - \begin{pmatrix} v(t) \\ a u_2 \end{pmatrix} = 0
$$

is replaced by

$$
\widetilde{A}_V^\top u - \begin{pmatrix} v(t) \\ 0 \end{pmatrix} = 0 \quad \text{with } \widetilde{A}_V^\top = \begin{pmatrix} 1 & 0 & 0 \\ 0 & 1 & 0 \end{pmatrix} \neq A_V^\top.
$$

We recognize that the *static* elements, whose element equations are independent of the output variables (here: The degenerated controlled source), are responsible for the higher index, because now the output variable is determined only via differential equations of the *dynamic* elements. We have $u_3 = -L \cdot G \cdot \dot{v}(t)$ for the differentiator circuit, i.e., the output voltage is the time derivative of the input voltage, and the problem is of index 2. This situation is typical for higher-index problems and is merely an electrical interpretation of the mathematical condition for index $\geqslant 2$.

Any extension of the differentiator circuit, which does not shortcut the inductor in Fig. 8.1, will keep the index $\geqslant 2$. By inserting LC-, LR- or RC-circuits into the feedback loop between ideal operational amplifier and inductor of the differentiator circuit, the index can be raised by 2, 1 or 1, respectively (REISSIG and FELDMANN [1996]).

Miller integrator. When we replace the inductor of the differentiator circuit by a capacitor then the circuit turns into an integrator. Fig. 8.3 shows such a circuit, which is called Miller integrator. The capacitor C_2 is mandatory for the circuit, while C_1 is added as a parasitic grounded capacitance, which may vanish.

We take this circuit to illustrate that due to the use of controlled sources the circuit parameters may have an impact on the structural properties of the network equations: Index, sensitivity of the solution with respect to input signals, and degree of freedom for assigning initial values. This is true even for linear circuits.

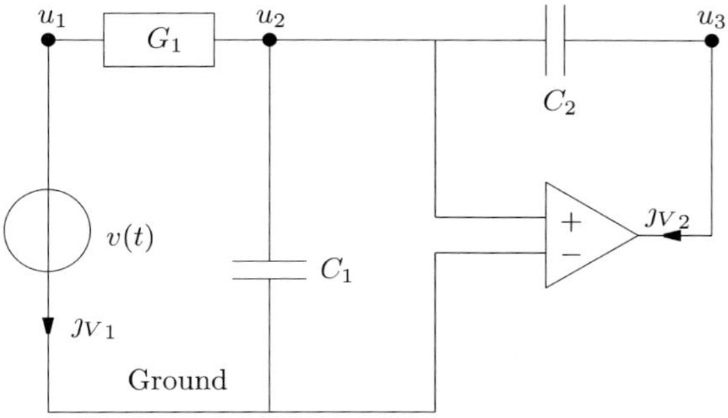

FIG. 8.3. Miller integrator circuit.

The function of this time-continuous version of an integrator is to integrate an input signal over time. Such integrators are important parts of integrated filter circuits, since they are used to substitute inductors of arbitrary inductance L, which are expensive to obtain otherwise in integrated technologies. Hereby the inductor relation is taken in admittance form

$$\jmath = \frac{1}{L} \int u \, \mathrm{d}t,$$

which requires the integration of the branch voltage u. For the sake of simplicity we use an ideal operational amplifier element with limited amplification a here.

Using Modified Nodal Analysis, the network equations read

$$\begin{aligned} &A_C C A_C^\top \cdot \dot{u} + A_R G A_R^\top \cdot u + A_V \jmath_V = 0, \\ &A_V^\top u - v(u,t) = 0, \end{aligned}$$

with

$$A_C = \begin{pmatrix} 0 & 0 \\ 1 & -1 \\ 0 & 1 \end{pmatrix}, \qquad A_R = \begin{pmatrix} 1 \\ -1 \\ 0 \end{pmatrix}, \qquad A_V = \begin{pmatrix} 1 & 0 \\ 0 & 0 \\ 0 & 1 \end{pmatrix},$$

$$C = \operatorname{diag}(C_1, C_2), \qquad G = G_1, \qquad v(u,t) = \begin{pmatrix} v(t) \\ a u_2 \end{pmatrix}.$$

If the amplification factor a tends to infinity then $u_2 = 0$, and for the capacitor current we get $C_2 \cdot \dot{u}_3 = -\jmath_{V2} = -G \cdot u_1 = -G \cdot v(t)$, from which follows the integrator function of the circuit:

$$u_3 = -\frac{G}{C_2} \int v(t) \, \mathrm{d}t.$$

For $a \neq 1 + C_1/C_2$, one can solve for $\dot{u}_2$ by inserting the last equation into the second: $\dot{u}_2 = G(v(t) - u_2)/C$ with $C = C_1 + C_2(1-a)$. All components are now fixed by u_2, the only degree of freedom:

$$u_1 = v(t), \qquad u_3 = a u_2, \qquad \jmath_{V1} = G(u_2 - v(t)),$$

$$\jmath_{V2} = \frac{C_2}{C} G(1-a)\big(v(t) - u_2\big).$$

TABLE 8.2
Miller integrator circuit: Impact of technical parameters on index, degree of freedom and sensitivity with respect to input signal

Technical parameter	Index	Degree of freedom	Sensitivity w.r.t. $v(t)$
$C_1 > 0$			
$a \neq 1 + C_1/C_2$	2	only u_2	only $v(t)$
$a = 1 + C_1/C_2$	3	–	$v(t)$ *and* $\dot{v}(t)$
$C_1 = 0$			
$a \neq 1$	1	only u_2	only $v(t)$
$a = 1$	2	–	only $v(t)$

For $a = 1 + C_1/C_2$, however, u_2 is fixed by the hidden algebraic relation $u_1 - u_2 = 0$. Now the solution is given at every time point by the input signal and its derivatives:

$$u_1 = u_2 = v(t), \qquad u_3 = av(t), \qquad \jmath_{V1} = 0, \qquad \jmath_{V2} = C_2(1-a)\dot{V}(t).$$

This reflects the impact of the technical parameters C_1, C_2 and a on the system w.r.t. input signals and degree of freedom for assigning initial values, see Table 8.2 for an overview. It remains to discuss the influence on the index:

First case: $C_1 > 0$. The derivative of the algebraic part with respect to the algebraic components $(u_1, \jmath_L, \jmath_V)^\top$ is singular, since the element relation for the amplifier $u_3 = au_2$ does not depend on $\jmath_{V2}$. After one differentiation we get by inserting the formulae for the differential variables u_2 and u_3 the linear algebraic relation

$$\begin{pmatrix} -(1-a)G_1/C_1 \\ (1-a)G_1/C_1 \\ (1-a)/C_1 + 1/C_2 \end{pmatrix}^\top \begin{pmatrix} u_1 \\ u_2 \\ \jmath_{V2} \end{pmatrix} = 0 \tag{8.1}$$

in $u_1, u_2, \jmath_{V2}$, which can be solved for $\jmath_{V2}$ iff $a \neq 1 + C_1/C_2$; in this case, the index is two. This is not surprising, since together with C_1 and C_2 the operational amplifier forms a loop of voltage source and capacitors.

For the exceptional case $a = 1 + C_1/C_2$ relation (8.1) reads

$$u_1 = u_2,$$

and a second differentiation is necessary to solve for $\jmath_{V2}$: The index is now three.

Second case: $C_1 = 0$. The partial derivative of the algebraic part with respect to the algebraic components $(u_1, u_2 + u_3, \jmath_{V1}, \jmath_{V2})^\top$ now reads

$$\begin{pmatrix} G & -G/2 & 1 & 0 \\ -G & G/2 & 0 & 1 \\ 1 & 0 & 0 & 0 \\ 0 & (1-a)/2 & 0 & 0 \end{pmatrix}.$$

The matrix is regular, and correspondingly the index is 1, iff $a \neq 1$ holds. For $a = 1$, however, the matrix has only rank 3. The last algebraic relation becomes $u_2 - u_3 = 0$, which fixes the differential component, too. To get the remaining differential relation from this equation, two differentiations are necessary. Hence the index is 2.

Conclusion. These results for controlled sources have an important practical consequence: It is not sufficient to rely only on structural aspects when trying to cope with higher-index problems in circuit simulation. This will be further elaborated when looking at stepwise refinements of a bipolar ringoscillator model in the following section. Possible solutions to this problem are discussed in Section 10.

9. Effects of refined modelling – a bipolar ringoscillator

The task of a ringoscillator is to generate autonomously an oscillating signal, which may be used for driving other parts of a circuit, but in many cases serves only for measuring the maximal clock rates which can be achieved with a given technology. The basic principle is to connect an odd number of inverter stages in a loop. Compared to standard MOS technologies, bipolar technologies are faster (by approximately an order of magnitude, such that frequencies of 10 GHz and higher are possible) due to a very small signal swing and high driving capabilities, but the circuits are not as compact and have a higher power consumption.

The basic model. A circuit diagram of our bipolar ringoscillator is shown in Fig. 9.1. Since it is simplified as far as possible it may look somewhat strange for an experienced circuit designer. On the other hand it has still its basic functionality, and can be extended in such a way that we observe the effects we want to discuss here.

Circuit description. The dashed box contains the core of the circuit and will be used as an icon in the extensions discussed later. It consists of three differential stages. The nodes between the resistors and the collector of the bipolar transistors (e.g., the nodes 1 and 2 for the left stage) are the outputs of each stage, while the nodes connected to the base of the bipolar transistors (e.g., the nodes 7 and 8 for the left stage) are its inputs. Each output of a stage is connected to the corresponding input of the next stage, thus forming a loop. Basically the circuit works in a *current mode*: The differential stages are driven by current sources, and due to the exponential characteristic of the bipolar transistor just that branch of each differential stage will take over almost all of the current, whose input node is at a higher voltage level. Since the Ohmic resistors cause a larger voltage drop for the branch carrying the larger current, its output will be at a lower voltage level, thus inverting the input signal.

In principle, one input of each differential stage may be fixed at a constant reference voltage. For speed advantages, often the complementary technique shown here is used, where both the original signal and its inverse are generated in each stage and propagated to the inputs of the next. Note that the circuit operates with negative voltages, which reduces the sensitivity of the signals with respect to perturbations of the power supply. First-order formulas for designing such an oscillator can be found in the textbooks (see e.g. HOFFMANN [1996]).

Network equations. With $u := (u_1, \dots, u_{10})^\top$, being the vector of node voltages at nodes $1, \dots, 10$, and $j_V = J_{SS}$ being the current through the only voltage source, the

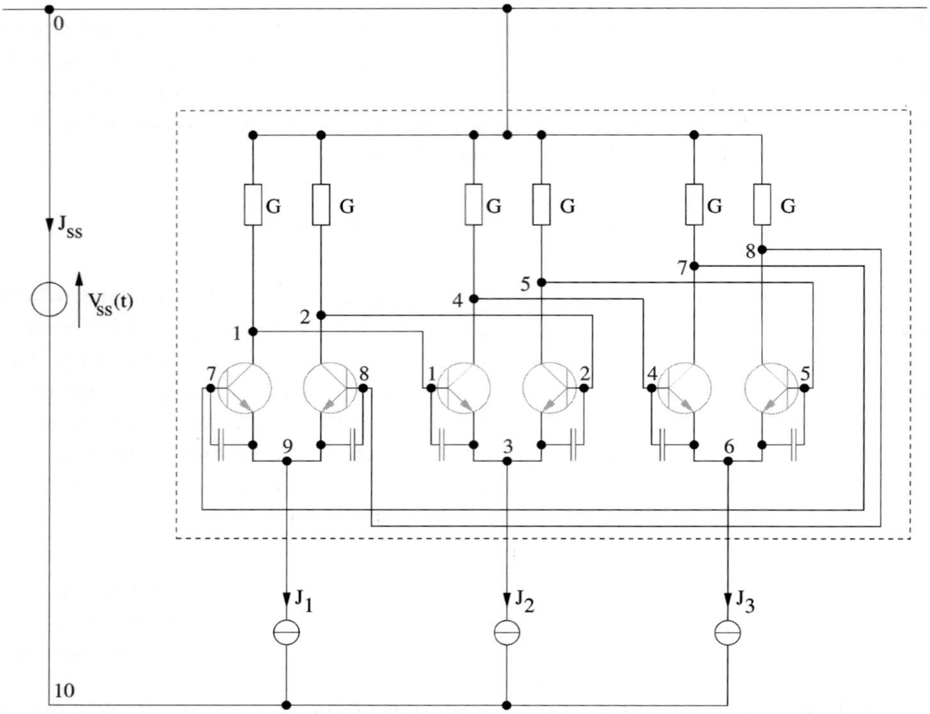

FIG. 9.1. Bipolar ringoscillator.

charge oriented MNA network equations read:

$$\begin{aligned}
&A_C\dot{q} + A_R\,\mathrm{diag}(G_1, G_2, G_3, G_4, G_5, G_6)A_R^\top u + A_V j_V + A_I\imath(A^\top u, t) = 0,\\
&A_V^\top u + V_{SS}(t) = 0,\\
&q - q_C(A_C^\top u) = 0,
\end{aligned}$$

with

$$A_C = \begin{pmatrix}
1 & 0 & 0 & 0 & 0 & 0\\
0 & 1 & 0 & 0 & 0 & 0\\
-1 & -1 & 0 & 0 & 0 & 0\\
0 & 0 & 1 & 0 & 0 & 0\\
0 & 0 & 0 & 1 & 0 & 0\\
0 & 0 & -1 & -1 & 0 & 0\\
0 & 0 & 0 & 0 & 1 & 0\\
0 & 0 & 0 & 0 & 0 & 1\\
0 & 0 & 0 & 0 & -1 & -1\\
0 & 0 & 0 & 0 & 0 & 0
\end{pmatrix},$$

$$
A_R = \begin{pmatrix}
1 & 0 & 0 & 0 & 0 & 0 \\
0 & 1 & 0 & 0 & 0 & 0 \\
0 & 0 & 0 & 0 & 0 & 0 \\
0 & 0 & 1 & 0 & 0 & 0 \\
0 & 0 & 0 & 1 & 0 & 0 \\
0 & 0 & 0 & 0 & 0 & 0 \\
0 & 0 & 0 & 0 & 1 & 0 \\
0 & 0 & 0 & 0 & 0 & 1 \\
0 & 0 & 0 & 0 & 0 & 0 \\
0 & 0 & 0 & 0 & 0 & 0
\end{pmatrix}, \quad
A_V = \begin{pmatrix} 0 \\ 0 \\ 0 \\ 0 \\ 0 \\ 0 \\ 0 \\ 0 \\ 0 \\ 1 \end{pmatrix},
$$

$$
A_I = \begin{pmatrix}
1 & 0 & 0 & 0 & 0 & 0 & 0 & 0 & 1 & 0 & 0 & 0 & 0 & 0 & 0 \\
0 & 1 & 0 & 0 & 0 & 0 & 0 & 0 & 0 & 1 & 0 & 0 & 0 & 0 & 0 \\
0 & 0 & -1 & -1 & 0 & 0 & 0 & 0 & -1 & -1 & 0 & 0 & 0 & 1 & 0 \\
0 & 0 & 1 & 0 & 0 & 0 & 0 & 0 & 0 & 0 & 1 & 0 & 0 & 0 & 0 \\
0 & 0 & 0 & 1 & 0 & 0 & 0 & 0 & 0 & 0 & 0 & 1 & 0 & 0 & 0 \\
0 & 0 & 0 & 0 & -1 & -1 & 0 & 0 & 0 & 0 & -1 & -1 & 0 & 0 & 1 \\
0 & 0 & 0 & 0 & 1 & 0 & 1 & 0 & 0 & 0 & 0 & 0 & 0 & 0 & 0 \\
0 & 0 & 0 & 0 & 0 & 1 & 0 & 1 & 0 & 0 & 0 & 0 & 0 & 0 & 0 \\
-1 & -1 & 0 & 0 & 0 & 0 & -1 & -1 & 0 & 0 & 0 & 0 & 1 & 0 & 0 \\
0 & 0 & 0 & 0 & 0 & 0 & 0 & 0 & 0 & 0 & 0 & 0 & -1 & -1 & -1
\end{pmatrix},
$$

$$
q_C(A_C^\top u) = \operatorname{diag}(c_{13}, c_{23}, c_{46}, c_{56}, c_{79}, c_{89}) \cdot A_C^\top u,
$$

$$
\imath(A^\top u, t) = \big(I_{C1}, \ldots, I_{C6}, I_{B1}, \ldots, I_{B6}, J_1(t), J_2(t), J_3(t)\big)^\top.
$$

Here I_{Cj} and I_{Bj} are the collector and base current of the bipolar transistor T_j ($j = 1, \ldots, 6$) introduced in Section 1. The capacitances c_{ij} between nodes i and j may be linear, or modelled in a nonlinear way: $c_{ij} = c_{ij}(A_C^\top u)$ (GÜNTHER and FELDMANN [1999b]).

The index. With the projector

$$
Q_C = \begin{pmatrix}
1/3 & 1/3 & 1/3 & 0 & 0 & 0 & 0 & 0 & 0 & 0 \\
1/3 & 1/3 & 1/3 & 0 & 0 & 0 & 0 & 0 & 0 & 0 \\
1/3 & 1/3 & 1/3 & 0 & 0 & 0 & 0 & 0 & 0 & 0 \\
0 & 0 & 0 & 1/3 & 1/3 & 1/3 & 0 & 0 & 0 & 0 \\
0 & 0 & 0 & 1/3 & 1/3 & 1/3 & 0 & 0 & 0 & 0 \\
0 & 0 & 0 & 1/3 & 1/3 & 1/3 & 0 & 0 & 0 & 0 \\
0 & 0 & 0 & 0 & 0 & 0 & 1/3 & 1/3 & 1/3 & 0 \\
0 & 0 & 0 & 0 & 0 & 0 & 1/3 & 1/3 & 1/3 & 0 \\
0 & 0 & 0 & 0 & 0 & 0 & 1/3 & 1/3 & 1/3 & 0 \\
0 & 0 & 0 & 0 & 0 & 0 & 0 & 0 & 0 & 1
\end{pmatrix}
$$

onto $\ker A_C^\top$ one shows that $\ker Q_C^\top A_V = \{0\}$ and $\ker(A_C A_R A_V)^\top = \{0\}$ hold. Since all sources are either independent or current sources that are not part of any VC loop and which are driven by branch voltages of capacitive paths, this model yields an index-1 problem. We only have to require that the charge model used for the capacitors in the nonlinear case yields a positive-definite generalized capacitance matrix.

Refined modelling. Eventually our basic circuit model has to be refined in order to get a higher degree of accuracy and to take nonideal operating conditions into account. Basically this is achieved by

- replacing idealized network elements by real circuits. As an example we will discuss the substitution of the current sources by transistor configurations,
- a more detailed modelling with respect to parasitic effects (see Table 9.1 for an overview).

The impact of a model refinement on the index is not a priori clear: Regularization to lower index, no change, and even an increase of the index may happen. In GÜNTHER and FELDMANN [1999b] some circuit configurations are reviewed which may yield higher-index problems. This will be illustrated in the following with some extensions of our basic ringoscillator model. Hereby it is sufficient for our purpose to modify only the circuit frame, while the core symbolized by a dashed box (see Fig. 9.1) remains unchanged.

Inductance of interconnect. Since power supply and ground line conduct a significant and rapidly changing current, it may be necessary to take their inductance into account (see Table 9.1). For the sake of simplicity we insert only an inductor with inductance L into the ground line of Fig. 9.1. The inclusion of an inductor into the power supply line gives no further insight here.

The differential index is raised from one to two, since the circuit now contains a cutset of an inductor and current sources. All node voltages in the cutset depend on the first derivatives of $J_1(t), \ldots, J_3(t)$, i.e., are index-2 variables. Numerically, this may not cause problems as far as the sources $J_i(t)$ are smooth. However it becomes apparent if the $J_i(t)$ are slightly perturbed with a 'noisy' signal of small amplitude and high frequency.

Realistic model for current sources. The sources providing the current for the differential stages in our basic ringoscillator model of Fig. 9.1 are in practice realized by bipolar transistors of npn-type, which are biased with a positive base-emitter voltage and a negative base-collector voltage. In this case the collector current is approximately given by

$$I_C \approx \beta \cdot \left(e^{\frac{U_{BE}}{U_T}} - 1\right)$$

(see Section 1), which defines $v_{Bias} = U_{BE}$ in order to get the same value for I_C as was provided by the current sources $J_1 \ldots J_3$ in the basic model.

Formally, the cutset of current sources and inductors is broken, and the index is reduced to 1. Numerically however, the bipolar transistors still act as current sources, and so one has to deal with a singularly perturbed index-2 problem if the regularizing capacitances c_{ij} at the three transistors acting as real current sources are small.

Modelling of crosstalk. If the interconnects are long parallel wires in the layout, then it may become necessary to take crosstalk between them into account (see Table 9.1). We restrict here to the simple case of crosstalk between the interconnect nodes 9 and 10 in our circuit of Fig. 9.1. Usually, crosstalk is modeled by adding a coupling capacitor between the nodes. However, sometimes also controlled sources are used for this purpose, especially in higher order models. We will focus here on the latter model since it may have a negative impact on the index, while the first one has a regularizing effect.

TABLE 9.1
Important parasitic effects in integrated circuit designs

Effect	Important for	Example	Impact on index
Nonideal element characteristics	high performance, analog circuits	limited output conductance of transistors	eventually decreasing
Resistance			
of diffusions	standard designs	emitter resistance of bipolar transistors	eventually decreasing
of interconnects	long interconnects, high currents	resistance of power supply, via holes	eventually decreasing
Capacitance			
of diffusions	standard designs	capacitive load	usually no change
of interconnects	large interconnects	signal cross-coupling	usually no change
Inductance			
of interconnects, package, etc.	high currents, fast switching	inductance of power supply	eventually increasing
Temperature effects	large temperature range	temperature dependence of mobility	no change
	high power	self-heating of power transistors	eventually increasing
Distributed (noncompact) elements	very fast switching signals	delay time of transmission lines	eventually increasing
	charge sensitive circuits	nonquasistationary element equations	usually no change
Parasitic semiconductor devices	compact design rules, unusual operating conditions	bipolar latchup in CMOS circuits	usually no change
External			
electromagnetic noise, radiation	flux or charge sensitive designs	α-radiation in dynamic memory cells	eventually increasing

Node 10 is split into a pair 10 and $10a$ which are connected by a voltage-controlled voltage source E_{Cross}. The controlling branch voltage is the voltage drop between nodes 9 and 10:

$$E_{Cross} = u_{10a} - u_{10} = \alpha_E \cdot (u_9 - u_{10}).$$

A reasonable value for the crosstalk factor α_E is between 1 and 10%. Note that here the mutual crosstalk from node 10 to node 9 is one order of magnitude smaller and can be neglected. Now the power supply voltage of node 10 is no longer constant. In our case, this will not have an effect on the oscillating waveforms, since the current provided by the sources J_1, J_2, J_3 is independent of u_{10}. But now the parasitic capacitor c_{10} of node

10 versus ground has to be included, since it is de- and upcharged simultaneously and such causes an additional load for the power supply current J_{SS}.

With J_L and J_E being the currents through the inductor and the controlled voltage source E_{Cross}, respectively, a first order approximation yields

$$\begin{aligned} J_{SS} = -J_E &= c_{10} \cdot \dot{u}_{10} - J_1 - J_2 - J_3 \\ &\approx \Delta J_{SS} - J_1 - J_2 - J_3, \end{aligned}$$

where

$$\Delta J_{SS} = -\alpha_E c_{10} \dot{u}_9$$

is caused by the crosstalk effect. The relative additional current

$$\left| \frac{\Delta J_{SS}}{J_1 + J_2 + J_3} \right|$$

is not very significant for smooth current sources, but it may increase dramatically if the current sources J_1, J_2, J_3 are somewhat noisy. The reason is, that u_9 is of index 2 due to the cutset of current sources/inductor. Since this variable controls the voltage source E_{Cross}, which is enclosed in a loop of voltage sources/capacitor anyway, its current J_E and therefore also the current J_{SS} of the power supply source V_{SS} are of index 3 (GÜNTHER and FELDMANN [1999b]). So J_{SS} depends on the second derivatives of the current sources $J_1(t)$, $J_2(t)$, $J_3(t)$.

Note that for the latter model the numerical integration schemes in standard simulation packages will fail in general if a noisy signal is applied to the input sources. Not even the startup behaviour of this circuit, where the power supply and input signals are ramped up to their final value, can be analysed in general due to the nonsmooth form of the ramp-up signals.

A detailed discussion of the bipolar ringoscillator and its refinement levels, including all technical parameters and models, derivation of network equations, and waveforms, can be found in GÜNTHER and FELDMANN [1999a].

After setup and analysis of the DAE network equations modelling electrical circuits in time domain, it remains to discuss the third step in circuit simulation: Numerical integration using DAE discretization schemes, which are tailored to the structure and index of the network equations.

CHAPTER III

Numerical Integration Schemes

The numerical integration of the network equations defines (at least from a mathematical point of view) the kernel of simulation packages in circuit design. This chapter does not aim at an introduction into numerical integration schemes for DAE systems: Neither in theory (convergence and stability) nor in general aspects of implementation (adaptivity, solution of nonlinear and linear systems). For this, the reader may consult a bunch of excellent textbooks (ASCHER and PETZOLD [1998], BRENAN, CAMPBELL and PETZOLD [1996], HAIRER and WANNER [1996]) or the survey article (RABIER and RHEINBOLDT [2002]).

In the following we first describe the conventional approach based on implicit linear multi-step methods, discuss the basic algorithms used, and how they are implemented and tailored to the needs of circuit simulation. Special care is demanded of index-2 systems. In addition, we introduce an alternative approach based on one-step methods. This recently developed scheme is compatible to the conventional one with respect to efficiency and robustness, and shows interesting numerical damping properties.

Throughout this chapter we will assume that the network equations correspond to RLC networks, and the only allowed controlled sources are those which keep the index between 1 and 2, depending on the network structure.

10. The conventional approach: Implicit linear multi-step formulas

To simplify notation, we first rewrite the network equations (4.1) in charge/flux oriented formulation

$$0=\underbrace{\begin{pmatrix} A_C & 0 \\ 0 & I \\ 0 & 0 \end{pmatrix}}_{A:=}\cdot\underbrace{\begin{pmatrix} \dot{q} \\ \dot{\phi} \end{pmatrix}}_{\dot{y}:=}+\underbrace{\begin{pmatrix} A_R r(A_R^\top u,t)+A_L j_L + A_V j_V + A_I \imath(u, j_L, j_V, t) \\ -A_L^\top u \\ v(u, j_L, j_V, t) - A_V^\top u \end{pmatrix}}_{f(x,t):=},$$

$$\underbrace{\begin{pmatrix} q \\ \phi \end{pmatrix}}_{y:=}=\underbrace{\begin{pmatrix} q_C(A_C^\top u) \\ \phi_L(j_L) \end{pmatrix}}_{g(x,t):=}$$

in a more compact linear-implicit form:

$$0=\mathcal{F}\big(\dot{y}(t), x(t), t\big) := A\cdot\dot{y}(t) + f\big(x(t), t\big), \tag{10.1a}$$

$$0 = y(t) - g\big(x(t)\big) \tag{10.1b}$$

with $x := (u, \jmath_L, \jmath_V)^\top$ being the vector of unknown network variables.

The basic algorithm. The conventional approach can be split into three main steps: Computation of consistent initial values, numerical integration of $\dot{y}$ based on multi-step schemes, transformation of the DAE into a nonlinear system and its numerical solution by Newton's procedure. Since the third step is usually performed with methods which are not very specific for circuit simulation, we will not discuss it further here.

Let us assume for the moment that the network equations are of index 1 – the index-2 case will be discussed later.

Consistent initial values. The first step in the transient analysis is to compute consistent initial values (x_0, y_0) for the initial time point t_0. In the index-1 case, this can be done by performing a steady state (DC operating point) analysis, i.e., to solve

$$\mathcal{F}(0, x_0, t_0) = 0 \tag{10.2}$$

for x_0 and then set $y_0 := g(x_0)$. If there are no controlled sources, the Jacobian $\partial\mathcal{F}/\partial x$ of (10.2) with respect to x_0 reads

$$\frac{\partial\mathcal{F}}{\partial x} = \begin{pmatrix} \widetilde{G}(A_R^\top u_0, t_0) & A_L & A_V \\ -A_L^\top & 0 & 0 \\ -A_V^\top & 0 & 0 \end{pmatrix}$$

with the definition $\widetilde{G}(A_R^\top u, t) := A_R G(A_R^\top u, t) A_R^\top$ already introduced in Section 4. Since $\ker(\partial\mathcal{F}/\partial x) = \ker(A_R, A_L, A_V)^\top \times \ker(A_L, A_V)$ holds, the matrix is only regular, if there are neither loops of independent voltage sources and/or inductors, nor cutsets of independent current sources and/or capacitors. If these topological conditions are violated, no steady state solution can be computed, and so most circuit analysis programs check and refuse these circuit configurations. Additional assumptions are implied in the case of controlled sources. But note that in the nonlinear case the Jacobian matrix also may become numerically singular, e.g., due to vanishing partial derivatives or in the case of bifurcation.

An approach always feasible in the index-1 case is to extract the algebraic constraints using the projector Q_C onto $\ker A_C^\top$:

$$Q_C^\top\big(A_R r(A_R^\top u, t) + A_L \jmath_L + A_V \jmath_V + A_I \imath(u, \jmath_L, \jmath_V, t)\big) = 0,$$
$$v(u, \jmath_L, \jmath_V, t) - A_V^\top u = 0.$$

If the index-1 topological conditions hold, this nonlinear system uniquely defines for $t = t_0$ the algebraic components $Q_C u_0$ and $\jmath_{V,0}$ for given (arbitrary) differential components $(I - Q_C)u_0$ and $\jmath_{L,0}$. The derivatives $\dot{y}_0$ have then to be chosen such that $A\dot{y}_0 + f(x_0, t_0) = 0$ holds.

Numerical integration. Starting from consistent initial values, the solution of the network equations is computed at discrete time points $t_1, t_2, \ldots,$ by numerical integration with implicit linear multi-step formulas.

The direct approach, which is shortly described here, was first proposed by GEAR [1971] for *backward differentiation formulas* (BDF methods): For a timestep h_k from t_{k-1} to $t_k = t_{k-1} + h_k$ the derivative $\dot{y}(t_k)$ in (10.1) is replaced by a linear ρ-step operator

ρ_k for the approximate $\dot{y}_k$, which is defined by

$$\rho_k = \frac{1}{h_k}\sum_{i=0}^{\rho} \gamma_{k,i} y_{k-i} - \sum_{i=1}^{\rho} \beta_{k,i} \dot{y}_{k-i} := \alpha_k y_k + r_k \tag{10.3}$$

with $y_{k-i} := g(x_{k-i})$, $i = 0, 1, \ldots, \rho$, and $\dot{y}_{k-i}$, $i = 1, \ldots, \rho$, already computed by previous operators ρ_{k-i}. The index k in the method coefficients $\beta_{k,i}$ and $\gamma_{k,i}$ indicate their dependence on the step size history in the case of variable step size implementations (see the paragraph about adaptivity below). The remainder r_k contains values of y and $\dot{y}$ for previous time points.

Transformation into a nonlinear system of equations. The numerical solution of the DAE system (10.1) is thus reduced to the solution of a system of nonlinear equations

$$\mathcal{F}\big(\alpha_k g(x_k) + r_k, x_k, t_k\big) = 0, \tag{10.4}$$

which is solved iteratively for x_k by applying Newton's method in a predictor-corrector scheme. Starting with a predictor step $x_k^{(0)}$ (x_{k-1} or some kind of extrapolated value from previous timepoint may be a reasonable choice), a new Newton correction $\Delta x_k^{(l)} := x_k^{(l)} - x_k^{(l-1)}$ is computed from a system of linear equations

$$D\mathcal{F}^{(l-1)} \Delta x_k^{(l)} = -\mathcal{F}^{(l-1)}, \qquad \mathcal{F}^{(l-1)} := \mathcal{F}(\alpha_k g(x_k^{(l-1)}) + r_k, x_k^{(l-1)}, t_k) \tag{10.5}$$

directly by sparse LU decomposition and forward backward substitution. Due to the structure of the nonlinear equations the Jacobian $D\mathcal{F}^{(l-1)}$ for Newton's scheme is

$$D\mathcal{F}^{(l-1)} = \alpha_k \cdot \mathcal{F}_{\dot{x}}^{(l-1)} + \mathcal{F}_x^{(l-1)}$$

$$\text{with} \quad \mathcal{F}_{\dot{x}}^{(l-1)} = A \cdot \frac{\partial g(x_k^{(l-1)})}{\partial x}, \qquad \mathcal{F}_x^{(l-1)} = \frac{\partial f(x_k^{(l-1)}, t_k)}{\partial x}.$$

If the step size h is sufficiently small, the regularity of $D\mathcal{F}^{(l-1)}$ follows from the regularity of the matrix pencil $\{A \cdot \partial g(x)/\partial x, \partial f/\partial x\}$ that is given at least for index-1 systems.

Implementation: Element stamps and cheap Jacobian. The implementation of the direct approach for one timestep into the analysis kernel of circuit simulation packages such as SPICE is outlined in Fig. 10.1.

In every Newton step (10.5), two main steps have to be performed:

- `LOAD` part: First the right-hand side $-\mathcal{F}^{(l-1)}$ of (10.5) and the Jacobian $D\mathcal{F}^{(l-1)}$ have to be computed;
- `SOLVE` part: The arising linear system is solved directly by sparse LU decomposition and forward backward substitution.

A characteristic feature of the implementation is that modelling and numerical integration (10.3) are interwoven in the `LOAD` part: First the arrays for right-hand side and Jacobian are zeroed. In a second step, these arrays are assembled by adding the contributions to $\mathcal{F}$ and $D\mathcal{F}$ element by element: So-called *element stamps* are used to evaluate the time-discretized models for network elements.

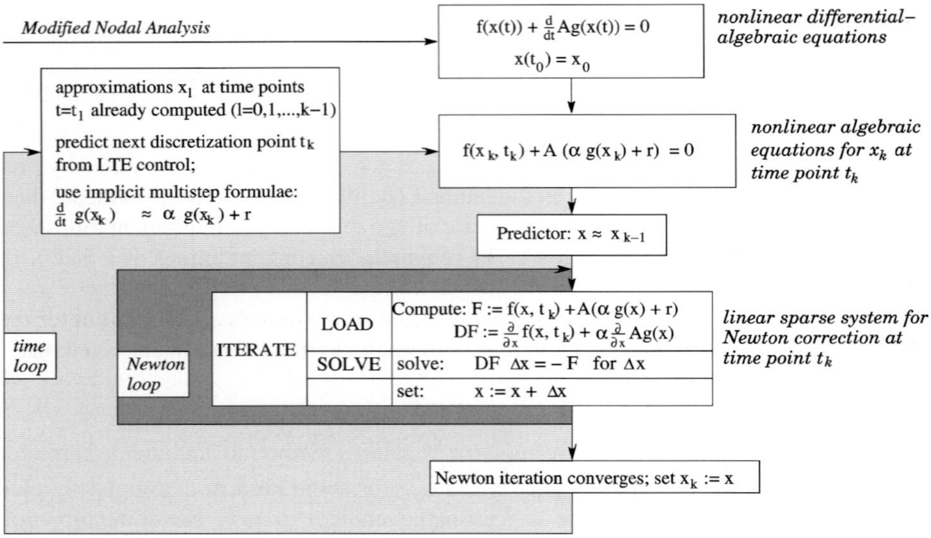

FIG. 10.1. The direct approach in SPICE like simulators.

Let us consider, for example, a linear capacitor with capacitance C between the nodes "+" and "−" with node potentials u_+ and u_- at time point t_k. Its characteristic equation reads $I_C(t_k) = \dot{q}_C(t_k)$, $q_C(t_k) = C \cdot (u_+ - u_-)$. After incorporating the approximation (10.3) for $\dot{q}_C$ one gets the approximate element relation

$$I_C = \alpha_k C \cdot (u_+ - u_-) + r_k,$$

which gives the following contributions to the Jacobian matrix for the rows corresponding to nodes "+" and "−" and columns corresponding to node potentials u_+ and u_-, and to the right-hand side (rhs) at nodes "+" and "−":

	u_+	u_-	rhs
+	$\alpha_k C$	$-\alpha_k C$	$-I_C$
−	$-\alpha_k C$	$\alpha_k C$	I_C

One consequence of using element stamps is the cheap availability of the Jacobian: For highly integrated circuits with a very sparse Jacobian, it is only slightly more expensive to evaluate both right-hand side and Jacobian than to evaluate only the right-hand side by its own. So, if not linear algebra aspects are dominant (which may happen for very large circuits) then the use of full rather than modified Newton may be appropriate in many cases.

BDF schemes and trapezoidal rule. It remains to answer the question which types of implicit linear multi-step formulae (10.3) are actually used. Since SPICE2 (NAGEL [1975]), most circuit simulators solve the network equations either with the *trapezoidal rule* (TR)

$$\rho_k = -\dot{y}_{k-1} + \frac{2}{h}(y_k - y_{k-1}) \quad (\rho = 1,\ \beta_{k,1} = 1,\ \gamma_{k,0} = -\gamma_{k,1} = 2) \tag{10.6}$$

or with BDF schemes:

$$\rho_k = \frac{1}{h_k} \sum_{i=0}^{\rho} \gamma_{k,i}\, y_{k-i} \quad (\beta_{k,1} = \cdots = \beta_{k,\rho} = 0). \tag{10.7}$$

For the BDF methods no derivatives of y at previous time points are needed. The first timestep is always performed by BDF1 (implicit Euler scheme) as starting procedure.

Why BDF schemes? The most appealing argument is to save function evaluations as much as possible, since they are extremely expensive in circuit simulation – see Gear's article (GEAR [1971]) which was explicitly dedicated for solving circuit equations, and consequently had been published in an electrical engineering journal. A second one is that the use of higher order methods does not require much extra cost. And the third one is a settled convergence and stability theory for fully-implicit (and not only semi-explicit) index-1 systems. Nonlinear index-1 network equations fit into this class of problems. For such systems the following convergence result for BDF schemes can be found in any textbook on DAEs: The ρ-step BDF method of fixed size h for $\rho < 7$ is feasible and converges to $\mathcal{O}(h^\rho)$ if all initial values are correct to $\mathcal{O}(h^\rho)$ and if the Newton process at each timestep is solved to accuracy $\mathcal{O}(h^{\rho+1})$. This convergence result has also been extended to variable stepsize BDF methods, provided that they are implemented in such a way that the method is stable for standard ODEs, i.e., the ratio of two succeeding stepsizes is bounded.

It should be noted that BDF schemes with order greater 3 are rarely used in practice because of the low smoothness properties of the transistor model equations. Stability properties give an additional argument for BDF1 and BDF2 schemes anyway: They are *A-stable*, i.e., for Dahlquist's linear test equation $\dot{x} = \lambda x$ the numerical solution with arbitrary stepsize h is bounded for all $\{\lambda;\ \mathrm{Re}(\lambda) < 0\}$ in the left half plane $\mathbb{C}^-$. In other words, no stability problems occur for stiff systems with decaying solutions. In contrast, convergent BDF schemes with higher order ($3 \leqslant \rho \leqslant 6$) cannot be A-stable because of the second Dahlquist barrier. However, they are $A(\alpha)$-stable with $0 < \alpha < \pi/2$, i.e., stable in the sectorial $\{\lambda;\ |\arg(-\lambda)| < \alpha, \lambda \neq 0\}$ of the left half plane; and at least for BDF3 $\alpha \approx 86 \cdot \frac{\pi}{180}$ is large enough to yield no serious stability problems in practice.

In addition to A-stability – and $A(\alpha)$-stability, respectively – the numerical solutions of BDF tend to zero (for fixed stepsize h) in the very stiff limit $\mathrm{Re}(\lambda) \to -\infty$. This *stiff decay* property (which is equivalent to the L-stability property for one-step methods) allows to skip rapidly varying solution details and still maintain a decent description of the solution on a coarse level in the very stiff case. Hence they are suitable for network equations that are generally very stiff because of the widely separated time constants in electrical circuits.

One consequence for A stable methods with stiff decay is numerical damping along the imaginary axis. This behaviour defines a serious shortcoming for BDF1 and BDF2 schemes: The solution is damped so strongly, that even for rather small timesteps oscillations may be damped out, and after some cycles a circuit seems to be quiescent even though it oscillates in reality.

A natural alternative to BDF2 is the trapezoidal rule TR, since it is the A-stable linear multi-step method of order 2 with smallest leading error coefficient. Due to its energy conserving property, it avoids the shortcoming of BDF methods: Oscillations are not damped at all – unfortunately, not even instabilities of highest frequency caused by numerical noise. This weak instability can be seen directly from (10.6): Errors of $\dot{y}_{k-1}$ are propagated to $\dot{y}_k = \rho_k$ without being damped, and errors of $\dot{y}_k$ propagate directly to the respective components of x_k.

One conclusion might be that TR would be a desirable integration rule, if it were damped sufficiently, but not as strongly as BDF. For this purpose several approaches are described to construct a combination of TR and BDF schemes (FELDMANN, WEVER, ZHENG, SCHULTZ and WRIEDT [1992]), so-called TR-BDF schemes. This name was first used in a paper by BANK, COUGHRAN, FICHTNER, GROSSE, ROSE and SMITH [1985]. The aim is to combine the advantages of both methods: Large timesteps and no loss of energy of the trapezoidal rule (TR) combined with the damping properties of BDF. An interesting interpretation of TR-BDF as a one-step method was presented in HOSEA and SHAMPINE [1996].

When looking for alternatives to TR-BDF, we will return in Section 12 for a more detailed discussion to the problem of preserving physical oscillations, while damping out artificial ones very efficiently.

Adaptivity: Stepsize selection and error control. Variable integration stepsizes are mandatory in circuit simulation since activity varies strongly over time. A simple criterion for timestep control can be obtained from the Newton process itself: The stepsize is reduced/increased, if the number of Newton iterations per timestep is larger/smaller than a given threshold (for example, 8 and 3); otherwise, the stepsize remains unchanged. This criterion is cheap to compute, but not very reliable: Linear problems converge with one single Newton step and hence would always be integrated with maximal stepsize.

The conventional strategy: *Estimating the local truncation error in $\dot{y}$*. A more reliable and still efficient stepsize prediction is based on estimating the local truncation error $\varepsilon_{\dot{y}} = \dot{y}(t_{k+1}) - \rho_{k+1}$ of the next step to be performed, i.e., the residual of the implicit linear multi-step formulas if the exact solution is inserted ($\beta_{k+1,0} = 1$):

$$\varepsilon_{\dot{y}} := \sum_{i=0}^{\rho} \beta_{k+1,i}\,\dot{y}_{(t_{k+1-i})} - \frac{1}{h_{k+1}} \sum_{i=0}^{\rho} \gamma_{k+1,i}\, g\big(x(t_{k+1-i})\big).$$

Usually the accuracy of $\dot{y}$ is controlled rather than that of y, because y itself is no quantity of interest for the user. This implies the loss of one integration order, as we will see now. After Taylor expansion around t_k the leading error term in $\varepsilon_{\dot{y}}$ turns out to be

$$\varepsilon_{\dot{y}} \approx \begin{cases} \frac{1}{2} h_{k+1} \frac{\mathrm{d}^2}{\mathrm{d}t^2} g(x(t_k)) & \text{for BDF1,} \\ \frac{1}{6} h_{k+1}(h_{k+1} + h_k) \frac{\mathrm{d}^3}{\mathrm{d}t^3} g(x(t_k)) & \text{for BDF2,} \\ \frac{1}{6} h_{k+1}^2 \frac{\mathrm{d}^3}{\mathrm{d}t^3} g(x(t_k)) & \text{for TR.} \end{cases}$$

The higher order time derivatives of g are usually estimated via divided differences based on $y_k, \ldots, y_{k-1-\rho}$. This is rather inaccurate, since only backward information is used to get the derivatives at the actual timepoint. So timestep control is always somewhat "behind" the actual timepoint, which makes it unstable and gives rise to overreactions, especially when the timesteps are large. This is another argument – besides that of low order smoothness of the element models – why BDF schemes of order greater than 3 are seldom used in practice. To improve the estimates for the higher order derivatives of g, it was suggested in KLAASSEN and PAAP [1987] to replace the divided differences by a higher order scheme – e.g., the trapezoidal rule – and to employ $\rho_k, \ldots, \rho_{k-p}$ for its evaluation. This improves accuracy, needs less backward stages, and is surely more consistent since the time derivatives entering the solution are either used for timestep control, and not any further approximations of them.

A new stepsize can be predicted by matching $\varepsilon_{\dot{y}}$ with a user defined error tolerance TOL. If h_{k+1} is not different from h_k, then TR allows due to its smaller error constant a timestep which is approximately 40% larger than for BDF2.

For an a-posteriori error check, the inequality $\|\varepsilon_{\dot{y}}\| \leqslant \mathrm{TOL}$ has to be evaluated with updated function evaluations for the higher order derivatives. Furthermore, an order control for variable order BDF schemes can be constructed very easily: The stepsize predictions for order $\rho - 1$, ρ and $\rho + 1$ are computed, and that order is chosen which gives the maximal timestep. In practice, the difference between the converged solution at $t = t_k$ and the initial value provided by a suitable predictor polynomial is a key value for the local truncation error estimation.

Modified timestep control. The main flaw of controlling $\varepsilon_{\dot{y}}$ is that the user has no direct control on the really interesting circuit variables, i.e., node potentials u and branch currents j_L, j_V. In order to overcome this disadvantage associated with charge/flux oriented integration, DENK [1990] used

$$\dot{g}\big(x(t)\big) = \frac{\partial g(x(t))}{\partial x}\dot{x}(t),$$

which means to assemble the terminal charges/branch fluxes in circuit nodes/branches and to perform classical integration on x rather than y. This method works well if Newton's procedure is started from a low order predictor. However, it requires the computation of the second derivatives of g, which are hard to get in practice or do not even exist due to poor smoothness properties of transistor models.

An alternative approach (SIEBER, FELDMANN, SCHULTZ and WRIEDT [1994]) is based on the idea not to transform the whole network equations as done by Denk, but only the local truncation error $\varepsilon_{\dot{y}}$ for $\dot{q}$ into a (cheap) estimate for the local error $\varepsilon_x := x(t_k) - x_k$ of $x(t)$. By expanding $\mathcal{F}(\dot{y}(t), x(t), t)$ at the actual time point t_k into a Taylor series around the approximate solution $(\dot{y}_k, x_k)$ and neglecting higher order terms, one obtains

$$\mathcal{F}\big(\dot{y}(t_k), x(t_k), t_k\big) \approx \mathcal{F}(\dot{y}_k, x_k, t_k) + \frac{\partial \mathcal{F}}{\partial \dot{y}}\big(\dot{y}(t) - \dot{y}_k\big) + \frac{\partial \mathcal{F}}{\partial x}\big(x(t) - x_k\big).$$

With the difference of exact and approximate value for $\dot{y}(t_k)$

$$\dot{g}\big(x(t_k)\big) - \dot{g}(x_k) = \alpha_k\big(g\big(x(t_k)\big) - g(x_k)\big) + \varepsilon_{\dot{y}} \approx \alpha_k \frac{\partial g}{\partial x}\big(x(t_k) - x_k\big) + \varepsilon_{\dot{y}}$$

example is the VC-loop of Fig. 7.1, where the current $\jmath_V$ depends on the time derivative $\dot{v}(t)$ of the input signal. This raises two questions for index-2 problems:

- How can we get consistent initial values?
- What happens when integration is started from nonconsistent initial values?

The standard method for the first problem consists of three steps (PANTELIDES [1988], ESTÉVEZ SCHWARZ and LAMOUR [1999]):

1. Select variables which can be given initial values, and initialize them;
2. setup equations for hidden constraints;
3. solve an augmented *nonlinear* system which includes the hidden constraints.

In ESTÉVEZ SCHWARZ [1999a] it was shown that the first and the second step can be done efficiently in circuit simulation by using the results of the previously described index monitor. However, a problem with this approach is that it is very much different from the handling of initial conditions in the lower index case. So an alternative was developed in ESTÉVEZ SCHWARZ [2000], which aims at being as near as possible to the standard algorithm for low index:

1. Find a solution without hidden constraints from solving Eq. (10.2);
2. setup and solve a *linear* system for corrections to this solution, such that the hidden constraints are fulfilled;
3. add the corrections to the initial values found in step 1 to get consistent ones.

Again the hidden constraints can be easily derived from the information provided by the index monitor. When the algorithm is applicable then the variables to be corrected turn out to be branch currents in VC-loops and node voltages in LI-cutsets; details can be found in ESTÉVEZ SCHWARZ [2000].

As an example we look at the circuit given in Fig. 10.2. It contains a VC-loop and is of index 2. The unknowns are the node voltages and the branch current of the voltage source, if an MNA formulation is used: $x = (u_1, u_2, u_3, \jmath_V)^\top$. The network equations are given by:

$$\text{KCL1:}\quad \jmath_V + C_1 \cdot (\dot{u}_1 - \dot{u}_2) + \frac{1}{R_1} u_1 = 0,$$

$$\text{KCL2:}\quad C_1 \cdot (\dot{u}_2 - \dot{u}_1) + \frac{1}{R_2} u_2 + C_2 \dot{u}_2 + \frac{1}{R_3} \cdot (u_2 - u_3) = 0,$$

$$\text{KCL3:}\quad \frac{1}{R_3} \cdot (u_3 - u_2) + C_3 \dot{u}_3 = 0,$$

$$\text{V-Source:}\quad u_1 = V(t).$$

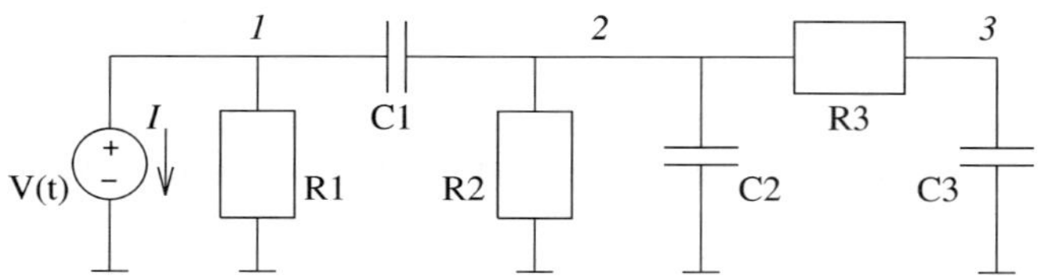

FIG. 10.2. An index-2 circuit.

The steady state DC solution

$$\dot{u}_1 = 0, \qquad \dot{u}_2 = 0, \qquad \dot{u}_3 = 0,$$
$$u_1 = V(0), \qquad u_2 = 0, \qquad u_3 = 0,$$
$$j_V = -\frac{1}{R_1} \cdot V(0)$$

solves the network equations, but violates the hidden constraint

$$\dot{u}_1 = \dot{V}(t).$$

To make it consistent, we need an additional current Δ_{jV} in the VC-loop, which we can compute from:

$$\text{KCL1:} \quad \Delta_{jV} + C_1 \cdot (\dot{u}_1 - \dot{u}_2) = 0,$$
$$\text{KCL2:} \quad C_1 \cdot (\dot{u}_2 - \dot{u}_1) + C_2 \dot{u}_2 = 0,$$
$$\text{V-Source:} \quad \dot{u}_1 = \dot{V}(0).$$

(Node 3 is not part of the VC-loop, and can be omitted here.) Its solution is added to the previous one to get the following consistent initial values:

$$\dot{u}_1 = \dot{V}(0), \qquad \dot{u}_2 = \frac{C_1}{C_1 + C_2} \cdot \dot{V}(0), \qquad \dot{u}_3 = 0,$$
$$u_1 = V(0), \qquad u_2 = 0, \qquad u_3 = 0,$$
$$j_V = -\frac{1}{R_1} \cdot V(0) - \frac{C_1 \cdot C_2}{C_1 + C_2} \cdot \dot{V}(0).$$

In case of a charge/flux oriented formulation the procedure is similar.

To answer the second question, we note that transient analysis may abort or yield wrong results if it is started from an inconsistent initial value. An example is given in ESTÉVEZ SCHWARZ [2000]. Fortunately, the multi-step methods are mostly started with a backward Euler step, and thanks to the special structure of the network equations this is sufficient in many cases to bring the solution back onto the right manifold, although integration was started from an inconsistent value (ESTÉVEZ SCHWARZ [2000]).[5] Note however that this is not true if integration is started with the trapezoidal rule; even with stiffly-accurate one-step methods it may take some timesteps to get back to the correct solution, if the initial values are not consistent.

Fixing the weak instability. The variable order, variable stepsize BDF for the index-2 network equations (10.1) reads

$$A \frac{1}{h_k} \sum_{i=0}^{\rho} \gamma_{k,i}\, g(x_{k-i}) + f(x_{k-i}, t_{k-i}) = \delta_k.$$

Here, the defect δ_k represents the perturbations in the kth step caused by the rounding errors and the defects arising when solving the nonlinear equations numerically. MÄRZ

[5] Some authors exploit this feature to get consistent initial values by performing some Eulersteps backward and then again forward in time (VALSA and VLACH [1995], BRACHTENDORF and LAUR [2001]).

and TISCHENDORF [1997] have shown that if the ratio of two succeeding stepsizes is bounded and the defect δ_k is small enough, then the BDF approach is feasible – i.e., the nonlinear equations to be solved per integration step are locally uniquely solvable with Newton's method – and convergent. However, a weakly instable term of the type

$$\max_{k \geqslant 0} \frac{1}{h_k} \|\mathcal{D}_k \delta_k\|$$

arises on the right-hand side for the error estimate of $\max_{k \geqslant \rho} \|x_k - x(t_k)\|$. Here $\mathcal{D}_k$ denotes a projector that filters out the higher-index components of the defect. In contrast to Hessenberg-type index-2 systems, this instability may affect all solution components, and may cause trouble for the timestep and error control. Remember, that the stepsize is decreased if the a-posteriori error check fails. For small stepsizes however, the weak instability is reflected by an error growth if the stepsize is decreased – the usual timestep and error control must fail!

Since all solution components may be affected, an appropriate error scaling – as done for Hessenberg systems – is no remedy. However, the instability can be fixed by reducing the most dangerous part of the defect δ_k, that is, those parts belonging to the range of $\mathcal{D}_k$. This defect correction can be done by generalizing the back propagation technique, since the projector can be computed very cheaply by pure graphical means with the use of an index monitor.

We finish this section with some remarks on a new BDF-based approach to integrate the network equations numerically, which shows some potential for the future: Modified Extended BDF.

In 1983, J. Cash proposed the Modified Extended BDF (MEBDF) method, which combines better stability properties and higher order of convergence than BDF, but requires more computations per step (CASH [1983, 2000]). One timestep with the MEBDF method consists of three BDF steps and an evaluation step. This results in more work compared to BDF, but the order of convergence increases with one for most circuits (BRUIN [2001]). This implies that for convergence order 3 we normally apply the 3-step BDF method, while with the MEBDF method a 2-step method suffices.

The k-step MEBDF-methods are A-stable (HAIRER, NØRSETT and WANNER [1987]) for $k \leqslant 3$, while for BDF this is restricted to the case $k \leqslant 2$ (CASH [1983]). Thus these MEBDF-methods 'break' Dahlquist's Law (HAIRER, NØRSETT and WANNER [1987]) that applies to real multistep methods: we have higher order methods with unconditional stability.

The approach looks attractive because implementation may re-use existing BDF-based datastructures efficiently. In the Modified version, also the number of needed LU-factorizations is reduced to only 1. Variants also allow parallelism (FRANK and VAN DER HOUWEN [2000]).

11. A second approach: One-step methods

Up to now, only multi-step methods have been used for the numerical discretization in professional packages. These conventional methods have achieved a high degree of maturity, and have proven to be efficient and very robust in an extremely large variety of applications. Nevertheless there is some motivation to look at alternative schemes also from an industrial point of view:

- The BDF methods are applicable to much more general classes of nonlinear DAEs; can methods be superior, which are definitely constructed for the special linear-implicit nonlinear form (10.1) of the circuit equations?
- In the charge/flux oriented form of conventional codes, timestep control is difficult, since charge/flux tolerances are not of interest for the user, and extra effort is necessary to derive charge/flux tolerances from the desirable user given node voltage or current tolerances.

  Are there methods with a more natural embedding of timestep control even in charge oriented formulation?
- The fully implicit methods used so far require in each timestep a nonlinear system to be solved. Can semi-implicit methods be employed, which need only linear systems to be solved?

Recently, a class of one-step methods was developed that give a positive answer to the three questions above. They are based on embedded Rosenbrock–Wanner (ROW) schemes, which have been used successfully for solving classical network equations (RENTROP [1990]), and

- are tailored to the special structure of charge/flux oriented network equations (10.1), and do not aim at solving arbitrary DAEs of non-Hessenberg type;
- enable a natural timestep control which applies directly on node potentials and branch currents;
- define linearly-implicit methods that need only linear systems to be solved.

Since these schemes turned out to be competitive with the standard multi-step methods even in an industrial environment, it seems worthwhile to introduce them in more detail here.

Charge/flux-oriented ROW schemes. In a first step, we apply a standard Rosenbrock–Wanner method to the linear-implicit DAE system (10.1) (HAIRE, LUBICH and ROCHE [1989], RENTROP, ROCHE and STEINEBACH [1989]). To simplify notation, we assume for the moment that the network equations do not explicitly depend on time, i.e., $f(x(t),t) \equiv f(x(t))$. For this homogeneous case, the numerical approximation for one ROW step reads

$$x_1 = x_0 + b^\top k, \tag{11.1a}$$

$$y_1 = y_0 + b^\top l, \tag{11.1b}$$

with weights $b := (b_1, \dots, b_s)^\top$ and increments $k := (k_1, \dots, k_s)^\top$, $l := (l_1, \dots, l_s)^\top$ defined by

$$\begin{pmatrix} A & \gamma h \frac{\partial f(x_0)}{\partial x} \\ -\gamma I & \gamma \frac{\partial g(x_0)}{\partial x} \end{pmatrix} \cdot \begin{pmatrix} l_i \\ k_i \end{pmatrix} = \begin{pmatrix} -hf(\sum_{j=1}^{i-1} \alpha_{ij} k_j) - h \frac{\partial f(x_0)}{\partial x} \sum_{j=1}^{i-1} \gamma_{ij} k_j \\ y_0 - g(\sum_{j=1}^{i-1} \alpha_{ij} k_j) + \sum_{j=1}^{i-1} (\alpha_{ij} + \gamma_{ij}) l_j - \frac{\partial g(x_0)}{\partial x} \sum_{j=1}^{i-1} \gamma_{ij} k_j \end{pmatrix} \tag{11.1c}$$

where $\alpha_{ij}=0$ for $i \geqslant j$, $\gamma_{ij}=0$ for $i>j$ and $\gamma_{ii}=\gamma\neq 0$, $i,j=1,\dots,s$. x_1 and y_1 are the approximations to the solution at time h with $x(0)=x_0$, $y(0)=y_0$. The increments are uniquely defined by the linear system (11.1c): The matrix

$$\begin{pmatrix} 0 & A\frac{\partial g(x_0)}{\partial x}+\gamma h\frac{\partial f(x_0)}{\partial x} \\ -\gamma I & \gamma\frac{\partial g(x_0)}{\partial x} \end{pmatrix}$$

obtained after one block Gaussian elimination step is nonsingular for sufficient small stepsizes h, since the matrix pencil $\{A\partial g(x)/\partial x, \partial f/\partial x\}$ is regular at least for index-1 systems.

In a second step, we use the special structure of (10.1) to eliminate the differential components y from the computation of x_1. The linear structure of the charge constraint (10.1b) allows for k_i to be computed independently from $l_1,\dots,l_{i-1}$. To fulfill charge conservation during integration, the differential variables y are projected at each grid point t_i in the integration interval $[0,T]$ on the charge constraint:

$$y_i := g(x_i), \quad \forall i \text{ with } t_i\in[0,T]. \tag{11.1d}$$

In the end, the computation of x_1 does only depend on x_0, and we have defined a class of charge/flux oriented ROW schemes by (11.1a), (11.1c) and (11.1d).

One notes that the same Jacobian information is needed in both multi-step schemes and charge/flux oriented ROW methods, but for different reasons: As iteration matrix for the multi-step schemes in the first case, and as system matrix of the linear equations which serve for getting the stage increments in the latter case. Note that the same Jacobian is used here for all stage equations. It is however possible to construct efficient higher order methods which exploit the fact that in circuit simulation the Jacobian is rather cheap to get (GÜNTHER and HOSCHEK [1997], GÜNTHER, HOSCHEK and WEINER [1999]); in this case the Jacobian would be different at each stage.

Convergence and order conditions. As shown in GÜNTHER [1998], classical convergence theory for semi-explicit index-1 problems can be applied to the ROW method (11.1a), (11.1c), (11.1d). Owing to the projection (11.1d), the local error $g(x_1)-g(x(h))$ must be $\mathcal{O}(h^{p+1})$ to obtain convergence order p. For arbitrary charge functions, this conditions leads to the requirement $x_1-x(h)=\mathcal{O}(h^{p+1})$, and we have the following convergence result: To obtain order p for the network equations (10.1) of index-1, the coefficients of the Rosenbrock method (11.1a), (11.1c), (11.1d) have to fulfill all order conditions for the algebraic variables up to order p in semi-explicit index-1 systems. This result applies also for a large class of index-2 network equations of the form (10.1).

The coefficients of the method are free to fulfil order conditions for a given method and to guarantee A- and L-stability, respectively. In contrast to multi-step methods, one can construct A- and L-stable methods of arbitrary order.

CHORAL – an embedded method of order (2)3. On account of the low smoothness properties of transistor models, as well as of the low accuracy demands usually required in practice, an embedded method of order (2)3 seems to be suitable. The corresponding scheme, CHORAL, has four stages and only three function evaluations. To avoid a constant term in the error estimate due to inconsistent initial values, both methods are chosen as stiffly accurate (HAIRER and WANNER [1996]), and in particular L-stable.

For the general nonhomogeneous case of (10.1), the numerical approximation x_k after one timestep from t_{k-1} to $t_k = t_{k-1} + h_k$, together with an embedded approximation $\hat{x}_k$ of lower order for error control and timestep prediction, is now given by

$$x_k = x_{k-1} + \sum_{i=1}^{s} d_i \kappa_i, \qquad \hat{x}_k = x_{k-1} + \sum_{i=1}^{s} \hat{d}_i \kappa_i,$$

where the increments κ_i are computed from linear systems

$$\left(\frac{1/\gamma}{h_k}\mathcal{F}^0_{\dot{x}} + \mathcal{F}^0_x\right)\kappa_i = \frac{1/\gamma}{h_k} A\big(g(x_{k-1}) - g(a_i)\big) - \sum_{j=1}^{i} \tilde{\beta}_{ij} f(a_j)$$
$$- \sum_{j=1}^{i-1} \tilde{\beta}_{ij} \frac{\partial f}{\partial x}(x_{k-1}, t_{k-1})\kappa_j - h\tilde{\tau}_i \frac{\partial f}{\partial t}(x_{k-1}, t_{k-1}), \tag{11.2}$$

whose right-hand sides can be setup after evaluating the functions $f(a_i)$ and $g(a_i)$ at internal stage values

$$a_i := x_{k-1} + \sum_{j=1}^{i-1} \sigma_{ij} \kappa_j.$$

The corresponding coefficient set of CHORAL with $\tilde{\beta}_{ij} := \beta_{ij}/\gamma$ and $\tilde{\tau}_i := \tau_i/\gamma$ is given in Table 11.1. Since the usual error estimate $\|x_1 - \hat{x}_1\| = \|\kappa_4\|$ for stiffly-accurate embedded ROW methods is used, a reliable error control and stepsize selection are offered that are based on node potentials and branch currents only. This makes timestep control very elegant, especially in comparison with the techniques discussed in the previous section for multi-step methods.

TABLE 11.1
Coefficients for CHORAL

$\gamma = 0.5728160624821349$	$\beta_{21} = -2.0302139317498051$
$d_1 = \hat{d}_1 = \sigma_{21} = \sigma_{31} = \sigma_{41} = 1/\gamma$	$\beta_{31} = 0.2707896390839690$
$d_2 = \hat{d}_2 = \sigma_{32} = \sigma_{42} = 0.0$	$\beta_{32} = 0.1563942984338961$
$d_3 = \hat{d}_3 = \sigma_{43} = 1.0$	$\beta_{41} = 2/3$
$d_4 = 1.0$	$\beta_{42} = 0.08757666432971973$
$\alpha_2 = 1.0$	$\beta_{43} = -0.3270593934785213$
$\alpha_3 = 1.0$	$\gamma_1 = \gamma$
$\alpha_4 = 1.0$	$\gamma_2 = -2.457397870$
$\tau_1 = 0.3281182414375370$	$\gamma_3 = 0$
$\tau_2 = -2.57057612180719$	$\gamma_4 = 0$
$\tau_3 = -0.229210360916031$	
$\tau_4 = 1/6$	

TABLE 11.2
CPU times: CHORAL versus BDF2 on HP workstation C200

Circuit	# transistors	# equations	CPU time	
			CHORAL	BDF2
LC oscillator	0	3	0.57s	0.33s
MOS ringoscillator	134	73	30.13s	27.61s
16 bit adder	544	283	2m41.32s	2m30.1s
1 Mbit DRAM	2005	1211	10m16.18s	8m29.15s
16 Mbit DRAM	5208	3500	23m37.18s	12m5.11s
ALU	13005	32639	97m31.64	82m21.03s

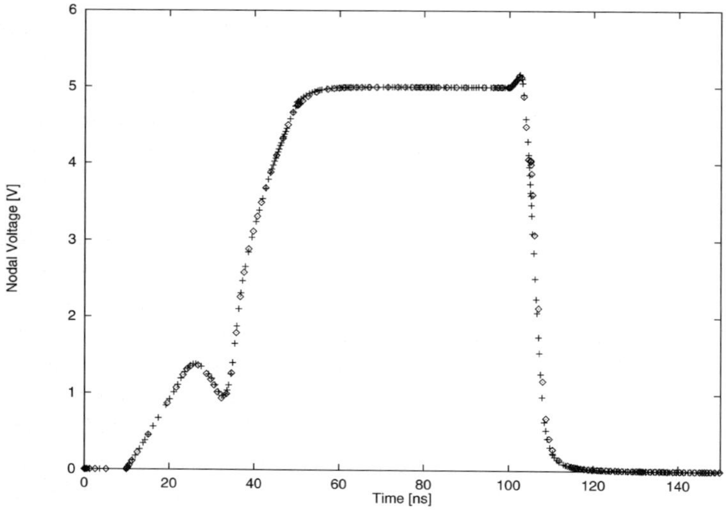

FIG. 11.1. One output nodal voltage for the 16 bit adder: Integration steps of CHORAL (◇) vs. BDF2 (+).

Practical experience. The implementation of CHORAL in an industrial circuit simulation package opened the possibility to gain experience not only with simple standard benchmark examples like an LC oscillator and MOS ringoscillator, but also for numerous real life problems (HOSCHEK [1999]). Some of them are included in Table 11.2: A 16 bit adder, critical path circuits of dynamic memory (DRAM) circuits, and an arithmetic logical unit ALU, which is the core of a central processing unit.

We see that CHORAL can cope even with large problems, and is competitive with BDF not only with respect to CPU times (Table 11.2) but also with respect to accuracy, see Fig. 11.1.

One reason for the efficiency of CHORAL seems to be the stepsize and error control that allow large stepsizes by only a few failures of the stepsize predictions. These results are confirmed by numerical tests reported in GÜNTHER [1998] for digital circuits, NAND gate and 2 bit adder: For nonstringent accuracy demands required in network analysis, CHORAL turned out to be as powerful and efficient as DASSL (BRENAN,

CAMPBELL and PETZOLD [1996]), the latter being a standard code for BDF integration of low index DAEs.

Particularly appealing are CHORAL's damping properties: Excitations and oscillations with physical significance are tracked, but perturbations are damped. This behaviour will be discussed in more detail in the following section.

12. Oscillatory circuits and numerical damping: A comparison

Dealing with oscillatory behaviour, we have to distinguish between two types of oscillations. The first type is given by oscillations of physical significance which reflect the behaviour of the mathematical model and the circuit, and should be preserved during numerical integration. The LC oscillator shown in Fig. 12.1 (left side) can serve as a basic example. This linear circuit consists of one capacitance $C = 4$ pF and inductance $L = 1$ nH in parallel driven by an initial current source $I_0 = 6$ A. Numerical approximations obtained by CHORAL and BDF2 are given in Fig. 12.1 (right side) for the branch current through the inductor. The current oscillates with the amplitude given by I_0 and frequency $\omega = 1/\sqrt{LC}$, which corresponds to a period of $T = 2\pi/\omega \approx 0.4$ nsec. While an error becomes visible both in phase and amplitude for BDF2, both phase and amplitude are preserved by CHORAL.

The second type is given by high frequent numerical noise, which should be attenuated by the integrator. It may be due to failures of stepsize and error control, or due to an inappropriate semidiscretization of a PDE model with respect to space (GÜNTHER [2001]). A third possible origin are discontinuities of the solution, which might be invoked by nonsmooth transistor models or input stimuli. In the latter case no problems should occur if integration is stopped at these points, and restarted with consistent initial values. Here the algorithms discussed in Section 10 for consistent initialization can be used efficiently for CHORAL, too. However, if integration is not stopped at these points, one may have to deal with inconsistent initial values. An example for such an

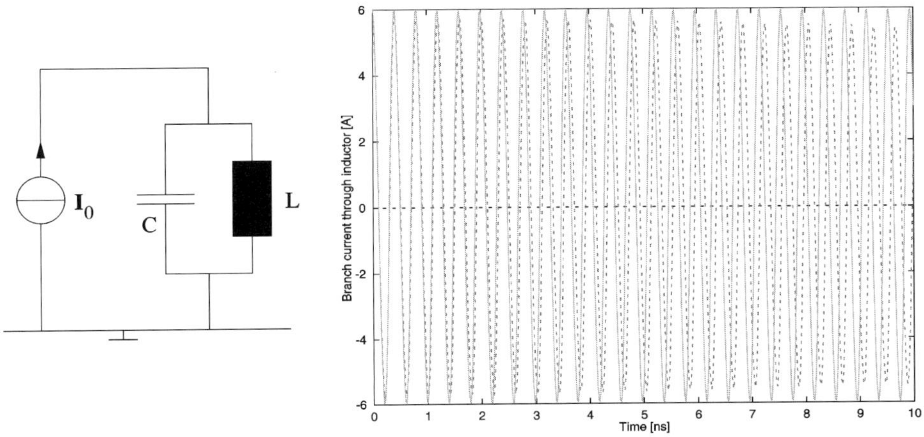

FIG. 12.1. LC oscillator (left) and simulation results (right) for BDF2 (- -) and CHORAL (–).

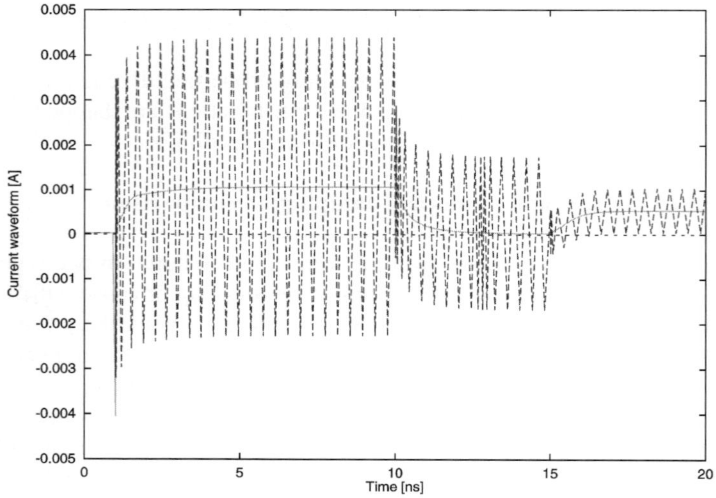

FIG. 12.2. Operational amplifier circuit: Simulation results for trapezoidal rule (- - -) and CHORAL (···).

effect is the current waveform of an operational amplifier circuit, for which the numerical results of the trapezoidal rule and CHORAL are shown in Fig. 12.2. At ≈ 1 nsec there is a sharp spike, which is invoked by traversing a discontinuity of a MOS capacitance model. Due to its energy conserving property, the trapezoidal rule maintains this perturbation, turning it into an oscillation with the actual timestep as period. This yields an impression that the circuit is unstable and oscillates.

In contrast, the perturbation is damped immediately by CHORAL, and only a few steps are necessary to get back to the smooth solution. The results with TR-BDF are not given here, but are similar to those of CHORAL.

Model equation: Harmonic oscillator. To explain these results for both physical and artificial oscillations, we investigate the model equation

$$\ddot{x} + \omega^2 x = 0 \tag{12.1}$$

of a harmonic oscillator with frequency ω over one period $[0, T := 2\pi/\omega]$. With initial values $x(0) = x_0$, $\dot{x}(0) = \dot{x}_0$, the solution reads

$$x(t) = r \cdot \operatorname{Re} \exp(i\varphi)$$

with

$$r := \sqrt{x_0^2 + \dot{x}_0^2/\omega^2}, \qquad \varphi := \omega t - \arctan\bigl(\dot{x}_0/(\omega x_0)\bigr).$$

Note that the LC oscillator discussed above corresponds to a harmonic oscillator with frequency $\omega = 1/\sqrt{LC}$.

The results obtained on model equation (12.1) with initial values $(x(0), \dot{x}(0))^\top = (1, 0)^\top$ for BDF2 and the trapezoidal rule, the integration schemes TR-BDF is based

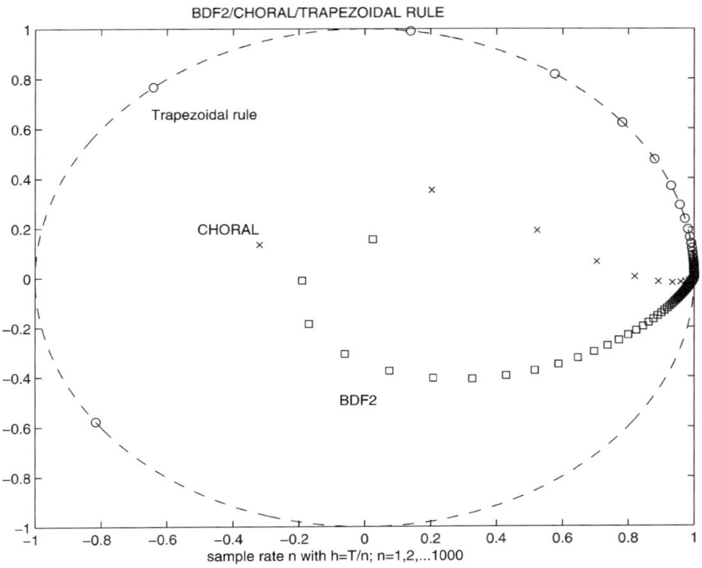

FIG. 12.3. Numerical approximation of BDF2 (□), trapezoidal rule (○) and CHORAL (×) for model equation (12.1) after one period $T = 2\pi/\omega$. The results are plotted for stepsizes $h = T/n$ ($n = 1, 2, \ldots, 1000$) in phase space $x = r\exp(i\varphi)$.

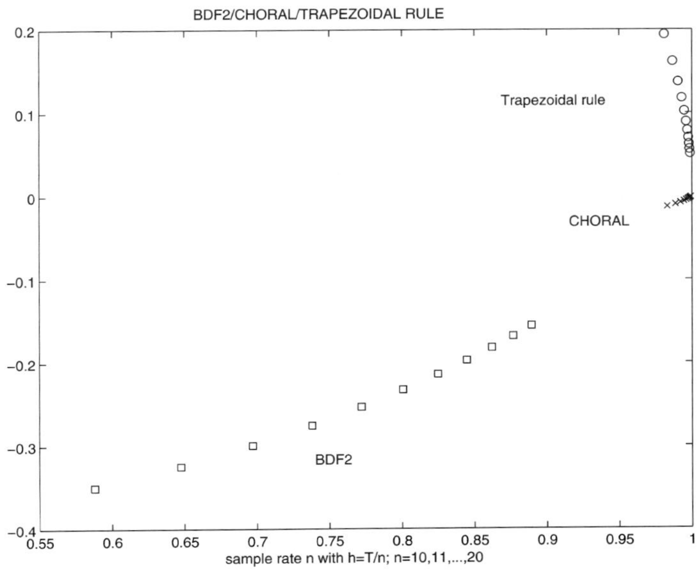

FIG. 12.4. Zoom into numerical approximation of BDF2 (□), trapezoidal rule (○) and CHORAL (×) on model equation (12.1) in phase space with sample rates $n = 10, 11, \ldots, 20$.

on, and CHORAL are given in Figs. 12.3 and 12.4. For each method one period was resolved with *sample rate* $n = 1, 2, \ldots, 1000$ steps of equidistant stepsize $h = T/n$.

Comparing the numerical approximations with the exact solution $(x(T), \dot{x}(T))^\top = (1, 0)^\top$ after one period, we see the following: Due to its energy conserving property, the trapezoidal rule generates no magnitude error for any n; however, for small sample rates one has to deal with rather large phase errors. BDF2 acts much worse: In addition to a phase error, one has to deal with amplitude errors, if one period is sampled too roughly. CHORAL, however, has only slight amplitude and phase errors even for rather small sample rates.

These results become more visible, if we zoom into the results for $n = 10, 11, \ldots, 20$. As a rule-of-thumb in circuit simulation, one has to sample one oscillation with approximately 10–20 points to get results which are accurate enough. Thus oscillations of physical significance which are approximated numerically using sample rates in the range of 10–20 yield rather large phase errors (trapezoidal rule) or both amplitude and phase errors (BDF2). CHORAL, however, is highlighted by only slight errors in phase and amplitude.

Analysis of one-step methods. These good properties of CHORAL applied to oscillatory circuits can be explained by investigating the model equation in more detail. Besides that, this analysis can illustrate its excellent damping properties as well. As a first step, we scale and rewrite (12.1) as an ODE system of first order. With $y := [x, \dot{x}/\omega]^\top$ we have

$$\dot{y} = Jy, \quad y(0) = \begin{pmatrix} x_0 \\ \dot{x}_0/\omega \end{pmatrix}, \tag{12.2}$$

where

$$J = \begin{pmatrix} 0 & \omega \\ -\omega & 0 \end{pmatrix}.$$

For one-step methods such as trapezoidal rule and CHORAL, the numerical solution y_n^h after one period with n equidistant steps of size $h = T/n$ reads

$$y_n^h = \left[R(hJ)\right]^n \begin{pmatrix} x_0 \\ \dot{x}_0/\omega \end{pmatrix}.$$

The stability matrix $R(hJ)$ has eigenvalues $R(\pm i\omega h)$ which are given by evaluating the scalar stability function $R(z)$ for imaginary arguments $z = \pm i\omega h$. Furthermore, its eigenvectors are $(1, i)^\top$ and $(1, -i)^\top$, and thus it holds

$$\left[R(hJ)\right]^n = U \begin{pmatrix} R(z)^n & 0 \\ 0 & R(-z)^n \end{pmatrix} U^{-1}, \quad U = \begin{pmatrix} 1 & 1 \\ i & -i \end{pmatrix}.$$

Therefore the numerical properties of a one-step method applied to the model equation is fixed by its stability function along the imaginary axis:

$$y_n^h = U \begin{pmatrix} R(z)^n & 0 \\ 0 & R(-z)^n \end{pmatrix} U^{-1} \begin{pmatrix} x_0 \\ \dot{x}_0/\omega \end{pmatrix},$$

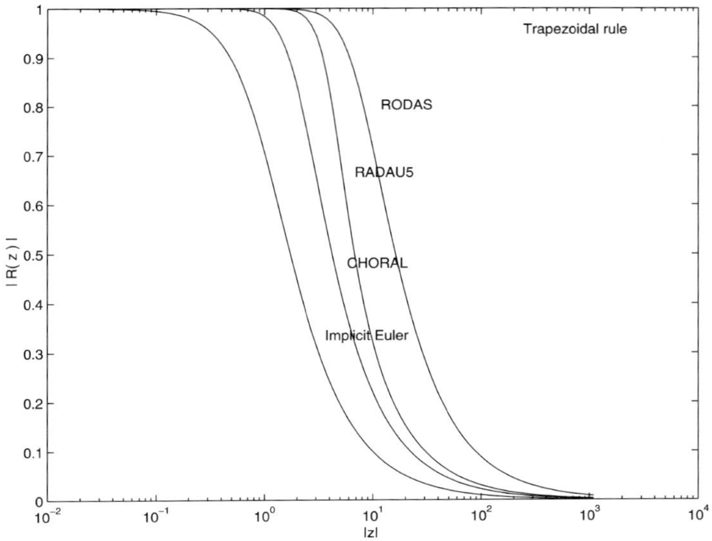

FIG. 12.5. Decay of stability functions along the imaginary axis (from left to right: Implicit Euler, CHORAL, RADAU5, RODAS, and trapezoidal rule with $|R(z)| \equiv 1$).

see Fig. 12.5. Note that we have

$$\lim_{z\to 0} R(z) = 1$$

for convergent methods, and

$$\lim_{z\to \pm\infty} R(z) = 0,$$

for L-stable methods (HAIRER and WANNER [1996]). Thus there is a range of small stepsizes where $|R(z)|$ is close to one and information is almost preserved, and another range where $|R(z)|$ tends to zero and strong damping prevails.

Depending on the type of oscillation, we demand different properties:

- *Oscillations of physical significance*. These should be preserved. Assuming a sample rate of 10–20 steps for oscillations of physical significance, we demand $|R(z)| \approx 1$ in the range of $|z| \in [0.1\pi, 0.2\pi]$.

  Having a look at Fig. 12.5, we see that this demand is fulfilled by all methods but the implicit Euler scheme.
- *Perturbations*. Such oscillations of high frequency, either numerical noise caused by timestep and error control, inconsistent initial values or by an inappropriate semidiscretization of a PDE model, should be damped as much and soon as possible. Hence a slight damping should already occur for $|z|$ larger than 0.2π, the limit for oscillations of physical significance, and $|R(z)| \approx 0$ for highly oscillatory signals, i.e., $|z| > 100$.

Except the trapezoidal rule, which is not L-stable, all methods show good damping properties for highly oscillatory signals ($|z| > 100$). But only the implicit Euler scheme and CHORAL damp already for $|z| > 0.2\pi$.

Summing up, CHORAL shows all the desired properties of a (nonideal) numerical low pass filter: Physical oscillations of low frequency are preserved, but highly oscillatory perturbations are efficiently damped.

The corresponding analysis for multi-step methods can be found in GÜNTHER, RENTROP and FELDMANN [2001].

CHAPTER IV

Numerical Treatment of Large Problems

Due to their reliability and robustness software codes employing the standard algorithms are established as workhorses, which are inevitable when designing electronic circuits. Especially for integrated circuit design one can distinguish two different steps in the design flow, where these tools are used:

- The *electrical design* stage comprises standard applications for characterization and optimization of functional building blocks, such as gates, operational amplifiers, oscillators etc. These analyses are run excessively in order to make sure that the functional units meet their specifications under a large variety of load and bias conditions and temperatures, and with different parameter sets representing the fluctuations of the technological processes in the fabrication lines.
- In the *verification stage* overall functionality of the circuit is checked. For this purpose the circuit parts containing the critical path inclusive parasitics – like capacitances and resistances of junctions and interconnects – are re-extracted from layout. This yields accurate but very large circuit models with many input nodes, which have to be biased with rather lengthy input stimuli in order to verify overall functionality of the circuit.

Typical data for these different kinds of application are given in Table 12.1. The dimension of the mathematical circuit model corresponds approximately to its number of transistors. Since the transistor models are fairly complex, most time in standard applications is spent for setting up the matrix and right hand side of the resulting linear system ('Load'). Due to the overlinear increase of the time spent for sparse Gaussian elimination, the computational expense for the linear solver becomes dominant for large

TABLE 12.1
Typical data for standard and large applications in circuit simulation

	Standard application	Large application
No of transistors	$10^1 \ldots 10^3$	$10^3 \ldots 10^5 (\ldots 10^6)$
No of equations	$10^1 \ldots 10^3$	$10^3 \ldots 10^5 (\ldots 10^6)$
No of timesteps	$10^2 \ldots 10^3$	$10^3 \ldots 10^6$
CPU times (on workstation or PC)	sec … min	hours … days
Load	85%	$85\% \ldots < 50\%$
Lin. solver	10%	$10\% \ldots > 50\%$
Overhead	5%	$5\% \ldots 2\%$

TABLE 12.2
How to speedup circuit simulation

Source of expense	Speedup possible by
Complexity of device models	– *higher level of abstraction*, using functional modelling with languages like VHDL-AMS – use of *table models* for devices or subblocks like gates
Overlinear expense for Gauss solver	*decomposition*
– typically $n^{1.2} \dots n^{1.8}$ (n: number of circuit nodes)	– decoupling into smaller blocks – use of iterative methods
Lack of adaptivity – global timestep control – global convergence control	*decomposition* – *higher degree of adaptivity* by exploiting different activity of different circuit parts at different times
Large number of devices	*parallelization*

applications. The overhead spent for timestep and convergence control etc. is usually below 5%.

Since the turnaround times for large applications are often beyond desirable limits, i.e., significantly more than $5 \dots 8$ hours, it is of a major interest to obtain speedups without sacrificing accuracy, universality and robustness too much.

Many attempts are pursued in the literature to overcome the computational limitations of circuit simulation. Table 12.2 shows the basic principles of these approaches, and which kind of problem they aim to improve.

The first row of Table 12.2 concerns modelling issues, which are not to be discussed here. The remaining rows of Table 12.2 roughly characterize the main aspects of our further discussion. First a glance at a simple MOS ringoscillator example will illustrate typical properties of the mathematical circuit models, which offer potentials for getting improvements.

13. Numerical properties of an MOS ringoscillator model

The task of a ringoscillator in bipolar technology and its basic principles were already explained in Section 9. Most of the very complex integrated circuits challenging circuit simulation however are fabricated in MOS technologies. As a typical representative we will now consider a simple ringoscillator in complementary MOS (CMOS) technology, and highlight some interesting properties of this circuit class.

Fig. 13.1 shows a circuit diagram of a CMOS ringoscillator consisting of 11 inverter stages, which are connected in a feedback loop. Each inverter is composed of a P-type MOS transistor – which is connected to power supply VDD – and of an N-type MOS transistor connected to ground. Furthermore a parasitic wiring capacitance to ground is added. When the input signal at the gate nodes of both transistors of an inverter is higher than a certain threshold voltage then the P-channel transistor is OFF, and the N-channel

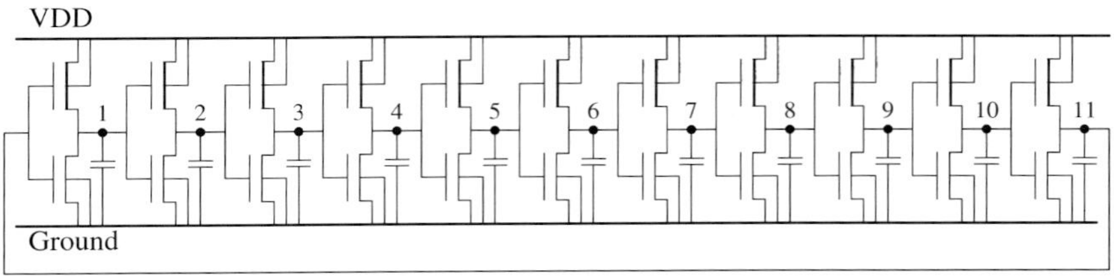

FIG. 13.1. CMOS ringsoscillator.

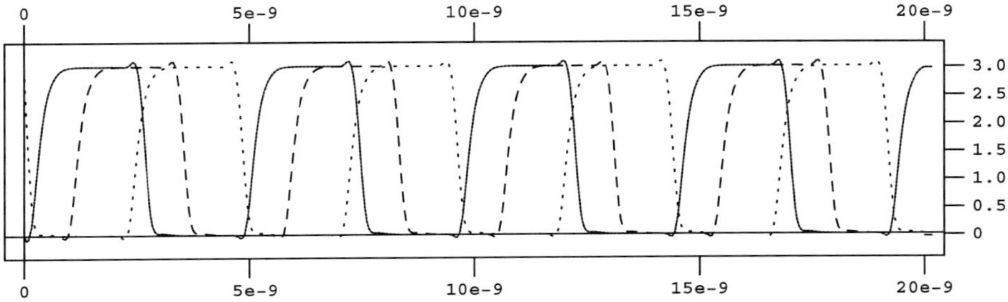

FIG. 13.2. CMOS ringoscillator – Waveforms (in Volt) over time (in sec): Nodes 1 (—), 6 (– – –) and 11 (···).

transistor is conducting, thus pulling the output signal at the common drain node to ground. Inversely, a low level input signal switches the N-channel transistor OFF and the P-channel transistor ON, such that the output node is loaded up to power supply voltage VDD. The inverted output signal drives the next inverter, and after passing all stages of the closed loop, it arrives with a certain time delay as input signal of the first one. This invokes an oscillation, and its period is usually just $2 \cdot 11$ times the average switching delay of one inverter stage. The waveforms of nodes 1, 6, and 11 are shown in Fig. 13.2; the other waveforms are identical – if the design is regular – but shifted in time.

Multirate. When looking at the waveforms of node 1 and 11 in Fig. 13.2 we recognize that the first inverter in the loop is only active in fairly small parts of an oscillation cycle, and more or less quiescent else. The same is true for all other inverters; but they are active at different time windows, since the signal is continuously propagating through the circuit. So the varying degree of activity for different circuit parts has usually no computational effect: Timestep control always has to take care about the smallest timestep in the whole circuit, unless multirate integration is being used.

In order to get an estimate for the potential benefit of multirate integration, we relate the global timestep h_{glob} to the timestep h_{loc} needed for accurate numerical integration of the first inverter, see Fig. 13.3. We see that h_{loc} determines the global timestep just when the first inverter is switching, and becomes much larger else. Obviously the relations are quite similar for all inverter stages. Using a slightly modified version of a

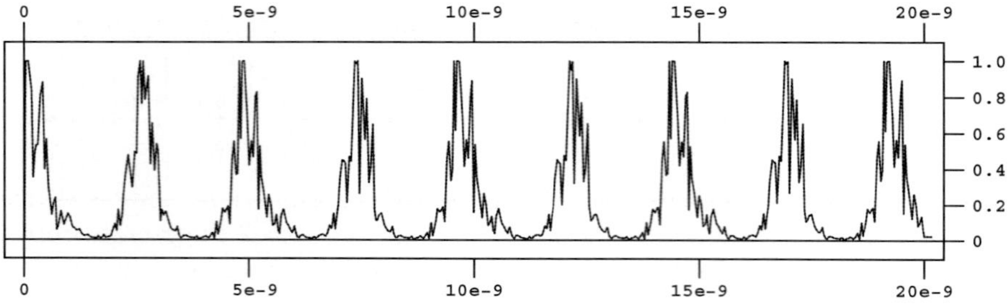

FIG. 13.3. CMOS ringoscillator – Timestep ratio h_{glob}/h_{loc} for the first inverter, over time.

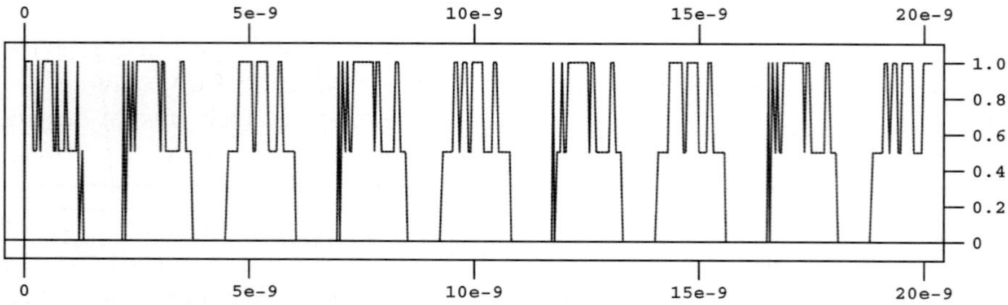

FIG. 13.4. CMOS ringoscillator – Iteration count ratio nc_{loc}/nc_{glob} for node 1, over time.

formula given in BARTEL, GÜNTHER and KVÆRNØ [2001], we can estimate the possible speedup in this case to be:

$$speedup = \frac{n \cdot m}{n_L + n_A \cdot m} \approx \frac{1}{\text{mean value}(h_{glob}/h_{loc})}$$

where n is the number of devices, n_A is the average number of active devices, $n_L = n - n_A$ the average number of inactive devices, and m is the average number of global timesteps within a timestep in case of no activity. Measuring the mean value from Fig. 13.3, we get

$$speedup \approx \frac{1}{0.24} \approx 4.2,$$

or with a more practical restriction of the local timesteps to $\leqslant 5 \cdot h_{glob}$ still a speedup factor of approximately 3.

In reality this figure will become smaller due to inevitable overhead; on the other hand it may further increase for larger circuits. So we conclude that multirate integration seemingly offers significant speedup potential for circuit simulation.

Latency. Another effect of the varying degree of activity is the different rate of convergence for different circuit parts, when applying fully implicit integration methods like BDF. Fig. 13.4 shows as an example the ratio nc_{loc}/nc_{glob} over time, where nc_{loc} counts the number of Newton iterations needed per timestep to get convergence of node 1, and

nc_{glob} is the global iteration count per timestep. Similar to the multirate formula we get a rough estimate

$$speedup = \frac{n}{n_L \cdot \mu + n_A}$$

for an algorithm which exploits the different rate of convergence for different circuit parts. Here, μ is an average ratio of iteration counts for the inactive and the active circuit parts. For our ringoscillator, an approximation follows which can be directly measured from Fig. 13.4:

$$speedup \approx \frac{1}{\text{mean value}(nc_{loc}/nc_{glob})} \approx \frac{1}{0.48} \approx 2.1.$$

A special case would be to omit re-evaluation of circuit parts which do not change from one timestep to the next ($\mu = 0 \rightarrow$ 'latency'). This gives an estimated $speedup = n/n_A$ in this case, making the exploitation of latency an interesting alternative to multirate integration.

Unidirectional signal flow. An inherent property of MOS transistors is to have – almost – no static current flow from the gate node into the device. So, when the signal flow in a circuit is passing the gate of an MOS transistor then it is mainly unidirectional in a local sense, and only dynamic effects can cause local feedback. This is illustrated in Fig. 13.5, where static and capacitive coupling in forward and backward direction is shown for the first inverter of the CMOS ringoscillator. Static backward coupling is negligible. Note that although capacitances are small, their coupling effect is comparable to static coupling, which is due to the high switching speed of $10^9 \ldots 10^{12}$ Volt per sec.

Unfortunately, those circuit configurations which propagate signals between source and drain node of the MOS transistors – like bus structures – do *not* exhibit unidirectional signal flow. Furthermore, *global* feedback coupling principles are extensively applied in circuit design.

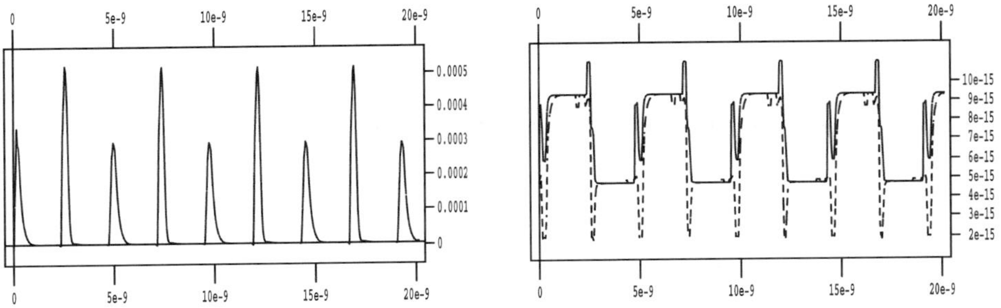

FIG. 13.5. CMOS ringoscillator – Forward (—) and backward (- - -) coupling coefficients for the first inverter, over time. Static/capacitive coupling is shown in the left/right diagram.

Parallelism. Finally we see that the circuit schematic consists of a large number of identical primitives (here: MOS transistors and capacitors), thus offering speedup potential by handling them in parallel. Since the model equations of transistors are usually very complex and contain many if-then-else branches, their evaluation is well suited for a medium or coarse grain type of parallelization.

14. Classification of existing methods

In the literature there is a rich variety of attempts to overcome the computational limitations of standard circuit simulation. A rough overview is given in Fig. 14.1. We see that decomposition techniques (HACHTEL and SANGIOVANNI-VINCENTELLI [1981], DE MICHELI, HSIEH and HAJJ [1987]) are applied at almost all stages of the standard algorithms, in order to

- apply relaxation methods,
- introduce a higher degree of adaptivity, and
- improve performance on parallel computers.

Single boxes (with larger fonts) in Fig. 14.1 represent single but large systems, while double boxes (with smaller fonts) indicate sets of decomposed, smaller subsystems.

The three columns on the left (ROW, MLN and standard) characterize algorithms which are sufficiently general to cope with any circuit of not too high DAE index. While standard TR-BDF as well as ROW integration has been discussed in Chapter III, the multi-level Newton method (MLN) will be described in more detail later on. A multi-level direct linear solver (block Gauss solver) is not included in the figure, since it can be seen as a special case of the multi-level Newton solver. Furthermore, modifications of the Newton method in the standard solver – as are described in ENGL, LAUR and DIRKS [1982], EICKHOFF and ENGL [1995] – are not included due to space limitations.

The right five columns (ITA, WR, WRN, PWL and Exp.Fit) describe approaches which can be efficiently used only for a restricted class of circuits, e.g., for more or less digital MOS circuits. These methods are shortly reviewed below; more details can be found in the survey papers (NEWTON and SANGIOVANNI-VINCENTELLI [1984], DE MICHELI, HSIEH and HAJJ [1987]) and in the book (WHITE and SANGIOVANNI-VINCENTELLI [1987]). Further developments are reviewed in URUHAMA [1988], SALEH and WHITE [1990], and their specific strengths and limitations are compared.

The formulas given below refer to network equations given in the compact form (10.1a). We assume that variables and equations are partitioned and reordered, such that each subblock i is characterized by just one entry in

$$x = (x_1, \dots x_i, \dots x_m)^\top, \qquad f = (f_1, \dots f_i, \dots f_m)^\top,$$
$$A = (A_1, \dots A_i, \dots A_m)^\top,$$

with m being the number of subblocks. Furthermore we assume that implicit multi-step methods

$$\dot{y} = \alpha y + \beta$$

are applied with α as leading integration coefficient and β giving the contributions of previous timepoints.

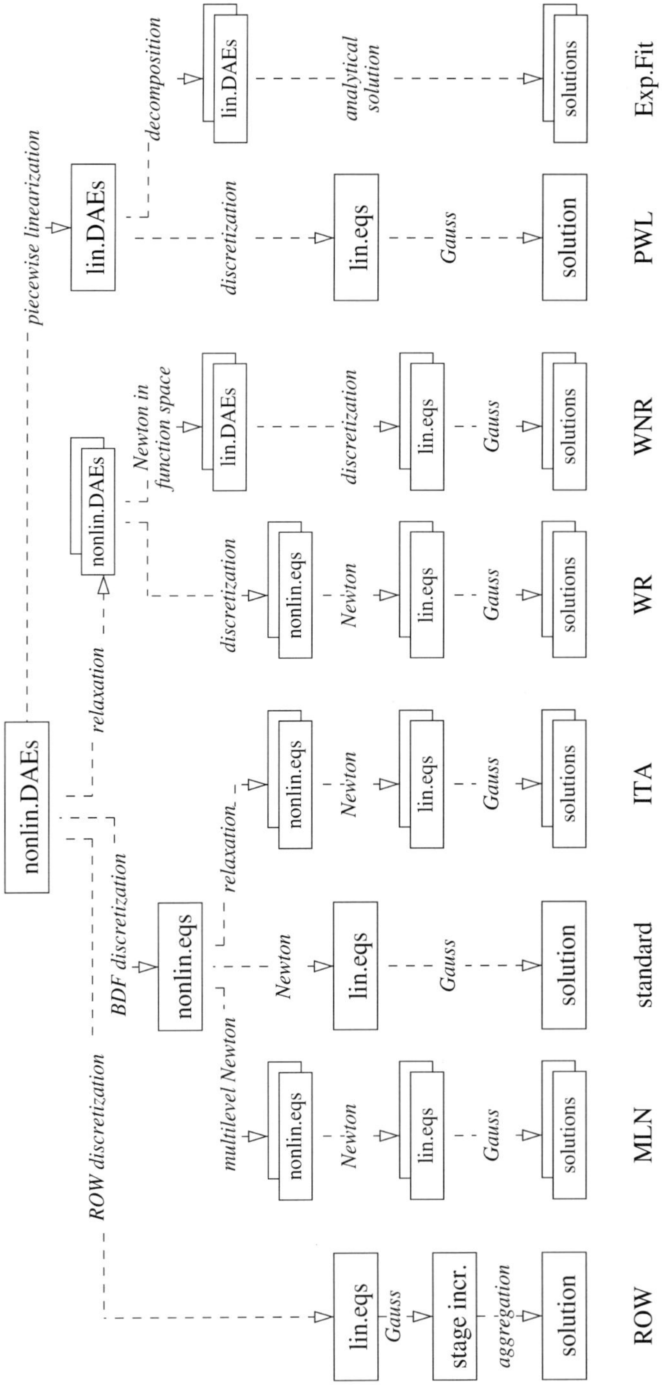

FIG. 14.1. Existing methods for circuit analysis in the time domain. (ROW: Rosenbrock Wanner method; MLN: multi-level Newton; ITA: iterated timing analysis; WR: waveform relaxation; WRN: waveform relaxation Newton; PWL: piecewise linear analysis; Exp.Fit: exponential fitting.)

ITA – iterated timing analysis. Historically, timing simulation was the first attempt to compute approximate waveforms for very large digital MOS circuits (CHAWLA, GUMMEL and KOZAK [1975]). Here the nonlinear systems were completely decomposed into single equations, which were approximately solved by performing one single Gauss–Jacobi or Gauss–Seidel step. In iterated timing analysis ITA the applicability of the method is significantly extended by blockwise decomposition, where the subblocks are solved with conventional methods (SALEH, KLECKNER and NEWTON [1983], DE MAN, ARNOUT and REYNAERT [1981]). Hence for each relaxation sweep j a sequence $i = 1, \dots, m$ of subsystems

$$A_i \cdot \big(\alpha y(x^j) + \beta\big) + f_i(x^j, t) = 0$$

has to be solved for x_i^j with

$$x^j = \begin{cases} (x_1^{j-1}, x_2^{j-1}, \dots, x_{i-1}^{j-1}, x_i^j, x_{i+1}^{j-1}, \dots, x_m^{j-1})^\top & \text{for Gauss–Jacobi relaxation,} \\ (x_1^j, x_2^j, \dots, x_i^j, x_{i+1}^{j-1}, \dots, x_m^{j-1})^\top & \text{for Gauss–Seidel relaxation.} \end{cases} \tag{14.1}$$

The lth iteration of the inner Newton process is described by

$$x_i^{j,l+1} = x_i^{j,l} + \Delta x_i^{j,l},$$

where $\Delta x_i^{j,l}$ is computed from

$$\left(\alpha A_i \cdot \frac{\partial y}{\partial x}\bigg|_{x=x^{j,l}} + \frac{\partial f}{\partial x}\bigg|_{x=x^{j,l}}\right) \Delta x_i^{j,l} = -A_i \cdot \big(\alpha y(x^{j,l}) + \beta\big) - f_i(x^{j,l}, t).$$

The particular Newton iterate reads, e.g., for Gauss–Seidel relaxation as

$$x^{j,l} = (x_1^j, \dots, x_{i-1}^j, x_i^{j,l}, x_{i+1}^{j-1}, \dots, x_m^{j-1})^\top.$$

A convergence proof of ITA is given in URUHAMA [1987] for circuits with strictly diagonal dominant capacitance matrix and Lipschitz continuous conductance matrix, provided that the timesteps are sufficiently small.

For efficiency reasons, often only one single Newton step is performed per relaxation sweep in ITA. This may cause some loss of accuracy and reliability, which is however acceptable in many cases. Adaptivity can be improved by exploiting the different activity of different circuit partitions. This is implemented in some kind of event control: Only those partitions of the system are scheduled for computation, which are activated by changing signals at their borders.

WR – waveform relaxation. This method is basically an application of Picard iteration to the network equations. The method can also be characterized as a block Gauss–Seidel–Newton or block Gauss–Jacobi–Newton method in the function space, since after decomposition of the circuit into subblocks the subsystems are solved for a whole waveform with standard methods, while their coupling is handled with relaxation methods. That means that for each relaxation sweep j a sequence of subsystems

$$A_i \cdot \dot{y}(x^j) + f_i(x^j, t) = 0, \quad i = 1, \dots, m$$

has to be solved for $x_i^j(t)$ in the time interval $0 \leqslant t \leqslant t_{end}$ with x^j being defined by (14.1).

WR was first discussed in LELARASMEE, RUEHLI and SANGIOVANNI-VINCENTELLI [1982], and a global convergence proof was given for circuits with a grounded capacitor at each node, provided that some Lipschitz conditions hold and the time windows used are sufficiently small.

The method has found much interest in the literature, since it offers a very natural way to improve adaptivity in form of multirate by integrating each subblock with its own timestep:

$$\dot{y} = \alpha_i^j y + \beta_i^j .$$

A survey including many practical aspects of WR methods is given in the book edited by DEBEVFE, ODEH and RUEHLI [1985]. Efficient parallelization of WR is described in ODENT, CLAESEN and DE MAN [1990]. Recent convergence theorems given in GRISTEDE, ZUKOWSKI and RUEHLI [1999] extend the class of feasible circuits and provide insight how to decompose the circuit for getting good convergence rates. The latter aspect has turned out to be a key issue for the performance of WR. Since simple partitioning schemes based on circuit topology sometimes give no satisfactory results, information about the entries of the Jacobian is often used for this purpose. This may even require repartitions from time to time, especially in case of strongly nonlinear circuits (ZECEVIC and GACIC [1999]).

Most of the literature published about WR deals with ordinary differential equations and therefore requires to have a grounded capacitor at each node, at least at the decoupling subblock borders. Extensions to DAEs with nondifferential – i.e., algebraic – coupling equations are discussed in ARNOLD and GÜNTHER [2001]; it is shown that certain contractivity conditions additionally must hold in order to ensure convergence in these cases.

WRN – waveform relaxation Newton. If the Newton process is applied directly in the function space, before time discretization, then we get the waveform Newton method WN:

Compute

$$x^{l+1}(t) = x^l(t) + \Delta x^l(t)$$

in the time interval $0 \leqslant t \leqslant t_{end}$ from solving

$$A \cdot \frac{\mathrm{d}}{\mathrm{d}t}\left(\left.\frac{\partial y}{\partial x}\right|_{x=x^l(t)} \cdot \Delta x^l\right) + \left.\frac{\partial f}{\partial x}\right|_{x=x^l(t)} \cdot \Delta x^l = -A \cdot \dot{y}(x^l) - f(x^l, t)$$

for $\Delta x^l(t)$. With

$$C^l(t) = A \cdot \left.\frac{\partial y}{\partial x}\right|_{x=x^l(t)}, \qquad G^l(t) = \left.\frac{\partial f}{\partial x}\right|_{x=x^l(t)}$$

this reads as

$$C^l(t) \cdot \frac{d\Delta x^l}{dt} + \left(G^l(t) + \dot{C}^l(t)\right)\Delta x^l = -A \cdot \dot{y}(x^l) - f(x^l, t).$$

Note that $\Delta x^l(0) = 0$ if $x^l(0) = x_0$.

A convergence proof of WN is given in SALEH and WHITE [1990] for very general circuit classes.

At a first glance, this method seems not to be very attractive for circuit simulation, since one cannot expect that initial waveforms are close to the solution. Its main advantage is that it yields systems of *linear* DAEs. These can eventually be solved with discretization methods which are more efficient than standard integration (PALUSINSKI, GUARINI and WRIGHT [1988]). The main motivation for presenting this method here is however, that it serves as a base for waveform relaxation Newton WRN.

If the nonlinear DAE subsystems of the WR method are solved with the WN method, then we get the WRN method:

Solving

$$A_i \cdot \dot{y}(x^j) + f_i(x^j, t) = 0, \quad i = 1, \ldots, m$$

with WN means to compute

$$x_i^{j,l+1}(t) = x_i^{j,l}(t) + \Delta x_i^{j,l}(t)$$

in the interval $0 \leqslant t \leqslant t_{end}$ from solving

$$C_i^{j,l}(t) \cdot \frac{d\Delta x_i^{j,l}}{dt} + \left(G_i^{j,l}(t) + \dot{C}_i^{j,l}(t)\right) \cdot \Delta x_i^{j,l} = -A_i \dot{y}(x^{j,l}) - f_i(x^{j,l}, t)$$

for $\Delta x_i^{j,l}(t)$, where $x^{j,l}$ is given by (14.1) and

$$C_i^{j,l}(t) = A_i \left.\frac{\partial y}{\partial x}\right|_{x=x^{j,l}}, \qquad G_i^{j,l}(t) = \left.\frac{\partial f_i}{\partial x}\right|_{x=x^{j,l}}.$$

We see again that here the DAEs are linear time variant.

A convergence proof for this method can be derived from the convergence of the WR and WN methods, from which it is composed (URUHAMA [1987], SALEH and WHITE [1990]). The adaptivity of this method is high due to its natural support of multirate integration. For efficiency reasons timesteps should be coarse in early stages of the relaxation process, and become finer only when it approaches convergence. The method is reported to be superior over WR for circuits which do not have a strongly unidirectional signal flow. For efficiency often only one Newton step is performed per relaxation; the price of slightly reduced accuracy and reliability seems to be acceptable in many applications. Finally, WRN allows for an efficient parallelization (SALEH, WEBBER, XIA and SANGIOVANNI-VINCENTELLI [1987]).

PWL – piecewise linear analysis. In an alternative approach, the nonlinear device characteristics are approximated by piecewise linear models. The resulting linear DAE systems are only piecewise valid. When they are solved with conventional time discretization schemes like BDF, then timestep control has to take care that their region of validity is not left within the timestep. This may slow down efficiency, if the model resolution is fine. For finding a solution, improved versions of Katzenelson's algorithm are used in general (KATZENELSON [1965], VLACH [1988a], YU and WING [1984]).

If the validity regions are explicitly included as constraints for the piecewise linearized network equations, then extra "state variables" have to be introduced, which define for which particular region a certain linear relation is valid. This *piecewise linear mapping* leads to systems of the following form:

$A \cdot \dot{y} + F \cdot x + B \cdot z + f_0(t) = 0$ piecewise linearization of $f(x, t)$;

$y = G \cdot x + E \cdot z + y_0$ piecewise linearization of $y(x)$;

$\bar{z} = H \cdot x + D \cdot z + z_0$ definition of region of validity;

$z \geqslant 0; \quad \bar{z} \geqslant 0;$

$z^{\top} \cdot \bar{z} = 0$ complementarity of state variables.

The dimension of the state variables $z, \bar{z} \in \mathbb{R}^{n_z}$ defines the maximal number of different regions of validity to be 2^{n_z}, since each component of z can be selected to be either $= 0$ or > 0, and the corresponding component of $\bar{z}$ is then defined from the complementarity condition. The crossing of a border between two regions of validity is characterized by just one component of z or $\bar{z}$ to become zero. Fortunately, this crossing can be performed by a rank one update of the system matrix; and if a hierarchical LU decomposition method is used for solving the linear system, this does not require much extra effort. A review of these techniques can be found in VAN BOKHOVEN [1987].

In LIN, KUH and MAREK-SADOWSKA [1993] piecewise linear circuit equations are obtained by mapping nonlinear conductances and capacitances into time variant linear conductances and capacitances, respectively.

The most appealing aspects of piecewise linear analysis PWL are, that no Newton iterations are necessary (LIN, KUH and MAREK-SADOWSKA [1993]), strong global convergence properties, and a uniform kind of modelling, based on tabulated data (VAN EIJNDHOVEN [1984]).

Exp.Fit – exponential fitting. If a PWL circuit model is decomposed into small subblocks then each subsystem can be solved analytically for a certain time interval, until the solution crosses the border of the particular linear model section. These techniques are known as exponential fitting methods (SARKANI and LINIGER [1974]). They have shown to offer high simulation speed, especially in timing simulators, where only one relaxation step is performed (ODRYNA and NASSIF [1986], VIDIGAL, NASSIF and DIRECTOR [1986], BAUER, FANG and BRAYTON [1988]). A mathematical analysis of exponential fitting methods in circuit simulation is presented in SILVEIRA, WHITE, NETO and VIDIGAL [1992]. It starts from the piecewise linearized version of (10.1a)

$$C \cdot \dot{x} + G \cdot x + f_0(t) = 0,$$

which is discretized with an explicit exponential formula of order 1:

$$x(t_{l+1}) = x(t_l) + D^{-1}\left(1 - e^{-Dh}\right)\dot{x}(t_l).$$

The solution is of the matrix exponential form

$$x(t) = e^{-Dt}x_0 - G^{-1}f_0(t),$$

where D is given by $D = C^{-1}G$. A standard method for solving this kind of equations is asymptotic waveform evaluation AWE (PILLAGE and ROHRER [1990], RATZLAFF and PILLAGE [1994]), which is based on moment matching methods. Unfortunately, the methods work only for regular

$$C = A \cdot \left.\frac{\partial y}{\partial x}\right|_{x=x_{t_l}},$$

i.e., ODEs. For the DAE case the Drazin inverse might come into play (WILKINSON [1982]). Since numerical evaluation of the matrix exponential containing the Drazin inverse turns out to be cumbersome (RENTROP [1990]), exponential fitting was only applicable to a restricted class of circuits up to now.

Concluding remarks.

- Surprisingly, iterative solvers for the linear equations of the standard or ROW approach are not included in Fig. 14.1. Although numerous attempts were made in the past to substitute direct Gaussian elimination by iterative solvers in general purpose circuit simulation programs, no one was really successful yet. Basically this is due to a lack of favourable numerical properties of the linearized circuit equations, in combination with restrictive accuracy requirements. Roughly speaking, the use of iterative linear solvers requires very good preconditioners, which are not much cheaper to get than direct LU factors. Furthermore, the widespread use of quasi Newton methods – taking Jacobians and their LU-factors from earlier iterations – alleviates the need for iterative solvers, if the circuit size is not too large, say less than 10^5 nodes. Only recently, promising approaches for iterative linear solvers were presented (SCHILDERS [2000], BOMHOF [2001]); both of them are particularly tailored to the specific structure of the network equations.
- As an interesting alternative to implicit integration adaptively controlled explicit integration has been suggested (DEVGAN and ROHRER [1994]) for fast timing analysis. Here explicit integration is stabilized by using the fact, that the signals saturate more or less rapidly, e.g., at the power supply or at the ground voltage level.
- Today software codes employing ITA or WR or exponential fitting algorithms or adaptively controlled explicit integration have obtained a high degree of maturity and robustness, which allows them to be successfully used even in industrial environments. Especially when exploiting hierarchical concepts on highly repetitive circuits, these codes can simulate several clocks of $10^7 \ldots 10^9$ transistor circuits on transistor level with reasonable accuracy in some hours.

  Note however that although these codes are often one or two orders of magnitude faster than standard circuit simulation packages, they cannot really substitute the latter. This is due to their restriction to time domain analysis as well as some lack of accuracy and universality and – even worse – reliability. So we will focus in the remaining part of this chapter on methods to speed up the standard transient analysis algorithms without sacrificing their universality and robustness. One is parallelization, and another one deals with multirate integration.

15. Parallelization

At a first glance, parallel circuit simulation offers a high speedup potential due to

- a large number of devices with fairly complex, but identical characteristic equations;
- large systems of linear equations to be solved;
- a small amount of purely serial overhead.

In practice however, the speedup of parallel versus serial simulation often saturates at a low level – say 2 to 4 – even on very powerful parallel computers. Further improvements can only be obtained by carefully adapting the granularity of parallelism to the particular computer architecture, which has an impact on both algorithms and coding.

A rough classification identifies three different granularity levels of parallelism (COX, BURCH, HOCEVAR, YANG and EPLER [1991]):

- *Fine grain* parallelism for single instruction multiple data or pipelining architectures like vector supercomputers, which were the workhorses for large industrial circuit simulation tasks in the past. Parallelization is basically achieved by vectorization.
- *Medium grain* parallelism for multiprocessor machines with shared memory. Such systems are presently often installed in industry for complex design tasks. Parallelization here is based on thread concepts.
- *Coarse grain* parallelism on loosely coupled clusters of workstations or PCs. Here it is essential to take care for a minimum data traffic over the local network. Parallelization is based on message passing systems like PVM or MPI. This level may also be useful for shared memory multiprocessors.

Due to reasons of cost effectiveness and flexibility, vector supercomputers are no longer used for circuit simulation. So vectorization will not be further considered here; literature can be found in (EICKHOFF [1991], FELDMANN, WEVER, ZHENG, SCHULTZ and WRIEDT [1992], EICKHOFF and ENGL [1995]).

While the levels of fine and medium grain parallelism directly aim at the classical circuit simulation algorithms, the coarse grain is best realized with a multi-level Newton method for solving the network equations at a particular timepoint. This method will be described in the following subsection, before we come back to parallel circuit simulation.

15.1. Multi-level Newton method

Originally, the multi-level Newton MLN method was developed to solve large nonlinear systems by decomposition without loosing the quadratic convergence of Newton's algorithm (RABBAT, SANGIOVANNI-VINCENTELLI and HSIEH [1979]). If a proper decomposition can be found then the method offers a good speedup potential by parallel execution on clusters of fast processors, but relatively slow interconnect network.

Our further discussion restricts on a two-level Newton method; an extension to more levels is possible, but not common practice.

We assume that the nonlinear system of equations $f(x) = 0$ with $x \in \mathbb{R}^n$, $f : \mathbb{R}^n \to \mathbb{R}^n$ has a regular Jacobian $\partial f/\partial x$. Further we assume that f is decomposed into m

subsystems f_i $(i = 1 \ldots m)$ and one master system f_{m+1}, and x is reordered such that

$$x^\top = (x_1, x_2, \ldots, x_m, x_{m+1}), \qquad f^\top = (f_1, f_2, \ldots, f_m, f_{m+1}),$$

where x_i $(i = 1 \ldots m)$ contain the inner variables of f_i and x_{m+1} contains the outer variables. Then

$$f_i = f_i(x_i, x_{m+1}) \quad (i = 1 \ldots m),$$
$$f_{m+1} = f_{m+1}(x_1, x_2, \ldots, x_m, x_{m+1}),$$

and the Jacobian of f has bordered block diagonal form:

$$\frac{\partial f}{\partial x} = \left(\begin{array}{cccc|c} \frac{\partial f_1}{\partial x_1} & & & & \frac{\partial f_1}{\partial x_{m+1}} \\ & \frac{\partial f_2}{\partial x_2} & & & \frac{\partial f_2}{\partial x_{m+1}} \\ & & \ddots & & \vdots \\ & & & \frac{\partial f_m}{\partial x_m} & \frac{\partial f_m}{\partial x_{m+1}} \\ \hline \frac{\partial f_{m+1}}{\partial x_1} & \frac{\partial f_{m+1}}{\partial x_2} & \cdots & \frac{\partial f_{m+1}}{\partial x_m} & \frac{\partial f_{m+1}}{\partial x_{m+1}} \end{array}\right).$$

Finally it is assumed that all submatrices $\partial f_i/\partial x_i$ $(i = 1 \ldots m)$ are regular; otherwise the decomposition has to be changed. Then the two-level Newton method contains a Newton loop for the master system, where for each outer iteration k the inner systems are solved for x_i with fixed outer variables x_{m+1}^k, see Fig. 15.1.[6]

Fig. 15.2 shows an example in two dimensions, i.e., $m = 1$, with the notation $S := 1$, $M := m + 1 = 2$. The inner Newton steps starting from the point (x_S^0, x_M^0) yield a solution on the curve $f_S = 0$ with fixed outer variable x_M^0. Then a Newton step is done into x_M direction to get the point (x_S^1, x_M^1); the latter may be further improved into x_S direction by adding the tangent correction, such that the next inner Newton cycle can start from the point (x_S^{T1}, x_M^{T1}).

The two-dimensional example illustrates how the two-level Newton scheme can be derived: The subsystem $f_S = 0$ defines an implicit relation $x_S = x_S(x_M)$, and the outer Newton method solves $f_M(x_S(x_M), x_M)$ for x_M.

The MLN approach can be characterized as follows:

- Quadratic convergence of the method is shown in RABBAT, SANGIOVANNI-VINCENTELLI and HSIEH [1979] under standard assumptions, if the inner nonlinear systems are solved with higher accuracy than the outer ones:

  $$\|\Delta x_i\| \leqslant \|\Delta x_{m+1}\|^2 \quad (i = 1 \ldots m).$$

  This may become difficult to achieve, especially for MLN methods with more than two levels. Methods for reducing the number of inner iterations without affecting quadratic convergence are described in ZHANG, BYRD and SCHNABEL [1992]. A simple practical rule to get sufficiently superlinear convergence is to solve the inner systems just somewhat more accurately than the actual norm of the outer

[6] In the linear case, the Schur matrix S_i and residuum R_i of Fig. 15.1 can be easily explained to be the Gauss updates for eliminating x_i in f_{m+1}, i.e., to transform the system into upper triangular form.

- *Initialization:*
  - get start vectors $x_1^0, x_2^0, \dots x_m^0, x_{m+1}^0$
  - set iteration indices $k = 0,\ j_i = 0\ (i = 1 \dots m)$
- *Outer Newton process*:
  do until convergence
  - do for all subsystems $i = 1 \dots m$
    * *Inner Newton process*:
      do until convergence
      · solve: $\partial f_i / \partial x_i \cdot \Delta x_i^{j_i} = -f_i$
      · add Newton correction: $x_i^{j_i+1} = x_i^{j_i} + \Delta x_i^{j_i}$
      · update inner iteration index: $j_i = j_i + 1$
      enddo
    * compute Schur matrix: $S_i = \frac{\partial f_{m+1}}{\partial x_i} \cdot (\frac{\partial f_i}{\partial x_i})^{-1} \cdot \frac{\partial f_i}{\partial x_{m+1}}$
    * compute residuum: $R_i = \frac{\partial f_{m+1}}{\partial x_i} \cdot (\frac{\partial f_i}{\partial x_i})^{-1} \cdot f_i$
    enddo
  - solve: $(\frac{\partial f_{m+1}}{\partial x_{m+1}} - \sum_{i=1}^{m} S_i) \cdot \Delta x_{m+1}^k = -f_{m+1} + \sum_{i=1}^{m} R_i$
  - add Newton correction: $x_{m+1}^{k+1} = x_{m+1}^k + \Delta x_{m+1}^k$
  - update master iteration index: $k = k + 1$
  - *Tangent correction*:
    do for all subsystems $i = 1 \dots m$
    * update inner variables: $x_i^{j_i+1} = x_i^{j_i} - (\frac{\partial f_i}{\partial x_i})^{-1} \cdot \frac{\partial f_i}{\partial x_{m+1}} \cdot \Delta x_{m+1}^k$
    * update inner iteration index: $j_i = j_i + 1$
    enddo
  enddo

FIG. 15.1. The two-level Newton method.

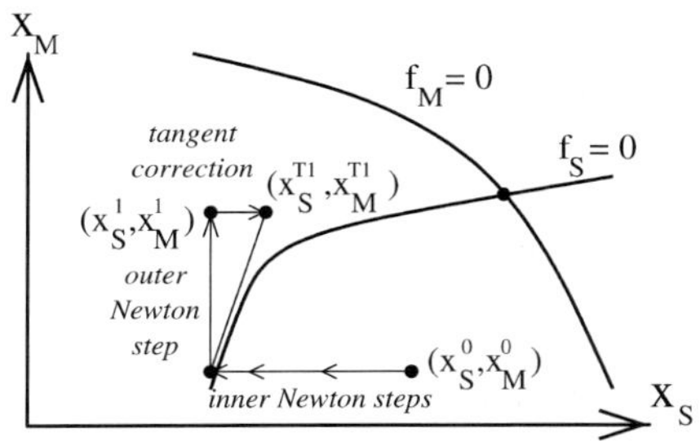

FIG. 15.2. Two-level Newton scheme for solving $f_S(x_S, x_M) = 0$, $f_M(x_S, x_M) = 0$. Index S: inner Newton, on subsystem; index M: outer Newton, on master system.

Newton process, e.g.:

$$\|\Delta x_i\| \leqslant \alpha \|\Delta x_{m+1}\| \quad (i = 1 \dots m) \text{ with } \alpha = 10^{-1} \dots 10^{-2}.$$

- A single-level Newton method is obtained, when only one inner iteration is performed and the solution is updated with the tangent correction (ZHANG [1989]).

This is useful for combining global and multi-level Newton steps, which may improve efficiency and robustness in certain cases (HONKALA, ROOS and VALTONEN [2001]).

- Due to additional inner Newton iterations one should expect that the multi-level method is more expensive than the standard Newton process. However in practice often nonlinearity can be shifted into the smaller subsystems, thus reducing the number of outer iterations and getting even better efficiency than with the single-level algorithm (FRÖHLICH, RIESS, WEVER and ZHENG [1998]).
- Originally, the tangent correction is not included in the MLN algorithm. Mainly it serves for getting a good start vector for the next inner iteration cycle (HOYER and SCHMIDT [1984], ZHANG [1989], WIEDL [1994]). In practice it turns out that the tangent correction should be omitted as long as the outer process is still far away from convergence.
- In case of sufficiently decreasing norms, quasi Newton steps may be employed for the outer iteration process by taking the Schur matrices of earlier iterations, and eventually avoiding expensive LU factorization of the outer system (FRÖHLICH, RIESS, WEVER and ZHENG [1998]).

In HONKALA, ROOS and VALTONEN [2001], HONKALA, KARANKO and ROOS [2002] a variant is described in which the Schur matrix actions and the tangent corrective actions are replaced by introducing a simple global Newton–Raphson step as outerloop action: in fact their method is Newton–Raphson in which each iteration is solved by m-parallel Newton–Raphson sub-processes, each of at most J-iterations (with $J \leqslant 5$) for the subsystems.

15.2. Parallel multi-level Newton: Loop over hierarchies

If the circuit is decomposed into a set of subblocks which are interconnected via a carrier network, then the MLN method described in Fig. 15.1 is a natural choice for parallelization at a coarse grain level: For each timestep, each subblock is solved in an inner Newton loop by a slave process on a separate processor unit, and then the master process performs an outer iteration for getting the carrier network solution. An ideal data flow for parallel MLN is shown in Fig. 15.3: The master sends the values of the carrier circuit variables x_{m+1} and the actual timestep to the slaves; then each slave performs its tangent correction, solves its subblock equations, computes Schur matrix S_i, residuum R_i and timestep control information, and sends all back to the master. The black boxes in Fig. 15.3 indicate when the particular process is busy. Of course, the relative time for data transmission should be much smaller than suggested in Fig. 15.3.

Decoupling. If the branch currents flowing from the slave subnets into the master carrier circuit are introduced as additional variables in the vector of unknowns, then the network equations are well decoupled, and the term $\partial f_{m+1}/\partial x_i$ arising in the Schur matrix

$$S_i = \frac{\partial f_{m+1}}{\partial x_i} \cdot \left(\frac{\partial f_i}{\partial x_i}\right)^{-1} \cdot \frac{\partial f_i}{\partial x_{m+1}}$$

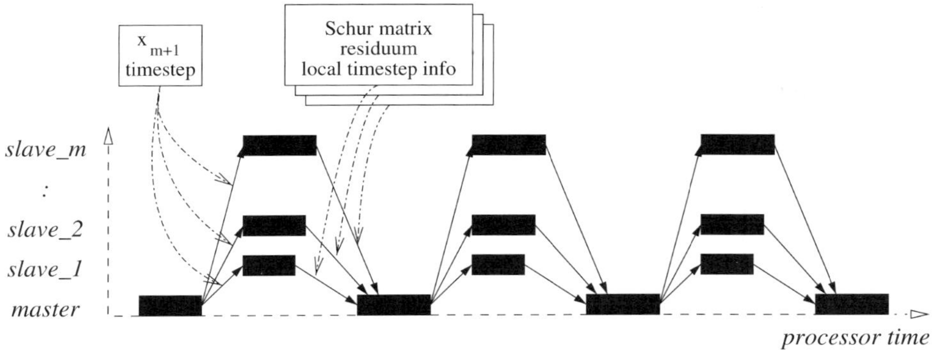

FIG. 15.3. Ideal data flow for parallel MLN.

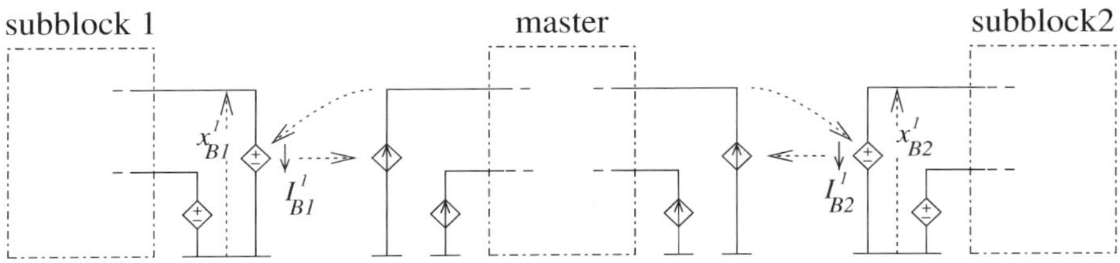

FIG. 15.4. Circuit decoupling with controlled sources. Pin currents I_{Bi} and pin voltages x_{Bi} are introduced for each subblock i.

and the residuum

$$R_i = \frac{\partial f_{m+1}}{\partial x_i} \cdot \left(\frac{\partial f_i}{\partial x_i}\right)^{-1} \cdot f_i$$

is simply a constant incidence matrix. Before showing more details, we make a further extension of the vector of unknowns: The pin voltages of the slave subblocks at the border to the master circuit are duplicated, and the new voltages are assigned to the slave subnets (FRÖHLICH, RIESS, WEVER and ZHENG [1998]). This is not necessary in principle for efficient parallelization; however there are two advantages:

- Circuits can be easily decoupled using controlled sources, as shown in Fig. 15.4 (WING and GIELCHINSKY [1972], WU [1976]). Since the latter are standard elements in any circuit simulator, no extra programming efforts are necessary for decoupling.[7]

[7] Decoupling by imposing pin voltages to the subblocks – as illustrated in Fig. 15.4 – is called "node tearing" in the literature (SANGIOVANNI-VINCENTELLI, CHEN and CHUA [1977]), and is mostly applied when simulating integrated circuits. This is surely adequate from a numerical aspect, as long as the circuit is voltage driven, i.e., its functionality is described in terms of voltage waveforms; an example is standard CMOS logic. For current driven circuits like some analog building blocks or power circuits with switches, "branch tearing" may be preferable (HACHTEL and SANGIOVANNI-VINCENTELLI [1981]). In this case the subcircuit pins are driven by controlled current sources, and the pin voltages are fed back into the master circuit.

- Numerical robustness of the MLN method can be improved by applying particular damping strategies for the controlled voltages sources, which drive the pins of the slave subcircuit blocks (WALLAT [1997]).

With these extensions, the vectors x_i and functions f_i, f_{m+1} of the standard MLN scheme (Fig. 15.1) have to be replaced by

$$x_i \Rightarrow \begin{pmatrix} x_{Si} \\ x_{Bi} \\ I_{Bi} \end{pmatrix}, \quad f_i(x_i, x_{m+1}) \Rightarrow \begin{pmatrix} f_{Si}(x_{Si}, x_{Bi}) \\ f_{Bi}(x_{Si}, x_{Bi}) + I_{Bi} \\ x_{Bi} - A_i \cdot x_{m+1} \end{pmatrix} \quad (i = 1, \ldots, m)$$

$$f_{m+1}(x_i, x_{m+1}) \Rightarrow f_{m+1}(x_{m+1}) - \sum_{i=1}^{m} A_i^\top \cdot I_{Bi},$$

where the variables $x_{Si} \in \mathbb{R}^{n_{Si}}$, $x_{Bi} \in \mathbb{R}^{n_{Bi}}$, $I_{Bi} \in \mathbb{R}^{n_{Bi}}$ and $x_{m+1} \in \mathbb{R}^{n_{m+1}}$ denote the inner network variables of subblock i, the pin voltages of subblock i, the pin currents leaving subblock i and the network variables of the master circuit, respectively. $A_i \in \{0, 1\}^{n_{Bi} \times n_{m+1}}$ are incidence matrices, projecting the nodes of the master system to the pin nodes of subblock i. Hereby are m the number of subblocks (slaves), n_{Si} and n_{Bi} the number of inner network variables and pins of subblock i, and n_{m+1} the dimension of the master system.

The network equations can be characterized as follows:

- $f_{Si} : \mathbb{R}^{n_{Si}} \times \mathbb{R}^{n_{Bi}} \to \mathbb{R}^{n_{Si}}$ are the inner network equations of subblock i;
- $f_{Bi} : \mathbb{R}^{n_{Si}} \times \mathbb{R}^{n_{Bi}} \to \mathbb{R}^{n_{Bi}}$ capture the currents flowing from the inner nodes of subblock i into its pins;
- $f_{m+1} : \mathbb{R}^{n_{m+1}} \to \mathbb{R}^{n_{m+1}}$ are the network equations of the master circuit without the slave contributions;

Consequently, the Jacobians $\partial f_i / \partial x_i$, $\partial f_i / \partial x_{m+1}$, and $\partial f_{m+1} / \partial x_i$ of the MLN scheme given in Fig. 15.1 have to be replaced by

$$\frac{\partial f_i}{\partial x_i} \Rightarrow J_i := \begin{pmatrix} \frac{\partial f_{Si}}{\partial x_{Si}} & \frac{\partial f_{Si}}{\partial x_{Bi}} & 0 \\ \frac{\partial f_{Bi}}{\partial x_{Si}} & \frac{\partial f_{Bi}}{\partial x_{Bi}} & I \\ 0 & I & 0 \end{pmatrix}, \quad \frac{\partial f_i}{\partial x_{m+1}} \Rightarrow \begin{pmatrix} 0 \\ 0 \\ -A_i \end{pmatrix},$$

$$\frac{\partial f_{m+1}}{\partial x_i} \Rightarrow (0 \quad 0 \quad -A_i^\top),$$

where I is a $n_{Bi} \times n_{Bi}$ unity matrix. With this decoupling, we get the following form for the Schur matrix:

$$S_i = A_i^\top \cdot \left(\frac{\partial f_{Bi}}{\partial x_{Si}} \cdot \left(\frac{\partial f_{Si}}{\partial x_{Si}} \right)^{-1} \cdot \frac{\partial f_{Si}}{\partial x_{Bi}} - \frac{\partial f_{Bi}}{\partial x_{Bi}} \right) \cdot A_i \quad (i = 1, \ldots, m). \tag{15.1}$$

The second factor can be shown to be just $\partial I_{Bi} / \partial x_{Bi}$, if the inner system is solved exactly. Since the latter is not possible in general, we conclude that the Schur matrix is a more or less good approximation for the admittance matrix of the particular subblock:

$$S_i \approx A_i^\top \cdot \frac{\partial I_{Bi}}{\partial x_{Bi}} \cdot A_i = A_i^\top \cdot \frac{\partial I_{Bi}}{\partial x_{m+1}}.$$

For its numerical computation we can use Eq. (15.1), which requires to calculate the inverse of $\partial f_{Si}/\partial x_{Si}$. This may become expensive since n_{Si} is large in general, and so it is more economic to exploit that the Schur matrix is just the lower right part of J_i^{-1}, where J_i is the inner iteration matrix:

$$J_i^{-1} = \begin{pmatrix} \left(\frac{\partial f_{Si}}{\partial x_{Si}}\right)^{-1} & 0 & -\left(\frac{\partial f_{Si}}{\partial x_{Si}}\right)^{-1} \frac{\partial f_{Si}}{\partial x_{Bi}} \\ 0 & 0 & I \\ -\frac{\partial f_{Bi}}{\partial x_{Si}} \left(\frac{\partial f_{Si}}{\partial x_{Si}}\right)^{-1} & I & \frac{\partial f_{Bi}}{\partial x_{Si}} \left(\frac{\partial f_{Si}}{\partial x_{Si}}\right)^{-1} \frac{\partial f_{Si}}{\partial x_{Bi}} - \frac{\partial f_{Bi}}{\partial x_{Bi}} \end{pmatrix}.$$

This submatrix can be computed columnwise, using the original system from the inner Newton process

$$J_i \cdot \begin{pmatrix} \Delta x_{Si} \\ \Delta x_{Bi} \\ \Delta I_{Bi} \end{pmatrix} = - \begin{pmatrix} f_{Si} \\ f_{Bi} + I_{Bi} \\ x_{Bi} - A_i \cdot x_{m+1} \end{pmatrix},$$

but taking different right hand sides: Solve

$$J_i \cdot \begin{pmatrix} \cdots \\ \cdots \\ s_i^k \end{pmatrix} = - \begin{pmatrix} 0 \\ 0 \\ e_i^k \end{pmatrix} \quad (k = 1, \ldots, n_{Bi})$$

for s_i^k, where e_i^k is the kth column of an $n_{Bi} \times n_{Bi}$ dimensional unity matrix. This requires one LU decomposition of J_i and only n_{Bi} forward backward substitutions, which can be done locally on each slave processor. The s_i^k then form the columns of the Schur matrix, which has to be transferred to the master processor for being assembled into the outer iteration matrix.

If the inner Newton loops are truncated after some iterations then there remains a defect of the subblock equations which enters the outer Newton process in form of the residuum R_i, see Fig. 15.1. In the decoupled formulation we get:

$$\begin{aligned} R_i = -A_i^\top \cdot \Bigg(& -\frac{\partial f_{Bi}}{\partial x_{Si}} \left(\frac{\partial f_{Si}}{\partial x_{Si}}\right)^{-1} \quad I \\ & -\frac{\partial f_{Bi}}{\partial x_{Si}} \left(\frac{\partial f_{Si}}{\partial x_{Si}}\right)^{-1} \frac{\partial f_{Si}}{\partial x_{Bi}} - \frac{\partial f_{Bi}}{\partial x_{Bi}} \Bigg) \cdot \begin{pmatrix} f_{Si} \\ f_{Bi} + I_{Bi} \\ x_{Bi} - A_i \cdot x_{m+1} \end{pmatrix} \\ = {} & A_i^\top \cdot \Delta I_{Bi} \quad (i = 1, \ldots, m). \end{aligned}$$

Here is ΔI_{Bi} an error term for the pin currents which is induced from truncating the inner Newton loop, and which can be computed from solving

$$J_i \cdot \begin{pmatrix} \cdots \\ \cdots \\ \Delta I_{Bi} \end{pmatrix} = - \begin{pmatrix} f_{Si} \\ f_{Bi} + I_{Bi} \\ x_{Bi} - A_i \cdot x_{m+1} \end{pmatrix}.$$

This can be done locally on each slave processor, and after being transferred to the master processor, the ΔI_{Bi} must be assembled into the right hand side for the next outer Newton step.

Finally the tangent correction has to be computed, before the next inner Newton loop is started. It is easy to see that this can be done locally either, as soon as the actual state of the master variables x_{m+1} is available on the slave processors.

Some remarks can be given on how to improve parallel performance:

- For the master circuit the linear solver is the most time critical part. Since the Schur matrices tend to be dense, a sparse block or even dense linear solver is adequate.
- We see from Fig. 15.3 that the master process is idle when the slaves are working, and vice versa. So, one slave process can be assigned to the master processor;[8] and on a shared memory machine a *parallel* linear solver utility should be used, which includes the slave processors for solving the large interconnect network of the master.
- A performance model for parallel MLN is described in GRÄB, GÜNTHER, WEVER and ZHENG [1996]. It can be used for dynamically adopting numerical parameters of the MLN method – like the maximal number of inner iterations – to the computer- and network-configuration and its actual load.

Partitioning. In an industrial environment an automatic tool is indispensable, which – by applying tearing methods – partitions a given circuit netlist into subnets, inserts the controlled sources for proper decoupling, and assigns the subnets for being solved in a particular process ($\rightarrow$ *static assignment*). The number of partitions is prescribed by the user. Hereby two different strategies can be applied: One makes the number of partitions just equal to the number of processors, which are actually available. In this case the partitions have to be equally balanced with respect to their computational load, and latency effects – i.e., the different rate of convergence for different partitions – cannot be exploited. The other strategy makes the number of partitions much larger than the number of processors, and assigns several subnets to one processor. This alleviates the needs to get partitions of equal workload, and opens a chance to exploit latency effects. Unfortunately, in real life applications it turned out, that a large number of partitions tends to increase the interconnect network significantly, which has to be solved in the master process DENK [2002]. This is a critical issue for the performance of MLN, and so in practice a small number of partitions is often more efficient than a large one.

The most essential requirements for the partitioner are (WALLAT [1997], FRÖHLICH, RIESS, WEVER and ZHENG [1998]):

1. prescribed number of partitions;
2. small number of interconnects between each partition and master;
3. small total number of interconnects;
4. equal computational weight (workload) for each partition;
5. solvability for each partition and carrier network;
6. small runtime;
7. all nonlinear elements put into partitions.

The second requirement is for keeping the dimension n_{Bi} of the Schur matrices small, which is desirable for reducing both the expense to calculate the Schur matrix and the

[8] More precisely: The master can start collecting data as soon as the fastest slave has finished its task; so the slave task with the smallest workload should be assigned to the master processor.

amount of data to be sent to the master. The third requirement is extremely important for the workload of the master process since the interconnect net mainly determines the dimension of the master system, which is *not* sparse in general. The 4th requirement takes care of a well balanced load of the slave processors, which is essential if each processor gets just one partition. The 5th requirement concerns that unsolvable circuit structures – like the loops of inductors and/or voltage sources discussed in Section 7 – may be generated due to insertion of controlled sources at the partition borders, which must be avoided. The runtime requirement is obvious, but hard to meet since partitioning problems are NP complete (GAREY and JOHNSON [1979], HACHTEL and SANGIOVANNI-VINCENTELLI [1981]). Hence heuristics have to be found which produce near optimal solutions in a short time. Finally, the last requirement is desirable to shift nonlinearity into the slaves, thus reducing the number of expensive outer iterations.

In Section 5 it was shown that decomposition may have an impact on the DAE index of the system. This is not a critical issue as long as the index does not get greater than 2 for the decomposed system, since state of the art integrators usually can cope with index-2 problems. However, in case of DAE index > 2 most integrators may run into severe numerical problems, and to avoid that should be a further requirement for the partitioner. Unfortunately, it may be difficult to find out for a certain circuit configuration if insertion of controlled sources would raise the index beyond 2, see ESTÉVEZ SCHWARZ and TISCHENDORF [2000]. So this requirement is not (yet) observed in any partitioning tool.

We will now shortly consider partitioning algorithms as far as they are suited for coarse grain parallelization of classical circuit simulation. Special partitioners for waveform relaxation or for placement tools, e.g., are not included here, since their objectives are different.

Since partitioning is closely related with the task to transform a given matrix into bordered block diagonal form (HACHTEL and SANGIOVANNI-VINCENTELLI [1981]), the methods suggested for the latter problem may be useful for partitioning as well. One of them is nested dissection. In its original form it provides partitions of decreasing size (HACHTEL and SANGIOVANNI-VINCENTELLI [1981]); it has to be checked if variants like that in GEORGE [1974] provide better balanced partitions (REISSIG [2001]).

In SANGIOVANNI-VINCENTELLI, CHEN and CHUA [1977] a *local* clustering algorithm was proposed, which turned out in practice to be efficient and to generate partitions of almost equal size. It starts from a vertex of a weighted network graph, and adds that neighbour vertex to the cluster, which provides the minimal number of edges to the vertices outside the cluster. If the cluster size is somewhat beyond its "optimal" value, then a backtracking step takes care that the number of edges crossing the border becomes minimal within a certain interval. This cluster is selected as a first partition, and the cluster process is restarted. The computational complexity of this method is $\mathcal{O}(n_V^2)$, where n_V is the number of vertices (SANGIOVANNI-VINCENTELLI, CHEN and CHUA [1977]).

A more accurate weight model for the computational cost of a partition is suggested in WALLAT [1997]. To this end, each circuit element is given a specific weight – depending on its computational complexity – and the cost of a partition is just the sum of the weights of its elements, plus a certain weight for each of its nodes, since the latter gives rise to one more circuit equation.

This model requires to formulate the cluster problem on hypergraphs which are weighted on both vertices and edges. It is however possible to extend the local clustering algorithm of SANGIOVANNI-VINCENTELLI, CHEN and CHUA [1977] properly, such that partitions with very well balanced computational cost can be generated even for large industrial designs in a reasonable time (WALLAT [1997]).

Partitions with even smaller numbers of interconnects can be expected from algorithms with a more *global* view. These operate usually in two steps: The first step provides an initial partitioning under global aspects, in order to meet the requirements 3 and 4. For this purpose bisection methods (COX, BURCH, HOCEVAR, YANG and EPLER [1991]), analytical placement (FRÖHLICH, RIESS, WEVER and ZHENG [1998]), or simply the hierarchy of the network description are used. In the second step the cut cost of each partition is reduced ($\rightarrow$ requirement 2) by shifting circuit nodes or branches between partitions. This is done using Fiduccia–Mattheyses like methods (COX, BURCH, HOCEVAR, YANG and EPLER [1991], ONOZUKA, KANH, MIZUTA, NAKATA and TANABE [1993]), minimizing ratio-cut (FRÖHLICH, RIESS, WEVER and ZHENG [1998]), or with some other heuristics (KAGE, KAWAFUJI and NIITSUMA [1994]).

Global partitioning methods often suffer from prohibitive runtimes when being applied to very large problems. As an alternative, a clustering algorithm was recently developed, which keeps global aspects in mind *and* aims at a very high computational efficiency (FRÖHLICH, GLÖCKEL and FLEISCHMANN [2000]). Basically it forms clusters by merging adjacent vertices (circuit elements) of an edge (circuit node) in a weighted modified network graph. For clustering, the simple edge weight criterion is replaced by a more sophisticated coupling measure, which takes care that adjacent vertices with only a few edges are preferred for clustering, and that the cluster size is well balanced. For each merging step the whole circuit is inspected; this brings the global aspects into account and finally enables excellent partitioning results in a reasonable time, as is demonstrated with a large number of actual designs from industry (FRÖHLICH [2002]). The method has a complexity of $\mathcal{O}(n_R \cdot n_V \cdot \log(n_R \cdot n_V))$, where n_R is an average number of edges per vertex, and n_V is the number of vertices.

Dynamic assignment techniques were explored in an experimental paper (COX, BURCH, HOCEVAR, YANG and EPLER [1991]), in order to check how far latency effects can be exploited, and which maximal degree of parallelism can be obtained. To this end the number of partitions was made quite large, and by using partial LU factorization and dynamic subtask allocation, $\approx 95\%$ of the total job could be executed in parallel, which is hard to obtain with static assignment. As a conclusion, dynamic resource allocation was recommended on shared memory machines, while static assignment fits better to clusters with distributed memory and slow interconnect network. It would be interesting to check if these results still hold for very large actual designs. Furthermore, it should be noted that dynamic allocation schemes require considerable programming efforts for the simulator itself, while partitioning requirements are less ambitious.

TABLE 15.1
Typical speedups with parallel MLN versus serial standard Newton

No of processors	Speedup
4	$2.5\ldots3$
8	$3.5\ldots5$
12	$5\ldots7$
16	>7

Results. The speedup obtainable with this kind of parallelization depends primarily on the circuit structure and on the quality of partitioning. In the best cases – with partitions providing equal workload and a small number of pins – almost linear speedups were reported, e.g., a factor 7.79 on 8 CPU's (FRÖHLICH, RIESS, WEVER and ZHENG [1998]). Sometimes even superlinear speedups can be observed, which is due to the shift of nonlinearity into smaller circuit blocks, or due to reduced memory needs. In the worst case of unbalanced partitions and large interconnect to the master circuit, the speedup may not be much larger than 1. Fortunately, the user can see in advance whether it makes sense to start parallel simulation with a given partitioning.

Typical speedups for real life applications are given in Table 15.1 (WEVER and ZHENG [1996], WALLAT [1997], FRÖHLICH, RIESS, WEVER and ZHENG [1998], FRÖHLICH, GLÖCKEL and FLEISCHMANN [2000]). Note that it often does not make much difference if the problem is run on a shared memory multiprocessor system or on a cluster of processors with relatively slow interconnect network. The latter aspect is of commercial interest, since it allows to set up a very cheap cluster of fast PCs for running large circuit simulations efficiently.

A reasonable number of processors is actually between 4 and 16. Each processor should have a fairly large load, since otherwise there is a risk to get a poor ratio of interconnect to partition size, making parallelization inefficient. Therefore scalability is limited: Increasing speedups can only be expected for an increasing number of processors, if the problem size is increasing as well (WEVER and ZHENG [1996]).

The runtime of an advanced partitioning tool is $1\ldots10$ min for circuits containing $15k\ldots150k$ transistors. This makes its possible to run several partitioning trials with different options, and to select the best partition found for performing the analysis.

Even more important than exact speedups is the chance to handle problems of a size which is almost one order of magnitude larger than with serial simulation. An actual example is a $500k$ transistor circuit including parasitics, which can be simulated over night on a 12 processor machine, giving full confidence in its functionality to the designer DENK [2002].

Note that the results reported here were obtained with a fully implicit integration method like BDF or TR-BDF. Semiimplicit numerical integration schemes – like the ROW method – do not require a parallel MLN method. However they can utilize the multi-level linear solver (VLACH [1988b]), which is naturally included in MLN, and so parallelization of the ROW method is achieved at almost no extra cost if a parallel MLN solver is available for fully implicit integration rules. Even partitioning is not affected by the particular choice for numerical integration.

15.3. Thread based parallelization: Loop over processes (threads)

This kind of parallelization is targeted for multiprocessor systems with a large shared memory. Hence interchange of data between processes is not of major concern. However cache effects are of great importance, and so it is essential to take care of data locality. We restrict on systems with a limited number of processors, as are commonly used in industrial environments.

From Table 12.1 we see that parallelization should focus on the load and on the linear solver part of a circuit simulator. Parallelization of the rest either does not impose any difficulties, or does not make sense due to its serial character and small runtimes.

Load. The load part consists of three tasks:

1. *Evaluation of the device characteristics* and their derivatives: This is an expensive, but perfectly parallelizable task, since all evaluations are independent from each other. Furthermore there are always many elements of the same kind (transistors, capacitors, ...); hence load balancing is trivial.

2. *Numerical integration* in case of BDF like methods; this is easy to parallelize as well.

3. *Stamping*, i.e., adding the element contributions to matrix and right hand side: This is some kind of protected operation, since different elements may stamp into the same entry of matrix or right hand side. Nevertheless, parallelization is necessary, since otherwise parallelization effects saturate already for 4...6 processors (SADAYAPPAN and VISVANATHAN [1988]). Possible solutions are (SADAYAPPAN and VISVANATHAN [1988], EICKHOFF [1991]):

- Edge colouring techniques can be applied: Circuit elements of the same kind sharing the same node are marked with different colours. Then all elements with identical colour can be stamped in parallel.
- The circuit is partitioned into equally sized subblocks with minimal number of interconnects between them. Elements of different subblocks not being at the border then can be stamped in parallel; the border elements can be stamped blockwise sequentially, partition per partition.
- The *stamp-into* operation is replaced by a *fetch-from* operation: Each entry of matrix and right hand side knows from which element it gets a contribution, and fetches it from there. All entries can act in parallel. If the number of elements connected to one node is very unbalanced, then some refinements of this method may be useful, e.g., the generation of subtasks.

In sum, parallelization of the load part can be done very efficiently, and good scalability can be expected.

Linear solver. The focus is here on parallel LU factorization, since forward backward substitution is far less expensive and easier to parallelize (FISCHER [2001]). Two main directions are pursued:

Tearing: The first approach is to partition the circuit with tearing methods into well balanced subblocks, and LU factorize the subblocks in parallel (SADAYAPPAN and VISVANATHAN [1988], VLACH [1988a], COX, BURCH, HOCEVAR, YANG and EPLER [1991]). This technique is closely related to the parallel multi-level Newton method described above. So we will not further discuss it here.

Clustering Gauss operations: The second approach is to cluster the Gauss operations of sparse LU factorization into sets of independent tasks, and perform all tasks of a set in parallel. These concepts are rapidly evolving at present due to their importance in a much more general framework. On overview can be found in Chapter 9 of this Handbook; an actual code is described in SCHENK [2000]. Note that in this framework the notation of parallel granularity (see, e.g., HEATH, NG and PEYTON [1990]) is different from what we have introduced at the beginning of this section.

We end with some comments on the second approach which directly concern circuit simulation aspects.

TABLE 15.2
Aspects of parallelization

Aspect	Thread based	Multi-level Newton
hardware	shared memory multiprocessor	– cluster of workstations or PCs – shared memory multiprocessor
hardware cost	moderate	cheap ... moderate
memory needs	large	small (eventually large for partitioner)
communication overhead	large	small
user handling	easy ... moderate	easy ... moderate ... difficult (depending on partitioner)
data flow	simple	complex on workstation cluster: – scatter partitions to slave processors – gather/merge resulting waveforms – restart and error management
scalability		
– small problems	good	no
– large problems	good ... moderate	moderate
spatial adaptivity (exploitation of latency)	low	– static assignment: low ... moderate – dynamic load balancing: high
algorithmic overhead	small	– Schur matrix – interconnect solver
algorithmic benefit	none	shift nonlinearity into subsystem
algorithmic challenge	partitioner: split Gauss operations	partitioner: split circuit
programming effort	moderate	– static assignment: low ... moderate – dynamic load balancing: high

Mixed direct/iterative linear solver: Parallelization of a mixed sparse direct/iterative linear solver was recently suggested in an interesting alternative, which directly aims at circuit simulation problems (BOMHOF [2001]), see Chapter 9 of this Handbook for details.

Adaptive partitioning: In the course of a transient analysis the linear solver is called quite often with an identical zero/nonzero pattern of the matrix. So it may be worthwhile to provide some learning phase in the algorithms, where partitioning is adapted until optimal speedups are obtained. In FISCHER [2001] such an adaptive partitioning method is described, which does not only provide significant speedup improvements, but also requires less CPU time than a conventional partitioner.

Ordering: The sparse LU factorization needs preordering of the matrix for minimal generation of fillins. Circuit simulation codes mostly employ the Markowitz method (MARKOWITZ [1957]), which is a variant of the minimum degree algorithms (DIRKS, FISCHER and RÜDIGER [2001]). For parallel LU factorization other methods may be better suited (BOMHOF [2001], REISSIG [2001]), although first experiments with nested dissection methods were not successful yet (FISCHER [2001]).

Thread based parallelization and parallel multi-level Newton MLN – a conclusion. Both schemes are realized in codes, which are used since some time in industrial environments, but there is no direct comparison available at present. As a first step, we try to compare them in Table 15.2 with respect to the most important aspects of parallelization, without giving numbers or assessments. If run on a shared memory system, MLN may be somewhat less efficient than a dedicated thread based version. The merits of MLN are an excellent performance/cost ratio and its flexibility, which even makes distributed simulation via Internet possible. One surprising fact is, that the most critical issue in both approaches is the partitioner, even if its objectives are somewhat different.

16. Hierarchical simulation

Very often the design of an electronic circuit is characterized by a hierarchical organization of models and submodels with active and passive components as final leafs. Compact transistor models (devices) are treated as building blocks in the modular design of the circuit. Submodels and devices are linked to the hierarchy by their terminal unknowns to the enclosing model. A device is a leaf of the tree. The top model is the circuit-level, which has only one terminal, the ground node (whose voltage is set to 0). Models and devices may have their own internal unknowns. The behaviour of the solution of a model at all nodes is completely determined by the values at the terminals (together with the internal sources) and the nonlinear interaction with its containing submodels and devices.

A hierarchically organized algorithm allows a datastructure of the circuit that is very close to the original design (RABBAT and HSIEH [1981], TER MATEN [1999], WEHRHAHN [1989], WEHRHAHN [1991]). A hierarchical formulation corresponds to a particular block partitioning of the problem that allows for parallelism in a natural way (BORCHARDT, GRUND and HORN [1997]). Even in a sequential approach particular algorithms can be pursued, starting with the observation that the overall matrix and

---

for all $i = 0, \ldots, I - 1$ (Time step iteration) **do**
  for $k = 0, \ldots, K - 1$ (Newton iteration) **do**
    Recursion I: Bottom-Up Matrix Assembly $[A\mathbf{x}^{n+1} = \mathbf{b} \equiv -\mathbf{F}(\mathbf{x}^n) + A\mathbf{x}^n]$
    and Decomposition $[A = UL,\ L\mathbf{x}^{n+1} = \mathbf{c} \equiv U^{-1}\mathbf{b}]$
    Recursion II: Top-Down Linear Solution $[\mathbf{x}^{n+1} = L^{-1}\mathbf{c}]$
    Recursion III: Bottom-Up error estimation
  end for
  Recursion III: Bottom-Up discretization-error estimation
end for

FIG. 16.1. The main hierarchical recursions.

the solution can be distributed over all hierarchical levels. Algorithms are usually defined recursively (see algorithm in Fig. 16.1). Depending on actions being done before or after the recursions and passing data to or lifting data from a submodel, or device, one can speak of Top-Down and of Bottom-Up recursions.

In the algorithm above Gaussian elimination is used because it nicely fits a hierarchical algorithm. This is in contrast to several iterative linear solvers in which needed preconditioners disturb the assumed block structure (however, for a collection of some recent results by some hierarchical-friendly methods, see SCHILDERS [2000]).

Bypass mechanisms. Each hierarchical branch normally depends continuously on the values of the terminals at some top-level, in addition to values of internal sources and values of time-derivatives. This allows for several forms of *bypassing* where we satisfy ourselves not to update results obtained previously in some part of a process.

- *Newton-level bypassing*: In a Newton–Raphson iteration one can decide to bypass a complete hierarchical branch starting from submodel S when its terminals do not change that much.

$$\mathbf{x}_{S,j}^{n+1} = \mathbf{x}_{S,j}^{n} \quad \text{if } \left\| \mathbf{x}_{M,i}^{n+1} - \mathbf{x}_{M,i}^{n} \right\| < \varepsilon \tag{16.1}$$

  where S denotes a submodel or device, and M the encompassing model. At this highest level i ranges from 1 to n_t^M (number of terminal unknowns of S at level M). At each sublevel S, j ranges from 1 to $n_t^S + n_i^S$ (where n_t^S, n_i^S are the number of terminal and internal unknowns at level S, respectively).

  Clearly, by this one can re-use matrix-contributions and right-hand side contributions from all submodels of which the tree starts at model M. Depending on the type of linear solver one also can re-use the local LU-decompositions.
- *Transient step bypassing*: The bypass approach may be extended to a transient step, when the extrapolated values (or the result of the predictor) indicate results close to the final one at the previous time level.
- *Cross-tree bypassing*: The above bypass approaches are examples of *in-tree bypassing*: one can only bypass a remainder of the hierarchical branch by comparing it to a previous approximation to the solution of the same branch.

A generalization to this might be called *cross-tree bypassing*. For this one identifies branches that formally have identical datastructures, for instance because each branch starts at different occurrences of a same model or device definition.

When one has determined the solution of one branch, its results may be copied to the other branch (when needed, one might postpone this). Note that this applies to the Newton as well as to the Transient step level (in Transient analysis one might also apply cross-tree bypassing during the stepsize determination).

The HSIM simulator, developed by Nassda Corporation (HSIM DATASHEET [2002]), exploits this type of bypassing. It efficiently stores repeated instances of the same subcircuit, providing by this "unlimited" design capacity (compared to flat-based simulators). It takes advantage of hierarchical structures operating under the same conditions in the design to dynamically reuse computed results (WANG and DENG [2001]). In mixed-signal analysis the bulk of the circuit is of a digital nature and only a minor part is a true analog part. In the digital part, a lot of branch matchings may occur, and also their boundary (terminal) values may be identical (in fact because of the digital nature, only very few different stages will be possible). Good speedups are reported when compared to conventional analog circuit simulators. In addition to bypassing, a hierarchical RC reduction algorithm compresses parasitic resistances and capacitances in the hierarchical database. Finally, a coupling decomposition algorithm efficiently models submodel couplings, such as crosstalk noises, as submodel interface currents and conductances.

17. Multirate integration

From our CMOS ringoscillator example in Section 13 we have seen that multirate integration offers significant speedup potential for circuit simulation. Waveform relaxation WR exploits the multirate behaviour in a very natural way: Subblocks are decoupled, and each of them can be integrated with its own local timestep. In standard circuit simulation however the subsystems are not decoupled. So when we solve a certain circuit part at a particular timepoint, we need information about the contribution of all other circuit parts, which is not available due to their different integration stepsize.

To get accurate, controllable, and cheap to compute estimates for these contributions has been a key problem in multirate integration.

Let us for the sake of simplicity assume that the circuit at a timepoint t can be separated into an active part x_A – which has to be integrated with a small timestep h – and a less active ("latent") part x_L, being integrated with a much larger timestep $H \gg h$. Then there are two different strategies to compute a solution in the time between t and $t + H$ (GEAR and WELLS [1984], SKELBOE [1984]):

- **Fastest first**: Integrate x_A with small stepsize h from t to $t + H$, using extrapolated values for x_L;
  then perform one integration step for x_L from t to $t + H$, taking interpolated values from x_A, if necessary.
- **Slowest first**: Integrate x_L with one step of size H, where $x_A(t + H)$ is extrapolated;
  then integrate x_A with small stepsizes h, taking interpolated values from x_L.

Both approaches are pursued in different implementations. While the first one seems to be straightforward – since it relies on the assumption that the slowly varying variables can be well extrapolated into future – offers the second computational advantages.

Roughly two directions can be recognized in the literature: One tries to extend standard circuit simulation techniques using multi-step integration methods; the second is oriented towards one-step methods. Both of them have in common, that for reducing overhead the circuit is partitioned into subblocks, each of which is handled with its own local timestep.

Multi-step methods. All multi-step methods known so far employ the fastest first principle, and make use of some kind of event control, see Fig. 17.1: Based on conventional timestep control, each subblock computes its next timepoint and puts it into an event list. The global timestep h is determined from the next entry of this event list, and depending on their own local stepsize h_i, the subblocks are marked to be active – if h_i is not much larger than h – or to be latent else. The active subblocks are evaluated in a conventional way, but the latent subblocks are replaced by some simple substitutes, which aim at extrapolating their terminal behaviour. With these substitutes, the reduced system is solved, and after getting convergence the latency assumption has to be verified a posteriori. The latter step is important to maintain the reliability of the standard algorithm. It may give rise to roll back the simulation for several timesteps, which is very critical since it degrades performance significantly.

- *Initialization*:
  - partition circuit into subblocks
  - compute initial values
  - setup and initialize event list
- *Transient simulation*:
  for each entry of event list do
  - mark subblocks to be active or latent
  - *Newton loop*:
    until convergence do
    * load matrix and right hand side for active subblocks as usual
    * load matrix and right hand side for substitute circuits of latent subblocks
    * solve reduced linear system
    * add Newton correction to active part of circuit variables
    * check convergence
  - *timestep control and verification step*:
    for all active subblocks do
    * perform timestep control
    enddo
    for all latent subblocks do
    * check latency assumption
    enddo
    update event list
  enddo

FIG. 17.1. Event controlled multirate integration with multi-step methods.

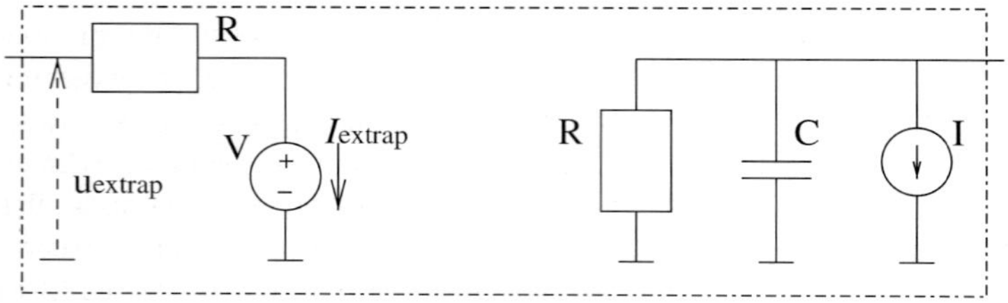

FIG. 17.2. Substitute circuits at the pins of latent subblocks (a) (FELDMANN, WEVER, ZHENG, SCHULTZ and WRIEDT [1992]) (left), (b) COX, BURCH, YANG and HOCEVAR [1989]) (right).

Alternatives for the substitute circuits are shown in Fig. 17.2. In (a) the value of R is fixed, and V is determined such that the extrapolated values for both pin current and voltage are consistent. (b) is just the Norton equivalent of the subblock at the pin node, i.e., $G = 1/R$ and C are its static and dynamic entry in the Jacobian, and I is the right hand side entry from the last iteration in active mode. Another approach suggests independent sources with extrapolated values for pin currents or voltages (SAKALLAH and DIRECTOR [1985]).

The speedup potential for this kind of multirate integration is mainly determined by savings of expensive device evaluations for the latent parts, and by solving a smaller system of equations. On the other hand there is slowdown due to roll back steps and due to overhead of event control. Overall speedup factors $2\ldots20$ have been reported (SAKALLAH and DIRECTOR [1985], FELDMANN, WEVER, ZHENG, SCHULTZ and WRIEDT [1992], COX, BURCH, YANG and HOCEVAR [1989]), but obviously methods and codes are not yet mature enough to be used in standard industrial environments.

One-step methods. Multirate one-step schemes so far have aimed at systems where the whole dynamics can be described by an initial value problem of ordinary differential equations

$$\dot{y} = f(t, y), \quad y(t_0) = y_0, \quad y \in \mathbb{R}^n. \tag{17.1}$$

For the sake of simplicity, we concentrate our investigations on autonomous initial value problems, whose state vector $y \in \mathbb{R}^n$ is partitioned into only two parts: Latent components $y_L \in \mathbb{R}^{n_L}$ and a small number of active components $y_A \in \mathbb{R}^{n_A}$ with $n_A + n_L = n$ and $n_A \ll n_L$,

$$\dot{y}_A = f_A(y_A, y_L), \quad y_A(t_0) = y_{A0}, \tag{17.2a}$$

$$\dot{y}_L = f_L(y_A, y_L), \quad y_L(t_0) = y_{L0}. \tag{17.2b}$$

The active components y_A are integrated with a small stepsize h, the latent components y_L with a large stepsize $H = mh$. The realisation of multirate one-step schemes now depends not only on the underlying numerical scheme, but also on which part is integrated first and, crucially, how the coupling is done.

One origin of these method are the split Runge–Kutta schemes by RICE [1960]. Multirate extrapolation (ENGSTLER and LUBICH [1997a]) and Runge–Kutta (ENGSTLER and LUBICH [1997b]) schemes are successfully used in stellar problems by Engstler and Lubich. In GÜNTHER and RENTROP [1993, 1994] multirate Rosenbrock–Wanner (MROW) methods are used for VLSI applications of electrical networks. One shortcoming of all these multirate methods derived so far is the coupling between active and latent components by interpolating and extrapolating state variables, which inevitably decompose the underlying one-step method in a two-step procedure, and thus makes their implementation very difficult into existing simulation packages. Recently, a new answer on how to realize the coupling, was given by KVÆRNØ and RENTROP [1999] for explicit Runge–Kutta schemes: The internal stages are used to compute the coupling terms, too. Meanwhile, this so-called *generalized multirate* approach was extended to implicit schemes, e.g., ROW- and W-methods, to manage also stiff problems as arise in network analysis. One should note that the coefficients of all these one-step schemes can be chosen such that one gets stable methods of any prescribed order of convergence.

We start to give the outline of a somewhat generic generalized multirate one-step method: The approximate solution $y_L^H(t_0+H)$ of y_L at time point t_0+H and $y_A^h(t_0+(\lambda+1)h)$ of y_A at time points $t_0+(\lambda+1)h$ are given by

$$y_L^H(t_0+H)=y_{L,0}+\sum_{i=1}^{s}\tilde{b}_i k_{L,i},$$

$$y_A^h\big(t_0+(\lambda+1)h\big)=y_A^h(t_0+\lambda h)+\sum_{i=1}^{s}b_i k_{A,i}^{\lambda}\quad(\lambda=0,\ldots,m-1),$$

where the increments are computed via the linear systems

$$\begin{aligned}k_{L,i}&=Hf_L\left(y_{L,0}+\sum_{j=1}^{i-1}\tilde{\alpha}_{ij}k_{L,j},\boxed{\bar{Y}_{A,i}}\right)\\&\quad+H\frac{\partial f_L}{\partial y_L}\bigg|_{y_0}\sum_{j=1}^{i}\tilde{\gamma}_{ij}k_{L,j}+mH\frac{\partial f_L}{\partial y_A}\bigg|_{y_0}\sum_{i=1}^{i}\tilde{\nu}_{ij}k_{A,j}^{0},\\k_{A,i}^{\lambda}&=hf_A\left(\boxed{\bar{Y}_{L,i}^{\lambda}},y_{A,\lambda}+\sum_{j=1}^{i-1}\alpha_{ij}k_{A,j}^{\lambda}\right)\\&\quad+h\frac{\partial f_A}{\partial y_A}\bigg|_{y_{0,\lambda}}\sum_{j=1}^{i}\gamma_{ij}k_{A,j}^{\lambda}+\frac{h}{m}\frac{\partial f_A}{\partial y_L}\bigg|_{y_{0,\lambda}}\sum_{j=1}^{i}\nu_{ij}k_{L,j}.\end{aligned}$$

Still, the coupling terms need to be defined, where we aim at

- active to latent: $\bar{Y}_{A,i}\approx y_A(t_0+\tilde{\alpha}_i\cdot H)$,
- latent to active: $\bar{Y}_{L,i}^{\lambda}\approx y_L(t_0+\lambda\cdot h+\alpha_i\cdot h)$,

with $\alpha_i:=\sum_{j=1}^{i-1}\alpha_{ij}$. Depending on the way how coupling terms are computed, we get different types of multirate formulae:

MROW (GÜNTHER and RENTROP [1993]): The coupling terms are defined by the usage of rational extrapolations y_A^{extra} and y_L^{extra}, respectively:

- active to latent: $\boxed{\bar{Y}_{A,i}}=y_A^{\text{extra}}(t_0+\tilde{\alpha}_i H)$;

- latent to active: $\boxed{\bar{Y}^{\lambda}_{L,i}} = y_L^{\text{extra}}(t_0 + (\lambda + \alpha_i)h)$.

Furthermore $N := (\nu_{ij})_{i,j=1,\dots,s} = 0$, $\tilde{N} := (\tilde{\nu}_{ij})_{i,j=1,\dots,s} = 0$; the coefficient matrices $A := (\alpha_{ij})_{i,j=1,\dots,s}$, $\tilde{A} := (\alpha_{ij})_{i,j=1,\dots,s}$ are strict lower triangular, and $G := (\gamma_{ij})_{i,j=1,\dots,s}$, $\tilde{G} := (\tilde{\gamma}_{ij})_{i,j=1,\dots,s}$ are lower diagonal with nonvanishing diagonals. Last, the Jacobian is evaluated at: $y_{0,\lambda} = (y_L^{\text{extra}}(t_0 + \lambda h), y_A^{\text{h}}(t_0 + \lambda h))$, $\lambda = 0, \dots, m-1$. Thus the computation over each macro step is decoupled, a kind of weakened slowest first strategy (GEAR and WELLS [1984]).

Generalized Multirate (KVÆRNØ and RENTROP [1999]): The coupling terms are computed by their 'own' RK-like methods,

$$\boxed{\bar{Y}_{A,i}} = y_{A,0} + m \sum_{j=1}^{i-1} \tilde{\delta}_{ij} k^0_{A,j}, \quad \text{and}$$

$$\boxed{\bar{Y}^{\lambda}_{L,i}} = y_{L,0} + \frac{1}{m} \sum_{j=1}^{i-1} \big(\delta_{ij} + F_j(\lambda)\big) k_{L,j},$$

which gives us a genuine one-step method. Fixing, where to evaluate the Jacobian and some finer structure of the coefficient matrices, yields different kinds of methods:

- *explicit Runge–Kutta* (KVÆRNØ and RENTROP [1999]): $G = N = \tilde{G} = \tilde{N} = 0$; thus no Jacobian is coupled.
- *partitioned Runge–Kutta* (GÜNTHER, KVÆRNØ and RENTROP [2001]): $G = N = \tilde{N} = 0$; $\tilde{G}$ with nonvanishing diagonal.
- *W-method* (BARTEL [2000]): G, $\tilde{G}$, N and $\tilde{N}$ have constant diagonals, which differ from zero at least for the first two matrices; in addition $y_{0,\lambda} = y_0$, i.e., the Jacobian is lagged over a single macro step in order to compute the micros.
- *ROW-method* (BARTEL [2000]): Conditions like W-method, plus evaluation of the Jacobian on the fine grid.

Generalized multirate schemes yield a compound step of macro and first micro step, and decouple all later micro steps. By the linear implicitness, we may sequentially compute the increments $k_{L,i}$ and $k^0_{A,i}$. If at least one diagonal element of $N, \tilde{N}$ vanishes, a block triangular form of the system matrix for the increments is obtained, such that the increments may be computed in an interleaved mode: $k_{L,1}, k^0_{A,1}, k_{L,2}, k^0_{A,2}, \dots$. Furthermore, we have ROW-type coefficients, i.e., we need just one decomposition per timestep.

Combining both coupling approaches leads to a hybrid scheme (BARTEL, GÜNTHER and KVÆRNØ [2001]): Whereas the latent and the first active step are computed simultaneously in a compound step, the remaining active steps within one macro step can be computed by an arbitrary stiff method, iff dense output formulae of enough accuracy are used for evaluating the latent part. First steps have now been made to generalize this idea of "mixed multirating" to the charge/flux oriented DAE network equations (STRIEBEL and GÜNTHER [2002]). Although multirate one-step schemes show promising features to gain speedup in circuit simulation, the reliability and robustness of these schemes does not yet allow to use them in standard packages.

CHAPTER V

Periodic Steady-State Problems

Periodic Steady-State (PSS) Problems have received special attention for simulating analog circuits. The aim was to efficiently study solutions of problems where a highly oscillating signal (carrier) was modulated by another signal. Due to nonlinear components the response to a single tone may give rise to higher harmonics, which in general is considered as (harmonic) distortion. When two tones are considered, intermodulation distortion may arise. Then an IC-designer is interested in detecting the (group of) components that contribute most to the distortion. The same analyses also allow study of Electromagnetic Compatibility Immunity.

The above problems were studied by techniques in the time domain, in the frequency domain, or by mixed time-frequency domain methods. In the last years, Radio Frequency (RF) simulation initiated renewed focussing on simulating PSS problems, especially in the time-domain (DUNLOP, DEMIR, FELDMANN, KAPUR, LONG, MELVILLE and ROYCHOWDHURY [1998], KUNDERT [1997], TELICHEVESKY, KUNDERT and WHITE [1996], TELICHEVESKY, KUNDERT, ELFADEL and WHITE [1996]).

This section describes several algorithms for simulating PSS problems. This will include forced problems (i.e., periodicity caused by external sources) as well as free oscillator problems. However, we will point out also that the separate algorithms are a step in a larger process: distortion analysis, immunity analysis, noise analysis. A complete algorithm shows a cascading sequence of basic simulation methods (for instance for providing initial approximative solutions). Another feature is that algorithms are favoured that exploit re-use of existing implementations.

18. RF simulation

In the past decade there has been an exponential growth in the consumer market for wireless products. Products like pagers, cordless and cellular phones are now common products for consumers all over the world. But also computers are no longer connected to other computers and their peripherals by copper wires only: wireless computer networks are used more and more. Not yet very common but growing steadily are the wireless home systems, connecting all kinds of equipment present in peoples homes. Furthermore there are promising markets in the automotive area in vehicular navigation and inter-vehicular communication.

The change from mainly professional wireless applications (military, private mobile radio, etc.) to a consumer market has severe implications for the total design process.

Where in the past there was time to build and measure several prototypes, nowadays the demands on time-to-market, time-to-quality, price, production volume, etc. are so severe that designers have to resort to simulation. In a marketing window of only a few months there clearly is no time for several iterations of these systems-on-silicon.

Although the RF part of these systems constitutes only a minor part of the total design area, it presents a major challenge in the total design cycle. This challenge is caused by the analogue/RF nature of the design but also by the lack of appropriate tools, models and design flows. Because the demand for RF simulation tools on this scale is relatively new, the developments of tools (the underlying principles and the commercial implementation there of) are lagging behind the designers needs. It is clear that we are only in the start-up phase of RF tooling and RF design flow development. Nevertheless, recently a lot of progress has been made in the research of mathematical principles for RF simulation. A number of these new ideas are already available in industrial and commercial software.

18.1. RF circuit and signal characteristics

An RF circuit forms the link between some baseband information signal and an antenna. A transmitter modulates the baseband signal on a high frequency carrier (sinusoid) and the task of the receiver is to retrieve the baseband signal from the modulated carrier. Thus, as compared to baseband circuits, RF circuits are special in the sense that they process modulated carriers. In the frequency domain a modulated carrier is a narrow band signal where the absolute bandwidth is related to the frequency of the carrier signal and the relative bandwidth is related to the modulating baseband signal. Practically, the ratio of the two frequencies is in the order of 100 or 1000.

Another major difference is that in RF systems, noise is a major issue. Noise consists of the (usually) small unwanted signals in a system. One can think of several forms of device noise (thermal noise, shot noise, flicker noise) but also of interferers like neighbouring channels, mirror frequencies, etc. All noise sources are of major importance because they directly translate to bit-error-rates of the transmitted data. Therefore it is imperative that RF designers can predict the overall noise quickly and accurately.

When dealing with narrow band signals in a noisy environment two mechanisms are of major importance. Firstly, if a narrow band signal is passed through a nonlinearity, the spectrum will be repeated about integer multiples of the carrier frequency resulting in a very wide but sparse spectrum. Secondly, the signal will interact with other signals in the circuit leading to wanted and unwanted frequency shifts. Although both mechanisms are always present and even interact, the first mechanism is less important for small signal levels (e.g., noise).

18.2. RF building blocks

RF systems are typically built from a limited number of different building blocks: oscillators, mixers, amplifiers/filters, dividers and power amplifiers. When building or discussing special RF circuit simulation functionality it is important to first determine the characteristics of each building block and the information which should be obtained during simulation:

- Oscillators are autonomous circuits which serve as a frequency reference signal often of very high accuracy. Therefore the frequency itself must be determined accurately but it is also important to be able to determine the frequency behaviour over time, i.e., the phase noise. Physically the phase noise is caused by the device noise of the oscillator's components.
- Mixers perform a frequency shift on the input spectrum. Because of unwanted nonlinearities the input signal will not only be shifted but also distorted. Furthermore, the mixer will add noise to the signal, again generated by the devices in the circuit.
- Amplifiers and filters also suffer from unwanted nonlinearities and add noise to the signal.
- Dividers are used to modify a frequency reference signal for example coming from an oscillator. They are strongly nonlinear and they add phase noise to the signal.
- Power amplifiers are much like small signal amplifiers. However, depending on the modulation type and efficiency requirements they may be strongly nonlinear. Assessing the nonlinearity, especially in the frequency domain is important.

18.3. Requirements for simulating RF circuits

As mentioned earlier, noise is of major importance in RF circuit design. Depending on the required accuracy and application area the noise can be seen as a small, independent signal in the circuit. Much more often, however, the small noise signals interact with the large signals in the circuit resulting in frequency shifts of the noise spectra (noise folding). In a few cases the noise can not even be considered as a small signal but interacts with the other (noise) signals in a nonlinear manner. The RF designer must be able to simulate all these different views on noise but the second one is considered the most important.

Nonlinearity (harmonic distortion and intermodulation distortion) is mainly a measure for the behaviour of a circuit under unwanted strong disturbances which enter the system.

RF designers must be able to extract this information by simulating a design with reasonable turn-around times. This has to do with the actual computing time required for a simulation job but also addresses the robustness of the software. Equally important, however, is that the results are accurate and hence reliable.

19. The basic two-step approach

From the above it is clear that conventional SPICE-like simulators are not sufficient: transient simulation of RF circuits suffers from excessive CPU times because they have to deal with the absolute bandwidth of the signals and will therefore only be used when no alternatives are available (e.g., full nonlinear noise simulation including time domain transient noise sources). AC analysis can easily deal with the high bandwidths but does neither take into account nonlinearities nor frequency shifts.

The newly developed RF simulation methods all somehow exploit the 'sparsity' of the signal spectra. The basic method is that of determining the periodic steady-state (PSS) solution of a circuit. Conceptually this can be seen as a generalisation of the well-known DC operating point: for baseband circuits the spectral content around 0 Hz (the DC point) is important. For RF circuits the (narrow) spectral content around specific frequencies (of the PSS solution) is of interest. This PSS solution can be obtained in the frequency domain (f.i. by applying the harmonic balance method) or in the time domain (by methods, like shooting, based on transient simulation methods). With baseband simulation, after determining the DC point, additional simulations like AC, noise, etc. can be done to obtain more information about the circuit. Similarly, based on the PSS solution several other simulations can be done like periodic AC, periodic noise, etc. In view of the RF circuit and signal characteristics, the PSS solution determines the nonlinear behaviour of the circuit while the periodic AC, etc. deals with the frequency shift.

The main difference between the time domain and frequency domain methods to obtain the PSS solution is that the former can easily deal with strongly nonlinear circuits and discontinuities and have good convergence properties while the latter deal naturally with components characterised in the frequency domain. Over the years combinations of both basic methods were developed resulting in mixed time-frequency domain approaches each with their own advantages and drawbacks.

A *two-step approach* appears to be powerful as well as practical for simulating *RF mixing noise*:

- Determine the *noiseless Periodic Steady-State* (*PSS*) *solution* as large-signal solution. This can be done in the time domain, the frequency domain or by using mixed time-frequency methods. The time-domain representation is a time-varying solution.

  Of course, a noiseless PSS-analysis (with or without determining the oscillation frequency), has value on its own for RF simulations.
- Apply a linearisation around the PSS-solution and study noise as a *small signal perturbation*. The noise sources may have frequencies that are different from the PSS-solution.

For simulating *RF phase noise*, or *timing jitter* (i.e., shifts in zero crossings of the solution) in the case of free oscillators, to apply as second step a linearisation around the PSS-solution and study noise as small signal perturbation is of limited use (DEMIR, MEHROTRA and ROYCHOWDHURY [2000]). In fact, the results are only usefull for small t, because the resulting perturbations may grow large with time. But it allows that the noise sources may have frequencies that are different from the PSS-solution.

The *nonlinear perturbation analysis*, proposed in DEMIR, MEHROTRA and ROYCHOWDHURY [2000], is an alternative to the second step. Also in this approach, the first step is necessary. The nonlinear perturbation analysis results in a correct phase deviation. For the orbital deviation, again a linearisation around the PSS-solution (but including phase deviation) can be used. This implies that periodicity of the coefficients of the linear time varying differential equation can not be assumed. It also implies that, in general, the phase deviation is a time varying function.

After defining the PSS problem mathematically, we describe in the next two sections some methods for these phases in some more detail.

20. The PSS problem

For the charge/flux oriented network equations (10.1) in compact form, the Periodic Steady-State (PSS) problem for one overall period $T > 0$ is defined as:

$$\frac{\mathrm{d}}{\mathrm{d}t} q(t, x) + j(t, x) = 0 \in \mathbb{R}^N, \tag{20.1}$$

$$x(0) - x(T) = 0 \tag{20.2}$$

with $q(t) := A \cdot g(x)$ denoting charges assembled at the respective nodes and fluxes, and $j(t, x) := f(x, t)$ including the static part and sources as well. This implies that for all $t \in \mathbb{R}$, $x(t) = x(t + T)$. A function $x : \mathbb{R} \to \mathbb{R}^n$ is called a *Periodic Steady-State Solution* if there is a $T > 0$ such that x satisfies (20.1)–(20.2). Note that according to this definition, a stationary solution (called the DC, direct current, solution), i.e., a solution of the form $x(t) \equiv x_0$, is also a PSS solution.

To define precisely the PSS problem, we have to introduce the concept of *limit cycles* and to define *stability* for PSS solutions and limit cycles.

> The *limit cycle* $\mathcal{C}(x)$ of a PSS solution x is the range of the function $x(t)$, i.e.,
>
> $$\mathcal{C}(x) = \{x(t) \mid t \in \mathbb{R}\}. \tag{20.3}$$
>
> A set $\mathcal{C}$ is called a limit cycle of (20.1) if there is a PSS solution x of (20.1) so that $\mathcal{C} = \mathcal{C}(x)$.
>
> A PSS solution x is called *stable* (some authors prefer the term *strongly stable*) if there is a $\delta > 0$ such that for every solution x^* to (20.1) which has the property that
>
> $$\exists_{\tau_1 > 0} \| x^*(0) - x(\tau_1) \| < \delta, \tag{20.4}$$
>
> there exists a $\tau_2 > 0$ such that
>
> $$\lim_{t \to \infty} \| x^*(t) - x(t + \tau_2) \| = 0. \tag{20.5}$$
>
> A limit cycle is called stable when all of its periodic steady-states are stable.

Periodic steady-states solutions that are not stable are not interesting for the IC designer, since they do not correspond to any physical behaviour of the modelled circuit. In fact, we want to actively avoid nonstable periodic steady-states solutions for this reason.

An exception to the above might be the DC solution, which is the most well-known unstable solution. Also numerically the DC solution is of interest because it provides a way to find (approximate, initial) solutions for finding stable solutions, by perturbing the DC solution.

For forced, or driven, (i.e., nonautonomous) problems all explicity time-dependent coefficients and sources are periodic with a common (known) period T. When dealing with autonomous circuits (also called free-running oscillator circuits) the functions q and j do not explicitly depend on time and j does not involve time-dependent external sources

$$\frac{\mathrm{d}}{\mathrm{d}t} q(x) + j(x) = 0 \in \mathbb{R}^N, \tag{20.6}$$

$$x(0) - x(T) = 0. \tag{20.7}$$

Despite this, a time-varying periodic steady-state solution may exist for some particular value of T. When this solution is nontrivial, i.e., different from the DC-solution, we will call this solution the oscillation solution and ω_{osc} and f_{osc}, given by $\omega_{\mathrm{osc}} = 2\pi f_{\mathrm{osc}} = \frac{2\pi}{T}$, the angular and 'normal' oscillation frequency, respectively. In the autonomous case, solution and oscillation frequency have to be determined both. Mathematically, the problem is a nonlinear eigenproblem.

In the autonomous case, it is clear that when $x(t)$ is a solution of (20.6)–(20.7), another solution can simply be constructed by making a time-shift: $\tilde{x}(t) = x(t - t_0)$. To make the problem unique, in practice one gauges the solution by requiring that

$$e_i^\top x(t_0) = c \tag{20.8}$$

(for some coordinate i and constant c) [clearly c should be determined in the range of x, but not equal to a DC-value], or by imposing a condition on the time-derivative[9]

$$e_i^\top x'(t_0) = c. \tag{20.9}$$

Now the system (20.6), (20.7) and (20.8), respectively, (20.9) defines a nonlinear problem with a "unique" solution for the unknowns x, T: i.e., small time shifts are excluded.

Rescaling the time by writing $t = sT$, with $s \in [0, 1]$, we have

$$\begin{aligned} \frac{\mathrm{d}}{\mathrm{d}t} q\big(x(t)\big) + j\big(x(t)\big) &= \frac{1}{T}\frac{\mathrm{d}}{\mathrm{d}s} q\big(x(sT)\big) + j\big(x(sT)\big) \\ &= \frac{1}{T}\frac{\mathrm{d}}{\mathrm{d}s} q\big(\hat{x}(s)\big) + j\big(\hat{x}(s)\big) \end{aligned} \tag{20.10}$$

where $\hat{x}(s) = x(sT)$. Note that $\hat{x}(1) = x(T)$. Hence, the problem (20.6)–(20.7) can also be studied on the unit interval for the function $\hat{x}(s)$ after scaling the s-derivative by a factor $1/T$.

In fact, T can be nicely added to the system as well

$$\frac{1}{T}\frac{\mathrm{d}}{\mathrm{d}s} q\big(\hat{x}(s)\big) + j\big(\hat{x}(s)\big) = 0, \tag{20.11}$$

$$\frac{\mathrm{d}}{\mathrm{d}s} T = 0, \tag{20.12}$$

$$\hat{x}(0) = \hat{x}(1), \tag{20.13}$$

$$e_i^\top \hat{x}(0) = c. \tag{20.14}$$

(Clearly, T automatically fulfills the periodicity condition.)

---

[9]In the following, the prime $'$ will denote differentiation w.r.t. time.

21. Perturbation analysis

Before describing algorithms for solving a PSS-problem, in this section we will consider the problem for a subsequent perturbation analysis. The PSS-solution of (20.1) will be denoted by x_{PSS}. It will also be called the noiseless time-varying large signal solution. Now we perturb the left-hand side of (20.1) by adding some small (noise) function n

$$\frac{\mathrm{d}}{\mathrm{d}t}q(x) + j(t,x) + n(t) = 0 \in \mathbb{R}^N, \tag{21.1}$$

which results in a solution

$$x(t) = x_{PSS}\big(t + \alpha(t)\big) + x_n(t), \tag{21.2}$$

in which the phase-shift function $\alpha(t)$ still has to be prescribed and $x_n(t)$ is small.

21.1. Linear perturbation analysis for forced systems

Linearising (21.1) around x_{PSS} (i.e., considering the case $\alpha(t) = 0$), results in a Linear Time Varying (LTV) differential equation for x_n

$$\frac{\mathrm{d}}{\mathrm{d}t}\big(C(t)x_n\big) + G(t)x_n + n(t) = 0 \in \mathbb{R}^N, \tag{21.3}$$

$$C(t) = \left.\frac{\partial q(x)}{\partial x}\right|_{x_{PSS}}, \qquad G(t) = \left.\frac{\partial j(t,x)}{\partial x}\right|_{x_{PSS}}. \tag{21.4}$$

In practical applications, a basic noise term has the form

$$n(t) = B(t)b(t), \tag{21.5}$$

$$B(t) = B\big(x_{PSS}(t)\big) \tag{21.6}$$

that consists of a normalized perturbation function $b(t)$, which is modulated by the periodical function $B(t) = B(x_{PSS}(t))$. Here $b(t)$ may be defined most conveniently in the frequency domain, while the $B(x_{PSS}(t))$ is defined by expressions in the time domain.

The validity of this approach has been discussed by DEMIR, MEHROTRA and ROYCHOWDHURY [2000]. For forced systems the perturbed solution $x(t)$ can be approximated by (21.2) with α being identically zero and x_n the solution of (21.3). However, when dealing with free oscillators a nontrivial choice for the phase-shift function $\alpha(t)$ has to be made too.

We note that the coefficients in (21.3) are periodic in t with period T. Thus, they can be expanded in exponentials $e^{i\omega_k t}$, in which $\omega_k = 2\pi k/T$. It is instructive to consider the case for a simple sine-wave source, i.e., when

$$n(t) = Ue^{i\nu t}, \tag{21.7}$$

in which U does not depend on time, and $\nu = 2\pi f_n$, where f_n may be interpreted as a noise frequency, that may be different from the ω_k. Introducing $y_n(t) \equiv e^{-i\nu t}x_n(t)$ results in a linear DAE of which source term and (complex) coefficients (that depend on the parameter ν) are periodical with period T

$$\frac{\mathrm{d}}{\mathrm{d}t}\big(C(t)y_n\big) + \big[G(t) + i\nu C(t)\big]y_n + U = 0 \in \mathbb{R}^N. \tag{21.8}$$

When $x_{PSS}(t) \equiv x_{DC}$, and $[G(t) + i\nu C(t)]$ is regular (and time-independent), the solution y_n is time-independent and simply equals the well-known AC-solution. For the general case, we find that y_n and x_n have expansions of the form (see also OKUMURA, TANIMOTO, ITAKURA and SUGAWARA [1993], TELICHEVESKY, KUNDERT and WHITE [1995])

$$y_n(t) = \sum_{k=-\infty}^{\infty} y_{n,k}^{(\nu)} e^{i\omega_k t}, \tag{21.9}$$

$$x_n(t) = \sum_{k=-\infty}^{\infty} y_{n,k}^{(\nu)} e^{i(\nu+\omega_k)t}. \tag{21.10}$$

Because of the periodic coefficients in (21.3) and (21.8), the determination of the $y_{n,k}^{(\nu)}$ is called Periodic AC (PAC) analysis. The expansion of $x_n(t)$ implies that

$$x_n(t+T) = \beta(\nu)x_n(t), \quad \text{or} \tag{21.11}$$

$$x_n(0) = \beta(-\nu)x_n(T), \quad \text{where} \tag{21.12}$$

$$\beta(\nu) = e^{i\nu T}. \tag{21.13}$$

It is clear that, for a single input frequency ν, the solution $x_n(t)$ contains frequencies of the form $(\nu + \omega_k)$, i.e., frequency folding occurs. If we allow for several input frequencies ν_i, we can also say that a certain output frequency might originate from a large number of possible input frequencies. Hence, noise components at a certain frequency might end up in a different frequency band. This is why, for example, $1/f$ noise which has its main energy at low frequencies, still plays an important role in RF circuits.

It is important to note that we described a *linear* perturbation analysis and we will not find contributions with frequencies, like $(\nu_1 + \nu_2 + \omega_k)$, $(\nu_1 + 2\nu_2 + \omega_k)$ etc. This assumption is in general not a severe limitation when simulating noise in RF circuits.

In OKUMURA, TANIMOTO, ITAKURA and SUGAWARA [1993], TELICHEVESKY, KUNDERT and WHITE [1995] one considers the integration of (21.3) in which case the factor β easily allows adaptive re-usage of linear algebra used for solving the PSS-problem (see also BOMHOF and VAN DER VORST [2001], BOMHOF [2001]). However, the integration of (21.8) also gives rise to elegant algorithms.

21.2. Floquet theory

When dealing with perturbed *oscillatory* systems

$$\frac{\mathrm{d}}{\mathrm{d}t}q(x) + j(x) + n(t) = 0 \in \mathbb{R}^N \tag{21.14}$$

it is no longer possible to assume that small perturbations $\mathbf{n}(t)$ lead to small deviations in $x_{PSS}(t)$ (an instructive example is provided by considering $y'(t) + \cos(t)y(t) - 1 = 0$, of which the inhomogeneous solution is not periodic at all; however, note that $y(t+2\pi)$ still satisfies the differential equation). The main reason is that the *period* of the large signal solution is influenced by $n(t)$. This can lead to large (momentary) frequency deviations such that the difference between the noiseless and noisy solution can no longer be considered to be small.

This section gives the necessary background of Floquet Theory when applied to oscillatory problems and which provides a way to a proper perturbation approach (DEMIR, MEHROTRA and ROYCHOWDHURY [2000], DEMIR [1998], LAMOUR, MÄRZ and WINKLER [1998a], LAMOUR [1998]). We start by noting that $x'_{PSS}(t)$ satisfies the homogeneous part of (21.3)

$$\frac{\mathrm{d}}{\mathrm{d}t}\big(C(t)x\big) + G(t)x = 0. \tag{21.15}$$

We assume the case of index 1 DAEs. At the end of the section the higher index case is considered.

Independent solutions. Let,

$$\mathbf{S}(t) = \left\{ z \in \mathbb{R}^N \,\middle|\, \left(G(t) + \frac{\mathrm{d}}{\mathrm{d}t}C(t)\right) z \in \mathrm{Im}\big(C(t)\big) \right\}, \tag{21.16}$$

$$\mathbf{N}(t) = \mathrm{Ker}\big(C(t)\big). \tag{21.17}$$

Then one has

$$\mathbf{S}(t) \cap \mathbf{N}(t) = 0, \tag{21.18}$$

$$\mathbf{S}(t) \oplus \mathbf{N}(t) = \mathbb{R}^N, \tag{21.19}$$

in the index-1 case. We assume that $\mathbf{S}(t)$ is m-dimensional. There are N independent solutions of the homogeneous problem: $u_1(t)e^{\mu_1 t}, \ldots, u_m(t)e^{\mu_m t}, u_{m+1}(t), \ldots, u_N(t)$. The first $u_1(t), \ldots, u_m(t)$ are a basis of $\mathbf{S}(t)$; the last, $u_{m+1}(t), \ldots, u_N(t)$, are a basis of $\mathbf{N}(t)$. The $\mu_1, \ldots, \mu_m$ are so-called Floquet exponents; the $e^{\mu_1 T}, \ldots, e^{\mu_m T}$ are Floquet multipliers. For a stable autonomous index 1 problem we can assume that $\mu_1 = 0$ and that $\mathrm{Re}(\mu_i) < 0$ for $i = 2, \ldots, m$. In this case we can choose $u_1(t) = x'_{PSS}(t)$.

Adjoint problem. The homogeneous adjoint (or dual) system corresponding to (21.3) is

$$C^\top(t)\frac{\mathrm{d}}{\mathrm{d}t}y - G^\top(t)y = 0. \tag{21.20}$$

Similar to the not-adjoint case we introduce

$$\mathbf{S}^\top(t) = \left\{ z \in \mathbb{R}^n \mid G^\top(t)z \in \mathrm{Im}\big(C^\top(t)\big) \right\}, \tag{21.21}$$

$$\mathbf{N}^\top(t) = \mathrm{Ker}\big(C^\top(t)\big), \tag{21.22}$$

which have the properties

$$\mathbf{S}^\top(t) \cap \mathbf{N}^\top(t) = 0, \tag{21.23}$$

$$\mathbf{S}^\top(t) \oplus \mathbf{N}^\top(t) = \mathbb{R}^N. \tag{21.24}$$

Also $\mathbf{S}^\top$ is m-dimensional. The adjoint problem has N independent solutions: $v_1(t)e^{-\mu_1 t}, \ldots, v_m(t)e^{-\mu_m t}, v_{m+1}(t), \ldots, v_N(t)$, where $v_1(t), \ldots, v_m(t)$ are a basis of $\mathbf{S}^\top(t)$ and the last, $v_{m+1}(t), \ldots, v_N(t)$, are a basis of $\mathbf{N}^\top(t)$.

Bi-orthogonality. It is easy to verify that if x and y are solutions of (21.15) and (21.20), respectively, the inner-products $y^\top(t)C(t)x(t)$ are constant, thus $y^\top(t)C(t)x(t) = y^\top(0)C(0)x(0)$, for all $t \geqslant 0$. More specifically, the bases $u_1(t), \ldots, u_N(t)$ and $v_1(t), \ldots, v_N(t)$ can be chosen such that, the $N \times N$ matrix $U(t)$ with as columns the $u_i(t)$ and the $N \times N$ matrix $v(t)$ with as rows the $v_i(t)$ satisfy a bi-orthogonality relation w.r.t. $C(t)$ and a nearly one w.r.t. $G(t)$

$$v(t)C(t)U(t) = \begin{pmatrix} I_m & 0 \\ 0 & 0 \end{pmatrix}, \tag{21.25}$$

$$v(t)G(t)U(t) = \begin{pmatrix} J_m^1 & 0 \\ J_m^2 & J_m^3 \end{pmatrix}. \tag{21.26}$$

Here I_m is a $m \times m$ identity matrix. J_m^1 is a $m \times m$ block matrix. J_m^2 and J_m^3 are just suitable block matrices.

State-transition matrix, monodromy matrix. Assuming a consistent initial condition $x(0) = x_0 \in \mathbf{S}(0)$, the solution $x_H(t)$ of (21.15) can be written as

$$x_H(t) = \sum_{i=1}^{m} u_i(t) \exp(\mu_i t) v_i^\top C(0) x_0, \tag{21.27}$$

$$= \Phi(t, 0) x_0, \tag{21.28}$$

$$\Phi(t, s) = \Theta(t, s) C(s), \tag{21.29}$$

$$\Theta(t, s) = U(t) D(t - s) v(s), \tag{21.30}$$

$$D(t - s) = \operatorname{diag}\bigl(\exp\bigl(\mu_1(t - s)\bigr), \ldots, \exp\bigl(\mu_m(t - s)\bigr), 0, \ldots, 0\bigr). \tag{21.31}$$

If x_0 is not a consistent initial value, one can write $x_0 = x_0^{(S)} + x_0^{(N)}$, where $x_0^{(S)} \in \mathbf{S}(0)$ and $x_0^{(N)} \in \mathbf{N}(0)$. Clearly $C(0)x_0 = C(0)x_0^{(S)}$, and $x_H(t)$ depends on $x_0^{(S)}$, rather then on x_0. For calculating consistent initial values we refer to BRACHTENDORF, WELSCH and LAUR [1995].

An inhomogeneous solution of (21.3) can be written as

$$x_P(t) = x_H(t) + \sum_{i=1}^{m} u_i(t) \int_0^t \exp\bigl(\mu_i(t - s)\bigr) v_i^\top(s) B(s) b(s)\, \mathrm{d}s$$
$$+ \Gamma(t) B(t) b(t), \tag{21.32}$$

$$= x_H(t) + \int_0^t \Theta(t, s) B(s) b(s)\, \mathrm{d}s + \Gamma(t) B(t) b(t). \tag{21.33}$$

Here $\Gamma(t)$ is a matrix with $\operatorname{Ker}(\Gamma(t)) = \operatorname{span}(C(t)u_1(t), \ldots, C(t)u_m(t))$.

The monodromy matrix is the matrix $\Phi(t, 0)$ after one period, i.e., $\Phi(T, 0)$ (this matrix one naturally studies when one considers shooting methods or applies Floquet theory for analyzing stability of a limit cycle). Because of the periodicity of the u_i, we see that the $u_i(0)$, for $i = 1, \ldots, m$ are eigenvectors of the monodromy matrix with corresponding eigenvalues $\exp(\mu_i T)$, and that the remaining $u_i(0)$, for $i = m+1, \ldots, N$, are eigenvectors for the $(N - (m-1))$-fold eigenvalue 0.

The adjoint problem (21.20) has the state-transition matrix

$$\Psi(t,s) = v^\top(t) D(s-t) U^\top(s) C^\top(s), \tag{21.34}$$

$$= \sum_{i=1}^{m} \exp\bigl(-\mu_i(t-s)\bigr) v_i(t) u_i^\top(s) C^\top(s). \tag{21.35}$$

Similar to the not-adjoint case, the $v_i(0)$ are eigenvectors of the associated monodromy matrix $\Psi(T,0)$.

The higher index case. The index-2 case is discussed in LAMOUR, MÄRZ and WINKLER [1998b] for quasilinear problems, which is sufficient here. It turns out that not only the algebraic but also the hidden constraints (see the discussion in Section 10) have to be observed when setting up the state transition and monodromy matrix. Especially they have to start from consistent initial values. The latter can either be computed with the methods sketched in Section 10, or columnwise as eigenvectors of a generalized eigenvalue problem. Of course the Floquet multipliers are only those of the independent part of the monodromy matrix. The stability criterion is again that all Floquet multipliers have magnitude < 1, except that one which has magnitude 1 in case of autonomous oscillation.

REMARK. Sometimes the monodromy matrix in the higher index case is defined to comprise only the linear independent parts of $\Phi(T,0)$ i.e., the basis vectors of $\mathbf{S}(t)$ SELTING and ZHENG [1997]. This delivers the same information as before, but may save some memory space and computational effort for its calculation.

21.3. Phase noise by nonlinear perturbation analysis

Phase-shift function $\alpha(t)$. We will take $u_1(t) = x'_{PSS}(t)$. Let $\alpha(t)$ be a (sufficiently smooth) phase- or time-shift function and let $s = t + \alpha(t)$ be the shifted time. If $x_{PSS}(t)$ is the PSS-solution of (20.6) then the phase-shifted function $y(t) \equiv x_{PSS}(s) = x_{PSS}(t + \alpha(t))$ satisfies

$$\begin{aligned}\frac{\mathrm{d}}{\mathrm{d}t} q(y) + j(y) &= \frac{\mathrm{d}}{\mathrm{d}s} q\bigl(x_{PSS}(s)\bigr) \cdot \frac{\mathrm{d}s}{\mathrm{d}t} + j\bigl(x_{PSS}(s)\bigr)\\ &= \frac{\mathrm{d}}{\mathrm{d}x_{PSS}} q\bigl(x_{PSS}(s)\bigr) \frac{\mathrm{d}x_{PSS}}{\mathrm{d}s} \alpha'(t)\\ &= C\bigl(t+\alpha(t)\bigr) u_1\bigl(t+\alpha(t)\bigr)\alpha'(t).\end{aligned} \tag{21.36}$$

Hence, the phase shifted function y satisfies a perturbed DAE in which the right-hand side has a particular form. Here u_1 is the tangent to the orbit.

We now consider perturbations of the form $B(x(t))b(t)$ (cf. also (21.3)) to the original DAE (20.6)

$$\frac{\mathrm{d}}{\mathrm{d}t}q(x) + j(x) + B\big(x(t)\big)b(t) = 0 \tag{21.37}$$

and express $B(x(t+\alpha(t)))b(t)$ into its components using the basis $\{C(t+\alpha(t))u_1(t+\alpha(t)), \ldots, C(t+\alpha(t))u_m(t+\alpha(t)), G(t+\alpha(t))u_{m+1}(t+\alpha(t)), \ldots, G(t+\alpha(t))u_N(t+\alpha(t))\}$

$$\begin{aligned} B\big(x\big(t+\alpha(t)\big)\big)b(t) = {} & \sum_{i=1}^{m} c_i\big(x, \alpha(t), t\big)C\big(t+\alpha(t)\big)u_i\big(t+\alpha(t)\big) \\ & + \sum_{i=m+1}^{N} c_i\big(x, \alpha(t), t\big)G\big(t+\alpha(t)\big)u_i\big(t+\alpha(t)\big), \end{aligned} \tag{21.38}$$

$$c_i\big(x, \alpha(t), t\big) = \tilde{v}_i\big(t+\alpha(t)\big)b(t), \tag{21.39}$$

$$\tilde{v}_i(t) = v_i^{\top}(t)B\big(x(t)\big). \tag{21.40}$$

Here the scalar functions $\tilde{v}_i(t)$ are periodical in t with period T.

The first component of $B(x(t+\alpha(t)))b(t)$ will be used to determine $\alpha(t)$. We define $\alpha(t)$ to satisfy the nonlinear, scalar, differential equation

$$\alpha'(t) = -v_1^{\top}\big(t+\alpha(t)\big)B\big(x_{PSS}\big(t+\alpha(t)\big)\big)b(t), \quad \alpha(0) = 0 \tag{21.41}$$

$$= -\tilde{v}_1\big(t+\alpha(t)\big)b(t), \quad \alpha(0) = 0. \tag{21.42}$$

(See also already KÄRTNER [1989, 1990] where a first start was made to treat the phase noise problem in the time-domain.) In DEMIR, MEHROTRA and ROYCHOWDHURY [2000], DEMIR [1998] it is argued that, in first order, (21.37) has a solution of the form $y(t) + z(t)$, with α determined by (21.42), and where the orbital deviation $z(t)$ satisfies $\|z\|_\infty <$ Const.$\|b\|_\infty$ (and even $z(t) \to 0$ $(t \to \infty)$). However, the phase shift function $\alpha(t)$ may increase with time (clearly, if $N = 1$, $B \equiv 1, b(t) \equiv \varepsilon$, and $v_1^{\top}(t) \equiv \kappa$, then $\alpha(t) = \kappa\varepsilon t$).

Determination of v_1. Note that for finding α, we clearly have to know v_1. In DEMIR, LONG and ROYCHOWDHURY [2000] this crucial vector is called *Perturbation Projection Vector*, or *PPV*. It represents a transfer between the perturbation of the DAE and the resulting phase shift.

In DEMIR, MEHROTRA and ROYCHOWDHURY [2000] v_1 is determined by performing first an eigenvalue/eigenvector analysis of the monodromy matrix of the adjoint problem to obtain $v_1(0)$, and followed by time integration (backward in time). For distincting the proper initial value $v_1(0)$ from other eigenvectors that have eigenvalues close to 0, one can exploit the bi-orthogonality relation (21.25), because $v_1(0)$ must have a nontrivial C-inner-product with $u_1(0)$.

Another, direct, approach is found in DEMIR, LONG and ROYCHOWDHURY [2000]. It nicely fits a Finite Difference Method approach and again exploits the bi-orthogonality relation (21.25) in an elegant way.

Phase noise analysis. For deterministic perturbations one has to integrate (21.42). Because of this action in the time domain, all Fourier components of $n(t)$ are treated in a combined way.

However, for stochastic noise, such a detail is not necessary. In DEMIR, MEHROTRA and ROYCHOWDHURY [2000], DEMIR [1998] expressions for the power due to the noise are derived that depend on the asymptotic behaviour (i.e., for large t) of the variance var$[\alpha(t)]$. The authors derive power spectrum expressions that depend on the Fourier components of the PSS-solution x_{PSS}, on the DC-component of $v_1(t)$, and on the power spectrum of b. The power of the jth harmonic of x_{PSS} is preserved in the power of the 'asymptotic' jth harmonic of y (i.e., the shifted x_{PSS}). Consequently, by summing over j, we see that also the total power is preserved.

Orbital deviation. In fact, the orbital deviation function z can be analysed by a proper linear perturbation analysis (but with linearised equations which now have nonperiodic coefficients!). Because n also affects the phase shifted function, around which one linearises for studying the orbital deviations, there is no simple summation formula known for cumulative noise contributions.

Other approaches. Finally, we briefly mention some alternative approaches for determining phase noise. In DE SMEDT and GIELEN [1997] a technique based on careful sampling is described to find phase noise effects due to specific noise sources. In DE SMEDT and GIELEN [1997] phase noise is considered from a parameter dependency point of view and an averaging technique is described that works well on (but is also restricted to) finite time intervals and is of interest in behavioural modeling. In HAJIMIRI and LEE [1998] a less accurate, but faster phase noise model is described that neglects the occurrence of α at the right-hand side in (21.42).

22. Algorithms for the PSS problem

In this section we describe some algorithms for solving PSS problems (i.e., for solving the noiseless, time varying, large signal). A general overview of numerical methods for highly oscillating problems can be found in PETZOLD, JAY and YEN [1997].

As time integrator we restrict ourselves to a θ-method ($0 \leqslant \theta \leqslant 1$): for the explicit ODE system $\dot{y}(t) = f(x(t), t)$, one step to compute the approximate y_{n+1} at time point $t_n + h$ from the previous approximate y_n at t_n reads

$$\frac{y_{n+1} - y_n}{h} = \theta f(y_{n+1}, t_n + h) + (1 - \theta) f(y_n, t_n).$$

This class of methods includes the explicit Euler-forward method ($\theta = 0$), the Trapezoidal Rule ($\theta = 0.5$) and the implicit Euler-backward scheme ($\theta = 1$). For other methods, f.i. BDF-like ones, see WELSCH [1998], WELSCH, BRACHTENDORF, SABELHAUS and LAUR [2001].

22.1. Direct time integration methods

Ordinary time integration usually starts from the DC solution. For forced (nonautonomous) problems the time integration usually is very slow, because the step-size will be determined by the highest oscillating component of the solution. For these problems, the Finite Difference Method (FDM), the Shooting Method (SM), the Harmonic Balance (HB), or the Envelope approach provide much more efficient alternatives. However, in analysing autonomous, free oscillating, problems, time integration also shows a nice property in securing to find stable limit cycles. For this reason, in this subsection, we will concentrate on finding a free oscillating solution, by exploiting time integration. With extrapolation techniques one can speed up convergence. In the past this approach has been applied by SKELBOE [1982] (even already for circuit problems) and PETZOLD [1981]. In SMITH, FORD and SIDI [1987] extrapolation techniques were generalized to sequences of vectors and has resulted in methods like Minimal Polynomial Extrapolation (MPE) and Reduced Rank Extrapolation (RRE). It is worth noting that all these methods can be implemented very elegantly within existing circuit simulators.

The basic Poincaré method. The basic method for solving (20.6)–(20.7) is called the Poincaré-map method. First we note that the length of the period can be estimated by looking for periodic recurring features in the computed circuit behaviour. A possible recurring feature is the point at which a specific condition is satisfied. This is equivalent to carrying out a Poincaré-map iteration, see HAIRER, NØRSETT and WANNER [1987], Section I.16. The idea is to cut the transient solution $x(t)$ by a hyperplane. The hyperplane is defined by an affine equation of the form $x^\top(t)n = \alpha$, for some vector n and scalar α. This equation is called the *switch equation*. The situation is visualised in Fig. 22.1. The basic Poincaré-map method can now be described as follows. Let an approximate solution x_0 and a required accuracy tolerance $\varepsilon > 0$ be given. The approximated solution $\tilde{x}$ and period $\widetilde{T}$ is computed by:

$$i := 0,\ t_0 := 0,\ x_0 := \text{some initial guess for } x$$

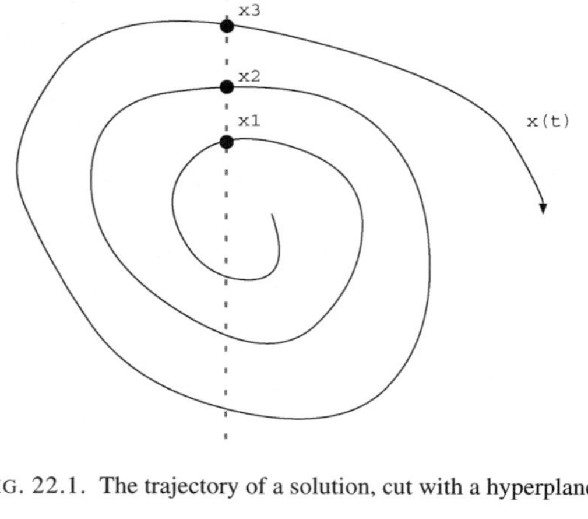

FIG. 22.1. The trajectory of a solution, cut with a hyperplane.

repeat

Starting with $t = t_i$, $x(t_i) = x_i$, integrate (19.6) until $(x(t), n) = \alpha$ and $\mathrm{d}(x(t), n)/\mathrm{d}t > 0$.

$x_{i+1} := x(t)$, $t_{i+1} := t$

$\delta := \|x_{i+1} - x_i\|$

$i := i + 1$

until $\delta \leqslant \varepsilon$

$\widetilde{T} := t_i - t_{i-1}$, $\tilde{x} := x_i$

The MPE accelerated Poincaré-map method. Let $x(t)$ be the solution of (20.6) with $x(0) = x_0$, and T_0 is the smallest $t > 0$ such that $(x(t), n) = \alpha$ and $\mathrm{d}(x(t), n)/\mathrm{d}t > 0$. Thus T_0 depends on x_0 as well.

Now we can define a function $F : \mathbb{R}^n \to \mathbb{R}^n$ by

$$F(x_0) := x(T_0). \tag{22.1}$$

The successive approximations of the Poincaré-map method satisfy the recursion relation

$$x_{n+1} = F(x_n). \tag{22.2}$$

This recursion is only in terms of the circuit state x; the period T does not appear explicitly in this iteration. Suppose that the sequence (22.2) converges linearly to some fixed point $\tilde{x}$ of F. A vector-extrapolation method to accelerate the basic method operates on the first k vectors of a sequence $\{x_n\}$, and produces an approximation y to the limit of $\{x_n\}$. This approximation is then used to restart (22.2) with $y_0 = y$ and the basic method generates a new sequence $y_0, y_1, y_2, \ldots$. Again, the acceleration method can be applied to this new sequence, resulting in a new approximation z of the limit. The sequence $x_0, y, z, \ldots$ converges much faster to the limit of $\{x_n\}$ than the sequence $\{x_n\}$ itself. Typically, if $\{x_n\}$ converges linearly, then $\{x_0, y, z, \ldots\}$ converges super-linearly.

A well-known acceleration method is minimal polynomial extrapolation (MPE). Rather than describing MPE here in detail, the reader is referred to SMITH, FORD and SIDI [1987]. For results with this approach we refer to HOUBEN and MAUBACH [2000, 2001], HOUBEN, TER MATEN, MAUBACH and PETERS [2001].

22.2. *Finite difference method*

The Finite Difference Method (FDM) solves the problem on a fixed time grid. Given is a number M, a series of M stepsizes $\{\Delta t_i\}$, implying a set of intermediate time-levels $\{t_i\}$ $(0 \leqslant i \leqslant M-1)$, where each t_i is the end point of the interval with length $\{\Delta t_i\}$. (See Fig. 22.2.) We assume that all t_i are contained in an interval of length T, that starts

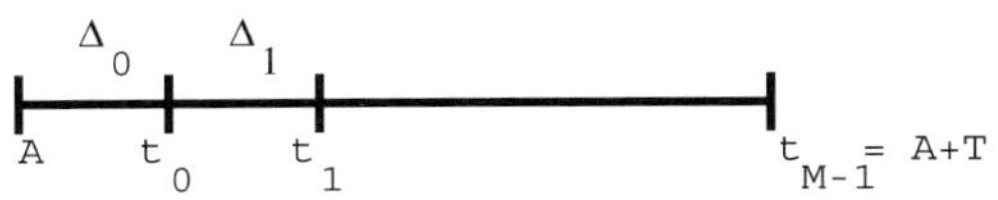

FIG. 22.2. Discretization of interval of length T, starting at A.

at $A = kT$ (for some $k \geqslant 0$). Thus

$$t_0 = A + \Delta t_0, \quad A = kT, \tag{22.3}$$

$$t_{M-1} = A + T, \tag{22.4}$$

$$t_i = A + \sum_{k=0}^{i} \Delta t_k, \quad i = 1, \ldots, M-1, \tag{22.5}$$

$$\Delta_k = \Delta t_k = t_k - t_{k-1}, \quad k = 1, \ldots, M-1, \tag{22.6}$$

$$\Delta_0 = \Delta t_0 = t_0 - (t_{M-1} - T). \tag{22.7}$$

Note that in general M will be available just at the start of the PSS-analysis. The periodicity is reflected in the definition of Δ_0.

We will write $\Delta_i = \Delta_i^s T$, for $\Delta_i^s \in [0, 1]$. Then $\sum_{k=0}^{M-1} \Delta_k^s = 1$. Clearly, with $t_i = s_i T$, solutions $\hat{x}(s)$ of the rescaled problem satisfy $\hat{x}(s_i) = x(t_i)$. Thus we will drop the ˆ and simply include the factor $1/T$ in the expressions when needed.

In the Finite Difference Method, M and the $\{\Delta_i^s\}$ will remain fixed during a complete (PSS-) Newton iteration.

For the next subsections we define

$$C(x) := \partial q(x)/\partial x, \qquad G(t, x) := \partial j(t, x)/\partial x \tag{22.8}$$

and we will write

$$C_i = C\big(x(t_i)\big), \qquad C_i^{(m)} = C\big(x^{(m)}(t_i)\big), \tag{22.9}$$

$$G_i = G\big(t_i, x(t_i)\big), \qquad G_i^{(m)} = G\big(t_i, x^{(m)}(t_i)\big). \tag{22.10}$$

22.2.1. *The basic FD-method*

The resulting discrete system of equations can be written as

$$F^D(x, T) = 0, \tag{22.11}$$

$$p^\top x - c = 0, \tag{22.12}$$

where $F^D : \mathbb{R}^{MN} \times R \to \mathbb{R}^{MN}$ is given by

$$\begin{aligned} F^D{}_0(x, T) &= \frac{q(x(t_0)) - q(x(t_{M-1}))}{\Delta_0} + \big[\theta j\big(x(t_0)\big) + (1-\theta) j\big(x(t_{M-1})\big)\big], \\ &= \frac{1}{T} \frac{q(x(t_0)) - q(x(t_{M-1}))}{\Delta_0^s} \\ &\quad + \big[\theta j\big(x(t_0)\big) + (1-\theta) j\big(x(t_{M-1})\big)\big], \end{aligned} \tag{22.13}$$

$$\begin{aligned} F^D{}_i(x, T) &= \frac{q(x(t_i)) - q(x(t_{i-1}))}{\Delta_i} + \big[\theta j\big(x(t_i)\big) + (1-\theta) j\big(x(t_{i-1})\big)\big], \\ &= \frac{1}{T} \frac{q(x(t_i)) - q(x(t_{i-1}))}{\Delta_i^s} \\ &\quad + \big[\theta j\big(x(t_i)\big) + (1-\theta) j\big(x(t_{i-1})\big)\big], \quad 1 \leqslant i \leqslant M-1. \end{aligned} \tag{22.14}$$

In (22.12), $p \in \mathbb{R}^{MN}$ is some given vector, usually a unit vector, in which case (22.12) only affects one time level. This last equation will only be imposed when dealing with free oscillators.

The above also offers the option to consider the frequency, $f = 1/T$, as natural unknown, rather than T. In that case

$$F^D(x, f) = 0, \tag{22.15}$$

$$p^\top x - c = 0, \tag{22.16}$$

where $F^D : \mathbb{R}^{MN} \to \mathbb{R}^{MN}$ is given by

$$F^D{}_0(x, f) = f\frac{q(x(t_0)) - q(x(t_{M-1}))}{\Delta_0^s} + \left[\theta j\big(x(t_0)\big) + (1-\theta) j\big(x(t_{M-1})\big)\right], \tag{22.17}$$

$$F^D{}_i(x, f) = f\frac{q(x(t_i)) - q(x(t_{i-1}))}{\Delta_i^s} + \left[\theta j\big(x(t_i)\big) + (1-\theta) j\big(x(t_{i-1})\big)\right], \quad 1 \leqslant i \leqslant M-1. \tag{22.18}$$

In the RF-case, one has $f \gg 1$ and thus $\frac{1}{T^2} \gg 1$. Hence, the f-variant behaves better scaled and we will restrict ourselve to that formulation.

Applying Newton–Raphson yields

$$\begin{pmatrix} Y^{(k)} & F^{(k)} \\ p^\top & 0 \end{pmatrix} \begin{pmatrix} x^{k+1} - x^k \\ f^{k+1} - f^k \end{pmatrix} = -\begin{pmatrix} F^D(x^k, f^k) \\ p^\top x^k - c \end{pmatrix} \tag{22.19}$$

in which

$$F^{(k)} = \frac{\partial}{\partial f} F^D\big(x^k, f^k\big), \tag{22.20}$$

$$Y^{(k)} = \frac{\partial}{\partial x} F^D\big(x^k, f^k\big) = L + B. \tag{22.21}$$

Here L and B are given by

$$L = \begin{bmatrix} \frac{C_0^{(k)}}{\Delta t_0} + \theta G_0^{(k)} & & & \\ -\frac{C_0^{(k)}}{\Delta t_1} + (1-\theta) G_0^{(k)} & \frac{C_1^{(k)}}{\Delta t_1} + \theta G_1^{(k)} & & \\ & \ddots & \ddots & \\ & & -\frac{C_{M-2}^{(k)}}{\Delta t_{M-1}} + (1-\theta) G_{M-2}^{(k)} & \frac{C_{M-1}^{(k)}}{\Delta t_{M-1}} + \theta G_{M-1}^{(k)} \end{bmatrix},$$

$$B = \begin{bmatrix} 0 \ldots & 0 & -\frac{C_{M-1}^{(k)}}{\Delta t_0} + (1-\theta) G_{M-1}^{(k)} \\ 0 \; 0 & & \\ \ddots & \ddots & \\ & 0 & 0 \end{bmatrix}. \tag{22.22}$$

For the fixed period problem, i.e., the nonautonomous case, we just drop the row for $p^\top$ and the column for F and come to

$$Y^{(k)}\left(x^{k+1} - x^k\right) = -F^D\left(x^k\right). \tag{22.23}$$

Some simple remarks apply:

- $C \equiv 0$ and $\theta = 1$: Then $B = 0$ and L is a block-diagonal matrix. The subsystems for each time level are decoupled and the solutions are the solutions obtained at an ordinary time integration, where the time dependent sources are evaluated at the proper time level. It is clear that when $C \equiv 0$, no oscillation can occur.
- In ordinary transient analysis the DAE character implies the necessary requirement $\theta \neq 0$. However, for the PSS-problem with the Finite Difference Method, $\theta = 0$ is a valid choice (because it is quite similar to $\theta = 1$, but viewed from the opposite time direction). For example: choose $M = 2$, $\Delta_i = T/2$, $C_i = \Delta_i C$, $G_i = G$, then the matrix $L + B$ looks like

$$L + B = \begin{bmatrix} C & -C + G \\ -C + G & C \end{bmatrix}. \tag{22.24}$$

  For commuting C, G (for instance $C = \mathrm{diag}(1, 0)$, $G = \mathrm{diag}(1, 1)$), the matrix $L + B$ has (nonzero) eigenvalues $\lambda_C + i\lambda_{-C+G}$.
- However, the DAE nature forbids to choose $\theta = 0.5$, because it makes the linear system singular! For seeing this, assume $C_i \equiv C$, $G_i \equiv G$ (constant), and an equidistant discretization with stepsize Δ (and with M odd). Define $C' = \frac{C}{\Delta}$, $G' = 0.5G$. Then

$$L = \begin{bmatrix} C' + G' & & & \\ -C' + G' & C' + G' & & \\ & \ddots & \ddots & \\ & & -C' + G' & C' + G' \end{bmatrix}, \tag{22.25}$$

$$B = \begin{bmatrix} 0 & \dots & 0 & -C' + G' \\ 0 & 0 & & \\ & \ddots & \ddots & \\ & & 0 & 0 \end{bmatrix}. \tag{22.26}$$

  When C' is singular there is a nontrivial vector v such that $C'v = 0$. Then also the large system is singular because $(L + B)w = 0$ for

$$w = (v, -v, v - v, \dots, v, -v)^\top. \tag{22.27}$$

  The trapezoidal rule looks to the mean of two subsequent function values and for this reason one can always add a zig-zag solution to such a "mean" value.

  Hence in practice one will have to take $\theta > 0.5$ and the choice is a trade-off between a better time-integration, but a nearly singular matrix, and more damping (and less order of time-integration), but with a better conditioned matrix.

We can rewrite the system (22.23) as

$$(L + \beta B)x = y, \tag{22.28}$$

in which $\beta = 1$. When studying linearizations around a PSS-solution responses to Fourier source terms give rise to linear systems in which β is complex, but satisfies $|\beta| = 1$ (see Section 21).

Block-Gaussian elimination allows to re-use direct solver modules from a circuit simulator. This way of decomposing the matrix meets a requirement that only memory for a limited number of full block matrices is used. In this way it is a sparse method. However, it may not be the most optimal LU-decomposition from the point of view of numerical stability, because not the most optimal pivots may be used. For several ideas we refer to ASCHER, MATTHEIJ and RUSSELL [1998] (Chapter 7), and BOMHOF and VAN DER VORST [2001], BOMHOF [2001] (for parallelizable algorithms).

The (block) lower-triangular matrix L is nonsingular and can be used as preconditioner for the matrix $(L + B)$ when using a Krylov space method (SAAD [1996]). For this case one needs to be able to determine $L^{-1}Bp$ for some vector p. For an iterative Krylov space method, the Krylov space can be extended by re-using the LU-decompositions of $\frac{C_i}{\Delta t} + G_i$ at each time-level.

For flat matrix circuit simulators, an efficient parallelizable GMRES-algorithm is described in BOMHOF and VAN DER VORST [2001].

22.2.2. FD for oscillator problem

For the oscillator problem, the sub-matrix Y in (22.21) is the same that one also encounters when applying the Finite-Difference Method to a forced Periodic Steady-State problem (with a fixed period T). From a software design point of view one would like to re-use software as much as possible. Indeed, when solving (22.19) a Block-Gaussian elimination procedure that uses Y^{-1} is attractive. Note that the complete Newton-matrix is nonsingular. In the limit, however, the sub-matrix Y in (22.21) becomes singular and one really needs (22.12) to gauge the complete problem.

In BRACHTENDORF, WELSCH and LAUR [1995], GOURARY, ULYANOV, ZHAROV, RUSAKOV, GULLAPALLI and MULVANEY [1998], WELSCH [1998] this problem was solved (in the frequency-domain) by introducing an artificial element in the circuit, a voltage source, of which the applied voltage E_{osc} had to be determined in such a way that the current through this source became 0. In that case the artificial element can be eliminated from the circuit and the solution on the remaining circuit gives the oscillator solution.

It is clear that such a voltage source can only be applied successfully at specific locations of the circuit. It is a requirement that for each value $E \neq E_{\text{osc}}$ a unique circuit solution results. When $E \to E_{\text{osc}}$, this unique circuit solution has to converge to the oscillator solution. In practice the user has to indicate where the oscillation will be perceived. This is not a drawback, because an IC-designer knows very well to choose a node where the oscillation occurs (as second node one can always use the ground node).

The approaches in BRACHTENDORF, WELSCH and LAUR [1995], WELSCH [1998] were considered more closely in BRACHTENDORF, LAMPE and LAUR [2000], LAMPE, BRACHTENDORF, TER MATEN, ONNEWEER and LAUR [2001]. Here also recommendations for increasing robustness were derived. In HOUBEN [1999], a similar approach was followed in the time-domain. We will consider these approaches more closely in the next subsections.

Artificial voltage source in the time-domain. The additional voltage source will be put between the nodes a and b. We assume the circuit unknowns to be ordered in such a way that at each time level $x(t) = (\ldots, x^a(t), x^b(t), i(E)(t))^\top$, where $x^a(t), x^b(t)$ are the voltage values at time level t at the nodes a and b respectively, and $i(E)(t)$ is the current through the artificial element $E(a,b)$ (thus $i(E)$ is the $(N+1)$th unknown). Let $E(a,b)(t_i) = \varepsilon_i$. The ε_i have to be determined in such a way that when the time profile of the voltage difference between the nodes a and b is identical to the time-varying voltage difference of the oscillator solution, the time profile of the current through the element is identically 0. In that case $x(t) = (\ldots, x^a(t), x^b(t), 0)^\top$, in which the part with the first N coordinates is identical to the oscillator solution at time level t. Because the artificial source E is added to the circuit, $i(E)$ does not occur as a controlling electrical variable in the user defined expressions. This implies that on each time level t_i

$$G_i = \begin{pmatrix} & & & \vdots \\ & 0 & & 1 \\ & & & -1 \\ \ldots & 1 & -1 & 0 \end{pmatrix}, \qquad C_i = \begin{pmatrix} & & & \vdots \\ & 0 & & 0 \\ & & & 0 \\ \ldots & 0 & 0 & 0 \end{pmatrix} \tag{22.29}$$

and

$$\frac{\partial j(x(t_i), \varepsilon_i)}{\partial \varepsilon_i} = -1, \qquad \frac{\partial q(x(t_i), \varepsilon_i)}{\partial \varepsilon_i} = 0. \tag{22.30}$$

In addition, the added equation on time level t_i is

$$i(E) = 0. \tag{22.31}$$

When the complete set of unknowns is written as $(x^\top(t_0), \ldots, x^\top(t_{M-1}), f, E^\top)^\top$, in which $E = (\varepsilon_0, \ldots, \varepsilon_{M-1})^\top$, Newton–Raphson can be formulated as

$$\begin{pmatrix} Y^{(k)} & F^{(k)} & \widetilde{\mathcal{E}} \\ p^\top & 0 & 0 \\ \mathcal{E} & 0 & 0 \end{pmatrix} \begin{pmatrix} x^{k+1} - x^k \\ f^{k+1} - f^k \\ E^{k+1} - E^k \end{pmatrix} = -\begin{pmatrix} F^D(x^k, f^k, \mathcal{E}^k) \\ p^\top x^k - c \\ \mathcal{E} x^k \end{pmatrix}. \tag{22.32}$$

Here

$$F^{(k)} = \left(\left(\frac{\partial}{\partial f} F_0^D(x^k, f^k) \right)^\top, \ldots, \left(\frac{\partial}{\partial f} F_{M-1}^D(x^k, f^k) \right)^\top \right)^\top, \tag{22.33}$$

$$\widetilde{\mathcal{E}} = \begin{pmatrix} -\theta e_{N+1} & 0 & \ldots & -(1-\theta) e_{N+1} \\ -(1-\theta) e_{N+1} & -\theta e_{N+1} & & 0 \\ \vdots & \ddots & \ddots & \vdots \\ 0 & & -(1-\theta) e_{N+1} & -\theta e_{N+1} \end{pmatrix}, \tag{22.34}$$

$$\mathcal{E} = \operatorname{diag}\left(e_{N+1}^\top, \ldots, e_{N+1}^\top\right). \tag{22.35}$$

Here x and p have length $M(N+1)$. Furthermore Y is a nonsingular $M(N+1) \times M(N+1)$ matrix that has a structure like in (22.21)–(22.22), but now based on the matrices G_i and C_i in (22.29), respectively. F is a vector of length $M(N+1)$, $\widetilde{\mathcal{E}}$ is a rectangular matrix of size $M(N+1) \times M$, and $\mathcal{E}$ is a rectangular matrix of size $M \times M(N+1)$ (M columns and M row-blocks of length $N+1$ each). Note that, for $\theta = 1$, one has $\widetilde{\mathcal{E}} = -\mathcal{E}^\top$.

Similar to (22.19), the linear system (22.32) can be solved using Block-Gaussian elimination that exploits the LU-decomposition of Y.

It seems natural to add the artificial voltage source to the same node as used for the gauge condition (20.8). This might indicate a possible source for conflicting requirements, because the gauge equation in its most simple form is a voltage or current condition (at a specific time level).

- Let us first consider the situation of a voltage condition. The basic point is that the (artificial) voltages ε_i are part of the Newton process and they will be tuned automatically such that in the limit they will not violate the gauge equation. More detailedly, the $p^\top$ in (22.32) might appear as a row m in Y. Then the corresponding entry F_m is zero. However, in $\widetilde{\mathcal{E}}$ we will find a (minus) one in the same row. This causes both rows to be independent. The corresponding ε_m will converge in one iteration, because of the linear dependency.
- Considering the situation of a current condition for the gauge equation, we remark that now $p^\top$ can not occur as row in Y. There is also no conflict with $\widetilde{\mathcal{E}}$, because $p^\top$ addresses real circuit unknowns known by the user, while $\mathcal{E}$ addresses the additional (artificial) circuit unknown $i(E)$. Hence $p^\top$ is also independent from the rows of $\mathcal{E}$.

In practice, one will put a resistor R in series with the artificial source E: the complete element $\widehat{E}(a,b)$ will act like a (linear) resistor $R(a,a')$ (of value R) and $E(a',b)$. Because of the linearity of $R(a,a')$, we can easily eliminate the unknown $x^{a'}(t)$ from the system. The effective Kirchhoff Voltage Law at time level t_i yields

$$x^a - x^b - Ri(E) - \varepsilon_i = 0. \tag{22.36}$$

The effect is that G_i in (22.29) simply changes into

$$G_i = \begin{pmatrix} & & & \vdots \\ & 0 & & 1 \\ & & & -1 \\ \dots & 1 & -1 & -R \end{pmatrix}. \tag{22.37}$$

The series resistance assures that no artificial voltage-shorts-inductor-loops are generated in the circuit. Note that the equation for $i(E)$ always has a nonzero diagonal element as pivot. In practice, $R = 1$.

Two-step approach. The two-step approach (BRACHTENDORF, WELSCH and LAUR [1995], GOURARY, ULYANOV, ZHAROV, RUSAKOV, GULLAPALLI and MULVANEY [1998], WELSCH [1998]) assumes that for given parameters f, E^k, the driven nonlinear problem $F^D(x(f,E^k)) = 0$ is solved. For updating f, E^k, Newton–Raphson can be used ("outer loop") in which one can exploit the Jacobian-matrix of the inner Newton–Raphson process for solving $F^D(x(f,E^k)) = 0$

$$\begin{pmatrix} p^\top \frac{\partial x}{\partial f} & p^\top \frac{\partial x}{\partial E} \\ \mathcal{E}\frac{\partial x}{\partial f} & \mathcal{E}\frac{\partial x}{\partial E} \end{pmatrix} \begin{pmatrix} f^{k+1} - f^k \\ E^{k+1} - E^k \end{pmatrix} = -\begin{pmatrix} p^\top x - c \\ \mathcal{E}x \end{pmatrix} \tag{22.38}$$

in which $\partial x/\partial f$ and $\partial x/\partial E$ are obtained by applying an ordinary sensitivity analysis to the inner, driven, problem. Here the Jacobian-matrix $Y = \partial F^D/\partial x$ of the inner Newton–

Raphson process is re-used in solving the systems

$$\frac{\partial F^D}{\partial f} + Y\frac{\partial x}{\partial f} = 0, \tag{22.39}$$

$$\frac{\partial F^D}{\partial E} + Y\frac{\partial x}{\partial E} = 0. \tag{22.40}$$

22.2.3. Normal projection of the Newton correction

In BRAMBILLA, D'AMORE and SANTOMAURO [1995] the idea is recalled to project the Newton correction for the circuit solution to become perpendicular to the orbit of the solution in some point t_0. Because the time derivative is the tangential derivative, this means that

$$\Delta x(t_0) \perp x'(t_0). \tag{22.41}$$

More generally, we may require that the overall inner product of Δx and $x'(t)$ is zero. This is similar to requiring

$$\sum_i \bigl(\Delta x(t_i), x'(t_i)\bigr) = 0. \tag{22.42}$$

In BRAMBILLA, D'AMORE and SANTOMAURO [1995] (22.41) is used to gauge the free oscillator problem rather than (22.12). Clearly, the algorithm has to find a t_0 where $x'(t_0) \neq 0$ (which will have to be approximated in practice). Near the limit solution this gauge excludes (small) time shifts. However, the algorithm does not exclude the DC-solution.

22.2.4. Initialization

For the driven problem, an initial timeprofile for the Finite Difference Method can be found by applying ordinary transient integration over several periods and collecting results at specific timepoints.

For the oscillator problem, the (Accelerated) Poincaré Method can be used to determine approximations for f and for a circuit solution as well – from these also initial values for the voltages as well as for the gauging value can be used. Note that FD uses a gauge value that may be different from the one used as switch value in Poincaré.

Alternatives can be found from pole-zero analysis and determining the eigenvector solutions of the dominant complex poles.

In Section 22.5, we will describe additional options when using Harmonic Balance.

22.3. Shooting Method

For the Shooting Method (SM), we define $F^S : \mathbb{R}^N \to \mathbb{R}^N$ by

$$F^S(x_0) = x(T),$$

with $x : [0, T] \to \mathbb{R}^N$ the solution of

$$\frac{\mathrm{d}}{\mathrm{d}t} q(x) + j(t, x) = 0, \quad \text{for } 0 \leqslant t \leqslant T, \tag{22.43}$$

$$x(0) = x_0. \tag{22.44}$$

For (single) shooting ('shooting-Newton' in TELICHEVESKY, KUNDERT, ELFADEL and WHITE [1996]) one has to solve

$$F^S(x_0) - x_0 = 0.$$

Using the Newton method (PSS Newton Process), one needs to evaluate $F^S(x_0)$ and the (monodromy) matrix $\Phi(x_0)$, defined by

$$\Phi(x_0) := \frac{\mathrm{d}F^S(x_0)}{\mathrm{d}x_0} \in \mathbb{R}^{N \times N}.$$

The calculation of $F^S(x_0)$ is in fact the result of a time integration. For instance, applying Euler-backward yields discrete equations at each time level, that are solved by an internal Newton method (*Time-Level Newton Process*):

$$\left(\frac{1}{\Delta t} C_{i+1}^{(m-1)} + G_{i+1}^{(m-1)}\right)\left(x_{i+1}^{(m)} - x_{i+1}^{(m-1)}\right)$$
$$= -\left\{\frac{q(t_{i+1}, x_{i+1}^{(m-1)}) - q(t_i, x_i)}{\Delta t} - j\left(t_{i+1}, x_{i+1}^{(m-1)}\right)\right\}. \tag{22.45}$$

Hence, this requires the solution of a system of linear equations with coefficient matrix $\frac{1}{\Delta t} C + G$. In fact this is a familiar process that is available in each conventional analog circuit simulator. The Newton matrix $\Phi(x_0)$ for the PSS Newton Process can be determined using a recursive procedure for the (matrix) quantities $\partial x_i / \partial x_0$

$$\left(\frac{1}{\Delta t} C(t_{i+1}, x_{i+1}) + G(t_{i+1}, x_{i+1})\right)\frac{\partial x_{i+1}}{\partial x_0} = \frac{1}{\Delta t} C(t_i, x_i)\frac{\partial x_i}{\partial x_0}, \tag{22.46}$$

$$\Phi(x_0) = \frac{\partial x(T)}{\partial x_0} = \frac{\partial x_M}{\partial x_0}. \tag{22.47}$$

The matrices C and G are rather sparse in contrast to the matrix $\Phi(x_0)$ that is rather full. In KUNDERT [1997], TELICHEVESKY, KUNDERT and WHITE [1995, 1996], TELICHEVESKY, KUNDERT, ELFADEL and WHITE [1996] the linear equations for the PSS-Newton are solved by means of a matrix-free method, by exploiting a Krylov-space method (GMRES or CGS). Here one needs to determine the result of $\Phi(x_0)p$, for some vector p, in order to extend the Krylov space. This can elegantly be done by a similar recursive procedure as above in (22.46), but now for a sequence of vectors. The charm of this recursion is that it re-uses the existing LU-decompositions of the Time-Level Newton Process; in addition the matrices C are needed. For GMRES a final least squares problem has to be solved. In fact, this has to be done in some flat-matrix structure. Assuming a k-dimensional Krylov space, the least squares problem is of order kN, where N is the number of unknowns in a flat circuit.

We collect some differences between the Shooting Method and the Finite Difference Method.

- In contrast to the Finite Difference Method, for the Shooting Method the time discretization can be chosen adaptively in a natural way, using the ordinary transient integration.

- The Shooting Method only needs an initial value to start from. But it may diverge rather fast in case of poles that cause instabilities.
- The FDM always assures periodicity for each iterand; in the limit also the discretized equations are satisfied. To contrast: each iterand of the Shooting Method satisfies the discretized equations, while reaching periodicity is the target of the method.
- In practice, the FDM is more stable than the Shooting Method, but – per iterand – it is much slower. The stability properties of shooting methods can be increased by applying multiple shooting that can be applied to the free oscillator problem as well (WELSCH [1998]).

The higher index case. From the remark concerning the higher index case at the end of Section 21.2, it follows that the shooting matrix should start from consistent initial values. Fortunately it can be shown (NEUBERT and SCHWARZ [2001]), that it is sufficient to start the calculation of $F^S(x_0)$ with $1(2)$ Backward Euler steps in case of DAE index $1(2)$. Alternatively, if the independent eigenvectors of the shooting matrix are known in the beginning, it is sufficient to calculate only the rectangular part comprising them (SELTING and ZHENG [1997], NEUBERT and SCHWARZ [2001], BAIZ [2003]).

Improving global convergence. Since the region of attraction for the PSS Newton Process is usually fairly small (see at the remarks concerning the differences between SM and FDM above), it is desirable to apply continuation methods which ensure global convergence under not too restrictive assumptions. A key issue here is that along the continuation path no bifurcations occur, which would make it difficult to track the "proper" solution branch. In BAIZ [2003] it is argued that this is best possible with an artificial homotopy

$$\rho(x, \lambda, a) := \lambda \cdot \big(F^S(x) - x\big) + (1 - \lambda) \cdot (x - a),$$

where λ, $0 \leqslant \lambda \leqslant 1$, is the homotopy parameter, and a is a start vector for the homotopy.

Using a theorem of Sard (see, e.g., CHOW, MALLET-PARET and YORKE [1978]) it is shown in BAIZ [2003] that under some reasonable assumptions for circuit models up to DAE index 2 the continuation path is smooth for almost any initial value a. So it can be traced with some kind of predictor-corrector techniques, starting from $\lambda = 0$ until the desired fixpoint $F^S(x) = x$ is obtained for $\lambda = 1$. The start vector a is obtained here from running a standard transient analysis over one cycle. For autonomous systems a gauging phase condition is added, while the frequency is an additional unknown.

22.4. *Waveform Newton*

In KEVENAAR [1994] the Waveform Newton Method has been described (for solving forced problems). Here one linearizes each time around a previously calculated periodic waveform $x^{(i)}$. This results in a linear DAE for the correction, in which the coefficients are periodic and depend on the last calculated waveform. From this we derive, that the next iterand $x^{(i+1)}$ satisfies

$$\begin{aligned}&\frac{\mathrm{d}}{\mathrm{d}t}\bigl[C\bigl(x^{(i)}\bigr)x^{(i+1)}\bigr]+G\bigl(t,x^{(i)}\bigr)x^{(i+1)}\\&\quad=-\left\{\frac{\mathrm{d}}{\mathrm{d}t}\bigl[q\bigl(x^{(i)}\bigr)-C\bigl(x^{(i)}\bigr)x^{(i)}\bigr]+\bigl[j\bigl(t,x^{(i)}\bigr)-G\bigl(t,x^{(i)}\bigr)x^{(i)}\bigr]\right\}.\end{aligned}\tag{22.48}$$

Similar to the Shooting Method case one can solve this linear DAE easily for an initial value of $x^{(i+1)}$ such that we have a periodic solution. Note that we can start with a nonperiodic waveform. All next iterands will automatically be periodic.

One can show, that, on a fixed grid, the Finite Difference Method and the above approach can generate the same solutions. However, the above approach, using the Shooting Method, allows to use adaptive integration. In this way both nice features of FDM (always periodic iterands) and of SM (adaptivity) are combined. As in FDM, each iterand is periodic, but only the limit satisfies the differential equations.

A nice feature is that the algorithm very elegantly extends to a Periodic AC analysis (see Section 21).

22.5. Harmonic Balance

Harmonic Balance (HB) is a nonlinear frequency-domain method for determining a periodic steady-state solution. The Fourier coefficients of the PSS are the solution of a nonlinear algebraic system of equations, that is usually solved by applying Newton's method. In the next we describe the method in some detail.

We assume d independent fundamental (angular) frequencies λ_j. Let $(.,.)$ denote the complex inner-product and Z be the set of integers. We write x (and similarly j and q) in an expansion of complex exponentials

$$x=\sum_{\omega_k\in\Lambda}X_k e^{\iota\omega_k t},\quad \text{with } \omega_k\in\Lambda\equiv\bigl\{\omega\mid\omega=(k,\lambda)\bigr\},\tag{22.49}$$

$$k\equiv(k_1,k_2,\ldots,k_d)^\top\in K\subset Z^d,\quad \lambda\equiv(\lambda_1,\lambda_2,\ldots,\lambda_d)^\top,\ \lambda_i>0,\tag{22.50}$$

where the (complex) X_k satisfies $X_{-k}=\overline{X}_k$.

Here λ and k are uniform for each component of x. The set K, containing integer tuples, is symmetrical about 0, while also $0\in K$. K is assumed to be finite. With K we denote the number of nonnegative (angular) frequencies (i.e., ω_k with $\omega_k\geqslant 0$). We also assume that all ω_k are different and $\omega_0=0$.

The choice of the fundamental frequencies λ_j will depend on the kinds of (modified) sine-wave sources used. We note that Λ should contain a sufficiently rich set of inter-distortion frequencies ω_k like $2\lambda_1, \lambda_1\pm\lambda_2$, that are required in a distortion analysis. In practice, too restrictive a choice of the finite set of K may give rise to aliasing problems when compared with the analytical problem. For some sine-wave sources, a 1-D set of frequencies will be sufficient.

Let $X=(X^1,X^2,\ldots,X^N)^\top$ be the Fourier transform of x. More specifically, using a real notation, $X^j=(X_0^{j,R},[X_1^{j,R},X_1^{j,I}],\ldots,[X_{K-1}^{j,R},X_{K-1}^{j,I}])^\top$, in which $X_k^j\equiv X_k^{j,R}+\iota X_k^{j,I}$ represents the kth Fourier coefficient of x^j.

With $\mathcal{F}$, we denote the mapping of the Fourier transform, thus $X = \mathcal{F}x$ and $x = \mathcal{F}^{-1}X$. The $\mathcal{F}$-transform of j (and similarly for q) is defined by $J(X) = \mathcal{F}j(\mathcal{F}^{-1}X) = \mathcal{F}j(x)$. By this Galerkin approach, the frequency-domain equivalent of (20.1) simply becomes

$$J(X) + \Omega Q(X) = 0, \quad \text{in which} \tag{22.51}$$

$$\Omega = \text{Block\_Diag}(\Omega_K, \ldots, \Omega_K), \tag{22.52}$$

$$\Omega_K = \text{Block\_Diag}\left(0, \begin{vmatrix} 0 & -\omega_1 \\ \omega_1 & 0 \end{vmatrix}, \ldots, \begin{vmatrix} 0 & -\omega_{K-1} \\ \omega_{K-1} & 0 \end{vmatrix}\right). \tag{22.53}$$

In the terminology of circuit analysis, the method of solving (20.1) by solving (22.51) is called the Harmonic Balance method. It is clear that the system given by (22.51) is a nonlinear algebraic set of equations in the frequency-domain.

In general, the system is solved by performing a Newton–Raphson iteration. A DC-analysis provides an initialization for the basic harmonic. For the other harmonics one can solve a set of AC-problems in parallel, each being linearised around the same DC-solution. Because each AC-problem is linear, this is very efficient. Note that this approach may be interpreted as the first iteration of a nonlinear block Gauss–Jacobi approach, using partitions between the components of different harmonics.

Clearly, in HB, a Newton–Raphson matrix is much larger than in ordinary DC or Transient Analysis. However, it still has a similar sparse structure as in the last two cases. Hence it is not surprisingly that quite some attention is made in literature concerning iterative methods applied to the linear system of equations arising in Harmonic Balance (BRACHTENDORF [1994], BRACHTENDORF, WELSCH and LAUR [1995], MELVILLE, FELDMANN and ROYCHOWDHURY [1995], RÖSCH [1992], RÖSCH and ANTREICH [1992], ROYCHOWDHURY and FELDMANN [1997]).

Sources. The choice of the fundamental frequencies λ_j depends on the kinds of sources used. For standard amplitude, frequency or phase modulated sources, a 2-D block of frequencies will usually be necessary, as explained below.

We assume voltage and current sources. The DC-sources are time-independent, the AC-sources may involve a simple sum of (co)sine-waves (SW-source). For Harmonic Balance the sources may also show amplitude modulation (SWAM), frequency modulation (SWFM) or phase modulation (SWPM) behaviour. Denoting a source by $s(t)$ and the carrier frequency and the signal frequency by ω_c and ω_s, respectively, the following cases can be distinguished (here θ simply denotes a phase shift).

Modulation	$x(t) = A(t)\cos(\psi(t) + \theta)$		$K_{\min} = \text{block}[n, m]$
	$A(t)$	$\psi(t)$	
AM	$a + b\sin(\omega_s t)$	$\omega_c t$	[1,1]
FM	a	$\int_0^t \omega_c + c\cos(\omega_s t)\,\mathrm{d}t$	$[1, m]$, $m \geqslant c/\omega_s$
PM	a	$\omega_c t + d\sin(\omega_s t)$	$[1, m]$, $m \geqslant d$

In the last column we have added the minimum rectangular subset of K in order to avoid obvious errors due to aliasing (assuming $\lambda_1 = \omega_c$, $\lambda_2 = \omega_s$). In general $\omega_c \gg \omega_s$. The following result, of which the proof is elementary, will be used in the sequel: *SWAM, SWFM and SWPM sources have a Fourier expansion with respect to the exponentials* $e^{\iota(n\omega_c + m\omega_s)t}$. *For SWAM and SWPM the coefficients are independent of* ω_c *and* ω_s. *For SWFM the coefficients depend on* c/ω_s.

A more careful evaluation of the coefficients reveals that the only nonzero harmonics are for $(k_1, k_2) = (1, m)$. For SWAM m is also restricted to $m \leqslant 1$, showing that a finite expansion is obtained. For SWFM and SWPM the dominant part of the infinite expansions depends on c/ω_s and d, respectively.

Discrete Fourier transform. Let $\mathcal{F}_{\eta,\mu}$ denote the Fourier transform using fundamental frequencies η, μ. We observe that in general, by the nonlinearity of i and q, $I(V)$ and $Q(V)$ depend on the specific λ_1, λ_2 mentioned previously. However, in practice, $I(V)$ and $Q(V)$ appear to be rather independent on λ_1, λ_2. This surprising phenomenon allows an efficient evaluation of $I(v)$ and $q(v)$. To be more specific, let λ_3, λ_4 be two other fundamental frequencies that satisfy the same assumptions as imposed on λ_1, λ_2, i.e. the corresponding set of frequencies generated by K and λ_3, λ_4 should not contain multiple values. In practice the nonlinearity in i (and similarly in q) with respect to v is only 'algebraic' in the following sense

$$I(V)_k = \left(\mathcal{F}_{\lambda_1,\lambda_2} i\left([\mathcal{F}_{\lambda_1,\lambda_2}]^{-1} V\right)\right)_k, \tag{22.54}$$

$$= \left(\mathcal{F}_{\lambda_3,\lambda_4} i\left([\mathcal{F}_{\lambda_3,\lambda_4}]^{-1} V\right)\right)_k, \tag{22.55}$$

for all $\lambda_1, \lambda_2, \lambda_3, \lambda_4$, which means that the Fourier coefficients are frequency independent. Nonlinear resistors are algebraic in the above sense when using the variables i and v; nonlinear capacitors when dealing with q and v (note that $i = \mathrm{d}q/\mathrm{d}t$); and nonlinear inductors when dealing with ϕ and i (note that $v = \mathrm{d}\phi/\mathrm{d}t$). The expansions in $e^{\iota(n\omega_c + m\omega_s)t}$ of the SWAM, SWFM, SWPM sources show that they also exhibit this algebraic behaviour.

The algebraic nonlinearity offers a way to exploit λ_3, λ_4 which are different from the fundamental analysis frequencies λ_1, λ_2, in determining $I(V)$ and its partial derivatives in an efficient and stable way using the Discrete Fourier Transform. For details we refer to BRACHTENDORF [1994], KUNDERT, WHITE and SANGIOVANNI-VINCENTELLI [1990], TER MATEN [1999], RÖSCH [1992]. In SEVAT [1994] several bijective mappings between (enveloping sets of) higher dimensional spectral sets K and a 1-dimensional equivalent (with no gaps) are considered that allow for proper usage of the DFT.

Numerical aspects of Harmonic Balance. Although HB is being successfully used for a wide range of applications, there are still some mathematical issues which have to be solved. One of them is error control and adaptivity, another one concerns DAE aspects.

- Accuracy of the HB solution is mainly determined by the sets k and λ in (22.50), which have to be provided by the user, or are determined from the type of sources, as is described above. In case of too few – or a not adequate set of – frequencies, alias effects may occur, or HB does not even converge. Harmonic Balance is useful only for mildly nonlinear problems, i.e., when all quantities have a Fourier expansion that can be well approximated by some finite one of limited length. Aliasing can be reduced by applying oversampling (TER MATEN [1999]). A rigorous mathematical adaptivity concept is not implemented, in general. A proposal is given in FRIESE [1996]; unfortunately, adaptivity here involves to reorganize the system matrix from time to time, which does not fit well into existing implementations.

- In practice, HB has been applied successfully even for index-2 problems, and no severe drawbacks or errors have been reported yet. There are however no theoretical investigations about the feasibility of this usage, and which impact may have a higher index on numerics.

HB oscillator algorithm. In BRACHTENDORF, WELSCH and LAUR [1995], GOURARY, ULYANOV, ZHAROV, RUSAKOV, GULLAPALLI and MULVANEY [1998], LAMPE, BRACHTENDORF, TER MATEN, ONNEWEER and LAUR [2001], WELSCH [1998] oscillator algorithms are given for Harmonic Balance that resemble the approach described in Section 22.2.2 for the time domain. However, there are some modifications:

- The gauge condition is replaced by the requirement that the imaginary part of the first harmonic at some predefined node has to be zero. Note that this allows the DC solution to be solution of the system. Indeed, the DC solution appears to be a strong attractor for the Newton process. Hence algorithms apply some additional deflation technique to exclude this solution.
- In practice the artificial element is defined directly in the frequency domain. For all harmonics but the first one the element acts as an 'open', i.e., the harmonic of the current is set to 0. For the first harmonic it acts as a voltage source in series with a resistor. For all analyses other than Harmonic Balance (that might be used for initialization), the current through the element is set to zero too.
- For initialization the equations are linearized around the DC-solution like in AC analysis. Kurokawa's method (KUROKAWA [1969]) calculates the response solution for an ordinary sinusoidal source with unit amplitude that replaces the artificial element and considers the admittance for the source element. The result is considered as a function of the frequency f. Where the imaginary part of the admittance becomes zero while the real part remains positive a good approximation for the oscillator can be found (the equation itself can be solved by applying for instance Newton–Raphson). For the circuit solution one uses the DC-solution plus the AC solution for the first harmonic. All other harmonics are set to 0.

  Alternatively one can solve a generalized eigenvalue problem (in practice one will consider the inverse eigenvalue problem) for the linearized equations (BRACHTENDORF, LAMPE and LAUR [2000]). Because an autonomous circuit can only start up oscillating when the DC-solution is unstable (Andronov–Hopf bifurcation theorem), one looks for eigenvalues $\lambda = \delta \pm j\omega$, where $\delta > 0$. The associated eigenvector also indicates where the artificial element may be attached. In addition it provides an estimate for the initial circuit solution. However, in practice, the f estimated by Kurokawa's method appears to be more accurate.

- The applied value of the artificial element is initialized by optimization techniques (GOURARY, ULYANOV, ZHAROV, RUSAKOV, GULLAPALLI and MULVANEY [1998], JOKUNEN [1997], WELSCH [1998]). In BRACHTENDORF, LAMPE and LAUR [2000], LAMPE, BRACHTENDORF, TER MATEN, ONNEWEER and LAUR [2001] techniques using affine damping improved convergence. Initial Global Optimization techniques improved robustness even more by providing much better initial estimates (LAMPE and LAUR [2002]). Note that the algorithm can be formulated as a full Newton process, but also as a two-step process. In the latter case for each applied value as internal step a driven Harmonic Balance process is executed until convergence. For updating the applied value and the frequency, the Jacobian matrix of the Harmonic Balance process can be reused for determining the sensitivities of the solution with respect to variations of the applied value and the frequency. In fact, this very elegantly reuses options for parameter sensitivity analysis.

Global convergence – a three stage approach. For getting global convergence properties, the application of a continuation method is adequate. In case of nonautonomous (forced) systems it is natural for this purpose to track a parameter dependent path in the frequency domain with some path-following methods, as is done in the DC domain by performing a DC transfer analysis. The parameter here may be a circuit parameter or the bias value of an independent source, either. In case of local parametrization along the solution path, even turning points in the parameter space can be tracked; and by watching the sign and magnitude of Floquet multipliers, circuit stability properties along the solution path can be analyzed (SELTING and ZHENG [1997]).

For autonomous oscillators the problem is more difficult since a good estimate for the frequency is important. So it is suggested in NEUBERT, SELTING and ZHENG [1998] for this case to start path-following from Hopf's bifurcation point. The latter is computed from a path-following procedure in the DC-domain, such that there is a three-stage approach for solving the whole problem:

1. Follow a path of DC steady states over a parameter λ – λ being a circuit parameter or the value of an independent source – until a Hopf bifurcation point is found. The latter is characterized by a sign change of the real part of a complex conjugate pair of generalized eigenvalues.
2. These eigenvalues and the corresponding eigenvectors provide first order information about frequency and Fourier coefficients of the oscillatory branch emanating from the DC path in Hopf's point. Since this information is not very accurate, an alternative method for getting the latter was developed ZHENG and NEUBERT [1997].
3. From this start point, follow the path of periodic steady states over λ, until the final value of λ is obtained.

Again it is worth to note that all these additional algorithmic steps can be implemented elegantly using Schur complement techniques, once the basic types of analysis are available.

Recently, homotopy approaches were considered in BRACHTENDORF, WELSCH and LAUR [1998], MA, TRAJKOVĆ and MAYARAM [2002].

22.6. *Envelope methods*

In TROYANOVSKY [1997] an envelope method is described in some detail. The envelope method is a generalisation of the Harmonic Balance method. It allows for multitone treatments, but in fact one of the periods may be infinite. The method mixes time-domain and frequency-domain approaches.

We write (22.49) as

$$x(t) = \sum_{\omega_k \in \Lambda} X_k(t) e^{\iota \omega_k t}, \quad \text{with } \omega_k \in \Lambda \equiv \{\omega \mid \omega = (k, \lambda)\}, \tag{22.56}$$

where the enveloping Fourier coefficients $X_k(t)$ represent modulation on top of the carrier sinusoids at frequencies ω_k. In order to describe the effect of $q(x)$ (and similarly of $j(x)$) we introduce

$$\tilde{x}(\tau_1, \tau_2) = \sum_{\omega_k \in \Lambda} X_k(\tau_1) e^{\iota \omega_k \tau_2}, \tag{22.57}$$

$$= \mathcal{F}_{\tau_2}^{-1} X(\tau_1) \tag{22.58}$$

(which in fact is a multivariate formulation; see also the next subsection). Here $X(\tau_1)$ is the Fourier transform in τ_2 of $\tilde{x}(\tau_1, \tau_2)$. Thus $x(t) = \tilde{x}(t, t)$. For fixed τ_1, the Fourier coefficients $Q_k(x(\tau_1))$ of q, when applied to $\tilde{x}(\tau_1, \tau_2)$, can be determined using the Fourier Transform in the τ_2 variable

$$q(\tilde{x}(\tau_1, \tau_2)) = \sum_{\omega_k \in \Lambda} Q_k(X(\tau_1)) e^{\iota \omega_k \tau_2}, \tag{22.59}$$

$$Q_k(X) = \mathcal{F}_{\tau_2} q(\mathcal{F}_{\tau_2}^{-1} X). \tag{22.60}$$

Clearly, $q(x(t)) = q(\tilde{x}(t, t))$. We will collect all $Q_k(X)$ in $q(x)$ (and similarly for $J(X)$). If we put the expansions of q and j in (20.1) we find a DAE for the $X(t)$

$$J(X(t)) + \frac{\mathrm{d}}{\mathrm{d}t} Q(X(t)) + \Omega Q(X)(t) = 0, \tag{22.61}$$

which can be solved using ordinary time integration methods. Stepping forward in time at each time level a nonlinear set of (complex-valued) algebraic equations has to be solved, that has the size of a Harmonic Balance problem. In RF applications, the envelope solution of the DAE (22.61) behaves much less oscillating (or is not oscillating at all) than that of (20.1). Hence, despite the larger nonlinear system of equations that has to be solved at each time level for (22.61), much larger time steps can be used than for (20.1).

Because of the separation of modes in τ_1 and τ_2 variables, different scaling effects can be separated. This is also the subject of the next subsection.

Analysis of high-quality oscillator circuits. Another kind of envelope following methods has been suggested for analysis of oscillatory circuits whose quality factor Q is so high that conventional methods like those described in PETZOLD [1981], SKELBOE

[1982] failed. These circuits exchange a very small amount of energy per cycle between the oscillator core and the driven, energy supplying circuit part, which makes the problem extremely stiff. In a state space diagram, the trajectories of one cycle are almost closed, even though the circuit is not yet in a steady state. So it seems reasonable to approximate the trajectory for this cycle by a really closed one, which can be computed by solving a PSS problem. Once this approximative trajectory is found, a transient analysis over a few cycles would correct this one into the real solution. From its dynamics a new estimate for a later cycle can be extrapolated, and the next step of this "envelope" method can be started ZHENG [1994]. In fact this method is again a mixed time-frequency approach.

A successful application of this idea is the startup analysis of quartz driven circuits (SCHMIDT-KREUSEL [1997], MATHIS [1998]). Since the quartz resonator oscillates very much like a harmonic oscillator, it can be substituted for one PSS step by a sinusoidal current source of a certain magnitude. Its phase can be arbitrarily set to zero, and a first guess for the frequency is just the resonator frequency of the quartz crystal. Once the PSS solution is found, and is "corrected" by a subsequent transient analysis over a few – 2, ..., 4, say, – cycles, the dynamic behaviour can be extrapolated over several hundred to thousand cycles, yielding a new value for the magnitude and the frequency of the substitute current source. So this method cannot only be seen as some kind of continuation method for the PSS problem, but also yields reasonable timing information about the startup process of the circuit.

22.7. *Multivariate extension*

In BRACHTENDORF, WELSCH, LAUR and BUNSE-GERSTNER [1996], BRACHTENDORF and LAUR [2000], ROYCHOWDHURY [1997], ROYCHOWDHURY, LONG and FELDMANN [1998], ROYCHOWDHURY [2001a, 2001b] multivariate extensions are described that apply to multitone situations. In fact, one introduces two or more independent time parameters, τ_1, τ_2 say. Then (20.1) is rewritten to

$$\frac{\mathrm{d}}{\mathrm{d}\tau_1}q(\hat{x}) + \frac{\mathrm{d}}{\mathrm{d}\tau_2}q(\hat{x}) + j(\hat{x}) = b(\tau_1, \tau_2) \in \mathbb{R}^N, \tag{22.62}$$

$$\hat{x}(0, \tau_2) = \hat{x}(T_1, \tau_2), \tag{22.63}$$

$$\hat{x}(\tau_1, 0) = \hat{x}(\tau_1, T_2). \tag{22.64}$$

After solving this partial differential problem (22.62) (hyperbolic for $\mathrm{d}q/\mathrm{d}x$ regular) on $[0, T_1] * [0, T_2]$ for $\hat{x}$, the solution $x(t)$ is found by $x(t) = \hat{x}(t(\mathrm{mod}\ T_1), t(\mathrm{mod}\ T_2))$.

It is clear that the above separation in two or more independent time parameters restricts one in formulating expressions. The aim is that on $[0, T_1] * [0, T_2]$ the solution $\hat{x}$ behaves smoothly and that only one period is met in each direction. In ROYCHOWDHURY [2001b] the problem of frequency modulation (FM) is considered more closely for the case of an oscillatory DAE

$$\omega(\tau_2)\frac{\mathrm{d}}{\mathrm{d}\tau_1}q(\hat{x}) + \frac{\mathrm{d}}{\mathrm{d}\tau_2}q(\hat{x}) + j(\hat{x}) = b(\tau_2), \tag{22.65}$$

$$\phi(t) = \int_0^t \omega(\tau_2)\, \mathrm{d}\tau_2, \tag{22.66}$$

$$x(t) = \hat{x}\big(\phi(t), t\big). \tag{22.67}$$

When (22.65) is solved, also the local frequency $\omega(\tau_2)$ is obtained (see also BRACHTENDORF and LAUR [2000]). The derivative, $\omega(\tau_2)$, of the 'warping' function ϕ, gives the extend of the stretch of the timescale in τ_2. For time integration methods of characteristics were studied recently (BRACHTENDORF and LAUR [2000], PULCH and GÜNTHER [2002]).

Optimal sweep following. An open question remains how $\omega(\tau_2)$ in (22.65) should be determined. One way to proceed is to observe that the differential equation (22.65) defines a two-dimensional manifold (called the *sweep*) in the state space $\mathbb{R}^N$. The choice of ω does not influence the sweep; however, it does influence the parametrisation of the sweep in terms of the coordinates τ_1 and τ_2.

In HOUBEN [2003], it is suggested to choose ω in such a way that

$$\int_0^T \left\| \frac{\mathrm{d}}{\mathrm{d}\tau_2} q(\hat{x}) \right\|^2 \mathrm{d}\tau_1 \tag{22.68}$$

becomes as small as possible. The rationale is that this will allow the largest stepsizes in the (slowly varying) τ_2-direction, thereby reducing computation time. It is shown in HOUBEN [2003] that this is the case for

$$\dot{\phi}(\tau_2) = \omega(\tau_2) = \frac{\int_0^T \big(b(\tau_2) - j(\hat{x}), \frac{\mathrm{d}}{\mathrm{d}\tau_2} q(\hat{x})\big)\, \mathrm{d}\tau_1}{\int_0^T \|\frac{\mathrm{d}}{\mathrm{d}\tau_2} q(\hat{x})\|^2\, \mathrm{d}\tau_1}. \tag{22.69}$$

Since this choice of ω is optimal with respect to the minimization of (22.68), the resulting method is called *Optimal Sweep Following*.

References

APPEL, T. (2000). A new timestep control in the circuit simulation package TITAN. Bachelor thesis (Technical University München, München).

ARNOLD, M. (1997). *Zur Theorie und zur numerischen Lösung von Anfangswertproblemen für differentiell-algebraische Systeme von höherem Index* (VDI-Verlag, Düsseldorf).

ARNOLD, M., GÜNTHER, M. (2001). Preconditioned dynamic iteration for coupled differential-algebraic systems. *BIT* **41**, 1–25.

ASCHER, U.M., MATTHEIJ, R.M.M., RUSSELL, R.D. (1998). *Numerical Solution of Boundary Value Problems for Ordinary Differential Equations* (Prentice Hall, Englewood Cliffs, NJ).

ASCHER, U.M., PETZOLD, L.R. (1998). *Computer Methods for Ordinary Differential Equations and Differential-Algebraic Equations* (SIAM, Philadelphia, PA).

BAIZ, A. (2003). Effiziente Lösung periodischer differential-algebraischer Gleichungssysteme in der Schaltungssimulation. PhD thesis (TU Darmstadt, Darmstadt).

BANK, R.E., COUGHRAN, W.M., FICHTNER, W., GROSSE, E.H., ROSE, D., SMITH, R.K. (1985). Transient simulation of silicon devices and circuits. *IEEE Trans. Comp. Aided Des. CAD* **4**, 436–451.

BARTEL, A. (2000). Generalised multirate — Two ROW-type versions for circuit simulation, Unclassified NatLab Report 804/2000 (Philips Research Laboratories, Eindhoven).

BARTEL, A., GÜNTHER, M., KVÆRNØ, A. (2001). Multirate methods in electrical circuit simulation. Numerics Report 2/2001 (Norwegian University of Science and Technology, Trondheim).

BAUER, R., FANG, A., BRAYTON, R. (1988). XPSim: A MOS VLSI simulator. In: *Proc. ICCAD'88, Santa Clara*, pp. 66–69.

BOMHOF, W. (2001). Iterative and parallel methods for linear systems with applications in circuit simulation. PhD thesis (Utrecht University, Utrecht).

BOMHOF, W., VAN DER VORST, H.A. (2001). A parallelizable GMRES-type method for p-cyclic matrices with applications in circuit simulation. In: Van Rienen, U., et al. (eds.), *Proc. SCEE-2000, Warnemünde*. In: Lecture Notes Comp. Sci. **18** (Springer, Berlin), pp. 293–300.

BORCHARDT, J., GRUND, F., HORN, D. (1997). Parallelized numerical methods for large systems of differential-algebraic equations in industrial applications. Preprint 382 (Weierstrass Institut für Angewandte Analysis und Stochastik, Berlin).

BOWERS, J.C., SEDORE, S.R. (1971). *SCEPTRE: A Computer Program for Circuit and System Analysis* (Prentice-Hall, Englewood Cliffs, NJ).

BRACHTENDORF, H.G. (1994). *Simulation des eingeschwungenen Verhaltens elektronischer Schaltungen*. PhD thesis (Universität Bremen, Bremen) (Shaker, Aachen) ISBN 3-8265-0226-4.

BRACHTENDORF, H.G., WELSCH, G., LAUR, R. (1995). A simulation tool for the analysis and verification of the steady state of circuit designs. *Int. J. Circ. Theory Appl.* **23**, 311–323.

BRACHTENDORF, H.G., WELSCH, G., LAUR, R., BUNSE-GERSTNER, A. (1996). Numerical steady state analysis of electronic circuits driven by multi-tone signals. *Electr. Eng.* **79**, 103–112.

BRACHTENDORF, H.G., WELSCH, G., LAUR, R. (1998). A time-frequency algorithm for the simulation of the initial transient response of oscillators. In: *Proc. ISCAS'98, Monterey, vol. I*, pp. 236–239.

BRACHTENDORF, H.G., LAMPE, S., LAUR, R. (2000). Comparison of the Philips and Bremen algorithm for calculating the steady state of autonomous oscillators. ITEM Report (Universität Bremen, Bremen).

BRACHTENDORF, H.G., LAUR, R. (2000). Analyse des transienten Verhaltens von Oszillatoren durch eine inverse Charakteristikenmethode. In: *Proc. ITG Workshop Mikroelektronik für die Informationstechnik, Darmstadt* (VDE, Berlin), pp. 29–34. ISBN 3-8007-2586-X.

BRACHTENDORF, H.G., LAUR, R. (2001). On consistent initial conditions for circuit's DAEs with higher index. *IEEE Trans. Circ. Syst. CAS I* **48**, 606–612.

BRENAN, K.E., CAMPBELL, S.L., PETZOLD, L.R. (1996). *Numerical Solution of Initial-Value Problems in Differential-Algebraic Equations* (SIAM, Philadelphia, PA).

BRAMBILLA, A., D'AMORE, D., SANTOMAURO, M. (1995). Simulation of autonomous circuits in the time domain. In: *Proc. ECCTD'95, Istanbul*, pp. 399–402.

BRUIN, S.M.A. (2001). Modified extended BDF applied to circuit equations. MSc-Thesis Free Univ. of Amsterdam. NatLab. Unclassified Report 2001/826 (Philips Research Laboratories, Eindhoven).

CALAHAN, D.A. (1968). A stable, accurate method of numerical integration for nonlinear systems. *Proc. IEEE* **56**, 744.

CALAHAN, D.A. (1972). *Computer Aided Network Design* (McGraw-Hill, New York).

CASH, J.R. (1983). The integration of stiff initial value problems in ODEs using modified extended backward differentiation formulae. *Comp. Math. Appl.* **9** (5), 645–657.

CASH, J.R. (2000). Modified extended backward differentiation formulae for the numerical solution of stiff initial value problems in ODEs and DAEs. *J. Comput. Appl. Math.* **125**, 117–130.

CHAWLA, B.R., GUMMEL, H.K., KOZAK, P. (1975). MOTIS – An MOS timing simulator. *IEEE Trans. Circ. Syst. CAS* **22**, 901–910.

CHOW, S., MALLET-PARET, J., YORKE, J. (1978). Finding zeros of maps: Homotopy methods that are constructive with probability one. *Math. Comput.* **32**, 887–889.

CHUA, L.O., LIN, P.-M. (1975). *Computer-Aided Analysis of Electronic Circuits: Algorithms & Computational Techniques* (Prentice-Hall, Englewood Cliffs, NJ).

CHUA, L.O. (1984). Nonlinear circuits. *IEEE Trans. Circ. Syst. CAS* **31**, 69–87.

COX, P.F., BURCH, R.G., YANG, P., HOCEVAR, D.E. (1989). New implicit integration method for efficient latency exploitation in circuit simulation. *IEEE Trans. Comp. Aided Des. CAD* **8**, 1051–1064.

COX, P.F., BURCH, R.G., HOCEVAR, D.E., YANG, P., EPLER, B.D. (1991). Direct circuit simulation algorithms for parallel processing. *IEEE Trans. Comp. Aided Des. CAD* **10**, 714–725.

DEBEVFE, P., ODEH, F., RUEHLI, A.E. (1985). Waveform techniques. In: Ruehli, A.E. (ed.), *Circuit Analysis, Simulation and Design, Part 2* (North Holland, Amsterdam), pp. 41–127.

DE MAN, H., ARNOUT, G., REYNAERT, P. (1981). Mixed mode circuit simulation techniques and their implementation in DIANA. In: Antognetti, P., et al. (eds.), *Computer Design Aids for VLSI Circuits*. In: NATO Advanced Study Institute series E **48** (Martinus Nijhoff, The Hague), pp. 113–174.

DE MICHELI, G., HSIEH, H.Y., HAJJ, I.N. (1987). Decomposition techniques for large scale circuit analysis and simulation. In: Ruehli, A.E. (ed.), *Circuit Analysis, Simulation and Design, Part II* (North Holland, Amsterdam), pp. 1–39.

DEMIR, A. (1998). Phase noise in oscillators: DAEs and coloured noise sources. In: *Proc. ICCAD'98, San Jose*, pp. 170–177.

DEMIR, A., LONG, D., ROYCHOWDHURY, J. (2000). Computing phase noise eigenfunctions directly from Harmonic Balance/shooting matrices. In: *Proc. ICCAD'2000, San Jose*, pp. 283–288.

DEMIR, A., MEHROTRA, A., ROYCHOWDHURY, J. (2000). Phase noise in oscillators: A unifying theory and numerical methods for characterization. In: *Proc. DAC'98, San Francisco*, pp. 26–31. Extended version. *IEEE Trans. Circ. Syst. CAS I* **47** (2000) 655–674.

DENK, G. (1990). An improved numerical integration method in the circuit simulator SPICE2-S. In: Bank, R.E., et al. (eds.), *Proc. Oberwolfach Conf.* In: Int. Ser. Num. Math. **93** (Birkhäuser, Basel), pp. 85–99.

DENK, G., RENTROP, P. (1991). Mathematical models in electric circuit simulation and their numerical treatment. In: Strehmel, K. (ed.), *Proc. NUMDIFF5* (Teubner, Leipzig), pp. 305–316.

DENK, G., (2002). Private communication.

DE SMEDT, B., GIELEN, G. (1997). Accurate simulation of phase noise in oscillators. In: *Proc. ESSCIRC'97, Southampton*, pp. 208–211.

DEUFLHARD, P., BORNEMANN, F. (2002). *Numerische Mathematik II* (De Gruyter, Berlin).

DEVGAN, A., ROHRER, R.A. (1994). Adaptively controlled explicit simulation. *IEEE Trans. Comp. Aided Des. CAD* **13**, 746–762.

DIRKS, H.K., FISCHER, M., RÜDIGER, J. (2001). Parallel algorithms for solving linear equations in VLSI circuit simulation. In: Van Rienen, U., et al. (eds.), *Proc. SCEE-2000, Warnemünde*. In: Lecture Notes Comp. Sci. Eng. **18** (Springer, Berlin), pp. 301–308.

DUNLOP, A., DEMIR, A., FELDMANN, P., KAPUR, S., LONG, D., MELVILLE, R., ROYCHOWDHURY, J. (1998). Tools and methodology for RF IC design. In: *Proc. DAC'98, San Francisco*, pp. 414–420.

EICKHOFF, K.M. (1991). Effiziente Methoden zur Simulation grosser MOS-Schaltungen. PhD thesis (Rheinisch-Westfälische Technische Hochschule, Aachen).

EICKHOFF, K.M., ENGL, W. (1995). Levelized incomplete LU-factorization and its application to large-scale circuit simulation. *IEEE Trans. Comp. Aided Des. CAD* **14**, 720–727.

ENGL, W.L., LAUR, R., DIRKS, H.K. (1982). MEDUSA – A simulator for modular circuits. *IEEE Trans. Comp. Aided Des. CAD* **1**, 85–93.

ENGSTLER, C., LUBICH, C. (1997a). Multirate extrapolation methods for differential equations with different time scales. *Computing* **58**, 173–185.

ENGSTLER, C., LUBICH, C. (1997b). MUR8: A multirate extension of the eighth-order Dormand-Prince method. *Appl. Numer. Math.* **25**, 185–192.

ESTÉVEZ SCHWARZ, D. (1999a). Topological analysis for consistent initialization in circuit simulation Preprint 99-3 (Institut für Mathematik, Humboldt-Universität zu Berlin, Berlin).

ESTÉVEZ SCHWARZ, D., LAMOUR, R. (1999). The computation of consistent initial values for nonlinear index-2 differential-algebraic equations. Preprint 99-13 (Humboldt-Universität zu Berlin, Berlin).

ESTÉVEZ SCHWARZ, D. (2000). Consistent initialization for differential-algebraic equations and its application to circuit simulation. PhD thesis (Humboldt Universität zu Berlin, Berlin) (Available at: http://dochost.rz.hu-berlin.de/dissertationen).

ESTÉVEZ SCHWARZ, D., TISCHENDORF, C. (2000). Structural analysis for electrical circuits and consequences for MNA. *Int. J. Circ. Theory Appl.* **28**, 131–162.

ESTÉVEZ SCHWARZ, D., FELDMANN, U., MÄRZ, R., STURTZEL, S., TISCHENDORF, C. (2003). Finding benificial DAE structures in circuit simulation. In: Jäger, W., Krebs, H.-J. (eds.), *Mathematics – Key Technology for the Future* (Springer, Berlin).

FELDMANN, U., WEVER, U., ZHENG, Q., SCHULTZ, R., WRIEDT, H. (1992). Algorithms for modern circuit simulation. *AEÜ* **46**, 274–285.

FELDMANN, U., GÜNTHER, M. (1999). Some remarks on regularization of circuit equations. In: *Proc. ISTET'99, Magdeburg*, pp. 343–348.

FISCHER, M. (2001). Multigranulare parallele Algorithmen zur Lösung von Gleichungssystemen der VLSI-Netzwerksimulation. PhD thesis (Christian-Albrechts-Universität zu Kiel, Kiel) (Shaker, Aachen) ISBN 3-8265-8565-8.

FRANK, J.E., VAN DER HOUWEN, P.J. (2000). Diagonalizable extended backward differential formulas. *BIT* **40** (3), 497–512.

FRIESE, T. (1996). Eine adaptive Spektralmethode zur Berechnung periodischer Orbits. In: Mathis, W., Noll, P. (eds.), *Proc. 2nd ITG Workshop 1995, Berlin* (VDE-Verlag, Berlin), pp. 21–26. ISBN 3-8007-2190-2.

FRÖHLICH, N., RIESS, B.M., WEVER, U.A., ZHENG, Q. (1998). A new approach for parallel simulation of VLSI circuits on a transistor level. *IEEE Trans. Circ. Syst. CAS I* **45**, 601–613.

FRÖHLICH, N., GLÖCKEL, V., FLEISCHMANN, J. (2000). A new partitioning method for parallel simulation of VLSI circuits on transistor level. In: *Proc. DATE'2000, Paris*, pp. 679–684.

FRÖHLICH, N. (2002). Verfahren zum Schaltungspartitionieren für die parallele Simulation auf Transistorebene. PhD thesis (TU München, München).

GAREY, M.R., JOHNSON, D.S. (1979). *Computers and Intractability: A Guide to the Theory of NP-Completeness* (Freeman, New York).

GEAR, C.W. (1971). Simultaneous numerical solution of differential-algebraic equations. *IEEE Trans. Circ. Theory CT* **18**, 89–95.

GEAR, C.W., WELLS, D.R. (1984). Multirate linear multistep methods. *BIT* **24**, 484–502.

GEAR, C.W. (1988). Differential-algebraic equation index transformations. *SIAM J. Sci. Stat. Comp.* **9**, 39–47.

GEAR, C.W. (1990). Differential-algebraic equations, indices, and integral algebraic equations. *SIAM J. Numer. Anal.* **27**, 1527–1534.

GEORGE, J.A. (1974). On block elimination for sparse linear systems. *SIAM J. Numer. Anal.*, 585–603.

GOURARY, M.M., ULYANOV, S.L., ZHAROV, M.M., RUSAKOV, S.G., GULLAPALLI, K.K., MULVANEY, B.J. (1998). Simulation of high-Q oscillators. In: *Proc. ICCAD'98, San Jose*, pp. 162–169.

GRÄB, R., GÜNTHER, M., WEVER, U., ZHENG, Q. (1996). Optimization of parallel multilevel Newton algorithms on workstation clusters. In: Bouge, L., et al. (eds.), *Proc. Euro-Par96*. In: Lecture Notes Comp. Sci. **1124** (Springer, Berlin), pp. 91–96.

GRIEPENTROG, E., MÄRZ, R. (1986). *Differential-Algebraic Equations and their Numerical Treatment* (Teubner, Leipzig).

GRISTEDE, G.D., ZUKOWSKI, C.A., RUEHLI, A.E. (1999). Measuring error propagation in waveform relaxation algorithms. *IEEE Trans. Circ. Syst. CAS I* **46**, 337–348.

GÜNTHER, M., RENTROP, P. (1993). Multirate ROW methods and latency of electric circuits. *Appl. Numer. Math.* **13**, 83–102.

GÜNTHER, M., RENTROP, P. (1994). Partitioning and multirate strategies in latent electric circuits. In: Bank, R.E., et al. (eds.), *Proc. Oberwolfach Conf.*. In: Int. Ser. Numer. Math. **117** (Birkhäuser, Basel), pp. 33–60.

GÜNTHER, M., HOSCHEK, M. (1997). ROW methods adapted to electric circuit simulation packages. *Comp. Appl. Math.* **82**, 159–170.

GÜNTHER, M. (1998). Simulating digital circuits numerically – A charge-oriented ROW approach. *Numer. Math.* **79**, 203–212.

GÜNTHER, M., FELDMANN, U. (1999a). CAD based electric circuit modeling in industry I: Mathematical structure and index of network equations. *Surv. Math. Ind.* **8**, 97–129.

GÜNTHER, M., FELDMANN, U. (1999b). CAD based electric circuit modeling in industry II: Impact of circuit configurations and parameters. *Surv. Math. Ind.* **8**, 131–157.

GÜNTHER, M., HOSCHEK, M., WEINER, R. (1999). ROW methods adapted to a cheap Jacobian. *Appl. Numer. Math.* **37**, 231–240.

GÜNTHER, M., HOSCHEK, M., RENTROP, P. (2000). Differential-algebraic equations in electric circuit simulation. *Int. J. Electron. Commun. (AEÜ)* **54**, 101–107.

GÜNTHER, M. (2001). *Partielle differential-algebraische Systeme in der numerischen Zeitbereichsanalyse elektrischer Schaltungen* (VDI-Verlag, Düsseldorf). ISBN 3-18-334320-7.

GÜNTHER, M., KVÆRNØ, A., RENTROP, P. (2001). Multirate partitioned Runge–Kutta methods. *BIT* **41**, 504–515.

GÜNTHER, M., RENTROP, P., FELDMANN, U. (2001). CHORAL – A one step method as numerical low pass filter in electrical network analysis. In: Van Rienen, U., et al. (eds.), *Proc. SCEE-2000, Warnemünde*. In: Lecture Notes Comp. Sci. Eng. **18** (Springer, Berlin), pp. 199–215.

GUPTA, G.K., GEAR, C.W., LEIMKUHLER, B.J. (1985). Implementing linear multistep formulas for solving DAEs. Rep. No. UIUCDCS-R-85-1205 (University of Illinois, Urbana).

GUSTAFSSON, K., LUNDH, M., SÖDERLIND, G. (1988). A PI stepsize control for the numerical solution of ordinary differential equations. *BIT* **28**, 270–287.

HACHTEL, G.D., BRAYTON, R.K., GUSTAVSON, F.G. (1971). The sparse tableau approach to network analysis and design. *IEEE Trans. Circ. Theory CT* **18**, 101–113.

HACHTEL, G.D., SANGIOVANNI-VINCENTELLI, A. (1981). A survey of third-generation simulation techniques. *Proc. IEEE* **69**, 1264–1280.

HAIRER, E., NØRSETT, S.P., WANNER, G. (1987). *Solving Ordinary Differential Equations I* (Springer, Berlin).

HAIRE, E., LUBICH, C., ROCHE, M. (1989). *The Numerical Solution of Differential-Algebraic Systems by Runge–Kutta Methods*, Lecture Notes in Mathematics **1409** (Springer, Berlin).

HAIRER, E., WANNER, G. (1996). *Solving Ordinary Differential Equations II*, second ed. (Springer, Berlin).

HAJIMIRI, A., LEE, T.H. (1998). A general theory of phase noise in electrical oscillators. *IEEE J. Solid-State Circ.* **33** (2), 179–194.

HEATH, M., NG, E., PEYTON, B. (1990). Parallel algorithms for sparse linear systems. In: Gallivan, K.A., et al. (eds.), *Parallel Algorithms for Matrix Computations* (SIAM, Philadelphia, PA), pp. 83–124.

HO, C.W., RUEHLI, A.E., BRENNAN, P.A. (1975). The modified nodal approach to network analysis. *IEEE Trans. Circ. Syst. CAS* **22**, 505–509.

HOFFMANN, K. (1996). *VLSI-Entwurf* (Oldenbourg, München).

HONKALA, M., ROOS, J., VALTONEN, M. (2001). New multilevel Newton–Raphson method for parallel circuit simulation. In: *Proc. ECCTD'01, Helsinki, vol. II*, pp. 113–116.

HONKALA, M., KARANKO, V., ROOS, J. (2002). Improving the convergence of combined Newton–Raphson and Gauss–Newton multilevel iteration method. In: *Proc. of ISCAS'02, Scottsdale, Arizona*, pp. 26–29.

HORNEBER, E.-H. (1985). *Simulation elektrischer Schaltungen auf dem Rechner* (Springer, Berlin).

HOSCHEK, M. (1999). *Einschrittverfahren zur numerischen Simulation elektrischer Schaltungen.* PhD thesis (TU Darmstadt, Darmstadt) (VDI-Verlag, Düsseldorf) ISBN 3-18-329320-X.

HOSEA, M.E., SHAMPINE, L.F. (1996). Analysis and implementation of TR-BDF2. *Appl. Numer. Math.* **20**, 21–37.

HOUBEN, S.H.M.J. (1999). Algorithms for periodic steady state analysis on electric circuits, Unclassified NatLab Report 804/99 (Philips Research Laboratories, Eindhoven).

HOUBEN, S.H.M.J. (2003). Circuits in motion. The numerical simulation of electrical circuits. PhD thesis (Eindhoven University of Technology, Eindhoven).

HOUBEN, S.H.M.J., MAUBACH, J.M. (2000). An accelerated Poincaré-map method for autonomous oscillators (Preprint TU Eindhoven, Eindhoven) (available at: http://www.win.tue.nl/~anwww/preprints/2000.html).

HOUBEN, S.H.M.J., MAUBACH, J.M. (2001). Periodic steady-state analysis of free-running oscillators. In: Van Rienen, U., et al. (eds.), *Proc. SCEE-2000, Warnemünde*. In: Lecture Notes Comp. Sci. Eng. **18** (Springer, Berlin), pp. 217–224.

HOUBEN, S.H.M.J., TER MATEN, E.J.W., MAUBACH, J.M., PETERS, J.M.F. (2001). Novel time-domain methods for free-running oscillators. In: *Proc. ECCTD'01, Helsinki, vol. III*, pp. 393–396.

HOYER, W., SCHMIDT, J.W. (1984). Newton-type decomposition methods for equations arising in network analysis. *ZAMM* **64**, 397–405.

HSIM DATASHEET (2002). The full-chip hierarchical circuit simulation and analysis tool. NASSDA; http://www.nassda.com.

JOKUNEN, H. (1997). Computation of the steady-state solution of nonlinear circuits with time-domain and large-signal-small-signal analysis methods. PhD thesis (Helsinki University of Technology, Espoo) (Acta Polytechnica Scandinavica, Electrical Engineering Series 87, 1–75).

KÄRTNER, F.X. (1989). Untersuchung des Rauschverhaltens von Oszillatoren. PhD thesis (TU München).

KÄRTNER, F.X. (1990). Analysis of white and $f^{-\alpha}$ noise in electrical oscillators. *Int. J. Circ. Theory Appl.* **18**, 485–519.

KAGE, T., KAWAFUJI, F., NIITSUMA, J. (1994). A circuit partitioning approach for parallel simulation, IEICE Trans. *Fundamentals E77-A* **3**, 461–466.

KAMPOWSKY, W., RENTROP, P., SCHMIDT, W. (1992). Classification and numerical simulation of electric circuits. *Surv. Math. Ind.* **2**, 23–65.

KATZENELSON, J. (1965). An algorithm for solving nonlinear resistor networks. *Bell Syst. Tech. J.* **44**, 1605–1620.

KEVENAAR, T.A.M. (1994). Periodic steady state analysis using shooting and waveform-Newton. *Int. J. Circ. Theory Appl.* **22**, 51–60.

KLAASSEN, B., PAAP, K.L. (1987). In: Proebster, W.E., Reiner, H. (eds.), *Proc. VLSI and Computers, CompEuro87* (IEEE Computer Society, Washington, DC), pp. 238–241. ISBN 0-8186-0773-4.

KUNDERT, K.S., WHITE, J.K., SANGIOVANNI-VINCENTELLI, A. (1990). *Steady-State Methods for Simulating Analog and Microwave Circuits* (Kluwer, Boston).

KUNDERT, K.S. (1995). *The Designer's Guide to Spice & Spectre* (Kluwer Academic, Boston).

KUNDERT, K.S. (1997). Simulation methods for RF integrated circuits. In: *Proc. ICCAD'97, San Jose*.

KUROKAWA, K. (1969). Some basic characteristics of broadband negative resistance oscillator circuits. *Bell System. Techn. J.* **48**, 1937–1955.

KVÆRNØ, A., RENTROP, P. (1999). Low order multirate Runge–Kutta methods in electric circuit simulation. Preprint 99/1 (IWRMM, University of Karlsruhe, Karlsruhe).

LAMOUR, R. (1998). Floquet theory for differential-algebraic equations (DAE). *ZAMM* **78** (3), S989–S990.

LAMOUR, R., MÄRZ, R., WINKLER, R. (1998a). How Floquet theory applies to index 1 differential-algebraic equations. *J. Math. Anal. Appl.* **217**, 372–394.

LAMOUR, R., MÄRZ, R., WINKLER, R. (1998b). Stability of periodic solutions of index-2 differential-algebraic systems. Preprint 98-23 (Humboldt Universität zu Berlin, Berlin).

LAMPE, S., BRACHTENDORF, H.G., TER MATEN, E.J.W., ONNEWEER, S.P., LAUR, R. (2001). Robust limit cycle calculations of oscillators. In: Van Rienen, U., et al. (eds.), *Proc. SCEE-2000, Warnemünde*. In: Lecture Notes Comp. Sci. Eng. **18** (Springer, Berlin), pp. 233–240.

LAMPE, S., LAUR, R. (2002). Initialisierungsprozedur für die Oszillatorsimulation basierend auf globalen Optimierungstechniken. In: *ANALOG 2002, Bremen*, pp. 81–86.

LEIMKUHLER, B.J. (1986). Error estimates for differential-algebraic equations. Report UIUCDCS-R-86-1287 (University of Illinois, Urbana).

LELARASMEE, E., RUEHLI, A.E., SANGIOVANNI-VINCENTELLI, A. (1982). The waveform relaxation method for time-domain analysis of large scale integrated circuits. *IEEE Trans. Comp. Aided Des. CAD* **1**, 131–145.

LIN, S., KUH, E.S., MAREK-SADOWSKA, M. (1993). Stepwise equivalent conductance circuit simulation technique. *IEEE Trans. Comp. Aided Des. CAD* **12**, 672–683.

MA, W., TRAJKOVĆ, L., MAYARAM, K. (2002). HomSSPICE: A homotopy-based circuit simulator for periodic-steady state analysis of oscillators. Paper presented at ISCAS-2002.

MÄRZ, R. (1992). Numerical methods for differential-algebraic equations. *Acta Numerica*, 141–198.

MÄRZ, R., TISCHENDORF, C. (1997). Recent results in solving index 2 differential-algebraic equations in circuit simulation. *SIAM J. Sci. Stat. Comp.* **18** (1), 139–159.

MARKOWITZ, H. (1957). The elimination form of the inverse and its application to linear programming. *Management Sci.* **3**, 255–269.

MATHIS, W. (1987). *Theorie nichtlinearer Netzwerke* (Springer, Berlin).

MATHIS, W., MAURITZ, H., ZHUANG, M. (1994). Entwurf von ODE- und DAE-Lösungsverfahren mit variabler Schrittweite nach regelungs-technischen Prinzipien: Möglichkeiten und Grenzen, 2. In: *Workshop "Identifizierungs-, Analyse- und Entwurfsmethoden für Deskriptorsysteme", Paderborn.*

MATHIS, W. (1998). An efficient method for the transient analysis of weakly damped crystal oscillators. In: *Proc. MTNS'98, Padova*, pp. 313–316.

MEETZ, K., ENGL, W.L. (1980). *Elektromagnetische Felder* (Springer, Berlin).

MELVILLE, R.C., FELDMANN, P., ROYCHOWDHURY, J. (1995). Efficient multi-tone distortion analysis of analog integrated circuits. In: *Proc. CICC'97, Santa Clara*, pp. 241–244.

MEYER, J.E. (1971). MOS models and circuit simulation. *RCA Rev.* **32**, 42–63.

MIURA-MATTAUSCH, M., FELDMANN, U., RAHM, A., BOLLU, M., SAVIGNAC, D. (1996). Unified complete MOSFET model for analysis of digital and analog circuits. *IEEE Trans. Comp. Aided Des. CAD* **15**, 1–7.

NAGEL, W. (1975). SPICE 2 – A computer program to simulate semiconductor circuits. MEMO ERL-M 520 (University of California Berkeley, Berkeley).

NEUBERT, R., SELTING, P., ZHENG, Q. (1998). Analysis of autonomous oscillators – A multistage approach. In: *Proc. MTNS'98, Padova*, pp. 1055–1058.

NEUBERT, A., SCHWARZ, A. (2001). Efficient analysis of oscillatory circuits. In: Van Rienen, U., et al. (eds.), *Proc. SCEE-2000, Warnemünde*. In: Lecture Notes Comp. Sci. Eng. **18** (Springer, Berlin), pp. 225–232.

NEWTON, A.R., SANGIOVANNI-VINCENTELLI, A. (1984). Relaxation-based electrical simulation. *IEEE Trans. Comp. Aided Des. CAD* **3**, 308–330.

ODENT, P., CLAESEN, L., DE MAN, H. (1990). Acceleration of relaxation-based circuit simulation using a multiprocesor system. *IEEE Trans. Comp. Aided Des. CAD* **9**, 1063–1072.

ODRYNA, P., NASSIF, S. (1986). The ADEPT timing simulation program. In: *VLSI Systems Design* (March), pp. 24–34.

OKUMURA, M., TANIMOTO, H., ITAKURA, T., SUGAWARA, T. (1993). Numerical noise analysis for nonlinear circuits with a periodic large signal excitation including cyclostationary noise sources. *IEEE Trans. Circ. Syst. CAS I* **40**, 581–590.

ONOZUKA, H., KANH, M., MIZUTA, C., NAKATA, T., TANABE, N. (1993). Development of parallelism for circuit simulation by tearing. In: *Proc. Europ. Des. Autom. Conf. EDAC'93*, pp. 12–17.

PALUSINSKI, O.A., GUARINI, M.W., WRIGHT, S.J. (1988). Spectral technique in electronic circuit analysis. *Int. J. Num. Modelling: Electr. Netw., Dev. Fields* **1**, 137–151.

PANTELIDES, C.C. (1988). The consistent initialization of differential-algebraic systems. *SIAM J. Sci. Stat. Comp.* **9**, 213–231.

PETZOLD, L.R. (1981). An efficient numerical method for highly oscillatory ordinary differential equations. *SIAM J. Numer. Anal.* **18** (3), 455–479.

PETZOLD, L.R., JAY, L., YEN, J. (1997). Numerical solution of highly oscillatory ordinary differential equations. *Acta Numerica* **6**, 437–484.

PILLAGE, L.T., ROHRER, R.A. (1990). Asymptotic waveform evaluation of timing analysis. *IEEE Trans. Comp. Aided Des. CAD* **9**, 352–366.

PULCH, R., GÜNTHER, M. (2002). A method of characteristics for solving multirate partial differential equations in radio frequency application. *Appl. Numer. Math.* **42**, 397–409.

RABBAT, G., HSIEH, H.-Y. (1981). Hierarchical computer-aided circuit design techniques. In: *Proc. of the 1981 European Conference on Circuit Theory and Design, The Hague* (Delft University Press, Delft), pp. 136–144.

RABBAT, N.B.G., SANGIOVANNI-VINCENTELLI, A.L., HSIEH, H.Y. (1979). A multilevel Newton algorithm with macromodeling and latency for the analysis of large-scale nonlinear circuits in the time domain. *IEEE Trans. Circ. Syst. CAS* **26**, 733–741.

RABIER, P., RHEINBOLDT, W. (1996). Time-dependent linear DAE's with discontinuous inputs. *J. Linear Algebra Appl.* **247**, 1–29.

RABIER, P.J., RHEINBOLDT, W.C. (2002). Theoretical and numerical analysis of differential-algebraic equations. In: Ciarlet, P.G., Lions, J.L. (eds.), *Handbook of Numerical Analysis, vol. VIII: Techniques of Scientific Computing (Part 4)* (Elsevier Science, Amsterdam), pp. 183–540.

RATZLAFF, C.L., PILLAGE, L.T. (1994). RICE: Rapid interconnect circuit evaluation using AWE. *IEEE Trans. Comp. Aided Des. CAD* **13**, 763–776.

REISSIG, G., FELDMANN, U. (1996). Computing the generic index of the circuit equations of linear active networks. In: *Proc. ISCAS'96, Atlanta, vol. III*, pp. 190–193.

REISSIG, G. (1998). Beiträge zu Theorie und Anwendung impliziter Differentialgleichungen. PhD thesis (TU Dresden, Dresden).

REISSIG, G. (2001). On the performance of minimum degree and minimum local fill heuristics in circuit simulation. Report Febr. 2001 (Dept. Chem. Eng., MIT, Cambridge).

RENTROP, P., ROCHE, M., STEINEBACH, G. (1989). The application of Rosenbrock–Wanner type methods with stepsize control in differential-algebraic equations. *Numer. Math.* **55**, 545–563.

RENTROP, P. (1990). ROW-type methods for the integration of electric circuits. In: Bank, R.E., et al. (eds.), *Proc. Oberwolfach Conf.* In: Int. Ser. Num. Math. **93** (Birkhäuser, Basel), pp. 59–71.

RICE, J.R. (1960). Split Runge–Kutta methods for simultaneous equations. *J. Res. Natl. Buv. Stand. B* **64**, 151–170.

RÖSCH, M. (1992). Schnelle Simulation des stationären Verhaltens nichtlinearer Schaltungen, PhD thesis (TU München, München).

RÖSCH, M., ANTREICH, K. (1992). Schnelle stationäre Simulation nichtlinearer Schaltungen im Frequenzbereich. *AEÜ* **46**, 168–176.

ROYCHOWDHURY, J. (1997). Efficient methods for simulating highly nonlinear multi-rate circuits. In: *Proc. DAC'97, Anaheim*, pp. 269–274.

ROYCHOWDHURY, J., FELDMANN, P. (1997). A new linear-time Harmonic Balance algorithm for cyclostationary noise analysis in RF. In: *Proc. ASP-DAC'97, Chiba*, pp. 483–492.

ROYCHOWDHURY, J., LONG, D., FELDMANN, P. (1998). Cyclostationary noise analysis of large RF circuits with multitone excitations. *IEEE J. Solid-State Circ.* **33** (3), 324–336.

ROYCHOWDHURY, J. (2001a). Analyzing circuits with widely-separated time scales using numerical PDE methods. *IEEE Trans. Circ. Syst. CAS I* **48**, 578–594.

ROYCHOWDHURY, J. (2001b). Multi-time PDEs for dynamical system analysis. In: Van Rienen, U., et al. (eds.), *Proc. SCEE-2000, Warnemünde*. In: Lecture Notes Comp. Sci. Eng. **18** (Springer, Berlin), pp. 3–14.

SAAD, Y. (1996). *Iterative Methods for Sparse Linear Systems* (PWS Publ., Boston).

SACKS-DAVIS, R. (1972). Error estimates for a stiff differential equation procedure. *SIAM J. Stat. Comp.* **3**, 367–384.

SADAYAPPAN, P., VISVANATHAN, V. (1988). Circuit simulation on shared-memory multiprocessors. *IEEE Trans. Computers* **37** (12), 1634–1642.

SAKALLAH, K.A., DIRECTOR, S.W. (1985). SAMSON2: An event driven VLSI circuit simulator. *IEEE Trans. Comp. Aided Des. CAD* **4**, 668–685.

SAKALLAH, K.A., YEN, Y., GREENBERG, S.S. (1990). A first-order charge conserving MOS capacitance model. *IEEE Trans. Comp. Aided Des. CAD* **9**, 99–108.

SALEH, R.A., KLECKNER, J.E., NEWTON, A.R. (1983). Iterated timing analysis in SPLICE1. In: *Proc. ICCAD'83, Santa Clara*, pp. 139–140.

SALEH, R., WEBBER, D., XIA, E., SANGIOVANNI-VINCENTELLI, A. (1987). Parallel waveform Newton algorithms for circuit simulation. In: *Proc. Int. Conf. Comp. Des.'87, Port Chester*, pp. 660–668.

SALEH, R.A., WHITE, J.K. (1990). Accelerating relaxation algorithms for circuit simulation using waveform-Newton and step-size refinement. *IEEE Trans. Comp. Aided Des. CAD* **9**, 951–958.

SANGIOVANNI-VINCENTELLI, A., CHEN, L.K., CHUA, L.O. (1977). A new tearing approach – Node-tearing nodal analysis. In: *Proc. ISCAS'77, Phoenix*, pp. 143–147.

SARKANI, E., LINIGER, W. (1974). Exponential fitting of matricial multistep methods for ordinary differential equations. *Math. Comput.* **28**, 1035–1052.

SCHENK, O. (2000). Scalable parallel sparse LU factorization methods on shared memory multiprocessors. PhD thesis (ETH Zürich, Zürich) (ISBN-3-89649-532-1, Series in Microelectronics 89, Hartung-Gorre, Konstanz).

SCHILDERS, W. (2000). Iterative Solution of linear systems in circuit simulation. In: *Progress in Industrial Mathematics at ECMI 2000, Mathematics in Industry, vol. 1* (Springer, Berlin), pp. 272–277.

SCHMIDT-KREUSEL, C. (1997). Zur rechnergestützten Analyse von Quarzoszillatoren. PhD thesis (Bergische Universität Wuppertal, Wuppertal).

SEVAT, M.F. (1994). Fourier transformation for harmonic balance. Nat. Lab. Report Nr. 6813/94 (Philips Research Laboratories).

SELTING, P., ZHENG, Q. (1997). Numerical stability analysis of oscillating integrated circuits. *J. Comp. Appl. Math.* **82**, 367–378.

SIEBER, E.-R., FELDMANN, U., SCHULTZ, R., WRIEDT, H. (1994). Timestep control for charge conserving integration in circuit simulation. In: Bank, R.E., et al. (eds.), *Proc. Oberwolfach Conf.* In: Int. Ser. Num. Math. **117** (Birkhäuser, Basel), pp. 103–113.

SILVEIRA, L.M., WHITE, J.K., NETO, H., VIDIGAL, L. (1992). On exponential fitting for circuit simulation. *IEEE Trans. Comp. Aided Des. CAD* **11**, 566–574.

SKELBOE, S. (1982). Time-domain steady-state analysis of nonlinear electrical systems. *Proc. IEEE* **70**, 1210–1228.

SKELBOE, S. (1984), Multirate integration methods. Report ECR-150 (University of Horsholm, Horsholm).

SMITH, D.A., FORD, W.F., SIDI, A. (1987). Extrapolation methods for vector sequences. *SIAM Rev.* **29** (2), 199–233.

SÖDERLIND, G. (2001). Automatic control and adaptive time-stepping. In: *Proc. ANODE 2001, Auckland*.

STRIEBEL, M., GÜNTHER, M. (2002). Towards distributed time integration in full chip design. *Appl. Numer. Math.*

TELICHEVESKY, R., KUNDERT, K.S., WHITE, J.K. (1995). Efficient steady-state analysis based on matrix-free Krylov-subspace methods. In: *Proc. DAC'95, San Francisco*, pp. 480–484.

TELICHEVESKY, R., KUNDERT, K.S., WHITE, J. (1996). Efficient AC and noise analysis of two-tone RF circuits. In: *Proc. DAC'96, Las Vegas*, pp. 292–297.

TELICHEVESKY, R., KUNDERT, K.S., ELFADEL, I., WHITE, J.K. (1996). Fast simulation algorithms for RF circuits. In: *Proc. CICC'96, San Diego*, pp. 437–444.

TER MATEN, E.J.W. (1999). Numerical methods for frequency domain analysis of electronic circuits. *Surv. Math. Ind.* **8**, 171–185.

TISCHENDORF, C. (1999). Topological index calculation of differential-algebraic equations in circuit simulation. *Surv. Math. Ind.* **8**, 187–199.

TROYANOVSKY, B. (1997). Frequency domain algorithms for simulating large signal distortion in semiconductor devices. PhD thesis (Stanford University, Stanford).

URUHAMA, K. (1987). Convergence of relaxation-based circuit simulation techniques. *IEICE Trans. E* **70**, 887–889.

URUHAMA, K. (1988). Relaxation based circuit simulation. *IEICE Trans. E* **71**, 1189–1194.

VALSA, J., VLACH, J. (1995). SWANN – A program for analysis of switched analog nonlinear networks. In: *Proc. ISCAS'95, Seattle*, pp. 1752–1755.

VAN BOKHOVEN, W.M.G. (1987). Piecewise linear solution techniques. In: Ruehli, A.E. (ed.), *Circuit Analysis, Simulation and Design, Part II* (North-Holland, Amsterdam), pp. 41–127.

VAN EIJNDHOVEN, J.T.J. (1984). A piecewise linear simulator for large scale integrated circuits. PhD thesis (Eindhoven University of Technology, Eindhoven).

VIDIGAL, L., NASSIF, S., DIRECTOR, S. (1986). CINNAMON: Coupled integration and nodal analysis of MOS networks. In: *Proc. DAC'86, Las Vegas*, pp. 179–185.

VLACH, M. (1988a). LU decomposition algorithms for parallel and vector computation. In: Ozawa, T. (ed.), *Analog Methods for Computer-Aided Circuit Analysis and Design* (Marcel Dekker, New York), pp. 37–64.

VLACH, M. (1988b). Decomposition techniques in circuit simulation. In: Ozawa, T. (ed.), *Analog Methods for Computer-Aided Circuit Analysis and Design* (Marcel Dekker, New York), pp. 93–115.

WALLAT, O. (1997). Partitionierung und Simulation elektrischer Netzwerke mit einem parallelen mehrstufigen Newton-Verfahren. PhD thesis (Universität Hamburg, Hamburg).

WANG, S., DENG, A.-C. (2001). Delivering a full-chip hierarchical circuit simulation & analysis solution for nanometer designs. White paper (Nassda Corporation).

WARD, D.E., DUTTON, R.W. (1978). A charge-oriented model for MOS transistor capacitances. *IEEE J. Solid-State Circ. SC* **13**, 703–708.

WEHRHAHN, E. (1989). Hierarchical circuit analysis. In: *1989 IEEE International Symposium on Circuits and Systems (Cat. No. 89CH2692-2), Portland, OR, USA, vol. 1*, pp. 701–704.

WEHRHAHN, E. (1991). Hierarchical sensitivity analysis of circuits. In: *1991 IEEE International Symposium on Circuits and Systems (Cat. No. 91CH3006-4), Singapore, vol. 2*, pp. 864–867.

WELSCH, G. (1998), Analyse des eingeschwungenen Zustands autonomer und nicht-autonomer elektronischer Schaltungen. PhD thesis (Universität Bremen, Bremen) (Shaker, Aachen) ISBN 3-8265-4233-9.

WELSCH, G., BRACHTENDORF, H.-G., SABELHAUS, C., LAUR, R. (2001). Minimization of the error in the calculation of the steady state by shooting methods. *IEEE Trans. Circ. Syst. I. Fund. Theory Appl.* **48** (10), 1252–1257.

WEVER, U., ZHENG, Q. (1996). Parallel transient analysis for circuit simulation. In: *Proc. 29th Annual Hawaii International Conference on System Sciences, Hawaii*, pp. 442–447.

WHITE, J.K., SANGIOVANNI-VINCENTELLI, A. (1987). *Relaxation Techniques for the Simulation of VLSI Circuits* (Kluwer, Boston).

WIEDL, W. (1994). Multilevel Newton Verfahren in der Transientenanalyse elektrischer Netzwerke. In: Bank, R.E., et al. (eds.), *Proc. Oberwolfach Conf.* In: Int. Ser. Numer. Math. **117** (Birkhäuser, Basel), pp. 103–113.

WILKINSON, J.H. (1982). Note on the practical significance of the Drazin inverse. In: Campbell, S.L. (ed.), *Recent Applications of Generalized Inverses* (Pitman, London), pp. 82–99.

WING, O., GIELCHINSKY, J. (1972). Computation of time response of large networks by partitioning. In: *Proc. Int. Symp. Circ. Theory*, pp. 125–129.

WU, F.F. (1976). Solution of large-scale networks by tearing. *IEEE Trans. Circ. Syst. CAS* **23**, 706–713.

YU, Q., WING, O. (1984). PLMAP: A piecewise linear MOS circuit analysis program. In: *Proc. ISCAS'84, Montreal*, pp. 530–533.

ZECEVIC, A.I., GACIC, N. (1999). A partitioning algorithm for the parallel solution of differential-algebraic equations by waveform relaxation. *IEEE Trans. Circ. Syst. CAS I* **46**, 421–434.

ZHANG, X., (1989). Parallel computation for the solution of block bordered nonlinear equations and their applications. PhD thesis (University of Colorado, Boulder).

ZHANG, X., BYRD, R.H., SCHNABEL, R.B. (1992). Parallel methods for solving nonlinear block bordered systems of equations. *SIAM J. Sci. Stat. Comp.* **13**, 841–859.

ZHENG, Q. (1994). The transient behavior of an oscillator. In: Bank, R.E., et al. (eds.), *Proc. Oberwolfach Conf.* In: Int. Ser. Numer. Math. **117** (Birkhäuser, Basel), pp. 143–154.

ZHENG, Q., NEUBERT, R. (1997). Computation of periodic solutions of differential-algebraic equations in the neighborhood of Hopf bifurcation points. *Int. J. Bifurcation and Chaos* **7**, 2773–2779.

Further reading

ANTOGNETTI, P., MASSOBRIO, G. (1987). *Semiconductor Device Modeling with SPICE* (McGraw-Hill, New York).

ANZILL, W., KÄRTNER, F.X., RUSSER, P. (1994). Simulation of the phase noise of oscillators in the frequency domain. *Int. J. Electron. Commun. (AEÜ)* **48**, 45–50.

BOMHOF, W., VAN DER VORST, H.A. (2000). A parallel linear system solver for circuit simulation problems. *Numer. Linear Algebra Appl.* **7**, 649–665.

BRACHTENDORF, H.G., MELVILLE, R., FELDMANN, P., LAMPE, S. (2002). Steady state calculation of oscillators using continuation methods. In: *Proc. Design Automation and Test in Europe (DATE 2002), Paris, France*, p. 1139.

BRAMBILLA, A., MAFFEZONI, P. (2000). Envelope following method for the transient analysis of electrical circuits. *IEEE Trans. Circ. Syst. CAS I* **47**, 999–1008.

CHANG, C.R., STEER, M.B., MARTIN, S., REESE JR, E. (1991). Computer-aided analysis of free-running microwave oscillators. *IEEE Trans. Microwave Theory Techniques* **39**, 1735–1745.

DEMIR, A., LIU, E.W.Y., SANGIOVANNI-VINCENTELLI, A. (1996). Time-domain non-Monte Carlo noise simulations for nonlinear dynamic circuits with arbitrary excitations. *IEEE Trans. Comp. Aided Des. CAD* **15**, 493–505.

DEMIR, A., SANGIOVANNI-VINCENTELLI, A. (1998). *Analysis and Simulation of Noise in Nonlinear Electronic Circuits and Systems* (Kluwer, Boston).

DUTTON, R.W., TROYANOVSKY, B., YU, Z., ARNBORG, T., ROTELLA, F., MA, G., SATO-IWANAGA, J. (1997). Device simulation for RF applications. In: *Proc. IEDM'97, Washington, DC*.

ESTÉVEZ SCHWARZ, D. (1999b). Consistent initialization for index-2 differential-algebraic equations and its application to circuit simulation. Preprint 99-5 (Institut für Mathematik, Humboldt-Universität zu Berlin, Berlin).

FELDMANN, P., MELVILLE, B., LONG, D. (1996). Efficient frequency domain analysis of large nonlinear analog circuits. In: *Proc. CICC'96, San Diego*, pp. 461–464.

FENG, D., PHILLIPS, J., NABORS, K., KUNDERT, K., WHITE, J. (1999). Efficient computation of quasi-periodic circuit operating conditions via a mixed frequency/time approach. In: *Proc. DAC'99, New Orleans*, pp. 635–640.

FILSETH, E.S., KUNDERT, K.S. (1996). Simulations strategies for RF design. *Electr. Eng.* **68** (832), 43–50.

GILMORE, R.J., STEER, M.B. (1991). Nonlinear circuit analysis using the method of Harmonic Balance – A review of the art: I. Introductory concepts, II. Advanced concepts. *Int. J. Microwave Millimeter-Wave Comp.-Aided Eng.* **1** (1), 22–37 (Part I), 2–1, 159–180 (Part II).

GÜNTHER, M. (1995). *Ladungsorientierte Rosenbrock–Wanner-Methoden zur numerischen Simulation digitaler Schaltungen*. PhD thesis (TU München, München) (VDI-Verlag, Düsseldorf) ISBN 3-18-316820-0.

GÜNTHER, M., RENTROP, P. (1996). The NAND-gate – A benchmark for the numerical simulation of digital circuits. In: Mathis, W., Noll, P. (eds.), *Proc. 2nd ITG Workshop 1995, Berlin* (VDE-Verlag, Berlin), pp. 27–33. ISBN 3-8007-2190-2.

GÜNTHER, M., RENTROP, P. (1999). PDAE-Netzwerkmodelle in der elektrischen Schaltungssimulation. In: *Proc. Analog'99, München*, pp. 31–38.

KANAGLEKAR, N., SIFRI, J. (1997). Integrate CAD methods for accurate RF IC simulation. *Microwaves RF*, 88–101.

KATO, T., TACHIBANA, W. (1998). Periodic steady-state analysis of an autonomous power electronic system by a modified shooting method. *IEE Trans. Power Electr.* **13** (3), 522–527.

LAMOUR, R. (1997). A shooting method for fully implicit index-2 differential-algebraic equations. *SIAM J. Sci. Stat. Comp.* **18**, 94–114.

MAYARAM, K., LEE, D.C., MOINIAN, S., RICH, D., ROYCHOWDHURY, J. (1997). Overview of computer-aided analysis tools for RFIC: Algorithms, features, and limitations. In: *Proc. CICC'97, Santa Clara*, pp. 505–512.

NGOYA, E., SUÁREZ, A., SOMMET, R., QUÉRÉ, R. (1995). Steady state analysis of free and forced oscillators by Harmonic Balance and stability investigation of periodic and quasi-periodic regimes. *Int. J. Microwave Millimeter-Wave Comp.-Aided Eng.* **5** (3), 210–223.

OKUMURA, M., SUGAWARA, T., TANIMOTO, H. (1990). An efficient small signal frequency analysis method for nonlinear circuits with two frequency excitations. *IEEE Trans. Comp. Aided Des. CAD* **9**, 225–235.

RENTROP, P., STREHMEL, K., WEINER, R. (1996). Ein Überblick über Einschrittverfahren zur numerischen Integration in der technischen Simulation. *GAMM Mitteilungen* **19** (1), 9–43.

SIMEON, B. (1998). Order reduction of stiff solvers at elastic multibody systems. *Appl. Numer. Math.* **28**, 459–475.

TER MATEN, E.J.W., VAN DE WIEL, M.C.J. (1999). Time-domain simulation of noise in dynamic nonlinear circuits. In: Arkeryd, L., et al. (eds.), *Proc. ECMI'98, Göteborg* (Teubner, Stuttgart), pp. 413–422.

TISCHENDORF, C. (1996). Solution of index-2 differential algebraic equations and its application in circuit simulation. PhD thesis (Humboldt Univ. zu Berlin, Logos Verlag Berlin).

VAN DER HOUWEN, P.J., SOMMEIJER, B.P. (1987). Explicit Runge–Kutta–Nyström methods with reduced phase errors for computing oscillating solutions. *SIAM J. Numer. Anal.* **24**, 595–617.

Simulation of EMC Behaviour

A.J.H. Wachters

Philips Research Laboratories, Prof. Holstlaan 4,
5656 AA, Eindhoven, The Netherlands
E-mail address: wachters@natlab.research.philips.com

W.H.A. Schilders

Philips Research Laboratories, IC Design, Prof. Holstlaan 4,
5656 AA, Eindhoven, The Netherlands
E-mail address: wil.schilders@philips.com

1. Introduction

In this chapter we describe methods that have been used to perform simulations of the electromagnetic behaviour of multilayer interconnection systems. Such a system consists of a number of planar conductors immersed in a configuration of homogeneous media of different permittivity bound by parallel planes. Examples of such systems are printed circuit boards, IC packages, filters and passive IC's. The program Fasterix, developed within Philips Research, has been used to simulate the electromagnetic behaviour of a variety of such systems (see DU CLOUX, MAAS and WACHTERS [1994]). One of the reasons for developing this programme were the strict regulations as far as electromagnetic compatibility are concerned. Electronic devices influence each other, but this influence should be kept to a minimum. This explains the large interest in simulations of EMC behaviour in the past 10 years. The present chapter is devoted to this subject, and gives a very detailed impression of how these simulations are enabled in practice. The development of numerical algorithms to solve the EMC problems is rather involved. A strong interplay is required between analytical and numerical techniques in order to obtain an efficient way of simulating devices. Evaluating fourfold integrals with singularities is not a trivial task, and requires a lot of tedious work. The chapter also shows that, even in rather elementary tasks like numerical integration, sophisticated

Numerical Methods in Electromagnetics
Special Volume (W.H.A. Schilders and E.J.W. ter Maten, Guest Editors) of
HANDBOOK OF NUMERICAL ANALYSIS, VOL. XIII
P.G. Ciarlet (Editor)

ISSN 1570-8659
DOI 10.1016/S1570-8659(04)13007-X

algorithms must be used to handle the complexity of the problem. This is characteristic for the present chapter: several methods that are not very well known will be discussed, such as the Kronrod and Patterson quadrature rules, and Orden's method for solving indefinite linear systems.

The structure of this chapter is as follows. Section 2 contains a derivation of the *Kirchhoff equations* from the *Maxwell equations* (also see Chapter 1 in this book), whereas Sections 3–7 contain the numerical methods required for the evaluation of the fourfold integrals that form the matrix elements in these equations. These matrix elements represent the resistors, inductors and capacitors in the equivalent circuit model for the interconnection system.

Section 8 presents an improvement of the method, treated in Section 5, for the analytical integration of the inner integrals for vector valued basis functions for a quadrilateral element. This improvement makes it possible to use non-planar conducting structures, such as bond wires, screens and boxes in a simulation. The analytical integration of the inner integrals over a triangular element for scalar and vector valued basis functions is discussed in Section 9.

Section 10 presents the solution methods used for the Kirchhoff equations. The linear algebra methods used for the solution of the linear system of equations and the generalized eigenvalue problems involved are discussed in Section 11. An efficient method for solving equations with large matrices is given in Section 12.

2. Derivation of Kirchhoff equations

In this section a derivation will be given of the Kirchhoff equations which describe the behaviour of an equivalent circuit of a PCB. For this purpose an equivalent boundary value problem will be derived from the Maxwell's equations. Next, we present a variational formulation of this problem and the function spaces that contain its weak solutions. To be able to compute these solutions the problem domain is subdivided into quadrilateral elements and the solutions are approximated by linear combinations of basis functions in finite dimensional subspaces of the original function spaces. We obtain a linear set of equations, which correspond to the *Kirchhoff equations*, the solutions of which represent the currents, charges and potentials in an electronic circuit. The matrix elements belonging to this set of equations are integrals. The fourfold integrals that represent the electromagnetic interaction between charges and currents in two elements of the discrete domain will be called *interaction integrals*.

Since in this chapter our attention is restricted to the evaluation methods for the interaction integrals, the derivation of the Kirchhoff equations will be given in a simplified form. For a more rigorous derivation see DU CLOUX, MAAS and WACHTERS [1994], or Chapter 1 in this volume.

2.1. *Introduction*

A PCB consists of a set of thin metal layers separated by dielectric layers. The metal layers are the conductors (see Fig. 2.1) that connect the external electronic components mounted on the PCB. By currents through the conductors electromagnetic fields will

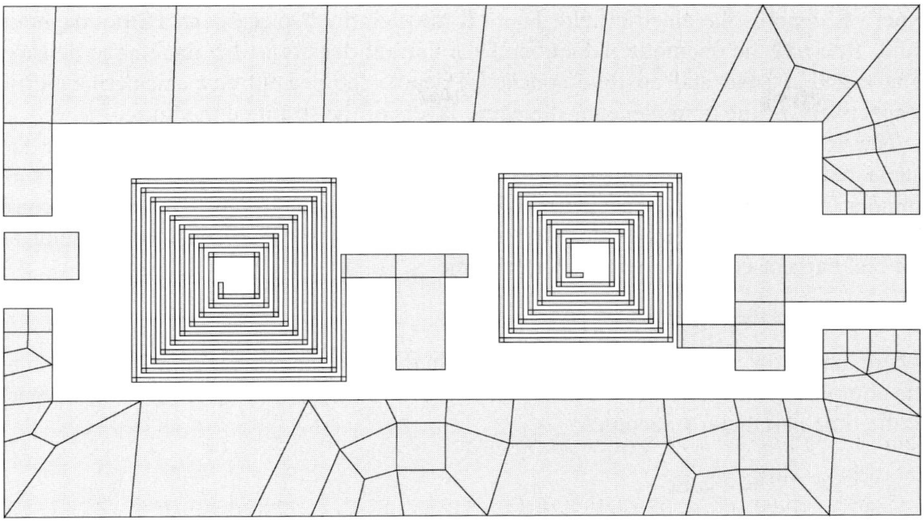

FIG. 2.1. A printed circuit board viewed from above. The patterns of the conductors are described by polygons, which are divided into quadrilaterals.

be generated that can cause crosstalk between the parts of an electronic system. This crosstalk often interferes with the desired signals and manifests itself as noise. For high frequencies this disturbance can sometimes be significantly large. A study of these electromagnetic fields may provide understanding of this disturbing interference between parts of an electronic system.

As will be demonstrated later in this chapter, the electromagnetic properties of a PCB can be translated into an equivalent circuit model (see also Chapter 9 in this volume for more general techniques). By putting such model, together with those for the external components, in a circuit analysis program, the potentials at the circuit nodes can be obtained. From these potentials the currents through the conductors and the radiated electromagnetic fields can be calculated. In fact, this provides a way of performing a coupled analysis of circuit behaviour and electromagnetics effects.

2.2. Maxwell's equations

The electromagnetic field in this electronic system can be described by the *Maxwell's equations*:

$$\nabla \times \mathbf{E} = -\frac{\partial \mathbf{B}}{\partial t},$$

$$\nabla \times \mathbf{H} = \mathbf{J} + \frac{\partial \mathbf{D}}{\partial t},$$

$$\nabla \cdot \mathbf{B} = 0,$$

$$\nabla \cdot \mathbf{D} = \rho,$$

where $\mathbf{E}$ denotes the electric field, $\mathbf{D} = \varepsilon\mathbf{E}$ the electric displacement, $\mathbf{H}$ the magnetic field, $\mathbf{B} = \mu\mathbf{H}$ the magnetic induction, $\mathbf{J}$ the current density and ρ the charge density. The ε and μ (real and positive constants) denote the permittivity and permeability, respectively, of the homogeneous dielectric layers of the stratified medium.

The vectors in the above system are real. For the time-periodic case, which we are interested in, the electromagnetic field is assumed to vary sinusoidally with time, with angular frequency $\omega \geqslant 0$. This periodic behaviour of the field can be expressed by complex vectors. It is customary (see RAMO [1984, Section 3.8]) to consider the fields as the real parts of complex vectors, for example

$$\mathbf{E}(\mathbf{x}, t) = \mathrm{Re}\big[\mathbf{E}(\mathbf{x}, \omega)\mathrm{e}^{-\mathrm{i}\omega t}\big],$$

where the *vector phasor* $\mathbf{E}(\mathbf{x}, \omega)$ is complex. In the following, the arguments $\mathbf{x}$ and ω of the complex quantities will be omitted, so that $\mathbf{E} = \mathbf{E}(\mathbf{x}, \omega)$. The derivative with respect to the time parameter t becomes

$$\frac{\partial}{\partial t}\mathbf{E}\mathrm{e}^{-\mathrm{i}\omega t} = -\mathrm{i}\omega\mathbf{E}.$$

Hence, for harmonic fields the Maxwell equations are given by

$$\nabla \times \mathbf{E} = \mathrm{i}\omega\mathbf{B}, \tag{2.1}$$

$$\nabla \times \mathbf{H} = \mathbf{J} - \mathrm{i}\omega\mathbf{D}, \tag{2.2}$$

$$\nabla \cdot \mathbf{B} = 0, \tag{2.3}$$

$$\nabla \cdot \mathbf{D} = \rho. \tag{2.4}$$

Since $\nabla \cdot (\nabla \times \mathbf{H}) = 0$, from (2.2) and (2.4) the *current continuity equation* follows:

$$\nabla \cdot \mathbf{J} - \mathrm{i}\omega\rho = 0. \tag{2.5}$$

Further the following relations hold:

$$\mathbf{J} = \sigma\mathbf{E} \quad \text{(Ohm's law)}, \tag{2.6}$$

$$\mathbf{D} = \varepsilon\mathbf{E}, \tag{2.7}$$

$$\mathbf{B} = \mu\mathbf{H}, \tag{2.8}$$

where the material properties σ and ε are assumed to be constant for each layer and μ is constant for the whole problem region.

2.3. *Equations to be solved*

The Maxwell equations form a set of coupled first order partial differential equations which give relations between electric and magnetic fields. In view of later applications it is convenient to introduce potentials (see JACKSON [1975, Section 6.4, pp. 219–220]) to obtain a smaller number of (second-order) equations, equivalent to the Maxwell equations.

The properties $\nabla \cdot (\nabla \times \mathbf{u}) = 0$ and $\nabla \times (\nabla\mathbf{u}) = 0$ for any $\mathbf{u} \in \mathbb{R}^3$ lead to the following observations.

Since $\nabla \cdot \mathbf{B} = 0$, the magnetic induction $\mathbf{B}$ can be defined in terms of a magnetic vector potential that satisfies

$$\mathbf{B} = \nabla \times \mathbf{A}. \tag{2.9}$$

Then, Eq. (2.1) can be rewritten as

$$\nabla \times (\mathbf{E} - \mathrm{i}\omega\mathbf{A}) = 0.$$

The argument of the rotation ($\nabla\times$) can be written as the gradient of some scalar function, namely the electric potential φ, so that

$$\mathbf{E} - \mathrm{i}\omega\mathbf{A} = -\nabla\varphi. \tag{2.10}$$

These definitions of the potentials are consistent with (2.1) and (2.3). By the other Maxwell's equations, (2.2) and (2.4), restrictions are imposed on these potentials. From (2.8) and (2.2) it follows that

$$\nabla \times \mathbf{B} = \mu(\nabla \times \mathbf{H}) = \mu\mathbf{J} - \mathrm{i}\omega\mu\mathbf{D}.$$

After substitution of (2.9), (2.7), (2.10) and property

$$\nabla \times (\nabla \times \mathbf{A}) = \nabla(\nabla \cdot \mathbf{A}) - \Delta\mathbf{A},$$

in this expression and

$$\mathbf{D} = \varepsilon\mathbf{E} = \varepsilon(-\nabla\varphi + \mathrm{i}\omega\mathbf{A}),$$

in (2.4) one obtains

$$\nabla(\nabla \cdot \mathbf{A}) - \Delta\mathbf{A} - \mathrm{i}\omega\mu\varepsilon\nabla\varphi - \omega^2\mu\varepsilon\mathbf{A} = \mu\mathbf{J}, \tag{2.11}$$

$$-\nabla \cdot (\varepsilon\nabla\varphi) + \mathrm{i}\omega\varepsilon\nabla \cdot \mathbf{A} = \rho. \tag{2.12}$$

Hence, the eight Maxwell equations have been reduced to four equations. However, the potentials $\mathbf{A}$ and φ are still not uniquely defined. If $(\mathbf{A}, \varphi)$ is a solution of (2.11) and (2.12), then $(\mathbf{A} + \nabla f, \varphi + \mathrm{i}\omega f)$ for an arbitrary scalar function f is also a solution. To express $\mathbf{A}$ and φ uniquely, an extra condition must be added. One possibility is to use the *Lorentz gauge condition*

$$\nabla \cdot \mathbf{A} - \mathrm{i}\omega\mu\varepsilon\varphi = 0. \tag{2.13}$$

After substitution of condition (2.13) in (2.11) and (2.12) one obtains the *Helmholtz equations*

$$(\Delta + k^2)\mathbf{A} = -\mu\mathbf{J}, \tag{2.14}$$

$$\nabla \cdot (\varepsilon\nabla\varphi) + \varepsilon k^2\varphi = -\rho, \tag{2.15}$$

where $k = \omega\sqrt{\varepsilon\mu}$. It can be shown that the solutions of Helmholtz equations are unique for appropriate Dirichlet and Neumann boundary conditions (see COLTON and KRESS [1992, Section 3]).

In summary, the total system of equations to be solved for $\mathbf{A}$, φ, $\mathbf{J}$ and ρ is as follows:

$$(\Delta + k^2)\mathbf{A} = -\mu\mathbf{J}, \tag{2.16}$$

$$\nabla \cdot (\varepsilon\nabla\varphi) + \varepsilon k^2\varphi = -\rho, \tag{2.17}$$

$$\mathbf{J} = \sigma\mathbf{E} = \sigma(-\nabla\varphi + \mathrm{i}\omega\mathbf{A}), \tag{2.18}$$

$$\nabla \cdot \mathbf{J} - \mathrm{i}\omega\rho = 0, \tag{2.19}$$

including appropriate boundary conditions.

2.4. *The boundary value problem*

In this subsection an equivalent boundary value problem will be posed. First, we will introduce the Green's functions, which are the solutions of an inhomogeneous Helmholtz equation for a homogeneous medium. Let the Green's function $G(\mathbf{x}', \mathbf{x}; k)$ be defined as the solution of the following equation

$$(\Delta + k^2)G(\mathbf{x}', \mathbf{x}; k) = -\delta(\mathbf{x}' - \mathbf{x}), \quad \mathbf{x} \in \Omega, \tag{2.20}$$

for a fixed point $\mathbf{x}'$ in a domain $\Omega \subset \mathbb{R}^3$ and the *Dirichlet* condition

$$G(\mathbf{x}', \mathbf{x}; k) = g(\mathbf{x}) \quad \text{on } \delta\Omega.$$

Here, $\delta(\mathbf{x}' - \mathbf{x})$ is the *Dirac delta function* with the properties:

$$\delta(\mathbf{x}' - \mathbf{x}) = 0, \quad \text{if } |\mathbf{x}' - \mathbf{x}| > 0,$$

$$\int_{B_R(\mathbf{x})} \delta(\mathbf{x}' - \mathbf{x})\,\mathrm{d}\mathbf{x}' = 1,$$

$$\int_{B_R(\mathbf{x})} \delta(\mathbf{x}' - \mathbf{x})\xi(\mathbf{x}')\,\mathrm{d}\mathbf{x}' = \xi(\mathbf{x}),$$

where the "sphere" $B_R(\mathbf{x}) = \{\mathbf{x}' \in \mathbb{R}^3;\ |\mathbf{x}' - \mathbf{x}| \leqslant R;\ \mathbf{x} \in \mathbb{R}^3\}$, and $\xi(\mathbf{x})$ is an arbitrary function over $B_R(\mathbf{x})$. The fundamental solution of Eq. (2.20) is

$$G_0(\mathbf{x}', \mathbf{x}; k) = \frac{\mathrm{e}^{\mathrm{i}k|\mathbf{x}' - \mathbf{x}|}}{4\pi|\mathbf{x}' - \mathbf{x}|}.$$

Thus, restricting the domain Ω to the conductors, the solutions of the Helmholtz equations (2.16) and (2.17) can be formulated as:

$$\mathbf{A}(\mathbf{x}, \omega) = \int_\Omega G_{\mathbf{A}}(\mathbf{x}', \mathbf{x}; k)\mu\mathbf{J}(\mathbf{x}', \omega)\,\mathrm{d}\mathbf{x}' + \mathbf{A}_0(\mathbf{x}, \omega), \tag{2.21}$$

$$\varphi(\mathbf{x}, \omega) = \int_\Omega G_\varphi(\mathbf{x}', \mathbf{x}; k)\frac{\rho(\mathbf{x}', \omega)}{\varepsilon}\,\mathrm{d}\mathbf{x}' + \varphi_0(\mathbf{x}, \omega), \tag{2.22}$$

where $\mathbf{A}_0$ and φ_0 are solutions of the homogeneous problem and

$$G_\varphi = \frac{\mathrm{e}^{\mathrm{i}k|\mathbf{x}' - \mathbf{x}|}}{4\pi|\mathbf{x}' - \mathbf{x}|}, \qquad G_{\mathbf{A}} = \begin{pmatrix} 1 & 0 & 0 \\ 0 & 1 & 0 \\ 0 & 0 & 1 \end{pmatrix} \frac{\mathrm{e}^{\mathrm{i}k|\mathbf{x}' - \mathbf{x}|}}{4\pi|\mathbf{x}' - \mathbf{x}|},$$

the fundamental solutions of the Helmholtz equations (2.16) and (2.17) for a homogeneous medium.

For a stratified inhomogeneous medium it is more difficult to obtain the Green's function. For $k|\mathbf{x}' - \mathbf{x}| \ll 1$, known as the *quasi-static* case, the method of images (see JACKSON [1975, Section 2.1, pp. 54–55]) can be used. Then the Green's function for the scalar potential becomes

$$G_\varphi = \sum_{j \in \mathrm{Images}(\varphi)} c_j \frac{\mathrm{e}^{\mathrm{i}k|\mathbf{x}'_j - \mathbf{x}|}}{4\pi |\mathbf{x}'_j - \mathbf{x}|}, \qquad \text{for } \mathbf{x}'_j = (x', y', z'_j).$$

The images for φ are due to reflections at the dielectric interfaces and the ground plane, all of which are perpendicular to the z-axis. The constants c_j only depend on the dielectric constants of the layers. If there are two or more of such reflection planes, the number of images is infinite. The Green's function for the vector potential has the form

$$G_\mathbf{A} = \sum_{j \in \mathrm{Images}(\mathbf{A})} M_j \frac{\mathrm{e}^{\mathrm{i}k|\mathbf{x}'_j - \mathbf{x}|}}{4\pi |\mathbf{x}'_j - \mathbf{x}|}, \qquad \text{for } \mathbf{x}'_j = (x', y', z'_j).$$

Since μ is constant, the images for $\mathbf{A}$ are only due to reflections at the groundplane, i.e., for each source point $\mathbf{x}'$ there is only one image point. If all metal layers are parallel to the ground layer and the ground plane is a perfect conductor, then

$$M_1 = -M_2 = \begin{pmatrix} 1 & 0 \\ 0 & 1 \end{pmatrix},$$

otherwise

$$M_1 = \begin{pmatrix} 1 & 0 & 0 \\ 0 & 1 & 0 \\ 0 & 0 & 1 \end{pmatrix}, \qquad M_2 = \begin{pmatrix} -1 & 0 & 0 \\ 0 & -1 & 0 \\ 0 & 0 & 1 \end{pmatrix}.$$

Now, we can formulate the following boundary value problem. Let Ω be the interior of the finite conductor regions, let $\Gamma = \delta\Omega$ be the boundary of these regions, and let Γ_V be that part of Γ that is restricted to the *connection ports*, where the external wires are connected to the conductors. For more details see DU CLOUX, MAAS and WACHTERS [1994]. After substitution of (2.21) in (2.18) the following boundary value problem can be formulated:

$$\frac{\mathbf{J}}{\sigma} + \nabla\varphi - \mathrm{i}\omega \int_\Omega G_\mathbf{A} \mu \mathbf{J} \, \mathrm{d}\mathbf{x}' = \mathbf{E}_0, \tag{2.23}$$

$$\nabla \cdot \mathbf{J} - \mathrm{i}\omega\rho = 0, \tag{2.24}$$

$$\varphi - \int_\Omega G_\varphi \frac{\rho}{\varepsilon} \, \mathrm{d}\mathbf{x}' = \varphi_0, \tag{2.25}$$

under the boundary conditions

$$\varphi(\mathbf{x}) = V_{\mathrm{fixed}}, \quad \mathbf{x} \in \Gamma_V,$$

$$\mathbf{J} \cdot \mathbf{n} = 0, \qquad \mathbf{x} \in \Gamma,$$

where φ, $\mathbf{J}$ and ρ are elements of the function spaces

$$\begin{aligned}
&\rho \in L^2(\Omega),\\
&\varphi \in H^1(\Omega) = \{u \in L^2(\Omega) \mid \nabla u \in L^2(\Omega)^3\},\\
&\mathbf{J} \in H^{\mathrm{div}}(\Omega) = \{\mathbf{v} \in L^2(\Omega)^3 \mid \nabla \cdot \mathbf{v} \in L^2(\Omega)\},
\end{aligned}$$

where $L^2(\Omega) = \{u \mid \int_\Omega u^2 \, d\mathbf{x} < \infty\}$. Further, $\mathbf{E}_0$ and φ_0 are due to irradiation from external sources, associated with the homogeneous solutions $\mathbf{A}_0$ and φ_0, respectively, of the Helmholtz equations, and $\mathbf{n}$ is the unit normal vector perpendicular to the boundary surface. The physical meaning of the boundary conditions is that no current will flow through the boundary, except through the connection ports.

2.5. *Variational formulation*

Assuming that the irradiation is zero, i.e., $\mathbf{E}_0 \equiv 0$ and $\varphi_0 \equiv 0$, the variational formulation of the boundary value problem is obtained by multiplying (2.23)–(2.25) with test functions (denoted by a tilde over the symbol) and integrating over the domain Ω of the conductors:

$$\int_\Omega \left\{ \frac{\mathbf{J}}{\sigma} + \nabla \varphi - \mathrm{i}\omega \int_\Omega G_{\mathbf{A}} \mu \mathbf{J} \, d\mathbf{x}' \right\} \cdot \tilde{\mathbf{J}} \, d\mathbf{x} = 0, \tag{2.26}$$

$$\int_\Omega \{\nabla \cdot \mathbf{J} - \mathrm{i}\omega\rho\} \tilde{\varphi} \, d\mathbf{x} = 0, \tag{2.27}$$

$$\int_\Omega \left\{ \varphi - \int_\Omega G_\varphi \frac{\rho}{\varepsilon} \, d\mathbf{x}' \right\} \tilde{\rho} \, d\mathbf{x} = 0, \tag{2.28}$$

where $\tilde{\mathbf{J}}$, $\tilde{\varphi}$ and $\tilde{\rho}$ are test functions in the infinite dimensional function spaces associated with $\mathbf{J}$, φ and ρ, respectively:

$$\begin{aligned}
&\rho, \tilde{\rho} \in L^2(\Omega),\\
&\varphi, \tilde{\varphi} \in H^1(\Omega) = \{u \in L^2(\Omega) \mid \nabla u \in L^2(\Omega)^3\},\\
&\mathbf{J}, \tilde{\mathbf{J}} \in H_0^{\mathrm{div}}(\Omega) = \{\mathbf{v} \in L^2(\Omega)^3 \mid \nabla \cdot \mathbf{v} \in L^2(\Omega);\ \mathbf{v} \cdot \mathbf{n} = 0 \text{ on } \Gamma\}.
\end{aligned}$$

After integration by parts, which is allowed since $\tilde{\mathbf{J}} \in H_0^{\mathrm{div}}(\Omega)$, and substitution of the boundary condition $\tilde{\mathbf{J}} \cdot \mathbf{n} = 0$, the following relation holds:

$$\int_\Omega \tilde{\mathbf{J}} \cdot \nabla \varphi \, d\mathbf{x} = -\int_\Omega \varphi \nabla \cdot \tilde{\mathbf{J}} \, d\mathbf{x}.$$

Substituting this expression in (2.26) gives the following weak formulation of the boundary value problem:

$$\rho(\mathbf{x}), \varphi(\mathbf{x}) \in L^2(\Omega), \quad \text{and} \quad \mathbf{J}(\mathbf{x}) \in H_0^{\mathrm{div}}(\Omega),$$

$$\int_\Omega \left\{ \frac{\mathbf{J}}{\sigma} \cdot \tilde{\mathbf{J}} - \varphi \nabla \cdot \tilde{\mathbf{J}} - \mathrm{i}\omega \int_\Omega G_{\mathbf{A}} \mu \mathbf{J} \, d\mathbf{x}' \cdot \tilde{\mathbf{J}} \right\} d\mathbf{x} = 0 \quad \text{for all } \tilde{\mathbf{J}} \in H_0^{\mathrm{div}}(\Omega), \tag{2.29}$$

$$\int_{\Omega} \{\nabla \cdot \mathbf{J} - \mathrm{i}\omega\rho\}\tilde{\varphi}\,\mathbf{dx} = 0 \quad \text{for all } \tilde{\varphi} \in L^2(\Omega), \tag{2.30}$$

$$\int_{\Omega} \left\{\varphi - \int_{\Omega} G_{\varphi}\frac{\rho}{\varepsilon}\,\mathbf{dx}'\right\}\tilde{\rho}\,\mathbf{dx} = 0 \quad \text{for all } \tilde{\rho} \in L^2(\Omega). \tag{2.31}$$

We assume that the conductors are planar and very thin so that, for the frequencies we are interested in, the quantities $\mathbf{J}$, φ and ρ are constant in the direction perpendicular to the conductors. Therefore, the dependence of the above expressions on the coordinate direction perpendicular to the layers may be separated from the dependence in parallel direction. Hence, the 3D integrals over the volume of the conductors may be replaced by 2D integrals over the surfaces, that result when the thickness of the conductor layers becomes zero. In the following Ω will be considered as a 2D manifold embedded in $\mathbb{R}^3$.

The system of Eqs. (2.23)–(2.25) is called the *operational formulation* of the problem and (2.29)–(2.31) is the *variational formulation*. It is easily seen that if $(\mathbf{J}, \varphi, \rho)$ is a solution of (2.23)–(2.25), it is also a solution of (2.29)–(2.31). Conversely, it can be shown (see AUBIN [1972, Section 1.5, p. 27]) that if the material constants σ^{-1}, ω, μ and ε^{-1} are bounded, and $(\mathbf{J}, \varphi, \rho)$ satisfy the variational formulation (2.29)–(2.31) for all $(\tilde{\mathbf{J}}, \tilde{\varphi}, \tilde{\rho})$ in the associated function spaces, the functions $(\mathbf{J}, \varphi, \rho)$ also satisfy the operational formulation of the boundary value problem.

2.6. *Discretisation*

To find an approximating solution of Eqs. (2.29)–(2.31), the function spaces are approximated by finite dimensional subspaces. Let us assume that the planar regions to which the conductors reduce when their thickness becomes zero consist of polygons, and let the domain of these regions be denoted by Ω_h. Then, the domain can be subdivided into convex quadrilaterals Ω_j as illustrated in Fig. 2.1. Since the planar conductor regions often have quadrilateral shapes with large aspect ratios, we have chosen quadrilateral elements instead of triangles. The set of quadrilaterals is referred to as the set of elements Ω_j, $j = 1, \ldots, N_{\text{elem}}$. The edges of the quadrilaterals inside the domain Ω_h, i.e., excluding the element edges in the boundary, are referred to as the set of edges $\mathcal{E}_l$, $l = 1, \ldots, N_{\text{edge}}$. On the domain Ω_h finite dimensional subspaces U_h, W_h and $H^{\text{div}}_{h,0}$ of the infinite dimensional function spaces L^2 and H^{div}_0 are taken. The *discrete formulation* associated with the problem (2.29)–(2.31) is to find the functions $(\varphi_h, \mathbf{J}_h, \rho_h)$ for which

$$\int_{\Omega_h} \left\{\frac{\mathbf{J}_h}{\sigma}\cdot\tilde{\mathbf{J}}_h - \varphi_h\nabla\cdot\tilde{\mathbf{J}}_h - \mathrm{i}\omega\int_{\Omega_h} G_{\mathbf{A}}\mu\mathbf{J}_h\,\mathbf{dx}'\cdot\tilde{\mathbf{J}}_h\right\}\mathbf{dx} = 0 \quad \text{for all } \tilde{\mathbf{J}}_h \in H^{\text{div}}_{h,0}, \tag{2.32}$$

$$\int_{\Omega_h} \{\nabla\cdot\mathbf{J}_h - \mathrm{i}\omega\rho_h\}\tilde{\varphi}_h\,\mathbf{dx} = 0 \quad \text{for all } \tilde{\varphi}_h \in U_h, \tag{2.33}$$

$$\int_{\Omega_h} \left\{\varphi_h - \int_{\Omega_h} G_{\varphi}\frac{\rho_h}{\varepsilon}\,\mathbf{dx}'\right\}\tilde{\rho}_h\,\mathbf{dx} = 0 \quad \text{for all } \tilde{\rho}_h \in W_h. \tag{2.34}$$

The functions φ_h, $\mathbf{J}_h$ and ρ_h are expanded in terms of basis functions, which span the finite dimensional subspaces defined above.

The scalar potential is expanded as

$$\varphi_h(\mathbf{x}) = \sum_{j=1}^{N_{\text{elem}}} V_j b_j(\mathbf{x}),$$

where V_j is the potential of element j, and $b_j(\mathbf{x})$ is defined by

$$b_j(\mathbf{x}) = \begin{cases} 1 & \text{for } \mathbf{x} \in \Omega_j, \\ 0 & \text{elsewhere.} \end{cases}$$

The surface charge density is expanded as

$$\rho_h(\mathbf{x}) = \sum_{j=1}^{N_{\text{elem}}} Q_j c_j(\mathbf{x}),$$

where Q_j is the charge of element j, while $c_j(\mathbf{x})$ are basis functions on the elements, adapted to include the singularity of the charge density (see Appendix A) near the conductor edge

$$c_j(\mathbf{x}) = f(\mathbf{x}) b_j(\mathbf{x}).$$

The function $f(\mathbf{x})$ is defined in Appendix A. It satisfies the following condition

$$\int_{\Omega_j} f(\mathbf{x})\,\mathrm{d}\mathbf{x} = |\Omega_j|, \tag{2.35}$$

where $|\Omega_j|$ is defined as the area of Ω_j.

Finally, the surface current density is expanded as

$$\mathbf{J}_h(\mathbf{x}) = \sum_{l=1}^{N_{\text{edge}}} I_l \tilde{\mathbf{w}}_l(\mathbf{x}),$$

where I_l is the current through edge l, and $\tilde{\mathbf{w}}_l(\mathbf{x})$ is defined by

$$\tilde{\mathbf{w}}_l(\mathbf{x}) = \begin{cases} f(\mathbf{x})\mathbf{w}_l(\mathbf{x}) & \text{for } \mathbf{x} \in \Omega_i \cup \Omega_j \text{ and } \mathcal{E}_l = \Omega_i \cap \Omega_j, \\ 0 & \text{otherwise.} \end{cases} \tag{2.36}$$

The form and properties of the basis functions $\mathbf{w}_l$ are defined in Appendix B. Note that the component of $\mathbf{w}_l$ normal to the edge is continuous at $\mathbf{x} \in \mathcal{E}_l$ when passing from Ω_i to Ω_j.

After substitution of the expansions of φ_h, $\mathbf{J}_h$ and ρ_h in (2.29)–(2.31) we obtain the following linear system of equations:

$$\sum_{l=1}^{N_{\text{edge}}} (R_{kl} - \mathrm{i}\omega L_{kl}) I_l - \sum_{j=1}^{N_{\text{elem}}} P_{kj} V_j = 0,$$

$$\mathrm{i}\omega \sum_{j=1}^{N_{\text{elem}}} M_{ij} Q_j - \sum_{l=1}^{N_{\text{edge}}} P_{li} I_l = 0,$$

$$\sum_{j=1}^{N_{\text{elem}}} (M_{ij} V_j - D_{ij} Q_j) = 0.$$

The matrix elements of **R**, **L**, **P**, **M** and **D** ($k, l = 1 \dots N_{\text{edge}}$ and $i, j = 1 \dots N_{\text{elem}}$) are given by

$$\begin{aligned}
R_{kl} &= \int_{\Omega_h} \frac{1}{\sigma} \tilde{\mathbf{w}}_l(\mathbf{x}) \cdot \tilde{\mathbf{w}}_k(\mathbf{x})\,\mathrm{d}\mathbf{x}, \\
L_{kl} &= \int_{\Omega_h} \tilde{\mathbf{w}}_l(\mathbf{x}) \cdot \left\{ \int_{\Omega_h} G_{\mathbf{A}}(\mathbf{x}, \mathbf{x}') \mu \tilde{\mathbf{w}}_k(\mathbf{x}')\,\mathrm{d}\mathbf{x}' \right\} \mathrm{d}\mathbf{x}, \qquad (2.37) \\
P_{kj} &= \int_{\Omega_j} b_j(\mathbf{x}) \nabla \cdot \tilde{\mathbf{w}}_k(\mathbf{x})\,\mathrm{d}\mathbf{x}, \\
M_{ij} &= \int_{\Omega_j} c_j(\mathbf{x}) b_i(\mathbf{x})\,\mathrm{d}\mathbf{x}, \\
D_{ij} &= \int_{\Omega_j} c_j(\mathbf{x}) \left\{ \int_{\Omega_i} G_\varphi(\mathbf{x}, \mathbf{x}') \frac{c_i(\mathbf{x}')}{\varepsilon}\,\mathrm{d}\mathbf{x}' \right\} \mathrm{d}\mathbf{x}. \qquad (2.38)
\end{aligned}$$

R is a sparse matrix and **L** and **D** are symmetrical, full matrices. From the definition of the basis functions b_i and c_j and Eq. (2.35) it follows that

$$M_{ij} = \delta_{ij} |\Omega_j|, \quad \delta_{ij} = \begin{cases} 1 & \text{if } i = j, \\ 0 & \text{if } i \neq j \end{cases}$$

so that **M** is a diagonal matrix of which the elements are the areas of the Ω_j.

LEMMA 2.1. *Let J be the Jacobian defined by the transformation* (B.1) *for* $\mathbf{x} \in \Omega_j$. *For one of the following conditions*

$$\begin{aligned}
&(1) \qquad f(\mathbf{x}) \equiv 1, \\
&(2) \qquad \nabla f \cdot \mathbf{w}_k = 0, \quad \textit{and} \quad J = |\Omega_j|,
\end{aligned}$$

the matrix **P** *has the form*

$$P_{kj} = \begin{cases} \pm 1 & \textit{if } \mathcal{E}_k \subset \Omega_j, \\ 0 & \textit{otherwise.} \end{cases}$$

PROOF. Let $\mathcal{E}_k \subset \Omega_j$, then Lemma B.2 shows that $\nabla \cdot \mathbf{w}_k = \frac{\pm 1}{J}$.
If condition (1) holds

$$P_{kj} = \int_{\Omega_j} b_j(\mathbf{x}) \nabla \cdot \tilde{\mathbf{w}}_k(\mathbf{x})\,\mathrm{d}\mathbf{x} = \int_{\Omega_j} \nabla \cdot \mathbf{w}_k(\mathbf{x})\,\mathrm{d}\mathbf{x} = \int_{\Omega_j} \frac{\pm 1}{J}\,\mathrm{d}\mathbf{x} = \pm 1.$$

If $\nabla f \cdot \mathbf{w}_k = 0$ of condition (2) holds

$$P_{kj} = \int_{\Omega_j} b_j(\mathbf{x}) \nabla \cdot \big(f(\mathbf{x}) \mathbf{w}_k(\mathbf{x})\big)\,\mathrm{d}\mathbf{x} = \int_{\Omega_j} f(\mathbf{x}) \nabla \cdot \mathbf{w}_k(\mathbf{x})\,\mathrm{d}\mathbf{x} = \pm 1 \int_{\Omega_j} \frac{f(\mathbf{x})}{J}\,\mathrm{d}\mathbf{x},$$

thus if $J = |\Omega_j|$ it follows from (2.35) that

$$P_{kj} = \frac{\pm 1}{|\Omega_j|} \int_{\Omega_j} f(\mathbf{x})\,\mathrm{d}\mathbf{x} = \pm 1.$$

If $\mathcal{E}_k \not\subset \Omega_j$, then the supports of b_j and $\nabla \cdot \tilde{\mathbf{w}}_k(\mathbf{x})$ are disjoint so that $P_{kj} = 0$. □

Condition (2) is fulfilled if the element Ω_j is rectangular and has two opposite edges lying in the boundary.

2.7. The Kirchhoff's equations

If Lemma 2.1 holds matrix **P** is an incidence matrix. Therefore, the elements and edges may be associated with the nodes and branches of a directed graph, so that our quasi-static electromagnetic model of a PCB is equivalent to a circuit of which the behaviour is described by the following set of $N_{\text{branches}} + 2N_{\text{nodes}}$ equations:

$$(\mathbf{R} - \mathrm{i}\omega\mathbf{L})I - \mathbf{P}V = 0,$$
$$-\mathbf{P}^{\mathrm{T}} I + \mathrm{i}\omega\mathbf{M}Q = 0,$$
$$\mathbf{M}^{\mathrm{T}} V - \mathbf{D}Q = 0,$$

and at particular nodes j, corresponding to elements $\Omega_j \subset \Gamma_V$, the excitation conditions

$$V_j = V_{\text{fixed},j}.$$

These equations are the *Kirchhoff's equations*, which are used in classical circuit theory. The meaning of the quantities in these equations is given below:

$V \sim$ Potentials at nodes,

$I \sim$ Currents over branches,

$\mathbf{P} \sim$ Incidence matrix between nodes and branches,

$\mathbf{R} \sim$ Resistance (of branches),

$\mathbf{L} \sim$ Inductance (of branches),

$\mathbf{M}\mathbf{D}^{-1}\mathbf{M}^{\mathrm{T}} \equiv \mathbf{C} \sim$ Capacitance (between nodes).

After elimination of Q we obtain a system of $N_{\text{branches}} + N_{\text{nodes}}$ equations for the unknown I and V:

$$(\mathbf{R} - \mathrm{i}\omega\mathbf{L})I - \mathbf{P}V = 0 \quad (\textit{Kirchhoff's voltage law}),$$
$$-\mathbf{P}^{\mathrm{T}} I + \mathrm{i}\omega\mathbf{C}V = 0 \quad (\textit{Kirchhoff's current law}).$$

If $N_{V,\text{fixed}}$ is the number of potentials for which

$$V_j = V_{\text{fixed},j},$$

and $N_V = N_{\text{nodes}} - N_{V,\text{fixed}}$ the final system of equations has $N_{\text{branches}} + N_V$ unknowns.

3. Interaction integrals

The Kirchhoff equations, which have been derived in Section 2, describe the behaviour of an equivalent circuit of a PCB. This system of equations is linear and the coefficients of the matrices associated with this system are integrals. The fourfold integrals that represent an electromagnetic interaction between charges and currents in two elements are called *interaction integrals*. These are the subject of the following sections.

There are two types of interaction integrals: the scalar-type interaction integral (2.38), representing the *capacitive* coupling between charges on the elements, and the vector-type interaction integral (2.37), representing the *inductive* coupling between currents flowing through the element edges. Since edges lie in two adjacent quadrilaterals the vector-type interaction integral (2.37) is an assemblage of four integrals over quadrilaterals. The integrals, defined on the quadrilaterals Ω_j and Ω_i, are of the following form

$$I = \int_{\Omega_j} \tilde{\psi}_j(\mathbf{x}) \cdot \mathbf{I}_i(\mathbf{x})\, \mathrm{d}\mathbf{x}, \tag{3.1}$$

where the inner integral $\mathbf{I}_i$, called the *source integral*, is

$$\mathbf{I}_i(\mathbf{x}) = \int_{\Omega_i} G(\mathbf{x}', \mathbf{x}) \tilde{\psi}_i(\mathbf{x}')\, \mathrm{d}\mathbf{x}'. \tag{3.2}$$

Here $\tilde{\psi}_i$ and $\tilde{\psi}_j$ are vector valued edge functions (cf. $\tilde{\mathbf{w}}_k$, $\tilde{\mathbf{w}}_l$ in Section 2.6) for vector-type integrals and scalar valued element functions (cf. c_i, c_j) for scalar-type integrals. The $\tilde{\psi}_i$ and $\tilde{\psi}_j$ contain a factor $f(\mathbf{x}) = \frac{1}{\sqrt{d(s_1,s_2)}}$, defined in Appendix A. This function is singular on the boundary of the domain of the conductors. G is the Green's function, which is singular if $|\mathbf{x}' - \mathbf{x}| = 0$, i.e., in the case of self-interaction ($\Omega_i = \Omega_j$) or if the integration elements are neighbours. In the following sections it will be assumed that $k|\mathbf{x}' - \mathbf{x}| \ll 1$ so that the expressions of Green's functions given in Section 2.4 reduce to the *quasi-static* form

$$G(\mathbf{x}', \mathbf{x}) = \sum_{k=1}^{N} c_k |\mathbf{x}_k' - \mathbf{x}|^{-1}, \tag{3.3}$$

where c_k is a scalar or matrix depending on the type of the integral, and N the number of images. Further, the factor 4π in the denominator of $G(\mathbf{x}', \mathbf{x})$ is omitted.

In some special cases (a part of) the integral can be evaluated analytically. For constant $\tilde{\psi}_j$ and $\tilde{\psi}_i$ and rectangular quadrilaterals Ω_j and Ω_i with corresponding edges in parallel the integral I can be evaluated analytically. For constant $\tilde{\psi}_j$ and $\tilde{\psi}_i$ and arbitrary, convex, quadrilaterals Ω_j and Ω_i the inner integral $\mathbf{I}_i$ can be evaluated analytically. For the vector valued $\tilde{\psi}_j$ and $\tilde{\psi}_i$ with constant f only the inner integral of $\mathbf{I}_i$ can be evaluated analytically. The analytical approach for these integrals is discussed in Sections 5 (scalar case) and 8 (vector-valued case).

In all other cases the integral I has to be evaluated numerically. For the numerical integration *quadrature rules* are needed. Several quadrature rules are discussed in

Section 4, in particular the *Patterson's quadrature rules*, which have the fastest convergence rates. However, as will be shown in that section, quadrature rules only reach fast convergence if the integrand is smooth enough, i.e., the integrand must be n times differentiable over the whole integration interval, for n large enough. Singularity of the integrand leads to very slow convergence. Therefore, methods to regularise or eliminate these singularities have to be investigated.

In Section 6 some methods are discussed by which the inner integrand is regularised such that it is smooth enough for integration by a quadrature process. The sources of the singularities in the inner integral are the Green's function and the factor for the boundary singularity. Regularisation of the Green's function is done by transformation to polar coordinates. For this purpose the element has to be divided into triangles, each of which has one edge of the element as base and the projection of the quadrature point of the outer integral I on the source element as vertex. The regularisation of the factor for the boundary singularity has to be treated in a special way. In particular, the treatment of the 'flat' triangles deserves special attention.

After numerical evaluation of the inner integral the outer integral only contains a factor for the boundary singularity. Regularisation of this singularity is done by a simple substitution, which is discussed in the last subsection of Section 6.

Throughout this chapter the integration domain of the outer integral, Ω_j, is referred to as the *object element* and the integration domain of the inner integral, Ω_i, as the *source element*. If the "distance" between the source and object element is large enough, then the Green's function can be approximated satisfactorily with a *Taylor expansion*. In Section 7 an error estimate of the Taylor expansion dependent on the distance will be given. In the same section the *moment integrals* are introduced. These are *twofold* integrals which, possibly, still contain the boundary singularity. After regularisation of this singularity, which can be done by an analogous substitution as discussed for the outer integral in the last subsection of Section 6, the moment integral can be evaluated by Patterson's quadrature process. The interaction integral I can be written as a linear combination of products of these moments.

This method has the advantage that, instead of a large number (quadratic with the number of elements or edges) of fourfold integrals, only a small number (linear with the number of elements or edges) of twofold moment integrals has to be evaluated. Moreover, these moment integrals can be evaluated in advance. Since it saves a lot of computer time, this method is preferred as an alternative for the numerical treatment of Section 6, if the distance between the source and object element is sufficiently large.

4. Numerical integration

Since most of the integrals, discussed in the previous chapters, cannot be evaluated analytically, we have to rely on numerical integration methods. For a detailed discussion of these methods see DAVIS and RABINOWITZ [1984].

In this chapter several numerical integration methods are discussed, in particular *Patterson's* quadrature formulae. Because special transformations are needed to regularise the singularity of the integrand, only one-dimensional Patterson's rules are used. First follows a short introduction to *quadrature formulae*.

4.1. Quadrature formulae

The essence of numerical quadrature is the approximation of an integral by a linear combination of the values of the integrand. Consider the integral

$$I(f)=\int_a^b f(x)\,\mathrm{d}x,$$

which has to be numerically evaluated. A quadrature formula is given by

$$K(f)=\sum_{i=0}^{n-1} w_i f(x_i), \tag{4.1}$$

where $f(x)$ is an arbitrary (smooth) function, x_i are different abscissae in the integration interval and w_i are the corresponding weights.

Often, we choose the weights and abscissae such that the rule is exact for polynomials up to a certain degree. Given an arbitrary set of n distinct abscissae $x_i \in [a,b]$, the corresponding weights can be determined by solving the linear system

$$\sum_{i=0}^{n-1} w_i {x_i}^k = \int_a^b x^k\,\mathrm{d}x, \qquad \text{for } k=0,\dots,n-1. \tag{4.2}$$

Note that the coefficient matrix (x_i^k) of the above system is a Vandermonde matrix, and therefore, non-singular if $x_i \neq x_j$ for $i \neq j$. Hence, there is always a unique solution.

If the nodes $a=x_0<\dots<x_{n-1}=b$ are equidistant, these quadrature formulae are called *Newton–Cotes* formulae.

For the w_i's obtained by solving the system (4.2) the integration formula (4.1) is at least *exact* for polynomials of degree $n-1$. If f is sufficiently smooth, say $f \in C^n[a,b]$, then the error is given by

$$\left|I(f)-K(f)\right| = Cf^{(n)}(\xi)(b-a)^{n+1}, \quad \xi \in (a,b),$$

where C is a constant and $f^{(n)}$ denotes the nth derivative of f.

When the behaviour of the function to be integrated is very distinct on different parts of the integration interval, it is advantageous to subdivide the interval. A Newton–Cotes formula can be applied to each subinterval. If the subdivision of the interval is done automatically it is called an *adaptive* subdivision. A *non-adaptive* subdivision is characterised by a predetermined choice of subdivision points.

Some examples of quadrature rules. Some integration formulae of the Newton–Cotes type, that approximate the integral $\int_a^b f(x)\,\mathrm{d}x$, and the corresponding error formulae are given below:

$$K_M=(b-a)f\left(\frac{b+a}{2}\right),$$

$$I-K_M=\frac{1}{24}(b-a)^3 f''(\xi_M) \quad (\textit{Midpoint rule}),$$

$$K_T = \frac{b-a}{2}\{f(a) + f(b)\},$$

$$I - K_T = \frac{-1}{12}(b-a)^3 f''(\xi_T) \quad (\textit{Trapezoidal rule}),$$

$$K_S = \frac{b-a}{6}\left\{f(a) + 4f\left(\frac{b+a}{2}\right) + f(b)\right\},$$

$$I - K_S = \frac{-1}{90}\left(\frac{b-a}{2}\right)^5 f^{(4)}(\xi_S) \quad (\textit{Simpson's rule}),$$

where the ξ_M, ξ_T, ξ_S are points in the interval (a, b) depending on the function f to be integrated.

For n-point Newton–Cotes formulae ($n \leqslant 8$) the associated weights are positive. The next theorem shows that these formulae are stable.

THEOREM 4.1 (Stability). *Consider the n-point quadrature formula*

$$K(f) = \sum_{i=1}^{n} w_i f(x_i),$$

to approximate the integral $\int_a^b f(x)\,\mathrm{d}x$. If the weights are all nonnegative, i.e., $w_i \geqslant 0$, the quadrature formula is stable.

PROOF. Let $\varepsilon(x)$ be a perturbation of $f(x)$ and let ε_f be a constant such that $|\varepsilon(x)| \leqslant \varepsilon_f$ for all $x \in [a, b]$. Then, from the positivity of the weights it follows that

$$\int_a^b \mathrm{d}x = (b-a) = \sum_{i=1}^{n} w_i = \sum_{i=1}^{n} |w_i|,$$

hence,

$$\sum_{i=1}^{n} w_i \varepsilon(x_i) \leqslant (b-a)\varepsilon_f. \qquad \square$$

A reason not to use higher order Newton–Cotes formulae ($n > 8$) is the possible instability due to negative weights. Instead one could use lower order formulae on subintervals of $[a, b]$.

Repeated quadrature. The successive application of a quadrature formula on ever smaller subintervals of $[a, b]$ to obtain an increasingly better approximation of the integral is called *repeated quadrature*. Here, we give an example for the trapezoidal rule.

Divide the interval $[a, b]$ in n equal subintervals $[x_{i-1}, x_i]$ of length h, such that $a = x_0 < x_1 < \cdots < x_{n-1} < x_n = b$ and where $h = \frac{b-a}{n}$. Repeated application of the trapezoidal rule gives

$$T_n = h\left\{\tfrac{1}{2} f(x_0) + f(x_1) + \cdots + f(x_{n-1}) + \tfrac{1}{2} f(x_n)\right\}.$$

Assuming that f is sufficiently differentiable, the error is given by (for a proof see DAVIS and RABINOWITZ [1984, Section 2.9])

$$I - T_n = C_1 h^2 + C_2 h^4 + \cdots + C_n h^{2n} + \mathcal{O}(h^{2n+2}),$$

where the constants C_i depend on f, but are independent of h. The following application of repeated quadrature is based on this error formula.

Romberg integration. Suppose one has a function $f \in C^{2(n+2)}[a, b]$, where $[a, b]$ is an interval of length h_0 in $\mathbb{R}$, and the approximations $T_0^{(0)}, T_1^{(0)}, \ldots, T_n^{(0)}$ of the integral of f over $[a, b]$. These $T_i^{(0)}$ have been obtained by applying repeated trapezoidal rules on the 2^i subintervals of length h_i, where $h_i = 2^{-i} h_0$. Again, the errors for the approximations are given by

$$\begin{aligned} I - T_i^{(0)} &= C_1 h_i^2 + C_2 h_i^4 + \cdots + \mathcal{O}\big(h_i^{2(n+2)}\big) \\ &= C_1 h_0^2 2^{-2i} + C_2 h_0^4 2^{-4i} + \cdots + \mathcal{O}\big(h_i^{2(n+2)}\big). \end{aligned}$$

Since the first terms in $I - T_{i-1}^{(0)}$ and $I - T_i^{(0)}$ are $C_1 h_{i-1}^2$ and $C_1 (2^{-1} h_{i-1})^2 = \frac{1}{4} C_1 h_{i-1}^2$, respectively, one can eliminate these terms by applying *Richardson extrapolation* to the sequence $T_0^{(0)}, \ldots, T_n^{(0)}$:

$$T_i^{(1)} = \frac{4T_i^{(0)} - T_{i-1}^{(0)}}{3}, \quad \text{for } i = 1, \ldots, n.$$

The errors for the newly obtained sequence of approximations are

$$\begin{aligned} I - T_i^{(1)} &= D_2 h_i^4 + D_3 h_i^6 + \cdots + \mathcal{O}\big(h_i^{2(n+2)}\big) \\ &= D_2 h_0^4 2^{-4i} + D_3 h_0^6 2^{-6i} + \cdots + \mathcal{O}\big(h_i^{2(n+2)}\big), \end{aligned}$$

where $D_i = \frac{2^2 - 2^{2i}}{2^2 - 1} C_i$. This process can be applied recursively on the sequences by

$$T_i^{(k)} = \frac{2^{2k} T_i^{(k-1)} - T_{i-1}^{(k-1)}}{2^{2k} - 1}, \quad \text{for } i = k, \ldots, n,$$

and the error for $T_i^{(n)}$ is

$$I - T_i^{(n)} = 2^{n(n+1)} C_{n+1} h_i^{2(n+1)} + \mathcal{O}\big(h_i^{2(n+2)}\big) = \mathcal{O}\bigg(\frac{2^{n(n+1)}}{4^{i(n+1)}} h_0^{2(n+1)}\bigg).$$

Thus, by *repeated* Richardson extrapolation, we get the *Romberg integration method*, so that the following theorem holds.

THEOREM 4.2. *Suppose $f \in C^\infty[a, b]$, then for $i \to \infty$ the sequences $T_i^{(k)}$, constructed as described above, converge towards $\int_a^b f(x)\,\mathrm{d}x$ for every $k = 0, 1, \ldots$ with error $\mathcal{O}(h_i^{2k+2})$.*

Moreover, the diagonal $T_n^{(n)}$ converges with error $\mathcal{O}(h_n^{2n+2})$.

TABLE 4.1
Romberg integration of the exponential function

i	$T_i^{(0)}$	$T_i^{(1)}$	$T_i^{(2)}$	$T_i^{(3)}$	$T_i^{(4)}$
0	2.03663128				
1	1.28898621	1.03977118			
2	1.06215961	0.98655075	0.98300272		
3	1.00205141	0.98201534	0.98171298	0.98169251	
4	0.98679198	0.98170551	0.98168485	0.98168441	0.98168437

EXAMPLE. Table 4.1 shows the results of the Romberg method for the integral $\int_0^4 \mathrm{e}^{-x}\,\mathrm{d}x$ (≈ 0.98168436). We can see that $T_n^{(n)}$ for $n > 0$ gives a much more accurate approximation than $T_n^{(0)}$. For the calculation of $T_0^{(0)}, \ldots, T_n^{(0)}$ a total of $2^n + 1$ function evaluations are needed.

Note, that these function evaluations are only necessary for the calculation of $T_i^{(0)}$, so that by repeated Richardson extrapolation with little extra cost an even better approximation can be obtained. It can also be proven that the sequence $(T_i^{(k)})_{i=k}^{\infty}$ converges faster towards $\int_a^b f(x)\,\mathrm{d}x$ than $(T_i^{(k-1)})_{i=k-1}^{\infty}$.

4.1.1. The Gaussian quadrature formulae

Gauss has proven that an n-point quadrature formula can be found which is exact for polynomials up to degree $2n-1$ and that this is the highest possible degree. This will be shown after the following definitions.

DEFINITION 4.1. Let $w(x)$ be a continuous function on (a, b), with $w(x) \geqslant 0$ on $[a, b]$. We define the *inner product* with respect to the weight function $w(x)$ as

$$\langle f, g\rangle = \int_a^b w(x) f(x) g(x)\,\mathrm{d}x, \qquad f, g \in C[a, b].$$

DEFINITION 4.2. Let $\{F_i(x), i = 0, \ldots, n\}$ be a set of nonzero polynomials with F_i of degree i. The polynomials are said to be *orthogonal* on $[a, b]$ with respect to the inner product $\langle\, ,\rangle$ if they satisfy

$$\langle F_i, F_j\rangle = 0 \quad \text{if } i \neq j.$$

Suppose $F_{2n-1}(x)$ is an arbitrary polynomial of degree $2n-1$. This can be expressed as

$$F_{2n-1}(x) = P_n(x) Q_{n-1}(x) + R_{n-1}(x),$$

where P_n is an nth degree polynomial with n distinct roots in $[a, b]$. Let these roots be the abscissae of a new quadrature formula. Then, this formula will integrate $P_n(x) Q_{n-1}(x)$ to zero and, by construction of the corresponding weights (see (4.2)), it will integrate $R_{n-1}(x)$ exactly, since this is a polynomial of degree $n-1$. Hence, if

$P_n(x)$ satisfies

$$\int_a^b P_n(x)Q_{n-1}(x)\,\mathrm{d}x = 0 \tag{4.3}$$

for any polynomial $Q_{n-1}(x)$ of degree $n-1$, $F_{2n-1}(x)$ will be integrated exactly by this quadrature formula. Since $P_n(x)^2 \geqslant 0$ is not integrated exactly, this is also the highest possible degree.

We must find a $P_n(x)$ such that condition (4.3) will be satisfied. This is the same as

$$\int_a^b P_n(x)\sum_{i=0}^{n-1} a_i x^i\,\mathrm{d}x = 0, \quad \text{for arbitrary } a_i,$$

so that for each individual term:

$$\int_a^b P_n(x)x^i\,\mathrm{d}x = 0, \quad \text{for } i = 0, \ldots, n-1. \tag{4.4}$$

If the ith degree polynomials $P_i(x)$, $i = 0, \ldots, n$, form an orthogonal set these conditions are satisfied, because any $(n-1)$th degree polynomial can be expressed as a linear combination of $P_i(x)$, $i = 0, \ldots, n-1$.

Consider the inner product

$$\langle f, g\rangle = \int_a^b f(x)g(x)\,\mathrm{d}x,$$

and take $a = -1$ and $b = 1$. The *Legendre*-polynomials (see Appendix C) are polynomials, that are mutually orthogonal with respect to this inner product, and therefore, they satisfy property (4.4). Hence, the quadrature formulae, of which the abscissae are the roots of the Legendre polynomials and the weights are constructed by solving Eqs. (4.2), have the property of integrating $F_{2n-1}(x)$ exactly on $[-1, 1]$. The following theorems hold. Theorem 4.4 proves the stability of Gaussian quadrature formulae.

THEOREM 4.3. *The n-point quadrature formula $K_G^{(n)}$, with abscissae the roots of the Legendre polynomial $P_n(x)$ (the Gaussian quadrature formula), is exact for all polynomials in $\Pi_{2n-1}[-1, 1]$. If $f(x) \in C^{2n}[-1, 1]$, the error incurred in integrating $f(x)$ is given by*

$$I(f) - K_G(f) = \frac{2^{2n+1}(n!)^4}{(2n+1)((2n)!)^3} f^{(2n)}(\xi), \quad \xi \in (-1, 1).$$

PROOF. For the error estimate see DAVIS and RABINOWITZ [1961, pp. 428–437]. □

THEOREM 4.4. *The weights w_i for a Gaussian formula $K_G^{(n)}$ are positive.*

PROOF. Let

$$l_i(x) = \prod_{\substack{j=1 \\ j\neq i}}^{n} (x - x_j) \quad (\in \Pi_{n-1}).$$

Then, $l_i(x_j) \neq 0$ for $i = j$ and $l_i(x_j) = 0$ for $i \neq j$.

The Gaussian formula certainly is exact for $l_i(x)^2$. Hence, from

$$0 < \int_{-1}^{1} l_i(x)^2\,\mathrm{d}x = \sum_{j=1}^{n} w_j l_i(x_j)^2 = w_i l_i(x_i)^2$$

it follows that $w_i > 0$ for all $i = 1, \ldots, n$. □

In a *quadrature process* quadrature rules of increasing order are applied successively, until the (estimated) relative error is smaller than a given tolerance. The absolute error can be estimated by taking the difference between the last two integral approximations, so that the relative error is estimated by the ratio of the absolute error to the last approximation. The following corollary holds.

COROLLARY 4.1. *The Gaussian quadrature process* $K_G^{(n)}$ *is convergent for every function* $f(x)$ *which is Riemann-integrable in* $[-1, 1]$, *i.e.,*

$$\lim_{n\to\infty} K_G^{(n)}(f) = \int_{-1}^{1} f(x)\,\mathrm{d}x.$$

PROOF. See DAVIS and RABINOWITZ [1984, Section 2.7.8]. □

Unfortunately, all the roots of the different Legendre polynomials, except zero, are different. Thus, the Gaussian quadrature process is rather inefficient, since in a step of the process no use is made of the integrands evaluated in the preceding steps. In the next subsection a more efficient method will be discussed.

4.2. *Kronrod's extension of quadrature formulae*

KRONROD [1965, p. 597] has suggested an extension of an n-point quadrature formula, by adding $n+1$ new abscissae to the original set, to yield a quadrature formula of degree $3n + 1$ (n even) or $3n + 2$ (n odd). This has the advantage that integrand evaluations needed for an n-point quadrature rule can be used again for the $(2n + 1)$-point quadrature rule. In the discussion of the Kronrod scheme we will restrict ourself to integrals with integration interval $[-1, 1]$. However, the results are applicable to integrals with an arbitrary finite interval $[a, b]$.

Let p be the number of points added to the original set of points and let F_{n+2p-1} be an arbitrary polynomial of degree $n + 2p - 1$. After division with remainder, this can be expressed as

$$F_{n+2p-1} = \tilde{P}_{n+p} Q_{p-1} + R_{n+p-1}.$$

Here $\tilde{P}_{n+p}$ is a polynomial whose roots are the $n + p$ abscissae of the new, extended quadrature formula. Since R_{n+p-1} is some polynomial of degree $n + p - 1$, it can always be exactly integrated by a $(n + p)$-point formula. Furthermore, Q_{p-1} can be expressed as

$$Q_{p-1} = \sum_{i=0}^{p-1} c_i x^i.$$

Therefore, if $\tilde{P}_{n+p}$ satisfies

$$\int_{-1}^{1} \tilde{P}_{n+p} x^i \,\mathrm{d}x = 0, \quad \text{for every } i = 0, \ldots, p-1,$$

then

$$\int_{-1}^{1} \tilde{P}_{n+p} Q_{p-1} \,\mathrm{d}x = 0.$$

Since $\int_{-1}^{1} F_{n+2p-1} \,\mathrm{d}x = \int_{-1}^{1} \tilde{P}_{n+p} Q_{p-1} \,\mathrm{d}x + \int_{-1}^{1} R_{n+p-1} \,\mathrm{d}x$, the quadrature formula is exact for all polynomials in Π_{n+2p-1}.

4.2.1. Application to Gauss–Legendre

Take $p = n + 1$ and P_n the nth degree Legendre polynomial. This choice of p yields the number of points required to subdivide the intervals spanned by the n original Gauss points and the boundaries (see next subsection). Let

$$\tilde{P}_{n+p} = K_{n+1} P_n,$$

then K_{n+1} can be determined by expanding it as a polynomial,

$$K_{n+1}(x) = x^{n+1} + \sum_{i=0}^{n} a_i x^i.$$

The coefficients a_i are calculated by solving the linear system

$$\int_{-1}^{1} K_{n+1}(x) P_n(x) x^k \,\mathrm{d}x = 0, \quad k = 0, \ldots, n.$$

As a result we can construct a quadrature formula, of which the abscissae are the n Gauss-points and the $n + 1$ roots of K_{n+1}. The corresponding weights are determined by solving the system (4.2), the method described earlier. The obtained formula is exact for F_{3n+1}. From an n'-point quadrature formula, with $n' = 2n + 1$, a new quadrature formula can be constructed by applying Kronrod's method to these n' points. The abscissae of this formula are the original n' points and the $n' + 1$ added points. The resulting quadrature formulae is exact for $F_{3n'+1}$. Since the formulae are symmetrical in the range interval $[-1, 1]$ (if x_i is a root, also $-x_i$ is a root) odd functions are always integrated exactly. Hence, the effective degree can be increased to $3n + 2$ when n is odd.

4.2.2. Patterson's quadrature formulae

PATTERSON [1968] has applied the Kronrod's method successively, starting with a 3-point Gaussian quadrature formula and developed a stable algorithm to calculate the nodes and corresponding weights of these quadrature formulae. Thus, he has derived a sequence of quadrature formulae of degrees $n = 7, 15, 31, 63, 127, 255$ and 511. The great advantage of these formulae compared to the Gaussian formulae is that all function evaluations of an n-point formula can be used in the extended $(2n + 1)$-point formula. The condition for Patterson's quadrature formulae to be stable is the positivity of the weights.

First we will give a justification for the choice of adding $n + 1$ points to the n original Gauss-points.

LEMMA 4.1. *Let $x_1^{(n)}, \ldots, x_n^{(n)}$ be the zeros of the Legendre polynomial $P_n(x)$ of degree n and let $y_1^{(n)}, \ldots, y_p^{(n)}$ be $p > \frac{n}{2}$ new points within the interval $(-1, 1)$. Let $K(f)$ be an extended quadrature formula*

$$K(f) = \sum_{j=1}^{n} w_j^{(n)} f\big(x_j^{(n)}\big) + \sum_{i=1}^{p} \tilde{w}_i^{(n)} f\big(y_i^{(n)}\big) \tag{4.5}$$

to approximate $I(f) = \int_{-1}^{1} f(x)\,\mathrm{d}x$. If this formula is exact for $f \in \Pi_{n+2p-1}$, then $p > n$.

PROOF. For $k \in \{1, \ldots, p\}$ define $P_n^*(x) = \prod_{j=1}^{n}(x - x_j^{(n)})$, the nth degree Legendre polynomial with leading coefficient 1, and $s_k(x) = \prod_{i=1, i \neq k}^{p} (x - y_i^{(n)})$. Let

$$g_k(x) = P_n^*(x) s_k(x), \quad P_n^* \in \Pi_n, \ s_k \in \Pi_{p-1}.$$

If $p \leqslant n$

$$I(g_k) = \int_{-1}^{1} g_k(x)\,\mathrm{d}x = \int_{-1}^{1} P_n^*(x) s_k(x)\,\mathrm{d}x = 0,$$

since $P_n^* \perp s_k \in \Pi_{n-1}$. But since $g_k \in \Pi_{n+p-1}$, it follows that

$$\begin{aligned} 0 = I(g_k) &= K(g_k) \\ &= \sum_{j=1}^{n} w_j^{(n)} g_k\big(x_j^{(n)}\big) + \sum_{i=1}^{p} \tilde{w}_i^{(n)} g_k\big(y_i^{(n)}\big) \\ &= 0 + \tilde{w}_k^{(n)} g_k\big(y_k^{(n)}\big). \end{aligned}$$

Since $g_k(y_k^{(n)}) \neq 0$ we find that $\tilde{w}_k^{(n)} = 0$. This holds for all $k \leqslant p$, which means that we have $K(f) = \sum_{j=1}^{n} w_j^{(n)} f(x_j^{(n)})$, which is the original Gauss formula. Therefore, $P_n(x)^2 \in \Pi_{2n}$ would be integrated to zero. However, if $p > \frac{n}{2}$, the formula should be exact for $f \in \Pi_{2n+1}$. But since $P_n(x)^2 \geqslant 0$ we have a contradiction. Hence $p > n$. □

From this lemma it follows that for exact integration of functions $f \in \Pi_{n+2p-1}$ the condition $p \geqslant n + 1$ is necessary. Let us consider $p = n + 1$. MONEGATO [1976a] has proven the following theorem.

THEOREM 4.5. *Let* (4.5) *be an extended Gauss–Legendre rule, then all* $\tilde{w}_i^{(n)} > 0$ *if and only if the nodes* $x_j^{(n)}$ *and* $y_i^{(n)}$ *interlace.*

SZEGÖ [1934] has proven that the nodes $x_j^{(n)}$ and $y_i^{(n)}$ interlace. Hence, the previous theorem shows that the weights $\tilde{w}_i^{(n)}$ are all positive. Furthermore, MONEGATO [1976b] has proven the next theorem.

THEOREM 4.6. *The weights* $w_j^{(n)}$ *of the extended Gauss–Legendre rules are positive.*

It should be noted that the weights $w_j^{(n)}$ are *not* the original Gauss weights. Although the weights of the extended Gauss–Legendre rules are all positive, the positivity of the weights of *all* successively obtained Patterson's rules has not been proven yet. However, from the tables it follows that the weights associated with the successive Patterson's quadrature rules (based on the 3-point Gauss–Legendre rule) until order 511 are all positive. Hence, the quadrature process by these quadrature rules is stable. Looking at Fig. 4.1 the conjecture might raise that the weights for higher order formulae are also positive.

Furthermore, tests have shown that the Patterson's rules converge *faster* to the true values of the integrals than the Gauss–Legendre rules for the same number of quadrature points. This was a reason for PATTERSON [1968] to state the following conjecture.

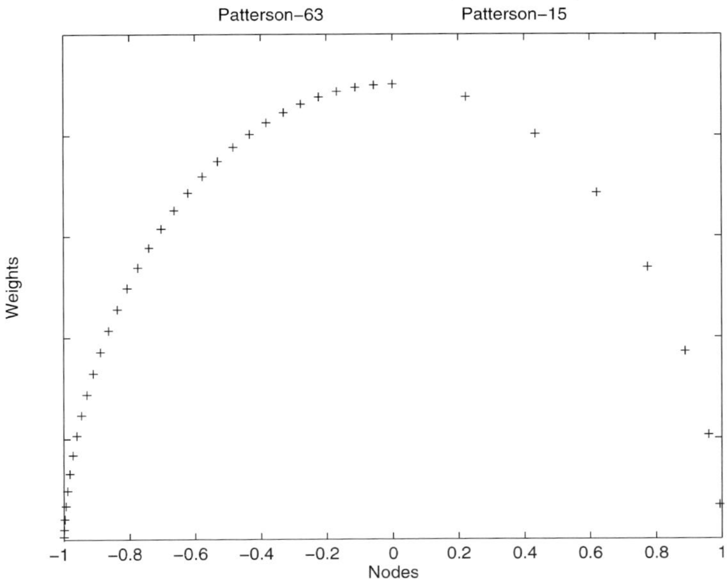

FIG. 4.1. The nodes x_i plotted against the weights w_i of the Patterson's 63-point formula in $(-1, 0]$ and the 15-point formulae in $[0, 1)$. The weights are scaled such that they can be compared.

CONJECTURE 4.1. *The quadrature process by Patterson's rules is stable and* uniformly *convergent for every function $f(x)$ which is Riemann-integrable in* $[-1, 1]$.

4.3. Comparison of integration methods

Next, we will compare the following four integration methods for some typical examples:

1. **Romberg's rule**
   Repeated trapezoidal rules and Richardson extrapolation until a certain accuracy has been reached. Error: $\mathcal{O}((\frac{b-a}{2^k})^{2(1+k)})$ for depth k, i.e., $2^k + 1$ function evaluations.
2. **Adaptive, recursive Simpson's rule**
   Repeated Simpson's rules on each subinterval recursively, until a certain tolerance level has been reached on each of the subintervals. Error: $\mathcal{O}((\frac{b-a}{2^k})^4)$ for depth k (the smallest subinterval has length $(b-a)2^{-k}$).
3. **Adaptive, recursive Newton–Cotes 8 panel rule**
   Repeated Newton–Cotes rules, where each interval is divided into 8 subintervals, recursively, until a certain accuracy has been reached on the subintervals. Error: $\mathcal{O}((\frac{b-a}{2^k})^9)$ for depth k.
4. **Patterson's quadrature rules**
   Successive extension of the Gauss–Legendre rule, up to degree 63.

The benefit of the Romberg's rule is its high theoretical convergence rate and the simplicity of the algorithm, and of an adaptive method its local refinement in the neighbourhood of a singularity. The first three methods use a 2-, 3- and 9-point Newton–Cotes formulae, and therefore, are only exact up to the corresponding degrees. For polynomial functions Patterson's rules of sufficiently high degree are exact. Therefore, we have chosen irregular integrands for comparison of the methods, so that none of these methods will be exact, but all converge to the exact values of integrals.

A measure for the convergence rate of the different methods is the number of necessary function evaluations to obtain a result with a given accuracy. In Table 4.2 the number of function evaluations is given for several tolerance levels for the relative error.

From the tables it is obvious that for the integrands chosen the Patterson's method is always the best choice. Compared to the other methods the number of function evaluations is very small. Since the Romberg's algorithm available to us makes use of global refinement for a singularity somewhere on the interval, the number of evaluation points becomes very large over the complete interval. Since Simpson's rule is a low order formula, it needs many subdivisions to obtain a sufficiently small error.

It appears that the available Romberg's algorithm and the adaptive Simpson's method are too expensive to obtain a satisfactory result. The adaptive Newton–Cotes 8 panel method is globally quite good, but the number of iterations, and therefore, the number of function evaluations is still quite large, because of its low order formula.

TABLE 4.2

Comparisons of the number of function evaluations needed for the different methods discussed, tested for several irregular functions. The tolerance is an upper bound for the absolute relative error

$I = \int_0^1 e^x \, dx$					$I = \int_0^1 \frac{1}{1+x^2} \, dx$				
Tol.	10^{-3}	10^{-6}	10^{-9}	10^{-12}	Tol.	10^{-3}	10^{-6}	10^{-9}	10^{-12}
Romb	5	9	17	33	Romb	9	33	65	129
AdSi	9	33	513	4097	AdSi	9	89	937	7885
ANC8	33	33	33	33	ANC8	33	33	49	113
Patt	7	7	15	15	Patt	7	15	31	31
$I = \int_0^1 x\sqrt{x} \, dx$					$I = \int_0^\pi \frac{1}{5+4\cos x} \, dx$				
Romb	9	129	2049	32769	Romb	17	129	129	513
AdSi	61	269	1965	6765	AdSi	37	321	3625	8193
ANC8	33	177	273	497	ANC8	33	65	129	289
Patt	7	31	63	63	Patt	15	31	63	63

4.3.1. Analysis of Patterson's quadrature rules

In this subsection two extra examples will be shown, that are of special interest for the study of the interaction integrals. The performance of the Patterson's quadrature rules for these integrals will be compared with two other quadrature rules.

The two integrals are:

$$I_1(\varepsilon) = \int_\varepsilon^1 \frac{1}{\sqrt{x}} \, dx,$$

$$I_2(\varepsilon) = \int_\varepsilon^{1-\varepsilon} \frac{1}{\sqrt{x(1-x)}} \, dx.$$

The smaller ε becomes, the better the integration interval approximates the interval [0, 1], and the more irregular the integrands become. Note that the singular behaviour near $x = 0$ for both integrals is about the same. Therefore, only the results for $I_2(\varepsilon)$ are shown. Those for $I_1(\varepsilon)$ are similar.

Table 4.3 shows the fast convergence rate of Patterson's quadrature rules compared to the other two methods. The results show that the number of function evaluations for the available Romberg's algorithm is too large to get a satisfactory result and that also the adaptive NC8 method needs too many evaluations. Although the number of evaluations for Patterson's rules is restricted to 63, the approximation of the integral is still quite accurate.

Next, we will compare the integrating power of Patterson's quadrature rules with that of the regular Gauss–Legendre rules. Fig. 4.2 shows the results obtained when these methods are applied to the integrand of $I_2(\varepsilon)$, which is not expected to be integrated exactly by these methods. The relative error is plotted against ε, which determines the integration bounds. For comparison this figure also shows the results for the Gauss–Legendre formula using the same number of points. Since an n-point Gauss rule is exact for polynomials of degree $2n - 1$ and an n-point Patterson rule is 'only' exact for

TABLE 4.3
(Absolute) relative errors $(|I_2 - K|)/|I_2|$ for different ε and the number of function evaluations needed. The maximum number of function evaluations for the different methods has been set to respectively 63, 8193 and 1048577

ε	Patterson		Adaptive NC8		Romberg	
10^{-10}	$3.3 \cdot 10^{-3}$	(63)	$2.6 \cdot 10^{-0}$	(8189)	$1.8 \cdot 10^{-2}$	(1048577)
10^{-7}	$3.0 \cdot 10^{-3}$	(63)	$7.3 \cdot 10^{-2}$	(8181)	$1.4 \cdot 10^{-4}$	(1048577)
10^{-5}	$7.7 \cdot 10^{-4}$	(63)	$2.6 \cdot 10^{-3}$	(8185)	$1.3 \cdot 10^{-11}$	(1048577)
10^{-4}	$2.1 \cdot 10^{-5}$	(63)	$5.5 \cdot 10^{-5}$	(8189)	$1.0 \cdot 10^{-14}$	(524289)
10^{-3}	$1.6 \cdot 10^{-8}$	(63)	$4.7 \cdot 10^{-8}$	(8185)	$7.4 \cdot 10^{-16}$	(65537)
10^{-2}	exact*	(63)	$1.6 \cdot 10^{-11}$	(8189)	$4.4 \cdot 10^{-15}$	(8193)

*Exact indicates an accuracy in excess of 16 digits.

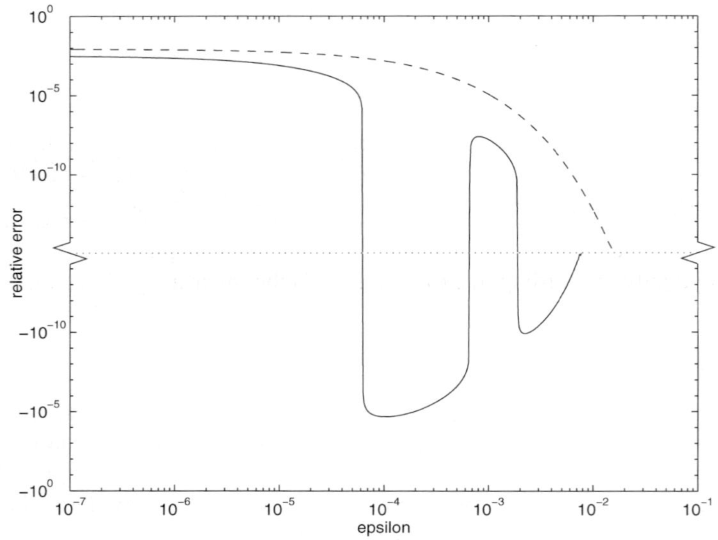

FIG. 4.2. Relative error in evaluating $I_2(\varepsilon)$ using Patterson-63. The dashed line shows the corresponding result for Gauss-63. The dotted line (rel. error = 0) indicates an accuracy in excess of 15 digits.

polynomials of $\frac{3}{2}n + 1$, one might expect that the Gauss–Legendre rules are better for higher degree polynomials. However, this is not the case. Tests have been performed, which show that integrals of high powers of x are better approximated by Patterson's rules than by Gauss–Legendre rules using the same number of points.

Apparently, the performance of Patterson's rules is superior to that of Gauss–Legendre rules. Tests on other almost singular integrands, show also that Patterson's rules are more accurate than the Gauss–Legendre rules for the same number of quadrature points. In summary, the advantages of the Patterson's rules above the Gauss–Legendre rules are

- All function evaluations of a quadrature rule can be used again in higher order quadrature rules,
- The relative error can be easily estimated, by taking the difference between two successive approximations.

5. Analytical integration

As we have already mentioned in Section 3, there are several cases where (a part of) the integral can be evaluated analytically. Basically, this is the case when the edges of the interaction domains are not a part of the boundary of the conductor region. A very special situation occurs when the quadrilateral elements are rectangular and parallel to the axes of the coordinate system. In that case the interaction integral with scalar valued basis functions can be evaluated completely analytically. For quadrilaterals which are not necessarily rectangular the partly analytical evaluation of the integrals is discussed in this section.

5.1. Analytical formula for scalar inner integral

In this subsection we consider the interaction integral with scalar valued basis functions. If none of the edges of the source element lie in the boundary of the conductor region, the factor for the boundary singularity is constant, therefore, the scalar basis functions are constant. In that case the inner part of the interaction integral can be evaluated completely analytically. According to expression (3.2) the inner integral has the form

$$I_i = \int_{\Omega_i} G(\mathbf{x}', \mathbf{x})\,\mathrm{d}\mathbf{x}',$$

where $\mathbf{x} = (x, y, z)$ is a fixed point of the interior of the object element. In this section only one term of the Green's function, $G(\mathbf{x}', \mathbf{x})$, of expression (3.3) will be taken into account.

The source quadrilateral Ω_i can be divided into triangles, of which the projection of $\mathbf{x}$ on the plane of the Ω_i is the common vertex (cf. Fig. 6.1). After transformation to polar coordinates, the inner integral I_i can be written as the sum of the four integrals over the triangles:

$$\int_{\Omega_i} \frac{1}{|\mathbf{x} - \mathbf{x}'|}\,\mathrm{d}\mathbf{x}' = \sum_{j=1}^{4} \int_{\varphi_{1j}}^{\varphi_{2j}} \int_0^{h_j/\cos\varphi} \frac{r}{\sqrt{r^2 + z^2}}\,\mathrm{d}r\,\mathrm{d}\varphi,$$

where φ_{1j} and φ_{2j} are the polar angles corresponding to the jth edge of Ω_i, $r^2 = (x - x')^2 + (y - y')^2$ and z stands for $z - z'$ (see Fig. 5.1).

Integration over r gives us

$$I_i = \sum_{j=1}^{4} \left\{ \int_{\varphi_{1j}}^{\varphi_{2j}} \sqrt{\frac{h_j^2}{\cos^2\varphi} + z^2}\,\mathrm{d}\varphi - |z|(\varphi_{2j} - \varphi_{1j}) \right\} \cdot \mathrm{sign}(h_j).$$

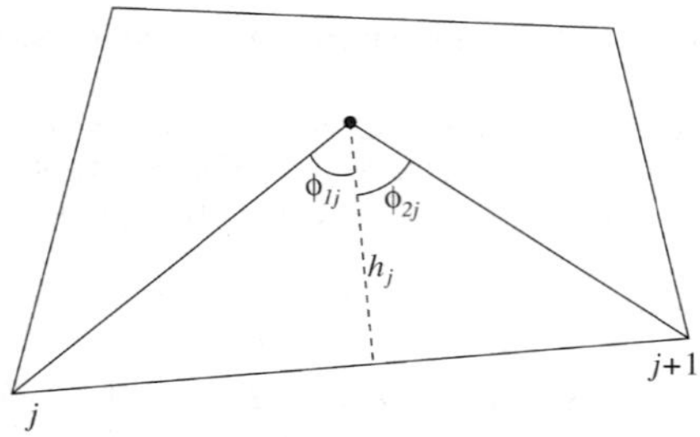

FIG. 5.1.

Finally, after the introduction of $p_j^2 = \frac{z^2}{h_j^2+z^2}$, $t_j = \sqrt{1 - p_j^2 \sin^2\varphi}$ and $q_j = \sqrt{1 - p_j^2}$, we obtain

$$I_i = \sum_{j=1}^{4} \left\{ \sqrt{h_j^2 + z^2} \int_{\varphi_{1j}}^{\varphi_{2j}} \frac{\sqrt{1 - p_j^2 \sin^2\varphi}}{\cos\varphi} \, \mathrm{d}\varphi - |z|(\varphi_{2j} - \varphi_{1j}) \right\} \cdot \operatorname{sign}(h_j)$$

$$= \sum_{j=1}^{4} \left\{ \sqrt{h_j^2 + z^2} \left[\frac{1}{2} q_j \ln \frac{t_j + q_j \sin\varphi}{t_j - q_j \sin\varphi} + p_j \arcsin(p_j \sin\varphi) \right]_{\varphi_{1j}}^{\varphi_{2j}} - |z|(\varphi_{2j} - \varphi_{1j}) \right\} \cdot \operatorname{sign}(h_j).$$

Note that if $h_j = 0$, then $\cos\varphi_{1j} = \cos\varphi_{2j} = 0$, $p_j = 1$, $q_j = 0$ and $t_{1j} = t_{2j} = 0$, so that

$$\left\{ |z| \arcsin(\sin\varphi) \Big|_{\varphi_{1j}}^{\varphi_{2j}} - |z|(\varphi_{2j} - \varphi_{1j}) \right\} \cdot \operatorname{sign}(h_j) = 0.$$

If $z = 0$, then

$$I_i = \sum_{j=1}^{4} \left[\frac{1}{2} h_j \ln \frac{1 + \sin\varphi}{1 - \sin\varphi} \right]_{\varphi_{1j}}^{\varphi_{2j}}.$$

5.2. *Analytical formula for vector valued inner integral*

Let us now consider the interaction integrals with vector valued basis functions, with a constant factor for the boundary singularity. In that case only the inner integral of $\mathbf{I}_i$ can be evaluated analytically. For this purpose we have to define some auxiliary quantities.

5.2.1. *Definitions of some auxiliary quantities*

Consider the quadrilateral $\mathbf{x}_1 \ldots \mathbf{x}_4$, with the vectors $\mathbf{v}_1 = \mathbf{x}_{12} - (\mathbf{x}_{12} + \mathbf{x}_{34})s_2$ and $\mathbf{v}_2 = -\mathbf{x}_{41} - (\mathbf{x}_{12} + \mathbf{x}_{34})s_1$ for $\mathbf{x}_{ij} = \mathbf{x}_j - \mathbf{x}_i$, and $\mathbf{w}_i$ as defined in Appendix B. After transformation to the isoparametric coordinates s_1 and s_2, the integrals over the edge functions $\mathbf{w}_i$ of the source element Ω_i, for a fixed object point $\mathbf{x}_m = (x_m, y_m, z_m)$, have the form

$$\int_0^1\int_0^1 \frac{\mathbf{w}_i(s_1,s_2)}{|\mathbf{x}_m-\mathbf{x}'|} J(s_1,s_2)\,\mathrm{d}s_1\,\mathrm{d}s_2,$$

where J is the Jacobian $|\mathbf{v}_1\times\mathbf{v}_2|$. This can be rewritten as

$$\int_0^1\int_0^1 \frac{\mathbf{q}_i(s_1,s_2)}{|\mathbf{x}_m-\mathbf{x}'|}\,\mathrm{d}s_1\,\mathrm{d}s_2, \tag{5.1}$$

where $\mathbf{q}_i$, for $j=1,\dots,4$ are the quadratic functions

$$\mathbf{q}_1=(1-s_2)\mathbf{v}_2,\quad \mathbf{q}_2=-s_1\mathbf{v}_1,\quad \mathbf{q}_3=-s_2\mathbf{v}_2,\quad \mathbf{q}_4=(1-s_1)\mathbf{v}_1. \tag{5.2}$$

Further, we introduce the vector $\mathbf{v}_0=(1-s_2)\mathbf{x}_1+s_2\mathbf{x}_4-\mathbf{x}_m$.

Since $\mathbf{x}'(s_1,s_2)=\mathbf{x}_1+\mathbf{x}_{12}s_1-\mathbf{x}_{41}s_2-(\mathbf{x}_{12}+\mathbf{x}_{34})s_1s_2$, the square of the denominator of the integrand of (5.1) can be rewritten as

$$|\mathbf{x}_m-\mathbf{x}'|^2=Q(s_1,s_2)=a(s_2)+b(s_2)s_1+c(s_2)s_1^2,$$

where a,b,c are quadratic functions of s_2:

$$\begin{aligned}
a(s_2)&=\mathbf{v}_0\cdot\mathbf{v}_0,\\
&=|\mathbf{x}_{41}|^2s_2^2-2\mathbf{x}_{41}\cdot\mathbf{x}_{m1}s_2+|\mathbf{x}_{m1}|^2,\\
b(s_2)&=2\mathbf{v}_1\cdot\mathbf{v}_0,\\
&=2\big\{(\mathbf{x}_{12}+\mathbf{x}_{34})\cdot\mathbf{x}_{41}s_2^2-\big\{(\mathbf{x}_{12}+\mathbf{x}_{34})\cdot\mathbf{x}_{m1}+\mathbf{x}_{12}\cdot\mathbf{x}_{41}\big\}s_2+\mathbf{x}_{12}\cdot\mathbf{x}_{m1}\big\},\\
c(s_2)&=\mathbf{v}_1\cdot\mathbf{v}_1,\\
&=|\mathbf{x}_{12}+\mathbf{x}_{34}|^2s_2^2-2(\mathbf{x}_{12}+\mathbf{x}_{34})\cdot\mathbf{x}_{12}s_2+|\mathbf{x}_{12}|^2.
\end{aligned}$$

5.2.2. *Form of the integrals*

Since the functions $\mathbf{q}_i$ in the expressions (5.2) are linear with respect to s_1 and s_2, the integral (5.1) is a linear combination of two integrals of the form:

$$\int_0^1 \frac{1}{\sqrt{Q(s_1,s_2)}}\,\mathrm{d}s_1=\int_0^1\frac{1}{\sqrt{a+bs_1+cs_1^2}}\,\mathrm{d}s_1,$$

$$\int_0^1 \frac{s_1}{\sqrt{Q(s_1,s_2)}}\,\mathrm{d}s_1=\int_0^1\frac{s_1}{\sqrt{a+bs_1+cs_1^2}}\,\mathrm{d}s_1,$$

where $Q(s_1,s_2)$ is quadratic with respect to s_1 and s_2 and $a(s_2)$, $b(s_2)$ and $c(s_2)$ are the quadratic functions as defined above. The integrals can be readily evaluated:

$$\frac{1}{\sqrt{c}}\ln\left(\frac{\sqrt{c}\sqrt{a+b+c}+c+\frac{1}{2}b}{\sqrt{c}\sqrt{a}+\frac{1}{2}b}\right),\qquad \frac{\sqrt{a+b+c}-\sqrt{a}}{c}-\frac{b}{2c}I_0.$$

After substitution of these integrals in (5.1) we obtain

$$\begin{aligned}
\int_0^1\int_0^1\frac{\mathbf{q}_i(s_1,s_2)}{|\mathbf{x}_m-\mathbf{x}'|}\,\mathrm{d}s_1\,\mathrm{d}s_2&=\int_0^1\overline{\mathcal{F}}_i\ln\left(\frac{\sqrt{c}\sqrt{a+b+c}+c+\frac{1}{2}b}{\sqrt{c}\sqrt{a}+\frac{1}{2}b}\right)\mathrm{d}s_2\\
&\quad+\int_0^1\overline{\mathcal{G}}_i\big(\sqrt{a+b+c}-\sqrt{a}\big)\,\mathrm{d}s_2,
\end{aligned}$$

where the vector valued functions $\overline{\mathcal{F}}_i$ and $\overline{\mathcal{G}}_i$ are defined by:

$$\overline{\mathcal{F}}_1(s_2) = (1-s_2)\frac{1}{\sqrt{c}}\left\{-\mathbf{x}_{41} + \frac{b}{2c}(\mathbf{x}_{12}+\mathbf{x}_{34})\right\},$$

$$\overline{\mathcal{G}}_1(s_2) = -\frac{1}{c}(1-s_2)(\mathbf{x}_{12}+\mathbf{x}_{34}),$$

$$\overline{\mathcal{F}}_2(s_2) = \frac{b}{2c\sqrt{c}}\{\mathbf{x}_{12} - s_2(\mathbf{x}_{12}+\mathbf{x}_{34})\},$$

$$\overline{\mathcal{G}}_2(s_2) = -\frac{1}{c}\{\mathbf{x}_{12} - s_2(\mathbf{x}_{12}+\mathbf{x}_{34})\},$$

$$\overline{\mathcal{F}}_3(s_2) = -s_2\frac{1}{\sqrt{c}}\left\{-\mathbf{x}_{41} + \frac{b}{2c}(\mathbf{x}_{12}+\mathbf{x}_{34})\right\},$$

$$\overline{\mathcal{G}}_3(s_2) = \frac{1}{c}s_2(\mathbf{x}_{12}+\mathbf{x}_{34}),$$

$$\overline{\mathcal{F}}_4(s_2) = \frac{1}{\sqrt{c}}\left(1+\frac{b}{2c}\right)\{\mathbf{x}_{12} - s_2(\mathbf{x}_{12}+\mathbf{x}_{34})\},$$

$$\overline{\mathcal{G}}_4(s_2) = -\frac{1}{c}\{\mathbf{x}_{12} - s_2(\mathbf{x}_{12}+\mathbf{x}_{34})\}.$$

After substitution of the expressions for a, b and c in the integrands, we obtain

$$\int_0^1\int_0^1 \frac{\mathbf{q}_i(s_1,s_2)}{|\mathbf{x}_m - \mathbf{x}'|}\,\mathrm{d}s_1\,\mathrm{d}s_2 = \int_0^1 \overline{\mathcal{F}}_i \ln\left(\frac{|\mathbf{v}_1||\mathbf{v}_1+\mathbf{v}_0| + \mathbf{v}_1\cdot(\mathbf{v}_1+\mathbf{v}_0)}{|\mathbf{v}_1||\mathbf{v}_0| + \mathbf{v}_1\cdot\mathbf{v}_0}\right)\mathrm{d}s_2 + \int_0^1 \overline{\mathcal{G}}_i\big(|\mathbf{v}_1+\mathbf{v}_0| - |\mathbf{v}_0|\big)\,\mathrm{d}s_2. \tag{5.3}$$

5.2.3. Evaluation for non-singular integrand

If the object point $\mathbf{x}_m$ is not in the source element, which means that either $z_m \neq 0$ or the projection point lies outside the element, the integrands of (5.3) are non-singular. Therefore, the integral (5.3) can be evaluated numerically with a satisfactory accuracy. However, if the distance, h, of $\mathbf{x}_m$ to the line $s_2 =$ constant through the integration point (s_1, s_2), which intersects the edges $\mathbf{x}_{23}$ and $\mathbf{x}_{41}$ of the source element, is almost zero, the integrand of the first integral of (5.3) is irregular and has to be analysed separately.

Behaviour of the integrand for $h \to 0$. Let d be defined as illustrated in Fig. 5.2. For $\alpha < \frac{\pi}{2}$ it holds that $d < 0$, and for $\alpha > \frac{\pi}{2}$ that $d > 0$. Note that $|\mathbf{v}_1| > 0$, since $\mathbf{v}_1$ is a convex combination of the edges $\mathbf{x}_{12}$ and $\mathbf{x}_{34}$, which are both nonzero.

An analysis of the argument L of the logarithm in the first integral of (5.3) shows that

$$L = \frac{|\mathbf{v}_1||\mathbf{v}_1+\mathbf{v}_0| + \mathbf{v}_1\cdot(\mathbf{v}_1+\mathbf{v}_0)}{|\mathbf{v}_1||\mathbf{v}_0| + \mathbf{v}_1\cdot\mathbf{v}_0} = \frac{|\mathbf{v}_1+\mathbf{v}_0| - d}{|\mathbf{v}_0| - |\mathbf{v}_1| - d} = \frac{\sqrt{h^2+d^2} - d}{\sqrt{h^2 + (|\mathbf{v}_1|+d)^2} - |\mathbf{v}_1| - d}. \tag{5.4}$$

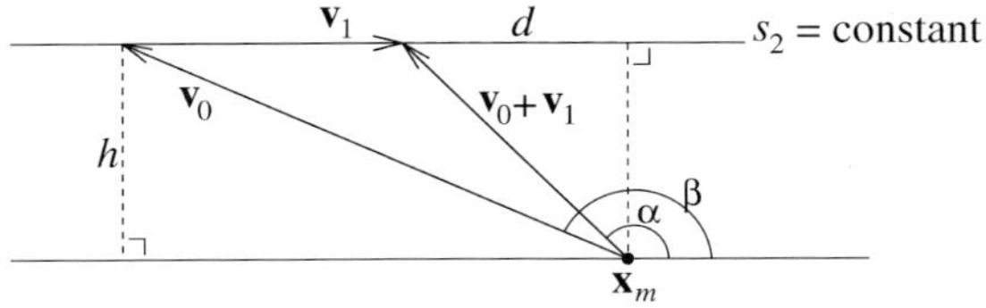

$$\sin\alpha = \frac{h}{|\mathbf{v}_1 + \mathbf{v}_0|}, \qquad \sin\beta = \frac{h}{|\mathbf{v}_0|}$$

$$h = \frac{|\mathbf{v}_0 \times \mathbf{v}_1|}{|\mathbf{v}_1|}, \qquad d = \frac{-\mathbf{v}_1 \cdot (\mathbf{v}_0 + \mathbf{v}_1)}{|\mathbf{v}_1|}$$

FIG. 5.2. For the analysis of the integrand of the first integral of (5.3) a plane has been drawn through $\mathbf{x}_m$ and the line $s_2 =$ constant. Note that this line is in the plane of the quadrilateral, but $\mathbf{x}_m$ is not, if $z_m \neq 0$.

For $d < 0$, L is singular if $\sqrt{h^2 + (|\mathbf{v}_1| - |d|)^2} = |\mathbf{v}_1| - |d|$, i.e., $h = 0$ and $|d| \leqslant |\mathbf{v}_1|$, so that for $h \to 0$ formula (5.4) can only be used for $d < -|\mathbf{v}_1|$. For $d = 0$, L is singular if $h = 0$.

If $d > 0$ and $h \to 0$, then $|\mathbf{v}_1 + \mathbf{v}_0| \to d$ and $|\mathbf{v}_0| - |\mathbf{v}_1| \to d$, so that the two terms in the numerator and the denominator of L have opposite signs and their value may become inaccurate due to cancellation of significant digits. This may affect the accuracy of L. However, L can be reformulated as follows:

$$L = \frac{|\mathbf{v}_1||\mathbf{v}_0| - \mathbf{v}_1 \cdot \mathbf{v}_0}{|\mathbf{v}_1||\mathbf{v}_1 + \mathbf{v}_0| - \mathbf{v}_1 \cdot (\mathbf{v}_1 + \mathbf{v}_0)} = \frac{|\mathbf{v}_0| + |\mathbf{v}_1| + d}{|\mathbf{v}_1 + \mathbf{v}_0| + d}$$
$$= \frac{\sqrt{h^2 + (|\mathbf{v}_1| + d)^2} + |\mathbf{v}_1| + d}{\sqrt{h^2 + d^2} + d}. \tag{5.5}$$

Since for $d > 0$ all terms in the numerator and in the denominator of the right-hand side are positive, both the numerator and denominator do not vanish, so that L can be evaluated accurately. If $d \leqslant 0$ and $h \to 0$, then $|\mathbf{v}_0| - |\mathbf{v}_1| \to -d$, so that L is singular.

In summary, for $d < -|\mathbf{v}_1|$ formula (5.4) must be used and for $d > 0$ formula (5.5). However, for $-|\mathbf{v}_1| \leqslant d \leqslant 0$, L is singular for $h = 0$. In this case the first integral of (5.3) has to be handled differently as will be shown in the next subsection.

5.2.4. *Evaluation for singular integrand*

If the integrand of the first integral of (5.3) is singular, some more provisions have to be made. Note that, if $z_m = 0$, $\mathbf{x}_m$ can also be written in isoparametric coordinates $s_i^{(m)}$.

The first integral of (5.3),

$$\int_0^1 \overline{\mathcal{F}}_i \ln\left(\frac{\sqrt{c}\sqrt{a+b+c} + c + \frac{1}{2}b}{\sqrt{c}\sqrt{a} + \frac{1}{2}b}\right) \mathrm{d}s_2,$$

can be rewritten as

$$\int_0^1 \overline{\mathcal{F}}_i \ln\left(\left(\sqrt{c}\sqrt{a+b+c} + c + \tfrac{1}{2}b\right)\left(\sqrt{c}\sqrt{a} - \tfrac{1}{2}b\right)\right) \mathrm{d}s_2$$
$$- \int_0^1 \overline{\mathcal{F}}_i \ln\left(ca - \left(\tfrac{1}{2}b\right)^2\right) \mathrm{d}s_2,$$

the last term of which is singular. After partial integration this integral becomes

$$\int_0^1 \mathcal{F}_i \ln\big(ca - \big(\tfrac{1}{2}b\big)^2\big)\,\mathrm{d}s_2 = \big[\mathcal{F}_i(s_2)\mathcal{L}(s_2)\big]_0^1 - \int_0^1 \mathcal{F}_i'(s_2)\mathcal{L}(s_2)\,\mathrm{d}s_2,$$

where

$$\mathcal{L}(s_2) = \int \ln\big(ca - \big(\tfrac{1}{2}b\big)^2\big)\,\mathrm{d}s_2. \tag{5.6}$$

After reformulation of the expressions for $\mathbf{x}_m$, $\mathbf{v}_0$ and $\mathbf{v}_1$:

$$\begin{aligned}
\mathbf{x}_m &= \mathbf{x}_1 + \mathbf{x}_{12}s_1^{(m)} - \mathbf{x}_{41}s_2^{(m)} - (\mathbf{x}_{12} + \mathbf{x}_{34})s_1^{(m)}s_2^{(m)},\\
\mathbf{v}_0 &= (1 - s_2)\mathbf{x}_1 + s_2\mathbf{x}_4 - \mathbf{x}_m = \mathbf{x}_1 - \mathbf{x}_{41}s_2 - \mathbf{x}_m\\
&= -\mathbf{x}_{12}s_1^{(m)} + \mathbf{x}_{41}\big(s_2^{(m)} - s_2\big) + (\mathbf{x}_{12} + \mathbf{x}_{34})s_1^{(m)}s_2^{(m)}\\
&= \mathbf{c}_0\big(s_2^{(m)} - s_2\big) - \mathbf{d}s_1^{(m)},\\
\mathbf{v}_1 &= \mathbf{x}_{12}(1 - s_2) - \mathbf{x}_{34}s_2\\
&= \big\{\mathbf{x}_{12}\big(1 - s_2^{(m)}\big) - \mathbf{x}_{34}s_2^{(m)}\big\} + (\mathbf{x}_{12} + \mathbf{x}_{34})\big(s_2^{(m)} - s_2\big)\\
&= \mathbf{d} + \mathbf{c}_1\big(s_2^{(m)} - s_2\big),
\end{aligned}$$

where $\mathbf{c}_0 = \mathbf{x}_{41}$, $\mathbf{c}_1 = \mathbf{x}_{12} + \mathbf{x}_{34}$ and $\mathbf{d} = \mathbf{x}_{12}(1 - s_2^{(m)}) - \mathbf{x}_{34}s_2^{(m)}$, we obtain for the argument of the logarithm of integral (5.6):

$$\begin{aligned}
ca - \big(\tfrac{1}{2}b\big)^2 &= |\mathbf{v}_0|^2|\mathbf{v}_1|^2 - (\mathbf{v}_0 \cdot \mathbf{v}_1)^2 = |\mathbf{v}_0|^2|\mathbf{v}_1|^2\big(1 - (\cos\beta)^2\big)\\
&= |\mathbf{v}_0|^2|\mathbf{v}_1|^2(\sin\beta)^2 = |\mathbf{v}_0 \times \mathbf{v}_1|^2\\
&= \big|(\mathbf{c}_0 \times \mathbf{c}_1)\big(s_2^{(m)} - s_2\big)^2 + \big(\mathbf{c}_0 + \mathbf{c}_1 s_1^{(m)}\big) \times \mathbf{d}\big(s_2^{(m)} - s_2\big)\big|^2\\
&= \big(s_2^{(m)} - s_2\big)^2\big\{C^2\big(s_2^{(m)} - s_2\big)^2 + B\big(s_2^{(m)} - s_2\big) + A^2\big\},
\end{aligned}$$

where

$$\begin{aligned}
C &= |\mathbf{c}_0 \times \mathbf{c}_1|, \qquad B = 2(\mathbf{c}_0 \times \mathbf{c}_1) \cdot \big(\big(\mathbf{c}_0 + \mathbf{c}_1 s_1^{(m)}\big) \times \mathbf{d}\big),\\
A &= \big|\big(\mathbf{c}_0 + \mathbf{c}_1 s_1^{(m)}\big) \times \mathbf{d}\big|.
\end{aligned}$$

Since the vectors $\mathbf{c}_0$, $\mathbf{c}_1$ and $\mathbf{d}$ all lie in the same plane, $\mathbf{c}_0 \times \mathbf{c}_1 \parallel (\mathbf{c}_0 + \mathbf{c}_1 s_1^{(m)}) \times \mathbf{d}$, so that

$$ca - \big(\tfrac{1}{2}b\big)^2 = y^2\big(Cy + \operatorname{sign}(B)A\big)^2,$$

where $y = s_2^{(m)} - s_2$. Therefore, the integral (5.6) becomes

$$\int \ln y^2\,\mathrm{d}s_2 + \int \ln\big(Cy + \operatorname{sign}(B)A\big)^2\,\mathrm{d}s_2.$$

Since $\mathrm{d}y = -\mathrm{d}s_2$, the first integral becomes

$$\int \ln y^2\,\mathrm{d}s_2 = -\int \ln y^2\,\mathrm{d}y = -y(\ln y^2 - 2),$$

and the second integral:

$$\int \ln\left(Cy + \mathrm{sign}(B)A\right)^2 \mathrm{d}s_2 = -\int \ln\left(Cy + \mathrm{sign}(B)A\right)^2 \mathrm{d}y$$
$$= -\frac{1}{C}\left(Cy + \mathrm{sign}(B)A\right)\left(\ln\left(Cy + \mathrm{sign}(B)A\right)^2 - 2\right).$$

Hence,

$$\mathcal{L}(s_2) = -y\left\{\ln y^2\left(Cy + \mathrm{sign}(B)A\right)^2 - 4\right\}$$
$$- \mathrm{sign}(B)\frac{A}{C}\left(\ln\left(Cy + \mathrm{sign}(B)A\right)^2 - 2\right).$$

If $C = 0$ the integral becomes:

$$\mathcal{L}(s_2) = -y\left(\ln(yA)^2 - 2\right).$$

6. Regularisations

In the previous section, integrands were discussed that did not contain a factor for the boundary singularity. These could be integrated partly analytically. For integrals with a factor for boundary singularity numerical methods must be used. As shown in Section 4 a very important condition for the numerical integration to give a satisfactory result is the smoothness of the integrand. So if the integrand contains singularities quadrature rules can give very inaccurate results. Since the interaction integral contains the Green's singularity as well as the boundary singularity, straightforward numerical integration is rather unreliable. In this section we will treat the elimination of the singularities by regularisation of the integrands. The Green's singularity and the boundary singularity will be treated separately.

6.1. The inner part of the interaction integral

As shown in Section 3 the interaction integral has the general form:

$$\int_{\Omega_j} \frac{\psi_j(\mathbf{x})}{\sqrt{d_j}} \left\{ \int_{\Omega_i} G(\mathbf{x}', \mathbf{x}) \frac{\psi_i(\mathbf{x}')}{\sqrt{d_i}} \,\mathrm{d}\mathbf{x}' \right\} \mathrm{d}\mathbf{x},$$

with d_j and d_i the distance functions $d_j(s_1, s_2)$ and $d_i(s_1', s_2')$, the specific form of which is given in Appendix A. Without any loss of generality we will assume that the z-coordinate of $\mathbf{x}'$ is always zero.

The first four subsections of this section deal with the *inner part* of the interaction integral

$$\mathbf{I}_i = \int_{\Omega_i} G(\mathbf{x}', \mathbf{x}) \frac{\psi_i(\mathbf{x}')}{\sqrt{d_i}} \,\mathrm{d}\mathbf{x}', \qquad (6.1)$$

the last subsection with the outer part. In this section only one term of the Green's function, $G(\mathbf{x}', \mathbf{x})$, of expression (3.3) will be taken into account.

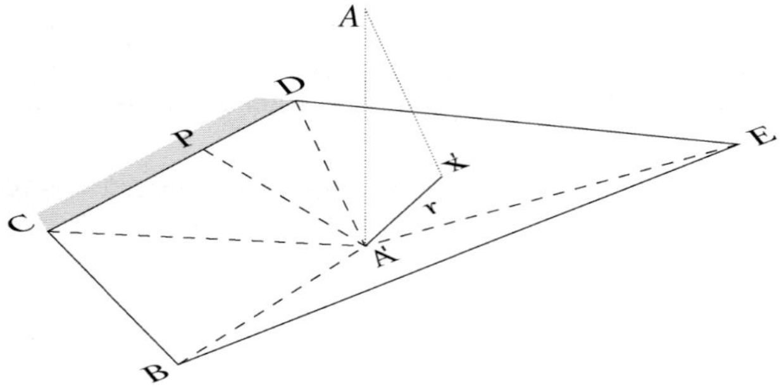

FIG. 6.1. Source quadrilateral divided into triangles, determined by the projection A' of the object point A.

Let A be the object point $\mathbf{x}$, and $A' = (x, y, 0)$, the projection of A on the source element plane (see Fig. 6.1). Let $r(x, y) = \sqrt{(x' - x)^2 + (y' - y)^2}$, then $G(\mathbf{x}', \mathbf{x})$ can be written as $|\mathbf{x}' - \mathbf{x}|^{-1} = 1/\sqrt{r^2 + z^2}$. Then (6.1) becomes

$$\iint_{\Omega_i} \frac{1}{\sqrt{r^2 + z^2}} \frac{\psi_i(\mathbf{x}')}{\sqrt{d_i}} \, \mathrm{d}\mathbf{x}'.$$

This integral has two sources of singularities: the Green's function $(r^2 + z^2)^{-1/2}$ and the boundary singularity $d_i^{-1/2} = d_i(s_1, s_2)^{-1/2}$. First we will concentrate on the former.

6.2. *The Green's singularity*

The expression $(r^2 + z^2)^{-1/2}$ depends on the distance z between the planes of the interacting quadrilaterals; for $z = 0$ it has a singularity in $r = 0$.

To remove the source of singularity, the quadrilateral will be divided into four triangles, each of which has an edge of the quadrilateral as base and midpoint A' as vertex (see Fig. 6.1). After transformation to polar coordinates the integral for one of four triangles becomes:

$$\int_{\varphi_1}^{\varphi_2} \int_0^{R(\varphi)} \frac{r}{\sqrt{r^2 + z^2}} \frac{\psi_i}{\sqrt{d}} \, \mathrm{d}r \, \mathrm{d}\varphi. \tag{6.2}$$

For example, for the triangle $\triangle A'CD$, φ_1 and φ_2 are the polar angles corresponding to $A'D$ and $A'C$, and $R(\varphi) = |A'P|$, with P the point on the boundary CD corresponding to the angle φ (see also Fig. 6.3). From the expression (6.2) one can see that if $z = 0$, i.e., $\mathbf{x}$ and $\mathbf{x}'$ are in the same plane, the Green's singularity has been completely eliminated. Even for $z \neq 0$ the function $r/\sqrt{r^2 + z^2}$ is regular.

6.3. *The boundary singularity*

To regularize the boundary singularity due to d in (6.2) is much more complicated. In Appendix A d is given as a function of the isoparametric coordinates (s_1, s_2). But the integrand of (6.2) must be integrated over the variables r and φ. Therefore, we need an expression of $\tilde{s}_i(r, \varphi)$ as function of r, φ, or an expression $s_i(r, \varphi)$, since the expressions $x'(r, \varphi)$ and $y'(r, \varphi)$ are well known. An unique expression $s_i(r, \varphi)$ only exists for points (x', y') inside and at the edges of the source quadrilateral, because an extra condition, $0 \leqslant s_i \leqslant 1$, must also be fulfilled.

Before deriving the expression $s_i(x', y')$ we will first define some auxiliary quantities. Having the quadrilateral $\mathbf{x}_1\mathbf{x}_2\mathbf{x}_3\mathbf{x}_4$ (in counterclockwise order), the edges can be defined as $\mathbf{x}_{ij} \equiv (\mathbf{x}_j - \mathbf{x}_i)$, for $j = i \bmod 4 + 1$.

The transformation can be given by

$$\mathbf{x}'(s_1, s_2) = \mathbf{x}_1 + \mathbf{x}_{12}s_1 - \mathbf{x}_{41}s_2 - (\mathbf{x}_{12} + \mathbf{x}_{34})s_1 s_2, \tag{6.3}$$

or

$$\mathbf{x}' - \mathbf{x}_1 - \mathbf{x}_{12}s_1 = -\big\{(\mathbf{x}_{12} + \mathbf{x}_{34})s_1 + \mathbf{x}_{41}\big\}s_2.$$

Taking the outer product of the left-hand side with the right-hand side, we obtain an implicit expression for s_1 in terms of $\mathbf{x}' = (x', y')$:

$$\begin{aligned} f(s_1)\mathbf{e}_z = (\mathbf{x}_{12} \times \mathbf{x}_{34})s_1^2 + \big\{(\mathbf{x}_{12} \times \mathbf{x}_{41}) - \big((\mathbf{x}' - \mathbf{x}_1) \times (\mathbf{x}_{12} + \mathbf{x}_{34})\big)\big\}s_1 \\ - \big((\mathbf{x}' - \mathbf{x}_1) \times \mathbf{x}_{41}\big) = 0, \end{aligned}$$

where $\mathbf{e}_z$ is the unit vector in the z-direction and $(\mathbf{a} \times \mathbf{b}) = (a_x b_y - a_y b_x)\mathbf{e}_z$, since the z-coordinates were assumed to be zero.

Note that $f(0)\mathbf{e}_z = -((\mathbf{x}' - \mathbf{x}_1) \times \mathbf{x}_{41})$ and $f(1)\mathbf{e}_z = ((\mathbf{x}' - \mathbf{x}_2) \times \mathbf{x}_{23})$. For a point $\mathbf{x}'$ inside or at the edges of the source quadrilateral Fig. 6.2 shows that $f(0) \leqslant 0$ and $f(1) \geqslant 0$, and that if $f(0) = 0$, then $f(1) \neq 0$, and if $f(1) = 0$, then $f(0) \neq 0$. Thus, f has exactly one root for $0 \leqslant s_1 \leqslant 1$. For a point $\mathbf{x}'$ outside the source quadrilateral there are generally two roots! From (6.3) it follows now

$$s_2 = -\frac{x' - x_1 - x_{12}s_1}{x_{41} + (x_{12} + x_{34})s_1} = -\frac{y' - y_1 - y_{12}s_1}{y_{41} + (y_{12} + y_{34})s_1}. \tag{6.4}$$

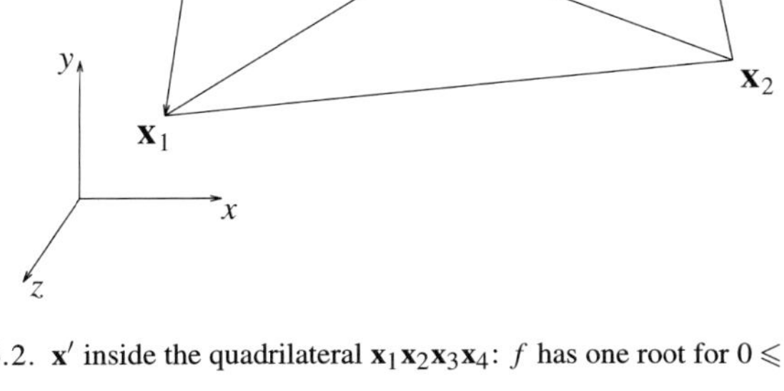

FIG. 6.2. $\mathbf{x}'$ inside the quadrilateral $\mathbf{x}_1\mathbf{x}_2\mathbf{x}_3\mathbf{x}_4$: f has one root for $0 \leqslant s_1 \leqslant 1$.

Hence we have an expression $s_i(x', y')$ for both isoparametric coordinates in terms of x' and y'. As a function of polar coordinates we have

$$\tilde{s}_i(r, \varphi) = s_i\big(x'(r, \varphi), y'(r, \varphi)\big),$$

so that the boundary singularity, in polar coordinates, has the form

$$d\big(\tilde{s}_1(r, \varphi), \tilde{s}_2(r, \varphi)\big)^{-1/2}.$$

The expression for d is too complicated for an analysis of the behaviour of the boundary singularity. Although d is a linear function of s_1 or s_2, generally, it is not the geometric distance from point $\mathbf{x}'$ to the boundary. Let $s' = |\mathbf{x}'(s_1, s_2) - \mathbf{x}'(0, \tilde{s}_2)|$ represent the distance from $\mathbf{x}'$ to the boundary of the conductor region. The following lemma holds:

LEMMA 6.1. *Let s' be defined as the distance $|\mathbf{x}'(s_1, s_2) - \mathbf{x}'(0, \tilde{s}_2)|$ to the boundary, where $\tilde{s}_2$ corresponds to the projection of $\mathbf{x}'$ to the boundary, and let d be the isoparametric distance to the boundary in the unit square. Then, if $\mathbf{x}'$ approaches the boundary, s' is approximately proportional to d, i.e., $s' \approx Cd$, where C is a constant.*

PROOF. Since the source for the boundary singularity is not constant, it may be assumed that the quadrilateral is rectangular. Suppose $d = s_2$. Let then $\mathbf{x}(s_1, s_2) = a(s_1)s_2 + b(s_1)$, for linear functions a and b. Since the line '$s_1 =$ constant' is perpendicular to the boundary, the point on the boundary with the shortest distance to $\mathbf{x}(s_1, s_2)$ is $\mathbf{x}(s_1, 0)$. So $s' = |\mathbf{x}(s_1, s_2) - \mathbf{x}(s_1, 0)| = |\{a(s_1)s_2 + b(s_1)\} - b(s_1)| = |a(s_1)s_2| = Cs_2$, since $a(s_1)$ is independent of s_2, hence constant for fixed s_1. The proof is analogous for other boundary singularities. □

The lemma says that, except for a constant, s' and d are equivalent functions. Therefore, in the following d will be represented as s'.

6.3.1. Transformations for one boundary

In order to evaluate the integrals over the triangles, we have to determine the integration bounds explicitly, in the particular cases. In Fig. 6.1 it can be seen that if one or more of the edges of the quadrilateral is a boundary of the domain, we have different situations for the integration triangles. We will first consider the case where the quadrilateral has only one boundary.

Triangle tangent to the boundary. The integrand with the boundary singularity has, after transformation to polar coordinates, the general form

$$\int_{\varphi_1}^{\varphi_2} \int_0^{R(\varphi)} \frac{\psi_i}{\sqrt{s'}} \, \mathrm{d}r \, \mathrm{d}\varphi,$$

where $R(\varphi) = |A'P|$, as can be seen in Fig. 6.3. Here we have the situation that one edge of the triangle is a boundary.

We want to regularise the boundary singularity by a certain substitution. So we have to find an expression for s' in terms of the integration variables. In the picture one

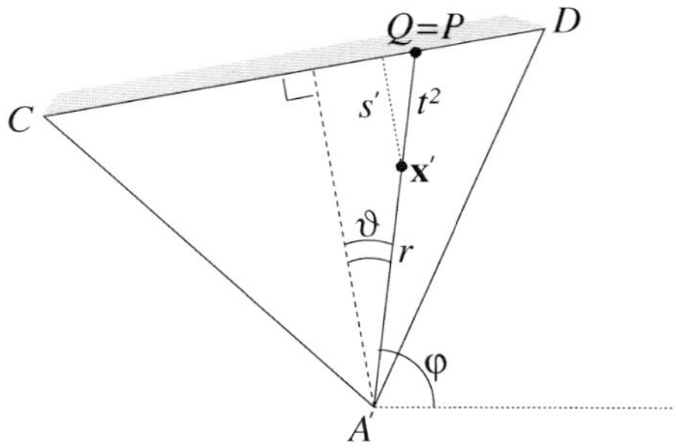

$$A' \equiv (x, y), \qquad CD = \text{boundary}$$
$$r \equiv |A'\mathbf{x}'|, \qquad R(\varphi) \equiv |A'Q|$$
$$t^2 \equiv |\mathbf{x}'Q|, \qquad s' = t^2 \cos\vartheta$$

FIG. 6.3. $\triangle A'CD$ with polar coordinates, where the edge CD is a boundary of the source domain.

can see that we have chosen t^2 such that $r = R(\varphi) - t^2$. To eliminate the $\sqrt{s'}$ singularity, which is now dependent on t and φ, we may substitute $s' = t^2 \cos\vartheta$, where $\vartheta = \frac{\pi}{2} - \angle A'QC$ is dependent on φ, since the position of Q is. Thus finally, with these substitutions the integral becomes

$$\int_{\varphi_1}^{\varphi_2} \int_0^{\sqrt{R(\varphi)}} \frac{\psi_i}{\sqrt{\cos\vartheta}} 2t \, \mathrm{d}t \, \mathrm{d}\varphi.$$

Since $\vartheta < \frac{\pi}{2}$, this is a regular integral, which could be evaluated rather accurate by Patterson, unless the triangle is very flat, i.e., $|\frac{\pi}{2} - \vartheta|$ small, then we will approach this integrand in a special way (see Section 6.4).

Triangle without tangent to the boundary. The latter case was a situation where one edge of the triangle is a boundary. The treatment changes a little if none of the edges is a boundary. Take, for instance, $\triangle A'DE$, where P, which is the intersubsection point of $A'\mathbf{x}'$ with the edge DE, is not a boundary point and Q is the intersubsection point of $A'P$ with the boundary CD. We will use the same kind of substitutions, but for that purpose some adaptions have to be made: let $q(\varphi) = |A'Q|$ and $R(\varphi) = |A'P|$, so we may substitute $r = q(\varphi) - t^2$. Then the inner integral becomes

$$\int_{\sqrt{q(\varphi)-R(\varphi)}}^{\sqrt{q(\varphi)}} \frac{\psi_i}{\sqrt{\cos\vartheta}} 2t \, \mathrm{d}t.$$

At first sight this seems satisfactory. However, if $A'P \| CD$ then ϑ becomes zero, or at least small if approximately $A'P \| CD$. This would introduce another singularity, namely $(\cos\vartheta)^{-1/2}$. Yet, if ϑ might become small then substitution won't be necessary anymore, for in that case $\sqrt{s'}$ changes hardly along AP, which means that it behaves very much like a constant. So here we would simply have the original integral

$$\int_0^{R(\varphi)} \frac{\psi_i}{\sqrt{s'}} \, \mathrm{d}r \, \mathrm{d}\varphi.$$

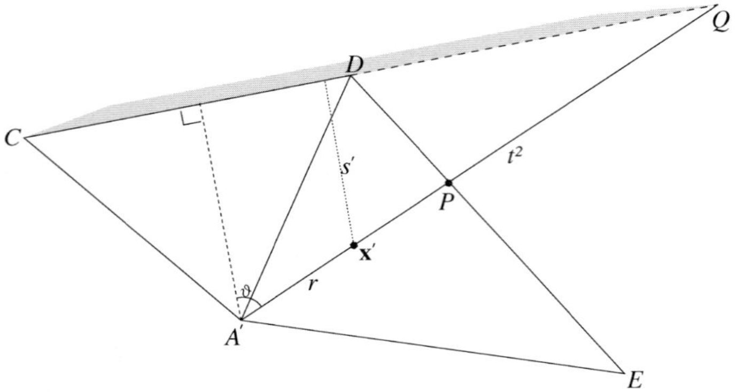

$A' \equiv (x, y)$, $CD = \text{boundary}$

$r \equiv |A'\mathbf{x}'|$, $R(\varphi) \equiv |A'P|$

$t^2 \equiv |\mathbf{x}'Q|$, $s' = t^2 \cos\vartheta$

$q(\varphi) \equiv |A'Q|$

FIG. 6.4. Integration over $\triangle A'DE$, with boundary edge CD.

Suppose (see Fig. 6.4) P comes in the neighbourhood of E, then Q lies at the other side of D so that A' lies in between P and Q. This means that the substitution becomes $r = t^2 - q(\varphi)$. So the integration bounds will be influenced differently as in the former case. The actual integral becomes then

$$\int_{\sqrt{q(\varphi)}}^{\sqrt{q(\varphi)+R(\varphi)}} \frac{\psi_i}{\sqrt{\cos\vartheta}} 2t\,\mathrm{d}t.$$

To distinguish the different cases we could split up the triangle in wedges, as illustrated in Fig. 6.5, such that we may use the former (or latter) substitution if the angle between $A'P$ and CD becomes larger than, say ε, or if $|\cos\vartheta| > \varepsilon$, and otherwise we will use no substitution. Hence for $\triangle A'DE$ we finally have the sum over the integrals.

For the remaining triangles we can use analogous substitutions to these.

Still leaves us the case where the projection A' of the object point lies outside the quadrilateral.

A' outside the quadrilateral. Suppose now that A is situated in such a way that A' lies outside the quadrilateral. The main difference from the previous cases is that, if we divide the quadrilateral into triangles, these triangles will overlap and we only want to integrate over the parts of the triangles that are inside the quadrilateral. If we consider the integral over $\triangle A'BC$, then even the whole triangle lies outside the quadrilateral, and so does $\triangle A'CD$. So we only have to integrate over the triangles $\triangle A'ED$ and $\triangle A'BE$. Let us consider the latter. In Fig. 6.6 we can see that a part of the triangle lies outside the quadrilateral. We only want to integrate over the part that intersects with the quadrilateral, so instead of the bounds 0 and $R(\varphi)$ for integration over r, we will use

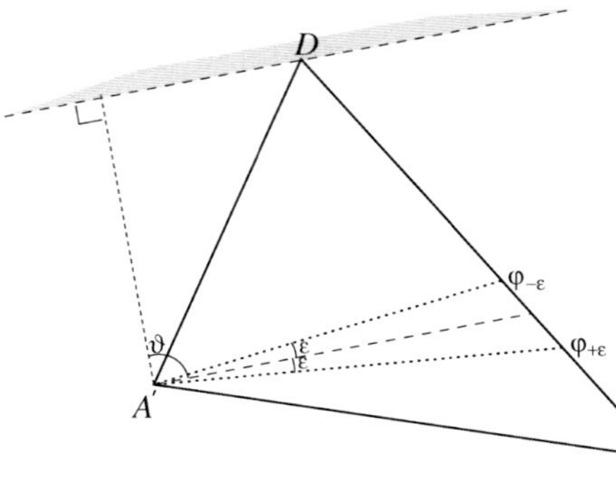

$$I_1 = \int_{\varphi_{-\varepsilon}}^{\varphi_D} \int_{\sqrt{q(\varphi) - R(\varphi)}}^{\sqrt{q(\varphi)}} \frac{\psi_i}{\sqrt{\cos\vartheta}} 2t \, \mathrm{d}t \, \mathrm{d}\varphi,$$

$$I_2 = \int_{\varphi_{+\varepsilon}}^{\varphi_{-\varepsilon}} \int_0^{R(\varphi)} \frac{\psi_i}{\sqrt{s'}} \, \mathrm{d}r \, \mathrm{d}\varphi,$$

$$I_3 = \int_{\varphi_E}^{\varphi_{+\varepsilon}} \int_{\sqrt{q(\varphi)}}^{\sqrt{q(\varphi) + R(\varphi)}} \frac{\psi_i}{\sqrt{\cos\vartheta}} 2t \, \mathrm{d}t \, \mathrm{d}\varphi.$$

FIG. 6.5.

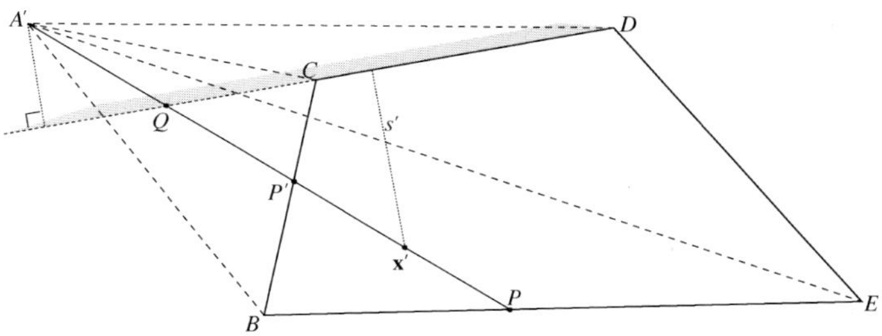

$A' \equiv (x, y)$, $CD =$ boundary

$r \equiv |A'\mathbf{x}'|$, $q(\varphi) \equiv |A'Q|$

$t^2 \equiv |\mathbf{x}'Q|$, $s' = t^2 \cos\vartheta$

$R_{\text{low}} \equiv |A'P'|$, $R_{\text{up}} \equiv |A'P|$

FIG. 6.6. A situation where A' lies outside the quadrilateral, where both $\triangle A'BC$ and $\triangle A'CD$ lie completely outside the quadrilateral, so only integration over $\triangle A'ED$ and $\triangle A'BE$, therefore the lower integration bound for r has to be calculated.

the lower bound R_{low} and the upper bound R_{up} (see the legend with Fig. 6.6), so that after substitution of $r = t^2 + q(\varphi)$ we obtain the integral

$$\int_{\varphi_B}^{\varphi_E} \int_{\sqrt{R_{\text{low}} - q(\varphi)}}^{\sqrt{R_{\text{up}} - q(\varphi)}} \frac{\psi_i}{\sqrt{\cos\vartheta}} 2t \, \mathrm{d}t \, \mathrm{d}\varphi.$$

The further approach is the same as in the previous subsections.

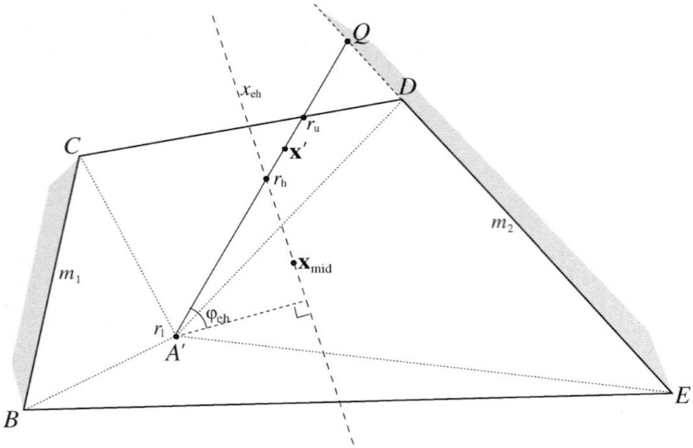

$$A' \equiv (x, y), \qquad r_l \sim A'$$

$$m_1 \equiv \vec{BC}, \qquad m_2 \equiv \vec{DE}$$

$$x_{eh} = \tfrac{1}{2}(m_2 - m_1)$$

$$a_{eh} = \mathrm{d}(A', x_{eh})$$

FIG. 6.7. Quadrilateral with two boundaries. On the left side of x_{eh} the singularity in m_1 dominates, on the right side m_2.

6.3.2. Transformations for two boundaries

In the case that the quadrilateral has two opposite boundaries m_1 and m_2, the approach for the boundary singularity $\{s_i(1-s_i)\}^{-1/2}$ is almost the same as for one boundary, except that we distinct the left-hand and the right-hand side of x_{eh}, where x_{eh} is the isoparametric midline between BC and DE. The picture shows a particular case, but we consider the general case, where A' can be anywhere, so also outside the quadrilateral. Consider $\triangle A'CD$ and we want to integrate over the line $A'Q$. Then we have the integration boundaries r_l ($\sim A'$) and r_u (this means that $\mathbf{x}'$ goes from 0 to r_u). In Fig. 6.7 it can be seen that the line $A'Q$ intersects with x_{eh} in the point r_h and that this intersubsection point lies in between r_l and r_u.

Over the line segment $r_l r_h$ the m_1-singularity dominates and over the line segment $r_h r_u$ the m_2-singularity dominates. So we can split the integral into a lower and upper part. For the lower part we use a substitution for $m = m_1$, which means: approach the integrand as if m_1 were the only boundary. For the upper part we will use $m = m_2$, which means: use the same treatment as with one boundary m_2. That is, if we let $s_1' \equiv \mathrm{dist}(\mathbf{x}', m_2)$ and $t_2^2 \equiv |\mathbf{x}'Q|$, then we use the substitution $s_1' = t_2^2 \cos\vartheta$. The substitution for t in r is completely analogue to the previous cases.

If a_{eh}, defined as the distance of A' to the midline x_{eh}, might become zero (A' lies on the midline) and the angle $\varphi_{eh} \approx \frac{\pi}{2}$, then we will not use a substitution, since the contribution of the singularity is nearly constant.

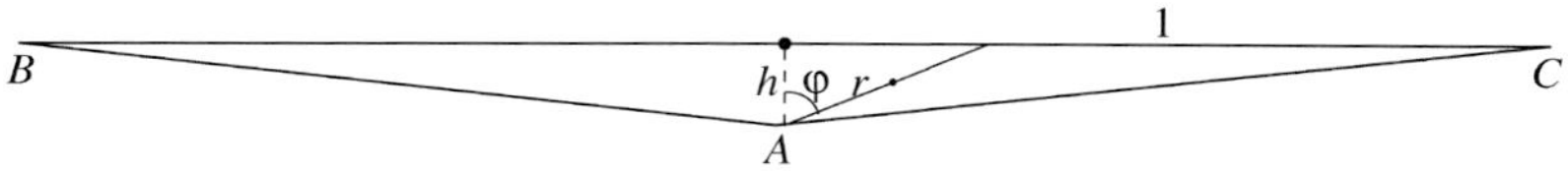

FIG. 6.8. Very flat triangle of height h.

6.4. *Flat triangles*

Sometimes we are dealing with flat triangles, which means that they have a very wide (half) top angle, or, which is the same, a small height compared to the base. Numerical integration over these kind of triangles (see e.g. Fig. 6.8) to the polar coordinates is not very accurate, since the inner integral of

$$\int_{-\varphi_t}^{\varphi_t} \int_0^{h/\cos\varphi} f(r,\varphi)\,\mathrm{d}r\,\mathrm{d}\varphi$$

varies very rapidly for $\varphi \approx \pm\frac{\pi}{2}$. First let us examine this problem in general.

It may be clear that for flat triangles the inner integral of

$$I = \int_{\varphi_1}^{\varphi_2} \int_{R_l}^{R_u} f(r,\varphi)\,\mathrm{d}r\,\mathrm{d}\varphi,$$

as function of φ, varies very rapidly. The reason for this is that the range for r, or the length of the radius over which is integrated, changes fast. A remedy for this would be adapting the inner integral, such that it becomes a function of φ expressing the average along the radius. Such an integral $F(\varphi)$ could be the following:

$$F(\varphi) = \left(\int_{R_l}^{R_u} f(r,\varphi)\,\mathrm{d}r\right) \Big/ (R_u - R_l),$$

so that we can rewrite

$$I = \int_{\varphi_1}^{\varphi_2} F(\varphi)(R_u - R_l)\,\mathrm{d}\varphi.$$

For the polar radii we have $R_l = \frac{k'}{\cos\varphi'}$ and $R_u = \frac{k}{\cos\varphi}$, where k' and k are the distances of the origin to the intersecting edges, φ' and φ are the angles with the normal of the corresponding edge (see Fig. 6.9). So the integral becomes

$$I = \int_{\varphi_1}^{\varphi_2} F(\varphi)\left(\frac{k}{\cos\varphi} - \frac{k'}{\cos\varphi}\right)\mathrm{d}\varphi.$$

Since $\int \frac{1}{\cos\varphi} = \frac{1}{2}\ln\frac{1+\sin\varphi}{1-\sin\varphi}$, we can eliminate the factors $\frac{1}{\cos\varphi}$ by the substitution

$$u = \tfrac{1}{2}k\ln\left(\frac{1+\sin\varphi}{1-\sin\varphi}\right) - \tfrac{1}{2}k'\ln\left(\frac{1+\sin\varphi'}{1-\sin\varphi'}\right), \tag{6.5}$$

for which $\frac{\mathrm{d}u}{\mathrm{d}\varphi} = \frac{k}{\cos\varphi} - \frac{k'}{\cos\varphi'}$. After this substitution the integral has the form

$$\int_{u(\varphi_1)}^{u(\varphi_2)} F(\varphi)\,\mathrm{d}u.$$

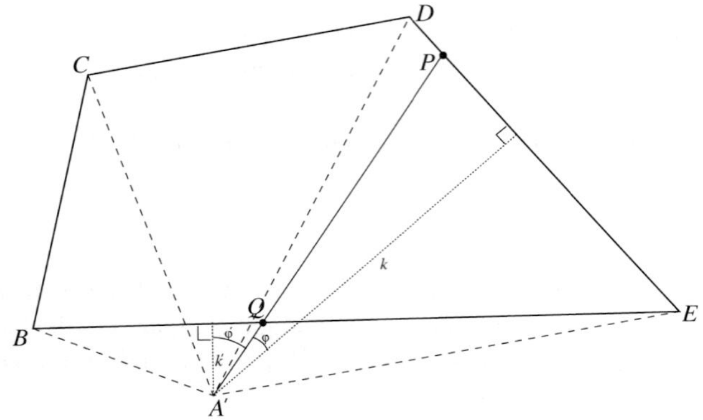

For integration over $A'P$:

$R_l \sim A'Q \qquad R_u \sim A'P$

$\varphi_1 \sim A'E \qquad \varphi_2 \sim A'D$

FIG. 6.9. Integration over $\triangle A'DE$ in a situation that A' near to the edge BE: if the polar angle φ goes from φ_1 to φ_2, the intersubsection point Q goes in the neighbourhood of E very rapidly along EB, so that the lower bound for the polar radius R_l also changes very rapidly.

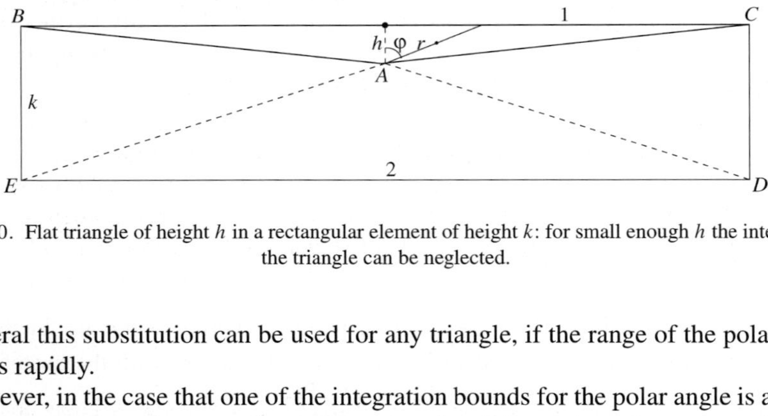

FIG. 6.10. Flat triangle of height h in a rectangular element of height k: for small enough h the integral over the triangle can be neglected.

In general this substitution can be used for any triangle, if the range of the polar radius changes rapidly.

However, in the case that one of the integration bounds for the polar angle is absolute approximately $\frac{\pi}{2}$, we have another tedious situation. When integrating over φ, the polar angle goes along the edge opposite to the origin. The intersubsection point with this edge goes very rapidly in the neighbourhood of the absolute angles $\frac{\pi}{2}$ (cf. Figs. 6.8 and 6.11). This will result in a function F, that does not depend very nicely on φ. We will discuss this situation in the following subsections.

6.4.1. Triangles inside the quadrilateral

If the origin (the integration point of the object domain) of the polar coordinate system lies inside the quadrilateral, we have the following situation: suppose the triangle ABC has base BC, a boundary of the quadrilateral $BCDE$ which is a rectangle of height k (cf. Fig. 6.10). Assuming that the base of the triangle has length 2, the height h depends on the size of the half top angle φ_t, for which holds: $|1 - \sin\varphi_t| = \varepsilon$. Then we can express $h(\varepsilon) = \tan(\arcsin(1-\varepsilon))^{-1}$ and for very small ε holds $h(\varepsilon) \ll 1$.

In particular we want to examine the singularity $s^{-1/2}$, where s is the normalized distance to the boundary BC. For the rectangle (size $2 \times k$) the integral becomes then

$$I_r = \int_0^k \int_{-1}^1 \frac{\sqrt{k}}{\sqrt{k-y}}\,\mathrm{d}x\,\mathrm{d}y = \int_0^k 2\frac{\sqrt{k}}{\sqrt{k-y}}\,\mathrm{d}y = \left[-4\sqrt{k}\sqrt{k-y}\right]_0^k = 4k,$$

and for the triangle (Fig. 6.8) the integral becomes, if r goes from A to a point on BC,

$$I_t = \int_{-\varphi_t}^{\varphi_t} \int_0^{h/\cos\varphi} \frac{r\sqrt{k}}{\sqrt{h - r\cos\varphi}}\,\mathrm{d}r\,\mathrm{d}\varphi = \int_{-\varphi_t}^{\varphi_t} \frac{4}{3}\frac{h\sqrt{hk}}{\cos(\varphi)^2}\,\mathrm{d}\varphi = \frac{8}{3}h\sqrt{hk}\tan\varphi_t,$$

where φ_t is the half top angle, so $\tan\varphi_t = \frac{1}{h}$.

Hence we can give the ratio of the two integrals, $\frac{I_t}{I_r}$, for different h, to get an idea of the contribution of the integral over the triangle to the integral over the complete quadrilateral:

$$\frac{I_t}{I_r} = \frac{\frac{8}{3}\sqrt{hk}}{4k} = \frac{2}{3}\sqrt{\frac{h}{k}}.$$

What we have here is a formula for the rate of contribution of the integral over the triangle to the integral over the rectangle. For h small enough the contribution will be significantly small and can therefore be neglected.

6.4.2. *Triangle partly outside the quadrilateral*

If the origin of the polar coordinate system lies outside the quadrilateral, but very near to the edge, we would have a situation that leads to a very unsmooth function $F(\varphi)$, to be considered as the average for a certain φ, as defined above. So here the approach will be different. Suppose we have a situation as illustrated in Fig. 6.11, where ε is small. For the new integration variable we will choose a point u that goes along $\overline{VM'}$. In order to determine such an u a few relations have to be looked at. Assume first that we have an u as described. Then we have

$$\overline{MS} = \overline{MV} + \overline{VS} = \overline{MV} + u\overline{VM'},$$

and we can give an expression for $\tan\varphi$ (for readability the vectors will now be denoted without an overline):

$$\tan\varphi = \frac{y_{MS}}{x_{MS}} = \frac{y_{MV} + u y_{VM'}}{x_{MV} + u x_{VM'}} = \frac{y_{MV} + u y_{VM'}}{x_{MV}(1-u)} = \frac{\tan\varphi_l - u\tan\beta}{1-u},$$

since $x_{VM'} = -x_{MV}$. From this we can determine u:

$$u = 1 - \frac{\tan\varphi_l - \tan(\varphi_l - \varepsilon)}{\tan\varphi - \tan(\varphi_l - \varepsilon)},$$

so that

$$\frac{\mathrm{d}u}{\mathrm{d}\varphi} = \frac{(1-u)^2 + (\tan\varphi_l - u\tan(\varphi_l - \varepsilon))^2}{\tan\varphi_l - \tan(\varphi_l - \varepsilon)}.$$

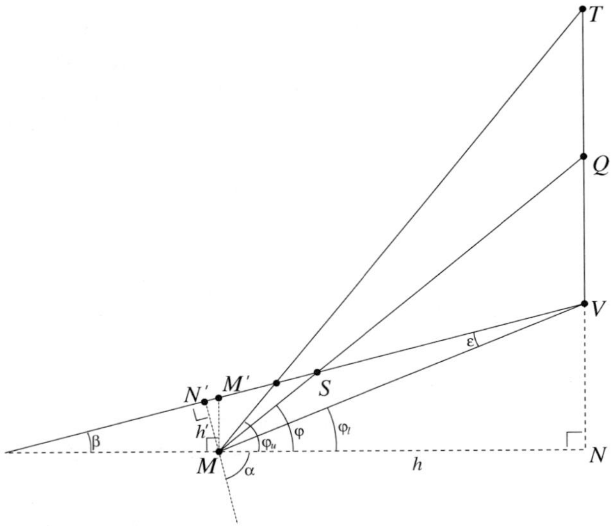

$$I = \int_{\varphi_l}^{\varphi^u} \int_{|MS|}^{|MQ|} f(r, \varphi)\, \mathrm{d}r\, \mathrm{d}\varphi$$

$$\varphi' \equiv \varphi + \alpha \qquad |MS| = \frac{h'}{\cos\varphi'}$$

$$\beta \equiv \varphi_l - \varepsilon \qquad |MQ| = \frac{h}{\cos\varphi}$$

FIG. 6.11. Flat triangle with origin M outside the quadrilateral and very small ε: for integration over $\triangle MVT$, the polar angle φ is substituted such that u becomes the new integration variable, that goes along the edge VM'.

If u goes along VM', and comes in the neighbourhood of M', we have the same difficulty again. To prevent this we can split up the triangle in wedges such that on the lower wedge (say $u \in [0, 0.9]$) we use the latter method and on the other wedge, the method discussed in Section 6.4, the substitution (6.5). Unless the lower integration bound of the polar angle of the upper wedge is still too flat; in that case we do again an analogue splitting up of the upper wedge, and so on.

6.5. The outer integral

Now that we have a method to evaluate the inner integral for an arbitrary object point $\mathbf{x}$, we are able to evaluate the outer part of interaction integral. This has the following form:

$$I = \int_{\Omega_j} \tilde{\psi}_j(\mathbf{x}) \cdot \mathbf{I}_i(\mathbf{x})\, \mathrm{d}\mathbf{x},$$

where the inner integral $\mathbf{I}_i$ is

$$\mathbf{I}_i(\mathbf{x}) = \int_{\Omega_i} G(\mathbf{x}', \mathbf{x}) \tilde{\psi}_i(\mathbf{x}')\, \mathrm{d}\mathbf{x}'.$$

After transformation to isoparametric coordinates the outer integral has the form

$$I = \int_0^1 \int_0^1 \tilde{\psi}_j(s_1, s_2) J(s_1, s_2) \cdot \mathbf{I}_i(s_1, s_2)\, ds_1\, ds_2$$
$$= \int_0^1 \int_0^1 \frac{\psi(s_1, s_2)}{\sqrt{d(s_1, s_2)}}\, ds_2\, ds_1.$$

The boundary singularities can be divided in three cases and are for each case regularised as follows:

If none of the edges of the corresponding element is a part of the boundary, then there is no boundary singularity. In that case we simply have:

$$I = \int_0^1 \int_0^1 \psi(s_1, s_2)\, ds_2\, ds_1,$$

so no substitution is used here.

The approach for an integrand with a factor for the boundary singularity is quite simple and divided into two cases:

With one edge in the boundary there are four possibilities, where $t = \sqrt{d}$ is substituted:

$$d = s_2 \implies I = 2\int_0^1 \int_0^1 \psi\big(s_1, s_2(t)\big)\, dt\, ds_1, \quad \text{where } s_2 = t^2,$$
$$d = 1 - s_1 \implies I = 2\int_0^1 \int_0^1 \psi\big(s_1(t), s_2\big)\, ds_2\, dt, \quad \text{where } s_1 = 1 - t^2,$$
$$d = 1 - s_2 \implies I = 2\int_0^1 \int_0^1 \psi\big(s_1, s_2(t)\big)\, dt\, ds_1, \quad \text{where } s_2 = 1 - t^2,$$
$$d = s_1 \implies I = 2\int_0^1 \int_0^1 \psi\big(s_1(t), s_2\big)\, ds_2\, dt, \quad \text{where } s_1 = t^2.$$

With two boundary edges, d has the form $s_i(1 - s_i)$, since the boundaries are always opposite to each other. Here we substitute $t = \arcsin(2s_i - 1)$:

$$d = s_2(1 - s_2) \implies I = \int_0^1 \int_{-1/2\pi}^{1/2\pi} \psi\big(s_1, s_2(t)\big)\, dt\, ds_1,$$

where $s_2 = \frac{1}{2}(1 + \sin t)$,

$$d = s_1(1 - s_1) \implies I = \int_{-1/2\pi}^{1/2\pi} \int_0^1 \psi\big(s_1(t), s_2\big)\, ds_2\, dt,$$

where $s_1 = \frac{1}{2}(1 + \sin t)$.

After these substitutions we have obtained non-singular integrals, which can be evaluated numerically by the method described in Section 4.

It is clear that the complete numerical integration takes a lot of function evaluations, namely $\mathcal{O}(k^4)$, if k is the average number of function evaluations needed for the quadrature with respect to each of the four integration variables. Totally, there are N^2 of such interaction integrals, where $N = n + m$ is the number of elements and edges, so for the evaluation of these integrals a total of $\mathcal{O}(N^2k^4)$ function evaluations would be needed. This number grows very rapidly if n is of order 10^3 or 10^4, and k might be 15 or 31 for irregular integrands. For a great part of the integrals an alternative method can be used. This is discussed in the next section.

7. Taylor expansion

In the previous section we have seen that the evaluation of the fourfold interaction integrals by numerical quadrature can take very much computer time. If the "distance" between two elements is larger than a given tolerance, the Green's function in the integrand of the interaction integral can be approximated by *Taylor expansion*. This method will be discussed in this section. In Section 7.7 a relation will be derived between the distance and the relative error made in the evaluation of the integral by this method.

7.1. *Interaction integral in general form*

The interaction integrals (2.37) and (2.38) belonging to the matrices $\mathbf{L}$ and $\mathbf{D}$ have the general form

$$I = \int_{\Omega_i} \tilde{\psi}_i(\mathbf{x}) \cdot \int_{\Omega_j} \tilde{\psi}_j(\mathbf{x}')G(\mathbf{x}' - \mathbf{x})\,\mathrm{d}\mathbf{x}'\,\mathrm{d}\mathbf{x}. \tag{7.1}$$

Here $\tilde{\psi}_i(\mathbf{x})$, $\tilde{\psi}_j(\mathbf{x})$ are vector valued basis functions (cf. $\tilde{\mathbf{w}}_k$, $\tilde{\mathbf{w}}_l$ in (2.37)) or scalar valued basis functions (cf. c_i, c_j in (2.38)) belonging to Ω_i and Ω_j, respectively, possibly containing a factor for singularity. Further, $\mathbf{x}$ and $\mathbf{x}'$ represent points in the object domain Ω_i and the source domain Ω_j, respectively.

The *distance* between two disjoint elements Ω_i and Ω_j (see Fig. 7.1) will be defined by

$$r_{\min}(\Omega_i, \Omega_j) = \min\bigl\{|\mathbf{x}' - \mathbf{x}|;\ \mathbf{x} \in \Omega_i,\ \mathbf{x}' \in \Omega_j\bigr\}.$$

If the distance is large enough we can apply Taylor expansion to the Green's function G with respect to $(\mathbf{x}' - \mathbf{x}'_m)$ and $(\mathbf{x} - \mathbf{x}_m)$, where $\mathbf{x}'_m$ and $\mathbf{x}_m$ are the midpoints of the source and object element, respectively.

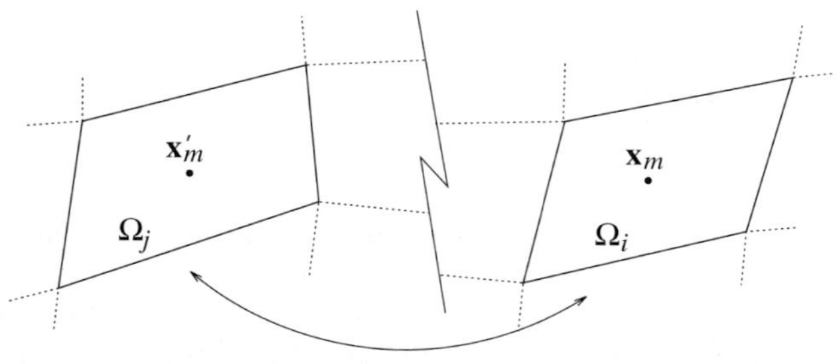

FIG. 7.1. Interaction between quadrilateral elements, $\Omega_i \ni \mathbf{x}$ and $\Omega_j \ni \mathbf{x}'$. These domains must be disjunct. $\mathbf{x}'_m$ and $\mathbf{x}_m$ are the midpoints of the elements.

7.2. Taylor expansion of a one-term Green's function

Consider the one-term Green's function

$$G(\mathbf{x}' - \mathbf{x}) = \frac{1}{|\mathbf{x}' - \mathbf{x}|}.$$

After the substitution $\mathbf{y} = \mathbf{x}' - \mathbf{x}$ and the second order Taylor expansion of $G(\mathbf{y})$ with respect to $\mathbf{y}_m = \mathbf{x}'_m - \mathbf{x}_m$:

$$\begin{aligned} G(\mathbf{y}) &= G(\mathbf{y}_m) + (\nabla G)(\mathbf{y}_m)^{\mathrm{T}}[\mathbf{y} - \mathbf{y}_m] \\ &\quad + \tfrac{1}{2}[\mathbf{y} - \mathbf{y}_m]^{\mathrm{T}}\big(\nabla(\nabla G)\big)(\mathbf{y}_m)[\mathbf{y} - \mathbf{y}_m] + \mathcal{O}\big(|\mathbf{y} - \mathbf{y}_m|^3\big), \end{aligned} \tag{7.2}$$

where the expressions for the gradient and the Hessian are given by

$$\begin{aligned} (\nabla G)(\mathbf{y}_m) &= -G^3(\mathbf{y}_m)\mathbf{y}_m, \\ \big(\nabla(\nabla G)\big)(\mathbf{y}_m) &= -G^3(\mathbf{y}_m)\mathbf{I} + 3G^5(\mathbf{y}_m)(\mathbf{y}_m \otimes \mathbf{y}_m). \end{aligned}$$

$\mathbf{I}$ is the identity matrix and $\mathbf{u} \otimes \mathbf{v}$ stands for $\mathbf{u}\mathbf{v}^{\mathrm{T}}$. Using the following shorthand notation

$$g_m = G(\mathbf{y}_m), \qquad \mathbf{g}_m = (\nabla G)(\mathbf{y}_m), \qquad \mathbf{G}_m = \big(\nabla(\nabla G)\big)(\mathbf{y}_m),$$

the Taylor approximation for the one-term Green's function becomes

$$G(\mathbf{y}) \approx g_m + \mathbf{g}_m^{\mathrm{T}}(\mathbf{y} - \mathbf{y}_m) + \tfrac{1}{2}(\mathbf{y} - \mathbf{y}_m)^{\mathrm{T}}\mathbf{G}_m(\mathbf{y} - \mathbf{y}_m). \tag{7.3}$$

7.3. Taylor expansion of a multiple-term Green's function

In this subsection we consider the multi-term Green's function

$$G(\mathbf{x}' - \mathbf{x}) = \sum_{i=0}^{N} \frac{c_i}{|\mathbf{x}'_i - \mathbf{x}|},$$

where N is the number of images. Analogously to Section 7.2 we substitute $\mathbf{y}_i = \mathbf{x}'_i - \mathbf{x}$ and $\mathbf{y}_{i\,m} = \mathbf{x}'_{i\,m} - \mathbf{x}_m$. If $\mathbf{y}_i - \mathbf{y}_{i\,m} = \mathbf{y} - \mathbf{y}_m$, the Taylor expansion for images becomes

$$G(\mathbf{y}) \approx g_m + \mathbf{g}_m^{\mathrm{T}}(\mathbf{y} - \mathbf{y}_m) + \tfrac{1}{2}(\mathbf{y} - \mathbf{y}_m)^{\mathrm{T}}\mathbf{G}_m(\mathbf{y} - \mathbf{y}_m), \tag{7.4}$$

where for $r_i = |\mathbf{y}_{i\,m}|$

$$g_m = \sum_{i=0}^{N} G_i(\mathbf{y}_{i\,m}) = \sum_{i=0}^{N} \frac{c_i}{r_i}, \tag{7.5}$$

$$\mathbf{g}_m = \sum_{i=0}^{N} (\nabla G_i)(\mathbf{y}_{i\,m}) = \sum_{i=0}^{N} \frac{-c_i}{r_i^3}\mathbf{y}_{i\,m}, \tag{7.6}$$

$$\mathbf{G}_m = \sum_{i=0}^{N} \big(\nabla(\nabla G_i)\big)(\mathbf{y}_{i\,m}) = \sum_{i=0}^{N} \frac{-c_i}{r_i^3}\mathbf{I} + 3\frac{c_i}{r_i^5}(\mathbf{y}_{i\,m} \otimes \mathbf{y}_{i\,m}). \tag{7.7}$$

Note, that the sum of images only contributes to the expansion coefficients.

Special attention has to be paid to the evaluation of the sums. If the value of a sum is relatively small compared to the individual terms, straightforward evaluation might result in a cancellation of significant digits. This occurs, for example, when $\sum c_i \approx 0$ and the r_i's are large, but their range is very small, i.e., $(r_i - r_0) \ll r_0$. By using the following identity for the sum in the expression of g_m

$$\sum \frac{c_i}{r_i} \quad \longrightarrow \quad \sum \frac{c_i(r_0 - r_i)}{r_i r_0} + \frac{1}{r_0} \sum c_i,$$

the terms in the first sum are much smaller than the terms in the original sum, so that less cancellation will occur. A similar treatment can be used for the derivatives.

7.4. *Substitution of the Taylor expansion in the interaction integral*

After substitution of the Taylor approximation (7.3) of $G(\mathbf{x}' - \mathbf{x})$ in (7.1) one obtains

$$\begin{aligned}
\tilde{I} &= \int_{\Omega_i} \tilde{\psi}_i(\mathbf{x}) \cdot \int_{\Omega_j} \tilde{\psi}_j(\mathbf{x}') g_m \, \mathbf{dx}' \, \mathbf{dx} \\
&+ \int_{\Omega_i} \tilde{\psi}_i(\mathbf{x}) \cdot \int_{\Omega_j} \tilde{\psi}_j(\mathbf{x}') \mathbf{g}_m^{\mathrm{T}} (\mathbf{y} - \mathbf{y}_m) \, \mathbf{dx}' \, \mathbf{dx} \\
&+ \frac{1}{2} \int_{\Omega_i} \tilde{\psi}_i(\mathbf{x}) \cdot \int_{\Omega_j} \tilde{\psi}_j(\mathbf{x}') (\mathbf{y} - \mathbf{y}_m)^{\mathrm{T}} \mathbf{G}_m (\mathbf{y} - \mathbf{y}_m) \, \mathbf{dx}' \, \mathbf{dx},
\end{aligned}$$

where $\mathbf{y} = \mathbf{x}' - \mathbf{x}$ and $\mathbf{y}_m = \mathbf{x}'_m - \mathbf{x}_m$. Since g_m, $\mathbf{g}_m$ and $\mathbf{G}_m$ are independent of $\mathbf{x}$ and $\mathbf{x}'$, they appear as constant terms in the integral, so that it may be rewritten as

$$\begin{aligned}
\tilde{I} &= g_m \int_{\Omega_i} \tilde{\psi}_i(\mathbf{x}) \cdot \int_{\Omega_j} \tilde{\psi}_j(\mathbf{x}') \, \mathbf{dx}' \, \mathbf{dx} \\
&- g_m^3 \int_{\Omega_i} \tilde{\psi}_i(\mathbf{x}) \cdot \int_{\Omega_j} \tilde{\psi}_j(\mathbf{x}') \{ \mathbf{y}_m^{\mathrm{T}} (\mathbf{y} - \mathbf{y}_m) \} \, \mathbf{dx}' \, \mathbf{dx} \\
&- \frac{1}{2} g_m^3 \int_{\Omega_i} \tilde{\psi}_i(\mathbf{x}) \cdot \int_{\Omega_j} \tilde{\psi}_j(\mathbf{x}') \{ \mathbf{I} \cdot [(\mathbf{y} - \mathbf{y}_m) \otimes (\mathbf{y} - \mathbf{y}_m)] \} \, \mathbf{dx}' \, \mathbf{dx} \\
&+ \frac{3}{2} g_m^5 \int_{\Omega_i} \tilde{\psi}_i(\mathbf{x}) \cdot \int_{\Omega_j} \tilde{\psi}_j(\mathbf{x}') \{ [\mathbf{y}_m \otimes \mathbf{y}_m] \cdot [(\mathbf{y} - \mathbf{y}_m) \otimes (\mathbf{y} - \mathbf{y}_m)] \} \, \mathbf{dx}' \, \mathbf{dx}.
\end{aligned}$$

The expressions between brackets $\{\ldots\}$ are scalar. For the $\otimes$-notation and the definition of inner products for matrices see Appendix D. The integral $\tilde{I}$ can be written in terms of moment integrals $M_{\alpha\beta}$

$$\begin{aligned}
\tilde{I} &= g_m M_{00} M'_{00} - g_m^3 \Big\{ M_{00} \cdot \sum_{\alpha \geqslant 1} \{\mathbf{y}_m\}_\alpha M'_{0\alpha} - M'_{00} \cdot \sum_{\alpha \geqslant 1} \{\mathbf{y}_m\}_\alpha M_{0\alpha} \Big\} \\
&- \frac{1}{2} g_m^3 \Big\{ M_{00} \cdot \sum_{\alpha \geqslant 1} M'_{\alpha\alpha} + M'_{00} \cdot \sum_{\alpha \geqslant 1} M_{\alpha\alpha} - 2 \sum_{\alpha \geqslant 1} M_{0\alpha} \cdot M'_{0\alpha} \Big\} \\
&+ \frac{3}{2} g_m^5 \Big\{ M_{00} \cdot \sum_{\alpha,\beta \geqslant 1} \{\mathbf{y}_m\}_\alpha \{\mathbf{y}_m\}_\beta M'_{\alpha\beta} + M'_{00} \cdot \sum_{\alpha,\beta \geqslant 1} \{\mathbf{y}_m\}_\alpha \{\mathbf{y}_m\}_\beta M_{\alpha\beta} \\
&- 2 \sum_{\alpha \geqslant 1} \{\mathbf{y}_m\}_\alpha M_{0\alpha} \cdot \sum_{\beta \geqslant 1} \{\mathbf{y}_m\}_\beta M'_{0\beta} \Big\}.
\end{aligned} \tag{7.8}$$

The moment integrals are defined by

$$M_{\alpha\beta} = \int_{\Omega_i} \tilde{\psi}_i(\mathbf{x})\{\mathbf{x}-\mathbf{x}_m\}_\alpha\{\mathbf{x}-\mathbf{x}_m\}_\beta \,\mathrm{d}\mathbf{x},$$

$$M'_{\alpha\beta} = \int_{\Omega_j} \tilde{\psi}_j(\mathbf{x}')\{\mathbf{x}'-\mathbf{x}'_m\}_\alpha\{\mathbf{x}'-\mathbf{x}'_m\}_\beta \,\mathrm{d}\mathbf{x}',$$

where $\{\mathbf{x}-\mathbf{x}_m\}_\alpha = 1$, $(x-x_m)$, $(y-y_m)$ or $(z-z_m)$ for $\alpha = 0$, 1, 2 or 3. They are scalars for scalar valued functions $\tilde{\psi}_i$, and vectors for vector valued functions $\tilde{\psi}_i$. Let n_x be the upper bound of α. In general, $n_x = 3$. However, if all elements are parallel to the x, y-plane then $z - z_m = 0$, so that $n_x = 2$. After transformation to isoparametric coordinates the moment integrals have the form

$$M = \int_0^1\int_0^1 \tilde{\mu}(s_1, s_2)\,\mathrm{d}s_2\,\mathrm{d}s_1.$$

If one or more of the edges of the integration domain are a part of a boundary, the function $\tilde{\mu}$ contains a factor for boundary singularity $d^{-1/2}$, and the general form of the moment integrals becomes

$$M = \int_0^1\int_0^1 \frac{\mu(s_1, s_2)}{\sqrt{d(s_1, s_2)}}\,\mathrm{d}s_2\,\mathrm{d}s_1,$$

where μ is a smooth function. For the regularisation of this boundary singularity a similar method can used as for that of the outer integral described in Section 6.5. The integrals obtained can be evaluated numerically by the Patterson's method described in Section 4.

If N is the number of elements and k is the average number of function evaluations for the quadrature with respect to each of the isoparametric parameters s_1 and s_2, the total number of function evaluations for calculating the moment integrals is of the order of Nk^2. Since the moment integrals can be evaluated in advance, the computer time for the evaluation of the interaction integrals is reduced considerably.

7.5. *Efficiency improvement of the algorithm*

In this subsection we concentrate on the efficiency of the algorithm to evaluate the interaction integral by expression (7.8) for a given set of moment integrals. Assuming that two identical terms are computed only once, the number of operations to evaluate the expression (7.8) with scalar moment integrals is

$$\begin{aligned} N_{\text{scalar}} &= n_s n_o\big[\{1\} + \big\{2(n_x+1)\big\} + \big\{2(n_x+1)\big\} + \big\{\tfrac{3}{2}n_x(n_x+1)+3\big\} + 4\big] \\ &= n_s n_o\big[(n_x+1)\big(\tfrac{3}{2}n_x+4\big)+4+4\big], \end{aligned}$$

where n_s and n_o are the number of different basis functions on the source and object quadrilaterals, i.e., $n_s = n_o = 1$. If $n_x = 2$, then $N_{\text{scalar}} = 29$.

For the expression (7.8) with vector moment integrals the number of operations is

$$N_{\text{vector}} = n_x n_s n_o\big\{(n_x+1)\big(\tfrac{3}{2}n_x+4\big)+5\big\} + 4n_s n_o,$$

where the first factor n_x is due to taking the inner product. The maximum value of n_s and n_o is 2. This is because for a vector-type interaction integral the source or object domain usually consists of two adjacent elements with an edge in common. If $n_x = 2$, then $N_{\text{vector}} = 56 n_s n_o$.

These numbers of operations can be reduced by calculating the inner products between the vectors and matrices which occur in expression (7.8) beforehand. Using the following shorthand notation for the moments due to the source element:

$$S_{00} = M'_{00}, \qquad \mathbf{s}_0 = \begin{bmatrix} M'_{01} \\ \vdots \\ M'_{0n_x} \end{bmatrix}, \qquad \mathbf{S} = \begin{bmatrix} M'_{11} & \cdots & M'_{1n_x} \\ \vdots & \ddots & \vdots \\ M'_{n_x 1} & \cdots & M'_{n_x n_x} \end{bmatrix},$$

and an analogous notation $\mathbf{m}_0$ and $\mathbf{M}$ for the object element, we obtain a compact expression for the integral of (7.8)

$$\begin{aligned} \tilde{I} &= M_{00}\big(g_m S_{00} + \mathbf{g}_m \cdot \mathbf{s}_0 + \tfrac{1}{2}\mathbf{G}_m \cdot \mathbf{S}\big) \\ &\quad + S_{00}\big(-\mathbf{g}_m \cdot \mathbf{m}_0 + \tfrac{1}{2}\mathbf{G}_m \cdot \mathbf{M}\big) - \mathbf{G}_m \cdot (\mathbf{m}_0 \otimes \mathbf{s}_0) \\ &= M_{00}\big(g_m S_{00} + \mathbf{g}_m \cdot \mathbf{s}_0 + \tfrac{1}{2}\mathbf{G}_m \cdot \mathbf{S}\big) \\ &\quad + \mathbf{m}_0 \cdot \big(-\mathbf{g}_m S_{00} - \mathbf{G}_m \mathbf{s}_0\big) + \mathbf{M} \cdot \tfrac{1}{2}\mathbf{G}_m S_{00}, \end{aligned} \tag{7.9}$$

where g_m, $\mathbf{g}_m$ and $\mathbf{G}_m$ are defined in Sections 7.2 and 7.3. For the definition of $\otimes$ see Appendix D.

The numbers of operations for the evaluation of expression (7.9) are:

$$\begin{aligned} N_{\text{scalar}} &= n_o n_s + n_s\big\{1 + n_x + \tfrac{1}{2}n_x(n_x+1)\big\} \\ &\quad + n_o n_s n_x + n_s\big(n_x + n_x^2\big) + \big(\tfrac{1}{2}n_x(n_x+1) + 1\big)n_o n_s \\ &= n_s\big\{(n_x+1)\big(\tfrac{3}{2}n_x + 1\big) + n_o(n_m+1)\big\}, \\ N_{\text{vector}} &= n_x n_s\big\{(n_x+1)\big(\tfrac{3}{2}n_x + 1\big) + n_o(n_m+1)\big\}, \end{aligned}$$

where $n_m = 1 + n_x + \frac{1}{2}n_x(n_x+1)$. Thus, for $n_x = 2, 3$

$$\begin{aligned} N_{\text{scalar}} &= 19, 33 \qquad (n_s = n_o = 1) \\ N_{\text{vector}} &\leqslant 104, 264 \quad (n_s = n_o = 2). \end{aligned}$$

7.6. *Moment integrals in local coordinate system*

So far, we have applied Taylor expansion with respect to midpoints of the elements in a global coordinate system. If all elements are on the same layer, or on parallel layers, the evaluation of the moment integrals is restricted to $n_x = 2$, otherwise $n_x = 3$. However, if the moment integral of each element is calculated in a local coordinate system for which the z-axis is perpendicular to the plane of that element, again $n_x = 2$. But for the evaluation of an interaction integral between an object and a source element, the moment integrals of the object element have to be transformed from the local coordinate system of the object element to the local coordinate system of the source element.

7.6.1. Transformation matrix

Let $\mathbf{Q}$ and $\mathbf{R}$ be the transformation matrices from a local to a global coordinate system of the source and object element, respectively, then

$$\mathbf{x} = \mathbf{x}_s + \mathbf{Q}\bar{\eta}, \qquad \mathbf{Q}\mathbf{Q}^{\mathrm{T}} = \mathbf{I},$$
$$\mathbf{x} = \mathbf{x}_o + \mathbf{R}\bar{\xi}, \qquad \mathbf{R}\mathbf{R}^{\mathrm{T}} = \mathbf{I},$$

so that

$$\bar{\eta} = \mathbf{Q}^{\mathrm{T}}(\mathbf{x}_o - \mathbf{x}_s + \mathbf{R}\bar{\xi}),$$

where $\bar{\eta} = (\eta_x, \eta_y, 0)$ and $\bar{\xi} = (\xi_x, \xi_y, 0)$ are the local coordinates in the source and object coordinate system, respectively, of point $\mathbf{x}$ with coordinates in the global coordinate system. Let $\bar{\xi}_m = (\xi_{m,x}, \xi_{m,y}, 0)$ and $\bar{\xi} = (\xi_x, \xi_y, 0)$ be points in the local object coordinate system, then

$$\begin{aligned}\bar{\eta}_m &= \mathbf{Q}^{\mathrm{T}}(\mathbf{x}_o - \mathbf{x}_s + \mathbf{R}\bar{\xi}_m),\\ \bar{\eta} &= \mathbf{Q}^{\mathrm{T}}(\mathbf{x}_o - \mathbf{x}_s + \mathbf{R}\bar{\xi}) = \bar{\eta}_m + \mathbf{Q}^{\mathrm{T}}\mathbf{R}(\bar{\xi} - \bar{\xi}_m)\\ &= \bar{\eta}_m + \mathbf{T}(\bar{\xi} - \bar{\xi}_m)\end{aligned}$$

are the coordinates of these points in the local source coordinate system, where $\mathbf{T} = \mathbf{Q}^{\mathrm{T}}\mathbf{R}$ is a rotation matrix.

7.6.2. Taylor expansion with transformation

Let $\bar{\eta}' = (\eta'_x, \eta'_y, 0)$ and $\bar{\eta} = \bar{\eta}_m + \mathbf{T}(\bar{\xi} - \bar{\xi}_m)$ be arbitrary points inside the source and object element, respectively, with coordinates in the local source coordinate system. Let $\Delta\bar{\eta}' = \bar{\eta}' - \bar{\eta}'_m$ and $\Delta\bar{\xi} = \bar{\xi} - \bar{\xi}_m$, then $\Delta\bar{\eta} = \mathbf{T}\Delta\bar{\xi}$, and

$$G(\bar{\eta}' - \bar{\eta}) = \sum_i c_i R_i^{-1},$$

$$\begin{aligned}R_i &= \left|\bar{\eta}'_i - \bar{\eta}_m - \mathbf{T}\Delta\bar{\xi}\right|\\ &= \big\{(\eta'_x - \eta_{m,x} - T_{11}\Delta\xi_x - T_{12}\Delta\xi_y)^2 + (\eta'_y - \eta_{m,y} - T_{21}\Delta\xi_x - T_{22}\Delta\xi_y)^2\\ &\quad + (\eta'_{i,z} - \eta'_{m,z} - T_{31}\Delta\xi_x - T_{32}\Delta\xi_y)^2\big\}^{-1/2}.\end{aligned}$$

Let $\mathbf{y} = \bar{\eta}' - \bar{\eta}$ and $\mathbf{y}_m = \bar{\eta}'_m - \bar{\eta}_m$. The Taylor expansion of $G(\mathbf{y})$ is similar to that of expression (7.2)

$$\begin{aligned}G(\mathbf{y}) &= G(\mathbf{y}_m) + (\nabla G)(\mathbf{y}_m)^{\mathrm{T}}[\mathbf{y} - \mathbf{y}_m]\\ &\quad + \tfrac{1}{2}[\mathbf{y} - \mathbf{y}_m]^{\mathrm{T}}\big(\nabla(\nabla G)\big)(\mathbf{y}_m)[\mathbf{y} - \mathbf{y}_m] + \mathcal{O}\big(|\mathbf{y} - \mathbf{y}_m|^3\big),\\ &= g_m + \mathbf{g}_m{}^{\mathrm{T}}(\Delta\bar{\eta}' - \mathbf{T}\Delta\bar{\xi}) + \tfrac{1}{2}(\Delta\bar{\eta}' - \mathbf{T}\Delta\bar{\xi})^{\mathrm{T}}\mathbf{G}_m(\Delta\bar{\eta}' - \mathbf{T}\Delta\bar{\xi}),\end{aligned}$$

where

$$g_m = G(\mathbf{y}_m), \qquad \mathbf{g}_m = (\nabla G)(\mathbf{y}_m), \qquad \mathbf{G}_m = \big(\nabla(\nabla G)\big)(\mathbf{y}_m).$$

∇ refers to derivatives with respect to the local source coordinates. Note that, in contrast to the expansion in Sections 7.2 and 7.3 for $n_x = 2$, all three elements of $\mathbf{g}_m$ and all nine elements of $\mathbf{G}_m$ are generally non-zero.

Let

$$\mathbf{s}_0 = (S_{01}, S_{02}, 0)^{\mathrm{T}}, \qquad \mathbf{m}_0 = (M_{01}, M_{02}, 0)^{\mathrm{T}},$$

$$\mathbf{S} = \begin{pmatrix} S_{11} & S_{12} & 0 \\ S_{21} & S_{22} & 0 \\ 0 & 0 & 0 \end{pmatrix}, \qquad \mathbf{M} = \begin{pmatrix} M_{11} & M_{12} & 0 \\ M_{21} & M_{22} & 0 \\ 0 & 0 & 0 \end{pmatrix}.$$

Then, similar to expression (7.9)

$$\begin{aligned} \tilde{I} &= M_{00}\big(g_m S_{00} + \mathbf{g}_m \cdot \mathbf{s}_0 + \tfrac{1}{2}\mathbf{G}_m \cdot \mathbf{S}\big) \\ &\quad + (-\mathbf{g}_m S_{00} - \mathbf{G}_m \mathbf{s}_0) \cdot \mathbf{T}\mathbf{m}_0 + (\mathbf{T}\mathbf{M}\mathbf{T}^{\mathrm{T}}) \cdot \tfrac{1}{2}\mathbf{G}_m S_{00}. \end{aligned}$$

For the evaluation of $\tilde{I}$, the quantities $\mathbf{m}_0$, $\mathbf{M}$ and $\mathbf{T}$ may be redefined as follows:

$$\mathbf{m}_0 = (M_{01}, M_{02})^{\mathrm{T}}, \qquad \mathbf{M} = \begin{pmatrix} M_{11} & M_{12} \\ M_{21} & M_{22} \end{pmatrix}, \qquad \mathbf{T} = \begin{pmatrix} T_{11} & T_{12} \\ T_{21} & T_{22} \\ T_{31} & T_{32} \end{pmatrix}.$$

Since $\mathbf{T}$ is a 3×2-matrix and $\mathbf{G}_m$ is a full 3×3-matrix, a complete reduction to two dimensions as in the previous subsection for $n_x = 2$ is not possible. An operation count for this transformed integral gives us:

$$\begin{aligned} N_{\text{scalar}} &= n_o n_s + n_s\big(1 + 2 + \tfrac{1}{2} \cdot 2 \cdot 3\big) + 3 n_o n_s + n_s(3 + 3 \cdot 2) \\ &\quad + n_o(3 \cdot 2) + n_o n_s\big(\tfrac{1}{2} \cdot 3 \cdot 4 + 1\big) + n_o(12 + 18) \\ &= 15 n_s + 36 n_o + 10 n_o n_s = 61, \\ N_{\text{vector}} &= 2(35 n_s + 72) = 284, \end{aligned}$$

which is larger than the number of operations for the untransformed case with $n_x = 3$ of Section 7.5. This is due to the large number of operations needed for the transformation of $\mathbf{m}_0$ and $\mathbf{M}$.

7.7. *Error estimates*

To estimate the error in the second order Taylor expansion of (7.2) we have derived the following expression for the third order terms:

$$\tfrac{1}{6}\overline{\mathbf{G}}_m \cdot [\mathbf{y} - \mathbf{y}_m]^3,$$

where $\mathbf{t}^\alpha = \underbrace{\mathbf{t} \otimes \cdots \otimes \mathbf{t}}_{\alpha}$, and

$$\overline{\mathbf{G}}_m = 3 g_m^5 \big[\mathbf{I} \otimes \mathbf{y}_m + \nabla \mathbf{H}\big] - 15 g_m^7 \overline{\mathbf{T}},$$

with the Hessian matrix $\mathbf{H} = \mathbf{y}_m \otimes \mathbf{y}_m$ and tensor $\overline{\mathbf{T}} = \mathbf{H} \otimes \mathbf{y}_m$. For the definition of $\otimes$ see Appendix D.

For any $\mathbf{y}$ there exists a point $\bar{\xi} = \mathbf{y}_m + \vartheta[\mathbf{y} - \mathbf{y}_m]$ for $0 < \vartheta < 1$ such that

$$G(\mathbf{y}) = g_m + \mathbf{g}_m \cdot [\mathbf{y} - \mathbf{y}_m] + \tfrac{1}{2}\mathbf{G}_m \cdot [\mathbf{y} - \mathbf{y}_m]^2 + \tfrac{1}{6}\overline{\mathbf{G}}(\bar{\xi}) \cdot [\mathbf{y} - \mathbf{y}_m]^3,$$

where

$$\overline{\mathbf{G}}(\bar{\xi}) = 3G^5(\bar{\xi})\big[\mathbf{I}\otimes\bar{\xi} + \nabla\mathbf{H}(\bar{\xi})\big] + 15G^7(\bar{\xi})\overline{\mathbf{T}}(\bar{\xi}),$$
$$\mathbf{H}(\bar{\xi}) = \bar{\xi}\otimes\bar{\xi},$$
$$\overline{\mathbf{T}}(\bar{\xi}) = \mathbf{H}(\bar{\xi})\otimes\bar{\xi}.$$

Let

$$r_{\min} = \min_{\substack{\mathbf{x}\in\Omega_i\\ \mathbf{x}'\in\Omega_j}} \|\mathbf{x}' - \mathbf{x}\|$$

be the smallest distance between two elements and

$$d = \max_{\mathbf{y}} \|\mathbf{y} - \mathbf{y}_m\|$$

the upper bound for $\|\mathbf{y}-\mathbf{y}_m\|$, where $\mathbf{y}\in\{\mathbf{x}'-\mathbf{x}\mid \mathbf{x}'\in\Omega_j, \mathbf{x}\in\Omega_i\}$. Note, that d depends on the choice of the midpoints, $\mathbf{x}_m$ and $\mathbf{x}'_m$. The following theorem states the relation that holds between d, $r_{\min}$ and the upper bound for the relative error in the second order Taylor expansion.

THEOREM 7.1. *Let $\tilde{G}$ be an approximation of G with the second order Taylor expansion* (7.3)*, then for every $\varepsilon > 0$ we have*

$$\frac{d}{r_{\min}} \leqslant \sqrt[3]{\frac{\varepsilon}{2}} \quad\Longrightarrow\quad \frac{\|\tilde{G}(\mathbf{x}'-\mathbf{x}) - G(\mathbf{x}'-\mathbf{x})\|_\infty}{|G(\mathbf{x}'-\mathbf{x})|} \leqslant \varepsilon,$$

for every $\mathbf{x}'\in\Omega_j, \mathbf{x}\in\Omega_i$.

PROOF. Let

$$\delta_G = |G-\tilde{G}| = \left|\tfrac{1}{6}\overline{\mathbf{G}}(\bar{\xi})\cdot[\mathbf{y}-\mathbf{y}_m]^3\right|$$

be the absolute error in the Taylor expansion and let further $\mathbf{y} = (\mathbf{x}'-\mathbf{x})$ and $\Delta\mathbf{y} = (\mathbf{y}-\mathbf{y}_m)$. We have

$$\mathbf{I}\otimes\bar{\xi} + \nabla\mathbf{H}(\bar{\xi}) = \begin{bmatrix} (3\xi_x, \xi_y, \xi_z) & (\xi_y, \xi_x, 0) & (\xi_z, 0, \xi_x) \\ (\xi_y, \xi_x, 0) & (\xi_x, 3\xi_y, \xi_z) & (0, \xi_z, \xi_y) \\ (\xi_z, 0, \xi_x) & (0, \xi_z, \xi_y) & (\xi_x, \xi_y, 3\xi_z) \end{bmatrix}.$$

If we multiply this by $\Delta\mathbf{y}^3 = \Delta\mathbf{y}\otimes\Delta\mathbf{y}\otimes\Delta\mathbf{y}$ and let $\mathbf{y} = (x, y, z)$, we obtain (see Appendix D)

$$\begin{aligned}\big(\mathbf{I}\otimes\bar{\xi} + \nabla\mathbf{H}(\bar{\xi})\big)\cdot\Delta\mathbf{y}^3 &= 3(\Delta x^2 + \Delta y^2 + \Delta z^2)(\xi_x\Delta x + \xi_y\Delta y + \xi_z\Delta z)\\ &= 3\|\Delta\mathbf{y}\|^2\langle\bar{\xi}, \Delta\mathbf{y}\rangle.\end{aligned}$$

Further, we have

$$\overline{\mathbf{T}}(\bar{\xi})\cdot\Delta\mathbf{y}^3 = (\xi_x\Delta x + \xi_y\Delta y + \xi_z\Delta z)^3 = \langle\bar{\xi}, \Delta\mathbf{y}\rangle^3.$$

An upper boundary for the error can be given by

$$\begin{aligned}
\delta_G &= \left|\tfrac{1}{6}\left\{3G^5(\bar{\xi})\left[\mathbf{I}\otimes\bar{\xi}+\nabla\mathbf{H}(\bar{\xi})\right]-15G^7(\bar{\xi})\overline{\mathbf{T}}(\bar{\xi})\right\}\cdot\Delta\mathbf{y}^3\right| \\
&\leqslant \max\left|\frac{9\|\Delta\mathbf{y}\|^2\langle\bar{\xi},\Delta\mathbf{y}\rangle}{6\|\bar{\xi}\|^5}-\frac{15\langle\bar{\xi},\Delta\mathbf{y}\rangle^3}{6\|\bar{\xi}\|^7}\right| \\
&= \max\left|\langle\bar{\xi},\Delta\mathbf{y}\rangle\left(\frac{9\|\Delta\mathbf{y}\|^2}{6\|\bar{\xi}\|^5}-\frac{15\langle\bar{\xi},\Delta\mathbf{y}\rangle^2}{6\|\bar{\xi}\|^7}\right)\right| \\
&= \max\left|\langle\bar{\xi},\Delta\mathbf{y}\rangle\frac{9\|\bar{\xi}\|^2\|\Delta\mathbf{y}\|^2-15\langle\bar{\xi},\Delta\mathbf{y}\rangle^2}{6\|\bar{\xi}\|^7}\right| \\
&= \max\left|\frac{\|\bar{\xi}\|^3\|\Delta\mathbf{y}\|^3}{\|\bar{\xi}\|^7}\cdot\cos\varphi\frac{9-15\cos^2\varphi}{6}\right| \\
&\leqslant \max\left|\frac{\|\Delta\mathbf{y}\|^3}{\|\bar{\xi}\|^4}\right|.
\end{aligned}$$

To compute the latter maximum we need to know the range of $\bar{\xi}$, which is the set $\Xi = \{\mathbf{x}'-\mathbf{x} \mid \mathbf{x}'\in\Omega_j, \mathbf{x}\in\Omega_i\}$. This region can be obtained as follows. Translate Ω_j over $-\mathbf{x}_m$, such that its midpoint is mapped onto $\mathbf{y}_m$. Move $-\Omega_i$ with its midpoint $\mathbf{x}_m$ along the edges of the translated Ω_j, as illustrated in Fig. 7.2.

From the definitions of $r_{\min}$ and d it follows that $r_{\min}$ is the lower bound of $\|\bar{\xi}\|$ and d the upper bound of $\|\Delta\mathbf{y}\|$, for $\bar{\xi},\mathbf{y}\in\Xi$, so that

$$\delta_G \leqslant \frac{d^3}{r_{\min}^4}.$$

Let $\mathbf{y}_{\min}$ be the point for which $\|\mathbf{y}_{\min}\| = r_{\min}$, then

$$\|\mathbf{y}\| = \|\mathbf{y}_{\min}+\mathbf{y}-\mathbf{y}_{\min}\| \leqslant \|\mathbf{y}_{\min}\| + \|\mathbf{y}-\mathbf{y}_{\min}\| \leqslant r_{\min}+d.$$

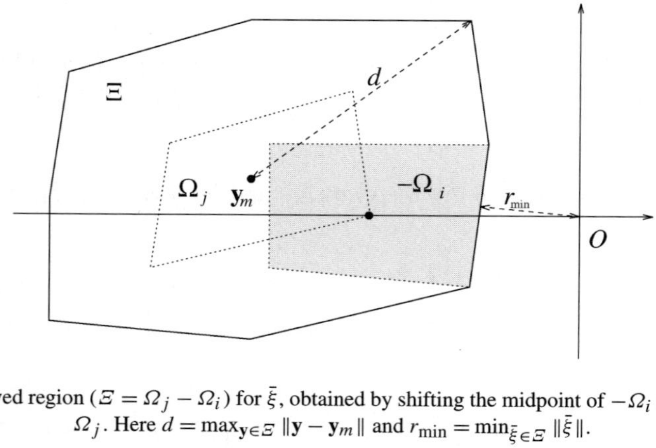

FIG. 7.2. Allowed region ($\Xi = \Omega_j - \Omega_i$) for $\bar{\xi}$, obtained by shifting the midpoint of $-\Omega_i$ along the edge of Ω_j. Here $d = \max_{\mathbf{y}\in\Xi}\|\mathbf{y}-\mathbf{y}_m\|$ and $r_{\min} = \min_{\bar{\xi}\in\Xi}\|\bar{\xi}\|$.

Since $G(\mathbf{y}) = \|\mathbf{y}\|^{-1}$, we have for the relative error $\rho_G = \frac{\delta_G}{G}$:

$$\rho_G \leqslant \alpha^3(1+\alpha) \quad \left(\alpha = \frac{d}{r_{\min}}\right).$$

Therefore, $\rho_G \leqslant \varepsilon$ if

$$\alpha^3(1+\alpha) \leqslant \varepsilon, \quad \text{or} \quad \alpha^3 \leqslant \frac{\varepsilon}{2}, \quad \text{or} \quad \frac{d}{r_{\min}} \leqslant \sqrt[3]{\frac{\varepsilon}{2}}, \quad \text{for } \varepsilon < 1. \qquad \square$$

With somewhat more effort it can be shown that for a third order Taylor expansion the condition $\alpha^4 \leqslant \frac{\varepsilon}{2}$ should be satisfied in order to keep the error small enough. Hence, we might state the following conjecture.

CONJECTURE 7.1. *Let $\tilde{G}$ be an approximation of G with the nth order Taylor expansion, then for every $\varepsilon > 0$ holds*

$$\alpha \leqslant \sqrt[n+1]{\frac{\varepsilon}{2}} \quad \Longrightarrow \quad \frac{\|\tilde{G} - G\|_\infty}{G} \leqslant \varepsilon.$$

8. Analytical integration of the inner integrals for vector valued basis functions

This section presents an improvement of the method, treated in Section 5 for the derivation of analytical expressions for the inner integral over a quadrilateral source element, of which the integrand is irregular and contains vector valued basis functions.

The integral is the sum of the integrals

$$\int_0^1 \overline{\mathcal{F}}_i(s_2) \ln\left(\frac{\sqrt{c}\sqrt{a+b+c}+c+\frac{1}{2}b}{\sqrt{c}\sqrt{a}+\frac{1}{2}b}\right) \mathrm{d}s_2, \tag{8.1}$$

and

$$\int_0^1 \overline{\mathcal{G}}_i(s_2)\left(\sqrt{a+b+c} - \sqrt{a}\right) \mathrm{d}s_2, \tag{8.2}$$

where a, b and c are quadratic functions of s_2. If the integration point of the outer integral, over the object element, lies in the interior or on the boundary of the source element, the integrand of the first integral is irregular. The integrand of the second integral has no singularity. In Section 5.2.4 the integral (8.1) is rewritten as

$$\int_0^1 \overline{\mathcal{F}}_i(s_2) \ln\left(\sqrt{c}\sqrt{a+b+c}+c+\tfrac{1}{2}b\right)\left(\sqrt{c}\sqrt{a}-\tfrac{1}{2}b\right) \mathrm{d}s_2$$
$$- \int_0^1 \overline{\mathcal{F}}_i \ln\left(ca - \left(\tfrac{1}{2}b\right)^2\right) \mathrm{d}s_2. \tag{8.3}$$

Only the last term has a singular integrand. After partial integration this integral becomes

$$\int_0^1 \overline{\mathcal{F}}_i(s_2) \ln\left(ca - \left(\tfrac{1}{2}b\right)^2\right) \mathrm{d}s_2 = \left[\overline{\mathcal{F}}_i(s_2)\mathcal{L}(s_2)\right]_0^1 - \int_0^1 \overline{\mathcal{F}}_i'(s_2)\mathcal{L}(s_2)\, \mathrm{d}s_2, \tag{8.4}$$

where

$$\mathcal{L}(s_2) = \int \ln\left(ca - \left(\tfrac{1}{2}b\right)^2\right) \mathrm{d}s_2. \tag{8.5}$$

In Section 5.2.4 an analytical expression for the integral (8.5) is derived. However, the integral in the right-hand side of (8.4) is evaluated numerically. Since the derivative of the integrand of (8.4) is singular, the numerical evaluation of this integral, by Patterson's rules, can become slowly convergent. Besides, the integral (8.5) can only be evaluated analytically if the integration point over the object element lies in the plane of the source element (see Section 5.2.4).

This section introduces another method, that allows for arbitrary orientation of object and source element. In this method the integration domain of the integrals (8.1) and (8.2) is split into at most three parts, for only one of which the integrand of (8.1) is singular. The domain of this singular part is chosen so that the approximation of (8.1) over it by an analytical expression is sufficiently accurate, and the integrals over the other parts can be evaluated by Patterson's rules. Let $\mathbf{v}_1 = \mathbf{x}_{12}(1 - s_2) - \mathbf{x}_{34}s_2$ for $\mathbf{x}_{ij} = \mathbf{x}_j - \mathbf{x}_i$, where $\mathbf{x}_1 \ldots \mathbf{x}_4$ are the vertices of the source element, and where s_2 is one of the iso-parametric coordinates of the integration point. It can be shown that only if the vector $\mathbf{v}_1$ is independent of s_2, the integral (8.1) can be calculated analytically over the full integration domain. This is the case for a source element with $\mathbf{x}_{12} + \mathbf{x}_{34} = 0$. In all other cases the integral (8.1) must be approximated. The approximation method makes use of the Taylor expansion of the vector $\mathbf{v}_1$ around the vector $\mathbf{d} = \mathbf{x}_{12}(1 - s_2^{(m)}) - \mathbf{x}_{34}s_2^{(m)}$, where $s_2^{(m)}$ is the iso-parametric coordinate, s_2, of the projection of the object integration point in the plane of the source element. If $s_2^{(m)} - s_2 = 0$, the integrand of (8.1) is singular.

8.1. *Definition of auxiliary quantities*

In this subsection the quantities are defined that will be used in the evaluation of integrals (8.1) and (8.2). Let us define the following quantities:

$$\begin{aligned}
&\mathbf{c}_0 = \mathbf{x}_{41}, && \hat{\mathbf{c}}_0 = -\mathbf{x}_{23}, && \mathbf{c}_1 = \mathbf{x}_{12} + \mathbf{x}_{34},\\
&\mathbf{d} = \mathbf{x}_{12} - \mathbf{c}_1 s_2^{(m)}, && y = s_2^{(m)} - s_2, && \mathbf{x} = \mathbf{c}_1 y,\\
&\mathbf{v}_0(y) = \tilde{\mathbf{v}}_0(y) + \mathbf{x}_{OM}, && \tilde{\mathbf{v}}_0(y) = \mathbf{c}_0 y - \mathbf{d}s_1^{(m)}, &&\\
&(\tilde{\mathbf{v}}_0 + \mathbf{v}_1)(y) = \hat{\mathbf{c}}_0 y + \mathbf{d}\left(1 - s_1^{(m)}\right), && \mathbf{v}_1(\mathbf{x}) = \mathbf{d} + \mathbf{x}, &&\\
&a(y) = \left|\mathbf{v}_0(y)\right|^2, && \tfrac{1}{2}b(y, \mathbf{x}) = \mathbf{v}_1(\mathbf{x})^{\mathrm{T}}\tilde{\mathbf{v}}_0(y), && c(\mathbf{x}) = \left|\mathbf{v}_1(\mathbf{x})\right|^2.
\end{aligned}$$

Note that for a element with parallel opposite edges $\mathbf{x} = 0$. The vector valued functions $\overline{\mathcal{F}}_i(y, \mathbf{x})$ and $\overline{\mathcal{G}}_i(y, \mathbf{x})$ can be written in terms of the above quantities as follows :

$$\overline{\mathcal{F}}_1(y, \mathbf{x}) = -\frac{\mathbf{c}_0}{|\mathbf{d}|}\left\{f_{000}(y, \mathbf{x}) + f_{010}(y, \mathbf{x})\right\} + \frac{\mathbf{c}_1}{|\mathbf{d}|}\left\{f_{001}(y, \mathbf{x}) + f_{011}(y, \mathbf{x})\right\},$$

$$\overline{\mathcal{F}}_2(y, \mathbf{x}) = \frac{\mathbf{d}}{|\mathbf{d}|} f_{001}(y, \mathbf{x}) + \frac{\mathbf{c}_1}{|\mathbf{d}|} f_{101}(y, \mathbf{x}),$$

$$\overline{\mathcal{F}}_3(y,\mathbf{x}) = -\frac{\mathbf{c}_0}{|\mathbf{d}|} f_{010}(y,\mathbf{x}) + \frac{\mathbf{c}_1}{|\mathbf{d}|} f_{011}(y,\mathbf{x}),$$

$$\overline{\mathcal{F}}_4(y,\mathbf{x}) = \frac{\mathbf{d}}{|\mathbf{d}|}\big\{f_{000}(y,\mathbf{x}) + f_{001}(y,\mathbf{x})\big\} + \frac{\mathbf{c}_1}{|\mathbf{d}|}\big\{f_{100}(y,\mathbf{x}) + f_{101}(y,\mathbf{x})\big\},$$

$$\overline{\mathcal{G}}_1(y,\mathbf{x}) = -\frac{\mathbf{c}_1}{|\mathbf{d}|}\big\{g_{00}(y,\mathbf{x}) + g_{01}(y,\mathbf{x})\big\},$$

$$\overline{\mathcal{G}}_2(y,\mathbf{x}) = -\frac{\mathbf{d}}{|\mathbf{d}|} g_{00}(y,\mathbf{x}) - \frac{\mathbf{c}_1}{|\mathbf{d}|} g_{10}(y,\mathbf{x}),$$

$$\overline{\mathcal{G}}_3(y,\mathbf{x}) = -\frac{\mathbf{c}_1}{|\mathbf{d}|} g_{01}(y,\mathbf{x}),$$

$$\overline{\mathcal{G}}_4(y,\mathbf{x}) = \overline{\mathcal{G}}_2(y,\mathbf{x}),$$

where

$$f_{jkl}(y,\mathbf{x}) = |\mathbf{d}|y^j\big(y - s_2^{(m)}\big)^k \tilde{f}_l(y,\mathbf{x}), \tag{8.6}$$

$$\tilde{f}_l(y,\mathbf{x}) = \frac{1}{\sqrt{c(\mathbf{x})}}\left(\frac{b(y,\mathbf{x})}{2c(\mathbf{x})}\right)^l = \frac{1}{|\mathbf{v}_1(\mathbf{x})|}\left(\frac{\mathbf{v}_1(\mathbf{x})^{\mathrm{T}}\tilde{\mathbf{v}}_0(y)}{|\mathbf{v}_1(\mathbf{x})|^2}\right)^l, \tag{8.7}$$

$$g_{jk}(y,\mathbf{x}) = |\mathbf{d}|y^j\big(y - s_2^{(m)}\big)^k \tilde{g}(y,\mathbf{x}), \tag{8.8}$$

$$\tilde{g}(y,\mathbf{x}) = \frac{1}{|\mathbf{v}_1(\mathbf{x})|^2}. \tag{8.9}$$

Further, the arguments of the logarithm in the integrand of the integral (8.1) and the arguments in the integrand of the integral (8.2) are

$$\big(\sqrt{c}\sqrt{a} + \tfrac{1}{2}b\big)(y,\mathbf{x}) = \big|\mathbf{v}_1(\mathbf{x})\big||\mathbf{d}|\mathcal{R}(y) + \mathbf{v}_1(\mathbf{x})^{\mathrm{T}}\tilde{\mathbf{v}}_0(y), \tag{8.10}$$

$$\big(\sqrt{c}\sqrt{a+b+c} + c + \tfrac{1}{2}b\big)(y,\mathbf{x}) = \big|\mathbf{v}_1(\mathbf{x})\big||\mathbf{d}|\hat{\mathcal{R}}(y) + \mathbf{v}_1(\mathbf{x})^{\mathrm{T}}(\tilde{\mathbf{v}}_0 + \mathbf{v}_1)(y), \tag{8.11}$$

$$\big(\sqrt{a}\big)(y) = |\mathbf{d}|\mathcal{R}(y), \tag{8.12}$$

$$\big(\sqrt{a+b+c}\big)(y) = |\mathbf{d}|\hat{\mathcal{R}}(y), \tag{8.13}$$

where

$$\mathcal{R}(y) = \sqrt{\frac{|\tilde{\mathbf{v}}_0(y)|^2}{|\mathbf{d}|^2} + \frac{|\mathbf{x}_{OM}|^2}{|\mathbf{d}|^2}},$$

$$\hat{\mathcal{R}}(y) = \sqrt{\frac{|(\tilde{\mathbf{v}}_0 + \mathbf{v}_1)(y)|^2}{|\mathbf{d}|^2} + \frac{|\mathbf{x}_{OM}|^2}{|\mathbf{d}|^2}}.$$

The derivatives with respect to $\mathbf{x}$ of the quantities depending on $\mathbf{x}$ are

$$\nabla\mathbf{v}_1(\mathbf{x}) = I,$$

$$\nabla\big|\mathbf{v}_1(\mathbf{x})\big|^n = n\big|\mathbf{v}_1(\mathbf{x})\big|^{n-2}\mathbf{v}_1(\mathbf{x}),$$

$$\nabla\big(\mathbf{v}_1(\mathbf{x})^{\mathrm{T}}\tilde{\mathbf{v}}_0(y)\big) = \tilde{\mathbf{v}}_0(y),$$

$$\nabla\big|\mathbf{v}_1(\mathbf{x})^{\mathrm{T}}\tilde{\mathbf{v}}_0(y)\big|^2 = 2\mathbf{v}_1(\mathbf{x})^{\mathrm{T}}\tilde{\mathbf{v}}_0(y)\tilde{\mathbf{v}}_0(y),$$

$$\nabla\big(\mathbf{v}_1(\mathbf{x})^{\mathrm{T}}(\tilde{\mathbf{v}}_0+\mathbf{v}_1)(y)\big)=(\tilde{\mathbf{v}}_0+\mathbf{v}_1)(y),$$
$$\nabla\big|\mathbf{v}_1(\mathbf{x})^{\mathrm{T}}(\tilde{\mathbf{v}}_0+\mathbf{v}_1)(y)\big|^2=2\mathbf{v}_1(\mathbf{x})^{\mathrm{T}}(\tilde{\mathbf{v}}_0+\mathbf{v}_1)(y)(\tilde{\mathbf{v}}_0+\mathbf{v}_1)(y).$$

The inner products of $\mathbf{x}$ with the first order derivatives with respect to $\mathbf{x}$ of the expressions in (8.7) for $l=0,1$ and (8.9) are

$$\mathbf{x}^{\mathrm{T}}(\nabla\tilde{f}_0)(y,\mathbf{x})=-\frac{\mathbf{x}^{\mathrm{T}}\mathbf{v}_1(\mathbf{x})}{|\mathbf{v}_1(\mathbf{x})|^3},$$
$$\mathbf{x}^{\mathrm{T}}(\nabla\tilde{f}_1)(y,\mathbf{x})=\frac{\mathbf{x}^{\mathrm{T}}\tilde{\mathbf{v}}_0(y)}{|\mathbf{v}_1(\mathbf{x})|^3}-3\frac{\mathbf{v}_1(\mathbf{x})^{\mathrm{T}}\tilde{\mathbf{v}}_0(y)}{|\mathbf{v}_1(\mathbf{x})|^2}\frac{\mathbf{x}^{\mathrm{T}}\mathbf{v}_1(\mathbf{x})}{|\mathbf{v}_1(\mathbf{x})|^3},$$
$$\mathbf{x}^{\mathrm{T}}(\nabla\tilde{g})(y,\mathbf{x})=-2\frac{\mathbf{x}^{\mathrm{T}}\mathbf{v}_1(\mathbf{x})}{|\mathbf{v}_1(\mathbf{x})|^4},$$

and with the second order derivatives with respect to $\mathbf{x}$ of the expressions in (8.7) for $l=0,1$ and (8.9) are

$$\mathbf{x}^{\mathrm{T}}\big(\nabla(\nabla\tilde{f}_0)\big)(y,\mathbf{x})\mathbf{x}=-\frac{1}{|\mathbf{v}_1(\mathbf{x})|^3}\left\{\mathbf{x}^{\mathrm{T}}\mathbf{x}-3\frac{(\mathbf{x}^{\mathrm{T}}\mathbf{v}_1(\mathbf{x}))^2}{|\mathbf{v}_1(\mathbf{x})|^2}\right\},$$
$$\begin{aligned}\mathbf{x}^{\mathrm{T}}\big(\nabla(\nabla\tilde{f}_1)\big)(y,\mathbf{x})\mathbf{x}=&-6\frac{(\mathbf{x}^{\mathrm{T}}\tilde{\mathbf{v}}_0(y))(\mathbf{x}^{\mathrm{T}}\mathbf{v}_1(\mathbf{x}))}{|\mathbf{v}_1(\mathbf{x})|^5}\\&-3\frac{\mathbf{v}_1(\mathbf{x})^{\mathrm{T}}\tilde{\mathbf{v}}_0(y)}{|\mathbf{v}_1(\mathbf{x})|^5}\left\{\mathbf{x}^{\mathrm{T}}\mathbf{x}-5\frac{(\mathbf{x}^{\mathrm{T}}\mathbf{v}_1(\mathbf{x}))^2}{|\mathbf{v}_1(\mathbf{x})|^2}\right\},\end{aligned}$$
$$\mathbf{x}^{\mathrm{T}}\big(\nabla(\nabla\tilde{g})\big)(y,\mathbf{x})\mathbf{x}=-\frac{2}{|\mathbf{v}_1(\mathbf{x})|^4}\left\{\mathbf{x}^{\mathrm{T}}\mathbf{x}-4\frac{(\mathbf{x}^{\mathrm{T}}\mathbf{v}_1(\mathbf{x}))^2}{|\mathbf{v}_1(\mathbf{x})|^2}\right\}.$$

Those with the first order derivatives with respect to $\mathbf{x}$ of the expressions in (8.10) and (8.11) are

$$\mathbf{x}^{\mathrm{T}}\big(\nabla\sqrt{c}\sqrt{a}+\tfrac{1}{2}b\big)(y,\mathbf{x})=|\mathbf{d}|\mathcal{R}(y)\frac{\mathbf{x}^{\mathrm{T}}\mathbf{v}_1(\mathbf{x})}{|\mathbf{v}_1(\mathbf{x})|}+\mathbf{x}^{\mathrm{T}}\tilde{\mathbf{v}}_0(y),$$
$$\mathbf{x}^{\mathrm{T}}\big(\nabla\sqrt{c}\sqrt{a+b+c}+c+\tfrac{1}{2}b\big)(y,\mathbf{x})=|\mathbf{d}|\hat{\mathcal{R}}(y)\frac{\mathbf{x}^{\mathrm{T}}\mathbf{v}_1(\mathbf{x})}{|\mathbf{v}_1(\mathbf{x})|}+\mathbf{x}^{\mathrm{T}}(\tilde{\mathbf{v}}_0+\mathbf{v}_1)(y),$$

and with the second order derivatives with respect to $\mathbf{x}$ of the expressions in (8.10) and (8.11) are

$$\mathbf{x}^{\mathrm{T}}\big(\nabla\big(\nabla\sqrt{c}\sqrt{a}+\tfrac{1}{2}b\big)\big)(y,\mathbf{x})\mathbf{x}=\frac{|\mathbf{d}|\mathcal{R}(y)}{|\mathbf{v}_1(\mathbf{x})|}\left\{\mathbf{x}^{\mathrm{T}}\mathbf{x}-\frac{(\mathbf{x}^{\mathrm{T}}\mathbf{v}_1(\mathbf{x}))^2}{|\mathbf{v}_1(\mathbf{x})|^2}\right\},$$
$$\mathbf{x}^{\mathrm{T}}\big(\nabla\big(\nabla\sqrt{c}\sqrt{a+b+c}+c+\tfrac{1}{2}b\big)\big)(y,\mathbf{x})\mathbf{x}=\frac{|\mathbf{d}|\hat{\mathcal{R}}(y)}{|\mathbf{v}_1(\mathbf{x})|}\left\{\mathbf{x}^{\mathrm{T}}\mathbf{x}-\frac{(\mathbf{x}^{\mathrm{T}}\mathbf{v}_1(\mathbf{x}))^2}{|\mathbf{v}_1(\mathbf{x})|^2}\right\}.$$

When $\mathbf{v}_1(\mathbf{x})$ approaches $\mathbf{d}$, the expressions in (8.6) and (8.8), and the inner products of $\mathbf{x}$ with their derivatives with respect to $\mathbf{x}$ become:

$$f_{jkl}(y)=y^j\big(y-s_2^{(m)}\big)^k\left(\frac{\mathbf{d}^{\mathrm{T}}\tilde{\mathbf{v}}_0(y)}{|\mathbf{d}|^2}\right)^l=y^j\big(y-s_2^{(m)}\big)^k\big(\alpha y-s_1^{(m)}\big)^l,$$

$$g_{jk}(y) = \frac{1}{|\mathbf{d}|} y^j \big(y - s_2^{(m)}\big)^k,$$

$$\mathbf{x}^{\mathrm{T}}(\nabla f_{jk0})(y) = -y^j \big(y - s_2^{(m)}\big)^k \frac{\mathbf{x}^{\mathrm{T}}\mathbf{d}}{|\mathbf{d}|^2} = -y^{j+1}\big(y - s_2^{(m)}\big)^k \varepsilon,$$

$$\mathbf{x}^{\mathrm{T}}(\nabla f_{jk1})(y) = y^j \big(y - s_2^{(m)}\big)^k \left\{ \frac{\mathbf{x}^{\mathrm{T}}\tilde{\mathbf{v}}_0(y)}{|\mathbf{d}|^2} - 3\frac{\mathbf{d}^{\mathrm{T}}\tilde{\mathbf{v}}_0(y)}{|\mathbf{d}|^2}\frac{\mathbf{x}^{\mathrm{T}}\mathbf{d}}{|\mathbf{d}|^2} \right\}$$
$$= y^{j+1}\big(y - s_2^{(m)}\big)^k \big\{ \big(\gamma y - \varepsilon s_1^{(m)}\big) - 3\big(\alpha y - s_1^{(m)}\big)\varepsilon \big\},$$

$$\mathbf{x}^{\mathrm{T}}(\nabla g_{jk})(y) = \frac{-2}{|\mathbf{d}|} y^j \big(y - s_2^{(m)}\big)^k \frac{\mathbf{x}^{\mathrm{T}}\mathbf{d}}{|\mathbf{d}|^2} = \frac{-2}{|\mathbf{d}|} y^{j+1}\big(y - s_2^{(m)}\big)^k \varepsilon,$$

$$\mathbf{x}^{\mathrm{T}}\big(\nabla(\nabla f_{jk0})\big)(y)\mathbf{x} = -y^j \big(y - s_2^{(m)}\big)^k \left(\frac{\mathbf{x}^{\mathrm{T}}\mathbf{x}}{|\mathbf{d}|^2} - 3\frac{(\mathbf{x}^{\mathrm{T}}\mathbf{d})^2}{|\mathbf{d}|^4} \right)$$
$$= -y^{j+2}\big(y - s_2^{(m)}\big)^k (\delta^2 - 3\varepsilon^2),$$

$$\mathbf{x}^{\mathrm{T}}\big(\nabla(\nabla f_{jk1})\big)(y)\mathbf{x}$$
$$= y^j \big(y - s_2^{(m)}\big)^k \left\{ -6\frac{\mathbf{x}^{\mathrm{T}}\tilde{\mathbf{v}}_0(y)}{|\mathbf{d}|^2}\frac{\mathbf{x}^{\mathrm{T}}\mathbf{d}}{|\mathbf{d}|^2} - 3\frac{\mathbf{d}^{\mathrm{T}}\tilde{\mathbf{v}}_0(y)}{|\mathbf{d}|^2}\left(\frac{\mathbf{x}^{\mathrm{T}}\mathbf{x}}{|\mathbf{d}|^2} - 5\frac{(\mathbf{x}^{\mathrm{T}}\mathbf{d})^2}{|\mathbf{d}|^4} \right) \right\}$$
$$= y^{j+2}\big(y - s_2^{(m)}\big)^k \big\{ -6\big(\gamma y - \varepsilon s_1^{(m)}\big)\varepsilon - 3\big(\alpha y - s_1^{(m)}\big)(\delta^2 - 5\varepsilon^2) \big\},$$

$$\mathbf{x}^{\mathrm{T}}\big(\nabla(\nabla g_{jk})\big)(y)\mathbf{x} = \frac{-2}{|\mathbf{d}|} y^j \big(y - s_2^{(m)}\big)^k \left(\frac{\mathbf{x}^{\mathrm{T}}\mathbf{x}}{|\mathbf{d}|^2} - 4\frac{(\mathbf{x}^{\mathrm{T}}\mathbf{d})^2}{|\mathbf{d}|^4} \right)$$
$$= \frac{-2}{|\mathbf{d}|} y^{j+2}\big(y - s_2^{(m)}\big)^k (\delta^2 - 4\varepsilon^2),$$

the expressions in (8.10) and (8.11), and the inner products of $\mathbf{x}$ with their derivatives with respect to $\mathbf{x}$ become:

$$\big(\sqrt{c}\sqrt{a} + \tfrac{1}{2}b\big)(y) = |\mathbf{d}|^2 \left\{ \mathcal{R}(y) + \frac{\mathbf{d}^{\mathrm{T}}\tilde{\mathbf{v}}_0(y)}{|\mathbf{d}|^2} \right\}$$
$$= |\mathbf{d}|^2 \big\{ \mathcal{R}(y) + \alpha y - s_1^{(m)} \big\},$$

$$\big(\sqrt{c}\sqrt{a+b+c} + c + \tfrac{1}{2}b\big)(y) = |\mathbf{d}|^2 \left\{ \hat{\mathcal{R}}(y) + \frac{\mathbf{d}^{\mathrm{T}}(\tilde{\mathbf{v}}_0 + \mathbf{v}_1)(y)}{|\mathbf{d}|^2} \right\}$$
$$= |\mathbf{d}|^2 \big\{ \hat{\mathcal{R}}(y) + \hat{\alpha} y + \big(1 - s_1^{(m)}\big) \big\},$$

$$\mathbf{x}^{\mathrm{T}}\big(\nabla\sqrt{c}\sqrt{a} + \tfrac{1}{2}b\big)(y) = |\mathbf{d}|^2 \left\{ \frac{\mathbf{x}^{\mathrm{T}}\mathbf{d}}{|\mathbf{d}|^2}\mathcal{R}(y) + \frac{\mathbf{x}^{\mathrm{T}}\tilde{\mathbf{v}}_0(y)}{|\mathbf{d}|^2} \right\}$$
$$= |\mathbf{d}|^2 y \big\{ \varepsilon\mathcal{R}(y) + \gamma y - \varepsilon s_1^{(m)} \big\},$$

$$\mathbf{x}^{\mathrm{T}}\big(\nabla\sqrt{c}\sqrt{a+b+c} + c + \tfrac{1}{2}b\big)(y) = |\mathbf{d}|^2 \left\{ \frac{\mathbf{x}^{\mathrm{T}}\mathbf{d}}{|\mathbf{d}|^2}\hat{\mathcal{R}}(y) + \frac{\mathbf{x}^{\mathrm{T}}(\tilde{\mathbf{v}}_0 + \mathbf{v}_1)(y)}{|\mathbf{d}|^2} \right\}$$
$$= |\mathbf{d}|^2 y \big\{ \varepsilon\hat{\mathcal{R}}(y) + \hat{\gamma} y + \varepsilon\big(1 - s_1^{(m)}\big) \big\},$$

$$\begin{aligned}\mathbf{x}^{\mathrm{T}}\big(\nabla\big(\nabla\sqrt{c}\sqrt{a}+\tfrac{1}{2}b\big)\big)(y)\mathbf{x} &= |\mathbf{d}|^2\mathcal{R}(y)\left\{\frac{\mathbf{x}^{\mathrm{T}}\mathbf{x}}{|\mathbf{d}|^2}-\frac{(\mathbf{x}^{\mathrm{T}}\mathbf{d})^2}{|\mathbf{d}|^4}\right\}\\ &= |\mathbf{d}|^2y^2\mathcal{R}(y)(\delta^2-\varepsilon^2),\\ \mathbf{x}^{\mathrm{T}}\big(\nabla\big(\nabla\sqrt{c}\sqrt{a+b+c}+c+\tfrac{1}{2}b\big)\big)(y)\mathbf{x} &= |\mathbf{d}|^2\hat{\mathcal{R}}(y)\left\{\frac{\mathbf{x}^{\mathrm{T}}\mathbf{x}}{|\mathbf{d}|^2}-\frac{(\mathbf{x}^{\mathrm{T}}\mathbf{d})^2}{|\mathbf{d}|^4}\right\}\\ &= |\mathbf{d}|^2y^2\hat{\mathcal{R}}(y)(\delta^2-\varepsilon^2),\end{aligned}$$

where

$$\begin{aligned}\mathcal{R}(y) &= \sqrt{\beta^2y^2-2\alpha s_1^{(m)}y+(s_1^{(m)})^2+\zeta^2},\\ \hat{\mathcal{R}}(y) &= \sqrt{\hat{\beta}^2y^2+2\hat{\alpha}(1-s_1^{(m)})y+(1-s_1^{(m)})^2+\zeta^2},\end{aligned}\tag{8.14}$$

and

$$\begin{aligned}\alpha &= \frac{\mathbf{d}^{\mathrm{T}}\mathbf{c}_0}{|\mathbf{d}|^2}, & \beta &= \frac{|\mathbf{c}_0|}{|\mathbf{d}|}, & \gamma &= \frac{\mathbf{c}_1^{\mathrm{T}}\mathbf{c}_0}{|\mathbf{d}|^2},\\ \hat{\alpha} &= \frac{\mathbf{d}^{\mathrm{T}}\hat{\mathbf{c}}_0}{|\mathbf{d}|^2}, & \hat{\beta} &= \frac{|\hat{\mathbf{c}}_0|}{|\mathbf{d}|}, & \hat{\gamma} &= \frac{\mathbf{c}_1^{\mathrm{T}}\hat{\mathbf{c}}_0}{|\mathbf{d}|^2},\\ \varepsilon &= \frac{\mathbf{d}^{\mathrm{T}}\mathbf{c}_1}{|\mathbf{d}|^2}, & \delta &= \frac{|\mathbf{c}_1|}{|\mathbf{d}|}, & \zeta &= \frac{|\mathbf{x}_{OM}|}{|\mathbf{d}|}.\end{aligned}$$

8.2. Taylor expansion of the integrals for the functions f_{jkl}, g_{jk}

The integrals for scalar functions f_{kl} and g_k are as follows:

$$I_{jkl}(y,\mathbf{x}) = \int f_{jkl}(y,\mathbf{x})\ln F(y,\mathbf{x})\,\mathrm{d}y,\tag{8.15}$$

$$I_{jk}(y,\mathbf{x}) = \int g_{jk}(y,\mathbf{x})G(y)\,\mathrm{d}y,\tag{8.16}$$

where

$$\begin{aligned}F(y,\mathbf{x}) &= \frac{(\sqrt{c}\sqrt{a}+\frac{1}{2}b)(y,\mathbf{x})}{(\sqrt{c}\sqrt{a+b+c}+c+\frac{1}{2}b)(y,\mathbf{x})},\\ G(y) &= \big(\sqrt{a}\big)(y)-\big(\sqrt{a+b+c}\big)(y) = |\mathbf{d}|\big(\mathcal{R}(y)-\hat{\mathcal{R}}(y)\big).\end{aligned}$$

The second order Taylor expansions of (8.15) and (8.16) with respect to $\mathbf{x}$ gives

$$\begin{aligned}I_{jkl}(y,\mathbf{x}) &= I_{jkl}(y)+\mathbf{x}^{\mathrm{T}}(\nabla I_{jkl})(y)+\tfrac{1}{2}\mathbf{x}^{\mathrm{T}}\big(\nabla(\nabla I_{jkl})\big)(y)\mathbf{x}+\mathcal{O}\big(|\mathbf{x}|^3\big),\\ I_{jk}(y,\mathbf{x}) &= I_{jk}(y)+\mathbf{x}^{\mathrm{T}}(\nabla I_{jk})(y)+\tfrac{1}{2}\mathbf{x}^{\mathrm{T}}\big(\nabla(\nabla I_{jk})\big)(y)\mathbf{x}+\mathcal{O}\big(|\mathbf{x}|^3\big),\end{aligned}$$

where

$$I_{jkl}(y) = \int f_{jkl}(y)\ln F(y)\,\mathrm{d}y,$$

$$\mathbf{x}^{\mathrm{T}}(\nabla I_{jkl})(y) = \int \mathbf{x}^{\mathrm{T}}(\nabla f_{jkl})(y) \ln F(y)\,\mathrm{d}y + \int f_{jkl}(y)\mathbf{x}^{\mathrm{T}}(\nabla \ln F)(y)\,\mathrm{d}y,$$

$$\begin{aligned}\mathbf{x}^{\mathrm{T}}\big(\nabla(\nabla I_{jkl})\big)(y)\mathbf{x} = &\int \mathbf{x}^{\mathrm{T}}\big(\nabla(\nabla f_{jkl})\big)(y)\mathbf{x}\ln F(y)\,\mathrm{d}y \\ &+ 2\int \mathbf{x}^{\mathrm{T}}(\nabla f_{jkl})(y)\mathbf{x}^{\mathrm{T}}(\nabla \ln F)(y)\,\mathrm{d}y \\ &+ \int f_{jkl}(y)\mathbf{x}^{\mathrm{T}}\big(\nabla(\nabla \ln F)\big)(y)\mathbf{x}\,\mathrm{d}y,\end{aligned}$$

$$I_{jk}(y) = \int g_{jk}(y)G(y)\,\mathrm{d}y,$$

$$\mathbf{x}^{\mathrm{T}}(\nabla I_{jk})(y) = \int \mathbf{x}^{\mathrm{T}}(\nabla g_{jk})(y)G(y)\,\mathrm{d}y,$$

$$\mathbf{x}^{\mathrm{T}}\big(\nabla(\nabla I_{jk})\big)(y)\mathbf{x} = \int \mathbf{x}^{\mathrm{T}}\big(\nabla(\nabla g_{jk})\big)(y)\mathbf{x}G(y)\,\mathrm{d}y,$$

and

$$\ln F(y) = \ln\big(\mathcal{R}(y) + \alpha y - s_1^{(m)}\big) - \ln\big(\hat{\mathcal{R}}(y) + \hat{\alpha} y + \big(1 - s_1^{(m)}\big)\big),$$

$$\mathbf{x}^{\mathrm{T}}(\nabla \ln F)(y) = y\left\{\frac{\varepsilon\mathcal{R}(y) + \gamma y - \varepsilon s_1^{(m)}}{\mathcal{R}(y) + \alpha y - s_1^{(m)}} - \frac{\varepsilon\hat{\mathcal{R}}(y) + \hat{\gamma} y + \varepsilon(1 - s_1^{(m)})}{\hat{\mathcal{R}}(y) + \hat{\alpha} y + (1 - s_1^{(m)})}\right\},$$

$$\begin{aligned}&\mathbf{x}^{\mathrm{T}}\big(\nabla(\nabla \ln F)\big)(y)\mathbf{x} \\ &\quad = y^2\left\{-\left(\frac{\varepsilon\mathcal{R}(y) + \gamma y - \varepsilon s_1^{(m)}}{\mathcal{R}(y) + \alpha y - s_1^{(m)}}\right)^2 + \left(\frac{\varepsilon\hat{\mathcal{R}}(y) + \hat{\gamma} y + \varepsilon(1 - s_1^{(m)})}{\hat{\mathcal{R}}(y) + \hat{\alpha} y + (1 - s_1^{(m)})}\right)^2\right. \\ &\qquad \left. + \frac{\mathcal{R}(y)(\delta^2 - \varepsilon^2)}{\mathcal{R}(y) + \alpha y - s_1^{(m)}} - \frac{\hat{\mathcal{R}}(y)(\delta^2 - \varepsilon^2)}{\hat{\mathcal{R}}(y) + \hat{\alpha} y + (1 - s_1^{(m)})}\right\}.\end{aligned}$$

Therefore, the integrals to be evaluated are

$$I_{j00}(y) = K_j(y),$$

$$I_{j10}(y) = I_{j+1,00}(y) - s_2^{(m)} I_{j00}(y),$$

$$I_{j01}(y) = \alpha K_{j+1}(y) - s_1^{(m)} K_j(y),$$

$$I_{j11}(y) = I_{j+1,01}(y) - s_2^{(m)} I_{j01}(y),$$

$$\mathbf{x}^{\mathrm{T}}(\nabla I_{j00})(y) = -\varepsilon\big(K_{j+1}(y) - L_{j+1}^{(1)}(y)\big),$$

$$\mathbf{x}^{\mathrm{T}}(\nabla I_{j10})(y) = \mathbf{x}^{\mathrm{T}}(\nabla I_{j+1,00})(y) - s_2^{(m)}\mathbf{x}^{\mathrm{T}}(\nabla I_{j00})(y),$$

$$\mathbf{x}^{\mathrm{T}}(\nabla I_{j01})(y) = (\gamma - 3\alpha\varepsilon)K_{j+2}(y) + \alpha\varepsilon L_{j+2}^{(1)}(y) + s_1^{(m)}\varepsilon\big(2K_{j+1} - L_{j+1}^{(1)}(y)\big),$$

$$\mathbf{x}^{\mathrm{T}}(\nabla I_{j11})(y) = \mathbf{x}^{\mathrm{T}}(\nabla I_{j+1,01})(y) - s_2^{(m)}\mathbf{x}^{\mathrm{T}}(\nabla I_{j01})(y),$$

$$
\begin{aligned}
&\mathbf{x}^{\mathrm{T}}\big(\nabla(\nabla I_{j00})\big)(y)\mathbf{x}\\
&\quad= -(\delta^2-3\varepsilon^2)K_{j+2}(y) - 2\varepsilon^2 L^{(1)}_{j+2}(y) - \varepsilon^2 L^{(2)}_{j+2}(y) + (\delta^2-\varepsilon^2)\Lambda_{j+2}(y),\\
&\mathbf{x}^{\mathrm{T}}\big(\nabla(\nabla I_{j10})\big)(y)\mathbf{x} = \mathbf{x}^{\mathrm{T}}\big(\nabla(\nabla I_{j+1,00})\big)(y)\mathbf{x} - s_2^{(m)}\mathbf{x}^{\mathrm{T}}\big(\nabla(\nabla I_{j00})\big)(y)\mathbf{x},\\
&\mathbf{x}^{\mathrm{T}}\big(\nabla(\nabla I_{j01})\big)(y)\mathbf{x} = -3\big\{2\gamma\varepsilon + \alpha(\delta^2-5\varepsilon^2)\big\}K_{j+3}(y) + 2(\gamma-3\alpha\varepsilon)\varepsilon L^{(1)}_{j+3}(y)\\
&\qquad\qquad + \alpha\big\{-\varepsilon^2 L^{(2)}_{j+3}(y) + (\delta^2-\varepsilon^2)\Lambda_{j+3}(y)\big\}\\
&\qquad\qquad + s_1^{(m)}\big\{3(\delta^2-3\varepsilon^2)K_{j+2}(y) + 4\varepsilon^2 L^{(1)}_{j+2}(y) + \varepsilon^2 L^{(2)}_{j+2}(y)\\
&\qquad\qquad - (\delta^2-\varepsilon^2)\Lambda_{j+2}(y)\big\},\\
&\mathbf{x}^{\mathrm{T}}\big(\nabla(\nabla I_{j11})\big)(y)\mathbf{x} = \mathbf{x}^{\mathrm{T}}\big(\nabla(\nabla I_{j+1,01})\big)(y)\mathbf{x} - s_2^{(m)}\mathbf{x}^{\mathrm{T}}\big(\nabla(\nabla I_{j01})\big)(y)\mathbf{x},\\
&I_{j0}(y) = M_j(y),\\
&I_{j1}(y) = I_{j+1,0}(y) - s_2^{(m)} I_{j0}(y),\\
&\mathbf{x}^{\mathrm{T}}(\nabla I_{j0})(y) = -2\varepsilon M_{j+1}(y),\\
&\mathbf{x}^{\mathrm{T}}(\nabla I_{j1})(y) = \mathbf{x}^{\mathrm{T}}(\nabla I_{j+1,0})(y) - s_2^{(m)}\mathbf{x}^{\mathrm{T}}(\nabla I_{j0})(y),\\
&\mathbf{x}^{\mathrm{T}}\big(\nabla(\nabla I_{j0})\big)(y)\mathbf{x} = -2(\delta^2-4\varepsilon^2)M_{j+2}(y),\\
&\mathbf{x}^{\mathrm{T}}\big(\nabla(\nabla I_{j1})\big)(y)\mathbf{x} = \mathbf{x}^{\mathrm{T}}\big(\nabla(\nabla I_{j+1,0})\big)(y)\mathbf{x} - s_2^{(m)}\mathbf{x}^{\mathrm{T}}\big(\nabla(\nabla I_{j0})\big)(y)\mathbf{x},
\end{aligned}
$$

where the auxiliary integrals $K_i(y)$, $L_i^{(n)}(y)$, $\Lambda_i(y)$, and $M_i(y)$ are

$$
\begin{aligned}
&K_i(y) = \mathcal{K}_i(y) - \hat{\mathcal{K}}_i(y),\\
&L_i^{(n)}(y) = \mathcal{L}_i^{(n)}\Big(y;\frac{\gamma}{\varepsilon},1\Big) - \hat{\mathcal{L}}_i^{(n)}\Big(y;\frac{\hat{\gamma}}{\varepsilon},1\Big),\\
&\Lambda_i(y) = \mathcal{L}_i^{(1)}(y;0,0) - \hat{\mathcal{L}}_i^{(1)}(y;0,0),\\
&M_i(y) = \mathcal{M}_i(y) - \hat{\mathcal{M}}_i(y),
\end{aligned}
$$

$$\mathcal{K}_i(y) = \int y^i \ln\big(\mathcal{R}(y) + \alpha y - s_1^{(m)}\big)\,\mathrm{d}y, \tag{8.17}$$

$$\hat{\mathcal{K}}_i(y) = \int y^i \ln\big(\hat{\mathcal{R}}(y) + \hat{\alpha} y + \big(1 - s_1^{(m)}\big)\big)\,\mathrm{d}y, \tag{8.18}$$

$$\mathcal{L}_i^{(n)}(y;\rho,\sigma) = \int y^i \left(\frac{\mathcal{R}(y) + \rho y - \sigma s_1^{(m)}}{\mathcal{R}(y) + \alpha y - s_1^{(m)}}\right)^n \mathrm{d}y, \tag{8.19}$$

$$\hat{\mathcal{L}}_i^{(n)}(y;\rho,\sigma) = \int y^i \left(\frac{\hat{\mathcal{R}}(y) + \rho y + \sigma(1 - s_1^{(m)})}{\hat{\mathcal{R}}(y) + \hat{\alpha} y + (1 - s_1^{(m)})}\right)^n \mathrm{d}y, \tag{8.20}$$

$$\mathcal{M}_i(y) = \int y^i \mathcal{R}(y)\,\mathrm{d}y, \tag{8.21}$$

$$\hat{\mathcal{M}}_i(y) = \int y^i \hat{\mathcal{R}}(y)\,\mathrm{d}y. \tag{8.22}$$

Note that the integral $L_i^{(n)}(y)$ need only to be calculated if $\varepsilon \neq 0$.

8.3. Analytical evaluation of the auxiliary integrals

In this subsection the analytical expressions are obtained for the auxiliary integrals defined by (8.17), (8.19) and (8.21). Those for the integrals defined by (8.18), (8.20) and (8.22) are similar.

The expression (8.14) can be rewritten as follows:

$$\mathcal{R}(y) = \sqrt{\beta^2 y^2 - 2\alpha s_1^{(m)} y + (s_1^{(m)})^2 + \zeta^2} = \sqrt{\tilde{z}^2 + f^2},$$

where

$$\tilde{z} = \beta y - \frac{\alpha}{\beta} s_1^{(m)},$$

$$f = \sqrt{\left(1 - \frac{\alpha^2}{\beta^2}\right)(s_1^{(m)})^2 + \zeta^2}.$$

Since $|\frac{\alpha}{\beta}| < 1$, and $s_1^{(m)}$ and ζ are not equal to zero simultaneously, $f > 0$, so that

$$y = \frac{f}{\beta}(z + h),$$

$$\mathcal{R}(y) + \alpha y - s_1^{(m)} = f\left(\sqrt{z^2 + 1} + pz + q\right),$$

where

$$z = \frac{\tilde{z}}{f}, \quad h = p\frac{s_1^{(m)}}{f}, \quad p = \frac{\alpha}{\beta}, \quad q = (p^2 - 1)\frac{s_1^{(m)}}{f},$$

and $|p| < 1$ and $|q| \leqslant 1$.

Similarly, one can write:

$$\mathcal{R}(y) + \rho y - \sigma s_1^{(m)} = f\left(\sqrt{z^2 + 1} + rz + s\right),$$

where

$$r = \frac{\rho}{\beta}, \quad s = (pr - \sigma)\frac{s_1^{(m)}}{f}.$$

Therefore, the auxiliary integrals defined by (8.17), (8.19) and (8.21) can be written as follows:

$$\mathcal{K}_i(y) = \left(\frac{f}{\beta}\right)^{i+1} \left(\frac{\ln(f)}{i+1}(z + h)^{i+1} + \sum_{k=0}^{i} \binom{i}{k} h^{i-k} \mathcal{K}_k(z)\right), \tag{8.23}$$

$$\mathcal{L}_i^{(n)}(y; \rho, \sigma) = \left(\frac{f}{\beta}\right)^{i+1} \sum_{k=0}^{i} \binom{i}{k} h^{i-k} \mathcal{L}_k^{(n)}(z; r, s), \tag{8.24}$$

$$\mathcal{M}_i(y) = f\left(\frac{f}{\beta}\right)^{i+1} \sum_{k=0}^{i} \binom{i}{k} h^{i-k} \mathcal{M}_k(z), \tag{8.25}$$

where

$$\mathcal{K}_k(z) = \int z^k \ln\bigl(pz + q + \sqrt{z^2+1}\bigr)\,dz, \tag{8.26}$$

$$\mathcal{L}_k^{(n)}(z; r, s) = \int z^k \left(\frac{rz + s + \sqrt{z^2+1}}{pz + q + \sqrt{z^2+1}}\right)^n dz, \tag{8.27}$$

$$\mathcal{M}_k(z) = \int z^k \sqrt{z^2+1}\,dz. \tag{8.28}$$

The expression for the integrals (8.26), (8.27) and (8.28) for $k = 0$ and the recursion formulae $k > 0$ are given in the following subsection.

8.4. Analytical expressions and recursion formulae

For $p = q = 0$ the formulae for the integrals (8.26) are

$$\mathcal{K}_0(z) = -z + \arctan z + \tfrac{1}{2} z \ln(z^2+1),$$

$$\mathcal{K}_1(z) = \tfrac{1}{4}(z^2+1)\bigl(\ln(z^2+1) - 1\bigr),$$

$$\mathcal{K}_k(z) = \frac{1}{k+1}\left\{-\frac{z^{k+1}}{k+1} + \tfrac{1}{2}(z^{k+1} + z^{k-1})\ln(z^2+1) - (k-1)\mathcal{K}_{k-2}(z)\right\}.$$

For $p \neq 0$ or $q \neq 0$ the formulae for the integrals (8.26) are

$$\mathcal{K}_k(z) = \frac{1}{(k+1)(p^2+q^2)}\left\{-\frac{q^2 z^{k+1}}{k+1} + (p^2+q^2) z^{k+1} \ln\bigl(pz + q + \sqrt{z^2+1}\bigr) - p\mathcal{I}_{k+1}(z) + q\mathcal{I}_k(z) - p(p^2-1)\mathcal{J}_{k+1}(z) + q(q^2-1)\mathcal{J}_k(z)\right\},$$

where

$$\mathcal{I}_0(z) = \int \frac{1}{\sqrt{z^2+1}}\,dz = \ln\bigl(z + \sqrt{z^2+1}\bigr),$$

$$\mathcal{I}_1(z) = \int \frac{z}{\sqrt{z^2+1}}\,dz = \sqrt{z^2+1}, \tag{8.29}$$

$$\mathcal{I}_k(z) = \int \frac{z^k}{\sqrt{z^2+1}}\,dz = \frac{1}{k}\bigl\{z^{k-1}\sqrt{z^2+1} - (k-1)\mathcal{I}_{k-2}(z)\bigr\},$$

and

$$\mathcal{J}_j(z) = \int \frac{z^j}{pz + q + \sqrt{z^2+1}}\,dz. \tag{8.30}$$

After the substitution $z = \frac{1}{2}(t - t^{-1})$, so that $\sqrt{z^2+1} = \frac{1}{2}(t + t^{-1})$, $dz = \frac{1}{2}(t + t^{-1}) \cdot t^{-1}\,dt$, and $t = z + \sqrt{z^2+1}$ the integral (8.30) can be written as follows:

$$\mathcal{J}_k(t) = \left(\frac{1}{2}\right)^k \int \frac{(t + t^{-1})(t - t^{-1})^k}{at^2 + bt + c}\,dt, = \left(\frac{1}{2}\right)^k \sum_{j=0}^{k} (-1)^j \binom{k}{j} T_{k-2j}(t),$$

$$T_i(t) = \tilde{\mathcal{J}}_{i+1}^{(1)}(t) + \tilde{\mathcal{J}}_{i-1}^{(1)}(t),$$

where

$$a = 1 + p, \quad b = 2q, \quad c = 1 - p, \quad -k \leqslant i \leqslant k,$$

and

$$\tilde{\mathcal{J}}_j^{(n)}(t) = \int \frac{t^j}{(at^2 + bt + c)^n}\,\mathrm{d}t. \tag{8.31}$$

Let constant $\tau \ll 1$, then the formulae for this integral are

$$\tilde{\mathcal{J}}_0^{(1)}(t) = \begin{cases} \frac{2}{\sqrt{4ac-b^2}}\left(\arctan\frac{2at+b}{\sqrt{4ac-b^2}} - \frac{\pi}{2}\right), & \text{for } b^2 - 4ac < -\tau(2at+b)^2, \\ \frac{1}{\sqrt{b^2-4ac}}\ln\left(\frac{2at+b-\sqrt{b^2-4ac}}{2at+b+\sqrt{b^2-4ac}}\right), & \text{for } b^2 - 4ac > \tau(2at+b)^2, \\ \frac{-2}{2at+b}\sum_{k=0}^{\infty}\frac{1}{2k+1}\left(\frac{b^2-4ac}{(2at+b)^2}\right)^k, & \text{for } |b^2 - 4ac| < \tau(2at+b)^2, \\ & \text{i.e., } p^2 + q^2 \approx 1, \end{cases}$$

$$\tilde{\mathcal{J}}_0^{(2)}(t) = \begin{cases} \frac{-1}{b^2-4ac}\left(\frac{2at+b}{at^2+bt+c} + 2a\tilde{\mathcal{J}}_0^{(1)}(t)\right), & \text{for } |b^2 - 4ac| > \tau(2at+b)^2, \\ \frac{-8a}{3(2at+b)^3}\sum_{k=0}^{\infty}\frac{k+1}{2k+3}\left(\frac{b^2-4ac}{(2at+b)^2}\right)^k, & \text{for } |b^2 - 4ac| < \tau(2at+b)^2, \end{cases}$$

$$\tilde{\mathcal{J}}_0^{(n)}(t) = \begin{cases} \frac{-1}{(n-1)(b^2-4ac)}\left(\frac{2at+b}{(at^2+bt+c)^{n-1}} + 2(2n-3)a\tilde{\mathcal{J}}_0^{(n-1)}(t)\right), & \text{for } b^2 \neq 4ac, \\ \frac{-2^{2n-1}a^{n-1}}{(2n-1)(2at+b)^{2n-1}}, & \text{for } b^2 = 4ac, \end{cases}$$

$$\tilde{\mathcal{J}}_{+1}^{(1)}(t) = \frac{1}{2a}\left(\ln(at^2 + bt + c) - b\tilde{\mathcal{J}}_0^{(1)}(t)\right),$$

$$\tilde{\mathcal{J}}_{-1}^{(1)}(t) = \frac{1}{2c}\left(\ln\frac{t^2}{at^2 + bt + c} - b\tilde{\mathcal{J}}_0^{(1)}(t)\right),$$

$$\tilde{\mathcal{J}}_{-1}^{(n)}(t) = \frac{1}{c}\left(\frac{1}{2(n-1)(at^2 + bt + c)^{n-1}} - \frac{b}{2}\tilde{\mathcal{J}}_0^{(n)}(t) + \tilde{\mathcal{J}}_{-1}^{(n-1)}(t)\right),$$

$$\tilde{\mathcal{J}}_{2n-1}^{(n)}(t) = \frac{1}{a}\left(\tilde{\mathcal{J}}_{2n-3}^{(n-1)}(t) - b\tilde{\mathcal{J}}_{2n-2}^{(n)}(t) - c\tilde{\mathcal{J}}_{2n-3}^{(n)}(t)\right),$$

$$\tilde{\mathcal{J}}_j^{(n)}(t) = \begin{cases} \frac{1}{(j-2n+1)a}\left(\frac{t^{j-1}}{(at^2+bt+c)^{n-1}} - (j-n)b\tilde{\mathcal{J}}_{j-1}^{(n)}(t) - (j-1)c\tilde{\mathcal{J}}_{j-2}^{(n)}(t)\right), \\ \quad \text{for } 0 < j \neq 2n-1, \\ \frac{1}{(j+1)c}\left(\frac{t^{j+1}}{(at^2+bt+c)^{n-1}} - (j-n+2)b\tilde{\mathcal{J}}_{j+1}^{(n)}(t) - (j-2n+3)a\tilde{\mathcal{J}}_{j+2}^{(n)}(t)\right), \\ \quad \text{for } j < -1, \end{cases}$$

where $n > 1$.

After the substitution $z = \frac{1}{2}(t - t^{-1})$, so that $\sqrt{z^2+1} = \frac{1}{2}(t + t^{-1})$, $\mathrm{d}z = \frac{1}{2}(t + t^{-1}) \cdot t^{-1}\,\mathrm{d}t$, and $t = z + \sqrt{z^2+1}$ the integral (8.27) can be written as follows:

$$\begin{aligned}
&\mathcal{L}_k^{(n)}(z; r, s) \\
&\quad = \left(\frac{1}{2}\right)^{k+1} \int (t + t^{-1}) t^{-1} (t - t^{-1})^k \left(\frac{r(t - t^{-1}) + 2s + (t + t^{-1})}{p(t - t^{-1}) + 2q + (t + t^{-1})}\right)^n \mathrm{d}t \\
&\quad = \left(\frac{1}{2}\right)^{k+1} \sum_{j=0}^{k} (-1)^j \binom{k}{j} T_{k-2j}^{(n)}(t; r, s),
\end{aligned}$$

$$T_i^{(n)}(t; r, s) = \int (t + t^{-1}) t^{i-1} \left(\frac{\breve{a}t^2 + \breve{b}t + \breve{c}}{at^2 + bt + c}\right)^n \mathrm{d}t, \tag{8.32}$$

where

$$\begin{aligned}
a &= 1 + p, & b &= 2q, & c &= 1 - p, & & \\
\breve{a} &= 1 + r, & \breve{b} &= 2s, & \breve{c} &= 1 - r, & & -k \leqslant i \leqslant k.
\end{aligned}$$

The integral (8.32) can be rewritten as follows:

$$T_i^{(n)}(t; r, s) = \int (t + t^{-1}) t^{i-1} \left(A + \frac{Bt + C}{at^2 + bt + c}\right)^n \mathrm{d}t,$$

where

$$A = \frac{\breve{a}}{a}, \quad B = \breve{b} - bA, \quad C = \breve{c} - cA.$$

For $n = 1, 2$ the integral (8.32) can be expressed in terms of the integrals (8.31) as follows:

$$\begin{aligned}
T_i^{(1)}(t; 0, 0) &= \tilde{\mathcal{J}}_{i+2}^{(1)}(t) + \tilde{\mathcal{J}}_{i-2}^{(1)}(t) + 2\tilde{\mathcal{J}}_i^{(1)}(t), \\
T_i^{(1)}(t; r, s) &= A\big(U_i(t) + U_{i-2}(t)\big) + B\big(\tilde{\mathcal{J}}_{i+1}^{(1)}(t) + \tilde{\mathcal{J}}_{i-1}^{(1)}(t)\big) \\
&\quad + C\big(\tilde{\mathcal{J}}_i^{(1)}(t) + \tilde{\mathcal{J}}_{i-2}^{(1)}(t)\big), \\
T_i^{(2)}(t; r, s) &= A^2\big(U_i(t) + U_{i-2}(t)\big) + 2AB\big(\tilde{\mathcal{J}}_{i+1}^{(1)}(t) + \tilde{\mathcal{J}}_{i-1}^{(1)}(t)\big) \\
&\quad + 2AC\big(\tilde{\mathcal{J}}_i^{(1)}(t) + \tilde{\mathcal{J}}_{i-2}^{(1)}(t)\big) + B^2\big(\tilde{\mathcal{J}}_{i+2}^{(2)}(t) + \tilde{\mathcal{J}}_i^{(2)}(t)\big) \\
&\quad + 2BC\big(\tilde{\mathcal{J}}_{i+1}^{(2)}(t) + \tilde{\mathcal{J}}_{i-1}^{(2)}(t)\big) + C^2\big(\tilde{\mathcal{J}}_i^{(2)}(t) + \tilde{\mathcal{J}}_{i-2}^{(2)}(t)\big),
\end{aligned}$$

where

$$U_i(t) = \int t^i\,\mathrm{d}t = \begin{cases} \ln(t), & \text{for } i = -1, \\ \frac{1}{i+1} t^{i+1}, & \text{for } i \neq -1. \end{cases}$$

The formulae for the integrals (8.28) are as follows:

$$\mathcal{M}_k(z) = \mathcal{I}_{k+2}(z) + \mathcal{I}_k(z),$$

where $\mathcal{I}_k(z)$ are the integrals (8.29).

9. Analytical integration of integrals over a triangle for scalar and vector valued basis functions

This section presents the analytical evaluation of the inner and moment integrals for a quadrilateral source element with scalar valued basis functions, and for a triangular source element with scalar or vector valued basis functions. For scalar valued basis functions the analytical evaluation of the inner integrals are already described in Section 5.2.4. For vector valued basis functions on a triangular element the integrals are decomposed into a sum of integrals, some of which are of the same type as those for the scalar valued basis functions. Therefore, in this section all the integrals over triangles will be treated.

Let $\mathbf{x}$ be the coordinates of the object point, O, the integration point of the outer integral. Let the source element with n edges be divided into triangles $\triangle_k$ for $k = 1, \ldots, n$, and let each $\triangle_k$ be the triangle with as top the point M and as base, $\mathbf{e}_k$, the kth edge of the element. For the inner integral the point M is the projection of point, O, in the plane of the element, for the moment integral the point M is the midpoint of the element.

The inner and moment integrals are, respectively:

$$I_j(\mathbf{x}) = \sum_k \int_{\triangle_k} \frac{\psi_j(\mathbf{x}')}{|\mathbf{x}' - \mathbf{x}|}\, \mathrm{d}\mathbf{x}', \tag{9.1}$$

$$M_{j,\alpha\beta} = \sum_k \int_{\triangle_k} \psi_j(\mathbf{x}')\{\mathbf{x}' - \mathbf{x}_M\}_\alpha \{\mathbf{x}' - \mathbf{x}_M\}_\beta\, \mathrm{d}\mathbf{x}'. \tag{9.2}$$

For scalar valued basis functions $\psi_j(\mathbf{x}') = 1$, and vector valued basis functions $\psi_j(\mathbf{x}') = (\mathbf{x}'_j - \mathbf{x}')/2J$, where $\mathbf{x}'_j$ is the jth vertex of the element, and J is the area of the element. The expression $\{\mathbf{x}' - \mathbf{x}_M\}_\alpha = 1$, $(x' - x_M)$, $(y' - y_M)$ or $(z' - z_M)$ for $\alpha = 0$, 1, 2 or 3.

Let $\mathbf{x}_{i,k}$ for $(i = 1, 2)$ be the vertices of $\mathbf{e}_k$, $\mathbf{n}$ be the normal to the plane of the element, $\mathbf{h}_k$ the perpendicular from M to $\mathbf{e}_k$, φ the angle between $\mathbf{x}' - \mathbf{x}_M$ and $\mathbf{h}_k$, and $\varphi_{i,k}$ the angle between $\mathbf{x}_{i,k} - \mathbf{x}_M$ and $\mathbf{h}_k$. Let P_k be the intersubsection of $\mathbf{x}' - \mathbf{x}_M$ and $\mathbf{e}_k$, and M_k the projection of M on $\mathbf{e}_k$. Let $\hat{\mathbf{a}}$ be the unit vector along vector $\mathbf{a}$. Then

$$\begin{aligned} &d = |OM|, \qquad h_k = |\mathbf{h}_k|,\\ &\mathbf{h}_k = MM_k = (\mathbf{x}_{1,k} - \mathbf{x}_M) - \big((\mathbf{x}_{1,k} - \mathbf{x}_M)\cdot \hat{\mathbf{e}}_k\big)\hat{\mathbf{e}}_k,\\ &\mathbf{x}' - \mathbf{x}_M = \big((\mathbf{x}' - \mathbf{x}_M)\cdot \hat{\mathbf{h}}_k\big)\hat{\mathbf{h}}_k + \big((\mathbf{x}' - \mathbf{x}_M)\cdot \hat{\mathbf{e}}_k\big)\hat{\mathbf{e}}_k. \end{aligned}$$

9.1. *Analytical formulae for the inner integrals*

After transformation to the polar coordinates $r = |\mathbf{x}' - \mathbf{x}_M|$ and φ, the inner integrals for scalar and vector valued basis functions become, respectively:

$$I(\mathbf{x}) = \sum_k I_k(\mathbf{x}), \tag{9.3}$$

$$\mathbf{I}_j(\mathbf{x}) = \frac{1}{2J}\left\{ I(\mathbf{x}) f x(\mathbf{x}'_j - \mathbf{x}_M) - \sum_k \big(\mathbf{I}_k^{(c)}(\mathbf{x}) + \mathbf{I}_k^{(s)}(\mathbf{x})\big)\right\}, \tag{9.4}$$

where

$$I_k(\mathbf{x}) = \mathcal{I}(\varphi_{2,k}, h_k) - \mathcal{I}(\varphi_{1,k}, h_k), \tag{9.5}$$

$$\mathbf{I}_k^{(c)}(\mathbf{x}) = \{\mathcal{I}_c(\varphi_{2,k}, h_k) - \mathcal{I}_c(\varphi_{1,k}, h_k)\}\hat{\mathbf{h}}_k, \tag{9.6}$$

$$\mathbf{I}_k^{(s)}(\mathbf{x}) = \{\mathcal{I}_s(\varphi_{2,k}, h_k) - \mathcal{I}_s(\varphi_{1,k}, h_k)\}\hat{\mathbf{e}}_k, \tag{9.7}$$

and $\varphi_{i,k} = \arctan\{\frac{1}{h_k}(\mathbf{x}_{i,k} - \mathbf{x}_M)\cdot\hat{\mathbf{e}}_k\}$.

Dropping the indices k the integral $I(\varphi, h)$ is defined by:

$$\begin{aligned}
\mathcal{I}(\varphi, h) &= \iint_0^{h/\cos\varphi} \frac{r}{\sqrt{r^2+d^2}}\,\mathrm{d}r\,\mathrm{d}\varphi \\
&= \int \big[\sqrt{r^2+d^2}\,\big]_0^{h/\cos\varphi}\,\mathrm{d}\varphi = \frac{h}{q}\int \frac{\sqrt{1+q^2x^2}}{1+x^2}\,\mathrm{d}x - \mathrm{d}\varphi \\
&= h\log\big(qx+\sqrt{1+q^2x^2}\big) + d\arctan\left(\frac{dq}{h\sqrt{1+q^2x^2}}\right) - \mathrm{d}\varphi \\
&= h\log\left(\frac{h\tan\varphi+s(\varphi)}{\sqrt{d^2+h^2}}\right) + d\arctan\left(\frac{d\tan\varphi}{s(\varphi)}\right) - \mathrm{d}\varphi \\
&= h\log\left(\frac{h\tan\varphi+s(\varphi)}{\sqrt{d^2+h^2}}\right) + d\arctan\left(\frac{(d-s(\varphi))\tan\varphi}{s(\varphi)+d\tan^2\varphi}\right),
\end{aligned} \tag{9.8}$$

where $x = \tan\varphi$, $q = \frac{h}{\sqrt{d^2+h^2}}$ and $s(\varphi) = |OP| = \sqrt{(\frac{h}{\cos\varphi})^2 + d^2} = \frac{h}{q}\sqrt{1+q^2x^2}$.

The integrals $I_c(\varphi, h)$ and $I_s(\varphi, h)$ are defined by:

$$\begin{aligned}
\mathcal{I}_c(\varphi, h) &= \iint_0^{h/\cos\varphi} \frac{r^2\cos\varphi}{\sqrt{r^2+d^2}}\,\mathrm{d}r\,\mathrm{d}\varphi \\
&= \frac{1}{2}\int \cos\varphi\big[r\sqrt{r^2+d^2} - d^2\log\big(\sqrt{r^2+d^2}+r\big)\big]_0^{h/\cos\varphi}\,\mathrm{d}\varphi \\
&= -\frac{1}{2}d^2\sin\varphi\log\left(\frac{s(\varphi)+h/\cos\varphi}{d}\right) + \frac{h^2}{2q}\int \frac{1}{\sqrt{1+q^2x^2}}\,\mathrm{d}x \\
&= -\frac{1}{2}d^2\sin\varphi\log\left(\frac{s(\varphi)+h/\cos\varphi}{d}\right) + \frac{1}{2}(h^2+d^2)\log\left(\frac{h\tan\varphi+s(\varphi)}{\sqrt{h^2+d^2}}\right),
\end{aligned} \tag{9.9}$$

$$\begin{aligned}
\mathcal{I}_s(\varphi, h) &= \iint_0^{h/\cos\varphi} \frac{r^2\sin\varphi}{\sqrt{r^2+d^2}}\,\mathrm{d}r\,\mathrm{d}\varphi \\
&= \frac{1}{2}\int \sin\varphi\big[r\sqrt{r^2+d^2} - d^2\log\big(\sqrt{r^2+d^2}+r\big)\big]_0^{h/\cos\varphi}\,\mathrm{d}\varphi \\
&= \frac{1}{2}d^2\cos\varphi\log\left(\frac{s(\varphi)+h/\cos\varphi}{d}\right) + \frac{1}{2}h^2q\int \frac{x}{\sqrt{1+q^2x^2}}\,\mathrm{d}x \\
&= \frac{1}{2}d^2\cos\varphi\log\left(\frac{s(\varphi)+h/\cos\varphi}{d}\right) + \frac{1}{2}hs(\varphi),
\end{aligned} \tag{9.10}$$

where $x = \tan\varphi$, $q = \frac{h}{\sqrt{d^2+h^2}}$ and $s(\varphi) = |OP| = \sqrt{(h/\cos\varphi)^2 + d^2} = \frac{h}{q}\sqrt{1+q^2x^2}$.

9.2. Analytical formulae for the moment integrals

After transformation to the polar coordinates $r = |\mathbf{x}' - \mathbf{x}_M|$ and φ, the moment integrals for scalar and vector valued basis functions become, respectively:

$$M_{\alpha\beta} = \sum_k M_{k,\alpha\beta}, \tag{9.11}$$

$$\mathbf{M}_{j,\alpha\beta} = \frac{1}{2J}\left\{ M_{\alpha\beta}\,(\mathbf{x}'_j - \mathbf{x}_M) - \sum_k \left(\mathbf{M}^{(c)}_{k,\alpha\beta} + \mathbf{M}^{(s)}_{k,\alpha\beta}\right)\right\}, \tag{9.12}$$

where

$$M_{k,\alpha\beta} = \mathcal{M}_{\alpha\beta}(\varphi_{2,k}, h_k) - \mathcal{M}_{\alpha\beta}(\varphi_{1,k}, h_k), \tag{9.13}$$

$$\mathbf{M}^{(c)}_{k,\alpha\beta} = \left\{\mathcal{M}^{(c)}_{\alpha\beta}(\varphi_{2,k}, h_k) - \mathcal{M}^{(c)}_{\alpha\beta}(\varphi_{1,k}, h_k)\right\}\hat{\mathbf{h}}_k, \tag{9.14}$$

$$\mathbf{M}^{(s)}_{k,\alpha\beta} = \left\{\mathcal{M}^{(s)}_{\alpha\beta}(\varphi_{2,k}, h_k) - \mathcal{M}^{(s)}_{\alpha\beta}(\varphi_{1,k}, h_k)\right\}\hat{\mathbf{e}}_k, \tag{9.15}$$

and $\varphi_{i,k} = \arctan\{\frac{1}{h_k}(\mathbf{x}_{i,k} - \mathbf{x}_M)\cdot\hat{\mathbf{e}}_k\}$.

Dropping the indices k the integrals $\mathcal{M}_{\alpha\beta}(\varphi, h)$, $\mathcal{M}^{(c)}_{\alpha\beta}(\varphi, h)$ and $\mathcal{M}^{(s)}_{\alpha\beta}(\varphi, h)$ are defined by:

$$\mathcal{M}_{00}(\varphi, h) = \iint_0^{h/\cos\varphi} r\,\mathrm{d}r\,\mathrm{d}\varphi = \tfrac{1}{2}h^2\int\frac{1}{\cos^2\varphi}\,\mathrm{d}\varphi = \tfrac{1}{2}h^2\tan\varphi, \tag{9.16}$$

$$\begin{aligned}\mathcal{M}_{0\alpha}(\varphi, h) &= \iint_0^{h/\cos\varphi}(\hat{h}_\alpha\cos\varphi + \hat{e}_\alpha\sin\varphi)r^2\,\mathrm{d}r\,\mathrm{d}\varphi\\ &= \tfrac{1}{3}h^3\int\frac{1}{\cos^3\varphi}(\hat{h}_\alpha\cos\varphi + \hat{e}_\alpha\sin\varphi)\,\mathrm{d}\varphi\\ &= \tfrac{1}{3}h^3\left(\hat{h}_\alpha\tan\varphi + \tfrac{1}{2}\hat{e}_\alpha\tan^2\varphi\right),\end{aligned} \tag{9.17}$$

$$\begin{aligned}\mathcal{M}_{\alpha\beta}(\varphi, h) &= \iint_0^{h/\cos\varphi}(\hat{h}_\alpha\cos\varphi + \hat{e}_\alpha\sin\varphi)(\hat{h}_\beta\cos\varphi + \hat{e}_\beta\sin\varphi)r^3\,\mathrm{d}r\,\mathrm{d}\varphi\\ &= \tfrac{1}{4}h^4\int\frac{1}{\cos^4\varphi}(\hat{h}_\alpha\cos\varphi + \hat{e}_\alpha\sin\varphi)(\hat{h}_\beta\cos\varphi + \hat{e}_\beta\sin\varphi)\,\mathrm{d}\varphi\\ &= \tfrac{1}{4}h^4\left\{\hat{h}_\beta\left(\hat{h}_\alpha\tan\varphi + \tfrac{1}{2}\hat{e}_\alpha\tan^2\varphi\right) + \hat{e}_\beta\left(\tfrac{1}{2}\hat{h}_\alpha\tan^2\varphi + \tfrac{1}{3}\hat{e}_\alpha\tan^3\varphi\right)\right\},\end{aligned} \tag{9.18}$$

$$\mathcal{M}^{(c)}_{00}(\varphi, h) = \iint_0^{h/\cos\varphi}\cos\varphi r^2\,\mathrm{d}r\,\mathrm{d}\varphi = \tfrac{1}{3}h^3\int\frac{1}{\cos^2\varphi}\,\mathrm{d}\varphi = \tfrac{1}{3}h^3\tan\varphi, \tag{9.19}$$

$$\begin{aligned}\mathcal{M}^{(c)}_{0\alpha}(\varphi, h) &= \iint_0^{h/\cos\varphi}\cos\varphi(\hat{h}_\alpha\cos\varphi + \hat{e}_\alpha\sin\varphi)r^3\,\mathrm{d}r\,\mathrm{d}\varphi\\ &= \tfrac{1}{4}h^4\int\frac{1}{\cos^3\varphi}(\hat{h}_\alpha\cos\varphi + \hat{e}_\alpha\sin\varphi)\,\mathrm{d}\varphi\\ &= \tfrac{1}{4}h^4\left(\hat{h}_\alpha\tan\varphi + \tfrac{1}{2}\hat{e}_\alpha\tan^2\varphi\right),\end{aligned} \tag{9.20}$$

$$\begin{aligned}\mathcal{M}^{(c)}_{\alpha\beta}(\varphi,h) &= \iint_0^{h/\cos\varphi} \cos\varphi\big(\hat{h}_\alpha\cos\varphi+\hat{e}_\alpha\sin\varphi\big)\big(\hat{h}_\beta\cos\varphi+\hat{e}_\beta\sin\varphi\big)r^4\,\mathrm{d}r\,\mathrm{d}\varphi\\ &= \tfrac{1}{5}h^5\int\frac{1}{\cos^4\varphi}\big(\hat{h}_\alpha\cos\varphi+\hat{e}_\alpha\sin\varphi\big)\big(\hat{h}_\beta\cos\varphi+\hat{e}_\beta\sin\varphi\big)\,\mathrm{d}\varphi\\ &= \tfrac{1}{5}h^5\big\{\hat{h}_\beta\big(\hat{h}_\alpha\tan\varphi+\tfrac{1}{2}\hat{e}_\alpha\tan^2\varphi\big)+\hat{e}_\beta\big(\tfrac{1}{2}\hat{h}_\alpha\tan^2\varphi+\tfrac{1}{3}\hat{e}_\alpha\tan^3\varphi\big)\big\},\end{aligned} \tag{9.21}$$

$$\mathcal{M}^{(s)}_{00}(\varphi,h) = \iint_0^{h/\cos\varphi}\sin\varphi r^2\,\mathrm{d}r\,\mathrm{d}\varphi = \tfrac{1}{3}h^3\int\frac{\sin\varphi}{\cos^3\varphi}\,\mathrm{d}\varphi = \tfrac{1}{6}h^3\tan^2\varphi, \tag{9.22}$$

$$\begin{aligned}\mathcal{M}^{(s)}_{0\alpha}(\varphi,h) &= \iint_0^{h/\cos\varphi}\sin\varphi\big(\hat{h}_\alpha\cos\varphi+\hat{e}_\alpha\sin\varphi\big)r^3\,\mathrm{d}r\,\mathrm{d}\varphi\\ &= \tfrac{1}{4}h^4\int\frac{\sin\varphi}{\cos^4\varphi}\big(\hat{h}_\alpha\cos\varphi+\hat{e}_\alpha\sin\varphi\big)\,\mathrm{d}\varphi\\ &= \tfrac{1}{4}h^4\big(\tfrac{1}{2}\hat{h}_\alpha\tan^2\varphi+\tfrac{1}{3}\hat{e}_\alpha\tan^3\varphi\big),\end{aligned} \tag{9.23}$$

$$\begin{aligned}\mathcal{M}^{(s)}_{\alpha\beta}(\varphi,h) &= \iint_0^{h/\cos\varphi}\sin\varphi\big(\hat{h}_\alpha\cos\varphi+\hat{e}_\alpha\sin\varphi\big)\big(\hat{h}_\beta\cos\varphi+\hat{e}_\beta\sin\varphi\big)r^4\,\mathrm{d}r\,\mathrm{d}\varphi\\ &= \tfrac{1}{5}h^5\int\frac{\sin\varphi}{\cos^5\varphi}\big(\hat{h}_\alpha\cos\varphi+\hat{e}_\alpha\sin\varphi\big)\big(\hat{h}_\beta\cos\varphi+\hat{e}_\beta\sin\varphi\big)\,\mathrm{d}\varphi\\ &= \tfrac{1}{5}h^5\big\{\hat{h}_\beta\big(\tfrac{1}{2}\hat{h}_\alpha\tan^2\varphi+\tfrac{1}{3}\hat{e}_\alpha\tan^3\varphi\big)+\hat{e}_\beta\big(\tfrac{1}{3}\hat{h}_\alpha\tan^3\varphi+\tfrac{1}{4}\hat{e}_\alpha\tan^4\varphi\big)\big\},\end{aligned} \tag{9.24}$$

where $\hat{h}_\alpha = \hat{\mathbf{h}}_x$, $\hat{\mathbf{h}}_y$ or $\hat{\mathbf{h}}_z$, and $\hat{e}_\alpha = \hat{\mathbf{e}}_x$, $\hat{\mathbf{e}}_y$ or $\hat{\mathbf{e}}_z$, for $\alpha = 1$, 2 or 3.

10. Solution of Kirchhoff's equations

10.1. Kirchhoff's equations

This section presents the solution methods for solving the *Kirchhoff's equations* describing the behaviour of a circuit which forms the electronic equivalent of an interconnection system consisting of a number of planar conductors immersed in a stratified medium. A derivation of Kirchhoff's equations from *Maxwell's equations* can be found in DU CLOUX, MAAS and WACHTERS [1994], and in Chapter 1 of the present volume. In these references, a weak formulation and discretisation of a mixed potential, boundary value problem is presented. Care is taken that, in the quasi-static approximation, the discretised equations admit an electronic circuit interpretation. The conductor surfaces are subdivided into a number of sufficiently small elements. The topology of the surfaces is described by the set of elements, the index set of which is denoted by $\mathcal{N}$, and a set of edges between adjacent elements, the index set of which is denoted by $\mathcal{E}$. The electric surface current, surface charge and scalar potential, defined on the conductor surfaces, are expanded in a number of basis functions, defined on the elements.

The Kirchhoff equations are:

$$(\mathbf{R}+s\mathbf{L})I-\mathbf{P}V=0, \tag{10.1}$$

$$\mathbf{P}^{\mathrm{T}}I+sQ=J, \tag{10.2}$$

$$\mathbf{D}Q=V, \tag{10.3}$$

where I collects the edge currents, Q the element charges, V the element potentials and J the external currents flowing into the interconnection system. Further, $\mathbf{R}$ denotes the resistance matrix, $\mathbf{L}$ the inductance matrix and $\mathbf{D}$ the elastance matrix. The matrix $\mathbf{P}$ denotes the incidence matrix. It consists of entries 0 and $\pm$ 1, and represents the topology. Finally, s denotes the complex frequency. Its imaginary part is $-\omega$. It is assumed that $|\omega| \leqslant \Omega$, where Ω is the maximum frequency for which the generated equivalent circuit should be valid. The matrices $\mathbf{R}$, $\mathbf{L}$, $\mathbf{D}$ and $\mathbf{P}$ are independent of s.

Elimination of the charges from (10.1)–(10.3) gives

$$(\mathbf{R}+s\mathbf{L})I-\mathbf{P}V=0, \tag{10.4}$$

$$\mathbf{P}^{\mathrm{T}}I+s\mathbf{C}V=J. \tag{10.5}$$

The charges are obtained from the potentials according to

$$\mathbf{Q}=\mathbf{C}V, \tag{10.6}$$

where $\mathbf{C}=\mathbf{D}^{-1}$ denotes the capacitance matrix.

The set of circuit nodes is defined to be a non-empty subset of the set of elements. Let N denote the index set of the set of circuit nodes, and N' the index set of its complement in $\mathcal{N}$. Introduction of this partitioning of the set of elements leads to the following equations

$$(\mathbf{R}+s\mathbf{L})\mathbf{I}-\mathbf{P}_{N'}\mathbf{V}_{N'}=\mathbf{P}_N\mathbf{V}_N, \tag{10.7}$$

$$-\mathbf{P}_{N'}^{\mathrm{T}}\mathbf{I}-s\mathbf{C}_{N'N'}\mathbf{V}_{N'}=s\mathbf{C}_{N'N}\mathbf{V}_N, \tag{10.8}$$

and

$$\mathbf{J}_N=\mathbf{P}_N^{\mathrm{T}}\mathbf{I}+s\mathbf{C}_{NN'}\mathbf{V}_{N'}+s\mathbf{C}_{NN}\mathbf{V}_N, \tag{10.9}$$

where $\mathbf{V}_N$ is the collection of prescribed vectors of circuit node voltages.

Let $\mathbb{R}$ and $\mathbb{C}$ be the sets of real and complex numbers, respectively, and $|.|$ denote the length of a set. From the discretisation it follows that the matrices

$$\begin{array}{llll}
\mathbf{R}\in\mathbb{R}^{|\mathcal{E}|\times|\mathcal{E}|}, & \mathbf{L}\in\mathbb{R}^{|\mathcal{E}|\times|\mathcal{E}|}, & \mathbf{P}_{N'}\in\mathbb{R}^{\mathcal{E}|\times|N'|}, & \mathbf{P}_N\in\mathbb{R}^{\mathcal{E}|\times|N|}, \\
\mathbf{C}_{N'N'}\in\mathbb{R}^{|N'|\times|N'|}, & \mathbf{C}_{N'N}\in\mathbb{R}^{|N'|\times|N|}, & \mathbf{C}_{NN'}\in\mathbb{R}^{|N|\times|N'|}, & \mathbf{C}_{NN}\in\mathbb{R}^{|N|\times|N|}, \\
\mathbf{I}\in\mathbb{C}^{|\mathcal{E}|\times|N|}, & \mathbf{V}_{N'}\in\mathbb{C}^{|N'|\times|N|}, & \mathbf{V}_N\in\mathbb{R}^{|N|\times|N|}, & \mathbf{J}_N\in\mathbb{C}^{|N|\times|N|}.
\end{array}$$

The matrices $\mathbf{R}$, $\mathbf{L}$ and $\mathbf{C}$ are symmetric and positive definite. The matrix $\mathbf{P}_{N'}$ has full column rank. From Eqs. (10.7)–(10.9) it follows that $\mathbf{J}_N$ is linearly related to $\mathbf{V}_N$, i.e.,

$$\mathbf{J}_N=\mathbf{Y}\,\mathbf{V}_N, \tag{10.10}$$

where $\mathbf{Y}$ is the admittance matrix of the interconnection system when observed from its circuit nodes.

10.2. Construction of the admittance matrix

Elimination of **I** from (10.7) and (10.8) gives

$$-\big(\mathbf{P}_{N'}^{\mathrm{T}}(\mathbf{R}+s\mathbf{L})^{-1}\mathbf{P}_{N'}+s\mathbf{C}_{N'N'}\big)\mathbf{V}_{N'}=\big(\mathbf{P}_{N'}^{\mathrm{T}}(\mathbf{R}+s\mathbf{L})^{-1}\mathbf{P}_{N}+s\mathbf{C}_{N'N}\big)\mathbf{V}_{N}, \tag{10.11}$$

$$\mathbf{I}=(\mathbf{R}+s\mathbf{L})^{-1}(\mathbf{P}_{N'}\mathbf{V}_{N'}+\mathbf{P}_{N}\mathbf{V}_{N}). \tag{10.12}$$

Let h be the mesh size, and $k_0=\omega\sqrt{\varepsilon_0\mu_0}$ the free-space wavenumber, where ε_0 and μ_0 denote the free-space permittivity and permeability, respectively. When $k_0h\ll 1$, it follows from the expressions for the matrix elements of **C**, **L** and **R** (see DU CLOUX, MAAS and WACHTERS [1994]), that the orders of the matrix elements are

$$\mathrm{i}\omega\mathbf{C}_{ij}=Z_0^{-1}\mathcal{O}(\mathrm{i}k_0h),\qquad \mathrm{i}\omega\mathbf{L}_{kl}=Z_0\mathcal{O}(\mathrm{i}k_0h),\qquad \mathbf{R}_{kl}=Z_s\mathcal{O}(1),$$

where $i,j\in\mathcal{N}$ and $k,l\in\mathcal{E}$, $Z_0=\sqrt{\mu_0/\varepsilon_0}$, and Z_s denotes the surface impedance of the conductors.

Therefore, the ratio between a matrix element of the second term and a corresponding matrix element of the first term in the left-hand side of (10.11) is $\mathcal{O}((\mathrm{i}k_0h)^2)$, if $Z=0$, and $\mathcal{O}(\mathrm{i}k_0h)$, if $Z\neq 0$. Returning to (10.11), $\mathbf{V}_{N'}$ may be expanded in powers of ik_0h, which is then substituted into (10.12) to obtain **I**. Neglecting higher order term in $\mathrm{i}k_0h$, it follows that

$$\mathbf{V}_{N'}=\mathbf{V}_0+\mathbf{V}_1,\qquad \mathbf{I}=\mathbf{I}_0+\mathbf{I}_1, \tag{10.13}$$

where $(\mathbf{V}_0,\mathbf{I}_0)$ and $(\mathbf{V}_1,\mathbf{V}_1)$ may be obtained from two sets of equations,

$$(\mathbf{R}+s\mathbf{L})\mathbf{I}_0-\mathbf{P}_{N'}\mathbf{V}_0=\mathbf{P}_N\mathbf{V}_N, \tag{10.14}$$

$$-\mathbf{P}_{N'}^{\mathrm{T}}\mathbf{I}_0=0, \tag{10.15}$$

$$(\mathbf{R}+s\mathbf{L})\mathbf{I}_1-\mathbf{P}_{N'}\mathbf{V}_1=0, \tag{10.16}$$

$$-\mathbf{P}_{N'}^{\mathrm{T}}\mathbf{I}_1=s(\mathbf{C}_{N'N'}\mathbf{V}_0+\mathbf{C}_{N'N}\mathbf{V}_N). \tag{10.17}$$

Let $\mathbf{V}_N$ be a unit matrix, then it follows from (10.10) that the admittance matrix $\mathbf{Y}=\mathbf{J}_N$. Substitution of (10.13) into (10.9) leads to

$$\mathbf{Y}=\mathbf{P}_N^{\mathrm{T}}(\mathbf{I}_0+\mathbf{I}_1)+s\mathbf{C}_{NN'}\mathbf{V}_0+s\mathbf{C}_{NN}+\mathcal{O}\big((\mathrm{i}k_0h)^2\big). \tag{10.18}$$

From (10.18) it follows that treating capacitive effects as a perturbation is consistent with the quasi-static modelling of the interconnection system (see DU CLOUX, MAAS and WACHTERS [1994]).

Depending on the frequency range of interest, four different methods can be distinguished to obtain a solution for these sets of equations. If one is only interested in the solution for a high (low) frequency range, it can be obtained by an expansion of $(\mathbf{V}_0,\mathbf{I}_0)$ and $(\mathbf{V}_1,\mathbf{I}_1)$ in s for relatively high (low) values of $|s|$.

However, if one is interested in the solution for the full frequency range there are two options. The first option is to solve Eqs. (10.14)–(10.17) for an appropriately chosen set of s values. The second option is to combine the solutions obtained for the high and low frequency ranges.

10.2.1. Admittance matrix for the high frequency range

In the high frequency range the **R** term in Eqs. (10.14) and (10.16) is considered as a perturbation of the $s\mathbf{L}$ term. Introducing the following expansions of $(\mathbf{V}_0, \mathbf{I}_0)$ and $(\mathbf{V}_1, \mathbf{V}_1)$

$$\mathbf{V}_0 = \mathbf{V}_{0,0} + s^{-1}\mathbf{V}_{0,1}, \qquad \mathbf{I}_0 = s^{-1}\mathbf{I}_{0,0} + s^{-2}\mathbf{I}_{0,1},$$
$$\mathbf{V}_1 = s^2\mathbf{V}_{1,0} + s\mathbf{V}_{1,1}, \qquad \mathbf{I}_1 = s\mathbf{I}_{1,0} + \mathbf{I}_{1,1},$$

and collecting the coefficients of s^0, s^{-1}, s^2 and s, respectively, the pairs $(\mathbf{V}_{i,j}, \mathbf{I}_{i,j})$ for $(i, j = 0, 1)$ may be obtained from the following four sets of equations

$$\mathbf{L}\mathbf{I}_{0,0} - \mathbf{P}_{N'}\mathbf{V}_{0,0} = \mathbf{P}_N\mathbf{V}_N, \tag{10.19}$$
$$-\mathbf{P}_{N'}^{\mathrm{T}}\mathbf{I}_{0,0} = 0, \tag{10.20}$$
$$\mathbf{L}\mathbf{I}_{0,1} - \mathbf{P}_{N'}\mathbf{V}_{0,1} = -\mathbf{R}\mathbf{I}_{0,0}, \tag{10.21}$$
$$-\mathbf{P}_{N'}^{\mathrm{T}}\mathbf{I}_{0,1} = 0, \tag{10.22}$$
$$\mathbf{L}\mathbf{I}_{1,0} - \mathbf{P}_{N'}\mathbf{V}_{1,0} = 0, \tag{10.23}$$
$$-\mathbf{P}_{N'}^{\mathrm{T}}\mathbf{I}_{1,0} = \mathbf{C}_{N'N'}\mathbf{V}_{0,0} + \mathbf{C}_{N'N}\mathbf{V}_N, \tag{10.24}$$
$$\mathbf{L}\mathbf{I}_{1,1} - \mathbf{P}_{N'}\mathbf{V}_{1,1} = -\mathbf{R}\mathbf{I}_{1,0}, \tag{10.25}$$
$$-\mathbf{P}_{N'}^{\mathrm{T}}\mathbf{I}_{1,1} = \mathbf{C}_{N'N'}\mathbf{V}_{0,1}. \tag{10.26}$$

The expansions of $(\mathbf{V}_0, \mathbf{I}_0)$ and $(\mathbf{V}_1, \mathbf{V}_1)$ are introduced to extend the validity of the high frequency range to lower frequencies. After substitution into the expression (10.18) and collection of the coefficients of powers of s one obtains

$$\mathbf{Y} = s^{-2}\mathbf{Y}_R + s^{-1}\mathbf{Y}_L + \mathbf{Y}_G + s\mathbf{Y}_C + \cdots, \tag{10.27}$$

where

$$\begin{aligned}\mathbf{Y}_L &= \mathbf{P}_N^{\mathrm{T}}\mathbf{I}_{0,0}, \qquad \mathbf{Y}_C = \mathbf{C}_{NN'}\mathbf{V}_{0,0} + \mathbf{C}_{NN} + \mathbf{P}_N^{\mathrm{T}}\mathbf{I}_{1,0},\\ \mathbf{Y}_R &= \mathbf{P}_N^{\mathrm{T}}\mathbf{I}_{0,1}, \qquad \mathbf{Y}_G = \mathbf{C}_{NN'}\mathbf{V}_{0,1} + \mathbf{P}_N^{\mathrm{T}}\mathbf{I}_{1,1}.\end{aligned} \tag{10.28}$$

An equivalent circuit that represents the admittance matrix consists of branches between every pair of circuit nodes. For a circuit with frequency independent components each branch can be approximated by a series resistor R and inductor L, in parallel with a capacitor C and a resistor of conductance G, so that for the branch between the circuit nodes i and j

$$R = -\mathbf{y}_{R,ij}\mathbf{y}_{L,ij}^{-2}, \quad L = \mathbf{y}_{L,ij}^{-1}, \quad C = \mathbf{y}_{C,ij}, \quad G = \mathbf{y}_{G,ij}, \tag{10.29}$$

where the branch admittance matrix element, $\mathbf{y}_{ij}$, is related to the admittance matrix elements $\mathbf{Y}_{ij}$ through $(i, j \in N)$

$$\mathbf{y}_{ij} = -\mathbf{Y}_{ij} \quad (i \neq j), \tag{10.30}$$
$$\mathbf{y}_{ii} = \sum_{j \in N} \mathbf{Y}_{ij}. \tag{10.31}$$

The diagonal element, $\mathbf{y}_{ii}$, represents the branch between the circuit node i and the ground plane or some reference at infinity. From Eqs. (10.19)–(10.22) it follows that $\mathbf{y}_{R,ii} = \mathbf{y}_{L,ii} = 0$.

If frequency dependent resistors are allowed each branch consists of a resistor $R = s^2\mathbf{y}_R^{-1}$ in parallel with an inductor L, a capacitor C and a resistor of conductance G given in (10.29). For passive IC's it can be shown that this is a good approximation for the frequency range of interest.

10.2.2. Admittance matrix for the low frequency range

In the low frequency range the $s\mathbf{L}$ term in Eqs. (10.14) and (10.16) is considered as a perturbation of the term $\mathbf{R}$. Introducing the following expansions of $(\mathbf{V}_0, \mathbf{I}_0)$ and $(\mathbf{V}_1, \mathbf{V}_1)$

$$\mathbf{V}_0 = \mathbf{V}_{0,0} + s\mathbf{V}_{0,1}, \qquad \mathbf{I}_0 = \mathbf{I}_{0,0} + s\mathbf{I}_{0,1},$$
$$\mathbf{V}_1 = s\mathbf{V}_{1,0}, \qquad \mathbf{I}_1 = s\mathbf{I}_{1,0},$$

and collecting the coefficients of powers of s, the pairs $(\mathbf{V}_{i,j}, \mathbf{I}_{i,j})$ for $(i, j = 0, 1)$ may be obtained from the following three sets of equations

$$\mathbf{R}\mathbf{I}_{0,0} - \mathbf{P}_{N'}\mathbf{V}_{0,0} = \mathbf{P}_N\mathbf{V}_N, \tag{10.32}$$
$$-\mathbf{P}_{N'}^{\mathrm{T}}\mathbf{I}_{0,0} = 0, \tag{10.33}$$
$$\mathbf{R}\mathbf{I}_{0,1} - \mathbf{P}_{N'}\mathbf{V}_{0,1} = -\mathbf{L}\mathbf{I}_{0,0}, \tag{10.34}$$
$$-\mathbf{P}_{N'}^{\mathrm{T}}\mathbf{I}_{0,1} = 0, \tag{10.35}$$
$$\mathbf{R}\mathbf{I}_{1,0} - \mathbf{P}_{N'}\mathbf{V}_{1,0} = 0, \tag{10.36}$$
$$-\mathbf{P}_{N'}^{\mathrm{T}}\mathbf{I}_{1,0} = \mathbf{C}_{N'N'}\mathbf{V}_{0,0} + \mathbf{C}_{N'N}\mathbf{V}_N, \tag{10.37}$$

The expansions of $(\mathbf{V}_0, \mathbf{I}_0)$ and $(\mathbf{V}_1, \mathbf{V}_1)$ are introduced to extend the validity of the low frequency range to higher frequencies. After substitution of them into the expression (10.18) and collection of the coefficients of powers of s one obtains

$$\mathbf{Y} = \mathbf{Y}_R + s\mathbf{Y}_C + \cdots, \tag{10.38}$$

where

$$\mathbf{Y}_R = \mathbf{P}_N^{\mathrm{T}}\mathbf{I}_{0,0}, \qquad \mathbf{Y}_C = \mathbf{C}_{NN'}\mathbf{V}_{0,0} + \mathbf{C}_{NN} + \mathbf{P}_N^{\mathrm{T}}(\mathbf{I}_{1,0} + \mathbf{I}_{1,0}).$$

An equivalent circuit that represents the admittance matrix consists of branches between every pair of circuit nodes. Each branch consists of a resistor R, in parallel with a capacitor C, so that for the branch between the circuit nodes i and j

$$R = \mathbf{y}_{R,ij}^{-1}, \qquad C = \mathbf{y}_{C,ij},$$

where the branch admittance matrix element $\mathbf{y}_{ij}$ is defined by the expressions (10.30) and (10.31) of Section 10.2.1. From Eqs. (10.32) and (10.33) it follows that $\mathbf{y}_{R,ii} = 0$.

10.2.3. Approximate admittance matrix for the full frequency range

Returning to (10.14) and (10.15), an expression for $\mathbf{I}_0$ can be obtained by introducing the null space of $\mathbf{P}_{N'}^{\mathrm{T}}$. Let $\mathcal{C} \in \mathbb{R}^{|\mathcal{E}|\times(|\mathcal{E}|-|N'|)}$ and

$$\mathbf{P}_{N'}^{\mathrm{T}}\mathcal{C} = 0, \qquad \mathcal{C}^{\mathrm{T}}\mathbf{P}_{N'} = 0, \tag{10.39}$$

then it follows from (10.14) that

$$\mathbf{I}_0 = \mathcal{C}\big(\mathcal{C}^{\mathrm{T}}(\mathbf{R} + s\mathbf{L})\mathcal{C}\big)^{-1}\mathcal{C}^{\mathrm{T}}\mathbf{P}_N\mathbf{V}_N. \tag{10.40}$$

Let $\mathbf{A} = \mathcal{C}^{\mathrm{T}}\mathbf{R}\mathcal{C}$, $\mathbf{B} = \mathcal{C}^{\mathrm{T}}\mathbf{L}\mathcal{C}$ and $\mathbf{A}, \mathbf{B} \in \mathbb{R}^{n\times n}$, where $n = |\mathcal{E}| - |N'|$. Consider the generalized eigenvalue problem $\mathbf{A}\mathbf{x} = \lambda\mathbf{B}\mathbf{x}$. Since $\mathbf{R}$ and $\mathbf{L}$ are symmetric and positive definite, and $\mathcal{C}$ has full column rank, $\mathbf{A}$ and $\mathbf{B}$ are symmetric and positive definite. Pencils $\mathbf{A} - \lambda\mathbf{B}$ of this variety are referred to as symmetric-definite pencils. For such pencils there exists a nonsingular matrix $\mathbf{X} = [\mathbf{x}_1, \ldots, \mathbf{x}_n]$ such that

$$\mathbf{X}^{\mathrm{T}}\mathbf{A}\mathbf{X} = \mathrm{diag}(a_1, \ldots, a_n), \qquad \mathbf{X}^{\mathrm{T}}\mathbf{B}\mathbf{X} = \mathrm{diag}(b_1, \ldots, b_n). \tag{10.41}$$

Moreover, $\mathbf{A}\mathbf{x}_i = \lambda_i\mathbf{x}_i$, for $i = 1, \ldots, n$, where $\lambda_i = \frac{a_i}{b_i} > 0$ (see GOLUB and VAN LOAN [1986], Section 8.6).

From this it follows that

$$(\mathbf{A} + s\mathbf{B})^{-1} = \sum_{i=1}^{n} \frac{\mathbf{x}_i\mathbf{x}_i^{\mathrm{T}}}{a_i + sb_i}. \tag{10.42}$$

Let $\mathbf{Y}_{RL} = \mathbf{P}_N^{\mathrm{T}}\mathbf{I}_0$ be the contribution of $\mathbf{I}_0$ to the admittance matrix $\mathbf{Y}$ of (10.18) then after substitution of (10.42) into (10.40) it follows that

$$\mathbf{Y}_{RL} = \sum_{i=1}^{n} \frac{\mathbf{H}_i}{\lambda_i + s}, \tag{10.43}$$

where

$$\mathbf{H}_i = b_i^{-1}\mathbf{P}_N^{\mathrm{T}}\mathcal{C}\mathbf{x}_i\mathbf{x}_i^{\mathrm{T}}\mathcal{C}^{\mathrm{T}}\mathbf{P}_N. \tag{10.44}$$

Let for the contributions of $\mathbf{V}_0$ and $\mathbf{I}_1$ to $\mathbf{Y}$ of (10.18) the high frequency approximation of (10.27) be taken, then

$$\mathbf{Y} = \mathbf{Y}_{RL} + \mathbf{Y}_G + s\mathbf{Y}_C + \cdots. \tag{10.45}$$

The numerical computation of all eigenvalues and eigenvectors of the generalized eigenvalue problem becomes prohibitively expensive as soon as n becomes larger than a few hundred. Therefore, the only practical way to obtain an expression for the admittance matrix $\mathbf{Y}$ is through approximation. In view of the expression (10.45), it is natural to look for an approximation of $\mathbf{Y}_{RL}$ with a number of terms, $m \ll n$.

In a computer program this is accomplished by calculating m, low and high, eigenvalues, λ_i, of the generalized eigenvalue problem, and some admittance matrices, Y_k, for an appropriately chosen set of $m + 2$, negative, real values of s. The set of these match frequencies, s_k, consists of some large negative values between $-\Omega$ and

$-\max(\lambda_1, \ldots, \lambda_m)$, and some small negative values between $-\min(\lambda_1, \ldots, \lambda_m)$ and 0. They are chosen to be real, so that the components of the equivalent circuit will be real.

There are two options to obtain the Y_k's. The first option, the sampling method, is to solve Eqs. (10.14)–(10.17) for each s_k. The second option, the perturbation method, is to calculate the Y_k's for the large negative s_k values by the high frequency approximation (10.27), and those for the small negative s_k values by the low frequency approximation (10.38).

An element of the branch admittance matrix, defined by the (10.30) and (10.31) of Section 10.2.1, is approximated by

$$\mathbf{y}_{ij}(s) = \mathbf{y}_{G,ij} + s\,\mathbf{y}_{C,ij} + \sum_{l=1}^{m} \frac{\mathbf{H}_{l,ij}}{\lambda_l + s}, \tag{10.46}$$

where the coefficients $\mathbf{y}_{G,ij}$, $\mathbf{y}_{C,ij}$ and $\mathbf{H}_{l,ij}$ are obtained by solving the following set of $m+2$ equations

$$\mathbf{y}_{G,ij} + s\mathbf{y}_{C,ij} + \sum_{l=1}^{m} \frac{\mathbf{H}_{l,ij}}{\lambda_l + s_k} = \mathbf{y}_{k,ij}, \quad \text{for } k = 1, \ldots, m+2. \tag{10.47}$$

An equivalent circuit which represents the admittance matrix consists of branches between every pair of circuit nodes. Each branch consists of m parallel connections of a series resistor R and inductor L, in parallel with a capacitor C, and a resistor of conductance G, so that for the branch between the circuit nodes i and j

$$R_l = \lambda_l \mathbf{H}_{l,ij}^{-1}, \qquad L_l = \mathbf{H}_{l,ij}^{-1}, \qquad C = \mathbf{y}_{C,ij}, \qquad G = \mathbf{y}_{G,ij}.$$

Since the components G are very small, and often introduce instabilities in the transient analysis of the equivalent circuit, in practice they are left out, so that instead of $m+2$, only $m+1$ match frequencies are needed.

10.3. Solution

The equivalent circuit for the interconnection system is submitted to a circuit analysis program, together with a description of the external components connected by the system, the bias conditions, and the frequency range or time domain for AC or transient analysis, respectively.

For AC analysis the output of the circuit analysis program is a list of nodal voltages for a number of frequencies. From these data the program can calculate the current density in the interconnection system by using the calculated transfer matrix. Next, the electromagnetic radiation can be calculated in the space around the system.

11. Linear algebra

This section presents linear algebra methods that can be used to solve the linear systems of equations obtained after discretisation. Particularly, the methods used for the solution of the linear system of equations and of the generalized eigenvalue problem will be discussed. The discussion is relatively brief, more details can be found in Chapter 8 of this volume.

Complex geometries of the interconnection system imply that a relatively large number of elements have to be used for a proper discretisation. As a result, the dimension of the coefficient matrices encountered in the linear system of equations is also large. This has a dramatic impact on the performance of the direct solution techniques: both the amount of storage needed and the time required to solve the linear system become prohibitively large. The former is often decisive, since it limits the size of problems which can be solved given the amount of memory space.

11.1. Solution of the linear systems of equations

In the solution of the Kirchhoff equations, discussed in Section 10, there are two types of linear system of equations.

In the first type of equation, e.g., $\mathbf{D}Q = V$ (see (10.3)), the matrix $\mathbf{D}$ is symmetric and positive definite. The solution of these systems can be performed by using the well known incomplete Cholesky conjugate gradient method, ICCG (see, e.g., BARRETT [1994]). For systems of a large dimension, n, it is often possible to perform the matrix vector multiplication in each iteration step in an efficient way. Therefore, the matrix $\mathbf{D}$ is approximated by the sum of a sparse matrix, $\mathbf{S}$, and a remainder matrix, $\mathbf{R} = \mathbf{V}\tilde{\mathbf{R}}\mathbf{V}$, where $\tilde{\mathbf{R}}$ is of dimension $m \ll n$, and $\mathbf{V}$ is an $n \times m$ prolongation matrix. This *matrix condensation* method will be discussed in Section 12. When there is a perfect ground plane present, as in the case of high frequency filter design, $\tilde{\mathbf{R}}$ is approximated by zero.

The second type of equations, e.g., (10.14) and (10.15), or (10.16) and (10.17), are of the form

$$\begin{pmatrix} \mathbf{A} & \mathbf{P} \\ \mathbf{P}^{\mathrm{T}} & 0 \end{pmatrix} \begin{pmatrix} I \\ V \end{pmatrix} = \begin{pmatrix} B_I \\ B_V \end{pmatrix}, \tag{11.1}$$

where $\mathbf{A} \in \mathbb{R}^{n\times n}$ is symmetric positive definite and $\mathbf{P} \in \mathbb{R}^{n\times m}$ such that

$$\mathbf{P}_{ij} \in \{-1, 0, 1\}, \quad \forall_{1\leqslant i\leqslant n, 1\leqslant j\leqslant m}.$$

Each row of $\mathbf{P}$ contains at most two non zero elements, which are of opposite sign:

$$\sum_{j=1}^{m} |\mathbf{P}_{ij}| \leqslant 2, \qquad -1 \leqslant \sum_{j=1}^{m} \mathbf{P}_{ij} \leqslant 1.$$

Finally, $\operatorname{rank}(\mathbf{P}) = m$.

The coefficient matrix in (11.1) can be decomposed into the following form:

$$\begin{pmatrix} \mathbf{A} & \mathbf{P} \\ \mathbf{P}^{\mathrm{T}} & 0 \end{pmatrix} = \begin{pmatrix} \mathbf{A} & 0 \\ \mathbf{P}^{\mathrm{T}} & \mathbf{I} \end{pmatrix} \begin{pmatrix} \mathbf{A}^{-1} & 0 \\ 0 & -\mathbf{P}^{\mathrm{T}}\mathbf{A}^{-1}\mathbf{P} \end{pmatrix} \begin{pmatrix} \mathbf{A} & \mathbf{P} \\ 0 & \mathbf{I} \end{pmatrix},$$

which shows that there are n positive and m negative eigenvalues. Unfortunately, most iterative techniques can only be applied to the solution of positive definite systems. For systems with both positive and negative eigenvalues, such methods may break down. Furthermore, convergence will often be extremely slow since the polynomials generated in the methods have to locate both positive and negative eigenvalues. Because of these problems, direct solution techniques are often preferred.

A solution to the problems sketched in the above is to transform (11.1) into a number of linear systems which can be solved using standard iterative techniques. The latter is often more attractive than using direct solution techniques, for a number of reasons. Firstly, the amount of memory space needed is much smaller. Secondly, approximations to the coefficient matrix can be used in the matrix vector products occurring in iterative methods rather than the entire matrix (see Section 12). Hence, there is no need to fully assemble the matrix.

As noted in the above, if iterative methods are to be used it is desirable that the coefficient matrices involved are positive definite. There are essentially two ways of achieving this, namely either using the *range space method* or the *null space*. Often, the use of the latter method is ruled out because of the need to construct a basis for the null space of a large matrix. However, for interconnection systems, the null space method appears to be extremely useful. In the following, the method will be described and advantages will be listed.

11.1.1. Null space method

The basis for the null space method is the observation that the solution of the second set of equations in (11.1), i.e., $\mathbf{P}^{\mathrm{T}}I = B_V$, can be cast into the form

$$I = \mathbf{P}\tilde{V} + \mathcal{C}X, \tag{11.2}$$

where $\tilde{V} \in \mathbb{R}^m$, $X \in \mathbb{R}^{n-m}$, $\mathcal{C} \in \mathbb{R}^{n\times(n-m)}$, and $\mathbf{P}\tilde{V}$ is a special solution of the second set of equations, satisfying

$$\mathbf{P}^{\mathrm{T}}\mathbf{P}\tilde{V} = B_V,$$

and $\mathcal{C}$ is a matrix whose columns form a basis for the null space of $\mathbf{P}^{\mathrm{T}}$. Note that (11.2) is the most general form of solutions of the second set of equations in (11.1).

After substitution of (11.2) into the first set of equations, one obtains

$$\mathbf{A}\mathcal{C}X + \mathbf{P}V = B_I - \mathbf{AP}X,$$

which, on multiplying by $\mathcal{C}^{\mathrm{T}}$ yields

$$\mathcal{C}^{\mathrm{T}}\mathbf{A}\mathcal{C}X + \mathcal{C}^{\mathrm{T}}\mathbf{P}V = \mathcal{C}^{\mathrm{T}}(B_I - \mathbf{AP}X).$$

Since the columns of $\mathcal{C}$ constitute a basis of the null space of $\mathbf{P}^{\mathrm{T}}$, it holds that $\mathbf{P}^{\mathrm{T}}\mathcal{C} = 0$ or, equivalently, $\mathcal{C}^{\mathrm{T}}\mathbf{P} = 0$. Hence, the equation just derived is actually equal to

$$\mathcal{C}^{\mathrm{T}}\mathbf{A}\mathcal{C}X = \mathcal{C}^{\mathrm{T}}(B_I - \mathbf{AP}\tilde{V}). \tag{11.3}$$

The conclusion is that there are three steps involved in solving the original system:

1. First solve the system $\mathbf{P}^{\mathrm{T}}\mathbf{P}\tilde{V} = B_V$ to obtain $\tilde{V}$ and, subsequently calculate $\mathbf{P}\tilde{V}$, which is a special solution of the second set of equations.
2. Next determine the unknown vector X by solving the system (11.3). Combining the result with the special solution obtained in step 1 leads to the vector of unknown currents I.

3. Having found the current vector I, determine the vector of unknown potentials, V, by solving $\mathbf{P}^{\mathrm{T}}\mathbf{P}V = \mathbf{P}^{\mathrm{T}}(B_I - \mathbf{A}I)$.

The first and third step involve solving systems with a coefficient matrix $\mathbf{P}^{\mathrm{T}}\mathbf{P}$ which, by the special structure of $\mathbf{P}$, is a positive definite symmetric matrix. In fact, it is an M-matrix, meaning that the diagonal entries are positive, the off-diagonal entries are non-positive and the inverse is positive. The solution of these systems can be performed by using the ICCG method. Note that $\mathbf{P}^{\mathrm{T}}\mathbf{P}$ is a sparse $m \times m$ matrix.

Crucial is the solution of the system in step 2 of the above procedure. Observe that, since $\mathbf{A}$ is a positive definite matrix, $\mathcal{C}^{\mathrm{T}}\mathbf{A}\mathcal{C}$, is also positive definite. Hence, standard numerical solution techniques can be applied to the system (11.3). Thus, the problem of indefiniteness is avoided by using this approach. Note that $\mathcal{C}^{\mathrm{T}}\mathbf{A}\mathcal{C}$ is an $(n-m) \times (n-m)$ matrix.

11.1.2. Construction of the null space matrix $\mathcal{C}$

The only problem is the construction of the null space matrix, $\mathcal{C}$. Fortunately, the matrix $\mathcal{C}$ only depends on the topology of the problem. Hence, the construction of the matrix $\mathcal{C}$ only has to be done once. A fortunate fact is that the elements of the null space can actually be interpreted physically. They are combinations of currents through branches constituting closed loops, the exterior of the problem area being considered as one node. This means that a considerable number of basis vectors can be constructed easily, by just finding all closed loops in the topology of the problem. Note that, in this way, a sparse basis is obtained. Most basis elements will consist of only a few non-zero entries. This is of importance when constructing the coefficient matrix $\mathcal{C}^{\mathrm{T}}\mathbf{A}\mathcal{C}$, because it saves computer time. Since, for most topologies, this procedure does not lead to all elements of the null space, the set of basis functions found needs to be completed. This can be done in a fairly simple way. Suppose the matrix $\mathbf{P}^{\mathrm{T}}$ is of the form

$$\mathbf{P}^{\mathrm{T}} = (FG),$$

where F is an $m \times m$ matrix and G an $m \times (n-m)$ matrix. Since rank($\mathbf{P}$) $= m$, it is possible to choose a suitable permutation of unknown currents and voltages, so that the matrix F is non singular. It is even possible to have an upper triangular F. Since every column of F contains at most two non-zeroes, a simple elimination process leads to the situation where F is the identity matrix.

Now assume that the matrix $\mathcal{C}$ is of the form

$$\mathcal{C} = \begin{pmatrix} M \\ N \end{pmatrix},$$

with M an $m \times (n-m)$ matrix and N an $(n-m) \times (n-m)$ matrix. Then the requirement $\mathbf{P}^{\mathrm{T}}\mathcal{C} = 0$ implies that

$$FM + GN = 0,$$

so that, if N is given, M follows from

$$M = -F^{-1}GN.$$

Orden (see ORDEN [1964]) already used this technique in 1964 to determine the null space. He chose N the identity matrix. Since a large number of basis elements are known, Orden's choice is not very efficient. Instead, we write

$$\mathcal{C} = \begin{pmatrix} M_1 & M_2 \\ N_1 & N_2 \end{pmatrix},$$

such that the first set of columns

$$\begin{pmatrix} M_1 \\ N_1 \end{pmatrix}$$

correspond to the basis elements already generated. Suppose that this sub-matrix contains k columns. The k columns of the matrix N_1 constitute a subspace of $\mathbb{R}^{n-m}$, which is at most k-dimensional. This means that there are at least $n-m-k$ unit vectors from $\mathbb{R}^{n-m}$ that are not in this subspace. In other words, it is possible to fill the matrix N_2 with unit vectors that are not in the span of the columns of N_1. Having constructed the matrix N_2, the matrix M_2 can be produced simply by

$$M_2 = -F^{-1}GN_2.$$

If F is the identity matrix, the entries of M_2 are all in the set $-1, 0, 1$ (since this also holds for the elements of G). In this way, a complete set of basis vectors for the null space can be found.

11.2. The calculation of a subset of the eigenvalues of a generalized eigenvalue problem

The smallest and largest eigenvalues, λ, of the generalized eigenvalue problem, $\mathbf{Ax} = \lambda\mathbf{Bx}$, mentioned in Section 10.2.3, can be calculated by dedicated routines available in many software libraries. An effective procedure is to use a generalization of an algorithm, developed by PARLETT and REID [1981] for large symmetric eigenvalue problems. This algorithm is a reliable and efficient method for finding all or part of the spectrum of a large symmetric matrix $\mathbf{M}$, based on the Lanczos algorithm, by tracking the progress of the eigenvalues of the Lanczos tridiagonal matrices towards the eigenvalues of $\mathbf{M}$.

VAN DER VORST [1982] has generalized this algorithm for the computations of eigenvalues of the product matrix, $\mathbf{M} = \mathbf{AB}^{-1}$, where $\mathbf{A}$ is symmetric, and $\mathbf{B}$ is symmetric positive definite. The method allows for the computation of the eigenvalues of $\mathbf{B}^{-1}\mathbf{A}$ which are equal to those $\mathbf{AB}^{-1}$, without the explicit need for an LL^{T}-factorization of the matrix $\mathbf{B}$. This makes the generalized scheme very attractive, especially if $\mathbf{B}$ has a sparse structure. The method is attractive if fast solvers are available for the solution of linear systems of the form $\mathbf{B}y = z$.

Since the small eigenvalues of the partially solved eigenvalue problem, $\mathbf{AB}^{-1}\mathbf{x} = \lambda\mathbf{x}$, obtained by the above method, are often not accurate enough, they are obtained by partially solving the inverse eigenvalue problem, $\mathbf{A}^{-1}\mathbf{Bx} = \mu\mathbf{x}$, where $\mu = \lambda^{-1}$.

In recent years, the computation of eigenvalues and the solution of generalized eigenvalue problems has received much attention. A very effective method has been described in the work by Van der Vorst and his co-workers: the Jacobi–Davidson method. For more details, we refer the reader to the recent literature on this subject.

12. Matrix condensation

This section presents an efficient method for solving equations with large matrices, A, of the type discussed in the previous sections. In Section 7 it was shown that in the integrand of an interaction integral between an object and a source element that are sufficiently apart, the Green function can be approximated by Taylor expansion. In that case, the integral can be expressed as a sum of the products of a moment integral, $M_{\alpha\beta}(\mathbf{x}_m)$, corresponding to the object element, and a moment integral, $M_{\alpha\beta}(\mathbf{x}'_m)$, corresponding to the source element, and a factor resulting from the Green function, $G(\mathbf{x}'_m - \mathbf{x}_m)$, where $\mathbf{x}_m$ and $\mathbf{x}'_m$ are the midpoints of the object and source element. The indices α and β refer to the terms in the Taylor expansion. For more details see also Section 12.1.

In the *matrix condensation* method treated in this section the elements are clustered into cells. Let n be the total number of elements and m be the number of cells ($m \ll n$). Let the interaction integrals between elements belonging to adjacent cells form a sparse $n \times n$ matrix S with N_s non-zero elements ($N_s \ll n^2$). Let the interaction between the non-adjacent cells form an $m \times m$ matrix, $\tilde{R}$, the non-zero coefficients of which describe some kind of averaged interaction between the elements of one cell with the elements of another cell.

The matrix, $\tilde{R}$, may be considered as the product matrix $WM^{\mathrm{T}} : G : MW^{\mathrm{T}}$. The matrix W is an $m \times n$ restriction matrix, the ith row of which has only non-zero coefficients for the elements belonging to the ith cell. The matrix $M^{\mathrm{T}} : G : M$ is an $n \times n$ matrix, where M is an $n_T \times n$ matrix (n_T is the number of Taylor terms), and G is an $n_T \times n_T$ matrix. Matrix G results from the Green function, $G(\mathbf{x}'_m - \mathbf{x}_m)$, and is different for each combination of a column of M^{T} and a row of M, which represent the moments of an object element with midpoint $\mathbf{x}_m$ and the moments of a source element with midpoint $\mathbf{x}'_m$, respectively. This special matrix product is denoted by ":".

In the matrix condensation method the matrix, A, is approximated by matrix $\tilde{A} = S + V\tilde{R}\,V^{\mathrm{T}}$, where V is an $n \times m$ prolongation matrix the jth column of which has only non-zero coefficients for the elements belonging to the jth cell. The non-zero coefficients of V are set equal to 1. Hence, the total number of matrix coefficients of matrix, A, to be calculated is reduced from n^2 to $N_s + m^2$.

For the solution of the equations with these large matrices an iterative method is used. In each iteration step a matrix vector multiplication $v = Au$ is performed. In the matrix condensation method the vector v after multiplication is $v1 + v2$, where $v1 = Su$ and $v2 = V\tilde{R}\,V^{\mathrm{T}}u$. Hence, the total number of operations in a matrix vector multiplication is reduced from $\mathcal{O}(n^2)$ to $\mathcal{O}(n)$.

There are two different methods to obtain the vector $v2$. One method is to construct the matrix, $\tilde{R}$, in advance. Therefore, the $n_T \times m$ cell moments MW^{T} are calculated from the $n_T \times n$ element moments M assuming that the non-zero elements in a row of W are all equal, and their sums equal to 1. Physically, this meaning that the elements belonging to a particular cell will all have the same charge. Since it is not always possible to compose the cells so that this is a good approximation, an alternative method is to use in each iteration step the vector u to construct an $m \times n$ restriction matrix U and to calculate the $n_T \times m$ source cell moments MU^{T}. Each row of matrix U contains the elements of u belonging to the corresponding cell completed with zeroes. Next, the matrix MU^{T} is multiplied by the $m \times m$ matrix $WM^{\mathrm{T}} : G : E$, where E is an $n_T \times m$ matrix with elements equal to 1. The matrix $WM^{\mathrm{T}} : G : E$ can be calculated in advance. This method allows the elements in a cell to have different charges. Note that the matrix, $\tilde{R}$, is no longer symmetric, so that instead of the ICCG method an iterative solution method must taken that can handle non-symmetric matrices, e.g., BICGSTAB (see BARRETT [1994]).

In the next subsection the coefficients of matrix $\tilde{R}$ will be derived.

12.1. Calculation of coefficients of matrix $\tilde{R}$

It has been shown that the interaction integrals belonging to the matrix $\mathbf{D}$ have the form:

$$I = \int_{\Omega_i} \tilde{\psi}_i(\mathbf{x}) \cdot \int_{\Omega_j} \tilde{\psi}_j(\mathbf{x}') G(\mathbf{x}' - \mathbf{x})\, \mathrm{d}\mathbf{x}'\, \mathrm{d}\mathbf{x}. \tag{12.1}$$

Here, $\tilde{\psi}_i(\mathbf{x})$ and $\tilde{\psi}_j(\mathbf{x})$ are basis functions belonging to the object domain Ω_i and the source domain Ω_j, respectively. Further, $\mathbf{x}$ and $\mathbf{x}'$ represent points in Ω_i and Ω_j, respectively.

Let the Green function be of the form:

$$G(\mathbf{x}' - \mathbf{x}) = \sum_{i=0}^{N} \frac{c_i}{|\mathbf{x}'_i - \mathbf{x}|},$$

where N is the number of images.

If the distance between two disjoint elements Ω_i and Ω_j, defined by

$$\min\bigl\{|\mathbf{x}' - \mathbf{x}|;\ \mathbf{x} \in \Omega_i,\ \mathbf{x}' \in \Omega_j\bigr\},$$

is large enough one can apply Taylor expansion to the Green function G with respect to $(\mathbf{x}' - \mathbf{x}'_m)$ and $(\mathbf{x} - \mathbf{x}_m)$, where $\mathbf{x}'_m$ and $\mathbf{x}_m$ are the midpoints of the source and object element, respectively. After the substitutions $\mathbf{y}_i = \mathbf{x}'_i - \mathbf{x}$, $\mathbf{y}_{im} = \mathbf{x}'_{im} - \mathbf{x}_m$ and $\mathbf{y}_i - \mathbf{y}_{im} = \mathbf{y} - \mathbf{y}_m$, the second order Taylor expansion becomes:

$$G(\mathbf{y}) = g_m + \mathbf{g}_m^{\mathrm{T}}(\mathbf{y} - \mathbf{y}_m) + \tfrac{1}{2}(\mathbf{y} - \mathbf{y}_m)^{\mathrm{T}} \mathbf{G}_m (\mathbf{y} - \mathbf{y}_m) + \mathcal{O}\bigl(|\mathbf{y} - \mathbf{y}_m|^3\bigr), \tag{12.2}$$

where for $r_i = |\mathbf{y}_{im}|$

$$g_m = \sum_{i=0}^{N} G_i(\mathbf{y}_{im}) = \sum_{i=0}^{N} \frac{c_i}{r_i}, \tag{12.3}$$

$$\mathbf{g}_m = \sum_{i=0}^{N} (\nabla G_i)(\mathbf{y}_{im}) = \sum_{i=0}^{N} \frac{-c_i}{r_i^3} \mathbf{y}_{im}, \tag{12.4}$$

$$\mathbf{G}_m = \sum_{i=0}^{N} \bigl(\nabla(\nabla G_i)\bigr)(\mathbf{y}_{im}) = \sum_{i=0}^{N} \frac{-c_i}{r_i^3} \mathbf{I} + 3\frac{c_i}{r_i^5}(\mathbf{y}_{im} \otimes \mathbf{y}_{im}). \tag{12.5}$$

For the definition of $\otimes$ see Appendix D. After substitution of the Taylor approximation (12.2) of $G(\mathbf{x}' - \mathbf{x})$ into the expression (12.1) one obtains

$$\tilde{I}(\mathbf{y}_m) = \int_{\Omega_i} \tilde{\psi}_i(\mathbf{x}) \cdot \int_{\Omega_j} \tilde{\psi}_j(\mathbf{x}') g_m \, d\mathbf{x}' \, d\mathbf{x} \tag{12.6}$$

$$+ \int_{\Omega_i} \tilde{\psi}_i(\mathbf{x}) \cdot \int_{\Omega_j} \tilde{\psi}_j(\mathbf{x}') \mathbf{g}_m^{\mathrm{T}} (\mathbf{y} - \mathbf{y}_m) \, d\mathbf{x}' \, d\mathbf{x} \tag{12.7}$$

$$+ \frac{1}{2} \int_{\Omega_i} \tilde{\psi}_i(\mathbf{x}) \cdot \int_{\Omega_j} \tilde{\psi}_j(\mathbf{x}') (\mathbf{y} - \mathbf{y}_m)^{\mathrm{T}} \mathbf{G}_m (\mathbf{y} - \mathbf{y}_m) \, d\mathbf{x}' \, d\mathbf{x}, \tag{12.8}$$

where $\mathbf{y} = \mathbf{x}' - \mathbf{x}$ and $\mathbf{y}_m = \mathbf{x}'_m - \mathbf{x}_m$.

Since g_m, $\mathbf{g}_m$ and $\mathbf{G}_m$ are independent of $\mathbf{x}$ and $\mathbf{x}'$, they appear as constant terms in the integral. Let the moment integrals be defined by

$$M_{\alpha\beta}(\mathbf{x}_m) = \int_{\Omega_i} \tilde{\psi}_i(\mathbf{x}) \{\mathbf{x} - \mathbf{x}_m\}_\alpha \{\mathbf{x} - \mathbf{x}_m\}_\beta \, d\mathbf{x},$$

where $\{\mathbf{x} - \mathbf{x}_m\}_\alpha = 1$, $(x - x_m)$, $(y - y_m)$ or $(z - z_m)$ for $\alpha = 0$, 1, 2 or 3.

After substitution into (12.8)

$$\tilde{I}(\mathbf{y}_m) = M^{\mathrm{T}} G M, \tag{12.9}$$

where the n_T-dimensional vector M contains the elements $M_{\alpha\beta}$ in the row-wise order $\alpha\beta = \{00, 01, \ldots, 33\}$. The $n_T \times n_T$ matrix G is of the form

$$G = \begin{bmatrix} g_m & \mathbf{g}_m^{\mathrm{T}} & \frac{1}{2}\tilde{\mathbf{G}}_m^{\mathrm{T}} \\ -\mathbf{g}_m & -\mathbf{G}_m & 0 \\ \frac{1}{2}\tilde{\mathbf{G}}_m & 0 & 0 \end{bmatrix},$$

where the scalar g_m, the 3-dimensional vector $\mathbf{g}_m$, and the 3×3 matrix $\mathbf{G}_m$ are defined by (12.3)–(12.5). $\tilde{\mathbf{G}}_m$ is a 9-dimensional vector that contains the matrix elements of $\mathbf{G}_m$ in row-wise order $\alpha\beta = \{11, 12, \ldots, 33\}$.

The moment integrals are calculated in advance. Next, the cell moments are calculated, defined by:

$$\mathcal{M}_{\alpha\beta}(\mathbf{x}_{m_{\mathrm{cell}}}) = \sum_{i \in \mathrm{cell}} w_i M_{\alpha\beta}(\mathbf{x}_{m_i}),$$

where $\mathbf{x}_{m_i}$ are the midpoints of the elements belonging to the cell with midpoint $\mathbf{x}_{m_{\mathrm{cell}}}$, and w_i are weight factors.

An expression for the coefficients of matrix $\tilde{R}$ is obtained after substituting in (12.9) the moments M by $\mathcal{M}$, and the matrix G by a similar matrix with the Green functions g_m, $\mathbf{g}_m$, $\mathbf{G}_m$ of $\mathbf{y}_{i\,m_{\mathrm{cell}}}$, where $\mathbf{y}_{m_{\mathrm{cell}}} = \mathbf{x}'_{m_{\mathrm{cell}}} - \mathbf{x}_{m_{\mathrm{cell}}}$, and $\mathbf{x}_{m_{\mathrm{cell}}}$ and $\mathbf{x}'_{m_{\mathrm{cell}}}$ are the midpoints of the object and source cell, respectively.

Appendix A. Boundary singularities

In this appendix we study the behavior of the potential, the fields and the surface charge densities in the neighbourhood of sharp "corners" or edges. We shall assume that they are infinitely sharp so that we can look at them closely enough that the behaviour of the fields is determined in functional form solely by the properties of the "corner" being considered and not by the details of the overall configuration.

The general situation in two dimensions is shown in Fig. A.1. The two conducting planes intersect at an angle β. The planes are assumed to be held at potential V. Since we are interested in the functional behaviour of the fields near the origin, we leave the "far away" behaviour unspecified as much as possible.

The geometry of Fig. A.1 suggests the use of polar coordinates. In terms of the polar coordinates (r, φ), the Laplace equation for the potential Φ, in two dimensions is

$$\frac{1}{r}\mathbf{d}r\left(r\frac{\mathrm{d}\Phi}{\mathrm{d}r}\right)+\frac{1}{r^2}\frac{\mathrm{d}^2\Phi}{\mathrm{d}\varphi^2}=0.$$

Using the separation of variables approach, we substitute

$$\Phi(r,\varphi)=R(r)F(\varphi).$$

This leads, upon multiplication by r^2/Φ, to

$$\frac{r}{R}\mathbf{d}r\left(r\frac{\mathrm{d}R}{\mathrm{d}r}\right)+\frac{1}{F}\frac{\mathrm{d}^2F}{\mathrm{d}\varphi^2}=0.$$

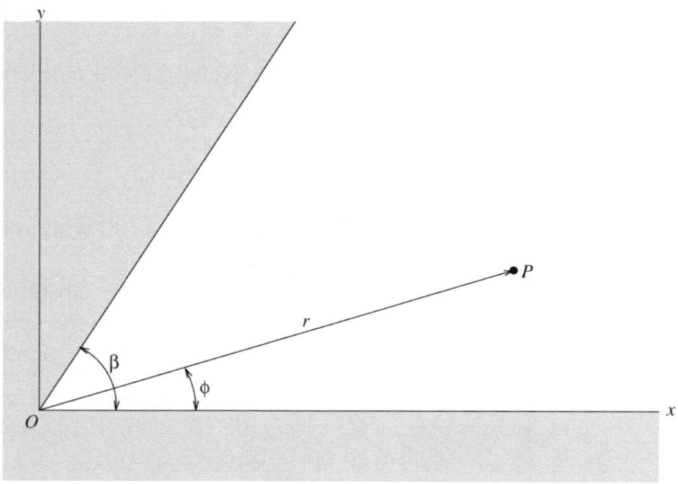

FIG. A.1. Intersubsection of two conducting planes, with potential V, defining a corner with opening angle β.

Since the two terms are separately function of r and φ, respectively, they each must be constant:

$$\frac{r}{R}\mathbf{d}r\left(r\frac{\mathrm{d}R}{\mathrm{d}r}\right)=\nu^2, \qquad \frac{1}{F}\frac{\mathrm{d}^2F}{\mathrm{d}\varphi^2}=-\nu^2.$$

The solutions to these equations are

$$\left.\begin{aligned} R(r) &= ar^{\nu}+br^{-\nu},\\ F(\varphi) &= A\cos(\nu\varphi)+B\sin(\nu\varphi) \end{aligned}\right\} \tag{A.1}$$

and for the special circumstance of $\nu = 0$, the solutions are

$$\left.\begin{aligned} R(r) &= a_0+b_0\ln r,\\ F(\varphi) &= A_0+B_0\varphi. \end{aligned}\right\} \tag{A.2}$$

These are the building blocks with which we construct the potential by linear superposition. For our situation the azimuthal angle is restricted to the range $0 \leqslant \varphi \leqslant \beta$. The boundary conditions are that $\Phi = V$ for all $r \geqslant 0$ when $\varphi = 0$ and $\varphi = \beta$. This requires that $b_0 = B_0 = 0$ in (A.2) and $b = A = 0$ in (A.1). Furthermore, it requires that ν be chosen to make $\sin(\nu\beta) = 0$. Hence

$$\nu=\frac{m\pi}{\beta}, \qquad m=1,2,\ldots$$

and the general solution becomes

$$\Phi(r,\varphi)=V+\sum_{m=1}^{\infty}a_m r^{m\pi/\beta}\sin(m\pi\varphi/\beta).$$

Since the series involves positive powers of $r^{\pi/\beta}$, for small enough r only the first term in the series will be important. Thus, near $r = 0$, the potential is approximately

$$\Phi(r,\varphi)\simeq V+a_1 r^{\pi/\beta}\sin(\pi\varphi/\beta).$$

The electric field components are

$$E_r(r,\varphi)=-\frac{\mathrm{d}\Phi}{\mathrm{d}r}\simeq-\frac{\pi a_1}{\beta}r^{(\pi/\beta)-1}\sin(\pi\varphi/\beta),$$

$$E_\varphi(r,\varphi)=-\frac{1}{r}\frac{\mathrm{d}\Phi}{\mathrm{d}\varphi}\simeq-\frac{\pi a_1}{\beta}r^{(\pi/\beta)-1}\cos(\pi\varphi/\beta).$$

The surface charge densities at $\varphi = 0$ and $\varphi = \beta$ are equal and are approximately

$$\rho(r)=\frac{E_\varphi(r,0)}{4\pi}\simeq-\frac{a_1}{4\beta}r^{(\pi/\beta)-1}.$$

The components of the field and the surface charge density near $r = 0$ all vary with distance as $r^{(\pi/\beta)-1}$. This dependence on r gives us for $\beta = 2\pi$ (the edge of a thin sheet) the singularity as $r^{-1/2}$. This is still integrable so that the charge within a finite distance from the edge is finite, but it implies that field strengths become very large at the edges of conducting sheets.

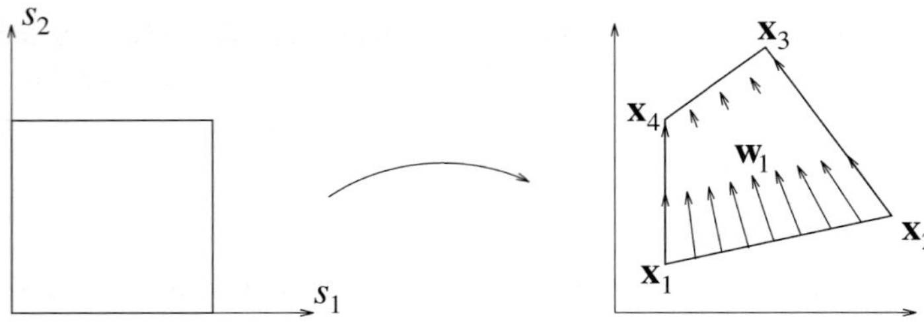

FIG. A.2. Transformation of the unit square to the quadrilateral in terms of the isoparametric coordinates s_1 and s_2, with the edge function $\mathbf{w}_1$.

To account for this boundary singularity the basis functions of the charge and the currents for the elements, of which one or more edges lie in the boundary of the 2D conductor region, will be adapted by the function $f(\mathbf{x})$ introduced in Section 2.6. Therefore, this function has the form $d^{-1/2}$. If one of the edges of the element lies in the boundary

$$d = s_i \quad \text{or} \quad d = 1 - s_i,$$

where s_i is one of the isoparametric coordinates (see Fig. A.2). If two opposite edges of the element lie in the boundary

$$d = s_i(1 - s_i).$$

Appendix B. Basis functions

In this appendix the vector valued basis functions for the current $\mathbf{J}$ on the edges of a quadrilateral element are defined. They span the function space H_h^{div}. For these basis functions some lemmas will be proven.

The mapping from a unit square with isoparametric coordinates s_1 and s_2 to the quadrilateral $\mathbf{x}_1 \ldots \mathbf{x}_4$, shown in Fig. A.2, is given by

$$\mathbf{x}(s_1, s_2) = (1 - s_2)\big[(1 - s_1)\mathbf{x}_1 + s_1\mathbf{x}_2\big] + s_2\big[(1 - s_1)\mathbf{x}_4 + s_1\mathbf{x}_3\big]. \tag{B.1}$$

For a particular element Ω_i the edge functions $\mathbf{w}_k$ are (see VAN WELIJ [1986, p. 371])

$$\begin{aligned}
&\mathbf{w}_1 = \frac{(1 - s_2)\mathbf{v}_2}{|\mathbf{v}_1 \times \mathbf{v}_2|}, \qquad \mathbf{w}_2 = \frac{-s_1\mathbf{v}_1}{|\mathbf{v}_1 \times \mathbf{v}_2|},\\
&\mathbf{w}_4 = \frac{(1 - s_1)\mathbf{v}_1}{|\mathbf{v}_1 \times \mathbf{v}_2|}, \qquad \mathbf{w}_3 = \frac{-s_2\mathbf{v}_2}{|\mathbf{v}_1 \times \mathbf{v}_2|},\\
&\mathbf{v}_1 = (\mathbf{x}_2 - \mathbf{x}_1) + s_2(\mathbf{x}_1 - \mathbf{x}_2 + \mathbf{x}_3 - \mathbf{x}_4),\\
&\mathbf{v}_2 = (\mathbf{x}_4 - \mathbf{x}_1) + s_1(\mathbf{x}_1 - \mathbf{x}_2 + \mathbf{x}_3 - \mathbf{x}_4).
\end{aligned} \tag{B.2}$$

For these basis functions the following lemmas hold:

LEMMA B.1. *Let* $\mathbf{w}_k$, $k = 1, \ldots, 4$, *be defined as in* (B.2) *and let* $\mathbf{J}_0(s_1, s_2) = \sum_{k=1}^{4} I_k \mathbf{w}_k(s_1, s_2)$ *for certain* $I_1, \ldots, I_4$. *Then, for* $(s_1, s_2) \in [0, 1] \times [0, 1]$:

$$\sum_{k=1}^{4} I_k = 0 \quad \Longrightarrow \quad \nabla \cdot \mathbf{J}_0 = 0.$$

PROOF. Consider the mapping (B.1) from the unit square to the quadrilateral. Let the vectors $\mathbf{v}_j$ and $\mathbf{v}_{12}$ be defined as $\mathbf{v}_j = \frac{\partial \mathbf{x}}{\partial s_j}$, and $\mathbf{v}_{12} = \frac{\partial^2 \mathbf{x}}{\partial s_1 \partial s_2}$. Then, we get the following relations:

$$\begin{aligned}
\mathbf{v}_1 &= (\mathbf{x}_2 - \mathbf{x}_1) + s_2 \mathbf{v}_{12}, \\
\mathbf{v}_2 &= (\mathbf{x}_4 - \mathbf{x}_1) + s_1 \mathbf{v}_{12}, \\
\mathbf{v}_{12} &= (\mathbf{x}_1 - \mathbf{x}_2 + \mathbf{x}_3 - \mathbf{x}_4).
\end{aligned}$$

In the following we will need the gradients ∇s_i $(i = 1, 2)$. These gradients are the rows of the inverse of the Jacobian matrix $\{\frac{\partial \mathbf{x}}{\partial s_i}\}$, which has $\mathbf{v}_j$ $(j = 1, 2)$ as columns. Hence $\mathbf{v}_j \cdot \nabla s_i = \delta_{ij}$. Let $J = (\mathbf{v}_1 \times \mathbf{v}_2) \cdot \mathbf{v}_3$, where $\mathbf{v}_3 = \frac{\mathbf{v}_1 \times \mathbf{v}_2}{|\mathbf{v}_1 \times \mathbf{v}_2|}$ is the unit normal vector to the quadrilateral, then the expressions for the gradients are:

$$\nabla s_1 = \frac{\mathbf{v}_2 \times \mathbf{v}_3}{J}, \qquad \nabla s_2 = \frac{\mathbf{v}_3 \times \mathbf{v}_1}{J}.$$

Now, let $f = I_4(1 - s_1) - I_2 s_1$ and $g = I_1(1 - s_2) - I_3 s_2$, then

$$\mathbf{J}_0 = \big(f \mathbf{v}_1 + g \mathbf{v}_2\big)/J,$$

and hence

$$\begin{aligned}
\nabla \cdot \mathbf{J}_0 = {} & \frac{1}{J}(\nabla f \cdot \mathbf{v}_1 + \nabla g \cdot \mathbf{v}_2) + \frac{1}{J^2}\big\{f(J\nabla \cdot \mathbf{v}_1 - \mathbf{v}_1 \cdot \nabla J) \\
& + g(J\nabla \cdot \mathbf{v}_2 - \mathbf{v}_2 \cdot \nabla J)\big\}.
\end{aligned}$$

Since $\nabla s_i \cdot \mathbf{v}_j = \delta_{ij}$,

$$\begin{aligned}
\nabla f \cdot \mathbf{v}_1 &= \frac{\mathrm{d}f}{\mathrm{d}s_1}(\nabla s_1 \cdot \mathbf{v}_1) = -I_2 - I_4, \\
\nabla g \cdot \mathbf{v}_2 &= \frac{\mathrm{d}g}{\mathrm{d}s_2}(\nabla s_2 \cdot \mathbf{v}_2) = -I_1 - I_3, \\
J\nabla \cdot \mathbf{v}_1 &= J(\nabla s_2 \cdot \mathbf{v}_{12}) = (\mathbf{v}_3 \times \mathbf{v}_1) \cdot \mathbf{v}_{12} = \mathbf{v}_3 \cdot (\mathbf{v}_1 \times \mathbf{v}_{12}) \\
&= \mathbf{v}_3 \cdot \big((\mathbf{x}_2 - \mathbf{x}_1) \times \mathbf{v}_{12}\big), \\
J\nabla \cdot \mathbf{v}_2 &= J(\nabla s_1 \cdot \mathbf{v}_{12}) = (\mathbf{v}_2 \times \mathbf{v}_3) \cdot \mathbf{v}_{12} = \mathbf{v}_3 \cdot (\mathbf{v}_{12} \times \mathbf{v}_2) \\
&= \mathbf{v}_3 \cdot \big(\mathbf{v}_{12} \times (\mathbf{x}_4 - \mathbf{x}_1)\big), \\
\mathbf{v}_1 \cdot \nabla J &= (\mathbf{v}_1 \cdot \nabla s_1)\big((\mathbf{x}_2 - \mathbf{x}_1) \times \mathbf{v}_{12}\big) \cdot \mathbf{v}_3 = \mathbf{v}_3 \cdot \big((\mathbf{x}_2 - \mathbf{x}_1) \times \mathbf{v}_{12}\big) = J\nabla \cdot \mathbf{v}_1, \\
\mathbf{v}_2 \cdot \nabla J &= (\mathbf{v}_2 \cdot \nabla s_2)\big(\mathbf{v}_{12} \times (\mathbf{x}_4 - \mathbf{x}_1)\big) \cdot \mathbf{v}_3 = \mathbf{v}_3 \cdot \big(\mathbf{v}_{12} \times (\mathbf{x}_4 - \mathbf{x}_1)\big) = J\nabla \cdot \mathbf{v}_2,
\end{aligned}$$

so that $\nabla \cdot \mathbf{J}_0 = -(I_1 + I_2 + I_3 + I_4)/J$. □

LEMMA B.2. *Let* $\mathbf{w}_1$ *for a particular element analogously be defined as in* (B.2):

$$\mathbf{w}_1 = \mp\frac{(1-s_2)\mathbf{v}_2}{J},$$

then

$$\nabla\cdot\mathbf{w}_1 = \frac{\pm 1}{J} = \frac{\pm 1}{|\mathbf{v}_1\times\mathbf{v}_2|}.$$

Further

$$\mathbf{w}_1\cdot\mathbf{n} = \frac{\pm 1}{|\mathbf{x}_2-\mathbf{x}_1|}.$$

PROOF.

$$\nabla\cdot\mathbf{w}_1 = \frac{\pm\nabla s_2\cdot\mathbf{v}_2}{J} \mp \frac{(1-s_2)}{J^2}(J\nabla\cdot\mathbf{v}_2 - \mathbf{v}_2\cdot\nabla J) = \frac{\pm 1}{J},$$

since in the proof of Lemma B.1 we have shown that $\nabla s_i\cdot\mathbf{v}_j = \delta_{ij}$ and $\mathbf{v}_l\cdot\nabla J = J\nabla\cdot\mathbf{v}_l$.

Let $\mathbf{n}$ be the outward normal on edge 1 in the plane of the quadrilateral, then for $s_2 = 0$

$$\mathbf{w}_1\cdot\mathbf{n} = \frac{\mp\mathbf{v}_2\cdot\mathbf{n}}{((\mathbf{x}_2-\mathbf{x}_1)\times\mathbf{v}_2)\cdot\mathbf{v}_3} = \frac{\mp\mathbf{v}_2\cdot\mathbf{n}}{(\mathbf{v}_3\times(\mathbf{x}_2-\mathbf{x}_1))\cdot\mathbf{v}_2}$$
$$= \frac{\mp\mathbf{v}_2\cdot\mathbf{n}}{-|\mathbf{x}_2-\mathbf{x}_1|\mathbf{v}_2\cdot\mathbf{n}} = \frac{\pm 1}{|\mathbf{x}_2-\mathbf{x}_1|},$$

i.e., $\mathbf{w}_1\cdot\mathbf{n}$ depends only on the length of edge 1. □

By similar reasoning it can be shown that the same holds for $\mathbf{w}_2$, $\mathbf{w}_3$ and $\mathbf{w}_4$.

In the following lemma it will be shown that, if $\mathbf{J}_h(\mathbf{x}) = \sum_{k=1}^4 I_k\tilde{\mathbf{w}}_k(\mathbf{x})$, where $\tilde{\mathbf{w}}_k(\mathbf{x})$ is defined by Eq. (2.36) and $\sum_{k=1}^4 I_k = 0$, which means that all the currents that enter an element Ω_i will leave the element, then $\nabla\cdot\mathbf{J}_h(\mathbf{x}) = 0$ will also hold for all $\mathbf{x}\in\Omega_i$.

LEMMA B.3. *If the following conditions hold*: $\nabla f(\mathbf{x}) = a\mathbf{n}$ *for some* $a\in\mathbb{R}$, $f(\mathbf{x})\neq 0$, *the Jacobian* $J(\mathbf{x})$ *is bounded for all* $\mathbf{x}\in\Omega_i$, *and* $\mathbf{J}_h\cdot\mathbf{n} = 0$, *then*

$$\sum_{k=1}^4 I_k = 0 \iff \nabla\cdot\mathbf{J}_h = 0.$$

PROOF. Define $\mathbf{J}_0(\mathbf{x}) = \sum_{k=1}^4 I_k\mathbf{w}_k(\mathbf{x})$. From Lemma B.1 it follows that $\mathbf{J}_0$ is divergence free, $\nabla\cdot\mathbf{J}_0 = 0$. Moreover, by the product rule

$$\nabla\cdot\mathbf{J}_h = \nabla\cdot(f\mathbf{J}_0) = \nabla f\cdot\mathbf{J}_0 + f\nabla\cdot\mathbf{J}_0$$
$$= \nabla f\cdot\mathbf{J}_0 + 0.$$

From $\mathbf{J}_0\cdot\mathbf{n} = 0$ and $\nabla f(\mathbf{x}) = a\mathbf{n}$ for some $a\in\mathbb{R}$, if follows that $\nabla f\cdot\mathbf{J}_0 = 0$. Hence $\nabla\cdot\mathbf{J}_h = 0$.

Conversely, if $\nabla \cdot \mathbf{J}_h = 0$, then

$$\nabla f \cdot \mathbf{J}_0 + f \sum_{k=1}^{4} I_k \nabla \mathbf{w}_k = 0.$$

From $\mathbf{J}_0 \cdot \mathbf{n} = 0$ and $\nabla f(\mathbf{x}) = a\mathbf{n}$, it follows that $\nabla f \cdot \mathbf{J}_0 = 0$. Using Lemma B.2, it follows that

$$\frac{-f(\mathbf{x})}{J(\mathbf{x})} \sum_{k=1}^{4} I_k = 0.$$

Since $\frac{f(\mathbf{x})}{J(\mathbf{x})} \neq 0$ for all $\mathbf{x} \in \Omega_i$, it follows that $\sum_{k=1}^{4} I_k = 0$. □

Appendix C. Legendre polynomials

DEFINITION C.1. By $\Pi_n[a,b]$ we denote the linear space of polynomials on $[a,b]$ of degree $\leqslant n$.

Consider the inner product $\langle\, ,\rangle$ on $[a,b]$ with respect to the continuous weight function $w(x) > 0$ on (a,b):

$$\langle f, g\rangle = \int_{-1}^{1} w(x) f(x) g(x)\,\mathrm{d}x.$$

Then, one can build up a system of orthogonal polynomials by the Gram–Schmidt process:

THEOREM C.1. *Suppose one has orthogonal polynomials $P_0, P_1, \ldots, P_{n-1}$ of degree $0, 1, \ldots, n-1$ respectively, then P_n, constructed by (Gram–Schmidt)*

$$P_n = x^n - \frac{\langle P_0, x^n\rangle}{\langle P_0, P_0\rangle} P_0 - \cdots - \frac{\langle P_{n-1}, x^n\rangle}{\langle P_{n-1}, P_{n-1}\rangle} P_{n-1},$$

is also orthogonal to $P_0, \ldots, P_{n-1}$. Moreover, the orthogonal polynomials P_n are unique apart from a multiplicative constant.

PROOF. From the orthogonality of $P_0, \ldots, P_{n-1}$ and Gram–Schmidt follows:

$$\langle P_k, P_n\rangle = \langle P_k, x^n\rangle - \frac{\langle P_k, x^n\rangle}{\langle P_k, P_k\rangle}\langle P_k, P_k\rangle = 0, \quad \forall k, 0 \leqslant k \leqslant n-1,$$

so that $P_0, \ldots, P_n$ form a system of orthogonal polynomials.

Uniqueness: Let $\tilde{P}_n, P_n \in \Pi_n$ and let $P_0, \ldots, P_{n-1}$ be an orthogonal system. Suppose furthermore that $\tilde{P}_n, P_n \perp P_0, \ldots, P_{n-1}$. Then, $\tilde{P}_n = \sum_{j=0}^{n} \beta_j P_j$ implies that $\beta_j = 0$, for $j \neq n$. □

THEOREM C.2. *All zeros of P_n are real, simple and contained in (a,b).*

PROOF. Let $P_n \neq 0$. P_n changes sign at the distinct points $x_1, \ldots, x_k$ only, while $k < n$ and $x_1, \ldots, x_k \in (a, b)$. Then, $s(x)$, defined as

$$s(x) = w(x) P_n(x) \prod_{i=1}^{k} (x - x_i),$$

does not change sign on (a, b). Hence $\int_a^b s(x)\,\mathrm{d}x \neq 0$.

Since $\prod_{i=1}^k (x - x_i) = \sum_{i=0}^k \beta_i P_i$ for some $\{\beta_i\}$, we also have, however,

$$\int_a^b s(x)\,\mathrm{d}x = \sum_{i=0}^{k} \beta_i \langle P_n, P_i \rangle = 0,$$

which is a contradiction. Since $P_n \in \Pi_n$ we have $k = n$. □

For the weight function $w(x) \equiv 1$ and $[a, b] = [-1, 1]$ the polynomials are called *Legendre polynomials*. These polynomials are uniquely determined by a multiplicative constant such that the leading coefficient is 1. All zeros of P_i $(i = 1, 2, \ldots)$ are simple, real and contained within $(-1, 1)$. Moreover, if $\tilde{x}$ is a zero of P_i, then $-\tilde{x}$ is also a zero.

THEOREM C.3. *A set of orthonormal polynomials $P_n^*(x)$, i.e., with leading coefficient 1, satisfy a three-term recurrence relationship*

$$P_n^*(x) = (a_n x + b_n) P_{n-1}^*(x) - c_n P_{n-2}^*(x), \quad n = 2, 3, \ldots.$$

PROOF. See DAVIS and RABINOWITZ [1961, pp.167–168, 234–255]. □

The Legendre polynomials (see Fig. C.1), P_n, can be defined by the three-term recursion

$$P_0(x) = 1,$$

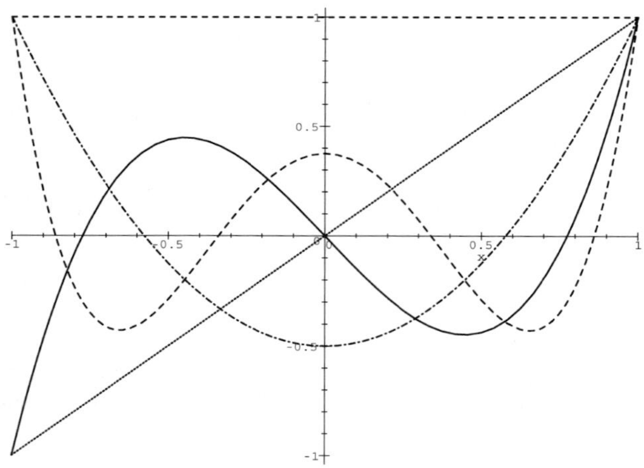

FIG. C.1. The Legendre polynomials $P_0(x), \ldots, P_4(x)$.

$$P_1(x) = x,$$

$$nP_n(x) = (2n-1)xP_{n-1}(x) - (n-1)P_{n-2}(x).$$

Appendix D. Inner products

We define the following inner products:

$$\mathbf{A}\cdot\mathbf{B} = \underbrace{\sum_{i_1=1}^{n}\cdots\sum_{i_d=1}^{n}}_{d} A_{i_1\ldots i_d}B_{i_1\ldots i_d}, \quad \text{where} \begin{cases} d=1 & \text{for vectors,}\\ d=2 & \text{for matrices,}\\ d=3 & \text{for tensors,}\\ \vdots \end{cases}$$

Let $\mathbf{x}_1 \otimes \cdots \otimes \mathbf{x}_\alpha$ be defined as

$$\mathbf{x}_1 \otimes \cdots \otimes \mathbf{x}_\alpha = \{(x_1)_{i_1}(x_2)_{i_2}\cdots(x_\alpha)_{i_\alpha}\}_{i_1,i_2,\ldots,i_\alpha},$$

so that, for example, in the two-dimensional case for $\mathbf{x} = \mathbf{x}_1$, $\mathbf{y} = \mathbf{x}_2$, and $i_1, i_2 = 1,\ldots,n$

$$\mathbf{x}\otimes\mathbf{y} = \begin{pmatrix} x_1\\ \vdots\\ x_n \end{pmatrix}(y_1\ldots y_n) = \begin{pmatrix} x_1y_1 & \ldots & x_1y_n\\ \vdots & \ddots & \vdots\\ x_ny_1 & \ldots & x_ny_n \end{pmatrix}.$$

LEMMA D.1. *If* $\mathbf{x}$ *and* $\mathbf{y}$ *are vectors of dimension* n *and* $\mathbf{A}$ *is a* $(n\times n)$*-matrix, then*

$$(\mathbf{Ay})^{\mathrm{T}}\mathbf{x} = \mathbf{A}\cdot(\mathbf{x}\otimes\mathbf{y}).$$

PROOF.

$$\mathbf{Ay} = \begin{pmatrix} A_{11}y_1 + \cdots + A_{1n}y_n\\ A_{21}y_1 + \cdots + A_{2n}y_n\\ \vdots\\ A_{n1}y_1 + \cdots + A_{nn}y_n \end{pmatrix},$$

$$\begin{aligned}(\mathbf{Ay})^{\mathrm{T}}\mathbf{x} = \mathbf{x}^{\mathrm{T}}(\mathbf{Ay}) &= (x_1A_{11}y_1 + \cdots + x_1A_{1n}y_n + x_2A_{21}y_1 + \cdots + x_nA_{nn}y_n)\\ &= \sum_{i=1}^{n}\sum_{j=1}^{n}A_{ij}x_iy_j\\ &= \mathbf{A}\cdot(\mathbf{x}\otimes\mathbf{y}).\end{aligned}$$ □

We have, analogously to the lemma, for tensors $\overline{\mathbf{T}}$

$$\overline{\mathbf{T}}\cdot(\mathbf{x}\otimes\mathbf{y}\otimes\mathbf{z}) = \big((\overline{\mathbf{T}}\mathbf{z})\mathbf{y}\big)^{\mathrm{T}}\mathbf{x}$$

where $\overline{\mathbf{T}}\mathbf{z} = \{T_{ij1}z_1 + T_{ij2}z_2 + \cdots + T_{ijn}z_n\}_{i\times j}$. This follows directly from the lemma.

In general, the following theorem holds.

THEOREM D.1. *Let* $\mathbf{M}_\alpha$ *be a* $\underbrace{n \times \cdots \times n}_{\alpha}$*-Tensor and let* $\mathbf{x}_1, \ldots, \mathbf{x}_\alpha$ *be vectors of dimension* n*. Then,*

$$\big(\ldots(\mathbf{M}_\alpha \mathbf{x}_\alpha)\ldots\mathbf{x}_3)\mathbf{x}_2\big)^{\mathrm{T}}\mathbf{x}_1 = \mathbf{M}_\alpha \cdot (\mathbf{x}_1 \otimes \cdots \otimes \mathbf{x}_\alpha).$$

PROOF. Analogously to the lemma. □

References

AUBIN, J.P. (1972). *Approximation of Elliptic Boundary Value Problems* (Wiley Interscience, New York).

BARRETT, R., et al. (1994). *Templates for the Solution of Linear Systems: Building Blocks for Iterative Methods*, second ed. (SIAM, Philadelphia, PA).

COLTON, D., KRESS, R. (1992). *Integral Equation Methods in Scattering Theory* (Krieger Publishing Company, Malabar, FL).

DAVIS, P.J., RABINOWITZ, P. (1961). Some geometrical theorems for abcissas and weights of Gauss type. *J. Math. Anal. Appl.* **2**, 167–168, 234–255, 428–437.

DAVIS, P.J., RABINOWITZ, P. (1984). *Methods of Numerical Integration*, second ed. (Academic Press, New York).

DU CLOUX, R., MAAS, G.P.J.F.M., WACHTERS, A.J.H. (1994). Quasi-static boundary element method for electromagnetic simulation of PCB's. *Philips J. Res.* **48**, 117–144.

GOLUB, G.H., VAN LOAN, C.F. (1986). *Matrix Computations* (North Oxford Academic Publishers Ltd, London).

JACKSON, J.D. (1975). *Classical Electrodynamics, vol. I* (John Wiley & Sons, New York).

KRONROD, A.S. (1965). *Nodes and Weights of Quadrature Formulas* (Consultants Bureau, New York), English transl. from Russian. MR 32:598.

MONEGATO, G. (1976a). A note on extended Gaussian quadrature rules. *Math. Comp.* **30**, 812–817.

MONEGATO, G. (1976b). Positivity of the weights of extended Gauss–Legendre quadrature rules. *Math. Comp.* **32**, 847–856.

ORDEN, A. (1964). Stationary points of quadratic functions under linear constraints. *Comput. J.* **7**, 238–242.

PARLETT, B.N., REID, J.K. (1981). Tracking the progress of the Lanczos algorithm for large symmetric eigenproblems. *IMA J. Numer. Anal.* **1**, 135–155.

PATTERSON, T.N.L. (1968). The optimum addition of points to quadrature formulae. *Math. Comp.* **22**, 847–856.

RAMO, S., WHINNERY, J.R., VAN DUZER, T. (1984). *Fields and Waves in Communication Electronics* (John Wiley and Sons, New York).

SZEGÖ, G. (1934). Über gewisse orthogonale Polynome, die zu einer oszillierenden Belegungsfunktion gehören. *Math. Ann.* **110**, 501–513.

VAN DER VORST, H.A. (1982). A generalized Lanczos scheme. *Math. Comp.* **39**, 559–561.

VAN WELIJ, J.S. (1986). Basis functions matching tangential components on element edges. In: *Proc. SISDEP-2 Conf., Swansea, UK*.

Solution of Linear Systems

O. Schenk

Integrated Systems Laboratory, ETHZ, Zürich, Switzerland

H.A. van der Vorst

Mathematical Institute, Utrecht University, Utrecht, The Netherlands

1. What to expect in this chapter?

Linear systems of large dimensions arise in various applications including circuit simulation, semiconductor problems, and electro magnetic modeling. In this chapter we will discuss state of the art numerical methods for the solution of these linear systems. These methods fall belong to the traditional two different classes: direct solution methods based on Gaussian elimination techniques and iterative methods.

Over the past two decades, impressive progress has been made in the design of efficient implementations for Gaussian elimination. These improvements are achieved by clever ordering techniques in order to keep nonzero fill in the usually sparse matrices of the linear systems limited. Furthermore, blocking techniques are exploited in order to realize high computational throughput on computers with a memory hierarchy. Section 2 gives an overview of the various techniques, with emphasis on those techniques that have been incorporated in the sparse direct solver PARDISO. This solver has been successfully and routinely used for large semiconductor device equations.

In Section 3 we consider the alternative for direct methods when these become too expensive, either in memory space requirements or in CPU time: iterative methods. The methods that we will discuss are the so-called Krylov subspace methods. This class of methods contains the currently most effective general purpose methods, like GMRES, CG, and Bi-CGSTAB. These iterative methods usually lead to inefficient computation when applied in a straight forward manner to the given linear system and the common remedy to this is *preconditioning*. Preconditioning is often based on some sort of incomplete Gaussian elimination and can be viewed as the bridge between the world of direct

Numerical Methods in Electromagnetics
Special Volume (W.H.A. Schilders and E.J.W. ter Maten, Guest Editors) of
HANDBOOK OF NUMERICAL ANALYSIS, VOL. XIII
P.G. Ciarlet (Editor)

ISSN 1570-8659
DOI 10.1016/S1570-8659(04)13008-1

methods and of iterative methods. Section 4 discusses some of the more effective forms of preconditioning. In that section we pay also attention to techniques for improving the degree of parallelism in preconditioning.

We conclude with a numerical example that gives some idea of the effectiveness of PARDISO (the direct approach) and preconditioned Bi-CGSTAB(ℓ) (a Krylov subspace iterative solver).

We have included many pointers to the literature for further details on the methods that we touch on in this chapter.

2. Direct solution method

2.1. Introduction

Developing an efficient parallel, or even serial, direct solver for sparse systems of linear equations is a challenging task that has been a subject of research for the past four decades. Several breakthroughs have been made during this time. Especially, the research on general unsymmetric sparse systems of linear equations was a very active area during the past few years. Recent algorithmic improvements alone have reduced the time required for the sequential direct solution of unsymmetric sparse systems of linear equations by almost an order of magnitude. Combined with significant advances in the performance to cost ratio of parallel computing hardware during this period, current sparse direct solver technology makes it possible to solve problems quickly and easily that might been considered impractically large until recently.

The main purpose of this chapter is to give a general overview over different sparse direct methods and the related literature. In particular the algorithms implemented in PARDISO – a high-performance and robust software for solving general sparse linear systems – will be described. The solver has been developed by SCHENK and GÄRTNER [2002a, 2002b], SCHENK, GÄRTNER and FICHTNER [2000], SCHENK, GÄRTNER, FICHTNER and STRICKER [2001] and successfully used for solving large semiconductor device equations. The algorithms employed in the package are also suitable for other application areas such as electromagnetic, circuit, fluid or structural engineering problems.

Typical direct solvers for general sparse systems of linear equations of the form $Ax = b$ have four distinct phases, namely, *Analysis* comprising ordering for fill-in reduction and symbolic factorization, *Numerical Factorization* of the sparse coefficient matrix A into triangular factors L and U using Gaussian elimination with or without partial pivoting, *Forward and Backward Elimination* to solve for x using the triangular factors L and U and the right-hand side vector b, and *Iterative Refinement* of the computed solution.

There is a vast variety of algorithms associated with each step. The review papers by DUFF [1998], GUPTA [2001], and HEATH, NG and PEYTON [1990] can serve as excellent reference of various algorithms. The two main books discussing direct solution of sparse linear systems are those by GEORGE and LIU [1981] and DUFF, ERISMAN and REID [1986]. The first focuses on the discussion of symmetric positive definite systems

TABLE 2.1
The serial computational complexity of the various phases of solving a sparse system of linear equations arising from two- and three-dimensional constant node-degree graphs with n vertices

Phase	Dense	2-D complexity	3-D complexity
Ordering:	–	$O(n)$	$O(n)$
Symbolic factorization	–	$O(n \log n)$	$O(n^{4/3})$
Numerical factorization	$O(n^3)$	$O(n^{3/2})$	$O(n^2)$
Forward/backward elimination	$O(n^2)$	$O(n \log n)$	$O(n^{4/3})$

and emphasizes graph theoretic aspects, while the latter considers both symmetric and unsymmetric systems.

Usually the analysis phase involve only graphs of the matrices, and hence only integer operations. The numerical factorization, the forward/backward elimination and iterative refinement involve floating-point operations. Nevertheless, as shown in Table 2.1, the numerical factorization is the most time consuming phase and the forward and backward elimination is about an order of magnitude faster.

Sparse direct solver technologies are important because of their generality and robustness. For many linear systems arising in semiconductor device and process simulations direct methods are often preferred because the effort involved in determining and computing a good preconditioner for an iterative solution may outweigh the cost of direct factorizations. Furthermore, direct methods provide an effective means for solving multiple systems with the same coefficient matrix and different right-hand side vectors because the factorization needs to be performed only once.

2.2. *Reordering matrices to reduce fill*

The process of elimination introduces fill into the factors of a sparse matrix.[1] The amount of fill can be controlled by reordering rows and columns of the matrix. The rows and columns of the matrix can be reordered to reduce the fill-in. Reducing fill reduces the amount of memory that the factorization uses and the number of floating-point operations that it performs.

Choosing an ordering of the unknowns x is equivalent to the columns of the coefficient matrix and then eliminating the unknowns in the natural order, and choosing an ordering of the equations is equivalent to permuting the rows of the matrix. For example, factoring the matrix

$$\begin{bmatrix} x & x & x & x & x & x \\ x & x & & & & \\ x & & x & & & \\ x & & & x & & \\ x & & & & x & \\ x & & & & & x \end{bmatrix}, \tag{2.1}$$

[1]When A is sparse, most of its elements are zero. The factorization tends to produce additional nonzero elements in the factors, which are called fill-in entries.

(where x's represent nonzeros) results in completely filled factors. Reversing the order of both rows and columns yields a matrix whose factors do not fill at all.

$$\begin{bmatrix} x & & & & & x \\ & x & & & & x \\ & & x & & & x \\ & & & x & & x \\ & & & & x & x \\ x & x & x & x & x & x \end{bmatrix}. \tag{2.2}$$

This section focuses on fill reduction for sparse symmetric matrices.[2] Therefore, only symmetric permutation are considered that preserve symmetry. Symmetry is preserved by applying the same permutation to both the rows and the columns of a symmetric matrix. Ordering the rows and columns in Eq. (2.1) corresponds to eliminating the center of the star first. The elimination adds edges to make its neighbors, which are all the remaining vertices, into a clique, thereby completely filling the graph. The reversed ordering shown in Eq. (2.2) eliminates the center of the star last, so no fill edges are introduced.

Factoring a permuted matrix PAP^T, where P is the permutation matrix, may dramatically reduce the amount of work required for the factorization (DUFF, ERISMAN and REID [1986]). There is no efficient algorithm for finding an optimal ordering. This problem has been shown by YANNAKAKIS [1981] to be NP-complete.[3] Furthermore, no algorithm that provides a provably good approximation has been discovered, although the existence if such an algorithm has not been ruled out. There are, however, several classes of algorithms that work well ll in practice and two of them will be sketched in the next section.

One of the most commonly applied heuristics for performing reorderings is the multiple minimum degree algorithm proposed by GEORGE and LIU [1989], LIU [1985]. This method has been almost exclusively used in direct methods and has been found to produce very good orderings. The alternative approach, vertex-separator based orderings, also called nested dissection, was first proposed by GEORGE [1973] for regular element meshes. The method is strongly related to graph partitioning and quality and runtime of the algorithm has been significantly improved recently for irregular problems, for example by BARNARD and SIMON [1995], GUPTA [1996], HENDRICKSON and ROTHBERG [1996], and KARYPIS and KUMAR [1998a].

The two most popular techniques for reordering sparse matrices, minimum degree and vertex based separator orderings, are reviewed next and their effectiveness of these reordering approaches will be evaluated in this section for a wide range of sparse matrices from real-world applications Integrated Systems Engineering AG [1998a, 1998b].

2.2.1. Minimum degree algorithms

The method of the minimum degree algorithm is first considered. The intuition behind the method is elementary. Since the elimination of a vertex x causes its neighbors to

[2]Fill reduction orderings for general unsymmetric matrices A is commonly applied to the structure symmetric extension $\tilde{A} = A + A^T$.

[3]The complexity of the algorithm is not bounded by a polynomial in n.

```
S = {}
while S ≠ V do
    for x ∈ V \ S do
        δ(x) = |adj(x)|
    end for
    pick z ∈ T = {y ∈ V \ S | δ(y) = min_{x∈V\S} δ(x)}
    order z next
    S = S ∪ {z}
    eliminate z and determine the resulting graph
end
```

FIG. 2.1. The minimum degree algorithm.

become adjacent,[4] minimum degree always chooses a vertex of minimum degree[5] from the elimination graph to be eliminated next. Unfortunately, this very simple method has historically proven to be quite difficult to implement efficiently GEORGE and LIU [1989]. Early implementations of the minimum degree algorithm required enormous runtime. But fortunately, several variants have since been developed, whose runtimes are quite reasonable in comparison to the cost of the factorization. The first, the multiple minimum degree algorithm (MMD) LIU [1985], reduces the runtime of the algorithm by eliminating a set of vertices of minimum degree simultaneously. This multiple elimination technique dramatically reduces the cost of updating the degrees of the neighbors of eliminated vertices, which is the main cost of the algorithm. Whereas the minimum degree algorithm must update the neighbors' degrees each time a vertex is eliminated, the multiple minimum degree will often eliminate many neighbors of a vertex before updating that vertex's degree. A second method, suggested by AMESTOY, DAVIS and DUFF [1996] to reduce the runtime, is a recent variant of minimum degree, called approximate minimum degree (AMD). AMD further reduces the runtime by computing an inexpensive upper bound on a vertex's degree rather than the true degree.

Fig. 2.1 implements the minimum degree algorithm. This algorithm mainly consists of a single loop which is executed n times where $n = |V|$.[6] First, the degree $\delta(x)$ of all nodes x in the current elimination graph is determined. Next, a node z is selected from the set of nodes with minimum degree in the current elimination graph. Once a node z with minimum degree is selected, it is added to the set S containing all reordered nodes, and z is eliminated from the current elimination graph. For the next step, it is necessary to recompute the degree of the remaining nodes and the new elimination graph.

Due to the degree approach, the minimum degree is a local algorithm. The algorithm adds a vertex at each step to the elimination tree (LIU [1990]) based on local degree information and it combines two or more subtrees together to form a larger subtree. As

[4]In other words, deleting a vertex/nodes x from the graph $G(A)$/the matrix A causes all incident edges of x to be removed from $G(A)$; new edges (fill-in) are introduced into the new elimination graph such that all adjacent vertices of x become pair-wise adjacent.

[5]The degree of a node x is defined by the number of adjacent nodes in the elimination graph $G(A)$.

[6]V is the set of all vertices/nodes in the graph $G(A)$/matrix A.

a result, in the minimum degree algorithm the elimination tree grows from the bottom up.

The elimination trees produced by the multiple minimum degree algorithm are high and unbalanced. These elimination trees exhibit little concurrency so that a subtree mapping leads to significant load imbalances. However, LIU [1988, 1989] considered a method that preserves the operation count and the fill-in of the graph of $G(PAP^T)$ and at the same time it is more appropriate for parallel elimination.

Due to the degree approach, the minimum degree is a greedy algorithm. The minimum degree heuristic is greedy in the sense that it makes reasonable local choices, but without considering their global impact.

For an example of the effects of this ordering, see Fig. 2.2.

2.2.2. *Vertex-separator-based orderings*

Another important family of ordering algorithms is based on vertex separators, or dissection of the graph. The basic idea is simple. A small subset of the vertices of the graph is chosen which splits the graph $G(A) = (V, E)$ of the matrix A into two disconnected components V_1 and V_2. The small subset of vertices is called separator and each connected component is called a domain. In the vertex-separator based orderings, V_1 is ordered first, V_2 second, while the separator in S are numbered last.

It has not yet been specified how to order the vertices within one domain or within the separator. The vertices of the domain are typically ordered using the same algorithm recursively, by finding a separator that breaks the domain into subdomains and ordering the separator last. The vertices of a separator are usually ordered using a variant of the minimum degree algorithm.

Dissection orderings use a more global view of the elimination process, compared to the minimum degree orderings. Ordering the separator last ensures that there are no fill edge created between a vertex v in one domain and a vertex u in another domain. The effect of ordering the separator last, therefore is to ensure that an entire block of zeros in the original matrix does not fill. By requiring that no domain is very large (does not contain more that $2/3$ of the number of vertices in the graph, for example), it is ensured that the zero blocks are large.

If the separator breaks the graph into two domains, then once the domains are eliminated, the separator's vertices form one large clique. If there are more domains, the graph of the trailing submatrix may remain sparse.

The best separator of an $\sqrt{n}$-by-$\sqrt{n}$ mesh consists of $\sqrt{n}$ vertices and dissects the mesh into two equal domains. Each domains can be dissected into $\sqrt{n}/2$-by-$\sqrt{n}/2$ subdomains using a $\sqrt{n}/2$ separator. Now four square subdomains have been obtained, similar to the original graph and the algorithm continues recursively. It can be shown that the number of nonzeros in the factors in $\Omega(n \log n)$, and that this is optimal for this class of graphs. In fact, there are algorithms that order any matrix whose graph is a planar graph (can be drawn on the place with no crossing edge) such that the factor has $\Omega(n \log n)$ nonzeros. Theses algorithms uses a dissection strategy similar the one that has been outlined for regular meshes.

In the vertex-separator based ordering algorithm, the vertex separator S can be computed from an edge separator of a 2-way graph partitioning. The vertex S separator can

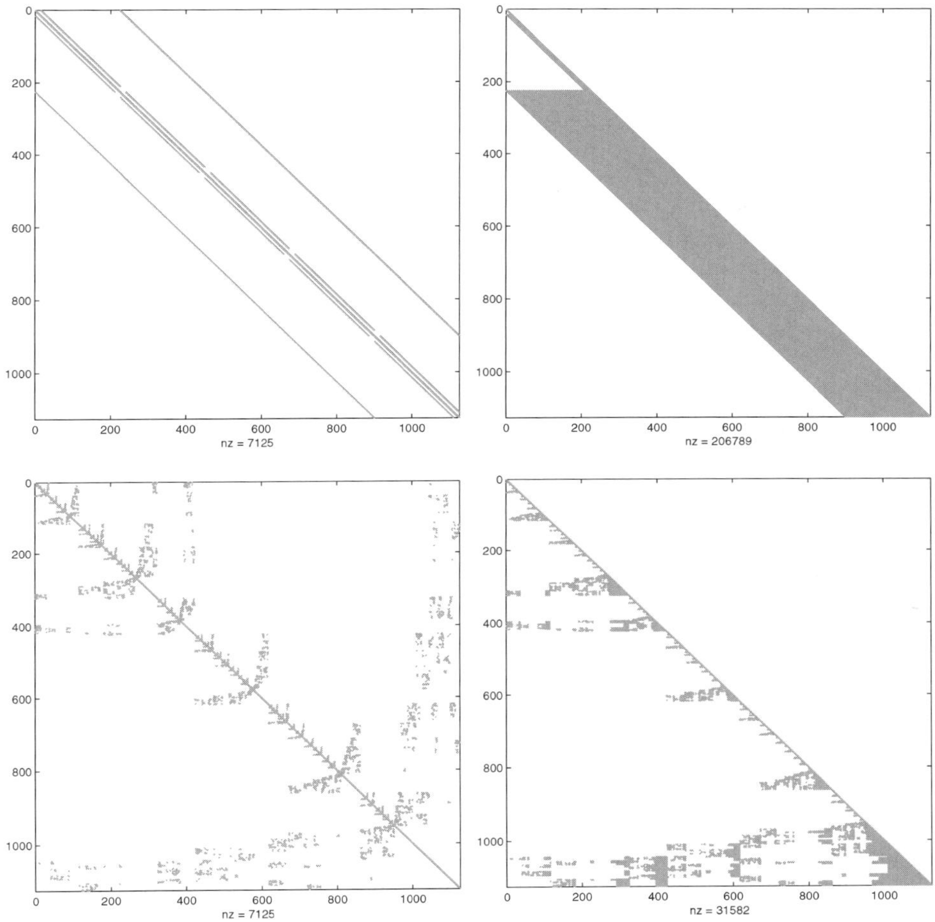

FIG. 2.2. The impact of ordering of a sparse matrix on fill in its Cholesky factor. The upper left figure shows the nonzero structure of a matrix whose graph is a 15-by-15-by-5 mesh. The ordering of the matrix corresponds to a natural row-by-row ordering of the mesh's vertices. The upper right figure shows the nonzero structure of the factors of the matrix. In the lower figures, the matrix (on the left) has been ordered using a minimum degree algorithm. Reordering reduces the number of nonzeros in the factor by 85%.

be computed from this edge separator by finding the minimum vertex cover (PAPADIMITRIOU and STEIGLITZ [1982], POTHEN and FAN [1990]) which has been found to compute small vertex separators from edge separators. Research on graph partitioning was a very active area in the last few years and has recently resulted into state-of-the art ordering codes, e.g., HENDRICKSON and LELAND [1995], HENDRICKSON and ROTHBERG [1996], KARYPIS and KUMAR [1998b], WALSHAW and CROSS [1999].

Graph partitioning. The 2-way graph partitioning problem is defined as follows: Given a graph $G = (V, E)$ with $|V| = n$, partition V into two subsets V_1, and V_2 such that $V_1 \cap V_2 = \emptyset$, $|V_1| = |V_2| = n/2$ and $V_1 \cup V_2 = V$, so that the number of edges of E whose incident vertices belong to different subsets is minimized. The effectiveness and the complexity of a nested dissection scheme depends on the quality/size of the separator and the effort to compute it. In general, small separators result in a small fill-in.

There are two classes of heuristics to solve the separator problem, depending on the information available about the graph G. The first class uses geometric information for each vertex v_i, such as $v_i = (x_i, y_i, z_i)$ coordinates (MILLER, TENG, THURSTON and VAVASIS [1998], GILBERT, MILLER and TENG [1998]). The second heuristic computes a partitioning without geometric information; only these methods are applicable for reordering sparse matrices. The Kernighan–Lin algorithm (KERNIGHAN and LIN [1970]) is one of the earliest graph partitioning methods without geometric information. It takes an initial partitioning and iteratively improves it by trying to swap groups of vertices between V_1 and V_2, greedily picking the group to swap that best minimizes the number of edge crossings at each step. In practice, it converges quickly to a local optimum if it has a good starting partition. Nowadays, this algorithm is used to improve partitions found by other methods. Due to the local view it is a nicely complement to algorithms which have a more global view of the problem but tend to ignore local characteristics. Until recently, one of the most prominent global algorithms has been the spectral partitioning method (BARNARD and SIMON [1995], FIEDLER [1973, 1975], POTHEN, SIMON and LIOU [1990]). However, these methods are expensive since they require the computation of the eigenvector corresponding to the smallest nonzero eigenvalue (Fiedler vector). Furthermore, it has recently turned out that variants of multilevel Kernighan–Lin algorithms have a smaller edge-cut compared with spectral methods KARYPIS and KUMAR [1998a]. The multilevel technique is used to accelerate the graph partitioning in HENDRICKSON and LELAND [1995], HENDRICKSON and ROTHBERG [1996], KARYPIS and KUMAR [1998b], WALSHAW and CROSS [1999].

2.2.3. *Multilevel nested dissection algorithms*

The various phases of the multilevel nested dissection (MLND) are shown in Fig. 2.3. During the coarsening phase, the size of the graph is successively decreased; during the initial partitioning phase, a bisection of the smaller graph is computed; and during the uncoarsening phase, the bisection is successively refined as it is projected to the larger graphs. During the uncoarsening phase the light lines indicate projected partitions, and dark lines indicate partitions that were produced after the refinement. Formally, the algorithm works on $G = (V_0, E_0)$ as follows (see also Fig. 2.4):

Coarsening phase. The graph is transformed in the coarsening phase into a sequence of smaller graphs $G_1, G_2, \ldots, G_m$ such that $|V_0| > |V_1| > |V_2| > \cdots > |V_m|$. Given the graph $G_i = (V_i, E_i)$, the coarser graph G_{i+1} can be obtained by collapsing adjacent vertices. Thus, the edge between two vertices is collapsed and a multinode consisting of these two vertices is created. This edge collapsing idea can be formally defined in terms of matchings. A matching of $G_i = (V_i, E_i)$ is a subset of edges no two of which share an endpoint. A matching is called maximal, if no other edge from E_i can be added. Thus, the next level coarser graph G_{i+1} is constructed from G_i by finding a maximal matching of G_i and collapsing the vertices being matched into multinodes. Since the

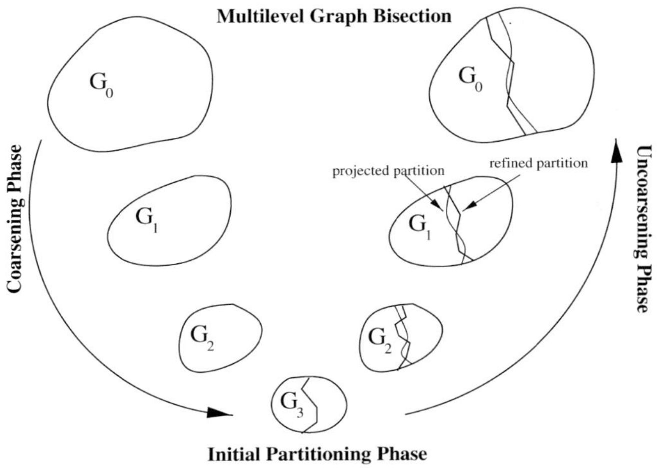

FIG. 2.3. The various phases of the multilevel graph bisection (KARYPIS and KUMAR [1998a]).

Function ML-Partition(G_0)
  if G_0 is small enough **then**
    Find partition (V_1, V_2) of G_0.
  else
    Coarsening Phase to obtain G_1.
    $(V_1', V_2') =$ ML-Partition(G_1).
    Uncoarsening Phase to obtain $(\widetilde{V}_1, \widetilde{V}_2)$.
    Kernighan-Lin Refinement to obtain (V_1, V_2).
  endif

FIG. 2.4. A multilevel graph bisection algorithm.

goal of collapsing vertices using matchings is to decrease the size of the graph G_i, the matching should be as large as possible.

The main difference between the various ordering packages (GUPTA [1996], HENDRICKSON and LELAND [1995], HENDRICKSON and ROTHBERG [1996], KARYPIS and KUMAR [1998b], WALSHAW and CROSS [1999]) is the construction of the maximal matching. One of the most popular methods are random matching (HENDRICKSON and LELAND [1993]), heavy edge matching (KARYPIS and KUMAR [1998a]), and heaviest edge matching (GUPTA [1996]).

Partitioning phase. A 2-way partition P_m of the graph $G_m = (V_m, E_m)$ is computed that splits V_m into two parts, each containing half the vertices of G_m. The partition of G_m can be obtained by using various algorithms such as spectral bisection (BARNARD and SIMON [1993], HENDRICKSON and LELAND [1995], POTHEN, SIMON and LIOU [1990]) or combinatorial methods (BARNES [1985], BUI, CHAUDHURI, LEIGHTON and SIPSER [1987], GEORGE [1973], GEORGE and LIU [1981], KERNIGHAN and LIN

[1970], GÖHRING and SAAD [1994], HAMMOND [1992]). It is shown in KARYPIS and KUMAR [1998a] that combinatorial methods generally finds smaller edge-cut separators compared with spectral bisection for partitioning the coarse graph. However, since the size of the coarsest graph G_m is small (i.e., $|V_m| < 100$), this step takes a small amount of time.

Uncoarsening phase. The partition P_m of G_m is projected back to G_0 by going through intermediate partitions $P_{m-1}, P_{m-2}, \ldots, P_1, P_0$. Each vertex v of G_{i+1} contains a distinct subset of vertices V_i^v of G_i. Obtaining P_i from P_{i+1} is done by simply assigning the set of vertices V_i^v collapsed to $v \in G_{i+1}$ to the appropriate partition in G_i. Although P_{i+1} is a local minimum partition of G_{i+1}, the projected partition P_i will not be, e.g., a local minimum with respect to G_i. Since G_i is finer, it has more degrees of freedom that can be used to improve P_i, and decrease the edge-cut of the partition. Hence, it may still be possible to improve the projected partition of G_i by local refinement heuristics. The refinement is usually done by using one of the variants of the Kernighan–Lin partition algorithm (FIDUCCIA and MATTHEYSES [1982], KERNIGHAN and LIN [1970]).

2.2.4. *The impact of the preprocessing algorithms*

The impact of the two preprocessing algorithms, minimum degree and vertex separator based orderings, is evaluated on a wide range of sparse matrices arising in two- and three-dimensional semiconductor process and device simulation problems (INTEGRATED SYSTEMS ENGINEERING AG [1998a, 1998b]). The characteristics of these matrices are described in Table 2.2. All matrices are structurally symmetric or have been extended to a structurally symmetric one. The purpose of the collection was to cover a wide range from small two-dimensional problems, up to larger three-dimensional matrices.

The quality of the ordering produced by the multilevel dissection algorithm from METIS (KARYPIS and KUMAR [1998a]) compared to that of multiple minimum degree

TABLE 2.2
Characteristics of the test matrices from semiconductor process and device simulation

#	Matrix	Rows	Nonzeros in A	Nonzeros/Row
1	2D eth-points	151’389	1’046’105	6.91
2	2D eth-load.motodop	35’804	221’938	6.19
3	2D eth-load.bic.hv15h	56’941	393’415	6.91
4	2D eth-big	13’209	91’465	6.92
5	2D ise-mosfet-1	12’024	250’740	20.85
6	2D ise-mosfet-2	24’123	504’765	20.92
7	3D eth-3d-eeprom	12’002	630’002	52.49
8	3D ise-igbt-coupled	18’668	412’674	22.10
9	3D eth-3d-eclt	25’170	1’236’312	49.11
10	3D ise-soir-coupled	29’907	2’004’323	67.01
11	3D eth-3d-mosfet	31’789	1’633’499	51.38
12	3D eth-3d-eclt-big	59’648	3’225’942	54.08

TABLE 2.3
The number of operations in Mflops required to factor the test matrices when ordered with multiple minimum degree MMD LIU [1985] and multilevel nested dissection MLND from the METIS package 4.0 (KARYPIS and KUMAR [1998a])

#	Matrix	MMD	MLND
s	2D eth-points	**239.7**	259.1
2	2D eth-load.motodop	**51.2**	52.1
3	2D eth-load.bic.hv15h	**123.1**	120.0
4	2D eth-big	36.3	**35.0**
5	2D ise-mosfet-1	**167.8**	169.2
6	2D ise-mosfet-2	**370.7**	380.4
7	3D eth-3d-eeprom	5'723.9	**3'105.4**
8	3D ise-igbt-coupled	8'048.3	**3'866.8**
9	3D eth-3d-eclt	27'042.3	**12'079.5**
10	3D ise-soir-coupled	55'476.8	**23'430.6**
11	3D eth-3d-mosfet	53'026.9	**22'339.8**
12	3D eth-3d-eclt-big	245'261.3	**75'102.6**

is shown in Table 2.3. It can be seen that MLND produces better orderings in terms of floating point performance for 7 out of the 12 sparse matrices. Interestingly, MLND performs consistently better than MMD if sparse matrices from three-dimensional applications are considered. When all six three-dimensional cases are considered, MMD produces orderings that require a total of 394'576 Mflops, whereas the produced orderings by MLND require only 139'921 Mflops. On the other hand, when the two-dimensional cases are considered, the factorization algorithms that use orderings from MMD require generally less fill-in and less floating point operations – but the difference in 2-D cases is a minor one.

The main conclusions that can be drawn from this study are: (1) the multilevel nested dissection algorithm used to find a fill-in reducing ordering is substantially better than multiple minimum degree for three-dimensional irregular matrices, and (2) the multiple minimum degree method performs better for most of the two-dimensional problems.

2.3. *Fast sparse matrix factorization on modern workstations*

2.3.1. *Two primary approaches to factorization: left-looking and multifrontal methods*

This section provides a brief description of the process of factorization of a sparse linear system. For simplicity of the notations, the Cholesky factorization is considered. The goal is to factor a matrix A into the form LL^T. The equations which govern the factorization are:

$$l_{j,j} = \left(a_{j,j} - \sum_{k=1}^{j-1} l_{j,k}^2\right)^{1/2}, \tag{2.3}$$

$$l_{i,j} = a_{i,j} - \sum_{k=1}^{j-1} l_{i,k} \cdot l_{j,k} / l_{j,j}. \tag{2.4}$$

```
1:  for j = 1 to n do
2:      f = A(j : n, j);
3:      for each k with L(j, k) ≠ 0 do
4:          f = f − L(j, k) · L(:, k);
5:      end for;
6:      L(j : n, j) = f;
7:      L(j, j) = √L(j, j);
8:      for each i with L(i, j) ≠ 0 do
9:          L(i, j) = L(i, j)/L(j, j);
10:     end for;
11: end for;
```

FIG. 2.5. Pseudo-code of the left-looking factorization method.

```
1:  set L = A;
2:  for k = 1 to n do
3:      L(k, k) = √L(k, k);
4:      for each i with L(i, k) ≠ 0 do
5:          L(i, k) = L(i, k)/L(k, k);
6:      end for;
7:      for each j with L(j, k) ≠ 0 do
8:          L(:, j) = L(:, j) − L(j, k) · L(:, k);
9:      end for;
10: end for;
```

FIG. 2.6. Pseudo-code of the multifrontal factorization method.

Since the matrix A is sparse, many of the entries in L will be zero. Therefore, it is only necessary to sum over those k for which $l_{j,k} \neq 0$. The above Eqs. (2.3) and (2.4) lead to two primary approaches to factorization: the left-looking method and the multifrontal method.

The left-looking method can be described by the pseudo-code[7] given in Fig. 2.5. In this method, a column j of L is computed by gathering all contributions to j from previously computed columns k. Since step 4 of the pseudo-code involves two columns, j and k, with potentially different nonzero structures, the problem of matching corresponding nonzeros must be resolved. In the left-looking method, the nonzeros are matched by scattering the contribution of each column k into a dense vector f. Once all k's have been processed, the net contribution is gathered from the dense vector f and added into column j. The classical form of this method is, e.g., employed in SPARSPAK (GEORGE and LIU [1980]).

The multifrontal method can be roughly described by the pseudo-code given in Fig. 2.6. In the multifrontal method, once a column k is completed it immediately generates all contributions which it will make to subsequent columns. In order to solve the

[7]Matlab notation is used for integer ranges: $(r : s)$ is the range of integers $(r, r+1, \ldots, s)$.

problem of matching nonzeros from column j and k in step 8, this set of contributions is collected into a dense lower triangular matrix, called the frontal matrix. This matrix is then stored in a separate storage area, called the update matrix stack. When a later column k of L is to be computed, all update matrices which affect k are removed from the stack and combined in a step called assembly. Column k is then completed, and its update matrices which affect k are removed from the stack. The update matrices are combined with the as yet unapplied updates matrices which modified k, and a new update matrix is placed on the stack. The columns are processed such that the needed update matrices are always at the top of the update matrix stack. This method was originally developed DUFF and REID [1983] to increase the percentage of vectorizable work in sparse factorization. Is has the disadvantage that it requires more storage than the left-looking method, since the update matrix stack must be maintained in addition to the storage for L.

2.3.2. Supernode sparse factorization: reducing the number of memory references

The concept of supernode elimination, which was first proposed in EISENSTAT, SCHULTZ and SHERMAN [1981] and successfully exploited by Ashcraft and others in ASHCRAFT, GRIMES, LEWIS, PEYTON and SIMON [1987], is important because it allows a high vector utilization and it is one key concept to decrease the number of memory references during sparse factorization.

In the process of sparse factorization[8] when column k of L modifies column j, the nonzeros of column k form nonzeros in corresponding positions of column j. As the factorization proceeds, this unioning of sparsity structures tends to create sets of columns with the same structures. These sets of columns are called supernodes. A supernode is a set of contiguous columns in the factor whose nonzero structure consists of a dense triangular block in the diagonal, and an identical set of nonzeros for each column below the diagonal block.[9] For example, in Fig. 2.7, the following supernodes can be identified: {A, B}, {D, E}, {G, H, I}, {K, L}, {M, N}, and {P, Q, R}. Supernodes arise in any sparse factor, and they are typically quite large.

Supernode elimination is a technique whereby the structure of a matrix's supernode is exploited in order to replace sparse vector operations (BLAS-1 operations) by dense matrix operations (BLAS-3 operations). When a column from a supernode is to update another column, then every column in that supernode will also update that column, since they all have the same structure. In the example matrix, the three columns {G, H, I} all update columns {K, L}, {M, N}, and {Q, R}.

Both sparse factorization methods, the left-looking and the multifrontal method, benefit from the supernode structure. The left-looking supernode method, e.g., exploits supernodes in the following way. Instead of scattering the contribution of each column of the supernodes into the dense vector, as it would ordinarily be done in general sparse factorization, the contribution of all columns in the supernode are first combined into a single dense vector, and that vector is then scattered. Since the storage of the nonze-

---

[8] A matrix A with a symmetric nonzero structure is considered.

[9] In EISENSTAT, GILBERT and LIU [1993] several possible ways have been considered to generalize the symmetric definition of supernodes to unsymmetric factorization.

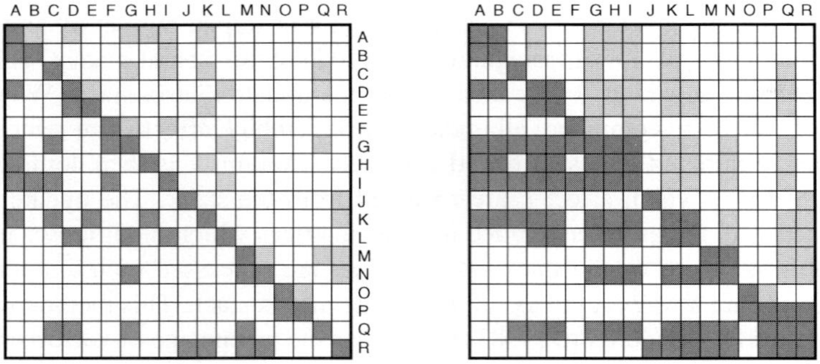

FIG. 2.7. The nonzero structure of a matrix A and its factors L and U.

ros of a single column is contiguous and all the columns have the same structure, this combination can be done as a series of dense vector operations.

Supernodes substantially decrease the number of memory references when performing sparse factorization. The supernode technique replaces a sequence of indirect vector operations with a sequence of direct vector operations followed by a single indirect operation. Each indirect operation requires the loading into the processor registers of both the index vector and the values vector. The pseudo-code and the supernode structure of a supernode-supernode Cholesky factorization is given in Fig. 2.8. In the next section it is shown how to organize a sparse left-looking LU factorization in order to compute the factors essentially by Level-3 BLAS (DONGARRA, DUCROZ, DUFF and HAMMARLING [1990]) and LAPACK (DONGARRA and DEMMEL [1991]) routines. A more detailed discussion of the advantages of supernode factorization on high-performance workstations can be found in SCHENK and GÄRTNER [2001].

2.3.3. Level-3 BLAS sparse factorization: the benefits of rectangular blocking

One key objective for LU factorization is to take advantage of the memory hierarchy on modern microprocessor workstations. These machines typically have multiple pipelined functional units, pipelined floating-point units, and a fast, but relatively small cache memory. Block algorithms are often used on these architectures, because positive cache effects and therefore a high floating-point performance can be expected. Basic concepts of block partitioning are adopted in the BLAS and LAPACK packages and many other standard benchmarks are designed with block partitioning in order to exploit the number crunching capabilities of these systems. A brief review of the block Gaussian elimination on dense linear systems is given in the next section and it is shown how a BLAS-3 supernode LU factorization can be organized on high-end workstations.

Block LU factorization of dense linear systems. Block LU or JIK-SDOT factorization is often used for the solution of a dense system of linear equations on modern workstations and vector supercomputers (DONGARRA [1998]). This is one feasible method on architectures with a hierarchical-memory system. Therefore, this factorization algorithm and some of its basic properties are reviewed in this section. A detailed descrip-

---

for each destination supernode $(r_2 : s_2)$ **do**
  for $j = r_2$ **to** s_2 **do**
    $f = A(j : n, j)$;
    for each supernode $(r_1 : s_1) < (r_2 : s_2)$ with $L(j, r_1 : s_1) \neq 0$ **do**
      for $k = r_1$ **to** s_1 **do**
        for $i = j$ **to** n with $L(i, k) \neq 0$ **do**
          $f = f - L(i, k) \cdot L(j, k)$;
        end for;
      end for;
    end for;
    $L(j : n, j) = f$;
  end for;
  Inner dense factorization for $L(r_2 : n, r_2 : s_2)$;
end for;

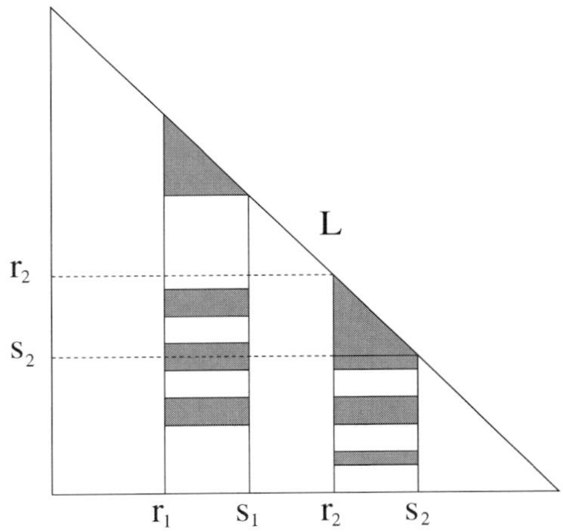

FIG. 2.8. Left-looking sparse Cholesky factorization with supernode-supernode updates.

tion of all possible forms of dense LU factorization can be found in DAYDÉ and DUFF [1989].

Each block contains nb columns and rows. At the kth step of the elimination process, one block column of L and one block row of U are computed. A block partitioning of A, L, and U is depicted in Fig. 2.9. The computation of one block column of L and one block row of U requires at each step the following operations:

1. The external modification of the block columns of L and the rows of U:

$$C_k \leftarrow C_k - \begin{pmatrix} A_k^1 \\ A_k^2 \end{pmatrix} B_k, \qquad U_k^2 \leftarrow U_k^2 - A_k^1 E_k. \tag{2.5}$$

2. The internal factorization of the diagonal block of C_k, using the factorization in the left part of Fig. 2.9, to obtain the factors L_k^1 and U_k^1.

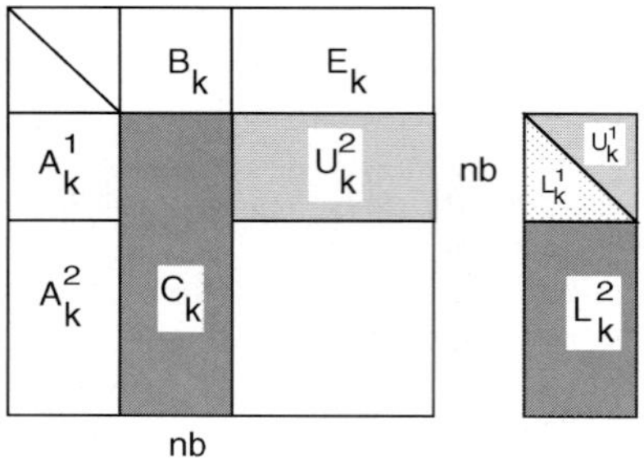

FIG. 2.9. Dense block LU factorization. The diagonal blocks L_k^1 and U_k^1 are stored together in one square block.

3. The internal factorization of the block columns of L and the rows of U:

$$L_k^2 \leftarrow L_k^2(U_k^1)^{-1}, \qquad U_k^2 \leftarrow (L_k^1)^{-1}U_k^2. \tag{2.6}$$

Having all external block updates processed in (2.5), the diagonal block can be factorized internally to obtain the factors L_k^1 and U_k^1. Finally, after the diagonal block is decomposed into the triangular matrices L_k^1 and U_k^1, the other blocks L_k^2 and U_k^2 can be determined.

In Section 2.3.4 the block technique is extended to sparse linear systems in order to compute the sparse LU factorization essentially by Level-3 BLAS routines.

2.3.4. Block LU factorization of sparse linear systems

The key idea on modern workstations deals with the data representation of the factors L and U. As it was discussed in the previous section, one important feature of the dense block LU factorization is the rectangular block structure of C_k in Fig. 2.9. The supernode diagonal portion U_k^1 of U is stored together with the supernode columns L_k^1 and L_k^2 of L. This supernode block numerical factorization involves mainly dense matrix-matrix multiplications resulting in high computational performance on modern computer architectures.

The impact of the rectangular supernode block structure is now discussed in detail. Fig. 2.7 depicts an example matrix A, the nonzero structure of the factors L and U and the left-looking factorization of an example supernode is represented in Fig. 2.10. The rectangular storage scheme of L and U is shown with two different types of shading in these figures.

In order to provide a more detailed picture of how the factorization schemes interact with the BLAS-3 routines, Figs. 2.11 and 2.12 present the external numerical factorization of the supernode {G, H, I} with the left-looking supernode-supernode approach. In the left-looking approach, the factorization of supernode {G, H, I} is performed by gathering all contributions to {G, H, I} from previously computed supernodes: {A, B}, {C}, {D, E}, and {F}. It is assumed that the nonzeros are stored continuously in an increasing order by columns on L and rows on U. Hence, the multiplication can be done without an index vector inside the DGEMM loop. Once one external supernode-supernode multiplication has been performed, the result is assembled in the destination supernode.

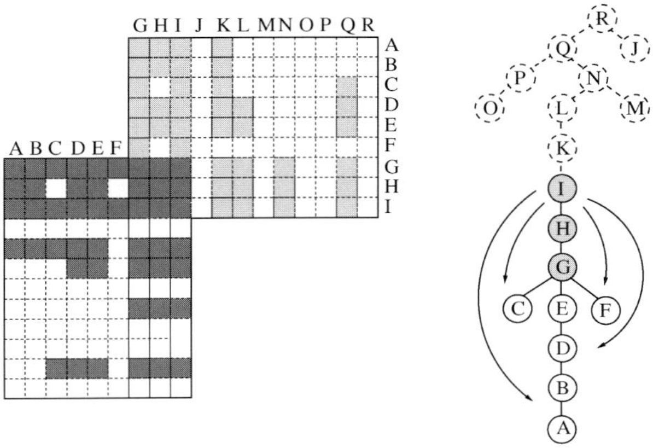

FIG. 2.10. The left-looking numerical factorization of supernode S(G, H, I). Nodes below node I in the elimination tree are involved in the left-looking factorization of supernode S(G, H, I).

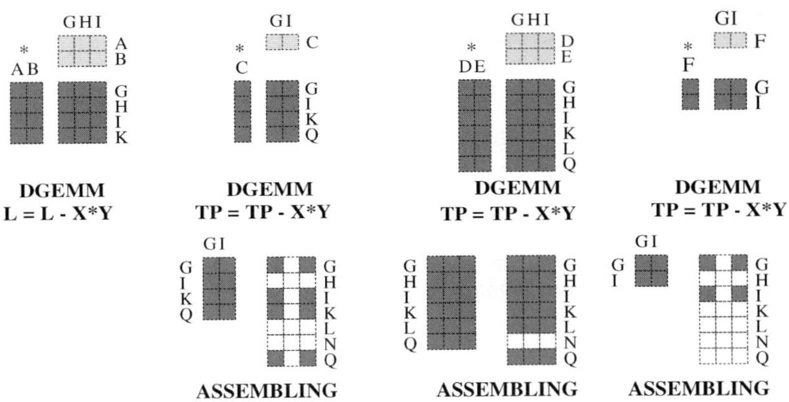

FIG. 2.11. The external numerical factorization of the factor L and the upper triangular dense diagonal block of the factor U.

It is important to note that the floating point operation phase is completely separated from the assembling phase and no indirectly accessed operands are required to perform the external supernode-supernode updates. The result is substantially fewer memory operations, since all elements are stored contiguously in the memory and $O(m^3)$ operations are performed with $O(m^2)$ operands if the two supernodes are $m \times m$ matrices. An assembly phase after the update with supernode {A, B} is not necessary since both supernodes share essentially the same nonzero structure. For the other three updates the temporary block TP contains the intermediate results of the supernode updates. The results are then scattered in the appropriate entries of L and U.

In examining Fig. 2.13, the question arises, which factorization method is used for the internal factorization of supernode {G, H, I}. In the example matrix, the three columns

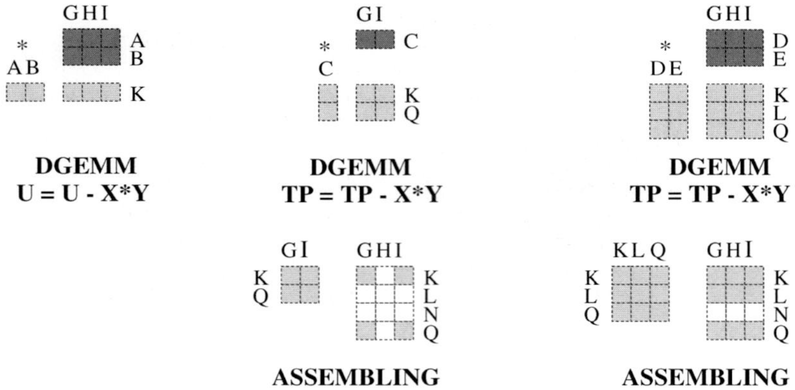

FIG. 2.12. The external numerical factorization of the remaining part of the factor U.

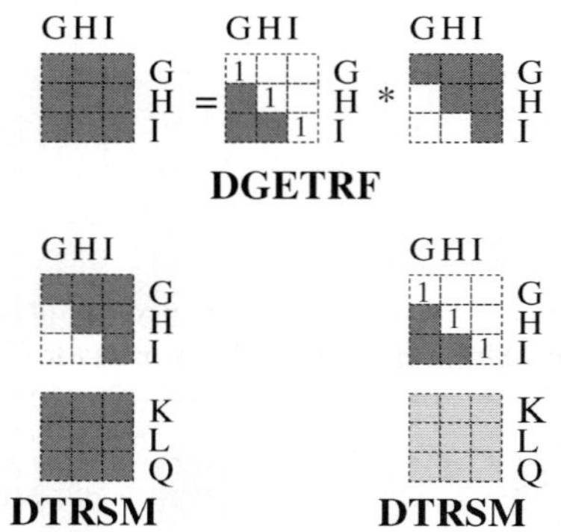

FIG. 2.13. The internal numerical factorization of the supernode dense diagonal block with Level-3 LAPACK routines.

and rows in supernode {G, H, I} are factorized by the LAPACK subroutines DGETRF. Finally, after the diagonal block is decomposed into the supernode triangular matrices, the remaining rows and columns of L and U can be determined by substitution with the routine DTRSM.[10]

2.3.5. *Block sparse factorization performance on modern workstations*

Any performance data given today will be invalid tomorrow – hence the typical patterns are illustrated by measurement data due to one main development platform: the COMPAQ Alpha workstation with an Alpha 21164 processor which is quite representative of the general class of modern high performance workstations. The 21164 is a four-way

[10]The LAPACK routines DGETRF computes an LU factorization of a general dense M-by-N matrix A using partial pivoting with row interchanges. DTRSM solves one of the dense matrix equations $AX = \alpha B$, $A^T X = \alpha B$, $XA = \alpha B$, or $XA^T = \alpha B$, where α is a scalar, X and B are M-by-N matrices, A is a unit, or nonunit, upper or lower triangular matrix.

TABLE 2.4
Performance of the numerical factorization for PARDISO on one CPU of a COMPAQ AlphaServer 4100

COMPAQ AlphaServer 4100, 600 MHz EV5.6 (21164A)					
Matrix	nnz(LU)	$\frac{nnz(LU)}{nnz(A)}$	# Mflops	Seconds	Mflop/s
7	7'614'492	18.45	3'866.8	9.11	424.36
9	17'914'354	14.49	12'079.5	25.96	465.29
10	28'342'865	14.14	23'430.6	46.21	507.03
11	26'732'577	16.36	22'339.8	44.89	497.63
12	63'785'130	19.77	75'102.6	146.45	512.99

superscalar RISC processor. The Alpha processor has an extremely fast clock rate of 600 MHz. It has an 8-KB first-level instruction cache, an 8-K first-level data cache, and a 96-K second-level cache on chip. Off the chip is a 8 MB direct-mapped third-level cache. The processor has one floating point add pipeline and one floating point multiply pipeline with a throughput of one floating point operation each per cycle. The peak floating rate is therefore 1'200 Mflop/s. By measurements, the DGEMM[11] routine from the DXML library achieves about 856 Mflop/s and the LINPACK[12] performance is reported to be 764 Mflop/s.

In order to evaluate the computational performance of the sparse block LU factorization algorithm, Table 2.4 gives the run-times in seconds and the Mflop/s rate for some matrices of Table 2.2 on an Alpha EV5.6 21164 processor. The described sparse Level-3 BLAS algorithm has been implemented into the PARDISO package SCHENK [2000]. The PARDISO performance ranges from 50% to 70% of the DGEMM performance showing that good use of the Level-3 BLAS can be obtained due to the rectangular supernode structure.

2.4. *Pivoting strategies to control accuracy*

So far only fill-in minimizing strategies and Level-3 BLAS numerical factorization methods with diagonal pivoting have been considered. It is well known that, e.g., for a special class of problems where A is symmetric and positive definite, pivots can be chosen down the diagonal (GOLUB and VAN LOAN [1996]). These diagonal pivots then are always nonzero and the element growth is limited. Symmetric permutations PAP^T can therefore be chosen solely on sparsity reasons.[13] Similar to the symmetric positive case, there exists another subclass of matrices for which Gaussian elimination without pivoting is stable. This subclass contains matrices that are diagonally dominant[14] but

[11]This is the Level-3 BLAS matrix-matrix multiplication routine $C = AB + C$, where A, B and C are rectangular dense matrices.

[12]This is a $1'000 \times 1'000$ matrix solution of $Ax = b$ using Gaussian elimination with code changes allowed for increased performance (such as Level-3 BLAS manufacturer-supplied numerical linear algebra libraries).

[13]The property of symmetric and positive definiteness is obviously preserved under symmetric permutations.

[14]A matrix $A \in \boldsymbol{R}^{nxn}$ is said to be diagonally dominant if $|a_{i,i}| \geqslant \sum_{j=1,j\neq i}^{n} |a_{i,j}|, i = 1, \ldots, n$.

are not positive definite. Permutations can also be chosen for such matrices with respect only to the sparsity.

However, there are also classes of sparse matrices that require pivoting in order to ensure numerical stability. Gaussian elimination, e.g., of a general sparse unsymmetric matrix or a sparse indefinite matrices, must be combined with pivoting. For these matrices it is possible to encounter arbitrarily small pivots on the diagonal. If still diagonal pivoting is used, large element growth may occur, yielding an unstable algorithm. This problem can be alleviated by pivoting on the element with the largest magnitude in each submatrix or column, interchanging rows and columns when needed. Generally, there are four pivoting strategies referred to as *complete and partial pivoting*, *Bunch and Kaufman*, *static*, and *supernode* pivoting that are widely used in sparse Gaussian elimination.

Complete and partial pivoting. Complete pivoting selects at each step k the element with the largest absolute value in the reduced submatrix (STOER and BULIRSCH [1983]). On the other hand, partial pivoting involves only a search for the element with the largest absolute value in either the row or the column. It is not as stable as complete pivoting, but in practice it gives good results for a lower computational cost (WILKINSON [1961]). However, complete pivoting can produce a stable factorization where partial pivoting fails (WRIGHT [1993]).

At each partial pivoting elimination step k a pivot element has to be chosen as the largest element in the column to be eliminated,

$$\left|a_{l,k}^{(k)}\right| = \max_{i=k,\dots,n} \left|a_{i,k}^{(k)}\right|. \tag{2.7}$$

Here, $a_{i,j}^{(k)}$ refers to elements in the matrix obtained after $(k-1)$ elimination steps.

It is fairly straightforward to implement a dense partial pivoting algorithm. For a sparse matrix, however, off-diagonal pivoting is tremendously difficult to implement mainly due to the following reason. If the element $a_{l,k}^{(k)}$ is chosen as a pivot element in the kth elimination step, it will modify a number of elements during the elimination. This is the reason why the nonzero pattern in L and U depends on the row interchanges and cannot be predetermined precisely from the structure of A. Consider, for example, the following unsymmetric sparse matrix given by GILBERT [1994]

$$A = \begin{pmatrix} 1 & & \\ & 2 & \\ \bullet & \bullet & 3 \end{pmatrix}. \tag{2.8}$$

Depending on the relative magnitudes of the nonzero entries, pivoting could cause the structure of U to be any of the four structures:

$$\begin{pmatrix} 1 & & \\ & 2 & \\ & & 3 \end{pmatrix}, \quad \begin{pmatrix} 1 & \bullet & \\ & 2 & \bullet \\ & & 3 \end{pmatrix}, \quad \begin{pmatrix} 1 & \bullet & \\ & 2 & \\ & & 3 \end{pmatrix}, \quad \begin{pmatrix} 1 & & \bullet \\ & 2 & \bullet \\ & & 3 \end{pmatrix}. \tag{2.9}$$

Consequently, one disadvantage of complete or partial pivoting during the numerical factorization is that the amount of fill-in cannot be determined a priori. Hence, the symbolic nonzero structure prediction cannot be treated as a separate process decoupled from the numerical factorization. Moreover, additional fill-in due to numerical pivoting occurs during the elimination. A threshold criterion is often used to restrict the partial pivoting rule and to balance sparsity and stability. Thus, a pivot candidate $a_{l,k}^{(k)}$ is only accepted if it satisfies the Markovitz inequality (MARKOWITZ [1957])

$$\left|a_{l,k}^{(k)}\right| \geqslant u \cdot \max_{i=k,\dots,n} \left|a_{i,k}^{(k)}\right|. \tag{2.10}$$

The threshold value u balances sparsity and numerical stability. Only sparsity considerations are made for $u = 0$, while $u = 1$ gives partial pivoting in (2.7). The value $u = 0.1$ has been recommended by many authors. It is remarkable that a small u sometimes introduces more fill-in than a larger one. An example is discussed in DUFF, ERISMAN and REID [1986], where the number of elements in the factor are almost the same for $u = 10^{-10}$ and $u = 1$, while a minimum is attained for $u = 0.1$.

The effect of threshold partial pivoting in LU numerical factorization is analyzed by AMESTOY [1990] and the number of operations with LU factorization are compared with the diagonal pivoting method. It is observed that the increase in the number of operations due to numerical threshold partial pivoting is significant and lies between 7% and 200% for sparse matrices from the Harwell–Boeing test collection (DUFF, GRIMES and LEWIS [1989]).

Bunch and Kaufman pivoting. An efficient strategy for symmetric matrices has been proposed by BUNCH and KAUFMANN [1977]. Let A be symmetric, but indefinite ($-c < x^T Ax < c, c > 0$). Although A may have an LDL^T factorization, the entries in the factors can have arbitrary magnitude. Pivoting along the diagonal[15] may, for general indefinite matrices, be insufficient. The small pivot candidates would cause excessive growth and rounding; they must therefore be rejected. Thus, LDL^T with symmetric pivoting cannot be recommended as a reliable approach to solve symmetric indefinite systems. The challenge is to involve the off-diagonal entries in the pivoting process while maintaining a large portion of the symmetry.

One way to control the element growth and to preserve symmetry at the same time is to use the combination of 1×1 and 2×2 pivots. These block pivots correspond to the simultaneous elimination of two columns in A. If there are no appropriate diagonal pivots, the hope is to find a suitable 2×2 submatrix with not too small determinants. The elimination procedure leads to the generalized LDL^T factorization, where the matrix D is a block diagonal with either 1×1 or 2×2 diagonal elements. In absence of large diagonal elements $a_{i,i}$, optimal pivot blocks

$$D = \begin{pmatrix} a_{i,i} & a_{i,j} \\ a_{j,i} & a_{j,j} \end{pmatrix},$$

are instead to be found and be permuted into pivot position. The large number of candidate blocks[16] makes it necessary to restrict the considered pivot blocks to smaller

[15]Symmetric pivoting PAP^T always selects the pivots from the diagonal and small diagonal entries cause large nonzero entries in the factors.

[16]In the kth elimination of a supernode with n_s rows and columns there are $(n_s - k + 1)(n_s - k)$ possible candidates.

subsets. For simplicity of notation, the first elimination step is considered. The procedure can then be described by the following steps:

1. Let $|a_{r,1}|$ be the largest element in magnitude in the first column. If $|a_{1,1}| \geqslant \rho|a_{r,1}|$, then choose $a_{1,1}$ as a 1×1 pivot block.
2. Otherwise, determine the largest off-diagonal element in magnitude $a_{r,s}$ in the rth row. If $|a_{1,1}a_{r,s}| \geqslant \rho a_{r,1}^2$, then choose $a_{1,1}$ as a 1×1 pivot block.
3. Otherwise, if $|a_{r,r}| \geqslant \rho|a_{r,s}|$ choose $a_{r,r}$ as a 1×1 pivot block. Else, choose

$$D = \begin{pmatrix} a_{1,1} & a_{1,r} \\ a_{r,1} & a_{r,r} \end{pmatrix},$$

as a 2×2 pivot block.

The parameter $\rho = (\sqrt{17}+1)/8$ is chosen to minimize the element growth. With this choice, the element growth after k steps is bounded by the factor $(2.57)^{k-1}$. Increasing ρ would decrease the growth but would also increase the cost for the pivot search. The choice of strategy must be a balance between stability and computational effort. A nice feature of the scheme is the fact that symmetric positive definite matrices pass the procedure without pivoting. The Cholesky factorization is therefore obtained in the form LDL^T.

Static pivoting. Static pivoting as an alternative to partial pivoting to stabilize sparse Gaussian elimination is proposed by LI and DEMMEL [1999]. The main advantage of static pivoting over partial pivoting is the possibility to permit a priori computation of the nonzero structure of the factors, which makes the factorization potentially more scalable on distributed-memory machines than factorizations in which the communications tasks only become apparent during the elimination process.

The original matrix A is permuted and scaled before the factorization to make the diagonal entries large compared to the off-diagonal entries by using the algorithm of DUFF and KOSTER [1997, 1999]. The magnitude of any tiny pivot, which is encountered during the factorization, is tested against a threshold of $\varepsilon^{1/2}\|A\|$, where ε is the machine precision and $\|A\|$ is the norm of A. If it is less than this value it is immediately set to this value (with the same sign) and the modified entry is used as pivot. This corresponds to a half-precision perturbation to the original matrix entry. As a result, the factorization is not exact and iterative refinement may be needed. However, numerical experiments demonstrate that the method is as stable as partial pivoting for a wide range of problems (AMESTOY, DUFF, L'EXCELLENT and LI [2000], SCHENK and GÄRTNER [2001]).

Supernode pivoting. In the previous Section 2.3 it was shown that significant gains in the execution times can be obtained by a Level-3 BLAS factorization method. The advantage of a Level-3 BLAS update is especially large for symmetric and structurally symmetric linear systems. Unfortunately, partial pivoting destroys the symmetric structure of the factors L and U. Due to the resulting unsymmetric structure of the factors L and U, an unsymmetric factorization concept has to be chosen. A supernode algorithm for these linear systems with partial pivoting has been developed in DEMMEL, GILBERT and LI [1999]. The kernel operation is based on a Level-2 BLAS supernode-column update and it is extended to a Level-2.5 BLAS algorithm. However, the computational performance of the Level-2.5 BLAS implementation achieves only 39% of the Level-3 performance on a COMPAQ Alpha 21164 (LI [1996]). To summarize, performance degradation with partial pivoting or threshold pivoting is due to:

1. additional fill-in during the numerical factorization,
2. the unsymmetric structure of L and U and a degradation from a Level-3 BLAS method to a Level-2.5 BLAS method,
3. the merging of symbolic and numerical factorization.

Consequently, one primary goal in Level-3 BLAS sparse LU factorization is to choose a pivot which will not create additional fill-in during the elimination and which allows to treat the symbolic factorization as a separate process. An alternative to Gaussian elimination with threshold partial pivoting is Gaussian elimination with complete pivoting in the diagonal block of the supernode (SCHENK and GÄRTNER [2000]). This strategy allows Level-3 BLAS updates and does not create any fill-in during the factorization. It chooses a tentative ordering for the nodes that reduces the fill-in without worrying about the possibility of instability. The tentative ordering may then be modified in the factorization phase to increase stability during the numerical factorization. This implies strong regularity assumptions on small supernodes and weak ones for the large supernodes. The supernode pivoting scheme has been successfully applied to unsymmetric matrices arising in semiconductor device and process simulations in SCHENK, GÄRTNER and FICHTNER [1999], SCHENK, GÄRTNER, SCHMIDTHÜSEN and FICHTNER [1999].

2.5. *Parallel strategies*

In this section, parallelism and granularity in the factorization process will be identified. In general, two important critical issues must be addressed in designing a parallel algorithm. It is necessary to exploit as much concurrency as possible and to maintain on the other hand, a sufficient level of per-processor efficiency by choosing an appropriate granularity for each task. For the parallel direct solution of sparse linear systems, the following four possible types of parallelism can be identified:

Node level parallelism. The first type of parallelism, called node level parallelism or type 0 parallelism, traverses the supernodes in the natural sequential ordering and solves each supernode factorization in parallel. The node level parallelism is obtained by simply running the uniprocessor algorithm under parallel multiplication routines form the BLAS and LAPACK library. Acceptable efficiency in parallel factorization requires sufficiently large matrices. However, for sparse matrix factorization, parallelism in the dense matrix kernels is quite limited, because the dense submatrices are typically small.

Elimination tree parallelism. The second source of parallelism, the elimination tree parallelism or type 1 parallelism, is generally exploited in all parallel sparse direct solver packages. Nodes in different subtrees correspond to independent tasks that can be executed in parallel. However, if only this type of parallelism is used, the speedups are very disappointing. Obviously it depends on the problem, but typically the maximum speedup is bounded by a factor of three or four. Is has been observed by AMESTOY and DUFF [1993] that often more than 75% of the computations are performed in the top three levels of the tree. It is thus necessary to obtain further parallelism within the large nodes near the root of the tree. The additional parallelism is based on parallel blocked versions of the numerical factorization.

1-d blocking/2-d blocking parallelism. Further parallelism can be obtained by a one-dimensional (1-d) blocking of the rows of the supernodes that are not nodes of type 1 – in the 1-d partition, each block column of L resides on only one process. Finally, if the supernode blocks close to the root are large enough, then a 2-d blocking of the rows and columns can be applied. The factors are decomposed into blocks of submatrices and these blocks are mapped onto a processor grid, in both row and column dimensions. Such a 2-d layout strikes a good balance among locality (by blocking), load balance (by cyclic mapping), and lower communication volume (by 2-d mapping). 2-d layouts were used in scalable implementation of sparse Cholesky factorization by GUPTA, KARYPIS and KUMAR [1997] and ROTHBERG [1996], and unsymmetric factorization by LI and DEMMEL [1999].

Pipelining parallelism. Another type of parallelism, called pipelining parallelism, is also suitable for a larger number of processors. Having studied the parallelism arising from different subtrees and a blocking of the supernodes, the relation between ancestors and descendants can be exploited with the pipelining parallelism. When the elimination process proceeds to a stage where there are more processors than independent subtrees, then the processors must work cooperatively on dependent columns.

Consider a left-looking algorithm and a simple situation with only two processors. Processor 1 gets a task 1 containing supernode j, processor 2 gets another task 2 containing supernode k, and node j is a descendant of node k in the tree. The dependency says that task 2 cannot finish its execution before task j finishes. However, processor 2 can start right away with the computations not involving supernode j – this includes the accumulation of the numerical updates using the already finished descendants in the elimination tree. Although a pipelining mechanism is complicated to implement, it is essential to achieve higher concurrency.

Demmel, Gilbert, and Li employed pipelining parallelism within a left-looking supernode algorithm in DEMMEL, GILBERT and LI [1999] while Schenk, Gärtner, and Fichtner exploited this type of parallelism in SCHENK, GÄRTNER and FICHTNER [2000] with a left-right looking strategy.

Parallel sparse direct solver packages. Parallel implementations of Cholesky and LU factorization have been treated by several authors. The benefit of using blocking techniques, higher level BLAS kernels, coupled with an increase in local cache memory and the communication speed of parallel processors, have made the parallel direct solution of sparse linear systems feasible on shared and distributed memory multiprocessing architectures.

Some recent performance results from several different parallel implementations are now reviewed. GUPTA, KARYPIS and KUMAR [1997] implemented a multifrontal algorithm using two-dimensional blocking and obtained a performance of 15 Gflop/s on 1024 nodes of the CRAY T3D for large sparse symmetric problems from structural analysis. A mixed 1-d/2-d distribution with static scheduling is used by HENON, RAMET and ROMAN [2002]. Their solver PaStiX performed sparse matrix factorization up to 33.3 Gflop/s on a 64 processor IBM SP3. Also recently, Amestoy, Duff, Excellent, and Li analyzed in an extensive comparison (AMESTOY, DUFF, L'EXCELLENT and LI [2000]) the performance characteristics of their two state-of-the art solvers for distributed memory machines. One is the multifrontal solver called MUMPS, the other is a supernodal solver called SuperLU, both targeted for message passing architectures. Schenk and Gärtner developed the supernode solver PARDISO based on a left-right looking approach to utilize shared memory multiprocessing systems with up to 64 processors and their experimental results show that the left-right looking algorithm is capable of using a moderate numbers of processors very efficiently (SCHENK and GÄRTNER [2000]). The algorithm delivers substantial speedup already for moderate problem sizes. Their approach has been tested on a wide range of architectures and they obtained, e.g., 117 Gflop/s on a 16 CPUs NEC SX5 (312 MHz) for an irregular sparse unsymmetric matrix from semiconductor laser device simulation at ETH Zurich.

In Table 2.5, the major characteristics of these full supported parallel sparse direct codes are summarized. The features and the methods of these packages have been presented at a minisymposium on parallel sparse direct methods at the tenth SIAM Conference on Parallel Processing for Scientific Computing. Some packages are targeted for special matrices such as symmetric and positive definite while others are targeted for the most general cases. This is reflected in column 3 ("scope") of the table. The matrices can be symmetric positive definite ("SPD"), symmetric and may be indefinite ("SYM") or unsymmetric ("UNS"). The sparse direct solver packages are further categorized in a parallel shared memory version ("OpenMP", "Threads") or a distributed memory version ("MPI"). The pivoting strategy is either diagonal pivoting ("No"), diagonal pivoting

TABLE 2.5
A selection of full supported parallel sparse direct solver packages

Code	Algorithm	Scope	Technique	Pivoting	Author
PARDISO	Left-right looking	SYM/ UNS	OpenMP	supernode	SCHENK, GÄRTNER and FICHTNER [2000]
SuperLU-MT	Left-looking	UNS	Threads	partial	DEMMEL, GILBERT and LI [1999]
MUMPS	Multifrontal	SYM/ UNS	MPI	partial	AMESTOY, DUFF and L'EXCELLENT [2000]
PaStiX	Multifrontal	SPD	MPI	No	HENON, RAMET and ROMAN [2002]
SuperLU-DIST	Right-looking	UNS	MPI	static	LI and DEMMEL [1999]

with a ordering ("static"), supernode pivoting ("supernode") or partial threshold pivoting ("partial"). The sixth column ("Author") reflects the contact person for the solver. It should be stressed that only fully supported packages are referred to. A comprehensive list of sparse direct solvers is also given in AMESTOY, DUFF, L'EXCELLENT and LI [2000].

3. Iterative solution methods

3.1. Introduction

The basic idea behind iterative methods is to replace the given system $Ax = b$ by a nearby simpler to solve system $Kx_0 = b$, and take x_0 as an approximation for x. The iteration comes from the systematic way in which the approximation can be improved. Indeed, we want the correction z that satisfies

$$A(x_0 + z) = b.$$

This leads to a new linear system

$$Az = b - Ax_0,$$

and we replace this system again by a nearby system, and often K is taken again:

$$Kz_0 = b - Ax_0.$$

This leads to the new approximation $x_1 = x_0 + z_0$. The correction procedure can be repeated for x_1, and so on, which gives an iterative method. In some iteration methods one selects a cycle of different approximations K, as, for instance, in ADI (VARGA [1962]) or SIP (STONE [1968]). In such cases one can regard the approximation for x after one cycle, as being obtained from the approximation prior to the cycle with an implicitly constructed K that represents the full cycle. This observation is of importance for the construction of preconditioners.

For the basic iteration, that we have introduced above, it follows that

$$\begin{aligned} x_{i+1} &= x_i + z_i \\ &= x_i + K^{-1}(b - Ax_i) \\ &= x_i + \tilde{b} - \widetilde{A}x_i, \end{aligned} \tag{3.1}$$

with $\tilde{b} = K^{-1}b$ and $\widetilde{A} = K^{-1}A$. We write K^{-1} for ease of notation; we (almost) never compute inverses of matrices explicitly. When we speak of $K^{-1}b$, we mean the vector $\tilde{b}$ that is solved from $K\tilde{b} = b$, and likewise for $K^{-1}Ax_i$.

The formulation in (3.1) can be interpreted as the basic iteration for the preconditioned linear system

$$\widetilde{A}x = \tilde{b}, \tag{3.2}$$

with approximation $K = I$ for $\widetilde{A} = K^{-1}A$.

In order to simplify the introduction of more advanced iteration methods, we will from now on assume that with some available preconditioner K the iterative schemes are applied to the (preconditioned) system (3.2), and we will skip the superscipt $\tilde{\cdot}$. This means that we iterate for $Ax = b$ with approximation $K = I$ for A. In some cases it will turn out to be more convenient to incorporate the preconditioner explicitly in the iteration scheme, but that will be clear from the context.

We have so arrived at the well-known Richardson iteration:

$$x_{i+1} = b + (I - A)x_i = x_i + r_i, \tag{3.3}$$

with the residual $r_i = b - Ax_i$.

Because relation (3.3) contains x_i as well as r_i, it cannot easily be analysed. Multiplication by $-A$ and adding b gives

$$b - Ax_{i+1} = b - Ax_i - Ar_i$$

or

$$\begin{aligned} r_{i+1} &= (I - A)r_i \qquad (3.4)\\ &= (I - A)^{i+1} r_0 \\ &= P_{i+1}(A) r_0. \qquad (3.5) \end{aligned}$$

In terms of the error, we get

$$A(x - x_{i+1}) = P_{i+1}(A)A(x - x_0),$$

so that, for nonsingular A:

$$x - x_{i+1} = P_{i+1}(A)(x - x_0).$$

In these expressions P_{i+1} is a (special) polynomial of degree $i + 1$. Note that $P_{i+1}(0) = 1$.

The expressions (3.4) and (3.5) lead to interesting observations. From (3.4) we conclude that

$$\|r_{i+1}\| \leqslant \|I - A\| \|r_i\|,$$

which shows we have guaranteed convergence for all initial r_0 if $\|I - A\| < 1$. This puts restrictions on the preconditioner (remember that A represents the preconditioned matrix). We will see later that it is not necessary that $\|I - A\| < 1$ for convergence of more advanced iterative schemes.

Eq. (3.5) is also of interest, because it shows that all residuals can be expressed in terms of powers of A times the initial residual. This observation will be crucial for the derivation of methods like the Conjugate Gradients method. It shows something more. Let us assume that A has n eigenvectors w_j, with corresponding eigenvectors λ_j. Then we can express r_0 in terms of the eigenvector basis as

$$r_0 = \sum_{j=1}^{n} \gamma_j w_j,$$

and we see that

$$r_i = P_i(A)r_0 = \sum_{j=1}^{n} \gamma_j P_i(\lambda_j) w_j.$$

This formula shows that the error reduction depends on how well the polynomial P_i damps the initial error components. It would be nice if we could construct iterative methods for which the corresponding error reduction polynomial P_i has better damping properties than for the standard iteration (3.3).

From now on we will also assume that $x_0 = 0$, which will help to make future formulas more simple. This does not mean a loss of generality, because the situation $x_0 \neq 0$ can be transformed, through the simple linear transformation $y = x - x_0$, to the system

$$Ay = b - Ax_0 = \bar{b}$$

for which obviously $y_0 = 0$.

With the simple Richardson iteration, we can proceed in different ways. One way is to include iteration parameters, for instance, by computing x_{i+1} as

$$x_{i+1} = x_i + \alpha_i r_i. \tag{3.6}$$

This leads to the error reduction formula

$$r_{i+1} = (I - \alpha_i A) r_i.$$

It follows that the error reduction polynomial P_i in this case can be expressed as

$$P_i = \prod_{j=1}^{i} (I - \alpha_j A).$$

An important consequence of this polynomial interpretation is that it is not longer necessary that $I - A$ has all its eigenvalues in the unit ball. The eigenvalues may, in principle, be anywhere as long as we see chance to construct iteration methods for which the corresponding iteration polynomials damp the unwanted error components. This is precisely what the modern Krylov subspace iteration methods attempt to do. As we will see, these methods proceed in an automatic manner and do not require user-specified parameters (such as the α_i).

We are now in the position to derive these advanced iterative methods. First we have to identify the subspace in which the successive approximate solutions are located. By repeating the simple Richardson iteration, we observe that

$$x_{i+1} = r_0 + r_1 + r_2 + \cdots + r_i \tag{3.7}$$

$$= \sum_{j=0}^{i} (I - A)^j r_0 \tag{3.8}$$

$$\in \operatorname{span}\{r_0, Ar_0, \ldots, A^i r_0\} \tag{3.9}$$

$$:= \mathcal{K}^{i+1}(A; r_0). \tag{3.10}$$

The m-dimensional space spanned by a given vector v, and increasing powers of A applied to v, up to the $(m-1)$th power, is called the m dimensional Krylov subspace, generated with A and v, denoted by $\mathcal{K}^m(A;v)$.

Apparently, the Richardson iteration, as it proceeds, delivers elements of Krylov subspaces of increasing dimension. This is also the case for the Richardson iteration (3.6) with parameters. Including local iteration parameters in the iteration would lead to other elements of the same Krylov subspaces. Let us write such an element still as x_{i+1}. Since $x_{i+1} \in K^{i+1}(A;r_0)$, we have that

$$x_{i+1} = Q_{i+1}(A)r_0,$$

with Q_{i+1} an arbitrary polynomial of degree $i+1$. It follows that

$$r_{i+1} = b - Ax_{i+1} = \big(I - AQ_{i+1}(A)\big)r_0 = \widetilde{P}_{i+1}(A)r_0, \tag{3.11}$$

with, just as in the standard Richardson iteration, $\widetilde{P}_{i+1}(0) = 1$.

The standard iteration (3.3) is characterized by the polynomial $P_{i+1}(A) = (I - A)^{i+1}$.

The consequence of this is, that if we want to make better combinations of the generated approximations, then we have to explore the Krylov subspace.

3.2. *The Krylov subspace approach*

Methods that attempt to generate better approximations from the Krylov subspace are often referred to as Krylov subspace methods. Because optimality usually refers to some sort of projection, they are also called Krylov projection methods. The most popular Krylov subspace methods, for identification of a good $x_k \in \mathcal{K}^k(A;r_0)$, can be distinguished in four different classes (we will still assume that $x_0 = 0$):

1. The *Ritz–Galerkin approach*: Construct the x_k for which the residual is orthogonal to the current subspace: $b - Ax_k \perp K^k(A;r_0)$.
2. The *minimum residual approach*: Identify the x_k for which the Euclidean norm $\|b - Ax_k\|_2$ is minimal over $K^k(A;r_0)$.
3. The *Petrov–Galerkin approach*: Find an x_k so that the residual $b - Ax_k$ is orthogonal to some other suitable k-dimensional subspace.
4. The *minimum error approach*: Compute the $x_k \in x_0 + A\mathcal{K}^k(A;r_0)$ such that $\|x - x_k\|$ is minimal.

The Ritz–Galerkin approach leads to well-known methods as Conjugate Gradients, the Lanczos method, FOM, and GENCG. The minimum residual approach leads to methods like GMRES, MINRES, and ORTHODIR. The main disadvantage of these two approaches is that, for most unsymmetric systems, they lead to long, and therefore expensive, recurrence relations for the approximate solutions. This can be relieved by selecting other subspaces for the orthogonality condition (the Galerkin condition). If we select the k-dimensional subspace in the third approach as $K^k(A^T;s_0)$, then we obtain the Bi-CG and QMR methods, and these methods work with short recurrences indeed. The SYMMLQ method (PAIGE and SAUNDERS [1975]) belongs to the fourth class. Also hybrids of these approaches have been proposed, like CGS, Bi-CGSTAB, BiCGSTAB(ℓ), FGMRES, and GMRESR.

The choice for a method is a delicate problem. If the matrix A is symmetric positive definite, then the choice is easy: Conjugate Gradients. For other types of matrices the situation is very diffuse. GMRES, proposed in 1986 by SAAD and SCHULTZ [1986], is the most robust method, but in terms of work per iteration step it is also relatively expensive. Bi-CG, which was suggested by Fletcher (FLETCHER [1976]), is a relatively inexpensive alternative, but it has problems with respect to convergence: the so-called breakdown situations. This aspect has received much attention. PARLETT, TAYLOR and LIU [1985] introduced the notion of look-ahead in order to overcome breakdowns and this was further perfected by FREUND, GUTKNECHT and NACHTIGAL [1993]. Other contributions to overcome specific breakdown situations were made by BANK and CHAN [1993], and FISCHER [1994]. We will discuss these approaches in Section 3.5.

The development of hybrid methods started with CGS, published in 1989 by SONNEVELD [1989], and was followed by Bi-CGSTAB, by VAN DER VORST [1992a], and others. The hybrid variants of GMRES: Flexible GMRES and GMRESR, in which GMRES is combined with some other iteration scheme, have been proposed in the mid-1990s.

Simple algorithms and unsophisticated software for some of these methods is provided in BARRETT, BERRY, CHAN, DEMMEL, DONATO, DONGARRA, EIJKHOUT, POZO, ROMINE and VAN DER VORST [1994]. This was complemented, with respect to theoretical aspects, by a very elegant textbook written by GREENBAUM [1997b]. Iterative methods with much attention to various forms of preconditioning have been described in AXELSSON [1994]. Another useful book on iterative methods was published by SAAD [1996]; it is very algorithm oriented, with, of course, a focus on GMRES and preconditioning techniques, like threshold ILU, ILU with pivoting, and incomplete LQ factorizations. A nice introduction for Krylov subspace methods, viewed from the standpoint of polynomial methods, can be found in FISCHER [1996].

An annotated entrance to the vast literature on preconditioned iterative methods is given in BRUASET [1995].

3.2.1. The Krylov subspace

In order to identify the approximations corresponding to the three different approaches, we need a suitable basis for the Krylov subspace; one that can be extended in a meaningful way for subspaces of increasing dimension. The obvious basis $r_0, Ar_0, \ldots, A^{i-1}r_0$, for $K^i(A; r_0)$, is not very attractive from a numerical point of view, since the vectors $A^j r_0$ for increasing j point more and more in the direction of the eigenvector corresponding to the in modulus largest eigenvalue. For that reason the basis vectors become dependent in finite precision arithmetic. It does not help to compute this nonorthogonal generic basis first and to orthogonalize it afterwards. The result would be that we have orthogonalized a very ill-conditioned set of basis vectors, which is numerically still not an attractive situation.

ARNOLDI [1951] has proposed to compute an orthogonal basis as follows. Start with $v_1 := r_0/\|r_0\|_2$. Then compute Av_1, make it orthogonal to v_1 and normalize the result, which gives v_2. The general procedure is as follows. Assume that we have already an orthonormal basis $v_1, \ldots, v_j$ for $K^j(A; r_0)$, then this basis is expanded by computing $t = Av_j$, and by orthonormalizing this vector t with respect to $v_1, \ldots, v_j$. In principle

$$
\begin{array}{l}
v_1 = r_0/\|r_0\|_2;\\
\text{for } j = 1, \ldots, m-1\\
\quad t = Av_j;\\
\quad \text{for } i = 1, \ldots, j\\
\quad\quad h_{i,j} = v_i^T t;\\
\quad\quad t = t - h_{i,j} v_i;\\
\quad \text{end};\\
\quad h_{j+1,j} = \|t\|_2;\\
\quad v_{j+1} = t/h_{j+1,j};\\
\text{end}
\end{array}
$$

FIG. 3.1. Arnoldi's method with modified Gram–Schmidt orthogonalization.

the orthonormalization process can be carried out in different ways, but the most commonly used approach is to do this by a modified Gram–Schmidt procedure (GOLUB and VAN LOAN [1996]).

This leads to an algorithm for the creation of an orthonormal basis for $K^m(A; r_0)$, as in Fig. 3.1. It is easily verified that $v_1, \ldots, v_m$ form an orthonormal basis for $\mathcal{K}^m(A; r_0)$ (that is, if the construction does not terminate at a vector $t = 0$). The orthogonalization leads to relations between the v_j, that can be formulated in a compact algebraic form. Let V_j denote the matrix with columns v_1 up to v_j, then it follows that

$$AV_{m-1} = V_m H_{m,m-1}. \tag{3.12}$$

The m by $m-1$ matrix $H_{m,m-1}$ is upper Hessenberg, and its elements $h_{i,j}$ are defined by the Arnoldi algorithm, with $h_{ij} = 0$ for $i > j+1$.

From a computational point of view, this construction is composed from three basic elements: a matrix vector product with A, inner products, and vector updates. We see that this orthogonalization becomes increasingly expensive for increasing dimension of the subspace, since the computation of each $h_{i,j}$ requires an inner product and a vector update.

Note that if A is symmetric, then so is $H_{m-1,m-1} = V_{m-1}^T A V_{m-1}$, so that in this situation $H_{m-1,m-1}$ is tridiagonal. This means that in the orthogonalization process, each new vector has to be orthogonalized with respect to the previous two vectors only, since all other inner products vanish. The resulting three term recurrence relation for the basis vectors of $K_m(A; r_0)$ is known as the *Lanczos method* (LANCZOS [1950]) and some very elegant methods are derived from it. In this symmetric case the orthogonalization process involves constant arithmetical costs per iteration step: one matrix vector product, two inner products, and two vector updates.

3.2.2. *The Ritz–Galerkin approach*

The Ritz–Galerkin conditions imply that $r_k \perp \mathcal{K}^k(A; r_0)$, and this is equivalent to

$$V_k^T(b - Ax_k) = 0.$$

Since $b = r_0 = \|r_0\|_2 v_1$, it follows that $V_k^T b = \|r_0\|_2 e_1$ with e_1 the first canonical unit vector in $\mathbb{R}^k$. With $x_k = V_k y$ we obtain

$$V_k^T A V_k y = \|r_0\|_2 e_1.$$

This system can be interpreted as the system $Ax = b$ projected onto the subspace $\mathcal{K}^k(A; r_0)$.

Obviously we have to construct the $k \times k$ matrix $V_k^T A V_k$, but this is, as we have seen, readily available from the orthogonalization process:

$$V_k^T A V_k = H_{k,k},$$

so that the x_k for which $r_k \perp \mathcal{K}^k(A; r_0)$ can be easily computed by first solving $H_{k,k} y = \|r_0\|_2 e_1$, and then forming $x_k = V_k y$. This algorithm is known as FOM or GENCG (SAAD and SCHULTZ [1986]).

When A is symmetric, then $H_{k,k}$ reduces to a tridiagonal matrix $T_{k,k}$, and the resulting method is known as the *Lanczos* method (LANCZOS [1952]). When A is in addition positive definite then we obtain, at least formally, the *Conjugate Gradients* method. In commonly used implementations of this method, one implicitly forms an LU factorization for $T_{k,k}$, without generating $T_{k,k}$ itself, and this leads to very elegant short recurrences for the x_j and the corresponding r_j, see Section 3.3.

The positive definiteness is necessary to guarantee the existence of the LU factorization, but it allows also for another useful interpretation. From the fact that $r_i \perp K^i(A; r_0)$, it follows that $A(x_i - x) \perp K^i(A; r_0)$, or $x_i - x \perp_A K^i(A; r_0)$. The latter observation expresses the fact that the error is A−orthogonal to the Krylov subspace and this is equivalent to the important observation that $\|x_i - x\|_A$ is minimal.[17] For an overview of the history of CG and main contributions on this subject, see GOLUB and O'LEARY [1989].

3.2.3. *The minimum residual approach*

The creation of an orthogonal basis for the Krylov subspace, with basis vectors $v_1, \ldots, v_{i+1}$, leads to

$$A V_i = V_{i+1} H_{i+1,i}, \tag{3.13}$$

where V_i is the matrix with columns v_1 to v_i. We look for an $x_i \in K^i(A; r_0)$, that is $x_i = V_i y$, for which $\|b - Ax_i\|_2$ is minimal. This norm can be rewritten as

$$\|b - Ax_i\|_2 = \|b - A V_i y\|_2 = \big\| \|r_0\|_2 V_{i+1} e_1 - V_{i+1} H_{i+1,i} y \big\|_2.$$

Now we exploit the fact that V_{i+1} is an orthonormal transformation with respect to the Krylov subspace $K^{i+1}(A; r_0)$:

$$\|b - Ax_i\|_2 = \big\| \|r_0\|_2 e_1 - H_{i+1,i} y \big\|_2,$$

and this final norm can simply be minimized by solving the minimum norm least squares problem for the $i + 1$ by i matrix $H_{i+1,i}$ and right-hand side $\|r_0\|_2 e_1$.

The GMRES method is based upon this approach, see Section 3.4.

---

[17] The A-norm is defined by $\|y\|_A^2 = (y, y)_A := (y, Ay)$, and we need the positive definiteness of A in order to get a proper inner product $(\cdot, \cdot)_A$.

3.2.4. The Petrov–Galerkin approach

For unsymmetric systems we can, in general, not reduce the matrix A to a symmetric system in a lower-dimensional subspace, by orthogonal projections. The reason is that we can not create an orthogonal basis for the Krylov subspace by a 3-term recurrence relation (FABER and MANTEUFFEL [1984]). We can, however, obtain a suitable nonorthogonal basis with a 3-term recurrence, by requiring that this basis is orthogonal with respect to some other basis.

We start by constructing an arbitrary basis for the Krylov subspace:

$$h_{i+1,i}v_{i+1} = Av_i - \sum_{j=1}^{i} h_{j,i}v_j, \tag{3.14}$$

which can be rewritten in matrix notation as $AV_i = V_{i+1}H_{i+1,i}$. The coefficients $h_{i+1,i}$ define the norm of v_{i+1}, and a natural choice would be to select them such that $\|v_{i+1}\|_2 = 1$. In Bi-CG implementations, a popular choice is to select $h_{i+1,i}$ such that $\|v_{i+1}\|_2 = \|r_{i+1}\|_2$.

Clearly, we cannot use V_i for the projection, but suppose we have a W_i for which $W_i^T V_i = D_i$ (an i by i diagonal matrix with diagonal entries d_i), and for which $W_i^T v_{i+1} = 0$.

Then

$$W_i^T AV_i = D_i H_{i,i}, \tag{3.15}$$

and now our goal is to find a W_i for which $H_{i,i}$ is tridiagonal. This means that $V_i^T A^T W_i$ should be tridiagonal too. This last expression has a similar structure as the right-hand side in (3.15), with only W_i and V_i reversed. This suggests to generate the w_i with A^T.

We start with an arbitrary $w_1 \neq 0$, such that $w_1^T v_1 \neq 0$. Then we generate v_2 with (3.14), and orthogonalize it with respect to w_1, which means that $h_{1,1} = w_1^T Av_1/(w_1^T v_1)$. Since $w_1^T Av_1 = (A^T w_1)^T v_1$, this implies that w_2, generated with

$$h_{2,1}w_2 = A^T w_1 - h_{1,1}w_1,$$

is also orthogonal to v_1.

This can be continued, and we see that we can create bi-orthogonal basis sets $\{v_j\}$, and $\{w_j\}$, by making the new v_i orthogonal to w_1 up to w_{i-1}, and then by generating w_i with the same recurrence coefficients, but with A^T instead of A.

Now we have that $W_i^T AV_i = D_i H_{i,i}$, and also that $V_i^T A^T W_i = D_i H_{i,i}$. This implies that $D_i H_{i,i}$ is symmetric, and hence $H_{i,i}$ is a tridiagonal matrix, which gives us the desired 3-term recurrence relation for the v_j's, and the w_j's. Note that $v_1, \ldots, v_i$ form a basis for $K^i(A; v_1)$, and $w_1, \ldots, w_i$ form a basis for $K^i(A^T; w_1)$.

3.3. The Conjugate Gradients method

As explained in Section 3.2.2, the conjugate gradient method can be viewed as a variant of the Lanczos method. The method is based on relation (3.12), which for symmetric A reduces to $AV_i = V_{i+1}H_{i+1,i}$ with tridiagonal $H_{i+1,i}$. For the kth column of V_k, we have that

$$Av_k = h_{k+1,k}v_{k+1} + h_{k,k}v_k + h_{k-1,k}v_{k-1}. \tag{3.16}$$

In the Galerkin approach, the new residual $b - Ax_{k+1}$ is orthogonal to the subspace spanned by $v_1, \ldots, v_k$, so that r_{k+1} is in the direction of v_{k+1}. Therefore, we can also select the scaling factor $h_{k+1,k}$ so that v_{k+1} coincides with r_{k+1}. This would be convenient, since the residual gives useful information on our solution, and we do not want to work with two sequences of auxiliary vectors.

From the consistency relation (3.11) we have that r_k can be written as

$$r_k = (I - AQ_{k-1}(A))r_0.$$

By inserting the polynomial expressions for the residuals in (3.16), and comparing the coefficient for r_0 in the new relation, we obtain

$$h_{k+1,k} + h_{k,k} + h_{k-1,k} = 0,$$

which defines $h_{k+1,k}$.

At the end of this section we will consider the situation where the recurrence relation terminates.

With R_i we denote the matrix with columns r_j:

$$R_i = (r_0, \ldots, r_{i-1}),$$

then we have

$$AR_i = R_{i+1}T_{i+1,i}, \tag{3.17}$$

where $T_{i+1,i}$ is a tridiagonal matrix (with $i+1$ rows and i columns) with elements $h_{i,j}$.

Since we are looking for a solution x_i in $K^i(A; r_0)$, that vector can be written as a combination of the basis vectors of the Krylov subspace, and hence

$$x_i = R_i y.$$

(Note that y has i components).

Furthermore, the Ritz–Galerkin condition says that the residual for x_i is orthogonal with respect to $r_0, \ldots, r_{i-1}$:

$$R_i^T(Ax_i - b) = 0,$$

and hence

$$R_i^T A R_i y - R_i^T b = 0.$$

Using Eq. (3.17), we obtain

$$R_i^T R_i T_{i,i} y = \|r_0\|_2^2 e_1.$$

Since $R_i^T R_i$ is a diagonal matrix with diagonal elements $\|r_0\|_2^2$ up to $\|r_{i-1}\|_2^2$ we find the desired solution by solving y from

$$T_{i,i} y = e_1 \quad \Rightarrow \quad y \quad \Rightarrow \quad x_i = R_i y.$$

So far we have only used the fact that A is symmetric and we have assumed that the matrix T_i is not singular. The Krylov subspace method that has been derived here is known as the Lanczos method for symmetric systems (LANCZOS [1952]).

Note that for some $j \leqslant n-1$ the construction of the orthogonal basis must terminate. In that case we have that $AR_{j+1} = R_{j+1} T_{j+1,j+1}$. Let y be the solution of the reduced system $T_{j+1,j+1} y = e_1$, and $x_{j+1} = R_{j+1} y$. Then it follows that $x_{j+1} = x$, i.e., we have arrived at the exact solution, since $Ax_{j+1} - b = AR_{j+1} y - b = R_{j+1} T_{j+1,j+1} y - b = R_{j+1} e_1 - b = 0$ (we have assumed that $x_0 = 0$).

The Conjugate Gradients method (HESTENES and STIEFEL [1952]), CG for short, is a clever variant on the above approach, which saves storage and computational effort. If we follow naively the above sketched approach, when solving the projected equations, then we see that we have to save all columns of R_i throughout the process in order to recover the current iteration vectors x_i. This can be done in a more memory friendly way. If we assume that the matrix A is in addition positive definite then, because of the relation

$$R_i^T A R_i = R_i^T R_i T_{i,i},$$

we conclude that $T_{i,i}$ can be transformed by a rowscaling matrix $R_i^T R_i$ into a positive definite symmetric tridiagonal matrix (note that $R_i^T A R_i$ is positive definite for $y \in \mathbb{R}^i$). This implies that $T_{i,i}$ can be LU decomposed without any pivoting:

$$T_{i,i} = L_i U_i,$$

with L_i lower bidiagonal, and U_i is upper bidiagonal with unit diagonal. This leads to two two-term recurrences for the update vector and for the residual vector.

It is not necessary to generate $T_{i,i}$ explicitly: we can obtain the required information in an easier way. For details on this see, for instance, GOLUB and VAN LOAN [1996], Chapter 10.2. The resulting method is known as the conjugate gradients method. The name stems from the property that the update vectors p_i, are A-orthogonal.

Note that the positive definiteness of A is only exploited for the flawless decomposition of the implicitly generated tridiagonal matrix $T_{i,i}$. This suggests that the conjugate gradients method may also work for certain nonpositive definite systems, but then at our own risk (PAIGE, PARLETT and VAN DER VORST [1995]).

3.3.1. Computational notes

The standard (unpreconditioned) Conjugate Gradients algorithm for the solution of $Ax = b$ can be represented by the following scheme:

CG is most often used in combination with a suitable approximation K for A; this K is called the preconditioner. We will assume that K is also positive definite. However, we cannot apply CG straight away for the explicitly preconditioned system $K^{-1}Ax = K^{-1}b$, as we suggested to do in the introduction, because $K^{-1}A$ is most likely not

x_0 is initial guess, $r_0 = b - Ax_0$
for $i = 1, 2, \ldots$
  $\rho_{i-1} = r_{i-1}^T r_{i-1}$
  if $i = 1$
    $p_i = r_{i-1}$
  else
    $\beta_{i-1} = \rho_{i-1}/\rho_{i-2}$
    $p_i = r_{i-1} + \beta_{i-1} p_{i-1}$
  endif
  $q_i = Ap_i$;
  $\alpha_i = \rho_{i-1}/p_i^T q_i$
  $x_i = x_{i-1} + \alpha_i p_i$
  $r_i = r_{i-1} - \alpha_i q_i$
  if x_i accurate enough **then quit**
end

FIG. 3.2. Conjugate Gradients without preconditioning.

symmetric. One way out is to apply the preconditioner differently. Assume that K is given in factored form:

$$K = LL^T,$$

as is the case for ILU preconditioners.

We then apply CG for the symmetrically preconditioned system

$$L^{-1}AL^{-T}y = L^{-1}b,$$

with $x = L^{-T}y$.

This approach has the disadvantage that K must be available in factored form and that we have to backtransform the approximate solution afterwards. There is a more elegant alternative. Note first that the CG method can be derived for any choice of the inner product. In our derivation we have used the standard inner product $(x, y) = \sum x_i y_i$, but we have not used any specific property of that inner product. Now we make a different choice:

$$[x, y] := (x, Ky).$$

It is easy to verify that $K^{-1}A$ is symmetric positive definite with respect to [,]:

$$\begin{aligned}[K^{-1}Ax, y] &= (K^{-1}Ax, Ky) = (Ax, y) \\ &= (x, Ay) = [x, K^{-1}Ay]. \end{aligned} \tag{3.18}$$

Hence, we can follow our CG procedure for solving the preconditioned system $K^{-1}Ax = K^{-1}b$, using the new [,]-inner product.

Apparently, we now are minimizing

$$\big[x_i - x, K^{-1}A(x_i - x)\big] = \big(x_i - x, A(x_i - x)\big),$$

x_0 is initial guess, $r_0 = b - Ax_0$
for $i = 1, 2, \ldots$
  Solve $Kw_{i-1} = r_{i-1}$
  $\rho_{i-1} = r_{i-1}^T w_{i-1}$
  if $i = 1$
    $p_i = w_{i-1}$
  else
    $\beta_{i-1} = \rho_{i-1}/\rho_{i-2}$
    $p_i = w_{i-1} + \beta_{i-1} p_{i-1}$
  endif
  $q_i = Ap_i$
  $\alpha_i = \rho_{i-1}/p_i^T q_i$
  $x_i = x_{i-1} + \alpha_i p_i$
  $r_i = r_{i-1} - \alpha_i q_i$
  if x_i accurate enough **then quit**
end

FIG. 3.3. Conjugate Gradients with preconditioning K.

which leads to the remarkable (and known) result that for this preconditioned system we still minimize the error in A-norm, but now over a Krylov subspace generated by $K^{-1}r_0$ and $K^{-1}A$.

In the computational scheme for preconditioned CG, in Fig. 3.3, for the solution of $Ax = b$ with preconditioner K, we have replaced the [,]-inner product again by the familiar standard inner product. E.g., note that with $\tilde{r}_{i+1} = K^{-1}Ax_{i+1} - K^{-1}b$ we have that

$$\begin{aligned}\rho_{i+1} &= [\tilde{r}_{i+1}, \tilde{r}_{i+1}]\\ &= [K^{-1}r_{i+1}, K^{-1}r_{i+1}] = [r_{i+1}, K^{-2}r_{i+1}]\\ &= (r_{i+1}, K^{-1}r_{i+1}),\end{aligned}$$

and $K^{-1}r_{i+1}$ is the residual corresponding to the preconditioned system $K^{-1}Ax = K^{-1}b$.

The coefficients α_j and β_j, generated by the Conjugate Gradients algorithms, as in Figs. 3.2 and 3.3, can be used to build the matrix $T_{i,i}$ in the following way:

$$T_{i,i} = \begin{pmatrix} \ddots & & \\ \ddots & -\frac{\beta_{j-1}}{\alpha_{j-1}} & \\ \ddots & \frac{1}{\alpha_j} + \frac{\beta_{j-1}}{\alpha_{j-1}} & \ddots \\ & -\frac{1}{\alpha_j} & \ddots \\ & & \ddots \end{pmatrix}. \tag{3.19}$$

Since $\alpha_j > 0$ and $\beta_j > 0$, we see that the above matrix is similar to the following symmetric tridiagonal matrix:

$$\widetilde{T}_{i,i} = \begin{pmatrix} \ddots & & \\ \ddots & -\frac{\sqrt{\beta_{j-1}}}{\alpha_{j-1}} & \\ \ddots & \frac{1}{\alpha_j} + \frac{\beta_{j-1}}{\alpha_{j-1}} & \ddots \\ & -\frac{\sqrt{\beta_j}}{\alpha_j} & \ddots \\ & & \ddots \end{pmatrix}.$$

The eigenvalues of the leading ith order minor of this matrix are the Ritz values of A (for Fig. 3.2) or the preconditioned matrix $K^{-1}A$ (for Fig. 3.3) with respect to the i-dimensional Krylov subspace spanned by the first i residual vectors. The Ritz values approximate the (extremal) eigenvalues of the (preconditioned) matrix increasingly well. These approximations can be used to get an impression of the relevant eigenvalues. They can also be used to construct upperbounds for the error in the delivered approximation with respect to the solution (KAASSCHIETER [1988], HAGEMAN and YOUNG [1981]). According to the results in VAN DER SLUIS and VAN DER VORST [1986], the eigenvalue information can also be used in order to understand or explain delays in the convergence behaviour.

The local convergence behavior of CG, and especially the occurrence of superlinear convergence, was first explained in a qualitative sense in CONCUS, GOLUB and O'LEARY [1976], and later in a quantitative sense in VAN DER SLUIS and VAN DER VORST [1986]. In both papers it was linked to the convergence of eigenvalues (Ritz values) of $T_{i,i}$ towards eigenvalues of A, for increasing i. The global convergence can be bounded with expressions that involve condition numbers, for details see for instance CONCUS, GOLUB and O'LEARY [1976], GOLUB and VAN LOAN [1996], AXELSSON [1977]. In AXELSSON [1977] the situation is analysed where the eigenvalues of $K^{-1}A$ are in disjunct intervals.

3.4. GMRES

As we have seen in Section 3.2.3, the minimal residual approach leads to a small minimum least squares problem that has to be solved:

$$H_{i+1,i}\,y = \|r_0\|_2 e_1.$$

In GMRES (SAAD and SCHULTZ [1986]) this is done efficiently with Givens rotations, that annihilate successively the subdiagonal elements in the upper Hessenberg matrix $H_{i+1,i}$.

In order to avoid excessive storage requirements and computational costs for the orthogonalization, GMRES is usually restarted after each m iteration steps. This algorithm is referred to as GMRES(m); the not-restarted version is often called 'full' GMRES. There is no simple rule to determine a suitable value for m; the speed of convergence may vary drastically for nearby values of m.

$r = b - Ax_0$, for a given initial guess x_0
for $j = 1, 2, \ldots$
  $\beta = ||r||_2$, $v_1 = r/\beta$; $\hat{b} = \beta e_1$;
  for $i = 1, 2, \ldots, m$
    $w = Av_i$;
    for $k = 1, \ldots, i$
      $h_{k,i} = v_k^T w$; $w = w - h_{k,i} v_k$;
    $h_{i+1,i} = \|w\|_2$; $v_{i+1} = w/h_{i+1,i}$;
    $r_{1,i} = h_{1,i}$;
    for $k = 2, \ldots, i$
      $\gamma = c_{k-1} r_{k-1,i} + s_{k-1} h_{k,i}$;
      $r_{k,i} = -s_{k-1} r_{k-1,i} + c_{k-1} h_{k,i}$;
      $r_{k-1,i} = \gamma$;
    $\delta = \sqrt{r_{i,i}^2 + h_{i+1,i}^2}$; $c_i = r_{i,i}/\delta$; $s_i = h_{i+1,i}/\delta$
    $r_{i,i} = c_i r_{i,i} + s_i h_{i+1,i}$
    $\hat{b}_{i+1} = -s_i \hat{b}_i$; $\hat{b}_i = c_i \hat{b}_i$;
    $\rho = |\hat{b}_{i+1}|$ $(= \|b - Ax_{(j-1)m+i}\|_2)$)
    if ρ is small enough **then**
      ($n_r = i$; goto *SOL*);
  $n_r = m$, $y_{n_r} = \hat{b}_{n_r}/r_{n_r,n_r}$
SOL: **for** $k = n_r - 1, \ldots, 1$
    $y_k = (\hat{b}_k - \sum_{i=k+1}^{n_r} r_{k,i} y_i)/r_{k,k}$
  $x = \sum_{i=1}^{n_r} y_i v_i$; **if** ρ small enough quit
  $r = b - Ax$

FIG. 3.4. unpreconditioned GMRES(m) with modified Gram–Schmidt.

We present in Fig. 3.4 the modified Gram–Schmidt version of GMRES(m) for the solution of the linear system $Ax = b$. The application to preconditioned systems, for instance, $K^{-1}Ax = K^{-1}b$ is straight-forward.

For complex valued systems, the scheme is as in Fig. 3.5. Note that the complex rotation is the only difference with respect to the real version.

The eigenvalues of $H_{i,i}$ are the Ritz values of A with respect to the Krylov subspace spanned by $v_1, \ldots, v_i$. They approximate eigenvalues of A increasingly well for increasing dimension i.

There is an interesting and simple relation between the Ritz–Galerkin approach (FOM and CG) and the minimum residual approach (GMRES and MINRES). In GMRES the projected system matrix $H_{i+1,i}$ is transformed by Givens rotations to an upper triangular matrix (with last row equal to zero). So, in fact, the major difference between FOM and GMRES is that in FOM the last $(i+1)$th row is simply discarded, while in GMRES this row is rotated to a zero vector. Let us characterize the Givens rotation, acting on rows i and $i+1$, in order to zero the element in position $(i+1, i)$, by the sine s_i and the cosine c_i. Let us further denote the residuals for FOM with an superscript F and those for GMRES with superscript G. Then we have the following relation between FOM and GMRES:

$r = b - Ax_0$, for a given initial guess x_0
for $j = 1, 2, \ldots$
  $\beta = ||r||_2$, $v_1 = r/\beta$; $\hat{b} = \beta e_1$;
  for $i = 1, 2, \ldots, m$
    $w = Av_i$;
    for $k = 1, \ldots, i$
      $h_{k,i} = v_k^* w$; $w = w - h_{k,i} v_k$;
    $h_{i+1,i} = ||w||_2$; $v_{i+1} = w/h_{i+1,i}$;
    $r_{1,i} = h_{1,i}$;
    for $k = 2, \ldots, i$
      $\gamma = c_{k-1} r_{k-1,i} + \bar{s}_{k-1} h_{k,i}$;
      $r_{k,i} = -s_{k-1} r_{k-1,i} + c_{k-1} h_{k,i}$;
      $r_{k-1,i} = \gamma$;
    $\delta = \sqrt{|r_{i,i}|^2 + |h_{i+1,i}|^2}$;
    if $|r_{i,i}| < |h_{i+1,i}|$
    then $\mu = r_{i,i}/h_{i+1,i}$; $\tau = \bar{\mu}/|\mu|$;
    else $\mu = h_{i+1,i}/r_{i,i}$; $\tau = \mu/|\mu|$;
    $c_i = |r_{i,i}|/\delta$; $s_i = |h_{i+1,i}|\tau/\delta$;
    $r_{i,i} = c_i r_{i,i} + \bar{s}_i h_{i+1,i}$;
    $\hat{b}_{i+1} = -s_i \hat{b}_i$; $\hat{b}_i = c_i \hat{b}_i$
    $\rho = |\hat{b}_{i+1}|$ $(= \|b - Ax_{(j-1)m+i}\|_2)$
    if ρ is small enough **then**
      ($n_r = i$; goto *SOL*);
  $n_r = m$, $y_{n_r} = \hat{b}_{n_r}/r_{n_r,n_r}$
SOL: **for** $k = n_r - 1, \ldots, 1$
      $y_k = (\hat{b}_k - \sum_{i=k+1}^{n_r} r_{k,i} y_i)/r_{k,k}$
  $x = \sum_{i=1}^{n_r} y_i v_i$; **if** ρ small enough quit
  $r = b - Ax$

FIG. 3.5. Unpreconditioned GMRES(m) for complex systems.

If $c_k \neq 0$ then the FOM and the GMRES residuals are related by

$$\|r_k^F\|_2 = \frac{\|r_k^G\|_2}{\sqrt{1 - (\|r_k^G\|_2/\|r_{k-1}^G\|_2)^2}} \tag{3.20}$$

(CULLUM and GREENBAUM [1996], Theorem 3.1). From this relation we see that when GMRES has a significant reduction at step k, in the norm of the residual (i.e., s_k is small, and $c_k \approx 1$), then FOM gives about the same result as GMRES. On the other hand when FOM has a breakdown ($c_k = 0$), then GMRES does not lead to an improvement in the same iteration step. Because of these relations we can link the convergence behaviour of GMRES with the convergence of Ritz values (the eigenvalues of the "FOM" part of the upper Hessenberg matrix). This has been exploited in VAN DER VORST and VUIK [1993], for the analysis and explanation of local effects in the convergence behaviour of GMRES.

There are various methods that are mathematically equivalent with FOM or GMRES. We will say that two methods are mathematically equivalent if they produce the same approximations $\{x_k\}$ in exact arithmetic. Among those that are equivalent to GMRES are: Orthomin (VINSOME [1976]), Orthodir (JEA and YOUNG [1980]), GENCR (ELMAN [1982]), and Axelsson's method (AXELSSON [1980]). These methods are often more expensive than GMRES per iteration step, and in some cases also less robust. Orthomin is still in use because this variant can be easily truncated (Orthomin(s)), in contrast to GMRES. The truncated and restarted versions of these algorithms are not necessarily mathematically equivalent.

Methods that are mathematically equivalent to FOM are: Orthores (JEA and YOUNG [1980]) and GENCG (CONCUS and GOLUB [1976], WIDLUND [1978]). In these methods the approximate solutions are constructed such that they lead to orthogonal residuals (which form a basis for the Krylov subspace; analogously to the CG method). A good overview of all these methods and their relations is given in SAAD [1996].

The GMRES method and FOM are closely related to vector extrapolation methods, when the latter are applied to linearly generated vector sequences. For a discussion on this, as well as for implementations for these matrix free methods, see SIDI [1991].

Note that when A is Hermitian (but not necessarily positive definite), the upper Hessenberg matrix $H_{i+1,i}$ reduces to a tridiagional system. This simplified structure can be exploited in order to avoid storage of all the basis vectors for the Krylov subspace, in a way similar as has been pointed out for CG. The resulting method is known as MINRES (PAIGE and SAUNDERS [1975]).

See Fig. 3.6 for an algorithmic formulation of MINRES. The formulation is derived from a MATLAB routine published in FISCHER [1996].

The usage of the 3-term recurrence relation for the columns of W_i makes MINRES very vulnerable for rounding errors, as has been shown in SLEIJPEN, VAN DER VORST and MODERSITZKI [2000]. It has been shown that rounding errors are propagated to the approximate solution with a factor proportional to the square of the condition number of A, whereas in GMRES these errors depend only on the condition number itself. Therefore, one should be careful with MINRES for ill-conditioned systems. If storage is no problem then GMRES should be preferred for ill-conditioned systems; if storage is a problem then one might consider the usage of SYMMLQ (PAIGE and SAUNDERS [1975]). SYMMLQ, however, may converge a good deal slower than MINRES for ill-conditioned systems. For more details on this, see SLEIJPEN, VAN DER VORST and MODERSITZKI [2000].

3.4.1. GMRESR and related approaches

In VAN DER VORST and VUIK [1994] it has been shown that the GMRES-method can be effectively combined (or rather preconditioned) with other iterative schemes. The iteration steps of GMRES (or GCR) are called outer iteration steps, while the iteration steps of the preconditioning iterative method are referred to as inner iterations. The combined method is called GMRES$\star$, where $\star$ stands for any given iterative scheme; in the case of GMRES as the inner iteration method, the combined scheme is called GMRESR (VAN DER VORST and VUIK [1994]).

Compute $v_1 = b - Ax_0$ for some initial guess x_0
$\beta_1 = \|v_1\|_2$; $\eta = \beta_1$;
$\gamma_1 = \gamma_0 = 1$; $\sigma_1 = \sigma_0 = 0$;
$v_0 = 0$; $w_0 = w_{-1} = 0$;
for $i = 1, 2, \ldots$
The Lanczos recurrence:
$v_i = \frac{1}{\beta_i} v_i$; $\alpha_i = v_i^T A v_i$;
$v_{i+1} = Av_i - \alpha_i v_i - \beta_i v_{i-1}$
$\beta_{i+1} = \|v_{i+1}\|_2$
QR part:
old Givens rot's on new column of T:
$\delta = \gamma_i \alpha_i - \gamma_{i-1}\sigma_i\beta_i$; $\rho_1 = \sqrt{\delta^2 + \beta_{i+1}^2}$
$\rho_2 = \sigma_i\alpha_i + \gamma_{i-1}\gamma_i\beta_i$; $\rho_3 = \sigma_{i-1}\beta_i$
New Givens rotation for subdiag elt:
$\gamma_{i+1} = \delta/\rho_1$; $\sigma_{i+1} = \beta_{i+1}/\rho_1$
Update of solution (with $W_i = V_i R_{i,i}^{-1}$)
$w_i = (v_i - \rho_3 w_{i-2} - \rho_2 w_{i-1})/\rho_1$
$x_i = x_{i-1} + \gamma_{i+1}\eta w_i$
$||r_i||_2 = |\sigma_{i+1}|\|r_{i-1}\|_2$
check convergence; continue if necessary
$\eta = -\sigma_{i+1}\eta$
end

FIG. 3.6. The unpreconditioned MINRES algorithm.

A similar approach has been followed for FGMRES (SAAD [1993]). In this method, the update directions for the approximate solution are preconditioned, whereas in GMRES⋆ the residuals are preconditioned. The latter approach offers more control over the reduction in the residual, in particular break-down situations can be easily detected and remedied.

In exact arithmetic GMRES⋆ is very close to the Generalized Conjugate Gradients method (AXELSSON and VASSILEVSKI [1991]); GMRES⋆, however, leads to a more efficient computational scheme.

The GMRES⋆ algorithm can be described by the computational scheme in Fig. 3.7.

A sufficient condition to avoid break-down in this method ($\|c\|_2 = 0$) is that the norm of the residual at the end of an inner iteration is smaller than the right-hand residual: $\|Az^{(m)} - r_i\|_2 < \|r_i\|_2$. This can easily be controlled during the inner iteration process. If stagnation occurs, i.e., no progress at all is made in the inner iteration, then it is suggested by VAN DER VORST and VUIK [1994] to do one (or more) steps of the LSQR method, which guarantees a reduction (although this reduction is often only small).

The idea behind these inner-outer iteration methods is that we explore parts of high-dimensional Krylov subspaces, hopefully localizing the same approximate solution that full GMRES would find over the entire subspace, but now at much lower computational costs. The alternatives for the inner iteration could be either one cycle of GMRES(m), since then we have also locally an optimal method, or some other iteration scheme,

---

x_0 is an initial guess; $r_0 = b - Ax_0$;
for $i = 0, 1, 2, 3, \ldots$
  Let $z^{(m)}$ be the approximate solution of $Az = r_i$
  obtained after m steps of an iterative method.
  $c = Az^{(m)}$ (often available from the iterative method)
  for $k = 0, \ldots, i - 1$
    $\alpha = (c_k, c)$
    $c = c - \alpha c_k$
    $z^{(m)} = z^{(m)} - \alpha u_k$
  $c_i = c/\|c\|_2$; $u_i = z^{(m)}/\|c\|_2$
  $x_{i+1} = x_i + (c_i, r_i)u_i$
  $r_{i+1} = r_i - (c_i, r_i)c_i$
  if x_{i+1} is accurate enough then quit
end

FIG. 3.7. The GMRES⋆ algorithm.

like for instance Bi-CGSTAB. As has been shown by VAN DER VORST [1992b] there are various situations for which we may expect stagnation or slow convergence for GMRES(m). In such cases it does not seem wise to use this method.

On the other hand it may also seem questionable whether a method like Bi-CGSTAB should lead to success in the inner iteration. This method does not satisfy a useful global minimization property and large part of its effectiveness comes from the underlying Bi-CG algorithm, which is based on bi-orthogonality relations. This means that for each outer iteration the inner iteration process has to build a bi-orthogonality relation again. It has been shown for the related Conjugate Gradients method that the orthogonality relations are determined largely by the distribution of the weights at the lower end of the spectrum and on the isolated eigenvalues at the upper end of the spectrum (VAN DER SLUIS and VAN DER VORST [1990]). By the nature of these kind of Krylov processes the largest eigenvalues and their corresponding eigenvector components quickly do enter the process after each restart, and hence it may be expected that much of the work is lost in rediscovering the same eigenvector components in the error over and over again, whereas these components may already be so small that further reduction in those directions in the outer iteration is waste of time, since it hardly contributes to a smaller norm of the residual.

This heuristic way of reasoning may explain in part our rather disappointing experiences with Bi-CGSTAB as the inner iteration process for GMRES⋆.

DE STURLER and FOKKEMA [1993] propose to prevent the outer search directions explicitly from being reinvestigated again in the inner process. This is done by keeping the Krylov subspace that is build in the inner iteration orthogonal with respect to the Krylov basis vectors generated in the outer iteration. The procedure works as follows.

In the outer iteration process the vectors $c_0, \ldots, c_{i-1}$ build an orthogonal basis for the Krylov subspace. Let C_i be the n by i matrix with columns $c_0, \ldots, c_{i-1}$. Then the inner iteration process at outer iteration i is carried out with the operator A_i instead of

A, and A_i is defined as

$$A_i = (I - C_i C_i^T)A. \tag{3.21}$$

It is easily verified that $A_i z \perp c_0, \ldots, c_{i-1}$ for all z, so that the inner iteration process takes place in a subspace orthogonal to these vectors. The additional costs, per iteration of the inner iteration process, are i inner products and i vector updates. In order to save on these costs, one should realize that it is not necessary to orthogonalize with respect to all previous c-vectors, and that "less effective" directions may be dropped, or combined with others. DE STURLER and FOKKEMA [1993] make suggestions for such strategies. Of course, these strategies are only effective in situations where we see too little residual reducing effect in the inner iteration process in comparison with the outer iterations of GMRES⋆.

3.5. *Bi-Conjugate Gradients*

We may proceed in a similar way as in the symmetric case:

$$AV_i = V_{i+1} T_{i+1,i}, \tag{3.22}$$

but here we use the matrix $W_i = [w_1, w_2, \ldots, w_i]$ for the projection of the system

$$W_i^T (b - Ax_i) = 0,$$

or

$$W_i^T A V_i y - W_i^T b = 0.$$

Using (3.22), we find that y_i satisfies

$$T_{i,i} y = \|r_0\|_2 e_1,$$

and $x_i = V_i y$. The resulting method is known as the Bi-Lanczos method LANCZOS [1952].

We have assumed that $d_i \neq 0$, that is $w_i^T v_i \neq 0$. The generation of the bi-orthogonal basis breaks down if for some i the value of $w_i^T v_i = 0$, this is referred to in literature as a *serious breakdown*. Likewise, when $w_i^T v_i \approx 0$, we have a near-breakdown. The way to get around this difficulty is the so-called Look-ahead strategy, which comes down to taking a number of successive basis vectors for the Krylov subspace together and to make them blockwise bi-orthogonal. This has been worked out in detail in PARLETT, TAYLOR and LIU [1985], FREUND, GUTKNECHT and NACHTIGAL [1993], FREUND and NACHTIGAL [1990], FREUND and NACHTIGAL [1991].

Another way to avoid breakdown is to restart as soon as a diagonal element gets small. Of course, this strategy looks surprisingly simple, but one should realise that at a restart the Krylov subspace, that has been built up so far, is thrown away, and this destroys the possibility of faster (i.e., superlinear) convergence. Moreover,the restarted process may suffer from break-down again. If this (rare) event happens then it is usually more effective to consider a look-ahead variant of the process.

We can try to construct an LU-decomposition, without pivoting, of $T_{i,i}$. If this decomposition exists, then, similar to CG, it can be updated from iteration to iteration and this leads to a recursive update of the solution vector, which avoids saving all intermediate r and w vectors. This variant of Bi-Lanczos is usually called Bi-Conjugate Gradients, or for short Bi-CG (FLETCHER [1976]). In Bi-CG, the d_i are chosen such that $v_i = r_{i-1}$, similarly to CG.

Of course one can in general not be certain that an LU decomposition (without pivoting) of the tridiagonal matrix $T_{i,i}$ exists, and this may lead also to breakdown (a breakdown of the *second kind*), of the Bi-CG algorithm. Note that this breakdown can be avoided in the Bi-Lanczos formulation of the iterative solution scheme, e.g., by making an LU-decomposition with 2 by 2 block diagonal elements (BANK and CHAN [1993]). It is also avoided in the QMR approach (see Section 3.5.1).

Note that for symmetric matrices Bi-Lanczos generates the same solution as Lanczos, provided that $w_1 = r_0$, and under the same condition Bi-CG delivers the same iterands as CG for positive definite symmetric matrices. However, the Bi-orthogonal variants do so at the cost of two matrix vector operations per iteration step.

For a computational scheme for Bi-CG, without provisions for breakdown, see BARRETT, BERRY, CHAN, DEMMEL, DONATO, DONGARRA, EIJKHOUT, POZO, ROMINE and VAN DER VORST [1994].

The scheme in Fig. 3.8 may be used for numerical experiments with the Bi-CG method. In the scheme the equation $Ax = b$ is solved with a suitable preconditioner K. The scheme has no provisions to prevent or cure break down.

As with conjugate gradients, the coefficients α_j and β_j, $j = 0, \ldots, i-1$, build the matrix T_i, as given in formula (3.19). This matrix is, for Bi-CG, in general not similar to a symmetric matrix. Its eigenvalues can be viewed as Petrov–Galerkin approximations, with respect to the spaces $\{\tilde{r}_j\}$ and $\{r_j\}$, of eigenvalues of A. For increasing values of i they tend to converge to eigenvalues of A. The convergence patterns, however, may be much more complicated and irregular than in the symmetric case.

3.5.1. QMR

The QMR method (FREUND and NACHTIGAL [1991]) relates to Bi-CG in a similar way as MINRES relates to CG. We start with the recurrence relations for the v_j:

$$AV_i = V_{i+1}T_{i+1,i}.$$

We would like to identify the x_i, with $x_i \in K^i(A; r_0)$, or $x_i = V_i y$, for which

$$\|b - Ax_i\|_2 = \|b - AV_i y\|_2 = \|b - V_{i+1}T_{i+1,i}y\|_2$$

is minimal, but the problem is that V_{i+1} is not orthogonal. However, we pretend that the columns of V_{i+1} are orthogonal. Then

$$\|b - Ax_i\|_2 = \left\| V_{i+1}\left(\|r_0\|_2 e_1 - T_{i+1,i}y\right)\right\|_2 = \left\|\left(\|r_0\|_2 e_1 - T_{i+1,i}y\right)\right\|_2,$$

and in FREUND and NACHTIGAL [1991] it is suggested to solve the projected minimum norm least squares problem $\|(\|r_0\|_2 e_1 - T_{i+1,i}y)\|_2$. The minimum value of this norm is called the quasi residual and will be denoted by $\|r_i^Q\|_2$.

x_0 is an initial guess; $r_0 = b - Ax_0$
Choose $\tilde{r}_0$ such that $(w_0, \tilde{r}_0) \neq 0$;
usually one chooses $\tilde{r}_0 = r_0$ or $\tilde{r}_0 = w_0$
for $i = 1, 2, \dots$
    Solve $K w_{i-1} = r_{i-1}$
    Solve $K^T \tilde{w}_{i-1} = \tilde{r}_{i-1}$
    $\rho_{i-1} = w_{i-1}^T \tilde{w}_{i-1}$
    if $\rho_{i-1} = 0$ **method fails**
    if $i = 1$
        $p_i = w_i$
        $\tilde{p}_i = \tilde{w}_i$
    else
        $\beta_{i-1} = \rho_{i-1}/\rho_{i-2}$
        $p_i = w_{i-1} + \beta_{i-1} p_{i-1}$
        $\tilde{p}_i = \tilde{w}_{i-1} + \beta_{i-1} \tilde{p}_{i-1}$
    endif
    $z_i = A p_i$
    $\tilde{z}_i = A^T \tilde{p}_i$
    $\alpha_i = \rho_{i-1}/(\tilde{p}_i^T z_i)$
    $x_i = x_{i-1} + \alpha_i p_i$
    $r_i = r_{i-1} - \alpha_i z_i$
    $\tilde{r}_i = \tilde{r}_{i-1} - \alpha_i \tilde{z}_i$
    if x_i is accurate enough **then quit**
end

FIG. 3.8. Bi-CG algorithm.

Since, in general, the columns of V_{i+1} are not orthogonal, the computed $x_i = V_i y$ does not solve the minimum residual problem, and therefore this approach is referred to as a Quasi-minimum residual approach (FREUND and NACHTIGAL [1991]). It can be shown that the norm of the residual r_i^{QMR} of QMR can be bounded in terms of the quasi residual

$$\|r_i^{\text{QMR}}\|_2 \leqslant \sqrt{i+1}\,\|r_i^Q\|_2.$$

The sketched approach leads to the simplest form of the QMR method. A more general form arises if the least squares problem is replaced by a weighted least squares problem (FREUND and NACHTIGAL [1991]). No strategies are yet known for optimal weights.

In FREUND and NACHTIGAL [1991] the QMR method is carried out on top of a look-ahead variant of the bi-orthogonal Lanczos method, which makes the method more robust. Experiments indicate that although QMR has a much smoother convergence behaviour than Bi-CG, it is not essentially faster than Bi-CG. This is confirmed explicitly by the following relation for the Bi-CG residual r_k^B and the quasi residual r_k^Q (in exact arithmetic):

$$\|r_k^B\|_2 = \frac{\|r_k^Q\|_2}{\sqrt{1 - (\|r_k^Q\|_2/\|r_{k-1}^Q\|_2)^2}}$$

(see CULLUM and GREENBAUM [1996], Theorem 4.1). This relation, which is similar to the relation for GMRES and FOM, shows that when QMR gives a significant reduction at step k, then Bi-CG and QMR have arrived at residuals of about the same norm (provided, of course, that the same set of starting vectors has been used).

It is tempting to compare QMR with GMRES, but this is difficult. GMRES really minimizes the 2-norm of the residual, but at the cost of increasing the work of keeping all residuals orthogonal and increasing demands for memory space. QMR does not minimize this norm, but often it has a convergence comparable to GMRES, at a cost of twice the amount of matrix vector products per iteration step. However, the generation of the basis vectors in QMR is relatively cheap and the memory requirements are limited and modest. QMR is preferred to Bi-CG in all cases because of its much smoother convergence behaviour, and also because QMR removes one break-down condition (even when implemented without look-ahead). Several variants of QMR, or rather Bi-CG, have been proposed, which increase the effectiveness of this class of methods in certain circumstances.

ZHOU and WALKER [1994] have shown that the Quasi-Minimum Residual approach can be followed for other methods, such as CGS and Bi-CGSTAB, as well. The main idea is that in these methods the approximate solution is updated as

$$x_{i+1} = x_i + \alpha_i p_i,$$

and the corresponding residual is updated as

$$r_{i+1} = r_i - \alpha_i A p_i.$$

This means that $AP_i = W_i R_{i+1}$, with W_i a lower bidiagonal matrix. The x_i are combinations of the p_i, so that we can try to find the combination $P_i y_i$ for which $\|b - AP_i y_i\|_2$ is minimal. If we insert the expression for AP_i, and ignore the fact that the r_i are not orthogonal, then we can minimize the norm of the residual in a quasi-minimum least squares sense, similar to QMR.

3.5.2. CGS

It is well known that the bi-conjugate gradient residual vector can be written as r_j $(= \rho_j v_j) = P_j(A)r_0$, and, likewise, the so-called shadow residual $\hat{r}_j$ $(= \rho_j w_j)$ can be written as $\hat{r}_j = P_j(A^T)\hat{r}_0$. Because of the bi-orthogonality relation we have that

$$\begin{aligned}(r_j, \hat{r}_i) &= (P_j(A)r_0, P_i(A^T)\hat{r}_0)\\ &= (P_i(A)P_j(A)r_0, \hat{r}_0) = 0,\end{aligned}$$

for $i < j$. The iteration parameters for bi-conjugate gradients are computed from innerproducts like the above. SONNEVELD [1989] observed that we can also construct the vectors $\tilde{r}_j = P_j^2(A)r_0$, using only the latter form of the innerproduct for recovering the bi-conjugate gradients parameters (which implicitly define the polynomial P_j). By doing so, the computation of the vectors $\hat{r}_j$ can be avoided and so can the multiplication by the matrix A^T.

The resulting CGS (SONNEVELD [1989]) method works in general very well for many unsymmetric linear problems. It converges often much faster than Bi-CG (about

---

x_0 is an initial guess; $r_0 = b - Ax_0$;
$\tilde{r}_0$ is an arbitrary vector, such that
$(r_0, \tilde{r}_0) \neq 0$,
e.g., $\tilde{r}_0 = r_0$; $\rho_0 = (r_0, \tilde{r}_0)$;
$\beta_{-1} = \rho_0$; $p_{-1} = q_0 = 0$;
for $i = 0, 1, 2, \ldots$
  $u_i = r_i + \beta_{i-1} q_i$;
  $p_i = u_i + \beta_{i-1}(q_i + \beta_{i-1} p_{i-1})$;
  solve $\hat{p}$ from $K\hat{p} = p_i$;
  $\hat{v} = A\hat{p}$;
  $\alpha_i = \frac{\rho_i}{(\tilde{r}_0, \hat{v})}$;
  $q_{i+1} = u_i - \alpha_i \hat{v}$;
  solve $\hat{u}$ from $K\hat{u} = u_i + q_{i+1}$
  $x_{i+1} = x_i + \alpha_i \hat{u}$;
  if x_{i+1} is accurate enough then quit;
  $r_{i+1} = r_i - \alpha_i A\hat{u}$;
  $\rho_{i+1} = (\tilde{r}_0, r_{i+1})$;
  if $\rho_{i+1} = 0$ then method fails to converge!;
  $\beta_i = \frac{\rho_{i+1}}{\rho_i}$;
end

FIG. 3.9. CGS algorithm.

twice as fast in some cases) and has the advantage that fewer vectors are stored than in GMRES. These three methods have been compared in many studies (see, e.g., RADICATI DI BROZOLO and ROBERT [1989], BRUSSINO and SONNAD [1989], POMMERELL and FICHTNER [1991], NACHTIGAL, REDDY and TREFETHEN [1992]).

CGS, however, usually shows a very irregular convergence behaviour. This behaviour can even lead to cancellation and a "spoiled" solution (VAN DER VORST [1992a]); see also Section 3.6. FREUND [1993] suggested a squared variant of QMR, which was called TFQMR. His experiments show that TFQMR is not necessarily faster than CGS, but it has certainly a much smoother convergence behavior.

The scheme in Fig. 3.9 represents the CGS process for the solution of $Ax = b$, with a given preconditioner K.

In exact arithmetic, the α_j and β_j are the same constants as those generated by BiCG. Therefore, they can be used to compute the Petrov–Galerkin approximations for eigenvalues of A.

CGS may be attractive in the context of Newton iterations for nonlinear systems, where the new iteration requires the solution of a linear system with the Jacobian of the nonlinear system.

3.5.3. Bi-CGSTAB

Bi-CGSTAB (VAN DER VORST [1992a]) is based on the following observation. Instead of squaring the Bi-CG iteration polynomial, as has been done in CGS, we can construct other iteration methods, by which x_i are generated so that $r_i = \widetilde{P}_i(A)P_i(A)r_0$ with other

---

x_0 is an initial guess; $r_0 = b - Ax_0$;
$\bar{r}_0$ is an arbitrary vector, such that
  $(\bar{r}_0, r_0) \neq 0$, e.g., $\bar{r}_0 = r_0$;
$\rho_{-1} = \alpha_{-1} = \omega_{-1} = 1$;
$v_{-1} = p_{-1} = 0$;
for $i = 0, 1, 2, \ldots$
  $\rho_i = (\bar{r}_0, r_i)$; $\beta_{i-1} = (\rho_i/\rho_{i-1})(\alpha_{i-1}/\omega_{i-1})$;
  $p_i = r_i + \beta_{i-1}(p_{i-1} - \omega_{i-1} v_{i-1})$;
  Solve $\hat{p}$ from $K\hat{p} = p_i$;
  $v_i = A\hat{p}$;
  $\alpha_i = \rho_i / (\bar{r}_0, v_i)$;
  $s = r_i - \alpha_i v_i$;
  if $\|s\|$ small enough **then**
    $x_{i+1} = x_i + \alpha_i \hat{p}$; quit;
  Solve z from $Kz = s$;
  $t = Az$;
  $\omega_i = (t, s)/(t, t)$;
  $x_{i+1} = x_i + \alpha_i \hat{p} + \omega_i z$;
  if x_{i+1} is accurate enough **then** quit;
  $r_{i+1} = s - \omega_i t$;
end

FIG. 3.10. The Bi-CGSTAB algorithm.

ith degree polynomials $\widetilde{P}$. An obvious possibility is to take for $\widetilde{P}_j$ a polynomial of the form

$$Q_i(x) = (1 - \omega_1 x)(1 - \omega_2 x) \ldots (1 - \omega_i x), \tag{3.23}$$

and to select suitable constants ω_j. This expression leads to an almost trivial recurrence relation for the Q_i.

In Bi-CGSTAB ω_j in the jth iteration step is chosen as to minimize r_j, with respect to ω_j, for residuals that can be written as $r_j = Q_j(A)P_j(A)r_0$.

The preconditioned Bi-CGSTAB algorithm for solving the linear system $Ax = b$, with preconditioning K reads as in Fig. 3.10.

The matrix K in this scheme represents the preconditioning matrix and the way of preconditioning (VAN DER VORST [1992a]). The above scheme in fact carries out the Bi-CGSTAB procedure for the explicitly postconditioned linear system

$$AK^{-1}y = b,$$

but the vectors y_i and the residual have been backtransformed to the vectors x_i and r_i corresponding to the original system $Ax = b$. Compared to CGS two extra innerproducts need to be calculated.

In exact arithmetic, the α_j and β_j have the same values as those generated by Bi-CG and CGS. Hence, they can be used to extract eigenvalue approximations for the eigenvalues of A (see Bi-CG).

Bi-CGSTAB can be viewed as the product of Bi-CG and GMRES(1). Of course, other product methods can be formulated as well. GUTKNECHT [1993] has proposed BiCGSTAB2, which is constructed as the product of Bi-CG and GMRES(2). A more general concept was described by SLEIJPEN and FOKKEMA [1993], under the name Bi-CGSTAB(ℓ).

3.5.4. *The Bi-CGSTAB(ℓ) methods*

In BiCGSTAB, the factor Q_k has only real roots by construction. It is well-known that optimal reduction polynomials for matrices with complex eigenvalues may have complex roots as well. If, for instance, the matrix A is real skew-symmetric, then GCR(1) stagnates forever, whereas a method like GCR(2) (or GMRES(2)), in which we minimize over two combined successive search directions, may lead to convergence, and this is mainly due to the fact that then complex eigenvalue components in the error can be effectively reduced.

This point of view was taken in GUTKNECHT [1993] for the construction of the BiCGSTAB2 method. SLEIJPEN and FOKKEMA [1993] generalized this idea to the combination of Bi-CG with GCR(ℓ), which leads to BI-CGSTAB(ℓ). In this approach one carries out ℓ successive steps with Bi-CG, and one uses the additional ℓ matrix vector products for the construction of a GCR(ℓ) factor.

There are variants of this approach in which more stable bases for the Krylov subspaces are generated (SLEIJPEN, VAN DER VORST and FOKKEMA [1994]), but for low values of ℓ a standard basis satisfies, together with a minimum norm solution obtained through solving the associated normal equations (which requires the solution of an ℓ by ℓ system. In most cases BiCGSTAB(2) will already give nice results for problems where BiCGSTAB fails.

Bi-CGSTAB(2) can be represented by the scheme in Fig. 3.11.

For the GCR(2) part we note that the 5 inner products can be taken together, in order to reduce start-up times for their global assembling. This gives the method BiCGSTAB(2) a (slight) advantage over BiCGSTAB. Furthermore we note that the updates in the GCR(2) may lead to more efficient code than for BiCGSTAB, since some of them can be combined.

3.6. *Accurate updating techniques*

Bi-CG and methods derived from Bi-CG can display rather irregular convergence behaviour. By irregular convergence we refer to the situation where successive residual vectors in the iterative process differ in orders of magnitude in norm, and some of these residuals may be even much bigger in norm than the starting residual. In particular the CGS method suffers from this phenomenon. We will show why this is a point of concern, even if eventually the (updated) residual satisfies a given tolerance.

In the Bi-CG algorithms, as well as in CG, we see in the algorithm typically a statement for the update of x_i, like

$$x_{i+1} = x_i + w_i \tag{3.24}$$

	x_0 is an initial guess; $r_0 = b - Ax_0$;
	$\hat{r}_0$ is an arbitrary vector, such that $(r, \hat{r}_0) \neq 0$,
	e.g., $\hat{r}_0 = r$;
	$\rho_0 = 1$; $u = 0$; $\alpha = 0$; $\omega_2 = 1$;
	for $i = 0, 2, 4, 6, \ldots$
	$\rho_0 = -\omega_2 \rho_0$
even BiCG step:	$\rho_1 = (\hat{r}_0, r_i)$; $\beta = \alpha\rho_1/\rho_0$; $\rho_0 = \rho_1$
	$u = r_i - \beta u$;
	$v = Au$
	$\gamma = (v, \hat{r}_0)$; $\alpha = \rho_0/\gamma$;
	$r = r_i - \alpha v$;
	$s = Ar$
	$x = x_i + \alpha u$;
odd BiCG step:	$\rho_1 = (\hat{r}_0, s)$; $\beta = \alpha\rho_1/\rho_0$; $\rho_0 = \rho_1$
	$v = s - \beta v$;
	$w = Av$
	$\gamma = (w, \hat{r}_0)$; $\alpha = \rho_0/\gamma$;
	$u = r - \beta u$
	$r = r - \alpha v$
	$s = s - \alpha w$
	$t = As$
GCR(2)-part:	$\omega_1 = (r, s)$; $\mu = (s, s)$; $\nu = (s, t)$; $\tau = (t, t)$;
	$\omega_2 = (r, t)$; $\tau = \tau - \nu^2/\mu$; $\omega_2 = (\omega_2 - \nu\omega_1/\mu)/\tau$;
	$\omega_1 = (\omega_1 - \nu\omega_2)/\mu$
	$x_{i+2} = x + \omega_1 r + \omega_2 s + \alpha u$
	$r_{i+2} = r - \omega_1 s - \omega_2 t$
	if x_{i+2} accurate enough **then** quit
	$u = u - \omega_1 v - \omega_2 w$
	end

FIG. 3.11. The Bi-CGSTAB(2) algorithm.

and a statement for the update of r_i, of the form

$$r_{i+1} = r_i - Aw_i. \tag{3.25}$$

We see that, in exact arithmetic, the relation $r_{i+1} = b - Ax_{i+1}$ holds, just as expected. A further inspection of these algorithms reveals that x_i is not used at other places in the basic algorithm, whereas the r_i is also used for the computation of the search direction and for iteration parameters. The important consequence of this is that rounding errors introduced by the actual evaluation of r_{i+1} through Eq. (3.25) will influence the further iteration process, but rounding errors in the evaluation of x_{i+1} by (3.24) will have no effect on the iteration. This would not be much of a problem if the rounding error

$$\delta r_{i+1} := fl(r_i - Aw_i) - (r_i - Aw_i)$$

would match the rounding error

$$\delta x_{i+1} := fl(x_i + w_i) - (x_i + w_i),$$

in the sense that $\delta r_{i+1} = -A\delta x_{i+1}$, since that would keep the desired relation $r_{i+1} = b - Ax_{i+1}$ intact. However, it will be obvious that this is idle hope, and the question remains how serious a possible deviation between r_j and $b - Ax_j$ can be.

Of course, we make rounding errors in (3.24) and (3.25) through the vector addition, but usually these errors will be small in comparison with the rounding errors introduced in the multiplication of w_i with A. Therefore, we will here consider only the effect of these errors. In this case, we can write the computed r_{i+1} as

$$r_{i+1} = r_j - Aw_i - \Delta_{Ai} w_i, \tag{3.26}$$

where Δ_A is an $n \times n$ matrix for which $|\Delta_{Ai}| \preceq n_A \bar{\xi} |A|$: n_A is the maximum number of nonzero matrix entries per row of A, $|B| := (|b_{ij}|)$ if $B = (b_{ij})$, $\bar{\xi}$ is the relative machine precision, the inequality $\preceq$ refers to element-wise $\leqslant$.

It then simply follows that

$$\begin{aligned} r_k - (b - Ax_k) &= \sum_{j=1}^{k} \Delta_{Aj} w_j \\ &= \sum_{j=1}^{k} \Delta_{Aj}(e_{j-1} - e_j), \end{aligned} \tag{3.27}$$

e_j is the approximation error in the jth approximation: $e_j := x - x_j$. Hence,

$$\begin{aligned} \big| \|r_k\| - \|b - Ax_k\| \big| &\leqslant 2kn_A\bar{\xi} \big\| |A| \big\| \max_j \|e_j\| \\ &\leqslant 2kn_A\bar{\xi} \big\| |A| \big\| \|A^{-1}\| \max_j \|r_j\|. \end{aligned} \tag{3.28}$$

Except for the factor k, the first upper–bound appears to be rather sharp. We see that an approximation with a large approximation error (and hence a large residual) may lead to inaccurate results in the remaining iteration process. Such large local approximation errors are typical for CGS, and VAN DER VORST [1992a] describes an example of the resulting numerical inaccuracy. If there are a number of approximations with comparable large approximation errors, then their multiplicity may replace the factor k, otherwise it will be only the largest approximation error that makes up virtually all of the bound for the deviation.

For more details we refer to SLEIJPEN and VAN DER VORST [1996], SLEIJPEN, VAN DER VORST and FOKKEMA [1994].

It is of course important to maintain a reasonable correspondence between r_k and $b - Ax_k$, and the easiest way to do this would be to replace the vector r_k by $b - Ax_k$. However, the vectors r_k steer the entire iterative process and their relation defines the projected matrix $T_{i,i}$. If we replace these vectors then we ignore the rounding errors to these vectors and it will be clear that the iteration process cannot compensate for these rounding errors. These rounding errors may be significant at iteration steps where the update to r_j is relatively large and the above sketched naive replacement strategy may then not be expected to work well. Indeed, if we replace r_{i+1} in CGS by $b - Ax_i$, instead

---

$z = x_0$; $\hat{x} = 0$; $r_{\min} = \|r_0\|_2$;
...
for $j = 0, 2, \ldots$ until convergence
    ...
    $\hat{x} = \hat{x} + w_j$ (instead of update of x_i)
    if $\|r_j\|_2 < r_{\min}$ (i.e., group update)
        $z = z + \hat{x}$
        $\hat{x} = 0$
        $r_j = b - Az$
        $r_{\min} = \|r_j\|_2$
    end if
end for

FIG. 3.12. The groupwise update strategy.

of updating it from r_i, then we observe stagnation in the convergence in many important situations. This means that we have to be more careful.

Neumaier (see SLEIJPEN and VAN DER VORST [1996] and references therein) suggested to replace r_j by $b - Ax_j$ in CGS only at places where $\|r_j\|_2$ is smaller than the smallest norm of the residual in the previous iteration history and to carry out a groupwise update for the iterates in between. Schematically, the groupwise update and residual replacement strategy of Neumaier can be described as in Fig. 3.12.

This scheme was further analysed and refined, in particular with a flying restart strategy, in SLEIJPEN and VAN DER VORST [1996]. Note that the errors in the evaluation of w_j itself are not so important: it is the different treatment of w_j in the updating of x_j and of r_j that causes the two vectors to loose their wanted relation. In this respect we may consider the vectors w_j as exact quantities.

At a replacement step we perturb the recurrence relation for the basis vectors of the Krylov subspace and we want these errors to be as small as possible. The updates w_j usually vary widely in norm in various stages of the iteration process, for instance in an early phase these norms may be larger than $\|r_0\|_2$, whereas they are small in the final phase of the iteration process. Specially in a phase between two successive smallest values of $\|r_j\|_2$, the norms of the updates may be a good deal larger than in the next interval between two smallest residual norms. Grouping the updates in groups of updates avoids that rounding errors within one group spoil the result for another group. More specifically, if we denote the sum of w_j's for the groups by S_i, and the total sum of updates by S, then groupwise updating leads to errors of the magnitude of $\xi|S_i|$, which can be much smaller than $\xi|S|$.

Now we have to determine how much we can perturb the recurrence relations for the Lanczos vectors r_j. This has been studied in much detail in TONG and YE [2000]. It has been observed by many authors that the driving recurrences $r_j = r_{j-1} - \alpha_{j-1}Aq_{j-1}$ and $q_j = r_j + \beta_{j-1}q_{j-1}$ are locally satisfied almost to machine precision and this is one of the main properties behind the convergence of the computed residuals (GREENBAUM [1997a], TONG and YE [2000], SLEIJPEN and VAN DER VORST [1996]). TONG and YE [2000] observed that these convergence is maintained even when we perturb the recur-

Input: an initial approximation $x = x_0$;
  a residual replacement threshold $\varepsilon = \sqrt{\xi}$; an estimate of $N\|A\|$;
Set $r_0 = b - Ax_0$; $\hat{x}_0 = 0$; $d_{\text{init}} = d_0 = \xi(\|r_0\| + N\|A\|\|x_0\|)$,
for $j = 1, 2, \ldots,$ until convergence
  Generate a correction vector q_j by the Iterative Method;
  $\hat{x}_j = \hat{x}_{j-1} + q_j$
  $r_j = r_{j-1} - Aq_j$
  $d_j = d_{j-1} + \xi N\|A\|\|\hat{x}_j\| + \xi\|r_j\|$
  if $d_{j-1} \leqslant \varepsilon\|r_{j-1}\|$, $d_j > \varepsilon\|r_j\|$ and $d_j > 1.1d_{\text{init}}$
    $z = z + \hat{x}_j$
    $\hat{x}_j = 0$
    $r_j = b - Az$
    $d_{\text{init}} = d_j = \xi(\|r_j\| + N\|A\|\|z\|)$
  end if
end for
$z = z + \hat{x}_j$

FIG. 3.13. The reliable updating strategy.

rence relations with perturbations that are significantly greater than machine precision, say of the order of the square root of the machine precision ξ, in a relative sense.

The idea, presented in VAN DER VORST and YE [2000], is to compute an upper bound for the deviation in r_j, with respect to $b - Ax_j$, in finite precision, and to replace r_j by $b - Ax_j$ as soon as this upper bound reaches the relative level of $\sqrt{\xi}$. This upper bound is denoted by d_j and it is computed from the recurrence

$$d_j = d_{j-1} + \xi N\|A\|\|\hat{x}_j\| + \xi\|r_j\|,$$

with N the maximal number of nonzero entries per row of A.

The replacement strategy for reliable updating is then implemented schematically as in Fig. 3.13.

REMARK. For this reliable implementation, we need to put a value for N (the maximal number of nonzero entries per row of A) and $\|A\|$. The number of nonzero entries may, in applications, vary from row to row, and selecting the maximum number may not be very realistic. In our experience with sparse matrices, the simple choice $N = 1$ still leads to a practical estimate d_n for $\|\delta_n\|$. For $\|A\|$, we suggest to take simply $\|A\|_\infty$.

In any case, we note that precise values are not essential, because the replacement threshold ε can be adjusted. We also need to choose this ε. Extensive numerical testing (see VAN DER VORST and YE [2000]) suggests that $\varepsilon \sim \sqrt{\xi}$ is a practical criterion. However, there are examples where this choice leads to stagnating residuals at some unacceptable level. In such cases, choosing a smaller ε will regain the convergence to $O(\xi)$.

The presented implementation requires one extra matrix-vector multiplication when a replacement is carried out. Since only a few steps with replacement are required, this extra cost is marginal relative to the other costs. However, some savings can be made by selecting a slightly smaller ε and carrying out residual replacement at the step next to the one for which the residual replacement criterion is satisfied (cf. SLEIJPEN and VAN DER VORST [1996]). It also requires one extra vector storage for the groupwise solution update (for z) and computation of a vector norm $\|\hat{x}_n\|$ for the update of d_n ($\|r_n\|$ is usually computed in the algorithm for stopping criteria).

4. Preconditioning

4.1. Introduction

There are many occasions and applications where iterative methods fail to converge or converge very slowly. The usual remedy is to apply a preconditioner, that is instead of $Ax = b$, one solves $KAx = Kb$ or a spectrally equivalent system, for example, $AKy = b$. The general problem of finding an efficient preconditioner, is to identify a linear operator K (the *preconditioner*) with the properties that:

1. K is a good approximation to A in some sense.
2. The cost of the construction of K is not prohibitive.
3. The system $Ky = z$ is much easier to solve than the original system.

By efficient, we mean that the iteration method converges much faster, in terms of CPU time, for the preconditioned system.

The choice of K varies from purely "black box" algebraic techniques which can be applied to general matrices to "problem dependent" preconditioners which exploit special features of a particular problem class. Although problem dependent preconditioners can be very powerful, there is still a practical need for efficient preconditioning techniques for large classes of problems. We refer the reader to AXELSSON [1994], CHAN and VAN DER VORST [1997], SAAD [1996] for further discussions on this. In this section, we will not go in details about preconditioning but rather give a sketch of important ideas for the construction of parallel preconditioners. For more details on implementation for high performance computers, see DONGARRA, DUFF, SORENSEN and VAN DER VORST [1998].

Originally, preconditioners were based on direct solution methods in which part of the computation is skipped. This leads to the notion of *Incomplete LU* (or *ILU*) *factorization* (MEIJERINK and VAN DER VORST [1977], AXELSSON [1994], SAAD [1996]).

4.2. Incomplete LU factorizations

Standard Gaussian elimination is equivalent to factoring the matrix A as $A = LU$, where L is lower triangular and U is upper triangular. In actual computations these factors are explicitly constructed. The main problem in sparse matrix computations is that the factors of A are often a good deal less sparse than A, which makes solution expensive, The basic idea in the point ILU preconditioner is to modify Gaussian elimination to allow fill-ins at only a restricted set of positions in the LU factors. Let the allowable fill-in positions be given by the index set S, i.e.,

$$\begin{aligned} l_{i,j} &= 0 \quad \text{if } j > i \quad \text{or} \quad (i,j) \notin S, \\ u_{i,j} &= 0 \quad \text{if } i > j \quad \text{or} \quad (i,j) \notin S. \end{aligned} \tag{4.1}$$

A commonly used strategy is to define S by:

$$S = \{(i,j) \mid a_{i,j} \neq 0\}. \tag{4.2}$$

That is, the only nonzeros allowed in the LU factors are those for which the corresponding entries in A are nonzero. Before we proceed with different strategies for the construction of effective incomplete factorizations we consider the question whether these factorizations exist. It can be shown that they exist for so-called M-matrices.[18]

The theory of M-matrices for iterative methods is very well covered by VARGA [1962]. These matrices occur frequently after discretisation of PDE's, and for M-matrices one can identify all sorts of approximating matrices K for which the basic splitting leads to a convergent iteration (3.1). For preconditioners for Krylov subspace methods it is not important that the basic iteration converges; primarily we want reduced condition numbers and/or better eigenvalue distributions for the preconditioned matrices. These latter properties are very difficult to prove. In fact, some of these effects have been proved only for very special model problems (VAN DER VORST and SLEIJPEN [1993], VAN DER VORST [1982]).

We now consider the actual construction of the incomplete decomposition. Let the preconditioner M be defined by the product of the incomplete LU factors, i.e., $M = LU$. For M to be a good preconditioner, it must be a good approximation to A in some measure. A typical strategy is to require the entries of M to match those of A on the set S:

$$m_{i,j} = a_{i,j} \quad \text{if } (i,j) \in S. \tag{4.3}$$

Even though the conditions (4.1) and (4.3) together are sufficient (for certain classes of matrices) to determine the nonzero entries of L and U directly, it is more natural and simpler to compute these entries based on a simple modification of the Gaussian elimination algorithm; see Fig. 4.1. The main difference from the usual Gaussian elimination algorithm is in the inner-most j-loop where an update to $a_{i,j}$ is computed only if it is allowed by the constraint set S.

After the completion of the algorithm, the incomplete LU factors are stored in the corresponding lower and upper triangular parts of the array A. It can be shown that the computed LU factors satisfy (4.3).

The incomplete factors $\widetilde{L}$ and $\widetilde{U}$ define the preconditioner $K = (\widetilde{L}\widetilde{U})^{-1}$. In the context of an iterative solver, this means that we have to evaluate expressions like $z = (\widetilde{L}\widetilde{U})^{-1}y$ for any given vector y. This is done in two steps: first obtain w from the solution of $\widetilde{L}w = y$ and then compute z from $\widetilde{U}z = w$. Straightforward implementation of these processes leads to recursions, for which vector and parallel computers are not ideally suited. This sort of observation has led to reformulations of the preconditioner, for example, with reordering techniques or with blocking techniques. It has

[18] The nonsingular matrix A is an M-matrix if $a_{i,j} \leqslant 0$, for $i \neq j$, and $A^{-1} \geqslant 0$.

ILU for an n by n matrix A (cf. AXELSSON [1994]):
for $k = 1, 2, \dots, n-1$
  $d1/a_{k,k}$
  for $i = k+1, k+2, \dots, n$
    if $(i, k) \in S$
      $e = da_{i,k};\ a_{i,k} = e$
      for $j = k+1, \dots, n$
        if $(i, j) \in S$ and $(k, j) \in S$
          $a_{i,j} = a_{i,j} - ea_{k,j}$
        end if
      end j
    end if
  end i
end k

FIG. 4.1. ILU for a general matrix A.

also led to different types of preconditioners, including diagonal scaling, polynomial preconditioning, and truncated Neumann series. These approaches may be useful in certain circumstances, but they tend to increase the computational complexity, because they often require more iteration steps or make each iteration step more expensive. Diagonal scaling can be done explicitly, without any further complications for parallel processing. This can be done before and in addition to the construction of another preconditioning.

4.3. An example of incomplete decompositions

We shall illustrate the above sketched process for a popular preconditioner for sparse positive definite symmetric matrices, namely, the *Incomplete Cholesky factorization* (GOLUB and VAN LOAN [1996], MEIJERINK and VAN DER VORST [1977, 1981], VARGA [1960]) with no fill-in. We will denote this preconditioner as IC(0). CG in combination with IC(0) is often referred to as ICCG(0). We shall consider IC(0) for the matrix with five nonzero diagonals, that arises after the 5-point finite-difference discretization of the two-dimensional Poisson equation over a rectangular region, using a grid of dimensions n_x by n_y. If the entries of the three nonzero diagonals in the upper triangular part of A are stored in three arrays $a(\cdot, 1)$ for the main diagonal, $a(\cdot, 2)$ for the first co-diagonal, and $a(\cdot, 3)$ for the n_xth co-diagonal, then the ith row of the symmetric matrix A can be represented as in (4.4).

$$A = \begin{pmatrix} \ddots & & \ddots & \ddots & \ddots & & \ddots \\ & a_{i-n_x,3} & & a_{i-1,2} & a_{i,1} & a_{i,2} & & a_{i,3} \\ & \ddots & & \ddots & \ddots & \ddots & & \ddots \end{pmatrix}. \tag{4.4}$$

This corresponds to the unknowns over a grid as shown below:

$$\begin{array}{ccccccccc}
\cdot & \cdot & \cdot & & \cdot & & & \cdot & \cdot \\
\cdot & \cdot & \cdot & & \cdot & & & \cdot & \cdot \\
 & & & & \downarrow & & & & \\
\cdot & \cdot & \cdot & \rightarrow & \cdot & \leftarrow & & \cdot & \cdot \\
 & & & & \uparrow & & & & \\
\cdot & \cdot & \cdot & & \cdot & & & \cdot & \cdot \\
\cdot & \cdot & \cdot & & \cdot & & & \cdot & \cdot \\
\cdot & \cdot & \cdot & & \cdot & & & \cdot & \cdot
\end{array}$$

If we write A as

$$A = L_A + \mathrm{diag}(A) + {L_A}^T,$$

in which L_A is the strictly lower triangular part of A, then the IC(0)-preconditioner can be written as

$$K = (L_A + D)D^{-1}(L_A^T + D).$$

This relation only holds if there are no corrections to off-diagonal nonzero entries in the incomplete elimination process for A and if we ignore all fill-in outside the nonzero structure of A. It is easy to do this for the 5-point Laplacian. For other matrices, we can force the relation to hold only if we ignore also Gaussian elimination corrections at places where A has nonzero entries. This may decrease the effectiveness of the preconditioner, because we then neglect more operations in the Gaussian elimination process.

For IC(0), the entries d_i of the diagonal matrix D can be computed from the relation

$$\mathrm{diag}(K) = \mathrm{diag}(A).$$

For the 5-diagonal A, this leads to the following relations for the d_i:

$$d_i = a_{i,1} - a_{i-1,2}^2/d_{i-1} - a_{i-n_x,3}^2/d_{i-n_x}. \tag{4.5}$$

Obviously this is a recursion in both directions over the grid. This aspect will be discussed later when considering the application of the preconditioner in the context of parallel and vector processing.

The so-called *modified incomplete decompositions* (DUPONT, KENDALL and RACHFORD JR [1968], GUSTAFSSON [1978]) follow from the requirement that

$$\mathrm{rowsum}(K) = \mathrm{rowsum}(A) + ch^2. \tag{4.6}$$

The term ch^2 is for grid-oriented problems with mesh-size h. Although in many applications this term is skipped (that is, one often takes $c = 0$), this may lead to ineffective preconditioning or even break-down of the preconditioner, see EIJKHOUT [1992]. In our context, the rowsum requirement in (4.6) amounts to an additional correction to the diagonal entries d_i, compared to those computed in (4.5).

For the solution of systems $Kw = r$, given by

$$K^{-1}r = (L_A^T + D)^{-1}D(L_A + D)^{-1}r,$$

it will almost never be advantageous to determine the matrices $(L_A^T + D)^{-1}$ and $(L_A + D)^{-1}$ explicitly, since these matrices are usually dense triangular matrices. Instead, for the computation of, say, $y = (L_A + D)^{-1}r$, y is solved from the linear lower triangular system $(L_A + D)y = r$. This step then leads typically to relations for the entries y_i, of the form

$$y_i = (r_i - a_{i-1,2}y_{i-1} - a_{i-n_x,3}y_{i-n_x})/d_i,$$

which again represents a recursion in both directions over the grid, of the same form as the recursion for the d_i.

For differently structured matrices, we can also perform incomplete LU factorizations. For efficient implementation, often many of the ideas, shown here for Incomplete Cholesky factorizations, apply. For more general matrices with the same nonzero structure as the 5-point Laplacian, some other well known approximations lead to precisely the same type of recurrence relations as for Incomplete LU and Incomplete Cholesky: for example, Gauss–Seidel, SOR, SSOR (HAGEMAN and YOUNG [1981]), and SIP (STONE [1968]). Hence these methods can often be made vectorizable or parallel in the same way as for Incomplete Cholesky preconditioning.

Since vector and parallel computers do not lend themselves well to recursions in a straightforward manner, the recursions just discussed may seriously degrade the effect of preconditioning on a vector or parallel computer, if carried out in the form given above. This sort of observation has led to different types of preconditioners, including diagonal scaling, polynomial preconditioning, and truncated Neumann series. Such approaches may be useful in certain circumstances, but they tend to increase the computational complexity (by requiring more iteration steps or by making each iteration step more expensive). On the other hand, various techniques have been proposed to vectorize the recursions, mainly based on reordering the unknowns or changing the order of computation. For regular grids, such approaches lead to highly vectorizable code for the standard incomplete factorizations (and consequently also for Gauss-Seidel, SOR, SSOR, and SIP). If our goal is to minimize computing time, there may thus be a trade-off between added complexity and increased vectorization. However, before discussing these techniques, we shall present a method of reducing the computational complexity of preconditioning.

4.4. Reordering the unknowns

A standard trick for exploiting parallelism is to select all unknowns that have no direct relationship with each other and to number them first. For the 5-point finite-difference discretization over rectangular grids, this approach is known as a *red-black ordering*. For elliptic PDEs, this leads to very parallel preconditioners. The performance of the preconditioning step is as high as the performance of the matrix-vector product. However, changing the order of the unknowns leads in general to a different preconditioner. DUFF and MEURANT [1989] report on experiments that show that most reordering schemes (for example, the red-black ordering) lead to a considerable increase in iteration steps (and hence in computing time) compared with the standard lexicographical ordering. For the red-black ordering associated with the discretized Poisson equation, it can be shown that the condition number of the preconditioned system is only about one quarter that of the unpreconditioned system for ILU, MILU and SSOR, with no asymptotic improvement as the gridsize h tends to zero (KUO and CHAN [1990]).

One way to obtain a better balance between parallelism and fast convergence, is to use more colours (DOI [1991]). In principle, since there is not necessarily any independence between different colours, using more colours decreases the parallelism but increases the global dependence and hence the convergence. In DOI and HOSHI [1992], up to 75 colours are used for a 76^2 grid on the NEC SX-3/14 resulting in a 2 Gflop/s performance, which is much better than for the wavefront ordering. With this large number of colours the speed of convergence for the preconditioned process is virtually the same as with a lexicographical ordering (DOI [1991]).

The concept of *multi-colouring* has been generalized to unstructured problems by JONES and PLASSMANN [1994]. They propose effective heuristics for the identification of large independent subblocks of a given matrix. For problems large enough to get sufficient parallelism in these subblocks, their approach leads to impressive speedups compared to the natural ordering on a single processor.

For a discussion of these techniques see DONGARRA, DUFF, SORENSEN and VAN DER VORST [1998].

4.5. *Hybrid techniques*

In the classical incomplete decompositions one ignores fill-in right from the start of the decomposition process. However, it might be a good idea to delay this until the matrix becomes too dense. This leads to a hybrid combination of direct and iterative techniques. One of such approaches has been described in BOMHOF and VAN DER VORST [2000]; we will describe it here in some detail.

We first permute the given matrix of the linear system $Ax = b$ to a doubly bordered block diagonal form:

$$\widetilde{A} = P^T A P = \begin{bmatrix} A_{00} & 0 & \cdots & 0 & A_{0m} \\ 0 & A_{11} & \ddots & \vdots & A_{1m} \\ \vdots & \ddots & \ddots & 0 & \vdots \\ 0 & \cdots & 0 & A_{m-1m-1} & \vdots \\ A_{m0} & A_{m1} & \cdots & \cdots & A_{mm} \end{bmatrix}. \tag{4.7}$$

Of course, the parallelism in the eventual method depends on the value of m, and some problems lend themselves more to this than others. Many circuit simulation problems can be rewritten in an effective way, as a circuit is often composed of components that are only locally coupled to others.

We permute the right-hand side b as well to $\tilde{b} = P^T b$, which leads to the system

$$\widetilde{A}\tilde{x} = \tilde{b}, \tag{4.8}$$

with $x = Px$.

The parts of $\tilde{b}$ and $\tilde{x}$ that correspond to the block ordering, will be denoted by $\tilde{b}_i$ and $\tilde{x}_i$. The first step in the (parallelizable) algorithm will be to eliminate the unknown parts $\tilde{x}_0, \ldots, \tilde{x}_{m-1}$, which is done by the algorithm in Fig. 4.2.

Parallel_for $i = 0, 1, \ldots, m-1$
  Decompose A_{ii}: $A_{ii} = L_{ii}U_{ii}$
  $L_{mi} = A_{mi}U_{ii}^{-1}$
  $U_{im} = L_{ii}^{-1}A_{im}$
  $y_i = L_{ii}^{-1}\tilde{b}_i$
  $S_i = L_{mi}U_{im}$
  $z_i = L_{mi}y_i$
end
$S = A_{mm} - \sum_{i=0}^{m-1} S_i$
$y_m = \tilde{b}_m - \sum_{i=0}^{m-1} z_i$
Solve $Sx_m = y_m$
Parallel_for $i = 0, 1, \ldots, m-1$
  $x_i = U_{ii}^{-1}(y_i - U_{im}x_m)$
end

FIG. 4.2. Parallel elimination.

Note that S in Fig. 4.2 denotes the Schur complement after the elimination of the blocks $0, 1, \ldots, m-1$. In many relevant situations, direct solution of the reduced system $Sx_m = y_m$ requires the dominating part of the total computational costs, and this is where we bring in the iterative component of the algorithm.

The next step is to construct a preconditioner for the reduced system. This is based on discarding small elements in S. The elements larger than some threshold value define the preconditioner C:

$$c_{ij} = \begin{cases} s_{ij} & \text{if } |s_{ij} > t|s_{ii}| \text{ or } |s_{ij}| > t|s_{jj}|, \\ 0 & \text{elsewhere} \end{cases} \tag{4.9}$$

with a parameter $0 \leqslant t < 1$. In the experiments, reported in BOMHOF and VAN DER VORST [2000] the value $t = 0.02$ turned out to be satisfactory, but this may need some experimentation for specific problems.

When we take C as the preconditioner, then we have to solve systems like $Cv = w$, and this requires decomposition of C. In order to prevent too much fill-in, it is suggested to reorder C with a minimum degree ordering. The system $Sx_m = y_m$ is then solved with, for instance, GMRES with preconditioner C. For the examples described in BOMHOF and VAN DER VORST [2000] it turns out that the convergence of GMRES was not very sensitive to the choice of t. The preconditioned iterative solution approach for the reduced system offers also opportunities for parallelism, although in BOMHOF and VAN DER VORST [2000] it is shown that even in serial mode the iterative solution (too sufficiently high precision) is often more efficient than direct solution of the reduced system.

In BOMHOF and VAN DER VORST [2000] heuristics are described for the decision on when the switch from direct to iterative should take place. These heuristics are based on mild assumptions on the speed of convergence of GMRES. The paper also reports on a number of experiments for linear systems, not only from circuit simulation, but also for some matrix problems taken from Matrix Market.[19] These experiments indicate that attractive savings in computational costs can be achieved, even in serial computation mode.

5. Example

In this section we compare the computational performance of the direct solver PARDISO and the iterative method BiCGSTAB(ℓ).[20] We have applied these methods for systems that arise in a Newton iteration (for the update with the Jacobian) of nonlinear systems. These nonlinear systems are related to the simulation of the behavior of a 3D MOS transistor with a doping profile and a mesh as is shown in Figs. 5.1 and 5.2, respectively.

The sizes of the various meshes for the comparisons are shown in Table 5.1. The systems were solved for a typical computation of a IdVg curve using the coupled Poisson, electron, and hole equations. The computations were carried out on a Compaq AlphaServer ES40, 4x667 MHz EV6.7, with 4GB of memory. The results are displayed

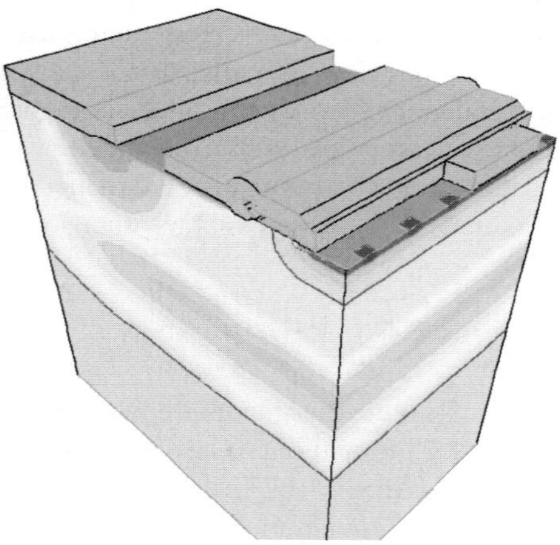

FIG. 5.1. 3D doping profile (red n+, blue p+) and geometry of the MOS transistor.

TABLE 5.1
Name and size of meshes used

Mesh name	Number of vertices
2D1	3836
2D2	37429
3D1	42302

TABLE 5.2
Name of mesh and Wall-clock time in seconds

Mesh	BiCGstab(ℓ)	PARDISO
2D1	170.79	171.76
2D2	9768	3677
3D1	9219	34512

[19] Collection of testmatrices available at ftp://ftp.cise.ufl.edu/cis/tech-reports/tr98/tr98-016.ps.

[20] The data and description for this example were provided by D. Fokkema, Philips Semiconductors, Nijmegen, The Netherlands.

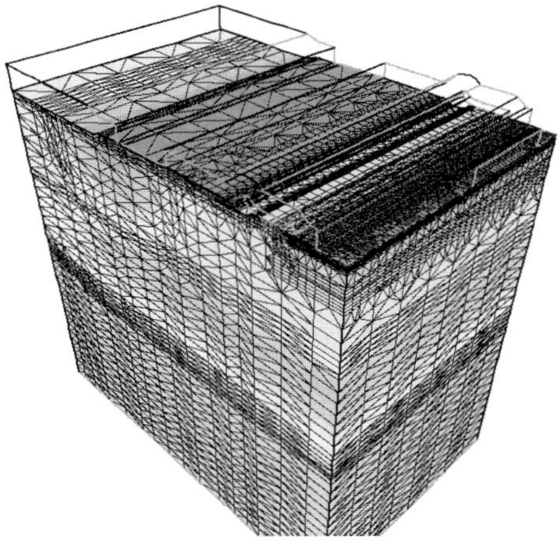

FIG. 5.2. The 3D discretization contains over 40'000 grid points.

in Table 5.2. This kind of performance behavior is typical for many problems in device simulation and the following observations can be made. For the small 2D1 mesh iterative and direct perform comparable, while for the larger 2D2 mesh the direct method is faster. For the 3D1 problem the iterative method has usually the advantage over the direct method.

References

AMESTOY, P.R., DUFF, I.S., L'EXCELLENT, J.-Y. (2000). Multifrontal parallel distributed symmetric and unsymmetric solvers. *Comput. Methods Appl. Mech. Engrg.* **184**, 501–520.

AMESTOY, P.R., DUFF, I.S., L'EXCELLENT, J.-Y., LI, X.S. (2000). Analysis and comparison of two general sparse solvers for distributed memory computers. Technical Report TR/PA/00/90 (CERFACS, Toulouse, France). ACM Trans. Math. Softw., submitted for publication.

AMESTOY, P.R., DAVIS, T.A., DUFF, I.S. (1996). An approximate minimum degree ordering algorithm. *SIAM J. Matrix Anal. Appl.* **17** (4), 886–905.

AMESTOY, P.R. (1990). Factorization of large unsymmetric sparse matrices based on a multifrontal approach in a multiprocessor environment. PhD thesis (Institut National Polytechnique de Toulouse) No. TH/PA/91/2.

AMESTOY, P.R., DUFF, I.S. (1993). Memory management issues in sparse multifrontal methods on multiprocessors. *Internat. J. Supercomputer Appl.* **7** (1), 64–82.

ARNOLDI, W.E. (1951). The principle of minimized iteration in the solution of the matrix eigenproblem. *Quart. Appl. Math.* **9**, 17–29.

ASHCRAFT, C.C., GRIMES, R.R., LEWIS, J.G., PEYTON, B.W., SIMON, H.D. (1987). Progress in sparse matrix methods for large linear systems on vector supercomputers. *Internat. J. Supercomputer Appl.* **1** (4), 10–30.

AXELSSON, O. (1977). Solution of linear systems of equations: iterative methods. In: Barker, V.A. (ed.), *Sparse Matrix Techniques* (Springer, Berlin), pp. 1–51.

AXELSSON, O. (1980). Conjugate gradient type methods for unsymmetric and inconsistent systems of equations. *Linear Algebra Appl.* **29**, 1–16.

AXELSSON, O. (1994). *Iterative Solution Methods* (Cambridge University Press, Cambridge).

AXELSSON, O., VASSILEVSKI, P.S. (1991). A black box generalized conjugate gradient solver with inner iterations and variable-step preconditioning. *SIAM J. Matrix Anal. Appl.* **12** (4), 625–644.

BANK, R.E., CHAN, T.F. (1993). An analysis of the composite step biconjugate gradient method. *Numer. Math.* **66**, 295–319.

BARNARD, S.T., SIMON, H. (1993). A fast multilevel implementation of recursive spectral bisection for partitioning unstructured problems. In: *Proceeding of the Sixth SIAM Conference on Parallel Processing for Scientific Computing*, pp. 711–718.

BARNARD, S.T., SIMON, H. (1995). A parallel implementation of multilevel recursive spectral bisection to adaptive unstructured meshes. In: *Proceeding of the Seventh SIAM Conference on Parallel Processing for Scientific Computing*, pp. 711–718.

BARNES, E.R. (1985). Partitioning the nodes of a graph. In: Alavi, Y., Chartrand, G., Lesniak, L., Lick, D.R., Wall, C.E. (eds.), *Graph Theory with Applications to Algorithms and Computer Science* (Wiley, New York), pp. 57–72.

BARRETT, B., BERRY, M., CHAN, T., DEMMEL, J., DONATO, J., DONGARRA, J., EIJKHOUT, V., POZO, R., ROMINE, C., VAN DER VORST, H. (1994). *Templates for the Solution of Linear Systems: Building Blocks for Iterative Methods* (SIAM, Philadelphia, PA).

BOMHOF, C.W., VAN DER VORST, H.A. (2000). A parallel linear system solver for circuit-simulation problems. *Numer. Linear Algebra Appl.* **7**, 649–665.

BRUASET, A.M. (1995). *A Survey of Preconditioned Iterative Methods* (Longman Scientific & and Technical, Harlow, UK).

BRUSSINO, G., SONNAD, V. (1989). A comparison of direct and preconditioned iterative techniques for sparse unsymmetric systems of linear equations. *Internat. J. Numer. Methods Eng.* **28**, 801–815.

BUI, T.N., CHAUDHURI, S., LEIGHTON, F.T., SIPSER, M. (1987). Graph bisection algorithms with good average case behaviour. *Combinatorica* **7**, 171–191.

BUNCH, J.R., KAUFMANN, L. (1977). Some stable methods for calculating inertia and solving symmetric linear systems. *Math. Comput.* **31**, 162–179.

CHAN, T.F., VAN DER VORST, H.A. (1997). Approximate and incomplete factorizations. In: Keyes, D.E., Sameh, A., Venkatakrishnan, V. (eds.), *Parallel Numerical Algorithms*. In: ICASE/LaRC Interdisciplinary Series in Science and Engineering (Kluwer, Dordrecht), pp. 167–202.

CONCUS, P., GOLUB, G.H. (1976). A generalized Conjugate Gradient method for nonsymmetric systems of linear equations. Technical Report STAN-CS-76-535 (Stanford University, Stanford, CA).

CONCUS, P., GOLUB, G.H., O'LEARY, D.P. (1976). A generalized conjugate gradient method for the numerical solution of elliptic partial differential equations. In: Bunch, J.R., Rose, D.J. (eds.), *Sparse Matrix Computations* (Academic Press, New York).

CULLUM, J., GREENBAUM, A. (1996). Relations between Galerkin and norm-minimizing iterative methods for solving linear systems. *SIAM J. Matrix Anal. Appl.* **17**, 223–247.

DAYDÉ, M.J., DUFF, I.S. (1989). Level 3 BLAS in *LU* factorization on the CRAY-2, ETA-10P, and IBM 3090-200/VF. *Internat. J. Supercomputer Appl. (3)*, 40–70.

DEMMEL, J., GILBERT, J., LI, X. (1999). An asynchronous parallel supernodal algorithm to sparse partial pivoting. *SIAM J. Matrix Anal. Appl.* **20** (4), 915–952.

DE STURLER, E., FOKKEMA, D.R. (1993). Nested Krylov methods and preserving the orthogonality. In: Duane Melson, N., Manteuffel, T.A., McCormick, S.F. (eds.), *Sixth Copper Mountain Conference on Multigrid Methods*. In: NASA Conference Publication **3324** (NASA), pp. 111–126.

DOI, S. (1991). On parallelism and convergence of incomplete LU factorizations. *Appl. Numer. Math.* **7**, 417–436.

DOI, S., HOSHI, A. (1992). Large numbered multicolor MILU preconditioning on SX-3/14. *Internat. J. Computer Math.* **44**, 143–152.

DONGARRA, J.J. (1998). Performance of various computers using standard linear equations software (Linpack benchmark report). Technical Report CS-89-85 (University of Tennessee Computer Science).

DONGARRA, J.J., DEMMEL, J. (1991). LAPACK: a portable high-performance numerical library for linear algebra. *Supercomputer* **8** (6), 33–38.

DONGARRA, J.J., DUCROZ, J., DUFF, I., HAMMARLING, S. (1990). A set of level 3 basic linear algebra subprograms. *ACM Trans. Math. Software* **16** (1), 1–28.

EISENSTAT, S.C., GILBERT, J.R., LIU, J.W. (1993). A supernodal approach to a sparse partial pivoting code. In: *Householder Symposium XII*.

DONGARRA, J.J., DUFF, I.S., SORENSEN, D.C., VAN DER VORST, H.A. (1998). *Numerical Linear Algebra for High-Performance Computers* (SIAM, Philadelphia, PA).

DUFF, I.S. (1998). Direct methods. Technical Report TR/PA/98/28 (CERFACS, Toulouse, France).

DUFF, I.S., KOSTER, J. (1997). The design and use of algorithms for permuting large entries to the diagonal of sparse matrices. Technical Report TR/PA/97/45 (CERFACS, Toulouse, France). Also appeared as Report RAL-TR-97-059 (Rutherford Appleton Laboratories, Oxfordshire).

DUFF, I.S., KOSTER, J. (1999). The design and use of algorithms for permuting large entries to the diagonal of sparse matrices. *SIAM J. Matrix Anal. Appl.* **20** (4), 889–901.

DUFF, I.S., ERISMAN, A.M., REID, J.K. (1986). *Direct Methods for Sparse Matrices* (Oxford University Press, Oxford).

DUFF, I.S., REID, J.K. (1983). The multifrontal solution of indefinite sparse symmetric linear systems. *ACM Trans. Math. Software* **9**, 302–325.

DUFF, I.S., ERISMAN, A.M., REID, J.K. (1986). *Direct Methods for Sparse Matrices* (Oxford Science Publications, Oxford).

DUFF, I.S., GRIMES, R., LEWIS, J. (1989). Sparse matrix test problems. *ACM Trans. Math. Software (15)*, 1–14.

DUFF, I.S., MEURANT, G.A. (1989). The effect of ordering on preconditioned conjugate gradient. *BIT* **29**, 635–657.

DUPONT, T., KENDALL, R.P., RACHFORD JR, H.H. (1968). An approximate factorization procedure for solving self-adjoint elliptic difference equations. *SIAM J. Numer. Anal.* **5** (3), 559–573.

EIJKHOUT, V. (1992). Beware of unperturbed modified incomplete point factorizations. In: Beauwens, R., de Groen, P. (eds.), *Iterative Methods in Linear Algebra, IMACS Int. Symp., Brussels, Belgium, 2–4 April, 1991* (North-Holland, Amsterdam), pp. 583–591.

EISENSTAT, S.C., SCHULTZ, M.H., SHERMAN, A.H. (1981). Algorithms and data structures for sparse symmetric Gaussian elimination. *SIAM J. Scientific Statist. Comput.* **2** (2), 225–237.

ELMAN, H.C. (1982). Iterative methods for large sparse nonsymmetric systems of linear equations. PhD thesis (Yale University, New Haven, CT).

FABER, V., MANTEUFFEL, T.A. (1984). Necessary and sufficient conditions for the existence of a conjugate gradient method. *SIAM J. Numer. Anal.* **21** (2), 352–362.

FIDUCCIA, C.M., MATTHEYSES, R.M. (1982). A linear-time heuristic for improving network partitions. In: *Proceedings of the 19th Design Automation Conference* (ACM), pp. 175–181.

FIEDLER, M. (1973). Algebraic connectivity of graphs. *Czecheslovak Math. J.* **23** (98), 298–305.

FIEDLER, M. (1975). A property of eigenvectors of nonnegative symmetric matrices and its application to graph theory. *Czecheslovak Math. J.* **25** (100), 619–633.

FISCHER, B. (1994). Orthogonal polynomials and polynomial based iteration methods for indefinite linear systems. PhD thesis (University of Hamburg, Hamburg, Germany).

FISCHER, B. (1996). *Polynomial Based Iteration Methods for Symmetric Linear Systems.* In: Advances in Numerical Mathematics (Wiley and Teubner, Chichester, Stuttgart).

FLETCHER, R. (1976). *Conjugate Gradient Methods for Indefinite Systems.* In: Lecture Notes Math. **506** (Springer, Berlin), pp. 73–89.

FREUND, R.W., GUTKNECHT, M.H., NACHTIGAL, N.M. (1993). An implementation of the look-ahead Lanczos algorithm for non-Hermitian matrices. *SIAM J. Sci. Comput.* **14**, 137–158.

FREUND, R.W., NACHTIGAL, N.M. (1990). An implementation of the look-ahead Lanczos algorithm for non-Hermitian matrices, part 2. Technical Report 90.46 (RIACS, NASA Ames Research Center).

FREUND, R.W., NACHTIGAL, N.M. (1991). QMR: a quasi-minimal residual method for non-Hermitian linear systems. *Numer. Math.* **60**, 315–339.

FREUND, R. (1993). A transpose-free quasi-minimal residual algorithm for non-Hermitian linear systems. *SIAM J. Sci. Comput.* **14**, 470–482.

GEORGE, A. (1973). Nested dissection of a regular finite element mesh. *SIAM J. Numer. Anal.* **10** (2), 345–363.

GEORGE, A., LIU, J.W.H. (1980). User guide for SPARSPAK: Waterloo sparse linear equations package Technical Report (Department of Computer Science, University of Waterloo).

GEORGE, A., LIU, J.W.H. (1981). *Computer Solution of Large Sparse Positive Definite Systems* (Prentice-Hall, Englewood Cliffs, NJ).

GEORGE, A., LIU, J.W.H. (1989). The evolution of the Minimum-Degree ordering algorithm. *SIAM Rev.* **31** (1), 1–19.

GILBERT, J.R. (1994). Predicting structures in sparse matrix computations. *SIAM J. Matrix Anal. Appl.* **15** (1), 62–79.

GILBERT, J.R., MILLER, G.L., TENG, S.-H. (1998). Geometric mesh partitioning: Implementation and experiments. *SIAM J. Scientific Statist. Comput.* **19** (6), 2091–2110.

GOLUB, G.H., O'LEARY, D.P. (1989). Some history of the conjugate gradient and Lanczos algorithms: 1948–1976. *SIAM Rev.* **31**, 50–102.

GOLUB, G.H., VAN LOAN, C.F. (1996). *Matrix Computations* (The Johns Hopkins University Press, Baltimore).

GÖHRING, T., SAAD, Y. (1994). Heuristics algorithms for automatic graph partitioning. Technical Report (University of Minnesota, Department of Computer Science, Minneapolis).

GREENBAUM, A. (1997a). Estimating the attainable accuracy of recursively computed residual methods. *SIAM J. Matrix Anal. Appl.* **18**, 535–551.

GREENBAUM, A. (1997b). *Iterative Methods for Solving Linear Systems* (SIAM, Philadelphia).

GUPTA, A. (1996). WGPP: Watson Graph Partitioning (and sparse matrix ordering) Package. Technical Report (IBM Research Division, T.J. Watson Research Center, Yorktown Heights).

GUPTA, A. (2001). Recent advances in direct methods for solving unsymmetric sparse systems of linear equations. Technical Report RC 22039 (98933) (IBM T.J. Watson Research Center, Yorktown Heights, NY).

GUPTA, A., KARYPIS, G., KUMAR, V. (1997). Highly scalable parallel algorithms for sparse matrix factorization. *IEEE Trans. Parallel Distribut. Syst.* **8** (5), 502–520.

GUSTAFSSON, I. (1978). A class of first order factorization methods. *BIT* **18**, 142–156.

GUTKNECHT, M.H. (1993). Variants of BICGSTAB for matrices with complex spectrum. *SIAM J. Sci. Comput.* **14**, 1020–1033.

HAGEMAN, L.A., YOUNG, D.M. (1981). *Applied Iterative Methods* (Academic Press, New York).

HAMMOND, S.W. (1992). Mapping unstructured grid problems to massively parallel computers. PhD thesis (Rensselaer Polytechnic Institute, Troy, NY).

HEATH, M.T., NG, F., PEYTON, B.M. (1990). Parallel algorithms for sparse linear systems. In: *Parallel Algorithms for Matrix Computations* (SIAM), pp. 83–124.

HENDRICKSON, B., LELAND, R. (1993). A multilevel algorithm for partitioning graphs. Technical Report SAND 93-1301 (Sandia National Laboratories, Albuquerque).

HENDRICKSON, B., LELAND, R. (1995). An improved spectral graph partitioning algorithm for mapping parallel computations. *SIAM J. Scientific Comput.* **16** (2), 452–469.

HENDRICKSON, B., ROTHBERG, E. (1996). Improving the runtime and quality of nested dissection orderings. Technical Report SAND 96-0868J (Sandia National Laboratories, Albuquerque).

HENON, P., RAMET, P., ROMAN, J. (2002). PaStiX: a high-performance parallel direct solver for sparse symmetric positive define systems. *Parallel Comput.* **28**, 301–321.

HESTENES, M.R., STIEFEL, E. (1952). Methods of conjugate gradients for solving linear systems. *J. Res. Natl. Bur. Stand.* **49**, 409–436.

INTEGRATED SYSTEMS ENGINEERING AG (1998a). DESSIS$_{-\mathrm{ISE}}$ Reference Manual. ISE Integrated Systems Engineering AG. http://www.ise.com.

INTEGRATED SYSTEMS ENGINEERING AG (1998b). DIOS$_{-\mathrm{ISE}}$ Reference Manual. ISE Integrated Systems Engineering AG. http://www.ise.com.

JEA, K.C., YOUNG, D.M. (1980). Generalized conjugate-gradient acceleration of nonsymmetrizable iterative methods. *Linear Algebra Appl.* **34**, 159–194.

JONES, M.T., PLASSMANN, P.E. (1994). The efficient parallel iterative solution of large sparse linear systems. In: George, A., Gilbert, J.R., Liu, J.W.H. (eds.), *Graph Theory and Sparse Matrix Computations*. In: IMA **56** (Springer, Berlin).

KAASSCHIETER, E.F. (1988). A practical termination criterion for the Conjugate Gradient method. *BIT* **28**, 308–322.

KARYPIS, G., KUMAR, V. (1998a). A fast and high quality multilevel scheme for partitioning irregular graphs. *SIAM J. Scientific Comput.* **20** (1), 359–392.

KARYPIS, G., KUMAR, V. (1998b). Multilevel algorithms for multi-constraint graph partitioning. Technical Report MN 98-019 (University of Minnesota, Department of Computer Science, Minneapolis, MN).

KERNIGHAN, B.W., LIN, S. (1970). An efficient heuristic procedure for partitioning graphs. *Bell System Technical J.* **49**, 291–307.

KUO, J.C.C., CHAN, T.F. (1990). Two-color Fourier analysis of iterative algorithms for elliptic problems with red/black ordering. *SIAM J. Sci. Statist. Comput.* **11**, 767–793.

LANCZOS, C. (1950). An iteration method for the solution of the eigenvalue problem of linear differential and integral operators. *J. Res. Natl. Bur. Stand* **45**, 225–280.

LANCZOS, C. (1952). Solution of systems of linear equations by minimized iterations. *J. Res. Natl. Bur. Stand* **49**, 33–53.

LI, X. (1996). Sparse Gaussian elimination on high performance computers. PhD thesis (University of California at Berkely, Department of Computer Science).

LI, X.S., DEMMEL, J.W. (1999). A scalable sparse direct solver using static pivoting. In: *Proceeding of the 9th SIAM Conference on Parallel Processing for Scientific Computing, San Antonio, Texas*.

LIU, J.W.H. (1985). Modification of the Minimum-Degree algorithm by multiple elimination. *ACM Trans. Math. Software* **11** (2), 141–153.

LIU, J.W.H. (1988). Equivalent sparse matrix reordering by elimination tree rotations. *SIAM J. Sci. Statist. Comput.* **9** (3), 424–444.

LIU, J.W.H. (1989). Reordering sparse matrices for parallel elimination. *Parallel Comput.* **11**, 73–91.

LIU, J.W.H. (1990). The role of elimination trees in sparse factorization. *SIAM J. Matrix Anal. Appl.* **11** (1), 134–172.

MARKOWITZ, H.M. (1957). The eliminaton form of the inverse and its application to linear programming. *Management Sci.* **3**, 255–269.

MEIJERINK, J.A., VAN DER VORST, H.A. (1977). An iterative solution method for linear systems of which the coefficient matrix is a symmetric M-matrix. *Math. Comp.* **31**, 148–162.

MEIJERINK, J.A., VAN DER VORST, H.A. (1981). Guidelines for the usage of incomplete decompositions in solving sets of linear equations as they occur in practical problems. *J. Comp. Physics* **44**, 134–155.

MILLER, G.L., TENG, S.-H., THURSTON, W., VAVASIS, S.A. (1998). Geometric separators for finite element meshes. *SIAM J. Sci. Statist. Comput.* **19** (2), 364–386.

NACHTIGAL, N.M., REDDY, S.C., TREFETHEN, L.N. (1992). How fast are nonsymmetric matrix iterations?. *SIAM J. Matrix Anal. Appl.* **13**, 778–795.

PAIGE, C.C., PARLETT, B.N., VAN DER VORST, H.A. (1995). Approximate solutions and eigenvalue bounds from Krylov subspaces. *Numer. Linear Algebra Appl.* **2** (2), 115–134.

PAIGE, C.C., SAUNDERS, M.A. (1975). Solution of sparse indefinite systems of linear equations. *SIAM J. Numer. Anal.* **12**, 617–629.

PAPADIMITRIOU, C.H., STEIGLITZ, K. (1982). *Combinatorial Optimization: Algorithms and Complexity* (Prentice-Hall, Englewood Cliffs, NJ).

PARLETT, B.N., TAYLOR, D.R., LIU, Z.A. (1985). A look-ahead Lanczos algorithm for unsymmetric matrices. *Math. Comp.* **44**, 105–124.

POMMERELL, C., FICHTNER, W. (1991). PILS: An iterative linear solver package for ill-conditioned systems. In: *Supercomputing '91* (IEEE Computer Society, Los Alamitos, CA), pp. 588–599.

POTHEN, A., FAN, C.J. (1990). Computing the block triangular form of a sparse matrix. *ACM Trans. Math. Software* **16** (4), 303–324.

POTHEN, A., SIMON, H.D., LIOU, K. (1990). Partitioning sparse matrices with eigenvectors of graphs. *SIAM J. Matrix Anal. Appl.* **11** (3), 430–452.

RADICATI DI BROZOLO, G., ROBERT, Y. (1989). Parallel conjugate gradient-like algorithms for solving sparse non-symmetric systems on a vector multiprocessor. *Parallel Comput.* **11**, 223–239.

ROTHBERG, E. (1996). Performance of panel and block approaches to sparse Cholesky factorization on the iPSC/860 and Paragon multicomputers. *SIAM J. Sci. Comput.* **17** (3), 699–711.

SAAD, Y. (1993). A flexible inner-outer preconditioned GMRES algorithm. *SIAM J. Sci. Comput.* **14**, 461–469.

SAAD, Y. (1996). *Iterative Methods for Sparse Linear Systems* (PWS Publishing Company, Boston).

SAAD, Y., SCHULTZ, M.H. (1986). GMRES: a generalized minimal residual algorithm for solving nonsymmetric linear systems. *SIAM J. Sci. Statist. Comput.* **7**, 856–869.

SCHENK, O. (2000). Scalable parallel sparse LU factorization methods on shared memory multiprocessors. PhD thesis (ETH Zürich).

SCHENK, O., GÄRTNER, K. (2000). Scalable parallel sparse LU factorization with a dynamical supernode pivoting approach in semiconductor device simulation. In: *Conference Proceedings of the 16th IMACS World Congress on Scientific Computation, Applied Mathematics and Simulation, Lausanne, Switzerland.*

SCHENK, O., GÄRTNER, K. (2001). Sparse factorization with two-level scheduling in PARDISO. In: *Proceedings of the 10th SIAM Conference on Parallel Processing for Scientific Computing, Portsmouth, Virgina* (SIAM).

SCHENK, O., GÄRTNER, K. (2002a). Solving unsymmetric sparse systems of linear equations with PARDISO. In: *Conference Proceedings of the Second International Conference on Computational Science ICCS2002, Amsterdam, The Netherlands.*

SCHENK, O., GÄRTNER, K. (2002b). Two-level dynamic scheduling in PARDISO: Improved scalability on shared memory multiprocessing systems. *Parallel Comput.* **28**, 187–197.

SCHENK, O., GÄRTNER, K., FICHTNER, W. (1999). Application of parallel sparse direct methods in semiconductor device and process simulation. In: Polychronopoulos, C., Joe, K., Fukuda, A., Tomita, S. (eds.), *High Performance Computing*. In: Lecture Notes in Computer Science **1615** (Springer, Berlin), pp. 206–219.

SCHENK, O., GÄRTNER, K., FICHTNER, W. (2000). Efficient sparse LU factorization with left-right looking strategy on shared memory multiprocessors. *BIT* **40** (1), 158–176.

SCHENK, O., GÄRTNER, K., FICHTNER, W., STRICKER, A. (2001). PARDISO: A high-performance serial and parallel sparse linear solver in semiconductor device simulation. *Future Generation of Computer Systems* **18**, 65–78.

SCHENK, O., GÄRTNER, K., SCHMIDTHÜSEN, B., FICHTNER, W. (1999). Numerical semiconductor device and process simulation on shared memory multiprocessors: algorithms, architectures, results. In: Yang, T. (ed.), Parallel Numerical Computations with Applications **515** (Kluwer Academic, Dordrecht), pp. 141–158.

SIDI, A. (1991). Efficient implementation of minimal polynomial and reduced rank extrapolation methods. *J. Comp. Appl. Math.* **36**, 305–337.

SLEIJPEN, G.L.G., FOKKEMA, D.R. (1993). BICGSTAB(ℓ) for linear equations involving unsymmetric matrices with complex spectrum. *ETNA* **1**, 11–32.

SLEIJPEN, G.L.G., VAN DER VORST, H.A. (1996). Reliable updated residuals in hybrid Bi-CG methods. *Computing* **56**, 141–163.

SLEIJPEN, G.L.G., VAN DER VORST, H.A., MODERSITZKI, J. (2000). Differences in the effects of rounding errors in Krylov solvers for symmetric indefinite linear systems. *SIAM J. Matrix Anal. Appl.* **22** (3), 726–751.

SLEIJPEN, G.L.G., VAN DER VORST, H.A., FOKKEMA, D.R. (1994). Bi-CGSTAB(ℓ) and other hybrid Bi-CG methods. *Numer. Algorithms* **7**, 75–109.

SONNEVELD, P. (1989). CGS: a fast Lanczos-type solver for nonsymmetric linear systems. *SIAM J. Sci. Statist. Comput.* **10**, 36–52.

STOER, J., BULIRSCH, R. (1983). *Introduction to Numerical Analysis* (Springer, Berlin).

STONE, H.S. (1968). Iterative solution of implicit approximations of multidimensional partial differential equations. *SIAM J. Numer. Anal.* **5**, 530–558.

TONG, C.H., YE, Q. (2000). Analysis of the finite precision Bi-Conjugate Gradient algorithm for nonsymmetric linear systems. *Math. Comp.* **69**, 1559–1575.

VAN DER SLUIS, A., VAN DER VORST, H.A. (1986). The rate of convergence of conjugate gradients. *Numer. Math.* **48**, 543–560.

VAN DER SLUIS, A., VAN DER VORST, H.A. (1990). SIRT- and CG-type methods for the iterative solution of sparse linear least-squares problems. *Linear Algebra Appl.* **130**, 257–302.

VAN DER VORST, H.A. (1992a). Bi-CGSTAB: A fast and smoothly converging variant of Bi-CG for the solution of non-symmetric linear systems. *SIAM J. Sci. Statist. Comput.* **13**, 631–644.

VAN DER VORST, H.A. (1992b). Conjugate gradient type methods for nonsymmetric linear systems. In: Beauwens, R., de Groen, P. (eds.), *Iterative Methods in Linear Algebra, IMACS Int. Symp., Brussels, Belgium, 2–4 April, 1991* (North-Holland, Amsterdam), pp. 67–76.

VAN DER VORST, H.A., SLEIJPEN, G.L.G. (1993). The effect of incomplete decomposition preconditioning on the convergence of Conjugate Gradients. In: Hackbusch, W., Wittum, G. (eds.), Incomplete Decompositions (ILU) – Algorithms, Theory, and Applications **41** (Braunschweig), pp. 179–187.

VAN DER VORST, H.A., VUIK, C. (1993). The superlinear convergence behaviour of GMRES. *J. Comp. Appl. Math.* **48**, 327–341.

VAN DER VORST, H.A., VUIK, C. (1994). GMRESR: A family of nested GMRES methods. *Numer. Linear Algebra Appl.* **1**, 369–386.

VAN DER VORST, H.A., YE, Q. (2000). Refined residual replacement techniques for subspace iterative methods for convergence of true residuals. *SIAM J. Sci. Comput.* **22**, 835–852.

VAN DER VORST, H.A. (1982). Preconditioning by incomplete decompositions. PhD thesis (Utrecht University, Utrecht, The Netherlands).

VARGA, R.S. (1960). Factorizations and normalized iterative methods. In: Langer, R.E. (ed.), *Boundary Problems in Differential Equations* (University of Wisconsin Press, Madison, WI), pp. 121–142.

VARGA, R.S. (1962). *Matrix Iterative Analysis* (Prentice-Hall, Englewood Cliffs NJ).

VINSOME, P.K.W. (1976). ORTHOMIN: an iterative method for solving sparse sets of simultaneous linear equations. In: *Proc. Fourth Symposium on Reservoir Simulation* (Society of Petroleum Engineers of AIME), pp. 149–159.

WALSHAW, C., CROSS, M., (1999). Parallel optimization algorithms for multilevel mesh partitioning. Technical Report 99/IM/44 (Univ. Greenwich, London SE18 6PF, UK).

WIDLUND, O. (1978). A Lanczos method for a class of nonsymmetric systems of linear equations. *SIAM J. Numer. Anal.* **15**, 801–812.

WILKINSON, J.H. (1961). Error analysis of direct methods of matrix inversion. *ACM Trans. Math. Software* **8**, 281–330.

WRIGHT, S.J. (1993). A collection of problems for which Gaussian elimination with partial pivoting is unstable. *SIAM J. Sci. Statist. Comput.* **14** (1), 231–238.

YANNAKAKIS, M. (1981). Computing the minimum fill-in is NP-complete. *SIAM J. Algebraic Discrete Methods* **2** (1), 77–79.

ZHOU, L., WALKER, H.F. (1994). Residual smoothing techniques for iterative methods. *SIAM J. Sci. Statist. Comput.* **15**, 297–312.

Reduced-Order Modeling

Zhaojun Bai

University of California, Department of Computer Science, One Shields Avenue, 3023 Engineering II, Davis, CA, USA
E-mail address: bai@cs.ucdavis.edu

Patrick M. Dewilde

TU Delft, Fac. Eletrotechniek Vakgroep CAS (Room L2.500), Mekelweg 4, 2628 CD, Delft, The Netherlands
E-mail address: p.dewile@dimes.tudelft.nl

Roland W. Freund

University of California, Davis, Department of Mathematics, One Shields Avenue, Davis, CA 95616, USA
E-mail address: freund@math.ucdavis.edu

Abstract

In recent years, reduced-order modeling techniques have proven to be powerful tools for various problems in circuit simulation. For example, today, reduction techniques are routinely used to replace the large RCL subcircuits that model the interconnect or the pin package of VLSI circuits by models of much smaller dimension. In this chapter, we review the reduced-order modeling techniques that are most widely employed in VLSI circuit simulation.

1. Introduction to the problem of model reduction

Roughly speaking, the problem of model reduction is to replace a given mathematical model of a system or process by a model that is much "smaller" than the original model, but still describes – at least approximately – certain aspects of the system or process. Clearly, model reduction involves a number of interesting issues. First and foremost

Numerical Methods in Electromagnetics
Special Volume (W.H.A. Schilders and E.J.W. ter Maten, Guest Editors) of
HANDBOOK OF NUMERICAL ANALYSIS, VOL. XIII
P.G. Ciarlet (Editor)

ISSN 1570-8659
DOI 10.1016/S1570-8659(04)13009-3

is the issue of selecting appropriate approximation schemes that allow the definition of suitable reduced-order models. In addition, it is often important that the reduced-order model preserves certain crucial properties of the original system, such as stability or passivity. Other issues include the characterization of the quality of the models, the extraction of the data from the original model that is needed to actually generate the reduced-order models, and the efficient and numerically stable computation of the models.

In this paper, we discuss reduced-order modeling techniques for large-scale linear dynamical systems, especially those that arise in the simulation of electronic circuits and of microelectromechanical systems.

We begin with a brief description of reduced-order modeling problems in circuit simulation. Electronic circuits are usually modeled as networks whose branches correspond to the circuit elements and whose nodes correspond to the interconnections of the circuit elements. Such networks are characterized by three types of equations. The *Kirchhoff's current law* (KCL) states that, for each node of the network, the currents flowing in and out of that node sum up to zero. The *Kirchhoff's voltage law* (KVL) states that, for each closed loop of the network, the voltage drops along that loop sum up to zero. The *branch constitutive relations* (BCRs) are equations that characterize the actual circuit elements. For example, the BCR of a linear resistor is Ohm's law. The BCRs are linear equations for simple devices, such as linear resistors, capacitors, and inductors, and they are nonlinear equations for more complex devices, such as diodes and transistors. Furthermore, in general, the BCRs involve time-derivatives of the unknowns, and thus they are ordinary differential equations. On the other hand, the KCLs and KVLs are linear algebraic equations that only depend on the topology of the circuit.

The KCLs, KVLs, and BCRs can be summarized as a system of first-order, in general nonlinear, *differential-algebraic equations* (DAEs) of the form

$$\frac{\mathrm{d}}{\mathrm{d}t}q(\hat{x},t)+f(\hat{x},t)=0, \tag{1.1}$$

together with suitable initial conditions. Here, $\hat{x}=\hat{x}(t)$ is the unknown vector of circuit variables at time t, the vector-valued function $f(\hat{x},t)$ represents the contributions of nonreactive elements such as resistors, sources, etc., and the vector-valued function $\frac{\mathrm{d}}{\mathrm{d}t}q(\hat{x},t)$ represents the contributions of reactive elements such as capacitors and inductors. There are a number of established methods, such as sparse tableau, nodal formulation, modified nodal analysis, etc. (see VLACH and SINGHAL [1994]), for generating a system of equations of the form (1.1) from a so-called *netlist* description of a given circuit. The vector functions $\hat{x}$, f, q, as well as their dimension, depend on the chosen formulation method. The most general method is sparse tableau, which consists of just listing all the KCLs, KVLs, and BCRs. The other formulation methods can be interpreted as starting from sparse tableau and eliminating some of the unknowns by using some of the KCL or KVL equations.

For all the standard formulation methods, the dimension of the system (1.1) is of the order of the number of elements in the circuit. Since today's VLSI circuits can have up to hundreds of millions of circuit elements, systems (1.1) describing such circuits can be of extremely large dimension. Reduced-order modeling allows to first replace large systems of the form (1.1) by systems of smaller dimension and then tackle these smaller systems by suitable DAE solvers. Ideally, one would like to apply nonlinear reduced-order modeling directly to the nonlinear system (1.1). However, since nonlinear reduction techniques are a lot less developed and less well-understood than linear ones, today, almost always linear reduced-order modeling is employed. To this end, one either linearizes the system (1.1) or decouples (1.1) into nonlinear and linear subsystems; see, e.g., FREUND [1999b] and the references given there.

For example, the first case arises in *small-signal analysis*; see, e.g., FREUND and FELDMANN [1996b]. Given a *DC operating point*, say $\hat{x}_0$, of the circuit described by (1.1), one linearizes the system (1.1) around $\hat{x}_0$. The resulting linearized version of (1.1) is of the following form:

$$E\frac{\mathrm{d}x}{\mathrm{d}t} = Ax + Bu(t), \tag{1.2}$$

$$y(t) = C^{\mathrm{T}}x(t). \tag{1.3}$$

Here, $A = \mathrm{D}_x f$ and $E = \mathrm{D}_x q$ are the Jacobian matrices of f and q, respectively, at the DC operating point $\hat{x}_0$, $x(t) = \hat{x}(t) - \hat{x}_0$, $u(t)$ is the vector of excitations applied to the sources of the circuit, and $y(t)$ is the vector of circuit variables of interest. Eqs. (1.2) and (1.3) represent a *time-invariant linear dynamical system*. Its *state-space dimension*, N, is the length of the vector x of circuit variables. For a circuit with many elements, the system (1.2) and (1.3) is thus of very high dimension. The idea of reduced-order modeling is then to replace the original system (1.2) and (1.3) by one the same form,

$$E_n\frac{\mathrm{d}z}{\mathrm{d}t} = A_n z + B_n u(t),$$

$$y(t) = C_n^{\mathrm{T}} z(t),$$

but of much smaller state-space dimension $n \ll N$.

Time-invariant linear dynamical systems of the form (1.2) and (1.3) also arise when equations describing linear subcircuits of a given circuit are decoupled from the system (1.1) that characterizes the whole circuit; see, e.g., [FREUND, 1999b]. For example, the interconnect or the pin package of VLSI circuits are often modeled as large linear RCL networks. Such linear subcircuits are described by systems of the form (1.2) and (1.3), where $x(t)$ is the vector of circuit variables associated with the subcircuit, and the vectors $u(t)$ and $y(t)$ contain the variables of the connections of the subcircuit to the, in general nonlinear, remainder of the whole circuit. By replacing, in the nonlinear system (1.1), the linear subsystem (1.2) and (1.3) by a reduced-order model of much smaller state-space dimension, the dimension of (1.1) can be reduced significantly before a DAE solver is then applied to such a smaller version of (1.1).

The remainder of this paper is organized as follows. In Section 2, we review some basic facts about time-invariant linear dynamical systems. In Section 3, we discuss reduced-order modeling of linear dynamical systems via Krylov-subspace techniques. In Section 4, we describe the use of Schur interpolation for various reduced-order modeling problems. In Section 5, we discuss Hankel-norm model reduction. Sections 6 and 7 are concerned with reduced-order modeling of second-order and semi-second-order dynamical systems. Finally, in Section 8, we make some concluding remarks.

2. Time-invariant linear dynamical systems

In this section, we review some basic facts about time-invariant linear dynamical systems and introduce reduced-order models defined by Padé or Padé-type approximants. We also discuss stability and passivity of linear dynamical systems.

2.1. State-space description

We consider m-input p-output time-invariant linear dynamical systems given by a *state-space description* of the form

$$E\frac{\mathrm{d}x}{\mathrm{d}t} = Ax + Bu(t), \tag{2.1}$$

$$y(t) = C^{\mathrm{T}}x(t) + Du(t), \tag{2.2}$$

together with suitable initial conditions. Here, $A, E \in \mathbb{R}^{N\times N}$, $B \in \mathbb{R}^{N\times m}$, $C \in \mathbb{R}^{N\times p}$, and $D \in \mathbb{R}^{p\times m}$ are given matrices, $x(t) \in \mathbb{R}^N$ is the vector of state variables, $u(t) \in \mathbb{R}^m$ is the vector of inputs, $y(t) \in \mathbb{R}^p$ the vector of outputs, N is the state-space dimension, and m and p are the number of inputs and outputs, respectively. Note that systems of the form (1.2) and (1.3) are just a special case of (2.1) and (2.2) with $D = 0$.

The linear system (2.1) and (2.2) is called *regular* if the matrix E in (2.1) is nonsingular, and it is called *singular* or a *descriptor system* if E is singular. Note that, in the regular case, the linear system (2.1) and (2.2) can always be re-written as

$$\frac{\mathrm{d}x}{\mathrm{d}t} = (E^{-1}A)x + (E^{-1}B)u(t),$$

$$y(t) = C^{\mathrm{T}}x(t) + Du(t),$$

which is just a system (2.1) and (2.2) with $E = I$.

The linear dynamical systems arising in circuit simulation are descriptor systems in general. Therefore, in the following, we allow $E \in \mathbb{R}^{N\times N}$ to be a general, possibly singular, matrix. The only assumption on the matrices $A, E \in \mathbb{R}^{N\times N}$ in (2.1) is that the matrix pencil $A - sE$ is *regular*, i.e., the matrix $A - sE$ is singular for only finitely many values of $s \in \mathbb{C}$.

In the case of singular E, Eq. (2.1) represents a system of DAEs. Solving DAEs is significantly more complex than solving systems of ordinary differential equations (ODEs). Moreover, there are constraints on the possible initial conditions that can be imposed on the solutions of (2.1). For a detailed discussion of DAEs and the structure of their solutions, we refer the reader to CAMPBELL [1980], CAMPBELL [1982], DAI [1989], VERGHESE, LÉVY and KAILATH [1981]. Here, we only present a brief glimpse of the issues arising in DAEs.

We start by bringing the matrices A and E in (2.1) to a certain normal form. For any regular pencil $A - sE$, there exist nonsingular matrices P and Q such that

$$P(A - sE)Q = \begin{bmatrix} A^{(1)} - sI & 0 \\ 0 & I - sJ \end{bmatrix}, \tag{2.3}$$

where the submatrix J is nilpotent. The matrix pencil on the right-hand side of (2.3) is called the *Weierstrass form* of $A - sE$. Assuming that the matrices A and E in (2.1) are already in Weierstrass form, the system (2.1) can be decoupled as follows:

$$\frac{dx^{(1)}}{dt} = A^{(1)}x^{(1)} + B^{(1)}u(t), \tag{2.4}$$

$$J\frac{dx^{(2)}}{dt} = x^{(2)} + B^{(2)}u(t). \tag{2.5}$$

The first subsystem, (2.4), is just a system of ODEs. Thus for any given initial condition $x^{(1)}(0) = \hat{x}^{(1)}$, there exits a unique solution of (2.4). Moreover, the so-called *free-response* of (2.4), i.e., the solutions $x(t)$ for $t \geqslant 0$ when $u \equiv 0$, consists of combinations of exponential modes at the eigenvalues of the matrix $A^{(1)}$. Note that, in view of (2.3), the eigenvalues of $A^{(1)}$ are just the finite eigenvalues of the pencil $A - sE$. The solutions of the second subsystem, (2.5), however, are of quite different nature. In particular, the free-response of (2.5) consists of $k_i - 1$ independent impulsive motions for each $k_i \times k_i$ Jordan block of the matrix J; see VERGHESE, LÉVY and KAILATH [1981].

For example, consider the case that the nilpotent matrix J in (2.5) is a single $k \times k$ Jordan block, i.e.,

$$J = \begin{bmatrix} 0 & 1 & 0 & \cdots & 0 \\ 0 & 0 & 1 & \ddots & \vdots \\ \vdots & \ddots & \ddots & \ddots & 0 \\ \vdots & \ddots & \ddots & \ddots & 1 \\ 0 & \cdots & \cdots & 0 & 0 \end{bmatrix} \in \mathbb{R}^{k \times k}.$$

The k components of the free-response $x^{(2)}(t)$ of (2.5) are then given by

$$\begin{aligned} x_1^{(2)}(t) &= -x_2^{(2)}(0-)\delta(t) - x_3^{(2)}(0-)\delta^{(1)}(t) - \cdots - x_k^{(2)}(0-)\delta^{(k-2)}(t), \\ x_2^{(2)}(t) &= -x_3^{(2)}(0-)\delta(t) - x_4^{(2)}(0-)\delta^{(1)}(t) - \cdots - x_k^{(2)}(0-)\delta^{(k-3)}(t), \\ \vdots &= \vdots \\ x_{k-1}^{(2)}(t) &= -x_k^{(2)}(0-)\delta(t), \\ x_k^{(2)}(t) &= 0. \end{aligned}$$

Here, $\delta(t)$ is the delta function and $\delta^{(i)}(t)$ is its ith derivative. Moreover, $x_i^{(2)}(0-)$, $i = 2, 3, \ldots, k$, are the components of the initial conditions that can be imposed on (2.4). Note that there are only $k-1$ degrees of freedom for the initial condition and that it is not possible to prescribe $x_1^{(2)}(0-)$. In particular, the free-response of (2.5) corresponding to an 1×1 Jordan blocks of J is just the zero solution, and there is no degree of freedom for the selection of an initial value corresponding to that block.

Finally, we remark that, in view of (2.3), the eigenvalues of the matrix pencil $A - sE$ corresponding to the subsystem (2.5) are just the infinite eigenvalues of $A - sE$.

2.2. *Reduced-order models and transfer functions*

The basic idea of reduced-order modeling is to replace a given system by a system of the same type, but with smaller state-space dimension. Thus, a *reduced-order model* of state-space dimension n of a given linear dynamical system (2.1) and (2.2) of dimension N is a system of the form

$$E_n \frac{\mathrm{d}z}{\mathrm{d}t} = A_n z + B_n u(t), \tag{2.6}$$

$$y(t) = C_n^{\mathrm{T}} z(t) + D_n u(t), \tag{2.7}$$

where $A_n, E_n \in \mathbb{R}^{n\times n}$, $B_n \in \mathbb{R}^{n\times m}$, $C_n \in \mathbb{R}^{n\times p}$, $D_n \in \mathbb{R}^{p\times m}$, and $n < N$.

The challenge then is to choose the matrices A_n, E_n, B_n, C_n, and D_n in (2.6) and (2.7) such that the reduced-order model in some sense approximates the original system. One possible measure of the approximation quality of a reduced-order model is based on the concept of transfer function.

If we assume zero initial conditions, then by applying the Laplace transform to the original system (2.1) and (2.2), we obtain the following algebraic equations:

$$sEX(s) = AX(s) + BU(s),$$

$$Y(s) = C^{\mathrm{T}} X(s) + DU(s).$$

Here, the frequency-domain variables $X(s)$, $U(s)$, and $Y(s)$ are the Laplace transforms of the time-domain variables of $x(t)$, $u(t)$, and $y(t)$, respectively. Note that $s \in \mathbb{C}$. Then, formally eliminating $X(s)$ in the above equations, we arrive at the frequency-domain input–output relation $Y(s) = H(s)U(s)$. Here,

$$H(s) := D + C^{\mathrm{T}}(sE - A)^{-1}B, \quad s \in \mathbb{C}, \tag{2.8}$$

is the so-called *transfer function* of the system (2.1) and (2.2). Note that

$$H : \mathbb{C} \mapsto (\mathbb{C} \cup \infty)^{p\times m}, \tag{2.9}$$

is an $p \times m$-matrix-valued rational function.

Similarly, the transfer function, H_n, of the reduced-order model (2.6) and (2.7) is given by

$$H_n(s) := D_n + C_n^{\mathrm{T}}(sE_n - A_n)^{-1}B_n, \quad s \in \mathbb{C}. \tag{2.10}$$

Note that H_n is a also an $p \times m$-matrix-valued rational function.

2.3. *Padé and Padé-type models*

The concept of transfer functions allows to define reduced-order models by means of Padé or Padé-type approximation.

Let $s_0 \in \mathbb{C}$ be any point such that s_0 is not a pole of the transfer function H given by (2.8). In practice, the point s_0 is chosen such that it is in some sense close to the frequency range of interest. We remark that the frequency range of interest is usually a subset of the imaginary axis in the complex s-plane. Since s_0 is not a pole of H, the function H admits the Taylor expansion

$$H(s) = M_0 + M_1(s - s_0) + M_2(s - s_0)^2 + \cdots + M_j(s - s_0)^j + \cdots \tag{2.11}$$

about s_0. The coefficients M_j, $j = 0, 1, \ldots$, in (2.11) are called the *moments* of H about the expansion point s_0. Note that the M_j's are $p \times m$ matrices.

A reduced-order model (2.6) and (2.7) of state-space dimension n is called an *nth Padé model* (at the expansion point s_0) of the original system (2.1) and (2.2) if the Taylor expansions about s_0 of the transfer functions H and H_n of the original system and the reduced-order system agree in as many leading terms as possible, i.e.,

$$H(s) = H_n(s) + \mathcal{O}\big((s - s_0)^{q(n)}\big), \tag{2.12}$$

where $q(n)$ is as large as possible. In FELDMANN and FREUND [1995b], FREUND [1995], it was shown that

$$q(n) \geqslant \left\lfloor \frac{n}{m} \right\rfloor + \left\lfloor \frac{n}{p} \right\rfloor,$$

with equality in the "generic" case. The meaning of "generic" will be described more precisely in Section 3.2.

Even though Padé models are defined via the local approximation property (2.12), in practice, they usually are excellent approximations over large frequency ranges. The following single-input single-output example illustrates this statement. The example is a circuit resulting from the so-called PEEC discretization RUEHLI [1974] of an electromagnetic problem. The circuit is an RCL network consisting of 2100 capacitors, 172 inductors, 6990 inductive couplings, and a single resistive source that drives the circuit. Modified nodal analysis is used to set up the circuit equations, resulting in a linear dynamical system of dimension $N = 306$. In turns out that a Padé model of dimension $n = 60$ is sufficient to produces an almost exact transfer function in the relevant frequency range $s = 2\pi\,\mathrm{i}\omega$, $0 \leqslant \omega \leqslant 5 \times 10^9$. The corresponding curves for $|H(s)|$ and $|H_{60}(s)|$ are shown in Fig. 2.1.

It is very tempting to compute Padé models directly via the definition (2.12). More precisely, one would first explicitly generate the $q(n)$ moments $M_0, M_1, \ldots, M_{q(n)-1}$, and then compute H_n and the system matrices in the reduced-order model (2.6) and (2.7) from these moments. However, computing Padé models directly from the moments is extremely ill-conditioned, and consequently, such a procedure is not viable; we refer the reader to FELDMANN and FREUND [1994], FELDMANN and FREUND [1995a] for a detailed discussion and numerical examples.

The preferred way to compute Padé models is to use Krylov-subspace techniques, such as a suitable Lanczos-type process, as we will describe in Section 3. This becomes possible after the transfer function (2.8) is rewritten in terms of a single matrix M, instead of the two matrices A and E. To this end, let

$$A - s_0 E = F_1 F_2, \quad \text{where } F_1,\ F_2 \in \mathbb{C}^{N \times N}, \tag{2.13}$$

be any factorization of $A - s_0 E$. For example, the matrices $A - s_0 E$ arising in circuit simulation are large, but sparse, and are such that a sparse LU factorization is feasible.

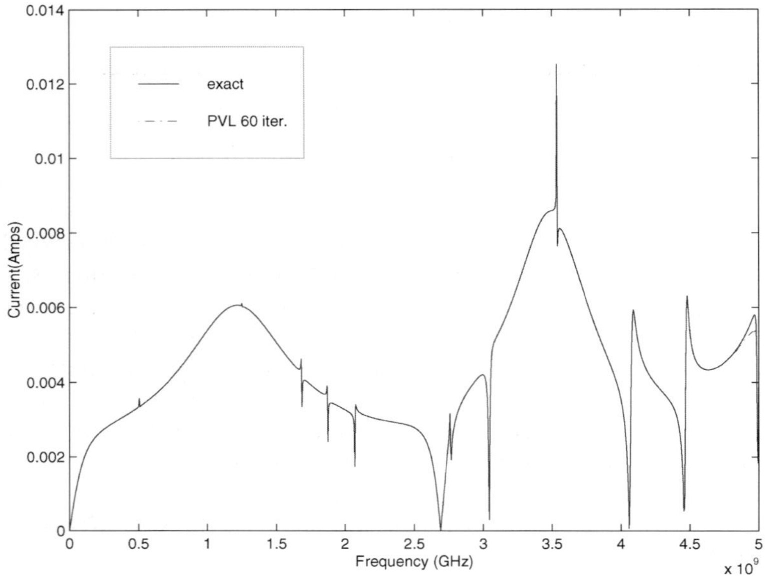

FIG. 2.1. The PEEC transfer function, exact and Padé model of dimension $n = 60$.

In this case, the matrices F_1 and F_2 in (2.13) are the lower and upper triangular factors, possibly with rows and columns permuted due to pivoting, of such a sparse LU factorization of $A - s_0E$. Using (2.13), the transfer function (2.8) can be rewritten as follows:

$$\begin{aligned} H(s) &= D + C^{\mathrm{T}}(sE - A)^{-1}B \\ &= D - C^{\mathrm{T}}\bigl(A - s_0E - (s - s_0)E\bigr)^{-1}B \\ &= D - L^{\mathrm{T}}\bigl(I - (s - s_0)M\bigr)^{-1}R, \end{aligned} \tag{2.14}$$

where

$$M := F_1^{-1}EF_2^{-1}, \quad R := F_1^{-1}B, \quad \text{and} \quad L := F_2^{-\mathrm{T}}C. \tag{2.15}$$

Note that (2.14) only involves one $N \times N$ matrix, namely M, instead of the two $N \times N$ matrices A and E in (2.8). This allows to apply Krylov-subspace methods to the single matrix M, with the $N \times m$ matrix R and the $N \times p$ matrix L as blocks of right and left starting vectors.

While Padé models often provide very good approximations in frequency domain, they also have undesirable properties. In particular, in general, Padé models do not preserve stability or passivity of the original system. However, by relaxing the Padé-approximation property (2.12), it is often possibly to obtain stable or passive models. More precisely, we call a reduced-order model (2.6) and (2.7) of state-space dimension n an *nth Padé-type model* (at the expansion point s_0) of the original system (2.1) and (2.2) if the Taylor expansions about s_0 of the transfer functions H and H_n of the original system and the reduced-order system agree in a number of leading terms, i.e.,

$$H(s) = H_n(s) + \mathcal{O}\big((s - s_0)^{q'}\big), \tag{2.16}$$

where $1 \leqslant q' < q(n)$.

2.4. *Stability*

An important property of linear dynamical systems is stability. An actual physical system needs to be stable in order to function properly. If a linear dynamical system (2.1) and (2.2) is used as a description of such a physical system, then clearly, it should also be stable. Moreover, when (2.1) and (2.2) is replaced by a reduced-order model that is then used in a time-domain analysis, the reduced-order model also needs to be stable.

In this subsection, we present a brief discussion of stability of linear descriptor systems. For a more general survey of the various concepts of stability of dynamical systems, we refer the reader to ANDERSON and VONGPANITLERD [1973], WILLEMS [1970].

A descriptor system of the form (2.1) and (2.2) is said to be *stable* if its free-response, i.e., the solutions $x(t)$, $t \geqslant 0$, of

$$E\frac{\mathrm{d}x}{\mathrm{d}t} = Ax, \qquad x(0) = x_0,$$

remain bounded as $t \to \infty$ for any possible initial vector x_0. Recall from the discussion in Section 2.1 that for singular E, there are certain restrictions on the possible initial vectors x_0.

Stability can easily be characterized in terms of the finite eigenvalues of the matrix pencil $A - sE$; see, e.g., MASUBUCHI, KAMITANE, OHARA and SUDA [1997]. More precisely, we have the following theorem.

THEOREM 2.1. *The descriptor system* (2.1) *and* (2.2) *is stable if, and only if, the following two conditions are satisfied*:
(i) *All finite eigenvalues* $\lambda \in \mathbb{C}$ *of the matrix pencil* $A - sE$ *satisfy* $\operatorname{Re}\lambda \leqslant 0$;
(ii) *All finite eigenvalues* λ *of the matrix pencil* $A - sE$ *with* $\operatorname{Re}\lambda = 0$ *are simple.*

We stress that, in view of Theorem 2.1, the infinite eigenvalues of the matrix pencil $A - sE$ have no effect on stability. The reason is that these infinite eigenvalues result only in impulsive motions, which go to zero as $t \to \infty$.

Recall that the transfer function H of the descriptor system (2.1) and (2.2) is of the form

$$H(s) = D + C^{\mathrm{T}}(sE - A)^{-1}B, \quad \text{where} \tag{2.17}$$

$$A, E \in \mathbb{R}^{N\times N}, \quad B \in \mathbb{R}^{N\times m}, \quad C \in \mathbb{R}^{N\times m}, \quad \text{and} \quad D \in \mathbb{R}^{p\times m}. \tag{2.18}$$

Note that any pole of H is necessarily an eigenvalue of the matrix pencil $A - sE$. Hence, it is tempting to determine stability via the poles of H. However, in general, not all eigenvalues of $A - sE$ are poles of H. For example, consider the following system

$$\frac{dx}{dt} = \begin{bmatrix} 1 & 0 \\ 0 & -1 \end{bmatrix} x + \begin{bmatrix} 0 \\ 1 \end{bmatrix} u(t),$$

$$y(t) = [1 \quad 1] x(t),$$

which is taken from p. 128 of ANDERSON and VONGPANITLERD [1973]. The pencil associated with this system is

$$A - sI = \begin{bmatrix} 1-s & 0 \\ 0 & -1-s \end{bmatrix}.$$

Its eigenvalues are ± 1, and hence this system is unstable. The transfer function $H(s) = 1/(s+1)$, however, only has the "stable" pole -1. Therefore, checking conditions (i) and (ii) of Theorem 2.1 only for the poles of H is, in general, not enough to guarantee stability. In order to infer stability of the system (2.1) and (2.2) from the poles of its transfer function, one needs an additional condition, which we formulate next.

Let H be a given $m \times p$-matrix-valued rational function. Any representation of H of the form (2.17) with matrices (2.18) is called a *realization* of H. Furthermore, a realization (2.17) of H is said to be *minimal* if the dimension N of the matrices (2.18) is as small as possible. We will also say that the state-space description (2.1) and (2.2) is a minimal realization if its transfer function (2.18) is a minimal realization.

The following theorem is the well-known characterization of minimal realizations in terms of conditions on the matrices (2.18); see, e.g., VERGHESE, LÉVY and KAILATH [1981]. We also refer the reader to the related results on controllability, observability, and minimal realizations of descriptor systems given in Chapter 2 of DAI [1989].

THEOREM 2.2. *Let H be a $m \times p$-matrix-valued rational function given by a realization* (2.17). *Then,* (2.17) *is a minimal realization of H if, and only if, the matrices* (2.18) *satisfy the following five conditions*:

(i) $\operatorname{rank}[A - sE \quad B] = N$ *for all* $s \in \mathbb{C}$;
(Finite controllability)

(ii) $\operatorname{rank}[E \quad B] = N$;
(Infinite controllability)

(iii) $\operatorname{rank}[A^{\mathrm{T}} - sE^{\mathrm{T}} \quad C] = N$ *for all* $s \in \mathbb{C}$;
(Finite observability)

(iv) $\operatorname{rank}[E^{\mathrm{T}} \quad C] = N$;
(Infinite observability)

(v) $A\ker(E) \subseteq \operatorname{Im}(E)$.
(Absence of nondynamic modes)

For descriptor systems given by a minimal realization, stability can indeed be checked via the poles of its transfer function.

THEOREM 2.3. *Let* (2.1) *and* (2.2) *be a minimal realization of a descriptor system, and let H be its transfer function* (2.17). *Then, the descriptor system* (2.1) *and* (2.2) *is stable* *if, and only if, all finite poles s_i of H satisfy* $\operatorname{Re} s_i \leqslant 0$ *and any pole with* $\operatorname{Re} s_i = 0$ *is simple.*

2.5. *Passivity*

In circuit simulation, reduced-order modeling is often applied to large passive linear subcircuits, such as RCL networks consisting of only resistors, inductors, and capacitors. When reduced-order models of such subcircuits are used within a simulation of the whole circuit, stability of the overall simulation can only be guaranteed if the reduced-order models preserve the passivity of the original subcircuits; see, e.g., CHIRLIAN [1967], ROHRER and NOSRATI [1981]. Therefore, it is important to have techniques to check passivity of a given reduced-order model.

Roughly speaking, a system is *passive* if it does not generate energy. For descriptor systems of the form (2.1) and (2.2), passivity is equivalent to positive realness of the transfer function. Moreover, such systems can only be passive if they have identical numbers of inputs and outputs. Thus, for the remainder of this subsection, we assume that $m = p$. Then, a system described by (2.1) and (2.2) is passive, i.e., it does not generate energy, if, and only if, its transfer function (2.17) is *positive real*; see, e.g., ANDERSON and VONGPANITLERD [1973]. A precise definition of positive realness is as follows.

DEFINITION 2.1. An $m \times m$-matrix-valued function $H : \mathbb{C} \mapsto (\mathbb{C} \cup \infty)^{m \times m}$ is called *positive real* if the following three conditions are satisfied:

(i) H is analytic in $\mathbb{C}_+ := \{s \in \mathbb{C} \mid \operatorname{Re} s > 0\}$;
(ii) $H(\bar{s}) = \overline{H(s)}$ for all $s \in \mathbb{C}$;
(iii) $H(s) + (H(s))^H \succeq 0$ for all $s \in \mathbb{C}_+$.

In Definition 2.1 and in the sequel, the notation $M \succeq 0$ means that the matrix M is Hermitian positive semi-definite. Similarly, $M \preceq 0$ means that M is Hermitian negative semi-definite.

For transfer functions H of the form (2.17), condition (ii) of Definition 2.1 is always satisfied since the matrices (2.18) are assumed to be real. Furthermore, condition (i) simply means that H cannot have poles in $\mathbb{C}_+$, and this can be checked easily. For the special case $m = 1$ of scalar-valued functions H, condition (iii) states that the real part of $H(s)$ is nonnegative for all s with nonnegative real part. In order to check this condition, it is sufficient to show that the real part of $H(s)$ is nonnegative for all purely imaginary s. This can be done by means of relatively elementary means. For example, in BAI and FREUND [2000], a procedure based on eigenvalue computations is proposed. For the general matrix-valued case, $m \geqslant 1$, however, checking condition (iii) is much more involved. One possibility is to employ a suitable extension of the classical positive real lemma (see, e.g. ANDERSON [1967], Chapter 5 of ANDERSON and VONGPANITLERD [1973], or Section 13.5 of ZHOU, DOYLE and GLOVER [1996]) that characterizes positive realness of regular linear systems via the solvability of certain linear matrix inequalities (LMIs). Such a version of the positive real lemma for general descriptor systems is stated in Theorem 2.4 below.

We remark that any matrix-valued rational function H has an expansion about $s = \infty$ of the form

$$H(s) = \sum_{j=-\infty}^{j_0} M_j s^j, \tag{2.19}$$

where $j_0 \geqslant 0$ is an integer. Moreover, the function H has a pole at $s = \infty$ if, and only if, $j_0 \geqslant 1$ and $M_{j_0} \neq 0$ in (2.19).

The positive real lemma for descriptor systems can now be stated as follows.

THEOREM 2.4 (Positive real lemma for descriptor systems (FREUND and JARRE [2004a])). *Let H be a real $m \times m$-matrix-valued rational function of the form* (2.17) *with matrices* (2.18).

(a) (Sufficient condition)
If the LMIs

$$\begin{bmatrix} A^{\mathrm{T}}X + X^{\mathrm{T}}A & X^{\mathrm{T}}B - C \\ B^{\mathrm{T}}X - C^{\mathrm{T}} & -D - D^{\mathrm{T}} \end{bmatrix} \preceq 0 \quad \textit{and} \quad E^{\mathrm{T}}X = X^{\mathrm{T}}E \succeq 0 \tag{2.20}$$

have a solution $X \in \mathbb{R}^{N\times N}$, then H is positive real.

(b) (Necessary condition)
Suppose that (2.17) *is a minimal realization of H and that the matrix M_0 in the expansion* (2.19) *satisfies*

$$(D - M_0) + (D - M_0)^{\mathrm{T}} \succeq 0. \tag{2.21}$$

If H is positive real, then there exists a solution $X \in \mathbb{R}^{N\times N}$ of the LMIs (2.20).

The result of Theorem 2.4 allows to check positive realness by solving the semi-definite programming problems of the form (2.20). Note that there are N^2 unknowns in (2.20), namely the entries of the $N \times N$ matrix X. Problems of the form (2.20) can be tackled with interior-point methods; see, e.g., BOYD, EL GHAOUI, FERON and BALAKRISHNAN [1994], FREUND and JARRE [2004a]. However, the computational complexity of these methods grows quickly with N, and thus, these methods are viable only for rather small values of N.

For the special case $E = I$, the result of Theorem 2.4 is just the classical positive real lemma. In this case, (2.20) reduces to the problem of finding a symmetric positive semi-definite matrix $X \in \mathbb{R}^{N\times N}$ such that

$$\begin{bmatrix} A^{\mathrm{T}}X + XA & XB - C \\ B^{\mathrm{T}}X - C^{\mathrm{T}} & -D - D^{\mathrm{T}} \end{bmatrix} \preceq 0.$$

Moreover, if $E = I$, the condition (2.21) is always satisfied, since in this case $M_0 = 0$ and $D + D^{\mathrm{T}} \succeq 0$.

2.6. *Linear RCL subcircuits*

In circuit simulation, an important special case of passive circuits is linear subcircuits that consist of only resistors, inductors, and capacitors. Such linear RCL subcircuits arise in the modeling of a circuit's interconnect and package; see, e.g., FREUND and FELDMANN [1997], FREUND and FELDMANN [1998], KIM, GOPAL and PILLAGE [1994], PILEGGI [1995].

The equations describing linear RCL subcircuits are of the form (2.1) and (2.2) with $D=0$ and $m=p$. Furthermore, the equations can be formulated such that the matrices $A,\ E \in \mathbb{R}^{N\times N}$ in (2.1) are symmetric and exhibit a block structure; see FREUND and FELDMANN [1996a], FREUND and FELDMANN [1998]. More precisely, we have

$$A = A^{\mathrm{T}} = \begin{bmatrix} -A_{11} & A_{12} \\ A_{12}^{\mathrm{T}} & 0 \end{bmatrix} \quad \text{and} \quad E = E^{\mathrm{T}} = \begin{bmatrix} E_{11} & 0 \\ 0 & -E_{22} \end{bmatrix}, \tag{2.22}$$

where the submatrices $A_{11},\ E_{11} \in \mathbb{R}^{N_1\times N_1}$ and $E_{22} \in \mathbb{R}^{N_2\times N_2}$ are symmetric positive semi-definite, and $N = N_1 + N_2$. Note that, except for the special case $N_2 = 0$, the matrices A and E are indefinite. The special case $N_2 = 0$ arises for RC subcircuits that contain only resistors and capacitors, but no inductors.

If the RCL subcircuit is viewed as an m-terminal component with $m = p$ inputs and outputs, then the matrices B and C in (2.1) and (2.2) are identical and of the form

$$B = C = \begin{bmatrix} B_1 \\ 0 \end{bmatrix} \quad \text{with } B_1 \in \mathbb{R}^{N_1\times m}. \tag{2.23}$$

In view of (2.22) and (2.23), the transfer function of such an m-terminal RCL subcircuit is given by

$$H(s) = B^{\mathrm{T}}(sE - A)^{-1}B, \quad \text{where } A = A^{\mathrm{T}},\ E = E^{\mathrm{T}}. \tag{2.24}$$

We call a transfer function H *symmetric* if it is of the form (2.24) with real matrices A, E, and B.

We will also use the following nonsymmetric formulation of (2.24). Let J be the block matrix

$$J = \begin{bmatrix} I_{N_1} & 0 \\ 0 & -I_{N_2} \end{bmatrix}, \tag{2.25}$$

where I_{N_1} and I_{N_2} is the $N_1 \times N_1$ and $N_2 \times N_2$ identity matrix, respectively.

Note that, by (2.23) and (2.25), we have $B = JB$. Using this relation, as well as (2.22), we can rewrite (2.24) as follows:

$$\begin{aligned} &H(s) = B^{\mathrm{T}}\big(s\tilde{E} - \tilde{A}\big)^{-1}B, \quad \text{where} \\ &\tilde{A} = \begin{bmatrix} -A_{11} & A_{12} \\ -A_{12}^{\mathrm{T}} & 0 \end{bmatrix}, \qquad \tilde{E} = \begin{bmatrix} E_{11} & 0 \\ 0 & E_{22} \end{bmatrix}. \end{aligned} \tag{2.26}$$

In this formulation, the matrix $\tilde{A}$ is no longer symmetric, but now

$$\tilde{A} + \tilde{A}^{\mathrm{T}} \preceq 0 \quad \text{and} \quad \tilde{E} \succeq 0. \tag{2.27}$$

It turns out that the properties are the key to ensure positive realness. Indeed, the following result was established as Theorem 13 in FREUND [2000b].

THEOREM 2.5. *Let $\tilde{A}$, $\tilde{E} \in \mathbb{R}^{N\times N}$, and $B \in \mathbb{R}^{N\times m}$. Assume that $\tilde{A}$ and $\tilde{E}$ satisfy (2.27), and that the matrix pencil $\tilde{A} - s\tilde{E}$ is regular. Then, the $m \times m$-matrix-valued function*

$$H(s) = B^{\mathrm{T}}\left(s\tilde{E} - \tilde{A}\right)^{-1} B$$

is positive real.

3. Krylov-subspace techniques

In this section, we discuss the use of Krylov-subspace methods for the construction of Padé and Padé-type reduced-order models of time-invariant linear dynamical systems. We also point the reader to FREUND [2003] for a more extended survey of Krylov-subspace methods for model reduction.

3.1. Block Krylov subspaces

We consider general descriptor systems of the form (2.1) and (2.2). The key to using Krylov-subspace techniques for reduced-order modeling of such systems is to first replace the matrix pair A and E by a single matrix M. To this end, let $s_0 \in \mathbb{C}$ be any given point such that the matrix $A - s_0E$ is nonsingular. Then, with M, R, and L denoting the matrices defined in (2.15), the linear system (2.1) and (2.2) can be rewritten in the following form:

$$M\frac{\mathrm{d}x}{\mathrm{d}t} = (I + s_0M)x + Ru(t), \tag{3.1}$$

$$y(t) = L^{\mathrm{T}}x(t) + Du(t). \tag{3.2}$$

Note that $M \in \mathbb{C}^{N\times N}$, $R \in \mathbb{C}^{N\times m}$, and $L \in \mathbb{C}^{N\times p}$, where N is the state-space dimension of the system, m is the number of inputs, and p is the number of outputs.

The transfer function H of the rewritten system (3.1) and (3.2) is given by (2.14). By expanding (2.14) about s_0, we obtain

$$H(s) = D - \sum_{j=0}^{\infty} L^{\mathrm{T}}M^jR(s - s_0)^j. \tag{3.3}$$

Recall from Section 2.3 that Padé and Padé-type reduced-order models are defined via the leading coefficients of an expansion of H about s_0. In view of (3.3), the jth coefficient of such an expansion can be expressed as follows:

$$-L^{\mathrm{T}}M^jR = -\left((M^{j-i})^{\mathrm{T}}L\right)^{\mathrm{T}}(M^iR), \quad i = 0, 1, \ldots, j. \tag{3.4}$$

Notice that the factors on the right-hand side of (3.4) are blocks of the *right* and *left block Krylov matrices*

$$\begin{aligned} &[R \quad MR \quad M^2R \quad \cdots \quad M^iR \quad \cdots] \quad \text{and} \\ &\left[L \quad M^{\mathrm{T}}L \quad (M^{\mathrm{T}})^2L \quad \cdots \quad (M^{\mathrm{T}})^kL \quad \cdots\right], \end{aligned} \tag{3.5}$$

respectively. As a result, all the information needed to generate Padé and Padé-type reduced-order models is contained in the block Krylov matrices (3.5). However, simply computing the blocks $M^i R$ and $(M^{\mathrm{T}})^i L$ in (3.5) and then generating the leading coefficients of the expansion (3.3) from these blocks is not a viable numerical procedure. The reason is that, in finite-precision arithmetic, as i increases, the blocks $M^i R$ and $(M^{\mathrm{T}})^i L$ quickly contain only information about the eigenspaces of the dominant eigenvalue of M. Instead, one needs to employ suitable Krylov-subspace methods that generate numerically better basis vectors for the subspaces associated with the block Krylov matrices (3.5).

Next, we give a formal definition of the subspaces induced by (3.5). Note that each block $M^i R$ consists of m column vectors of length N. By scanning these column vectors of the right block Krylov matrix in (3.5) from left to right and by deleting any column that is linearly dependent on columns to its left, we obtain the *deflated* right block Krylov matrix

$$\begin{bmatrix} R_1 & MR_2 & M^2R_3 & \cdots & M^{i_{\max}-1}R_{i_{\max}} \end{bmatrix}. \tag{3.6}$$

This process of detecting and deleting the linearly dependent columns is called *exact deflation*. We remark that the matrix (3.6) is finite, since at most N of the column vectors can be linearly independent. Furthermore, a column $M^i r$ being linearly dependent on columns to its left in (3.5) implies that any column $M^{i'} r$, $i' \geqslant i$, is linearly dependent on columns to its right. Therefore, in (3.6), for each $i = 1, 2, \ldots, i_{\max}$, the matrix R_i is a submatrix of R_{i-1}, where, for $i = 1$, we set $R_0 = R$.

Let m_i denote the number of columns of R_i. The matrix (3.6) thus has

$$n_{\max}^{(\mathrm{r})} :=\leqslant m_1 + m_2 + \cdots + m_{i_{\max}},$$

columns. For each integer n with $1 \leqslant n \leqslant n_{\max}^{(\mathrm{r})}$, we define the *nth right block Krylov subspace* $\mathcal{K}_n(M, R)$ (induced by M and R) as the subspace spanned by the first n columns of the deflated right block Krylov matrix (3.6).

Analogously, by deleting the linearly independent columns of the left block Krylov matrix in (3.5), we obtain a deflated left block Krylov matrix of the form

$$\begin{bmatrix} L_1 & M^{\mathrm{T}}L_2 & (M^{\mathrm{T}})^2L_3 & \cdots & (M^{\mathrm{T}})^{i_{\max}-1}L_{k_{\max}} \end{bmatrix}. \tag{3.7}$$

Let $n_{\max}^{(\mathrm{l})}$ be the number of columns of the matrix (3.7). Then for each integer n with $1 \leqslant n \leqslant n_{\max}^{(\mathrm{l})}$, we define the *nth left block Krylov subspace* $\mathcal{K}_n(M^{\mathrm{T}}, L)$ (induced by M^{T} and L) as the subspace spanned by the first n columns of the deflated left block Krylov matrix (3.7).

For a more detailed discussion of block Krylov subspaces and deflation, we refer the reader to ALIAGA, BOLEY, FREUND and HERNÁNDEZ [2000], FREUND [2000b].

3.2. *Approaches based on Lanczos and Lanczos-type methods*

In this section, we discuss reduced-order modeling approaches that employ Lanczos and Lanczos-type methods for the construction of suitable basis vectors for the right and left block Krylov subspaces $\mathcal{K}_n(M, R)$ and $\mathcal{K}_n(M^{\mathrm{T}}, L)$.

3.2.1. The MPVL algorithm

For the special case $m = p = 1$ of single-input single-output linear dynamical systems, each of the "blocks" R and L only consists of a single vector, say r and l, and $\mathcal{K}_n(M, r)$ and $\mathcal{K}_n(M^T, l)$ are just the standard nth right and left Krylov subspaces induced by single vectors. The classical Lanczos process (LANCZOS [1950]) is a well-known procedure for computing two sets of bi-orthogonal basis vectors for $\mathcal{K}_n(M, r)$ and $\mathcal{K}_n(M^T, l)$. Moreover, these vectors are generated by means of three-term recurrences the coefficients of which define a tridiagonal matrix T_n. It turns out that T_n contains all the information that is needed to set up an nth Padé reduced-order model of a given single-input single-output time-invariant linear dynamical system. The associated computational procedure is called the *Padé via Lanczos* (PVL) algorithm in FELDMANN and FREUND [1994], FELDMANN and FREUND [1995a].

Here, we describe in some detail an extension of the PVL algorithm to the case of general m-input p-output time-invariant linear dynamical systems. The underlying block Krylov subspace method is the *nonsymmetric band Lanczos algorithm* (FREUND [2000a]) for constructing two sets of right and left Lanczos vectors

$$v_1, v_2, \ldots, v_n \quad \text{and} \quad w_1, w_2, \ldots, w_n, \tag{3.8}$$

respectively. These vectors span the nth right and left block Krylov subspaces (induced by M and R, and M^T and L, respectively):

$$\begin{aligned} &\operatorname{span}\{v_1, v_2, \ldots, v_n\} = \mathcal{K}_n(M, R) \quad \text{and} \\ &\operatorname{span}\{w_1, w_2, \ldots, w_n\} = \mathcal{K}_n(M^T, L). \end{aligned} \tag{3.9}$$

Moreover, the vectors (3.8) are constructed to be bi-orthogonal:

$$w_j^T v_k = \begin{cases} 0 & \text{if } j \neq k, \\ \delta_j & \text{if } j = k, \end{cases} \quad \text{for all } j, k = 1, 2, \ldots, n. \tag{3.10}$$

It turns out that the Lanczos vectors (3.8) can be constructed by means of recurrence relations of length at most $m + p + 1$. The recurrence coefficients for the first n right Lanczos vectors define an $n \times n$ matrix $T_n^{(pr)}$ that is "essentially" a band matrix with total bandwidth $m + p + 1$. Similarly, the recurrence coefficients for the first n left Lanczos vectors define an $n \times n$ band matrix $\widetilde{T}_n^{(pr)}$ with total bandwidth $m + p + 1$. For a more detailed discussion of the structure of $T_n^{(pr)}$ and $\widetilde{T}_n^{(pr)}$, we refer the reader to ALIAGA, BOLEY, FREUND and HERNÁNDEZ [2000], FREUND [2000a].

Algorithm 3.1 below gives a complete description of the numerical procedure that generates the Lanczos vectors (3.8) with properties (3.9) and (3.10). In order to obtain a Padé reduced-order model based on this algorithm, one does not need the Lanczos vectors themselves, but rather the matrix of right recurrence coefficients $T_n^{(pr)}$, the matrices $\rho_n^{(pr)}$ and $\eta_n^{(pr)}$ that contain the recurrence coefficients from processing the starting blocks R and L, respectively, and the diagonal matrix

$$\Delta_n = \operatorname{diag}(\delta_1, \delta_2, \ldots, \delta_n),$$

whose diagonal entries are the δ_j's from (3.10). The following algorithm produces the matrices $T_n^{(pr)}$, $\rho_n^{(pr)}$, $\eta_n^{(pr)}$, and Δ_n as output.

ALGORITHM 3.1 (*Nonsymmetric band Lanczos algorithm*).

INPUT: A matrix $M \in \mathbb{C}^{N \times N}$;
A block of m right starting vectors $R = [\, r_1 \quad r_2 \quad \cdots \quad r_m \,] \in \mathbb{C}^{N \times m}$;
A block of p left starting vectors $L = [\, l_1 \quad l_2 \quad \cdots \quad l_p \,] \in \mathbb{C}^{N \times p}$.

OUTPUT: The $n \times n$ Lanczos matrix $T_n^{(\mathrm{pr})}$, and the matrices $\rho_n^{(\mathrm{pr})}$, $\eta_n^{(\mathrm{pr})}$, and Δ_n.

(0) For $k = 1, 2, \ldots, m$, set $\hat{v}_k = r_k$.
For $k = 1, 2, \ldots, p$, set $\hat{w}_k = l_k$.
Set $m_\mathrm{c} = m$, $p_\mathrm{c} = p$, and $\mathcal{I}_v = \mathcal{I}_w = \emptyset$.

For $n = 1, 2, \ldots,$ until convergence or $m_\mathrm{c} = 0$ or $p_\mathrm{c} = 0$ or $\delta_n = 0$ do:

(1) (If necessary, deflate $\hat{v}_n$.)
Compute $\|\hat{v}_n\|_2$.
Decide if $\hat{v}_n$ should be deflated. If yes, do the following:
(a) Set $\hat{v}_{n-m_\mathrm{c}}^{\mathrm{defl}} = \hat{v}_n$ and store this vector. Set $\mathcal{I}_v = \mathcal{I}_v \cup \{n - m_\mathrm{c}\}$.
(b) Set $m_\mathrm{c} = m_\mathrm{c} - 1$.
If $m_\mathrm{c} = 0$, set $n = n - 1$ and stop.
(c) For $k = n, n+1, \ldots, n + m_\mathrm{c} - 1$, set $\hat{v}_k = \hat{v}_{k+1}$.
(d) Repeat all of Step (1).

(2) (If necessary, deflate $\hat{w}_n$.)
Compute $\|\hat{w}_n\|_2$.
Decide if $\hat{w}_n$ should be deflated. If yes, do the following:
(a) Set $\hat{w}_{n-p_\mathrm{c}}^{\mathrm{defl}} = \hat{w}_n$ and store this vector. Set $\mathcal{I}_w = \mathcal{I}_w \cup \{n - p_\mathrm{c}\}$.
(b) Set $p_\mathrm{c} = p_\mathrm{c} - 1$.
If $p_\mathrm{c} = 0$, set $n = n - 1$ and stop.
(c) For $k = n, n+1, \ldots, n + p_\mathrm{c} - 1$, set $\hat{w}_k = \hat{w}_{k+1}$.
(d) Repeat all of Step (2).

(3) (Normalize $\hat{v}_n$ and $\hat{w}_n$ to obtain v_n and w_n.)
Set

$$t_{n,n-m_\mathrm{c}} = \|\hat{v}_n\|_2, \qquad \tilde{t}_{n,n-p_\mathrm{c}} = \|\hat{w}_n\|_2,$$

$$v_n = \frac{\hat{v}_n}{t_{n,n-m_\mathrm{c}}}, \quad \text{and} \quad w_n = \frac{\hat{w}_n}{\tilde{t}_{n,n-p_\mathrm{c}}}.$$

(4) (Compute δ_n and check for possible breakdown.)
Set $\delta_n = w_n^\mathrm{T} v_n$. If $\delta_n = 0$, set $n = n - 1$ and stop.

(5) (Orthogonalize the right candidate vectors against w_n.)
For $k = n+1, n+2, \ldots, n + m_\mathrm{c} - 1$, set

$$t_{n,k-m_\mathrm{c}} = \frac{w_n^\mathrm{T} \hat{v}_k}{\delta_n} \quad \text{and} \quad \hat{v}_k = \hat{v}_k - v_n t_{n,k-m_\mathrm{c}}.$$

(6) (Orthogonalize the left candidate vectors against v_n.)
For $k = n+1, n+2, \ldots, n + p_\mathrm{c} - 1$, set

$$\tilde{t}_{n,k-p_\mathrm{c}} = \frac{\hat{w}_k^\mathrm{T} v_n}{\delta_n} \quad \text{and} \quad \hat{w}_k = \hat{w}_k - w_n \tilde{t}_{n,k-p_\mathrm{c}}.$$

(7) (Advance the right block Krylov subspace to get $\hat{v}_{n+m_\mathrm{c}}$.)

(a) Set $\hat{v}_{n+m_c} = M v_n$.
(b) For $k \in \mathcal{I}_w$ (in ascending order), set

$$\tilde{\sigma} = \left(\hat{w}_k^{\text{defl}}\right)^{\mathrm{T}} v_n, \qquad \tilde{t}_{n,k} = \frac{\tilde{\sigma}}{\delta_n},$$

and, if $k > 0$, set

$$t_{k,n} = \frac{\tilde{\sigma}}{\delta_k} \quad \text{and} \quad \hat{v}_{n+m_c} = \hat{v}_{n+m_c} - v_k t_{k,n}.$$

(c) Set $k_v = \max\{1,\ n - p_c\}$.
(d) For $k = k_v, k_v + 1, \ldots, n-1$, set

$$t_{k,n} = \tilde{t}_{n,k} \frac{\delta_n}{\delta_k} \quad \text{and} \quad \hat{v}_{n+m_c} = \hat{v}_{n+m_c} - v_k\, t_{k,n}.$$

(e) Set

$$t_{n,n} = \frac{w_n^{\mathrm{T}} \hat{v}_{n+m_c}}{\delta_n} \quad \text{and} \quad \hat{v}_{n+m_c} = \hat{v}_{n+m_c} - v_n\, t_{n,n}.$$

(8) (Advance the left block Krylov subspace to get $\hat{w}_{n+p_c}$.)
(a) Set $\hat{w}_{n+p_c} = M^{\mathrm{T}} w_n$.
(b) For $k \in \mathcal{I}_v$ (in ascending order), set

$$\sigma = w_n^{\mathrm{T}} \hat{v}_k^{\text{defl}}, \qquad t_{n,k} = \frac{\sigma}{\delta_n},$$

and, if $k > 0$, set

$$\tilde{t}_{k,n} = \frac{\sigma}{\delta_k} \quad \text{and} \quad \hat{w}_{n+p_c} = \hat{w}_{n+p_c} - w_k \tilde{t}_{k,n}.$$

(c) Set $k_w = \max\{1, n - m_c\}$.
(d) For $k = k_w, k_w + 1, \ldots, n-1$, set

$$\tilde{t}_{k,n} = t_{n,k} \frac{\delta_n}{\delta_k} \quad \text{and} \quad \hat{w}_{n+p_c} = \hat{w}_{n+p_c} - w_k \tilde{t}_{k,n}.$$

(e) Set

$$\tilde{t}_{n,n} = t_{n,n} \quad \text{and} \quad \hat{w}_{n+p_c} = \hat{w}_{n+p_c} - w_n\, \tilde{t}_{n,n}.$$

(9) Set

$$\begin{aligned}
T_n^{(\text{pr})} &= [\, t_{i,k} \,]_{i,k=1,2,\ldots,n}\,, \\
\rho_n^{(\text{pr})} &= [\, t_{i,k-m} \,]_{i=1,2,\ldots,n;k=1,2,\ldots,k_\rho}\,, \qquad \text{where } k_\rho = m + \min\{0, n - m_c\}, \\
\eta_n^{(\text{pr})} &= [\, \tilde{t}_{i,k-p} \,]_{i=1,2,\ldots,n;k=1,2,\ldots,k_\eta}\,, \qquad \text{where } k_\eta = p + \min\{0, n - p_c\}, \\
\Delta_n &= \operatorname{diag}(\delta_1, \delta_2, \ldots, \delta_n).
\end{aligned}$$

(10) Check if n is large enough. If yes, stop.

REMARK 3.1. When applied to single starting vectors, i.e., for the special case $m = p = 1$, Algorithm 3.1 reduces to the classical nonsymmetric Lanczos process (LANCZOS [1950]).

REMARK 3.2. It can be shown that, at step n of Algorithm 3.1, exact deflation of a vector in the right, respectively left, block Krylov matrix (3.5) occurs if, and only if, $\hat{v}_n = 0$, respectively $\hat{w}_n = 0$, in Step (1), respectively Step (2). Therefore, to run Algorithm 3.1 with exact deflation only, one deflates $\hat{v}_n$ if $\|\hat{v}_n\|_2 = 0$ in Step (1), and one deflates $\hat{w}_n$ if $\|\hat{w}_n\|_2 = 0$ in Step (2). In finite-precision arithmetic, however, so-called *inexact deflation* is employed. This means that in Step (1), $\hat{v}_n$ is deflated if $\|\hat{v}_n\|_2 \leqslant \varepsilon$, and in Step (2), $\hat{w}_n$ is deflated if $\|\hat{w}_n\|_2 \leqslant \varepsilon$, where $\varepsilon = \varepsilon(M) > 0$ is a suitably chosen small constant.

REMARK 3.3. The occurrence of $\delta_n = 0$ in Step (4) of Algorithm 3.1 is called a *breakdown*. In finite-precision arithmetic, in Step (4) one should also check for *near-breakdowns*, i.e., if $\delta_n \approx 0$. In general, it cannot be excluded that breakdowns or near-breakdowns occur, although they are very unlikely. Furthermore, by using so-called *look-ahead* techniques, it is possible to remedy the problem of possible breakdowns or near-breakdowns. For the sake of simplicity, we have stated the band Lanczos algorithm without look-ahead only. A look-ahead version of Algorithm 3.1 is described in ALIAGA, BOLEY, FREUND and HERNÁNDEZ [2000].

The *matrix-Padé via Lanczos* (MPVL) algorithm, which was first introduced in FELDMANN and FREUND [1995b], FREUND [1995], consists of applying Algorithm 3.1 to the matrices M, R, and L defined in (2.15), and running it for n steps. The matrices $T_n^{(\mathrm{pr})}$, $\rho_n^{(\mathrm{pr})}$, $\eta_n^{(\mathrm{pr})}$, and Δ_n produced by Algorithm 3.1 are then used to set up a reduced-order model of the original linear dynamical system (2.1) and (2.2) as follows:

$$T_n^{(\mathrm{pr})}\frac{\mathrm{d}z}{\mathrm{d}t} = \left(s_0 T_n^{(\mathrm{pr})} - I\right)z + \rho_n^{(\mathrm{pr})}u(t), \tag{3.11}$$

$$y(t) = \left(\eta_n^{(\mathrm{pr})}\right)^{\mathrm{T}}\Delta_n z(t) + Du(t). \tag{3.12}$$

Note that the transfer function of this reduced-order model is given by

$$H_n(s) = D + \left(\eta_n^{(\mathrm{pr})}\right)^{\mathrm{T}}\Delta_n\left(I - (s - s_0)T_n^{(\mathrm{pr})}\right)^{-1}\rho_n^{(\mathrm{pr})}. \tag{3.13}$$

The reduced-order model (3.11) and (3.12) is indeed a matrix-Padé model of the original system.

THEOREM 3.1 (Matrix-Padé model (FELDMANN and FREUND [1995b], FREUND [1995])). *Suppose that Algorithm* 3.1 *is run with exact deflation only and that* $n \geqslant \max\{m, p\}$. *Then, the reduced-order model* (3.11) *and* (3.12) *is a matrix-Padé model of the linear dynamical system* (2.1) *and* (2.2). *More precisely, the Taylor expansions about* s_0 *of the transfer functions,* H, (2.8) *and,* H_n, (3.13) *agree in as many leading coefficients as possible, i.e.,*

$$H(s) = H_n(s) + \mathcal{O}\left((s - s_0)^{q(n)}\right),$$

where $q(n)$ is as large as possible. In particular,

$$q(n) \geqslant \left\lfloor \frac{n}{m} \right\rfloor + \left\lfloor \frac{n}{p} \right\rfloor.$$

A disadvantage of Padé models is that, in general, they do not preserve the stability and possibly passivity of the original linear dynamical system. In part, these problems can be overcome by means of suitable post-processing techniques, such as the ones described in BAI, FELDMANN and FREUND [1998], BAI and FREUND [2001a]. However, the reduced-order models obtained by post-processing of Padé models are necessarily no longer optimal in the sense of Padé approximation. Furthermore, post-processing techniques are not guaranteed to always result in stable and possibly passive reduced-order models.

For special cases, however, Padé models can be shown to be stable and passive. In particular, this is the case for linear dynamical systems describing RC subcircuits, RL subcircuits, and LC subcircuits; see BAI and FREUND [2001b], FREUND and FELDMANN [1996a], FREUND and FELDMANN [1997], FREUND and FELDMANN [1998].

Next, we describe the SyMPVL algorithm (FREUND and FELDMANN [1996a], FREUND and FELDMANN [1997], FREUND and FELDMANN [1998]), which is a special version of MPVL tailored to linear RCL subcircuits.

3.2.2. *The SyMPVL algorithm*

Recall from Section 2.6 that linear RCL subcircuits can be described by linear dynamical systems (2.1) and (2.2) with $D = 0$, symmetric matrices A and E of the form (2.22), and matrices $B = C$ of the form (2.23). Furthermore, the transfer function, H, (2.24) is symmetric.

We now assume that the expansion point s_0 for the Padé approximation is chosen to be real and nonnegative, i.e., $s_0 \geqslant 0$. Together with (2.22) it follows that the matrix $A - s_0 E$ is symmetric indefinite, with N_1 nonpositive and N_2 nonnegative eigenvalues. Thus, $A - s_0 E$ admits a factorization of the following form:

$$A - s_0 E = -F_1 J F_1^{\mathrm{T}}, \tag{3.14}$$

where J is the block matrix defined in (2.25). Instead of the general factorization (2.13), we now use (3.14). By (3.14) and (2.15), the matrices M, R, and L, are then of the following form:

$$M = F_1^{-1} E F_1^{-\mathrm{T}} J, \quad R = F_1^{-1} B, \quad \text{and} \quad L = -J F_1^{-1} C.$$

Since $E = E^{\mathrm{T}}$ and $B = C$, it follows that

$$J M = M^{\mathrm{T}} J \quad \text{and} \quad L = -J R.$$

This means that M is J-symmetric and the left starting block L is (up to its sign) the J-multiple of the right starting block R. These two properties imply that all the right and left Lanczos vectors generated by the band Lanczos Algorithm 3.1 are J-multiples of each other:

$$w_j = J v_j \quad \text{for all } j = 1, 2, \ldots, n.$$

Consequently, Algorithm 3.1 simplifies in that only the right Lanczos vectors need to be computed. The resulting version of MPVL for computing matrix-Padé models of RCL subcircuits is just the SyMPVL algorithm. The computational costs of SyMPVL are half of that of the general MPVL algorithm.

Let $H_n^{(1)}$ denote the matrix-Padé model generated by SyMPVL after n Lanczos steps. For general RCL subcircuits, however, $H_n^{(1)}$ will not preserve the passivity of the original system.

An additional reduced-order model that is guaranteed to be passive can be obtained as follows, provided that all right Lanczos vectors are stored. Let

$$V_n = [\, v_1 \quad v_2 \quad \cdots \quad v_n \,]$$

denote the matrix that contains the first n right Lanczos vectors as columns. Then, by projecting the matrices in the representation (2.26) of the transfer function H of the original RCL subcircuit onto the columns of V_n, we obtain the following reduced-order transfer function:

$$H_n^{(2)}(s) = \left(V_n^{\mathrm{T}} B\right)^{\mathrm{T}} \left(s V_n^{\mathrm{T}} \tilde{E} V_n - V_n^{\mathrm{T}} \tilde{A} V_n\right)^{-1} V_n^{\mathrm{T}} B. \tag{3.15}$$

The passivity of the original RCL subcircuit, together with Theorem 2.5 implies that the reduced-order model defined by $H_n^{(2)}$ is indeed passive. Furthermore, in FREUND [2000b], it is shown that $H_n^{(2)}$ is a matrix-Padé-type approximation of the original transfer function and that, at the expansion point s_0, $H_n^{(2)}$ matches half as many leading coefficients of H as the matrix-Padé approximant $H_n^{(1)}$.

Next, we illustrate the behavior of SyMPVL with two circuit examples.

3.2.3. *A package model*

The first example arises is the analysis of a 64-pin package model used for an RF integrated circuit. Only eight of the package pins carry signals, the rest being either unused or carrying supply voltages. The package is characterized as a passive linear dynamical system with $m = p = 16$ inputs and outputs, representing 8 exterior and 8 interior terminals. The package model is described by approximately 4000 circuit elements, resistors, capacitors, inductors, and inductive couplings, resulting in a linear dynamical system with a state-space dimension of about 2000.

In FREUND and FELDMANN [1997], SyMPVL was used to compute a Padé-based reduced-order model of the package, and it was found that a model $H_n^{(1)}$ of order $n = 80$ is sufficient to match the transfer-function components of interest. However, the model $H_n^{(1)}$ has a few poles in the right half of the complex plane, and therefore, it is not passive.

In order to obtain a passive reduced-order model, we ran SyMPVL again on the package example, and this time, also generated the projected reduced-order model $H_n^{(2)}$ given by (3.15). The expansion point $s_0 = 5\pi \times 10^9$ was used. Recall that $H_n^{(2)}$ is only a Padé-type approximant and thus less accurate than the Padé approximant $H_n^{(2)}$. Therefore, one now has to go to order $n = 112$ to obtain a projected reduced-order model $H_n^{(2)}$ that matches the transfer-function components of interest. Figs. 3.1 and 3.2 show

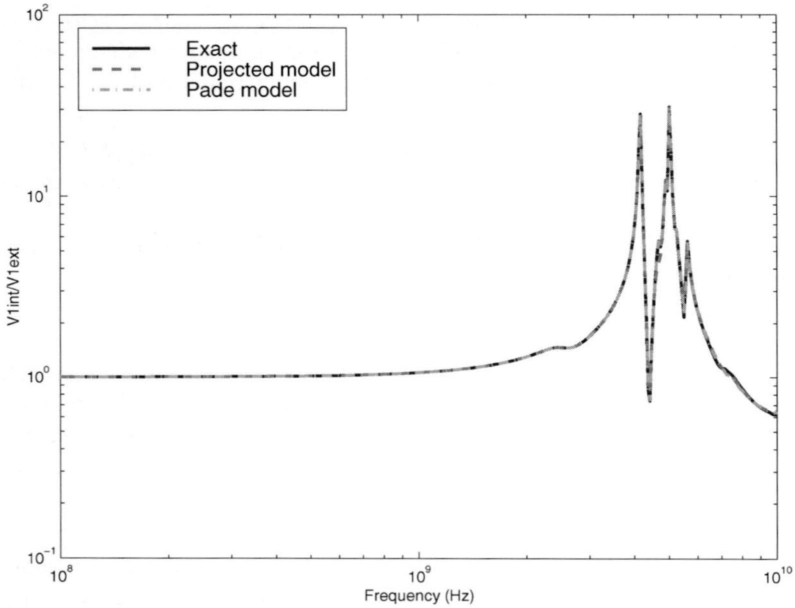

FIG. 3.1. Package: Pin no. 1 external to Pin no. 1 internal, exact, projected model, and Padé model.

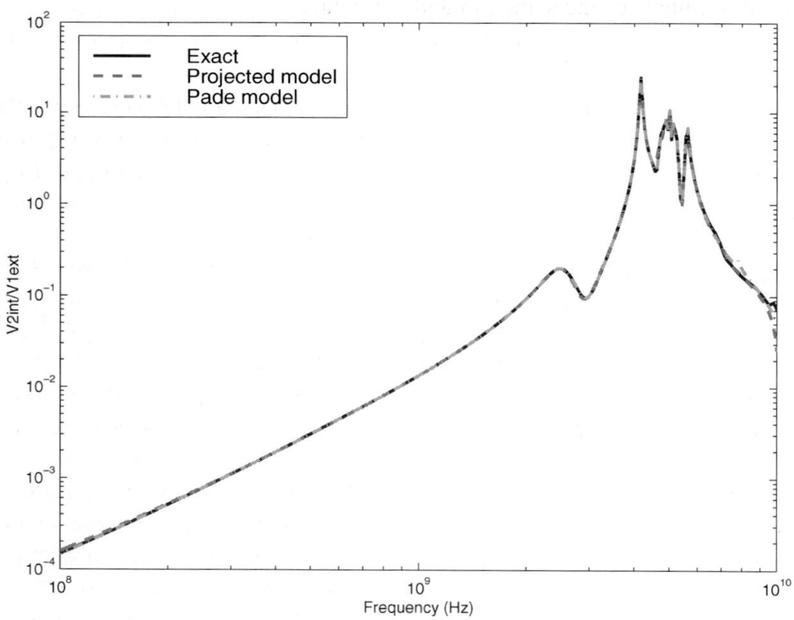

FIG. 3.2. Package: Pin no. 1 external to Pin no. 2 internal, exact, projected model, and Padé model.

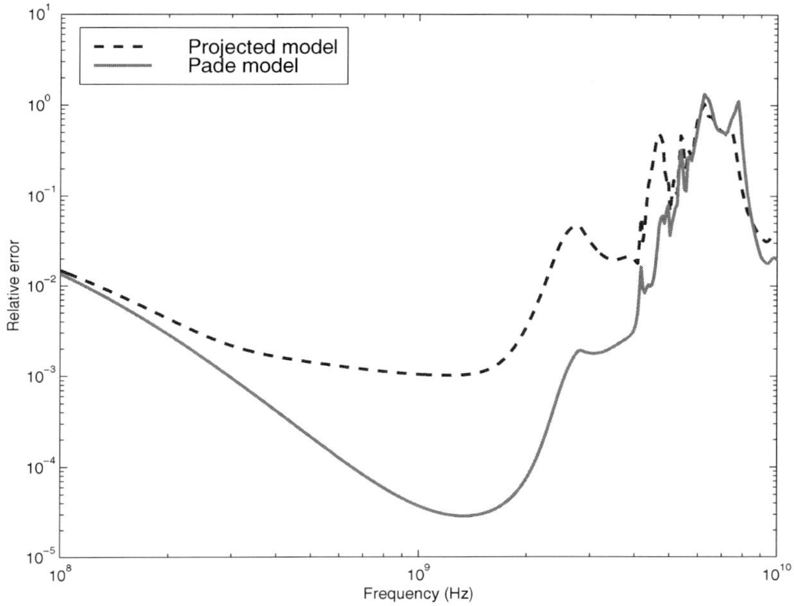

FIG. 3.3. Relative error of projected model and Padé model.

the voltage-to-voltage transfer function between the external terminal of Pin no. 1 and the internal terminals of the same pin and the neighboring Pin no. 2, respectively. The plots show results with the projected model $H_n^{(2)}$ and the Padé model $H_n^{(2)}$, both of order $n = 112$, compared with an exact analysis.

In Fig. 3.3, we compare the relative error of the projected model $H_{112}^{(2)}$ and the Padé model $H_{112}^{(1)}$ of the same size. Clearly, the Padé model is more accurate. However, out of the 112 poles of $H_{112}^{(1)}$, 22 have positive real parts, violating the passivity of the Padé model. On the other hand, the projected model is passive.

3.2.4. An extracted RC circuit

This is an extracted RC circuit with about 4000 elements and $m = 20$ ports. The expansion point $s_0 = 0$ was used. Since the projected model and the Padé model are identical for RC circuits, we only computed the Padé model via SyMPVL.

The point of this example is to illustrate the usefulness of the deflation procedure built into SyMPVL. It turned out that sweeps through the first two Krylov blocks, R and MR, of the block Krylov matrix (3.5) were sufficient to obtain a reduced-order model that matches the transfer function in the frequency range of interest. During the sweep through the second block, 6 almost linearly dependent vectors were discovered and deflated. As a result, the reduced-order model obtained with deflation is only of size $n = 2m - 6 = 34$. When SyMPVL was rerun on this example, with deflation turned off, a reduced-order model of size $n = 40$ was needed to match the transfer function. In Fig. 3.4, we show the $H_{1,11}$ component of the reduced-order model obtained with de-

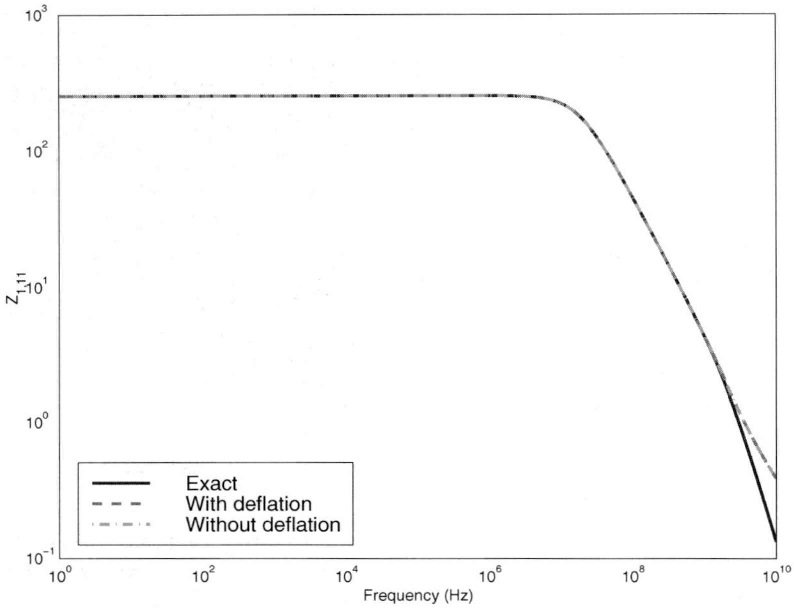

FIG. 3.4. Impedance $H_{1,11}$.

flation and without deflation, compared to the exact transfer function. Clearly, deflation leads to a significantly smaller reduced-order model that is as accurate as the bigger one generated without deflation.

3.3. Approaches based on the Arnoldi process

The Arnoldi process (ARNOLDI [1951]) is another widely-used Krylov-subspace method. A band version of the Arnoldi process that is suitable for multiple starting vectors can also be used for reduced-order modeling. However, the models generated from the band Arnoldi process are only Padé-type models.

In contrast to the band Lanczos algorithm, the band Arnoldi process only involves one of the starting blocks, namely R, and it only uses matrix–vector products with M. Moreover, the band Arnoldi process only generates one set of vectors, $v_1, v_2, \ldots, v_n$, instead of the two sequences of right and left vectors produced by the band Lanczos algorithm. The Arnoldi vectors span the nth right block Krylov subspace (induced by M and R):

$$\operatorname{span}\{v_1, v_2, \ldots, v_n\} = \mathcal{K}_n(M, R).$$

The Arnoldi vectors are constructed to be orthonormal:

$$V_n^H V_n = I, \quad \text{where } V_n := [\, v_1 \quad v_2 \quad \cdots \quad v_n \,].$$

After n iterations, the Arnoldi process has generated the first n Arnoldi vectors, namely the n columns of the matrix V_n, as well as an $n \times n$ matrix $G_n^{(\mathrm{pr})}$ of recurrence coefficients, and, provided that $n \geqslant m$, an $n \times m$ matrix $\rho_n^{(\mathrm{pr})}$. The matrices $G_n^{(\mathrm{pr})}$ and $\rho_n^{(\mathrm{pr})}$ are projections of the matrices M and R onto the subspace spanned by the columns of V_n, which is just the block Krylov subspace $\mathcal{K}_n(M, R)$. More precisely, we have

$$G_n^{(\mathrm{pr})} = V_n^H M V_n \quad \text{and} \quad \rho_n^{(\mathrm{pr})} = V_n^H R. \tag{3.16}$$

The band Arnoldi process can be stated as follows.

ALGORITHM 3.2 (*Band Arnoldi process*).

INPUT: A matrix $M \in \mathbb{C}^{n \times n}$;
A block of m right starting vectors $R = [\, r_1 \quad r_2 \quad \cdots \quad r_m \,] \in \mathbb{C}^{n \times m}$.

OUTPUT: The $n \times n$ Arnoldi matrix $G_n^{(\mathrm{pr})}$.
The matrix $V_n = [\, v_1 \quad v_2 \quad \cdots \quad v_n \,]$ containing the first n Arnoldi vectors, and the matrix $\rho_n^{(\mathrm{pr})}$.

(0) For $k = 1, 2, \ldots, m$, set $\hat{v}_k = r_k$.
Set $m_c = m$ and $\mathcal{I} = \emptyset$.

For $n = 1, 2, \ldots,$ until convergence or $m_c = 0$ do:

(1) (If necessary, deflate $\hat{v}_n$.)
Compute $\|\hat{v}_n\|_2$.
Decide if $\hat{v}_n$ should be deflated. If yes, do the following:
(a) Set $\hat{v}_{n-m_c}^{\mathrm{defl}} = \hat{v}_n$ and store this vector. Set $\mathcal{I} = \mathcal{I} \cup \{n - m_c\}$.
(b) Set $m_c = m_c - 1$. If $m_c = 0$, set $n = n - 1$ and stop.
(c) For $k = n, n+1, \ldots, n + m_c - 1$, set $\hat{v}_k = \hat{v}_{k+1}$.
(d) Repeat all of Step (1).

(2) (Normalize $\hat{v}_n$ to obtain v_n.)
Set

$$g_{n,n-m_c} = \|\hat{v}_n\|_2 \quad \text{and} \quad v_n = \frac{\hat{v}_n}{g_{n,n-m_c}}.$$

(3) (Orthogonalize the candidate vectors against v_n.)
For $k = n+1, n+2, \ldots, n + m_c - 1$, set

$$g_{n,k-m_c} = v_n^H \hat{v}_k \quad \text{and} \quad \hat{v}_k = \hat{v}_k - v_n\, g_{n,k-m_c}.$$

(4) (Advance the block Krylov subspace to get $\hat{v}_{n+m_c}$.)
(a) Set $\hat{v}_{n+m_c} = M v_n$.
(b) For $k = 1, 2, \ldots, n$, set

$$g_{k,n} = v_k^H \hat{v}_{n+m_c} \quad \text{and} \quad \hat{v}_{n+m_c} = \hat{v}_{n+m_c} - v_k\, g_{k,n}.$$

(5) (a) For $k \in \mathcal{I}$, set $g_{n,k} = v_n^H \hat{v}_k^{\mathrm{defl}}$.
(b) Set

$$G_n^{(\mathrm{pr})} = [\, g_{i,k} \,]_{i,k=1,2,\ldots,n},$$
$$\rho_n^{(\mathrm{pr})} = [\, g_{i,k-m} \,]_{i=1,2,\ldots,n;\, k=1,2,\ldots,k_\rho},$$
where $k_\rho = m + \min\{0, n - m_c\}$.

(6) Check if n is large enough. If yes, stop.

Note that, in contrast to the band Lanczos algorithm, the band Arnoldi process requires the storage of all previously computed Arnoldi vectors.

Like the band Lanczos algorithm, the band Arnoldi process can also be employed to reduced-order modeling. Let M, R, and L be the matrices defined in (2.15). After running Algorithm 3.2 (applied to M and R) for n steps, we have obtained the matrices $G_n^{(\mathrm{pr})}$ and $\rho_n^{(\mathrm{pr})}$, as well as the matrix V_n of Arnoldi vectors. The transfer function H_n of a reduced-order model H_n can now be defined as follows:

$$H_n(s) = \left(V_n^H L\right)^H \left(I - (s - s_0) V_n^H M V_n\right)^{-1} \left(V_n^H R\right).$$

Using the relations (3.16) for $G_n^{(\mathrm{pr})}$ and $\rho_n^{(\mathrm{pr})}$, the formula for H_n reduces to

$$H_n(s) = \left(V_n^H L\right)^H \left(I - (s - s_0) G_n^{(\mathrm{pr})}\right)^{-1} \rho_n^{(\mathrm{pr})}. \tag{3.17}$$

The matrices $G_n^{(\mathrm{pr})}$ and $\rho_n^{(\mathrm{pr})}$ are directly available from Algorithm 3.2. In addition, one also needs to compute the matrix

$$\eta_n^{(\mathrm{pr})} = V_n^H L.$$

It turns out that the transfer function (3.17) defines a matrix-Padé-type reduced-order model.

THEOREM 3.2 (Matrix-Padé-type model (FREUND [2000b], ODABASIOGLU [1996])). *Suppose that Algorithm 3.2 is run with exact deflation only and that $n \geqslant m$. Then, the reduced-order model associated with the reduced-order transfer function* (3.17) *is a matrix-Padé-type model of the linear dynamical system* (2.1) *and* (2.2). *More precisely, the Taylor expansions about s_0 of the transfer functions, H,* (2.8) *and, H_n,* (3.17) *agree in at least*

$$q'(n) \geqslant \left\lfloor \frac{n}{m} \right\rfloor$$

leading coefficients:

$$H(s) = H_n(s) + \mathcal{O}\left((s - s_0)^{q'(n)}\right). \tag{3.18}$$

REMARK 3.4. The number $q'(n)$ is the exact number of terms matched in the expansion (3.18) provided that no exact deflations occur in Algorithm 3.2. In the case of exact deflations, the number of matching terms is somewhat higher, but so is the number of matching terms for the matrix-Padé model of Theorem 3.1; see FREUND [2000b]. In particular, the matrix-Padé model is always more accurate than the matrix-Padé-type model obtained from Algorithm 3.2. On the other hand, the band Arnoldi process is certainly simpler than the band Lanczos process. Furthermore, the true orthogonality of the Arnoldi vectors in general results in better numerical behavior than the bi-orthogonality of the Lanczos vectors.

REMARK 3.5. For the special case of RCL subcircuits, the algorithm PRIMA proposed in ODABASIOGLU [1996], ODABASIOGLU, CELIK and PILEGGI [1997] can be interpreted as a special case of the Arnoldi reduced-order modeling procedure described here. Furthermore, in FREUND [1999a], FREUND [2000b] it is shown that the reduced-order model produced by PRIMA is mathematically equivalent to the additional passive model produced by SyMPVL. In contrast to PRIMA, however, SyMPVL also produces a true matrix-Padé model, and thus PRIMA does not appear to have any real advantage over or be even competitive with SyMPVL.

REMARK 3.6. An improved variant of PRIMA is the SPRIM reduction algorithm, which was recently proposed by FREUND [2004c]. While PRIMA generates provably passive reduced-order models, it does not preserve other structures, such as reciprocity or the block structure of the circuit matrices, inherent to RCL circuits. This has motivated the development of algorithms such as ENOR (SHEEHAN [1999]) and its variants (CHEN, LUK and CHEN [2003]) that generate passive and reciprocal reduced-order models, yet still match as many moments as PRIMA. However, the moment-matching property of the PRIMA models is not optimal. SPRIM overcomes these disadvantages of PRIMA. In particular, SPRIM generates provably passive and reciprocal macromodels of multi-port RCL circuits, and the SPRIM models match twice as many moments as the corresponding PRIMA models obtained with identical computational work. For a detailed description of SPRIM and its properties, we refer the reader to FREUND [2004c]. Here, we only present one example, which is taken from FREUND [2004c]. The example is a circuit resulting from the so-called PEEC discretization (RUEHLI [1974]) of an electromagnetic problem. The circuit is an RCL network consisting of

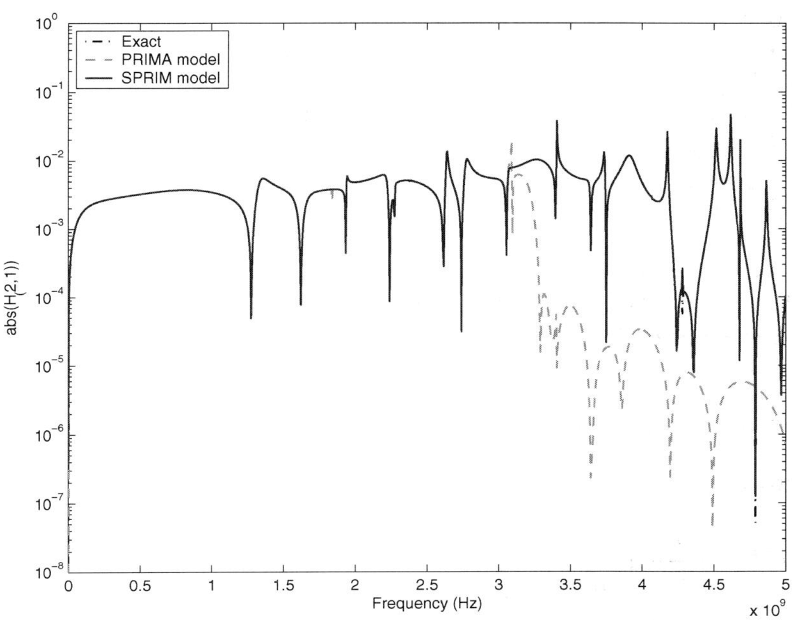

FIG. 3.5. $|H_{2,1}|$ for PEEC circuit.

2100 capacitors, 172 inductors, 6990 inductive couplings, and a single resistive source that drives the circuit. The circuit is formulated as a 2-port. We compare the PRIMA and SPRIM models corresponding to the same dimension n of the underlying block Krylov subspace. The expansion point $s_0 = 2\pi \times 10^9$ was used. In Fig. 3.5, we plot the absolute value of the (2, 1) component, $H_{2,1}$, of the 2×2-matrix-valued transfer function over the frequency range of interest. The dimension $n = 120$ was sufficient for SPRIM to match the exact transfer function. The corresponding PRIMA model of the same dimension, however, has not yet converged to the exact transfer function in large parts of the frequency range of interest. Fig. 3.5 clearly illustrates the better approximation properties of SPRIM due to the matching of twice as many moments as PRIMA.

4. Schur interpolation

4.1. The setting

The modeling of physical effects often produces large, positive definite Hermitian matrices. For example, the modeling of interconnects in an integrated circuit produces in first instance a full elastance matrix G from which a sparse approximating capacitance matrix C has to be derived. Likewise, the behavior of the substrate of an integrated circuit is modeled by a conductivity matrix, and the inductive behavior of the interconnects by an inductance matrix. These matrices are positive definite, because they express either conservation of energy or dissipation. It is a non-trivial problem to find low-complexity approximations to a positive definite matrix, which are positive definite in their own right. For example, if $G = [G_{i,j}]$ is positive definite, then the matrix G_a obtained by putting elements outside a given band equal to zero, i.e., $(G_a)_{i,j} = G_{i,j}$ for $|i - j| < n$ some n, and zero otherwise, will not necessarily be positive definite. If a matrix is diagonally dominant, then putting some off-diagonal elements equal to zero while keeping the Hermitian property would preserve the dominance and hence also the positive definiteness. We shall analyze some of the properties of such schemes soon. An important observation is that properties such as "banded" and "diagonally dominant" are not preserved under inversion: the inverse of a banded matrix is not banded (except when the matrix is block diagonal) and the inverse of a diagonally dominant matrix is not diagonally dominant. Consider for example the matrix (for real a)

$$M_a = \begin{bmatrix} 1 & a & a^2 \\ a & 1 & a \\ a^2 & a & 1 \end{bmatrix}.$$

It is positive definite for $|a| < 1$ with inverse

$$M_a^{-1} = \frac{1}{1-a^2} \begin{bmatrix} 1 & -a & 0 \\ -a & 1+a^2 & -a \\ 0 & -a & 1 \end{bmatrix}.$$

If we truncate M_a by putting $(M_a)_{1,3} = (M_a)_{3,1} = 0$, then the resulting matrix will be positive definite only when in addition $a \leqslant 1/\sqrt{2}$. We see that the inverse of M_a is diagonally dominant for $|a| < 1$ while that is only the case for M_a when $a < (\sqrt{5} - 1)/2$. So, why would it be better to truncate a matrix rather than its inverse? A related issue is whether the inverse of a banded matrix has the same computational complexity as the original. Further in this section we shall develop a nice theory that is capable of answering such questions.

Another approach would be to perform the approximation on a Cholesky factor R where $G = R^H R$, R is upper triangular and R^H represents the Hermitian conjugate of R, rather than on the original matrix. Assuming that the off-diagonal elements of R become small the farther they are located from the main diagonal, it makes sense to approximate R by a banded matrix. Also, approximating R by some approximant R_a will produce automatically an approximant $G_a = R_a^H R_a$ that is positive definite. At first sight it would appear that it is not any better to approximate the square root than the original – an ε relative error on the square root of a scalar quantity would roughly produce a 2ε error on the square. The situation with matrices is, however, vastly different, since the condition number of the square root of a (positive definite) matrix, or of its Cholesky factor is just the square root of the original. Still the question arises whether a direct, element-wise approximation of the square root would be a "good" approximation technique, in the sense of either strong norms or complexity? What we need is a theory to gauge both complexity and approximation error. In addition, we would like the approximation procedure to be as simple as possible, for example, it should use a minimal amount of computations in its own right.

We start out this section with the celebrated theory of maximum-entropy interpolation of positive definite matrices. It gives a good stronghold on low-complexity approximation when "low-complexity" is understood as minimizing the number of independent algebraic parameters, e.g., by putting a sufficient number of elements in the matrix or its inverse zero. Immediately the question arises when the sparsity pattern of a positive definite matrix is preserved in its Cholesky factors. This question also has a very neat answer, namely when the matrix entries exhibit a "chordal pattern". In that case, the maximum-entropy interpolant can be found directly, in a minimal number of computations equal to the number of non-zero entries in the matrix, by a matrix interpolation algorithm that is a matrix version of the celebrated Schur interpolation algorithm of complex function analysis. The approximating properties of Schur's algorithm are known and we shall spend a few words explaining them. Finally, we shall show ways of generalizing Schur's algorithm to a more complex situation, namely the so-called "multiple band case".

4.2. Maximum-entropy interpolation of strictly positive definite matrices

Suppose that the following information on an otherwise unknown strictly positive definite (and of course Hermitian) matrix G of size $N \times N$ is given:

- The diagonal elements $G_{k,k}$ for all $k = 1, 2, \ldots, N$;
- Some off-diagonal elements, characterized by a set $\mathcal{S}$: if $(i, j) \in \mathcal{S}$ then $G_{i,j}$ is known. Since G is Hermitian, we restrict elements of $\mathcal{S}$ to be in the strictly upper triangular zone where $i < j$.

This information is known as "interpolating conditions". The question we ask is: *is it possible to find a positive definite matrix G_a which has the assigned element values on the main diagonal and the set $\mathcal{S}$, and is otherwise in some sense "of minimal complexity"?*

It turns out that this question has a nice definite answer if "complexity" here is understood to mean: "the value of the off-diagonal elements $(G^{-1})_{i,j}$ is zero for (i, j) not in $\mathcal{S}$". A comfortable treatment of the theory leading to this result requires the introduction of the notion of "entropy of a strictly positive matrix H", originating from stochastic system theory and which is given by the (finite) quantity:

$$\mathcal{E}(H) = \log \det H.$$

The following theorem is valid.

THEOREM 4.1. *Suppose that the diagonal elements $G_{k,k}$ and some off-diagonal elements belonging to an off-diagonal set of indices $\mathcal{S}$ of a strictly positive definite matrix G are given. Then, there exists a unique strictly positive definite matrix G_a such that G_a interpolates the given entries, i.e., $(G_a)_{i,j} = G_{i,j}$ for $i = j$ and $(i, j) \in \mathcal{S}$, and which is such that $(G_a^{-1})_{i,j} = 0$ for (i, j) not in $\mathcal{S}$. This G_a also maximizes the entropy $\mathcal{E}(H) = \log \det H$ over all H that meet the interpolation conditions.*

SKETCH OF PROOF. Suppose that H is a strictly positive definite matrix depending on some parameter ξ. The differential of the entropy with respect to ξ is then given by

$$\frac{\partial}{\partial \xi} \log \det H = \frac{1}{\det H} \frac{\partial \det H}{\partial \xi}.$$

Let us observe that the dependency of $\det H$ on a given entry $H_{i,j}$ can be expressed using the Cramer minor expansion based on the row i:

$$\det H = \sum_{k=1}^{N} H_{i,k} M_{i,k},$$

where $M_{i,k}$ is the minor corresponding to the element at the position (i, k). The minor $M_{i,k}$ does not depend on any element in the ith row of H, in particular it does not depend on $H_{i,j}$ – the determinant is linear in that element. Let now $\xi = G_{i,j}$ for some (i, j) not in $\mathcal{S}$, corresponding to the position of an element that must be determined. Since the $\log \det H$ surface is smooth over the space of parameters to be determined, an extremum will only occur if each possible ξ is chosen so that the variation of the entropy with respect to ξ is zero (or else at the border of feasibility, but that situation cannot lead to a maximum since the border corresponds to matrices whose determinant is zero). The variation for $\xi = G_{i,j}$ on G is now given by:

$$\frac{\partial}{\partial \xi} \log \det G = \frac{1}{\det G} \frac{\partial \det G}{\partial \xi} = \frac{M_{i,j}}{\det G} = (G^{-1})_{j,i}.$$

Hence the top of the entropy surface in the parameter space of the unknown entries of the matrix G_a, i.e., the entries not in $\mathcal{S}$, must correspond to a strictly positive definite extension G_a of G for which $(G_a^{-1})_{i,j} = 0$. The proof now terminates by showing that this top exists and is unique. This must be reasonable in view of the fact that there is a uniform upper bound on the entropy, namely

$$\sum_{k=1}^{N} \log G_{k,k}.$$

This bound can be obtained through recursive evaluation via Cholesky decomposition, and the fact that the interpolating set is convex, if H_1 and H_2 are strictly positive definite and interpolating, so is $kH_1 + (1-k)H_2$ for $0 \leqslant k \leqslant 1$. □

Hence the maximum-entropy extension of entries of a strictly positive definite matrix does exist, and it produces a sparse inverse matrix! This is already a very useful result for model reduction of, for example, capacitive models of IC interconnects, as we shall soon see. However, it is a theoretical result in that the proof of existence does not produce a direct algorithm to compute the result. One may resort to dynamic optimization, and, indeed, that should lead to a solution, but maybe a problematic one, first because it leads to complex computations involving all the elements outside the interpolating set, and second because the entropy surface is most likely very flat, making the optimum hard to find even though there are very good algorithms for convex optimization. Hence it pays to find a way of computing the solution directly on the basis of the known data, if possible. This question is related to the question whether a sparsity pattern in an original, strictly positive definite matrix G is preserved in the Cholesky factor L, where $G = LL^H$, a question which we now address.

4.3. Chordal systems

Assume that we are given a strictly positive definite matrix G whose diagonal elements are known and which is otherwise sparse with upper triangular sparsity pattern $\mathcal{S}$, i.e., $G_{i,j} = 0$ for (i, j) with $i < j$ not belonging to $\mathcal{S}$ (G is of course Hermitian). Connected to $\mathcal{S}$ there is a *sparsity graph* defined as follows:

- Nodes: there are N nodes corresponding to the N rows of the matrix;
- Edges: there is an edge between node i and node j iff $(i, j) \in \mathcal{S}$, assuming $i < j$.

For example, a matrix with fillings

$$\begin{bmatrix} * & * & \cdot & * & \cdot \\ * & * & * & \cdot & * \\ \cdot & * & * & * & \cdot \\ * & \cdot & * & * & * \\ \cdot & * & \cdot & * & * \end{bmatrix} \tag{4.1}$$

has the sparsity graph shown in Fig. 4.1.

We say that a sparsity graph is *chordal* when there is no loop of more than three nodes that has no chord in the graph, a chord being a direct connection between two nodes (with reference to a polygone). The graph shown in Fig. 4.1 is non-chordal, the loop 1-2-3-4-1 has no chords (and there are more such loops). It turns out that the Cholesky

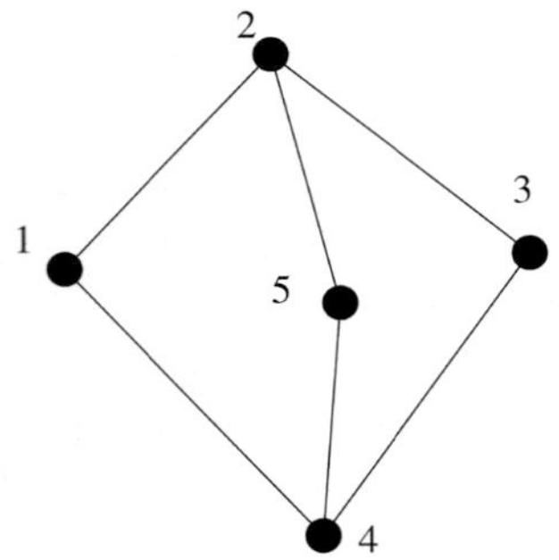

FIG. 4.1. Sparsity graph of the matrix template (4.1).

factorization of a positive definite matrix with chordal sparsity graph will suffer no fill-ins provided it is executed in the right order. To find that order we need another property of chordal graphs.

We shall say that a node of a graph has an *adjacent clique* if the subgraph consisting of that node and the nodes directly connected to it together with the edges connecting these nodes form a clique, i.e., are fully connected. A chordal graph now has the two following properties:

- The graph obtained by deleting one node with the edges connected to it is chordal;
- It has at least one node which has an adjacent clique.

The first property is almost evident, while the second property can be proven recursively on the number of nodes. Hence, a reordering and peeling off of the nodes of a chordal graph is possible whereby each node in turn has an adjacent clique in the remaining graph: start with such a node in the original graph, remove it with its connecting edges and continue recursively. Finding a node with an adjacent clique can be done in less than N^2 steps, hence the complexity of the reordering is certainly polynomial in N.

With this reordering of nodes, performing the Cholesky factorization in the order of peeling will not produce any fill-ins, exactly because of the adjacent clique property at each step. The converse is "generically" true as well, if a Cholesky factorization does not result in fill-ins *generically* (an element might accidentally become zero), then the sparsity graph must be chordal as well. It turns out that the maximum-entropy interpolant of a matrix with chordal sparsity pattern can be computed directly on the given entries, the famous algorithm to do so is the generalized Schur algorithm described in the next subsection. Unfortunately, many problems in modeling or reduced modeling of integrated circuits involve strictly positive definite matrices that do not have chordal sparsity patterns. In particular, multiband patterns are almost essentially non chordal and hence will need additional, non-exact techniques for reduced modeling. This question is treated in the section on multiband generalization. A special case of a chordal graph is a graph representing a staircase filling, i.e., a filling corresponding to a non-regular band. One would obtain such a graph if in the order of nodes with adjacent cliques, each node in turn belongs to the adjacency set of its predecessor.

4.4. *Schur's algorithm in the chordal case*

We are now ready to introduce the generalized matrix Schur algorithm, originally presented as an estimation algorithm in DEPRETTERE [1981], and whose matrix properties were analyzed in DEWILDE and DEPRETTERE [1987]. The application of the algorithm to reduced modeling of integrated circuits was given in DEWILDE [1988]. We utilize the algebraic framework of the latter paper, slightly generalizing it to cover chordal sparsity in addition to staircases. Let the original, $N \times N$ strictly positive definite matrix be $G = [G_{i,j}]$ and let D be its main diagonal:

$$D = \text{diag}(G_{1,1}, G_{2,2}, \dots, G_{N,N}).$$

It is advantageous to work with a normalized version of G, for theoretical purposes if not for numerical ones. Hence, let

$$g = D^{-1/2} G D^{-1/2}.$$

The matrix g will have all its diagonal elements equal to one (the situation could be generalized to the case where all the entries in G are in fact matrices, the block case, but for simplicity of explanation we keep the procedures scalar and shall indicate later on how to handle the block-matrix case). Let us assume, moreover, that the nodes are put in a correct adjacent-clique order, the staircase order will do if available.

4.4.1. *A side excursion: the classical Schur parametrization case*

Before engaging in the description of the matrix Schur algorithm, let us make a brief side excursion to the original algorithm involved in Schur's parametrization of a contractive, analytic function on the unit disc $\mathbf{D} = \{z\colon |z| < 1\}$ of the complex plane. Suppose that

$$s(z) = s_0 + s_1 z + s_2 z^2 + \cdots$$

is such a function, represented by its MacLaurin series. The question answered by the Schur parametrization is whether the given MacLaurin series does indeed correspond to a contractive function. To start, either $|s_0| = 1$ and $s(z)$ reduces to a constant of modulus one (by the maximum modulus theorem of complex analysis), or $|s_0| < 1$ and then a new contractive function which is analytic in $\mathbf{D}$ may be derived from $s(z)$ via the recipe:

$$s^{(1)}(z) = \frac{s(z) - s_0}{z(1 - \overline{s_0} s(z))} = s_0^{(1)} + s_1^{(1)} z + \cdots.$$

Notice that the transformation

$$s \mapsto \frac{s - s_0}{1 - \overline{s_0} s}$$

maps the unit disc onto itself. The procedure may be repeated on $s^{(1)}(z)$, yielding a criterion on $s_0^{(0)}$ and a new $s^{(2)}(z)$, and then recursively continued further. Let $\rho_0 = s_0$, $\rho_1 = s_0^{(1)}, \dots$ be the so-called "Schur parameters" for $s(z)$. In an inverse scattering context where they often appear, the ρ_k's are also called reflection coefficients. The sequence of Schur parameters of a contractive function that is analytic in $\mathbf{D}$ is either finite, in which case the last coefficient is of unit modulus, or infinite, and then all Schur parameters are less than one in modulus. The Schur parameters determine $s(z)$ uniquely, just as the s_k's do, one series can be converted into the other and vice versa. In his famous 1917 paper, SCHUR [1917] demonstrates that $s(z)$ is contractive in the unit disc iff the Schur parametrization satisfies one of these two properties – this is the Schur criterion for contractivity (the proof is in fact pretty straightforward). The transformation that leads from $s^{(k)}(z)$ to $s^{(k+1)}(z)$ is obviously bilinear. It can be linearized if it is put in matrix form. Let us write for that purpose

$$s^{(n)}(z) = \frac{\delta^{(n)}(z)}{\gamma^{(n)}(z)}.$$

Then the following linear recursion produces the same effect as the original Schur parametrization

$$z\left[\gamma^{(n+1)}(z)\ \delta^{(n+1)}(z)\right] = \left[\gamma^{(n)}(z)\ \delta^{(n)}(z)\right]\frac{1}{\sqrt{1-|\rho_n|^2}}\begin{bmatrix} z & -\rho_n \\ -\overline{\rho_n} & 1 \end{bmatrix}$$

when the Schur parameter chosen as

$$\rho_n = \frac{\delta^{(n)}(0)}{\gamma^{(n)}(0)}$$

is less than one in magnitude (the square roots are included for normalization purposes, they may be dispensed with in practical computations). The recursion is started with $[\gamma^{(0)}(z)\ \delta^{(0)}(z)] = [1\ s(z)]$. Aside from a shift represented by z, the Schur recursion involves transformations with a hyperbolic matrix, sometimes called a Halmos transformation and defined as

$$H(\rho) = \frac{1}{\sqrt{1-|\rho|^2}}\begin{bmatrix} 1 & -\rho \\ -\overline{\rho} & 1 \end{bmatrix}.$$

Let us define the signature matrix

$$J = \begin{bmatrix} 1 & \\ & -1 \end{bmatrix}.$$

Then, we compute easily that $H(\rho)J(H(\rho))^H = (H(\rho))^H J H(\rho) = J$, which represents the hyperbolic property.

The original Schur theory works on a contractive function $s(z)$. Alternatively, one could start from what is known as a *positive real function* $\phi(z)$, i.e., a function that is analytic in $\mathbf{D}$ and such that $\mathrm{Re}(\phi(z)) = (\phi(z) + \overline{\phi(z)})/2 \geqslant 0$ in $\mathbf{D}$. The Cayley transformation relates a contractive function $s(z)$ to a *positive real function* $\phi(z)$ (i.e., a function with positive real part $\mathrm{Re}(\phi(z))$ for all z in the unit disc):

$$s(z) = \frac{\phi(z) - 1}{\phi(z) + 1}.$$

Schur's parametrization provides a test for positive reality on the sequence defined by the MacLaurin expansion of ϕ, the linearized recursion can now be started with

$$\left[\gamma^{(0)}(z)\ \delta^{(0)}(z)\right] = \tfrac{1}{2}\left[\phi(z)+1\ \phi(z)-1\right].$$

After $n+1$ steps it will yield

$$\tfrac{1}{2}\left[\phi(z)+1\ \phi(z)-1\right]\theta_0(z)\theta_2(z)\cdots\theta_n(z) = z^n\left[\gamma^{(n)}(z)\ \delta^{(n)}(z)\right]$$

with each $\theta_i(z)$ representing an elementary Schur step. Let us introduce the para-Hermitian conjugate of a function of z as $f^*(z) = \overline{f(1/\overline{z})}$. In the Schur parametrization theory (see, e.g., DEWILDE, VIEIRA and KAILATH [1978]), one deduces that the overall Schur matrix $\Theta_n(z) = \theta_0(z)\theta_1(z)\cdots\theta_n(z)$ has the form

$$\Theta_n(z) = \frac{1}{2}\begin{bmatrix} (1+\phi_n^*(z))T_{rn}^{-*}(z) & (1-\phi_n(z))T_{fn}^{-1}(z) \\ (1-\phi_n^*(z))T_{rn}^{-*}(z) & (1-\phi_n(z))T_{fn}^{-1}(z) \end{bmatrix}$$

in which $\phi_n(z)$ is also PR in $\mathbf{D}$, $T_{rn}(z)$ and $T_{fn}(z)$ are analytic in $\mathbf{D}$ and

$$\frac{\phi_n(z)+\phi_n^*(z)}{2} = T_{rn}(z)T_{rn}^*(z) = T_{fn}^*(z)T_{fn}(z).$$

(Notice that the para-Hermitian conjugate is equal to the Hermitian conjugate only on the unit circle.) Outside the unit circle it is its analytic continuation, when definable. Often in the engineering literature, the para-Hermitian conjugate is denoted by a sub-star, in contrast to the upper star, which is often interpreted as equal to complex conjugation. Here we use upper star, to indicate that the upper-stared quantity corresponds in fact to the analytic continuation of the adjoint in the Fourier domain on the unit circle). One of the central properties of $\phi_n(z)$, resulting from the Schur parametrization, is that it interpolates the original $\phi(z)$ to order n:

$$\phi(z) = \phi_n(z) + z^{n+1}r(z)$$

in which $r(z)$ is analytic in $\mathbf{D}$. Remark also that $\phi_n^{-1}(z)$ is polynomial hence $\phi_n(z)$ is of the "autoregressive type". The theory of maximum entropy interpolation is well developed in complex function theory, and it is satisfied by $\phi_n(z)$ as a maximum entropy interpolant of order n for $\phi(z)$, whereby the entropy measure now must be taken as

$$\int_{-\pi}^{\pi} \log \mathrm{Re}\big(\phi(\xi)\big)\frac{\mathrm{d}\xi}{2\pi}.$$

4.4.2. The matrix case

In the matrix case, the hyperbolic transformation will play a role similar to the complex case. We embed the Halmos transformation in an otherwise unitary matrix and index its position, much as is done in the classical QR algorithm based on Jacobi transformations.

This leads to $2N \times 2N$ hyperbolic matrices $\theta_{i,j}(\rho)$ of the form

$$
\left[\begin{array}{c|c}
\begin{array}{cccccccccc}
1 &&&&&&&&& \\
& \ddots &&&&&&&& \\
&& 1 &&&&&&& \\
&&& \frac{1}{\sqrt{1-|\rho|^2}} & 0 \dots &&&&& \\
&&& 0 & 1 &&&&& \\
&&& \vdots && \ddots &&&& \\
&&&&&& 1 &&& \\
&&&&&&& 1 && \\
&&&&&&&& 1 & \\
&&&&&&&&& \ddots \\
&&&&&&&&&& 1
\end{array}
&
\begin{array}{cccccccccc}
0 &&&&&&&&& \\
& \ddots &&&&&&&& \\
&& 0 &&&&&&& \\
&&& 0 & \cdots 0 & -\frac{\rho}{\sqrt{1-|\rho|^2}} &&&& \\
&&&& 0 & 0 &&&& \\
&&&&& \ddots \quad \vdots &&&& \\
&&&&& 0 &&&& \\
&&&&&& 0 &&& \\
&&&&&&& 0 && \\
&&&&&&&& \ddots & \\
&&&&&&&&& 0
\end{array}
\\ \hline
\begin{array}{cccccccccc}
0 &&&&&&&&& \\
& \ddots &&&&&&&& \\
&& 0 &&&&&&& \\
&&& 0 &&&&&& \\
&&&& 0 &&&&& \\
&&& \vdots && \ddots &&&& \\
&&& 0 && 0 &&&& \\
&&& -\frac{\rho^*}{\sqrt{1-|\rho|^2}} & 0 \cdots 0 & 0 &&&& \\
&&&&&& 0 &&& \\
&&&&&&& \ddots && \\
&&&&&&&& 0 &
\end{array}
&
\begin{array}{cccccccccc}
1 &&&&&&&&& \\
& \ddots &&&&&&&& \\
&& 1 &&&&&&& \\
&&& 1 &&&&&& \\
&&&& 1 &&&&& \\
&&&&& \ddots & \vdots &&& \\
&&&&& 1 & 0 &&& \\
&&&&& \cdots 0 & \frac{1}{\sqrt{1-|\rho|^2}} &&& \\
&&&&&&& 1 && \\
&&&&&&&& \ddots & \\
&&&&&&&&& 1
\end{array}
\end{array}\right]
$$

in which the elements of the elementary hyperbolic transformation are on the intersections of rows i and $N + j$ with columns i and $N + j$. Similar as in the complex case we will use it to eliminate entries in an appropriate linearization. We give the analog for Schur's algorithm in the matrix case first before motivating it. This simplifies the discussion, but an alert reader will recognize the similarities with the complex case. First we define the equivalent of the function $\phi(z)$ (as announced earlier we work on a normalized version of G, although this is not strictly necessary):

$$
\Phi = \begin{bmatrix}
1 & 2g_{1,2} & 2g_{1,3} & \cdots & 2g_{1,N} \\
0 & 1 & 2g_{2,3} & \cdots & 2g_{2,N} \\
0 & 0 & 1 & \cdots & 2g_{3,N} \\
\vdots & \vdots & \vdots & \ddots & \vdots \\
0 & 0 & 0 & \cdots & 1
\end{bmatrix}.
$$

Hence $g = \frac{1}{2}(\Phi + \Phi^H)$, where A^H indicates the Hermitian conjugate of the matrix A. We define as initial data

$$\Gamma_0 = \begin{bmatrix} 1 & g_{1,1} & \cdots & g_{1,N} \\ 0 & 1 & \cdots & g_{2,N} \\ \vdots & \vdots & \ddots & \vdots \\ 0 & 0 & \cdots & 1 \end{bmatrix}, \qquad \Delta_0 = \begin{bmatrix} 0 & g_{1,1} & \cdots & g_{1,N} \\ 0 & 0 & \cdots & g_{2,N} \\ \vdots & \vdots & \ddots & \vdots \\ 0 & 0 & \cdots & 0 \end{bmatrix}.$$

Hence we have

$$[\,\Gamma_0 \quad \Delta_0\,] = \tfrac{1}{2}[\,\Phi + I \quad \Phi - I\,].$$

Let $S_0 = \Gamma_0^{-1}\Delta_0$. We see that

$$g = \frac{\Phi + \Phi^H}{2} = \tfrac{1}{4}(\Phi + I)\big(I - S_0 S_0^H\big)\big(\Phi^H + I\big),$$

and hence S_0 is a contractive matrix in the sense that $S_0 S_0^H \preceq I$. We shall say that a couple of $N \times N$ upper triangular matrices $[\Gamma \;\; \Delta]$ are (*strictly*) *admissible* if Γ is invertible and $\Gamma^{-1}\Delta$ is (strictly) contractive. Define the $2N \times 2N$ signature matrix

$$J = \begin{bmatrix} I_N & \\ & -I_N \end{bmatrix},$$

where I_N is the unit matrix of dimension N. If Θ is a $2N \times 2N$ is a J-unitary matrix, i.e.,

$$\Theta J \Theta^H = \Theta^H J \Theta = J,$$

then any transformation of an admissible $[\Gamma \;\; \Delta]$ on the right with Θ will yield a new matrix

$$[\,\Gamma' \quad \Delta'\,] = [\,\Gamma \quad \Delta\,]\Theta,$$

which is (strictly) admissible when the original is (strictly) admissible. A product of J-unitary matrices will itself be J-unitary as well.

The Schur elimination procedure based on the chordal set $\mathcal{S}$ will consist in applying a sequence of elementary Halmos transformations on recursively computed admissible matrices, starting with $[\Gamma_0 \;\; \Delta_0]$, in the adjacent-clique order on the interpolation data. Each Halmos transformation is intended to eliminate one off-diagonal entry corresponding to a position in the set $\mathcal{S}$. Let the matrices G, g, Γ_0, Δ_0 be ordered in the adjacent-clique order, and suppose that the elements of $\mathcal{S}$ in row i are given by $(i, n_{i,1}), (i, n_{i,2}), \ldots, (i, n_{i,m_i})$ where $i < n_{i,1} < \cdots < n_{i,m_i}$ (the set may even be empty of course). We shall perform the elimination procedure in the strict order $(1, n_{1,1}), (1, n_{1,2}), \ldots, (2, n_{2,1}), \ldots$. Let us number these steps by the integer K. At step K corresponding to, say, the predecessor of $(i, n_{i,k})$, we have available an admissible pair $[\Gamma_K \;\; \Delta_K]$, which is such that the elements $(\Delta_K)_{i,j}$ have been annihilated for all pairs (i, j)'s in the elimination list preceding $(i, n_{i,k})$. The new step will annihilate $(\Delta_K)_{i,n_{i,k}}$ and use for that purpose an elimination matrix of the Halmos type, namely

$H_{i,n_{i,k}}(\rho_{i,n_{i,k}})$ with

$$\rho_{i,n_{i,k}} = (\Gamma_K)^{-1}_{i,n_{i,k}}(\Delta_K)_{i,n_{i,k}}.$$

At least three remarks are important here:

- The element $(\Delta_{K+1})_{i,n_{i,k}}$ is set equal to zero by the elimination procedure;
- The elements that were put to zero in previous steps remain zero in all the subsequent eliminations because of the adjacent-clique order;
- There are no fill-ins, also due to the adjacent-clique property at each step.

After completion of all the steps, an overall elimination matrix Θ_t results given by

$$\Theta_t = \theta_{1,n_{1,1}}\theta_{1,n_{1,2}}\cdots\theta_{N,n_{N,m_N}}$$

and finally

$$[\,\Gamma_t \quad \Delta_t\,] = [\,\Gamma_0 \quad \Delta_0\,]\,\Theta_t$$

are obtained, in which all the elements belonging to the set $\mathcal{S}$ in Δ_t have been annihilated (as well as all the diagonal elements due to the initial normalization). In parallel, the entries in Θ_t are essentially constrained to the diagonal, the set $\mathcal{S}$ and its reflection. To make this statement more precise, let

$$\Theta_t = \begin{bmatrix} \Theta_{1,1} & \Theta_{1,2} \\ \Theta_{2,1} & \Theta_{2,2} \end{bmatrix}.$$

Then, the non-zero entries of $\Theta_{1,1}$ are restricted to diagonals and $\mathcal{S}^*$, those of $\Theta_{2,2}$ to diagonals and $\mathcal{S}$ while the non-zero entries of $\Theta_{1,2}$ are restricted to $\mathcal{S}$ and those of $\Theta_{2,1}$ are restricted to $\mathcal{S}^*$. This follows also from the special structure of $\mathcal{S}$ and the order in which the eliminations have been done. We shall call such a J-unitary matrix "$\mathcal{S}$-based". The following theorem from DEWILDE and DEPRETTERE [1987] holds.

THEOREM 4.2. *An $\mathcal{S}$-based J-unitary matrix Θ_t has the form*

$$\frac{1}{2}\begin{bmatrix} (I+\Phi_t^H)L_t^{-H} & (I-\Phi_t)M_t^{-1} \\ (I-\Phi_t^H)L_t^{-H} & (I+\Phi_t)M_t^{-1} \end{bmatrix}$$

in which Φ_t, L_t and M_t are upper triangular matrices, Φ_t has unit main diagonal, L_t and M_t are invertible, and in addition

$$\frac{\Phi_t+\Phi_t^H}{2} = L_t L_t^H = M_t^H M_t.$$

The Schur procedure executed as detailed above yields the following interpolation result.

THEOREM 4.3 (DEWILDE and DEPRETTERE [1987]). *Let $g_t = \frac{1}{2}(\Phi_t + \Phi_t^H)$ be the result of the Schur elimination procedure based on the chordal set $\mathcal{S}$. Then,*

$$(g - g_t)_{i,j} = 0 \quad \textit{for } (i,j) \in \mathcal{S}$$

and, in particular,

$$\Phi - \Phi_t = 2\Delta_t M_t,$$

where Δ_t *is defined by* $[\Gamma_0\ \Delta_0]\Theta_t = [\Gamma_t\ \Delta_t]$.

Given the theory developed so far, the two theorems are not too hard to prove. The Schur recursion necessitates a number of elementary Halmos transformations precisely equal to the number of elements in the interpolation set $\mathcal{S}$, and it produces the desired maximum-entropy interpolant, due to the fact that the appropriate entries in the inverse matrix are zero. Notice also that L_t^{-1} and M_t^{-1} have supports on $\mathcal{S}$ and the diagonal, while L_t, M_t and Φ_t are full matrices, which in practical calculations will never be computed – a banded computational scheme exists for vector–matrix multiplication with both L_t and L_t^{-1}, see DEWILDE and DEPRETTERE [1987].

4.5. *Generalizations*

The preceding theory works only for matrices with a chordal sparsity pattern. Can the theory be extended to more general types of matrices, in particular to matrices with multiple bands, as often occur in 2D or 3D finite element or finite difference problems. We give an indication on how an approximate technique may yield satisfactory results. We refer the reader to the literature (NELIS, DEWILDE and DEPRETTERE [1989]) for further information. A first remark is that in some, quite common cases, a double banded (or even multibanded) matrix can be chordal. For example, a $2n \times 2n$ matrix with four $n \times n$ blocks with filling pattern as in

$$\left[\begin{array}{cccc|cccc}
* & * & & & * & * & & \\
* & * & * & & * & * & * & \\
& * & * & * & & * & * & * \\
& & * & * & & & * & * \\
\hline
* & * & & & * & * & & \\
* & * & * & & * & * & * & \\
& * & * & * & & * & * & * \\
& & * & * & & & * & *
\end{array}\right]$$

is actually of chordal type and can be solved exactly using Schur matrix interpolation (more general forms can easily be derived using the theory of adjacent cliques described above). This result can be used to factorize more general matrices approximatively. For example, a (positive definite) block matrix of the type

$$\begin{bmatrix} A_{11} & A_{12} & \\ A_{21} & A_{22} & A_{23} \\ & A_{32} & A_{33} \end{bmatrix}$$

in which all the non-zero blocks are only sparsely specified and which is such that the two submatrices

$$A_1 := \begin{bmatrix} A_{11} & A_{12} \\ A_{21} & A_{22} \end{bmatrix}, \qquad A_2 := \begin{bmatrix} A_{22} & A_{23} \\ A_{32} & A_{33} \end{bmatrix}$$

have chordal filling specifications has a sparse approximant for its inverse which can be constructed from sparse approximants of A_1, A_2, and A_{22} as follows. Let A_{ME} indicate the maximum-entropy approximant of a sparsely specified matrix A, then A_{ME}^{-1} has corresponding sparse fillings according to the theory developed above. In addition, let us introduce one more bit of notation: by "$\Box A$" we mean the operation of fitting the matrix A in a larger matrix that extends its range of indices while padding it with zeros. A "good" approximant for the ME inverse of A is then given by

$$A_{ME}^{-1} \approx \Box(A_1)_{ME}^{-1} + \Box(A_2)_{ME}^{-1} - \Box(A_{22})_{ME}^{-1}. \tag{4.2}$$

The significance of this formula is that the inverse of the maximum-entropy interpolant for the matrix A based on the given non-chordal definition pattern is expressed in terms of maximum interpolants of submatrices whose definition pattern is presumably chordal and which can hence be computed by a fast algorithm such as the Schur parametrization given in the previous section. We give a short motivation for this result, a complete theory with proofs is given in NELIS [1989]. The main property used is the fact that for reasonably well-conditioned positive definite matrices with entries specified on a given pattern, the inverse of the ME approximant of a principal submatrix is actually a good approximation of the restriction of the inverse of the ME approximant to the same indices as the submatrix – in matrix notation, let $A(i, j)$ be the principal submatrix obtained by restraining A to the index range $i \cdots j$ then, utilizing the same pattern of specified entries,

$$\big(A(i, j)\big)_{ME}^{-1} \approx \big(A_{ME}^{-1}\big)(i, j). \tag{4.3}$$

Notice that the two matrices now have the same sparsity pattern corresponding to the pattern given, but they are not numerically the same. This opens the way for a "calculus of sparse inverse matrices" of the ME type. The formula (4.2) can now be interpreted as defining block-wise approximations on the ME inverse of the original matrix, whereby the middle matrix (corresponding to the "22" block) is repeated trice, each time with a different approximant. There is no guarantee that (4.2) actually defines a positive definite matrix, but since the approximants are assumed close, the approximation should be good when the original matrix is well conditioned, a detailed error analysis can be found in the already cited thesis by NELIS [1989]. The reason why (4.3) holds is the fact that ME approximants actually define strong norm approximants on the Cholesky factors. This seems to have been remarked first in DEWILDE and DYM [1981]. Formula (4.2) generalizes to large matrices with intricate block sparsity patterns and has been used successfully in the modern finite-element modeling program for interconnects of integrated circuits SPACE (see VAN DER MEIJS [1992]).

5. Hankel-norm model reduction

5.1. The setting

In this section we are interested in linear operators – of the type T where T induces a linear map $y = Tu$ – and where T is represented by a "model", more precisely a model that represents the linear computations the computer actually executes, based on its sequential intake of data, use of memory and sequential production of results. Such a model is called a "state-space model" because it is an instance of a classical model for a time-varying dynamical system adapted to the computational context. We start out with a simple but computationally intensive representation of the desired function and we shall proceed to reduce that representation to another one with much lower computational complexity. The model we start with will be a direct derivative of all the known data and will therefore be of much too high complexity, called a "model of high complexity". Our goal will be to reduce that model to one of smallest possible computational complexity, given a specified and acceptable tolerance on the accuracy. Here, T may be a matrix, but it may also be an infinite-dimensional operator, the theory that we shall present is not restricted to finite operators. In our basic framework, T will be a lower triangular operator, it represents a "causal" transfer between the vectors u and y viewed as time series. If it so happens that T does not satisfies this property, e.g., when T is a full matrix, then we would decompose T first either additively or multiplicatively into lower/upper operators: $T = L + U$ or $T = LU$ and then approximate L and dually U separately, but it may also be useful to move the main diagonal up so that the whole matrix becomes lower triangular, see the special case treated later. Our theory does not really become more complicated if we assume that T is in fact a block matrix, i.e., that the entries in T are actually matrices themselves, provided dimensions in rows and columns match. Hence T will look as follows:

$$T = \begin{bmatrix} \ddots & & & & \\ \ddots & 0 & & 0 & \\ \ddots & T_{-1,-1} & 0 & & \\ \ddots & T_{0,-1} & \boxed{T_{0,0}} & 0 & \\ \ddots & T_{1,-1} & T_{1,0} & T_{1,1} & \ddots \\ \iddots & & \ddots & \ddots & \ddots \end{bmatrix}.$$

Here, the $T_{i,j}$ block has dimension $n_i \times m_j$ and represents a partial map of the vector entry u_j to an additive component of y_i in the output vector in the map:

$$\begin{bmatrix} \ddots & & & & \\ \ddots & 0 & & 0 & \\ \ddots & T_{-1,-1} & 0 & & \\ \ddots & T_{0,-1} & \boxed{T_{0,0}} & 0 & \\ \ddots & T_{1,-1} & T_{1,0} & T_{1,1} & \ddots \\ \iddots & & \ddots & \ddots & \ddots \end{bmatrix} \begin{bmatrix} \vdots \\ u_{-2} \\ u_{-1} \\ \boxed{u_0} \\ u_1 \\ u_2 \\ \vdots \end{bmatrix} = \begin{bmatrix} \vdots \\ y_{-2} \\ y_{-1} \\ \boxed{y_0} \\ y_1 \\ y_2 \\ \vdots \end{bmatrix}.$$

We identify the 0th (block-)element in a matrix by putting a square around it, and similarly for the $(0, 0)$th element of a matrix of operator for orientation purposes. The (linear) computation defined by $y = Tu$ as executed by a computer that takes the input data

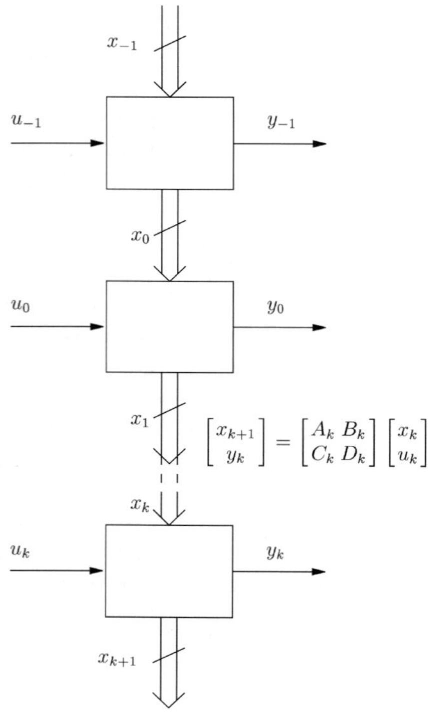

FIG. 5.1. A causal state-space realization of an operator T: the state represents the data available for computation at a given stage.

sequence u and produces the output sequence y can be represented by a "causal model" for T. The transfer from the input vector u to the output vector y can indeed be written in terms of an intermediate sequence $\{x_k\}$ of data which the computer stores in memory, and so-called *realization matrices* representing the computations at the sequence point k, as:

$$\begin{aligned} x_{k+1} &= A_k x_k + B_k u_k, \\ y_k &= C_k x_k + D_k u_k. \end{aligned}$$

This is called a "time-varying state-space representation" of the computation. The dimension δ_k of the vector x_k is called the state dimension at point k, and the dimensions of the realization matrices A_k, B_k, C_k, D_k are respectively $\delta_{k+1} \times \delta_k$, $\delta_{k+1} \times m_k$, $n_k \times \delta_k$, $n_k \times m_k$. A graphical representation of the state representation is shown in Fig. 5.1.

We call A_k the *state transition matrix* at point k, while the other matrices B_k, C_k, and D_k stand for partial local maps input–state, state–output, and input–output, respectively, at point k. The system will be strictly causal when $D_k = 0$. It may happen that some of the vectors and matrices are not present. For example, if a matrix is represented by a state model, then the initial state in the representation (e.g., x_0) will not be present. In that case we say that the dimension of the respective vector is zero, it is represented

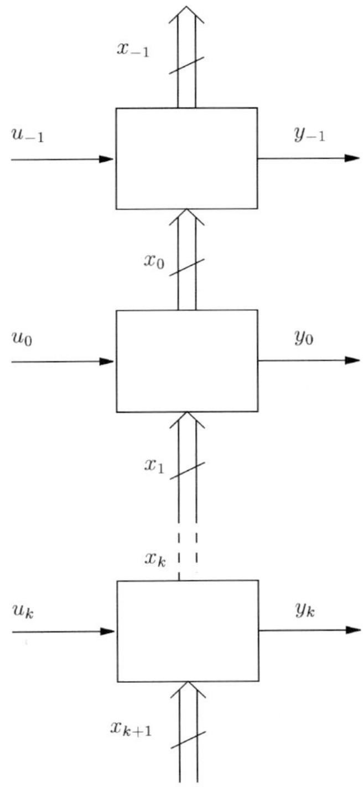

FIG. 5.2. The signal flow in an anticausal system.

by a place holder "·", but there is no numerical value present (not even zero). Similarly, the last state will also not be present when a finite matrix is represented, and disappears accordingly.

An anticausal system – represented by an upper block matrix – may similarly have an anticausal state realization as follows:

$$\begin{aligned} x'_k &= A'_k x'_{k+1} + B'_k u_k, \\ y_k &= C'_k X'_{k+1} + D'_k u_k. \end{aligned}$$

Here, we have chosen to make the realization strictly anticausal by putting $D'_k = 0$. A graphical representation of an anticausal linear system is shown in Fig. 5.2.

It is convenient for notational purposes to assemble the realization matrices in diagonal matrices or operators with appropriate dimensions. So we define:

$$A = \begin{bmatrix} \ddots & & & \\ & A_{-1} & & \\ & & \boxed{A_0} & \\ & & & A_1 & \\ & & & & \ddots \end{bmatrix}, \qquad B = \begin{bmatrix} \ddots & & & \\ & B_{-1} & & \\ & & \boxed{B_0} & \\ & & & B_1 \\ \ddots & & & \end{bmatrix},$$

and so forth. Introducing also the causal shift matrix:

$$Z = \begin{bmatrix} \ddots & & & & \\ & \ddots & 0 & & \\ & & I & \boxed{0} & \\ & & & I & 0 \\ & & & & \ddots & \ddots \end{bmatrix},$$

we see that the original operator T can be expressed in terms of these diagonal matrices as follows:

$$T = D + C(I - ZA)^{-1}ZB.$$

This generalizes the classical representation matrix for stationary discrete time systems, where Z now replaces the classical causal shift z (notice, however, that the unit matrices in Z may have different dimensions and that Z does not commute with matrices in the sense that $ZA \neq AZ$, as the scalar shift "z" would). Some care must be exercised when one interprets these formulas in the case of finite matrices. The block diagonals, A, B, C, D, and Z are all block matrices or operators with appropriate dimensions. So, A will map a state sequence of dimensions $\ldots, \delta_{-1}, \boxed{\delta_0}, \delta_1, \ldots$ to $\ldots, \delta_0, \boxed{\delta_1}, \delta_2, \ldots$, and Z applied to the same state sequence will map to $\ldots, \delta_{-2}, \boxed{\delta_{-1}}, \delta_0, \ldots$. Numerically, Z will be a perfect unit matrix in the finite case, but its block decomposition will make it look like the shift matrix that it is, for a shift on a finite sequence will keep the numerical values of that sequence, but will shift their indices! The inverse in the formula for T can be interpreted in a purely formal sense as meaning $(I - AZ)^{-1} = I + AZ + (AZ)^2 + \cdots$, but the series will of course also converge in the operator sense, if AZ is idempotent (which would be the case with finite matrices) or if $(AZ)^k$ converges to zero quickly enough. To make the notion of convergence more precise, we introduce a norm on the input and output spaces, namely the ℓ_2 or quadratic norm:

$$\|u\| = \sqrt{\sum_j \|u_j\|^2},$$

where the $\|u_j\|$ is the usual Euclidean norm on a vector (square root of the sum of magnitudes square of the entries). In this paper we treat operators T that are bounded as maps between input and output spaces endowed with the quadratic norm (this corresponds to the L_∞ norm on the unit circle in the classical case). A sufficient condition for this is that the spectral radius of AZ is strictly less than one, in which case the Neumann series $I + AZ + (AZ)^2 + \cdots$ converges in norm. If that is the case we say that the realization for T is *uniformly exponentially stable* or *ues* (exponential stability of time-varying systems is extensively treated in the time-varying literature). To characterize this case further we define:

$$\ell_A = \sigma(ZA) = \lim_{n\to\infty} \left\|(ZA)^n\right\|^{1/n}$$

and a system realization will be ues if $\ell_A < 1$. The "Z" can be taken out of the formula for ℓ_A if we define the South-East diagonal shift (with "$*$" indicating the adjoint operator):

$$A^{(1)} = ZAZ^*$$

so that

$$\ell_A = \lim_{n\to\infty} \|AA^{(1)}A^{(2)}\cdots A^{(n-1)}\|^{1/n}.$$

The "continuous product" that appears in the formula is useful for other purposes. In particular, if we express the block entries in T in terms of a realization, we obtain, for $i > j$, $T_{i,j} = C_i A_{i-1}\cdots A_{j+1}B_j$, and we see that the entries become (uniformly) exponentially small for large $i - j$ when $\ell_A < 1$. In a later section, we shall see how we can recover a realization from the entries in T, but before doing so we turn to some more definitions and properties in the basic framework.

5.1.1. Lyapunov transformations

A state realization for an operator or matrix is not unique, even when it is minimal. In fact, we can permit ourselves a state transformation that introduces at each point k a transformed state x'_k related to the original via $x_k = R_k x'_k$ where the state transformation matrix R_k is non-singular for each k. In the case of infinite systems we usually require even more, namely R_k and R_k^{-1} should be uniformly bounded over k. Such transformations we call "Lyapunov transformations". They have the nice property that they are preserving the exponential stability of the realization. Under the state transformation, a causal realization transforms as follows:

$$\begin{bmatrix} A_k & B_k \\ C_k & D_k \end{bmatrix} \mapsto \begin{bmatrix} R_{k+1}^{-1}A_kR_k & R_{k+1}^{-1}B_k \\ C_kR_k & D_k \end{bmatrix}, \tag{5.1}$$

or, when expressed in the global diagonal notation:

$$\begin{bmatrix} A & B \\ C & D \end{bmatrix} \mapsto \begin{bmatrix} (R^{(-1)})^{-1}AR & (R^{(-1)})^{-1}B \\ CR & D_k \end{bmatrix}.$$

State transformations are very important not only to achieve canonical representations discussed below, but also to obtain algebraically minimal calculations – see Chapter 14 of DEWILDE and VAN DER VEEN [1998].

5.1.2. Input/output normal forms

We say that a realization is in *output normal form* when

$$A^*A + C^*C = I$$

i.e., $A_k^*A_k + C_k^*C_k = I$ for each k. From (5.1) and putting $M_k = R_k^{-*}R_k^{-1}$, we see that a realization can be brought to output normal form if a bounded and invertible solution exists to the recursive set of *Lyapunov–Stein* equations

$$A_k^*M_{k+1}A_k + C_k^*C_k = M_k,$$

or, equivalently, if $A^*M^{(-1)}A + C^*C = M$ has a boundedly invertible diagonal operator M as a solution. The existence of the solution has been much studied in Lyapunov stability theory, we suffice here with some facts. If the original realization is ues (i.e., if $\ell_A < 1$), then the Lyapunov–Stein equation always has a bounded solution M. The solution M can be expressed as the so-called observability Gramian:

$$M = \sum_{k=0}^{\infty} (A^{\{k\}})^* (C^* C)^{(-k)} (A^{\{k\}}),$$

where we have put

$$A^{\{k\}} = A^{(-k+1)} \cdots A^{(-1)} A$$

and the sum converges in norm because of the ues assumption. The state transformation needed to bring the system in output normal is then obtained from $M^{-1} = RR^*$. The problem with its existence is whether M is boundedly invertible. We shall say that the system is *strictly observable* if that is the case. In the sequel we shall normally assume this property to be valid.

5.1.3. Realization theory and canonical spaces

One may wonder when a causal transfer operator T has a finite-dimensional realization at each time point k. It turns out (see DEWILDE and VAN DER VEEN [1998]) that this will be the case iff each kth order operator

$$H_k := \begin{bmatrix} T_{k+1,k} & T_{k+1,k-1} & T_{k+1,k-2} & \cdots \\ T_{k+2,k} & T_{k+2,k-1} & T_{k+2,k-2} & \cdots \\ T_{k+3,k} & T_{k+3,k-1} & T_{k+3,k-2} & \cdots \\ \vdots & \vdots & \vdots & \end{bmatrix}$$

has finite rank δ_k. We call these operators local Hankel matrices, and their rank δ_k actually gives the minimal state dimension needed at point k. The here defined Hankel operators do not have the classical Hankel structure (elements equal along anti-diagonals), but they do fit the general functional definition of Hankel operators as exemplified in Fig. 5.3, where the matrices are shown in a graphical way (notice that the columns in the picture are in reverse order, the definition of the H_k fits the classical matrix representation).

Realization theory shows that any collection of minimal factorizations of all H_k will produce a minimal realization. If we express the Hankel operators in terms of a state

$$\left[\begin{array}{ccc|cc} \ddots & \ddots & & & \\ & \ddots & T_{k,k} & & \\ \hline & T_{k+1,k-1} & T_{k+1,k} & T_{k+1,k+1} & \\ & T_{k+2,k-1} & T_{k+2,k} & T_{k+2,k+1} & T_{k+2,k+2} \\ & & & \ddots & \ddots \end{array}\right]$$

FIG. 5.3. Generalized Hankel operators in a matrix or operator.

space representation we have:

$$H_k = \mathcal{O}_k \mathcal{R}_k = \begin{bmatrix} C_k \\ C_{k+1}A_k \\ \vdots \end{bmatrix} [\, B_{k-1} \quad A_{k-1}B_{k-2} \quad \cdots \,],$$

and the "realization theory" is reduced to reading the A_k, B_k, C_k, D_k from the factorization. The columns of $\mathcal{O}_k$ form a basis for the columns of the Hankel matrix H_k while the rows of $\mathcal{R}_k$ form a basis for its rows. We shall obtain a realization in output normal form iff the columns of $\mathcal{O}_k$ have been chosen orthonormal for each k. The realization derived from the factorization is then given by:

$$B_{k-1} = (\mathcal{R}_k)_1, \qquad C_k = (\mathcal{O}_k)_0, \qquad A_k = \mathcal{O}_{k+1}^{\dagger}\mathcal{O}_k^{\downarrow},$$

where $(\mathcal{R}_k)_1$ is the first element of the "reachability" matrix $\mathcal{R}_k$, $(\mathcal{O}_k)_0$ the top element of the "observability" matrix $\mathcal{O}_k$, the "$\dagger$" indicates the Moore–Penrose inverse, and the "downarrow" on $\mathcal{O}_k$ indicates a matrix equal to $\mathcal{O}_k$ except for its first block-element, which has been deleted. The matrix A_k is uniquely defined because of the minimality of the factorization, even when any general inverse is used.

5.1.4. Balanced realization

It is also possible to define a balanced realization, by using a factorization based on a singular value decomposition of the Hankel operator:

$$H_k = U_k \begin{bmatrix} \sqrt{\sigma_1} & & \\ & \ddots & \\ & & \sqrt{\sigma_k} \end{bmatrix} \cdot \begin{bmatrix} \sqrt{\sigma_1} & & \\ & \ddots & \\ & & \sqrt{\sigma_k} \end{bmatrix} V_k.$$

However, balanced realizations and approximations are only of limited use in time-varying theory, they are unable to handle transfer operators of low rank with sparse entries far from the main diagonal adequately (see DEWILDE and VAN DER VEEN [1998]). We give them here for the sake of completeness.

5.1.5. Reachability/observability bases in terms of realizations

It is easy to produce a direct relation between realizations and reachability or controllability bases, in particular we find:

$$\mathbf{F}_0 = C(I - ZA)^{-1} = \begin{bmatrix} \ddots & & & & \\ \ddots & C_{-1} & & & \\ \ddots & C_0A_{-1} & \boxed{C_0} & & \cdots \\ \ddots & C_1A_0A_{-1} & C_1A_0 & C_1 & \\ \ddots & \ddots & \ddots & \ddots & \ddots \end{bmatrix}$$

and dually

$$\mathbf{F} = B^*Z^*(I - A^*Z^*)^{-1}.$$

Each block column of $\mathbf{F}_0$ or $\mathbf{F}$ forms the basis for a local observability or controllability space.

5.2. *Hankel-norm model reduction*

We are given a lower (block-)operator T (we write: $T \in \mathcal{L}$) that we wish to approximate by a lower operator T_a of minimal complexity and that meets a certain pre-assigned complexity. First we make the notion "complexity" and "meeting a pre-assigned norm" more concrete.

5.2.1. *Complexity*

We identify "complexity" with "local state dimension". Suppose indeed that at stage k the state dimension (the total number of floating-point numbers the system has stored in memory from its past) is δ_k. Then it can be shown that number, together with the dimensions of the local input and output space determines the local computational complexity. It turns out that the number of floating point operations needed at stage k is given by $\frac{1}{2}(m_k + n_k + \delta_k)(m_k + n_k + \delta_{k+1} + 1)$ (see DEWILDE and VAN DER VEEN [1998]), exactly equal to the number of "algebraically free parameters" at that stage.

5.2.2. *Norm*

What is an adequate approximating norm? In the classical model reduction context an L_∞-type norm is known to be too strong (because the polynomials or rationals are not dense in such a space), while an L_2 norm is usually too weak, because it gives rise to undesirable phenomena like the Gibbs phenomenon. A good compromise, one that also offers quite a bit of flexibility, is provided by the Hankel norm, i.e., the supremum of the norms of the local Hankel operators we defined before. This is the norm we shall be using, hence we define

$$\|T\|_H = \sup_k \|H_k\|.$$

We still need to characterize the approximation accuracy needed. We take as measure for precision a Hermitian, strictly positive diagonal operator Γ – in fact it could be taken as $\Gamma = \varepsilon \cdot I$ for some small epsilon, but we may need the extra freedom of accommodating the precision at each time point.

5.2.3. *High-order model*

As described earlier, we start out our model reduction by selecting an appropriate representation of the desired computation as a high-complexity or high-order model that can be used computationally. An example of such a high-order model is given by a truncated Taylor-like series of high-enough order so that the truncation error has hardly any impact, but other, more convenient high-order representations may be adequate as well. If

$$T \approx T_0 + ZT_1 + Z^2 T_2 + \cdots + Z^n T_n$$

(with n sufficiently large and where each T_k represents a shifted diagonal of T), then a simple but high-complexity realization for T is given by the generalized companion form (in formal output normal form)

$$\begin{bmatrix} A & B \\ C & D \end{bmatrix} = \left[\begin{array}{cccc|c} 0 & & & & T_n \\ I & 0 & & & T_{n-1} \\ & \ddots & \ddots & & \vdots \\ & & I & 0 & T_1 \\ \hline 0 & \cdots & 0 & I & T_0 \end{array}\right]. \tag{5.2}$$

Expression (5.2) should be interpreted as a matrix consisting of block diagonals. At time point k the local realization has the form

$$\begin{bmatrix} A_k & B_k \\ C_k & D_k \end{bmatrix} = \left[\begin{array}{cccc|c} 0 & & & & T_{n,k} \\ I & 0 & & & T_{n-1,k} \\ & \ddots & \ddots & & \vdots \\ & & I & 0 & T_{1,k} \\ \hline 0 & \cdots & 0 & I & T_{0,k} \end{array}\right].$$

Also the shift matrix Z must be interpreted in a block fashion and now has the form

$$\begin{bmatrix} Z & & & \\ & Z & & \\ & & \ddots & \\ & & & Z \end{bmatrix}$$

conformal with the block-diagonal decomposition of A. Given the higher model for T and the precision Γ, the model reduction problem becomes:

Find a causal operator T_a of minimal state complexity such that

$$\left\| (T - T_a)\Gamma^{-1} \right\|_H \leqslant 1,$$

i.e., T_a approximates T up to a precision given by Γ. It is customary to take the higher model T so that it is strictly causal, i.e., $T_0 = 0$ and to require the same of the low-order approximation. We follow that habit since it does not impair generality and simplifies some properties. Before embarking on the solution and its properties, we introduce the main ingredients needed.

5.2.4. Ingredient #1: Nehari reduction

The Nehari theorem adapted to our context is as follows.

THEOREM 5.1. *For any bounded, strictly causal operator T,*

$$\|T\|_H = \min_{T'' \in \mathcal{U}} \|T + T''\|,$$

where the norm in the second member is the operator norm and T'' is a bounded, anticausal operator.

A proof of the Nehari theorem in the general context of nest algebras (to which our setup conforms) goes back to the work of ARVESON [1975]. For a proof restricted to our specific context, see DEWILDE and VAN DER VEEN [1998]. Application of the Nehari theorem reduces the problem to: *Find a (general) bounded operator T' so that its causal part $T_a = \mathbf{P}T'$ is of minimal complexity and*

$$\|(T - T')\Gamma^{-1}\| \leqslant 1.$$

5.2.5. Ingredient #2: external factorization

We are given $T \in \mathcal{L}$. An "external factorization" consists of finding $\Delta \in \mathcal{L}$ and $U \in \mathcal{L}$ unitary such that $T = U\Delta^*$ (a more general type relaxes the requirement on U, see further). This type of factorization is reminiscent of the coprime factorization of classical system theory, where U is an all-pass function that collects the "poles" of T and Δ^* is obtained as $U^*T - U^*$ pushes the poles of T to anticausality. It is easy to perform an external factorization on the state-space representation of T, especially when it is given in output normal form. So suppose that the realizations are given as (we use the $\approx$ sign to represent realizations).

$$\mathbf{T}_k \approx \begin{bmatrix} A_k & B_k \\ C_k & D_k \end{bmatrix}$$

in which, for all k,

$$A_k^* A_k + C_k^* C_k = I.$$

Then, the kth realization matrix for U is found by completing the first block column to form unitary matrices:

$$\begin{bmatrix} A_k & B_{Uk} \\ C_k & D_{Uk} \end{bmatrix}$$

thereby producing B_{Uk} and D_{Uk} as completing matrices. The "remainder" Δ_k is then given by

$$\Delta_k \approx \begin{bmatrix} A_k & B_{Uk} \\ B_k^* A_k + D_k^* C_k & \\ & B_k^* B_{Uk} + D_k^* D_{Uk} \end{bmatrix}.$$

Algorithmically, a simplified "Householder-type" algorithm will provide the missing data. Numerical analysts would write, somewhat equivocally

$$\begin{bmatrix} B_{Uk} \\ D_{Uk} \end{bmatrix} = \begin{bmatrix} A_k \\ C_k \end{bmatrix}^{\perp}.$$

5.2.6. Ingredient #3: J-unitary operators

In interpolation and approximation theory, J-unitary operators of various types play a central, if not crucial role. Causal J-unitary operators map input spaces of the type $\ell_2^{\mathcal{M}_1} \times \ell_2^{\mathcal{M}_2}$ to output spaces of the type $\ell_2^{\mathcal{N}_1} \times \ell_2^{\mathcal{N}_2}$, hence they are of the block type:

$$\Theta = \begin{bmatrix} \Theta_{11} & \Theta_{12} \\ \Theta_{21} & \Theta_{22} \end{bmatrix}.$$

These spaces are endowed with a non-definite metric. We denote

$$J_{\mathcal{M}} = \begin{bmatrix} I_{\mathcal{M}_1} & \\ & -I_{\mathcal{M}_2} \end{bmatrix}, \qquad J_{\mathcal{N}} = \begin{bmatrix} I_{\mathcal{N}_1} & \\ & -I_{\mathcal{N}_2} \end{bmatrix}.$$

The J-unitary operators that we shall use will all be bounded and causal. The J-unitarity means

$$\Theta J_{\mathcal{N}} \Theta^* = J_{\mathcal{M}}, \qquad \Theta^* J_{\mathcal{M}} \Theta = J_{\mathcal{N}}.$$

It has important consequences for the block entries of Θ:

- Θ_{22} is boundedly invertible and $\|\Theta_{22}^{-1}\| \ll 1$;
- $\|\Theta_{22}^{-1}\Theta_{21}\| \ll 1$.

The operator Θ_{22}^{-1} turns out to be of great importance in model-reduction theory. It is most likely of mixed type (causal/anticausal). We return later to its state-space analysis.

5.2.7. *Method of solution*

With the ingredients previously detailed, the actual method to generate the solution appears very straightforward. It consists of two steps:

Step 1: Perform a coprime external factorization:

$$T = U \Delta^* \tag{5.3}$$

with $\Delta \in \mathcal{L}$ and $U \in \mathcal{L}$.

Step 2: Perform an external factorization of the type:

$$\Theta \begin{bmatrix} U^* \\ -\Gamma^{-1}T^* \end{bmatrix} = \begin{bmatrix} A' \\ -B' \end{bmatrix}. \tag{5.4}$$

Here, Θ is a block lower-triangular J-unitary operator of dimensions conforming to $\left[\begin{smallmatrix} U^* \\ -\Gamma^{-1}T^* \end{smallmatrix}\right]$, $A' \in \mathcal{L}$, and $B' \in \mathcal{L}$. The solution of the interpolation problem is now given by

$$\begin{aligned} T' &= \mathcal{B}'^* \Theta_{22}^{-*} \Gamma, \\ T_a &= \text{strictly lower part of } T'. \end{aligned} \tag{5.5}$$

Before embarking on computational issues, we show first that this recipe indeed produces a T' and a T_a that satisfies the norm and the minimality conditions. The norm condition is easy to treat directly. As to the study of complexity, it will be based on the state-space properties of the operator Θ appearing in the special J-unitary external factorization that have to be studied first.

5.2.8. *The norm condition*

From the second block row in (5.5), we obtain

$$\Theta_{21} U^* - \Gamma^{-1} \Theta_{22} T^* = -B',$$

and since Θ_{22} is invertible, it follows immediately by reordering of terms that

$$(T - T')\Gamma^{-1} = \left[\Theta_{22}^{-1}\Theta_{21}U^*\right]^*,$$

where we have put $T' = B'^* \Theta_{22}^{-*} \Gamma$. Hence,

$$\|(T - T')\Gamma^{-1}\| < 1$$

since $\|U^*\| = 1$ and $\|\Theta_{22}^{-1}\Theta_{21}\| < 1$.

5.2.9. The construction of the special J-external factorization

We are looking for a minimal Θ that meets the factorization condition expressed in (5.4). As is the case of the regular external factorization, it will be based on the completion of appropriate reachability operators. A realization for $[U \;\; -T\Gamma^{-1}]$ is given by

$$\begin{bmatrix} A & B_U & -B\Gamma^{-1} \\ C & D_U & 0 \end{bmatrix}$$

whose reachability part is given by $[A \;\; B_U \;\; -B\Gamma^{-1}]$, based on the realization

$$\begin{bmatrix} A & B_U \\ C & D_U \end{bmatrix}$$

for U (notice that in case one starts out with a companion form as detailed above, this part of the procedure is actually trivial, we simply have $U = Z^n$). From the realization theory we can deduce next that a bounded, causal ues J-unitary operator has the property that it possesses a realization which is J-unitary for some, still to be determined state signature

$$J_{\mathcal{B}} = \begin{bmatrix} I_{\mathcal{B}_+} & \\ & -I_{\mathcal{B}_-} \end{bmatrix}. \tag{5.6}$$

Hence, an appropriate state transformation should be able to produce the desired $J_{\mathcal{B}}$ and J-unitarity based on such a signature matrix on the state. As is the case for the regular external factorization, a somewhat special reachability Gramian will play a central role in finding this transformation. Indeed, let $\{R_k\}$ be the set of state-transformation matrices needed. Then the reachability matrices transform as

$$\begin{bmatrix} R_{k+1}^{-1} A_k R_k & R_{k+1}^{-1}(B_U)_k & -R_{k+1}^{-1} B\Gamma^{-1} \end{bmatrix},$$

and we wish each of these matrices to be part of a J-unitary matrix, i.e., they have each to be J-isometric for an adequate local signature matrix. Suppose that we already have the signature matrices $(J_{\mathcal{B}})_k$, and let $\Lambda_k = R_k (J_{\mathcal{B}})_k R_k^*$, then the J-unitarity of the Gramian can be expressed as follows:

$$A_k \Lambda_k A_k^* + (B_U)_k (B_U)_k^* - B_k \Gamma_k^{-2} B_k^* = \Lambda_{k+1}. \tag{5.7}$$

A solution for Λ will exist if this Lyapunov–Stein equation has a definite solution that is also boundedly invertible. Note that because of the ues condition on A, the equation has a unique bounded solution; the question is whether the solution is also boundedly invertible. The existence of the solution can be studied directly in terms of the original data by eliminating B_U, since

$$AA^* + B_U B_U^* = I.$$

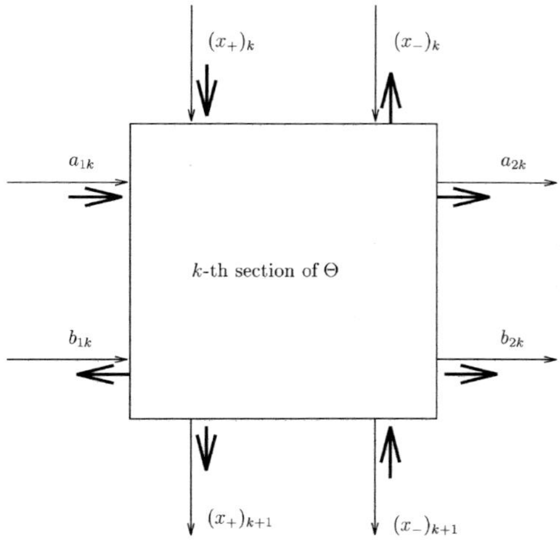

FIG. 5.4. The dataflow in a Theta section is shown with normal arrows, the "energy flow" indicating the sign of the quadratic norms is indicated with fat arrows.

Setting $M = I - \Lambda$ the equation turns into

$$M_{k+1} = A_k M_k A_k^* + B_k \Gamma_k^{-2} B_k^*.$$

Here, M is the reachability Gramian of $T\Gamma^{-1}$, and we find that a solution to the J-unitary embedding problem exists iff $(I - M)^{-1}$ exists and is bounded, i.e., iff the eigenvalues of M_k are bounded away from 1, uniformly over k. In the case the solution is not definite, a "borderline" solution may exist, and thus the case becomes singular. Although that singular case is beyond the present treatment, we shall devote some words to it in the discussion at the end of this section. Let us now assume that a strictly definite solution does exist and analyze it further. Let the inertia of Λ_k be given by

$$\Lambda_k = R_k \begin{bmatrix} (I_{\mathcal{B}_+})_k & \\ & -(I_{\mathcal{B}_-})_k \end{bmatrix} R_k^*.$$

After application of the state transformation $R_{k+1}^{-1} \cdots R_k$, the dataflow for Θ looks as in Fig. 5.4.

Associated with the various signature matrices, we can also imagine an "energy flow" representing the conservation of quadratic norm or energy which follows from the J-unitarity imposed on Θ. The energy flow corresponding to the signature is shown by fat arrows in Fig. 5.4.

5.2.10. Complexity analysis

We have as proposed solution

$$T_a = \text{strictly causal part of } B'^* \Theta_{22}^{-*} \Gamma.$$

In this expression, B'^* is anticausal while Θ_{22}^{-*} is of mixed causality. We first establish that the complexity of T_a is essentially determined by the (strictly) causal part of Θ_{22}^{-*}.

Next we shall analyze the complexity of the latter. Let

$$B' = d + cZ(I - aZ)^{-1}b, \quad \text{causal part of}$$
$$\Theta_{22}^{-*} = D_2 + C_2 Z(I - A_2 Z)^{-1} B_2,$$

be minimal realizations for B' and the causal part of Θ_{22}^{-*}, respectively (for the existence of the latter, see further). The computation of the causal part for the product is straightforward:

$$\begin{aligned}&\text{causal part of } B'^* \Theta_{22}^{-*} \Gamma\\ &\quad = d^* D_2 \Gamma + d^* C_2 Z(I - A_2 Z)^{-1} B_2 \Gamma\\ &\qquad + \text{causal part of } b^*(I - Z^* a^*)^{-1}(c^* C_2)^{(-1)}(I - A_2 Z)^{-1} B_2 \Gamma.\end{aligned}$$

The computation reduces to the "generalized partial-fraction decomposition" of the last part. This is handled in the following generic lemma.

LEMMA 5.1. *Let a and A_2 be transition operators with $\ell_a \leqslant 1$, $\ell_{A_2} \leqslant 1$ and at least one less than one, then*

$$\begin{aligned}&(I - Z^* a^*)^{-1}(c^* C_2)^{(-1)}(A - A_2 Z)^{-1}\\ &\quad = (I - Z^* a^*)^{-1} Z^* a^* m + m + m A_2 Z(I - A_2 Z)^{-1},\end{aligned}$$

where m is the unique bounded solution of the Lyapunov–Stein equation

$$m^{(1)} = c^* C_2 + a^* m A_2.$$

PROOF. The proof of the lemma is by direct computation, after chasing the denominators and identifying the entries. □

Applying the lemma to the product that defines T_a, we obtain

$$T_a = \left(d^* D_2 \Gamma + b^* m B_2 \Gamma\right) + \left(d^* C_2 + b^* m\right) Z(I - A_2 Z)^{-1} B_2 \Gamma.$$

We see that T_a inherits the complexity of Θ_{22}^{-*}, at least essentially (further cancellations are theoretically possible but not very likely). In fact, they have the same reachability space based on $\{A_2, B_2\Gamma\}$. The complexity analysis hence proceeds with the analysis of the complexity of Θ_{22}^{-*}. This can be done in a particularly elegant way by studying the strict-past/future decomposition of the operator Θ. We decompose an arbitrary signal (say a belonging to some ℓ_2-space) in its strict-past part and its future part ($a_k = a_{pk} + a_{fk}$). Let the corresponding operators be denoted by $\mathbf{P}_k$ for the projection on the future and $\mathbf{P}'_k = I - \mathbf{P}_k$ for the projection on the strict past, then the splitting of the operator Θ happens as shown in Fig. 5.5, where we also have indicated the sign decomposition of the state discussed earlier. The arrows in Fig. 5.5 indicate flow of energy in the sense that each block satisfies the energy balance with respect to incoming and outgoing energetic contributions (isometric or J-isometric depending on whether a signal is considered an input or an output in the formulation at hand). The causal part of Θ_{22}^{-*} will of course

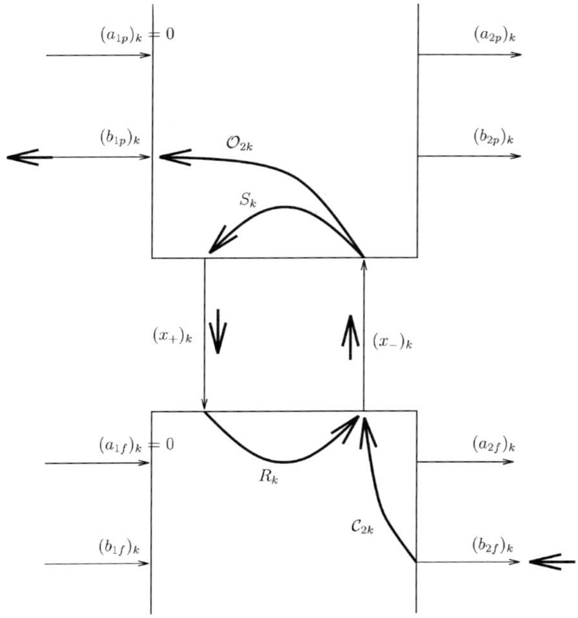

FIG. 5.5. The figure shows the signal flow for $(\Theta_{22}^{-1})_k$. The energy flow of Fig. 5.4 applies, here the relevant signal propagation is indicated with fat arrows.

correspond to the anticausal part of Θ_{22}^{-1}. Writing out

$$\Theta_{22}^{-1} = B_2^* Z^* \left(I - A_2^* Z^*\right)^{-1} C_2^* + D_2^* + \text{causal part},$$

we see that the relevant state dimension is given by the state dimension needed for the operator represented by the first term that produces the map b_{2f} to b_{1p} with $a_1 = 0$ and $b_{2p} = 0$ since Θ_{22}^{-1} maps b_2 to b_1 under the assumption $a_1 = 0$ and the portion b_{1f} in b_1 is to be neglected by the restriction to the lower part of the result (with a slight abuse of notation we can handle all time points k in the same global formula – see DEWILDE and VAN DER VEEN [1998] for details). With reference to the situation in Fig. 5.5, let us define two new diagonal operators $S : x_- \mapsto x_+$ (in the past) and $R : x_+ \mapsto x_-$ (in the future). It is not hard to see (and a more detailed analysis would show) that both these operators are causal and strictly contractive. With b_{2f} as only non-zero input in this configuration, and with energy conservation in vigor, we see that both b_{1p} and x_+ are solely dependent on x_-. In fact, we have

$$\begin{aligned} x_- &= (I - RS)^{-1} \mathcal{C}_2 b_{2f}, \\ b_{1p} &= \mathcal{O}_2 x_-, \\ x_+ &= S x_-, \end{aligned}$$

where $\mathcal{C}_2$ and $\mathcal{O}_2$ are appropriate reachability and observability maps derived from the anticausal part of Θ_{22}^{-1} (and which we do not detail any further here). The map from

b_{2f} to b_{1p} then factors as

$$b_{2p} = \mathcal{O}_2 \cdot (I - RS)^{-1} \mathcal{R}_2 b_{2f},$$

and its state complexity is determined by the dimension of the "anticausal" state x_-. Hence, T_a has the same complexity as the strict lower part of Θ_{22}^{-*}, which is locally equal to the dimension δ_{k-} of x_-. This dimension is now easy to gauge from the original construction of Θ and is given by the following theorem.

THEOREM 5.2. *Assuming that there exists an ε so that all singular values of all H_k, Hankel matrices of $T\Gamma^{-1}$, are at least ε distant from 1, the dimension δ_{k-} is given by the number of singular values of H_k larger than one. This is also the minimal dimension of any strictly causal approximant T_a satisfying $\|(T - T_a)\Gamma^{-1}\| < 1$.*

PROOF. Recall that the dimension of x_{k-} is given by the number of eigenvalues of M_k larger than one, where M_k satisfies

$$M_{k+1} = A_k M_k A_k^* + B_k \Gamma_k^{-2} B_k^*.$$

Since we started out with a system in output normal form, and $H_k = \mathcal{O}_k \mathcal{R}_k$, we have

$$H_k^* H_k = \mathcal{R}_k^* \mathcal{O}_k^* \mathcal{O}_k \mathcal{R}_k = \mathcal{R}_k^* \mathcal{R}_k = M_k,$$

where $\mathcal{O}_k^* \mathcal{O}_k = I$ since we assumed the system in output normal form, and the singular values of H_k equal the eigenvalues of M_k. This proves the first statement. As for the second assertion, its proof is much more complex, and based on the fact that all approximants which meet the norm condition can be generated by loading Θ in a contractive and causal operator S_L, more precisely, all T' have the form

$$T' = T + US^*\Gamma.$$

Here,

$$S = (S_L\Theta_{21} + \Theta_{22})^{-1}(S_L\Theta_{11} + \Theta_{12}),$$

and U is as defined earlier. It turns out that the complexity of its lower part is then at least equal to the complexity of the lower part of Θ_{22}^{-*}. For a complete treatment, see Chapter 10 of DEWILDE and VAN DER VEEN [1998], in particular Theorem 10.18. □

These are the basic results on Hankel-norm approximation of a lower operator. Many more properties can be derived on this new and interesting method, in particular, state-space representations for the approximants are relatively easy to derive, for details we refer to the literature cited.

5.3. The recursive Schur algorithm for Hankel-norm approximation

A low-complexity Hankel-norm approximation to a strictly upper but otherwise general matrix can be derived from an elementary Schur-type elimination algorithm using both orthogonal and hyperbolic elementary matrices. It is a direct application of the previous theory to finite matrices and was first presented in DEWILDE and VAN DER VEEN [1998]. Here, we present the result without proof.

Suppose that the original matrix to be approximated is given by

$$T = \begin{bmatrix} \boxed{0} & & & \\ t_{21} & \underline{0} & & \\ \vdots & & \ddots & \\ t_{n1} & t_{n2} & \cdots & \underline{0} \end{bmatrix},$$

then a trivial external factorization for $T = U\Delta^*$ is given by

$$U = \begin{bmatrix} \boxed{1\,0\,0\,0} \\ 0\,1\,0\,0 \\ 0\,0\,1\,0 \\ 0\,0\,0\,1 \end{bmatrix}, \qquad \Delta = \begin{bmatrix} \boxed{0\; t_{21}^* \cdots t_{n1}^*} \\ 0 \cdots t_{n2}^* \\ \ddots \vdots \\ 0 \end{bmatrix}.$$

According to the theory in the previous sections, the Θ matrix necessary for the Hankel-norm approximation must now have the following three properties:

1. It must be J-unitary for appropriate signature matrices;
2. It must be block lower;
3. It must make the product

$$\Theta \begin{bmatrix} U^* \\ -T^* \end{bmatrix}$$

lower. (Point 1 may seem cryptic but will be partly justified in the sequel.)

The right-hand side signature of Θ is certainly given by $J_2 = I_n \oplus -I_n$, in accordance with the right factor, the left-hand side signature will follow from the construction and will differ case by case. It is possible at this point to determine the local arrow dimensions of Θ but not yet the signs of the state and output arrows. To illustrate the point, let us assume that the entries in T are scalar. Because of the structure of U^* and T^*, the first block in a realization for Θ will have n positive inputs (from U^*) and one negative input (from $-T^*$), and it will have $n-1$ states going to the next stage. This means that this first stage must have two outputs (the signs of the outgoing states and outputs are yet to be determined – see Fig. 5.5 for extra information).

The matrix to be block lowered using elementary operations is given by:

$$\begin{bmatrix} U^* \\ -T^* \end{bmatrix} = \begin{array}{c} + \\ + \\ \vdots \\ + \\ \hline - \\ - \\ \vdots \\ - \end{array} \left[\begin{array}{cccc} \boxed{\begin{matrix}1\\0\\\vdots\\0\end{matrix}} & \begin{matrix}0\\1\\\vdots\\0\end{matrix} & \begin{matrix}\\\\\ddots\\\cdots\end{matrix} & \begin{matrix}\\\\\\1\end{matrix} \\ \hline \boxed{-t_{11}^*} & -t_{21}^* & \cdots & -t_{n1}^* \\ 0 & -t_{22}^* & \cdots & -t_{n2}^* \\ \vdots & & \ddots & \vdots \\ 0 & 0 & \cdots & -t_{nn}^* \end{array}\right].$$

The elimination procedure now starts with the elimination of $-t_{n1}^*$ in the first row of the second block, using the last row in the first block. We indicate this state of affairs with the pair of indices $\langle n, 1\rangle$. Since the sign of the last row of the first block is positive, and that of the first row of the second block negative, a hyperbolic rotation must be used, which can be of two forms, depending on the magnitude of $-t_{n1}^*$. One possibility is

$$\frac{1}{\sqrt{1-|\rho_{n1}|^2}}\begin{bmatrix} 1 & \overline{\rho_{n1}} \\ \rho_{n1} & 1 \end{bmatrix},$$

in which case $\rho_{n1} = t_{n1}$ has to be smaller than one in magnitude and the target signature is $\langle +, -\rangle$, while the other possibility, when $|t_{n1}| > 1$, is

$$\frac{1}{\sqrt{1-|\rho_{n1}|^2}}\begin{bmatrix} \rho_{n1} & 1 \\ 1 & \overline{\rho_{n1}} \end{bmatrix},$$

in which case $\rho_{n1} = 1/t_{n1}$, again of magnitude smaller than one. The case where $|t_{n1}| = 1$ is not allowable in the present state of the theory (for an extension, see DEWILDE [1995]), the respective coefficient in Γ then has to be adapted (the condition on the singular values of the Hankel operator is not satisfied). It may happen that in the course of the elimination procedure, a signature of the type $\langle +, +\rangle$ or $\langle -, -\rangle$ is encountered. In that case a regular (unitary) Jacobi rotation will do, and if $\langle -, +\rangle$ as initial signature is found, then the mirror case of the case detailed above holds. The type of rotations used in the scheme will determine the actual flow of energy between the stages of the realization for Θ. The resulting complexity can also be deduced directly from the signature resulting at the output. For example, if the output sequence is $\langle +, -\rangle$, $\langle +, -\rangle$, ..., then all state transitions have positive signs and Θ_{22} is causally invertible. The low-complexity approximant then reduces to a diagonal matrix. At the opposite side, and taking for example the 4×4 case, the output sequence $\langle +, +\rangle$, $\langle +, +\rangle$, $\langle -, -\rangle$, $\langle -, -\rangle$ will result in a state sequence given by $\langle +, +, -\rangle$, $\langle -, -\rangle$, $\langle -\rangle$, resulting in an "approximant" of maximal complexity. The principle involved is that at each state there must be an equal number of incoming and outgoing arrows on the one hand, and an equal number of incoming and outgoing energy arrows as well. A connection will bear a "+" sign if the two arrows point in the same direction and a "−" sign in the opposite case. From the resulting diagram, a realization for Θ_{22}^{-*} can be derived, and from there a realization for the approximant T_a, we refer to the literature cited for details. Although the algorithm does provide for an optimal solution, the computational details are still somewhat extensive.

6. Second-order linear dynamical systems

Second-order models arise naturally in the study of many types of physical systems, such as electrical and mechanical systems; see, e.g., BAI [2002] and the references given there. A *time-invariant multi-input multi-output second-order system* is described by equations of the form

$$M\frac{d^2q}{dt^2} + D\frac{dq}{dt} + Kq = Pu(t), \tag{6.1}$$

$$y(t) = L^{\mathrm{T}}q(t), \tag{6.2}$$

together with initial conditions $q(0) = q_0$ and $\frac{dq}{dt}(0) = \dot{q}_0$. Here, $q(t) \in \mathbb{R}^N$ is the vector of state variables, $u(t) \in \mathbb{R}^m$ is the input force vector, and $y(t) \in \mathbb{R}^p$ is the output measurement vector. Moreover, $M, D, K \in \mathbb{R}^{N\times N}$ are system matrices, such as mass, damping, and stiffness matrices in structural dynamics, $P \in \mathbb{R}^{N\times m}$ is the input distribution matrix, and $L \in \mathbb{R}^{N\times p}$ is the output measurement matrix. Finally, N is the state-space dimension, and m and p are the number of inputs and outputs, respectively. In most practical cases, m and p are much smaller than N.

The second-order system (6.1) and (6.2) can be reformulated as an equivalent linear first-order system in many different ways. We will use the following equivalent linear system:

$$E\frac{dx}{dt} = Ax + Bu(t), \tag{6.3}$$

$$y(t) = C^{\mathrm{T}}x(t), \tag{6.4}$$

where

$$x = \begin{bmatrix} q \\ \frac{dq}{dt} \end{bmatrix}, \qquad A = \begin{bmatrix} -K & 0 \\ 0 & W \end{bmatrix}, \qquad E = \begin{bmatrix} D & M \\ W & 0 \end{bmatrix},$$

$$B = \begin{bmatrix} P \\ 0 \end{bmatrix}, \qquad C = \begin{bmatrix} L \\ 0 \end{bmatrix}.$$

Here, $W \in \mathbb{R}^{N\times N}$ can be any non-singular matrix. A common choice is the identity matrix, $W = I$. If the matrices M, D, and K are all symmetric and M is nonsingular, as it is often the case in structural dynamics, we can choose $W = M$. The resulting matrices A and E in the linearized system (6.3) are then symmetric, and thus preserve the symmetry of the original second-order system.

Assume that, for simplicity, we have zero initial conditions, i.e., $q(0) = q_0$, $\frac{dq}{dt}(0) = 0$, and $u(0) = 0$ in (6.1) and (6.2). Then, by taking the Laplace transform of (6.1) and (6.2), we obtain the following system:

$$s^2MQ(s) + DQ(s) + KQ(s) = PU(s),$$

$$Y(s) = L^{\mathrm{T}}Q(s).$$

Eliminating $Q(s)$ results in the frequency-domain input–output relation $Y(s) = H(s)U(s)$, where

$$H(s) := L^{\mathrm{T}}\left(s^2M + sD + K\right)^{-1}P$$

is the transfer function. In view of the equivalent linearized system (6.3) and (6.4), the transfer function can also be written as

$$H(s) = C^{\mathrm{T}}(sE - A)^{-1}B.$$

If the matrix K in (6.1) is nonsingular, then $s_0 = 0$ is guaranteed not to be a pole of H. In this case, H can be expanded about $s_0 = 0$ as follows:

$$H(s) = M_0 + M_1s + M_2s^2 + \cdots,$$

where the matrices M_j are the so-called *low-frequency moments*. In terms of the matrices of the linearized system (6.3) and (6.4), the moments are given by

$$M_j = -C^{\mathrm{T}}\left(A^{-1}E\right)^j A^{-1}B, \quad j = 0, 1, 2, \ldots.$$

6.1. *Frequency-response analysis methods*

In this subsection, we describe the use of eigensystem analysis to tackle the second-order system (6.1) and (6.2) directly.

We assume that the input force vector $u(t)$ of (6.1) is time-harmonic:

$$u(t) = \tilde{u}(\omega)\,\mathrm{e}^{\mathrm{i}\omega t},$$

where ω is the frequency of the system. Correspondingly, we assume that the state variables of the second-order system can be represented as follows:

$$q(t) = \tilde{q}(\omega)\,\mathrm{e}^{\mathrm{i}\omega t}.$$

The problem of solving the system of second-order differential equations (6.1) then reduces to solving the parameterized linear system of equations

$$\left(-\omega^2 M + \mathrm{i}\omega D + K\right)\tilde{q}(\omega) = P\tilde{u}(\omega) \tag{6.5}$$

for $\tilde{q}(\omega)$. This approach is called the *direct frequency-response analysis method*. For a given frequency ω_0, one can use a linear system solver, either direct or iterative, to obtain the desired vector $\tilde{q}(\omega_0)$.

Alternatively, we can try to reduce the cost of solving the large-scale parameterized linear system of Eq. (6.5) by first applying an eigensystem analysis. This approach is called the *modal frequency-response analysis* in structural dynamics. The basic idea is to first transfer the coordinates $\tilde{q}(\omega)$ of the state vector $q(t)$ to new coordinates $p(\omega)$ as follows:

$$q(t) \cong W_k\, p(\omega)\,\mathrm{e}^{\mathrm{i}\omega t}.$$

Here, W_k consists of k selected modal shapes to retain the modes whose resonant frequencies lie within the range of forcing frequencies. More precisely, W_k consists of k selected eigenvectors of the underlying quadratic eigenvalue problem $(\lambda^2 M + \lambda D + K)w = 0$. Eq. (6.5) is then approximated by

$$\left(-\omega^2 M W_k + \mathrm{i}\omega D W_k + K W_k\right)p(\omega) = P\tilde{u}(\omega).$$

Multiplying this equation from the left by W_k^{T}, we obtain a $k \times k$ parameterized linear system of equations for $p(\omega)$:

$$\left(-\omega^2 (W_k^{\mathrm{T}} M W_k) + \mathrm{i}\omega (W_k^{\mathrm{T}} D W_k) + (W_k^{\mathrm{T}} K W_k)\right)p(\omega) = W_k^{\mathrm{T}} P(\omega).$$

Typically, $k \ll n$. The main question now is how to obtain the desired modal shapes W_k. One possibility is to simply extract W_k from the matrix pair (M, K) by ignoring the contribution of the damping term. This is called the *modal superposition method* in structural dynamics. This approach is applicable under the assumption that the damping term is of a certain form. For example, this is the case for so-called Rayleigh damping $D = \alpha M + \beta K$, where α and β are scalars (see CLOUGH and PENZIEN [1975]). In general, however, one may need to solve the full quadratic eigenvalue problem $(\lambda^2 M + \lambda D + K)w = 0$ in order to obtain the desired modal shapes W_k. Some of these techniques have been reviewed in the recent survey paper by TISSEUR and MEERBERGEN [2001] on the quadratic eigenvalue problem.

6.2. *Reduced-order modeling based on linearization*

An obvious approach to constructing reduced-order models of the second-order system (6.1) and (6.2) is to apply any of the model-reduction techniques for linear systems to the linearized system (6.3) and (6.4). In particular, we can employ the Krylov-subspace techniques discussed in Section 3.

The resulting approach can be summarized as follows:

(1) Linearize the second-order system (6.1) and (6.2) by properly defining the $2N \times 2N$ matrices A and E of the equivalent linear system (6.3) and (6.4). Select an expansion point s_0 "close" to the frequency range of interest and such that the matrix $A - s_0 E$ is nonsingular.

(2) Apply a suitable Krylov process, such as the nonsymmetric band Lanczos algorithm described in Section 3.2, to the matrix $M := (A - s_0 E)^{-1} E$ and the blocks of right and left starting vectors $R := (A - s_0 E)^{-1} B$ and $L := C$ to obtain bi-orthogonal Lanczos basis matrices V_n and W_n for the nth right and left block-Krylov subspaces $\mathcal{K}_n(M, R)$ and $\mathcal{K}_n(M^{\mathrm{T}}, L)$.

(3) Approximate the state vector $x(t)$ by $V_n z(t)$ where $z(t)$ is determined by the following linear reduced-order model of the linear system (6.3) and (6.4):

$$E_n \frac{\mathrm{d}z}{\mathrm{d}t} = A_n z + B_n u(t), \qquad y(t) = C_n^{\mathrm{T}} z(t).$$

Here, $E_n = T_n$, $A_n = \Delta_n + s_0 T_n$, $B_n = \rho_n^{(\mathrm{pr})}$, $C_n = \eta_n^{(\mathrm{pr})}$, and T_n, Δ_n, $\rho_n^{(\mathrm{pr})}$, $\eta_n^{(\mathrm{pr})}$ are the matrices generated by the nonsymmetric band Lanczos algorithm.

In Fig. 6.1, we show the results of this approach applied to the linear-drive multimode resonator structure described in CLARK, ZHOU and PISTER [1998]. The solid lines are the Bode plots of the frequency response of the original second-order system, which is of dimension $N = 63$. The dashed line in the left, respectively right, plot is the Bode plot of the frequency response of the reduced-order model of dimension $n = 8$, respectively $n = 12$. The relative error between the transfer functions of the original system and the reduced-order model of dimension $n = 12$ is less than 10^{-4} over the frequency range shown in Fig. 6.1.

There are a couple of advantages of the linearization approach. First, one can directly employ existing reduced-order modeling techniques developed for linear systems. Second, one can also exploit the structures of the linearized system matrices A and E in a Krylov process to reduce the computational cost. However, the linearization approach also has disadvantages. In particular, it ignores the physical meaning of the original system matrices, and more importantly, the reduced-order models are no longer in a second-order form. For engineering design and control of structural systems, it is often

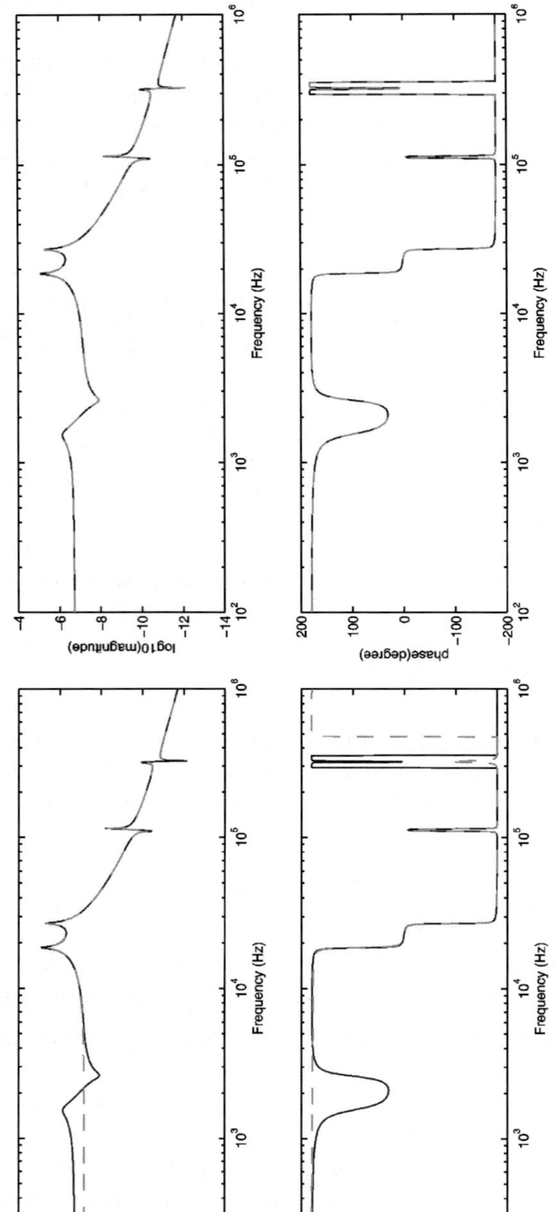

FIG. 6.1. Bode plots for the original system and the reduced-order model of dimension $n = 8$ (left) and $n = 12$ (right).

desirable to have reduced-order models that preserve the second-order form; see, e.g., SU and CRAIG [1991].

While the straightforward linearization approach has the above disadvantages, it is possible to exploit the inherent structure of the Krylov subspaces associated with the linearized system to construct reduced-order models of second-order, or even higher-order, systems that preserve the higher-order structure. Such structure-preserving linearization approaches are described in FREUND [2004a], FREUND [2004b].

6.3. Reduced-order modeling based on second-order systems

In this section, we discuss a Krylov-subspace technique that produces a reduced-order model of second-order form. This approach is based on the work of SU and CRAIG [1991].

The key observation is the following. In view of the linearization (6.3) and (6.4) of the second-order system (6.1) and (6.2), the desired Krylov subspace for reduced-order modeling is

$$\operatorname{span}\{\widetilde{B}, (A^{-1}E)\widetilde{B}, (A^{-1}E)^2\widetilde{B}, \dots, (A^{-1}E)^{n-1}\widetilde{B}\}.$$

Here, $\widetilde{B} := -A^{-1}[B\ C]$. Moreover, we have assumed that the matrix A in (6.3) is nonsingular. Let us set

$$R_j = \begin{bmatrix} R_j^d \\ R_j^v \end{bmatrix} := (-A^{-1}E)^j \widetilde{B},$$

where R_j^d is the vector of length N corresponding to the displacement portion of the vector R_j, and R_j^v is the vector of length N corresponding to the velocity portion of the vector R_j, see SU and CRAIG [1991]. Then, in view of the structure of the matrices A and E, we have

$$\begin{bmatrix} R_j^d \\ R_j^v \end{bmatrix} = (-A^{-1}E)\begin{bmatrix} R_{j-1}^d \\ R_{j-1}^v \end{bmatrix} = \begin{bmatrix} K^{-1}DR_{j-1}^d + K^{-1}MR_{j-1}^d \\ -R_{j-1}^d \end{bmatrix}.$$

Note that the jth velocity-portion vector R_j^v is the same (up to its sign) as the $(j-1)$st displacement-portion vector R_{j-1}^d. In other words, the second portion R_j^v of R_j is the "one-step" delay of the first portion R_{j-1}^d of R_j. This suggests that one may simply choose

$$\operatorname{span}\{R_0^d, R_1^d, R_2^d, \dots, R_{n-1}^d\} \tag{6.6}$$

as the projection subspace used for reduced-order modeling.

In practice, for numerical stability, one may opt to employ the Arnoldi process to generate an orthonormal basis Q_n of the subspace (6.6). The resulting procedure can be summarized as follows.

ALGORITHM 6.1 (*Algorithm by Su and Craig Jr.*).

(0) (Initialization)

Set $R_0^d = K^{-1}[P\ L]$, $R_0^v = 0$, $U_0 S_0 V_0^{\mathrm{T}} = (R_0^d)^{\mathrm{T}} K R_0^d$ (by computing an SVD),

$$Q_1^d = R_0^d U_0 S_0^{-1/2}, \text{ and } Q_1^v = 0.$$

(1) (Arnoldi loop)
For $j = 1, 2, \ldots, n-1$ do:
Set $R_j^d = K^{-1}(DQ_{j-1}^d + MQ_{j-1}^v)$ and $R_j^v = -Q_{j-1}^d$.
(2) (Orthogonalization)
For $i = 1, 2, \ldots, j$ do:
Set $T_i = (Q_i^d)^{\mathrm{T}} K R_j^d$, $R_j^d = R_j^d - Q_i^d T_i$, and $R_j^v = R_j^v - Q_i^v T_i$.
(3) (Normalization)
Set $U_0 S_0 V_0^{\mathrm{T}} = (R_j^d)^{\mathrm{T}} K R_j^d$ (by computing an SVD),

$$Q_{j+1}^d = R_j^d U_0 S_0^{-1/2}, \text{ and } Q_{j+1}^v = R_j^v U_0 S_0^{-1/2}.$$

An approximation of the state vector $q(t)$ can then be obtained by constraining $q(t)$ to the subspace spanned by the columns of Q_n, i.e., $q(t) \approx Q_n z(t)$. Moreover, the reduced-order state vector $z(t)$ is defined as the solution of the following second-order system:

$$M_n \frac{\mathrm{d}^2 q}{\mathrm{d}t^2} + D_n \frac{\mathrm{d}q}{\mathrm{d}t} + K_n q = P_n u(t), \tag{6.7}$$

$$y(t) = L_n^{\mathrm{T}} q(t), \tag{6.8}$$

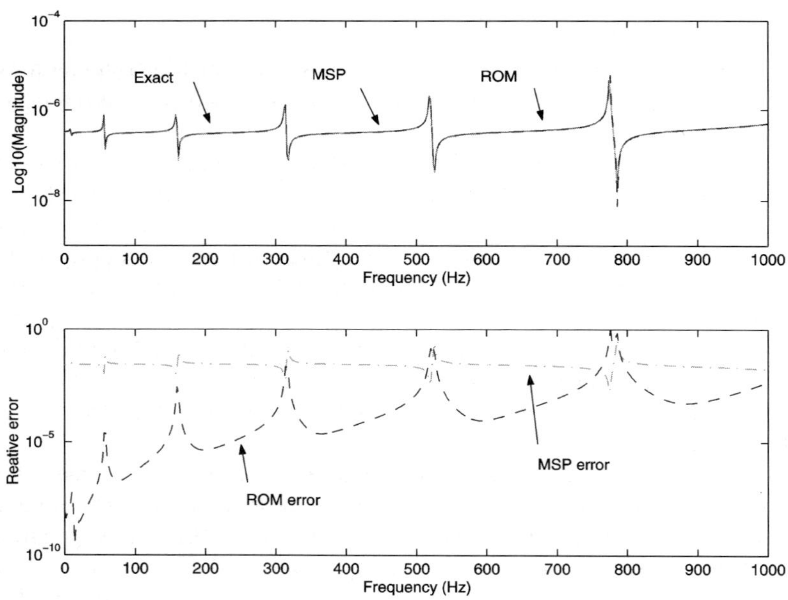

FIG. 6.2. Frequency-response analysis (top plot) and relative errors (bottom plot) of a finite-element model of a shaft.

where $M_n := Q_n^{\mathrm{T}} M Q_n$, $D_n := Q_n^{\mathrm{T}} D Q_n$, $K_n := Q_n^{\mathrm{T}} K Q_n$, $P_n := Q_n^{\mathrm{T}} P$, and $L_n := Q_n^{\mathrm{T}} L$. Note that (6.7) and (6.8) is a reduced-order model in second-order form of the original second-order system (6.1) and (6.2).

In SU and CRAIG [1991], a number of advantages of this approach are described. Here, we present some numerical results of a frequency-response analysis of a second-order system of order $N = 400$, which arises from a finite-element model of a shaft on bearing support with a damper. In the top of Fig. 6.2, we plot the magnitudes of the transfer function H computed exactly, approximated by the model-superposition (MSP) method, and approximated by the Krylov-subspace technique (ROM). For the MSP method, we used the 80 modal shapes W_{80} from the matrix pencil (M, K). The reduced-order model (6.7) and (6.8) is also of dimension $n = 80$. The bottom plot of Fig. 6.2 shows the relative errors between the exact transfer function and its approximations based on the MSP method (dash-dotted line) and the ROM method (dashed line). The plots indicate that no accuracy has been lost by the Krylov subspace-based method.

7. Semi-second-order dynamical systems

In some applications, in particular in the simulation of MEMS devices (SENTURIA, ALURU and WHITE [1997]), the underlying mathematical models are second-order systems with nonlinear excitation forces of the following type:

$$\begin{aligned} M\frac{\mathrm{d}^2 q}{\mathrm{d}t^2} + D\frac{\mathrm{d}q}{\mathrm{d}t} + Kq &= Pu\left(q, \frac{\mathrm{d}q}{\mathrm{d}t}, t\right), \\ y(t) &= L^{\mathrm{T}} q(t). \end{aligned} \tag{7.1}$$

Here, the system matrices M, D, K, P, and L have the same interpretation as in the standard second-order system (6.1) and (6.2). However, excitation force u is now a nonlinear function of q, and possibly $\frac{\mathrm{d}q}{\mathrm{d}t}$.

Systems of the form (7.1) and (7.1) are called *semi-second-order* time-invariant multi-input multi-output linear dynamical systems. Such systems are used as the underlying mathematical models in SUGAR [2001], which is a system-level simulation package for MEMS devices. For example, Fig. 7.1 shows a simple electrostatic gap-closing actuator, which is used as a demo in SUGAR. In this case, the excitation force u includes the electrostatic potential between the plates and is proportional to $(v(t)/\operatorname{gap}(q))^2$, where $v(t)$ is the voltage between electrodes and $\operatorname{gap}(q)$ is a scalar function of q for the distance between the two place electrodes. For mode details about the model used for the electrostatic gap-closing actuator, see BAI, BINDEL, CLARK, DEMMEL, PISTER and ZHOU [2000].

Instead of treating the semi-second-order system (7.1) and (7.1) as a general nonlinear system, we can exploit the structure of the system and apply the idea of "nonlinear dynamics using linear modes". This approach is suggested in ANANTHASURESH, GUPTA and SENTURIA [1996], where a non-damped system, i.e., $D = 0$ is considered and the eigenmodes of M and K are used to extract a reduced-order model. In BAI, BINDEL, CLARK, DEMMEL, PISTER and ZHOU [2000], we described a Krylov-subspace based reduced-order modeling technique for systems (7.1) and (7.1). The idea is to first ignore

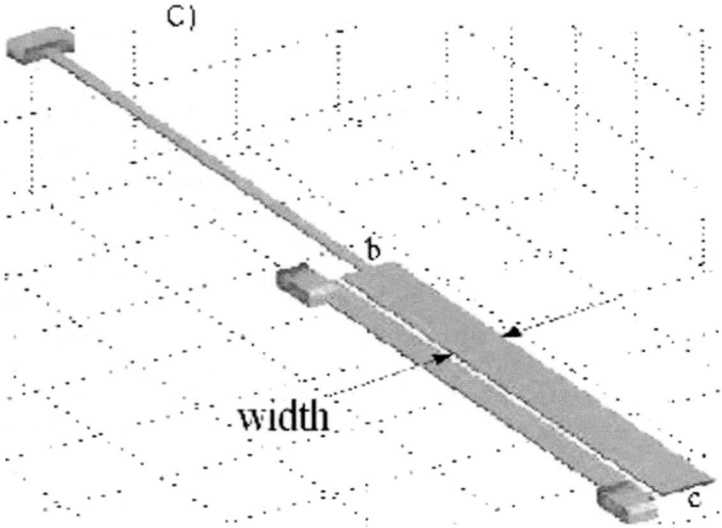

FIG. 7.1. Electrostatic gap-closing actuator.

the nonlinearity in the force term u, and treat the system as a second-order system. Using the approach discussed in Section 6.2, a projection space V_n is constructed, which may be regarded as the *linear Krylov modes*. The vector q is then expanded in terms of the constructed subspace, namely $q(t) \approx V_n z(t)$, and we obtain the following reduced-order model in terms of the vector $z(t)$:

$$E_n \frac{\mathrm{d}z}{\mathrm{d}t} = A_n x + B_n u\big(V_n z(t), t\big),$$
$$y(t) = C_n^{\mathrm{T}} z(t).$$

Here, the definitions of E_n, A_n, B_n, and C_n are the same as in Section 6.2. Note that the excitation force term $u(q,t)$ of the full-order system is replaced by $u(V_n z(t), t)$ in the reduced-order model. When the reduce-order model is solved by a numerical method, it is necessary that $u(V_n z_j, t)$ can be evaluated for the given z_j, which may be regarded as the approximation of $z(t)$ at time step $t = t_j$.

In Fig. 7.2, we illustrate this approach for the transient analysis of the electrostatic gap-closing actuator shown in Fig. 7.1. The first plot shows the output $y(t)$ of the original system and the output $\tilde{y}(t)$ of the reduced-order system of dimension $n = 6$. The original systems has dimension $N = 30$. The second plot shows the accuracy of the reduced-order model of dimension $n = 6$ in terms of the relative error $\|y(t) - \tilde{y}(t)\|/\|y(t)\|$.

We remark that, as indicated in GABBAY, MEHNER and SENTURIA [2000], the use of linear (eigen or Krylov) modes may not adequately capture all the features of nonlinear behavior. It is the subject of current research to further understand the approach sketched in this section and its limitations.

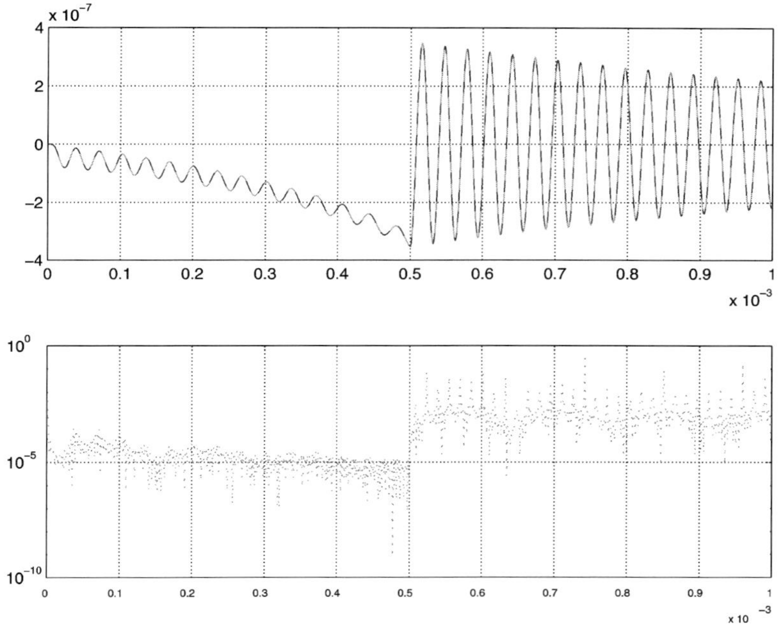

FIG. 7.2. Transient responses of the gap-closing actuator.

8. Concluding remarks

We presented a survey of the most common techniques for reduced-order modeling of large-scale linear dynamical systems. By and large, the area of linear reduced-order modeling is fairly well explored, and we have a number of efficient techniques at our disposal. Still, some open problems remain. One such problem is the construction of reduced-order models that preserve stability or passivity and at the same time, have optimal approximation properties. In particular in circuit simulation, reduced-order modeling is used to substitute large linear subsystems within the simulation of even larger, in general nonlinear systems. It would be important to better understand the effects of these substitutions on the overall nonlinear simulation.

Finally, the systems arising in the simulation of electronic circuits are nonlinear in general, and it would be highly desirable to apply nonlinear reduced-order modeling techniques directly to these nonlinear systems. However, the area of nonlinear reduced-order modeling is in its infancy compared to the state-of-the-art of linear reduced-order modeling. We expect that further progress in model reduction will mainly occur in the area of nonlinear reduced-order modeling.

References

ALIAGA, J.I., BOLEY, D.L., FREUND, R.W., HERNÁNDEZ, V. (2000). A Lanczos-type method for multiple starting vectors. *Math. Comp.* **69**, 1577–1601.

ANANTHASURESH, G.K., GUPTA, R.K., SENTURIA, S.D. (1996). An approach to macromodeling of MEMS for nonlinear dynamic simulation. In: *Microelectromechanical Systems (MEMS)*. In: ASME Dynamics Systems & Control (DSC) Ser. **59**, pp. 401–407.

ANDERSON, B.D.O. (1967). A system theory criterion for positive real matrices. *SIAM J. Control* **5**, 171–182.

ANDERSON, B.D.O., VONGPANITLERD, S. (1973). *Network Analysis and Synthesis* (Prentice-Hall, Englewood Cliffs, NJ).

ARNOLDI, W.E. (1951). The principle of minimized iterations in the solution of the matrix eigenvalue problem. *Quart. Appl. Math.* **9**, 17–29.

ARVESON, W. (1975). Interpolation problems in nest algebras. *J. Funct. Anal.* **20**, 208–233.

BAI, Z. (2002). Krylov subspace techniques for reduced-order modeling of large-scale dynamical systems. *Appl. Numer. Math.* **43** (1–2), 9–44.

BAI, Z., BINDEL, D., CLARK, J., DEMMEL, J., PISTER, K.S.J., ZHOU, N. (2000). New numerical techniques and tools in SUGAR for 3D MEMS simulation. In: *Technical Proc. 4th Internat. Conf. on Modeling and Simulation of Microsystems*, pp. 31–34.

BAI, Z., FELDMANN, P., FREUND, R.W. (1998). How to make theoretically passive reduced-order models passive in practice. In: *Proc. IEEE 1998 Custom Integrated Circuits Conference* (IEEE, Piscataway, NJ), pp. 207–210.

BAI, Z., FREUND, R.W. (2000). Eigenvalue-based characterization and test for positive realness of scalar transfer functions. *IEEE Trans. Automat. Control* **45** (12), 2396–2402.

BAI, Z., FREUND, R.W. (2001a). A partial Padé-via-Lanczos method for reduced-order modeling. *Linear Algebra Appl.* **332**, 139–164.

BAI, Z., FREUND, R.W. (2001b). A symmetric band Lanczos process based on coupled recurrences and some applications. *SIAM J. Sci. Comput.* **23** (2), 542–562.

BOYD, S., EL GHAOUI, L., FERON, E., BALAKRISHNAN, V. (1994). *Linear Matrix Inequalities in System and Control Theory* (SIAM Publications, Philadelphia, PA).

CAMPBELL, S.L. (1980). *Singular Systems of Differential Equations* (Pitman, London, UK).

CAMPBELL, S.L. (1982). *Singular Systems of Differential Equations II* (Pitman, London, UK).

CHEN, T.-H., LUK, C., CHEN, C.C.-P. (2003). In: *Technical Digest of the 2003 IEEE/ACM Int. Conf. on Computer-Aided Design* (IEEE Computer Society Press, Los Alamitos, CA), pp. 786–792.

CHIRLIAN, P.M. (1967). *Integrated and Active Network Analysis and Synthesis* (Prentice-Hall, Englewood Cliffs, NJ).

CLARK, J.V., ZHOU, N., PISTER, K.S.J. (1998). MEMS simulation using SUGAR v0.5. In: *Proc. Solid-State Sensors and Actuators Workshop* (Hilton Head Island, SC), pp. 191–196.

CLOUGH, R.W., PENZIEN, J. (1975). *Dynamics of Structures* (McGraw-Hill, New York).

DAI, L. (1989). *Singular Control Systems*, Lecture Notes in Control and Information Sciences **118** (Springer-Verlag, Berlin, Germany).

DEPRETTERE, E. (1981). Mixed-form time-variant lattice recursions. In: *Outils et Modèles Mathématiques pour l'Automatique, l'Analyse de Systèmes et le Traitement du Signal* (CNRS, Paris), pp. 545–562.

DEWILDE, P. (1988). New algebraic methods for modeling large-scale integrated circuits. *Circuit Theory Appl.* **16**, 473–503.

DEWILDE, P. (1995). J-unitary matrices for algebraic approximation and interpolation – the singular case. In: Moonen, M., Moor, B.D. (eds.), *SVD and Signal Processing, III, Algorithms, Architectures and Applications* (Elsevier, Amsterdam), pp. 209–223.

DEWILDE, P., DEPRETTERE, E.F. (1987). Approximate inversion of positive matrices with applications to modelling. In: Curtain, R.F. (ed.), *Modelling, Robustness and Sensitivity Reduction in Control Systems*. In: NATO ASI Series **F34** (Springer-Verlag, Berlin), pp. 211–238.

DEWILDE, P., DYM, H. (1981). Schur recursions, error formulas, and convergence of rational estimators for stationary stochastic sequences. *IEEE Trans. Inf. Theory* **27** (4), 446–461.

DEWILDE, P., VAN DER VEEN, A.-J. (1998). *Time-Varying Systems and Computations* (Kluwer, Dordrecht).

DEWILDE, P., VIEIRA, A., KAILATH, T. (1978). On a generalized Szegö–Levinson realization algorithm for optimal linear predictors based on a network synthesis approach. *IEEE Trans. Circuits Syst.* **25** (9), 663–675.

FELDMANN, P., FREUND, R.W. (1994). Efficient linear circuit analysis by Padé approximation via the Lanczos process. In: *Proceedings of EURO-DAC'94 with EURO-VHDL'94* (IEEE Computer Society Press, Los Alamitos, CA), pp. 170–175.

FELDMANN, P., FREUND, R.W. (1995a). Efficient linear circuit analysis by Padé approximation via the Lanczos process. *IEEE Trans. Computer-Aided Design* **14**, 639–649.

FELDMANN, P., FREUND, R.W. (1995b). Reduced-order modeling of large linear subcircuits via a block Lanczos algorithm. In: *Proc. 32nd ACM/IEEE Design Automation Conference* (ACM, New York, NY), pp. 474–479.

FREUND, R.W. (1995). Computation of matrix Padé approximations of transfer functions via a Lanczos-type process. In: Chui, C., Schumaker, L. (eds.), *Approximation and Interpolation*. In: Approximation Theory VIII **1** (World Scientific Publishing Co Inc., Singapore), pp. 215–222.

FREUND, R.W. (1999a). Passive reduced-order models for interconnect simulation and their computation via Krylov-subspace algorithms. In: *Proc. 36th ACM/IEEE Design Automation Conference* (ACM, New York, NY), pp. 195–200.

FREUND, R.W. (1999b). Reduced-order modeling techniques based on Krylov subspaces and their use in circuit simulation. In: Datta, B.N. (ed.), Applied and Computational Control, Signals, and Circuits **1** (Birkhäuser, Boston), pp. 435–498.

FREUND, R.W. (2000a). Band Lanczos method (Section 7.10). In: Bai, Z., Demmel, J., Dongarra, J., Ruhe, A., van der Vorst, H. (eds.), *Templates for the Solution of Algebraic Eigenvalue Problems: A Practical Guide* (SIAM Publications, Philadelphia, PA), pp. 205–216. Also available online from http://cm.bell-labs.com/cs/doc/99.

FREUND, R.W. (2000b). Krylov-subspace methods for reduced-order modeling in circuit simulation. *J. Comput. Appl. Math.* **123** (1–2), 395–421.

FREUND, R.W. (2003). Model reduction methods based on Krylov subspaces. *Acta Numer.* **12**, 267–319.

FREUND, R.W. (2004a). Krylov subspaces associated with higher-order linear dynamical systems. Technical report, Department of Mathematics, University of California, Davis, CA, submitted for publication.

FREUND, R.W. (2004b). Padé-type model reduction of second-order and higher-order linear dynamical systems. Technical report, Department of Mathematics, University of California, Davis, CA, USA, submitted for publication.

FREUND, R.W. (2004c). SPRIM: structure-preserving reduced-order interconnect macromodeling. In: *Tech. Dig. 2004 IEEE/ACM International Conference on Computer-Aided Design* (IEEE Computer Society Press, Los Alamitos, CA), pp. 80–87.

FREUND, R.W., FELDMANN, P. (1996a). Reduced-order modeling of large passive linear circuits by means of the SyPVL algorithm. In: *Tech. Dig. 1996 IEEE/ACM International Conference on Computer-Aided Design* (IEEE Computer Society Press, Los Alamitos, CA), pp. 280–287.

FREUND, R.W., FELDMANN, P. (1996b). Small-signal circuit analysis and sensitivity computations with the PVL algorithm. In: IEEE Trans. Circuits and Systems – II: Analog and Digital Signal Processing **43**, pp. 577–585.

FREUND, R.W., FELDMANN, P. (1997). The SyMPVL algorithm and its applications to interconnect simulation. In: *Proc. 1997 Internat. Conf. on Simulation of Semiconductor Processes and Devices* (IEEE, Piscataway, NJ), pp. 113–116.

FREUND, R.W., FELDMANN, P. (1998). Reduced-order modeling of large linear passive multi-terminal circuits using matrix-Padé approximation. In: *Proc. Design, Automation and Test in Europe Conference 1998* (IEEE Computer Society Press, Los Alamitos, CA), pp. 530–537.

FREUND, R.W., JARRE, F. (2004a). An extension of the positive real lemma to descriptor systems. *Optim. Methods Softw.* **18** (1), 69–87.

FREUND R.W., JARRE, F. (2004b). Numerical computation of nearby positive real systems in the descriptor case. Technical Report, Department of Mathematics, University of California, Davis, California, USA, in preparation.

GABBAY, L.D., MEHNER, J.E., SENTURIA, S.D. (2000). Computer-aided generation of nonlinear reduced-order dynamic macromodels – I: Non-stress-stiffened case. *J. Microelectromech. Syst.* **9** (2), 262–269.

KIM, S.-Y., GOPAL, N., PILLAGE, L.T. (1994). Time-domain macromodels for VLSI interconnect analysis. *IEEE Trans. Computer-Aided Design* **13**, 1257–1270.

LANCZOS, C. (1950). An iteration method for the solution of the eigenvalue problem of linear differential and integral operators. *J. Res. Natl. Bur. Stand.* **45**, 255–282.

MASUBUCHI, I., KAMITANE, Y., OHARA, A., SUDA, N. (1997). H_∞ control for descriptor systems; a matrix inequalities approach. *Automatica J. IFAC* **33** (4), 669–673.

NELIS, H. (1989). Sparse approximations of inverse matrices. Ph.D. thesis, Delft Univ. Techn., The Netherlands.

NELIS, H., DEWILDE, P., DEPRETTERE, E. (1989). Inversion of partially specified positive definite matrices by inverse scattering. In: The Gohberg Anniversary Collection. Operator Theory: Advances and Applications **40** (Birkhäuser Verlag, Basel), pp. 325–357.

ODABASIOGLU, A. (1996). Provably passive RLC circuit reduction. M.S. thesis, Department of Electrical and Computer Engineering, Carnegie Mellon University.

ODABASIOGLU, A., CELIK, M., PILEGGI, L.T. (1997). PRIMA: passive reduced-order interconnect macromodeling algorithm. In: *Tech. Dig. 1997 IEEE/ACM International Conference on Computer-Aided Design* (IEEE Computer Society Press, Los Alamitos, CA), pp. 58–65.

PILEGGI, L.T. (1995). Coping with RC(L) interconnect design headaches. In: *Tech. Dig. 1995 IEEE/ACM International Conference on Computer-Aided Design* (IEEE Computer Society Press, Los Alamitos, CA), pp. 246–253.

ROHRER, R.A., NOSRATI, H. (1981). Passivity considerations in stability studies of numerical integration algorithms. *IEEE Trans. Circuits Syst.* **28**, 857–866.

RUEHLI, A.E. (1974). Equivalent circuit models for three-dimensional multiconductor systems. *IEEE Trans. Microwave Theory Tech.* **22**, 216–221.

SCHUR, I. (1917). Über Potenzreihen, die im Innern des Einheitskreises beschränkt sind, I. *J. Reine Angew. Math.* **147**, 205–232. English transl. In: Operator Theory: Advances and Applications **18** (Birkhäuser Verlag, Basel), 1986, pp. 31–59.

SENTURIA, S.D., ALURU, N., WHITE, J. (1997). Simulating the behavior of MEMS devices: Computational methods and needs. *IEEE Comput. Sci. Eng. Mag.* **4**, 30–43.

SHEEHAN, B.N. (1999). ENOR: model order reduction of RLC circuits using nodal equations for efficient factorization. In: *Proc. 36th ACM/IEEE Design Automation Conference* (ACM, New York, NY), pp. 17–21.

SU, T.-J., CRAIG JR., R.R. (1991). Model reduction and control of flexible structures using Krylov vectors. *J. Guidance Control Dynam.* **14**, 260–267.

SUGAR (2001). A MEMS simulation program. Available at 2001; http://www-bsac.eecs.berkeley.edu/cadtools/sugar.

TISSEUR, F., MEERBERGEN, K. (2001). The quadratic eigenvalue problem. *SIAM Rev.* **43** (2), 235–286.

VAN DER MEIJS, N. (1992). Accurate and efficient layout extraction. Ph.D. thesis, Delft Univ. Techn., The Netherlands.

VERGHESE, G.C., LÉVY, B.C., KAILATH, T. (1981). A generalized state-space for singular systems. *IEEE Trans. Automat. Control* **26** (4), 811–831.

VLACH, J., SINGHAL, K. (1994). *Computer Methods for Circuit Analysis and Design*, second ed. (Van Nostrand Reinhold, New York, NY).

WILLEMS, J.L. (1970). *Stability Theory of Dynamical Systems* (John Wiley & Sons, Inc., New York, NY).

ZHOU, K., DOYLE, J.C., GLOVER, K. (1996). *Robust and Optimal Control* (Prentice-Hall, Upper Saddle River, NJ).

Subject Index

16 bit adder, 578
ΦC-loop, 552

A- and L-stability, 576
A- and L-stable, 576
a-posteriori error check, 569
A-stability, 567
A-stable, 567, 568
$A(\alpha)$-stability, 567
absorbing boundary conditions (ABCs), 201, 203, 204, 278, 279, 286
acceptor, 33
acceptor concentration, 36
accumulation layer, 27, 37
accumulation mode, 36
acoustic phonon scattering, 448, 458, 460
across (quantities), 129
action functional, 14, 15, 17, 19
action integral, 17, 60, 61, 76
active, 613, 614
active networks, 550
adaptive grids, 500
adaptive mesh refinement (AMR), 501, 502, 504–507, 509, 511, 513, 515
adaptive partitioning, 610
adaptively controlled explicit integration, 596
adaptivity, 568, 586, 590
adjoint (operator), 125
adjoint system, 625
admittance form, 530, 533, 555
admittance matrix, 731–736
affine
– connection, 70, 74
– – gravitational, 74
– equivalence, 124, 186
– map, 112
– space, 110, 111
algebraic
– components, 550
– constraint, 547
– – hidden, 547
– index, 547
– variables, 546
alternating-direction-implicit (ADI), 509
– finite-difference time-domain (FDTD) method, 271
alternative finite-difference grids, 226
ALU, 578
amber, 68
ambient (space), 117
Ampère, 138
– law, 10, 14, 205
– rule, 116
amplitude modulation, 642
AMR, 501, 502, 504–507, 509, 511, 513, 515
analysis, 564
analytical ABCs, 204
analytical boundary conditions, 278
analytical placement, 606
analytical properties of DAEs, 545
angular frequency, 664
angular momentum, 13
approach, conventional, 535
area, 129
artificial element, 644
artificial voltage source, 635, 636, 644
asymptotic waveform evaluation (AWE), 596
averaging technique, 629
axial gauge, 66
axial vector, 115, 137

backward differentiation formulas (BDF methods), 564
backward Euler, 573
ballistic
– channel, 93

– conductor, 86, 87
– silicon diode, 429
– transport, 64, 85, 86
band gap, 27, 31
bandwidth, 618, 619
barrier height, 36
barycenter, 112
barycentric dual, 153
– mesh, 154, 181
baseband signal, 618
basic elements, 532
basis function, 669, 746
basis, contravariant, 96
basis, covariant, 96
battery region, 90
BBW model, 450, 453, 514
BDF, 571, 629
– Modified Extended, 574
BDF schemes, 564, 567
BDF1, 567
BDF2, 567, 568, 579, 580
Berenger, 279
Berenger's perfectly matched layer, 279
Bernoulli function, 30, 79, 385, 398
Bianci identities, 44
bias conditions, 551, 552
BICGSTAB, 742
bifurcation, 564
Biot–Savart's law, 68
bipolar ringoscillator, 557
bipolar transistor, 530, 557
bisection methods, 606
block-centered scheme, 161
block-Gaussian elimination, 576, 635
Boltzmann constant, 21
Boltzmann transport equation, 21, 23, 24, 27–29
Boltzmann transport theory, 27
bordered block diagonal form, 598
Bose–Einstein statistics, 448
boson, 67
bound frame, 116
bound state, 33, 36
bound vector, 112
boundary
– of a cell, 114
– of a chain, 122, 151
– of a manifold, 113
– singularity, 744, 746
boundary conditions, 7, 19, 62, 495
box/surface-integration method, 78
branch, 50
– current, 530, 532
– fluxes, 533
– tearing, 601
– voltage, 530, 532
Breit–Wigner expansion, 38
Brezzi–Douglas–Marini elements, 345
Brillouin zone, 27
built-in potential, 323
bulk, 86
bypass, 611

CAD, 529
canonical momentum, 61, 66
canonical quantization, 14, 93
capacitance, 17, 94, 530, 672, 731
– differential, 534
– general nonlinear, 534
– generalized matrix of, 539, 551
capacitance model, 536
capacitance-oriented formulation, 535
capacitive coupling, 673
capacitive paths, 559
capacitor, 55, 93, 530, 532, 662
– linear, 530
carrier, 618
– frequency, 642
– heating, 24, 30
– mean energy, 29
– network, 600
– temperature, 28
carrier transport
– quantum mechanical features, 83
– quantum theory of, 83
Cartesian coordinates, 100
Cartesian grid, 74, 77
categories, 115
Caughey–Thomas formula, 452
causality, 63
causality principle, 25
cell, 113, 114
cell variable, 40
center (of a cell), 153
centered finite-differences, 211
CFL condition, 469, 474
chain, 121
chainlet, 125
characteristic equations, 529, 532, 537
characteristic impedance, 49
charge, 662
– accumulation, 26
– conservation, 44, 148, 535, 536, 576
– constraint, 576
– density, 27, 60, 664
– distribution, 15, 39, 62
– neutrality, 532, 535
– oriented network equations, 558

– storing element, 535
charge-oriented models, 535
charge/flux oriented
– formulation, 533, 536, 563
– – of MNA, 537
– integration, 569
– ROW schemes, 575
chart, 113
chemical potential, 22, 26, 32, 86, 87
– difference, 87
– position dependent, 88
CHORAL, 576, 579, 580, 582
circuit, 49
– analysis program, 736
– configurations, 541, 550, 552, 564
– modeling, 8
– topology, 93
circulation, 10, 47, 99, 128
classical
– approach, 533
– electrodynamics, 13, 84
– electromagnetism, 83
– field, 68
– Hall effect, 64
CLF condition, 471
closed
– cell, 114
– curve, 95, 96, 99
– electric circuit, 88, 89
– form, 134
– loops, 739
– paving, 150
– surface, 95
closure relations, 457
clustering, 605
– Gauss operations, 609
CMOS, 586
– technology, 31
coarse graining, 21, 22
coaxial cable, 17
– square, 81
coboundary, 134
cocycle, 134
codomain (of a map), 111
coherence length, 91
coherent transport, 93
cohomology, 134, 151, 183
commutative diagram, 174
commuting diagram property, 347–349
compact model, 530
complement (of subspace), 116
complementary operator methods (COM), 204
complementary technique, 557
complex energy eigenvalues, 36
complex-valued numerical wavenumbers, 241
components
– algebraic, 550
– differential, 550
compound step, 616
computational physics, 7
computer, 74
computer calculations, 68
computer-aided analysis, 529
conditions, topological, 549
conductance, 529, 530
– formula, 86
– generalized matrix of, 539, 551
– matrices, 551
– quantization, 84, 86, 88, 91
conducting
– medium, 23
conduction
– band, 26, 35
– – edge, 32
– channel, 85, 88
– – state, 89
conductivity, 12, 45
– tensor, 25
conductor, 20, 45, 47, 78, 95
confinement potential, 65
confining potential, 64
conjugacy (of operators), 169
conjugated variables, 26
connected regions, 95
connection ports, 667
conservation
– laws, 11, 13
– of angular momentum, 13
– of electric charge, 11
– of linear momentum, 12
conservative schemes, 467
consistency of orientations, 118
consistency of scheme, 170
consistent, 626
– initial condition, 626
– initial values, 564
constitutive equations, 12, 20, 21, 29, 31
constitutive relation, 41, 43, 532
constraint, 61
– algebraic, 627
– hidden, 571, 572, 627
constriction, 84
contact resistance, 86, 88, 89
continuation method, 640, 645
continuity equation, 11, 25, 396
continuous energy spectrum, 34
contravariant vector, 69

convective currents, 28
conventional formulation of MNA, 539
convergence, 567, 576
– of scheme, 160, 172, 174
Cooper pair, 67, 91
coordinate
– covariance, 68, 70
– curve, 98
– differentials, 69
– surfaces, 96
– system, 69
– – Euclidean, 73
– transformation, 69, 70
Coulomb blockade, 54
Coulomb gauge, 42, 61, 63, 75, 76
Coulomb potential, 62
counting measure, 127
coupled analysis, 663
coupling of higher-index configurations, 552
coupling of index-2 problems, 552
Courant stability factor, 232
Courant stability limit, 263
covariance, 68, 70
covariant derivative, 74
covariant vector, 69
covector, 133
CPU time, 500
cross product, 124
crossing wires, 80
crosstalk, 560, 663
cryogenic temperatures, 64
crystal defects, 22
crystal lattice, 24
curl, 135
curl–curl operator, 77
current, 130, 662
– carrying state, 88
– conservation, 384
– continuity equation, 379, 385, 664
– density, 17, 18, 27, 35, 42, 60, 92, 133, 320, 321, 664
– – solenoidal, 92
– density operator, 35
– distribution, 15
– in metals, 23
– limiting mechanism, 87
– mode, 557
– operator, 94
– paradox, 35, 36
– source, 533
current-voltage characteristics, 25, 37
curvature, 68, 74, 83
– local, 71
– of space–time, 69
curvilinear coordinates, 96, 98, 100
cutsets, 539
– of inductors, 541
cycle, 122
cyclic coordinate, 88, 96, 97
cyclotron frequency, 65
cylindrical symmetry, 66

DAE, 529, 531, 540
– approach, 540
– index, 605
– models, 542
Dahlquist barrier, second, 567
Dahlquist's linear test equation, 567
damping properties, 579
Debye length, 325, 326
decay time, 25
decaying subband state, 35
decoherence, 86
decomposition, 586, 590
decoupling, 600
defect correction, 574
degeneracy (of elements), 187
degrees of freedom, 8, 13, 555
– field, 61
– of a link, 75
Delaunay triangulation, 365, 366, 372, 386, 388
delocalized state, 88
delta function, 40
delta-normalization, 33
dense output, 616
density, 142
depletion layer, 27
depletion region, 326, 429
device lifetime, 38
dielectric, 78
– constant, 36, 667
– material, 41
– media, 38, 40
– medium, 45
– polarization, 38
diffeomorphism, 115
differential
– capacitance, 534
– components, 550
– forms, 125, 127, 130
– – of higher degree, 181
– geometry, 83, 95
– index, 547, 560
– of a function, 134
– operator, 62, 74
– stages, 557
– variables, 546, 550

differential-algebraic equation (DAE), 529, 531, 540
differential-algebraic network equations, 529
differentiator circuit, 552, 553
diffusion, 26
– equation, 30
dipole charges, 38
dipole moment, 40
Dirac delta function, 666
direct frame, 115, 116
direct linear transform, 124
direct tunneling, 31
director, 116
Dirichlet condition, 141
discrete energy, 160, 169
discrete formulation, 669
discrete Fourier transform (DFT), 258
discretization, 8, 78
– grid, 74
– Scharfetter–Gummel scheme, 79
disordered materials, electron propagation through, 85
dispersion, 160
displaced Maxwellian distribution, 24, 29
displacement, 41
– current, 10
dissipative function, 543
dissipative processes, 12
distance, between near-by points, 70
distortion, 617
distribution function, 22, 24
distributions, 547
divergence, 135
divergence-free basis function, 395, 396
divided differences, 568
domain decomposition methods, 542
domain (of a map), 111
dot product, 127
DRAM, 578
Drazin inverse, 596
drift-diffusion, 317
Drude's model, 23–25, 86
dual
– mesh, 152
– of a cell, 152
– of a face, 152
– of operator, 134
– system, 625
– Whitney forms, 190
duality (of cells and W. forms), 182
duality product, 130
dynamic assignment, 606
dynamic elements, 554
edge state, 64, 65
effective group action, 111
effective mass, 26, 32
– approximation, 32
efficiency, 570
eigenproblem, 622
eigenvalue, 735
eigenvector, 735
Einstein, 70
– general theory of relativity, 68
– relation, 28, 452
– relations, 320
– theory of gravity, 68
elastance, 731
elastic collision, 23
elastic scattering, 12
electric
– charge, 7, 9, 141
– – density, 9, 12, 20
– circuit, 10, 14, 19, 84, 87, 90, 93
– – topology of, 83, 84, 87
– conductance
– – quantization of, 84
– current, 7, 9–11, 83
– – density, 9, 10, 12, 20, 24, 30
– displacement, 40, 93, 664
– field, 9, 10, 14, 30, 66, 664
– – circulation, 88
– – conservative, 87–89
– – external, 92
– – irrotational, 87, 88, 92
– – localized, 93
– – non-conservative, 89
– flux, 10, 41
– monopole, 39
– permittivity, 9
– potential, 665
– susceptibility, 40, 41
electrical
– charges, 532
– conductance, 25
– conductivity, 23, 29
– polarization, 43
electrodynamics, 44, 60, 69
– geometrical interpretation, 73
– geometry of, 68
electromagnetic
– compatibility, 661
– effects, 663
– energy, 12, 44
– field, 9, 11–15, 67, 75, 83, 84, 95
– – quantization of, 93
– – spatial localization of, 83

– – tensor, 12, 74
– potentials, 67
– radiation, 14
– waves, 63
electromagnetism, 7, 15, 83, 94
electromotive force, 10, 51, 87, 91, 94, 132
electron diffusivity, 28
electron mobility, 28
electron–electron interaction, 32, 36
electrostatic
– confinement, 90
– potential, 87
– – drop, 87
element stamps, 565
element, charge storing, 535
elements
– dynamic, 554
– energy storing, 533
– one-port, 532
– static, 554
– two-terminal, 532
embedded method, 576
embedding, 114
embraced flux, 131
EMC, 661
energy, 142, 143
– balance equation, 29
– barrier, 33, 35
– conservation of, 160
– conserving, 568, 580
– dissipation, 83, 86, 90, 91
– flux, 29
– norms, 158
– rate equation, 92
– relaxation time, 29
– spectrum, 33
– – continuous, 88
– – discrete, 88
– storing elements, 533
energy-balance model, 328, 409
energy-transport model, 30, 327, 404
Engquist–Osher flux, 470
ensemble average, 94
entropy, 21, 456
envelope, 630
– method, 646
– wave function, 32, 35
equations of motion, 61, 63
equipotential volume, 90
equivalent circuit, 49, 733, 736
– model, 663
error
– control, 568, 577
– estimate, 343, 347, 355, 360, 375, 394, 401, 570
– scaling, 574
– tolerance, 569
Euclidean
– coordinate system, 73
– geometry, 70
– space, 109, 123
Euler–Lagrange equations, 15, 17
Euler-backward method, 629
Euler-forward method, 629
event control, 592, 613
exact form, 134
exact sequence, 180, 183
explicit integration
– adaptively controlled, 596
exponential fitting, 387, 406, 595
extended hydrodynamical model, 454
exterior derivative, 134
external orientation, 117
extrapolation, 630

face, 149
Faraday, 138
Faraday's law, 10, 13, 44, 68, 205, 530
fast Fourier transform (FFT), 225, 258
Fasterix, 661
fastest first, 612
FDM, 630, 631, 641
feedback loop, 554
Fermi's Golden Rule, 23
Fiduccia–Mattheyses, 606
field equations, 74, 76
finite difference method (FDM), 630, 631, 641
– θ-method, 629
finite difference schemes, 467
finite differences, 211
finite-difference time-domain (FDTD) method, 199, 200
– alternating-direction-implicit (ADI), 271
first Maxwell equation, 15, 42
first order approximation, 562
Floquet exponents, 625
Floquet multipliers, 625
Floquet theory, 624, 625
flux, 128, 131, 132
– conservation, 535
– quantization, 91–93
formulation
– capacitance-oriented, 535
– charge/flux-oriented, 533, 536
formulation of energy storing elements
– charge/flux-oriented, 534
– conventional, 534

forward backward substitution, 565
forward biased, 429
Fourier expansion, 62
Fourier law, 451
fourth Maxwell equation, 15, 94
fourth-order-accurate finite-difference scheme, 251, 252
fractional step method, 464
frame, 115
free group action, 111
free vector, 112
frequency folding, 624
frequency modulation, 642
fully implicit methods, 575
fully-implicit index-1 systems, 567
functional, 15
– differentiation, 76
– modelling, 586

Galerkin, 642
Galerkin Hodge, 157
Gamma function, 461
gate, 84
– arms, 84
– current, 34–36
– – density, 36
– electrode, 31, 33
– leakage current, 35–37
– length, 31
– stack, 31, 33–36
– tunneling current, 36
– voltage, 31, 64, 85
gauge, 13, 60, 62, 64, 67, 73, 75
– condition, 60, 61, 67, 75, 76, 665
– covariance, 68
– covariant variables, 74
– field, 14, 79
– invariance, 14, 60, 61, 68
– – principle of, 69
– theories, 68, 72
– transformation, 14, 42, 60, 67
gauging, 162
Gauss' law, 10, 25, 40, 44, 61, 63, 68, 93
– for the electric field, 206
– for the magnetic field, 206
Gauss' theorem, 52, 95, 97, 98
Gauss–Legendre quadrature rule, 681
Gauss–Legendre rule, 683, 685, 686
Gaussian quadrature, 680, 682
– formula, 678, 679
general nonlinear capacitance, 534
general relativity, 68
general theory of relativity, 68, 69
generalized coordinates, 21, 61
generalized eigenvalue problem, 662, 735, 740
generalized momenta, 21
generalized multirate, 615
generic function, 536
geometric grading, 293
geometrical interpretation, 68, 74
geometry, 68, 83
– non-Euclidean, 68
ghost field, 63, 80
ghost modes, 165
Gibbs' ensemble, 25
Godunov flux, 470
Godunov method, 470
Godunov scheme, 466, 468, 469
grain (of mesh), 167
granularity of parallelism, 597
gravitational field, 69
gravitational forces, 7
gravity, 69, 83
Green's function, 7, 62, 666, 667, 687, 706, 741
– free space, 62
– poles, 63
Green's theorem, 96
grid
– Cartesian, 74, 77
– generation, 8
– links of a, 74, 75, 77, 78
– nodes, 74, 77
– sampling density, 232
– staggered, collocated, 227
– staggered, uncollocated, 209, 226
gridding methods, 202, 226
ground line, 560
group action, 110
gyrators, 550

Hall bar, 83
Hall resistance, 64
Hall voltage, 64
Hamiltonian, 14, 15, 94
– of a closed electric circuit, 94
harmonic
– average, 378, 381, 385
– balance (HB), 620, 630, 638, 641
– modes, 45
– oscillator, 580
– – functions, 65
Hartree approximation, 32, 33, 36
heat flow, 451
Heisenberg equations of motion, 94
Helmholtz
– decomposition, 165
– equation, 46, 48, 665–668

– operator, 45
– theorem, 13, 42, 87, 95, 99
Hermite functions, 66
Hessenberg type, 571
Hessenberg-type index-2 systems, 574
heterojunction, 64, 84
hexagonal grids, 228, 256
hexahedral elements, 184
hidden constraints, 571, 572
– algebraic, 547
hierarchical simulation, 610
high-quality oscillator circuits, 646
higher index, 552, 627, 640, 644
higher-index case, 546
higher-index components, 574
higher-index problems, 547, 554, 560
highly oscillatory perturbations, 584
Hilbert space, 25
Hodge operator, 139, 140
hole, 26
– diffusivity, 28
– mobility, 28
homogeneous magnetic field, 64
homogeneous space, 111, 112
homology, 123
hybrid analysis, 537
hydrodynamic model, 28, 29
hyperbolic system, 461, 464, 465
hypergraphs, 606
hysteresis, 44

ICCG, 742
icon, 117
ideal operational amplifier, 554, 555
idealized network elements, 560
ill-posed problems, 529
impedance, 49
– form, 533
implicit
– Euler scheme, 567
– linear multi-step methods, 563
– numerical integration schemes, 540
impurities, 36
inappropriate regularization, 543
incidence matrix, 150, 672, 731
incidence number, 150
incompressible stationary flow, 41
inconsistent initial values, 579
indefinite linear systems, 662
index, 545
– algebraic, 547
– differential, 547, 560
– perturbation, 547, 549
– tractability, 547
index monitor, 571, 572, 574
index-1
– case, 546
– problem, 551
– systems
– – fully-implicit, 567
index-2, 644
– configurations, 549
– problem
– – singularly perturbed, 560
– systems
– – of Hessenberg-type, 574
index-3 variables, 552
induced orientation, 119
inductance, 17, 19, 20, 81, 83, 92, 94, 551, 672, 731
– generalized matrix of, 539, 551
induction, 136
inductive coupling, 673
inductor, 56, 532, 662
inelastic scattering, 12
inf-sup condition, 338, 339, 341, 347, 354, 358
initial values
– consistent, 564
– inconsistent, 579
inner orientation, 116, 117
insulator, 19–21, 26, 31, 38
integral, 128
integral theorem, 95
interacting electromagnetic field, 15
interaction
– Hamiltonian, 23
– integral, 662, 673, 693, 705, 706
– Lagrangian, 67
interconnection system, 731, 732, 736, 737
interconnects, 552, 560
interface, 27
– state, 34
intermediate state, 35
internal energy, 26
internal orientation, 116
internal resistance, 90
intrinsic concentration, 320
intrinsic grid velocity anisotropy, 238
intrinsic semiconductors, 26
inversion layer, 31–35
inverter stages, 557
irradiation, 668
irreversible processes, 21
irrotational field, 87
isoparametric elements, 184
isotropy, 124
– group, 111

iterated timing analysis, 592
iteration matrix, 570
iterative linear solvers, 596

J · **E** theorem, 51, 53, 58, 92, 96
Jacobian, 98, 565, 566, 570
– sparse, 566
Jacobian matrix, 539, 540, 564
JL cutsets, 552
Jordan decomposition, 546

Kane dispersion relation, 446
Kane model, 429
Katzenelson, 594
Kirchhoff, 8
– current law (KCL), 532, 672
– equations, 662, 672, 673, 730, 731, 737
– laws, 50, 532, 533, 537
– voltage law (KVL), 532, 672
Klein bottle, 121
Kramers–Kronig relations, 25
Kronrod quadrature rule, 662, 680, 681
Kubo's theory, 25
Kurokawa's method, 644

L–R-circuit, 91
L-stability, 567
L-stable, 577, 583
Lagrange equations of the first kind, 542
Lagrange multiplier, 76, 353, 355, 381, 456
Lagrangian, 13, 15, 60, 66, 542, 543
– density, 15, 67
– electromagnetic field, 15
– multipliers, 542, 543
Lanczos algorithm, 740
Landau gauge, 64, 65
Landauer–Büttiker formula, 85, 86, 88, 93, 94
Laplace equation, 744
Laplace transformation, 173
Laplace's equation, 76
Laplacian, 78, 100
large signal solution, 623
latency, 588
latent, 613, 614
lattice temperature, 29
Lax theorem, 172
Lax–Friedrichs flux, 470, 471
LC oscillator, 579
lead, 88–90
– resistance, 90
Legendre polynomial, 39, 679, 682, 749, 750
Lenz' law, 91
LI-cutsets, 549, 572
limit cycle, 621
line integral, 10
linear
– algebra, 736
– capacitor, 530
– momentum, 12
– perturbation analysis, 624
– PI-controller, 570
– resistors, 529
– response, 12
– system solution, 736, 737
Linear Time Varying (LTV), 623
linearly-implicit methods, 575
link, 79
– variables, 75
Liouville's theorem, 22
little group, 111
local error, 569
local truncation error, 568, 569
longitudinal
– component, 99
– current, 64
– electric field, 48
– magnetic field, 47
– polarization, 63
– resistance, 64
loops, 539
– of capacitors, 541
Lorentz force, 12, 16, 64
Lorentz gauge condition, 665
Lorenz gauge, 63
low pass filter
– numerical, 584
low-dispersion algorithms, 204
low-dispersion FDTD algorithms, 250
LU decomposition, sparse, 565

M-matrix, 363, 372, 382, 383, 385, 392, 409, 416, 739
macro models, 552
macroscopic
– field equations, 44
– leads, 83, 86
– Maxwell equations, 20, 44
magnetic
– charge, 144
– energy, 19, 92
– field, 14, 18, 43, 44, 65, 136, 664
– – external, 91
– – induced, 87, 91
– – lines, 91
– flux, induced, 10, 44, 91, 93
– fluxe, 532
– induction, 9, 14, 136, 664, 665

– media, 42
– moment, 43
– – density, 43
– monopoles, 10, 44
– permeability, 9
– susceptibility, 43
– vector potential, 665
magnetization, 43
magnetohydrodynamics, 21
magnetomotive force, 138
magnets, 68
manifold, 113
– with boundary, 113
mass, 127
massive scalar particles, 67
Maszstab Invarianz, 69
matching orientations, 122
matrix
– condensation, 741
– exponential form, 595
– nilpotent, 546
– pencil, 565, 576
matrix-free method, 639
matter, 68
maximum entropy distribution, 457
Maximum Entropy Principle (MEP), 456
Maxwell
– equations, 7–9, 13, 15–17, 20, 44–47, 60, 84, 95, 140, 205, 662–665, 730
– – differential form of, 11
– – integral form, 9
– laws, 68
– stress tensor, 12
– theory, 83
Maxwell–Ampère law, 75
Maxwell–Boltzmann statistics, 320, 323, 330
measure, 127
mechanical energy, 11
MESFET, 485, 493, 513, 514
mesoscopic active area, 86, 89
mesoscopic device, 83, 87
mesoscopic ring, 83, 89
Metal Semiconductor Field Effect Transistor (MESFET), 485, 493, 513, 514
metal-oxide-semiconductor field-effect transistor, 64
metals, 21, 25, 29
method of images, 667
methods
– fully implicit, 575
– linearly-implicit, 575
– semi-implicit, 575
metric tensor, 70, 71
microcircuit, 89
Miller integrator, 552, 554
Minimal Polynomial Extrapolation (MPE), 630, 631
Minkowski space, 83
MinMod limiter, 472
MIS capacitor, 31, 32, 35
MIS transistor, 36
mixed scheme, 162
mixed-hybrid scheme, 164
MLN, 597
MNA, 537, 538, 555
– charge/flux oriented formulation of, 537
– conventional formulation of, 539
mobile charges, 38
mobility, 24
Möbius band, 118, 120, 121
model parameters, 551
model refinement, 560
modelling, functional, 586
models, charge-oriented, 535
modified Bessel functions of second kind, 461
Modified Extended BDF, 574
Modified Nodal Analysis (MNA), 537, 538, 555
modified timestep control, 569
molecular charge distribution, 40
moment
– equations, 449, 454
– expansion, 27
– integral, 674, 709, 710, 743
– matching, 596
momentum flux, 29
momentum relaxation time, 28
momentum representation, 63
monodromy, 639
– matrix, 626
morphisms, 115
MOS transistor, 321, 432
MPE, 630, 631
MROW, 615
multi-level Newton, 597
multi-ports, 533
multi-resolution time-domain (MRTD) technique, 204
multi-step methods, 573
– implicit linear, 563
multiply connected, 96
– region, 92, 97
– surface, 95
multipole moments, 39
multirate, 587, 593, 612
– generalized, 615
multirate extrapolation, 614

multirate Rosenbrock–Wanner, 615
multivector, 176

NAND gate, 578
Nessyahu–Tadmor scheme, 472, 475, 476, 478, 485
nested dissection, 605
network approach, 531
network elements, 529
– idealized, 560
network equations, 529, 531, 563
– charge oriented, 558
– charge/flux oriented formulation of, 537, 563
– differential-algebraic, 529
– in charge/flux oriented formulation, 537
– stiff, 567
network topology, 529, 532, 545
– laws, 532
networks, active, 550
Neumann condition, 141
Newton correction, 565
Newton–Cotes quadrature rules, 675, 676, 684
Newton–Raphson matrix, 75, 76
Newton–Raphson method, 8, 75
Newton's method, 565
Newton's procedure, 564
nilpotency, 546
nilpotent matrix, 546
nilpotent part, 546
Nodal Analysis (NA), 530, 531, 537
node tearing, 601
node voltages, 532
noise, 617
– $1/f$, 624
– device, 618
– flicker, 618
– frequency, 623
– phase, 619, 620
– shot, 618
– thermal, 618
– timing jitter, 620
noiseless, 623, 625
noiseless solution, 620
noisy solution, 625
non-commuting operators, 88
non-equilibrium state, 27
non-interacting Liouville equation, 36
non-linear response, 12
non-local resistance, 87
non-singular Lagrangian density, 61
nonconformity, 343, 352
nonlinear perturbation analysis, 620
nonpolar optical phonon
– interaction, 448
– scattering, 460
nonpolar phonon scattering, 458
nonsplitting schemes, 478
nonuniform Yee grid, 222
norator, 553, 554
normal continuity, 132
normal field, 128
normal projection, 638
normal trees, 552
normalized perturbation function, 623
nuclear decay, 35, 36
null space, 739, 740
– method, 738
nullator, 554
numerical
– accuracy, 536
– damping, 567
– dispersion, 231, 250
– flux, 468
– instability, 261
– integration, 563, 674, 705
– integration schemes
– – implicit, 540
– low pass filter, 584
– noise, 568, 579
– phase velocity, 234
– regularization, 543
– simulation, 95
– stability, 75, 261
numerically singular, 540, 564

ODE model, 540, 542
ohmic contacts, 322
ohmic loss, 45
ohmic response, 84
Ohm's law, 12, 23, 24, 79, 87, 140, 530, 664
one-electron Hamiltonian, 64
one-particle Schrödinger equation, 14, 36
one-particle wave function, 36
one-port elements, 532
one-step methods, 563
– stiffly-accurate, 573
open cell, 114
open-ended conductor, 88
open-ended region, 87, 88
operational amplifier, 553
– circuit, 580
– ideal, 554, 555
operational formulation, 669
Optimal Sweep Following, 648
orbit, 627
– of group action, 111, 186

Orden's method, 662, 740
order conditions, 576
ordinary differential equations (ODEs), 529, 531
orientable
– manifold, 118
– surface, 95
orientation, 95, 115
– of vector space, 115
orthogonal construction (of dual mesh), 153
orthogonal polynomials, 678, 749
oscillation frequency, 622
oscillation solution, 622
oscillations, physical, 584
outer orientation, 117
oxide thickness, 31, 34, 37

PAC, 620, 624
parabolic band, 446
– limit, 459
parallel, 635
parallel transport, 69, 74
parallel-plate capacitor, 41
parallelism, 589
parallelization, 586, 597
– thread based, 608
parasitic effects, 560
partition of unity, 180, 181
partitioning, 604
– adaptive, 610
– dynamic, 606
– of circuits, 542
– requirements, 604
– static, 604
passive IC, 734
passive, strictly, 548
path integral, 14
Patterson quadrature
– formulae, 674
– rule, 662, 674, 681–686, 709
Pauli's exclusion principle, 23, 24
paving, 149
PCB, 662, 663
Péclet number, 413
perfect conductor, 54
perfectly matched layer (PML), 204, 279
– absorbing boundary conditions, 278
periodic
– AC, 620, 624
– boundary conditions, 64
– noise, 620
– steady-state, 617, 620, 621
– – solution, 621
permeability, 43, 664, 732
permittivity, 33, 664, 732
persistent current, 83, 89
perturbation index, 547, 549
Perturbation Projection Vector, 628
perturbations, highly oscillatory, 584
perturbative methods, 45
perturbed oscillatory systems, 624
Petrov–Galerkin formulation, 402
phase
– coherence, 84, 87
– modulation, 642
– noise, 620
– – analysis, 629
– space, 21, 25
phase-coherent transport, 87
phase-shift, 623
– function, 627
phenomenological theory, 68
phonons, 22, 36
photon modes, 14
physical oscillations, 584
PI-controller, linear, 570
Picard iteration, 592
piecewise linear analysis, 594
piecewise linear mapping, 595
piecewise smooth manifold, 114
piecewise-polynomial reconstruction, 473
placement, analytical, 606
plane waves, 64
plasma physics, 21
pn-diode, 421, 428
Poincaré, 638
Poincaré lemma, 134
Poincaré-map, 630
Poisson equation, 33, 36, 447, 480, 487, 507
polar coordinates, 744
polar vector, 115
polarization, 40, 41
poly-depletion, 38
polynomial grading, 293
positive-definite, 539, 731, 735
positive-definiteness, 551
potential, 13, 144, 662
– barrier, 89, 90
– difference, 14, 87
– hill, 90
– well, 33, 35, 64, 65
power supply, 560
Poynting theorem, 11, 12, 144
Poynting vector, 11, 12
predictor step, 565
predictor-corrector scheme, 565
principle of least action, 19
printed circuit boards, 661

prismatic elements, 186
probes (as modelled by chains), 133
problems
– autonomous, 622
– driven, 621
– forced, 621
– free-running, 622
– ill-posed, 529, 541
– nonautonomous, 621
– oscillator, 622
projective system, 184
properties of DAE
– analytical, 545
– structural, 545
properties, structural, 552
proxy (vector, field), 132
pseudo-spectral time-domain (PSTD)
– method, 258
– technique, 204
PSS, 617, 620, 621
pyramidal elements, 190
Pythagoras' theorem, 70

QL-cutsets, 552
quadrature
– formula, 368, 412, 675
– repeated, 676
– rule, 673
quadrilateral element, 662, 669
quadrupole moment, 39
quality factor, 646
quantized conductance, 87
quantum
– circuit theory, 93
– dot, 83, 86
– dynamics, 94
– electrodynamics, 83, 84, 93
– Hall effect, 63
– point contact, 83, 84, 86
– wire, 83
quantum mechanical
– probability density, 14
– reflection, 87
quantum mechanics, 7, 8, 69, 83, 84, 88, 94
quantum-Liouville equation, 25
quasi-Fermi levels, 320
quasi-neutral limit, 326
quasi-static, 667, 732
– approximation, 730
– form, 673
quasistationary behaviour, 532

radiation, 60, 63
– boundary conditions (RBCs), 278
– field, 14
– loss, 91
– resistance, 91
Radio Frequency (RF), 617, 618
range (of a map), 111
range space method, 738
ratio-cut, 606
Raviart–Thomas element, 344, 380
recombination-generation, 324, 334
Reduced Rank Extrapolation (RRE), 630
refinement, 155
– criteria, 504
regular group action, 111
regularization, 543, 560, 693
– by including parasitic effects, 543
– inappropriate, 543
– numerical, 543
– of curl–curl system, 164
regularizing effect, 560
relative boundary, 152
relaxation methods, 590
repeated quadrature, 676, 677
requirements, 604
reservoir, 86–88
resistance, 25, 26, 84, 86, 91, 672, 731
resistivity, 81
resistor, 530, 532, 662
– linear, 529
resonance, 31, 33–35, 89
– energies, 34, 36
– lifetime, 35, 36, 38
– width, 36
resonant
– bound state, 36
– energies, 33, 34
– state, 35
retarded Green function, 63
reverse biased, 428
RF, 617, 618
Richardson extrapolation, 677, 678
Riemann geometry, 69–71
Riemann problem, 468, 469
ringoscillator, 557, 586
– bipolar, 557
RLC-network, 548, 550, 551
robustness, 570
roll back, 613
Romberg integration, 677, 678, 684
Rosenbrock–Wanner (ROW) schemes, 575
– charge/flux-oriented, 575
Runge–Kutta, 614
– split, 614
running wave, 35

S parameters, 48
sample rate, 582
sampling, 629
scalability, 607
scalar, 69
– field, 14, 61, 67, 80, 96
– – charged, 67
– potential, 13–15, 48, 62, 63, 68, 87, 96, 99
– product (of forms), 142
scattering, 22, 36
– in a semiconductor, 448
scattering mechanisms
– elastic, 86
– electron–electron, 86
– electron–phonon, 86
– inelastic, 86, 90
scattering processes, 23
Scharfetter–Gummel scheme, 30
Schmitt trigger, 529, 538, 551
Schottky contact, 496
Schrödinger equation, 7, 32–36, 65, 94
Schur matrix, 599, 602
second Dahlquist barrier, 567
second-order accurate central differencing, 211
self-interaction, 67
semi-implicit approximation, 212
semi-implicit methods, 575
semiclassical Boltzmann equation, 447
semiconducting materials, 28
semiconductor, 20, 21, 26, 27, 32, 78, 446, 447, 503
– structure
– – high-mobility, 84
– – nanometer-sized, 83
series impedance, 49
shielding, 18
shock capturing methods, 465
shooting, 620
– method, 626, 630, 638, 641
shunt admittance, 49
signal flow, unidirectional, 589
signal frequency, 642
silicon, 31
– diode, 482, 484, 509–511
– dioxide, 31
– substrate, 31
simplices, 154
simplifying assumptions, 529
simply connected region, 9, 13, 87, 88
Simpson's quadrature rule, 684
single-electron Schrödinger equation, 86
singular Lagrangian, 61
singular matrix, 75
singular operator, 61
singularly perturbed index-2 problem, 560
singularly-perturbed ODE, 543
skew frame, 115, 116
skew linear transform, 124
skin effect, 17
SLIC method, 508, 511
Slotboom variable, 321, 330, 379, 391
slowest first, 612
SM, 630
small signal perturbation, 620
smooth manifold, 113
solenoid, 66
source integral, 673
span, 116
sparse, 539
sparse Jacobian, 566
sparse LU decomposition, 565
Sparse Tableau Approach (STA), 537
spatial confinement, 89
spectral width, 33, 34
speed of light, 63
spherical harmonics, 38
spiral inductor, 83
split Runge–Kutta, 614
splitting approach, 506
splitting strategy, 476
spurious modes, 165
stability, 567, 621
– function, 582
– matrix, 582
– of scheme, 172, 174
stabilizer, 111
stable, 621
– strongly, 621
staggered, collocated grid, 227
staggered, uncollocated grid, 209, 226
stamping, 608
standing waves, 36
star construction (of dual mesh), 153
star (of cell), 180
star-shaped, 153
state equations, 541
State Variable approach, 552
state-space model, 541
static assignment, 604
static condensation, 361, 381, 408
static elements, 554
statistical operator, 36
statistical physics, 21
steady state, 564
step function, 63
step size, 565
stepsize selection, 568, 577

stiff decay, 567
stiffly accurate, 577
stiffly-accurate one-step methods, 573
Stokes theorem, 95, 99, 134
– multiply connected regions, 87
straight form, 130
stratified inhomogeneous medium, 667
streamline diffusion, 402
stressing conditions, 38
stretched-coordinate formulation of Berenger's PML, 284
strictly passive, 548
strongly stable, 621
structural properties, 552
– of the DAE network equations, 545
structural symmetry, 539
subband energies, 33, 34
subband state, 31, 33, 35
subcircuit partitioning, 542, 550
substitute circuits, 614
substrate, 34, 35
superconducting ring, 91
superconductor, 54, 67, 91
– type-I, 91
supraconvergent FDTD algorithm, 222
surface charge, 34
surface element, 10, 98
surface integral, 97
symmetric, 735
symmetry (of Hodge map), 142

't Hooft gauge, 67
table model, 586
tangent, 627
– correction, 599
– map, 113
– space, 112
– vector, 98
tangential continuity, 132
Taylor expansion, 674, 706, 710–712, 715, 720, 741
TE modes, 48
telegraph equation, 49
TEM modes, 47
temperature, 22, 26
temporal gauge, 66
tensor, 69
– equations, 70
terminal charges, 533
test function, 668
tetradecahedron/dual-tetrahedron mesh, 230
theory of relativity, 83
thermal
– conductance, 28
– conductivity, 28, 29
– equilibrium, 22, 26, 27, 35
– – local, 87
– flux, 28
– voltage, 320
third Maxwell equation, 87
thread based parallelization, 608
through (quantity), 129, 137
time reversal invariance, 21
time-reversal symmetry breaking, 36
timestep control, 568, 569, 577
– modified, 569
timing jitter, 620
timing simulation, 592
TM modes, 47
topological
– conditions, 549, 552, 564
– criteria, 571
– structure, 529
topology, 88, 96
toroidal circuit, 87
toroidal region, 93, 96
TR-BDF, 580
– schemes, 568
trace (of a form), 141
tractability index, 547
transfer matrix, 33
transistor, 31
transitive function, 113
transitive group action, 111
translational invariance, 32, 45, 46, 64, 65
translational symmetry, 17
transmission
– coefficient, 36, 86, 89
– conditions, 131, 141, 142
– line, 44, 49
– – theory, 48
– matrix, 34
– probability, 86
transverse
– component, 99
– potential, 48
– subspaces, 116
Trapezoidal Rule (TR), 567, 568, 580, 629, 676
traveling state, 36
trial function, 19
triangular element, 662
tunneling, 34–36, 54
– currents, 31, 38
twisted chain, 122, 129
twisted form, 130, 133
two-dimensional electron gas, 64, 84
two-step approach, 637

two-terminal device, 86
two-terminal elements, 532

unconditional stability, 271
uniaxial perfectly matched layer (UPML), 204, 279, 286
unidirectional signal flow, 589
uniform electric field, 29
uniform mesh, 167
uniform nonoscillatory reconstruction, 480
uniform refinement, 167
uniformly accurate central scheme of order 2, 479
unifying theory, 7
unisolvence, 183
UNO limiter, 472
unstable, 621
unstaggered, collocated grid, 227
upwind, 384, 416
upwind schemes, 468

vacuum, 9, 67
– expectation value, 67
– impedance of, 91
valence band, 26
valley, 32
– index, 32
variable step size implementations, 565
variables
– algebraic, 546
– differential, 546, 550
variational calculus, 16
variational formulation, 668, 669
variational principle, 15, 16
VC loops, 549, 551, 552, 572, 573
vector at a point, 112
vector calculus, 83, 100
vector extrapolation, 630
vector field, 66, 95, 96, 99
– conservative, 97
– irrotational, 96
– longitudinal part, 95
– non-conservative, 97
– transverse part, 95
vector phasor, 664
vector potential, 13–15, 43, 63, 64, 66, 68, 79, 80, 91, 99
– irrotational, 91
vector-extrapolation, 631
vectorial area, 124, 181
velocity field, 41
voltage sources, 533
volume (form), 124
volume integral, 97
von Klitzing resistance, 64, 85, 91
Voronoi–Delaunay dual, 153

warping function, 648
wave equation, 13
wave function, 14, 32, 33, 35, 74, 89, 92
wave guide, 44
waveform evaluation, asymptotic, 596
waveform Newton, 640
waveform relaxation, 592
– Newton, 593
weak decay, 7
weak instability, 568, 573
wedge product, 142
Whitney complex, 174
Whitney forms, 174, 176
Whitney map, 172
Wiedemann–Franz law, 29, 453

Yee algorithm, 208
– divergence-free nature, 221
Yee scheme, 159